ROBUST ADAPTIVE CONTROL

Petros A. Ioannou
Electrical Engineering-Systems
University of Southern California

Jing Sun
Naval Architecture and Marine Engineering
University of Michigan

Dover Publications
Garden City, New York

Copyright

Copyright © 1996, 2012 by Petros A. Ioannou and Jing Sun
All rights reserved.

Bibliographical Note

This Dover edition, first published in 2012, is a slightly altered republication of the work originally published in 1996 by Prentice-Hall, Inc., Upper Saddle River, New Jersey. This edition does not include the personal computer floppy disk that came with the original, and all references to it have been omitted.

Library of Congress Cataloging-in-Publication Data

Ioannou, P. A. (Petros A.), 1953–.
   Robust adaptive control / Petros A. Ioannou and Jin Sun.—Dover edition.
     p. cm.
   Summary: "Presented in a tutorial style, this text reduces the confusion and difficulty in grasping the design, analysis and robustness of a wide class of adaptive controls for continuous-time plants. The treatment unifies, simplifies, and explains most of the techniques for designing and analyzing adaptive control systems. Excellent text and authoritative reference"—Provided by publisher.
   Includes bibliographical references and index.
   ISBN-13: 978-0-486-49817-1 (pbk.)
   ISBN-10: 0-486-49817-4 (pbk.)
   1. Adaptive control systems. I. Sun, Jing, 1961–. II. Title.
TJ217.I66 2012
629.8'36—dc23

2012022707

Manufactured in the United States by LSC Communications Book LLC
49817405   2021
www.doverpublications.com

To Kira, Andreas, and Alexandros
　　　　　　　　　　P. A. I.

To Bing, Alex, and Alan
　　　　　　　　J. S.

Contents

Preface   xiii

List of Acronyms   xvii

1 Introduction   1
    1.1 Control System Design Steps 1
    1.2 Adaptive Control 5
       1.2.1 Robust Control 6
       1.2.2 Gain Scheduling 7
       1.2.3 Direct and Indirect Adaptive Control 8
       1.2.4 Model Reference Adaptive Control 12
       1.2.5 Adaptive Pole Placement Control 14
       1.2.6 Design of On-Line Parameter Estimators 16
    1.3 A Brief History 23

2 Models for Dynamic Systems   26
    2.1 Introduction 26
    2.2 State-Space Models 27
       2.2.1 General Description 27
       2.2.2 Canonical State-Space Forms 29
    2.3 Input/Output Models 34
       2.3.1 Transfer Functions 34
       2.3.2 Coprime Polynomials 39
    2.4 Plant Parametric Models 47
       2.4.1 Linear Parametric Models 49
       2.4.2 Bilinear Parametric Models 58

2.5 Problems 61

3 Stability — 66
3.1 Introduction 66
3.2 Preliminaries 67
3.2.1 Norms and $\mathcal{L}_p$ Spaces 67
3.2.2 Properties of Functions 72
3.2.3 Positive Definite Matrices 78
3.3 Input/Output Stability 79
3.3.1 $\mathcal{L}_p$ Stability 79
3.3.2 The $\mathcal{L}_{2\delta}$ Norm and I/O Stability 85
3.3.3 Small Gain Theorem 96
3.3.4 Bellman-Gronwall Lemma 101
3.4 Lyapunov Stability 105
3.4.1 Definition of Stability 105
3.4.2 Lyapunov's Direct Method 108
3.4.3 Lyapunov-Like Functions 117
3.4.4 Lyapunov's Indirect Method 118
3.4.5 Stability of Linear Systems 120
3.5 Positive Real Functions and Stability 126
3.5.1 Positive Real and Strictly Positive Real Transfer Functions 126
3.5.2 PR and SPR Transfer Function Matrices 132
3.6 Stability of LTI Feedback Systems 134
3.6.1 A General LTI Feedback System 134
3.6.2 Internal Stability 135
3.6.3 Sensitivity and Complementary Sensitivity Functions . 136
3.6.4 Internal Model Principle 137
3.7 Problems 139

4 On-Line Parameter Estimation — 144
4.1 Introduction 144
4.2 Simple Examples 146
4.2.1 Scalar Example: One Unknown Parameter 146
4.2.2 First-Order Example: Two Unknowns 151
4.2.3 Vector Case 156

	4.2.4	Remarks	161
4.3	Adaptive Laws with Normalization		162
	4.3.1	Scalar Example	162
	4.3.2	First-Order Example	165
	4.3.3	General Plant	169
	4.3.4	SPR-Lyapunov Design Approach	171
	4.3.5	Gradient Method	180
	4.3.6	Least-Squares	192
	4.3.7	Effect of Initial Conditions	200
4.4	Adaptive Laws with Projection		203
	4.4.1	Gradient Algorithms with Projection	203
	4.4.2	Least-Squares with Projection	206
4.5	Bilinear Parametric Model		208
	4.5.1	Known Sign of ρ^*	208
	4.5.2	Sign of ρ^* and Lower Bound ρ_0 Are Known	212
	4.5.3	Unknown Sign of ρ^*	215
4.6	Hybrid Adaptive Laws		217
4.7	Summary of Adaptive Laws		220
4.8	Parameter Convergence Proofs		220
	4.8.1	Useful Lemmas	220
	4.8.2	Proof of Corollary 4.3.1	235
	4.8.3	Proof of Theorem 4.3.2 (iii)	236
	4.8.4	Proof of Theorem 4.3.3 (iv)	239
	4.8.5	Proof of Theorem 4.3.4 (iv)	240
	4.8.6	Proof of Corollary 4.3.2	241
	4.8.7	Proof of Theorem 4.5.1(iii)	242
	4.8.8	Proof of Theorem 4.6.1 (iii)	243
4.9	Problems		245

5 Parameter Identifiers and Adaptive Observers — 250

5.1	Introduction		250
5.2	Parameter Identifiers		251
	5.2.1	Sufficiently Rich Signals	252
	5.2.2	Parameter Identifiers with Full-State Measurements	258
	5.2.3	Parameter Identifiers with Partial-State Measurements	260
5.3	Adaptive Observers		267

		5.3.1 The Luenberger Observer 267

        5.3.1 The Luenberger Observer 267
        5.3.2 The Adaptive Luenberger Observer 269
        5.3.3 Hybrid Adaptive Luenberger Observer 276
  5.4 Adaptive Observer with Auxiliary Input 279
  5.5 Adaptive Observers for Nonminimal Plant Models 287
        5.5.1 Adaptive Observer Based on Realization 1 287
        5.5.2 Adaptive Observer Based on Realization 2 292
  5.6 Parameter Convergence Proofs 297
        5.6.1 Useful Lemmas 297
        5.6.2 Proof of Theorem 5.2.1 301
        5.6.3 Proof of Theorem 5.2.2 302
        5.6.4 Proof of Theorem 5.2.3 306
        5.6.5 Proof of Theorem 5.2.5 309
  5.7 Problems 310

6 Model Reference Adaptive Control   313

  6.1 Introduction 313
  6.2 Simple Direct MRAC Schemes 315
        6.2.1 Scalar Example: Adaptive Regulation 315
        6.2.2 Scalar Example: Adaptive Tracking 320
        6.2.3 Vector Case: Full-State Measurement 325
        6.2.4 Nonlinear Plant 327
  6.3 MRC for SISO Plants 330
        6.3.1 Problem Statement 331
        6.3.2 MRC Schemes: Known Plant Parameters 333
  6.4 Direct MRAC with Unnormalized Adaptive Laws 343
        6.4.1 Relative Degree $n^* = 1$ 345
        6.4.2 Relative Degree $n^* = 2$ 355
        6.4.3 Relative Degree $n^* = 3$ 362
  6.5 Direct MRAC with Normalized Adaptive Laws 372
        6.5.1 Example: Adaptive Regulation 372
        6.5.2 Example: Adaptive Tracking 379
        6.5.3 MRAC for SISO Plants 383
        6.5.4 Effect of Initial Conditions 395
  6.6 Indirect MRAC 396
        6.6.1 Scalar Example 397

    6.6.2 Indirect MRAC with Unnormalized Adaptive Laws . . 401
    6.6.3 Indirect MRAC with Normalized Adaptive Law 407
6.7 Relaxation of Assumptions in MRAC 412
    6.7.1 Assumption P1: Minimum Phase 412
    6.7.2 Assumption P2: Upper Bound for the Plant Order . . 413
    6.7.3 Assumption P3: Known Relative Degree n^* 414
    6.7.4 Tunability 415
6.8 Stability Proofs of MRAC Schemes 417
    6.8.1 Normalizing Properties of Signal m_f 417
    6.8.2 Proof of Theorem 6.5.1: Direct MRAC 418
    6.8.3 Proof of Theorem 6.6.2: Indirect MRAC 424
6.9 Problems 429

7 Adaptive Pole Placement Control 434
7.1 Introduction 434
7.2 Simple APPC Schemes 436
    7.2.1 Scalar Example: Adaptive Regulation 436
    7.2.2 Modified Indirect Adaptive Regulation 440
    7.2.3 Scalar Example: Adaptive Tracking 442
7.3 PPC: Known Plant Parameters 447
    7.3.1 Problem Statement 448
    7.3.2 Polynomial Approach 449
    7.3.3 State-Variable Approach 454
    7.3.4 Linear Quadratic Control 459
7.4 Indirect APPC Schemes 466
    7.4.1 Parametric Model and Adaptive Laws 466
    7.4.2 APPC Scheme: The Polynomial Approach 468
    7.4.3 APPC Schemes: State-Variable Approach 478
    7.4.4 Adaptive Linear Quadratic Control (ALQC) 486
7.5 Hybrid APPC Schemes 494
7.6 Stabilizability Issues and Modified APPC 498
    7.6.1 Loss of Stabilizability: A Simple Example 499
    7.6.2 Modified APPC Schemes 502
    7.6.3 Switched-Excitation Approach 506
7.7 Stability Proofs 513
    7.7.1 Proof of Theorem 7.4.1 513

		7.7.2 Proof of Theorem 7.4.2	519

       7.7.3 Proof of Theorem 7.5.1 523
  7.8 Problems 527

8 Robust Adaptive Laws   530
  8.1 Introduction 530
  8.2 Plant Uncertainties and Robust Control 531
       8.2.1 Unstructured Uncertainties 532
       8.2.2 Structured Uncertainties: Singular Perturbations ... 536
       8.2.3 Examples of Uncertainty Representations 539
       8.2.4 Robust Control 541
  8.3 Instability Phenomena in Adaptive Systems 544
       8.3.1 Parameter Drift 545
       8.3.2 High-Gain Instability 548
       8.3.3 Instability Resulting from Fast Adaptation 549
       8.3.4 High-Frequency Instability 551
       8.3.5 Effect of Parameter Variations 552
  8.4 Modifications for Robustness: Simple Examples 554
       8.4.1 Leakage 556
       8.4.2 Parameter Projection 565
       8.4.3 Dead Zone 566
       8.4.4 Dynamic Normalization 571
  8.5 Robust Adaptive Laws 575
       8.5.1 Parametric Models with Modeling Error 576
       8.5.2 SPR-Lyapunov Design Approach with Leakage ... 582
       8.5.3 Gradient Algorithms with Leakage 592
       8.5.4 Least-Squares with Leakage 602
       8.5.5 Projection 603
       8.5.6 Dead Zone 606
       8.5.7 Bilinear Parametric Model 613
       8.5.8 Hybrid Adaptive Laws 616
       8.5.9 Effect of Initial Conditions 623
  8.6 Summary of Robust Adaptive Laws 623
  8.7 Problems 625

CONTENTS

9 Robust Adaptive Control Schemes — 634
- 9.1 Introduction — 634
- 9.2 Robust Identifiers and Adaptive Observers — 635
  - 9.2.1 Dominantly Rich Signals — 638
  - 9.2.2 Robust Parameter Identifiers — 643
  - 9.2.3 Robust Adaptive Observers — 648
- 9.3 Robust MRAC — 650
  - 9.3.1 MRC: Known Plant Parameters — 651
  - 9.3.2 Direct MRAC with Unnormalized Adaptive Laws — 656
  - 9.3.3 Direct MRAC with Normalized Adaptive Laws — 666
  - 9.3.4 Robust Indirect MRAC — 687
- 9.4 Performance Improvement of MRAC — 693
  - 9.4.1 Modified MRAC with Unnormalized Adaptive Laws — 697
  - 9.4.2 Modified MRAC with Normalized Adaptive Laws — 703
- 9.5 Robust APPC Schemes — 709
  - 9.5.1 PPC: Known Parameters — 710
  - 9.5.2 Robust Adaptive Laws for APPC Schemes — 713
  - 9.5.3 Robust APPC: Polynomial Approach — 715
  - 9.5.4 Robust APPC: State Feedback Law — 722
  - 9.5.5 Robust LQ Adaptive Control — 730
- 9.6 Adaptive Control of LTV Plants — 732
- 9.7 Adaptive Control for Multivariable Plants — 734
  - 9.7.1 Decentralized Adaptive Control — 735
  - 9.7.2 The Command Generator Tracker Approach — 736
  - 9.7.3 Multivariable MRAC — 739
- 9.8 Stability Proofs of Robust MRAC Schemes — 744
  - 9.8.1 Properties of Fictitious Normalizing Signal — 744
  - 9.8.2 Proof of Theorem 9.3.2 — 748
- 9.9 Stability Proofs of Robust APPC Schemes — 759
  - 9.9.1 Proof of Theorem 9.5.2 — 759
  - 9.9.2 Proof of Theorem 9.5.3 — 763
- 9.10 Problems — 768
- A Swapping Lemmas — 774
- B Optimization Techniques — 783
  - B.1 Notation and Mathematical Background — 783
  - B.2 The Method of Steepest Descent (Gradient Method) — 785

	B.3	Newton's Method 786
	B.4	Gradient Projection Method 788
	B.5	Example 791

Bibliography   **795**

Index   **819**

Preface

The area of adaptive control has grown to be one of the richest in terms of algorithms, design techniques, analytical tools, and applications. Several books and research monographs already exist on the topics of parameter estimation and adaptive control.

Despite this rich literature, the field of adaptive control may easily appear to an outsider as a collection of unrelated tricks and modifications. Students are often overwhelmed and sometimes confused by the vast number of what appear to be unrelated designs and analytical methods achieving similar results. Researchers concentrating on different approaches in adaptive control often find it difficult to relate their techniques with others without additional research efforts.

The purpose of this book is to alleviate some of the confusion and difficulty in understanding the design, analysis, and robustness of a wide class of adaptive control for continuous-time plants. The book is the outcome of several years of research, whose main purpose was not to generate new results, but rather unify, simplify, and present in a tutorial manner most of the existing techniques for designing and analyzing adaptive control systems.

The book is written in a self-contained fashion to be used as a textbook on adaptive systems at the senior undergraduate, or first and second graduate level. It is assumed that the reader is familiar with the materials taught in undergraduate courses on linear systems, differential equations, and automatic control. The book is also useful for an industrial audience where the interest is to implement adaptive control rather than analyze its stability properties. Tables with descriptions of adaptive control schemes presented in the book are meant to serve this audience. In addition, the first author coauthored the adaptive control toolbox (see http://www.adaptivecontroltoolbox.com/ACT/Comand-Center.html) based on Matlab/

Simulink which can be used to simulate a wide range of identification and adaptive control algorithims presented in the book.

A significant part of the book, devoted to parameter estimation and learning in general, provides techniques and algorithms for on-line fitting of dynamic or static models to data generated by real systems. The tools for design and analysis presented in the book are very valuable in understanding and analyzing similar parameter estimation problems that appear in neural networks, fuzzy systems, and other universal approximators. The book will be of great interest to the neural and fuzzy logic audience who will benefit from the strong similarity that exists between adaptive systems, whose stability properties are well established, and neural networks, fuzzy logic systems where stability and convergence issues are yet to be resolved.

The book is organized as follows: Chapter 1 is used to introduce adaptive control as a method for controlling plants with parametric uncertainty. it also provides some background and a brief history of the development of adaptive control. Chapter 2 presents a review of various plant model representations that are useful for parameter identification and control. A considerable number of stability results that are useful in analyzing and un≠derstanding the properties of adaptive and nonlinear systems in general are presented in Chapter 3. Chapter 4 deals with the design and analysis of on-line parameter estimators or adaptive laws that form the backbone of every adaptive control scheme presented in the chapters to follow. The design of parameter identifiers and adaptive observers for stable plants is presented in Chapter 5. Chapter 6 is devoted to the design and analysis of a wide class of model reference adaptive controllers for minimum phase plants. The design of adaptive control for plants that are not necessarily minimum phase is presented in Chapter 7. These schemes are based on pole placement control strategies and are referred to as adaptive pole placement control. While Chapters 4 through 7 deal with plant models that are free of disturbances, unmodeled dynamics and noise, Chapters 8 and 9 deal with the robustness issues in adaptive control when plant model uncertainties, such as bounded disturbances and unmodeled dynamics, are present.

The book can be used in various ways. The reader who is familiar with stability and linear systems may start from Chapter 4. An introductory course in adaptive control could be covered in Chapters 1, 2, and 4 to 9, by excluding the more elaborate and difficult proofs of theorems that are presented either in the last section of chapters or in the appendices. Chapter 3 could be used for reference and for covering relevant stability results that arise during the course. A higher-level course intended for graduate students that are interested in a deeper understanding of adaptive control could cover all chapters with more emphasis on the design and stability proofs. A course for an industrial audience could contain Chapters 1, 2, and 4 to 9 with emphasis on the design of adaptive control algorithms rather than stability proofs and convergence.

Acknowledgments

The writing of this book has been surprisingly difficult and took a long time to evolve to its present form. Several versions of the book were completed only to be put aside after realizing that new results and techniques would lead to a better version. This book was published by Prentice-Hall in 1996 and went out of print some years later due to the change of policy of the publishing company toward supporting specialized control textbooks. The book, however, became popular and was used widely by many universities in teaching adaptive control. The authors made the book available in electronic form shortly after it went out of print and as a result it remained as an uninterrupted source for studying adaptive control.

A long list of friends and colleagues have helped us in the preparation of the book in many different ways. We are especially grateful to Petar Kokotovic who introduced the first author to the field of adaptive control back in 1979. Since then he has been a great advisor, friend, and colleague. His continuous enthusiasm and hard work for research has been the strongest driving force behind our research and that of our students. We thank Brian Anderson, Karl Åström, Mike Athans, Bo Egardt, Graham Goodwin, Rick Johnson, Gerhard Kreisselmeier, Yoan Landau, Lennart Ljung, David Mayne, late R. Monopoli, Bob Narendra, and Steve Morse for their work, interactions, and continuous enthusiasm in adaptive control that helped us lay the foundations of most parts of the book.

We would especially like to express our deepest appreciation to Laurent Praly and Kostas Tsakalis. Laurent was the first researcher to recognize and publicize the beneficial effects of dynamic normalization on robustness that opened the way to a wide class of robust adaptive control algorithms addressed in the book. His interactions with us and our students is highly appreciated. Kostas, a former student of the first author, is responsible for many mathematical tools and stability arguments used in Chapters 6 and 9. His continuous interactions helped us to decipher many of the cryptic concepts and robustness properties of model reference adaptive control.

We are thankful to our former and current students and visitors who collaborated with us in research and contributed to this work: Farid Ahrned-Zaid, C. C. Chien, Aniruddha Datta, Marios Polycarpou, Houmair Raza, Alex Stotsky, Tim Sun, Hualin Tan, Gang Tao, Hui Wang, Tom Xu, and Youping Zhang. We are grateful to many colleagues for stimulating discussions at conferences workshops, and meetings. They have helped us broaden our understanding of the field. In particular, we would like to mention Anu Annaswamy, Erwei Bai, Bob Bitmead, Marc Bodson, Stephen Boyd, Sara Dasgupta, the late Howard Elliot, Li-chen Fu, Fouad Giri, David Hill, Ioannis Kanellakopoulos, Pramod Khargonekar, Hassan Khalil, Bob Kosut, Jim Krause, Miroslav Krstić, Rogelio Lozano-Leal, Iven Mareels, Rick Middleton, David Mudget, Romeo Ortega, Brad Riedle, Charles Rohrs, Ali Saberi, Shankar Sastry, Lena Valavani, Jim Winkelman, Erik Ydstie, Jessy Grizzle and Anna Stefanopoulou.

Finally, we acknowledge the support of several organizationrs including Ford Motor Company, General Motors Project Trilby, National Science Foundation, Rockwell International, and Lockheed. Special thanks are due to Bob Borcherts, Roger Fruechte, Neil Schilke, and James Rillings of former Project Trilby; Bill Powers, Mike Shulman, and Steve Eckert of Ford Motor Company; and Bob Rooney and Houssein Youseff of Lockheed whose support of our research made this book possible.

Petros A. Ioannou
Jing Sun

List of Acronyms

ALQC	Adaptive linear quadratic control
APPC	Adaptive pole placement control
B-G	Bellman Gronwall (lemma)
BIBO	Bounded-input bounded-output
CEC	Certainty equivalence control
I/O	Input/output
LKY	Lefschetz-Kalman-Yakubovich (lemma)
LQ	Linear quadratic
LTI	Linear time invariant
LTV	Linear time varying
MIMO	Multi-input multi-output
MKY	Meyer-Kalman-Yakubovich (lemma)
MRAC	Model reference adaptive control
MRC	Model reference control
PE	Persistently exciting
PI	Proportional plus integral
PPC	Pole placement control
PR	Positive real
SISO	Single input single output
SPR	Strictly positive real
TV	Time varying
UCO	Uniformly completely observable
a.s.	Asymptotically stable
e.s.	Exponentially stable
m.s.s.	(In the) mean square sense
u.a.s.	Uniformly asymptotically stable
u.b.	Uniformly bounded
u.s.	Uniformly stable
u.u.b.	Uniformly ultimately bounded
w.r.t.	With respect to

Chapter 1

Introduction

1.1 Control System Design Steps

The design of a controller that can alter or modify the behavior and response of an unknown plant to meet certain performance requirements can be a tedious and challenging problem in many control applications. By plant, we mean any process characterized by a certain number of inputs u and outputs y, as shown in Figure 1.1.

The plant inputs u are processed to produce several plant outputs y that represent the measured output response of the plant. The control design task is to choose the input u so that the output response $y(t)$ satisfies certain given performance requirements. Because the plant process is usually complex, i.e., it may consist of various mechanical, electronic, hydraulic parts, etc., the appropriate choice of u is in general not straightforward. The control design steps often followed by most control engineers in choosing the input u are shown in Figure 1.2 and are explained below.

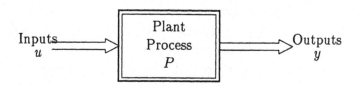

Figure 1.1 Plant representation.

Step 1. Modeling

The task of the control engineer in this step is to understand the processing mechanism of the plant, which takes a given input signal $u(t)$ and produces the output response $y(t)$, to the point that he or she can describe it in the form of some mathematical equations. These equations constitute the mathematical model of the plant. An exact plant model should produce the same output response as the plant, provided the input to the model and initial conditions are exactly the same as those of the plant. The complexity of most physical plants, however, makes the development of such an exact model unwarranted or even impossible. But even if the exact plant model becomes available, its dimension is likely to be infinite, and its description nonlinear or time varying to the point that its usefulness from the control design viewpoint is minimal or none. This makes the task of modeling even more difficult and challenging, because the control engineer has to come up with a mathematical model that describes accurately the input/output behavior of the plant and yet is simple enough to be used for control design purposes. A simple model usually leads to a simple controller that is easier to understand and implement, and often more reliable for practical purposes.

A plant model may be developed by using physical laws or by processing the plant input/output (I/O) data obtained by performing various experiments. Such a model, however, may still be complicated enough from the control design viewpoint and further simplifications may be necessary. Some of the approaches often used to obtain a simplified model are

(i)   Linearization around operating points
(ii) Model order reduction techniques

In approach (i) the plant is approximated by a linear model that is valid around a given operating point. Different operating points may lead to several different linear models that are used as plant models. Linearization is achieved by using Taylor's series expansion and approximation, fitting of experimental data to a linear model, etc.

In approach (ii) small effects and phenomena outside the frequency range of interest are neglected leading to a lower order and simpler plant model. The reader is referred to references [67, 106] for more details on model reduction techniques and approximations.

1.1. CONTROL SYSTEM DESIGN STEPS

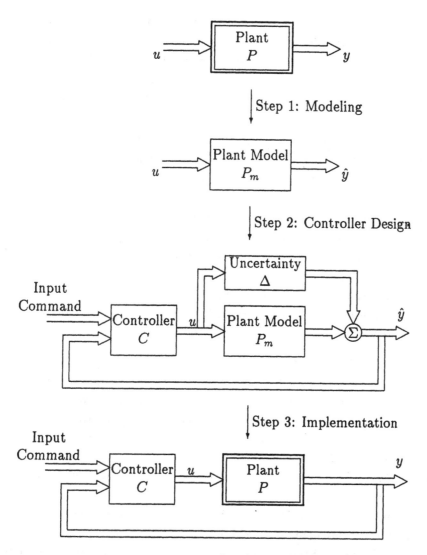

Figure 1.2 Control system design steps.

In general, the task of modeling involves a good understanding of the plant process and performance requirements, and may require some experience from the part of the control engineer.

Step 2. Controller Design

Once a model of the plant is available, one can proceed with the controller design. The controller is designed to meet the performance requirements for the plant model. If the model is a good approximation of the plant, then one would hope that the controller performance for the plant model would be close to that achieved when the same controller is applied to the plant.

Because the plant model is always an approximation of the plant, the effect of any discrepancy between the plant and the model on the performance of the controller will not be known until the controller is applied to the plant in Step 3. One, however, can take an intermediate step and analyze the properties of the designed controller for a plant model that includes a class of plant model uncertainties denoted by Δ that are likely to appear in the plant. If Δ represents most of the unmodeled plant phenomena, its representation in terms of mathematical equations is not possible. Its characterization, however, in terms of some known bounds may be possible in many applications. By considering the existence of a general class of uncertainties Δ that are likely to be present in the plant, the control engineer may be able to modify or redesign the controller to be less sensitive to uncertainties, i.e., to be more robust with respect to Δ. This robustness analysis and redesign improves the potential for a successful implementation in Step 3.

Step 3. Implementation

In this step, a controller designed in Step 2, which is shown to meet the performance requirements for the plant model and is robust with respect to possible plant model uncertainties Δ, is ready to be applied to the unknown plant. The implementation can be done using a digital computer, even though in some applications analog computers may be used too. Issues, such as the type of computer available, the type of interface devices between the computer and the plant, software tools, etc., need to be considered a priori. Computer speed and accuracy limitations may put constraints on the complexity of the controller that may force the control engineer to go back to Step 2 or even Step 1 to come up with a simpler controller without violating the performance requirements.

Another important aspect of implementation is the final adjustment, or as often called the tuning, of the controller to improve performance by compensating for the plant model uncertainties that are not accounted for during the design process. Tuning is often done by trial and error, and depends very much on the experience and intuition of the control engineer.

In this book we will concentrate on Step 2. We will be dealing with the design of control algorithms for a class of plant models described by the linear differential equation

$$\dot{x} = Ax + Bu, \quad x(0) = x_0$$
$$y = C^\top x + Du \qquad (1.1.1)$$

In (1.1.1) $x \in \mathcal{R}^n$ is the state of the model, $u \in \mathcal{R}^r$ the plant input, and $y \in \mathcal{R}^l$ the plant model output. The matrices $A \in \mathcal{R}^{n \times n}$, $B \in \mathcal{R}^{n \times r}$, $C \in \mathcal{R}^{n \times l}$, and $D \in \mathcal{R}^{l \times r}$ could be constant or time varying. This class of plant models is quite general because it can serve as an approximation of nonlinear plants around operating points. A controller based on the linear model (1.1.1) is expected to be simpler and easier to understand than a controller based on a possibly more accurate but nonlinear plant model.

The class of plant models given by (1.1.1) can be generalized further if we allow the elements of A, B, and C to be completely unknown and changing with time or operating conditions. The control of plant models (1.1.1) with A, B, C, and D unknown or partially known is covered under the area of adaptive systems and is the main topic of this book.

1.2 Adaptive Control

According to Webster's dictionary, to adapt means "to change (oneself) so that one's behavior will conform to new or changed circumstances." The words "adaptive systems" and "adaptive control" have been used as early as 1950 [10, 27].

The design of autopilots for high-performance aircraft was one of the primary motivations for active research on adaptive control in the early 1950s. Aircraft operate over a wide range of speeds and altitudes, and their dynamics are nonlinear and conceptually time varying. For a given operating point, specified by the aircraft speed (Mach number) and altitude, the complex aircraft dynamics can be approximated by a linear model of the same form as (1.1.1). For example, for an operating point i, the linear aircraft model has the following form [140]:

$$\dot{x} = A_i x + B_i u, \quad x(0) = x_0$$
$$y = C_i^\top x + D_i u \qquad (1.2.1)$$

where A_i, B_i, C_i, and D_i are functions of the operating point i. As the aircraft goes through different flight conditions, the operating point changes

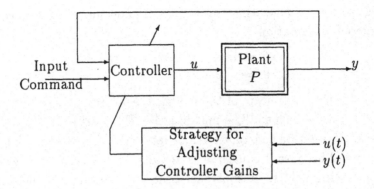

Figure 1.3 Controller structure with adjustable controller gains.

leading to different values for A_i, B_i, C_i, and D_i. Because the output response $y(t)$ carries information about the state x as well as the parameters, one may argue that in principle, a sophisticated feedback controller should be able to learn about parameter changes by processing $y(t)$ and use the appropriate gains to accommodate them. This argument led to a feedback control structure on which adaptive control is based. The controller structure consists of a feedback loop and a controller with adjustable gains as shown in Figure 1.3. The way of changing the controller gains in response to changes in the plant and disturbance dynamics distinguishes one scheme from another.

1.2.1 Robust Control

A constant gain feedback controller may be designed to cope with parameter changes provided that such changes are within certain bounds. A block diagram of such a controller is shown in Figure 1.4 where $G(s)$ is the transfer function of the plant and $C(s)$ is the transfer function of the controller. The transfer function from y^* to y is

$$\frac{y}{y^*} = \frac{C(s)G(s)}{1+C(s)G(s)} \qquad (1.2.2)$$

where $C(s)$ is to be chosen so that the closed-loop plant is stable, despite parameter changes or uncertainties in $G(s)$, and $y \approx y^*$ within the frequency range of interest. This latter condition can be achieved if we choose $C(s)$

1.2. ADAPTIVE CONTROL

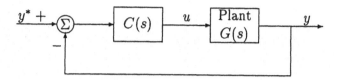

Figure 1.4 Constant gain feedback controller.

so that the loop gain $|C(jw)G(jw)|$ is as large as possible in the frequency spectrum of y^* provided, of course, that large loop gain does not violate closed-loop stability requirements. The tracking and stability objectives can be achieved through the design of $C(s)$ provided the changes within $G(s)$ are within certain bounds. More details about robust control will be given in Chapter 8.

Robust control is not considered to be an adaptive system even though it can handle certain classes of parametric and dynamic uncertainties.

1.2.2 Gain Scheduling

Let us consider the aircraft model (1.2.1) where for each operating point i, $i = 1, 2, \ldots, N$, the parameters $A_i, B_i, C_i,$ and D_i are known. For a given operating point i, a feedback controller with constant gains, say θ_i, can be designed to meet the performance requirements for the corresponding linear model. This leads to a controller, say $C(\theta)$, with a set of gains $\{\theta_1, \theta_2, \ldots, \theta_i, \ldots, \theta_N\}$ covering N operating points. Once the operating point, say i, is detected the controller gains can be changed to the appropriate value of θ_i obtained from the precomputed gain set. Transitions between different operating points that lead to significant parameter changes may be handled by interpolation or by increasing the number of operating points. The two elements that are essential in implementing this approach is a look-up table to store the values of θ_i and the plant auxiliary measurements that correlate well with changes in the operating points. The approach is called *gain scheduling* and is illustrated in Figure 1.5.

The gain scheduler consists of a look-up table and the appropriate logic for detecting the operating point and choosing the corresponding value of θ_i from the table. In the case of aircraft, the auxiliary measurements are the Mach number and the dynamic pressure. With this approach plant

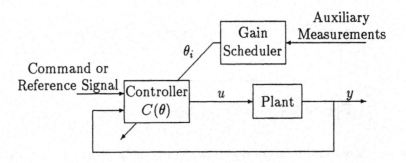

Figure 1.5 Gain scheduling.

parameter variations can be compensated by changing the controller gains as functions of the auxiliary measurements.

The advantage of gain scheduling is that the controller gains can be changed as quickly as the auxiliary measurements respond to parameter changes. Frequent and rapid changes of the controller gains, however, may lead to instability [226]; therefore, there is a limit as to how often and how fast the controller gains can be changed.

One of the disadvantages of gain scheduling is that the adjustment mechanism of the controller gains is precomputed off-line and, therefore, provides no feedback to compensate for incorrect schedules. Unpredictable changes in the plant dynamics may lead to deterioration of performance or even to complete failure. Another possible drawback of gain scheduling is the high design and implementation costs that increase with the number of operating points.

Despite its limitations, gain scheduling is a popular method for handling parameter variations in flight control [140, 210] and other systems [8].

1.2.3 Direct and Indirect Adaptive Control

An adaptive controller is formed by combining an on-line parameter estimator, which provides estimates of unknown parameters at each instant, with a control law that is motivated from the known parameter case. The way the parameter estimator, also referred to as *adaptive law* in the book, is combined with the control law gives rise to two different approaches. In the first approach, referred to as *indirect adaptive control*, the plant parameters are estimated on-line and used to calculate the controller parameters. This

1.2. ADAPTIVE CONTROL

approach has also been referred to as *explicit adaptive control*, because the design is based on an explicit plant model.

In the second approach, referred to as *direct adaptive control*, the plant model is parameterized in terms of the controller parameters that are estimated directly without intermediate calculations involving plant parameter estimates. This approach has also been referred to as *implicit adaptive control* because the design is based on the estimation of an implicit plant model.

In indirect adaptive control, the plant model $P(\theta^*)$ is parameterized with respect to some unknown parameter vector θ^*. For example, for a linear time invariant (LTI) single-input single-output (SISO) plant model, θ^* may represent the unknown coefficients of the numerator and denominator of the plant model transfer function. An on-line parameter estimator generates an estimate $\theta(t)$ of θ^* at each time t by processing the plant input u and output y. The parameter estimate $\theta(t)$ specifies an estimated plant model characterized by $\hat{P}(\theta(t))$ that for control design purposes is treated as the "true" plant model and is used to calculate the controller parameter or gain vector $\theta_c(t)$ by solving a certain algebraic equation $\theta_c(t) = F(\theta(t))$ at each time t. The form of the control law $C(\theta_c)$ and algebraic equation $\theta_c = F(\theta)$ is chosen to be the same as that of the control law $C(\theta_c^*)$ and equation $\theta_c^* = F(\theta^*)$ that could be used to meet the performance requirements for the plant model $P(\theta^*)$ if θ^* was known. It is, therefore, clear that with this approach, $C(\theta_c(t))$ is designed at each time t to satisfy the performance requirements for the estimated plant model $\hat{P}(\theta(t))$, which may be different from the unknown plant model $P(\theta^*)$. Therefore, the principal problem in indirect adaptive control is to choose the class of control laws $C(\theta_c)$ and the class of parameter estimators that generate $\theta(t)$ as well as the algebraic equation $\theta_c(t) = F(\theta(t))$ so that $C(\theta_c(t))$ meets the performance requirements for the plant model $P(\theta^*)$ with unknown θ^*. We will study this problem in great detail in Chapters 6 and 7, and consider the robustness properties of indirect adaptive control in Chapters 8 and 9. The block diagram of an indirect adaptive control scheme is shown in Figure 1.6.

In direct adaptive control, the plant model $P(\theta^*)$ is parameterized in terms of the unknown controller parameter vector θ_c^*, for which $C(\theta_c^*)$ meets the performance requirements, to obtain the plant model $P_c(\theta_c^*)$ with exactly the same input/output characteristics as $P(\theta^*)$.

The on-line parameter estimator is designed based on $P_c(\theta_c^*)$ instead of

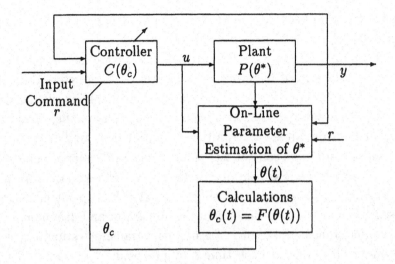

Figure 1.6 Indirect adaptive control.

$P(\theta^*)$ to provide direct estimates $\theta_c(t)$ of θ_c^* at each time t by processing the plant input u and output y. The estimate $\theta_c(t)$ is then used to update the controller parameter vector θ_c without intermediate calculations. The choice of the class of control laws $C(\theta_c)$ and parameter estimators generating $\theta_c(t)$ for which $C(\theta_c(t))$ meets the performance requirements for the plant model $P(\theta^*)$ is the fundamental problem in direct adaptive control. The properties of the plant model $P(\theta^*)$ are crucial in obtaining the parameterized plant model $P_c(\theta_c^*)$ that is convenient for on-line estimation. As a result, direct adaptive control is restricted to a certain class of plant models. As we will show in Chapter 6, a class of plant models that is suitable for direct adaptive control consists of all SISO LTI plant models that are minimum-phase, i.e., their zeros are located in Re $[s] < 0$. The block diagram of direct adaptive control is shown in Figure 1.7.

The principle behind the design of direct and indirect adaptive control shown in Figures 1.6 and 1.7 is conceptually simple. The design of $C(\theta_c)$ treats the estimates $\theta_c(t)$ (in the case of direct adaptive control) or the estimates $\theta(t)$ (in the case of indirect adaptive control) as if they were the true parameters. This design approach is called *certainty equivalence* and can be used to generate a wide class of adaptive control schemes by combining different on-line parameter estimators with different control laws.

1.2. ADAPTIVE CONTROL

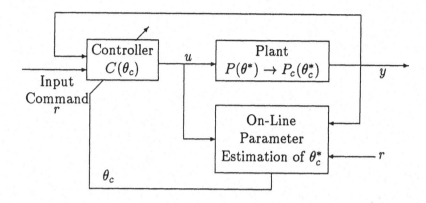

Figure 1.7 Direct adaptive control.

The idea behind the certainty equivalence approach is that as the parameter estimates $\theta_c(t)$ and $\theta(t)$ converge to the true ones θ_c^* and θ^*, respectively, the performance of the adaptive controller $C(\theta_c)$ tends to that achieved by $C(\theta_c^*)$ in the case of known parameters.

The distinction between direct and indirect adaptive control may be confusing to most readers for the following reasons: The direct adaptive control structure shown in Figure 1.7 can be made identical to that of the indirect adaptive control by including a block for calculations with an identity transformation between updated parameters and controller parameters. In general, for a given plant model the distinction between the direct and indirect approach becomes clear if we go into the details of design and analysis. For example, direct adaptive control can be shown to meet the performance requirements, which involve stability and asymptotic tracking, for a minimum-phase plant. It is still not clear how to design direct schemes for nonminimum-phase plants. The difficulty arises from the fact that, in general, a convenient (for the purpose of estimation) parameterization of the plant model in terms of the desired controller parameters is not possible for nonminimum-phase plant models.

Indirect adaptive control, on the other hand, is applicable to both minimum- and nonminimum-phase plants. In general, however, the mapping between $\theta(t)$ and $\theta_c(t)$, defined by the algebraic equation $\theta_c(t) \triangleq F(\theta(t))$, cannot be guaranteed to exist at each time t giving rise to the so-called *stabilizability* problem that is discussed in Chapter 7. As we will show in

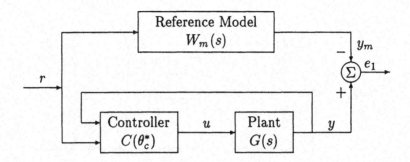

Figure 1.8 Model reference control.

Chapter 7, solutions to the stabilizability problem are possible at the expense of additional complexity.

Efforts to relax the minimum-phase assumption in direct adaptive control and resolve the stabilizability problem in indirect adaptive control led to adaptive control schemes where both the controller and plant parameters are estimated on-line, leading to combined direct/indirect schemes that are usually more complex [112].

1.2.4 Model Reference Adaptive Control

Model reference adaptive control (MRAC) is derived from the model following problem or model reference control (MRC) problem. In MRC, a good understanding of the plant and the performance requirements it has to meet allow the designer to come up with a model, referred to as the *reference model*, that describes the desired I/O properties of the closed-loop plant. The objective of MRC is to find the feedback control law that changes the structure and dynamics of the plant so that its I/O properties are exactly the same as those of the reference model. The structure of an MRC scheme for a LTI, SISO plant is shown in Figure 1.8. The transfer function $W_m(s)$ of the reference model is designed so that for a given reference input signal $r(t)$ the output $y_m(t)$ of the reference model represents the desired response the plant output $y(t)$ should follow. The feedback controller denoted by $C(\theta_c^*)$ is designed so that all signals are bounded and the closed-loop plant transfer function from r to y is equal to $W_m(s)$. This transfer function matching guarantees that for any given reference input $r(t)$, the tracking error

1.2. ADAPTIVE CONTROL

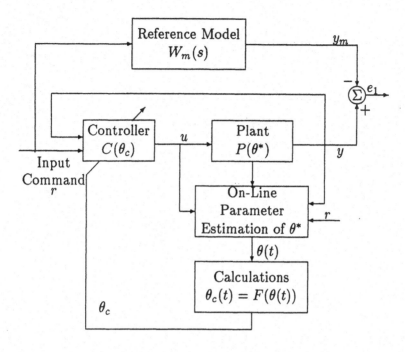

Figure 1.9 Indirect MRAC.

$e_1 \triangleq y - y_m$, which represents the deviation of the plant output from the desired trajectory y_m, converges to zero with time. The transfer function matching is achieved by canceling the zeros of the plant transfer function $G(s)$ and replacing them with those of $W_m(s)$ through the use of the feedback controller $C(\theta_c^*)$. The cancellation of the plant zeros puts a restriction on the plant to be minimum phase, i.e., have stable zeros. If any plant zero is unstable, its cancellation may easily lead to unbounded signals.

The design of $C(\theta_c^*)$ requires the knowledge of the coefficients of the plant transfer function $G(s)$. If θ^* is a vector containing all the coefficients of $G(s) = G(s, \theta^*)$, then the parameter vector θ_c^* may be computed by solving an algebraic equation of the form

$$\theta_c^* = F(\theta^*) \tag{1.2.3}$$

It is, therefore, clear that for the MRC objective to be achieved the plant model has to be minimum phase and its parameter vector θ^* has to be known exactly.

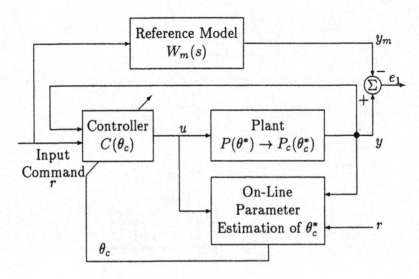

Figure 1.10 Direct MRAC.

When θ^* is unknown the MRC scheme of Figure 1.8 cannot be implemented because θ_c^* cannot be calculated using (1.2.3) and is, therefore, unknown. One way of dealing with the unknown parameter case is to use the certainty equivalence approach to replace the unknown θ_c^* in the control law with its estimate $\theta_c(t)$ obtained using the direct or the indirect approach. The resulting control schemes are known as MRAC and can be classified as indirect MRAC shown in Figure 1.9 and direct MRAC shown in Figure 1.10.

Different choices of on-line parameter estimators lead to further classifications of MRAC. These classifications and the stability properties of both direct and indirect MRAC will be studied in detail in Chapter 6.

Other approaches similar to the certainty equivalence approach may be used to design direct and indirect MRAC schemes. The structure of these schemes is a modification of those in Figures 1.9 and 1.10 and will be studied in Chapter 6.

1.2.5 Adaptive Pole Placement Control

Adaptive pole placement control (APPC) is derived from the pole placement control (PPC) and regulation problems used in the case of LTI plants with known parameters.

1.2. ADAPTIVE CONTROL

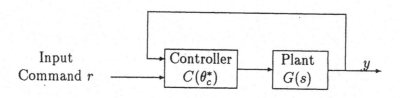

Figure 1.11 Pole placement control.

In PPC, the performance requirements are translated into desired locations of the poles of the closed-loop plant. A feedback control law is then developed that places the poles of the closed-loop plant at the desired locations. A typical structure of a PPC scheme for a LTI, SISO plant is shown in Figure 1.11.

The structure of the controller $C(\theta_c^*)$ and the parameter vector θ_c^* are chosen so that the poles of the closed-loop plant transfer function from r to y are equal to the desired ones. The vector θ_c^* is usually calculated using an algebraic equation of the form

$$\theta_c^* = F(\theta^*) \tag{1.2.4}$$

where θ^* is a vector with the coefficients of the plant transfer function $G(s)$.

If θ^* is known, then θ_c^* is calculated from (1.2.4) and used in the control law. When θ^* is unknown, θ_c^* is also unknown, and the PPC scheme of Figure 1.11 cannot be implemented. As in the case of MRC, we can deal with the unknown parameter case by using the certainty equivalence approach to replace the unknown vector θ_c^* with its estimate $\theta_c(t)$. The resulting scheme is referred to as *adaptive pole placement control* (APPC). If $\theta_c(t)$ is updated directly using an on-line parameter estimator, the scheme is referred to as *direct APPC*. If $\theta_c(t)$ is calculated using the equation

$$\theta_c(t) = F(\theta(t)) \tag{1.2.5}$$

where $\theta(t)$ is the estimate of θ^* generated by an on-line estimator, the scheme is referred to as *indirect APPC*. The structure of direct and indirect APPC is the same as that shown in Figures 1.6 and 1.7 respectively for the general case.

The design of APPC schemes is very flexible with respect to the choice of the form of the controller $C(\theta_c)$ and of the on-line parameter estimator.

For example, the control law may be based on the linear quadratic design technique, frequency domain design techniques, or any other PPC method used in the known parameter case. Various combinations of on-line estimators and control laws lead to a wide class of APPC schemes that are studied in detail in Chapter 7.

APPC schemes are often referred to as *self-tuning regulators* in the literature of adaptive control and are distinguished from MRAC. The distinction between APPC and MRAC is more historical than conceptual because as we will show in Chapter 7, MRAC can be considered as a special class of APPC. MRAC was first developed for continuous-time plants for model following, whereas APPC was initially developed for discrete-time plants in a stochastic environment using minimization techniques.

1.2.6 Design of On-Line Parameter Estimators

As we mentioned in the previous sections, an adaptive controller may be considered as a combination of an on-line parameter estimator with a control law that is derived from the known parameter case. The way this combination occurs and the type of estimator and control law used gives rise to a wide class of different adaptive controllers with different properties. In the literature of adaptive control the on-line parameter estimator has often been referred to as the *adaptive law*, *update law*, or *adjustment mechanism*. In this book we will often refer to it as the adaptive law. The design of the adaptive law is crucial for the stability properties of the adaptive controller. As we will see in this book the adaptive law introduces a multiplicative nonlinearity that makes the closed-loop plant nonlinear and often time varying. Because of this, the analysis and understanding of the stability and robustness of adaptive control schemes are more challenging.

Some of the basic methods used to design adaptive laws are

(i) Sensitivity methods
(ii) Positivity and Lyapunov design
(iii) Gradient method and least-squares methods based on estimation error cost criteria

The last three methods are used in Chapters 4 and 8 to design a wide class of adaptive laws. The sensitivity method is one of the oldest methods used in the design of adaptive laws and will be briefly explained in this section together with the other three methods for the sake of completeness. It will not be used elsewhere in this book for the simple reason that in theory the adaptive laws based on the last three methods can be shown to have better stability properties than those based on the sensitivity method.

(i) Sensitivity methods

This method became very popular in the 1960s [34, 104], and it is still used in many industrial applications for controlling plants with uncertainties. In adaptive control, the sensitivity method is used to design the adaptive law so that the estimated parameters are adjusted in a direction that minimizes a certain performance function. The adaptive law is driven by the partial derivative of the performance function with respect to the estimated parameters multiplied by an error signal that characterizes the mismatch between the actual and desired behavior. This derivative is called *sensitivity function* and if it can be generated on-line then the adaptive law is implementable. In most formulations of adaptive control, the sensitivity function cannot be generated on-line, and this constitutes one of the main drawbacks of the method. The use of approximate sensitivity functions that are implementable leads to adaptive control schemes whose stability properties are either weak or cannot be established.

As an example let us consider the design of an adaptive law for updating the controller parameter vector θ_c of the direct MRAC scheme of Figure 1.10.

The tracking error e_1 represents the deviation of the plant output y from that of the reference model, i.e., $e_1 \triangleq y - y_m$. Because $\theta_c = \theta_c^*$ implies that $e_1 = 0$ at steady state, a nonzero value of e_1 may be taken to imply that $\theta_c \neq \theta_c^*$. Because y depends on θ_c, i.e., $y = y(\theta_c)$ we have $e_1 = e_1(\theta_c)$ and, therefore, one way of reducing e_1 to zero is to adjust θ_c in a direction that minimizes a certain cost function of e_1. A simple cost function for e_1 is the quadratic function

$$J(\theta_c) = \frac{e_1^2(\theta_c)}{2} \qquad (1.2.6)$$

A simple method for adjusting θ_c to minimize $J(\theta_c)$ is the method of steepest descent or gradient method (see Appendix B) that gives us the adaptive law

$$\dot{\theta}_c = -\gamma \nabla J(\theta_c) = -\gamma e_1 \nabla e_1(\theta_c) \qquad (1.2.7)$$

where
$$\nabla e_1(\theta_c) \triangleq \left[\frac{\partial e_1}{\partial \theta_{c1}}, \frac{\partial e_1}{\partial \theta_{c2}}, ..., \frac{\partial e_1}{\partial \theta_{cn}}\right]^T \quad (1.2.8)$$
is the gradient of e_1 with respect to
$$\theta_c = [\theta_{c1}, \theta_{c2}, ..., \theta_{cn}]^T$$
Because
$$\nabla e_1(\theta_c) = \nabla y(\theta_c)$$
we have
$$\dot{\theta}_c = -\gamma e_1 \nabla y(\theta_c) \quad (1.2.9)$$
where $\gamma > 0$ is an arbitrary design constant referred to as the *adaptive gain* and $\frac{\partial y}{\partial \theta_{ci}}, i = 1, 2, ..., n$ are the sensitivity functions of y with respect to the elements of the controller parameter vector θ_c. The sensitivity functions $\frac{\partial y}{\partial \theta_{ci}}$ represent the sensitivity of the plant output to changes in the controller parameter θ_c.

In (1.2.7) the parameter vector θ_c is adjusted in the direction of steepest descent that decreases $J(\theta_c) = \frac{e_1^2(\theta_c)}{2}$. If $J(\theta_c)$ is a convex function, then it has a global minimum that satisfies $\nabla y(\theta_c) = 0$, i.e., at the minimum $\dot{\theta}_c = 0$ and adaptation stops.

The implementation of (1.2.9) requires the on-line generation of the sensitivity functions ∇y that usually depend on the unknown plant parameters and are, therefore, unavailable. In these cases, approximate values of the sensitivity functions are used instead of the actual ones. One type of approximation is to use some a priori knowledge about the plant parameters to compute the sensitivity functions.

A popular method for computing the approximate sensitivity functions is the so-called MIT rule. With this rule the unknown parameters that are needed to generate the sensitivity functions are replaced by their on-line estimates. Unfortunately, with the use of approximate sensitivity functions, it is not possible, in general, to prove global closed-loop stability and convergence of the tracking error to zero. In simulations, however, it was observed that the MIT rule and other approximation techniques performed well when the adaptive gain γ and the magnitude of the reference input signal are small. Averaging techniques are used in [135] to confirm these observations and establish local stability for a certain class of reference input signals. Globally,

1.2. ADAPTIVE CONTROL

however, the schemes based on the MIT rule and other approximations may go unstable. Examples of instability are presented in [93, 187, 202].

We illustrate the use of the MIT rule for the design of an MRAC scheme for the plant

$$\ddot{y} = -a_1\dot{y} - a_2 y + u \tag{1.2.10}$$

where a_1 and a_2 are the unknown plant parameters, and $\dot{y}$ and y are available for measurement.

The reference model to be matched by the closed loop plant is given by

$$\ddot{y}_m = -2\dot{y}_m - y_m + r \tag{1.2.11}$$

The control law

$$u = \theta_1^* \dot{y} + \theta_2^* y + r \tag{1.2.12}$$

where

$$\theta_1^* = a_1 - 2, \quad \theta_2^* = a_2 - 1 \tag{1.2.13}$$

will achieve perfect model following. The equation (1.2.13) is referred to as the matching equation. Because a_1 and a_2 are unknown, the desired values of the controller parameters θ_1^* and θ_2^* cannot be calculated from (1.2.13). Therefore, instead of (1.2.12) we use the control law

$$u = \theta_1 \dot{y} + \theta_2 y + r \tag{1.2.14}$$

where θ_1 and θ_2 are adjusted using the MIT rule as

$$\dot{\theta}_1 = -\gamma e_1 \frac{\partial y}{\partial \theta_1}, \quad \dot{\theta}_2 = -\gamma e_1 \frac{\partial y}{\partial \theta_2} \tag{1.2.15}$$

where $e_1 = y - y_m$. To implement (1.2.15), we need to generate the sensitivity functions $\frac{\partial y}{\partial \theta_1}, \frac{\partial y}{\partial \theta_2}$ on-line.

Using (1.2.10) and (1.2.14) we obtain

$$\frac{\partial \ddot{y}}{\partial \theta_1} = -a_1 \frac{\partial \dot{y}}{\partial \theta_1} - a_2 \frac{\partial y}{\partial \theta_1} + \dot{y} + \theta_1 \frac{\partial \dot{y}}{\partial \theta_1} + \theta_2 \frac{\partial y}{\partial \theta_1} \tag{1.2.16}$$

$$\frac{\partial \ddot{y}}{\partial \theta_2} = -a_1 \frac{\partial \dot{y}}{\partial \theta_2} - a_2 \frac{\partial y}{\partial \theta_2} + y + \theta_1 \frac{\partial \dot{y}}{\partial \theta_2} + \theta_2 \frac{\partial y}{\partial \theta_2} \tag{1.2.17}$$

If we now assume that the rate of adaptation is slow, i.e., $\dot{\theta}_1$ and $\dot{\theta}_2$ are small, and the changes of $\ddot{y}$ and $\dot{y}$ with respect to θ_1 and θ_2 are also small, we can interchange the order of differentiation to obtain

$$\frac{d^2}{dt^2}\frac{\partial y}{\partial \theta_1} = (\theta_1 - a_1)\frac{d}{dt}\frac{\partial y}{\partial \theta_1} + (\theta_2 - a_2)\frac{\partial y}{\partial \theta_1} + \dot{y} \qquad (1.2.18)$$

$$\frac{d^2}{dt^2}\frac{\partial y}{\partial \theta_2} = (\theta_1 - a_1)\frac{d}{dt}\frac{\partial y}{\partial \theta_2} + (\theta_2 - a_2)\frac{\partial y}{\partial \theta_2} + y \qquad (1.2.19)$$

which we may rewrite as

$$\frac{\partial y}{\partial \theta_1} = \frac{1}{p^2 - (\theta_1 - a_1)p - (\theta_2 - a_2)}\dot{y} \qquad (1.2.20)$$

$$\frac{\partial y}{\partial \theta_2} = \frac{1}{p^2 - (\theta_1 - a_1)p - (\theta_2 - a_2)}y \qquad (1.2.21)$$

where $p(\cdot) \triangleq \frac{d}{dt}(\cdot)$ is the differential operator.

Because a_1 and a_2 are unknown, the above sensitivity functions cannot be used. Using the MIT rule, we replace a_1 and a_2 with their estimates $\hat{a}_1$ and $\hat{a}_2$ in the matching equation (1.2.13), i.e., we relate the estimates $\hat{a}_1$ and $\hat{a}_2$ with θ_1 and θ_2 using

$$\hat{a}_1 = \theta_1 + 2, \ \hat{a}_2 = \theta_2 + 1 \qquad (1.2.22)$$

and obtain the approximate sensitivity functions

$$\frac{\partial y}{\partial \theta_1} \simeq \frac{1}{p^2 + 2p + 1}\dot{y}, \ \frac{\partial y}{\partial \theta_2} \simeq \frac{1}{p^2 + 2p + 1}y \qquad (1.2.23)$$

The equations given by (1.2.23) are known as the *sensitivity filters or models*, and can be easily implemented to generate the approximate sensitivity functions for the adaptive law (1.2.15).

As shown in [93, 135], the MRAC scheme based on the MIT rule is locally stable provided the adaptive gain γ is small, the reference input signal has a small amplitude and sufficient number of frequencies, and the initial conditions $\theta_1(0)$ and $\theta_2(0)$ are close to θ_1^* and θ_2^* respectively.

For larger γ and $\theta_1(0)$ and $\theta_2(0)$ away from θ_1^* and θ_2^*, the MIT rule may lead to instability and unbounded signal response.

1.2. ADAPTIVE CONTROL

The lack of stability of MIT rule based adaptive control schemes prompted several researchers to look for different methods of designing adaptive laws. These methods include the positivity and Lyapunov design approach, and the gradient and least-squares methods that are based on the minimization of certain estimation error criteria. These methods are studied in detail in Chapters 4 and 8, and are briefly described below.

(ii) Positivity and Lyapunov design

This method of developing adaptive laws is based on the direct method of Lyapunov and its relationship with *positive real* functions. In this approach, the problem of designing an adaptive law is formulated as a stability problem where the differential equation of the adaptive law is chosen so that certain stability conditions based on Lyapunov theory are satisfied. The adaptive law developed is very similar to that based on the sensitivity method. The only difference is that the sensitivity functions in the approach (i) are replaced with ones that can be generated on-line. In addition, the Lyapunov-based adaptive control schemes have none of the drawbacks of the MIT rule-based schemes.

The design of adaptive laws using Lyapunov's direct method was suggested by Grayson [76], Parks [187], and Shackcloth and Butchart [202] in the early 1960s. The method was subsequently advanced and generalized to a wider class of plants by Phillipson [188], Monopoli [149], Narendra [172], and others.

A significant part of Chapters 4 and 8 will be devoted to developing adaptive laws using the Lyapunov design approach.

(iii) Gradient and least-squares methods based on estimation error cost criteria

The main drawback of the sensitivity methods used in the 1960s is that the minimization of the performance cost function led to sensitivity functions that are not implementable. One way to avoid this drawback is to choose a cost function criterion that leads to sensitivity functions that are available for measurement. A class of such cost criteria is based on an error referred to as the *estimation error* that provides a measure of the discrepancy between the estimated and actual parameters. The relationship of the estimation error with the estimated parameters is chosen so that the cost function is convex, and its gradient with respect to the estimated parameters is implementable.

Several different cost criteria may be used, and methods, such as the gradient and least-squares, may be adopted to generate the appropriate sensitivity functions.

As an example, let us design the adaptive law for the direct MRAC law (1.2.14) for the plant (1.2.10).

We first rewrite the plant equation in terms of the desired controller parameters given by (1.2.13), i.e., we substitute for $a_1 = 2 + \theta_1^*$, $a_2 = 1 + \theta_2^*$ in (1.2.10) to obtain

$$\ddot{y} = -2\dot{y} - y - \theta_1^* \dot{y} - \theta_2^* y + u \qquad (1.2.24)$$

which may be rewritten as

$$y = \theta_1^* \dot{y}_f + \theta_2^* y_f + u_f \qquad (1.2.25)$$

where

$$\dot{y}_f = -\frac{1}{s^2 + 2s + 1}\dot{y}, \quad y_f = -\frac{1}{s^2 + 2s + 1}y, \quad u_f = \frac{1}{s^2 + 2s + 1}u \qquad (1.2.26)$$

are signals that can be generated by filtering.

If we now replace θ_1^* and θ_2^* with their estimates θ_1 and θ_2 in equation (1.2.25), we will obtain,

$$\hat{y} = \theta_1 \dot{y}_f + \theta_2 y_f + u_f \qquad (1.2.27)$$

where $\hat{y}$ is the estimate of y based on the estimate θ_1 and θ_2 of θ_1^* and θ_2^*. The error

$$\varepsilon_1 \stackrel{\Delta}{=} y - \hat{y} = y - \theta_1 \dot{y}_f - \theta_2 y_f - u_f \qquad (1.2.28)$$

is, therefore, a measure of the discrepancy between θ_1, θ_2 and θ_1^*, θ_2^*, respectively. We refer to it as the *estimation error*. The estimates θ_1 and θ_2 can now be adjusted in a direction that minimizes a certain cost criterion that involves ε_1. A simple such criterion is

$$J(\theta_1, \theta_2) = \frac{\varepsilon_1^2}{2} = \frac{1}{2}(y - \theta_1 \dot{y}_f - \theta_2 y_f - u_f)^2 \qquad (1.2.29)$$

which is to be minimized with respect to θ_1, θ_2. It is clear that $J(\theta_1, \theta_2)$ is a convex function of θ_1, θ_2 and, therefore, the minimum is given by $\nabla J = 0$.

If we now use the gradient method to minimize $J(\theta_1, \theta_2)$, we obtain the adaptive laws

$$\dot{\theta}_1 = -\gamma_1 \frac{\partial J}{\partial \theta_1} = \gamma_1 \varepsilon_1 \dot{y}_f, \quad \dot{\theta}_2 = -\gamma_2 \frac{\partial J}{\partial \theta_2} = \gamma_2 \varepsilon_1 y_f \qquad (1.2.30)$$

where $\gamma_1, \gamma_2 > 0$ are the adaptive gains and $\varepsilon_1, \dot{y}_f, y_f$ are all implementable signals.

Instead of (1.2.29), one may use a different cost criterion for ε_1 and a different minimization method leading to a wide class of adaptive laws. In Chapters 4 to 9 we will examine the stability properties of a wide class of adaptive control schemes that are based on the use of estimation error criteria, and gradient and least-squares type of optimization techniques.

1.3 A Brief History

Research in adaptive control has a long history of intense activities that involved debates about the precise definition of adaptive control, examples of instabilities, stability and robustness proofs, and applications.

Starting in the early 1950s, the design of autopilots for high-performance aircraft motivated an intense research activity in adaptive control. High-performance aircraft undergo drastic changes in their dynamics when they fly from one operating point to another that cannot be handled by constant-gain feedback control. A sophisticated controller, such as an adaptive controller, that could learn and accommodate changes in the aircraft dynamics was needed. Model reference adaptive control was suggested by Whitaker et al. in [184, 235] to solve the autopilot control problem. The sensitivity method and the MIT rule was used to design the adaptive laws of the various proposed adaptive control schemes. An adaptive pole placement scheme based on the optimal linear quadratic problem was suggested by Kalman in [96].

The work on adaptive flight control was characterized by "a lot of enthusiasm, bad hardware and non-existing theory" [11]. The lack of stability proofs and the lack of understanding of the properties of the proposed adaptive control schemes coupled with a disaster in a flight test [219] caused the interest in adaptive control to diminish.

The 1960s became the most important period for the development of control theory and adaptive control in particular. State space techniques and stability theory based on Lyapunov were introduced. Developments in dynamic programming [19, 20], dual control [53] and stochastic control in general, and in system identification and parameter estimation [13, 229] played a crucial role in the reformulation and redesign of adaptive control. By 1966 Parks and others found a way of redesigning the MIT rule-based adaptive laws used in the MRAC schemes of the 1950s by applying the Lyapunov design approach. Their work, even though applicable to a special class of LTI plants, set the stage for further rigorous stability proofs in adaptive control for more general classes of plant models.

The advances in stability theory and the progress in control theory in the 1960s improved the understanding of adaptive control and contributed to a strong renewed interest in the field in the 1970s. On the other hand, the simultaneous development and progress in computers and electronics that made the implementation of complex controllers, such as the adaptive ones, feasible contributed to an increased interest in applications of adaptive control. The 1970s witnessed several breakthrough results in the design of adaptive control. MRAC schemes using the Lyapunov design approach were designed and analyzed in [48, 153, 174]. The concepts of positivity and hyperstability were used in [123] to develop a wide class of MRAC schemes with well-established stability properties. At the same time parallel efforts for discrete-time plants in a deterministic and stochastic environment produced several classes of adaptive control schemes with rigorous stability proofs [72, 73]. The excitement of the 1970s and the development of a wide class of adaptive control schemes with well established stability properties was accompanied by several successful applications [80, 176, 230].

The successes of the 1970s, however, were soon followed by controversies over the practicality of adaptive control. As early as 1979 it was pointed out that the adaptive schemes of the 1970s could easily go unstable in the presence of small disturbances [48]. The nonrobust behavior of adaptive control became very controversial in the early 1980s when more examples of instabilities were published demonstrating lack of robustness in the presence of unmodeled dynamics or bounded disturbances [85, 197]. This stimulated many researchers, whose objective was to understand the mechanisms of instabilities and find ways to counteract them. By the mid 1980s, several

1.3. A BRIEF HISTORY

new redesigns and modifications were proposed and analyzed, leading to a body of work known as *robust adaptive control*. An adaptive controller is defined to be robust if it guarantees signal boundedness in the presence of "reasonable" classes of unmodeled dynamics and bounded disturbances as well as performance error bounds that are of the order of the modeling error.

The work on robust adaptive control continued throughout the 1980s and involved the understanding of the various robustness modifications and their unification under a more general framework [48, 87, 84].

The solution of the robustness problem in adaptive control led to the solution of the long-standing problem of controlling a linear plant whose parameters are unknown and changing with time. By the end of the 1980s several breakthrough results were published in the area of adaptive control for linear time-varying plants [226].

The focus of adaptive control research in the late 1980s to early 1990s was on performance properties and on extending the results of the 1980s to certain classes of nonlinear plants with unknown parameters. These efforts led to new classes of adaptive schemes, motivated from nonlinear system theory [98, 99] as well as to adaptive control schemes with improved transient and steady-state performance [39, 211].

Adaptive control has a rich literature full with different techniques for design, analysis, performance, and applications. Several survey papers [56, 183], and books and monographs [3, 15, 23, 29, 48, 55, 61, 73, 77, 80, 85, 94, 105, 123, 144, 169, 172, 201, 226, 229, 230] have already been published. Despite the vast literature on the subject, there is still a general feeling that adaptive control is a collection of unrelated technical tools and tricks. The purpose of this book is to unify the various approaches and explain them in a systematic and tutorial manner.

Chapter 2

Models for Dynamic Systems

2.1 Introduction

In this chapter, we give a brief account of various models and parameterizations of LTI systems. Emphasis is on those ideas that are useful in studying the parameter identification and adaptive control problems considered in subsequent chapters.

We begin by giving a summary of some canonical state space models for LTI systems and of their characteristics. Next we study I/O descriptions for the same class of systems by using transfer functions and differential operators. We express transfer functions as ratios of two polynomials and present some of the basic properties of polynomials that are useful for control design and system modeling.

We then examine several nonminimal state-space representations of LTI systems that we express in a form in which parameters, such as coefficients of polynomials in the transfer function description, are separated from signals formed by filtering the system inputs and outputs. These parametric models and their properties are crucial in parameter identification and adaptive control problems to be studied in subsequent chapters.

The intention of this chapter is not to give a complete picture of all aspects of LTI system modeling and representation, but rather to present a summary of those ideas that are used in subsequent chapters. For further

2.2 State-Space Models

2.2.1 General Description

discussion on the topic of modeling and properties of linear systems, we refer the reader to several standard books on the subject starting with the elementary ones [25, 41, 44, 57, 121, 180] and moving to the more advanced ones [30, 42, 95, 198, 237, 238].

2.2 State-Space Models

2.2.1 General Description

Many systems are described by a set of differential equations of the form

$$\begin{aligned} \dot{x}(t) &= f(x(t), u(t), t), \quad x(t_0) = x_0 \\ y(t) &= g(x(t), u(t), t) \end{aligned} \tag{2.2.1}$$

where

- t is the time variable
- $x(t)$ is an n-dimensional vector with real elements that denotes the state of the system
- $u(t)$ is an r-dimensional vector with real elements that denotes the input variable or control input of the system
- $y(t)$ is an l-dimensional vector with real elements that denotes the output variables that can be measured
- f, g are real vector valued functions
- n is the dimension of the state x called the order of the system
- $x(t_0)$ denotes the value of $x(t)$ at the initial time $t = t_0 \geq 0$

When f, g are linear functions of x, u, (2.2.1) takes the form

$$\begin{aligned} \dot{x} &= A(t)x + B(t)u, \quad x(t_0) = x_0 \\ y &= C^\top(t)x + D(t)u \end{aligned} \tag{2.2.2}$$

where $A(t) \in \mathcal{R}^{n \times n}, B(t) \in \mathcal{R}^{n \times r}, C(t) \in \mathcal{R}^{n \times l}$, and $D(t) \in \mathcal{R}^{l \times r}$ are matrices with time-varying elements. If in addition to being linear, f, g do not depend on time t, we have

$$\begin{aligned} \dot{x} &= Ax + Bu, \quad x(t_0) = x_0 \\ y &= C^\top x + Du \end{aligned} \tag{2.2.3}$$

where $A, B, C,$ and D are matrices of the same dimension as in (2.2.2) but with constant elements.

We refer to (2.2.2) as the finite-dimensional linear time-varying (LTV) system and to (2.2.3) as the finite dimensional LTI system.

The solution $x(t), y(t)$ of (2.2.2) is given by

$$\begin{aligned} x(t) &= \Phi(t, t_0)x(t_0) + \int_{t_0}^{t} \Phi(t, \tau)B(\tau)u(\tau)d\tau \\ y(t) &= C^{\top}(t)x(t) + D(t)u(t) \end{aligned} \quad (2.2.4)$$

where $\Phi(t, t_0)$ is the state transition matrix defined as a matrix that satisfies the linear homogeneous matrix equation

$$\frac{\partial \Phi(t, t_0)}{\partial t} = A(t)\Phi(t, t_0), \quad \Phi(t_0, t_0) = I$$

For the LTI system (2.2.3), $\Phi(t, t_0)$ depends only on the difference $t - t_0$, i.e.,

$$\Phi(t, t_0) = \Phi(t - t_0) = e^{A(t-t_0)}$$

and the solution $x(t), y(t)$ of (2.2.3) is given by

$$\begin{aligned} x(t) &= e^{A(t-t_0)}x_0 + \int_{t_0}^{t} e^{A(t-\tau)}Bu(\tau)d\tau \\ y(t) &= C^{\top}x(t) + Du(t) \end{aligned} \quad (2.2.5)$$

where e^{At} can be identified to be

$$e^{At} = \mathcal{L}^{-1}[(sI - A)^{-1}]$$

where $\mathcal{L}^{-1}$ denotes the inverse Laplace transform and s is the Laplace variable.

Usually the matrix D in (2.2.2), (2.2.3) is zero, because in most physical systems there is no direct path of nonzero gain between the inputs and outputs.

In this book, we are concerned mainly with LTI, SISO systems with $D = 0$. In some chapters and sections, we will also briefly discuss systems of the form (2.2.2) and (2.2.3).

2.2. STATE-SPACE MODELS

2.2.2 Canonical State-Space Forms

Let us consider the SISO, LTI system

$$\dot{x} = Ax + Bu, \quad x(t_0) = x_0$$
$$y = C^T x \qquad (2.2.6)$$

where $x \in \mathcal{R}^n$. The *controllability matrix* P_c of (2.2.6) is defined by

$$P_c \triangleq [B, AB, \ldots, A^{n-1}B]$$

A necessary and sufficient condition for the system (2.2.6) to be *completely controllable* is that P_c is nonsingular. If (2.2.6) is completely controllable, the linear transformation

$$x_c = P_c^{-1} x \qquad (2.2.7)$$

transforms (2.2.6) into the *controllability canonical form*

$$\dot{x}_c = \begin{bmatrix} 0 & 0 & \cdots & 0 & -a_0 \\ 1 & 0 & \cdots & 0 & -a_1 \\ 0 & 1 & \cdots & 0 & -a_2 \\ \vdots & & \ddots & & \vdots \\ 0 & 0 & \cdots & 1 & -a_{n-1} \end{bmatrix} x_c + \begin{bmatrix} 1 \\ 0 \\ 0 \\ \vdots \\ 0 \end{bmatrix} u \qquad (2.2.8)$$

$$y = C_c^T x_c$$

where the a_i's are the coefficients of the characteristic equation of A, i.e., $\det(sI - A) = s^n + a_{n-1} s^{n-1} + \cdots + a_0$ and $C_c^T = C^T P_c$.

If instead of (2.2.7), we use the transformation

$$x_c = M^{-1} P_c^{-1} x \qquad (2.2.9)$$

where

$$M = \begin{bmatrix} 1 & a_{n-1} & \cdots & a_2 & a_1 \\ 0 & 1 & \cdots & a_3 & a_2 \\ \vdots & \vdots & \ddots & \vdots & \vdots \\ 0 & 0 & \cdots & 1 & a_{n-1} \\ 0 & 0 & \cdots & 0 & 1 \end{bmatrix}$$

we obtain the following *controller canonical form*

$$\dot{x}_c = \begin{bmatrix} -a_{n-1} & -a_{n-2} & \cdots & -a_1 & -a_0 \\ 1 & 0 & \cdots & 0 & 0 \\ 0 & 1 & \cdots & 0 & 0 \\ \vdots & \vdots & \ddots & & \vdots \\ 0 & 0 & \cdots & 1 & 0 \end{bmatrix} x_c + \begin{bmatrix} 1 \\ 0 \\ 0 \\ \vdots \\ 0 \end{bmatrix} u \quad (2.2.10)$$

$$y = C_0^T x_c$$

where $C_0^T = C^T P_c M$. By rearranging the elements of the state vector x_c, (2.2.10) may be written in the following form that often appears in books on linear system theory

$$\dot{x}_c = \begin{bmatrix} 0 & 1 & 0 & \cdots & 0 \\ 0 & 0 & 1 & \cdots & 0 \\ \vdots & \vdots & & \ddots & \\ 0 & 0 & & \cdots & 1 \\ -a_0 & -a_1 & \cdots & -a_{n-2} & -a_{n-1} \end{bmatrix} x_c + \begin{bmatrix} 0 \\ 0 \\ \vdots \\ 0 \\ 1 \end{bmatrix} u \quad (2.2.11)$$

$$y = C_1^T x_c$$

where C_1 is defined appropriately.

The *observability* matrix P_o of (2.2.6) is defined by

$$P_o \triangleq \begin{bmatrix} C^T \\ C^T A \\ \vdots \\ C^T A^{n-1} \end{bmatrix} \quad (2.2.12)$$

A necessary and sufficient condition for the system (2.2.6) to be *completely observable* is that P_o is nonsingular. By following the dual of the arguments presented earlier for the *controllability* and *controller canonical forms*, we arrive at *observability and observer forms* provided P_o is nonsingular [95], i.e., the *observability canonical form* of (2.2.6) obtained by using the trans-

2.2. STATE-SPACE MODELS

formation $x_o = P_o x$ is

$$\dot{x}_o = \begin{bmatrix} 0 & 1 & 0 & \cdots & 0 \\ 0 & 0 & 1 & \cdots & 0 \\ \vdots & \vdots & & \ddots & \vdots \\ 0 & 0 & & \cdots & 1 \\ -a_0 & -a_1 & \cdots & -a_{n-2} & -a_{n-1} \end{bmatrix} x_o + B_o u \quad (2.2.13)$$

$$y = [1, 0, \ldots, 0] x_o$$

and the *observer canonical form* is

$$\dot{x}_o = \begin{bmatrix} -a_{n-1} & 1 & 0 & \cdots & 0 \\ -a_{n-2} & 0 & 1 & \cdots & 0 \\ \vdots & \vdots & & \ddots & \\ -a_1 & 0 & 0 & \cdots & 1 \\ -a_0 & 0 & 0 & \cdots & 0 \end{bmatrix} x_o + B_1 u \quad (2.2.14)$$

$$y = [1, 0, \ldots, 0] x_o$$

where B_o, B_1 may be different.

If the rank of the controllability matrix P_c for the nth-order system (2.2.6) is less than n, then (2.2.6) is said to be *uncontrollable*. Similarly, if the rank of the observability matrix P_o is less than n, then (2.2.6) is *unobservable*.

The system represented by (2.2.8) or (2.2.10) or (2.2.11) is completely controllable but not necessarily observable. Similarly, the system represented by (2.2.13) or (2.2.14) is completely observable but not necessarily controllable.

If the nth-order system (2.2.6) is either unobservable or uncontrollable then its I/O properties for zero initial state, i.e., $x_0 = 0$ are completely characterized by a lower order completely controllable and observable system

$$\begin{aligned} \dot{x}_{co} &= A_{co} x_{co} + B_{co} u, \quad x_{co}(t_0) = 0 \\ y &= C_{co}^\mathsf{T} x_{co} \end{aligned} \quad (2.2.15)$$

where $x_{co} \in \mathcal{R}^{n_r}$ and $n_r < n$. It turns out that no further reduction in the order of (2.2.15) is possible without affecting the I/O properties for all inputs.

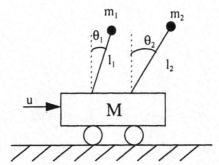

Figure 2.1 Cart with two inverted pendulums

For this reason (2.2.15) is referred to as the *minimal state-space representation* of the system to be distinguished from the *nonminimal state-space representation* that corresponds to either an uncontrollable or unobservable system.

A minimal state space model does not describe the uncontrollable or unobservable parts of the system. These parts may lead to some unbounded states in the nonminimal state-space representation of the system if any initial condition associated with these parts is nonzero. If, however, the uncontrollable or unobservable parts are asymptotically stable [95], they will decay to zero exponentially fast, and their effect may be ignored in most applications. A system whose uncontrollable parts are asymptotically stable is referred to as *stabilizable*, and the system whose unobservable parts are asymptotically stable is referred to as *detectable* [95].

Example 2.2.1 Let us consider the cart with the two inverted pendulums shown in Figure 2.1, where M is the mass of the cart, m_1 and m_2 are the masses of the bobs, and l_1 and l_2 are the lengths of the pendulums, respectively. Using Newton's law and assuming small angular deviations of $|\theta_1|, |\theta_2|$, the equations of motions are given by

$$M\dot{v} = -m_1 g \theta_1 - m_2 g \theta_2 + u$$
$$m_1(\dot{v} + l_1 \ddot{\theta}_1) = m_1 g \theta_1$$
$$m_2(\dot{v} + l_2 \ddot{\theta}_2) = m_2 g \theta_2$$

where v is the velocity of the cart, u is an external force, and g is the acceleration due to gravity. To simplify the algebra, let us assume that $m_1 = m_2 = 1kg$ and $M = 10m_1$. If we now let $x_1 = \theta_1, x_2 = \dot{\theta}_1, x_3 = \theta_1 - \theta_2, x_4 = \dot{\theta}_1 - \dot{\theta}_2$ be the state variables, we obtain the following state-space representation for the system:

$$\dot{x} = Ax + Bu$$

2.2. STATE-SPACE MODELS

where $x = [x_1, x_2, x_3, x_4]^T$

$$A = \begin{bmatrix} 0 & 1 & 0 & 0 \\ 1.2\alpha_1 & 0 & -0.1\alpha_1 & 0 \\ 0 & 0 & 0 & 1 \\ 1.2(\alpha_1 - \alpha_2) & 0 & \alpha_2 - 0.1(\alpha_1 - \alpha_2) & 0 \end{bmatrix}, \quad B = \begin{bmatrix} 0 \\ \beta_1 \\ 0 \\ \beta_1 - \beta_2 \end{bmatrix}$$

and $\alpha_1 = \frac{g}{l_1}, \alpha_2 = \frac{g}{l_2}, \beta_1 = -\frac{0.1}{l_1}$, and $\beta_2 = -\frac{0.1}{l_2}$.
The controllability matrix of the system is given by

$$P_c = [B, AB, A^2B, A^3B]$$

We can verify that

$$\det P_c = \frac{(0.011)^2 g^2 (l_1 - l_2)^2}{l_1^4 l_2^4}$$

which implies that the system is controllable if and only if $l_1 \neq l_2$.

Let us now assume that θ_1 is the only variable that we measure, i.e., the measured output of the system is

$$y = C^T x$$

where $C = [1, 0, 0, 0]^T$. The observability matrix of the system based on this output is given by

$$P_o = \begin{bmatrix} C^T \\ C^T A \\ C^T A^2 \\ C^T A^3 \end{bmatrix}$$

By performing the calculations, we verify that

$$\det P_o = 0.01 \frac{g^2}{l_1^2}$$

which implies that the system is always observable from $y = \theta_1$.

When $l_1 = l_2$, the system is uncontrollable. In this case, $\alpha_1 = \alpha_2, \beta_1 = \beta_2$, and the matrix A and vector B become

$$A = \begin{bmatrix} 0 & 1 & 0 & 0 \\ 1.2\alpha_1 & 0 & -0.1\alpha_1 & 0 \\ 0 & 0 & 0 & 1 \\ 0 & 0 & \alpha_1 & 0 \end{bmatrix}, \quad B = \begin{bmatrix} 0 \\ \beta_1 \\ 0 \\ 0 \end{bmatrix}$$

indicating that the control input u cannot influence the states x_3, x_4. It can be verified that for $x_3(0), x_4(0) \neq 0$, all the states will grow to infinity for all possible inputs u. For $l_1 = l_2$, the control of the two identical pendulums is possible provided the initial angles and angular velocities are identical, i.e., $\theta_1(0) = \theta_2(0)$ and $\dot\theta_1(0) = \dot\theta_2(0)$, which imply that $x_3(0) = x_4(0) = 0$. ▽

2.3 Input/Output Models

2.3.1 Transfer Functions

Transfer functions play an important role in the characterization of the I/O properties of LTI systems and are widely used in classical control theory.

We define the transfer function of an LTI system by starting with the differential equation that describes the dynamic system. Consider a system described by the nth-order differential equation

$$y^{(n)}(t) + a_{n-1}y^{(n-1)}(t) + \cdots + a_0 y(t) = b_m u^{(m)}(t) + b_{m-1}u^{(m-1)}(t) + \cdots + b_0 u(t) \quad (2.3.1)$$

where $y^{(i)}(t) \triangleq \frac{d^i}{dt^i} y(t)$, and $u^{(i)}(t) \triangleq \frac{d^i}{dt^i} u(t)$; $u(t)$ is the input variable, and $y(t)$ is the output variable; the coefficients $a_i, b_j, i = 0, 1\ldots, n-1, j = 0, 1, \ldots, m$ are constants, and n and m are constant integers. To obtain the transfer function of the system (2.3.1), we take the Laplace transform on both sides of the equation and assume *zero initial conditions*, i.e.,

$$(s^n + a_{n-1}s^{n-1} + \cdots + a_0)Y(s) = (b_m s^m + b_{m-1}s^{m-1} + \cdots + b_0)U(s)$$

where s is the Laplace variable. The transfer function $G(s)$ of (2.3.1) is defined as

$$G(s) \triangleq \frac{Y(s)}{U(s)} = \frac{b_m s^m + b_{m-1}s^{m-1} + \cdots + b_0}{s^n + a_{n-1}s^{n-1} + \cdots + a_0} \quad (2.3.2)$$

The inverse Laplace $g(t)$ of $G(s)$, i.e.,

$$g(t) \triangleq \mathcal{L}^{-1}[G(s)]$$

is known as the impulse response of the system (2.3.1) and

$$y(t) = g(t) * u(t)$$

where $*$ denotes convolution. When $u(t) = \delta_\Delta(t)$ where $\delta_\Delta(t)$ is the *delta function* defined as

$$\delta_\Delta(t) = \lim_{\epsilon \to 0} \frac{\mathcal{I}(t) - \mathcal{I}(t-\epsilon)}{\epsilon}$$

where $\mathcal{I}(t)$ is the unit step function, then

$$y(t) = g(t) * \delta_\Delta(t) = g(t)$$

2.3. INPUT/OUTPUT MODELS

Therefore, when the input to the LTI system is a delta function (often referred to as a unit impulse) at $t = 0$, the output of the system is equal to $g(t)$, the impulse response.

We say that $G(s)$ is *proper* if $G(\infty)$ is finite i.e., $n \geq m$; *strictly proper* if $G(\infty) = 0$, i.e., $n > m$; and *biproper* if $n = m$.

The *relative degree* n^* of $G(s)$ is defined as $n^* = n - m$, i.e., $n^* =$ degree of denominator - degree of numerator of $G(s)$.

The *characteristic equation* of the system (2.3.1) is defined as the equation $s^n + a_{n-1}s^{n-1} + \cdots + a_0 = 0$.

In a similar way, the transfer function may be defined for the LTI system in the state space form (2.2.3), i.e., taking the Laplace transform on each side of (2.2.3) we obtain

$$sX(s) - x(0) = AX(s) + BU(s)$$
$$Y(s) = C^T X(s) + DU(s) \qquad (2.3.3)$$

or

$$Y(s) = \left(C^T(sI - A)^{-1}B + D\right)U(s) + C^T(sI - A)^{-1}x(0)$$

Setting the initial conditions to zero, i.e., $x(0) = 0$ we get

$$Y(s) = G(s)U(s) \qquad (2.3.4)$$

where

$$G(s) = C^T(sI - A)^{-1}B + D$$

is referred to as the *transfer function matrix* in the case of multiple inputs and outputs and simply as the transfer function in the case of SISO systems. We may also represent $G(s)$ as

$$G(s) = \frac{C^T\{\mathrm{adj}(sI - A)\}B}{\det(sI - A)} + D \qquad (2.3.5)$$

where adj$Q$ denotes the adjoint of the square matrix $Q \in \mathcal{R}^{n \times n}$. The (i, j) element q_{ij} of adj$Q$ is given by

$$q_{ij} = (-1)^{i+j}\det(Q_{ji}); \qquad i, j = 1, 2, \ldots n$$

where $Q_{ji} \in \mathcal{R}^{(n-1)\times(n-1)}$ is a submatrix of Q obtained by eliminating the jth row and the ith column of the matrix Q.

It is obvious from (2.3.5) that the poles of $G(s)$ are included in the eigenvalues of A. We say that A is *stable* if all its eigenvalues lie in $\text{Re}[s] < 0$ in which case $G(s)$ is a *stable transfer function*. It follows that $\det(sI - A) = 0$ is the characteristic equation of the system with transfer function given by (2.3.5).

In (2.3.3) and (2.3.4) we went from a state-space representation to a transfer function description in a straightforward manner. The other way, i.e., from a proper transfer function description to a state-space representation, is not as straightforward. It is true, however, that for every proper transfer function $G(s)$ there exists matrices A, B, C, and D such that

$$G(s) = C^T(sI - A)^{-1}B + D$$

As an example, consider a system with the transfer function

$$G(s) = \frac{b_m s^m + b_{m-1} s^{m-1} + \cdots + b_0}{s^n + a_{n-1} s^{n-1} + \cdots + a_0} = \frac{Y(s)}{U(s)}$$

where $n > m$. Then the system may be represented in the controller form

$$\dot{x} = \begin{bmatrix} -a_{n-1} & -a_{n-2} & \cdots & -a_1 & -a_0 \\ 1 & 0 & \cdots & 0 & 0 \\ 0 & 1 & \cdots & 0 & 0 \\ \vdots & \vdots & \ddots & & \vdots \\ 0 & 0 & \cdots & 1 & 0 \end{bmatrix} x + \begin{bmatrix} 1 \\ 0 \\ \vdots \\ 0 \\ 0 \end{bmatrix} u \quad (2.3.6)$$

$$y = [0, 0, \ldots, b_m, \ldots, b_1, b_0] x$$

or in the observer form

$$\dot{x} = \begin{bmatrix} -a_{n-1} & 1 & 0 & \cdots & 0 \\ -a_{n-2} & 0 & 1 & \cdots & 0 \\ \vdots & \vdots & & \ddots & \vdots \\ -a_1 & 0 & 0 & \cdots & 1 \\ -a_0 & 0 & 0 & \cdots & 0 \end{bmatrix} x + \begin{bmatrix} 0 \\ \vdots \\ b_m \\ \vdots \\ b_0 \end{bmatrix} u \quad (2.3.7)$$

$$y = [1, 0, \ldots, 0] x$$

2.3. INPUT/OUTPUT MODELS

One can go on and generate many different state-space representations describing the I/O properties of the same system. The canonical forms in (2.3.6) and (2.3.7), however, have some important properties that we will use in later chapters. For example, if we denote by (A_c, B_c, C_c) and (A_o, B_o, C_o) the corresponding matrices in the controller form (2.3.6) and observer form (2.3.7), respectively, we establish the relations

$$[\text{adj}(sI - A_c)]B_c = [s^{n-1}, \ldots, s, 1]^T \triangleq \alpha_{n-1}(s) \quad (2.3.8)$$

$$C_o^T \text{adj}(sI - A_o) = [s^{n-1}, \ldots, s, 1] = \alpha_{n-1}^T(s) \quad (2.3.9)$$

whose right-hand sides are independent of the coefficients of $G(s)$. Another important property is that in the triples (A_c, B_c, C_c) and (A_o, B_o, C_o), the $n+m+1$ coefficients of $G(s)$ appear explicitly, i.e., (A_c, B_c, C_c) (respectively (A_o, B_o, C_o)) is completely characterized by $n+m+1$ parameters, which are equal to the corresponding coefficients of $G(s)$.

If $G(s)$ has no zero-pole cancellations then both (2.3.6) and (2.3.7) are minimal state-space representations of the same system. If $G(s)$ has zero-pole cancellations, then (2.3.6) is unobservable, and (2.3.7) is uncontrollable. If the zero-pole cancellations of $G(s)$ occur in $\text{Re}[s] < 0$, i.e., stable poles are cancelled by stable zeros, then (2.3.6) is detectable, and (2.3.7) is stabilizable. Similarly, a system described by a state-space representation is unobservable or uncontrollable, if and only if the transfer function of the system has zero-pole cancellations. If the unobservable or uncontrollable parts of the system are asymptotically stable, then the zero-pole cancellations occur in $\text{Re}[s] < 0$.

An alternative approach for representing the differential equation (2.3.1) is by using the differential operator

$$p(\cdot) \triangleq \frac{d(\cdot)}{dt}$$

which has the following properties:

$$(i) \quad p(x) = \dot{x}; \quad (ii) \quad p(xy) = \dot{x}y + x\dot{y}$$

where x and y are any differentiable functions of time and $\dot{x} \triangleq \frac{dx(t)}{dt}$.

The inverse of the operator p denoted by p^{-1} or simply by $\frac{1}{p}$ is defined as

$$\frac{1}{p}(x) \triangleq \int_0^t x(\tau)d\tau + x(0) \quad \forall t \geq 0$$

where $x(t)$ is an integrable function of time. The operators $p, \frac{1}{p}$ are related to the Laplace operator s by the following equations

$$\mathcal{L}\{p(x)\}|_{x(0)=0} = sX(s)$$

$$\mathcal{L}\{\frac{1}{p}(x)\}|_{x(0)=0} = \frac{1}{s}X(s)$$

where $\mathcal{L}$ is the Laplace transform and $x(t)$ is any differentiable function of time. Using the definition of the differential operator, (2.3.1) may be written in the compact form

$$R(p)(y) = Z(p)(u) \tag{2.3.10}$$

where

$$R(p) = p^n + a_{n-1}p^{n-1} + \cdots + a_0$$

$$Z(p) = b_m p^m + b_{m-1}p^{m-1} + \cdots + b_0$$

are referred to as the *polynomial differential operators* [226].

Equation (2.3.10) has the same form as

$$R(s)Y(s) = Z(s)U(s) \tag{2.3.11}$$

obtained by taking the Laplace transform on both sides of (2.3.1) and assuming zero initial conditions. Therefore, for zero initial conditions one can go from representation (2.3.10) to (2.3.11) and vice versa by simply replacing s with p or p with s appropriately. For example, the system

$$Y(s) = \frac{s+b_0}{s^2+a_0}U(s)$$

may be written as

$$(p^2 + a_0)(y) = (p + b_0)(u)$$

with $y(0) = \dot{y}(0) = 0$, $u(0) = 0$ or by abusing notation (because we never defined the operator $(p^2 + a_0)^{-1}$) as

$$y(t) = \frac{p+b_0}{p^2+a_0}u(t)$$

Because of the similarities of the forms of (2.3.11) and (2.3.10), we will use s to denote both the differential operator and Laplace variable and express the system (2.3.1) with zero initial conditions as

$$y = \frac{Z(s)}{R(s)}u \tag{2.3.12}$$

2.3. INPUT/OUTPUT MODELS

where y and u denote $Y(s)$ and $U(s)$, respectively, when s is taken to be the Laplace operator, and y and u denote $y(t)$ and $u(t)$, respectively, when s is taken to be the differential operator.

We will often refer to $G(s) = \frac{Z(s)}{R(s)}$ in (2.3.12) as the filter with input $u(t)$ and output $y(t)$.

Example 2.3.1 Consider the system of equations describing the motion of the cart with the two pendulums given in Example 2.2.1, where $y = \theta_1$ is the only measured output. Eliminating the variables θ_1, θ_2, and $\dot{\theta}_2$ by substitution, we obtain the fourth order differential equation

$$y^{(4)} - 1.1(\alpha_1 + \alpha_2)y^{(2)} + 1.2\alpha_1\alpha_2 y = \beta_1 u^{(2)} - \alpha_1\beta_2 u$$

where $\alpha_i, \beta_i, i = 1, 2$ are as defined in Example 2.2.1, which relates the input u with the measured output y.

Taking the Laplace transform on each side of the equation and assuming zero initial conditions, we obtain

$$[s^4 - 1.1(\alpha_1 + \alpha_2)s^2 + 1.2\alpha_1\alpha_2]Y(s) = (\beta_1 s^2 - \alpha_1\beta_2)U(s)$$

Therefore, the transfer function of the system from u to y is given by

$$\frac{Y(s)}{U(s)} = \frac{\beta_1 s^2 - \alpha_1\beta_2}{s^4 - 1.1(\alpha_1 + \alpha_2)s^2 + 1.2\alpha_1\alpha_2} = G(s)$$

For $l_1 = l_2$, we have $\alpha_1 = \alpha_2, \beta_1 = \beta_2$, and

$$G(s) = \frac{\beta_1(s^2 - \alpha_1)}{s^4 - 2.2\alpha_1 s^2 + 1.2\alpha_1^2} = \frac{\beta_1(s^2 - \alpha_1)}{(s^2 - \alpha_1)(s^2 - 1.2\alpha_1)}$$

has two zero-pole cancellations. Because $\alpha_1 > 0$, one of the zero-pole cancellations occurs in $\text{Re}[s] > 0$ which indicates that any fourth-order state representation of the system with the above transfer function is not stabilizable. ▽

2.3.2 Coprime Polynomials

The I/O properties of most of the systems studied in this book are represented by proper transfer functions expressed as the ratio of two polynomials in s with real coefficients, i.e.,

$$G(s) = \frac{Z(s)}{R(s)} \qquad (2.3.13)$$

where $Z(s) = b_m s^m + b_{m-1} s^{m-1} + \cdots + b_0$, $R(s) = s^n + a_{n-1} s^{n-1} + \cdots + a_0$ and $n \geq m$.

The properties of the system associated with $G(s)$ depend very much on the properties of $Z(s)$ and $R(s)$. In this section, we review some of the general properties of polynomials that are used for analysis and control design in subsequent chapters.

Definition 2.3.1 *Consider the polynomial $X(s) = \alpha_n s^n + \alpha_{n-1} s^{n-1} + \cdots + \alpha_0$. We say that $X(s)$ is* **monic** *if $\alpha_n = 1$ and $X(s)$ is* **Hurwitz** *if all the roots of $X(s) = 0$ are located in $\operatorname{Re}[s] < 0$. We say that the* **degree** *of $X(s)$ is n if the coefficient α_n of s^n satisfies $\alpha_n \neq 0$.*

Definition 2.3.2 *A system with a transfer function given by (2.3.13) is referred to as* **minimum phase** *if $Z(s)$ is Hurwitz; it is referred to as* **stable** *if $R(s)$ is Hurwitz.*

As we mentioned in Section 2.3.1, a system representation is minimal if the corresponding transfer function has no zero-pole cancellations, i.e., if the numerator and denominator polynomials of the transfer function have no common factors other than a constant. The following definition is widely used in control theory to characterize polynomials with no common factors.

Definition 2.3.3 *Two polynomials $a(s)$ and $b(s)$ are said to be* **coprime** *(or* **relatively prime***) if they have no common factors other than a constant.*

An important characterization of coprimeness of two polynomials is given by the following Lemma.

Lemma 2.3.1 (Bezout Identity) *Two polynomials $a(s)$ and $b(s)$ are coprime if and only if there exist polynomials $c(s)$ and $d(s)$ such that*

$$c(s)a(s) + d(s)b(s) = 1$$

For a proof of Lemma 2.3.1, see [73, 237].

The Bezout identity may have infinite number of solutions $c(s)$ and $d(s)$ for a given pair of coprime polynomials $a(s)$ and $b(s)$ as illustrated by the following example.

2.3. INPUT/OUTPUT MODELS

Example 2.3.2 Consider the coprime polynomials $a(s) = s+1, b(s) = s+2$. Then the Bezout identity is satisfied for

$$c(s) = s^n + 2s^{n-1} - 1, \quad d(s) = -s^n - s^{n-1} + 1$$

and any $n \geq 1$. ▽

Coprimeness is an important property that is often exploited in control theory for the design of control schemes for LTI systems. An important theorem that is very often used for control design and analysis is the following.

Theorem 2.3.1 *If $a(s)$ and $b(s)$ are coprime and of degree n_a and n_b, respectively, where $n_a > n_b$, then for any given arbitrary polynomial $a^*(s)$ of degree $n_{a^*} \geq n_a$, the polynomial equation*

$$a(s)l(s) + b(s)p(s) = a^*(s) \qquad (2.3.14)$$

has a unique solution $l(s)$ and $p(s)$ whose degrees n_l and n_p, respectively, satisfy the constraints $n_p < n_a, n_l \leq \max(n_{a^*} - n_a, n_b - 1)$.

Proof From Lemma 2.3.1, there exist polynomials $c(s)$ and $d(s)$ such that

$$a(s)c(s) + b(s)d(s) = 1 \qquad (2.3.15)$$

Multiplying Equation (2.3.15) on both sides by the polynomial $a^*(s)$, we obtain

$$a^*(s)a(s)c(s) + a^*(s)b(s)d(s) = a^*(s) \qquad (2.3.16)$$

Let us divide $a^*(s)d(s)$ by $a(s)$, i.e.,

$$\frac{a^*(s)d(s)}{a(s)} = r(s) + \frac{p(s)}{a(s)}$$

where $r(s)$ is the quotient of degree $n_{a^*} + n_d - n_a$; n_{a^*}, n_a, and n_d are the degrees of $a^*(s), a(s)$, and $d(s)$, respectively, and $p(s)$ is the remainder of degree $n_p < n_a$. We now use

$$a^*(s)d(s) = r(s)a(s) + p(s)$$

to express the left-hand side of (2.3.16) as

$$a^*(s)a(s)c(s) + r(s)a(s)b(s) + p(s)b(s) = [a^*(s)c(s) + r(s)b(s)]a(s) + p(s)b(s)$$

and rewrite (2.3.16) as
$$l(s)a(s) + p(s)b(s) = a^*(s) \qquad (2.3.17)$$
where $l(s) = a^*(s)c(s) + r(s)b(s)$. The above equation implies that the degree of $l(s)a(s) =$ degree of $(a^*(s) - p(s)b(s)) \leq \max\{n_{a^*}, n_p + n_b\}$. Hence, the degree of $l(s)$, denoted by n_l, satisfies $n_l \leq \max\{n_{a^*} - n_a, n_p + n_b - n_a\}$. We, therefore, established that polynomials $l(s)$ and $p(s)$ of degree $n_l \leq \max\{n_{a^*} - n_a, n_p + n_b - n_a\}$ and $n_p < n_a$ respectively exist that satisfy (2.3.17). Because $n_p < n_a$ implies that $n_p \leq n_a - 1$, the degree n_l also satisfies $n_l \leq \max\{n_a^* - n_a, n_b - 1\}$. We show the uniqueness of $l(s)$ and $p(s)$ by proceeding as follows: We suppose that $(l_1(s), p_1(s)), (l_2(s), p_2(s))$ are two solutions of (2.3.17) that satisfy the degree constraints $n_p < n_a, n_l \leq \max\{n_{a^*} - n_a, n_b - 1\}$, i.e.,
$$a(s)l_1(s) + b(s)p_1(s) = a^*(s), \quad a(s)l_2(s) + b(s)p_2(s) = a^*(s)$$
Subtracting one equation from another, we have
$$a(s)(l_1(s) - l_2(s)) + b(s)(p_1(s) - p_2(s)) = 0 \qquad (2.3.18)$$
which implies that
$$\frac{b(s)}{a(s)} = \frac{l_2(s) - l_1(s)}{p_1(s) - p_2(s)} \qquad (2.3.19)$$
Because $n_p < n_a$, equation (2.3.19) implies that $b(s), a(s)$ have common factors that contradicts with the assumption that $a(s)$ and $b(s)$ are coprime. Thus, $l_1(s) = l_2(s)$ and $p_1(s) = p_2(s)$, which implies that the solution $l(s)$ and $p(s)$ of (2.3.17) is unique, and the proof is complete. □

If no constraints are imposed on the degrees of $l(s)$ and $p(s)$, (2.3.14) has an infinite number of solutions. Equations of the form (2.3.14) are referred to as *Diophantine* equations and are widely used in the algebraic design of controllers for LTI plants. The following example illustrates the use of Theorem 2.3.1 for designing a stable control system.

Example 2.3.3 Let us consider the following plant
$$y = \frac{s-1}{s^3}u \qquad (2.3.20)$$
We would like to choose the input $u(t)$ so that the closed-loop characteristic equation of the plant is given by $a^*(s) = (s+1)^5$, i.e., u is to be chosen so that the closed-loop plant is described by
$$(s+1)^5 y = 0 \qquad (2.3.21)$$

2.3. INPUT/OUTPUT MODELS

Let us consider the control input in the form of

$$u = -\frac{p(s)}{l(s)}y \qquad (2.3.22)$$

where $l(s)$ and $p(s)$ are polynomials with real coefficients whose degrees and coefficients are to be determined. Using (2.3.22) in (2.3.20), we have the closed-loop plant

$$s^3 l(s) y = -(s-1)p(s)y$$

i.e.,

$$[l(s)s^3 + p(s)(s-1)]y = 0$$

If we now choose $l(s)$ and $p(s)$ to satisfy the Diophantine equation

$$l(s)s^3 + p(s)(s-1) = (s+1)^5 \qquad (2.3.23)$$

then the closed-loop plant becomes the same as the desired one given by (2.3.21).

Because (2.3.23) may have an infinite number of solutions for $l(s), p(s)$, we use Theorem 2.3.1 to choose $l(s)$ and $p(s)$ with the lowest degree. According to Theorem 2.3.1, Equation (2.3.23) has a unique solution $l(s), p(s)$ of degree equal to at most 2. Therefore, we assume that $l(s), p(s)$ have the form

$$l(s) = l_2 s^2 + l_1 s + l_0$$

$$p(s) = p_2 s^2 + p_1 s + p_0$$

which we use in (2.3.23) to obtain the following polynomial equation

$$l_2 s^5 + l_1 s^4 + (l_0 + p_2)s^3 + (p_1 - p_2)s^2 + (p_0 - p_1)s - p_0 = s^5 + 5s^4 + 10s^3 + 10s^2 + 5s + 1$$

Equating the coefficients of the same powers of s on each side of the above equation, we obtain the algebraic equations

$$\begin{cases} l_2 = 1 \\ l_1 = 5 \\ l_0 + p_2 = 10 \\ p_1 - p_2 = 10 \\ p_0 - p_1 = 5 \\ -p_0 = 1 \end{cases}$$

which have the unique solution of $l_2 = 1, l_1 = 5, l_0 = 26, p_2 = -16, p_1 = -6, p_0 = -1$. Hence,

$$l(s) = s^2 + 5s + 26, \quad p(s) = -16s^2 - 6s - 1$$

and from (2.3.22) the control input is given by

$$u = \frac{16s^2 + 6s + 1}{s^2 + 5s + 26} y \qquad \triangledown$$

Another characterization of coprimeness that we use in subsequent chapters is given by the following theorem:

Theorem 2.3.2 (Sylvester's Theorem) *Two polynomials $a(s) = a_n s^n + a_{n-1} s^{n-1} + \cdots + a_0$, $b(s) = b_n s^n + b_{n-1} s^{n-1} + \cdots + b_0$ are coprime if and only if their Sylvester matrix S_e is nonsingular, where S_e is defined to be the following $2n \times 2n$ matrix:*

$$S_e \triangleq \begin{bmatrix}
a_n & 0 & 0 & \cdots & 0 & 0 & b_n & 0 & 0 & \cdots & 0 & 0 \\
a_{n-1} & a_n & 0 & & 0 & 0 & b_{n-1} & b_n & 0 & & 0 & 0 \\
\cdot & a_{n-1} & a_n & \ddots & & \vdots & \cdot & b_{n-1} & b_n & \ddots & & \vdots \\
\cdot & \cdot & \cdot & \ddots & & \vdots & \cdot & \cdot & \cdot & \ddots & & \vdots \\
\cdot & \cdot & \cdot & \cdot & & 0 & \cdot & \cdot & \cdot & \cdot & & 0 \\
a_1 & \cdot & \cdot & \cdot & a_n & b_1 & \cdot & \cdot & \cdot & \cdot & & b_n \\
a_0 & a_1 & \cdot & \cdot & a_{n-1} & b_0 & b_1 & \cdot & \cdot & \cdot & & b_{n-1} \\
0 & a_0 & \cdot & \cdot & \cdot & 0 & b_0 & \cdot & \cdot & \cdot & & \cdot \\
0 & 0 & \cdot & \cdot & \cdot & 0 & 0 & \cdot & \cdot & \cdot & & \cdot \\
\vdots & & \ddots & \cdot & \cdot & \vdots & & 0 & \ddots & \cdot & & \cdot \\
\vdots & & \ddots & a_0 & a_1 & \vdots & & & \ddots & & b_0 & b_1 \\
0 & 0 & \cdots & 0 & 0 & a_0 & 0 & 0 & \cdots & & 0 & 0 & b_0
\end{bmatrix}$$
(2.3.24)

Proof If Consider the following polynomial equation

$$a(s)c(s) + b(s)d(s) = 1 \qquad (2.3.25)$$

where $c(s) = c_{n-1} s^{n-1} + c_{n-2} s^{n-2} + \cdots + c_0$, $d(s) = d_{n-1} s^{n-1} + d_{n-2} s^{n-2} + \cdots + d_0$ are some polynomials. Equating the coefficients of equal powers of s on both sides of (2.3.25), we obtain the algebraic equation

$$S_e p = e_{2n} \qquad (2.3.26)$$

where $e_{2n} = [0, 0, \ldots, 0, 1]^T$ and

$$p = [c_{n-1}, c_{n-2}, \ldots, c_0, d_{n-1}, d_{n-2}, \ldots, d_0]^T$$

2.3. INPUT/OUTPUT MODELS

Equations (2.3.25) and (2.3.26) are equivalent in the sense that any solution of (2.3.26) satisfies (2.3.25) and vice versa. Because S_e is nonsingular, equation (2.3.26) has a unique solution for p. It follows that (2.3.25) also has a unique solution for $c(s)$ and $d(s)$ which according to Lemma 2.3.1 implies that $a(s), b(s)$ are coprime.

Only if We claim that if $a(s)$ and $b(s)$ are coprime, then for all nonzero polynomials $p(s)$ and $q(s)$ of degree $n_p < n$ and $n_q < n$, respectively, we have

$$a(s)p(s) + b(s)q(s) \not\equiv 0 \qquad (2.3.27)$$

If the claim is not true, there exists nonzero polynomials $p_1(s)$ and $q_1(s)$ of degree $n_{p_1} < n$ and $n_{q_1} < n$, respectively, such that

$$a(s)p_1(s) + b(s)q_1(s) \equiv 0 \qquad (2.3.28)$$

Equation (2.3.28) implies that $b(s)/a(s)$ can be expressed as

$$\frac{b(s)}{a(s)} = -\frac{p_1(s)}{q_1(s)}$$

which, because $n_{p_1} < n$ and $n_{q_1} < n$, implies that $a(s)$ and $b(s)$ have common factors, thereby contradicting the assumption that $a(s), b(s)$ are coprime. Hence, our claim is true and (2.3.27) holds.

Now (2.3.27) may be written as

$$S_e x \neq 0 \qquad (2.3.29)$$

where $x \in \mathcal{R}^{2n}$ contains the coefficients of $p(s), q(s)$. Because (2.3.27) holds for all nonzero $p(s)$ and $q(s)$ of degree $n_p < n$ and $n_q < n$, respectively, then (2.3.29) holds for all vectors $x \in \mathcal{R}^{2n}$ with $x \neq 0$, which implies that S_e is nonsingular. □

The determinant of S_e is known as the *Sylvester resultant* and may be used to examine the coprimeness of a given pair of polynomials. If the polynomials $a(s)$ and $b(s)$ in Theorem 2.3.2 have different degrees—say $n_b < n_a$—then $b(s)$ is expressed as a polynomial of degree n_a by augmenting it with the additional powers in s whose coefficients are taken to be equal to zero.

Example 2.3.4 Consider the polynomials

$$a(s) = s^2 + 2s + 1, \quad b(s) = s - 1 = 0s^2 + s - 1$$

Their Sylvester matrix is given by

$$S_e = \begin{bmatrix} 1 & 0 & 0 & 0 \\ 2 & 1 & 1 & 0 \\ 1 & 2 & -1 & 1 \\ 0 & 1 & 0 & -1 \end{bmatrix}$$

Because $\det S_e = 4 \neq 0$, $a(s)$ and $b(s)$ are coprime polynomials. $\triangledown$

The properties of the Sylvester matrix are useful in solving a class of Diophantine equations of the form

$$l(s)a(s) + p(s)b(s) = a^*(s)$$

for $l(s)$ and $p(s)$ where $a(s), b(s)$, and $a^*(s)$ are given polynomials.

For example, equation $a(s)l(s) + b(s)p(s) = a^*(s)$ with $n_a = n, n_{a^*} = 2n - 1$, and $n_b = m < n$ implies the algebraic equation

$$S_e x = f \qquad (2.3.30)$$

where $S_e \in \mathcal{R}^{2n \times 2n}$ is the Sylvester matrix of $a(s), b(s)$, and $x \in \mathcal{R}^{2n}$ is a vector with the coefficients of the polynomials $l(s)$ and $p(s)$ whose degree according to Theorem 2.3.1 is at most $n - 1$ and $f \in \mathcal{R}^{2n}$ contains the coefficients of $a^*(s)$. Therefore, given $a^*(s), a(s)$, and $b(s)$, one can solve (2.3.30) for x, the coefficient vector of $l(s)$ and $p(s)$. If $a(s), b(s)$ are coprime, S_e^{-1} exists and, therefore, the solution of (2.3.30) is unique and is given by

$$x = S_e^{-1} f$$

If $a(s), b(s)$ are not coprime, then S_e is not invertible, and (2.3.30) has a solution if and only if the vector f is in the range of S_e. One can show through algebraic manipulations that this condition is equivalent to the condition that $a^*(s)$ contains the common factors of $a(s)$ and $b(s)$.

Example 2.3.5 Consider the same control design problem as in Example 2.3.3, where the control input $u = -\frac{p(s)}{l(s)} y$ is used to force the plant $y = \frac{s-1}{s^3} u$ to satisfy $(s + 1)^5 y = 0$. We have shown that the polynomials $l(s)$ and $p(s)$ satisfy the Diophantine equation

$$l(s)a(s) + p(s)b(s) = (s+1)^5 \qquad (2.3.31)$$

where $a(s) = s^3$ and $b(s) = s - 1$. The corresponding Sylvester matrix S_e of $a(s)$ and $b(s)$ is

$$S_e = \begin{bmatrix} 1 & 0 & 0 & 0 & 0 & 0 \\ 0 & 1 & 0 & 0 & 0 & 0 \\ 0 & 0 & 1 & 1 & 0 & 0 \\ 0 & 0 & 0 & -1 & 1 & 0 \\ 0 & 0 & 0 & 0 & -1 & 1 \\ 0 & 0 & 0 & 0 & 0 & -1 \end{bmatrix}$$

Because $\det S_e = -1$, we verify that $a(s), b(s)$ are coprime.

As in Example 2.3.3, we like to solve (2.3.31) for the unknown coefficients $l_i, p_i, i = 0, 1, 2$ of the polynomials $l(s) = l_2 s^2 + l_1 s + l_0$ and $p(s) = p_2 s^2 + p_1 s + p_0$. By equating the coefficients of equal powers of s on each side of (2.3.31), we obtain the algebraic equation

$$S_e x = f \tag{2.3.32}$$

where $f = [1, 5, 10, 10, 5, 1]^T$ and $x = [l_2, l_1, l_0, p_2, p_1, p_0]^T$. Because S_e is a nonsingular matrix, the solution of (2.3.32) is given by

$$x = S_e^{-1} f = [1, 5, 26, -16, -6, -1,]^T$$

which is the same as the solution we obtained in Example 2.3.3 (verify!). ▽

2.4 Plant Parametric Models

Let us consider the plant represented by the following minimal state-space form:

$$\begin{aligned} \dot{x} &= Ax + Bu, \quad x(0) = x_0 \\ y &= C^T x \end{aligned} \tag{2.4.1}$$

where $x \in \mathcal{R}^n, u \in \mathcal{R}^1$, and $y \in \mathcal{R}^1$ and A, B, and C have the appropriate dimensions. The triple (A, B, C) consists of $n^2 + 2n$ elements that are referred to as the plant parameters. If (2.4.1) is in one of the canonical forms studied in Section 2.2.2, then n^2 elements of (A, B, C) are fixed to be 0 or 1 and at most $2n$ elements are required to specify the properties of the plant. These $2n$ elements are the coefficients of the numerator and denominator of the transfer function $\frac{Y(s)}{U(s)}$. For example, using the Laplace transform in (2.4.1), we obtain

$$Y(s) = C^T (sI - A)^{-1} BU(s) + C^T (sI - A)^{-1} x_0$$

which implies that

$$Y(s) = \frac{Z(s)}{R(s)}U(s) + \frac{C^\top\{\text{adj}(sI - A)\}}{R(s)}x_0 \qquad (2.4.2)$$

where $R(s)$ is a polynomial of degree n and $Z(s)$ of degree at most $n - 1$. Setting $x_0 = 0$, we obtain the transfer function description

$$y = \frac{Z(s)}{R(s)}u \qquad (2.4.3)$$

where without loss of generality, we can assume $Z(s)$ and $R(s)$ to be of the form

$$\begin{align} Z(s) &= b_{n-1}s^{n-1} + b_{n-2}s^{n-2} + \cdots + b_1 s + b_0 \qquad (2.4.4)\\ R(s) &= s^n + a_{n-1}s^{n-1} + \cdots + a_1 s + a_0 \end{align}$$

If $Z(s)$ is of degree $m < n-1$, then the coefficients $b_i, i = n-1, n-2, \ldots, m+1$ are equal to zero. Equations (2.4.3) and (2.4.4) indicate that at most $2n$ parameters are required to uniquely specify the I/O properties of (2.4.1). When more than $2n$ parameters in (2.4.3) are used to specify the same I/O properties, we say that the plant is *overparameterized*. For example, the plant

$$y = \frac{Z(s)}{R(s)}\frac{\Lambda(s)}{\Lambda(s)}u \qquad (2.4.5)$$

where $\Lambda(s)$ is Hurwitz of arbitrary degree $r > 0$, has the same I/O properties as the plant described by (2.4.3), and it is, therefore, overparameterized. In addition, any state representation of order $n+r > n$ of (2.4.5) is nonminimal.

For some estimation and control problems, certain plant parameterizations are more convenient than others. A plant parameterization that is useful for parameter estimation and some control problems is the one where parameters are lumped together and separated from signals. In parameter estimation, the parameters are the unknown constants to be estimated from the measurements of the I/O signals of the plant.

In the following sections, we present various parameterizations of the same plant that are useful for parameter estimation to be studied in later chapters.

2.4. PLANT PARAMETRIC MODELS

2.4.1 Linear Parametric Models

Parameterization 1

The plant equation (2.4.3) may be expressed as an nth-order differential equation given by

$$y^{(n)} + a_{n-1}y^{(n-1)} + \cdots + a_0 y = b_{n-1}u^{(n-1)} + b_{n-2}u^{(n-2)} + \cdots + b_0 u \quad (2.4.6)$$

If we lump all the parameters in (2.4.6) in the parameter vector

$$\theta^* = [b_{n-1}, b_{n-2}, \ldots, b_0, a_{n-1}, a_{n-2}, \ldots, a_0]^\top$$

and all I/O signals and their derivatives in the signal vector

$$\begin{aligned} Y &= [u^{(n-1)}, u^{(n-2)}, \ldots, u, -y^{(n-1)}, -y^{(n-2)}, \ldots, -y]^\top \\ &= [\alpha_{n-1}^\top(s)u, -\alpha_{n-1}^\top(s)y]^\top \end{aligned}$$

where $\alpha_i(s) \triangleq [s^i, s^{i-1}, \ldots, 1]^\top$, we can express (2.4.6) and, therefore, (2.4.3) in the compact form

$$y^{(n)} = \theta^{*\top} Y \quad (2.4.7)$$

Equation (2.4.7) is linear in θ^*, which, as we show in Chapters 4 and 5, is crucial for designing parameter estimators to estimate θ^* from the measurements of $y^{(n)}$ and Y. Because in most applications the only signals available for measurement is the input u and output y and the use of differentiation is not desirable, the use of the signals $y^{(n)}$ and Y should be avoided. One way to avoid them is to filter each side of (2.4.7) with an nth-order stable filter $\frac{1}{\Lambda(s)}$ to obtain

$$z = \theta^{*\top}\phi \quad (2.4.8)$$

where

$$z \triangleq \frac{1}{\Lambda(s)} y^{(n)} = \frac{s^n}{\Lambda(s)} y$$

$$\phi \triangleq \left[\frac{\alpha_{n-1}^\top(s)}{\Lambda(s)} u, -\frac{\alpha_{n-1}^\top(s)}{\Lambda(s)} y \right]^\top$$

and

$$\Lambda(s) = s^n + \lambda_{n-1}s^{n-1} + \cdots + \lambda_0$$

is an arbitrary Hurwitz polynomial in s. It is clear that the scalar signal z and vector signal ϕ can be generated, without the use of differentiators, by simply filtering the input u and output y with stable proper filters $\frac{s^i}{\Lambda(s)}, i = 0, 1, \ldots n$. If we now express $\Lambda(s)$ as

$$\Lambda(s) = s^n + \lambda^\top \alpha_{n-1}(s)$$

where $\lambda = [\lambda_{n-1}, \lambda_{n-2}, \ldots, \lambda_0]^\top$, we can write

$$z = \frac{s^n}{\Lambda(s)} y = \frac{\Lambda(s) - \lambda^\top \alpha_{n-1}(s)}{\Lambda(s)} y = y - \lambda^\top \frac{\alpha_{n-1}(s)}{\Lambda(s)} y$$

Therefore,

$$y = z + \lambda^\top \frac{\alpha_{n-1}(s)}{\Lambda(s)} y$$

Because $z = \theta^{*\top} \phi = \theta_1^{*\top} \phi_1 + \theta_2^{*\top} \phi_2$, where

$$\theta_1^* \triangleq [b_{n-1}, b_{n-2}, \ldots, b_0]^\top, \quad \theta_2^* \triangleq [a_{n-1}, a_{n-2}, \ldots, a_0]^\top$$

$$\phi_1 \triangleq \frac{\alpha_{n-1}(s)}{\Lambda(s)} u, \quad \phi_2 \triangleq -\frac{\alpha_{n-1}(s)}{\Lambda(s)} y$$

it follows that

$$y = \theta_1^{*\top} \phi_1 + \theta_2^{*\top} \phi_2 - \lambda^\top \phi_2$$

Hence,

$$y = \theta_\lambda^{*\top} \phi \qquad (2.4.9)$$

where $\theta_\lambda^* = [\theta_1^{*\top}, \theta_2^{*\top} - \lambda^\top]^\top$. Equations (2.4.8) and (2.4.9) are represented by the block diagram shown in Figure 2.2.

A state-space representation for generating the signals in (2.4.8) and (2.4.9) may be obtained by using the identity

$$[\mathrm{adj}(sI - \Lambda_c)] l = \alpha_{n-1}(s)$$

where Λ_c, l are given by

$$\Lambda_c = \begin{bmatrix} -\lambda_{n-1} & -\lambda_{n-2} & \cdots & -\lambda_0 \\ 1 & 0 & \cdots & 0 \\ \vdots & & \ddots & \vdots \\ 0 & \cdots & 1 & 0 \end{bmatrix}, \quad l = \begin{bmatrix} 1 \\ 0 \\ \vdots \\ 0 \end{bmatrix}$$

2.4. PLANT PARAMETRIC MODELS

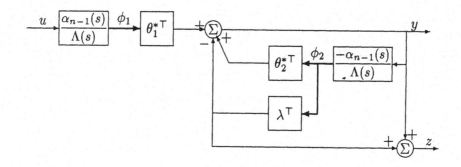

Figure 2.2 Plant Parameterization 1.

which implies that

$$\det(sI - \Lambda_c) = \Lambda(s), \quad (sI - \Lambda_c)^{-1}l = \frac{\alpha_{n-1}(s)}{\Lambda(s)}$$

Therefore, it follows from (2.4.8) and Figure 2.2 that

$$\begin{aligned}
\dot{\phi}_1 &= \Lambda_c \phi_1 + lu, \quad \phi_1 \in \mathcal{R}^n \\
\dot{\phi}_2 &= \Lambda_c \phi_2 - ly, \quad \phi_2 \in \mathcal{R}^n \\
y &= \theta_\lambda^{*T} \phi \\
z &= y + \lambda^T \phi_2 = \theta^{*T} \phi
\end{aligned} \quad (2.4.10)$$

Because $\Lambda(s) = \det(sI - \Lambda_c)$ and $\Lambda(s)$ is Hurwitz, it follows that Λ_c is a stable matrix.

The parametric model (2.4.10) is a nonminimal state-space representation of the plant (2.4.3). It is nonminimal because $2n$ integrators are used to represent an nth-order system. Indeed, the transfer function $Y(s)/U(s)$ computed using (2.4.10) or Figure 2.2, i.e.,

$$\frac{Y(s)}{U(s)} = \frac{Z(s)}{R(s)} \frac{\Lambda(s)}{\Lambda(s)} = \frac{Z(s)}{R(s)}$$

involves n stable zero-pole cancellations.

The plant (2.4.10) has the same I/O response as (2.4.3) and (2.4.1) provided that all state initial conditions are equal to zero, i.e., $x_0 = 0$, $\phi_1(0) = \phi_2(0) = 0$. In an actual plant, the state x in (2.4.1) may represent physical variables and the initial state x_0 may be different from zero. The effect of the initial state x_0 may be accounted for in the model (2.4.10) by applying the same procedure to equation (2.4.2) instead of equation (2.4.3). We can verify (see Problem 2.9) that if we consider the effect of initial condition x_0, we will obtain the following representation

$$\begin{aligned}
\dot\phi_1 &= \Lambda_c\phi_1 + lu, \quad \phi_1(0) = 0 \\
\dot\phi_2 &= \Lambda_c\phi_2 - ly, \quad \phi_2(0) = 0 \\
y &= \theta_\lambda^{*T}\phi + \eta_0 \\
z &= y + \lambda^T\phi_2 = \theta^{*T}\phi + \eta_0
\end{aligned} \qquad (2.4.11)$$

where η_0 is the output of the system

$$\begin{aligned}
\dot\omega &= \Lambda_c\omega, \quad \omega(0) = \omega_0 \\
\eta_0 &= C_0^T\omega
\end{aligned} \qquad (2.4.12)$$

where $\omega \in \mathcal{R}^n, \omega_0 = B_0 x_0$ and $C_0 \in \mathcal{R}^n, B_0 \in \mathcal{R}^{n\times n}$ are constant matrices that satisfy $C_0^T\{\operatorname{adj}(sI - \Lambda_c)\}B_0 = C^T\{\operatorname{adj}(sI - A)\}$.

Because Λ_c is a stable matrix, it follows from (2.4.12) that ω, η_0 converge to zero exponentially fast. Therefore, the effect of the nonzero initial condition x_0 is the appearance of the exponentially decaying to zero term η_0 in the output y and z.

Parameterization 2

Let us now consider the parametric model (2.4.9)

$$y = \theta_\lambda^{*T}\phi$$

and the identity $W_m(s)W_m^{-1}(s) = 1$, where $W_m(s) = Z_m(s)/R_m(s)$ is a transfer function with relative degree one, and $Z_m(s)$ and $R_m(s)$ are Hurwitz polynomials. Because θ_λ^* is a constant vector, we can express (2.4.9) as

$$y = W_m(s)\theta_\lambda^{*T}W_m^{-1}(s)\phi$$

If we let

$$\psi \triangleq \frac{1}{W_m(s)}\phi = \left[\frac{\alpha_{n-1}^T(s)}{W_m(s)\Lambda(s)}u, -\frac{\alpha_{n-1}^T(s)}{W_m(s)\Lambda(s)}y\right]^T$$

2.4. PLANT PARAMETRIC MODELS

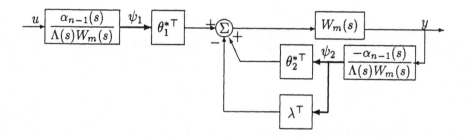

Figure 2.3 Plant Parameterization 2.

we have
$$y = W_m(s)\theta_\lambda^{*T}\psi \qquad (2.4.13)$$

Because all the elements of $\frac{\alpha_{n-1}(s)}{\Lambda(s)W_m(s)}$ are proper transfer functions with stable poles, the state $\psi = [\psi_1^T, \psi_2^T]^T$, where

$$\psi_1 = \frac{\alpha_{n-1}(s)}{W_m(s)\Lambda(s)}u, \quad \psi_2 = -\frac{\alpha_{n-1}(s)}{W_m(s)\Lambda(s)}y$$

can be generated without differentiating y or u. The dimension of ψ depends on the order n of $\Lambda(s)$ and the order of $Z_m(s)$. Because $Z_m(s)$ can be arbitrary, the dimension of ψ can be also arbitrary.

Figure 2.3 shows the block diagram of the parameterization of the plant given by (2.4.13). We refer to (2.4.13) as *Parameterization 2*. In [201], Parameterization 2 is referred to as the *model reference representation* and is used to design parameter estimators for estimating θ_λ^* when $W_m(s)$ is a strictly positive real transfer function (see definition in Chapter 3).

A special case of (2.4.13) is the one shown in Figure 2.4 where
$$W_m(s) = \frac{1}{s + \lambda_0}$$

and $(s + \lambda_0)$ is a factor of $\Lambda(s)$, i.e.,
$$\Lambda(s) = (s + \lambda_0)\Lambda_q(s) = s^n + \lambda_{n-1}s^{n-1} + \cdots + \lambda_0$$

where
$$\Lambda_q(s) = s^{n-1} + q_{n-2}s^{n-2} + \cdots + q_1 s + 1$$

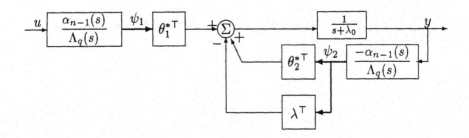

Figure 2.4 Plant Parameterization 2 with $\Lambda(s) = (s + \lambda_0)\Lambda_q(s)$ and $W_m(s) = \frac{1}{s+\lambda_0}$.

The plant Parameterization 2 of Figure 2.4 was first suggested in [131], where it was used to develop stable adaptive observers. An alternative parametric model of the plant of Figure 2.4 can be obtained by first separating the biproper elements of $\frac{\alpha_{n-1}(s)}{\Lambda_q(s)}$ as follows:

For any vector $c \triangleq [c_{n-1}, c_{n-2}, \ldots, c_1, c_0]^T \in \mathcal{R}^n$, we have

$$\frac{c^T \alpha_{n-1}(s)}{\Lambda_q(s)} = \frac{c_{n-1}s^{n-1}}{\Lambda_q(s)} + \frac{\bar{c}^T \alpha_{n-2}(s)}{\Lambda_q(s)} \qquad (2.4.14)$$

where $\bar{c} \triangleq [c_{n-2}, \ldots, c_1, c_0]^T, \alpha_{n-2} \triangleq [s^{n-2}, \ldots, s, 1]^T$. Because $\Lambda_q(s) = s^{n-1} + \bar{q}^T \alpha_{n-2}(s)$, where $\bar{q} = [q_{n-2}, \ldots, q_1, 1]^T$, we have $s^{n-1} = \Lambda_q(s) - \bar{q}^T \alpha_{n-2}$, which after substitution we obtain

$$\frac{c^T \alpha_{n-1}(s)}{\Lambda_q(s)} = c_{n-1} + \frac{(\bar{c} - c_{n-1}\bar{q})^T \alpha_{n-2}(s)}{\Lambda_q(s)} \qquad (2.4.15)$$

We use (2.4.15) to obtain the following expressions:

$$\theta_1^{*T} \frac{\alpha_{n-1}(s)}{\Lambda_q(s)} u = b_{n-1}u + \bar{\theta}_1^{*T} \frac{\alpha_{n-2}(s)}{\Lambda_q(s)} u$$

$$-(\theta_2^{*T} - \lambda^T)\frac{\alpha_{n-1}(s)}{\Lambda_q(s)}y = (\lambda_{n-1} - a_{n-1})y - \bar{\theta}_2^{*T}\frac{\alpha_{n-2}(s)}{\Lambda_q(s)}y \quad (2.4.16)$$

where $\bar{\theta}_1^{*T} = \bar{b} - b_{n-1}\bar{q}$, $\bar{\theta}_2^{*T} = \bar{a} - \bar{\lambda} - (a_{n-1} - \lambda_{n-1})\bar{q}$ and $\bar{a} = [a_{n-2}, \ldots, a_1, a_0]^T$, $\bar{b} = [b_{n-1}, \ldots, b_1, b_0]^T$, $\bar{\lambda} = [\lambda_{n-2}, \ldots, \lambda_1, \lambda_0]^T$. Using (2.4.16), Figure 2.4 can be reconfigured as shown in Figure 2.5.

2.4. PLANT PARAMETRIC MODELS

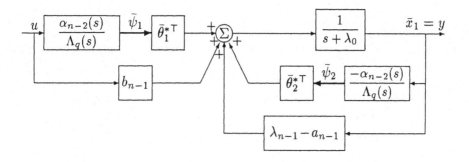

Figure 2.5 Equivalent plant Parameterization 2.

A nonminimal state space representation of the plant follows from Figure 2.5, i.e.,

$$\begin{aligned}
\dot{\bar{x}}_1 &= -\lambda_0 \bar{x}_1 + \bar{\theta}^{*T}\bar{\psi}, & \bar{x}_1 &\in \mathcal{R}^1 \\
\dot{\bar{\psi}}_1 &= \bar{\Lambda}_c \bar{\psi}_1 + \bar{l}u, & \bar{\psi}_1 &\in \mathcal{R}^{n-1} \\
\dot{\bar{\psi}}_2 &= \bar{\Lambda}_c \bar{\psi}_2 - \bar{l}y, & \bar{\psi}_2 &\in \mathcal{R}^{n-1} \\
y &= \bar{x}_1
\end{aligned} \qquad (2.4.17)$$

where $\bar{\theta}^* = [b_{n-1}, \bar{\theta}_1^{*T}, \lambda_{n-1} - a_{n-1}, \bar{\theta}_2^{*T}]^T$, $\bar{\psi} = [u, \bar{\psi}_1^T, y, \bar{\psi}_2^T]^T$ and

$$\bar{\Lambda}_c = \begin{bmatrix} -q_{n-2} & -q_{n-3} & \cdots & -q_0 \\ 1 & 0 & \cdots & 0 \\ \vdots & \ddots & & \vdots \\ 0 & \cdots & 1 & 0 \end{bmatrix}, \quad \bar{l} = \begin{bmatrix} 1 \\ 0 \\ \vdots \\ 0 \end{bmatrix}$$

As with Parameterization 1, if we account for the initial condition $x(0) = x_0 \neq 0$, we obtain

$$\begin{aligned}
\dot{\bar{x}}_1 &= -\lambda_0 \bar{x}_1 + \bar{\theta}^{*T}\bar{\psi}, & \bar{x}_1(0) &= 0 \\
\dot{\bar{\psi}}_1 &= \bar{\Lambda}_c \bar{\psi}_1 + \bar{l}u, & \bar{\psi}_1(0) &= 0 \\
\dot{\bar{\psi}}_2 &= \bar{\Lambda}_c \bar{\psi}_2 - \bar{l}y, & \bar{\psi}_2(0) &= 0 \\
y &= \bar{x}_1 + \eta_0
\end{aligned} \qquad (2.4.18)$$

where η_0 is the output of the system

$$\begin{aligned}
\dot{\omega} &= \Lambda_c \omega, \quad \omega(0) = \omega_0, \quad \omega \in \mathcal{R}^n \\
\eta_0 &= C_0^T \omega
\end{aligned}$$

Example 2.4.1 (**Parameterization 1**) Let us consider the differential equation

$$y^{(4)} + a_2 y^{(2)} + a_0 y = b_2 u^{(2)} + b_0 u \qquad (2.4.19)$$

that describes the motion of the cart with the two pendulums considered in Examples 2.2.1, 2.3.1, where

$$a_2 = -1.1(\alpha_1 + \alpha_2), \quad a_0 = 1.2\alpha_1\alpha_2, \quad b_2 = \beta_1, \quad b_0 = -\alpha_1\beta_2$$

Equation (2.4.19) is of the same form as (2.4.6) with $n = 4$ and coefficients $a_3 = a_1 = b_3 = b_1 = 0$. Following (2.4.7), we may rewrite (2.4.19) in the compact form

$$y^{(4)} = \theta_0^{*T} Y_0 \qquad (2.4.20)$$

where $\theta_0^* = [b_2, b_0, a_2, a_0]^T$, $Y_0 = [u^{(2)}, u, -y^{(2)}, -y]^T$. Because y and u are the only signals we can measure, $y^{(4)}, Y_0$ are not available for measurement.

If we filter each side of (2.4.20) with the filter $\frac{1}{\Lambda(s)}$, where $\Lambda(s) = (s+2)^4 = s^4 + 8s^3 + 24s^2 + 32s + 16$, we have

$$z = \theta_0^{*T} \phi_0 \qquad (2.4.21)$$

where $z = \frac{s^4}{(s+2)^4} y$, $\phi_0 = \left[\frac{s^2}{(s+2)^4} u, \frac{1}{(s+2)^4} u, -\frac{s^2}{(s+2)^4} y, -\frac{1}{(s+2)^4} y \right]^T$ are now signals that can be generated from the measurements of y and u by filtering. Because in (2.4.19) the elements $a_3 = a_1 = b_3 = b_1 = 0$, the dimension of θ_0^*, ϕ_0 is 4 instead of 8, which is implied by (2.4.8).

Similarly, following (2.4.9) we have

$$y = \theta_\lambda^{*T} \phi \qquad (2.4.22)$$

where

$$\theta_\lambda^* = [0, b_2, 0, b_0, -8, a_2 - 24, -32, a_0 - 16]^T$$

$$\phi = \left[\frac{\alpha_3^T(s)}{(s+2)^4} u, -\frac{\alpha_3^T(s)}{(s+2)^4} y \right]^T, \quad \alpha_3(s) = [s^3, s^2, s, 1]^T$$

We may rewrite (2.4.22) by separating the elements of θ_λ^* that do not depend on the parameters of (2.4.19) to obtain

$$y = \theta_{0\lambda}^{*T} \phi_0 + h_0^T \phi$$

where $\theta_{0\lambda}^* = [b_2, b_0, a_2 - 24, a_0 - 16]^T$, $h_0 = [0, 0, 0, 0, -8, 0, -32, 0]^T$. We obtain a state-space representation of (2.4.21) and (2.4.22) by using (2.4.10), i.e.,

$$\begin{aligned}
\dot{\phi}_1 &= \Lambda_c \phi_1 + lu, & \phi_1 &\in \mathcal{R}^4 \\
\dot{\phi}_2 &= \Lambda_c \phi_2 - ly, & \phi_2 &\in \mathcal{R}^4 \\
y &= \theta_\lambda^{*T} \phi = \theta_{0\lambda}^{*T} \phi_0 + h_0^T \phi \\
z &= \theta_0^{*T} \phi_0
\end{aligned}$$

2.4. PLANT PARAMETRIC MODELS

where
$$\Lambda_c = \begin{bmatrix} -8 & -24 & -32 & -16 \\ 1 & 0 & 0 & 0 \\ 0 & 1 & 0 & 0 \\ 0 & 0 & 1 & 0 \end{bmatrix}, \quad l = \begin{bmatrix} 1 \\ 0 \\ 0 \\ 0 \end{bmatrix}$$

$$\phi_0 = \begin{bmatrix} 0 & 1 & 0 & 0 & 0 & 0 & 0 & 0 \\ 0 & 0 & 0 & 1 & 0 & 0 & 0 & 0 \\ 0 & 0 & 0 & 0 & 0 & 1 & 0 & 0 \\ 0 & 0 & 0 & 0 & 0 & 0 & 0 & 1 \end{bmatrix} \phi$$

and $\phi = [\phi_1^T, \phi_2^T]^T$. Instead of (2.4.22), we can also write

$$y = \theta_0^{*T} \phi_0 - \lambda^T \phi_2$$

where $\lambda = [8, 24, 32, 16]^T$. ▽

Example 2.4.2 (**Parameterization 2**) Consider the same plant as in Example 2.4.1, i.e.,
$$y = \theta_\lambda^{*T} \phi$$

where $\theta_\lambda^{*T} = [0, b_2, 0, b_0, -8, a_2 - 24, -32, a_0 - 16]$,

$$\phi = \left[\frac{\alpha_3^T(s)}{(s+2)^4} u, \; -\frac{\alpha_3^T(s)}{(s+2)^4} y \right]^T$$

Now we write
$$y = \frac{1}{s+2} \theta_\lambda^{*T} \psi$$

where
$$\psi \triangleq \left[\frac{\alpha_3^T(s)}{(s+2)^3} u, \; -\frac{\alpha_3^T(s)}{(s+2)^3} y \right]^T$$

Using simple algebra, we have

$$\frac{\alpha_3(s)}{(s+2)^3} = \frac{1}{(s+2)^3} \begin{bmatrix} s^3 \\ s^2 \\ s \\ 1 \end{bmatrix} = \begin{bmatrix} 1 \\ 0 \\ 0 \\ 0 \end{bmatrix} + \frac{1}{(s+2)^3} \begin{bmatrix} -6 & -12 & -8 \\ 1 & 0 & 0 \\ 0 & 1 & 0 \\ 0 & 0 & 1 \end{bmatrix} \alpha_2(s)$$

where $\alpha_2(s) = [s^2, s, 1]^T$. Therefore, ψ can be expressed as

$$\psi = \left[u - \bar{\lambda}^T \frac{\alpha_2(s)}{(s+2)^3} u, \; \frac{\alpha_2^T(s)}{(s+2)^3} u, \; -y + \bar{\lambda}^T \frac{\alpha_2(s)}{(s+2)^3} y, \; -\frac{\alpha_2^T(s)}{(s+2)^3} y \right]^T$$

where $\bar{\lambda} = [6, 12, 8]^T$, and $\theta_\lambda^{*T}\psi$ can be expressed as

$$\theta_\lambda^{*T}\psi = \bar{\theta}^{*T}\bar{\psi} \qquad (2.4.23)$$

where

$$\bar{\theta}^* = [b_2, 0, b_0, 8, a_2 + 24, 64, a_0 + 48]^T$$

$$\bar{\psi} = \left[\frac{\alpha_2^T(s)}{(s+2)^3}u, y, -\frac{\alpha_2^T(s)}{(s+2)^3}y\right]^T$$

Therefore,

$$y = \frac{1}{s+2}\bar{\theta}^{*T}\bar{\psi} \qquad (2.4.24)$$

A state-space realization of (2.4.24) is

$$\begin{aligned}
\dot{\bar{x}}_1 &= -2\bar{x}_1 + \bar{\theta}^{*T}\bar{\psi}, & \bar{x}_1 &\in \mathcal{R}^1 \\
\dot{\bar{\psi}}_1 &= \bar{\Lambda}_c\bar{\psi}_1 + \bar{l}u, & \bar{\psi}_1 &\in \mathcal{R}^3 \\
\dot{\bar{\psi}}_2 &= \bar{\Lambda}_c\bar{\psi}_2 - \bar{l}y, & \bar{\psi}_2 &\in \mathcal{R}^3 \\
y &= \bar{x}_1
\end{aligned}$$

where $\bar{\psi} = [\bar{\psi}_1^T, y, \bar{\psi}_2^T]^T$,

$$\bar{\Lambda}_c = \begin{bmatrix} -6 & -12 & -8 \\ 1 & 0 & 0 \\ 0 & 1 & 0 \end{bmatrix}, \quad \bar{l} = \begin{bmatrix} 1 \\ 0 \\ 0 \end{bmatrix}$$

$\triangledown$

2.4.2 Bilinear Parametric Models

Let us now consider the parameterization of a special class of systems expressed as

$$y = k_0 \frac{Z_0(s)}{R_0(s)} u \qquad (2.4.25)$$

where k_0 is a scalar, $R_0(s)$ is monic of degree n, and $Z_0(s)$ is monic and Hurwitz of degree $m < n$. In addition, let $Z_0(s)$ and $R_0(s)$ satisfy the Diophantine equation

$$k_0 Z_0(s) P(s) + R_0(s) Q(s) = Z_0(s) A(s) \qquad (2.4.26)$$

where

$$Q(s) = s^{n-1} + q^T \alpha_{n-2}(s)$$

2.4. PLANT PARAMETRIC MODELS

$$P(s) = p^\top \alpha_{n-1}(s)$$

$$\alpha_i(s) \triangleq [s^i, s^{i-1}, \ldots, s, 1]^\top$$

$q \in \mathcal{R}^{n-1}, p \in \mathcal{R}^n$ are the coefficient vectors of $Q(s) - s^{n-1}$, $P(s)$, respectively, and $A(s)$ is a monic Hurwitz polynomial of degree $2n - m - 1$. The Diophantine equation (2.4.26) relating $Z_0(s), R_0(s), k_0$ to $P(s), Q(s)$, and $A(s)$ arises in control designs, such as model reference control, to be discussed in later chapters. The polynomials $P(s)$ and $Q(s)$ are usually the controller polynomials to be calculated by solving (2.4.26) for a given $A(s)$. Our objective here is to obtain a parameterization of (2.4.25), in terms of the coefficients of $P(s)$ and $Q(s)$, that is independent of the coefficients of $Z_0(s)$ and $R_0(s)$. We achieve this objective by using (2.4.26) to eliminate the dependence of (2.4.25) on $Z_0(s)$ and $R_0(s)$ as follows:

From (2.4.25), we obtain

$$Q(s)R_0(s)y = k_0 Z_0(s)Q(s)u \quad (2.4.27)$$

by rewriting (2.4.25) as $R_0(s)y = k_0 Z_0(s)u$ and operating on each side by $Q(s)$. Using $Q(s)R_0(s) = Z_0(s)(A(s) - k_0 P(s))$ obtained from (2.4.26) in (2.4.27), we have

$$Z_0(s)(A(s) - k_0 P(s))y = k_0 Z_0(s)Q(s)u \quad (2.4.28)$$

Because $Z_0(s)$ is Hurwitz, we filter each side of (2.4.28) by $\frac{1}{Z_0(s)}$ to obtain

$$A(s)y = k_0 P(s)y + k_0 Q(s)u \quad (2.4.29)$$

and write (2.4.29) as

$$A(s)y = k_0[p^\top \alpha_{n-1}(s)y + q^\top \alpha_{n-2}(s)u + s^{n-1}u] \quad (2.4.30)$$

We now have various choices to make. We can filter each side of (2.4.30) with the stable filter $\frac{1}{A(s)}$ and obtain

$$y = k_0 \left[p^\top \frac{\alpha_{n-1}}{A(s)} y + q^\top \frac{\alpha_{n-2}(s)}{A(s)} u + \frac{s^{n-1}}{A(s)} u \right]$$

which may be written in the compact form

$$y = k_0(\theta^{*\top}\phi + z_0) \quad (2.4.31)$$

where $\theta^* = [q^\top, p^\top]^\top$, $\phi = \left[\frac{\alpha_{n-2}^\top(s)}{A(s)}u, \frac{\alpha_{n-1}^\top(s)}{A(s)}y\right]^\top$, and $z_0 = \frac{s^{n-1}}{A(s)}u$. We can also filter each side of (2.4.30) using an arbitrary stable filter $\frac{1}{\Lambda(s)}$ whose order n_λ satisfies $2n - m - 1 \geq n_\lambda \geq n - 1$ to obtain

$$y = W(s)k_0(\theta^{*\top}\phi + z_0) \tag{2.4.32}$$

where now $\phi = \left[\frac{\alpha_{n-2}^\top(s)}{\Lambda(s)}u, \frac{\alpha_{n-1}^\top(s)}{\Lambda(s)}y\right]^\top$, $z_0 = \frac{s^{n-1}}{\Lambda(s)}u$, and $W(s) = \frac{\Lambda(s)}{A(s)}$ is a proper transfer function.

In (2.4.31) and (2.4.32), ϕ and z_0 may be generated by filtering the input u and output y of the system. Therefore, if u and y are measurable, then all signals in (2.4.31) and (2.4.32) can be generated, and the only possible unknowns are k_0 and θ^*. If k_0 is known, it can be absorbed in the signals ϕ and z_0, leading to models that are affine in θ^* of the form

$$\bar{y} = W(s)\theta^{*\top}\bar{\phi} \tag{2.4.33}$$

where $\bar{y} = y - W(s)k_0 z_0$, $\bar{\phi} = k_0\phi$. If k_0, however, is unknown and is part of the parameters of interest, then (2.4.31) and (2.4.32) are not affine with respect to the parameters k_0 and θ^*, but instead, k_0 and θ^* appear in a special bilinear form. For this reason, we refer to (2.4.31) and (2.4.32) as *bilinear parametric models* to distinguish them from (2.4.7) to (2.4.9) and (2.4.33), which we refer to as *linear parametric* or affine parametric models. The forms of the linear and bilinear parametric models are general enough to include parameterizations of some systems with dynamics that are not necessarily linear, as illustrated by the following example.

Example 2.4.3 Let us consider the nonlinear scalar system

$$\dot{x} = a_0 f(x,t) + b_0 g(x,t) + c_0 u \tag{2.4.34}$$

where a_0, b_0, and c_0 are constant scalars; $f(x,t)$ and $g(x,t)$ are known nonlinear functions that can be calculated at each time t; and u, x is the input and state of the system, respectively. We assume that f, g, and u are such that for each initial condition $x(0) = x_0$, (2.4.34) has only one solution defined for all $t \in [0, \infty)$. If x and u are measured, (2.4.34) can be expressed in the form of parametric model (2.4.33) by filtering each side of (2.4.34) with a stable strictly proper transfer function $W_f(s)$, i.e.,

$$z = W_f(s)\theta^{*\top}\phi \tag{2.4.35}$$

2.5. PROBLEMS

where $z = sW_f(s)x$, $\theta^* = [a_0, b_0, c_0]^T$, and $\phi = [f(x,t), g(x,t), u]^T$. Instead of (2.4.35), we may also write (2.4.34) in the form

$$\dot{x} = -a_m x + a_m x + \theta^{*T} \phi$$

for some $a_m > 0$, or

$$x = \frac{1}{s + a_m}[a_m x + \theta^{*T} \phi]$$

Then

$$z \triangleq x - \frac{a_m}{s + a_m} x = \frac{1}{s + a_m} \theta^{*T} \phi \qquad (2.4.36)$$

which is in the form of (2.4.35) with $W_f(s) = \frac{1}{s + a_m}$. We may continue and rewrite (2.4.35) (respectively (2.4.36)) as

$$z = \theta^{*T} \phi_f, \quad \phi_f = W_f(s)\phi \qquad (2.4.37)$$

which is in the form of (2.4.8). ▽

The nonlinear example demonstrates the fact that the parameter θ^* appears linearly in (2.4.35) and (2.4.37) does not mean that the dynamics are linear.

2.5 Problems

2.1 Verify that $x(t)$ and $y(t)$ given by (2.2.4) satisfy the differential equation (2.2.2).

2.2 Check the controllability and observability of the following systems:

(a)

$$\dot{x} = \begin{bmatrix} -0.2 & 0 \\ -1 & 0.8 \end{bmatrix} x + \begin{bmatrix} 1 \\ 1 \end{bmatrix} u$$

$$y = [-1, 1]x$$

(b)

$$\dot{x} = \begin{bmatrix} -1 & 1 & 0 \\ 0 & -1 & 0 \\ 0 & 0 & -2 \end{bmatrix} x + \begin{bmatrix} 0 \\ 1 \\ 1 \end{bmatrix} u$$

$$y = [1, 1, 1]x$$

(c)
$$\dot{x} = \begin{bmatrix} -5 & 1 \\ -6 & 0 \end{bmatrix} x + \begin{bmatrix} 1 \\ 1 \end{bmatrix} u$$
$$y = [1, 1]x$$

2.3 Show that (A, B) is controllable if and only if the augmented matrix $[sI - A, B]$ is of full rank for all $s \in \mathcal{C}$.

2.4 The following state equation describes approximately the motion of a hot air balloon:
$$\begin{bmatrix} \dot{x}_1 \\ \dot{x}_2 \\ \dot{x}_3 \end{bmatrix} = \begin{bmatrix} -\frac{1}{\tau_1} & 0 & 0 \\ \sigma & -\frac{1}{\tau_2} & 0 \\ 0 & 1 & 0 \end{bmatrix} \begin{bmatrix} x_1 \\ x_2 \\ x_3 \end{bmatrix} + \begin{bmatrix} 1 \\ 0 \\ 0 \end{bmatrix} u + \begin{bmatrix} 0 \\ \frac{1}{\tau_2} \\ 0 \end{bmatrix} w$$
$$y = [0 \ 0 \ 1]x$$

where x_1: the temperature change of air in the balloon away from the equilibrium temperature; x_2: vertical velocity of the balloon; x_3: change in altitude from equilibrium altitude; u: control input that is proportional to the change in heat added to air in the balloon; w: vertical wind speed; and σ, τ_1, τ_2 are parameters determined by the design of the balloon.

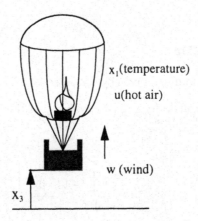

(a) Let $w = 0$. Is the system completely controllable? Is it completely observable?

(b) If it is completely controllable, transform the state-space representation into the controller canonical form.

(c) If it is completely observable, transform the state-space representation into the observer canonical form.

2.5. PROBLEMS

(d) Assume $w = $ constant. Can the augmented state $x_a \triangleq [x^T, w]^T$ be observed from y?

(e) Assume $u = 0$. Can the states be controlled by w?

2.5 Derive the following transfer functions for the system described in Problem 2.4:

(a) $G_1(s) \triangleq \frac{Y(s)}{U(s)}$ when $w = 0$ and $y = x_3$.

(b) $G_2(s) \triangleq \frac{Y(s)}{W(s)}$ when $u = 0$ and $y = x_3$.

(c) $G_3(s) \triangleq \frac{Y_1(s)}{U(s)}$ when $w = 0$ and $y_1 = x_1$.

(d) $G_4(s) \triangleq \frac{Y_1(s)}{W(s)}$ when $u = 0$ and $y_1 = x_1$.

2.6 Let $a(s) = (s + \alpha)^3, b(s) = \beta$, where α, β are constants with $\beta \neq 0$.

(a) Write the Sylvester matrix of $a(s)$ and $b(s)$.

(b) Suppose $p_0(s), l_0(s)$ is a solution of the polynomial equation

$$a(s)l(s) + b(s)p(s) = 1 \tag{2.5.1}$$

Show that $(p_1(s), l_1(s))$ is a solution of (2.5.1) if and only if $p_1(s), l_1(s)$ can be expressed as

$$p_1(s) = p_0(s) + r(s)a(s)$$
$$l_1(s) = l_0(s) - r(s)b(s)$$

for any polynomial $r(s)$.

(c) Find the solution of (2.5.1) for which $p(s)$ has the lowest degree and $l(s)/p(s)$ is a proper rational function.

2.7 Consider the third order plant

$$y = G(s)u$$

where

$$G(s) = \frac{b_2 s^2 + b_1 s + b_0}{s^3 + a_2 s^2 + a_1 s + a_0}$$

(a) Write the parametric model of the plant in the form of (2.4.8) or (2.4.13) when $\theta^* = [b_2, b_1, b_0, a_2, a_1, a_0]^T$.

(b) If a_0, a_1, and a_2 are known, i.e., $a_0 = 2, a_1 = 1$, and $a_2 = 3$, write a parametric model for the plant in terms of $\theta^* = [b_2, b_1, b_0]^T$.

(c) If b_0, b_1, and b_2 are known, i.e., $b_0 = 1, b_1 = b_2 = 0$, develop a parametric model in terms of $\theta^* = [a_2, a_1, a_0]^T$.

2.8 Consider the spring-mass-dashpot system shown below:

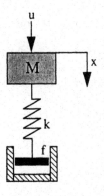

where k is the spring constant, f the viscous-friction or damping coefficient, m the mass of the system, u the forcing input, and x the displacement of the mass M. If we assume a "linear" spring, i.e., the force acting on the spring is proportional to the displacement, and a friction force proportional to velocity, i.e., $\dot{x}$, we obtain, using Newton's law, the differential equation

$$M\ddot{x} = u - kx - f\dot{x}$$

that describes the dynamic system.

(a) Give a state-space representation of the system.

(b) Calculate the transfer function that relates x with u.

(c) Obtain a linear parametric model of the form

$$z = \theta^{*\top}\phi$$

where $\theta^* = [M, k, f]^\top$ and z, ϕ are signals that can be generated from the measurements of u, x without the use of differentiators.

2.9 Verify that (2.4.11) and (2.4.12) are nonminimal state-space representations of the system described by (2.4.1). Show that for the same input $u(t)$, the output response $y(t)$ is exactly the same for both systems. (Hint: Verify that $C_0^\top[adj(sI - \Lambda_c)]B_0 = C^\top[adj(sI - A)]$ for some $C_0 \in \mathcal{R}^n, B_0 \in \mathcal{R}^{n \times n}$ by using the identity

$$[adj(sI - A)] = s^{n-1}I + s^{n-2}(A + a_{n-1}I) + s^{n-3}(A^2 + a_{n-1}A + a_{n-2}I)$$
$$+ \cdots + (A^{n-1} + a_{n-1}A^{n-2} + \cdots + a_1 I)$$

and choosing C_0 such that (C_0, Λ_c) is an observable pair.)

2.10 Write a state-space representation for the following systems:

2.5. PROBLEMS

(a) $\phi = \frac{\alpha_{n-1}(s)}{\Lambda(s)} u$, $\Lambda(s)$ is monic of order n.

(b) $\phi = \frac{\alpha_{n-1}(s)}{\Lambda_1(s)} u$, $\Lambda_1(s)$ is monic of order $n-1$.

(c) $\phi = \frac{\alpha_m(s)}{\Lambda_1(s)} u$, $m \leq n-1$, $\Lambda_1(s)$ is monic of order $n-1$.

2.11 Show that
$$(sI - \Lambda_c)^{-1} l = \left(C_o^T (sI - \Lambda_o)^{-1} \right)^T = \frac{\alpha_{n-1}(s)}{\Lambda(s)}$$
where (Λ_c, l) is in the controller form and (C_o, Λ_o) is in the observer form.

2.12 Show that there exists constant matrices $Q_i \in \mathcal{R}^{(n-1) \times (n-1)}$ such that
$$(sI - \Lambda_0)^{-1} d_i = Q_i \frac{\alpha_{n-2}(s)}{\Lambda(s)}, \quad i = 1, 2, \ldots, n$$
where $d_1 = -\lambda$; $\Lambda(s) = s^{n-1} + \lambda^T \alpha_{n-2}(s) = \det(sI - \Lambda_0)$, $\Lambda_0 = \begin{bmatrix} -\lambda & \begin{array}{c} I_{n-2} \\ \hline 0 \end{array} \end{bmatrix}$;
$d_i = [0, \ldots, 0, 1, 0, \ldots, 0]^T \in \mathcal{R}^{n-1}$ whose $(i-1)$th element is equal to 1, and $i = 2, 3, \ldots, n$.

Chapter 3

Stability

3.1 Introduction

The concept of stability is concerned with the investigation and characterization of the behavior of dynamic systems.

Stability plays a crucial role in system theory and control engineering, and has been investigated extensively in the past century. Some of the most fundamental concepts of stability were introduced by the Russian mathematician and engineer Alexandr Lyapunov in [133]. The work of Lyapunov was extended and brought to the attention of the larger control engineering and applied mathematics community by LaSalle and Lefschetz [124, 125, 126], Krasovskii [107], Hahn [78], Massera [139], Malkin [134], Kalman and Bertram [97], and many others.

In control systems, we are concerned with changing the properties of dynamic systems so that they can exhibit acceptable behavior when perturbed from their operating point by external forces. The purpose of this chapter is to present some basic definitions and results on stability that are useful for the design and analysis of control systems. Most of the results presented are general and can be found in standard textbooks. Others are more specific and are developed for adaptive systems. The proofs for most of the general results are omitted, and appropriate references are provided. Those that are very relevant to the understanding of the material presented in later chapters are given in detail.

In Section 3.2, we present the definitions and properties of various norms

3.2. PRELIMINARIES

and functions that are used in the remainder of the book. The concept of I/O stability and some standard results from functional analysis are presented in Section 3.3. These include useful results on the I/O properties of linear systems, the small gain theorem that is widely used in robust control design and analysis, and the $\mathcal{L}_{2\delta}$-norm and Bellman-Gronwall (B-G) Lemma that are important tools in the analysis of adaptive systems. The definitions of Lyapunov stability and related theorems for linear and nonlinear systems are presented in Section 3.4. The concept of passivity, in particular of positive real and strictly positive real transfer functions, and its relation to Lyapunov stability play an important role in the design of stable adaptive systems. Section 3.5 contains some basic results on positive real functions, and their connections to Lyapunov functions and stability that are relevant to adaptive systems.

In Section 3.6, the focus is on some elementary results and principles that are used in the design and analysis of LTI feedback systems. We concentrate on the notion of internal stability that we use to motivate the correct way of computing the characteristic equation of a feedback system and determining its stability properties. The use of sensitivity and complementary sensitivity functions and some fundamental trade-offs in LTI feedback systems are briefly mentioned to refresh the memory of the reader. The internal model principle and its use to reject the effects of external disturbances in feedback systems is presented.

A reader who is somewhat familiar with Lyapunov stability and the basic properties of norms may skip this chapter. He or she may use it as reference and come back to it whenever necessary. For the reader who is unfamiliar with Lyapunov stability and I/O properties of linear systems, the chapter offers a complete tutorial coverage of all the notions and results that are relevant to the understanding of the rest of the book.

3.2 Preliminaries

3.2.1 Norms and $\mathcal{L}_p$ Spaces

For many of the arguments for scalar equations to be extended and remain valid for vector equations, we need an analog for vectors of the absolute value of a scalar. This is provided by the norm of a vector.

Definition 3.2.1 *The* **norm** $|x|$ *of a vector x is a real valued function with the following properties:*

(i) $|x| \geq 0$ *with* $|x| = 0$ *if and only if* $x = 0$
(ii) $|\alpha x| = |\alpha||x|$ *for any scalar* α
(iii) $|x + y| \leq |x| + |y|$ *(triangle inequality)*

The norm $|x|$ of a vector x can be thought of as the size or length of the vector x. Similarly, $|x - y|$ can be thought of as the distance between the vectors x and y.

An $m \times n$ matrix A represents a linear mapping from n-dimensional space $\mathcal{R}^n$ into m-dimensional space $\mathcal{R}^m$. We define the *induced norm* of A as follows:

Definition 3.2.2 *Let $|\cdot|$ be a given vector norm. Then for each matrix $A \in \mathcal{R}^{m \times n}$, the quantity $\|A\|$ defined by*

$$\|A\| \triangleq \sup_{\substack{x \neq 0 \\ x \in \mathcal{R}^n}} \frac{|Ax|}{|x|} = \sup_{|x| \leq 1} |Ax| = \sup_{|x| = 1} |Ax|$$

is called the **induced (matrix) norm** *of A corresponding to the vector norm $|\cdot|$.*

The induced matrix norm satisfies the properties (i) to (iii) of Definition 3.2.1.

Some of the properties of the induced norm that we will often use in this book are summarized as follows:

(i) $|Ax| \leq \|A\||x|, \quad \forall x \in \mathcal{R}^n$
(ii) $\|A + B\| \leq \|A\| + \|B\|$
(iii) $\|AB\| \leq \|A\|\|B\|$

where A, B are arbitrary matrices of compatible dimensions. Table 3.1 shows some of the most commonly used norms on $\mathcal{R}^n$.

It should be noted that the function $\|A\|_s \triangleq \max_{ij} |a_{ij}|$, where $A \in \mathcal{R}^{m \times n}$ and a_{ij} is the (i, j) element of A satisfies the properties (i) to (iii) of Definition 3.2.1. It is not, however, an induced matrix norm because no vector norm exists such that $\|\cdot\|_s$ is the corresponding induced norm. Note that $\|\cdot\|_s$ does not satisfy property (c).

3.2. PRELIMINARIES

Table 3.1 Commonly used norms

Norm on $\mathcal{R}^n$	Induced norm on $\mathcal{R}^{m \times n}$
$\|x\|_\infty = \max_i \|x_i\|$ (infinity norm)	$\|A\|_\infty = \max_i \sum_j \|a_{ij}\|$ (row sum)
$\|x\|_1 = \sum_i \|x_i\|$	$\|A\|_1 = \max_j \sum_i \|a_{ij}\|$ (column sum)
$\|x\|_2 = (\sum_i \|x_i\|^2)^{1/2}$	$\|A\|_2 = [\lambda_m(A^T A)]^{1/2}$, where $\lambda_m(M)$
(Euclidean norm)	is the maximum eigenvalue of M

Example 3.2.1
(i) Let $x = [1, 2, -10, 0]^T$. Using Table 3.1, we have

$$|x|_\infty = 10, \quad |x|_1 = 13, \quad |x|_2 = \sqrt{105}$$

(ii) Let

$$A = \begin{bmatrix} 0 & 5 \\ 1 & 0 \\ 0 & -10 \end{bmatrix}, \quad B = \begin{bmatrix} -1 & 5 \\ 0 & 2 \end{bmatrix}$$

Using Table 3.1, we have

$\|A\|_1 = 15, \quad \|A\|_2 = 11.18, \quad \|A\|_\infty = 10$
$\|B\|_1 = 7, \quad \|B\|_2 = 5.465, \quad \|B\|_\infty = 6$
$\|AB\|_1 = 35, \quad \|AB\|_2 = 22.91, \quad \|AB\|_\infty = 20$

which can be used to verify property (iii) of the induced norm. ▽

For functions of time, we define the $\mathcal{L}_p$ norm

$$\|x\|_p \triangleq \left(\int_0^\infty |x(\tau)|^p d\tau \right)^{1/p}$$

for $p \in [1, \infty)$ and say that $x \in \mathcal{L}_p$ when $\|x\|_p$ exists (i.e., when $\|x\|_p$ is finite). The $\mathcal{L}_\infty$ norm is defined as

$$\|x\|_\infty \triangleq \sup_{t \geq 0} |x(t)|$$

and we say that $x \in \mathcal{L}_\infty$ when $\|x\|_\infty$ exists.

In the above $\mathcal{L}_p, \mathcal{L}_\infty$ norm definitions, $x(t)$ can be a scalar or a vector function. If x is a scalar function, then $|\cdot|$ denotes the absolute value. If x is a vector function in $\mathcal{R}^n$ then $|\cdot|$ denotes any norm in $\mathcal{R}^n$.

Similarly, for sequences we define the l_p norm as

$$\|x\|_p \triangleq \left(\sum_{i=1}^{\infty}|x_i|^p\right)^{1/p}, \quad 1 \leq p < \infty$$

and the l_∞ norm as

$$\|x\|_\infty \triangleq \sup_{i \geq 1}|x_i|$$

where $x = (x_1, x_2, \ldots)$ and $x_i \in \mathcal{R}$. We say $x \in l_p$ (respectively $x \in l_\infty$) if $\|x\|_p$ (respectively $\|x\|_\infty$) exists.

We are usually concerned with classes of functions of time that do not belong to $\mathcal{L}_p$. To handle such functions we define the $\mathcal{L}_{pe}$ norm

$$\|x_t\|_p \triangleq \left(\int_0^t |x(\tau)|^p d\tau\right)^{\frac{1}{p}}$$

for $p \in [1, \infty)$ and say that $x \in \mathcal{L}_{pe}$ when $\|x_t\|_p$ exists for any finite t. Similarly, the $\mathcal{L}_{\infty e}$ norm is defined as

$$\|x_t\|_\infty \triangleq \sup_{0 \leq \tau \leq t}|x(\tau)|$$

The function t^2 does not belong to $\mathcal{L}_p$ but $t^2 \in \mathcal{L}_{pe}$. Similarly, any continuous function of time belongs to $\mathcal{L}_{pe}$ but it may not belong to $\mathcal{L}_p$.

For each $p \in [1, \infty]$, the set of functions that belong to $\mathcal{L}_p$ (respectively $\mathcal{L}_{pe}$) form a *linear vector space* called $\mathcal{L}_p$ *space* (respectively $\mathcal{L}_{pe}$ *space*) [42]. If we define the truncated function f_t as

$$f_t(\tau) \triangleq \begin{cases} f(\tau) & 0 \leq \tau \leq t \\ 0 & \tau > t \end{cases}$$

for all $t \in [0, \infty)$, then it is clear that for any $p \in [1, \infty]$, $f \in \mathcal{L}_{pe}$ implies that $f_t \in \mathcal{L}_p$ for any finite t. The $\mathcal{L}_{pe}$ space is called the *extended $\mathcal{L}_p$ space* and is defined as the set of all functions f such that $f_t \in \mathcal{L}_p$.

It can be easily verified that the $\mathcal{L}_p$ and $\mathcal{L}_{pe}$ norms satisfy the properties of the norm given by Definition 3.2.1. It should be understood, however, that elements of $\mathcal{L}_p$ and $\mathcal{L}_{pe}$ are *equivalent classes* [42], i.e., if $f, g \in \mathcal{L}_p$ and $\|f - g\|_p = 0$, the functions f and g are considered to be the same element of $\mathcal{L}_p$ even though $f(t) \neq g(t)$ for some values of t. The following lemmas give some of the properties of $\mathcal{L}_p$ and $\mathcal{L}_{pe}$ spaces that we use later.

3.2. PRELIMINARIES

Lemma 3.2.1 (Hölder's Inequality) *If $p, q \in [1, \infty]$ and $\frac{1}{p} + \frac{1}{q} = 1$, then $f \in \mathcal{L}_p, g \in \mathcal{L}_q$ imply that $fg \in \mathcal{L}_1$ and*

$$\|fg\|_1 \leq \|f\|_p \|g\|_q$$

When $p = q = 2$, the Hölder's inequality becomes the *Schwartz inequality*, i.e.,

$$\|fg\|_1 \leq \|f\|_2 \|g\|_2 \tag{3.2.1}$$

Lemma 3.2.2 (Minkowski Inequality) *For $p \in [1, \infty]$, $f, g \in \mathcal{L}_p$ imply that $f + g \in \mathcal{L}_p$ and*

$$\|f + g\|_p \leq \|f\|_p + \|g\|_p \tag{3.2.2}$$

The proofs of Lemma 3.2.1 and 3.2.2 can be found in any standard book on real analysis such as [199, 200].

We should note that the above lemmas also hold for the truncated functions f_t, g_t of f, g, respectively, provided $f, g \in \mathcal{L}_{pe}$. For example, if f and g are continuous functions, then $f, g \in \mathcal{L}_{pe}$, i.e., $f_t, g_t \in \mathcal{L}_p$ for any finite $t \in [0, \infty)$ and from (3.2.1) we have $\|(fg)_t\|_1 \leq \|f_t\|_2 \|g_t\|_2$, i.e.,

$$\int_0^t |f(\tau) g(\tau)| d\tau \leq \left(\int_0^t |f(\tau)|^2 d\tau \right)^{\frac{1}{2}} \left(\int_0^t |g(\tau)|^2 d\tau \right)^{\frac{1}{2}} \tag{3.2.3}$$

which holds for any finite $t \geq 0$. We use the above Schwartz inequality extensively throughout this book.

Example 3.2.2 Consider the function $f(t) = \frac{1}{1+t}$. Then,

$$\|f\|_\infty = \sup_{t \geq 0} \left| \frac{1}{1+t} \right| = 1, \quad \|f\|_2 = \left(\int_0^\infty \frac{1}{(1+t)^2} dt \right)^{\frac{1}{2}} = 1$$

$$\|f\|_1 = \int_0^\infty \frac{1}{1+t} dt = \lim_{t \to \infty} \ln(1+t) \to \infty$$

Hence, $f \in \mathcal{L}_2 \bigcap \mathcal{L}_\infty$ but $f \notin \mathcal{L}_1$; f, however, belongs to $\mathcal{L}_{1e}$, i.e., for any finite $t \geq 0$, we have

$$\int_0^t \frac{1}{1+\tau} d\tau = \ln(1+t) < \infty \qquad \triangledown$$

Example 3.2.3 Consider the functions

$$f(t) = 1+t, \quad g(t) = \frac{1}{1+t}, \quad \text{for } t \geq 0$$

It is clear that $f \notin \mathcal{L}_p$ for any $p \in [1, \infty]$ and $g \notin \mathcal{L}_1$. Both functions, however, belong to $\mathcal{L}_{pe}$; and can be used to verify the Schwartz inequality (3.2.3)

$$\|(fg)_t\|_1 \leq \|f_t\|_2 \|g_t\|_2$$

i.e.,

$$\int_0^t 1 d\tau \leq \left(\int_0^t (1+\tau)^2 d\tau\right)^{\frac{1}{2}} \left(\int_0^t \frac{1}{(1+\tau)^2} d\tau\right)^{\frac{1}{2}}$$

for any $t \in [0, \infty)$ or equivalently

$$t \leq \left(\frac{t(t^2 + 3t + 3)}{3}\right)^{\frac{1}{2}} \left(\frac{t}{1+t}\right)^{\frac{1}{2}}$$

which is true for any $t \geq 0$. $\triangledown$

In the remaining chapters of the book, we adopt the following notation regarding norms unless stated otherwise. We will drop the subscript 2 from $|\cdot|_2, \|\cdot\|_2$ when dealing with the Euclidean norm, the induced Euclidean norm, and the $\mathcal{L}_2$ norm. If $x : \mathcal{R}^+ \mapsto \mathcal{R}^n$, then

$|x(t)|$ represents the vector norm in $\mathcal{R}^n$ at each time t

$\|x_t\|_p$ represents the $\mathcal{L}_{pe}$ norm of the function $|x(t)|$

$\|x\|_p$ represents the $\mathcal{L}_p$ norm of the function $|x(t)|$

If $A \in \mathcal{R}^{m \times n}$, then

$\|A\|_i$ represents the induced matrix norm corresponding to the vector norm $|\cdot|_i$.

If $A : \mathcal{R}^+ \mapsto \mathcal{R}^{m \times n}$ has elements that are functions of time t, then

$\|A(t)\|_i$ represents the induced matrix norm corresponding to the vector norm $|\cdot|_i$ at time t.

3.2.2 Properties of Functions

Let us start with some definitions.

Definition 3.2.3 (Continuity) *A function $f : [0, \infty) \mapsto \mathcal{R}$ is **continuous** on $[0, \infty)$ if for any given $\epsilon_0 > 0$ there exists a $\delta(\epsilon_0, t_0)$ such that $\forall t_0, t \in [0, \infty)$ for which $|t - t_0| < \delta(\epsilon_0, t_0)$ we have $|f(t) - f(t_0)| < \epsilon_0$.*

3.2. PRELIMINARIES

Definition 3.2.4 (Uniform Continuity) *A function $f : [0, \infty) \mapsto \mathcal{R}$ is* **uniformly continuous** *on $[0, \infty)$ if for any given $\epsilon_0 > 0$ there exists a $\delta(\epsilon_0)$ such that $\forall t_0, t \in [0, \infty)$ for which $|t - t_0| < \delta(\epsilon_0)$ we have $|f(t) - f(t_0)| < \epsilon_0$.*

Definition 3.2.5 (Piecewise Continuity) *A function $f : [0, \infty) \mapsto \mathcal{R}$ is* **piecewise continuous** *on $[0, \infty)$ if f is continuous on any finite interval $[t_0, t_1] \subset [0, \infty)$ except for a finite number of points.*

Definition 3.2.6 (Absolute Continuity) *A function $f : [a, b] \mapsto \mathcal{R}$ is* **absolutely continuous** *on $[a, b]$ iff, for any $\epsilon_0 > 0$, there is a $\delta > 0$ such that*

$$\sum_{i=1}^{n} |f(\alpha_i) - f(\beta_i)| < \epsilon_0$$

for any finite collection of subintervals (α_i, β_i) of $[a, b]$ with $\sum_{i=1}^{n} |\alpha_i - \beta_i| < \delta$.

Definition 3.2.7 (Lipschitz) *A function $f : [a, b] \to \mathcal{R}$ is* **Lipschitz** *on $[a, b]$ if $|f(x_1) - f(x_2)| \leq k|x_1 - x_2| \; \forall x_1, x_2 \in [a, b]$, where $k \geq 0$ is a constant referred to as the* **Lipschitz constant**.

The function $f(t) = \sin(\frac{1}{t})$ is continuous on $(0, \infty)$, but is not uniformly continuous (verify!).

A function defined by a square wave of finite frequency is not continuous on $[0, \infty)$, but it is piecewise continuous.

Note that a uniformly continuous function is also continuous. A function f with $\dot{f} \in \mathcal{L}_\infty$ is uniformly continuous on $[0, \infty)$. Therefore, an easy way of checking the uniform continuity of $f(t)$ is to check the boundedness of $\dot{f}$. If f is Lipschitz on $[a, b]$, then it is absolutely continuous.

The following facts about functions are important in understanding some of the stability arguments which are often made in the analysis of adaptive systems.

Fact 1 $\lim_{t \to \infty} \dot{f}(t) = 0$ *does not* imply that $f(t)$ has a limit as $t \to \infty$.

For example, consider the function $f(t) = \sin(\sqrt{1+t})$. We have

$$\dot{f} = \frac{\cos \sqrt{1+t}}{2\sqrt{1+t}} \to 0 \text{ as } t \to \infty$$

but $f(t)$ has no limit. Another example is $f(t) = \sqrt{1+t}\sin(\ln(1+t))$, which is an unbounded function of time. Yet

$$\dot f(t) = \frac{\sin(\ln(1+t))}{2\sqrt{1+t}} + \frac{\cos(\ln(1+t))}{\sqrt{1+t}} \to 0 \text{ as } t \to \infty$$

Fact 2 $\lim_{t \to \infty} f(t) = c$ for some constant $c \in \mathcal{R}$ *does not* imply that $\dot f(t) \to 0$ as $t \to \infty$.

For example, the function $f(t) = \frac{\sin(1+t)^n}{1+t}$ tends to zero as $t \to \infty$ for any finite integer n but

$$\dot f = -\frac{\sin(1+t)^n}{(1+t)^2} + n(1+t)^{n-2}\cos(1+t)^n$$

has no limit for $n \geq 2$ and becomes unbounded as $t \to \infty$ for $n > 2$.

Some important lemmas that we frequently use in the analysis of adaptive schemes are the following:

Lemma 3.2.3 *The following is ture for scalar-valued functions:*
(i) *A function $f(t)$ that is bounded from below and is nonincreasing has a limit as $t \to \infty$.*
(ii) *Consider the nonnegative scalar functions $f(t), g(t)$ defined for all $t \geq 0$. If $f(t) \leq g(t)$ $\forall t \geq 0$ and $g \in \mathcal{L}_p$, then $f \in \mathcal{L}_p$ for all $p \in [1, \infty]$.*

Proof (i) Because f is bounded from below, its infimum f_m exists, i.e.,

$$f_m = \inf_{0 \leq t \leq \infty} f(t)$$

which implies that there exists a sequence $\{t_n\} \in \mathcal{R}^+$ such that $\lim_{n \to \infty} f(t_n) = f_m$. This, in turn, implies that given any $\epsilon_0 > 0$ there exists an integer $N > 0$ such that

$$|f(t_n) - f_m| < \epsilon_0, \quad \forall n \geq N$$

Because f is nonincreasing, there exists an $n_0 \geq N$ such that for any $t \geq t_{n_0}$ and some $n_0 \geq N$ we have

$$f(t) \leq f(t_{n_0})$$

and

$$|f(t) - f_m| \leq |f(t_{n_0}) - f_m| < \epsilon_0$$

for any $t \geq t_{n_0}$. Because $\epsilon_0 > 0$ is any given number, it follows that $\lim_{t \to \infty} f(t) = f_m$.

3.2. PRELIMINARIES

(ii) We have

$$z(t) \triangleq \left(\int_0^t f^p(\tau) d\tau \right)^{\frac{1}{p}} \leq \left(\int_0^\infty g^p(\tau) d\tau \right)^{\frac{1}{p}} < \infty, \quad \forall t \geq 0$$

Because $0 \leq z(t) < \infty$ and $z(t)$ is nondecreasing, we can establish, as in (i), that $z(t)$ has a limit, i.e., $\lim_{t \to \infty} z(t) = \bar{z} < \infty$, which implies that $f \in \mathcal{L}_p$. For $p = \infty$, the proof is straightforward. □

Lemma 3.2.3 (i) does not imply that $f \in \mathcal{L}_\infty$. For example, the function $f(t) = \frac{1}{t}$ with $t \in (0, \infty)$ is bounded from below, i.e., $f(t) \geq 0$ and is nonincreasing, but it becomes unbounded as $t \to 0$. If, however, $f(0)$ is finite, then it follows from the nonincreasing property $f(t) \leq f(0)$ $\forall t \geq 0$ that $f \in \mathcal{L}_\infty$. A special case of Lemma 3.2.3 that we often use in this book is when $f \geq 0$ and $\dot{f} \leq 0$.

Lemma 3.2.4 *Let $f, V : [0, \infty) \mapsto \mathcal{R}$. Then*

$$\dot{V} \leq -\alpha V + f, \quad \forall t \geq t_0 \geq 0$$

implies that

$$V(t) \leq e^{-\alpha(t-t_0)} V(t_0) + \int_{t_0}^t e^{-\alpha(t-\tau)} f(\tau) d\tau, \quad \forall t \geq t_0 \geq 0$$

for any finite constant α.

Proof Let $w(t) \triangleq \dot{V} + \alpha V - f$. We have $w(t) \leq 0$ and

$$\dot{V} = -\alpha V + f + w$$

implies that

$$V(t) = e^{-\alpha(t-t_0)} V(t_0) + \int_{t_0}^t e^{-\alpha(t-\tau)} f(\tau) d\tau + \int_{t_0}^t e^{-\alpha(t-\tau)} w(\tau) d\tau$$

Because $w(t) \leq 0$ $\forall t \geq t_0 \geq 0$, we have

$$V(t) \leq e^{-\alpha(t-t_0)} V(t_0) + \int_{t_0}^t e^{-\alpha(t-\tau)} f(\tau) d\tau$$

□

Lemma 3.2.5 *If $f, \dot{f} \in \mathcal{L}_\infty$ and $f \in \mathcal{L}_p$ for some $p \in [1, \infty)$, then $f(t) \to 0$ as $t \to \infty$.*

The result of Lemma 3.2.5 is a special case of a more general result given by Barbălat's Lemma stated below.

Lemma 3.2.6 (Barbălat's Lemma [192]) *If $\lim_{t\to\infty} \int_0^t f(\tau)d\tau$ exists and is finite, and $f(t)$ is a uniformly continuous function, then $\lim_{t\to\infty} f(t) = 0$.*

Proof Assume that $\lim_{t\to\infty} f(t) = 0$ does not hold, i.e., either the limit does not exist or it is not equal to zero. This implies that there exists an $\epsilon_0 > 0$ such that for every $T > 0$, one can find a sequence of numbers $t_i > T$ such that $|f(t_i)| > \epsilon_0$ for all i.

Because f is uniformly continuous, there exists a number $\delta(\epsilon_0) > 0$ such that

$$|f(t) - f(t_i)| < \frac{\epsilon_0}{2} \text{ for every } t \in [t_i, t_i + \delta(\epsilon_0)]$$

Hence, for every $t \in [t_i, t_i + \delta(\epsilon_0)]$, we have

$$\begin{aligned} |f(t)| &= |f(t) - f(t_i) + f(t_i)| \geq |f(t_i)| - |f(t) - f(t_i)| \\ &\geq \epsilon_0 - \frac{\epsilon_0}{2} = \frac{\epsilon_0}{2} \end{aligned}$$

which implies that

$$\left| \int_{t_i}^{t_i+\delta(\epsilon_0)} f(\tau)d\tau \right| = \int_{t_i}^{t_i+\delta(\epsilon_0)} |f(\tau)|\, d\tau > \frac{\epsilon_0 \delta(\epsilon_0)}{2} \qquad (3.2.4)$$

where the first equality holds because $f(t)$ retains the same sign for $t \in [t_i, t_i+\delta(\epsilon_0)]$. On the other hand, $g(t) \triangleq \int_0^t f(\tau)d\tau$ has a limit as $t \to \infty$ implies that there exists a $T(\epsilon_0) > 0$ such that for any $t_2 > t_1 > T(\epsilon_0)$ we have

$$|g(t_1) - g(t_2)| < \frac{\epsilon_0 \delta(\epsilon_0)}{2}$$

i.e.,

$$\left| \int_{t_1}^{t_2} f(\tau)d\tau \right| < \frac{\epsilon_0 \delta(\epsilon_0)}{2}$$

which for $t_2 = t_i + \delta(\epsilon_0), t_1 = t_i$ contradicts (3.2.4), and, therefore, $\lim_{t\to\infty} f(t) = 0$. □

3.2. PRELIMINARIES

The proof of Lemma 3.2.5 follows directly from that of Lemma 3.2.6 by noting that the function $f^p(t)$ is uniformly continuous for any $p \in [1, \infty)$ because $f, \dot{f} \in \mathcal{L}_\infty$.

The condition that $f(t)$ is uniformly continuous is crucial for the results of Lemma 3.2.6 to hold as demonstrated by the following example.

Example 3.2.4 Consider the following function described by a sequence of isosceles triangles of base length $\frac{1}{n^2}$ and height equal to 1 centered at n where $n = 1, 2, \ldots \infty$ as shown in the figure below:

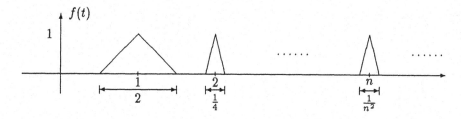

This function is continuous but not uniformly continuous. It satisfies

$$\lim_{t \to \infty} \int_0^t f(\tau) d\tau = \frac{1}{2} \sum_{n=1}^{\infty} \frac{1}{n^2} = \frac{\pi^2}{12}$$

but $\lim_{t \to \infty} f(t)$ does not exist. ▽

The above example also serves as a counter example to the following situation that arises in the analysis of adaptive systems: We have a function $V(t)$ with the following properties: $V(t) \geq 0$, $\dot{V} \leq 0$. As shown by Lemma 3.2.3 these properties imply that $\lim_{t \to \infty} V(t) = V_\infty$ exists. However, there is no guarantee that $\dot{V}(t) \to 0$ as $t \to \infty$. For example consider the function

$$V(t) = \pi - \int_0^t f(\tau) d\tau$$

where $f(t)$ is as defined in Example 3.2.4. Clearly,

$$V(t) \geq 0, \quad \dot{V} = -f(t) \leq 0, \quad \forall t \geq 0$$

and

$$\lim_{t \to \infty} V(t) = V_\infty = \pi - \frac{\pi^2}{12}$$

but $\lim_{t\to\infty} \dot{V}(t) = -\lim_{t\to\infty} f(t)$ does not exist. According to Barbălat's lemma, a sufficient condition for $\dot{V}(t) \to 0$ as $t \to \infty$ is that $\dot{V}$ is uniformly continuous.

3.2.3 Positive Definite Matrices

A square matrix $A \in \mathcal{R}^{n \times n}$ is called *symmetric* if $A = A^\top$. A symmetric matrix A is called *positive semidefinite* if for every $x \in \mathcal{R}^n$, $x^\top A x \geq 0$ and *positive definite* if $x^\top A x > 0$ $\forall x \in \mathcal{R}^n$ with $|x| \neq 0$. It is called *negative semidefinite* (*negative definite*) if $-A$ is positive semidefinite (positive definite).

The definition of a positive definite matrix can be generalized to nonsymmetric matrices. In this book we will always assume that the matrix is symmetric when we consider positive or negative definite or semidefinite properties.

We write $A \geq 0$ if A is positive semidefinite, and $A > 0$ if A is positive definite. We write $A \geq B$ and $A > B$ if $A - B \geq 0$ and $A - B > 0$, respectively.

A symmetric matrix $A \in \mathcal{R}^{n \times n}$ is positive definite if and only if any one of the following conditions holds:

(i) $\lambda_i(A) > 0, i = 1, 2, \ldots, n$ where $\lambda_i(A)$ denotes the ith eigenvalue of A, which is real because $A = A^\top$.

(ii) There exists a nonsingular matrix A_1 such that $A = A_1 A_1^\top$.

(iii) Every principal minor of A is positive.

(iv) $x^\top A x \geq \alpha |x|^2$ for some $\alpha > 0$ and $\forall x \in \mathcal{R}^n$.

The decomposition $A = A_1 A_1^\top$ in (ii) is unique when A_1 is also symmetric. In this case, A_1 is positive definite, it has the same eigenvectors as A, and its eigenvalues are equal to the square roots of the corresponding eigenvalues of A. We specify this unique decomposition of A by denoting A_1 as $A^{\frac{1}{2}}$, i.e., $A = A^{\frac{1}{2}} A^{\frac{\top}{2}}$ where $A^{\frac{1}{2}}$ is a positive definite matrix and $A^{\top/2}$ denotes the transpose of $A^{1/2}$.

A symmetric matrix $A \in \mathcal{R}^{n \times n}$ has n orthogonal eigenvectors and can be decomposed as

$$A = U^\top \Lambda U \qquad (3.2.5)$$

3.3. INPUT/OUTPUT STABILITY

where U is a unitary (orthogonal) matrix (i.e., $U^\top U = I$) with the eigenvectors of A, and Λ is a diagonal matrix composed of the eigenvalues of A. Using (3.2.5), it follows that if $A \geq 0$, then for any vector $x \in \mathcal{R}^n$

$$\lambda_{min}(A)|x|^2 \leq x^\top A x \leq \lambda_{max}(A)|x|^2$$

Furthermore, if $A \geq 0$ then

$$\|A\|_2 = \lambda_{max}(A)$$

and if $A > 0$ we also have

$$\|A^{-1}\|_2 = \frac{1}{\lambda_{min}(A)}$$

where $\lambda_{max}(A), \lambda_{min}(A)$ is the maximum and minimum eigenvalue of A, respectively.

We should note that if $A > 0$ and $B \geq 0$, then $A + B > 0$, but it is not true in general that $AB \geq 0$.

3.3 Input/Output Stability

The systems encountered in this book can be described by an I/O mapping that assigns to each input a corresponding output, or by a state variable representation. In this section we shall present some basic results concerning I/O stability. These results are based on techniques from functional analysis [42], and most of them can be applied to both continuous- and discrete-time systems. Similar results are developed in Section 3.4 by using the state variable approach and Lyapunov theory.

3.3.1 $\mathcal{L}_p$ Stability

We consider an LTI system described by the convolution of two functions $u, h : \mathcal{R}^+ \to \mathcal{R}$ defined as

$$y(t) = u * h \triangleq \int_0^t h(t-\tau)u(\tau)d\tau = \int_0^t u(t-\tau)h(\tau)d\tau \qquad (3.3.1)$$

where u, y is the input and output of the system, respectively. Let $H(s)$ be the Laplace transform of the I/O operator $h(\cdot)$. $H(s)$ is called the transfer function and $h(t)$ the impulse response of the system (3.3.1). The system (3.3.1) may also be represented in the form

$$Y(s) = H(s)U(s) \qquad (3.3.2)$$

where $Y(s), U(s)$ is the Laplace transform of y, u respectively.

We say that the system represented by (3.3.1) or (3.3.2) is $\mathcal{L}_p$ *stable* if $u \in \mathcal{L}_p \Rightarrow y \in \mathcal{L}_p$ and $\|y\|_p \leq c\|u\|_p$ for some constant $c \geq 0$ and any $u \in \mathcal{L}_p$. When $p = \infty$, $\mathcal{L}_p$ stability, i.e., $\mathcal{L}_\infty$ stability, is also referred to as bounded-input bounded-output *(BIBO) stability*.

The following results hold for the system (3.3.1).

Theorem 3.3.1 *If $u \in \mathcal{L}_p$ and $h \in \mathcal{L}_1$ then*

$$\|y\|_p \leq \|h\|_1 \|u\|_p \qquad (3.3.3)$$

where $p \in [1, \infty]$.

When $p = 2$ we have a sharper bound for $\|y\|_p$ than that of (3.3.3) given by the following Lemma.

Lemma 3.3.1 *If $u \in \mathcal{L}_2$ and $h \in \mathcal{L}_1$, then*

$$\|y\|_2 \leq \sup_{\omega} |H(j\omega)| \|u\|_2 \qquad (3.3.4)$$

For the proofs of Theorem 3.3.1, Lemma 3.3.1 see [42].

Remark 3.3.1 It can be shown that (3.3.4) also holds [232] when $h(\cdot)$ is of the form

$$h(t) = \begin{cases} 0 & t < 0 \\ \sum_{i=0}^{\infty} f_i \delta(t - t_i) + f_a(t) & t \geq 0 \end{cases}$$

where $f_a \in \mathcal{L}_1, \sum_{i=0}^{\infty} |f_i| < \infty$ and t_i are nonnegative finite constants. The Laplace transform of $h(t)$ is now given by

$$H(s) = \sum_{i=0}^{\infty} f_i e^{-st_i} + H_a(s)$$

which is not a rational function of s. The biproper transfer functions that are of interest in this book belong to the above class.

3.3. INPUT/OUTPUT STABILITY

Remark 3.3.2 We should also note that (3.3.3) and (3.3.4) hold for the truncated functions of u, y, i.e.,

$$\|y_t\|_p \leq \|h\|_1 \|u_t\|_p$$

for any $t \in [0, \infty)$ provided $u \in \mathcal{L}_{pe}$. Similarly,

$$\|y_t\|_2 \leq \sup_\omega |H(j\omega)| \|u_t\|_2$$

for any $t \in [0, \infty)$ provided $u \in \mathcal{L}_{2e}$. This is clearly seen by noticing that $u \in \mathcal{L}_{pe} \Rightarrow u_t \in \mathcal{L}_p$ for any finite $t \geq 0$.

It can be shown [42] that inequality (3.3.3) is sharp for $p = \infty$ because $\|h\|_1$ is the induced norm of the map $T: u \mapsto Tu \triangleq y$ from $\mathcal{L}_\infty$ into $\mathcal{L}_\infty$, i.e., $\|T\|_\infty = \|h\|_1$. Similarly for (3.3.4) it can be shown that the induced norm of the linear map $T: \mathcal{L}_2 \mapsto \mathcal{L}_2$ is given by

$$\|T\|_2 = \sup_{\omega \in \mathcal{R}} |H(j\omega)| \qquad (3.3.5)$$

i.e., the bound (3.3.4) is also sharp.

The induced $\mathcal{L}_2$ norm in (3.3.5) is referred to as the H_∞ norm for the transfer function $H(s)$ and is denoted by

$$\|H(s)\|_\infty \triangleq \sup_{\omega \in \mathcal{R}} |H(j\omega)|$$

Let us consider the simple case where $h(t)$ in (3.3.1) is the impulse response of an LTI system whose transfer function $H(s)$ is a rational function of s. The following theorem and corollaries hold.

Theorem 3.3.2 *Let $H(s)$ be a strictly proper rational function of s. Then $H(s)$ is analytic in $\text{Re}[s] \geq 0$ if and only if $h \in \mathcal{L}_1$.*

Corollary 3.3.1 *If $h \in \mathcal{L}_1$, then*
(i) *h decays exponentially, i.e., $|h(t)| \leq \alpha_1 e^{-\alpha_0 t}$ for some $\alpha_1, \alpha_0 > 0$*
(ii) *$u \in \mathcal{L}_1 \Rightarrow y \in \mathcal{L}_1 \cap \mathcal{L}_\infty, \dot{y} \in \mathcal{L}_1$, y is continuous and $\lim_{t \to \infty} |y(t)| = 0$*
(iii) *$u \in \mathcal{L}_2 \Rightarrow y \in \mathcal{L}_2 \cap \mathcal{L}_\infty, \dot{y} \in \mathcal{L}_2$, y is continuous and $\lim_{t \to \infty} |y(t)| = 0$*
(iv) *For $p \in [1, \infty]$, $u \in \mathcal{L}_p \Rightarrow y, \dot{y} \in \mathcal{L}_p$ and y is continuous*

For proofs of Theorem 3.3.2 and Corollary 3.3.1, see [42].

Corollary 3.3.2 *Let $H(s)$ be biproper and analytic in $\mathrm{Re}[s] \geq 0$. Then $u \in \mathcal{L}_2 \cap \mathcal{L}_\infty$ and $\lim_{t \to \infty} |u(t)| = 0$ imply that $y \in \mathcal{L}_2 \cap \mathcal{L}_\infty$ and $\lim_{t \to \infty} |y(t)| = 0$.*

Proof $H(s)$ may be expressed as

$$H(s) = d + H_a(s)$$

where d is a constant and $H_a(s)$ is strictly proper and analytic in $Re[s] \geq 0$. We have

$$y = du + y_a, \quad y_a = H_a(s)u$$

where, by Corollary 3.3.1, $y_a \in \mathcal{L}_2 \cap \mathcal{L}_\infty$ and $|y_a(t)| \to 0$ as $t \to \infty$. Because $u \in \mathcal{L}_2 \cap \mathcal{L}_\infty$ and $u(t) \to 0$ as $t \to \infty$, it follows that $y \in \mathcal{L}_2 \cap \mathcal{L}_\infty$ and $|y(t)| \to 0$ as $t \to \infty$. $\square$

Example 3.3.1 Consider the system described by

$$y = H(s)u, \quad H(s) = \frac{e^{-\alpha s}}{s + \beta}$$

for some constant $\alpha > 0$. For $\beta > 0$, $H(s)$ is analytic in $Re[s] \geq 0$. The impulse response of the system is given by

$$h(t) = \begin{cases} e^{-\beta(t-\alpha)} & t \geq \alpha \\ 0 & t < \alpha \end{cases}$$

and $h \in \mathcal{L}_1$ if and only if $\beta > 0$. We have

$$\|h\|_1 = \int_0^\infty |h(t)| dt = \int_\alpha^\infty e^{-\beta(t-\alpha)} dt = \frac{1}{\beta}$$

and

$$\|H(s)\|_\infty = \sup_\omega \left| \frac{e^{-\alpha j\omega}}{j\omega + \beta} \right| = \frac{1}{\beta}$$

$\triangledown$

Example 3.3.2 Consider the system described by

$$y = H(s)u, \quad H(s) = \frac{2s+1}{s+5}$$

3.3. INPUT/OUTPUT STABILITY

The impulse response of the system is given by

$$h(t) = \begin{cases} 2\delta_\Delta(t) - 9e^{-5t} & t \geq 0 \\ 0 & t < 0 \end{cases}$$

where $h_a = -9e^{-5t} \in \mathcal{L}_1$. This system belongs to the class described in Remark 3.3.1. We have

$$\|H(s)\|_\infty = \sup_\omega \left|\frac{1+2j\omega}{5+j\omega}\right| = \sup_\omega \left(\frac{1+4\omega^2}{25+\omega^2}\right)^{\frac{1}{2}} = 2$$

Hence, according to (3.3.4) and Remarks 3.3.1 and 3.3.2, for any $u \in \mathcal{L}_{2e}$, we have

$$\|y_t\|_2 \leq 2\|u_t\|_2$$

for any $t \in [0, \infty)$. $\triangledown$

Definition 3.3.1 (μ–small in the mean square sense (m.s.s.)) Let $x : [0, \infty) \mapsto \mathcal{R}^n$, where $x \in \mathcal{L}_{2e}$, and consider the set

$$S(\mu) = \left\{ x : [0, \infty) \mapsto \mathcal{R}^n \, \Big| \, \int_t^{t+T} x^\top(\tau)x(\tau)d\tau \leq c_0\mu T + c_1, \quad \forall t, T \geq 0 \right\}$$

for a given constant $\mu \geq 0$, where $c_0, c_1 \geq 0$ are some finite constants, and c_0 is independent of μ. We say that x is μ–small in the m.s.s. if $x \in S(\mu)$.

Using the proceeding definition, we can obtain a result similar to that of Corollary 3.3.1 (iii) in the case where $u \notin \mathcal{L}_2$ but $u \in S(\mu)$ for some constant $\mu \geq 0$.

Corollary 3.3.3 Consider the system (3.3.1). If $h \in \mathcal{L}_1$, then $u \in S(\mu)$ implies that $y \in S(\mu)$ and $y \in \mathcal{L}_\infty$ for any finite $\mu \geq 0$. Furthermore

$$|y(t)|^2 \leq \frac{\alpha_1^2}{\alpha_0} \frac{e^{\alpha_0}}{(1-e^{-\alpha_0})}(c_0\mu + c_1), \qquad \forall t \geq t_0 \geq 0$$

where α_0, α_1 are the parameters in the bound for h in Corollary 3.3.1 (i).

Proof Using Corollary 3.3.1 (i), we have

$$|y(t)| \leq \int_{t_0}^t |h(t-\tau)u(\tau)|d\tau \leq \int_{t_0}^t \alpha_1 e^{-\alpha_0(t-\tau)}|u(\tau)|d\tau, \qquad \forall t \geq t_0 \geq 0$$

for some constants $\alpha_1, \alpha_0 > 0$. Using the Schwartz inequality we obtain

$$|y(t)|^2 \leq \alpha_1^2 \int_{t_0}^t e^{-\alpha_0(t-\tau)}d\tau \int_{t_0}^t e^{-\alpha_0(t-\tau)}|u(\tau)|^2 d\tau$$

$$\leq \frac{\alpha_1^2}{\alpha_0} \int_{t_0}^t e^{-\alpha_0(t-\tau)}|u(\tau)|^2 d\tau \qquad (3.3.6)$$

Therefore, for any $t \geq t_0 \geq 0$ and $T \geq 0$ we have

$$\int_t^{t+T} |y(\tau)|^2 d\tau \leq \frac{\alpha_1^2}{\alpha_0} \int_t^{t+T} \int_{t_0}^\tau e^{-\alpha_0(\tau-s)}|u(s)|^2 ds d\tau$$

$$= \frac{\alpha_1^2}{\alpha_0} \int_t^{t+T} \left(\int_{t_0}^t e^{-\alpha_0(\tau-s)}|u(s)|^2 ds + \int_t^\tau e^{-\alpha_0(\tau-s)}|u(s)|^2 ds \right) d\tau \qquad (3.3.7)$$

Using the identity involving the change of the sequence of integration, i.e.,

$$\int_t^{t+T} f(\tau) \int_t^\tau g(s) ds d\tau = \int_t^{t+T} g(s) \int_s^{t+T} f(\tau) d\tau ds \qquad (3.3.8)$$

for the second term on the right-hand side of (3.3.7), we have

$$\int_t^{t+T} |y(\tau)|^2 d\tau \leq \frac{\alpha_1^2}{\alpha_0} \int_t^{t+T} e^{-\alpha_0 \tau} d\tau \int_{t_0}^t e^{\alpha_0 s}|u(s)|^2 ds$$

$$+ \frac{\alpha_1^2}{\alpha_0} \int_t^{t+T} e^{\alpha_0 s}|u(s)|^2 \left(\int_s^{t+T} e^{-\alpha_0 \tau} d\tau \right) ds$$

$$\leq \frac{\alpha_1^2}{\alpha_0^2} \left(\int_{t_0}^t e^{-\alpha_0(t-s)}|u(s)|^2 ds + \int_t^{t+T} |u(s)|^2 ds \right)$$

where the last inequality is obtained by using $e^{-\alpha_0 t} - e^{-\alpha_0(t+T)} \leq e^{-\alpha_0 t}$. Because $u \in \mathcal{S}(\mu)$ it follows that

$$\int_t^{t+T} |y(\tau)|^2 d\tau \leq \frac{\alpha_1^2}{\alpha_0^2} [\Delta(t,t_0) + c_0 \mu T + c_1] \qquad (3.3.9)$$

where $\Delta(t,t_0)) \triangleq \int_{t_0}^t e^{-\alpha_0(t-s)}|u(s)|^2 ds$. If we establish that $\Delta(t,t_0) \leq c$ for some constant c independent of t, t_0 then we can conclude from (3.3.9) that $y \in \mathcal{S}(\mu)$. We start with

$$\Delta(t,t_0) = \int_{t_0}^t e^{-\alpha_0(t-s)}|u(s)|^2 ds$$

$$\leq e^{-\alpha_0 t} \sum_{i=0}^{n_t} \int_{i+t_0}^{i+1+t_0} e^{\alpha_0 s} |u(s)|^2 ds \qquad (3.3.10)$$

$$\leq e^{-\alpha_0 t} \sum_{i=0}^{n_t} e^{\alpha_0 (i+1+t_0)} \int_{i+t_0}^{i+1+t_0} |u(s)|^2 ds$$

where n_t is an integer that depends on t and satisfies $n_t + t_0 \leq t < n_t + 1 + t_0$. Because $u \in S(\mu)$, we have

$$\Delta(t, t_0) \leq e^{-\alpha_0 t}(c_0 \mu + c_1) \sum_{i=0}^{n_t} e^{\alpha_0(i+1+t_0)} \leq \frac{c_0 \mu + c_1}{1 - e^{-\alpha_0}} e^{\alpha_0} \qquad (3.3.11)$$

Using (3.3.11) in (3.3.9) we have

$$\int_t^{t+T} |y(\tau)|^2 d\tau \leq \frac{\alpha_1^2}{\alpha_0^2} \left(c_0 \mu T + c_1 + \frac{c_0 \mu + c_1}{1 - e^{-\alpha_0}} e^{\alpha_0} \right)$$

for any $t \geq t_0 \geq 0$. Setting $\hat{c}_0 = \frac{c_0 \alpha_1^2}{\alpha_0^2}$ and $\hat{c}_1 = \left(c_1 + \frac{c_0 \mu + c_1}{1 - e^{-\alpha_0}} e^{\alpha_0} \right) \frac{\alpha_1^2}{\alpha_0^2}$, it follows that $y \in S(\mu)$.

From (3.3.6), (3.3.10), and (3.3.11), we can calculate the upper bound for $|y(t)|^2$. □

Definition 3.3.1 may be generalized to the case where μ is not necessarily a constant as follows.

Definition 3.3.2 Let $x : [0, \infty) \mapsto \mathcal{R}^n$, $w : [0, \infty) \mapsto \mathcal{R}^+$ where $x \in \mathcal{L}_{2e}$, $w \in \mathcal{L}_{1e}$ and consider the set

$$S(w) = \left\{ x, w \,\bigg|\, \int_t^{t+T} x^\top(\tau) x(\tau) d\tau \leq c_0 \int_t^{t+T} w(\tau) d\tau + c_1, \forall t, T \geq 0 \right\}$$

where $c_0, c_1 \geq 0$ are some finite constants. We say that x is w-small in the m.s.s. if $x \in S(w)$.

We employ Corollary 3.3.3, and Definitions 3.3.1 and 3.3.2 repeatedly in Chapters 8 and 9 for the analysis of the robustness properties of adaptive control systems.

3.3.2 The $\mathcal{L}_{2\delta}$ Norm and I/O Stability

The definitions and results of the previous sections are very helpful in developing I/O stability results based on a different norm that are particularly useful in the analysis of adaptive schemes.

In this section we consider the *exponentially weighted $\mathcal{L}_2$ norm* defined as
$$\|x_t\|_{2\delta} \triangleq \left(\int_0^t e^{-\delta(t-\tau)} x^\top(\tau) x(\tau) d\tau \right)^{\frac{1}{2}}$$
where $\delta \geq 0$ is a constant. We say that $x \in \mathcal{L}_{2\delta}$ if $\|x_t\|_{2\delta}$ exists. When $\delta = 0$ we omit it from the subscript and use the notation $x \in \mathcal{L}_{2e}$.

We refer to $\|(\cdot)\|_{2\delta}$ as the $\mathcal{L}_{2\delta}$ norm. For any finite time t, the $\mathcal{L}_{2\delta}$ norm satisfies the properties of the norm given by Definition 3.2.1, i.e.,

(i) $\|x_t\|_{2\delta} \geq 0$
(ii) $\|\alpha x_t\|_{2\delta} = |\alpha| \|x_t\|_{2\delta}$ for any constant scalar α
(iii) $\|(x+y)_t\|_{2\delta} \leq \|x_t\|_{2\delta} + \|y_t\|_{2\delta}$

It also follows that

(iv) $\|\alpha x_t\|_{2\delta} \leq \|x_t\|_{2\delta} \sup_t |\alpha(t)|$ for any $\alpha \in \mathcal{L}_\infty$

The notion of $\mathcal{L}_{2\delta}$ norm has been introduced mainly to simplify the stability and robustness analysis of adaptive systems. To avoid confusion, we should point out that the $\mathcal{L}_{2\delta}$ norm defined here is different from the exponentially weighted norm used in many functional analysis books that is defined as $\left\{ \int_0^t e^{\delta \tau} x^\top(\tau) x(\tau) d\tau \right\}^{\frac{1}{2}}$. The main difference is that this exponentially weighted norm is a nondecreasing function of t, whereas the $\mathcal{L}_{2\delta}$ norm may not be.

Let us consider the LTI system given by
$$y = H(s)u \qquad (3.3.12)$$
where $H(s)$ is a rational function of s and examine $\mathcal{L}_{2\delta}$ *stability*, i.e., given $u \in \mathcal{L}_{2\delta}$, what can we say about the $\mathcal{L}_p, \mathcal{L}_{2\delta}$ properties of the output $y(t)$ and its upper bounds.

Lemma 3.3.2 *Let $H(s)$ in (3.3.12) be proper. If $H(s)$ is analytic in $\mathrm{Re}[s] \geq -\delta/2$ for some $\delta \geq 0$ and $u \in \mathcal{L}_{2e}$ then*

(i)
$$\|y_t\|_{2\delta} \leq \|H(s)\|_{\infty\delta} \|u_t\|_{2\delta}$$

where
$$\|H(s)\|_{\infty\delta} \triangleq \sup_\omega \left| H\left(j\omega - \frac{\delta}{2}\right) \right|$$

3.3. INPUT/OUTPUT STABILITY

(ii) *Furthermore, when $H(s)$ is strictly proper, we have*

$$|y(t)| \leq \|H(s)\|_{2\delta}\|u_t\|_{2\delta}$$

where

$$\|H(s)\|_{2\delta} \triangleq \frac{1}{\sqrt{2\pi}} \left\{ \int_{-\infty}^{\infty} \left| H\left(j\omega - \frac{\delta}{2}\right) \right|^2 d\omega \right\}^{\frac{1}{2}}$$

The norms $\|H(s)\|_{2\delta}$, $\|H(s)\|_{\infty\delta}$ are related by the inequality

$$\|H(s)\|_{2\delta} \leq \frac{1}{\sqrt{2p-\delta}} \|(s+p)H(s)\|_{\infty\delta}$$

for any $p > \delta/2 \geq 0$.

Proof The transfer function $H(s)$ can be expressed as $H(s) = d + H_a(s)$ with

$$h(t) = \begin{cases} 0 & t < 0 \\ d\delta_\Delta(t) + h_a(t) & t \geq 0 \end{cases}$$

Because d is a finite constant, $H(s)$ being analytic in $\text{Re}[s] \geq -\delta/2$ implies that $h_a \in \mathcal{L}_1$, i.e., the pair $\{H(s), h(t)\}$ belongs to the class of functions considered in Remark 3.3.1.

If we define

$$h_\delta(t) = \begin{cases} 0 & t < 0 \\ d\delta_\Delta(t) + e^{\frac{\delta}{2}t} h_a(t) & t \geq 0 \end{cases}$$

$y_\delta(t) \triangleq e^{\frac{\delta}{2}t} y(t)$ and $u_\delta(t) \triangleq e^{\frac{\delta}{2}t} u(t)$, it follows from (3.3.1) that

$$y_\delta(t) = \int_0^t e^{\frac{\delta}{2}(t-\tau)} h(t-\tau) e^{\frac{\delta}{2}\tau} u(\tau) d\tau = h_\delta * u_\delta$$

Now $u \in \mathcal{L}_{2e} \Rightarrow u_\delta \in \mathcal{L}_{2e}$. Therefore, applying Lemma 3.3.1 and Remark 3.3.1 for the truncated signals $y_{\delta t}, u_{\delta t}$ at time t and noting that $H(s - \delta/2)$ is the Laplace transform of h_δ we have

$$\|y_{\delta t}\|_2 \leq \|H(s - \delta/2)\|_\infty \|u_{\delta t}\|_2 \qquad (3.3.13)$$

Because $e^{-\frac{\delta}{2}t}\|y_{\delta t}\|_2 = \|y_t\|_{2\delta}$, $e^{-\frac{\delta}{2}t}\|u_{\delta t}\|_2 = \|u_t\|_{2\delta}$, and $\|H(s-\delta/2)\|_\infty = \|H(s)\|_{\infty\delta}$, (i) follows directly from (3.3.13).

For $d = 0$, i.e., $H(s)$ is strictly proper, we have

$$|y(t)| \leq \left| \int_0^t e^{\frac{\delta}{2}(t-\tau)} h(t-\tau) e^{-\frac{\delta}{2}(t-\tau)} u(\tau) d\tau \right|$$

$$\leq \left(\int_0^t e^{\delta(t-\tau)} |h(t-\tau)|^2 d\tau \right)^{\frac{1}{2}} \|u_t\|_{2\delta}$$

where the second inequality is obtained by applying the Schwartz inequality. Then,

$$|y(t)| \leq \left(\int_0^\infty e^{\delta(t-\tau)}|h(t-\tau)|^2 d\tau\right)^{\frac{1}{2}} \|u_t\|_{2\delta}$$

$$= \frac{1}{\sqrt{2\pi}}\left(\int_{-\infty}^\infty |H(j\omega - \delta/2)|^2 d\omega\right)^{\frac{1}{2}} \|u_t\|_{2\delta} \quad (3.3.14)$$

where the equality is obtained by assuming that $H(s)$ is strictly proper and applying Parseval's Theorem [42](p. 236), implies (ii).

Because $H(s)$ is strictly proper, we can write

$$\|H(s)\|_{2\delta} = \frac{1}{\sqrt{2\pi}}\left(\int_{-\infty}^\infty |(j\omega+p_0)H(j\omega-\delta/2)|^2 \frac{1}{|j\omega+p_0|^2}d\omega\right)^{\frac{1}{2}}$$

$$\leq \frac{1}{\sqrt{2\pi}}\left(\int_{-\infty}^\infty \frac{1}{|j\omega+p_0|^2}d\omega\right)^{\frac{1}{2}} \sup_\omega(|(j\omega+p_0)H(j\omega-\delta/2)|)$$

$$= \frac{1}{\sqrt{2p_0}}\|(s+p_0+\delta/2)H(s)\|_{\infty\delta}$$

for any $p_0 > 0$. Setting $p_0 = p - \delta/2$, the result follows. □

Remark 3.3.3 Lemma 3.3.2 can be extended to the case where $H(s)$ is not rational in s but belongs to the general class of transfer functions described in Remark 3.3.1.

We refer to $\|H(s)\|_{2\delta}, \|H(s)\|_{\infty\delta}$ defined in Lemma 3.3.2 as the δ-shifted H_2 and H_∞ norms, respectively.

Lemma 3.3.3 *Consider the linear time-varying system given by*

$$\begin{aligned}\dot{x} &= A(t)x + B(t)u, \quad x(0) = x_0 \\ y &= C^\top(t)x + D(t)u\end{aligned} \quad (3.3.15)$$

where $x \in \mathcal{R}^n, y \in \mathcal{R}^r, u \in \mathcal{R}^m$, and the elements of the matrices A, B, C, and D are bounded continuous functions of time. If the state transition matrix $\Phi(t,\tau)$ of (3.3.15) satisfies

$$\|\Phi(t,\tau)\| \leq \lambda_0 e^{-\alpha_0(t-\tau)} \quad (3.3.16)$$

for some $\lambda_0, \alpha_0 > 0$ and $u \in \mathcal{L}_{2e}$, then for any $\delta \in [0,\delta_1)$ where $0 < \delta_1 < 2\alpha_0$ is arbitrary, we have

3.3. INPUT/OUTPUT STABILITY

(i) $|x(t)| \leq \frac{c\lambda_0}{\sqrt{2\alpha_0-\delta}}\|u_t\|_{2\delta} + \epsilon_t$

(ii) $\|x_t\|_{2\delta} \leq \frac{c\lambda_0}{\sqrt{(\delta_1-\delta)(2\alpha_0-\delta_1)}}\|u_t\|_{2\delta} + \epsilon_t$

(iii) $\|y_t\|_{2\delta} \leq c_0\|u_t\|_{2\delta} + \epsilon_t$

where

$$c_0 = \frac{c\lambda_0}{\sqrt{(\delta_1-\delta)(2\alpha_0-\delta_1)}} \sup_t \|C^T(t)\| + \sup_t \|D(t)\|, \quad c = \sup_t \|B(t)\|$$

and ϵ_t is an exponentially decaying to zero term because $x_0 \neq 0$.

Proof The solution $x(t)$ of (3.3.15) can be expressed as

$$x(t) = \Phi(t,0)x_0 + \int_0^t \Phi(t,\tau)B(\tau)u(\tau)d\tau$$

Therefore,

$$|x(t)| \leq \|\Phi(t,0)\|\|x_0\| + \int_0^t \|\Phi(t,\tau)\|\,\|B(\tau)\|\,|u(\tau)|d\tau$$

Using (3.3.16) we have

$$|x(t)| \leq \epsilon_t + c\lambda_0 \int_0^t e^{-\alpha_0(t-\tau)}|u(\tau)|d\tau \qquad (3.3.17)$$

where c and λ_0 are as defined in the statement of the lemma. Expressing $e^{-\alpha_0(t-\tau)}$ as $e^{-(\alpha_0-\frac{\delta}{2})(t-\tau)}e^{-\frac{\delta}{2}(t-\tau)}$ and applying the Schwartz inequality, we have

$$\begin{aligned}|x(t)| &\leq \epsilon_t + c\lambda_0 \left(\int_0^t e^{-(2\alpha_0-\delta)(t-\tau)}d\tau\right)^{\frac{1}{2}} \left(\int_0^t e^{-\delta(t-\tau)}|u(\tau)|^2 d\tau\right)^{\frac{1}{2}} \\ &\leq \epsilon_t + \frac{c\lambda_0}{\sqrt{2\alpha_0-\delta}}\|u_t\|_{2\delta}\end{aligned}$$

which completes the proof of (i). Using property (iii) of Definition 3.2.1 for the $\mathcal{L}_{2\delta}$ norm, it follows from (3.3.17) that

$$\|x_t\|_{2\delta} \leq \|\epsilon_t\|_{2\delta} + c\lambda_0\|Q_t\|_{2\delta} \qquad (3.3.18)$$

where

$$\|Q_t\|_{2\delta} \triangleq \left\|\left(\int_0^t e^{-\alpha_0(t-\tau)}|u(\tau)|d\tau\right)\right\|_{t_{2\delta}} = \left[\int_0^t e^{-\delta(t-\tau)}\left(\int_0^\tau e^{-\alpha_0(\tau-s)}|u(s)|ds\right)^2 d\tau\right]^{\frac{1}{2}}$$

Using the Schwartz inequality we have

$$\left(\int_0^\tau e^{-\alpha_0(\tau-s)}|u(s)|ds\right)^2 = \left(\int_0^\tau e^{-(\alpha_0-\frac{\delta_1}{2})(\tau-s)} e^{-\frac{\delta_1}{2}(\tau-s)}|u(s)|ds\right)^2$$

$$\leq \int_0^\tau e^{-(2\alpha_0-\delta_1)(\tau-s)}ds \int_0^\tau e^{-\delta_1(\tau-s)}|u(s)|^2 ds$$

$$\leq \frac{1}{2\alpha_0-\delta_1}\int_0^\tau e^{-\delta_1(\tau-s)}|u(s)|^2 ds$$

i.e.,

$$\|Q_t\|_{2\delta} \leq \frac{1}{\sqrt{2\alpha_0-\delta_1}}\left(\int_0^t e^{-\delta(t-\tau)}\int_0^\tau e^{-\delta_1(\tau-s)}|u(s)|^2 ds d\tau\right)^{\frac{1}{2}} \quad (3.3.19)$$

Interchanging the sequence of integration, (3.3.19) becomes

$$\|Q_t\|_{2\delta} \leq \frac{1}{\sqrt{2\alpha_0-\delta_1}}\left(\int_0^t e^{-\delta t+\delta_1 s}|u(s)|^2 \int_s^t e^{-(\delta_1-\delta)\tau}d\tau ds\right)^{\frac{1}{2}}$$

$$= \frac{1}{\sqrt{2\alpha_0-\delta_1}}\left(\int_0^t e^{-\delta t+\delta_1 s}|u(s)|^2 \frac{e^{-(\delta_1-\delta)s}-e^{-(\delta_1-\delta)t}}{\delta_1-\delta}ds\right)^{\frac{1}{2}}$$

$$= \frac{1}{\sqrt{2\alpha_0-\delta_1}}\left(\int_0^t \frac{e^{-\delta(t-s)}-e^{-\delta_1(t-s)}}{\delta_1-\delta}|u(s)|^2 ds\right)^{\frac{1}{2}}$$

$$\leq \frac{1}{\sqrt{(2\alpha_0-\delta_1)(\delta_1-\delta)}}\left(\int_0^t e^{-\delta(t-s)}|u(s)|^2 ds\right)^{\frac{1}{2}}$$

for any $\delta < \delta_1 < 2\alpha_0$. Because $\|\epsilon_t\|_{2\delta} \leq \epsilon_t$, the proof of (ii) follows.

The proof of (iii) follows directly by noting that

$$\|y_t\|_{2\delta} \leq \|(C^\top x)_t\|_{2\delta} + \|(Du)_t\|_{2\delta} \leq \|x_t\|_{2\delta} \sup_t \|C^\top(t)\| + \|u_t\|_{2\delta} \sup_t \|D(t)\|$$

□

A useful extension of Lemma 3.3.3, applicable to the case where $A(t)$ is not necessarily stable and $\delta = \delta_0 > 0$ is a given fixed constant, is given by the following Lemma that makes use of the following definition.

Definition 3.3.3 *The pair* $(C(t), A(t))$ *in (3.3.15) is* **uniformly completely observable** *(UCO) if there exist constants* $\beta_1, \beta_2, \nu > 0$ *such that for all* $t_0 \geq 0$,

$$\beta_2 I \geq N(t_0, t_0+\nu) \geq \beta_1 I$$

3.3. INPUT/OUTPUT STABILITY

where $N(t_0, t_0 + \nu) \triangleq \int_{t_0}^{t_0+\nu} \Phi^T(\tau, t_0) C(\tau) C^T(\tau) \Phi(\tau, t_0) d\tau$ is the so-called observability grammian [1, 201] and $\Phi(t, \tau)$ is the state transition matrix associated with $A(t)$.

Lemma 3.3.4 *Consider a linear time-varying system of the same form as (3.3.15) where $(C(t), A(t))$ is UCO, and the elements of $A, B, C,$ and D are bounded continuous functions of time. For any given finite constant $\delta_0 > 0$, we have*

(i) $|x(t)| \leq \frac{\lambda_1}{\sqrt{2\alpha_1 - \delta_0}} (c_1 \|u_t\|_{2\delta_0} + c_2 \|y_t\|_{2\delta_0}) + \epsilon_t$

(ii) $\|x(t)\|_{2\delta_0} \leq \frac{\lambda_1}{\sqrt{(\delta_1 - \delta_0)(2\alpha_1 - \delta_1)}} (c_1 \|u_t\|_{2\delta_0} + c_2 \|y_t\|_{2\delta_0}) + \epsilon_1$

(iii) $\|y_t\|_{2\delta_0} \leq \|x_t\|_{2\delta_0} \sup_t \|C^T(t)\| + \|u_t\|_{2\delta_0} \sup_t \|D(t)\|$

where $c_1, c_2 \geq 0$ are some finite constants; δ_1, α_1 satisfy $\delta_0 < \delta_1 < 2\alpha_1$, and ϵ_t is an exponentially decaying to zero term because $x_0 \neq 0$.

Proof Because (C, A) is uniformly completely observable, there exists a matrix $K(t)$ with bounded elements such that the state transition matrix $\Phi_c(t, \tau)$ of $A_c(t) \triangleq A(t) - K(t) C^T(t)$ satisfies

$$\|\Phi_c(t, \tau)\| \leq \lambda_1 e^{-\alpha_1 (t - \tau)}$$

for some constants $\alpha_1, \delta_1, \lambda_1$ that satisfy $\alpha_1 > \frac{\delta_1}{2} > \frac{\delta_0}{2}$, $\lambda_1 > 0$. Let us now rewrite (3.3.15), by using what is called "output injection," as

$$\dot{x} = (A - KC^T)x + Bu + KC^T x$$

Because $C^T x = y - Du$, we have

$$\dot{x} = A_c(t)x + \bar{B}u + Ky$$

where $\bar{B} = B - KD$. Following exactly the same procedure as in the proof of Lemma 3.3.3, we obtain

$$|x(t)| \leq \frac{\lambda_1}{\sqrt{2\alpha_1 - \delta_0}} (c_1 \|u_t\|_{2\delta_0} + c_2 \|y_t\|_{2\delta_0}) + \epsilon_t$$

where $c_1 = \sup_t \|\bar{B}(t)\|$, $c_2 = \sup_t \|K(t)\|$ and ϵ_t is an exponentially decaying to zero term due to $x(0) = x_0$. Similarly,

$$\|x_t\|_{2\delta_0} \leq \frac{\lambda_1}{\sqrt{(\delta_1 - \delta_0)(2\alpha_1 - \delta_1)}} (c_1 \|u_t\|_{2\delta_0} + c_2 \|y_t\|_{2\delta_0}) + \epsilon_t$$

by following exactly the same steps as in the proof of Lemma 3.3.3. The proof of (iii) follows directly from the expression of y. □

Instead of the interval $[0, t)$, the $\mathcal{L}_{2\delta}$ norm can be defined over any arbitrary interval of time as follows:

$$\|x_{t,t_1}\|_{2\delta} \triangleq \left(\int_{t_1}^{t} e^{-\delta(t-\tau)} x^T(\tau) x(\tau) d\tau\right)^{\frac{1}{2}}$$

for any $t_1 \geq 0$ and $t \geq t_1$. This definition allow us to use the properties of the $\mathcal{L}_{2\delta}$ norm over certain intervals of time that are of interest. We develop some of these properties for the LTI, SISO system

$$\begin{aligned} \dot{x} &= Ax + Bu, \quad x(0) = x_0 \\ y &= C^T x + Du \end{aligned} \quad (3.3.20)$$

whose transfer function is given by

$$y = [C^T(sI - A)^{-1}B + D]u = H(s)u \quad (3.3.21)$$

Lemma 3.3.5 *Consider the LTI system (3.3.20), where A is a stable matrix and $u \in \mathcal{L}_{2e}$. Let α_0, λ_0 be the positive constants that satisfy $\|e^{A(t-\tau)}\| \leq \lambda_0 e^{-\alpha_0(t-\tau)}$. Then for any constant $\delta \in [0, \delta_1)$ where $0 < \delta_1 < 2\alpha_0$ is arbitrary, for any finite $t_1 \geq 0$ and $t \geq t_1$ we have*

(i)
 (a) $|x(t)| \leq \lambda_0 e^{-\alpha_0(t-t_1)} |x(t_1)| + c_1 \|u_{t,t_1}\|_{2\delta}$
 (b) $\|x_{t,t_1}\|_{2\delta} \leq c_0 e^{-\frac{\delta}{2}(t-t_1)} |x(t_1)| + c_2 \|u_{t,t_1}\|_{2\delta}$

(ii) $\|y_{t,t_1}\|_{2\delta} \leq c_3 e^{-\frac{\delta}{2}(t-t_1)} |x(t_1)| + \|H(s)\|_{\infty\delta} \|u_{t,t_1}\|_{2\delta}$

(iii) *Furthermore if $D = 0$, i.e., $H(s)$ is strictly proper, then*

$$|y(t)| \leq c_4 e^{-\alpha_0(t-t_1)} |x(t_1)| + \|H(s)\|_{2\delta} \|u_{t,t_1}\|_{2\delta}$$

where

$$c_1 = \|B\| c_0, \quad c_0 = \frac{\lambda_0}{\sqrt{2\alpha_0 - \delta}}, \quad c_2 = \frac{\|B\| \lambda_0}{\sqrt{(\delta_1 - \delta)(2\alpha_0 - \delta_1)}}$$

$$c_3 = \|C^T\| c_0, \quad c_4 = \|C^T\| \lambda_0$$

3.3. INPUT/OUTPUT STABILITY

Proof Define $v(\tau)$ as
$$v(\tau) = \begin{cases} 0 & \text{if } \tau < t_1 \\ u(\tau) & \text{if } \tau \geq t_1 \end{cases}$$

From (3.3.20) we have
$$x(t) = e^{A(t-t_1)}x(t_1) + \bar{x}(t) \qquad \forall t \geq t_1 \qquad (3.3.22)$$

where
$$\bar{x}(t) = \int_{t_1}^{t} e^{A(t-\tau)}Bu(\tau)d\tau \qquad \forall t \geq t_1$$

We can now rewrite $\bar{x}(t)$ as
$$\bar{x}(t) = \int_0^t e^{A(t-\tau)}Bv(\tau)d\tau \qquad \forall t \geq 0 \qquad (3.3.23)$$

Similarly
$$y(t) = C^T e^{A(t-t_1)}x(t_1) + \bar{y}(t) \qquad \forall t \geq t_1 \qquad (3.3.24)$$

$$\bar{y}(t) = \int_0^t C^T e^{A(t-\tau)}Bv(\tau)d\tau + Dv(t) \qquad \forall t \geq 0 \qquad (3.3.25)$$

It is clear that $\bar{x}$ in (3.3.23) and $\bar{y}$ in (3.3.25) are the solutions of the system
$$\begin{aligned} \dot{\bar{x}} &= A\bar{x} + Bv, \quad \bar{x}(0) = 0 \\ \bar{y} &= C^T \bar{x} + Dv \end{aligned} \qquad (3.3.26)$$

whose transfer function is $C^T(sI - A)^{-1}B + D = H(s)$.

Because A is a stable matrix, there exists constants λ_0, $\alpha_0 > 0$ such that
$$\|e^{A(t-\tau)}\| \leq \lambda_0 e^{-\alpha_0(t-\tau)}$$

which also implies that $H(s)$ is analytic in $\text{Re}[s] \geq -\alpha_0$.

Let us now apply the results of Lemma 3.3.3 to (3.3.26). We have
$$|\bar{x}(t)| \leq \frac{\|B\|\lambda_0}{\sqrt{2\alpha_0 - \delta}}\|v_t\|_{2\delta} = c_1\|v_t\|_{2\delta}$$

$$\|\bar{x}_t\|_{2\delta} \leq \frac{\|B\|\lambda_0}{\sqrt{(\delta_1 - \delta)(2\alpha_0 - \delta_1)}}\|v_t\|_{2\delta} = c_2\|v_t\|_{2\delta}$$

for some $\delta_1 > 0$, $\delta > 0$ such that $0 < \delta < \delta_1 < 2\alpha_0$. Because $\|v_t\|_{2\delta} = \|u_{t,t_1}\|_{2\delta}$ and $\|\bar{x}_{t,t_1}\|_{2\delta} \leq \|\bar{x}_t\|_{2\delta}$, it follows that for all $t \geq t_1$

$$|\bar{x}(t)| \leq c_1\|u_{t,t_1}\|_{2\delta}, \quad \|\bar{x}_{t,t_1}\|_{2\delta} \leq c_2\|u_{t,t_1}\|_{2\delta} \qquad (3.3.27)$$

From (3.3.22) we have
$$|x(t)| \leq \lambda_0 e^{-\alpha_0(t-t_1)}|x(t_1)| + |\bar{x}(t)| \qquad \forall t \geq t_1$$

which together with (3.3.27) imply (i)(a). Using (3.3.22) we have

$$\|x_{t,t_1}\|_{2\delta} \leq \|(e^{A(t-t_1)}x(t_1))_{t,t_1}\|_{2\delta} + \|\bar{x}_{t,t_1}\|_{2\delta}$$

which implies that

$$\begin{aligned}\|x_{t,t_1}\|_{2\delta} &\leq \left(\int_{t_1}^{t} e^{-\delta(t-\tau)} e^{-2\alpha_0(\tau-t_1)} d\tau\right)^{\frac{1}{2}} \lambda_0 |x(t_1)| + \|\bar{x}_{t,t_1}\|_{2\delta} \\ &\leq \frac{\lambda_0 e^{-\frac{\delta}{2}(t-t_1)}}{\sqrt{2\alpha_0 - \delta}} |x(t_1)| + \|\bar{x}_{t,t_1}\|_{2\delta} \end{aligned} \quad (3.3.28)$$

From (3.3.27) and (3.3.28), (i)(b) follows.

Let us now apply the results of Lemma 3.3.2 to the system (3.3.26), also described by

$$\bar{y} = H(s)v$$

we have

$$\|\bar{y}_t\|_{2\delta} \leq \|H(s)\|_{\infty\delta} \|v_t\|_{2\delta}$$

and for $H(s)$ strictly proper

$$|\bar{y}(t)| \leq \|H(s)\|_{2\delta} \|v_t\|_{2\delta}$$

for any $0 \leq \delta < 2\alpha_0$. Since $\|v_t\|_{2\delta} = \|u_{t,t_1}\|_{2\delta}$ and $\|\bar{y}_{t,t_1}\|_{2\delta} \leq \|\bar{y}_t\|_{2\delta}$, we have

$$\|\bar{y}_{t,t_1}\|_{2\delta} \leq \|H(s)\|_{\infty\delta} \|u_{t,t_1}\|_{2\delta} \quad (3.3.29)$$

and

$$|\bar{y}(t)| \leq \|H(s)\|_{2\delta} \|u_{t,t_1}\|_{2\delta}, \quad \forall t \geq t_1 \quad (3.3.30)$$

From (3.3.24) we have

$$|y(t)| \leq \|C^\top\| \lambda_0 e^{-\alpha_0(t-t_1)} |x(t_1)| + |\bar{y}(t)|, \quad \forall t \geq t_1 \quad (3.3.31)$$

which implies, after performing some calculations, that

$$\|y_{t,t_1}\|_{2\delta} \leq \|C^\top\| \frac{\lambda_0}{\sqrt{2\alpha_0 - \delta}} e^{-\frac{\delta}{2}(t-t_1)} |x(t_1)| + \|\bar{y}_{t,t_1}\|_{2\delta}, \quad \forall t \geq t_1 \quad (3.3.32)$$

Using (3.3.29) in (3.3.32) we establish (ii) and from (3.3.30) and (3.3.31) we establish (iii). □

By taking $t_1 = 0$, Lemma 3.3.5 also shows the effect of the initial condition $x(0) = x_0$ of the system (3.3.20) on the bounds for $|y(t)|$ and $\|y_t\|_{2\delta}$.

We can obtain a similar result as in Lemma 3.3.4 over the interval $[t_1, t]$ by extending Lemma 3.3.5 to the case where A is not necessarily a stable matrix and $\delta = \delta_0 > 0$ is a given fixed constant, provided (C, A) is an observable pair.

3.3. INPUT/OUTPUT STABILITY

Lemma 3.3.6 *Consider the LTV system (3.3.15) where the elements of $A(t)$, $B(t)$, $C(t)$, and $D(t)$ are bounded continuous functions of time and whose state transition matrix $\Phi(t,\tau)$ satisfies*

$$\|\Phi(t,\tau)\| \leq \lambda_0 e^{-\alpha_0(t-\tau)}$$

$\forall t \geq \tau$ *and* $t, \tau \in [t_1, t_2)$ *for some* $t_2 > t_1 \geq 0$ *and* $\alpha_0, \lambda_0 > 0$. *Then for any* $\delta \in [0, \delta_1)$ *where* $0 < \delta_1 < 2\alpha_0$ *is arbitrary, we have*

(i) $|x(t)| \leq \lambda_0 e^{-\alpha_0(t-t_1)}|x(t_1)| + \frac{c\lambda_0}{\sqrt{2\alpha_0-\delta}}\|u_{t,t_1}\|_{2\delta}$

(ii) $\|x_{t,t_1}\|_{2\delta} \leq \frac{\lambda_0}{\sqrt{2\alpha_0-\delta}}e^{-\frac{\delta}{2}(t-t_1)}|x(t_1)| + \frac{c\lambda_0}{\sqrt{(\delta_1-\delta)(2\alpha_0-\delta_1)}}\|u_{t,t_1}\|_{2\delta}, \forall t \in [t_1, t_2)$

where $c = \sup_t \|B(t)\|$.

Proof The solution $x(t)$ of (3.3.15) is given by

$$x(t) = \Phi(t, t_1)x(t_1) + \int_{t_1}^{t} \Phi(t,\tau)B(\tau)u(\tau)d\tau$$

Hence,

$$|x(t)| \leq \lambda_0 e^{-\alpha_0(t-t_1)}|x(t_1)| + c\lambda_0 \int_{t_1}^{t} e^{-\alpha_0(t-\tau)}|u(\tau)|d\tau$$

Proceeding as in the proof of Lemma 3.3.3 we establish (i). Now

$$\|x_{t,t_1}\|_{2\delta} \leq \lambda_0 |x(t_1)| \left(\int_{t_1}^{t} e^{-\delta(t-\tau)}e^{-2\alpha_0(\tau-t_1)}d\tau \right)^{\frac{1}{2}} + c\lambda_0 \|Q_{t,t_1}\|_{2\delta}$$

$$\|Q_{t,t_1}\|_{2\delta} \triangleq \left\| \left(\int_{t_1}^{t} e^{-\alpha_0(t-\tau)}|u(\tau)|d\tau \right)_{t,t_1} \right\|_{2\delta}$$

Following exactly the same step as in the proof of Lemma 3.3.3 we establish that

$$\|Q_{t,t_1}\|_{2\delta} \leq \frac{1}{\sqrt{(2\alpha_0-\delta_1)(\delta_1-\delta)}}\|u_{t,t_1}\|_{2\delta}$$

Because

$$\|x_{t,t_1}\|_{2\delta} \leq \frac{\lambda_0|x(t_1)|}{\sqrt{2\alpha_0-\delta}}e^{-\frac{\delta}{2}(t-t_1)} + \|Q_{t,t_1}\|_{2\delta}$$

the proof of (ii) follows. $\square$

Example 3.3.3 (i) Consider the system described by

$$y = H(s)u$$

where $H(s) = \frac{2}{s+3}$. We have

$$\|H(s)\|_{\infty\delta} = \sup_{\omega}\left|\frac{2}{j\omega + 3 - \frac{\delta}{2}}\right| = \frac{4}{6-\delta}, \quad \forall \delta \in [0,6)$$

and

$$\|H(s)\|_{2\delta} = \frac{1}{\sqrt{2\pi}}\left(\int_{-\infty}^{\infty}\frac{4}{\omega^2 + \frac{(6-\delta)^2}{4}}d\omega\right)^{\frac{1}{2}} = \frac{2}{\sqrt{6-\delta}}, \quad \forall \delta \in [0,6)$$

For $u(t) = 1, \forall t \geq 0$, we have $y(t) = \frac{2}{3}(1 - e^{-3t})$, which we can use to verify inequality (ii) of Lemma 3.3.2, i.e.,

$$|y(t)| = \frac{2}{3}|1 - e^{-3t}| \leq \frac{2}{\sqrt{6-\delta}}\left(\frac{1-e^{-\delta t}}{\delta}\right)^{\frac{1}{2}}$$

holds $\forall t \in [0,\infty)$ and $\delta \in (0,6)$.

(ii) The system in (i) may also be expressed as

$$\dot{y} = -3y + 2u, \quad y(0) = 0$$

Its transition matrix $\Phi(t,0) = e^{-3t}$ and from Lemma 3.3.3, we have

$$|y(t)| \leq \frac{2}{\sqrt{6-\delta}}\|u_t\|_{2\delta}, \quad \forall \delta \in [0,6)$$

For $u(t) = 1, \forall t \geq 0$, the above inequality implies

$$|y(t)| = \frac{2}{3}|1 - e^{-3t}| \leq \frac{2}{\sqrt{6-\delta}}\left(\frac{1-e^{-\delta t}}{\delta}\right)^{\frac{1}{2}}$$

which holds for all $\delta \in (0,6)$. $\triangledown$

3.3.3 Small Gain Theorem

Many feedback systems, including adaptive control systems, can be put in the form shown in Figure 3.1. The operators H_1, H_2 act on e_1, e_2 to produce the outputs y_1, y_2; u_1, u_2 are external inputs. Sufficient conditions for H_1, H_2 to guarantee existence and uniqueness of solutions e_1, y_1, e_2, y_2 for given inputs u_1, u_2 in $\mathcal{L}_{pe}$ are discussed in [42]. Here we assume that H_1, H_2 are such that the existence and uniqueness of solutions are guaranteed. The

3.3. INPUT/OUTPUT STABILITY

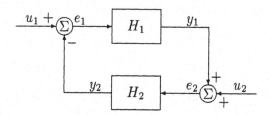

Figure 3.1 Feedback system.

problem is to determine conditions on H_1, H_2 so that if u_1, u_2 are bounded in some sense, then e_1, e_2, y_1, y_2 are also bounded in the same sense.

Let $\mathcal{L}$ be a normed linear space defined by

$$\mathcal{L} \triangleq \{f : \mathcal{R}^+ \mapsto \mathcal{R}^n \mid \|f\| < \infty\}$$

where $\|\cdot\|$ corresponds to any of the norms introduced earlier. Let $\mathcal{L}_e$ be the extended normed space associated with $\mathcal{L}$, i.e.,

$$\mathcal{L}_e = \{f : \mathcal{R}^+ \mapsto \mathcal{R}^n \mid \|f_t\| < \infty, \forall t \in \mathcal{R}^+\}$$

where

$$f_t(\tau) = \begin{cases} f(\tau) & \tau \leq t \\ 0 & \tau > t \end{cases}$$

The following theorem known as the *small gain theorem* [42] gives sufficient conditions under which bounded inputs produce bounded outputs in the feedback system of Figure 3.1.

Theorem 3.3.3 *Consider the system shown in Figure 3.1. Suppose H_1, H_2: $\mathcal{L}_e \mapsto \mathcal{L}_e$; $e_1, e_2 \in \mathcal{L}_e$. Suppose that for some constants $\gamma_1, \gamma_2 \geq 0$ and β_1, β_2, the operators H_1, H_2 satisfy*

$$\|(H_1 e_1)_t\| \leq \gamma_1 \|e_{1t}\| + \beta_1$$

$$\|(H_2 e_2)_t\| \leq \gamma_2 \|e_{2t}\| + \beta_2$$

$\forall t \in \mathcal{R}^+$. *If*

$$\gamma_1 \gamma_2 < 1$$

then

(i)
$$\|e_{1t}\| \leq (1-\gamma_1\gamma_2)^{-1}(\|u_{1t}\| + \gamma_2\|u_{2t}\| + \beta_2 + \gamma_2\beta_1)$$
$$\|e_{2t}\| \leq (1-\gamma_1\gamma_2)^{-1}(\|u_{2t}\| + \gamma_1\|u_{1t}\| + \beta_1 + \gamma_1\beta_2) \quad (3.3.33)$$

for any $t \geq 0$.

(ii) *If in addition, $\|u_1\|, \|u_2\| < \infty$, then e_1, e_2, y_1, y_2 have finite norms, and the norms of e_1, e_2 are bounded by the right-hand sides of (3.3.33) with all subscripts t dropped.*

The constants γ_1, γ_2 are referred to as the gains of H_1, H_2 respectively. When $u_2 \equiv 0$, there is no need to separate the gain of H_1 and H_2. In this case, one can consider the "loop gain" H_2H_1 as illustrated by the following corollary:

Corollary 3.3.4 *Consider the system of Figure 3.1 with $u_2 \equiv 0$. Suppose that*
$$\|(H_2H_1e_1)_t\| \leq \gamma_{21}\|e_{1t}\| + \beta_{21}$$
$$\|(H_1e_1)_t\| \leq \gamma_1\|e_{1t}\| + \beta_1$$

$\forall t \in \mathcal{R}^+$ *for some constants $\gamma_{21}, \gamma_1 \geq 0$ and β_{21}, β_1. If $\gamma_{21} < 1$, then*

(i)
$$\|e_{1t}\| \leq (1-\gamma_{21})^{-1}(\|u_{1t}\| + \beta_{21})$$
$$\|y_{1t}\| \leq \gamma_1(1-\gamma_{21})^{-1}(\|u_{1t}\| + \beta_{21}) + \beta_1 \quad (3.3.34)$$

for any $t \geq 0$.

(ii) *If in addition $\|u_1\| < \infty$, then e_1, e_2, y_1, y_2 have finite norms and (3.3.34) holds without the subscript t.*

The proofs of Theorem 3.3.3 and Corollary 3.3.4 follow by using the properties of the norm [42].

The small gain theorem is a very general theorem that applies to both continuous and discrete-time systems with multiple inputs and outputs.

As we mentioned earlier, Theorem 3.3.3 and Corollary 3.3.4 assume the existence of solutions $e_1, e_2 \in \mathcal{L}_e$. In practice, u_1, u_2 are given external inputs and e_1, e_2 are calculated using the operators H_1, H_2. Therefore, the existence of $e_1, e_2 \in \mathcal{L}_e$ depends on the properties of H_1, H_2.

Example 3.3.4 Let us consider the feedback system

3.3. INPUT/OUTPUT STABILITY

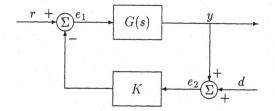

where $G(s) = \frac{e^{-\alpha s}}{s+2}$, $\alpha \geq 0$ is a constant time delay, and K is a constant feedback gain. The external input r is an input command, and d is a noise disturbance. We are interested in finding conditions on the gain K such that

(i) $r, d \in \mathcal{L}_\infty \Longrightarrow e_1, e_2, y \in \mathcal{L}_\infty$
(ii) $r, d \in \mathcal{L}_2 \Longrightarrow e_1, e_2, y \in \mathcal{L}_2$

The system is in the form of the general feedback system given in Figure 3.1, i.e.,

$$u_1 = r, \; u_2 = d$$

$$H_1 e_1(t) = \int_0^{t-\alpha} e^{2\alpha} e^{-2(t-\tau)} e_1(\tau) d\tau$$

$$H_2 e_2(t) = K e_2(t)$$

where $e^{-2(t-\alpha)}$ for $t \geq \alpha$ comes from the impulse response $g(t)$ of $G(s)$, i.e., $g(t) = e^{-2(t-\alpha)}$ for $t \geq \alpha$ and $g(t) = 0$ for $t < \alpha$.

(i) Because

$$\begin{aligned}
|H_1 e_1(t)| &\leq e^{2\alpha} \int_0^{t-\alpha} e^{-2(t-\tau)} |e_1(\tau)| d\tau \\
&\leq e^{2\alpha} \int_0^{t-\alpha} e^{-2(t-\tau)} d\tau \|e_{1t}\|_\infty \\
&\leq \frac{1}{2} \|e_{1t}\|_\infty
\end{aligned}$$

we have $\gamma_1 = \frac{1}{2}$. Similarly, the $\mathcal{L}_\infty$-gain of H_2 is $\gamma_2 = |K|$. Therefore, for $\mathcal{L}_\infty$-stability the small gain theorem requires

$$\frac{|K|}{2} < 1, \quad \text{i.e.,} \quad |K| < 2$$

(ii) From Lemma 3.3.1, we have

$$\|(H_1 e_1)_t\|_2 \leq \sup_\omega \left| \frac{e^{-\alpha j\omega}}{2 + j\omega} \right| \|e_{1t}\|_2 = \frac{1}{2} \|e_{1t}\|_2$$

which implies that the $\mathcal{L}_2$ gain of H_1 is $\gamma_1 = \frac{1}{2}$. Similarly the $\mathcal{L}_2$ gain of H_2 is $\gamma_2 = |K|$, and the condition for $\mathcal{L}_2$-stability is $|K| < 2$.

For this simple system, however, with $\alpha = 0$, we can verify that $r, d \in \mathcal{L}_\infty \Longrightarrow e_1, e_2, y \in \mathcal{L}_\infty$ if and only if $K > -2$, which indicates that the condition given by the small gain theorem (for $\alpha = 0$) is conservative. $\triangledown$

Example 3.3.5 Consider the system

$$\dot{x} = A_c x, \quad A_c = A + B$$

where $x \in \mathcal{R}^n$, A is a stable matrix, i.e., all the eigenvalues of A are in $\text{Re}[s] < 0$ and B is a constant matrix. We are interested in obtaining an upper bound for B such that A_c is a stable matrix. Let us represent the system in the form of Figure 3.1 as the following:

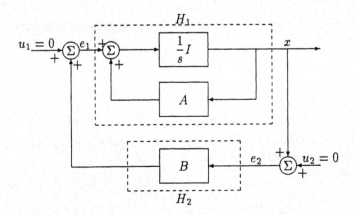

We can verify that the $\mathcal{L}_\infty$ gain of H_1 is $\gamma_1 = \frac{\alpha_1}{\alpha_0}$ where $\alpha_1, \alpha_0 > 0$ are the constants in the bound $\|e^{A(t-\tau)}\| \leq \alpha_1 e^{-\alpha_0(t-\tau)}$ that follows from the stability of A. The $\mathcal{L}_\infty$ gain of H_2 is $\gamma_2 = \|B\|$. Therefore for $\mathcal{L}_\infty$ stability, we should have

$$\|B\| \frac{\alpha_1}{\alpha_0} < 1$$

or

$$\|B\| < \frac{\alpha_0}{\alpha_1}$$

Now $\mathcal{L}_\infty$ stability implies that $A_c = A + B$ is a stable matrix. (Note that the initial condition for x is taken to be zero, i.e., $x(0) = 0$.) $\triangledown$

Despite its conservatism, the small gain theorem is widely used to design robust controllers for uncertain systems. In many applications, certain loop transformations are needed to transform a given feedback system to the form of the feedback system of Figure 3.1 where H_1, H_2 have finite gains [42].

3.3.4 Bellman-Gronwall Lemma

A key lemma for analysis of adaptive control schemes is the following.

Lemma 3.3.7 (Bellman-Gronwall Lemma I) [232] *Let $\lambda(t), g(t), k(t)$ be nonnegative piecewise continuous functions of time t. If the function $y(t)$ satisfies the inequality*

$$y(t) \leq \lambda(t) + g(t) \int_{t_0}^{t} k(s)y(s)ds, \quad \forall t \geq t_0 \geq 0 \qquad (3.3.35)$$

then

$$y(t) \leq \lambda(t) + g(t) \int_{t_0}^{t} \lambda(s)k(s) \left[exp\left(\int_{s}^{t} k(\tau)g(\tau)d\tau \right) \right] ds \quad \forall t \geq t_0 \geq 0 \qquad (3.3.36)$$

In particular, if $\lambda(t) \equiv \lambda$ is a constant and $g(t) \equiv 1$, then

$$y(t) \leq \lambda exp\left(\int_{t_0}^{t} k(s)ds \right) \quad \forall t \geq t_0 \geq 0$$

Proof Let us define

$$q(t) \triangleq k(t) e^{-\int_{t_0}^{t} g(\tau)k(\tau)d\tau}$$

Because $k(t)$ is nonnegative, we have $q(t) \geq 0 \ \forall t \geq t_0$. Multiplying both sides of (3.3.35) by $q(t)$, and rearranging the inequality we obtain

$$q(t)y(t) - q(t)g(t) \int_{t_0}^{t} k(s)y(s)ds \leq \lambda(t)q(t) \qquad (3.3.37)$$

From the expression of $q(t)$, one can verify that

$$q(t)y(t) - q(t)g(t) \int_{t_0}^{t} k(s)y(s)ds = \frac{d}{dt}\left(e^{-\int_{t_0}^{t} g(\tau)k(\tau)d\tau} \int_{t_0}^{t} k(s)y(s)ds \right) \qquad (3.3.38)$$

Using (3.3.38) in (3.3.37) and integrating both sides of (3.3.37), we obtain

$$e^{-\int_{t_0}^{t} g(\tau)k(\tau)d\tau} \int_{t_0}^{t} k(s)y(s)ds \leq \int_{t_0}^{t} \lambda(s)q(s)ds$$

Therefore,

$$\begin{aligned}
\int_{t_0}^{t} k(s)y(s)ds &\leq e^{\int_{t_0}^{t} g(\tau)k(\tau)d\tau} \int_{t_0}^{t} \lambda(s)q(s)ds \\
&= e^{\int_{t_0}^{t} g(\tau)k(\tau)d\tau} \int_{t_0}^{t} \lambda(s)k(s) e^{-\int_{t_0}^{s} g(\tau)k(\tau)d\tau} ds \\
&= \int_{t_0}^{t} \lambda(s)k(s) e^{\int_{s}^{t} g(\tau)k(\tau)d\tau} ds \qquad (3.3.39)
\end{aligned}$$

Using (3.3.39) in (3.3.35), the proof for the inequality (3.3.36) is complete.

Consider the special case where λ is a constant and $g = 1$. Define

$$q_1 \triangleq \lambda + \int_{t_0}^{t} k(s)y(s)ds$$

From (3.3.35), we have

$$y(t) \leq q_1(t)$$

Now

$$\dot{q}_1 = ky$$

Because $k \geq 0$, we have

$$\dot{q}_1 \leq kq_1$$

Let $w = \dot{q}_1 - kq_1$. Clearly, $w \leq 0$ and

$$\dot{q}_1 = kq_1 + w$$

which implies

$$q_1(t) = e^{\int_{t_0}^{t} k(\tau)d\tau} q_1(t_0) + \int_{t_0}^{t} e^{\int_{\tau}^{t} k(s)ds} w(\tau)d\tau \qquad (3.3.40)$$

Because $k \geq 0, w \leq 0 \ \forall t \geq t_0$ and $q_1(t_0) = \lambda$, it follows from (3.3.40) that

$$y(t) \leq q_1(t) \leq \lambda e^{\int_{t_0}^{t} k(\tau)d\tau}$$

and the proof is complete. □

The reader can refer to [32, 232] for alternative proofs of the B-G Lemma. Other useful forms of the B-G lemma are given by Lemmas 3.3.8 and 3.3.9.

Lemma 3.3.8 (B-G Lemma II) *Let $\lambda(t), k(t)$ be nonnegative piecewise continuous function of time t and let $\lambda(t)$ be differentiable. If the function $y(t)$ satisfies the inequality*

$$y(t) \leq \lambda(t) + \int_{t_0}^{t} k(s)y(s)ds, \quad \forall t \geq t_0 \geq 0$$

then

$$y(t) \leq \lambda(t_0)e^{\int_{t_0}^{t} k(s)ds} + \int_{t_0}^{t} \dot{\lambda}(s)e^{\int_{s}^{t} k(\tau)d\tau}ds, \quad \forall t \geq t_0 \geq 0.$$

3.3. INPUT/OUTPUT STABILITY

Proof Let
$$z(t) = \lambda(t) + \int_{t_0}^{t} k(s)y(s)ds$$
it follows that z is differentiable and $z \geq y$. We have
$$\dot{z} = \dot{\lambda} + ky, \quad z(t_0) = \lambda(t_0)$$
Let $v = z - y$, then
$$\dot{z} = \dot{\lambda} + kz - kv$$
whose state transition matrix is
$$\Phi(t,\tau) = \exp \int_{\tau}^{t} k(s)ds$$
Therefore,
$$z(t) = \Phi(t,t_0)z(t_0) + \int_{t_0}^{t} \Phi(t,\tau)[\dot{\lambda}(\tau) - k(\tau)v(\tau)]d\tau$$
Because
$$\int_{t_0}^{t} \Phi(t,\tau)k(\tau)v(\tau)d\tau \geq 0$$
resulting from $\Phi(t,\tau), k(\tau), v(\tau)$ being nonnegative, we have
$$z(t) \leq \Phi(t,t_0)z(t_0) + \int_{t_0}^{t} \Phi(t,\tau)\dot{\lambda}(\tau)d\tau$$
Using the expression for $\Phi(t,t_0)$ in the above inequality, we have
$$y(t) \leq z(t) \leq \lambda(t_0)e^{\int_{t_0}^{t} k(s)ds} + \int_{t_0}^{t} \dot{\lambda}(s)e^{\int_{s}^{t} k(\tau)d\tau}ds$$
and the proof is complete. □

Lemma 3.3.9 (B-G Lemma III) *Let c_0, c_1, c_2, α be nonnegative constants and $k(t)$ a nonnegative piecewise continuous function of time. If $y(t)$ satisfies the inequality*
$$y(t) \leq c_0 e^{-\alpha(t-t_0)} + c_1 + c_2 \int_{t_0}^{t} e^{-\alpha(t-\tau)}k(\tau)y(\tau)d\tau, \quad \forall t \geq t_0$$
then
$$y(t) \leq (c_0 + c_1)e^{-\alpha(t-t_0)}e^{c_2 \int_{t_0}^{t} k(s)ds} + c_1\alpha \int_{t_0}^{t} e^{-\alpha(t-\tau)}e^{c_2 \int_{\tau}^{t} k(s)ds}d\tau, \quad \forall t \geq t_0$$

Proof The proof follows directly from Lemma 3.3.8 by rewriting the given inequality of y as

$$\bar{y}(t) \leq \lambda(t) + \int_{t_0}^{t} \bar{k}(\tau)\bar{y}(\tau)d\tau$$

where $\bar{y}(t) = e^{\alpha t}y(t), \bar{k}(t) = c_2 k(t), \lambda(t) = c_0 e^{\alpha t_0} + c_1 e^{\alpha t}$. Applying Lemma 3.3.8, we obtain

$$e^{\alpha t}y(t) \leq (c_0 + c_1)e^{\alpha t_0}e^{c_2 \int_{t_0}^{t} k(s)ds} + c_1\alpha \int_{t_0}^{t} e^{\alpha \tau} e^{c_2 \int_{\tau}^{t} k(s)ds} d\tau$$

The result follows by multiplying each side of the above inequality by $e^{-\alpha t}$. $\square$

The B-G Lemma allows us to obtain an explicit bound for $y(t)$ from the implicit bound of $y(t)$ given by the integral inequality (3.3.35). Notice that if $y(t) \geq 0$ and $\lambda(t) = 0 \; \forall t \geq 0$, (3.3.36) implies that $y(t) \equiv 0 \; \forall t \geq 0$.

In many cases, the B-G Lemma may be used in place of the small gain theorem to analyze a class of feedback systems in the form of Figure 3.1 as illustrated by the following example.

Example 3.3.6 Consider the same system as in Example 3.3.5. We have

$$x(t) = e^{At}x(0) + \int_{0}^{t} e^{A(t-\tau)}Bx(\tau)d\tau$$

Hence,

$$|x(t)| \leq \alpha_1 e^{-\alpha_0 t}|x(0)| + \int_{0}^{t} \alpha_1 e^{-\alpha_0(t-\tau)}\|B\||x(\tau)|d\tau$$

i.e.,

$$|x(t)| \leq \alpha_1 e^{-\alpha_0 t}|x(0)| + \alpha_1 e^{-\alpha_0 t}\|B\| \int_{0}^{t} e^{\alpha_0 \tau}|x(\tau)|d\tau$$

Applying the B-G Lemma I with $\lambda = \alpha_1 e^{-\alpha_0 t}|x(0)|, g(t) = \alpha_1\|B\|e^{-\alpha_0 t}, k(t) = e^{\alpha_0 t}$, we have

$$|x(t)| \leq \alpha_1 e^{-\alpha_0 t}|x(0)| + \alpha_1 |x(0)|e^{-\gamma t}$$

where

$$\gamma = \alpha_0 - \alpha_1 \|B\|$$

Therefore, for $|x(t)|$ to be bounded from above by a decaying exponential (which implies that $A_c = A + B$ is a stable matrix), B has to satisfy

$$\|B\| < \frac{\alpha_0}{\alpha_1}$$

3.4. LYAPUNOV STABILITY

which is the same condition we obtained in Example 3.3.5 using the small gain theorem. In this case, we assume that $|x(0)| \neq 0$, otherwise for $x(0) = 0$ we would have $\lambda(t) = 0$ and $|x(t)| = 0 \; \forall t \geq 0$ which tells us nothing about the stability of A_c. The reader may like to verify the same result using B-G Lemmas II and III. $\triangledown$

3.4 Lyapunov Stability

3.4.1 Definition of Stability

We consider systems described by ordinary differential equations of the form

$$\dot{x} = f(t, x), \quad x(t_0) = x_0 \tag{3.4.1}$$

where $x \in \mathcal{R}^n$, $f : \mathcal{J} \times \mathcal{B}(r) \mapsto \mathcal{R}$, $\mathcal{J} = [t_0, \infty)$ and $\mathcal{B}(r) = \{x \in \mathcal{R}^n \mid |x| < r\}$. We assume that f is of such nature that for every $x_0 \in \mathcal{B}(r)$ and every $t_0 \in \mathcal{R}^+$, (3.4.1) possesses one and only one solution $x(t; t_0, x_0)$.

Definition 3.4.1 *A state x_e is said to be an* **equilibrium state** *of the system described by (3.4.1) if*

$$f(t, x_e) \equiv 0 \quad for\; all\; t \geq t_0$$

Definition 3.4.2 *An equilibrium state x_e is called an* **isolated equilibrium state** *if there exists a constant $r > 0$ such that $\mathcal{B}(x_e, r) \stackrel{\triangle}{=} \{x \mid |x - x_e| < r\} \subset \mathcal{R}^n$ contains no equilibrium state of (3.4.1) other than x_e.*

The equilibrium state $x_{1e} = 0, x_{2e} = 0$ of

$$\dot{x}_1 = x_1 x_2, \quad \dot{x}_2 = x_1^2$$

is not isolated because any point $x_1 = 0, x_2 = $ constant is an equilibrium state. The differential equation

$$\dot{x} = (x - 1)^2 x$$

has two isolated equilibrium states $x_e = 1$ and $x_e = 0$.

Definition 3.4.3 *The equilibrium state x_e is said to be* **stable** *(in the sense of Lyapunov) if for arbitrary t_0 and $\epsilon > 0$ there exists a $\delta(\epsilon, t_0)$ such that $|x_0 - x_e| < \delta$ implies $|x(t; t_0, x_0) - x_e| < \epsilon$ for all $t \geq t_0$.*

Definition 3.4.4 *The equilibrium state x_e is said to be* **uniformly stable (u.s.)** *if it is stable and if $\delta(\epsilon, t_0)$ in Definition 3.4.3 does not depend on t_0.*

Definition 3.4.5 *The equilibrium state x_e is said to be* **asymptotically stable (a.s.)** *if (i) it is stable, and (ii) there exists a $\delta(t_0)$ such that $|x_0 - x_e| < \delta(t_0)$ implies $\lim_{t \to \infty} |x(t; t_0, x_0) - x_e| = 0$.*

Definition 3.4.6 *The set of all $x_0 \in \mathcal{R}^n$ such that $x(t; t_0, x_0) \to x_e$ as $t \to \infty$ for some $t_0 \geq 0$ is called the* **region of attraction** *of the equilibrium state x_e. If condition (ii) of Definition 3.4.5 is satisfied, then the equilibrium state x_e is said to be* **attractive**.

Definition 3.4.7 *The equilibrium state x_e is said to be* **uniformly asymptotically stable (u.a.s.)** *if (i) it is uniformly stable, (ii) for every $\epsilon > 0$ and any $t_0 \in \mathcal{R}^+$, there exist a $\delta_0 > 0$ independent of t_0 and ϵ and a $T(\epsilon) > 0$ independent of t_0 such that $|x(t; t_0, x_0) - x_e| < \epsilon$ for all $t \geq t_0 + T(\epsilon)$ whenever $|x_0 - x_e| < \delta_0$.*

Definition 3.4.8 *The equilibrium state x_e is* **exponentially stable (e.s.)** *if there exists an $\alpha > 0$, and for every $\epsilon > 0$ there exists a $\delta(\epsilon) > 0$ such that*

$$|x(t; t_0, x_0) - x_e| \leq \epsilon e^{-\alpha(t - t_0)} \quad \text{for all } t \geq t_0$$

whenever $|x_0 - x_e| < \delta(\epsilon)$.

Definition 3.4.9 *The equilibrium state x_e is said to be* **unstable** *if it is not stable.*

When (3.4.1) possesses a unique solution for each $x_0 \in \mathcal{R}^n$ and $t_0 \in \mathcal{R}^+$, we need the following definitions for the global characterization of solutions.

Definition 3.4.10 *A solution $x(t; t_0, x_0)$ of (3.4.1) is* **bounded** *if there exists a $\beta > 0$ such that $|x(t; t_0, x_0)| < \beta$ for all $t \geq t_0$, where β may depend on each solution.*

3.4. LYAPUNOV STABILITY

Definition 3.4.11 *The solutions of (3.4.1) are* **uniformly bounded** *(u.b.) if for any $\alpha > 0$ and $t_0 \in \mathcal{R}^+$, there exists a $\beta = \beta(\alpha)$ independent of t_0 such that if $|x_0| < \alpha$, then $|x(t; t_0, x_0)| < \beta$ for all $t \geq t_0$.*

Definition 3.4.12 *The solutions of (3.4.1) are* **uniformly ultimately bounded (u.u.b.)** *(with bound B) if there exists a $B > 0$ and if corresponding to any $\alpha > 0$ and $t_0 \in \mathcal{R}^+$, there exists a $T = T(\alpha) > 0$ (independent of t_0) such that $|x_0| < \alpha$ implies $|x(t; t_0, x_0)| < B$ for all $t \geq t_0 + T$.*

Definition 3.4.13 *The equilibrium point x_e of (3.4.1) is* **asymptotically stable in the large (a.s. in the large)** *if it is stable and every solution of (3.4.1) tends to x_e as $t \to \infty$ (i.e., the region of attraction of x_e is all of $\mathcal{R}^n$).*

Definition 3.4.14 *The equilibrium point x_e of (3.4.1) is* **uniformly asymptotically stable in the large (u.a.s. in the large)** *if (i) it is uniformly stable, (ii) the solutions of (3.4.1) are uniformly bounded, and (iii) for any $\alpha > 0$, any $\epsilon > 0$ and $t_0 \in \mathcal{R}^+$, there exists $T(\epsilon, \alpha) > 0$ independent of t_0 such that if $|x_0 - x_e| < \alpha$ then $|x(t; t_0, x_0) - x_e| < \epsilon$ for all $t \geq t_0 + T(\epsilon, \alpha)$.*

Definition 3.4.15 *The equilibrium point x_e of (3.4.1) is* **exponentially stable in the large (e.s. in the large)** *if there exists $\alpha > 0$ and for any $\beta > 0$, there exists $k(\beta) > 0$ such that*

$$|x(t; t_0, x_0)| \leq k(\beta) e^{-\alpha(t-t_0)} \quad for\ all\ t \geq t_0$$

whenever $|x_0| < \beta$.

Definition 3.4.16 *If $x(t; t_0, x_0)$ is a solution of $\dot{x} = f(t, x)$, then the trajectory $x(t; t_0, x_0)$ is said to be* **stable (u.s., a.s., u.a.s., e.s., unstable)** *if the equilibrium point $z_e = 0$ of the differential equation*

$$\dot{z} = f(t, z + x(t; t_0, x_0)) - f(t, x(t; t_0, x_0))$$

is stable (u.s., a.s., u.a.s., e.s., unstable, respectively).

The above stability concepts and definitions are illustrated by the following example:

Example 3.4.1

(i) $\dot{x} = 0$ has the equilibrium state $x_e = c$, where c is any constant, which is not an isolated equilibrium state. It can be easily verified that $x_e = c$ is stable, u.s. but not a.s.

(ii) $\dot{x} = -x^3$ has an isolated equilibrium state $x_e = 0$. Its solution is given by

$$x(t) = x(t; t_0, x_0) = \left(\frac{x_0^2}{1 + 2x_0^2(t - t_0)} \right)^{\frac{1}{2}}$$

Now given any $\epsilon > 0$, $|x_0| < \delta = \epsilon$ implies that

$$|x(t)| = \sqrt{\frac{x_0^2}{1 + 2x_0^2(t - t_0)}} \leq |x_0| < \epsilon \quad \forall t \geq t_0 \geq 0 \qquad (3.4.2)$$

Hence, according to Definition 3.4.3, $x_e = 0$ is stable. Because $\delta = \epsilon$ is independent of t_0, $x_e = 0$ is also u.s. Furthermore, because $x_e = 0$ is stable and $x(t) \to x_e = 0$ as $t \to \infty$ for all $x_0 \in \mathcal{R}$, we have a.s. in the large. Let us now check whether $x_e = 0$ is u.a.s. in the large by using Definition 3.4.14. We have already shown u.s. From (3.4.2) we have that $x(t)$ is u.b. To satisfy condition (iii) of Definition 3.4.14, we need to find a $T > 0$ independent of t_0 such that for any $\alpha > 0$ and $\epsilon > 0$, $|x_0| < \alpha$ implies $|x(t)| < \epsilon$ for all $t \geq t_0 + T$. From (3.4.2) we have

$$|x(t)| \leq |x(t_0 + T)| = \sqrt{\frac{x_0^2}{1 + 2x_0^2 T}} < \sqrt{\frac{1}{2T}}, \quad \forall t \geq t_0 + T$$

Choosing $T = \frac{1}{2\epsilon^2}$, it follows that $|x(t)| < \epsilon \ \forall t \geq t_0 + T$. Hence, $x_e = 0$ is u.a.s. in the large. Using Definition 3.4.15, we can conclude that $x_e = 0$ is not e.s.

(iii) $\dot{x} = (x - 2)x$ has two isolated equilibrium states $x_e = 0$ and $x_e = 2$. It can be shown that $x_e = 0$ is e.s. with the region of attraction $R_a = \{x \mid x < 2\}$ and $x_e = 2$ is unstable.

(iv) $\dot{x} = -\frac{1}{1+t}x$ has an equilibrium state $x_e = 0$ that is stable, u.s., a.s. in the large but is not u.a.s. (verify).

(v) $\dot{x} = (t \sin t - \cos t - 2)x$ has an isolated equilibrium state $x_e = 0$ that is stable, a.s. in the large but not u.s. (verify). $\qquad \triangledown$

3.4.2 Lyapunov's Direct Method

The stability properties of the equilibrium state or solution of (3.4.1) can be studied by using the so-called direct method of Lyapunov (also known as Lyapunov's second method) [124, 125]. The objective of this method is

3.4. LYAPUNOV STABILITY

to answer questions of stability by using the form of $f(t,x)$ in (3.4.1) rather than the explicit knowledge of the solutions. We start with the following definitions [143].

Definition 3.4.17 *A continuous function $\varphi : [0,r] \mapsto \mathcal{R}^+$ (or a continuous function $\varphi : [0,\infty) \mapsto \mathcal{R}^+$) is said to belong to* **class $\mathcal{K}$**, *i.e., $\varphi \in \mathcal{K}$ if*

(i) $\varphi(0) = 0$
(ii) φ *is strictly increasing on $[0,r]$ (or on $[0,\infty)$).*

Definition 3.4.18 *A continuous function $\varphi : [0,\infty) \mapsto \mathcal{R}^+$ is said to belong to* **class $\mathcal{KR}$**, *i.e., $\varphi \in \mathcal{KR}$ if*

(i) $\varphi(0) = 0$
(ii) φ *is strictly increasing on $[0,\infty)$*
(iii) $\lim_{r \to \infty} \varphi(r) = \infty$.

The function $\varphi(|x|) = \frac{x^2}{1+x^2}$ belongs to class $\mathcal{K}$ defined on $[0,\infty)$ but not to class $\mathcal{KR}$. The function $\varphi(|x|) = |x|$ belongs to class $\mathcal{K}$ and class $\mathcal{KR}$. It is clear that $\varphi \in \mathcal{KR}$ implies $\varphi \in \mathcal{K}$, but not the other way.

Definition 3.4.19 *Two functions $\varphi_1, \varphi_2 \in \mathcal{K}$ defined on $[0,r]$ (or on $[0,\infty)$) are said to be* **of the same order of magnitude** *if there exist positive constants k_1, k_2 such that*

$$k_1 \varphi_1(r_1) \leq \varphi_2(r_1) \leq k_2 \varphi_1(r_1), \quad \forall r_1 \in [0,r] \ (\text{ or } \forall r_1 \in [0,\infty))$$

The function $\varphi_1(|x|) = \frac{x^2}{1+2x^2}$ and $\varphi_2 = \frac{x^2}{1+x^2}$ are of the same order of magnitude (verify!).

Definition 3.4.20 *A function $V(t,x) : \mathcal{R}^+ \times \mathcal{B}(r) \mapsto \mathcal{R}$ with $V(t,0) = 0 \ \forall t \in \mathcal{R}^+$ is* **positive definite** *if there exists a continuous function $\varphi \in \mathcal{K}$ such that $V(t,x) \geq \varphi(|x|) \ \forall t \in \mathcal{R}^+$, $x \in \mathcal{B}(r)$ and some $r > 0$. $V(t,x)$ is called* **negative definite** *if $-V(t,x)$ is positive definite.*

The function $V(t,x) = \frac{x^2}{1-x^2}$ with $x \in \mathcal{B}(1)$ is positive definite, whereas $V(t,x) = \frac{1}{1+t}x^2$ is not. The function $V(t,x) = \frac{x^2}{1+x^2}$ is positive definite for all $x \in \mathcal{R}$.

Definition 3.4.21 *A function* $V(t,x) : \mathcal{R}^+ \times \mathcal{B}(r) \mapsto \mathcal{R}$ *with* $V(t,0) = 0$ $\forall t \in \mathcal{R}^+$ *is said to be* **positive (negative) semidefinite** *if* $V(t,x) \geq 0$ $(V(t,x) \leq 0)$ *for all* $t \in \mathcal{R}^+$ *and* $x \in \mathcal{B}(r)$ *for some* $r > 0$.

Definition 3.4.22 *A function* $V(t,x) : \mathcal{R}^+ \times \mathcal{B}(r) \mapsto \mathcal{R}$ *with* $V(t,0) = 0$ $\forall t \in \mathcal{R}^+$ *is said to be* **decrescent** *if there exists* $\varphi \in \mathcal{K}$ *such that* $|V(t,x)| \leq \varphi(|x|)$ $\forall t \geq 0$ *and* $\forall x \in \mathcal{B}(r)$ *for some* $r > 0$.

The function $V(t,x) = \frac{1}{1+t}x^2$ is decrescent because $V(t,x) = \frac{1}{1+t}x^2 \leq x^2$ $\forall t \in \mathcal{R}^+$ but $V(t,x) = tx^2$ is not.

Definition 3.4.23 *A function* $V(t,x) : \mathcal{R}^+ \times \mathcal{R}^n \mapsto \mathcal{R}$ *with* $V(t,0) = 0$ $\forall t \in \mathcal{R}^+$ *is said to be* **radially unbounded** *if there exists* $\varphi \in \mathcal{KR}$ *such that* $V(t,x) \geq \varphi(|x|)$ *for all* $x \in \mathcal{R}^n$ *and* $t \in \mathcal{R}^+$.

The function $V(x) = \frac{x^2}{1+x^2}$ satisfies conditions (i) and (ii) of Definition 3.4.23 (i.e., choose $\varphi(|x|) = \frac{|x|^2}{1+|x|^2}$). However, because $V(x) \leq 1$, one cannot find a function $\varphi(|x|) \in \mathcal{KR}$ to satisfy $V(x) \geq \varphi(|x|)$ for all $x \in \mathcal{R}^n$. Hence, V is not radially unbounded.

It is clear from Definition 3.4.23 that if $V(t,x)$ is radially unbounded, it is also positive definite for all $x \in \mathcal{R}^n$ but the converse is not true. The reader should be aware that in some textbooks "positive definite" is used for radially unbounded functions, and "locally positive definite" is used for our definition of positive definite functions.

Let us assume (without loss of generality) that $x_e = 0$ is an equilibrium point of (3.4.1) and define $\dot{V}$ to be the time derivative of the function $V(t,x)$ along the solution of (3.4.1), i.e.,

$$\dot{V} = \frac{\partial V}{\partial t} + (\nabla V)^\top f(t,x) \tag{3.4.3}$$

where $\nabla V = [\frac{\partial V}{\partial x_1}, \frac{\partial V}{\partial x_2}, \ldots, \frac{\partial V}{\partial x_n}]^\top$ is the gradient of V with respect to x. The second method of Lyapunov is summarized by the following theorem.

Theorem 3.4.1 *Suppose there exists a positive definite function* $V(t,x) : \mathcal{R}^+ \times \mathcal{B}(r) \mapsto \mathcal{R}$ *for some* $r > 0$ *with continuous first-order partial derivatives with respect to* x, t, *and* $V(t,0) = 0$ $\forall t \in \mathcal{R}^+$. *Then the following statements are true:*

3.4. LYAPUNOV STABILITY

(i) If $\dot{V} \leq 0$, then $x_e = 0$ is **stable**.
(ii) If V is decrescent and $\dot{V} \leq 0$, then $x_e = 0$ is **u.s.**
(iii) If V is decrescent and $\dot{V} < 0$, then x_e is **u.a.s.**
(iv) If V is decrescent and there exist $\varphi_1, \varphi_2, \varphi_3 \in \mathcal{K}$ of the same order of magnitude such that

$$\varphi_1(|x|) \leq V(t,x) \leq \varphi_2(|x|), \quad \dot{V}(t,x) \leq -\varphi_3(|x|)$$

for all $x \in \mathcal{B}(r)$ and $t \in \mathcal{R}^+$, then $x_e = 0$ is **e.s.**

In the above theorem, the state x is restricted to be inside the ball $\mathcal{B}(r)$ for some $r > 0$. Therefore, the results (i) to (iv) of Theorem 3.4.1 are referred to as local results. Statement (iii) is equivalent to that there exist $\varphi_1, \varphi_2, \varphi_3 \in \mathcal{K}$, where $\varphi_1, \varphi_2, \varphi_3$ do **not** have to be of the same order of magnitude, such that $\varphi_1(|x|) \leq V(t,x) \leq \varphi_2(|x|), \dot{V}(t,x) \leq -\varphi_3(|x|)$.

Theorem 3.4.2 *Assume that (3.4.1) possesses unique solutions for all $x_0 \in \mathcal{R}^n$. Suppose there exists a positive definite, decrescent and radially unbounded function $V(t,x) : \mathcal{R}^+ \times \mathcal{R}^n \mapsto \mathcal{R}^+$ with continuous first-order partial derivatives with respect to t, x and $V(t,0) = 0 \ \forall t \in \mathcal{R}^+$. Then the following statements are true:*

(i) If $\dot{V} < 0$, then $x_e = 0$ is **u.a.s. in the large**.
(ii) If there exist $\varphi_1, \varphi_2, \varphi_3 \in \mathcal{KR}$ of the same order of magnitude such that

$$\varphi_1(|x|) \leq V(t,x) \leq \varphi_2(|x|), \quad \dot{V}(t,x) \leq -\varphi_3(|x|)$$

then $x_e = 0$ is **e.s. in the large**.

Statement (i) of Theorem 3.4.2 is also equivalent to that there exist $\varphi_1, \varphi_2 \in \mathcal{K}$ and $\varphi_3 \in \mathcal{KR}$ such that

$$\varphi_1(|x|) \leq V(t,x) \leq \varphi_2(|x|), \quad \dot{V}(t,x) \leq -\varphi_3(|x|), \ \forall x \in \mathcal{R}^n$$

For a proof of Theorem 3.4.1, 3.4.2, the reader is referred to [32, 78, 79, 97, 124].

Theorem 3.4.3 *Assume that (3.4.1) possesses unique solutions for all $x_0 \in \mathcal{R}^n$. If there exists a function $V(t,x)$ defined on $|x| \geq R$ (where R may be large) and $t \in [0, \infty)$ with continuous first-order partial derivatives with respect to x, t and if there exist $\varphi_1, \varphi_2 \in \mathcal{KR}$ such that*

(i) $\varphi_1(|x|) \leq V(t,x) \leq \varphi_2(|x|)$
(ii) $\dot{V}(t,x) \leq 0$

for all $|x| \geq R$ and $t \in [0,\infty)$, then, the solutions of (3.4.1) are u.b. If in addition there exists $\varphi_3 \in \mathcal{K}$ defined on $[0,\infty)$ and

(iii) $\dot{V}(t,x) \leq -\varphi_3(|x|)$ *for all $|x| \geq R$ and $t \in [0,\infty)$*

then, the solutions of (3.4.1) are u.u.b.

Let us examine statement (ii) of Theorem 3.4.1 where V decrescent and $\dot{V} \leq 0$ imply $x_e = 0$ is u.s. If we remove the restriction of V being decrescent in (ii), we obtain statement (i), i.e., $\dot{V} \leq 0$ implies $x_e = 0$ is stable but not necessarily u.s. Therefore, one might tempted to expect that by removing the condition of V being decrescent in statement (iii), we obtain $x_e = 0$ is a.s., i.e., $\dot{V} < 0$ alone implies $x_e = 0$ is a.s. This intuitive conclusion is not true, as demonstrated by a counter example in [206] where a first-order differential equation and a positive definite, nondecrescent function $V(t,x)$ are used to show that $\dot{V} < 0$ does not imply a.s.

The system (3.4.1) is referred to as *nonautonomous*. When the function f in (3.4.1) does not depend explicitly on time t, the system is referred to as *autonomous*. In this case, we write

$$\dot{x} = f(x) \qquad (3.4.4)$$

Theorem 3.4.1 to 3.4.3 also hold for (3.4.4) because it is a special case of (3.4.1). In the case of (3.4.4), however, $V(t,x) = V(x)$, i.e., it does not depend explicitly on time t, and all references to the word "decrescent" and "uniform" could be deleted. This is because $V(x)$ is always decrescent and the stability (respectively a.s.) of the equilibrium $x_e = 0$ of (3.4.4) implies u.s. (respectively u.a.s.).

For the system (3.4.4), we can obtain a stronger result than Theorem 3.4.2 for a.s. as indicated below.

Definition 3.4.24 *A set Ω in $\mathcal{R}^n$ is* **invariant** *with respect to equation (3.4.4) if every solution of (3.4.4) starting in Ω remains in Ω for all t.*

Theorem 3.4.4 *Assume that (3.4.4) possesses unique solutions for all $x_0 \in \mathcal{R}^n$. Suppose there exists a positive definite and radially unbounded function $V(x): \mathcal{R}^n \mapsto \mathcal{R}^+$ with continuous first-order derivative with respect to x and $V(0) = 0$. If*

3.4. LYAPUNOV STABILITY

(i) $\dot{V} \leq 0 \; \forall x \in \mathcal{R}^n$

(ii) *The origin $x = 0$ is the only invariant subset of the set*

$$\Omega = \left\{ x \in \mathcal{R}^n \; \middle| \; \dot{V} = 0 \right\}$$

then the equilibrium $x_e = 0$ of (3.4.4) is a.s. in the large.

Theorems 3.4.1 to 3.4.4 are referred to as Lyapunov-type theorems. The function $V(t, x)$ or $V(x)$ that satisfies any Lyapunov-type theorem is referred to as Lyapunov function.

Lyapunov functions can be also used to predict the instability properties of the equilibrium state x_e. Several instability theorems based on the second method of Lyapunov are given in [232].

The following examples demonstrate the use of Lyapunov's direct method to analyze the stability of nonlinear systems.

Example 3.4.2 Consider the system

$$\begin{aligned} \dot{x}_1 &= x_2 + cx_1(x_1^2 + x_2^2) \\ \dot{x}_2 &= -x_1 + cx_2(x_1^2 + x_2^2) \end{aligned} \tag{3.4.5}$$

where c is a constant. Note that $x_e = 0$ is the only equilibrium state. Let us choose

$$V(x) = x_1^2 + x_2^2$$

as a candidate for a Lyapunov function. $V(x)$ is positive definite, decrescent, and radially unbounded. Its time derivative along the solution of (3.4.5) is

$$\dot{V} = 2c(x_1^2 + x_2^2)^2 \tag{3.4.6}$$

If $c = 0$, then $\dot{V} = 0$, and, therefore, $x_e = 0$ is u.s.. If $c < 0$, then $\dot{V} = -2|c|(x_1^2+x_2^2)^2$ is negative definite, and, therefore, $x_e = 0$ is u.a.s. in the large. If $c > 0$, $x_e = 0$ is unstable (because in this case V is strictly increasing $\forall t \geq 0$), and, therefore, the solutions of (3.4.5) are unbounded [232]. ▽

Example 3.4.3 Consider the following system describing the motion of a simple pendulum

$$\begin{aligned} \dot{x}_1 &= x_2 \\ \dot{x}_2 &= -k \sin x_1 \end{aligned} \tag{3.4.7}$$

where $k > 0$ is a constant, x_1 is the angle, and x_2 the angular velocity. We consider a candidate for a Lyapunov function, the function $V(x)$ representing the total energy of the pendulum given as the sum of the kinetic and potential energy, i.e.,

$$V(x) = \frac{1}{2}x_2^2 + k\int_0^{x_1} \sin\eta \, d\eta = \frac{1}{2}x_2^2 + k(1 - \cos x_1)$$

$V(x)$ is positive definite and decrescent $\forall x \in \mathcal{B}(\pi)$ but not radially unbounded. Along the solution of (3.4.7) we have

$$\dot{V} = 0$$

Therefore, the equilibrium state $x_e = 0$ is u.s. $\triangledown$

Example 3.4.4 Consider the system

$$\begin{aligned} \dot{x}_1 &= x_2 \\ \dot{x}_2 &= -x_2 - e^{-t}x_1 \end{aligned} \qquad (3.4.8)$$

Let us choose the positive definite, decrescent, and radially unbounded function

$$V(x) = x_1^2 + x_2^2$$

as a Lyapunov candidate. We have

$$\dot{V} = -2x_2^2 + 2x_1x_2(1 - e^{-t})$$

Because for this choice of V function neither of the preceding Lyapunov theorems is applicable, we can reach no conclusion. So let us choose another V function

$$V(t, x) = x_1^2 + e^t x_2^2$$

In this case, we obtain

$$\dot{V}(t, x) = -e^t x_2^2$$

This V function is positive definite, and $\dot{V}$ is negative semidefinite. Therefore, Theorem 3.4.1 is applicable, and we conclude that the equilibrium state $x_e = 0$ is stable. However, because V is not decrescent, we cannot conclude that the equilibrium state $x_e = 0$ is u.s. $\triangledown$

Example 3.4.5 Let us consider the following differential equations that arise quite often in the analysis of adaptive systems

$$\begin{aligned} \dot{x} &= -x + \phi x \\ \dot{\phi} &= -x^2 \end{aligned} \qquad (3.4.9)$$

3.4. LYAPUNOV STABILITY

The equilibrium state is $x_e = 0, \phi_e = c$, where c is any constant, and, therefore, the equilibrium state is not isolated.

Let us define $\bar{\phi} = \phi - c$, so that (3.4.9) is transformed into

$$\dot{x} = -(1-c)x + \bar{\phi}x$$
$$\dot{\bar{\phi}} = -x^2 \tag{3.4.10}$$

We are interested in the stability of the equilibrium point $x_e = 0, \phi_e = c$ of (3.4.9), which is equivalent to the stability of $x_e = 0, \bar{\phi}_e = 0$ of (3.4.10). We choose the positive definite, decrescent, radially unbounded function

$$V(x, \bar{\phi}) = \frac{x^2}{2} + \frac{\bar{\phi}^2}{2} \tag{3.4.11}$$

Then,

$$\dot{V}(x, \bar{\phi}) = -(1-c)x^2$$

If $c > 1$, then $\dot{V} > 0$ for $x \neq 0$; therefore, $x_e = 0, \bar{\phi}_e = 0$ is unstable. If, however, $c \leq 1$, then $x_e = 0, \bar{\phi}_e = 0$ is u.s. For $c < 1$ we have

$$\dot{V}(x, \bar{\phi}) = -c_0 x^2 \leq 0 \tag{3.4.12}$$

where $c_0 = 1 - c > 0$. From Theorem 3.4.3 we can also conclude that the solutions $x(t), \bar{\phi}(t)$ are u.b. but nothing more. We can exploit the properties of V and $\dot{V}$, however, and conclude that $x(t) \to 0$ as $t \to \infty$ as follows.

From (3.4.11) and (3.4.12) we conclude that because $V(t) = V(x(t), \bar{\phi}(t))$ is bounded from below and is nonincreasing with time, it has a limit, i.e., $\lim_{t \to \infty} V(t) = V_\infty$. Now from (3.4.12) we have

$$\lim_{t \to \infty} \int_0^t x^2 d\tau = \int_0^\infty x^2 d\tau = \frac{V(0) - V_\infty}{c_0} < \infty$$

i.e., $x \in \mathcal{L}_2$. Because the solution $x(t), \bar{\phi}(t)$ is u.b., it follows from (3.4.10) that $\dot{x} \in \mathcal{L}_\infty$, which together with $x \in \mathcal{L}_2$ imply (see Lemma 3.2.5) that $x(t) \to 0$ as $t \to \infty$. $\triangledown$

Example 3.4.6 Consider the differential equation

$$\dot{x}_1 = -2x_1 + x_1 x_2 + x_2$$
$$\dot{x}_2 = -x_1^2 - x_1$$

Consider $V(x) = \frac{x_1^2}{2} + \frac{x_2^2}{2}$. We have $\dot{V} = -2x_1^2 \leq 0$ and the equilibrium $x_{1e} = 0, x_{2e} = 0$ is u.s. The set defined in Theorem 3.4.4 is given by

$$\Omega = \{x_1, x_2 \mid x_1 = 0\}$$

Because $\dot{x}_1 = x_2$ on Ω, any solution that starts from Ω with $x_2 \neq 0$ leaves Ω. Hence, $x_1 = 0, x_2 = 0$ is the only invariant subset of Ω. Therefore the equilibrium $x_{1e} = 0, x_{2e} = 0$ is a.s. in the large. $\triangledown$

In the proceeding examples, we assume implicitly that the differential equations considered have unique solutions. As indicated in Section 3.4.1, this property is assumed for the general differential equation (3.4.1) on which all definitions and theorems are based. The following example illustrates that if the property of existence of solution is overlooked, an erroneous stability result may be obtained when some of the Lyapunov theorems of this section are used.

Example 3.4.7 Consider the second-order differential equation

$$\begin{aligned} \dot{x}_1 &= -2x_1 - x_2 \mathrm{sgn}(x_1), & x_1(0) &= 1 \\ \dot{x}_2 &= |x_1|, & x_2(0) &= 0 \end{aligned}$$

where

$$\mathrm{sgn}(x_1) = \begin{cases} 1 & \text{if } x_1 \geq 0 \\ -1 & \text{if } x_1 < 0 \end{cases}$$

The function

$$V(x_1, x_2) = x_1^2 + x_2^2$$

has a time derivative $\dot{V}$ along the solution of the differential equation that satisfies

$$\dot{V}(x_1, x_2) = -4x_1^2 \leq 0$$

Hence, according to Theorem 3.4.1, $x_1, x_2 \in \mathcal{L}_\infty$ and the equilibrium $x_{1e} = 0, x_{2e} = 0$ is u.s. Furthermore, we can show that $x_1(t) \to 0$ as $t \to \infty$ as follows: Because $V \geq 0$ and $\dot{V} \leq 0$ we have that $\lim_{t \to \infty} V(x_1(t), x_2(t)) = V_\infty$ for some $V_\infty \in \mathcal{R}^+$. Hence,

$$\int_0^t 4x_1^2(\tau) d\tau = V(x_1(0), x_2(0)) - V(x_1(t), x_2(t))$$

which implies that

$$4 \int_0^\infty x_1^2(\tau) d\tau < \infty$$

i.e., $x_1 \in \mathcal{L}_2$. From $x_1, x_2 \in \mathcal{L}_\infty$ we have that $\dot{x}_1 \in \mathcal{L}_\infty$, which together with $x_1 \in \mathcal{L}_2$ and Lemma 3.2.5 imply that $x_1(t) \to 0$ as $t \to \infty$.

The above conclusions are true provided continuous functions $x_1(t), x_2(t)$ with $x_1(0) = 1$ and $x_2(0) = 0$ satisfying the differential equation for all $t \in [0, \infty)$ exist. However, the solution of the above differential equation exists only in the time

3.4. LYAPUNOV STABILITY

interval $t \in [0, 1]$, where it is given by $x_1(t) = (1-t)e^{-t}$, $x_2(t) = te^{-t}$. The difficulty in continuing the solution beyond $t = 1$ originates from the fact that in a small neighborhood of the point $(x_1(1), x_2(1)) = (0, e^{-1})$, $\dot{x}_1 = -2x_1 - x_2\text{sgn}(x_1) < 0$ if $x_1 > 0$ and $\dot{x}_1 > 0$ if $x_1 < 0$. This causes $\dot{x}_1$ to change sign infinitely many times around the point $(0, e^{-1})$, which implies that no continuous functions $x_1(t), x_2(t)$ exist to satisfy the given differential equation past the point $t = 1$. ▽

The main drawback of the Lyapunov's direct method is that, in general, there is no procedure for finding the appropriate Lyapunov function that satisfies the conditions of Theorems 3.4.1 to 3.4.4 except in the case where (3.4.1) represents a LTI system. If, however, the equilibrium state $x_e = 0$ of (3.4.1) is u.a.s. the existence of a Lyapunov function is assured as shown in [139].

3.4.3 Lyapunov-Like Functions

The choice of an appropriate Lyapunov function to establish stability by using Theorems 3.4.1 to 3.4.4 in the analysis of a large class of adaptive control schemes may not be obvious or possible in many cases. However, a function that resembles a Lyapunov function, but does not possess all the properties that are needed to apply Theorems 3.4.1 to 3.4.4, can be used to establish some properties of adaptive systems that are related to stability and boundedness. We refer to such a function as the *Lyapunov-like* function. The following example illustrates the use of Lyapunov-like functions.

Example 3.4.8 Consider the third-order differential equation

$$\begin{aligned} \dot{x}_1 &= -x_1 - x_2 x_3, & x_1(0) &= x_{10} \\ \dot{x}_2 &= x_1 x_3, & x_2(0) &= x_{20} \\ \dot{x}_3 &= x_1^2, & x_3(0) &= x_{30} \end{aligned} \quad (3.4.13)$$

which has the nonisolated equilibrium points in $\mathcal{R}^3$ defined by $x_1 = 0, x_2 = constant$, $x_3 = 0$ or $x_1 = 0, x_2 = 0, x_3 = constant$. We would like to analyze the stability properties of the solutions of (3.4.13) by using an appropriate Lyapunov function and applying Theorems 3.4.1 to 3.4.4. If we follow Theorems 3.4.1 to 3.4.4, then we should start with a function $V(x_1, x_2, x_3)$ that is positive definite in $\mathcal{R}^3$. Instead of doing so let us consider the simple quadratic function

$$V(x_1, x_2) = \frac{x_1^2}{2} + \frac{x_2^2}{2}$$

which is positive semidefinite in $\mathcal{R}^3$ and, therefore, does not satisfy the positive definite condition in $\mathcal{R}^3$ of Theorems 3.4.1 to 3.4.4. The time derivative of V along the solution of the differential equation (3.4.13) satisfies

$$\dot{V} = -x_1^2 \leq 0 \tag{3.4.14}$$

which implies that V is a nonincreasing function of time. Therefore,

$$V(x_1(t), x_2(t)) \leq V(x_1(0), x_2(0)) \triangleq V_0$$

and $V, x_1, x_2 \in \mathcal{L}_\infty$. Furthermore, V has a limit as $t \to \infty$, i.e.,

$$\lim_{t \to \infty} V(x_1(t), x_2(t)) = V_\infty$$

and (3.4.14) implies that

$$\int_0^t x_1^2(\tau) d\tau = V_0 - V(t), \quad \forall t \geq 0$$

and

$$\int_0^\infty x_1^2(\tau) d\tau = V_0 - V_\infty < \infty$$

i.e., $x_1 \in \mathcal{L}_2$. From $x_1 \in \mathcal{L}_2$ we have from (3.3.13) that $x_3 \in \mathcal{L}_\infty$ and from $x_1, x_2, x_3 \in \mathcal{L}_\infty$ that $\dot{x}_1 \in \mathcal{L}_\infty$. Using $\dot{x}_1 \in \mathcal{L}_\infty$, $x_1 \in \mathcal{L}_2$ and applying Lemma 3.2.5 we have $x_1(t) \to 0$ as $t \to \infty$. By using the properties of the positive semidefinite function $V(x_1, x_2)$, we have established that the solution of (3.4.13) is uniformly bounded and $x_1(t) \to 0$ as $t \to \infty$ for any finite initial condition $x_1(0)$, $x_2(0)$, $x_3(0)$. Because the approach we follow resembles the Lyapunov function approach, we are motivated to refer to $V(x_1, x_2)$ as the Lyapunov-like function. In the above analysis we also assumed that (3.4.13) has a unique solution. For discussion and analysis on existence and uniqueness of solutions of (3.4.13) the reader is referred to [191]. ▽

We use Lyapunov-like functions and similar arguments as in the example above to analyze the stability of a wide class of adaptive schemes considered throughout this book.

3.4.4 Lyapunov's Indirect Method

Under certain conditions, conclusions can be drawn about the stability of the equilibrium of a nonlinear system by studying the behavior of a certain linear

3.4. LYAPUNOV STABILITY

system obtained by linearizing (3.4.1) around its equilibrium state. This method is known as the *first method of Lyapunov* or as *Lyapunov's indirect method* and is given as follows [32, 232]: Let $x_e = 0$ be an equilibrium state of (3.4.1) and assume that $f(t, x)$ is continuously differentiable with respect to x for each $t \geq 0$. Then in the neighborhood of $x_e = 0$, f has a Taylor series expansion that can be written as

$$\dot{x} = f(t, x) = A(t)x + f_1(t, x) \qquad (3.4.15)$$

where $A(t) = \nabla f|_{x=0}$ is referred to as the Jacobian matrix of f evaluated at $x = 0$ and $f_1(t, x)$ represents the remaining terms in the series expansion.

Theorem 3.4.5 *Assume that $A(t)$ is uniformly bounded and that*

$$\lim_{|x| \to 0} \sup_{t \geq 0} \frac{|f_1(t, x)|}{|x|} = 0$$

Let $z_e = 0$ be the equilibrium of

$$\dot{z}(t) = A(t)z(t)$$

The following statements are true for the equilibrium $x_e = 0$ of (3.4.15):

(i) *If $z_e = 0$ is u.a.s. then $x_e = 0$ is u.a.s.*
(ii) *If $z_e = 0$ is unstable then $x_e = 0$ is unstable*
(iii) *If $z_e = 0$ is u.s. or stable, no conclusions can be drawn about the stability of $x_e = 0$.*

For a proof of Theorem 3.4.5 see [232].

Example 3.4.9 Consider the second-order differential equation

$$m\ddot{x} = -2\mu(x^2 - 1)\dot{x} - kx$$

where m, μ, and k are positive constants, which is known as the Van der Pol oscillator. It describes the motion of a mass-spring-damper with damping coefficient $2\mu(x^2 - 1)$ and spring constant k, where x is the position of the mass. If we define the states $x_1 = x, x_2 = \dot{x}$, we obtain the equation

$$\begin{aligned} \dot{x}_1 &= x_2 \\ \dot{x}_2 &= -\frac{k}{m}x_1 - \frac{2\mu}{m}(x_1^2 - 1)x_2 \end{aligned}$$

which has an equilibrium at $x_{1e} = 0, x_{2e} = 0$. The linearization of this system around $(0,0)$ gives us

$$\begin{bmatrix} \dot{z}_1 \\ \dot{z}_2 \end{bmatrix} = \begin{bmatrix} 0 & 1 \\ -\frac{k}{m} & \frac{2\mu}{m} \end{bmatrix} \begin{bmatrix} z_1 \\ z_2 \end{bmatrix}$$

Because $m, \mu > 0$ at least one of the eigenvalues of the matrix A is positive and therefore the equilibrium $(0,0)$ is unstable. ∇

3.4.5 Stability of Linear Systems

Equation (3.4.15) indicates that certain classes of nonlinear systems may be approximated by linear ones in the neighborhood of an equilibrium point or, as often called in practice, operating point. For this reason we are interested in studying the stability of linear systems of the form

$$\dot{x}(t) = A(t)x(t) \tag{3.4.16}$$

where the elements of $A(t)$ are piecewise continuous for all $t \geq t_0 \geq 0$, as a special class of the nonlinear system (3.4.1) or as an approximation of the linearized system (3.4.15). The solution of (3.4.16) is given by [95]

$$x(t; t_0, x_0) = \Phi(t, t_0) x_0$$

for all $t \geq t_0$, where $\Phi(t, t_0)$ is the *state transition matrix* and satisfies the matrix differential equation

$$\frac{\partial}{\partial t} \Phi(t, t_0) = A(t) \Phi(t, t_0), \quad \forall t \geq t_0$$

$$\Phi(t_0, t_0) = I$$

Some additional useful properties of $\Phi(t, t_0)$ are

(i) $\Phi(t, t_0) = \Phi(t, \tau) \Phi(\tau, t_0) \ \forall t \geq \tau \geq t_0$ (semigroup property)
(ii) $\Phi(t, t_0)^{-1} = \Phi(t_0, t)$
(iii) $\frac{\partial}{\partial t_0} \Phi(t, t_0) = -\Phi(t, t_0) A(t_0)$

Necessary and sufficient conditions for the stability of the equilibrium state $x_e = 0$ of (3.4.16) are given by the following theorems.

3.4. LYAPUNOV STABILITY

Theorem 3.4.6 *Let $\|\Phi(t,\tau)\|$ denote the induced matrix norm of $\Phi(t,\tau)$ at each time $t \geq \tau$. The equilibrium state $x_e = 0$ of (3.4.16) is*

(i) *stable if and only if the solutions of (3.4.16) are bounded or equivalently*

$$c(t_0) \stackrel{\triangle}{=} \sup_{t \geq t_0} \|\Phi(t, t_0)\| < \infty$$

(ii) *u.s. if and only if*

$$c_0 \stackrel{\triangle}{=} \sup_{t_0 \geq 0} c(t_0) = \sup_{t_0 \geq 0} \left(\sup_{t \geq t_0} \|\Phi(t, t_0)\| \right) < \infty$$

(iii) *a.s. if and only if*

$$\lim_{t \to \infty} \|\Phi(t, t_0)\| = 0$$

for any $t_0 \in \mathcal{R}^+$

(iv) *u.a.s. if and only if there exist positive constants α and β such that*

$$\|\Phi(t, t_0)\| \leq \alpha e^{-\beta(t-t_0)}, \quad \forall t \geq t_0 \geq 0$$

(v) *e.s. if and only if it is u.a.s.*

(vi) *a.s., u.a.s., e.s. in the large if and only if it is a.s., u.a.s., e.s., respectively.*

Theorem 3.4.7 [1] *Assume that the elements of $A(t)$ are u.b. for all $t \in \mathcal{R}^+$. The equilibrium state $x_e = 0$ of the linear system (3.4.16) is u.a.s. if and only if, given any positive definite matrix $Q(t)$, which is continuous in t and satisfies*

$$0 < c_1 I \leq Q(t) \leq c_2 I < \infty$$

for all $t \geq t_0$, the scalar function defined by

$$V(t, x) = x^T \int_t^\infty \Phi^T(\tau, t) Q(\tau) \Phi(\tau, t) \, d\tau \, x \qquad (3.4.17)$$

exists (i.e., the integral defined by (3.4.17) is finite for finite values of x and t) and is a Lyapunov function of (3.4.16) with

$$\dot{V}(t, x) = -x^T Q(t) x$$

It follows using the properties of $\Phi(t, t_0)$ that $P(t) \triangleq \int_t^\infty \Phi^\top(\tau, t) Q(\tau) \Phi(\tau, t) d\tau$ satisfies the equation

$$\dot{P}(t) = -Q(t) - A^\top(t) P(t) - P(t) A(t) \qquad (3.4.18)$$

i.e., the Lyapunov function (3.4.17) can be rewritten as $V(t, x) = x^\top P(t) x$, where $P(t) = P^\top(t)$ satisfies (3.4.18).

Theorem 3.4.8 *A necessary and sufficient condition for the u.a.s of the equilibrium $x_e = 0$ of (3.4.16) is that there exists a symmetric matrix $P(t)$ such that*

$$\gamma_1 I \leq P(t) \leq \gamma_2 I$$
$$\dot{P}(t) + A^\top(t) P(t) + P(t) A(t) + \nu C(t) C^\top(t) \leq O$$

are satisfied $\forall t \geq 0$ and some constant $\nu > 0$, where $\gamma_1 > 0, \gamma_2 > 0$ are constants and $C(t)$ is such that $(C(t), A(t))$ is a UCO pair (see Definition 3.3.3).

When $A(t) = A$ is a constant matrix, the conditions for stability of the equilibrium $x_e = 0$ of

$$\dot{x} = Ax \qquad (3.4.19)$$

are given by the following theorem.

Theorem 3.4.9 *The equilibrium state $x_e = 0$ of (3.4.19) is stable if and only if*

(i) *All the eigenvalues of A have nonpositive real parts.*
(ii) *For each eigenvalue λ_i with $\text{Re}\{\lambda_i\} = 0$, λ_i is a simple zero of the minimal polynomial of A (i.e., of the monic polynomial $\psi(\lambda)$ of least degree such that $\psi(A) = O$).*

Theorem 3.4.10 *A necessary and sufficient condition for $x_e = 0$ to be a.s. in the large is that any one of the following conditions is satisfied* [1]:

(i) *All the eigenvalues of A have negative real parts*

---

[1] Note that (iii) includes (ii). Because (ii) is used very often in this book, we list it separately for easy reference.

3.4. LYAPUNOV STABILITY

(ii) *For every positive definite matrix Q, the following Lyapunov matrix equation*
$$A^\top P + PA = -Q$$
has a unique solution P that is also positive definite.

(iii) *For any given matrix C with (C, A) observable, the equation*
$$A^\top P + PA = -C^\top C$$
has a unique solution P that is positive definite.

It is easy to verify that for the LTI system given by (3.4.19), if $x_e = 0$ is stable, it is also u.s. If $x_e = 0$ is a.s., it is also u.a.s. and e.s. in the large.

In the rest of the book we will abuse the notation and call the matrix A in (3.4.19) *stable* when the equilibrium $x_e = 0$ is a.s., i.e., when all the eigenvalues of A have negative real parts and *marginally stable* when $x_e = 0$ is stable, i.e., A satisfies (i) and (ii) of Theorem 3.4.9.

Let us consider again the linear time-varying system (3.4.16) and suppose that for each fixed t all the eigenvalues of the matrix $A(t)$ have negative real parts. In view of Theorem 3.4.10, one may ask whether this condition for $A(t)$ can ensure some form of stability for the equilibrium $x_e = 0$ of (3.4.16). The answer is unfortunately no in general, as demonstrated by the following example given in [232].

Example 3.4.10 Let

$$A(t) = \begin{bmatrix} -1 + 1.5\cos^2 t & 1 - 1.5\sin t \cos t \\ -1 - 1.5\sin t \cos t & -1 + 1.5\sin^2 t \end{bmatrix}$$

The eigenvalues of $A(t)$ for each fixed t,

$$\lambda(A(t)) = -.25 \pm j.5\sqrt{1.75}$$

have negative real parts and are also independent of t. Despite this the equilibrium $x_e = 0$ of (3.4.16) is unstable because

$$\Phi(t, 0) = \begin{bmatrix} e^{.5t}\cos t & e^{-t}\sin t \\ -e^{.5t}\sin t & e^{-t}\cos t \end{bmatrix}$$

is unbounded w.r.t. time t. ▽

Despite Example 3.4.10, Theorem 3.4.10 may be used to obtain some sufficient conditions for a class of $A(t)$, which guarantee that $x_e = 0$ of (3.4.16) is u.a.s. as indicated by the following theorem.

Theorem 3.4.11 *Let the elements of $A(t)$ in (3.4.16) be differentiable[2] and bounded functions of time and assume that*

(A1) $Re\{\lambda_i(A(t))\} \leq -\sigma_s \ \forall t \geq 0$ *and for* $i = 1, 2, \ldots, n$ *where* $\sigma_s > 0$ *is some constant.*

(i) *If $\|\dot{A}\| \in \mathcal{L}_2$, then the equilibrium state $x_e = 0$ of (3.4.16) is u.a.s. in the large.*

(ii) *If any one of the following conditions:*

   (a) $\int_t^{t+T} \|\dot{A}(\tau)\| d\tau \leq \mu T + \alpha_0$, *i.e.*, $(\|\dot{A}\|)^{\frac{1}{2}} \in \mathcal{S}(\mu)$

   (b) $\int_t^{t+T} \|\dot{A}(\tau)\|^2 d\tau \leq \mu^2 T + \alpha_0$, *i.e.*, $\|\dot{A}\| \in \mathcal{S}(\mu^2)$

   (c) $\|\dot{A}(t)\| \leq \mu$

is satisfied for some $\alpha_0, \mu \in \mathcal{R}^+$ and $\forall t \geq 0, T \geq 0$, then there exists a $\mu^* > 0$ such that if $\mu \in [0, \mu^*)$, the equilibrium state x_e of (3.4.16) is u.a.s. in the large.

Proof Using (A1), it follows from Theorem 3.4.10 that the Lyapunov equation

$$A^T(t)P(t) + P(t)A(t) = -I \qquad (3.4.20)$$

has a unique bounded solution $P(t)$ for each fixed t. We consider the following Lyapunov function:

$$V(t, x) = x^T P(t) x$$

Then along the solution of (3.4.16) we have

$$\dot{V} = -|x(t)|^2 + x^T(t)\dot{P}(t)x(t) \qquad (3.4.21)$$

From (3.4.20), $\dot{P}$ satisfies

$$A^T(t)\dot{P}(t) + \dot{P}(t)A(t) = -Q(t), \quad \forall t \geq 0 \qquad (3.4.22)$$

where $Q(t) = \dot{A}^T(t)P(t) + P(t)\dot{A}(t)$. Because of (A1), it can be verified [95] that

$$\dot{P}(t) = \int_0^\infty e^{A^T(t)\tau} Q(t) e^{A(t)\tau} d\tau$$

---

[2] The condition of differentiability can be relaxed to Lipschitz continuity.

3.4. LYAPUNOV STABILITY

satisfies (3.4.22) for each $t \geq 0$, therefore,

$$\|\dot{P}(t)\| \leq \|Q(t)\| \int_0^\infty \|e^{A^\top(t)\tau}\| \|e^{A(t)\tau}\| d\tau$$

Because (A1) implies that $\|e^{A(t)\tau}\| \leq \alpha_1 e^{-\alpha_0 \tau}$ for some $\alpha_1, \alpha_0 > 0$ it follows that

$$\|\dot{P}(t)\| \leq c\|Q(t)\|$$

for some $c \geq 0$. Then,

$$\|Q(t)\| \leq 2\|P(t)\|\|\dot{A}(t)\|$$

together with $P \in \mathcal{L}_\infty$ imply that

$$\|\dot{P}(t)\| \leq \beta \|\dot{A}(t)\|, \quad \forall t \geq 0 \tag{3.4.23}$$

for some constant $\beta \geq 0$. Using (3.4.23) in (3.4.21) and noting that P satisfies $0 < \beta_1 \leq \lambda_{min}(P) \leq \lambda_{max}(P) \leq \beta_2$ for some $\beta_1, \beta_2 > 0$, we have that

$$\dot{V}(t) \leq -|x(t)|^2 + \beta\|\dot{A}(t)\||x(t)|^2 \leq -\beta_2^{-1}V(t) + \beta\beta_1^{-1}\|\dot{A}(t)\|V(t)$$

therefore,

$$V(t) \leq e^{-\int_{t_0}^t (\beta_2^{-1} - \beta\beta_1^{-1}\|\dot{A}(\tau)\|)d\tau} V(t_0) \tag{3.4.24}$$

Let us prove (ii) first. Using condition (a) in (3.4.24) we have

$$V(t) \leq e^{-(\beta_2^{-1} - \beta\beta_1^{-1}\mu)(t-t_0)} e^{\beta\beta_1^{-1}\alpha_0} V(t_0)$$

Therefore, for $\mu^* = \frac{\beta_1}{\beta_2 \beta}$ and $\forall \mu \in [0, \mu^*)$, $V(t) \to 0$ exponentially fast, which implies that $x_e = 0$ is u.a.s. in the large.

To be able to use (b), we rewrite (3.4.24) as

$$V(t) \leq e^{-\beta_2^{-1}(t-t_0)} e^{\beta\beta_1^{-1} \int_{t_0}^t \|\dot{A}(\tau)\|d\tau} V(t_0)$$

Using the Schwartz inequality and (b) we have

$$\int_{t_0}^t \|\dot{A}(\tau)\|d\tau \leq \left(\int_{t_0}^t \|\dot{A}(\tau)\|^2 d\tau\right)^{\frac{1}{2}} \sqrt{t-t_0}$$

$$\leq [\mu^2(t-t_0)^2 + \alpha_0(t-t_0)]^{\frac{1}{2}}$$

$$\leq \mu(t-t_0) + \sqrt{\alpha_0}\sqrt{t-t_0}$$

Therefore,

$$V(t) \leq e^{-\alpha(t-t_0)} y(t) V(t_0)$$

where $\alpha = (1-\gamma)\beta_2^{-1} - \beta\beta_1^{-1}\mu$,

$$\begin{aligned} y(t) &= \exp\left[-\gamma\beta_2^{-1}(t-t_0) + \beta\beta_1^{-1}\sqrt{\alpha_0}\sqrt{t-t_0}\right] \\ &= \exp\left[-\gamma\beta_2^{-1}\left(\sqrt{t-t_0} - \frac{\beta\beta_1^{-1}\sqrt{\alpha_0}}{2\gamma\beta_2^{-1}}\right)^2 + \frac{\alpha_0\beta^2\beta_2}{4\gamma\beta_1^2}\right] \end{aligned}$$

and γ is an arbitrary constant that satisfies $0 < \gamma < 1$. It can be shown that

$$y(t) \leq \exp\left[\alpha_0\frac{\beta^2\beta_2}{4\gamma\beta_1^2}\right] \triangleq c \quad \forall t \geq t_0$$

hence,

$$V(t) \leq ce^{-\alpha(t-t_0)}V(t_0)$$

Choosing $\mu^* = \frac{\beta_1(1-\gamma)}{\beta_2\beta}$, we have that $\forall \mu \in [0, \mu^*)$, $\alpha > 0$ and, therefore, $V(t) \to 0$ exponentially fast, which implies that $x_e = 0$ is u.a.s in the large.

Since (c) implies (a), the proof of (c) follows directly from that of (a).

The proof of (i) follows from that of (ii)(b), because $\|\dot{A}\| \in \mathcal{L}_2$ implies (b) with $\mu = 0$. $\square$

In simple words, Theorem 3.4.11 states that if the eigenvalues of $A(t)$ for each fixed time t have negative real parts and if $A(t)$ varies sufficiently slowly most of the time, then the equilibrium state $x_e = 0$ of (3.4.16) is u.a.s.

3.5 Positive Real Functions and Stability

3.5.1 Positive Real and Strictly Positive Real Transfer Functions

The concept of PR and SPR transfer functions plays an important role in the stability analysis of a large class of nonlinear systems, which also includes adaptive systems.

The definition of PR and SPR transfer functions is derived from network theory. That is, a PR (SPR) rational transfer function can be realized as the driving point impedance of a passive (dissipative) network. Conversely, a passive (dissipative) network has a driving point impedance that is rational and PR (SPR). A passive network is one that does not generate energy, i.e., a network consisting only of resistors, capacitors, and inductors. A dissipative

3.5. POSITIVE REAL FUNCTIONS AND STABILITY

network dissipates energy, which implies that it is made up of resistors and capacitors, inductors that are connected in parallel with resistors.

In [177, 204], the following equivalent definitions have been given for PR transfer functions by an appeal to network theory.

Definition 3.5.1 *A rational function $G(s)$ of the complex variable $s = \sigma + j\omega$ is called PR if*

(i) $G(s)$ *is real for real s.*
(ii) $\text{Re}[G(s)] \geq 0$ *for all $\text{Re}[s] > 0$.*

Lemma 3.5.1 *A rational proper transfer function $G(s)$ is PR if and only if*
(i) $G(s)$ *is real for real s.*
(ii) $G(s)$ *is analytic in $\text{Re}[s] > 0$, and the poles on the $j\omega$-axis are simple and such that the associated residues are real and positive.*
(iii) *For all real value ω for which $s = j\omega$ is not a pole of $G(s)$, one has $\text{Re}[G(j\omega)] \geq 0$.*

For SPR transfer functions we have the following Definition.

Definition 3.5.2 [177] *Assume that $G(s)$ is not identically zero for all s. Then $G(s)$ is SPR if $G(s - \epsilon)$ is PR for some $\epsilon > 0$.*

The following theorem gives necessary and sufficient conditions in the frequency domain for a transfer function to be SPR:

Theorem 3.5.1 [89] *Assume that a rational function $G(s)$ of the complex variable $s = \sigma + j\omega$ is real for real s and is not identically zero for all s. Let n^* be the relative degree of $G(s) = Z(s)/R(s)$ with $|n^*| \leq 1$. Then, $G(s)$ is SPR if and only if*

(i) $G(s)$ *is analytic in $\text{Re}[s] \geq 0$*

(ii) $\text{Re}[G(j\omega)] > 0, \ \forall \omega \in (-\infty, \infty)$

(iii) (a) *When $n^* = 1$, $\lim_{|\omega| \to \infty} \omega^2 \text{Re}[G(j\omega)] > 0$.*
 (b) *When $n^* = -1$, $\lim_{|\omega| \to \infty} \frac{G(j\omega)}{j\omega} > 0$.*

It should be noted that when $n^* = 0$, (i) and (ii) in Theorem 3.5.1 are necessary and sufficient for $G(s)$ to be SPR. This, however, is not true for $n^* = 1$ or -1. For example,

$$G(s) = (s + \alpha + \beta)/[(s + \alpha)(s + \beta)]$$

$\alpha, \beta > 0$ satisfies (i) and (ii) of Theorem 3.5.1, but is not SPR because it does not satisfy (iiia). It is, however, PR.

Some useful properties of SPR functions are given by the following corollary.

Corollary 3.5.1 (i) $G(s)$ is PR (SPR) if and only if $1/G(s)$ is PR (SPR)
(ii) If $G(s)$ is SPR, then, $|n^*| \leq 1$, and the zeros and poles of $G(s)$ lie in $\text{Re}[s] < 0$
(iii) If $|n^*| > 1$, then $G(s)$ is not PR.

A necessary condition for $G(s)$ to be PR is that the Nyquist plot of $G(j\omega)$ lies in the right half complex plane, which implies that the phase shift in the output of a system with transfer function $G(s)$ in response to a sinusoidal input is less than 90°.

Example 3.5.1 Consider the following transfer functions:

(i) $\quad G_1(s) = \dfrac{s-1}{(s+2)^2}$, $\quad$ (ii) $G_2(s) = \dfrac{1}{(s+2)^2}$
(iii) $\quad G_3(s) = \dfrac{s+3}{(s+1)(s+2)}$, $\quad$ (iv) $G_4(s) = \dfrac{1}{s+\alpha}$

Using Corollary 3.5.1 we conclude that $G_1(s)$ is not PR, because $1/G_1(s)$ is not PR. We also have that $G_2(s)$ is not PR because $n^* > 1$.

For $G_3(s)$ we have that

$$\text{Re}[G_3(j\omega)] = \frac{6}{(2-\omega^2)^2 + 9\omega^2} > 0, \quad \forall \omega \in (-\infty, \infty)$$

which together with the stability of $G_3(s)$ implies that $G_3(s)$ is PR. Because $G_3(s)$ violates (iii)(a) of Theorem 3.5.1, it is not SPR.

The function $G_4(s)$ is SPR for any $\alpha > 0$ and PR, but not SPR for $\alpha = 0$. $\quad\nabla$

The relationship between PR, SPR transfer functions, and Lyapunov stability of corresponding dynamic systems leads to the development of several

3.5. POSITIVE REAL FUNCTIONS AND STABILITY

stability criteria for feedback systems with LTI and nonlinear parts. These criteria include the celebrated Popov's criterion and its variations [192]. The vital link between PR, SPR transfer functions or matrices and the existence of a Lyapunov function for establishing stability is given by the following lemmas.

Lemma 3.5.2 (Kalman-Yakubovich-Popov (KYP) Lemma)[7, 192]
Given a square matrix A with all eigenvalues in the closed left half complex plane, a vector B such that (A, B) is controllable, a vector C and a scalar $d \geq 0$, the transfer function defined by

$$G(s) = d + C^\top (sI - A)^{-1} B$$

is PR if and only if there exist a symmetric positive definite matrix P and a vector q such that

$$A^\top P + PA = -qq^\top$$
$$PB - C = \pm(\sqrt{2d})q$$

Lemma 3.5.3 (Lefschetz-Kalman-Yakubovich (LKY) Lemma) [89, 126] *Given a stable matrix A, a vector B such that (A, B) is controllable, a vector C and a scalar $d \geq 0$, the transfer function defined by*

$$G(s) = d + C^\top (sI - A)^{-1} B$$

is SPR if and only if for any positive definite matrix L, there exist a symmetric positive definite matrix P, a scalar $\nu > 0$ and a vector q such that

$$A^\top P + PA = -qq^\top - \nu L$$
$$PB - C = \pm q\sqrt{2d}$$

The lemmas above are applicable to LTI systems that are controllable. This controllability assumption is relaxed in [142, 172].

Lemma 3.5.4 (Meyer-Kalman-Yakubovich (MKY) Lemma) *Given a stable matrix A, vectors B, C and a scalar $d \geq 0$, we have the following: If*

$$G(s) = d + C^\top (sI - A)^{-1} B$$

is SPR, then for any given $L = L^\top > 0$, there exists a scalar $\nu > 0$, a vector q and a $P = P^\top > 0$ such that

$$A^\top P + PA = -qq^\top - \nu L$$

$$PB - C = \pm q\sqrt{2d}$$

In many applications of SPR concepts to adaptive systems, the transfer function $G(s)$ involves stable zero-pole cancellations, which implies that the system associated with the triple (A, B, C) is uncontrollable or unobservable. In these situations the MKY lemma is the appropriate lemma to use.

Example 3.5.2 Consider the system

$$y = G(s)u$$

where $G(s) = \frac{s+3}{(s+1)(s+2)}$. We would like to verify whether $G(s)$ is PR or SPR by using the above lemmas. The system has the state space representation

$$\begin{aligned} \dot{x} &= Ax + Bu \\ y &= C^\top x \end{aligned}$$

where

$$A = \begin{bmatrix} 0 & 1 \\ -2 & -3 \end{bmatrix}, \quad B = \begin{bmatrix} 0 \\ 1 \end{bmatrix}, \quad C = \begin{bmatrix} 3 \\ 1 \end{bmatrix}$$

According to the above lemmas, if $G(s)$ is PR, we have

$$PB = C$$

which implies that

$$P = \begin{bmatrix} p_1 & 3 \\ 3 & 1 \end{bmatrix}$$

for some $p_1 > 0$ that should satisfy $p_1 - 9 > 0$ for P to be positive definite. We need to calculate $p_1, \nu > 0$ and q such that

$$A^\top P + PA = -qq^\top - \nu L \tag{3.5.1}$$

is satisfied for any $L = L^\top > 0$. We have

$$-\begin{bmatrix} 12 & 11 - p_1 \\ 11 - p_1 & 0 \end{bmatrix} = -qq^\top - \nu L$$

Now $Q = \begin{bmatrix} 12 & 11 - p_1 \\ 11 - p_1 & 0 \end{bmatrix}$ is positive semidefinite for $p_1 = 11$ but is indefinite for $p_1 \neq 11$. Because no $p_1 > 9$ exists for which Q is positive definite, no $\nu > 0$ can

3.5. POSITIVE REAL FUNCTIONS AND STABILITY

be found to satisfy (3.5.1) for any given $L = L^\top > 0$. Therefore, $G(s)$ is not SPR, something we have already verified in Example 3.5.1. For $p_1 = 11$, we can select $\nu = 0$ and $q = [\sqrt{12}, 0]^\top$ to satisfy (3.5.1). Therefore, $G(s)$ is PR. $\triangledown$

The KYP and MKY Lemmas are useful in choosing appropriate Lyapunov or Lyapunov-like functions to analyze the stability properties of a wide class of continuous-time adaptive schemes for LTI plants. We illustrate the use of MKY Lemma in adaptive systems by considering the stability properties of the system

$$\begin{aligned} \dot{e} &= A_c e + B_c \theta^\top \omega \\ \dot{\theta} &= -\Gamma e_1 \omega \\ e_1 &= C^\top e \end{aligned} \quad (3.5.2)$$

that often arises in the analysis of a certain class of adaptive schemes. In (3.5.2), $\Gamma = \Gamma^\top > 0$ is constant, $e \in \mathcal{R}^m, \theta \in \mathcal{R}^n, e_1 \in \mathcal{R}^1$ and $\omega = C_0^\top e + C_1^\top e_m$, where e_m is continuous and $e_m \in \mathcal{L}_\infty$. The stability properties of (3.5.2) are given by the following theorem.

Theorem 3.5.2 *If A_c is a stable matrix and $G(s) = C^\top(sI - A_c)^{-1} B_c$ is SPR, then $e, \theta, \omega \in \mathcal{L}_\infty$; $e, \dot{\theta} \in \mathcal{L}_\infty \bigcap \mathcal{L}_2$ and $e(t), e_1(t), \dot{\theta}(t) \to 0$ as $t \to \infty$.*

Proof Because we made no assumptions about the controllability or observability of (A_c, B_c, C_c), we concentrate on the use of the MKY Lemma. We choose the function

$$V(e, \theta) = e^\top P e + \theta^\top \Gamma^{-1} \theta$$

where $P = P^\top > 0$ satisfies

$$\begin{aligned} A_c^\top P + P A_c &= -qq^\top - \nu L \\ P B_c &= C \end{aligned}$$

for some vector q, scalar $\nu > 0$ and $L = L^\top > 0$ guaranteed by the SPR property of $G(s)$ and the MKY Lemma. Because $\omega = C_0^\top e + C_1^\top e_m$ and $e_m \in \mathcal{L}_\infty$ can be treated as an arbitrary continuous function of time in $\mathcal{L}_\infty$, we can establish that $V(e, \theta)$ is positive definite in $\mathcal{R}^{n+m}$, and V is a Lyapunov function candidate.

The time derivative of V along the solution of (3.5.2) is given by

$$\dot{V} = -e^\top qq^\top e - \nu e^\top L e \leq 0$$

which by Theorem 3.4.1 implies that the equilibrium $e_e, \theta_e = 0$ is u.s. and $e_1, e, \theta \in \mathcal{L}_\infty$. Because $\omega = C_0^T e + C_1^T e_m$ and $e_m \in \mathcal{L}_\infty$, we also have $\omega \in \mathcal{L}_\infty$. By exploiting the properties of $\dot{V}, V$ further, we can obtain additional properties about the solution of (3.5.2) as follows: From $V \geq 0$ and $\dot{V} \leq 0$ we have that

$$\lim_{t \to \infty} V(e(t), \theta(t)) = V_\infty$$

and, therefore,

$$\nu \int_0^\infty e^T L e d\tau \leq V_0 - V_\infty$$

where $V_0 = V(e(0), \theta(0))$. Because $\nu \lambda_{min}(L)|e|^2 \leq \nu e^T L e$, it follows that $e \in \mathcal{L}_2$. From (3.5.2) we have $|\dot{\theta}(t)| \leq \|\Gamma\| \|e_1\| |\omega|$. Since $\omega \in \mathcal{L}_\infty$ and $e_1 \in \mathcal{L}_\infty \bigcap \mathcal{L}_2$ it follows from Lemma 3.2.3 (ii) that $\dot{\theta} \in \mathcal{L}_\infty \bigcap \mathcal{L}_2$. Using $e, \theta, \omega \in \mathcal{L}_\infty$ we obtain $\dot{e} \in \mathcal{L}_\infty$, which together with $e \in \mathcal{L}_2$, implies that $e(t) \to 0$ as $t \to \infty$. Hence, $e_1, \dot{\theta}, e \to 0$ as $t \to \infty$, and the proof is complete. □

The arguments we use to prove Theorem 3.5.2 are standard in adaptive systems and will be repeated in subsequent chapters.

3.5.2 PR and SPR Transfer Function Matrices

The concept of PR transfer functions can be extended to PR transfer function matrices as follows.

Definition 3.5.3 [7, 172] *An $n \times n$ matrix $G(s)$ whose elements are rational functions of the complex variable s is called PR if*

(i) $G(s)$ *has elements that are analytic for* $\text{Re}[s] > 0$
(ii) $G^*(s) = G(s^*)$ *for* $\text{Re}[s] > 0$
(iii) $G^T(s^*) + G(s)$ *is positive semidefinite for* $\text{Re}[s] > 0$

where $*$ denotes complex conjugation.

Definition 3.5.4 [7] *An $n \times n$ matrix $G(s)$ is SPR if $G(s - \epsilon)$ is PR for some $\epsilon > 0$.*

Necessary and sufficient conditions in the frequency domain for $G(s)$ to be SPR are given by the following theorem.

3.5. POSITIVE REAL FUNCTIONS AND STABILITY

Theorem 3.5.3 [215] *Consider the $n \times n$ rational transfer matrix*

$$G(s) = C^T(sI - A)^{-1}B + D \qquad (3.5.3)$$

where $A, B, C,$ and D are real matrices with appropriate dimensions. Assume that $G(s) + G^T(-s)$ has rank n almost everywhere in the complex plane. Then $G(s)$ is SPR if and only if

(i) *all elements of $G(s)$ are analytic in $\mathrm{Re}[s] \geq 0$.*

(ii) $G(j\omega) + G^T(-j\omega) > 0 \; \forall \omega \in \mathcal{R}.$

(iii) (a) $\lim_{\omega \to \infty} \omega^2 [G(j\omega) + G^T(-j\omega)] > 0, \; D + D^T \geq 0 \; \text{if} \; \det[D + D^T] = 0.$

(b) $\lim_{\omega \to \infty} [G(j\omega) + G^T(-j\omega)] > 0 \; \text{if} \; \det[D + D^T] \neq 0.$

Necessary and sufficient conditions on the matrices $A, B, C,$ and D in (3.5.3) for $G(s)$ to be PR, SPR are given by the following lemmas which are generalizations of the lemmas in the SISO case to the MIMO case.

Theorem 3.5.4 [215] *Assume that $G(s)$ given by (3.5.3) is such that $G(s) + G^T(-s)$ has rank n almost everywhere in the complex plane, $\det(sI - A)$ has all its zeros in the open left half plane and (A, B) is completely controllable. Then $G(s)$ is SPR if and only if for any real symmetric positive definite matrix L, there exist a real symmetric positive definite matrix P, a scalar $\nu > 0$, real matrices Q and K such that*

$$A^T P + PA = -QQ^T - \nu L$$

$$PB = C \pm QK$$

$$K^T K = D^T + D$$

Lemma 3.5.5 [177] *Assume that the transfer matrix $G(s)$ has poles that lie in $\mathrm{Re}[s] < -\gamma$ where $\gamma > 0$ and (A, B, C, D) is a minimal realization of $G(s)$. Then, $G(s)$ is SPR if and only if a matrix $P = P^T > 0$, and matrices Q, K exist such that*

$$A^T P + PA = -QQ^T - 2\gamma P$$

$$PB = C \pm QK$$

$$K^T K = D + D^T$$

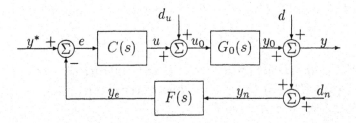

Figure 3.2 General feedback system.

The SPR properties of transfer function matrices are used in the analysis of a class of adaptive schemes designed for multi-input multi-output (MIMO) plants in a manner similar to that of SISO plants.

3.6 Stability of LTI Feedback Systems

3.6.1 A General LTI Feedback System

The block diagram of a typical feedback system is shown in Figure 3.2. Here $G_0(s)$ represents the transfer function of the plant model and $C(s), F(s)$ are the cascade and feedback compensators, respectively. The control input u generated from feedback is corrupted by an input disturbance d_u to form the plant input u_0. Similarly, the output of the plant y_0 is corrupted by an output disturbance d to form the actual plant output y. The measured output y_n, corrupted by the measurement noise d_n, is the input to the compensator $F(s)$ whose output y_e is subtracted from the reference (set point) y^* to form the error signal e.

The transfer functions $G_0(s), C(s), F(s)$ are generally proper and causal, and are allowed to be rational or irrational, which means they may include time delays.

The feedback system can be described in a compact I/O matrix form by treating $R = [y^*, d_u, d, d_n]^T$ as the input vector, and $E = [e, u_0, y, y_n]^T$ and $Y = [y_0, y_e, u]^T$ as the output vectors, i.e.,

$$E = H(s)R, \quad Y = I_1 E + I_2 R \tag{3.6.1}$$

3.6. STABILITY OF LTI FEEDBACK SYSTEMS

where

$$H(s) = \frac{1}{1 + FCG_0} \begin{bmatrix} 1 & -FG_0 & -F & -F \\ C & 1 & -FC & -FC \\ CG_0 & G_0 & 1 & -FCG_0 \\ CG_0 & G_0 & 1 & 1 \end{bmatrix}$$

$$I_1 = \begin{bmatrix} 0 & 0 & 1 & 0 \\ -1 & 0 & 0 & 0 \\ 0 & 1 & 0 & 0 \end{bmatrix}, \quad I_2 = \begin{bmatrix} 0 & 0 & -1 & 0 \\ 1 & 0 & 0 & 0 \\ 0 & -1 & 0 & 0 \end{bmatrix}$$

3.6.2 Internal Stability

Equation (3.6.1) relates all the signals in the system with the external inputs y^*, d_u, d, d_n. From a practical viewpoint it is important to guarantee that for any bounded input vector R, all the signals at any point in the feedback system are bounded. This motivates the definition of the so-called *internal stability* [152, 231].

Definition 3.6.1 *The feedback system is internally stable if for any bounded external input R, the signal vectors Y, E are bounded and in addition*

$$\|Y\|_\infty \leq c_1 \|R\|_\infty, \quad \|E\|_\infty \leq c_2 \|R\|_\infty \tag{3.6.2}$$

for some constants $c_1, c_2 \geq 0$ that are independent of R.

A necessary and sufficient condition for the feedback system (3.6.1) to be internally stable is that each element of $H(s)$ has stable poles, i.e., poles in the open left half s-plane [231].

The concept of internal stability may be confusing to some readers because of the fact that in most undergraduate books the stability of the feedback system is checked by examining the roots of the characteristic equation

$$1 + FCG_0 = 0$$

The following example is used to illustrate such confusions.

Example 3.6.1 Consider

$$G_0(s) = \frac{1}{s-2}, \quad C(s) = \frac{s-2}{s+5}, \quad F(s) = 1$$

for which the characteristic equation

$$1 + FCG_0 = 1 + \frac{(s-2)}{(s+5)(s-2)} = 1 + \frac{1}{s+5} = 0 \qquad (3.6.3)$$

has a single root at $s = -6$, indicating stability. On the other hand, the transfer matrix $H(s)$ calculated from this example yields

$$\begin{bmatrix} e \\ u_0 \\ y \\ y_n \end{bmatrix} = \frac{1}{s+6} \begin{bmatrix} s+5 & -(s+5)/(s-2) & -(s+5) & -(s+5) \\ s-2 & s+5 & -(s-2) & -(s-2) \\ 1 & (s+5)/(s-2) & s+5 & 1 \\ 1 & (s+5)/(s-2) & s+5 & s+5 \end{bmatrix} \begin{bmatrix} y^* \\ d_u \\ d \\ d_n \end{bmatrix} \qquad (3.6.4)$$

indicating that a bounded d_u will produce unbounded e, y, y_n, i.e., the feedback system is not internally stable. We should note that in calculating (3.6.3) and (3.6.4), we assume that $\frac{s+\alpha}{s+\alpha}g(s)$ and $g(s)$ are the same transfer functions for any constant α. In (3.6.3) the exact cancellation of the pole at $s = 2$ of $G_0(s)$ by the zero at $s = 2$ of $C(s)$ led to the wrong stability result, whereas in (3.6.4) such cancellations have no effect on internal stability. This example indicates that internal stability is more complete than the usual stability derived from the roots of $1 + FCG_0 = 0$. If, however, G_0, C, F are expressed as the ratio of coprime polynomials, i.e., $G_0(s) = \frac{n_0(s)}{p_0(s)}, C(s) = \frac{n_c(s)}{p_c(s)}, F(s) = \frac{n_f(s)}{p_f(s)}$ then a necessary and sufficient condition for internal stability [231] is that the roots of the characteristic equation

$$p_0 p_c p_f + n_0 n_c n_f = 0 \qquad (3.6.5)$$

are in the open left-half s-plane. For the example under consideration, $n_0 = 1, p_0 = s - 2, n_c = s - 2, p_c = s + 5, n_f = 1, p_f = 1$ we have

$$(s-2)(s+5) + (s-2) = (s-2)(s+6) = 0$$

which has one unstable root indicating that the feedback is not internally stable.

$\triangledown$

3.6.3 Sensitivity and Complementary Sensitivity Functions

Although internal stability guarantees that all signals in the feedback system are bounded for any bounded external inputs, performance requirements put restrictions on the size of some of the signal bounds. For example, one of the main objectives of a feedback controller is to keep the error between the plant output y and the reference signal y^* small in the presence of external inputs, such as reference inputs, bounded disturbances, and noise. Let us consider

3.6. STABILITY OF LTI FEEDBACK SYSTEMS

the case where in the feedback system of Figure 3.2, $F(s) = 1, d_u = 0$. Using (3.6.1), we can derive the following relationships between the plant output y and external inputs y^*, d, d_n:

$$y = T_0 y^* + S_0 d - T_0 d_n \qquad (3.6.6)$$

where

$$S_0 \triangleq \frac{1}{1 + CG_0}, \quad T_0 \triangleq \frac{CG_0}{1 + CG_0}$$

are referred to as the sensitivity function and complementary sensitivity function, respectively. It follows that S_0, T_0 satisfy

$$S_0 + T_0 = 1 \qquad (3.6.7)$$

It is clear from (3.6.6) that for good tracking and output disturbance rejection, the loop gain $L_0 \triangleq CG_0$ has to be chosen large so that $S_0 \approx 0$ and $T_0 \approx 1$. On the other hand, the suppression of the effect of the measurement noise d_n on y requires L_0 to be small so that $T_0 \approx 0$, which from (3.6.7) implies that $S_0 \approx 1$. This illustrates one of the basic trade-offs in feedback design, which is good reference tracking and disturbance rejection ($|L_0| \gg 1, S_0 \approx 0, T_0 \approx 1$) has to be traded off against suppression of measurement noise ($|L_0| \ll 1, S_0 \approx 1, T_0 \approx 0$). In a wide class of control problems, y^*, d are usually low frequency signals, and d_n is dominant only at high frequencies. In this case, $C(s)$ can be designed so that at low frequencies the loop gain L_0 is large, i.e., $S_0 \approx 0, T_0 \approx 1$, and at high frequencies L_0 is small, i.e., $S_0 \approx 1, T_0 \approx 0$. Another reason for requiring the loop gain L_0 to be small at high frequencies is the presence of dynamic plant uncertainties whose effect is discussed in later chapters.

3.6.4 Internal Model Principle

In many control problems, the reference input or setpoint y^* can be modeled as

$$Q_r(s) y^* = 0 \qquad (3.6.8)$$

where $Q_r(s)$ is a known polynomial, and $s \triangleq \frac{d}{dt}$ is the differential operator. For example, when $y^* =$ constant, $Q_r(s) = s$. When $y^* = t$, $Q_r(s) = s^2$ and when $y^* = A\sin\omega_0 t$ for some constants A and ω_0, then $Q_r(s) = s^2 + \omega_0^2$, etc. Similarly, a deterministic disturbance d can be modeled as

$$Q_d(s)d = 0 \tag{3.6.9}$$

for some known $Q_d(s)$, in cases where sufficient information about d is available. For example, if d is a sinusoidal signal with unknown amplitude and phase but with known frequency ω_d then it can be modeled by (3.6.9) with $Q_d(s) = s^2 + \omega_d^2$.

The idea behind the internal model principle is that by including the factor $\frac{1}{Q_r(s)Q_d(s)}$ in the compensator $C(s)$, we can null the effect of y^*, d on the tracking error $e = y^* - y$. To see how this works, consider the feedback system in Figure 3.2 with $F(s) = 1, d_u = d_n = 0$ and with the reference input y^* and disturbance d satisfying (3.6.8) and (3.6.9) respectively for some known polynomials $Q_r(s), Q_d(s)$. Let us now replace $C(s)$ in Figure 3.2 with

$$\bar{C}(s) = \frac{C(s)}{Q(s)}, \quad Q(s) = Q_r(s)Q_d(s) \tag{3.6.10}$$

where $C(s)$ in (3.6.10) is now chosen so that the poles of each element of $H(s)$ in (3.6.1) with $C(s)$ replaced by $\bar{C}(s)$ are stable. From (3.6.6) with $d_u = d_n = 0$ and C replaced by C/Q, we have

$$e = y^* - y = \frac{1}{1 + \frac{CG_0}{Q}}y^* - \frac{1}{1 + \frac{CG_0}{Q}}d = \frac{1}{Q + CG_0}Q(y^* - d)$$

Because $Q = Q_r Q_d$ nulls d, y^*, i.e., $Q(s)d = 0, Q(s)y^* = 0$, it follows that

$$e = \frac{1}{Q + CG_0}[0]$$

which together with the stability of $1/(Q + CG_0)$ guaranteed by the choice of $C(s)$ imply that $e(t) = y^*(t) - y(t)$ tends to zero exponentially fast.

The property of exact tracking guaranteed by the internal model principle can also be derived from the values of the sensitivity and complementary sensitivity functions S_0, T_0 at the frequencies of y^*, d. For example, if $y^* = \sin\omega_0 t$ and $d = \sin\omega_1 t$, i.e., $Q(s) = (s^2 + \omega_0^2)(s^2 + \omega_1^2)$ we have

$$S_0 = \frac{Q}{Q + CG_0}, \quad T_0 = \frac{CG_0}{Q + CG_0}$$

3.7. PROBLEMS

and $S_0(j\omega_0) = S_0(j\omega_1) = 0$, $T_0(j\omega_0) = T_0(j\omega_1) = 1$.

A special case of the internal model principle is integral control where $Q(s) = s$ which is widely used in industrial control to null the effect of constant set points and disturbances on the tracking error.

3.7 Problems

3.1 (a) Sketch the unit disk defined by the set $\{x \,|\, x \in \mathcal{R}^2, |x|_p \leq 1\}$ for (i) $p = 1$, (ii) $p = 2$, (iii) $p = 3$, and (iv) $p = \infty$.

(b) Calculate the $\mathcal{L}_2$ norm of the vector function

$$x(t) = [e^{-2t}, e^{-t}]^T$$

by (i) using the $|\cdot|_2$-norm in $\mathcal{R}^2$ and (ii) using the $|\cdot|_\infty$ norm in $\mathcal{R}^2$.

3.2 Let $y = G(s)u$ and $g(t)$ be the impulse response of $G(s)$. Consider the induced norm of the operator T defined as $\|T\|_\infty = \sup_{\|u\|_\infty = 1} \frac{\|y\|_\infty}{\|u\|_\infty}$, where $T : u \in \mathcal{L}_\infty \mapsto y \in \mathcal{L}_\infty$. Show that $\|T\|_\infty = \|g\|_1$.

3.3 Take $u(t) = f(t)$ where $f(t)$, shown in the figure, is a sequence of pulses centered at n with width $\frac{1}{n^3}$ and amplitude n, where $n = 1, 2, \ldots, \infty$.

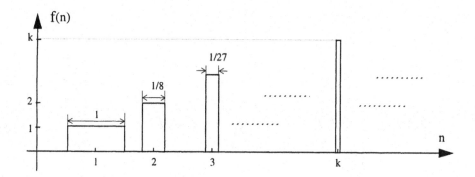

(a) Show that $u \in \mathcal{L}_1$ but $u \notin \mathcal{L}_2$ and $u \notin \mathcal{L}_\infty$.

(b) If $y = G(s)u$ where $G(s) = \frac{1}{s+1}$ and $u = f$, show that $y \in \mathcal{L}_1 \cap \mathcal{L}_\infty$ and $|y(t)| \to 0$ as $t \to \infty$.

3.4 Consider the system depicted in the following figure:

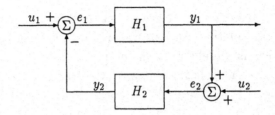

Let $H_1, H_2: \mathcal{L}_{\infty e} \mapsto \mathcal{L}_{\infty e}$ satisfy

$$\|(H_1 e_1)_t\| \leq \gamma_1 \|e_{1t}\| + \beta_1$$

$$\|(H_2 e_2)_t\| \leq \gamma_2 \|e_{2t}\| + \int_0^t e^{-\alpha(t-\tau)} \gamma(\tau) \|e_{2\tau}\| d\tau + \beta_2$$

for all $t \geq 0$, where $\gamma_1 \geq 0, \gamma_2 \geq 0$, $\alpha > 0$, β_1, β_2 are constants, $\gamma(t)$ is a nonnegative continuous function and $\|(\cdot)_t\|, \|(\cdot)\|$ denote the $\mathcal{L}_{\infty e}, \mathcal{L}_\infty$ norm respectively. Let $\|u_1\| \leq c, \|u_2\| \leq c$ for some constant $c \geq 0$, and $\gamma_1 \gamma_2 < 1$ (small gain). Show that

(a) If $\gamma \in \mathcal{L}_2$, then $e_1, e_2, y_1, y_2 \in \mathcal{L}_\infty$.

(b) If $\gamma \in \mathcal{S}(\mu)$, then $e_1, e_2, y_1, y_2 \in \mathcal{L}_\infty$ for any $\mu \in [0, \mu^*)$, where $\mu^* = \frac{\alpha^2(1-\gamma_1\gamma_2)^2}{2c_0\gamma_1^2}$.

3.5 Consider the system described by the following equation:

$$\dot{x} = A(t)x + f(t, x) + u \tag{3.7.1}$$

where $f(t, x)$ satisfies

$$|f(t, x)| \leq \gamma(t)|x| + \gamma_0(t)$$

for all $x \in \mathcal{R}^n$, $t \geq 0$ and $f(t, 0) = 0$, where $\gamma(t) \geq 0, \gamma_0(t)$ are continuous functions. If the equilibrium $y_e = 0$ of $\dot{y} = A(t)y$ is u.a.s and $\gamma \in \mathcal{S}(\mu)$, show that the following statements hold for some $\mu^* > 0$ and any $\mu \in [0, \mu^*)$:

(a) $u \in \mathcal{L}_\infty$ and $\gamma_0 \in \mathcal{S}(\nu)$ for any $\nu \geq 0$ implies $x \in \mathcal{L}_\infty$

(b) $u \equiv 0, \gamma_0 \equiv 0$ implies that the equilibrium $x_e = 0$ of (3.7.1) is u.a.s in the large

(c) $u, \gamma_0 \in \mathcal{L}_2$ implies that $x \in \mathcal{L}_\infty$ and $\lim_{t \to \infty} x(t) = 0$

(d) If $u \equiv \gamma_0 \equiv 0$ then the solution $x(t; t_0, x_0)$ of (3.7.1) satisfies

$$|x(t; t_0, x_0)| \leq K e^{-\beta(t-t_0)} |x(t_0)| \quad \text{for } t \geq t_0 \geq 0$$

where $\beta = \alpha - c_0 \mu K > 0$ for some constant c_0 and $K, \alpha > 0$ are constants in the bound for the state transition matrix $\Phi(t, \tau)$ of $\dot{y} = A(t)y$, i.e.,

$$\|\Phi(t, \tau)\| \leq K e^{-\alpha(t-\tau)}, \quad \text{for } t \geq \tau \geq t_0$$

3.7. PROBLEMS

3.6 Consider the LTI system
$$\dot{x} = (A + \epsilon B)x$$
where $\epsilon > 0$ is a scalar. Calculate $\epsilon^* > 0$ such that for all $\epsilon \in [0, \epsilon^*)$, the equilibrium state $x_e = 0$ is e.s. in the large when

(a)
$$A = \begin{bmatrix} -1 & 10 \\ 0 & -2 \end{bmatrix}, \quad B = \begin{bmatrix} 1 & 0 \\ 2 & 5 \end{bmatrix}$$

(b)
$$A = \begin{bmatrix} 0 & 10 \\ 0 & -1 \end{bmatrix}, \quad B = \begin{bmatrix} 5 & -8 \\ 0 & 2 \end{bmatrix}$$

using (i) the Euclidean norm and (ii) the infinity norm.

3.7 Consider the system given by the following block diagram:

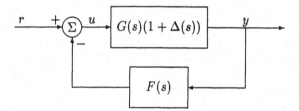

where $F(s)$ is designed such that the closed-loop system is internally stable when $\Delta(s) \equiv 0$, i.e., $\frac{G}{1+FG}, \frac{1}{1+FG}$ are stable transfer functions. Derive conditions on G, F using the small gain theorem such that the mapping $T : r \mapsto y$ is bounded in $\mathcal{L}_2$ for any $\Delta(s)$ that satisfies $\|\Delta(s)\|_\infty \leq \delta$, where $\delta > 0$ is a given constant.

3.8 Consider the LTI system
$$\dot{x} = (A + B)x, \quad x \in \mathcal{R}^n$$
where $Re\lambda_i(A) < 0$ and B is an arbitrary constant matrix. Find a bound on $\|B\|$ for $x_e = 0$ to be e.s. in the large by

(a) Using an appropriate Lyapunov function

(b) Without the use of a Lyapunov function

3.9 Examine the stability of the equilibrium states of the following differential equations:

(a) $\dot{x} = \sin t \, x$

(b) $\dot{x} = (3t \sin t - t)x$

(c) $\dot{x} = a(t)x$, where $a(t)$ is a continuous function with $a(t) < 0 \ \forall t \geq t_0 \geq 0$

3.10 Use Lyapunov's direct method to analyze the stability of the following systems:

(a) $\begin{cases} \dot{x}_1 = -x_1 + x_1 x_2 \\ \dot{x}_2 = -\gamma x_1^2 \end{cases}$

(b) $\ddot{x} + 2\dot{x}^3 + 2x = 0$

(c) $\dot{x} = \begin{bmatrix} 1 & -3 \\ 2 & -5 \end{bmatrix} x$

(d) $\begin{cases} \dot{x}_1 = -2x_1 + x_1 x_2 \\ \dot{x}_2 = -x_1^2 - \sigma x_2 \end{cases}$

3.11 Find the equilibrium state of the scalar differential equation

$$\dot{x} = -(x-1)(x-2)^2$$

and examine their stability properties using

(a) Linearization

(b) Appropriate Lyapunov functions

3.12 Consider the system described by

$$\dot{x}_1 = x_2$$
$$\dot{x}_2 = -x_2 - (k_0 + \sin t)x_1$$

where $k_0 > 2$ is a constant. Use an appropriate Lyapunov function to investigate stability of the equilibrium states.

3.13 Check the PR and SPR properties of the systems given by the following transfer functions:

(a) $G_1(s) = \frac{s+5}{(s+1)(s+4)}$

(b) $G_2(s) = \frac{s}{(s+2)^2}$

(c) $G_3(s) = \frac{s-2}{(s+3)(s+5)}$

(d) $G_4(s) = \frac{1}{s^2+2s+2}$

3.14 Assume that the transfer function $G_1(s)$ and $G_2(s)$ are PR. Check whether the following transfer functions are PR in general:

(a) $G_a(s) = G_1(s) + G_2(s)$

(b) $G_m(s) = G_1(s)G_2(s)$

(c) $G_f(s) = G_1(s)(1 + G_1(s)G_2(s))^{-1}$

3.7. PROBLEMS

Repeat (a), (b), and (c) when $G_1(s)$ and $G_2(s)$ are SPR.

3.15 (a) Let
$$G_1(s) = \frac{1}{s+1}, \quad L(s) = \frac{s-1}{(s+1)^2}$$

Find a bound[3] on $\epsilon > 0$ for the transfer function
$$G_\epsilon(s) = G_1(s) + \epsilon L(s)$$
to be PR, SPR.

(b) Repeat (a) when
$$G_1(s) = \frac{s+5}{(s+2)(s+3)}, \quad L(s) = -\frac{1}{s+1}$$

Comment on your results.

3.16 Consider the following feedback system:

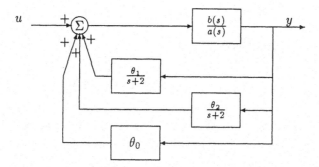

where $a(s) = s^2 - 3s + 2$ and $b(s) = s + \alpha$. Choose the constants $\theta_0, \theta_1, \theta_2$ such that for $\alpha = 5$, the transfer function
$$\frac{y(s)}{u(s)} = W_m(s) = \frac{1}{s+1}$$

What if $\alpha = -5$? Explain.

3.17 Consider the following system:
$$\dot{e} = Ae + B\phi \sin t, \quad e_1 = C^\top e$$
$$\dot{\phi} = -e_1 \sin t$$

where $\phi, e_1 \in \mathcal{R}^1, e \in \mathcal{R}^n$, (A, B) is controllable and $W_m(s) = C^\top (sI - A)^{-1} B$ is SPR. Use Lemma 3.5.3 to find an appropriate Lyapunov function to study the stability properties of the system (i.e., of the equilibrium state $\phi_e = 0, e_e = 0$).

---
[3] Find an $\epsilon^* > 0$ such that for all $\epsilon \in [0, \epsilon^*)$, $G_\epsilon(s)$ is PR, SPR.

Chapter 4

On-Line Parameter Estimation

4.1 Introduction

In Chapter 2, we discussed various types of model representations and parameterizations that describe the behavior of a wide class of dynamic systems. Given the structure of the model, the model response is determined by the values of certain constants referred to as plant or model parameters. In some applications these parameters may be measured or calculated using the laws of physics, properties of materials, etc. In many other applications, this is not possible, and the parameters have to be deduced by observing the system's response to certain inputs. If the parameters are fixed for all time, their determination is easier, especially when the system is linear and stable. In such a case, simple frequency or time domain techniques may be used to deduce the unknown parameters by processing the measured response data off-line. For this reason, these techniques are often referred to as off-line parameter estimation techniques. In many applications, the structure of the model of the plant may be known, but its parameters may be unknown and changing with time because of changes in operating conditions, aging of equipment, etc., rendering off-line parameter estimation techniques ineffective. The appropriate estimation schemes to use in this case are the ones that provide frequent estimates of the parameters of the plant model by properly processing the plant I/O data on-line. We refer to these scheme

4.1. INTRODUCTION

as *on-line estimation* schemes.

The purpose of this chapter is to present the design and analysis of a wide class of schemes that can be used for on-line parameter estimation. The essential idea behind on-line estimation is the comparison of the observed system response $y(t)$, with the output of a parameterized model $\hat{y}(\theta,t)$ whose structure is the same as that of the plant model. The parameter vector $\theta(t)$ is adjusted continuously so that $\hat{y}(\theta,t)$ approaches $y(t)$ as t increases. Under certain input conditions, $\hat{y}(\theta,t)$ being close to $y(t)$ implies that $\theta(t)$ is close to the unknown parameter vector θ^* of the plant model. The on-line estimation procedure, therefore, involves three steps: In the first step, an appropriate parameterization of the plant model is selected. This is an important step because some plant models are more convenient than others. The second step involves the selection of the adjustment law referred to as adaptive law for generating or updating $\theta(t)$. The adaptive law is usually a differential equation whose state is $\theta(t)$ and is designed using stability considerations or simple optimization techniques to minimize the difference between $y(t)$ and $\hat{y}(\theta,t)$ with respect to $\theta(t)$ at each time t. The third and final step is the design of the plant input so that the properties of the adaptive law imply that $\theta(t)$ approaches the unknown plant parameter vector θ^* as $t \to \infty$. This step is more important in problems where the estimation of θ^* is one of the objectives and is treated separately in Chapter 5. In adaptive control, where the convergence of $\theta(t)$ to θ^* is usually not one of the objectives, the first two steps are the most important ones. In Chapters 6 and 7, a wide class of adaptive controllers are designed by combining the on-line estimation schemes of this chapter with appropriate control laws. Unlike the off-line estimation schemes, the on-line ones are designed to be used with either stable or unstable plants. This is important in adaptive control where stabilization of the unknown plant is one of the immediate objectives.

The chapter is organized as follows: We begin with simple examples presented in Section 4.2, which we use to illustrate the basic ideas behind the design and analysis of adaptive laws for on-line parameter estimation. These examples involve plants that are stable and whose states are available for measurement. In Section 4.3, we extend the results of Section 4.2 to plants that may be unstable, and only part of the plant states is available for measurement. We develop a wide class of adaptive laws using a Lyapunov design approach and simple optimization techniques based on the gradient method and least squares. The on-line estimation problem where the unknown parameters are constrained to lie inside known convex sets is discussed in Section 4.4. The estimation of parameters that appear in a special bilinear form is a problem that often appears in model reference adaptive control. It is treated in Section 4.5. In the adaptive laws of Sections 4.2 to 4.5, the parameter estimates are generated continuously with time. In Section 4.6, we discuss a class of adaptive laws, referred to as *hybrid adaptive laws*, where the parameter estimates are generated at finite intervals of time. A summary of the adaptive laws developed is given in Section 4.7 where the main equations of the adaptive laws and their properties are organized in tables. Section 4.8 contains most of the involved proofs dealing with parameter convergence.

4.2 Simple Examples

In this section, we use simple examples to illustrate the derivation and properties of simple on-line parameter estimation schemes. The simplicity of these examples allows the reader to understand the design methodologies and stability issues in parameter estimation without having to deal with the more complex differential equations that arise in the general case.

4.2.1 Scalar Example: One Unknown Parameter

Let us consider the following plant described by the algebraic equation

$$y(t) = \theta^* u(t) \tag{4.2.1}$$

where $u \in \mathcal{L}_\infty$ is the scalar input, $y(t)$ is the output, and θ^* is an unknown scalar. Assuming that $u(t), y(t)$ are measured, it is desired to obtain an estimate of θ^* at each time t. If the measurements of y, u were noise free, one could simply calculate $\theta(t)$, the estimate of θ^*, as

$$\theta(t) = \frac{y(t)}{u(t)} \tag{4.2.2}$$

whenever $u(t) \neq 0$. The division in (4.2.2), however, may not be desirable because $u(t)$ may assume values arbitrarily close to zero. Furthermore, the

4.2. SIMPLE EXAMPLES

effect of noise on the measurement of u, y may lead to an erroneous estimate of θ^*. The noise and computational error effects in (4.2.2) may be reduced by using various other nonrecursive or off-line methods especially when θ^* is a constant for all t.

In our case, we are interested in a recursive or on-line method to generate $\theta(t)$. We are looking for a differential equation, which depends on signals that are measured, whose solution is $\theta(t)$ and its equilibrium state is $\theta_e = \theta^*$. The procedure for developing such a differential equation is given below.

Using $\theta(t)$ as the estimate of θ^* at time t, we generate the estimated or predicted value $\hat{y}(t)$ of the output $y(t)$ as

$$\hat{y}(t) = \theta(t)u(t) \qquad (4.2.3)$$

The prediction or estimation error ϵ_1, which reflects the parameter uncertainty because $\theta(t)$ is different from θ^*, is formed as the difference between $\hat{y}$ and y, i.e.,

$$\epsilon_1 = y - \hat{y} = y - \theta u \qquad (4.2.4)$$

The dependence of ϵ_1 on the parameter estimation error $\tilde{\theta} \stackrel{\Delta}{=} \theta - \theta^*$ becomes obvious if we use (4.2.1) to substitute for y in (4.2.4), i.e.,

$$\epsilon_1 = \theta^* u - \theta u = -\tilde{\theta} u \qquad (4.2.5)$$

The differential equation for generating $\theta(t)$ is now developed by minimizing various cost criteria of ϵ_1 with respect to θ using the gradient or Newton's method. Such criteria are discussed in great detail in Section 4.3. For this example, we concentrate on the simple cost criterion

$$J(\theta) = \frac{\epsilon_1^2}{2} = \frac{(y - \theta u)^2}{2}$$

which we minimize with respect to θ. For each time t, the function $J(\theta)$ is convex over $\mathcal{R}^1$; therefore, any local minimum of J is also global and satisfies $\nabla J(\theta) = 0$. One can solve $\nabla J(\theta) = -(y - \theta u)u = 0$ for θ and obtain the nonrecursive scheme (4.2.2) or use the gradient method (see Appendix B) to form the recursive scheme

$$\dot{\theta} = -\gamma \nabla J(\theta) = \gamma(y - \theta u)u = \gamma \epsilon_1 u, \quad \theta(0) = \theta_0 \qquad (4.2.6)$$

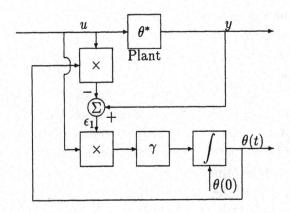

Figure 4.1 Implementation of the scalar adaptive law (4.2.6).

where $\gamma > 0$ is a scaling constant, which we refer to as the *adaptive gain*. In the literature, the differential equation (4.2.6) is referred to as the *update law* or the *adaptive law* or the *estimator*, to name a few. In this book, we refer to (4.2.6) as the *adaptive law* for updating $\theta(t)$ or estimating θ^* on-line.

The implementation of the adaptive law (4.2.6) is shown in Figure 4.1. The stability properties of (4.2.6) are analyzed by rewriting (4.2.6) in terms of the parameter error $\tilde{\theta} = \theta - \theta^*$, i.e.,

$$\dot{\tilde{\theta}} = \dot{\theta} - \dot{\theta}^* = \gamma \epsilon_1 u - \dot{\theta}^*$$

Because $\epsilon_1 = \theta^* u - \theta u = -\tilde{\theta} u$ and θ^* is constant, i.e., $\dot{\theta}^* = 0$, we have

$$\dot{\tilde{\theta}} = -\gamma u^2 \tilde{\theta}, \quad \tilde{\theta}(0) = \theta(0) - \theta^* \qquad (4.2.7)$$

We should emphasize that (4.2.7) is used only for analysis. It cannot be used to generate $\theta(t)$ because given an initial estimate $\theta(0)$ of θ^*, the initial value $\tilde{\theta}(0) = \theta(0) - \theta^*$, which is required for implementing (4.2.7) is unknown due to the unknown θ^*.

Let us analyze (4.2.7) by choosing the Lyapunov function

$$V(\tilde{\theta}) = \frac{\tilde{\theta}^2}{2\gamma}$$

The time derivative $\dot{V}$ of V along the solution of (4.2.7) is given by

$$\dot{V} = \frac{\tilde{\theta}^\top \dot{\tilde{\theta}}}{\gamma}$$

4.2. SIMPLE EXAMPLES

which after substitution of $\dot{\tilde{\theta}}$ from (4.2.7) becomes

$$\dot{V} = -u^2\tilde{\theta}^2 = -\epsilon_1^2 \leq 0 \qquad (4.2.8)$$

which implies that the equilibrium $\tilde{\theta}_e = 0$ of (4.2.7) is u.s.

Because no further information about $u(t)$ is assumed other than $u \in \mathcal{L}_\infty$, we cannot guarantee that $\dot{V} < 0$ (e.g., take $u(t) = 0$) and, therefore, cannot establish that $\tilde{\theta}_e = 0$ is a.s. or e.s. We can, however, use the properties of $V, \dot{V}$ to establish convergence for the estimation error and other signals in (4.2.6). For example, because $V \geq 0$ is a nonincreasing function of time, the $\lim_{t \to \infty} V(\tilde{\theta}(t)) = V_\infty$ exists. Therefore, from (4.2.8) we have

$$\int_0^\infty \epsilon_1^2(\tau)d\tau = -\int_0^\infty \dot{V}(\tau)d\tau = V_0 - V_\infty$$

where $V_0 = V(\tilde{\theta}(0))$, which implies that $\epsilon_1 \in \mathcal{L}_2$. Now from (4.2.6) and $u \in \mathcal{L}_\infty$, we also have that $\dot{\theta} \in \mathcal{L}_\infty \cap \mathcal{L}_2$. Because, as we have shown in Chapter 3, a square integrable function may not have a limit, let alone tend to zero with time, we cannot establish that $\epsilon_1(t), \dot{\theta}(t) \to 0$ as $t \to \infty$ without additional conditions. If, however, we assume that $\dot{u} \in \mathcal{L}_\infty$, then it follows that $\dot{\epsilon}_1 = -\dot{\tilde{\theta}}u - \tilde{\theta}\dot{u} \in \mathcal{L}_\infty$; therefore, from Lemma 3.2.5 we have $\epsilon_1(t) \to 0$ as $t \to \infty$, which is implied by $\epsilon_1 \in \mathcal{L}_2, \dot{\epsilon}_1 \in \mathcal{L}_\infty$. This, in turn, leads to

$$\lim_{t \to \infty} \dot{\tilde{\theta}}(t) = \lim_{t \to \infty} \dot{\theta}(t) = 0 \qquad (4.2.9)$$

The conclusion of this analysis is that for any $u, \dot{u} \in \mathcal{L}_\infty$, the adaptive law (4.2.6) guarantees that the estimated output $\hat{y}(t)$ converges to the actual output $y(t)$ and the speed of adaptation (i.e., the rate of change of the parameters $\dot{\theta}$) decreases with time and converges to zero asymptotically.

One important question to ask at this stage is whether $\theta(t)$ converges as $t \to \infty$ and, if it does, is the $\lim_{t \to \infty} \theta(t) = \theta^*$?

A quick look at (4.2.9) may lead some readers to conclude that $\dot{\theta}(t) \to 0$ as $t \to \infty$ implies that $\theta(t)$ does converge to a constant. This conclusion is obviously false because the function $\theta(t) = \sin(\ln(2+t))$ satisfies (4.2.9) but has no limit. We have established, however, that $V(\tilde{\theta}(t)) = \frac{\tilde{\theta}^2}{2\gamma}$ converges to V_∞ as $t \to \infty$, i.e., $\lim_{t \to \infty} \tilde{\theta}^2(t) = 2\gamma V_\infty$, which implies that $\tilde{\theta}(t)$ and, therefore, $\theta(t)$ does converge to a constant, i.e., $\lim_{t \to \infty} \theta(t) = \pm\sqrt{2\gamma V_\infty} + \theta^* \triangleq \bar{\theta}$.

Hence the question which still remains unanswered is whether $\bar{\theta} = \theta^*$. It is clear from (4.2.1) that for $u(t) = 0$, a valid member of the class of input signals considered, $y(t) = 0 \ \forall t \geq 0$, which provides absolutely no information about the unknown θ^*. It is, therefore, obvious that without additional conditions on the input $u(t), y(t)$ may not contain sufficient information about θ^* for the identification of θ^* to be possible.

For this simple example, we can derive explicit conditions on $u(t)$ that guarantee parameter convergence by considering the closed-form solution of (4.2.7), i.e.,

$$\tilde{\theta}(t) = e^{-\gamma \int_0^t u^2(\tau)d\tau} \tilde{\theta}(0) \quad (4.2.10)$$

For inputs $u(t) = 0$ or $u(t) = e^{-t}$, $\tilde{\theta}(t)$ does not tend to zero, whereas for $u^2(t) = \frac{1}{1+t}$, $\tilde{\theta}(t)$ tends to zero asymptotically but not exponentially. A necessary and sufficient condition for $\tilde{\theta}(t)$ to converge to zero exponentially fast is that $u(t)$ satisfies

$$\int_t^{t+T_0} u^2(\tau)d\tau \geq \alpha_0 T_0 \quad (4.2.11)$$

$\forall t \geq 0$ and for some $\alpha_0, T_0 > 0$ (see Problem 4.1). It is clear that $u(t) = 1$ satisfies (4.2.11), whereas $u(t) = 0$ or e^{-t} or $\frac{1}{1+t}$ does not. The property of u given by (4.2.11) is referred to as *persistent excitation* (PE) and is crucial in many adaptive schemes where parameter convergence is one of the objectives. The signal u, which satisfies (4.2.11), is referred to be *persistently exciting* (PE). The PE property of signals is discussed in more detail in Section 4.3.

It is clear from (4.2.10), (4.2.11) that the rate of exponential convergence of $\tilde{\theta}(t)$ to zero is proportional to the adaptive gain γ and the constant α_0 in (4.2.11), referred to as the *level of excitation*. Increasing the value of γ will speed up the convergence of $\theta(t)$ to θ^*. A large γ, however, may make the differential equation (4.2.6) "stiff" and, therefore, more difficult to solve numerically.

The same methodology as above may be used for the identification of an n-dimensional vector θ^* that satisfies the algebraic equation

$$y = \theta^{*\mathsf{T}} \phi \quad (4.2.12)$$

where $y \in \mathcal{R}^1, \phi \in \mathcal{R}^n$ are bounded signals available for measurement. We will deal with this general case in subsequent sections.

4.2.2 First-Order Example: Two Unknowns

Consider the following first-order plant

$$\dot{x} = -ax + bu, \quad x(0) = x_0 \qquad (4.2.13)$$

where the parameters a and b are constant but unknown, and the input u and state x are available for measurement. We assume that $a > 0$ and $u \in \mathcal{L}_\infty$ so that $x \in \mathcal{L}_\infty$. The objective is to generate an adaptive law for estimating a and b on-line by using the observed signals $u(t)$ and $x(t)$.

As in Section 4.2.1, the adaptive law for generating the estimates $\hat{a}$ and $\hat{b}$ of a and b, respectively, is to be driven by the estimation error

$$\epsilon_1 = x - \hat{x} \qquad (4.2.14)$$

where $\hat{x}$ is the estimated value of x formed by using the estimates $\hat{a}$ and $\hat{b}$. The state $\hat{x}$ is usually generated by an equation that has the same form as the plant but with a and b replaced by $\hat{a}$ and $\hat{b}$, respectively. For example, considering the plant equation (4.2.13), we can generate $\hat{x}$ from

$$\dot{\hat{x}} = -\hat{a}\hat{x} + \hat{b}u, \quad \hat{x}(0) = \hat{x}_0 \qquad \text{(P)}$$

Equation (P) is known as the *parallel model* configuration [123] and the estimation method based on (P) as the *output error method* [123, 172]. The plant equation, however, may be rewritten in various different forms giving rise to different equations for generating $\hat{x}$. For example, we can add and subtract the term $a_m x$, where $a_m > 0$ is an arbitrary design constant, in (4.2.13) and rewrite the plant equation as

$$\dot{x} = -a_m x + (a_m - a)x + bu$$

i.e.,

$$x = \frac{1}{s + a_m}[(a_m - a)x + bu] \qquad (4.2.15)$$

Furthermore, we can proceed and rewrite (4.2.15) as

$$x = \theta^{*\mathsf{T}} \phi$$

where $\theta^* = [b, a_m - a]^\mathsf{T}, \phi = \left[\frac{1}{s+a_m}u, \frac{1}{s+a_m}x\right]^\mathsf{T}$, which is in the form of the algebraic equation considered in Section 4.2.1. Therefore, instead of using

(P), we may also generate $\hat{x}$ from

$$\dot{\hat{x}} = -a_m\hat{x} + (a_m - \hat{a})x + \hat{b}u$$

that is

$$\hat{x} = \frac{1}{s+a_m}[(a_m - \hat{a})x + \hat{b}u] \qquad \text{(SP)}$$

by considering the parameterization of the plant given by (4.2.15). Equation (SP) is widely used for parameter estimation and is known as the *series-parallel model* [123]. The estimation method based on (SP) is called the *equation error method* [123, 172]. Various other models that are a combination of (P) and (SP) are generated [123] by considering different parameterizations for the plant (4.2.13).

The estimation error $\epsilon_1 = x - \hat{x}$ satisfies the differential equation

$$\dot{\epsilon}_1 = -a\epsilon_1 + \tilde{a}\hat{x} - \tilde{b}u \qquad \text{(P1)}$$

for model (P) and

$$\dot{\epsilon}_1 = -a_m\epsilon_1 + \tilde{a}x - \tilde{b}u \qquad \text{(SP1)}$$

for model (SP) where

$$\tilde{a} \triangleq \hat{a} - a, \quad \tilde{b} \triangleq \hat{b} - b$$

are the parameter errors. Equations (P1) and (SP1) indicate how the parameter error affects the estimation error ϵ_1. Because $a, a_m > 0$, zero parameter error, i.e., $\tilde{a} = \tilde{b} = 0$, implies that ϵ_1 converges to zero exponentially. Because $\tilde{a}, \tilde{b}$ are unknown, ϵ_1 is the only measured signal that we can monitor in practice to check the success of estimation. We should emphasize, however, that $\epsilon_1 \to 0$ does not imply that $\tilde{a}, \tilde{b} \to 0$ unless some PE properties are satisfied by $\hat{x}, x, u$ as we will demonstrate later on in this section. We should also note that ϵ_1 cannot be generated from (P1) and (SP1) because $\tilde{a}$ and $\tilde{b}$ are unknown. Equations (P1) and (SP1) are, therefore, only used for the purpose of analysis.

Let us now use the error equation (SP1) to derive the adaptive laws for estimating a and b. We assume that the adaptive laws are of the form

$$\dot{\hat{a}} = f_1(\epsilon_1, \hat{x}, x, u), \quad \dot{\hat{b}} = f_2(\epsilon_1, \hat{x}, x, u) \qquad (4.2.16)$$

where f_1 and f_2 are functions of measured signals, and are to be chosen so that the equilibrium state

$$\hat{a}_e = a, \quad \hat{b}_e = b, \quad \epsilon_{1e} = 0 \qquad (4.2.17)$$

4.2. SIMPLE EXAMPLES

of the third-order differential equation described by (SP1) (where $x \in \mathcal{L}_\infty$ is treated as an independent function of time) and (4.2.16) is u.s., or, if possible, u.a.s., or, even better, e.s.

We choose f_1, f_2 so that a certain function $V(\epsilon_1, \tilde{a}, \tilde{b})$ and its time derivative $\dot{V}$ along the solution of (SP1), (4.2.16) are such that V qualifies as a Lyapunov function that satisfies some of the conditions given by Theorems 3.4.1 to 3.4.4 in Chapter 3. We start by considering the quadratic function

$$V(\epsilon_1, \tilde{a}, \tilde{b}) = \frac{1}{2}(\epsilon_1^2 + \tilde{a}^2 + \tilde{b}^2) \qquad (4.2.18)$$

which is positive definite, decrescent, and radially unbounded in $\mathcal{R}^3$. The time derivative of V along any trajectory of (SP1), (4.2.16) is given by

$$\dot{V} = -a_m \epsilon_1^2 + \tilde{a} x \epsilon_1 - \tilde{b} u \epsilon_1 + \tilde{a} f_1 + \tilde{b} f_2 \qquad (4.2.19)$$

and is evaluated by using the identities $\dot{\tilde{a}} = \dot{\hat{a}}$, $\dot{\tilde{b}} = \dot{\hat{b}}$, which hold because a and b are assumed to be constant.

If we choose $f_1 = -\epsilon_1 x$, $f_2 = \epsilon_1 u$, we have

$$\dot{V} = -a_m \epsilon_1^2 \leq 0 \qquad (4.2.20)$$

and (4.2.16) becomes

$$\dot{\hat{a}} = -\epsilon_1 x, \quad \dot{\hat{b}} = \epsilon_1 u \qquad (4.2.21)$$

where $\epsilon_1 = x - \hat{x}$ and $\hat{x}$ is generated by (SP).

Applying Theorem 3.4.1 to (4.2.18) and (4.2.20), we conclude that V is a Lyapunov function for the system (SP1), (4.2.16) where x and u are treated as independent bounded functions of time and the equilibrium given by (4.2.17) is u.s. Furthermore, the trajectory $\epsilon_1(t), \hat{a}(t), \hat{b}(t)$ is bounded for all $t \geq 0$. Because $\epsilon_1 = x - \hat{x}$ and $x \in \mathcal{L}_\infty$ we also have that $\hat{x} \in \mathcal{L}_\infty$; therefore, all signals in (SP1) and (4.2.21) are uniformly bounded. As in the example given in Section 4.2.1, (4.2.18) and (4.2.20) imply that

$$\lim_{t \to \infty} V(\epsilon_1(t), \tilde{a}(t), \tilde{b}(t)) = V_\infty < \infty$$

and, therefore,

$$\int_0^\infty \epsilon_1^2(\tau) d\tau = -\frac{1}{a_m} \int_0^\infty \dot{V} d\tau = \frac{1}{a_m}(V_0 - V_\infty)$$

where $V_0 = V(\epsilon_1(0), \tilde{a}(0), \tilde{b}(0))$, i.e., $\epsilon_1 \in \mathcal{L}_2$. Because $u, \tilde{a}, \tilde{b}, x, \epsilon_1 \in \mathcal{L}_\infty$, it follows from (SP1) that $\dot{\epsilon}_1 \in \mathcal{L}_\infty$, which, together with $\epsilon_1 \in \mathcal{L}_2$, implies that $\epsilon_1(t) \to 0$ as $t \to \infty$, which, in turn, implies that $\dot{\tilde{a}}(t), \dot{\tilde{b}}(t) \to 0$ as $t \to \infty$.

It is worth noting that $\epsilon_1(t), \dot{\tilde{a}}(t), \dot{\tilde{b}}(t) \to 0$ as $t \to \infty$ do not imply that $\tilde{a}$ and $\tilde{b}$ converge to any constant let alone to zero. As in the example of Section 4.2.1, we can use (4.2.18) and (4.2.20) and establish that

$$\lim_{t \to \infty} (\tilde{a}^2(t) + \tilde{b}^2(t)) = 2V_\infty$$

which again does not imply that $\tilde{a}$ and $\tilde{b}$ have a limit, e.g., take

$$\tilde{a}(t) = \sqrt{2V_\infty} \sin \sqrt{1+t}, \quad \tilde{b}(t) = \sqrt{2V_\infty} \cos \sqrt{1+t}$$

The failure to establish parameter convergence may motivate the reader to question the choice of the Lyapunov function given by (4.2.18) and of the functions f_1, f_2 in (4.2.19). The reader may argue that perhaps for some other choices of V and f_1, f_2, u.a.s could be established for the equilibrium (4.2.17) that will automatically imply that $\tilde{a}, \tilde{b} \to 0$ as $t \to \infty$. Since given a differential equation, there is no procedure for finding the appropriate Lyapunov function to establish stability in general, this argument appears to be quite valid. We can counteract this argument, however, by applying simple intuition to the plant equation (4.2.13). In our analysis, we put no restriction on the input signal u, apart from $u \in \mathcal{L}_\infty$, and no assumption is made about the initial state x_0. For $u = 0$, an allowable input in our analysis, and $x_0 = 0$, no information can be extracted about the unknown parameters a, b from the measurements of $x(t) = 0, u(t) = 0, \forall t \geq 0$. Therefore, no matter how intelligent an adaptive law is, parameter error convergence to zero cannot be achieved when $u = 0 \; \forall t \geq 0$. This simplistic explanation demonstrates that additional conditions have to be imposed on the input signal u to establish parameter error convergence to zero. Therefore, no matter what V and f_1, f_2 we choose, we can not establish u.a.s. without imposing conditions on the input u. These conditions are similar to those imposed on the input u in Section 4.2.1, and will be discussed and analyzed in Chapter 5.

In the adaptive law (4.2.21), the adaptive gains are set equal to 1. A similar adaptive law with arbitrary adaptive gains $\gamma_1, \gamma_2 > 0$ is derived by

4.2. SIMPLE EXAMPLES

considering

$$V(\epsilon_1, \tilde{a}, \tilde{b}) = \frac{1}{2}\left(\epsilon_1^2 + \frac{\tilde{a}^2}{\gamma_1} + \frac{\tilde{b}^2}{\gamma_2}\right)$$

instead of (4.2.18). Following the same procedure as before we obtain

$$\dot{\hat{a}} = -\gamma_1 \epsilon_1 x, \quad \dot{\hat{b}} = \gamma_2 \epsilon_1 u$$

where $\gamma_1, \gamma_2 > 0$ are chosen appropriately to slow down or speed up adaptation.

Using (4.2.18) with model (P1) and following the same analysis as with model (SP1), we obtain

$$\dot{\hat{a}} = -\epsilon_1 \hat{x}, \quad \dot{\hat{b}} = \epsilon_1 u \qquad (4.2.22)$$

and

$$\dot{V} = -a\epsilon_1^2 \leq 0$$

Hence, the same conclusions as with (4.2.21) are drawn for (4.2.22).

We should note that $\dot{V}$ for (P1) depends on the unknown a, whereas for (SP1) it depends on the known design scalar a_m. Another crucial difference between model (P) and (SP) is their performance in the presence of noise, which becomes clear after rewriting the adaptive law for $\hat{a}$ in (4.2.21), (4.2.22) as

$$\dot{\hat{a}} = -(x - \hat{x})\hat{x} = \hat{x}^2 - x\hat{x} \qquad \text{(P)}$$

$$\dot{\hat{a}} = -(x - \hat{x})x = -x^2 + x\hat{x} \qquad \text{(SP)}$$

If the measured plant state x is corrupted by some noise signal v, i.e., x is replaced by $x + v$ in the adaptive law, it is clear that for the model (SP), $\dot{\hat{a}}$ will depend on v^2 and v, whereas for model (P) only on v. The effect of noise (v^2) may result in biased estimates in the case of model (SP), whereas the quality of estimation will be less affected in the case of model (P). The difference between the two models led some researchers to the development of more complicated models that combine the good noise properties of the parallel model (P) with the design flexibility of the series-parallel model (SP) [47, 123].

Simulations

We simulate the parallel and series-parallel estimators and examine the effects of the input signal u, the adaptive gain and noise disturbance on their performance. For simplicity, we consider a first-order example $y = \frac{b}{s+a}u$ with two unknown parameters a and b. Two adaptive estimators

$$\dot{\hat{a}} = -\epsilon_1 \hat{x}, \quad \dot{\hat{b}} = \epsilon_1 u$$
$$\dot{\hat{x}} = -\hat{a}\hat{x} + \hat{b}u, \quad \epsilon_1 = x - \hat{x}$$

and

$$\dot{\hat{a}} = -\epsilon_1 x, \quad \dot{\hat{b}} = \epsilon_1 u$$
$$\dot{\hat{x}} = -a_m \hat{x} + (a_m - \hat{a})x + \hat{b}u, \quad \epsilon_1 = x - \hat{x}$$

based on the parallel and series-parallel model, respectively, are simulated with $a = 2$ and $b = 1$. The results are given in Figures 4.2 and Figure 4.3, respectively. Plots (a) and (b) in Figure 4.2 and 4.3 give the time response of the estimated parameters when the input $u = \sin 5t$, and the adaptive gain $\gamma = 1$ for (a) and $\gamma = 5$ for (b). Plots (c) in both figures give the results of estimation for a step input, where persistent excitation and, therefore, parameter convergence are not guaranteed. Plots (d) show the performance of the estimator when the measurement $x(t)$ is corrupted by $d(t) = 0.1n(t)$, where $n(t)$ is a normally distributed white noise.

It is clear from Figures 4.2 (a,b) and Figure 4.3 (a,b) that the use of a larger value of the adaptive gain γ led to a faster convergence of $\hat{a}$ and $\hat{b}$ to their true values. The lack of parameter convergence to the true values in Figure 4.2 (c), 4.3 (c) is due to the use of a non-PE input signal. As expected, the parameter estimates are more biased in the case of the series-parallel estimator shown in Figure 4.3 (d) than those of the parallel one shown in Figure 4.2 (d).

4.2.3 Vector Case

Let us extend the example of Section 4.2.2 to the higher-order case where the plant is described by the vector differential equation

$$\dot{x} = A_p x + B_p u \tag{4.2.23}$$

4.2. SIMPLE EXAMPLES

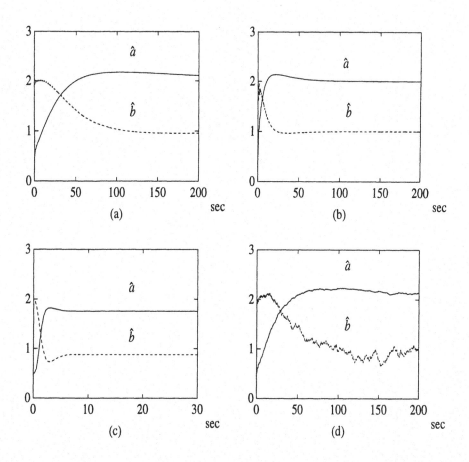

Figure 4.2 Simulation results of the parallel estimator. (a) $u = \sin 5t$, $\gamma = 1$, no measurement noise; (b) $u = \sin 5t$, $\gamma = 5$, no measurement noise; (c) $u =$ unit step function, $\gamma = 1$, no measurement noise; (d) $u = \sin 5t, \gamma = 1$, output x is corrupted by $d(t) = 0.1n(t)$, where $n(t)$ is a normally distributed white noise.

where the state $x \in \mathcal{R}^n$ and input $u \in \mathcal{R}^r$ are available for measurement, $A_p \in \mathcal{R}^{n \times n}, B_p \in \mathcal{R}^{n \times r}$ are unknown, A_p is stable, and $u \in \mathcal{L}_\infty$. As in the scalar case, we form the parallel model

$$\dot{\hat{x}} = \hat{A}_p \hat{x} + \hat{B}_p u, \quad \hat{x} \in \mathcal{R}^n \tag{P}$$

where $\hat{A}_p(t), \hat{B}_p(t)$ are the estimates of A_p, B_p at time t to be generated by

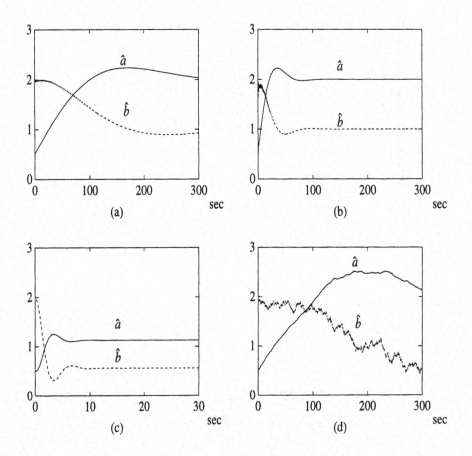

Figure 4.3 Simulation results of the series-parallel estimator. (a) $u = \sin 5t$, $\gamma = 1$, no measurement noise; (b) $u = \sin 5t, \gamma = 5$, no measurement noise; (c) $u =$ unit step function, $\gamma = 1$, no measurement noise; (d) $u = \sin 5t, \gamma = 1$, output x is corrupted by $d(t) = 0.1n(t)$, where $n(t)$ is the normally distributed white noise.

an adaptive law, and $\hat{x}(t)$ is the estimate of the vector $x(t)$. Similarly, by considering the plant parameterization

$$\dot{x} = A_m x + (A_p - A_m)x + B_p u$$

where A_m is an arbitrary stable matrix, we define the series-parallel model as

$$\dot{\hat{x}} = A_m \hat{x} + (\hat{A}_p - A_m)x + \hat{B}_p u \qquad \text{(SP)}$$

4.2. SIMPLE EXAMPLES

The estimation error vector ϵ_1 defined as

$$\epsilon_1 \triangleq x - \hat{x}$$

satisfies

$$\dot{\epsilon}_1 = A_p \epsilon_1 - \tilde{A}_p \hat{x} - \tilde{B}_p u \qquad (P1)$$

for model (P) and

$$\dot{\epsilon}_1 = A_m \epsilon_1 - \tilde{A}_p x - \tilde{B}_p u \qquad (SP1)$$

for model (SP), where $\tilde{A}_p \triangleq \hat{A}_p - A_p$, $\tilde{B}_p \triangleq \hat{B}_p - B_p$.

Let us consider the parallel model design and use (P1) to derive the adaptive law for estimating the elements of A_p, B_p. We assume that the adaptive law has the general structure

$$\dot{\hat{A}}_p = F_1(\epsilon_1, x, \hat{x}, u), \quad \dot{\hat{B}}_p = F_2(\epsilon_1, x, \hat{x}, u) \qquad (4.2.24)$$

where F_1 and F_2 are functions of known signals that are to be chosen so that the equilibrium

$$\hat{A}_{pe} = A_p, \quad \hat{B}_{pe} = B_p, \quad \epsilon_{1e} = 0$$

of (P1), (4.2.24) has some desired stability properties. We start by considering the function

$$V(\epsilon_1, \tilde{A}_p, \tilde{B}_p) = \epsilon_1^T P \epsilon_1 + \text{tr}\left(\frac{\tilde{A}_p^T P \tilde{A}_p}{\gamma_1}\right) + \text{tr}\left(\frac{\tilde{B}_p^T P \tilde{B}_p}{\gamma_2}\right) \qquad (4.2.25)$$

where $\text{tr}(A)$ denotes the trace of a matrix A, $\gamma_1, \gamma_2 > 0$ are constant scalars, and $P = P^T > 0$ is chosen as the solution of the Lyapunov equation

$$P A_p + A_p^T P = -I \qquad (4.2.26)$$

whose existence is guaranteed by the stability of A_p (see Theorem 3.4.10). The time derivative $\dot{V}$ of V along the trajectory of (P1), (4.2.24) is given by

$$\begin{aligned} \dot{V} &= \dot{\epsilon}_1^T P \epsilon_1 + \epsilon_1^T P \dot{\epsilon}_1 \\ &+ \text{tr}\left(\frac{\dot{\tilde{A}}_p^T P \tilde{A}_p}{\gamma_1} + \frac{\tilde{A}_p^T P \dot{\tilde{A}}_p}{\gamma_1}\right) + \text{tr}\left(\frac{\dot{\tilde{B}}_p^T P \tilde{B}_p}{\gamma_2} + \frac{\tilde{B}_p^T P \dot{\tilde{B}}_p}{\gamma_2}\right) \end{aligned}$$

which after substituting for $\dot{\epsilon}_1, \dot{\tilde{A}}_p, \dot{\tilde{B}}_p$ becomes

$$\dot{V} = \epsilon_1^T(PA_p + A_p^T P)\epsilon_1 - 2\epsilon_1^T P\tilde{A}_p\hat{x} - 2\epsilon_1^T P\tilde{B}_p u + \text{tr}\left(2\frac{\tilde{A}_p^T PF_1}{\gamma_1} + 2\frac{\tilde{B}_p^T PF_2}{\gamma_2}\right) \quad (4.2.27)$$

We use the following properties of trace to manipulate (4.2.27):

(i) $\text{tr}(AB) = \text{tr}(BA)$
(ii) $\text{tr}(A+B) = \text{tr}(A) + \text{tr}(B)$ for any $A, B \in \mathcal{R}^{n \times n}$
(iii) $\text{tr}(yx^T) = x^T y$ for any $x, y \in \mathcal{R}^{n \times 1}$

We have

$$\epsilon_1^T P\tilde{A}_p\hat{x} = \hat{x}^T \tilde{A}_p^T P\epsilon_1 = \text{tr}(\tilde{A}_p^T P\epsilon_1\hat{x}^T)$$

$$\epsilon_1^T P\tilde{B}_p u = \text{tr}(\tilde{B}_p^T P\epsilon_1 u^T)$$

and, therefore,

$$\dot{V} = -\epsilon_1^T \epsilon_1 + 2\text{tr}\left(\frac{\tilde{A}_p^T PF_1}{\gamma_1} - \tilde{A}_p^T P\epsilon_1\hat{x}^T + \frac{\tilde{B}_p PF_2}{\gamma_2} - \tilde{B}_p^T P\epsilon_1 u^T\right) \quad (4.2.28)$$

The obvious choice for F_1, F_2 to make $\dot{V}$ negative is

$$\dot{\tilde{A}}_p = F_1 = \gamma_1 \epsilon_1 \hat{x}^T, \quad \dot{\tilde{B}}_p = F_2 = \gamma_2 \epsilon_1 u^T \quad (4.2.29)$$

In the case of the series-parallel model, we choose

$$V(\epsilon_1, \tilde{A}_p, \tilde{B}_p) = \epsilon_1^T P\epsilon_1 + \text{tr}\left(\frac{\tilde{A}_p^T P\tilde{A}_p}{\gamma_1}\right) + \text{tr}\left(\frac{\tilde{B}_p^T P\tilde{B}_p}{\gamma_2}\right) \quad (4.2.30)$$

where $P = P^T > 0$ is the solution of the Lyapunov equation

$$A_m^T P + PA_m = -I \quad (4.2.31)$$

By following the same procedure as in the case of the parallel model, we obtain

$$\dot{\tilde{A}}_p = \gamma_1 \epsilon_1 x^T, \quad \dot{\tilde{B}}_p = \gamma_2 \epsilon_1 u^T \quad (4.2.32)$$

4.2. SIMPLE EXAMPLES

Remark 4.2.1 If instead of (4.2.30), we choose

$$V(\epsilon_1, \tilde{A}_p, \tilde{B}_p) = \epsilon_1^\top P \epsilon_1 + \operatorname{tr}\left(\tilde{A}_p^\top \tilde{A}_p\right) + \operatorname{tr}\left(\tilde{B}_p^\top \tilde{B}_p\right) \qquad (4.2.33)$$

where $P = P^\top > 0$ satisfies (4.2.31), we obtain

$$\dot{\hat{A}}_p = P\epsilon_1 x^\top, \quad \dot{\hat{B}}_p = P\epsilon_1 u^\top \qquad (4.2.34)$$

In this case P is the adaptive gain matrix and is calculated using (4.2.31). Because A_m is a known stable matrix, the calculation of P is possible. It should be noted that if we use the same procedure for the parallel model, we will end up with an adaptive law that depends on P that satisfies the Lyapunov equation (4.2.26) for A_p. Because A_p is unknown, P cannot be calculated; therefore, the adaptive laws corresponding to (4.2.34) for the parallel model are not implementable.

The time derivative $\dot{V}$ of V in both the parallel and series-parallel estimators satisfies

$$\dot{V} = -\epsilon_1^\top \epsilon_1 \leq 0$$

which implies that the equilibrium $\hat{A}_{pe} = A_p, \hat{B}_{pe} = B_p, \epsilon_{1e} = 0$ of the respective equations is u.s. Using arguments similar to those used in Section 4.2.2 we establish that $\epsilon_1 \in \mathcal{L}_2, \dot{\epsilon}_1 \in \mathcal{L}_\infty$ and that

$$|\epsilon_1(t)| \to 0, \quad \|\dot{\hat{A}}_p(t)\| \to 0, \quad \|\dot{\hat{B}}_p(t)\| \to 0 \quad as\ t \to \infty$$

The convergence properties of $\hat{A}_p, \hat{B}_p$ to their true values A_p, B_p, respectively, depend on the properties of the input u. As we will discuss in Chapter 5, if u belongs to the class of sufficiently rich inputs, i.e., u has enough frequencies to excite all the modes of the plant, then the vector $[x^\top, u^\top]^\top$ is PE and guarantees that $\hat{A}_p, \hat{B}_p$ converge to A_p, B_p, respectively, exponentially fast.

4.2.4 Remarks

(i) In this section we consider the design of on-line parameter estimators for simple plants that are stable, whose states are accessible for measurement and whose input u is bounded. Because no feedback is used and the plant is not disturbed by any signal other than u, the stability of the plant is not an issue. The main concern, therefore, is the stability properties of the estimator or adaptive law that generates the on-line estimates for the unknown plant parameters.

(ii) For the examples considered, we are able to design on-line parameter estimation schemes that guarantee that the estimation error ϵ_1 converges to zero as $t \to \infty$, i.e., the predicted state $\hat{x}$ approaches that of the plant as $t \to \infty$ and the estimated parameters change more and more slowly as time increases. This result, however, is not sufficient to establish parameter convergence to the true parameter values unless the input signal u is sufficiently rich. To be sufficiently rich, u has to have enough frequencies to excite all the modes of the plant.

(iii) The properties of the adaptive schemes developed in this section rely on the stability of the plant and the boundedness of the plant input u. Consequently, they may not be appropriate for use in connection with control problems where u is the result of feedback and is, therefore, no longer guaranteed to be bounded a priori. In the following sections, we develop on-line parameter estimation schemes that do not rely on the stability of the plant and the boundedness of the plant input.

4.3 Adaptive Laws with Normalization

In Section 4.2, we used several simple examples to illustrate the design of various adaptive laws under the assumption that the full state of the plant is available for measurement, the plant is stable, and the plant input is bounded. In this section, we develop adaptive laws that do not require the plant to be stable or the plant input to be bounded a priori. Such adaptive laws are essential in adaptive control, to be considered in later chapters, where the stability of the plant and the boundedness of the plant input are properties to be proven and, therefore, cannot be assumed to hold a priori.

We begin with simple examples of plants whose inputs and states are not restricted to be bounded and develop adaptive laws using various approaches. These results are then generalized to a wider class of higher-order plants whose output rather than the full state vector is available for measurement.

4.3.1 Scalar Example

Let us consider the simple plant given by the algebraic equation

$$y(t) = \theta^* u(t) \tag{4.3.1}$$

4.3. ADAPTIVE LAWS WITH NORMALIZATION

where u and, therefore, y are piecewise continuous signals but not necessarily bounded, and θ^* is to be estimated by using the measurements of y and u. As in Section 4.2.1, we can generate the estimate $\hat{y}$ of y and the estimation error ϵ_1 as

$$\hat{y} = \theta u$$

$$\epsilon_1 = y - \hat{y} = y - \theta u$$

where $\theta(t)$ is the estimate of θ^* at time t.

Because u and y are not guaranteed to be bounded, the minimization problem

$$\min_\theta J(\theta) = \min_\theta \frac{(y - \theta u)^2}{2}$$

is ill posed and, therefore, the procedure of Section 4.2.1 where $u, y \in \mathcal{L}_\infty$ does not extend to the case where $u, y \notin \mathcal{L}_\infty$. This obstacle is avoided by dividing each side of (4.3.1) with some function referred to as the *normalizing signal* $m > 0$ to obtain

$$\bar{y} = \theta^* \bar{u} \qquad (4.3.2)$$

where

$$\bar{y} = \frac{y}{m}, \quad \bar{u} = \frac{u}{m}$$

are the normalized values of y and u, respectively, and $m^2 = 1 + n_s^2$. The signal n_s is chosen so that $\frac{u}{m} \in \mathcal{L}_\infty$. A straightforward choice for n_s is $n_s = u$, i.e., $m^2 = 1 + u^2$.

Because $\bar{u}, \bar{y} \in \mathcal{L}_\infty$, we can follow the procedure of Section 4.2.1 and develop an adaptive law based on (4.3.2) rather than on (4.3.1) as follows:

The estimated value $\hat{\bar{y}}$ of $\bar{y}$ is generated as

$$\hat{\bar{y}} = \theta \bar{u}$$

and the estimation error as

$$\bar{\epsilon}_1 = \bar{y} - \hat{\bar{y}} = \bar{y} - \theta \bar{u}$$

It is clear that $\bar{\epsilon}_1 = \frac{\epsilon_1}{m} = \frac{y - \theta u}{m}$ and $\hat{\bar{y}} = \hat{y}/m$. The adaptive law for θ is now developed by solving the well-posed minimization problem

$$\min_\theta J(\theta) = \min_\theta \frac{(\bar{y} - \theta \bar{u})^2}{2} = \min_\theta \frac{(y - \theta u)^2}{2m^2}$$

Using the gradient method we obtain

$$\dot{\theta} = \gamma \bar{\epsilon}_1 \bar{u}, \quad \gamma > 0 \tag{4.3.3}$$

or in terms of the unnormalized signals

$$\dot{\theta} = \gamma \frac{\epsilon_1 u}{m^2} \tag{4.3.4}$$

where m may be chosen as $m^2 = 1 + u^2$.

For clarity of presentation, we rewrite (4.3.3) and (4.3.4) as

$$\dot{\theta} = \gamma \epsilon u \tag{4.3.5}$$

where

$$\epsilon = \frac{\epsilon_1}{m^2}$$

We refer to ϵ as the *normalized estimation error*. This notation enables us to unify results based on different approaches and is adopted throughout the book.

Let us now analyze (4.3.5) by rewriting it in terms of the parameter error $\tilde{\theta} = \theta - \theta^*$. Using $\dot{\tilde{\theta}} = \dot{\theta}$ and $\epsilon = \epsilon_1/m^2 = -\frac{\tilde{\theta}u}{m^2} = -\frac{\tilde{\theta}\bar{u}}{m}$, we have

$$\dot{\tilde{\theta}} = -\gamma \bar{u}^2 \tilde{\theta} \tag{4.3.6}$$

We propose the Lyapunov function

$$V(\tilde{\theta}) = \frac{\tilde{\theta}^2}{2\gamma}$$

whose time derivative along the solution of (4.3.6) is given by

$$\dot{V} = -\tilde{\theta}^2 \bar{u}^2 = -\epsilon^2 m^2 \leq 0$$

Hence, $\tilde{\theta}, \theta \in \mathcal{L}_\infty$ and $\epsilon m \in \mathcal{L}_2$. Because $\tilde{\theta}, \bar{u} \in \mathcal{L}_\infty$, it follows from $\epsilon = -\tilde{\theta}\frac{\bar{u}}{m}$ that $\epsilon, \epsilon m \in \mathcal{L}_\infty$. If we now rewrite (4.3.5) as $\dot{\theta} = \gamma \epsilon m \bar{u}$, it follows from $\epsilon m \in \mathcal{L}_\infty \cap \mathcal{L}_2$ and $\bar{u} \in \mathcal{L}_\infty$ that $\dot{\theta} \in \mathcal{L}_\infty \cap \mathcal{L}_2$. Since $\frac{d}{dt}\epsilon m = -\dot{\tilde{\theta}}\bar{u} - \tilde{\theta}\dot{\bar{u}}$ and $\dot{\tilde{\theta}}, \tilde{\theta}, \bar{u} \in \mathcal{L}_\infty$, it follows that for $\dot{\bar{u}} \in \mathcal{L}_\infty$ we have $\frac{d}{dt}(\epsilon m) \in \mathcal{L}_\infty$, which, together with $\epsilon m \in \mathcal{L}_2$, implies that $\epsilon m \to 0$ as $t \to \infty$. This, in turn, implies that $\dot{\tilde{\theta}} \to 0$ as $t \to \infty$.

4.3. ADAPTIVE LAWS WITH NORMALIZATION

The significance of this example is that even in the case of unbounded y, u we are able to develop an adaptive law that guarantees bounded parameter estimates and a speed of adaptation that is bounded in an $\mathcal{L}_2$ and $\mathcal{L}_\infty$ sense (i.e., $\dot{\theta} \in \mathcal{L}_\infty \cap \mathcal{L}_2$).

When $n_s = 0$, i.e., $m = 1$ the adaptive law (4.3.5) becomes the *unnormalized* one considered in Section 4.2. It is obvious that for $m = 1$, in (4.3.5), i.e., $\epsilon = \epsilon_1$, we can still establish parameter boundedness, but we can not guarantee boundedness for $\dot{\theta}$ in an $\mathcal{L}_p$ sense unless $u \in \mathcal{L}_\infty$. As we will demonstrate in Chapter 6, the property that $\dot{\theta} \in \mathcal{L}_2$ is crucial for stability when adaptive laws of the form (4.3.5) are used with control laws based on the certainty equivalence principle to stabilize unknown plants.

4.3.2 First-Order Example

Let us now consider the same plant (4.2.15) given in Section 4.2.2, i.e.,

$$x = \frac{1}{s + a_m}[(a_m - a)x + u] \qquad (4.3.7)$$

where for simplicity we assume that $b = 1$ is known, u is piecewise continuous but not necessarily bounded, and a may be positive or negative. Unlike the example in Section 4.2.2, we make no assumptions about the boundedness of x and u. Our objective is to develop an adaptive law for estimating a on-line using the measurements of x, u. If we adopt the approach of the example in Section 4.2.2, we will have

$$\hat{x} = \frac{1}{s + a_m}[(a_m - \hat{a})x + u], \quad \epsilon_1 = x - \hat{x} \qquad (4.3.8)$$

and

$$\dot{\hat{a}} = -\epsilon_1 x. \qquad (4.3.9)$$

Let us now analyze (4.3.8) and (4.3.9) without using any assumption about the boundedness of x, u. We consider the estimation error equation

$$\dot{\epsilon}_1 = -a_m \epsilon_1 + \tilde{a} x \qquad (4.3.10)$$

where $\tilde{a} \stackrel{\Delta}{=} \hat{a} - a$ and propose the same function

$$V = \frac{\epsilon_1^2}{2} + \frac{\tilde{a}^2}{2} \qquad (4.3.11)$$

as in Section 4.2.2. The time derivative of V along (4.3.9) and (4.3.10) is given by

$$\dot{V} = -a_m \epsilon_1^2 \leq 0 \qquad (4.3.12)$$

Because x is not necessarily bounded, it cannot be treated as an independent bounded function of time in (4.3.10) and, therefore, (4.3.10) cannot be decoupled from (4.3.8). Consequently, (4.3.8) to (4.3.10) have to be considered and analyzed together in $\mathcal{R}^3$, the space of $\epsilon_1, \tilde{a}, \hat{x}$. The chosen function V in (4.3.11) is only positive semidefinite in $\mathcal{R}^3$, which implies that V is not a Lyapunov function; therefore, Theorems 3.4.1 to 3.4.4 cannot be applied. V is, therefore, a Lyapunov-like function, and the properties of V, $\dot{V}$ allow us to draw some conclusions about the behavior of the solution $\epsilon_1(t), \tilde{a}(t)$ without having to apply the Lyapunov Theorems 3.4.1 to 3.4.4. From $V \geq 0$ and $\dot{V} = -a_m \epsilon_1^2 \leq 0$ we conclude that $V \in \mathcal{L}_\infty$, which implies that $\epsilon_1, \tilde{a} \in \mathcal{L}_\infty$, and $\epsilon_1 \in \mathcal{L}_2$. Without assuming $x \in \mathcal{L}_\infty$, however, we cannot establish any bound for $\dot{\tilde{a}}$ in an $\mathcal{L}_p$ sense.

As in Section 4.3.1, let us attempt to use normalization and modify (4.3.9) to achieve bounded speed of adaptation in some sense. The use of normalization is not straightforward in this case because of the dynamics introduced by the transfer function $\frac{1}{s+a_m}$, i.e., dividing each side of (4.3.7) by m may not help because

$$\frac{x}{m} \neq \frac{1}{s+a_m}\left[(a_m - a)\frac{x}{m} + \frac{u}{m}\right]$$

For this case, we propose the error signal

$$\epsilon = x - \hat{x} - \frac{1}{s+a_m}\epsilon n_s^2 = \frac{1}{s+a_m}(\tilde{a}x - \epsilon n_s^2) \qquad (4.3.13)$$

i.e.,

$$\dot{\epsilon} = -a_m \epsilon + \tilde{a}x - \epsilon n_s^2$$

where n_s is a normalizing signal to be designed.

Let us now use the error equation (4.3.13) to develop an adaptive law for $\hat{a}$. We consider the Lyapunov-like function

$$V = \frac{\epsilon^2}{2} + \frac{\tilde{a}^2}{2} \qquad (4.3.14)$$

4.3. ADAPTIVE LAWS WITH NORMALIZATION

whose time derivative along the solution of (4.3.13) is given by

$$\dot{V} = -a_m \epsilon^2 - \epsilon^2 n_s^2 + \tilde{a}\epsilon x + \tilde{a}\dot{\tilde{a}}$$

Choosing

$$\dot{\tilde{a}} = \dot{\hat{a}} = -\epsilon x \qquad (4.3.15)$$

we have

$$\dot{V} = -a_m \epsilon^2 - \epsilon^2 n_s^2 \leq 0$$

which together with (4.3.14) imply $V, \epsilon, \tilde{a} \in \mathcal{L}_\infty$ and $\epsilon, \epsilon n_s \in \mathcal{L}_2$. If we now write (4.3.15) as

$$\dot{\tilde{a}} = -\epsilon m \frac{x}{m}$$

where $m^2 = 1 + n_s^2$ and choose n_s so that $\frac{x}{m} \in \mathcal{L}_\infty$, then $\epsilon m \in \mathcal{L}_2$ (because $\epsilon, \epsilon n_s \in \mathcal{L}_2$) implies that $\dot{\tilde{a}} \in \mathcal{L}_2$. A straightforward choice for n_s is $n_s = x$, i.e., $m^2 = 1 + x^2$.

The effect of n_s can be roughly seen by rewriting (4.3.13) as

$$\dot{\epsilon} = -a_m \epsilon - \epsilon n_s^2 + \tilde{a}x \qquad (4.3.16)$$

and solving for the "quasi" steady-state response

$$\epsilon_s = \frac{\tilde{a}x}{a_m + n_s^2} \qquad (4.3.17)$$

obtained by setting $\dot{\epsilon} \approx 0$ in (4.3.16) and solving for ϵ. Obviously, for $n_s^2 = x^2$, large ϵ_s implies large $\tilde{a}$ independent of the boundedness of x, which, in turn, implies that large ϵ_s carries information about the parameter error $\tilde{a}$ even when $x \notin \mathcal{L}_\infty$. This indicates that n_s may be used to normalize the effect of the possible unbounded signal x and is, therefore, referred to as the *normalizing signal*. Because of the similarity of ϵ_s with the normalized estimation error defined in (4.3.5), we refer to ϵ in (4.3.13), (4.3.16) as the *normalized estimation error* too.

Remark 4.3.1 The normalizing term ϵn_s^2 in (4.3.16) is similar to the nonlinear "damping" term used in the control of nonlinear systems [99]. It makes $\dot{V}$ more negative by introducing the negative term $-\epsilon^2 n_s^2$ in the expression for $\dot{V}$ and helps establish that $\epsilon n_s \in \mathcal{L}_2$. Because $\dot{\hat{a}}$ is

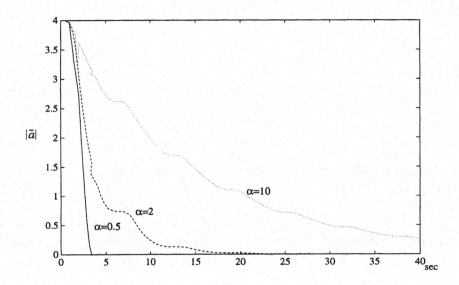

Figure 4.4 Effect of normalization on the convergence and performance of the adaptive law (4.3.15).

bounded from above by $\epsilon\sqrt{n_s^2 + 1} = \epsilon m$ and $\epsilon m \in \mathcal{L}_2$, we can conclude that $\dot{\hat{a}} \in \mathcal{L}_2$, which is a desired property of the adaptive law. Note, however, that $\dot{\hat{a}} \in \mathcal{L}_2$ does not imply that $\dot{\hat{a}} \in \mathcal{L}_\infty$. In contrast to the example in Section 4.3.1, we have not been able to establish that $\dot{\hat{a}} \in \mathcal{L}_\infty$. As we will show in Chapter 6 and 7, the $\mathcal{L}_2$ property of the derivative of the estimated parameters is sufficient to establish stability in the adaptive control case.

Simulations

Let us simulate the effect of normalization on the convergence and performance of the adaptive law (4.3.15) when $a = 0$ is unknown, $u = \sin t$, and $a_m = 2$. We use $n_s^2 = \alpha x^2$ and consider different values of $\alpha \geq 0$. The simulation results are shown in Figure 4.4. It is clear that large values of α lead to a large normalizing signal that slows down the speed of convergence.

4.3. ADAPTIVE LAWS WITH NORMALIZATION

4.3.3 General Plant

Let us now consider the SISO plant

$$\dot{x} = Ax + Bu, \quad x(0) = x_0 \\ y = C^\top x \qquad (4.3.18)$$

where $x \in \mathcal{R}^n$ and only y, u are available for measurement. Equation (4.3.18) may also be written as

$$y = C^\top (sI - A)^{-1} Bu + C^\top (sI - A)^{-1} x_0$$

or as

$$y = \frac{Z(s)}{R(s)} u + \frac{C^\top \operatorname{adj}(sI - A) x_0}{R(s)} \qquad (4.3.19)$$

where $Z(s), R(s)$ are in the form

$$Z(s) = b_{n-1} s^{n-1} + b_{n-2} s^{n-2} + \cdots + b_1 s + b_0$$

$$R(s) = s^n + a_{n-1} s^{n-1} + \cdots + a_1 s + a_0$$

The constants a_i, b_i for $i = 0, 1, \ldots, n-1$ are the plant parameters. A convenient parameterization of the plant that allows us to extend the results of the previous sections to this general case is the one where the unknown parameters are separated from signals and expressed in the form of a linear equation. Several such parameterizations have already been explored and presented in Chapter 2. We summarize them here and refer to Chapter 2 for the details of their derivation.

Let

$$\theta^* = [b_{n-1}, b_{n-2}, \ldots, b_1, b_0, a_{n-1}, a_{n-2}, \ldots, a_1, a_0]^\top$$

be the vector with the unknown plant parameters. The vector θ^* is of dimension $2n$. If some of the coefficients of $Z(s)$ are zero and known, i.e., $Z(s)$ is of degree $m < n - 1$ where m is known, the dimension of θ^* may be reduced. Following the results of Chapter 2, the plant (4.3.19) may take any one of the following parameterizations:

$$z = \theta^{*\top} \phi + \eta_0 \qquad (4.3.20)$$

$$y = \theta_\lambda^{*\top} \phi + \eta_0 \qquad (4.3.21)$$

$$y = W(s)\theta_\lambda^{*\mathsf{T}}\psi + \eta_0 \qquad (4.3.22)$$

where

$$z = W_1(s)y, \quad \phi = H(s)\begin{bmatrix} u \\ y \end{bmatrix}, \quad \psi = H_1(s)\begin{bmatrix} u \\ y \end{bmatrix}, \quad \eta_0 = c_0^{\mathsf{T}} e^{\Lambda_c t} B_0 x_0$$

$$\theta_\lambda^* = \theta^* - b_\lambda$$

$W_1(s), H(s), H_1(s)$ are some known proper transfer function matrices with stable poles, $b_\lambda = [0, \lambda^{\mathsf{T}}]^{\mathsf{T}}$ is a known vector, and Λ_c is a stable matrix which makes η_0 to be an exponentially decaying to zero term that is due to nonzero initial conditions. The transfer function $W(s)$ is a known strictly proper transfer function with relative degree 1, stable poles, and stable zeros.

Instead of dealing with each parametric model separately, we consider the general model

$$z = W(s)\theta^{*\mathsf{T}}\psi + \eta_0 \qquad (4.3.23)$$

where $W(s)$ is a proper transfer function with stable poles, $z \in \mathcal{R}^1, \psi \in \mathcal{R}^{2n}$ are signal vectors available for measurement and $\eta_0 = c_0^{\mathsf{T}} e^{\Lambda_c t} B_0 x_0$. Initially we will assume that $\eta_0 = 0$, i.e.,

$$z = W(s)\theta^{*\mathsf{T}}\psi \qquad (4.3.24)$$

and use (4.3.24) to develop adaptive laws for estimating θ^* on-line. The effect of η_0 and, therefore, of the initial conditions will be treated in Section 4.3.7.

Because θ^* is a constant vector, going from form (4.3.23) to form (4.3.20) is trivial, i.e., rewrite (4.3.23) as $z = \theta^{*\mathsf{T}} W(s)\psi + \eta_0$ and define $\phi = W(s)\psi$. As illustrated in Chapter 2, the parametric model (4.3.24) may also be a parameterization of plants other than the LTI one given by (4.3.18). What is crucial about (4.3.24) is that the unknown vector θ^* appears linearly in an equation where all other signals and parameters are known exactly. For this reason we will refer to (4.3.24) as the *linear parametric model*. In the literature, (3.4.24) has also been referred to as the *linear regression model*.

In the following section we use different techniques to develop adaptive laws for estimating θ^* on-line by assuming that $W(s)$ is a known, proper transfer function with stable poles, and z, ψ are available for measurement.

4.3.4 SPR-Lyapunov Design Approach

This approach dominated the literature of continuous adaptive schemes [48, 149, 150, 153, 172, 178, 187]. It involves the development of a differential equation that relates the estimation or normalized estimation error with the parameter error through an SPR transfer function. Once in this form the KYP or the MKY Lemma is used to choose an appropriate Lyapunov function V whose time derivative $\dot{V}$ is made nonpositive, i.e., $\dot{V} \leq 0$ by properly choosing the differential equation of the adaptive law.

The development of such an error SPR equation had been a challenging problem in the early days of adaptive control [48, 150, 153, 178]. The efforts in those days were concentrated on finding the appropriate transformation or generating the appropriate signals that allow the expression of the estimation/parameter error equation in the desired form.

In this section we use the SPR-Lyapunov design approach to design adaptive laws for estimating θ^* in the parametric model (4.3.24). The connection of the parametric model (4.3.24) with the adaptive control problem is discussed in later chapters. By treating parameter estimation independently of the control design, we manage to separate the complexity of the estimation part from that of the control part. We believe this approach simplifies the design and analysis of adaptive control schemes, to be discussed in later chapters, and helps clarify some of the earlier approaches that appear tricky and complicated to the nonspecialist.

Let us start with the linear parametric model

$$z = W(s)\theta^{*\top}\psi \qquad (4.3.25)$$

Because θ^* is a constant vector, we can rewrite (4.3.25) in the form

$$z = W(s)L(s)\theta^{*\top}\phi \qquad (4.3.26)$$

where

$$\phi = L^{-1}(s)\psi$$

and $L(s)$ is chosen so that $L^{-1}(s)$ is a proper stable transfer function and $W(s)L(s)$ is a proper SPR transfer function.

Remark 4.3.2 For some $W(s)$ it is possible that no $L(s)$ exists such that $W(s)L(s)$ is proper and SPR. In such cases, (4.3.25) could be properly manipulated and put in the form of (4.3.26). For example, when

$W(s) = \frac{s-1}{s+2}$, no $L(s)$ can be found to make $W(s)L(s)$ SPR. In this case, we write (4.3.25) as $\bar{z} = \frac{s+1}{(s+2)(s+3)}\theta^{*T}\phi$ where $\phi = \frac{s-1}{s+1}\psi$ and $\bar{z} = \frac{1}{s+3}z$. The new $W(s)$ in this case is $W(s) = \frac{s+1}{(s+2)(s+3)}$ and a wide class of $L(s)$ can be found so that WL is SPR.

The significance of the SPR property of $W(s)L(s)$ is explained as we proceed with the design of the adaptive law.

Let $\theta(t)$ be the estimate of θ^* at time t. Then the estimate $\hat{z}$ of z at time t is constructed as

$$\hat{z} = W(s)L(s)\theta^T\phi \qquad (4.3.27)$$

As with the examples in the previous section, the estimation error ϵ_1 is generated as

$$\epsilon_1 = z - \hat{z}$$

and the normalized estimation error as

$$\epsilon = z - \hat{z} - W(s)L(s)\epsilon n_s^2 = \epsilon_1 - W(s)L(s)\epsilon n_s^2 \qquad (4.3.28)$$

where n_s is the normalizing signal which we design to satisfy

$$\frac{\phi}{m} \in \mathcal{L}_\infty, \quad m^2 = 1 + n_s^2 \qquad (A1)$$

Typical choices for n_s that satisfy (A1) are $n_s^2 = \phi^T\phi$, $n_s^2 = \phi^T P\phi$ for any $P = P^T > 0$, etc. When $\phi \in \mathcal{L}_\infty$, (A1) is satisfied with $m = 1$, i.e., $n_s = 0$ in which case $\epsilon = \epsilon_1$.

We examine the properties of ϵ by expressing (4.3.28) in terms of the parameter error $\tilde{\theta} \triangleq \theta - \theta^*$, i.e., substituting for z, $\hat{z}$ in (4.3.28) we obtain

$$\epsilon = WL(-\tilde{\theta}^T\phi - \epsilon n_s^2) \qquad (4.3.29)$$

For simplicity, let us assume that $L(s)$ is chosen so that WL is strictly proper and consider the following state space representation of (4.3.29):

$$\begin{aligned} \dot{e} &= A_c e + B_c(-\tilde{\theta}^T\phi - \epsilon n_s^2) \\ \epsilon &= C_c^T e \end{aligned} \qquad (4.3.30)$$

where A_c, B_c, and C_c are the matrices associated with a state space representation that has a transfer function $W(s)L(s) = C_c^T(sI - A_c)^{-1}B_c$.

4.3. ADAPTIVE LAWS WITH NORMALIZATION

The error equation (4.3.30) relates ϵ with the parameter error $\tilde{\theta}$ and is used to construct an appropriate Lyapunov type function for designing the adaptive law of θ. Before we proceed with such a design, let us examine (4.3.30) more closely by introducing the following remark.

Remark 4.3.3 The normalized estimation error ϵ and the parameters A_c, B_c, and C_c in (4.3.30) can be calculated from (4.3.28) and the knowledge of WL, respectively. However, the state error e cannot be measured or generated because of the unknown input $\tilde{\theta}^\top \phi$.

Let us now consider the following Lyapunov-like function for the differential equation (4.3.30):

$$V(\tilde{\theta}, e) = \frac{e^\top P_c e}{2} + \frac{\tilde{\theta}^\top \Gamma^{-1} \tilde{\theta}}{2} \qquad (4.3.31)$$

where $\Gamma = \Gamma^\top > 0$ is a constant matrix and $P_c = P_c^\top > 0$ satisfies the algebraic equations

$$\begin{aligned} P_c A_c + A_c^\top P_c &= -q q^\top - \nu L_c \\ P_c B_c &= C_c \end{aligned} \qquad (4.3.32)$$

for some vector q, matrix $L_c = L_c^\top > 0$ and a small constant $\nu > 0$. Equation (4.3.32) is guaranteed by the SPR property of $W(s)L(s) = C_c^\top(sI - A_c)^{-1} B_c$ and the KYL Lemma if (A_c, B_c, C_c) is minimal or the MKY Lemma if (A_c, B_c, C_c) is nonminimal.

Remark 4.3.4 Because the signal vector ϕ in (4.3.30) is arbitrary and could easily be the state of another differential equation, the function (4.3.31) is not guaranteed to be positive definite in a space that includes ϕ. Hence, V is a Lyapunov-like function.

The time derivative $\dot{V}$ along the solution of (4.3.30) is given by

$$\dot{V}(\tilde{\theta}, e) = -\frac{1}{2} e^\top q q^\top e - \frac{\nu}{2} e^\top L_c e + e^\top P_c B_c [-\tilde{\theta}^\top \phi - \epsilon n_s^2] + \tilde{\theta}^\top \Gamma^{-1} \dot{\tilde{\theta}} \qquad (4.3.33)$$

We now need to choose $\dot{\tilde{\theta}} = \dot{\theta}$ as a function of signals that can be measured so that the indefinite terms in $\dot{V}$ are canceled out. Because e is not available for measurement, $\dot{\theta}$ cannot depend on ϵ explicitly. Therefore, at first glance, it seems that the indefinite term $-e^\top P_c B_c \tilde{\theta}^\top \phi = -\tilde{\theta}^\top \phi e^\top P_c B_c$ cannot be cancelled because the choice $\dot{\theta} = \dot{\tilde{\theta}} = \Gamma \phi e^\top P_c B_c$ is not acceptable due to the presence of the unknown signal e.

Here, however, is where the SPR property of WL becomes handy. We know from (4.3.32) that $P_c B_c = C_c$ which implies that $e^\top P_c B_c = e^\top C_c = \epsilon$. Therefore, (4.3.33) can be written as

$$\dot{V}(\tilde{\theta}, e) = -\frac{1}{2} e^\top q q^\top e - \frac{\nu}{2} e^\top L_c e - \epsilon \tilde{\theta}^\top \phi - \epsilon^2 n_s^2 + \tilde{\theta}^\top \Gamma^{-1} \dot{\tilde{\theta}} \qquad (4.3.34)$$

The choice for $\dot{\tilde{\theta}} = \dot{\theta}$ to make $\dot{V} \leq 0$ is now obvious, i.e., for

$$\dot{\theta} = \dot{\tilde{\theta}} = \Gamma \epsilon \phi \qquad (4.3.35)$$

we have

$$\dot{V}(\tilde{\theta}, e) = -\frac{1}{2} e^\top q q^\top e - \frac{\nu}{2} e^\top L_c e - \epsilon^2 n_s^2 \leq 0 \qquad (4.3.36)$$

which together with (4.3.31) implies that $V, e, \epsilon, \theta, \tilde{\theta} \in \mathcal{L}_\infty$ and that

$$\lim_{t \to \infty} V(\tilde{\theta}(t), e(t)) = V_\infty < \infty$$

Furthermore, it follows from (4.3.36) that

$$\int_0^\infty \epsilon^2 n_s^2 d\tau + \frac{\nu}{2} \int_0^\infty e^\top L_c e d\tau \leq V(\tilde{\theta}(0), e(0)) - V_\infty \qquad (4.3.37)$$

Because $\lambda_{min}(L_c)|e|^2 \leq e^\top L_c e$ and $V(\tilde{\theta}(0), e(0))$ is finite for any finite initial condition, (4.3.37) implies that $\epsilon n_s, e \in \mathcal{L}_2$ and therefore $\epsilon = C_c^\top e \in \mathcal{L}_2$. From the adaptive law (4.3.35), we have

$$|\dot{\theta}| \leq \|\Gamma\| |\epsilon m| \frac{|\phi|}{m}$$

where $m^2 = 1 + n_s^2$. Since $\epsilon^2 m^2 = \epsilon^2 + \epsilon^2 n_s^2$ and $\epsilon, \epsilon n_s \in \mathcal{L}_2$ we have that $\epsilon m \in \mathcal{L}_2$, which together with $\frac{|\phi|}{m} \in \mathcal{L}_\infty$ implies that $\dot{\theta} = \dot{\tilde{\theta}} \in \mathcal{L}_2$.

We summarize the properties of (4.3.35) by the following theorem.

Theorem 4.3.1 *The adaptive law (4.3.35) guarantees that*

4.3. ADAPTIVE LAWS WITH NORMALIZATION

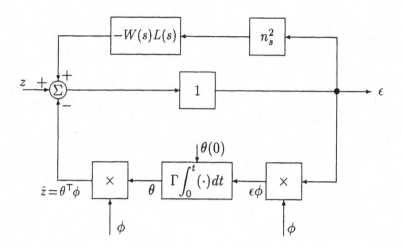

Figure 4.5 Block diagram for implementing adaptive algorithm (4.3.35) with normalized estimation error.

(i) $\theta, \epsilon \in \mathcal{L}_\infty$
(ii) $\epsilon, \epsilon n_s, \dot\theta \in \mathcal{L}_2$

independent of the boundedness properties of ϕ.

Remark 4.3.5 Conditions (i) and (ii) of Theorem 4.3.1 specify the quality of estimation guaranteed by the adaptive law (4.3.35). In Chapters 6 and 7, we will combine (4.3.35) with appropriate control laws to form adaptive control schemes. The stability properties of these schemes depend on the properties (i) and (ii) of the adaptive law.

Remark 4.3.6 The adaptive law (4.3.35) using the normalized estimation error generated by (4.3.28) can be implemented using the block diagram shown in Figure 4.5. When $W(s)L(s) = 1$, the normalized estimation error becomes $\epsilon = \epsilon_1/(1 + n_s^2)$ with $\epsilon_1 = z - \hat{z}$, which is the same normalization used in the gradient algorithm. This result can be obtained using simple block diagram transformation, as illustrated in Figure 4.6.

Remark 4.3.7 The normalizing effect of the signal n_s can be explained by setting $\dot\epsilon = 0$ in (4.3.30) and solving for the "quasi" steady-state

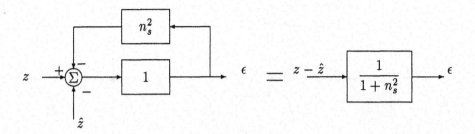

Figure 4.6 Two equivalent block diagrams for generating the normalized estimation error when $W(s)L(s) = 1$.

response ϵ_{ss} of ϵ, i.e.,

$$\epsilon_{ss} = \frac{\alpha(-\tilde{\theta}^\top \phi)}{1 + \alpha n_s^2} = \frac{\epsilon_{1ss}}{1 + \alpha n_s^2} \qquad (4.3.38)$$

where $\alpha = -C_c^\top A_c^{-1} B_c$ is positive, i.e., $\alpha > 0$, because of the SPR property of WL and ϵ_{1ss} is the "quasi" steady state response of ϵ_1. Because of n_s, ϵ_{ss} cannot become unbounded as a result of a possibly unbounded signal ϕ. Large ϵ_{ss} implies that $\tilde{\theta}$ is large; therefore, large ϵ carries information about $\tilde{\theta}$, which is less affected by ϕ.

Remark 4.3.8 The normalizing signal n_s may be chosen as $n_s^2 = \phi^\top \phi$ or as $n_s^2 = \phi^\top P(t) \phi$ where $P(t) = P^\top(t) > 0$ has continuous bounded elements. In general, if we set $n_s = 0$ we cannot establish that $\dot{\theta} \in \mathcal{L}_2$ which, as we show in later chapters, is a crucial property for establishing stability in the adaptive control case. In some special cases, we can afford to set $n_s = 0$ and still establish that $\dot{\theta} \in \mathcal{L}_2$. For example, if $\phi \in \mathcal{L}_\infty$ or if $\theta, \epsilon \in \mathcal{L}_\infty$ implies that $\phi \in \mathcal{L}_\infty$, then it follows from (4.3.35) that $\epsilon \in \mathcal{L}_2 \Rightarrow \dot{\theta} \in \mathcal{L}_2$.

When $n_s^2 = 0$, i.e., $m = 1$, we refer to (4.3.35) as the *unnormalized adaptive law*. In this case $\epsilon = \epsilon_1$ leading to the type of adaptive laws considered in Section 4.2. In later chapters, we show how to use both the normalized and unnormalized adaptive laws in the design of adaptive control schemes.

Another desired property of the adaptive law (4.3.35) is the convergence

4.3. ADAPTIVE LAWS WITH NORMALIZATION

of $\theta(t)$ to the unknown vector θ^*. Such a property is achieved for a special class of vector signals ϕ described by the following definition:

Definition 4.3.1 (Persistence of Excitation (PE)) *A piecewise continuous signal vector $\phi : \mathcal{R}^+ \mapsto \mathcal{R}^n$ is PE in $\mathcal{R}^n$ with a level of excitation $\alpha_0 > 0$ if there exist constants $\alpha_1, T_0 > 0$ such that*

$$\alpha_1 I \geq \frac{1}{T_0} \int_t^{t+T_0} \phi(\tau)\phi^\top(\tau) d\tau \geq \alpha_0 I, \quad \forall t \geq 0 \tag{4.3.39}$$

Although the matrix $\phi(\tau)\phi^\top(\tau)$ is singular for each τ, (4.3.39) requires that $\phi(t)$ varies in such a way with time that the integral of the matrix $\phi(\tau)\phi^\top(\tau)$ is uniformly positive definite over any time interval $[t, t+T_0]$.

If we express (4.3.39) in the scalar form, i.e.,

$$\alpha_1 \geq \frac{1}{T_0} \int_t^{t+T_0} (q^\top \phi(\tau))^2 d\tau \geq \alpha_0, \quad \forall t \geq 0 \tag{4.3.40}$$

where q is any constant vector in $\mathcal{R}^n$ with $|q| = 1$, then the condition can be interpreted as a condition on the energy of ϕ in all directions. The properties of PE signals as well as various other equivalent definitions and interpretations are given in the literature [1, 12, 22, 24, 52, 75, 127, 141, 171, 172, 201, 242].

Corollary 4.3.1 *If $n_s, \phi, \dot{\phi} \in \mathcal{L}_\infty$ and ϕ is PE, then (4.3.35) guarantees that $\theta(t) \to \theta^*$ exponentially fast.*

The proof of Corollary 4.3.1 is long and is given in Section 4.8.

Corollary 4.3.1 is important in the case where parameter convergence is one of the primary objectives of the adaptive system. We use Corollary 4.3.1 in Chapter 5 to establish parameter convergence in parameter identifiers and adaptive observers for stable plants.

The condition that $\dot{\phi}$ appears only in the case of the adaptive laws based on the SPR-Lyapunov approach with $W(s)L(s)$ strictly proper. It is a condition in Lemma 4.8.3 (iii) that is used in the proof of Corollary 4.3.1 in Section 4.8.

The results of Theorem 4.3.1 are also valid when $W(s)L(s)$ is biproper (see Problem 4.4). In fact if $W(s)$ is minimum phase, one may choose $L(s) =$

$W^{-1}(s)$ leading to $W(s)L(s) = 1$. For $WL = 1$, (4.3.28), (4.3.29) become

$$\epsilon = \frac{z - \hat{z}}{m^2} = -\frac{\tilde{\theta}^T \phi}{m^2}$$

where $m^2 = 1 + n_s^2$. In this case we do not need to employ the KYP or MKY Lemma because the Lyapunov-like function

$$V(\tilde{\theta}) = \frac{\tilde{\theta}^T \Gamma^{-1} \tilde{\theta}}{2}$$

leads to

$$\dot{V} = -\epsilon^2 m^2$$

by choosing

$$\dot{\theta} = \Gamma \epsilon \phi \qquad (4.3.41)$$

The same adaptive law as (4.3.41) can be developed by using the gradient method to minimize a certain cost function of ϵ with respect to θ. We discuss this method in the next section.

Example 4.3.1 Let us consider the following signal

$$y = A \sin(\omega t + \varphi)$$

that is broadcasted with a known frequency ω but an unknown phase φ and unknown amplitude A. The signal y is observed through a device with transfer function $W(s)$ that is designed to attenuate any possible noise present in the measurements of y, i.e.,

$$z = W(s)y = W(s)A\sin(\omega t + \varphi) \qquad (4.3.42)$$

For simplicity let us assume that $W(s)$ is an SPR transfer function. Our objective is to use the knowledge of the frequency ω and the measurements of z to estimate A, φ. Because A, φ may assume different constant values at different times we would like to use the results of this section and generate an on-line estimation scheme that provides continuous estimates for A, φ. The first step in our approach is to transform (4.3.42) in the form of the linear parametric model (4.3.24). This is done by using the identity

$$A\sin(\omega t + \varphi) = A_1 \sin \omega t + A_2 \cos \omega t$$

where

$$A_1 = A \cos \varphi, \quad A_2 = A \sin \varphi \qquad (4.3.43)$$

to express (4.3.42) in the form

$$z = W(s)\theta^{*T}\phi$$

4.3. ADAPTIVE LAWS WITH NORMALIZATION

where $\theta^* = [A_1, A_2]^T$ and $\phi = [\sin\omega t, \cos\omega t]^T$.

From the estimate of θ^*, i.e., A_1, A_2, we can calculate the estimate of A, φ by using the relationship (4.3.43). Using the results of this section the estimate $\theta(t)$ of θ^* at each time t is given by

$$\dot{\theta} = \Gamma\epsilon\phi$$

$$\epsilon = z - \hat{z} - W(s)\epsilon n_s^2, \quad \hat{z} = W(s)\theta^T\phi, \quad n_s^2 = \alpha\phi^T\phi$$

where $\theta = [\hat{A}_1, \hat{A}_2]^T$ and $\hat{A}_1, \hat{A}_2$ is the estimate of A_1, A_2, respectively. Since $\phi \in \mathcal{L}_\infty$, the normalizing signal may be taken to be equal to zero, i.e., $\alpha = 0$. The adaptive gain Γ may be chosen as $\Gamma = diag(\gamma)$ for some $\gamma > 0$, leading to

$$\dot{\hat{A}}_1 = \gamma\epsilon\sin\omega t, \quad \dot{\hat{A}}_2 = \gamma\epsilon\cos\omega t \qquad (4.3.44)$$

The above adaptive law guarantees that $\hat{A}_1, \hat{A}_2, \epsilon \in \mathcal{L}_\infty$ and $\epsilon, \dot{\hat{A}}_1, \dot{\hat{A}}_2 \in \mathcal{L}_2$. Since $\phi \in \mathcal{L}_\infty$ we also have that $\dot{\hat{A}}_1, \dot{\hat{A}}_2 \in \mathcal{L}_\infty$. As we mentioned earlier the convergence of $\hat{A}_1, \hat{A}_2$ to A_1, A_2 respectively is guaranteed provided ϕ is PE. We check the PE property of ϕ by using (4.3.39). We have

$$\frac{1}{T_0}\int_t^{t+T_0} \phi(\tau)\phi^T(\tau)d\tau = \frac{1}{T_0}S(t, T_0)$$

where

$$S(t, T_0) \triangleq \begin{bmatrix} \dfrac{T_0}{2} - \dfrac{\sin 2\omega(t+T_0) - \sin 2\omega t}{4\omega} & -\dfrac{\cos 2\omega(t+T_0) - \cos 2\omega t}{4\omega} \\ -\dfrac{\cos 2\omega(t+T_0) - \cos 2\omega t}{4\omega} & \dfrac{T_0}{2} + \dfrac{\sin 2\omega(t+T_0) - \sin 2\omega t}{4\omega} \end{bmatrix}$$

For $T_0 = \frac{2\pi}{\omega}$, we have

$$\frac{1}{T_0}\int_t^{t+T_0} \phi(\tau)\phi^T(\tau)d\tau = \begin{bmatrix} \frac{\pi}{\omega} & 0 \\ 0 & \frac{\pi}{\omega} \end{bmatrix}$$

Hence, the PE condition (4.3.39) is satisfied with $T_0 = 2\pi/\omega$, $0 < \alpha_0 \leq \pi/\omega$, $\alpha_1 \geq \frac{\pi}{\omega}$; therefore, ϕ is PE, which implies that $\hat{A}_1, \hat{A}_2$ converge to A_1, A_2 exponentially fast.

Using (4.3.43), the estimate $\hat{A}, \hat{\varphi}$ of A, φ, respectively, is calculated as follows:

$$\hat{A}(t) = \sqrt{\hat{A}_1^2(t) + \hat{A}_2^2(t)}, \quad \hat{\varphi} = \cos^{-1}\left(\frac{\hat{A}_1(t)}{\hat{A}(t)}\right) \qquad (4.3.45)$$

The calculation of $\hat{\varphi}$ at each time t is possible provided $\hat{A}(t) \neq 0$. This implies that $\hat{A}_1, \hat{A}_2$ should not go through zero at the same time, which is something that cannot be guaranteed by the adaptive law (4.3.44). We know, however, that $\hat{A}_1, \hat{A}_2$

converge to A_1, A_2 exponentially fast, and A_1, A_2 cannot be both equal to zero (otherwise $y \equiv 0\ \forall t \geq 0$). Hence, after some finite time T, $\hat{A}_1, \hat{A}_2$ will be close enough to A_1, A_2 for $\bar{\hat{A}}(T) \neq 0$ to imply $\hat{A}(t) \neq 0\ \forall t \geq T$.

Because $\hat{A}_1, \hat{A}_2 \to A_1, A_2$ exponentially fast, it follows from (4.3.45) that $\hat{A}, \hat{\varphi}$ converge to A, φ exponentially fast.

Let us simulate the above estimation scheme when $W(s) = \frac{2}{s+2}$, $\omega = 2$ rad/sec and the unknown A, φ are taken as $A = 10, \varphi = 16° = 0.279$ rad for $0 \leq t \leq 20$ sec and $A = 7, \varphi = 25° = 0.463$ rad for $t > 20$ sec for simulation purposes. The results are shown in Figure 4.7, where $\gamma = 1$ is used. ▽

Example 4.3.2 Consider the following plant:

$$y = \frac{b_1 s + b_0}{s^2 + 3s + 2} u$$

where b_1, b_0 are the only unknown parameters to be estimated. We rewrite the plant in the form of the parametric model (4.3.24) by first expressing it as

$$y = \frac{1}{(s+1)(s+2)} \theta^{*T} \psi \qquad (4.3.46)$$

where $\theta^* = [b_1, b_0]^T$, $\psi = [\dot{u}, u]^T$. We then choose $L(s) = s + 2$ so that $W(s)L(s) = \frac{1}{s+1}$ is SPR and rewrite (4.3.46) as

$$y = \frac{1}{s+1} \theta^{*T} \phi \qquad (4.3.47)$$

where $\phi = \left[\frac{s}{s+2} u, \frac{1}{s+2} u\right]^T$ can be generated by filtering u. Because (4.3.47) is in the form of parametric model (4.3.24), we can apply the results of this section to obtain the adaptive law

$$\dot{\theta} = \Gamma \epsilon \phi$$

$$\epsilon = y - \frac{1}{s+1}(\theta^T \phi + \epsilon n_s^2), \quad n_s = \alpha \phi^T \phi$$

where $\alpha > 0$ and $\theta = [\hat{b}_1, \hat{b}_0]^T$ is the on-line estimate of θ^*. This example illustrates that the dimensionality of θ, ϕ may be reduced if some of the plant parameters are known. ▽

4.3.5 Gradient Method

Some of the earlier approaches to adaptive control in the early 1960s [20, 34, 96, 104, 115, 123, 175, 220] involved the use of simple optimization techniques

4.3. ADAPTIVE LAWS WITH NORMALIZATION

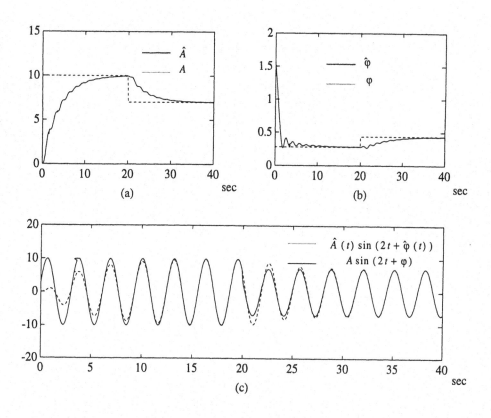

Figure 4.7 Simulation results for Example 4.3.1.

such as the gradient or steepest descent method to minimize a certain performance cost with respect to some adjustable parameters. These approaches led to the development of a wide class of adaptive algorithms that had found wide applications in industry. Despite their success in applications, the schemes of the 1960s lost their popularity because of the lack of stability in a global sense. As a result, starting from the late 1960s and early 1970s, the schemes of the 1960s have been replaced by new schemes that are based on Lyapunov theory. The gradient method, however, as a tool for designing adaptive laws retained its popularity and has been widely used in discrete-time [73] and, to a less extent, continuous-time adaptive systems. In contrast to the schemes of the 1960s, the schemes of the 1970s and 1980s that are based on gradient methods are shown to have global stability properties.

What made the difference with the newer schemes were new formulations of the parameter estimation problem and the selection of different cost functions for minimization.

In this section, we use the gradient method and two different cost functions to develop adaptive laws for estimating θ^* in the parametric model

$$z = W(s)\theta^{*\top}\psi \tag{4.3.24}$$

The use of the gradient method involves the development of an algebraic estimation error equation that motivates the selection of an appropriate cost function $J(\theta)$ that is convex over the space of $\theta(t)$, the estimate of θ^* at time t, for each time t. The function $J(\theta)$ is then minimized with respect to θ for each time t by using the gradient method described in Appendix B. The algebraic error equation is developed as follows:

Because θ^* is constant, the parametric model (4.3.24) can be written in the form

$$z = \theta^{*\top}\phi \tag{4.3.48}$$

where $\phi = W(s)\psi$.

The parametric model (4.3.48) has been the most popular one in discrete time adaptive control. At each time t, (4.3.48) is an algebraic equation where the unknown θ^* appears linearly. Because of the simplicity of (4.3.48), a wide class of recursive adaptive laws may be developed.

Using (4.3.48) the estimate $\hat{z}$ of z at time t is generated as

$$\hat{z} = \theta^\top \phi$$

where $\theta(t)$ is the estimate of θ^* at time t. The normalized estimation error ϵ is then constructed as

$$\epsilon = \frac{z - \hat{z}}{m^2} = \frac{z - \theta^\top \phi}{m^2} \tag{4.3.49}$$

where $m^2 = 1 + n_s^2$ and n_s is the normalizing signal designed so that

$$\frac{\phi}{m} \in \mathcal{L}_\infty \tag{A1}$$

As in Section 4.3.4, typical choices for n_s are $n_s^2 = \phi^\top \phi$, $n_s^2 = \phi^\top P \phi$ for $P = P^\top > 0$, etc.

4.3. ADAPTIVE LAWS WITH NORMALIZATION

For analysis purposes we express ϵ as a function of the parameter error $\tilde{\theta} \triangleq \theta - \theta^*$, i.e., substituting for z in (4.3.49) we obtain

$$\epsilon = -\frac{\tilde{\theta}^T \phi}{m^2} \quad (4.3.50)$$

Clearly the signal $\epsilon m = -\tilde{\theta}^T \frac{\phi}{m}$ is a reasonable measure of the parameter error $\tilde{\theta}$ because for any piecewise continuous signal vector ϕ (not necessarily bounded), large ϵm implies large $\tilde{\theta}$. Several adaptive laws for θ can be generated by using the gradient method to minimize a wide class of cost functions of ϵ with respect to θ. In this section we concentrate on two different cost functions that attracted considerable interest in the adaptive control community.

Instantaneous Cost Function

Let us consider the simple quadratic cost function

$$J(\theta) = \frac{\epsilon^2 m^2}{2} = \frac{(z - \theta^T \phi)^2}{2m^2} \quad (4.3.51)$$

motivated from (4.3.49), (4.3.50), that we like to minimize with respect to θ. Because of the property (A1) of m, $J(\theta)$ is convex over the space of θ at each time t; therefore, the minimization problem is well posed. Applying the gradient method, the minimizing trajectory $\theta(t)$ is generated by the differential equation

$$\dot{\theta} = -\Gamma \nabla J(\theta)$$

where $\Gamma = \Gamma^T > 0$ is a scaling matrix that we refer to as the adaptive gain. From (4.3.51) we have

$$\nabla J(\theta) = -\frac{(z - \theta^T \phi)\phi}{m^2} = -\epsilon \phi$$

and, therefore, the adaptive law for generating $\theta(t)$ is given by

$$\dot{\theta} = \Gamma \epsilon \phi \quad (4.3.52)$$

We refer to (4.3.52) as the *gradient algorithm*.

Remark 4.3.9 The adaptive law (4.3.52) has the same form as (4.3.35) developed using the Lyapunov design approach. As shown in Section 4.3.4, (4.3.52) follows directly from the Lyapunov design method by taking $L(s) = W^{-1}(s)$.

Remark 4.3.10 The convexity of $J(\theta)$ (as explained in Appendix B) guarantees the existence of a single global minimum defined by $\nabla J(\theta) = 0$. Solving $\nabla J(\theta) = -\epsilon\phi = -\frac{z-\theta^T\phi}{m^2}\phi = 0$, i.e., $\phi z = \phi\phi^T \theta$, for θ will give us the nonrecursive gradient algorithm

$$\theta(t) = (\phi\phi^T)^{-1}\phi z$$

provided that $\phi\phi^T$ is nonsingular. For $\phi \in \mathcal{R}^{n\times 1}$ and $n > 1$, $\phi\phi^T$ is always singular, the following nonrecursive algorithm based on N data points could be used:

$$\theta(t) = \left(\sum_{i=1}^{N} \phi(t_i)\phi^T(t_i)\right)^{-1} \sum_{i=1}^{N} \phi(t_i) z(t_i)$$

where $t_i \leq t$, $i = 1, \ldots, N$ are the points in time where the measurements of ϕ and z are taken.

Remark 4.3.11 The minimum of $J(\theta)$ corresponds to $\epsilon = 0$, which implies $\dot{\theta} = 0$ and the end of adaptation. The proof that $\theta(t)$ will converge to a trajectory that corresponds to ϵ being small in some sense is not directly guaranteed by the gradient method. A Lyapunov type of analysis is used to establish such a result as shown in the proof of Theorem 4.3.2 that follows.

Theorem 4.3.2 *The adaptive law (4.3.52) guarantees that*

(i) $\epsilon, \epsilon n_s, \theta, \dot{\theta} \in \mathcal{L}_\infty$
(ii) $\epsilon, \epsilon n_s, \dot{\theta} \in \mathcal{L}_2$

independent of the boundedness of the signal vector ϕ and

(iii) *if $n_s, \phi \in \mathcal{L}_\infty$ and ϕ is PE, then $\theta(t)$ converges exponentially to θ^**

4.3. ADAPTIVE LAWS WITH NORMALIZATION

Proof Because θ^* is constant, $\dot{\tilde{\theta}} = \dot{\theta}$ and from (4.3.52) we have

$$\dot{\tilde{\theta}} = \Gamma \epsilon \phi \qquad (4.3.53)$$

We choose the Lyapunov-like function

$$V(\tilde{\theta}) = \frac{\tilde{\theta}^\top \Gamma^{-1} \tilde{\theta}}{2}$$

Then along the solution of (4.3.53), we have

$$\dot{V} = \tilde{\theta}^\top \phi \epsilon = -\epsilon^2 m^2 \leq 0 \qquad (4.3.54)$$

where the second equality is obtained by substituting $\tilde{\theta}^\top \phi = -\epsilon m^2$ from (4.3.50). Hence, $V, \tilde{\theta} \in \mathcal{L}_\infty$, which, together with (4.3.50), implies that $\epsilon, \epsilon m \in \mathcal{L}_\infty$. In addition, we establish from the properties of $V, \dot{V}$, by applying the same argument as in the previous sections, that $\epsilon m \in \mathcal{L}_2$, which implies that $\epsilon, \epsilon n_s \in \mathcal{L}_2$. Now from (4.3.53) we have

$$|\dot{\tilde{\theta}}| = |\dot{\theta}| \leq \|\Gamma\| |\epsilon m| \frac{|\phi|}{m} \qquad (4.3.55)$$

which together with $\frac{|\phi|}{m} \in \mathcal{L}_\infty$ and $\epsilon m \in \mathcal{L}_2 \cap \mathcal{L}_\infty$ implies that $\dot{\theta} \in \mathcal{L}_2 \cap \mathcal{L}_\infty$ and the proof for (i) and (ii) is complete.

The proof for (iii) is long and more complicated and is given in Section 4.8. □

Remark 4.3.12 The property $V(\tilde{\theta}) \geq 0$ and $\dot{V} \leq 0$ of the Lyapunov-like function implies that $\lim_{t \to \infty} V(\tilde{\theta}(t)) = V_\infty$. This, however, does not imply that $\dot{V}(t)$ goes to zero as $t \to \infty$. Consequently, we cannot conclude that ϵ or ϵm go to zero as $t \to \infty$, i.e., that the steepest descent reaches the global minimum that corresponds to $\nabla J(\theta) = -\epsilon \phi = 0$. If however, $\dot{\phi}/m, \dot{m}/m \in \mathcal{L}_\infty$, we can establish that $\frac{d}{dt}(\epsilon m) \in \mathcal{L}_\infty$, which, together with $\epsilon m \in \mathcal{L}_2$, implies that $\epsilon(t)m(t) \to 0$ as $t \to \infty$. Because $m^2 = 1 + n_s^2$ we have $\epsilon(t) \to 0$ as $t \to \infty$ and from (4.3.55) that $\dot{\theta}(t) \to 0$ as $t \to \infty$. Now $|\nabla J(\theta)| \leq |\epsilon \phi| \leq |\epsilon m| \frac{|\phi|}{m}$, which implies that $|\nabla J(\theta(t))| \to 0$ as $t \to \infty$, i.e., $\theta(t)$ converges to a trajectory that corresponds to a global minimum of $J(\theta)$ asymptotically with time provided $\frac{\dot{\phi}}{m}, \frac{\dot{m}}{m} \in \mathcal{L}_\infty$.

Remark 4.3.13 Even though the form of the gradient algorithm (4.3.52) is the same as that of the adaptive law (4.3.35) based on the SPR-Lyapunov design approach, their properties are different. For example,

(4.3.52) guarantees that $\dot{\theta} \in \mathcal{L}_\infty$, whereas such property has not been shown for (4.3.35).

The speed of convergence of the estimated parameters to their true values, when $n_s, \phi \in \mathcal{L}_\infty$ and ϕ is PE, is characterized in the proof of Theorem 4.3.2 (iii) in Section 4.8. It is shown that

$$\tilde{\theta}^\top(t)\Gamma^{-1}\tilde{\theta}(t) \leq \gamma^n \tilde{\theta}^\top(0)\Gamma^{-1}\tilde{\theta}(0)$$

where $0 \leq t \leq nT_0$, n is an integer and

$$\gamma = 1 - \gamma_1, \quad \gamma_1 = \frac{2\alpha_0 T_0 \lambda_{min}(\Gamma)}{2m_0 + \beta^4 T_0^2 \lambda_{max}^2(\Gamma)}$$

where α_0 is the level of excitation of ϕ, $T_0 > 0$ is the size of the time interval in the PE definition of ϕ, $m_0 = \sup_{t \geq 0} m^2(t)$ and $\beta = \sup_{t \geq 0} |\phi(t)|$. We established that $0 < \gamma < 1$. The smaller the γ, i.e., the larger the γ_1, the faster the parameter error converges to zero. The constants α_0, T_0, β and possibly m_0 are all interdependent because they all depend on $\phi(t)$. It is, therefore, not very clear how to choose $\phi(t)$, if we can, to increase the size of γ_1.

Integral Cost Function

A cost function that attracted some interest in the literature of adaptive systems [108] is the integral cost function

$$J(\theta) = \frac{1}{2}\int_0^t e^{-\beta(t-\tau)}\epsilon^2(t,\tau)m^2(\tau)d\tau \qquad (4.3.56)$$

where $\beta > 0$ is a design constant and

$$\epsilon(t,\tau) = \frac{z(\tau) - \theta^\top(t)\phi(\tau)}{m^2(\tau)}, \quad \epsilon(t,t) = \epsilon \qquad (4.3.57)$$

is the normalized estimation error at time τ based on the estimate $\theta(t)$ of θ^* at time $t \geq \tau$. The design constant β acts as a forgetting factor, i.e., as time t increases the effect of the old data at time $\tau < t$ is discarded exponentially. The parameter $\theta(t)$ is to be chosen at each time t to minimize the integral square of the error on all past data that are discounted exponentially.

4.3. ADAPTIVE LAWS WITH NORMALIZATION

Using (4.3.57), we express (4.3.56) in terms of the parameter θ, i.e.,

$$J(\theta) = \frac{1}{2}\int_0^t e^{-\beta(t-\tau)} \frac{(z(\tau) - \theta^\top(t)\phi(\tau))^2}{m^2(\tau)} d\tau \qquad (4.3.58)$$

Clearly, $J(\theta)$ is convex over the space of θ for each time t and the application of the gradient method for minimizing $J(\theta)$ w.r.t. θ yields

$$\dot{\theta} = -\Gamma \nabla J = \Gamma \int_0^t e^{-\beta(t-\tau)} \frac{(z(\tau) - \theta^\top(t)\phi(\tau))}{m^2(\tau)} \phi(\tau) d\tau \qquad (4.3.59)$$

where $\Gamma = \Gamma^\top > 0$ is a scaling matrix that we refer to as the adaptive gain. Equation (4.3.59) is implemented as

$$\begin{aligned} \dot{\theta} &= -\Gamma(R(t)\theta + Q(t)) \\ \dot{R} &= -\beta R + \frac{\phi\phi^\top}{m^2}, \quad R(0) = 0 \\ \dot{Q} &= -\beta Q - \frac{z\phi}{m^2}, \quad Q(0) = 0 \end{aligned} \qquad (4.3.60)$$

where $R \in \mathcal{R}^{n\times n}, Q \in \mathcal{R}^{n\times 1}$. We refer to (4.3.59) or (4.3.60) as the *integral adaptive law*. Its form is different from that of the previous adaptive laws we developed. The properties of (4.3.60) are also different and are given by the following theorem.

Theorem 4.3.3 *The integral adaptive law (4.3.60) guarantees that*

(i) $\epsilon, \epsilon n_s, \theta, \dot{\theta} \in \mathcal{L}_\infty$
(ii) $\epsilon, \epsilon n_s, \dot{\theta} \in \mathcal{L}_2$
(iii) $\lim_{t\to\infty} |\dot{\theta}(t)| = 0$
(iv) *if $n_s, \phi \in \mathcal{L}_\infty$ and ϕ is PE then $\theta(t)$ converges exponentially to θ^*. Furthermore, for $\Gamma = \gamma I$ the rate of convergence can be made arbitrarily large by increasing the value of the adaptive gain γ.*

Proof Because $\frac{\phi}{m} \in \mathcal{L}_\infty$, it follows that $R, Q \in \mathcal{L}_\infty$ and, therefore, the differential equation for θ behaves as a linear time-varying differential equation with a bounded input. Substituting for $z = \phi^\top \theta^*$ in the differential equation for Q we verify that

$$Q(t) = -\int_0^t e^{-\beta(t-\tau)} \frac{\phi(\tau)\phi^\top(\tau)}{m^2} d\tau \theta^* = -R(t)\theta^* \qquad (4.3.61)$$

and, therefore,
$$\dot{\theta} = \dot{\tilde{\theta}} = -\Gamma R(t)\tilde{\theta} \tag{4.3.62}$$

We analyze (4.3.62) by using the Lyapunov-like function
$$V(\tilde{\theta}) = \frac{\tilde{\theta}^\top \Gamma^{-1} \tilde{\theta}}{2} \tag{4.3.63}$$

whose time derivative along the solution of (4.3.62) is given by
$$\dot{V} = -\tilde{\theta}^\top R(t) \tilde{\theta} \tag{4.3.64}$$

Because $R(t) = R^\top(t) \geq 0 \ \forall t \geq 0$ it follows that $\dot{V} \leq 0$; therefore, $V, \tilde{\theta}, \theta \in \mathcal{L}_\infty$, $(\tilde{\theta}^\top R \tilde{\theta})^{\frac{1}{2}} = |R^{\frac{1}{2}} \tilde{\theta}| \in \mathcal{L}_2$. From $\epsilon = -\frac{\tilde{\theta}^\top \phi}{m^2}$ and $\tilde{\theta}, \frac{\phi}{m} \in \mathcal{L}_\infty$ we conclude that $\epsilon, \epsilon m$ and, therefore, $\epsilon n_s \in \mathcal{L}_\infty$.

From (4.3.62) we have
$$|\dot{\theta}| \leq \|\Gamma R^{\frac{\top}{2}}\| |R^{\frac{1}{2}} \tilde{\theta}| \tag{4.3.65}$$

which together with $R \in \mathcal{L}_\infty$ and $|R^{\frac{1}{2}} \tilde{\theta}| \in \mathcal{L}_\infty \bigcap \mathcal{L}_2$ imply that $\dot{\theta} \in \mathcal{L}_\infty \bigcap \mathcal{L}_2$. Since $\dot{\tilde{\theta}}, \dot{R} \in \mathcal{L}_\infty$, it follows from (4.3.62) that $\ddot{\tilde{\theta}} \in \mathcal{L}_\infty$, which, together with $\dot{\tilde{\theta}} \in \mathcal{L}_2$, implies $\lim_{t\to\infty} |\dot{\theta}(t)| = \lim_{t\to\infty} |\Gamma R(t)\tilde{\theta}(t)| = 0$.

To show that $\epsilon m \in \mathcal{L}_2$ we proceed as follows. We have
$$\frac{d}{dt} \tilde{\theta}^\top R \tilde{\theta} = \epsilon^2 m^2 - 2\tilde{\theta}^\top R \Gamma R \tilde{\theta} - \beta \tilde{\theta}^\top R \tilde{\theta}$$

Therefore,
$$\int_0^t \epsilon^2 m^2 d\tau = \tilde{\theta}^\top R \tilde{\theta} + 2 \int_0^t \tilde{\theta}^\top R \Gamma R \tilde{\theta} d\tau + \beta \int_0^t \tilde{\theta}^\top R \tilde{\theta} d\tau$$

Because $\lim_{t\to\infty}[\tilde{\theta}^\top(t) R(t) \tilde{\theta}(t)] = 0$ and $|R^{\frac{1}{2}} \tilde{\theta}| \in \mathcal{L}_2$ it follows that
$$\lim_{t\to\infty} \int_0^t \epsilon^2 m^2 d\tau = \int_0^\infty \epsilon^2 m^2 d\tau < \infty$$

i.e., $\epsilon m \in \mathcal{L}_2$.

Hence, the proof for (i) to (iii) is complete. The proof for (iv) is given in Section 4.8. □

Remark 4.3.14 In contrast to the adaptive law based on the instantaneous cost, the integral adaptive law guarantees that $\dot{\theta}(t) = -\Gamma \nabla J(\theta(t)) \to 0$ as $t \to \infty$ without any additional conditions on the signal vector ϕ

4.3. ADAPTIVE LAWS WITH NORMALIZATION

and m. In this case, $\theta(t)$ converges to a trajectory that minimizes the integral cost asymptotically with time. As we demonstrated in Chapter 3 using simple examples, the convergence of $\dot\theta(t)$ to a zero vector does not imply that $\theta(t)$ converges to a constant vector.

In the proof of Theorem 4.3.3 (iv) given in Section 4.8, we have established that when $n_s, \phi \in \mathcal{L}_\infty$ and ϕ is PE, the parameter error $\tilde\theta$ satisfies

$$|\tilde\theta(t)| \leq \sqrt{\frac{\lambda_{max}(\Gamma)}{\lambda_{min}(\Gamma)}} |\tilde\theta(T_0)| e^{\frac{-\alpha}{2}(t-T_0)}, \quad \forall t \geq T_0$$

where

$$\alpha = 2\beta_1 e^{-\beta T_0} \lambda_{min}(\Gamma), \quad \beta_1 = \alpha_0 T_0 \alpha_0', \quad \alpha_0' = \sup_t \frac{1}{m^2(t)}$$

and α_0, T_0 are constants in the definition of the PE property of ϕ, i.e., $\alpha_0 > 0$ is the level of excitation and $T_0 > 0$ is the length of the time interval. The size of the constant $\alpha > 0$ indicates the speed with which $|\tilde\theta(t)|$ is guaranteed to converge to zero. The larger the level of excitation α_0, the larger the α is. A large normalizing signal decreases the value of α and may have a negative effect on the speed of convergence. If Γ is chosen as $\Gamma = \gamma I$, then it becomes clear that a larger γ guarantees a faster convergence of $|\tilde\theta(t)|$ to zero.

Example 4.3.3 Let us consider the same problem as in Example 4.3.1. We consider the equation

$$z = W(s)A\sin(\omega t + \varphi) = W(s)\theta^{*T}\phi \qquad (4.3.66)$$

where $\theta^* = [A_1, A_2]^T, \phi = [\sin\omega t, \cos\omega t]^T, A_1 = A\cos\varphi, A_2 = A\sin\varphi$. We need to estimate A, φ using the knowledge of $\phi, \omega, W(s)$ and the measurement of z.

We first express (4.3.66) in the form of the linear parametric model (4.3.48) by filtering ϕ with $W(s)$, i.e.,

$$z = \theta^{*T}\phi_0, \quad \phi_0 = W(s)\phi \qquad (4.3.67)$$

and then obtain the adaptive law for estimating θ^* by applying the results of this section. The gradient algorithm based on the instantaneous cost is given by

$$\dot\theta = \Gamma\epsilon\phi_0, \quad \epsilon = \frac{z - \theta^T\phi_0}{m^2}, \quad m^2 = 1 + \alpha\phi_0^T\phi_0 \qquad (4.3.68)$$

where $\theta = [\hat A_1, \hat A_2]^T$ is the estimate of θ^* and $\alpha \geq 0$. Because $\phi \in \mathcal{L}_\infty$ and $W(s)$ has stable poles, $\phi_0 \in \mathcal{L}_\infty$ and α can be taken to be equal to zero.

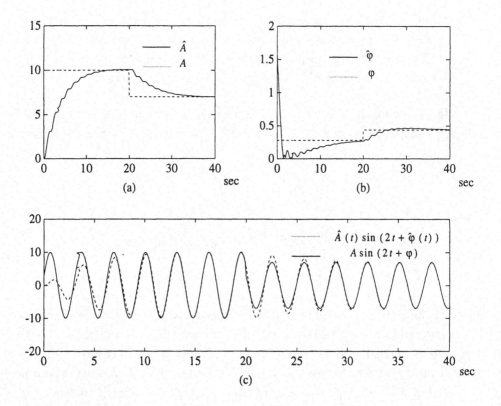

Figure 4.8 Simulation results for Example 4.3.3: Performance of the gradient adaptive law (4.3.68) based on the instantaneous cost.

The gradient algorithm based on the integral cost is given by

$$\begin{aligned}
\dot{\theta} &= -\Gamma(R(t)\theta + Q), \quad \theta(0) = \theta_0 \\
\dot{R} &= -\beta R + \frac{\phi_0 \phi_0^\mathsf{T}}{m^2}, \quad R(0) = 0 \\
\dot{Q} &= -\beta Q - \frac{z\phi_0}{m^2}, \quad Q(0) = 0
\end{aligned} \quad (4.3.69)$$

where $R \in \mathcal{R}^{2\times 2}, Q \in \mathcal{R}^{2\times 1}$. The estimate $\hat{A}, \hat{\varphi}$ of the unknown constants A, φ is calculated from the estimates $\hat{A}_1, \hat{A}_2$ in the same way as in Example 4.3.1. We can establish, as shown in Example 4.3.1, that ϕ_0 satisfies the PE conditions; therefore, both adaptive laws (4.3.68) and (4.3.69) guarantee that θ converges to θ^* exponentially fast.

Let us now simulate (4.3.68) and (4.3.69). We choose $\Gamma = diag(\gamma)$ with $\gamma = 10$ for both algorithms, and $\beta = 0.1$ for (4.3.69). We also use $\omega = 2$ rad/sec and

4.3. ADAPTIVE LAWS WITH NORMALIZATION

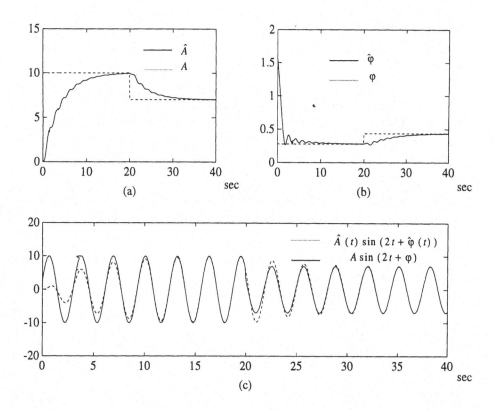

Figure 4.9 Simulation results for Example 4.3.3: Performance of the gradient adaptive law (4.3.69) based on the integral cost.

$W(s) = \frac{2}{s+2}$. Figure 4.8 shows the performance of the adaptive law (4.3.68) based on the instantaneous cost, and Figure 4.9 shows that of (4.3.69) based on the integral cost. ▽

Remark 4.3.15 This example illustrates that for the same estimation problem the gradient method leads to adaptive laws that require more integrators than those required by adaptive laws based on the SPR-Lyapunov design approach. Furthermore, the gradient algorithm based on the integral cost is far more complicated than that based on the instantaneous cost. This complexity is traded off by the better convergence properties of the integral algorithm described in Theorem 4.3.3.

4.3.6 Least-Squares

The least-squares is an old method dating back to Gauss in the eighteenth century where he used it to determine the orbit of planets. The basic idea behind the least-squares is fitting a mathematical model to a sequence of observed data by minimizing the sum of the squares of the difference between the observed and computed data. In doing so, any noise or inaccuracies in the observed data are expected to have less effect on the accuracy of the mathematical model.

The method of least-squares has been widely used in parameter estimation both in a recursive and nonrecursive form mainly for discrete-time systems [15, 52, 73, 80, 127, 144]. The method is simple to apply and analyze in the case where the unknown parameters appear in a linear form, such as in the linear parametric model

$$z = \theta^{*\top} \phi \qquad (4.3.70)$$

Before embarking on the use of least-squares to estimate θ^* in (4.3.70), let us illustrate its use and properties by considering the simple scalar plant

$$y = \theta^* u + d_n \qquad (4.3.71)$$

where d_n is a noise disturbance; $y, u \in \mathcal{R}^+$ and $u \in \mathcal{L}_\infty$. We examine the following estimation problem: Given the measurements of $y(\tau), u(\tau)$ for $0 \leq \tau < t$, find a "good" estimate $\theta(t)$ of θ^* at time t. One possible solution is to calculate $\theta(t)$ from

$$\theta(t) = \frac{y(\tau)}{u(\tau)} = \theta^* + \frac{d_n(\tau)}{u(\tau)}$$

for some $\tau < t$ for which $u(\tau) \neq 0$. Because of the noise disturbance, however, such an estimate may be far off from θ^*. A more natural approach is to generate θ by minimizing the cost function

$$J(\theta) = \frac{1}{2} \int_0^t (y(\tau) - \theta(t) u(\tau))^2 d\tau \qquad (4.3.72)$$

with respect to θ at any given time t. The cost $J(\theta)$ penalizes all the past errors from $\tau = 0$ to t that are due to $\theta(t) \neq \theta^*$. Because $J(\theta)$ is a convex

4.3. ADAPTIVE LAWS WITH NORMALIZATION

function over $\mathcal{R}^1$ at each time t, its minimum satisfies

$$\nabla J(\theta) = -\int_0^t y(\tau)u(\tau)d\tau + \theta(t)\int_0^t u^2(\tau)d\tau = 0 \qquad (4.3.73)$$

for any given time t, which gives

$$\theta(t) = \left(\int_0^t u^2(\tau)d\tau\right)^{-1}\int_0^t y(\tau)u(\tau)d\tau \qquad (4.3.74)$$

provided of course the inverse exists. This is the celebrated *least-squares estimate*. The least-squares method considers all past data in an effort to provide a good estimate for θ^* in the presence of noise d_n. For example, when $u(t) = 1 \; \forall t \geq 0$ and d_n has a zero average value, we have

$$\lim_{t \to \infty} \theta(t) = \lim_{t \to \infty} \frac{1}{t}\int_0^t y(\tau)u(\tau)d\tau = \theta^* + \lim_{t \to \infty} \frac{1}{t}\int_0^t d_n(\tau)d\tau = \theta^*$$

i.e., $\theta(t)$ converges to the exact parameter value despite the presence of the noise disturbance d_n.

Let us now extend this problem to the linear model (4.3.70). As in Section 4.3.5, the estimate $\hat{z}$ of z and the normalized estimation error are generated as

$$\hat{z} = \theta^\top \phi, \qquad \epsilon = \frac{z - \hat{z}}{m^2} = \frac{z - \theta^\top \phi}{m^2} \qquad (4.3.49)$$

where $m^2 = 1 + n_s^2$, $\theta(t)$ is the estimate of θ^* at time t, and m satisfies $\phi/m \in \mathcal{L}_\infty$.

We consider the following cost function

$$J(\theta) = \frac{1}{2}\int_0^t e^{-\beta(t-\tau)}\frac{[z(\tau) - \theta^\top(t)\phi(\tau)]^2}{m^2(\tau)}d\tau + \frac{1}{2}e^{-\beta t}(\theta - \theta_0)^\top Q_0(\theta - \theta_0)$$

(4.3.75)

where $Q_0 = Q_0^\top > 0$, $\beta \geq 0$, $\theta_0 = \theta(0)$, which is a generalization of (4.3.72) to include discounting of past data and a penalty on the initial estimate θ_0 of θ^*. The cost (4.3.75), apart from the additional term that penalizes the initial parameter error, is identical to the integral cost (4.3.58) considered in Section 4.3.5. The method, however, for developing the estimate $\theta(t)$ for θ^* is different. Because $z/m, \phi/m \in \mathcal{L}_\infty$, $J(\theta)$ is a convex function of θ over $\mathcal{R}^n$ at each time t. Hence, any local minimum is also global and satisfies

$$\nabla J(\theta(t)) = 0, \quad \forall t \geq 0$$

i.e.,

$$\nabla J(\theta) = e^{-\beta t} Q_0 (\theta(t) - \theta_0) - \int_0^t e^{-\beta(t-\tau)} \frac{z(\tau) - \theta^\top(t)\phi(\tau)}{m^2(\tau)} \phi(\tau) d\tau = 0$$

which yields the so-called *nonrecursive least-squares* algorithm

$$\theta(t) = P(t) \left[e^{-\beta t} Q_0 \theta_0 + \int_0^t e^{-\beta(t-\tau)} \frac{z(\tau)\phi(\tau)}{m^2(\tau)} d\tau \right] \quad (4.3.76)$$

where

$$P(t) = \left[e^{-\beta t} Q_0 + \int_0^t e^{-\beta(t-\tau)} \frac{\phi(\tau)\phi^\top(\tau)}{m^2(\tau)} d\tau \right]^{-1} \quad (4.3.77)$$

Because $Q_0 = Q_0^\top > 0$ and $\phi \phi^\top$ is positive semidefinite, $P(t)$ exists at each time t. Using the identity

$$\frac{d}{dt} P P^{-1} = \dot{P} P^{-1} + P \frac{d}{dt} P^{-1} = 0$$

we can show that P satisfies the differential equation

$$\dot{P} = \beta P - P \frac{\phi \phi^\top}{m^2} P, \quad P(0) = P_0 = Q_0^{-1} \quad (4.3.78)$$

Therefore, the calculation of the inverse in (4.3.77) is avoided by generating P as the solution of the differential equation (4.3.78). Similarly, differentiating $\theta(t)$ w.r.t. t and using (4.3.78) and $\epsilon m^2 = z - \theta^\top \phi$, we obtain

$$\dot{\theta} = P \epsilon \phi \quad (4.3.79)$$

We refer to (4.3.79) and (4.3.78) as the *continuous-time recursive least-squares algorithm with forgetting factor*.

The stability properties of the least-squares algorithm depend on the value of the forgetting factor β as discussed below.

Pure Least-Squares

In the identification literature, (4.3.79) and (4.3.78) with $\beta = 0$ is referred to as the *"pure" least-squares algorithm* and has a very similar form as the Kalman filter. For this reason, the matrix P is usually called the *covariance matrix*.

4.3. ADAPTIVE LAWS WITH NORMALIZATION

Setting $\beta = 0$, (4.3.78), (4.3.79) become

$$\dot{\theta} = P\epsilon\phi$$
$$\dot{P} = -\frac{P\phi\phi^T P}{m^2}, \quad P(0) = P_0 \qquad (4.3.80)$$

In terms of the P^{-1} we have

$$\frac{d}{dt}P^{-1} = \frac{\phi\phi^T}{m^2}$$

which implies that $\frac{d(P^{-1})}{dt} \geq 0$, and, therefore, P^{-1} may grow without bound. In the matrix case, this means that P may become arbitrarily small and slow down adaptation in some directions. This is the so-called *covariance wind-up* problem that constitutes one of the main drawbacks of the pure least-squares algorithm.

Despite its deficiency, the pure least-squares algorithm has the unique property of guaranteeing parameter convergence to constant values as described by the following theorem:

Theorem 4.3.4 *The pure least-squares algorithm (4.3.80) guarantees that*

(i) $\epsilon, \epsilon n_s, \theta, \dot{\theta}, P \in \mathcal{L}_\infty$.
(ii) $\epsilon, \epsilon n_s, \dot{\theta} \in \mathcal{L}_2$.
(iii) $\lim_{t\to\infty} \theta(t) = \bar{\theta}$, *where* $\bar{\theta}$ *is a constant vector.*
(iv) *If* $n_s, \phi \in \mathcal{L}_\infty$ *and* ϕ *is PE, then* $\theta(t)$ *converges to* θ^* *as* $t \to \infty$.

Proof From (4.3.80) we have that $\dot{P} \leq 0$, i.e., $P(t) \leq P_0$. Because $P(t)$ is nonincreasing and bounded from below (i.e., $P(t) = P^T(t) \geq 0, \forall t \geq 0$) it has a limit, i.e.,

$$\lim_{t\to\infty} P(t) = \bar{P}$$

where $\bar{P} = \bar{P}^T \geq 0$ is a constant matrix. Let us now consider

$$\frac{d}{dt}(P^{-1}\tilde{\theta}) = -P^{-1}\dot{P}P^{-1}\tilde{\theta} + P^{-1}\dot{\tilde{\theta}} = \frac{\phi\phi^T\tilde{\theta}}{m^2} + \epsilon\phi = 0$$

where the last two equalities are obtained by using $\dot{\theta} = \dot{\tilde{\theta}}, \frac{d}{dt}P^{-1} = -P^{-1}\dot{P}P^{-1}$ and $\epsilon = -\frac{\tilde{\theta}^T\phi}{m^2} = -\frac{\phi^T\tilde{\theta}}{m^2}$. Hence, $P^{-1}(t)\tilde{\theta}(t) = P_0^{-1}\tilde{\theta}(0)$, and, therefore, $\tilde{\theta}(t) = P(t)P_0^{-1}\tilde{\theta}(0)$ and $\lim_{t\to\infty}\tilde{\theta}(t) = \bar{P}P_0^{-1}\tilde{\theta}(0)$, which implies that $\lim_{t\to\infty}\theta(t) = \theta^* + \bar{P}P_0^{-1}\tilde{\theta}(0) \triangleq \bar{\theta}$.

Because $P(t) \leq P_0$ and $\tilde{\theta}(t) = P(t)P_0^{-1}\tilde{\theta}(0)$ we have $\theta, \tilde{\theta} \in \mathcal{L}_\infty$, which, together with $\frac{\phi}{m} \in \mathcal{L}_\infty$, implies that $\epsilon m = -\frac{\tilde{\theta}^\top \phi}{m}$ and $\epsilon, \epsilon n_s \in \mathcal{L}_\infty$. Let us now consider the function

$$V(\tilde{\theta}, t) = \frac{\tilde{\theta}^\top P^{-1}(t)\tilde{\theta}}{2}$$

The time derivative $\dot{V}$ of V along the solution of (4.3.80) is given by

$$\dot{V} = \epsilon\tilde{\theta}^\top \phi + \frac{\tilde{\theta}^\top \phi\phi^\top \tilde{\theta}}{2m^2} = -\epsilon^2 m^2 + \frac{\epsilon^2 m^2}{2} = -\frac{\epsilon^2 m^2}{2} \leq 0$$

which implies that $V \in \mathcal{L}_\infty$, $\epsilon m \in \mathcal{L}_2$; therefore, $\epsilon, \epsilon n_s \in \mathcal{L}_2$. From (4.3.80) we have

$$|\dot{\theta}| \leq \|P\|\frac{|\phi|}{m}|\epsilon m|$$

Because $P, \frac{\phi}{m}, \epsilon m \in \mathcal{L}_\infty$ and $\epsilon m \in \mathcal{L}_2$, we have $\dot{\theta} \in \mathcal{L}_\infty \bigcap \mathcal{L}_2$, which completes the proof for (i), (ii), and (iii). The proof of (iv) is given in Section 4.8. $\square$

Remark 4.3.16

(i) We should note that the convergence rate of $\theta(t)$ to θ^* in Theorem 4.3.4 is not guaranteed to be exponential even when ϕ is PE. As shown in the proof of Theorem 4.3.4 (iv) in Section 4.8, $P(t), \tilde{\theta}(t)$ satisfy

$$P(t) \leq \frac{\bar{m}}{(t-T_0)\alpha_0}I, \quad |\tilde{\theta}(t)| \leq \frac{P_0^{-1}\bar{m}}{(t-T_0)\alpha_0}|\tilde{\theta}(0)|, \quad \forall t > T_0$$

where $\bar{m} = \sup_t m^2(t)$, i.e., $|\tilde{\theta}(t)|$ is guaranteed to converge to zero with a speed of $\frac{1}{t}$.

(ii) The convergence of $\theta(t)$ to $\bar{\theta}$ as $t \to \infty$ does not imply that $\dot{\theta}(t) \to 0$ as $t \to \infty$ (see examples in Chapter 3).

(iii) We can establish that $\epsilon, \dot{\theta} \to 0$ as $t \to \infty$ if we assume that $\dot{\phi}/m, \dot{m}/m \in \mathcal{L}_\infty$ as in the case of the gradient algorithm based on the instantaneous cost.

Pure Least-Squares with Covariance Resetting

The so called wind-up problem of the pure least-squares algorithm is avoided by using various modifications that prevent $P(t)$ from becoming singular.

4.3. ADAPTIVE LAWS WITH NORMALIZATION

One such modification is the so-called *covariance resetting* described by

$$\dot{\theta} = P\epsilon\phi$$
$$\dot{P} = -\frac{P\phi\phi^\top P}{m^2}, \quad P(t_r^+) = P_0 = \rho_0 I \quad (4.3.81)$$

where t_r is the time for which $\lambda_{min}(P(t)) \leq \rho_1$ and $\rho_0 > \rho_1 > 0$ are some design scalars. Because of (4.3.81), $P(t) \geq \rho_1 I \ \forall t \geq 0$; therefore, P is guaranteed to be positive definite for all $t \geq 0$.

Strictly speaking, (4.3.81) is no longer the least-squares algorithm that we developed by setting $\nabla J(\theta) = 0$ and $\beta = 0$. It does, however, behave as a pure least-squares algorithm between resetting points. The properties of (4.3.81) are similar to those of the gradient algorithm based on the instantaneous cost. In fact, (4.3.81) may be viewed as a gradient algorithm with time-varying adaptive gain P.

Theorem 4.3.5 *The pure least-squares with covariance resetting algorithm (4.3.81) has the following properties:*

(i) $\epsilon, \epsilon n_s, \theta, \dot{\theta} \in \mathcal{L}_\infty$.
(ii) $\epsilon, \epsilon n_s, \dot{\theta} \in \mathcal{L}_2$.
(iii) *If $n_s, \phi \in \mathcal{L}_\infty$ and ϕ is PE then $\theta(t)$ converges exponentially to θ^*.*

Proof The covariance matrix $P(t)$ has elements that are discontinuous functions of time whose values between discontinuities are defined by the differential equation (4.3.81). At the discontinuity or resetting point t_r, $P(t_r^+) = P_0 = \rho_0 I$; therefore, $P^{-1}(t_r^+) = \rho_0^{-1} I$. Between discontinuities $\frac{d}{dt}P^{-1}(t) \geq 0$, i.e., $P^{-1}(t_2) - P^{-1}(t_1) \geq 0$ $\forall t_2 \geq t_1 \geq 0$ such that $t_r \notin [t_1, t_2]$, which implies that $P^{-1}(t) \geq \rho_0^{-1} I, \forall t \geq 0$. Because of the resetting, $P(t) \geq \rho_1 I, \forall t \geq 0$. Therefore, (4.3.81) guarantees that

$$\rho_0 I \geq P(t) \geq \rho_1 I, \quad \rho_1^{-1} I \geq P^{-1}(t) \geq \rho_0^{-1} I, \quad \forall t \geq 0$$

Let us now consider the function

$$V(\tilde{\theta}) = \frac{\tilde{\theta}^\top P^{-1} \tilde{\theta}}{2} \quad (4.3.82)$$

where P is given by (4.3.81). Because P^{-1} is a bounded positive definite symmetric matrix, it follows that V is decrescent and radially unbounded in the space of $\tilde{\theta}$. Along the solution of (4.3.81) we have

$$\dot{V} = \frac{1}{2}\tilde{\theta}^\top \frac{d(P^{-1})}{dt}\tilde{\theta} + \tilde{\theta}^\top P^{-1}\dot{\tilde{\theta}} = -\epsilon^2 m^2 + \frac{1}{2}\tilde{\theta}^\top \frac{d(P^{-1})}{dt}\tilde{\theta}$$

Between resetting points we have from (4.3.81) that $\frac{d(P^{-1})}{dt} = \frac{\phi\phi^\top}{m^2}$; therefore,

$$\dot{V} = -\epsilon^2 m^2 + \frac{1}{2}\frac{(\tilde{\theta}^\top \phi)^2}{m^2} = -\frac{\epsilon^2 m^2}{2} \leq 0 \qquad (4.3.83)$$

$\forall t \in [t_1, t_2]$ where $[t_1, t_2]$ is any interval in $[0, \infty)$ for which $t_r \notin [t_1, t_2]$.

At the points of discontinuity of P, we have

$$V(t_r^+) - V(t_r) = \frac{1}{2}\tilde{\theta}^\top (P^{-1}(t_r^+) - P^{-1}(t_r))\tilde{\theta}$$

Because $P^{-1}(t_r^+) = \frac{1}{\rho_0}I, P^{-1}(t_r) \geq \frac{1}{\rho_0}I$, it follows that $V(t_r^+) - V(t_r) \leq 0$, which implies that $V \geq 0$ is a nonincreasing function of time for all $t \geq 0$. Hence, $V \in \mathcal{L}_\infty$ and $\lim_{t\to\infty} V(t) = V_\infty < \infty$. Because the points of discontinuities t_r form a set of measure zero, it follows from (4.3.83) that $\epsilon m, \epsilon \in \mathcal{L}_2$. From $V \in \mathcal{L}_\infty$ and $\rho_1^{-1}I \geq P^{-1}(t) \geq \rho_0^{-1}I$ we have $\tilde{\theta} \in \mathcal{L}_\infty$, which implies that $\epsilon, \epsilon m \in \mathcal{L}_\infty$. Using $\epsilon m \in \mathcal{L}_\infty \bigcap \mathcal{L}_2$ and $\rho_0 I \geq P \geq \rho_1 I$ we have $\dot{\theta} \in \mathcal{L}_\infty \bigcap \mathcal{L}_2$ and the proof of (i) and (ii) is, therefore, complete.

The proof of (iii) is very similar to the proof of Theorem 4.3.2 (iii) and is omitted. □

Modified Least-Squares with Forgetting Factor

When $\beta > 0$, the problem of $P(t)$ becoming arbitrarily small in some directions no longer exists. In this case, however, $P(t)$ may grow without bound since $\dot{P}$ may satisfy $\dot{P} > 0$ because $\beta P > 0$ and the fact that $\frac{P\phi\phi^\top P}{m^2}$ is only positive semidefinite.

One way to avoid this complication is to modify the least-squares algorithm as follows:

$$\begin{aligned}\dot{\theta} &= P\epsilon\phi \\ \dot{P} &= \begin{cases} \beta P - \frac{P\phi\phi^\top P}{m^2} & \text{if } \|P(t)\| \leq R_0 \\ 0 & \text{otherwise} \end{cases}\end{aligned} \qquad (4.3.84)$$

where $P(0) = P_0 = P_0^\top > 0$, $\|P_0\| \leq R_0$ and R_0 is a constant that serves as an upper bound for $\|P\|$. This modification guarantees that $P \in \mathcal{L}_\infty$ and is referred to as the *modified least-squares with forgetting factor*. The above algorithm guarantees the same properties as the pure least-squares with covariance resetting given by Theorem 4.3.5. They can be established

4.3. ADAPTIVE LAWS WITH NORMALIZATION

by choosing the same Lyapunov-like function as in (4.3.82) and using the identity $\frac{d P^{-1}}{dt} = -P^{-1}\dot{P}P^{-1}$ to establish

$$\frac{dP^{-1}}{dt} = \begin{cases} -\beta P^{-1} + \frac{\phi\phi^\top}{m^2} & \text{if } \|P\| \leq R_0 \\ 0 & \text{otherwise} \end{cases}$$

where $P^{-1}(0) = P_0^{-1}$, which leads to

$$\dot{V} = \begin{cases} -\frac{\epsilon^2 m^2}{2} - \frac{\beta}{2}\tilde{\theta}^\top P^{-1}\tilde{\theta} & \text{if } \|P\| \leq R_0 \\ -\frac{\epsilon^2 m^2}{2} & \text{otherwise} \end{cases}$$

Because $\dot{V} \leq -\frac{\epsilon^2 m^2}{2} \leq 0$ and $P(t)$ is bounded and positive definite $\forall t \geq 0$, the rest of the analysis is exactly the same as in the proof of Theorem 4.3.5.

Least-Squares with Forgetting Factor and PE

The covariance modifications described above are not necessary when n_s, $\phi \in \mathcal{L}_\infty$ and ϕ is PE. The PE property of ϕ guarantees that over an interval of time, the integral of $-P\frac{\phi\phi^\top}{m^2}P$ is a negative definite matrix that counteracts the effect of the positive definite term βP with $\beta > 0$ in the covariance equation and guarantees that $P \in \mathcal{L}_\infty$. This property is made precise by the following corollary:

Corollary 4.3.2 *If $n_s, \phi \in \mathcal{L}_\infty$ and ϕ is PE then the recursive least-squares algorithm with forgetting factor $\beta > 0$ given by (4.3.78) and (4.3.79) guarantees that $P, P^{-1} \in \mathcal{L}_\infty$ and that $\theta(t)$ converges exponentially to θ^*.*

The proof is presented in Section 4.8.

The use of the recursive least-squares algorithm with forgetting factor with $\phi \in \mathcal{L}_\infty$ and ϕ PE is appropriate in parameter estimation of stable plants where parameter convergence is the main objective. We will address such cases in Chapter 5.

Let us illustrate the design of a least-squares algorithm for the same system considered in Example 4.3.1.

Example 4.3.4 The system

$$z = W(s)A\sin(\omega t + \varphi)$$

where A, φ are to be estimated on-line is rewritten in the form

$$z = \theta^{*\mathsf{T}} \phi_0$$

where $\theta^* = [A_1, A_2]^\mathsf{T}, \phi_0 = W(s)\phi, \phi = [\sin \omega t, \cos \omega t]^\mathsf{T}$. The least-squares algorithm for estimating θ^* is given by

$$\begin{aligned} \dot{\theta} &= P\epsilon\phi_0 \\ \dot{P} &= \beta P - P\frac{\phi_0 \phi_0^\mathsf{T}}{m^2} P, \quad P(0) = \rho_0 I \end{aligned}$$

where $\epsilon = \frac{z - \theta^\mathsf{T} \phi_0}{m^2}, m^2 = 1 + \phi_0^\mathsf{T} \phi_0$ and $\beta \geq 0, \rho_0 > 0$ are design constants. Because ϕ_0 is PE, no modifications are required. Let us simulate the above scheme when $A = 10, \varphi = 16° = 0.279$ rad, $\omega = 2$ rad/sec, $W(s) = \frac{2}{s+2}$. Figure 4.10 gives the time response of $\hat{A}$ and $\hat{\varphi}$, the estimate of A and φ, respectively, for different values of β. The simulation results indicate that the rate of convergence depends on the choice of the forgetting factor β. Larger β leads to faster convergence of $\hat{A}, \hat{\varphi}$ to $A = 10, \varphi = 0.279$, respectively. $\triangledown$

4.3.7 Effect of Initial Conditions

In the previous sections, we developed a wide class of on-line parameter estimators for the linear parametric model

$$z = W(s)\theta^{*\mathsf{T}}\psi + \eta_0 \qquad (4.3.85)$$

where η_0, the exponentially decaying to zero term that is due to initial conditions, is assumed to be equal to zero. As shown in Chapter 2 and in Section 4.3.3, η_0 satisfies the equation

$$\begin{aligned} \dot{\omega}_0 &= \Lambda_c \omega_0, \quad \omega_0(0) = B_0 x_0 \\ \eta_0 &= C_0^\mathsf{T} \omega_0 \end{aligned} \qquad (4.3.86)$$

where Λ_c is a stable matrix, and x_0 is the initial value of the plant state at $t = 0$.

4.3. ADAPTIVE LAWS WITH NORMALIZATION

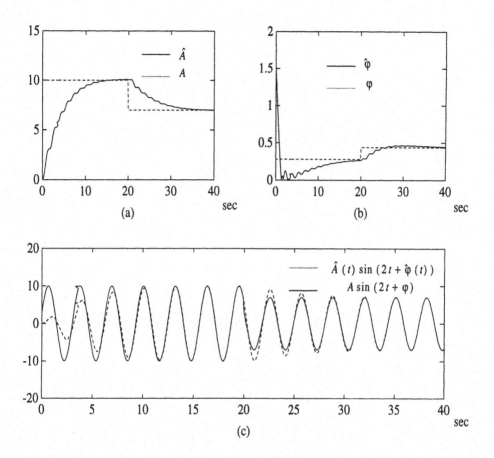

Figure 4.10 Simulation results of Example 4.3.4 for the least-squares algorithm with forgetting factor and PE signals.

Let us analyze the effect of η_0 on the gradient algorithm

$$\begin{aligned}
\dot{\theta} &= \Gamma \epsilon \phi \\
\epsilon &= \frac{z - \hat{z}}{m^2}, \quad \hat{z} = \theta^\top \phi \\
\phi &= W(s)\psi
\end{aligned} \quad (4.3.87)$$

that is developed for the model (4.3.85) with $\eta_0 = 0$ in Section 4.3.5.

We first express (4.3.87) in terms of the parameter error $\tilde{\theta} = \theta - \theta^*$, i.e.,

$$\dot{\tilde{\theta}} = \Gamma \epsilon \phi$$

$$\epsilon = \frac{z - \hat{z}}{m^2} = \frac{-\tilde{\theta}^\top \phi + \eta_0}{m^2} \qquad (4.3.88)$$

It is clear that η_0 acts as a disturbance in the normalized estimation error and, therefore, in the adaptive law for θ. The question that arises now is whether η_0 will affect the properties of (4.3.87) as described by Theorem 4.3.2. We answer this question as follows.

Instead of the Lyapunov-like function

$$V(\tilde{\theta}) = \frac{\tilde{\theta}^\top \Gamma^{-1} \tilde{\theta}}{2}$$

used in the case of $\eta_0 = 0$, we propose the function

$$V(\tilde{\theta}, \omega_0) = \frac{\tilde{\theta}^\top \Gamma^{-1} \tilde{\theta}}{2} + \omega_0^\top P_0 \omega_0$$

where $P_0 = P_0^\top > 0$ satisfies the Lyapunov equation

$$P_0 \Lambda_c + \Lambda_c^\top P_0 = -\gamma_0 I$$

for some $\gamma_0 > 0$ to be chosen. Then along the solution of (4.3.87) we have

$$\dot{V} = \tilde{\theta}^\top \phi \epsilon - \gamma_0 |\omega_0|^2 = -\epsilon^2 m^2 + \epsilon \eta_0 - \gamma_0 |\omega_0|^2$$

Because $\eta_0 = C_0^\top \omega_0$ we have

$$\dot{V} \leq -\epsilon^2 m^2 + |\epsilon| \|C_0^\top\| |\omega_0| - \gamma_0 |\omega_0|^2$$

$$\leq -\frac{\epsilon^2 m^2}{2} - \frac{1}{2}\left(\epsilon m - |C_0^\top| \frac{|\omega_0|}{m}\right)^2 - |\omega_0|^2 \left(\gamma_0 - \frac{|C_0^\top|^2}{2m^2}\right)$$

By choosing $\gamma_0 \geq \frac{|C_0^\top|^2}{2}$ we have

$$\dot{V} \leq -\frac{\epsilon^2 m^2}{2} \leq 0$$

which implies that $\tilde{\theta} \in \mathcal{L}_\infty, \epsilon m \in \mathcal{L}_2$. Because $\eta_0 \in \mathcal{L}_\infty \cap \mathcal{L}_2$ and $\frac{\phi}{m} \in \mathcal{L}_\infty$ we have $\epsilon, \epsilon m, \dot{\theta} \in \mathcal{L}_\infty \cap \mathcal{L}_2$. Hence, (i) and (ii) of Theorem 4.3.2 also hold when $\eta_0 \neq 0$. In a similar manner we can show that $\eta_0 \neq 0$ does not affect (iii) of Theorem 4.3.2. As in every dynamic system, $\eta_0 \neq 0$ will affect the transient response of $\theta(t)$ depending on how fast $\eta_0(t) \to 0$ as $t \to \infty$.

The above procedure can be applied to all the results of the previous sections to establish that initial conditions do not affect the established properties of the adaptive laws developed under the assumption of zero initial conditions.

4.4 Adaptive Laws with Projection

In Section 4.3 we developed a wide class of adaptive laws for estimating the constant vector θ^* that satisfies the linear parametric model

$$z = W(s){\theta^*}^T \psi \qquad (4.4.1)$$

by allowing θ^* to lie anywhere in $\mathcal{R}^n$. In many practical problems where θ^* represents the parameters of a physical plant, we may have some a priori knowledge as to where θ^* is located in $\mathcal{R}^n$. This knowledge usually comes in terms of upper or lower bounds for the elements of θ^* or in terms of a well-defined subset of $\mathcal{R}^n$, etc. One would like to use such a priori information and design adaptive laws that are constrained to search for estimates of θ^* in the set where θ^* is located. Intuitively such a procedure may speed up convergence and reduce large transients that may occur when $\theta(0)$ is chosen to be far away from the unknown θ^*. Another possible reason for constraining $\theta(t)$ to lie in a certain set that contains θ^* arises in cases where $\theta(t)$ is required to have certain properties satisfied by all members of the set so that certain desired calculations that involve $\theta(t)$ are possible. Such cases do arise in adaptive control and will be discussed in later chapters.

We examine how to modify the adaptive laws of Section 4.3 to handle the case of constrained parameter estimation in the following sections.

4.4.1 Gradient Algorithms with Projection

Let us start with the gradient method where the unconstrained minimization of $J(\theta)$ considered in Section 4.3.5 is extended to

$$\text{minimize } J(\theta)$$
$$\text{subject to } \theta \in \mathcal{S} \qquad (4.4.2)$$

where $\mathcal{S}$ is a *convex set with a smooth boundary* almost everywhere. Let $\mathcal{S}$ be given by

$$\mathcal{S} = \{\theta \in \mathcal{R}^n \mid g(\theta) \le 0\} \qquad (4.4.3)$$

where $g : \mathcal{R}^n \mapsto R$ is a smooth function. The solution of the constrained minimization problem follows from the *gradient projection method* discussed in Appendix B and is given by

$$\dot{\theta} = \Pr(-\Gamma \nabla J) \triangleq \begin{cases} -\Gamma \nabla J & \text{if } \theta \in \mathcal{S}^0 \\ & \text{or } \theta \in \delta(\mathcal{S}) \text{ and } -(\Gamma \nabla J)^\top \nabla g \leq 0 \\ -\Gamma \nabla J + \Gamma \frac{\nabla g \nabla g^\top}{\nabla g^\top \Gamma \nabla g} \Gamma \nabla J & \text{otherwise} \end{cases}$$
(4.4.4)

where $\mathcal{S}^0$ is the interior of $\mathcal{S}$, $\delta(\mathcal{S})$ is the boundary of $\mathcal{S}$ and $\theta(0)$ is chosen to be in $\mathcal{S}$, i.e., $\theta(0) \in \mathcal{S}$.

Let us now use (4.4.4) to modify the gradient algorithm $\dot{\theta} = -\Gamma \nabla J(\theta) = \Gamma \epsilon \phi$ given by (4.3.52). Because $\nabla J = -\epsilon \phi$, (4.4.4) becomes

$$\dot{\theta} = \Pr(\Gamma \epsilon \phi) = \begin{cases} \Gamma \epsilon \phi & \text{if } \theta \in \mathcal{S}^0 \\ & \text{or if } \theta \in \delta(\mathcal{S}) \text{ and } (\Gamma \epsilon \phi)^\top \nabla g \leq 0 \\ \Gamma \epsilon \phi - \Gamma \frac{\nabla g \nabla g^\top}{\nabla g^\top \Gamma \nabla g} \Gamma \epsilon \phi & \text{otherwise} \end{cases}$$
(4.4.5)

where $\theta(0) \in \mathcal{S}$.

In a similar manner we can modify the integral adaptive law (4.3.60) by substituting $\nabla J = R(t)\theta + Q(t)$ in (4.4.4).

The principal question we need to ask ourselves at this stage is whether projection will destroy the properties of the unconstrained adaptive laws developed in Section 4.3.5. This question is answered by the following Theorem.

Theorem 4.4.1 *The gradient adaptive laws of Section 4.3.5 with the projection modification given by (4.4.4) retain all their properties that are established in the absence of projection and in addition guarantee that $\theta \in \mathcal{S}$ $\forall t \geq 0$ provided $\theta(0) = \theta_0 \in \mathcal{S}$ and $\theta^* \in \mathcal{S}$.*

Proof It follows from (4.4.4) that whenever $\theta \in \delta(\mathcal{S})$ we have $\dot{\theta}^\top \nabla g \leq 0$, which implies that the vector $\dot{\theta}$ points either inside $\mathcal{S}$ or along the tangent plane of $\delta(\mathcal{S})$ at point θ. Because $\theta(0) = \theta_0 \in \mathcal{S}$, it follows that $\theta(t)$ will never leave $\mathcal{S}$, i.e., $\theta(t) \in \mathcal{S}$ $\forall t \geq 0$.

The adaptive law (4.4.4) has the same form as the one without projection except for the additional term

$$Q = \begin{cases} \Gamma \frac{\nabla g \nabla g^\top}{\nabla g^\top \Gamma \nabla g} \Gamma \nabla J & \text{if } \theta \in \delta(\mathcal{S}) \text{ and } -(\Gamma \nabla J)^\top \nabla g > 0 \\ 0 & \text{otherwise} \end{cases}$$

in the expression for $\dot{\theta}$. If we use the same function V as in the unconstrained case to analyze the adaptive law with projection, the time derivative $\dot{V}$ of V will have

4.4. ADAPTIVE LAWS WITH PROJECTION

the additional term

$$\tilde{\theta}^\top \Gamma^{-1} Q = \begin{cases} \tilde{\theta}^\top \frac{\nabla g \nabla g^\top}{\nabla g^\top \Gamma \nabla g} \Gamma \nabla J & \text{if } \theta \in \delta(\mathcal{S}) \text{ and } -(\Gamma \nabla J)^\top \nabla g > 0 \\ 0 & \text{otherwise} \end{cases}$$

Because of the convex property of $\mathcal{S}$ and the assumption that $\theta^* \in \mathcal{S}$, we have $\tilde{\theta}^\top \nabla g = (\theta - \theta^*)^\top \nabla g \geq 0$ when $\theta \in \delta(\mathcal{S})$. Because $\nabla g^\top \Gamma \nabla J = (\Gamma \nabla J)^\top \nabla g < 0$ for $\theta \in \delta(\mathcal{S})$ and $-(\Gamma \nabla J)^\top \nabla g > 0$, it follows that $\tilde{\theta}^\top \Gamma^{-1} Q \leq 0$. Therefore, the term $\tilde{\theta}^\top \Gamma^{-1} Q$ introduced by the projection can only make $\dot{V}$ more negative and does not affect the results developed from the properties of $V, \dot{V}$. Furthermore, the $\mathcal{L}_2$ properties of $\dot{\theta}$ will not be affected by projection because, with or without projection, $\dot{\theta}$ can be shown to satisfy

$$|\dot{\theta}|^2 \leq c |\Gamma \nabla J|^2$$

for some constant $c \in \mathcal{R}^+$. □

The projection modification (4.4.4) holds also for the adaptive laws based on the SPR-Lyapunov design approach even though these adaptive laws are not derived from the constrained optimization problem defined in (4.4.2). The reason is that the adaptive law (4.3.35) based on the SPR-Lyapunov approach and the gradient algorithm based on the instantaneous cost have the same form, i.e.,

$$\dot{\theta} = \Gamma \epsilon \phi$$

where ϵ, ϕ are, of course, not the same signals in general. Therefore, by substituting for $-\Gamma \nabla J = \Gamma \epsilon \phi$ in (4.4.4), we can obtain the SPR-Lyapunov based adaptive law with projection.

We can establish, as done in Theorem 4.4.1, that the adaptive laws based on the SPR-Lyapunov design approach with the projection modification retain their original properties established in the absence of projection.

Remark 4.4.1 We should emphasize that the set $\mathcal{S}$ is required to be convex. The convexity of $\mathcal{S}$ helps in establishing the fact that the projection does not alter the properties of the adaptive laws established without projection. Projection, however, may affect the transient behavior of the adaptive law.

Let us now consider some examples of constrained parameter estimation.

Example 4.4.1 Consider the linear parametric model

$$z = \theta^{*T}\phi$$

where $\theta^* = [\theta_1^*, \theta_2^*, \ldots, \theta_n^*]^T$ is an unknown constant vector, θ_1^* is known to satisfy $|\theta_1^*| \geq \rho_0 > 0$ for some known constant ρ_0, $\text{sgn}(\theta_1^*)$ is known, and z, ϕ can be measured. We would like to use this *a priori* information about θ_1^* and constrain its estimation to always be inside a convex set $\mathcal{S}$ which contains θ^* and is defined as

$$\mathcal{S} \triangleq \{\theta \in \mathcal{R}^n \mid g(\theta) = \rho_0 - \theta_1 \text{sgn}(\theta_1^*) \leq 0\}$$

The gradient algorithm with projection becomes

$$\dot{\theta} = \begin{cases} \Gamma\epsilon\phi & \text{if } \rho_0 - \theta_1\text{sgn}(\theta_1^*) < 0 \\ & \text{or if } \rho_0 - \theta_1\text{sgn}(\theta_1^*) = 0 \text{ and } (\Gamma\epsilon\phi)^T \nabla g \leq 0 \\ \Gamma\epsilon\phi - \Gamma\frac{\nabla g \nabla g^T}{\nabla g^T \Gamma \nabla g}\Gamma\epsilon\phi & \text{otherwise} \end{cases} \quad (4.4.6)$$

where $\epsilon = \frac{z - \theta^T \phi}{m^2}$ and $\theta_1(0)$ satisfies $\rho_0 - \theta_1(0)\text{sgn}(\theta_1^*) < 0$. For simplicity, let us choose $\Gamma = \text{diag}\{\gamma_1, \Gamma_0\}$ where $\gamma_1 > 0$ is a scalar and $\Gamma_0 = \Gamma_0^T > 0$, and partition ϕ, θ as $\phi = [\phi_1, \phi_0^T]^T$, $\theta = [\theta_1, \theta_0^T]^T$ where $\phi_1, \theta_1 \in \mathcal{R}^1$. Because

$$\nabla g = [-\text{sgn}(\theta_1^*), 0, \ldots, 0]^T$$

it follows from (4.4.6) that

$$\begin{aligned} \dot{\theta}_1 &= \begin{cases} \gamma_1\epsilon\phi_1 & \text{if } \theta_1\text{sgn}(\theta_1^*) > \rho_0 \text{ or} \\ & \text{if } \theta_1\text{sgn}(\theta_1^*) = \rho_0 \text{ and } \gamma_1\phi_1\epsilon\text{sgn}(\theta_1^*) \geq 0 \\ 0 & \text{otherwise} \end{cases} \\ \dot{\theta}_0 &= \Gamma_0\epsilon\phi_0 \end{aligned} \quad (4.4.7)$$

where $\theta_0(0)$ is arbitrary and $\theta_1(0)$ satisfies $\theta_1(0)\text{sgn}(\theta_1^*) > \rho_0$. $\triangledown$

4.4.2 Least-Squares with Projection

The gradient projection method can also be adopted in the case of the least squares algorithm

$$\begin{aligned} \dot{\theta} &= P\epsilon\phi \\ \dot{P} &= \beta P - P\frac{\phi\phi^T}{m^2}P, \quad P(0) = P_0 = Q_0^{-1} \end{aligned} \quad (4.4.8)$$

developed in Section 4.3.6 by viewing (4.4.8) as a gradient algorithm with time varying scaling matrix P and $\epsilon\phi$ as the gradient of some cost function

4.4. ADAPTIVE LAWS WITH PROJECTION

J. If $\mathcal{S} = \{\theta \in \mathcal{R}^n \mid g(\theta) \leq 0\}$ is the convex set for constrained estimation, then (4.4.8) is modified as

$$\dot{\theta} = \Pr(P\epsilon\phi) = \begin{cases} P\epsilon\phi & \text{if } \theta \in \mathcal{S}^0 \\ & \text{or if } \theta \in \delta(\mathcal{S}) \text{ and } (P\epsilon\phi)^\top \nabla g \leq 0 \\ P\epsilon\phi - P\dfrac{\nabla g \nabla g^\top}{\nabla g^\top P \nabla g} P\epsilon\phi & \text{otherwise} \end{cases} \quad (4.4.9)$$

where $\theta(0) \in \mathcal{S}$ and

$$\dot{P} = \begin{cases} \beta P - P\dfrac{\phi\phi^\top}{m^2}P & \text{if } \theta \in \mathcal{S}^0 \\ & \text{or if } \theta \in \delta(\mathcal{S}) \text{ and } (P\epsilon\phi)^\top \nabla g \leq 0 \\ 0 & \text{otherwise} \end{cases} \quad (4.4.10)$$

where $P(0) = P_0 = P_0^\top > 0$.

It can be shown as in Section 4.4.1 that the least-squares with projection has the same properties as the corresponding least-squares without projection.

The equation for the covariance matrix P is modified so that at the point of projection on the boundary of $\mathcal{S}$, P is a constant matrix, and, therefore, the adaptive law at that point is a gradient algorithm with constant scaling that justifies the use of the gradient projection method explained in Appendix B.

Example 4.4.2 Let us now consider a case that often arises in adaptive control in the context of robustness. We would like to constrain the estimates $\theta(t)$ of θ^* to remain inside a bounded convex set $\mathcal{S}$. Let us choose $\mathcal{S}$ as

$$\mathcal{S} = \{\theta \in \mathcal{R}^n \mid \theta^\top \theta - M_0^2 \leq 0\}$$

for some known constant M_0 such that $|\theta^*| \leq M_0$. The set $\mathcal{S}$ represents a sphere in $\mathcal{R}^n$ centered at $\theta = 0$ and of radius M_0. We have

$$\nabla g = 2\theta$$

and the least-squares algorithm with projection becomes

$$\dot{\theta} = \begin{cases} P\epsilon\phi & \text{if } \theta^\top \theta < M_0^2 \\ & \text{or if } \theta^\top \theta = M_0^2 \text{ and } (P\epsilon\phi)^\top \theta \leq 0 \\ \left(I - \dfrac{P\theta\theta^\top}{\theta^\top P\theta}\right) P\epsilon\phi & \text{otherwise} \end{cases} \quad (4.4.11)$$

where $\theta(0)$ satisfies $\theta^\top(0)\theta(0) \leq M_0^2$ and P is given by

$$\dot{P} = \begin{cases} \beta P - P\frac{\phi\phi^\top}{m^2}P & \text{if } \theta^\top\theta < M_0^2 \\ & \text{or if } \theta^\top\theta = M_0^2 \text{ and } (P\epsilon\phi)^\top\theta \leq 0 \\ 0 & \text{otherwise} \end{cases} \quad (4.4.12)$$

Because $P = P^\top > 0$ and $\theta^\top P\theta > 0$ when $\theta^\top\theta = M_0^2$, no division by zero occurs in (4.4.11). $\triangledown$

4.5 Bilinear Parametric Model

As shown in Chapter 2, a certain class of plants can be parameterized in terms of their desired controller parameters that are related to the plant parameters via a Diophantine equation. Such parameterizations and their related estimation problem arise in direct adaptive control, and in particular, direct MRAC, which is discussed in Chapter 6.

In these cases, θ^*, as shown in Chapter 2, appears in the form

$$z = W(s)[\rho^*(\theta^{*\top}\psi + z_0)] \quad (4.5.1)$$

where ρ^* is an unknown constant; z, ψ, z_0 are signals that can be measured and $W(s)$ is a known proper transfer function with stable poles. Because the unknown parameters ρ^*, θ^* appear in a special bilinear form, we refer to (4.5.1) as the *bilinear parametric model*.

The procedure of Section 4.3 for estimating θ^* in a linear model extends to (4.5.1) with minor modifications when the $\text{sgn}(\rho^*)$ is known or when $\text{sgn}(\rho^*)$ and a lower bound ρ_0 of $|\rho^*|$ are known. When the $\text{sgn}(\rho^*)$ is unknown the design and analysis of the adaptive laws require some additional modifications and stability arguments. We treat each case of known and unknown $\text{sgn}(\rho^*), \rho_0$ separately.

4.5.1 Known Sign of ρ^*

The SPR-Lyapunov design approach and the gradient method with an instantaneous cost function discussed in the linear parametric case extend to the bilinear one in a rather straightforward manner.

4.5. BILINEAR PARAMETRIC MODEL

Let us start with the SPR-Lyapunov design approach. We rewrite (4.5.1) in the form

$$z = W(s)L(s)\rho^*(\theta^{*\mathsf{T}}\phi + z_1) \tag{4.5.2}$$

where $z_1 = L^{-1}(s)z_0, \phi = L^{-1}(s)\psi$ and $L(s)$ is chosen so that $L^{-1}(s)$ is proper and stable and WL is proper and SPR. The estimate $\hat{z}$ of z and the normalized estimation error are generated as

$$\hat{z} = W(s)L(s)\rho(\theta^{\mathsf{T}}\phi + z_1) \tag{4.5.3}$$

$$\epsilon = z - \hat{z} - W(s)L(s)\epsilon\, n_s^2 \tag{4.5.4}$$

where n_s is designed to satisfy

$$\frac{\phi}{m}, \frac{z_1}{m} \in \mathcal{L}_\infty, \quad m^2 = 1 + n_s^2 \tag{A2}$$

and $\rho(t), \theta(t)$ are the estimates of ρ^*, θ^* at time t, respectively. Letting $\tilde{\rho} \triangleq \rho - \rho^*, \tilde{\theta} \triangleq \theta - \theta^*$, it follows from (4.5.2) to (4.5.4) that

$$\epsilon = W(s)L(s)[\rho^*\theta^{*\mathsf{T}}\phi - \tilde{\rho}z_1 - \rho\theta^{\mathsf{T}}\phi - \epsilon n_s^2]$$

Now $\rho^*\theta^{*\mathsf{T}}\phi - \rho\theta^{\mathsf{T}}\phi = \rho^*\theta^{*\mathsf{T}}\phi - \rho^*\theta^{\mathsf{T}}\phi + \rho^*\theta^{\mathsf{T}}\phi - \rho\theta^{\mathsf{T}}\phi = -\rho^*\tilde{\theta}^{\mathsf{T}}\phi - \tilde{\rho}\theta^{\mathsf{T}}\phi$ and, therefore,

$$\epsilon = W(s)L(s)[-\rho^*\tilde{\theta}^{\mathsf{T}}\phi - \tilde{\rho}\xi - \epsilon n_s^2], \quad \xi = \theta^{\mathsf{T}}\phi + z_1 \tag{4.5.5}$$

A minimal state representation of (4.5.5) is given by

$$\begin{aligned} \dot{e} &= A_c e + B_c(-\rho^*\tilde{\theta}^{\mathsf{T}}\phi - \tilde{\rho}\xi - \epsilon n_s^2) \\ \epsilon &= C_c^{\mathsf{T}} e \end{aligned} \tag{4.5.6}$$

where $C_c^{\mathsf{T}}(sI - A_c)^{-1}B_c = W(s)L(s)$ is SPR. The adaptive law is now developed by considering the Lyapunov-like function

$$V(\tilde{\theta}, \tilde{\rho}) = \frac{e^{\mathsf{T}} P_c e}{2} + |\rho^*|\frac{\tilde{\theta}^{\mathsf{T}}\Gamma^{-1}\tilde{\theta}}{2} + \frac{\tilde{\rho}^2}{2\gamma}$$

where $P_c = P_c^{\mathsf{T}} > 0$ satisfies the algebraic equations given by (4.3.32) that are implied by the KYL Lemma, and $\Gamma = \Gamma^{\mathsf{T}} > 0, \gamma > 0$. Along the solution of (4.5.6), we have

$$\dot{V} = -\frac{e^{\mathsf{T}} qq^{\mathsf{T}} e}{2} - \frac{\nu}{2}e^{\mathsf{T}} L_c e - \rho^*\epsilon\tilde{\theta}^{\mathsf{T}}\phi - \epsilon\tilde{\rho}\xi - \epsilon^2 n_s^2 + |\rho^*|\tilde{\theta}^{\mathsf{T}}\Gamma^{-1}\dot{\tilde{\theta}} + \frac{\tilde{\rho}\dot{\tilde{\rho}}}{\gamma}$$

where $\nu > 0, L_c = L_c^T > 0$. Because $\rho^* = |\rho^*|\text{sgn}(\rho^*)$ it follows that by choosing

$$\dot{\tilde{\theta}} = \dot{\theta} = \Gamma\epsilon\phi\text{sgn}(\rho^*)$$
$$\dot{\tilde{\rho}} = \dot{\rho} = \gamma\epsilon\xi \qquad (4.5.7)$$

we have

$$\dot{V} = -\frac{e^T qq^T e}{2} - \frac{\nu}{2}e^T L_c e - \epsilon^2 n_s^2 \leq 0$$

The rest of the analysis continues as in the case of the linear model. We summarize the properties of the bilinear adaptive law by the following theorem.

Theorem 4.5.1 *The adaptive law (4.5.7) guarantees that*

(i) $\epsilon, \theta, \rho \in \mathcal{L}_\infty$.
(ii) $\epsilon, \epsilon n_s, \dot{\theta}, \dot{\rho} \in \mathcal{L}_2$.
(iii) *If $\phi, \dot{\phi} \in \mathcal{L}_\infty$, ϕ is PE and $\xi \in \mathcal{L}_2$, then $\theta(t)$ converges to θ^* as $t \to \infty$.*
(iv) *If $\xi \in \mathcal{L}_2$, the estimate ρ converges to a constant $\bar{\rho}$ independent of the properties of ϕ.*

Proof The proof of (i) and (ii) follows directly from the properties of $V, \dot{V}$ by following the same procedure as in the linear parametric model case and is left as an exercise for the reader. The proof of (iii) is established by using the results of Corollary 4.3.1 to show that the homogeneous part of (4.5.6) with $\tilde{\rho}\xi$ treated as an external input together with the equation of $\dot{\tilde{\theta}}$ in (4.5.7) form an e.s. system. Because $\tilde{\rho}\xi \in \mathcal{L}_2$ and A_c is stable, it follows that $e, \tilde{\theta} \to 0$ as $t \to \infty$. The details of the proof are given in Section 4.8. The proof of (iv) follows from $\epsilon, \xi \in \mathcal{L}_2$ and the inequality

$$\int_0^t |\dot{\rho}|d\tau \leq \gamma \int_0^t |\epsilon\xi|d\tau \leq \gamma \left(\int_0^\infty \epsilon^2 d\tau\right)^{\frac{1}{2}} \left(\int_0^\infty \xi^2 d\tau\right)^{\frac{1}{2}} < \infty$$

which implies that $\dot{\rho} \in \mathcal{L}_1$. Therefore, we conclude that $\rho(t)$ has a limit $\bar{\rho}$, i.e., $\lim_{t\to\infty} \rho(t) = \bar{\rho}$. □

The lack of convergence of ρ to ρ^* is due to $\xi \in \mathcal{L}_2$. If, however, ϕ, ξ are such that $\phi_\alpha \triangleq [\phi^T, \xi]^T$ is PE, then we can establish by following the same approach as in the proof of Corollary 4.3.1 that $\tilde{\theta}, \tilde{\rho}$ converge to zero

4.5. BILINEAR PARAMETRIC MODEL

exponentially fast. For $\xi \in \mathcal{L}_2$, the vector ϕ_α cannot be PE even when ϕ is PE.

For the gradient method we rewrite (4.5.1) as

$$z = \rho^*(\theta^{*\top}\phi + z_1) \tag{4.5.8}$$

where $z_1 = W(s)z_0, \phi = W(s)\psi$. Then the estimate $\hat{z}$ of z and the normalized estimation error ϵ are given by

$$\begin{aligned}\hat{z} &= \rho(\theta^\top \phi + z_1) \\ \epsilon &= \frac{z - \hat{z}}{m^2} = \frac{z - \rho(\theta^\top \phi + z_1)}{m^2}\end{aligned} \tag{4.5.9}$$

where n_s^2 is chosen so that

$$\frac{\phi}{m}, \frac{z_1}{m} \in \mathcal{L}_\infty, \quad m^2 = 1 + n_s^2 \tag{A2}$$

As in the case of the linear model, we consider the cost function

$$J(\rho, \theta) = \frac{\epsilon^2 m^2}{2} = \frac{(z - \rho^*\theta^\top \phi - \rho\xi + \rho^*\xi - \rho^*z_1)^2}{2m^2}$$

where $\xi = \theta^\top \phi + z_1$ and the second equality is obtained by using the identity $-\rho(\theta^\top \phi + z_1) = -\rho\xi - \rho^*\theta^\top \phi + \rho^*\xi - \rho^*z_1$. Strictly speaking $J(\rho, \theta)$ is not a convex function of ρ, θ over $\mathcal{R}^{n+1}$ because of the dependence of ξ on θ. Let us, however, ignore this dependence and treat ξ as an independent function of time. Using the gradient method and treating ξ as an arbitrary function of time, we obtain

$$\dot{\theta} = \Gamma_1 \rho^* \epsilon \phi, \quad \dot{\rho} = \gamma \epsilon \xi \tag{4.5.10}$$

where $\Gamma_1 = \Gamma_1^\top > 0, \gamma > 0$ are the adaptive gains. The adaptive law (4.5.10) cannot be implemented due to the unknown ρ^*. We go around this difficulty as follows: Because Γ_1 is arbitrary, we assume that $\Gamma_1 = \frac{\Gamma}{|\rho^*|}$ for some other arbitrary matrix $\Gamma = \Gamma^\top > 0$ and use it together with $\rho^* = |\rho^*|\text{sgn}(\rho^*)$ to get rid of the unknown parameter ρ^*, i.e., $\Gamma_1 \rho^* = \frac{\Gamma}{|\rho^*|}\rho^* = \Gamma\text{sgn}(\rho^*)$ leading to

$$\dot{\theta} = \Gamma\epsilon\phi\,\text{sgn}(\rho^*), \quad \dot{\rho} = \gamma\epsilon\xi \tag{4.5.11}$$

which is implementable. The properties of (4.5.11) are given by the following theorem.

Theorem 4.5.2 *The adaptive law (4.5.11) guarantees that*

(i) $\epsilon, \epsilon n_s, \theta, \dot\theta, \rho, \dot\rho \in \mathcal{L}_\infty$.
(ii) $\epsilon, \epsilon n_s, \dot\theta, \dot\rho \in \mathcal{L}_2$.
(iii) *If* $n_s, \phi \in \mathcal{L}_\infty$, ϕ *is PE and* $\xi \in \mathcal{L}_2$, *then* $\theta(t)$ *converges to* θ^* *as* $t \to \infty$.
(iv) *If* $\xi \in \mathcal{L}_2$, *then* ρ *converges to a constant* $\bar\rho$ *as* $t \to \infty$ *independent of the properties of* ϕ.

The proof follows from that of the linear parametric model and of Theorem 4.5.1, and is left as an exercise for the reader.

The extension of the integral adaptive law and least-squares algorithm to the bilinear parametric model is more complicated and difficult to implement due to the appearance of the unknown ρ^* in the adaptive laws. This problem is avoided by assuming the knowledge of a lower bound for $|\rho^*|$ in addition to $\text{sgn}(\rho^*)$ as discussed in the next section.

4.5.2 Sign of ρ^* and Lower Bound ρ_0 Are Known

The complications with the bilinearity in (4.5.1) are avoided if we rewrite (4.5.1) in the form of the linear parametric model

$$z = \bar\theta^{*\top} \bar\phi \qquad (4.5.12)$$

where $\bar\theta^* = [\bar\theta_1^*, \bar\theta_2^{*\top}]^\top, \bar\phi = [z_1, \phi^\top]^\top$, and $\bar\theta_1^* = \rho^*, \bar\theta_2^* = \rho^*\theta^*$. We can now use the methods of Section 4.3 to generate the estimate $\bar\theta(t)$ of $\bar\theta^*$ at each time t. From the estimate $\bar\theta = [\bar\theta_1, \bar\theta_2^\top]^\top$ of $\bar\theta^*$, we calculate the estimate ρ, θ of ρ^*, θ^* as follows:

$$\rho(t) = \bar\theta_1(t), \quad \theta(t) = \frac{\bar\theta_2(t)}{\bar\theta_1(t)} \qquad (4.5.13)$$

The possibility of division by zero or a small number in (4.5.13) is avoided by constraining the estimate of $\bar\theta_1$ to satisfy $|\bar\theta_1(t)| \geq \rho_0 > 0$ for some $\rho_0 \leq |\rho^*|$. This is achieved by using the gradient projection method and assuming that ρ_0 and $\text{sgn}(\rho^*)$ are known. We illustrate the design of such a gradient algorithm as follows:

By considering (4.5.12) and following the procedure of Section 4.3, we generate

$$\hat z = \bar\theta^\top \bar\phi, \quad \epsilon = \frac{z - \hat z}{m^2} \qquad (4.5.14)$$

4.5. BILINEAR PARAMETRIC MODEL

where $m^2 = 1 + n_s^2$ and n_s is chosen so that $\bar{\phi}/m \in \mathcal{L}_\infty$, e.g. $n_s^2 = \bar{\phi}^\top \bar{\phi}$. The adaptive law is developed by using the gradient projection method to solve the constrained minimization problem

$$\min_{\bar{\theta}} J(\bar{\theta}) = \min_{\bar{\theta}} \frac{(z - \bar{\theta}^\top \bar{\phi})^2}{2m^2}$$

subject to $\rho_0 - \bar{\theta}_1 \text{sgn}(\rho^*) \leq 0$

i.e.,

$$\dot{\bar{\theta}} = \begin{cases} \Gamma \epsilon \bar{\phi} & \text{if } \rho_0 - \bar{\theta}_1 \text{sgn}(\rho^*) < 0 \\ & \text{or if } \rho_0 - \bar{\theta}_1 \text{sgn}(\rho^*) = 0 \text{ and } (\Gamma \epsilon \bar{\phi})^\top \nabla g \leq 0 \\ \Gamma \epsilon \bar{\phi} - \Gamma \frac{\nabla g \nabla g^\top}{\nabla g^\top \Gamma \nabla g} \Gamma \epsilon \bar{\phi} & \text{otherwise} \end{cases}$$

(4.5.15)

where $g(\bar{\theta}) = \rho_0 - \bar{\theta}_1 \text{sgn}(\rho^*)$. For simplicity, let us assume that $\Gamma = diag\{\gamma_1, \Gamma_2\}$ where $\gamma_1 > 0$ is a scalar and $\Gamma_2 = \Gamma_2^\top > 0$ and simplify the expressions in (4.5.15). Because

$$\nabla g = [-\text{sgn}(\rho^*), 0, \ldots, 0]^\top$$

it follows from (4.5.15) that

$$\dot{\bar{\theta}}_1 = \begin{cases} \gamma_1 \epsilon \bar{\phi}_1 & \text{if } \bar{\theta}_1 \text{sgn}(\rho^*) > \rho_0 \\ & \text{or if } \bar{\theta}_1 \text{sgn}(\rho^*) = \rho_0 \text{ and } -\gamma_1 \bar{\phi}_1 \epsilon \text{sgn}(\rho^*) \leq 0 \\ 0 & \text{otherwise} \end{cases}$$

(4.5.16)

where $\bar{\theta}_1(0)$ satisfies $\bar{\theta}_1(0) \text{sgn}(\rho^*) \geq \rho_0$, and

$$\dot{\bar{\theta}}_2 = \Gamma_2 \epsilon \bar{\phi}_2 \qquad (4.5.17)$$

where $\bar{\phi}_1 \stackrel{\triangle}{=} z_1, \bar{\phi}_2 \stackrel{\triangle}{=} \phi$.

Because $\bar{\theta}_1(t)$ is guaranteed by the projection to satisfy $|\bar{\theta}_1(t)| \geq \rho_0 > 0$, the estimate $\rho(t), \theta(t)$ can be calculated using (4.5.13) without the possibility of division by zero. The properties of the adaptive law (4.5.16), (4.5.17) with (4.5.13) are summarized by the following theorem.

Theorem 4.5.3 *The adaptive law described by (4.5.13), (4.5.16), (4.5.17) guarantees that*

(i) $\epsilon, \epsilon n_s, \rho, \theta, \dot{\rho}, \dot{\theta} \in \mathcal{L}_\infty$.

(ii) $\epsilon, \epsilon n_s, \dot{\rho}, \dot{\theta} \in \mathcal{L}_2$.

(iii) If $n_s, \bar{\phi} \in \mathcal{L}_\infty$ and $\bar{\phi}$ is PE, then $\bar{\theta}, \theta, \rho$ converge to $\bar{\theta}^*, \theta^*, \rho^*$, respectively, exponentially fast.

Proof Consider the Lyapunov-like function

$$V = \frac{\tilde{\bar{\theta}}_1^2}{2\gamma_1} + \frac{\tilde{\bar{\theta}}_2^T \Gamma_2^{-1} \tilde{\bar{\theta}}_2}{2}$$

where $\tilde{\bar{\theta}}_1 \triangleq \bar{\theta}_1 - \bar{\theta}_1^*$, $\tilde{\bar{\theta}}_2 \triangleq \bar{\theta}_2 - \bar{\theta}_2^*$. Then along the solution of (4.5.16), (4.5.17), we have

$$\dot{V} = \begin{cases} -\epsilon^2 m^2 & \text{if } \bar{\theta}_1 \text{sgn}(\rho^*) > \rho_0 \\ & \text{or if } \bar{\theta}_1 \text{sgn}(\rho^*) = \rho_0 \text{ and } -\gamma_1 \bar{\phi}_1 \epsilon \text{sgn}(\rho^*) \le 0 \\ \tilde{\bar{\theta}}_2^T \bar{\phi}_2 \epsilon & \text{if } \bar{\theta}_1 \text{sgn}(\rho^*) = \rho_0 \text{ and } -\gamma_1 \bar{\phi}_1 \epsilon \text{sgn}(\rho^*) > 0 \end{cases} \quad (4.5.18)$$

Because $\epsilon m^2 = -\tilde{\bar{\theta}}^T \bar{\phi} = -\tilde{\bar{\theta}}_1 \bar{\phi}_1 - \tilde{\bar{\theta}}_2^T \bar{\phi}_2$, we have $\tilde{\bar{\theta}}_2^T \bar{\phi}_2 \epsilon = -\epsilon^2 m^2 - \tilde{\bar{\theta}}_1 \epsilon \bar{\phi}_1$. For $\bar{\theta}_1 \text{sgn}(\rho^*) = \rho_0$ (i.e., $\bar{\theta}_1 = \rho_0 \text{sgn}(\rho^*)$) and $-\gamma_1 \bar{\phi}_1 \epsilon \text{sgn}(\rho^*) > 0$, we have $\tilde{\bar{\theta}}_1 \bar{\phi}_1 \epsilon = (\rho_0 \text{sgn}(\rho^*) - |\rho^*| \text{sgn}(\rho^*)) \bar{\phi}_1 \epsilon = (\rho_0 - |\rho^*|) \text{sgn}(\rho^*) \bar{\phi}_1 \epsilon > 0$ (because $\rho_0 - |\rho^*| < 0$ and $\text{sgn}(\rho^*) \bar{\phi}_1 \epsilon < 0$), which implies that

$$\epsilon \tilde{\bar{\theta}}_2^T \bar{\phi}_2 = -\epsilon^2 m^2 - \epsilon \bar{\phi}_1 (\rho_0 - |\rho^*|) \text{sgn}(\rho^*) < -\epsilon^2 m^2$$

Therefore, projection introduces the additional term $-\epsilon \bar{\phi}_1 (\rho_0 - |\rho^*|) \text{sgn}(\rho^*)$ that can only make $\dot{V}$ more negative. Substituting for $\epsilon \tilde{\bar{\theta}}_2^T \bar{\phi}_2 < -\epsilon^2 m^2$ in (4.5.18) we have

$$\dot{V} \le -\epsilon^2 m^2$$

which implies that $\bar{\theta}_1, \bar{\theta}_2 \in \mathcal{L}_\infty$; $\epsilon, \epsilon n_s \in \mathcal{L}_\infty \cap \mathcal{L}_2$.

Because $\bar{\phi}/m \in \mathcal{L}_\infty$ and $\epsilon m \in \mathcal{L}_\infty \cap \mathcal{L}_2$, it follows from (4.5.16), (4.5.17) that $\dot{\bar{\theta}}_i \in \mathcal{L}_\infty \cap \mathcal{L}_2$, $i = 1, 2$.

Using (4.5.13), we have $\dot{\rho} = \dot{\bar{\theta}}_1$, $\dot{\theta} = \frac{\dot{\bar{\theta}}_2}{\bar{\theta}_1} - \frac{\dot{\bar{\theta}}_1 \bar{\theta}_2}{\bar{\theta}_1^2}$, which, together with $\dot{\bar{\theta}}_i \in \mathcal{L}_\infty \cap \mathcal{L}_2, i = 1, 2, |\bar{\theta}_1| \ge \rho_0 > 0$, imply that $\dot{\rho}, \dot{\theta} \in \mathcal{L}_\infty \cap \mathcal{L}_2$.

The convergence of $\bar{\theta}$ to $\bar{\theta}^*$ follows from that of Theorem 4.3.2 (iii) and the fact that projection can only make $\dot{V}$ more negative. The convergence of θ, ρ to θ^*, ρ^* follows from that of $\bar{\theta}^*$, equation (4.5.13), assumption $|\bar{\theta}_1^*| \ge \rho_0$ and $|\bar{\theta}_1(t)| > \rho_0$ $\forall t \ge 0$. $\square$

In a similar manner one may use (4.5.12) and (4.5.13) to derive adaptive laws using the integral cost function and least-squares with $\bar{\theta}_1$ constrained to satisfy $|\bar{\theta}_1(t)| \ge \rho_0 > 0$ $\forall t \ge 0$.

4.5.3 Unknown Sign of ρ^*

The problem of designing adaptive laws for the bilinear model (4.5.1) with sgn(ρ^*) unknown was motivated by Morse [155] in the context of MRAC where such an estimation problem arises. Morse conjectured that the sgn(ρ^*) is necessary for designing appropriate adaptive control laws used to solve the stabilization problem in MRAC. Nussbaum [179] used a simple example to show that although Morse's conjecture was valid for a class of adaptive control laws, the sgn(ρ^*) is not necessary for stabilization in MRAC if a different class of adaptive or control laws is employed. This led to a series of results on MRAC [137, 156, 157, 167, 236] where the sgn(ρ^*) was no longer required to be known.

In our case, we use the techniques of [179] to develop adaptive laws for the bilinear model (4.5.1) that do not require the knowledge of sgn(ρ^*). The design and analysis of these adaptive laws is motivated purely from stability arguments that differ from those used when sgn(ρ^*) is known.

We start with the parametric model

$$z = \rho^*(\theta^{*\top}\phi + z_1)$$

and generate $\hat{z}, \epsilon$ as

$$\hat{z} = N(x)\rho(\theta^\top\phi + z_1), \quad \epsilon = \frac{z - \hat{z}}{m^2}$$

where $m^2 = 1 + n_s^2$ is designed so that $\frac{\phi}{m}, \frac{z_1}{m} \in \mathcal{L}_\infty$,

$$N(x) = x^2 \cos x \qquad (4.5.19)$$

x is generated from

$$x = w + \frac{\rho^2}{2\gamma}; \quad \dot{w} = \epsilon^2 m^2, \quad w(0) = 0 \qquad (4.5.20)$$

where $\gamma > 0$. The following adaptive laws are proposed for generating ρ, θ:

$$\dot{\theta} = N(x)\Gamma\epsilon\phi, \quad \dot{\rho} = N(x)\gamma\epsilon\xi \qquad (4.5.21)$$

where $\xi = \theta^\top\phi + z_1$. The function $N(x)$ plays the role of an adaptive gain in (4.5.21) and has often been referred to as the *Nussbaum gain*. Roughly speaking, $N(x)$ accounts for the unknown sign of ρ^* by changing the sign of the vector field of θ, ρ periodically with respect to the signal x.

The design of (4.5.19) to (4.5.21) is motivated from the analysis that is given in the proof of the following theorem, which states the properties of (4.5.19) to (4.5.21).

Theorem 4.5.4 *The adaptive law (4.5.19) to (4.5.21) guarantees that*

(i) $\quad x, w, \theta, \rho \in \mathcal{L}_\infty$

(ii) $\quad \epsilon, \epsilon n_s, \dot{\theta}, \dot{\rho} \in \mathcal{L}_\infty \cap \mathcal{L}_2$

Proof We start by expressing the normalized estimation error ϵ in terms of the parameter error $\tilde{\theta}$, i.e.,

$$\begin{aligned}
\epsilon m^2 &= z - \hat{z} \\
&= \rho^* \theta^{*\top}\phi + \rho^* z_1 - \rho^* \theta^\top \phi - \rho^* z_1 + \rho^* \xi - N(x)\rho\xi \\
&= -\rho^* \tilde{\theta}^\top \phi + \rho^* \xi - N(x)\rho\xi \quad (4.5.22)
\end{aligned}$$

We choose the function

$$V = \frac{\tilde{\theta}^\top \Gamma^{-1} \tilde{\theta}}{2}$$

whose time derivative along the solution of (4.5.21) is given by

$$\dot{V} = \tilde{\theta}^\top \phi \epsilon N(x) \quad (4.5.23)$$

Substituting $\tilde{\theta}^\top \phi = \frac{1}{\rho^*}[-\epsilon m^2 + \rho^* \xi - N(x)\rho\xi]$ from (4.5.22) into (4.5.23), we obtain

$$\dot{V} = -\frac{N(x)}{\rho^*}[\epsilon^2 m^2 - \rho^* \epsilon \xi + N(x)\rho\epsilon\xi]$$

or

$$\dot{V} = -\frac{N(x)}{\rho^*}\left[\epsilon^2 m^2 + \frac{\rho\dot{\rho}}{\gamma}\right] + \frac{\dot{\rho}}{\gamma}$$

From (4.5.20), we have $\dot{x} = \epsilon^2 m^2 + \frac{\rho\dot{\rho}}{\gamma}$ which we use to rewrite $\dot{V}$ as

$$\dot{V} = -\frac{N(x)\dot{x}}{\rho^*} + \frac{\dot{\rho}}{\gamma} \quad (4.5.24)$$

Integrating (4.5.24) on both sides we obtain

$$V(t) - V(0) = \frac{\rho(t) - \rho(0)}{\gamma} - \frac{1}{\rho^*}\int_0^{x(t)} N(\sigma)d\sigma$$

Because $N(x) = x^2 \cos x$ and $\int_0^{x(t)} \sigma^2 \cos \sigma d\sigma = 2x \cos x + (x^2 - 2) \sin x$, it follows that

$$V(t) - V(0) = \frac{\rho(t) - \rho(0)}{\gamma} - \frac{1}{\rho^*}[2x \cos x + (x^2 - 2) \sin x] \qquad (4.5.25)$$

From $x = w + \frac{\rho^2}{2\gamma}$ and $w \geq 0$, we conclude that $x(t) \geq 0$. Examination of (4.5.25) shows that for large x, the term $x^2 \sin x$ dominates the right-hand side of (4.5.25) and oscillates between $-x^2$ and x^2. Because

$$V(t) = V(0) + \frac{\rho(t) - \rho(0)}{\gamma} - \frac{1}{\rho^*}[2x \cos x - 2 \sin x] - \frac{x^2 \sin x}{\rho^*}$$

and $V(t) \geq 0$, it follows that x has to be bounded, otherwise the inequality $V \geq 0$ will be violated for large x. Bounded x implies that $V, w, \rho, \theta \in \mathcal{L}_\infty$. Because $w(t) = \int_0^t \epsilon^2 m^2 dt$ is a nondecreasing function bounded from above, the $\lim_{t \to \infty} w(t) = \int_0^\infty \epsilon^2 m^2 dt$ exists and is finite which implies that $\epsilon m \in \mathcal{L}_2$, i.e., $\epsilon, \epsilon n_s \in \mathcal{L}_2$. The rest of the proof follows directly as in the case of the linear parametric model and is omitted. □

4.6 Hybrid Adaptive Laws

The adaptive laws developed in Sections 4.3 to 4.5 update the estimate $\theta(t)$ of the unknown parameter vector θ^* continuously with time, i.e., at each time t we have a new estimate. For computational and robustness reasons it may be desirable to update the estimates only at specific instants of time t_k where $\{t_k\}$ is an unbounded monotonically increasing sequence in $\mathcal{R}^+$. Let $t_k = kT_s$ where $T_s = t_{k+1} - t_k$ is the "sampling" period and $k = 0, 1, 2, \ldots,$. Consider the design of an adaptive law that generates the estimate of the unknown θ^* at the discrete instances of time $t = 0, T_s, 2T_s, \cdots$.

We can develop such an adaptive law for the gradient algorithm

$$\dot{\theta} = \Gamma \epsilon \phi \qquad (4.6.1)$$

$$\epsilon = \frac{z - \hat{z}}{m^2} \qquad (4.6.2)$$

where z is the output of the linear parametric model

$$z = \theta^{*T} \phi$$

and
$$\hat{z} = \theta^T \phi \tag{4.6.3}$$

Integrating (4.6.1) from $t_k = kT_s$ to $t_{k+1} = (k+1)T_s$ we have

$$\theta_{k+1} = \theta_k + \Gamma \int_{t_k}^{t_{k+1}} \epsilon(\tau)\phi(\tau)d\tau, \quad \theta_0 = \theta(0), \quad k = 0, 1, 2, \ldots \tag{4.6.4}$$

where $\theta_k \triangleq \theta(t_k)$. Equation (4.6.4) generates a sequence of estimates, i.e., $\theta_0 = \theta(0), \theta_1 = \theta(T_s), \theta_2 = \theta(2T_s), \ldots, \theta_k = \theta(kT_s)$ of θ^*.

If we now use (4.6.4) instead of (4.6.1) to estimate θ^*, the error ϵ and $\hat{z}$ have to be generated using θ_k instead of $\theta(t)$, i.e.,

$$\hat{z}(t) = \theta_k^T \phi(t), \quad \epsilon(t) = \frac{z(t) - \hat{z}(t)}{m^2(t)}, \quad \forall t \in [t_k, t_{k+1}] \tag{4.6.5}$$

We refer to (4.6.4), (4.6.5) as the *hybrid adaptive law*. The following theorem establishes its stability properties.

Theorem 4.6.1 *Let m, T_s, Γ be chosen so that*

(a) $\frac{\phi^T \phi}{m^2} \leq 1$, $m \geq 1$
(b) $2 - T_s \lambda_m \geq \gamma$ *for some* $\gamma > 0$

where $\lambda_m = \lambda_{\max}(\Gamma)$. Then the hybrid adaptive law (4.6.4), (4.6.5) guarantees that

(i) $\theta_k \in l_\infty$.
(ii) $\Delta \theta_k \in l_2$; $\epsilon, \epsilon m \in \mathcal{L}_\infty \cap \mathcal{L}_2$, *where* $\Delta \theta_k = \theta_{k+1} - \theta_k$.
(iii) *If $m, \phi \in \mathcal{L}_\infty$ and ϕ is PE, then $\theta_k \to \theta^*$ as $k \to \infty$ exponentially fast.*

Proof As in the stability proof for the continuous adaptive laws, we evaluate the rate of change of the Lyapunov-like function

$$V(k) = \tilde{\theta}_k^T \Gamma^{-1} \tilde{\theta}_k \tag{4.6.6}$$

along the trajectory generated by (4.6.4) and (4.6.5), where $\tilde{\theta}_k \triangleq \theta_k - \theta^*$. Notice that

$$\Delta V(k) = (2\tilde{\theta}_k + \Delta \theta_k)^T \Gamma^{-1} \Delta \theta_k$$

4.6. HYBRID ADAPTIVE LAWS

where $\Delta V(k) = V(k+1) - V(k)$. Using (4.6.4) we obtain

$$\Delta V(k) = 2\tilde{\theta}_k^\top \int_{t_k}^{t_{k+1}} \epsilon(\tau)\phi(\tau)d\tau + \left(\int_{t_k}^{t_{k+1}} \epsilon(\tau)\phi(\tau)d\tau\right)^\top \Gamma \int_{t_k}^{t_{k+1}} \epsilon(\tau)\phi(\tau)d\tau$$

Because $\epsilon m^2 = -\tilde{\theta}_k^\top \phi(t)$ and $\lambda_m \geq \|\Gamma\|$, we have

$$\Delta V(k) \leq -2\int_{t_k}^{t_{k+1}} \epsilon^2(\tau)m^2(\tau)d\tau + \lambda_m \left(\int_{t_k}^{t_{k+1}} |\epsilon(\tau)m(\tau)|\frac{|\phi(\tau)|}{m(\tau)}d\tau\right)^2 \quad (4.6.7)$$

Using the Schwartz inequality, we can establish that

$$\left(\int_{t_k}^{t_{k+1}} |\epsilon(\tau)m(\tau)|\frac{|\phi(\tau)|}{m(\tau)}d\tau\right)^2 \leq \int_{t_k}^{t_{k+1}} \epsilon^2(\tau)m^2(\tau)d\tau \int_{t_k}^{t_{k+1}} \left(\frac{|\phi(\tau)|}{m(\tau)}\right)^2 d\tau$$

$$\leq T_s \int_{t_k}^{t_{k+1}} \epsilon^2(\tau)m^2(\tau)d\tau \quad (4.6.8)$$

where the last inequality follows from assumption (a). Using (4.6.8) in (4.6.7), it follows that

$$\Delta V(k) \leq -(2 - T_s\lambda_m)\int_{t_k}^{t_{k+1}} \epsilon^2(\tau)m^2(\tau)d\tau \quad (4.6.9)$$

Therefore, if $2 - T_s\lambda_m > \gamma$ for some $\gamma > 0$, we have $\Delta V(k) \leq 0$, which implies that $V(k)$ is a nonincreasing function and thus the boundedness of $V(k), \tilde{\theta}_k$ and θ_k follows. From (4.6.9), one can easily see that

$$\int_0^{t_{k+1}} \epsilon^2(\tau)m^2(\tau)d\tau \leq \frac{V(0) - V(k+1)}{(2 - T_s\lambda_m)} \quad (4.6.10)$$

We can establish that $\lim_{k\to\infty} V(k+1)$ exists and, from (4.6.10), that $\epsilon m \in \mathcal{L}_\infty \bigcap \mathcal{L}_2$. Because $m \geq 1$, it follows immediately that $\epsilon \in \mathcal{L}_\infty \bigcap \mathcal{L}_2$.

Similarly, we can obtain that

$$\Delta\theta_k^\top \Delta\theta_k \leq T_s\lambda_m^2 \int_{t_k}^{t_{k+1}} \epsilon^2(\tau)m^2(\tau)d\tau \quad (4.6.11)$$

by using the Schwartz inequality and condition (a) of the theorem. Therefore,

$$\sum_{k=1}^\infty \Delta\theta_k^\top \Delta\theta_k \leq T_s\lambda_m^2 \int_0^\infty \epsilon^2(\tau)m^2(\tau)d\tau < \infty$$

which implies $\Delta\theta_k \in l_2$ and, thus, completes the proof for (i) and (ii).

The proof for (iii) is relatively involved and is given in Section 4.8. □

For additional reading on hybrid adaptation the reader is referred to [60] where the term *hybrid* was first introduced and to [172, 173] where different hybrid algorithms are introduced and analyzed.

The hybrid adaptive law (4.6.4) may be modified to guarantee that $\theta_k \in \mathcal{R}^n$ belongs to a certain convex subset $\mathcal{S}$ of $\mathcal{R}^n$ by using the gradient projection method. That is, if $\mathcal{S}$ is defined as

$$\mathcal{S} = \left\{ \theta \in \mathcal{R}^n | \theta^\top \theta \leq M_0^2 \right\}$$

then the hybrid adaptive law (4.6.4) with projection becomes

$$\bar{\theta}_{k+1} = \theta_k + \Gamma \int_{t_k}^{t_{k+1}} \epsilon(\tau) \phi(\tau) d\tau$$

$$\theta_{k+1} = \begin{cases} \bar{\theta}_{k+1} & \text{if } \bar{\theta}_{k+1} \in \mathcal{S} \\ \frac{\bar{\theta}_{k+1}}{|\bar{\theta}_{k+1}|} M_0 & \text{if } \bar{\theta}_{k+1} \notin \mathcal{S} \end{cases}$$

and $\theta_0 \in \mathcal{S}$. As in the continuous-time case, it can be shown that the hybrid adaptive law with projection has the same properties as those of (4.6.4). In addition it guarantees that $\theta_k \in \mathcal{S}$, $\forall k \geq 0$. The details of this analysis are left as an exercise for the reader.

4.7 Summary of Adaptive Laws

In this section, we present tables with the adaptive laws developed in the previous sections together with their properties.

4.8 Parameter Convergence Proofs

In this section, we present the proofs of the theorems and corollaries of the previous sections that deal with parameter convergence. These proofs are useful for the reader who is interested in studying the behavior and convergence properties of the parameter estimates. They can be omitted by the reader whose interest is mainly on adaptive control where parameter convergence is not part of the control objective.

4.8.1 Useful Lemmas

The following two lemmas are used in the proofs of corollaries and theorems presented in this sections.

4.8. PARAMETER CONVERGENCE PROOFS

Table 4.1 Adaptive law based on SPR-Lyapunov design approach

Parametric model	$z = W(s)\theta^{*\top}\psi$
Parametric model rewritten	$z = W(s)L(s)\theta^{*\top}\phi, \quad \phi = L^{-1}(s)\psi$
Estimation model	$\hat{z} = W(s)L(s)\theta^{\top}\phi$
Normalized estimation error	$\epsilon = z - \hat{z} - W(s)L(s)\epsilon n_s^2$
Adaptive law	$\dot{\theta} = \Gamma\epsilon\phi$
Design variables	$L^{-1}(s)$ proper and stable; $W(s)L(s)$ proper and SPR; $m^2 = 1 + n_s^2$ and n_s chosen so that $\frac{\phi}{m} \in \mathcal{L}_\infty$ (e. g., $n_s^2 = \alpha\phi^{\top}\phi$ for some $\alpha > 0$)
Properties	(i) $\epsilon, \theta \in \mathcal{L}_\infty$; (ii) $\epsilon, \epsilon n_s, \dot{\theta} \in \mathcal{L}_2$

Lemma 4.8.1 (Uniform Complete Observability (UCO) with Output Injection). *Assume that there exists constants $\nu > 0, k_\nu \geq 0$ such that for all $t_0 \geq 0$, $K(t) \in \mathcal{R}^{n \times l}$ satisfies the inequality*

$$\int_{t_0}^{t_0+\nu} |K(\tau)|^2 d\tau \leq k_\nu \tag{4.8.1}$$

$\forall t \geq 0$ *and some constants $k_0, \nu > 0$. Then (C, A), where $C \in \mathcal{R}^{n \times l}, A \in \mathcal{R}^{n \times n}$, is a UCO pair if and only if $(C, A + KC^{\top})$ is a UCO pair.*

Proof We show that if there exist positive constants $\beta_1, \beta_2 > 0$ such that the observability grammian $N(t_0, t_0 + \nu)$ of the system (C, A) satisfies

$$\beta_1 I \leq N(t_0, t_0 + \nu) \leq \beta_2 I \tag{4.8.2}$$

then the observability grammian $N_1(t_0, t_0 + \nu)$ of $(C, A + KC^{\top})$ satisfies

$$\beta_1' I \leq N_1(t_0, t_0 + \nu) \leq \beta_2' I \tag{4.8.3}$$

for some constant $\beta_1', \beta_2' > 0$. From the definition of the observability grammian matrix, (4.8.3) is equivalent to

Table 4.2 Gradient algorithms

Parametric model	$z = \theta^{*T}\phi$
Estimation model	$\hat{z} = \theta^T \phi$
Normalized estimation error	$\epsilon = \dfrac{z - \hat{z}}{m^2}$
A. Based on instantaneous cost	
Adaptive law	$\dot{\theta} = \Gamma\epsilon\phi$
Design variables	$m^2 = 1 + n_s^2,\, n_s^2 = \alpha\phi^T\phi,\, \alpha > 0,\, \Gamma = \Gamma^T > 0$
Properties	(i) $\epsilon, \epsilon n_s, \theta, \dot{\theta} \in \mathcal{L}_\infty$; (ii) $\epsilon, \epsilon n_s, \dot{\theta} \in \mathcal{L}_2$
B. Based on the integral cost	
Adaptive law	$\dot{\theta} = -\Gamma(R\theta + Q)$ $\dot{R} = -\beta R + \dfrac{\phi\phi^T}{m^2},\; R(0) = 0$ $\dot{Q} = -\beta Q - \dfrac{z\phi}{m^2},\; Q(0) = 0$
Design variables	$m^2 = 1 + n_s^2$, n_s chosen so that $\phi/m \in \mathcal{L}_\infty$ (e.g., $n_s^2 = \alpha\phi^T\phi,\, \alpha > 0$); $\beta > 0,\, \Gamma = \Gamma^T > 0$
Properties	(i) $\epsilon, \epsilon n_s, \theta, \dot{\theta}, R, Q \in \mathcal{L}_\infty$; (ii) $\epsilon, \epsilon n_s, \dot{\theta} \in \mathcal{L}_2$; (iii) $\lim_{t\to\infty} \dot{\theta} = 0$

$$\beta_1'|x_1(t_0)|^2 \leq \int_{t_0}^{t_0+\nu} |C^T(t)x_1(t)|^2 dt \leq \beta_2'|x_1(t_0)|^2 \qquad (4.8.4)$$

where x_1 is the state of the system

$$\begin{aligned}\dot{x}_1 &= (A + KC^T)x_1 \\ y_1 &= C^T x_1\end{aligned} \qquad (4.8.5)$$

which is obtained, using output injection, from the system

4.8. PARAMETER CONVERGENCE PROOFS

Table 4.3 Least-squares algorithms

Parametric model	$z = \theta^{*\top}\phi$
Estimation model	$\hat{z} = \theta^{\top}\phi$
Normalized estimation error	$\epsilon = (z - \hat{z})/m^2$
A. Pure least-squares	
Adaptive law	$\dot{\theta} = P\epsilon\phi$ $\dot{P} = -P\frac{\phi\phi^{\top}}{m^2}P, \quad P(0) = P_0$
Design variables	$P_0 = P_0^{\top} > 0;\ m^2 = 1 + n_s^2,\ n_s$ chosen so that $\phi/m \in \mathcal{L}_\infty$ (e.g., $n_s^2 = \alpha\phi^{\top}\phi, \alpha > 0$ or $n_s^2 = \phi^{\top}P\phi$)
Properties	(i) $\epsilon, \epsilon n_s, \theta, \dot{\theta}, P \in \mathcal{L}_\infty$; (ii) $\epsilon, \epsilon n_s, \dot{\theta} \in \mathcal{L}_2$; (iii) $\lim_{t\to\infty} \theta(t) = \bar{\theta}$
B. Least-squares with covariance resetting	
Adaptive law	$\dot{\theta} = P\epsilon\phi$ $\dot{P} = -P\frac{\phi\phi^{\top}}{m^2}P, \quad P(t_r^+) = P_0 = \rho_0 I,$ where t_r is the time for which $\lambda_{min}(P) \leq \rho_1$
Design variables	$\rho_0 > \rho_1 > 0;\ m^2 = 1 + n_s^2,\ n_s$ chosen so that $\phi/m \in \mathcal{L}_\infty$ (e.g., $n_s^2 = \alpha\phi^{\top}\phi, \alpha > 0$)
Properties	(i) $\epsilon, \epsilon n_s, \theta, \dot{\theta}, P \in \mathcal{L}_\infty$; (ii) $\epsilon, \epsilon n_s, \dot{\theta} \in \mathcal{L}_2$
C. Least-squares with forgetting factor	
Adaptive law	$\dot{\theta} = P\epsilon\phi$ $\dot{P} = \begin{cases} \beta P - P\frac{\phi\phi^{\top}}{m^2}P, & if\ \|P(t)\| \leq R_0 \\ 0 & otherwise \end{cases}$ $P(0) = P_0$
Design variables	$m^2 = 1 + n_s^2,\ n_s^2 = \alpha\phi^{\top}\phi$ or $\phi^{\top}P\phi;\ \beta > 0, R_0 > 0$ scalars; $P_0 = P_0^{\top} > 0,\ \|P_0\| \leq R_0$
Properties	(i) $\epsilon, \epsilon n_s, \theta, \dot{\theta}, P \in \mathcal{L}_\infty$; (ii) $\epsilon, \epsilon n_s, \dot{\theta} \in \mathcal{L}_2$

Table 4.4 Adaptive laws for the bilinear model

Parametric model : $z = W(s)\rho^*(\theta^{*\top}\psi + z_0)$	
A. SPR-Lyapunov design: sign of ρ^* known	
Parametric model rewritten	$z = W(s)L(s)\rho^*(\theta^{*\top}\phi + z_1)$ $\phi = L^{-1}(s)\psi,\ z_1 = L^{-1}(s)z_0$
Estimation model	$\hat{z} = W(s)L(s)\rho(\theta^\top\phi + z_1)$
Normalized estimation error	$\epsilon = z - \hat{z} - W(s)L(s)\epsilon n_s^2$
Adaptive law	$\dot{\theta} = \Gamma\epsilon\phi\,\mathrm{sgn}(\rho^*)$ $\dot{\rho} = \gamma\epsilon\xi,\ \xi = \theta^\top\phi + z_1$
Design variables	$L^{-1}(s)$ proper and stable; $W(s)L(s)$ proper and SPR; $m^2 = 1 + n_s^2$; n_s chosen so that $\frac{\phi}{m}, \frac{z_1}{m} \in \mathcal{L}_\infty$ (e.g. $n_s^2 = \alpha(\phi^\top\phi + z_1^2),\ \alpha > 0$); $\Gamma = \Gamma^\top > 0, \gamma > 0$
Properties	(i) $\epsilon, \theta, \rho \in \mathcal{L}_\infty$; (ii) $\epsilon, \epsilon n_s, \dot\theta, \dot\rho \in \mathcal{L}_2$
B. Gradient algorithm: $\mathrm{sgn}(\rho^*)$ known	
Parametric model rewritten	$z = \rho^*(\theta^{*\top}\phi + z_1)$ $\phi = W(s)\psi,\ z_1 = W(s)z_0$
Estimation model	$\hat{z} = \rho(\theta^\top\phi + z_1)$
Normalized estimation error	$\epsilon = \dfrac{z - \hat{z}}{m^2}$
Adaptive law	$\dot{\theta} = \Gamma\epsilon\phi\,\mathrm{sgn}(\rho^*)$ $\dot{\rho} = \gamma\epsilon\xi,\ \xi = \theta^\top\phi + z_1$
Design variables	$m^2 = 1 + n_s^2$; n_s chosen so that $\frac{\phi}{m}, \frac{z_1}{m} \in \mathcal{L}_\infty$ (e.g., $n_s^2 = \phi^\top\phi + z_1^2$); $\Gamma = \Gamma^\top > 0, \gamma > 0$
Properties	(i) $\epsilon, \epsilon n_s, \theta, \rho, \dot\theta, \dot\rho \in \mathcal{L}_\infty$; (ii) $\epsilon, \epsilon n_s, \dot\theta, \dot\rho \in \mathcal{L}_2$

$$\dot{x} = Ax$$
$$y = C^\top x \qquad (4.8.6)$$

Form (4.8.5) and (4.8.6), it follows that $e \triangleq x_1 - x$ satisfies

$$\dot{e} = Ae + KC^\top x_1$$

4.8. PARAMETER CONVERGENCE PROOFS

Table 4.4 (Continued)

	C. Gradient algorithm with projection **Sign (ρ^*) and lower bound $0 < \rho_0 \leq |\rho^*|$ known**
Parametric model rewritten	$z = \bar{\theta}^{*\top}\bar{\phi}$ $\bar{\theta}^* = [\bar{\theta}_1^*, \bar{\theta}_2^{*\top}]^\top, \bar{\theta}_1^* = \rho^*, \bar{\theta}_2^* = \rho^*\theta^*$ $\bar{\phi} = [z_1, \phi^\top]^\top$
Estimation model	$\hat{z} = \bar{\theta}^\top \bar{\phi}$
Normalized estimation error	$\epsilon = \dfrac{z - \hat{z}}{m^2}$
Adaptive law	$\dot{\bar{\theta}}_1 = \begin{cases} \gamma_1 \epsilon z_1 & \text{if } \bar{\theta}_1 \text{sgn}(\rho^*) > \rho_0 \text{ or} \\ & \text{if } \bar{\theta}_1 \text{sgn}(\rho^*) = \rho_0 \text{ and } -\gamma_1 z_1 \epsilon \text{sgn}(\rho^*) \leq 0 \\ 0 & \text{otherwise} \end{cases}$ $\dot{\bar{\theta}}_2 = \Gamma_2 \epsilon \phi$ $\rho = \bar{\theta}_1, \theta = \dfrac{\bar{\theta}_2}{\bar{\theta}_1}$
Design variables	$m^2 = 1 + n_s^2$; n_s chosen so that $\dfrac{\phi}{m} \in \mathcal{L}_\infty$ (e.g., $n_s^2 = \alpha \bar{\phi}^\top \bar{\phi}$, $\alpha > 0$); $\gamma_1 > 0$; $\bar{\theta}_1(0)$ satisfies $|\bar{\theta}_1(0)| \geq \rho_0$; $\Gamma_2 = \Gamma_2^\top > 0, \gamma > 0$
Properties	(i) $\epsilon, \epsilon n_s, \theta, \rho, \dot{\theta}, \dot{\rho} \in \mathcal{L}_\infty$; (ii) $\epsilon, \epsilon n_s, \dot{\theta}, \dot{\rho} \in \mathcal{L}_2$

	D. Gradient algorithm without projection **Unknown sign (ρ^*)**
Parametric model	$z = \rho^*(\theta^{*\top}\phi + z_1)$
Estimation model	$\hat{z} = N(x)\rho(\theta^\top \phi + z_1)$ $N(x) = x^2 \cos x$ $x = w + \dfrac{\rho^2}{2\gamma}, \dot{w} = \epsilon^2 m^2, w(0) = 0$
Normalized estimation error	$\epsilon = \dfrac{z - \hat{z}}{m^2}$
Adaptive law	$\dot{\theta} = N(x)\Gamma \epsilon \phi$ $\dot{\rho} = N(x)\gamma \epsilon \xi, \ \xi = \theta^\top \phi + z_1$
Design variables	$m^2 = 1 + n_s^2$; n_s chosen so that $\dfrac{\phi}{m}, \dfrac{z_1}{m} \in \mathcal{L}_\infty$; (e.g., $n_s^2 = \phi^\top \phi + z_1^2$); $\gamma > 0, \Gamma = \Gamma^\top > 0$
Properties	(i) $\epsilon, \epsilon n_s, \theta, \rho, \dot{\theta}, \ \dot{\rho}, x, w \in \mathcal{L}_\infty$; (ii) $\epsilon, \epsilon n_s, \dot{\theta}, \dot{\rho} \in \mathcal{L}_2$

Table 4.5. Hybrid adaptive law

Parametric model	$z = \theta^{*\top}\phi$
Estimation model	$\hat{z} = \theta_k^\top \phi, t \in [t_k, t_{k+1})$
Normalized estimation error	$\epsilon = \dfrac{z - \hat{z}}{m^2}$
Adaptive law	$\theta_{k+1} = \theta_k + \Gamma \int_{t_k}^{t_{k+1}} \epsilon(\tau)\phi(\tau)d\tau, k = 0, 1, 2, \ldots,$
Design variables	Sampling period $T_s = t_{k+1} - t_k > 0, t_k = kT_s$; $m^2 = 1 + n_s^2$ and n_s chosen so that $\|\phi\|/m \leq 1$ (e.g., $n_s^2 = \alpha\phi^\top\phi, \alpha \geq 1$) $\Gamma = \Gamma^\top > 0$ $2 - T_s\lambda_{max}(\Gamma) > \gamma$ for some constant $\gamma > 0$
Properties	(i) $\theta_k \in l_\infty, \epsilon, \epsilon n_s \in \mathcal{L}_\infty$ (ii) $\|\theta_{k+1} - \theta_k\| \in l_2$; $\epsilon, \epsilon n_s \in \mathcal{L}_2$

Consider the trajectories $x(t)$ and $x_1(t)$ with the same initial conditions. We have

$$e(t) = \int_{t_0}^{t} \Phi(t,\tau)K(\tau)C^\top(\tau)x_1(\tau)d\tau \qquad (4.8.7)$$

where Φ is the state transition matrix of (4.8.6). Defining

$$\bar{x}_1 \triangleq \begin{cases} KC^\top x_1/|KC^\top x_1| & \text{if } |C^\top x_1| \neq 0 \\ K/|K| & \text{if } |C^\top x_1| = 0 \end{cases}$$

we obtain, using the Schwartz inequality, that

$$|C^\top(t)e(t)|^2 \leq \int_{t_0}^{t} \left|C^\top(t)\Phi(t,\tau)K(\tau)C^\top(\tau)x_1(\tau)\right|^2 d\tau$$

$$\leq \int_{t_0}^{t} \left|C^\top(t)\Phi(t,\tau)\bar{x}_1(\tau)\right|^2 |K(\tau)|^2 d\tau \int_{t_0}^{t} \left|C^\top(\tau)x_1(\tau)\right|^2 d\tau \qquad (4.8.8)$$

4.8. PARAMETER CONVERGENCE PROOFS

Using the triangular inequality $(a+b)^2 \leq 2a^2 + 2b^2$ and (4.8.8), we have

$$\int_{t_0}^{t_0+\nu} |C^\top(t)x_1(t)|^2 dt \leq 2\int_{t_0}^{t_0+\nu} |C^\top(t)x(t)|^2 dt + 2\int_{t_0}^{t_0+\nu} |C^\top(t)e(t)|^2 dt$$

$$\leq 2\int_{t_0}^{t_0+\nu} |C^\top(t)x(t)|^2 dt$$

$$+ 2\int_{t_0}^{t_0+\nu} \int_{t_0}^{t} \left|C^\top(t)\Phi(t,\tau)\bar{x}_1(\tau)\right|^2 |K(\tau)|^2 d\tau \int_{t_0}^{t} \left|C^\top(\tau)x_1(\tau)\right|^2 d\tau dt$$

$$\leq 2\beta_2 |x_1(t_0)|^2$$

$$+ 2\int_{t_0}^{t_0+\nu} \int_{t_0}^{t} \left|C^\top(t)\Phi(t,\tau)\bar{x}_1(\tau)\right|^2 |K(\tau)|^2 d\tau \int_{t_0}^{t} \left|C^\top(\tau)x_1(\tau)\right|^2 d\tau dt$$

where the last inequality is obtained using the UCO property of (C, A) and the condition that $x(t_0) = x_1(t_0)$. Applying the B-G Lemma, we obtain

$$\int_{t_0}^{t_0+\nu} |C^\top(t)x_1(t)|^2 dt \leq 2\beta_2 |x_1(t_0)|^2 e^{\left\{\int_{t_0}^{t_0+\nu} \int_{t_0}^{t} 2|C^\top(t)\Phi(t,\tau)\bar{x}_1(\tau)|^2 |K(\tau)|^2 d\tau dt\right\}} \quad (4.8.9)$$

By interchanging the sequence of integration, we have

$$\int_{t_0}^{t_0+\nu}\int_{t_0}^{t} |C^\top(t)\Phi(t,\tau)\bar{x}_1(\tau)|^2 |K(\tau)|^2 d\tau dt = \int_{t_0}^{t_0+\nu}\int_{\tau}^{t_0+\nu} |C^\top(t)\Phi(t,\tau)\bar{x}_1(\tau)|^2 dt |K(\tau)|^2 d\tau$$

Because (C, A) being UCO and $|\bar{x}_1| = 1$ imply that

$$\int_{\tau}^{t_0+\nu} \left|C^\top(t)\Phi(t,\tau)\bar{x}_1(\tau)\right|^2 dt \leq \int_{t_0}^{t_0+\nu} \left|C^\top(t)\Phi(t,\tau)\right|^2 dt \leq \beta_2$$

for any $t_0 \leq \tau \leq t_0 + \nu$, it follows from (4.8.1) and the above two equations that

$$\int_{t_0}^{t_0+\nu}\int_{t_0}^{t} \left|C^\top(t)\Phi(t,\tau)\bar{x}_1(\tau)\right|^2 |K(\tau)|^2 d\tau dt \leq k_\nu \beta_2 \quad (4.8.10)$$

and, therefore, (4.8.9) leads to

$$\int_{t_0}^{t_0+\nu} |C^\top(t)x_1(t)|^2 dt \leq 2\beta_2 e^{2\beta_2 k_\nu} |x_1(t_0)|^2 \quad (4.8.11)$$

On the other hand, using $x_1 = x + e$ and the triangular inequality $(a+b)^2 \geq \frac{1}{2}a^2 - b^2$, we have

$$\int_{t_0}^{t_0+\nu} |C^\top(t)x_1(t)|^2 dt \geq \frac{1}{2}\int_{t_0}^{t_0+\nu} |C^\top(t)x(t)|^2 dt - \int_{t_0}^{t_0+\nu} |C^\top(t)e(t)|^2 dt$$

$$\geq \frac{\beta_1}{2}|x_1(t_0)|^2 - \int_{t_0}^{t_0+\nu} |C^\top(t)e(t)|^2 dt \quad (4.8.12)$$

where the last inequality is obtained using the UCO property of (C, A) and the fact $x(t_0) = x_1(t_0)$. Substituting (4.8.8) for $|C^T e|^2$, we obtain

$$\int_{t_0}^{t_0+\nu} |C^T(t)x_1(t)|^2 dt \geq \frac{\beta_1}{2}|x_1(t_0)|^2$$
$$- \left(\int_{t_0}^{t_0+\nu} |C^T(t)x_1(t)|^2 dt\right) \int_{t_0}^{t_0+\nu} \int_{t_0}^{t} |C^T(t)\Phi(t,\tau)\bar{x}_1(\tau)|^2 |K(\tau)|^2 d\tau dt$$

by using the fact that

$$\int_{t_0}^{t} |C^T(\tau)x_1(\tau)|^2 d\tau \leq \int_{t_0}^{t_0+\nu} |C^T(\tau)x_1(\tau)|^2 d\tau$$

for any $t \leq t_0 + \nu$. Using (4.8.10), we have

$$\int_{t_0}^{t_0+\nu} |C^T(t)x_1(t)|^2 dt \geq \frac{\beta_1}{2}|x_1(t_0)|^2 - \beta_2 k_\nu \int_{t_0}^{t_0+\nu} |C^T(t)x_1(t)|^2 dt$$

or

$$\int_{t_0}^{t_0+\nu} |C^T(t)x_1(t)|^2 dt \geq \frac{\beta_1}{2(1+\beta_2 k_\nu)}|x_1(t_0)|^2 \qquad (4.8.13)$$

Setting $\beta_1' = \frac{\beta_1}{2(1+\beta_2 k_\nu)}, \beta_2' = 2\beta_2 e^{2\beta_2 k_\nu}$, we have shown that (4.8.4) holds for $\beta_1', \beta_2' > 0$ and therefore $(C, A + KC^T)$ is UCO, and, hence, the if part of the lemma is proved.

The proof for the only if part is exactly the same as the if part since (C, A) can be obtained from $(C, A + KC^T)$ using output injection. $\square$

Lemma 4.8.2 *Let $H(s)$ be a proper stable transfer function and $y = H(s)u$. If $u \in \mathcal{L}_\infty$, then*

$$\int_{t}^{t+T} y^2(\tau) d\tau \leq k_1 \int_{t}^{t+T} u^2(\tau) d\tau + k_2$$

for some constants $k_1, k_2 \geq 0$ and any $t \geq 0, T > 0$. Furthermore, if $H(s)$ is strictly proper, then $k_1 \leq \alpha \|H(s)\|_\infty^2$ for some constant $\alpha > 0$.

Proof Let us define
$$f_S(\tau) = \begin{cases} f(\tau) & \text{if } \tau \in \mathcal{S} \\ 0 & \text{otherwise} \end{cases}$$

where $\mathcal{S} \subset \mathcal{R}$ is a subset of $\mathcal{R}$ that can be an open or closed interval. Then we can write

$$u(\tau) = u_{[0,t)}(\tau) + u_{[t,t+T]}(\tau), \qquad \forall 0 \leq \tau \leq t + T$$

4.8. PARAMETER CONVERGENCE PROOFS

Because $H(s)$ can be decomposed as $H(s) = h_0 + H_1(s)$ where h_0 is a constant and $H_1(s)$ is strictly proper, we can express $y = H(s)u$ as

$$y(\tau) = h_0 u(\tau) + H_1(s)\left\{u_{[0,t)} + u_{[t,t+T]}\right\} \triangleq h_0 u(\tau) + y_1(\tau) + y_2(\tau), \quad \forall 0 \leq \tau \leq t+T$$

where $y_1 \triangleq H_1(s)u_{[0,t)}$, $y_2 \triangleq H_1(s)u_{[t,t+T]}$. Therefore, using the inequality

$$(a+b+c)^2 \leq 3a^2 + 3b^2 + 3c^2$$

we have

$$\int_t^{t+T} y^2(\tau)d\tau \leq 3h_0^2 \int_t^{t+T} u^2(\tau)d\tau + 3\int_t^{t+T} y_1^2(\tau)d\tau + 3\int_t^{t+T} y_2^2(\tau)d\tau \quad (4.8.14)$$

Using Lemma 3.3.1, Remark 3.3.2, and noting that $y_2(\tau) = 0$ for all $\tau < t$, we have

$$\int_t^{t+T} y_2^2(\tau)d\tau = \int_0^{t+T} y_2^2(\tau)d\tau \triangleq \|y_{2(t+T)}\|_2^2 \leq \|H_1(s)\|_\infty^2 \|u_{[t,t+T]}\|_2^2$$

where

$$\|u_{[t,t+T]}\|_2^2 = \int_0^\infty u_{[t,t+T]}^2(\tau)d\tau = \int_t^{t+T} u^2(\tau)d\tau$$

Hence,

$$\int_t^{t+T} y_2^2(\tau)d\tau \leq \|H_1(s)\|_\infty^2 \int_t^{t+T} u^2(\tau)d\tau \quad (4.8.15)$$

To evaluate the second term in (4.8.14), we write

$$y_1(\tau) = h_1(\tau) * u_{[0,t)}(\tau) = \int_0^\tau h_1(\tau-\sigma)u_{[0,t)}(\sigma)d\sigma$$

$$= \begin{cases} \int_0^\tau h_1(\tau-\sigma)u(\sigma)d\sigma & \text{if } \tau < t \\ \int_0^t h_1(\tau-\sigma)u(\sigma)d\sigma & \text{if } \tau \geq t \end{cases}$$

where $h_1(t)$ is the impulse response of $H_1(s)$. Because $H_1(s)$ is a strictly proper stable transfer function, $|h_1(\tau-\sigma)| \leq \alpha_1 e^{-\alpha_2(\tau-\sigma)}$ for some constants $\alpha_1, \alpha_2 > 0$, then using the Schwartz inequality and the boundedness of $u(\tau)$, we have

$$\int_t^{t+T} y_1^2(\tau)d\tau = \int_t^{t+T} \left(\int_0^t h_1(\tau-\sigma)u(\sigma)d\sigma\right)^2 d\tau$$

$$\leq \int_t^{t+T} \left(\int_0^t |h_1(\tau-\sigma)|d\sigma \int_0^t |h_1(\tau-\sigma)|u^2(\sigma)d\sigma\right) d\tau$$

$$\leq \alpha_1^2 \int_t^{t+T} \left(\int_0^t e^{-\alpha_2(\tau-\sigma)}d\sigma \int_0^t e^{-\alpha_2(\tau-\sigma)}u^2(\sigma)d\sigma\right) d\tau$$

$$\leq \frac{\alpha_1^2 \bar{\alpha}}{\alpha_2^2} \int_t^{t+T} e^{-2\alpha_2(\tau-t)}d\tau$$

$$\leq \frac{\alpha_1^2 \bar{\alpha}}{2\alpha_2^3} \quad (4.8.16)$$

where $\bar{\alpha} = sup_\sigma u^2(\sigma)$. Using (4.8.15) and (4.8.16) in (4.8.14) and defining $k_1 = 3(h_0^2 + \|H_1(s)\|_\infty^2)$, $k_2 = \frac{3\alpha_1^2 \bar{\alpha}}{2\alpha_2^3}$, it follows that

$$\int_t^{t+T} y^2(\tau) d\tau \leq k_1 \int_t^{t+T} u^2(\tau) d\tau + k_2$$

and the first part of the Lemma is proved.

If $H(s)$ is strictly proper, then $h_0 = 0$, $H_1(s) = H(s)$ and $k_1 = 3\|H(s)\|_\infty^2$, therefore the proof is complete. □

Lemma 4.8.3 (Properties of PE Signals) *If $w : \mathcal{R}^+ \mapsto \mathcal{R}^n$ is PE and $w \in \mathcal{L}_\infty$, then the following results hold:*

(i) $w_1 \triangleq Fw$, *where $F \in \mathcal{R}^{m \times n}$ with $m \leq n$ is a constant matrix, is PE if and only if F is of rank m.*

(ii) *Let (a) $e \in \mathcal{L}_2$ or (b) $e \in \mathcal{L}_\infty$ and $e \to 0$ as $t \to \infty$, then $w_2 = w + e$ is PE.*

(iii) *Let $e \in \mathcal{S}(\mu)$ and $e \in \mathcal{L}_\infty$. There exists a $\mu^* > 0$ such that for all $\mu \in [0, \mu^*)$, $\omega_\mu = w + e$ is PE.*

(iv) *If in addition $\dot{w} \in \mathcal{L}_\infty$ and $H(s)$ is a stable, minimum phase, proper rational transfer function, then $w_3 = H(s)w$ is PE.*

Proof The proof of (i) and (iii) is quite trivial, and is left as an exercise for the reader.

To prove (iv), we need to establish that the inequality

$$\beta_2 T \geq \int_t^{t+T} (q^\top w_3(\tau))^2 d\tau \geq \beta_1 T \qquad (4.8.17)$$

holds for some constants $\beta_1, \beta_2, T > 0$, any $t \geq 0$ and any $q \in \mathcal{R}^n$ with $|q| = 1$.

The existence of an upper bound in (4.8.17) is implied by the assumptions that $w \in \mathcal{L}_\infty$, $H(s)$ has stable poles and therefore $w_3 \in \mathcal{L}_\infty$. To establish the lower bound, we define

$$z(t) \triangleq \frac{a^r}{(s+a)^r} q^\top w$$

and write

$$z(t) = \frac{a^r}{(s+a)^r} q^\top H^{-1}(s) w_3 = \frac{a^r}{(s+a)^r} H^{-1}(s) q^\top w_3$$

where $a > 0$ is arbitrary at this moment and $r > 0$ is an integer that is chosen to be equal to the relative degree of $H(s)$ so that $\frac{a^r}{(s+a)^r} H^{-1}(s)$ is a proper stable transfer function.

According to Lemma 4.8.2, we have

$$\int_t^{t+T} z^2(\tau) d\tau \leq k_1 \int_t^{t+T} \left(q^\top w_3(\tau)\right)^2 d\tau + k_2$$

4.8. PARAMETER CONVERGENCE PROOFS

for any $t, T > 0$ and some $k_1, k_2 > 0$ which may depend on a, or equivalently

$$\int_t^{t+T} (q^\top w_3(\tau))^2 d\tau \geq \frac{1}{k_1}\left(\int_t^{t+T} z^2(\tau) d\tau - k_2\right) \tag{4.8.18}$$

On the other hand, we have

$$\begin{aligned} z(t) &= \frac{a^r}{(s+a)^r} q^\top w \\ &= q^\top w + \left(\frac{a^r - (s+a)^r}{s(s+a)^r}\right) q^\top \dot{w} \\ &= q^\top w + z_1 \end{aligned}$$

where

$$z_1 = \frac{a^r - (s+a)^r}{s(s+a)^r} q^\top \dot{w}$$

It is shown in Appendix A (see Lemma A.2) that

$$\left\|\frac{a^r - (s+a)^r}{s(s+a)^r}\right\|_\infty \leq \frac{k}{a}$$

for some constant $k > 0$ that is independent of a. Therefore, applying Lemma 4.8.2, we have

$$\begin{aligned} \int_t^{t+T} z_1^2(\tau) d\tau &\leq \frac{k_3}{a^2} \int_t^{t+T} (q^\top \dot{w})^2 d\tau + \bar{k}_3 \\ &\leq \frac{k_4 T}{a^2} + \bar{k}_3 \end{aligned}$$

where $k_3 = 3k^2, k_4 = k_3 \sup_t |\dot{w}|^2$ and $\bar{k}_3 \geq 0$ is independent of α, and the second inequality is obtained using the boundedness of $\dot{w}$. Using the inequality $(x+y)^2 \geq \frac{1}{2}x^2 - y^2$, we have

$$\begin{aligned} \int_t^{t+T} z^2(\tau) d\tau &= \int_t^{t+T} (q^\top w(\tau) + z_1(\tau))^2 d\tau \\ &\geq \frac{1}{2} \int_t^{t+T} (q^\top w)^2 d\tau - \int_t^{t+T} z_1^2(\tau) d\tau \\ &\geq \frac{1}{2} \int_t^{t+T} (q^\top w)^2 d\tau - \frac{k_4 T}{a^2} - \bar{k}_3 \end{aligned} \tag{4.8.19}$$

Because w is PE, i.e.,

$$\int_t^{t+T_0} (q^\top w)^2 d\tau \geq \alpha_0 T_0$$

for some $T_0, \alpha_0 > 0$ and $\forall t \geq 0$, we can divide the interval $[t, t+T]$ into subintervals of length T_0 and write

$$\int_t^{t+T} (q^\top w)^2 d\tau \geq \sum_{i=1}^{n_0} \int_{t+(i-1)T_0}^{t+iT_0} (q^\top w)^2 d\tau \geq n_0 \alpha_0 T_0$$

where n_0 is the largest integer that satisfies $n_0 \leq \frac{T}{T_0}$. From the definition of n_0, we have $n_0 \geq \frac{T}{T_0} - 1$; therefore, we can establish the following inequality

$$\int_t^{t+T} (q^\top w)^2 d\tau \geq \alpha_0 \left(\frac{T}{T_0} - 1\right) T_0 = \alpha_0 (T - T_0) \qquad (4.8.20)$$

Because the above analysis holds for any $T > 0$, we assume that $T > T_0$. Using (4.8.19) and (4.8.20) in (4.8.18), we have

$$\begin{aligned}
\int_t^{t+T} (q^\top w_3)^2 d\tau &\geq \frac{1}{k_1}\left(\frac{\alpha_0}{2}(T - T_0) - \frac{k_4}{a^2}T - \bar{k}_3 - k_2\right) \\
&= \frac{1}{k_1}\left\{\left(\frac{\alpha_0}{2} - \frac{k_4}{a^2}\right)T - \left(\frac{\alpha_0 T_0}{2} + \bar{k}_3 + k_2\right)\right\}
\end{aligned} \qquad (4.8.21)$$

Since (4.8.21) holds for any $a, T > 0$ and α_0, k_4 are independent of a, we can first choose a to satisfy

$$\frac{\alpha_0}{2} - \frac{k_4}{a^2} \geq \frac{\alpha_0}{4}$$

and then fix T so that

$$\frac{\alpha_0}{4}T - \left(\frac{\alpha_0 T_0}{2} + \bar{k}_3 + k_2\right) > \beta_1 k_1 T$$

for a fixed $\beta_1 > 0$. It follows that

$$\int_t^{t+T} (q^\top w_3(\tau))^2 d\tau \geq \beta_1 T > 0$$

and the lower bound in (4.8.17) is established. $\square$

Lemma 4.8.4 *Consider the system*

$$\begin{aligned}
\dot{Y}_1 &= A_c Y_1 - B_c \phi^\top Y_2 \\
\dot{Y}_2 &= 0 \\
y_0 &= C_c^\top Y_1
\end{aligned} \qquad (4.8.22)$$

where A_c is a stable matrix, (C_c, A_c) is observable, and $\phi \in \mathcal{L}_\infty$. If ϕ_f defined as

$$\phi_f \triangleq C_c^\top (sI - A_c)^{-1} B_c \phi$$

4.8. PARAMETER CONVERGENCE PROOFS

satisfies

$$\alpha_1 I \leq \frac{1}{T_0} \int_t^{t+T_0} \phi_f(\tau)\phi_f^\top(\tau)d\tau \leq \alpha_2 I, \quad \forall t \geq 0 \qquad (4.8.23)$$

for some constants $\alpha_1, \alpha_2, T_0 > 0$, *then (4.8.22) is UCO.*

Proof The UCO of (4.8.22) follows if we establish that the observability grammian $\mathcal{N}(t, t+T)$ of (4.8.22) defined as

$$\mathcal{N}(t, t+T) \triangleq \int_t^{t+T} \Phi^\top(\tau,t) C C^\top \Phi(\tau,t) d\tau$$

where $C = [C_c^\top \; 0]^\top$ satisfies

$$\beta I \geq \mathcal{N}(t, t+T) \geq \alpha I$$

for some constant $\alpha, \beta > 0$, where $\Phi(t, t_0)$ is the state transition matrix of (4.8.22). The upper bound βI follows from the boundedness of $\Phi(t, t_0)$ that is implied by $\phi \in \mathcal{L}_\infty$ and the fact that A_c is a stable matrix. The lower bound will follow if we establish the following inequality:

$$\int_t^{t+T} y_0^2(\tau) d\tau \geq \alpha \left(|Y_1(t)|^2 + |Y_2|^2 \right)$$

where Y_2 is independent of t due to $\dot{Y}_2 = 0$. From (4.8.22), we can write

$$y_0(\tau) = C_c^\top Y_1(\tau) = C_c^\top e^{A_c(\tau-t)} Y_1(t) - \int_t^\tau C_c^\top e^{A_c(\tau-\sigma)} B_c \phi^\top(\sigma) d\sigma Y_2$$
$$\triangleq y_1(\tau) + y_2(\tau)$$

for all $\tau \geq t$, where $y_1(\tau) \triangleq C_c^\top e^{A_c(\tau-t)} Y_1(t)$, $y_2(\tau) \triangleq -\int_t^\tau C_c^\top e^{A_c(\tau-\sigma)} B_c \phi^\top(\sigma) d\sigma Y_2$. Using the inequalities $(x+y)^2 \geq \frac{x^2}{2} - y^2$ and $(x+y)^2 \geq \frac{y^2}{2} - x^2$ with $x = y_1, y = y_2$ over the intervals $[t, t+T']$, $[t+T', t+T]$, respectively, we have

$$\int_t^{t+T} y_0^2(\tau) d\tau \geq \int_t^{t+T'} \frac{y_1^2(\tau)}{2} d\tau - \int_t^{t+T'} y_2^2(\tau) d\tau$$
$$+ \int_{t+T'}^{t+T} \frac{y_2^2(\tau)}{2} d\tau - \int_{t+T'}^{t+T} y_1^2(\tau) d\tau \qquad (4.8.24)$$

for any $0 < T' < T$. We now evaluate each term on the right-hand side of (4.8.24). Because A_c is a stable matrix, it follows that

$$|y_1(\tau)| \leq k_1 e^{-\gamma_1(\tau-t)} |Y_1(t)|$$

for some $k_1, \gamma_1 > 0$, and, therefore,

$$\int_{t+T'}^{t+T} y_1^2(\tau)d\tau \leq \frac{k_1^2}{2\gamma_1}e^{-2\gamma_1 T'}|Y_1(t)|^2 \quad (4.8.25)$$

On the other hand, since (C_c, A_c) is observable, we have

$$\int_t^{t+T'} e^{A_c^\top(t-\tau)}C_c C_c^\top e^{A_c(t-\tau)}d\tau \geq k_2 I$$

for any $T' > T_1$ and some constants $k_2, T_1 > 0$. Hence,

$$\int_t^{t+T'} y_1^2(\tau)d\tau \geq k_2|Y_1(t)|^2 \quad (4.8.26)$$

Using $y_2(\tau) = -\phi_f^\top(\tau)Y_2$ and the fact that

$$\alpha_2 n_1 T_0 I \geq \int_{t+T'}^{t+T} \phi_f \phi_f^\top d\tau \geq n_0 \alpha_1 T_0 I$$

where n_0, n_1 is the largest and smallest integer respectively that satisfy

$$n_0 \leq \frac{T - T'}{T_0} \leq n_1$$

i.e., $n_0 \geq \frac{T-T'}{T_0} - 1$, $n_1 \leq \frac{T-T'}{T_0} + 1$, we can establish the following inequalities satisfied by y_2:

$$\int_t^{t+T'} y_2^2(\tau)d\tau \leq \alpha_2 T_0\left(\frac{T'}{T_0} + 1\right)|Y_2|^2$$

$$\int_{t+T'}^{t+T} y_2^2(\tau)d\tau \geq \alpha_1 T_0\left(\frac{T-T'}{T_0} - 1\right)|Y_2|^2 \quad (4.8.27)$$

Using (4.8.25), (4.8.26), (4.8.27) in (4.8.24), we have

$$\int_t^{t+T} y_0^2(\tau)d\tau \geq \left(\frac{k_2}{2} - \frac{k_1^2}{2\gamma_1}e^{-2\gamma_1 T'}\right)|Y_1(t)|^2$$
$$+ \left(\frac{\alpha_1 T_0}{2}\left(\frac{T-T'}{T_0} - 1\right) - \alpha_2 T_0\left(\frac{T'}{T_0} + 1\right)\right)|Y_2|^2 \quad (4.8.28)$$

Because the inequality (4.8.28) is satisfied for all T, T' with $T' > T_1$, let us first choose T' such that $T' > T_1$ and

$$\frac{k_2}{2} - \frac{k_1^2}{2\gamma_1}e^{-2\gamma_1 T'} \geq \frac{k_2}{4}$$

4.8. PARAMETER CONVERGENCE PROOFS

Now choose T to satisfy

$$\frac{\alpha_1 T_0}{2}\left(\frac{T-T'}{T_0} - 1\right) - \alpha_2 T_0 \left(\frac{T'}{T_0} + 1\right) \geq \beta_1$$

for a fixed β_1. We then have

$$\int_t^{t+T} y_0^2(\tau) d\tau \geq \alpha \left(|Y_1(t)|^2 + |Y_2(t)|^2\right)$$

where $\alpha = min\{\beta_1, \frac{k_2}{4}\}$. Hence, (4.8.22) is UCO. □

4.8.2 Proof of Corollary 4.3.1

Consider equations (4.3.30), (4.3.35), i.e.,

$$\begin{aligned} \dot{e} &= A_c e + B_c(-\tilde{\theta}^\top \phi - \epsilon n_s^2) \\ \dot{\tilde{\theta}} &= \Gamma \phi \epsilon \\ \epsilon &= C_c^\top e \end{aligned} \quad (4.8.29)$$

that describe the stability properties of the adaptive law. In proving Theorem 4.3.1, we have also shown that the time derivative $\dot{V}$ of

$$V = \frac{e^\top P_c e}{2} + \frac{\tilde{\theta}^\top \Gamma^{-1} \tilde{\theta}}{2}$$

where $\Gamma = \Gamma^\top > 0$ and $P_c = P_c^\top > 0$, satisfies

$$\dot{V} \leq -\nu' \epsilon^2 \quad (4.8.30)$$

for some constant $\nu' > 0$. Defining

$$A(t) = \begin{bmatrix} A_c - B_c C_c^\top n_s^2 & -B_c \phi^\top \\ \Gamma \phi C_c^\top & 0 \end{bmatrix}, \quad C = [C_c^\top \ 0]^\top, \quad P = \frac{1}{2}\begin{bmatrix} P_c & 0 \\ 0 & \Gamma^{-1} \end{bmatrix}$$

and $x = [e^\top, \tilde{\theta}^\top]^\top$, we rewrite (4.8.29) as

$$\dot{x} = A(t)x, \quad \epsilon = C^\top x$$

and express the above Lyapunov-like function V and its derivative $\dot{V}$ as

$$\begin{aligned} V &= x^\top P x \\ \dot{V} &= 2x^\top P A x + x^\top \dot{P} x \\ &= x^\top (PA + A^\top P + \dot{P})x \leq -\nu' x^\top C C^\top x = -\nu' \epsilon^2 \end{aligned}$$

where $\dot{P} = 0$. It therefore follows that P, as defined above, satisfies the inequality

$$A^\top(t)P + PA(t) + \nu' CC^\top \leq O$$

Using Theorem 3.4.8, we can establish that the equilibrium $e_e = 0, \tilde{\theta}_e = 0$ (i.e., $x_e = 0$) of (4.8.29) is u.a.s, equivalently e.s., provided (C, A) is a UCO pair.

According to Lemma 4.8.1, (C, A) and $(C, A + KC^\top)$ have the same UCO property, where

$$K \triangleq \begin{bmatrix} B_c n_s^2 \\ -\Gamma\phi \end{bmatrix}$$

is bounded. We can therefore establish that (4.8.29) is UCO by showing that $(C, A+KC^\top)$ is a UCO pair. We write the system corresponding to $(C, A+KC^\top)$ as

$$\begin{aligned} \dot{Y}_1 &= A_c Y_1 - B_c \phi^\top Y_2 \\ \dot{Y}_2 &= 0 \\ y_0 &= C_c^\top Y_1 \end{aligned} \quad (4.8.31)$$

Because ϕ is PE and $C_c^\top(sI - A_c)^{-1}B_c$ is stable and minimum phase (which is implied by $C_c^\top(sI - A_c)^{-1}B_c$ being SPR) and $\dot{\phi} \in \mathcal{L}_\infty$, it follows from Lemma 4.8.3 (iii) that

$$\phi_f(\tau) \triangleq \int_t^\tau C_c^\top e^{A_c(\tau-\sigma)} B_c \phi(\sigma) d\sigma$$

is also PE; therefore, there exist constants $\alpha_1, \alpha_2, T_0 > 0$ such that

$$\alpha_2 I \geq \frac{1}{T_0} \int_t^{t+T_0} \phi_f(\tau)\phi_f^\top(\tau) d\tau \geq \alpha_1 I, \quad \forall t \geq 0$$

Hence, applying Lemma 4.8.4 to the system (4.8.31), we conclude that $(C, A + KC^\top)$ is UCO which implies that (C, A) is UCO. Therefore, we conclude that the equilibrium $\tilde{\theta}_e = 0, e_e = 0$ of (4.8.29) is e.s. in the large. $\square$

4.8.3 Proof of Theorem 4.3.2 (iii)

The parameter error equation (4.3.53) may be written as

$$\begin{aligned} \dot{\tilde{\theta}} &= A(t)\tilde{\theta} \\ y_0 &= C^\top(t)\tilde{\theta} \end{aligned} \quad (4.8.32)$$

where $A(t) = -\Gamma\frac{\phi\phi^\top}{m^2}, C^\top(t) = -\frac{\phi^\top}{m}$, $y_0 = \epsilon m$. The system (4.8.32) is analyzed using the Lyapunov-like function

$$V = \frac{\tilde{\theta}^\top \Gamma^{-1} \tilde{\theta}}{2}$$

4.8. PARAMETER CONVERGENCE PROOFS

that led to

$$\dot V = -\frac{(\tilde\theta^\top \phi)^2}{m^2} = -\epsilon^2 m^2$$

along the solution of (4.8.32). We need to establish that the equilibrium $\tilde\theta_e = 0$ of (4.8.32) is e.s. We achieve that by using Theorem 3.4.8 as follows: Let $P = \Gamma^{-1}$, then $V = \frac{\tilde\theta^\top P \tilde\theta}{2}$ and

$$\dot V = \frac{1}{2}\tilde\theta^\top [PA(t) + A^\top(t)P + \dot P]\tilde\theta = -\tilde\theta^\top C(t)C^\top(t)\tilde\theta$$

where $\dot P = 0$. This implies that

$$\dot P + PA(t) + A^\top P + 2C(t)C^\top(t) \le 0$$

and according to Theorem 3.4.8, $\tilde\theta_e = 0$ is e.s. provided (C, A) is UCO. Using Lemma 4.8.1, we have that (C, A) is UCO if $(C, A + KC^\top)$ is UCO for some K that satisfies the condition of Lemma 4.8.1. We choose

$$K = -\Gamma \frac{\phi}{m}$$

leading to $A + KC^\top = 0$. We consider the following system that corresponds to the pair $(C, A + KC^\top)$, i.e.,

$$\begin{aligned}\dot Y &= 0 \\ y_0 &= C^\top Y = -\frac{\phi^\top}{m}Y\end{aligned} \qquad (4.8.33)$$

The observability grammian of (4.8.33) is given by

$$\mathcal{N}(t, t+T) = \int_t^{t+T} \frac{\phi(\tau)\phi^\top(\tau)}{m^2(\tau)} d\tau$$

Because ϕ is PE and $m \ge 1$ is bounded, it follows immediately that the grammian matrix $\mathcal{N}(t, t+T)$ is positive definite for some $T > 0$ and for all $t \ge 0$, which implies that (4.8.33) is UCO which in turn implies that (C, A) is UCO; thus, the proof is complete. □

In the following, we give an alternative proof of Theorem 4.3.2 (iii), which does not make use of the UCO property.

From (4.3.54), we have

$$V(t+T) = V(t) - \int_t^{t+T} \epsilon^2 m^2 d\tau = V(t) - \int_t^{t+T} \frac{(\tilde\theta^\top(\tau)\phi(\tau))^2}{m^2} d\tau \qquad (4.8.34)$$

for any $t, T > 0$. Expressing $\tilde{\theta}^\top(\tau)\phi(\tau)$ as $\tilde{\theta}^\top(\tau)\phi(\tau) = \tilde{\theta}^\top(t)\phi(\tau) + (\tilde{\theta}(\tau) - \tilde{\theta}(t))^\top \phi(\tau)$ and using the inequality $(x+y)^2 \geq \frac{1}{2}x^2 - y^2$, it follows that

$$\int_t^{t+T} \frac{(\tilde{\theta}^\top(\tau)\phi(\tau))^2}{m^2} d\tau \geq \frac{1}{m_0} \int_t^{t+T} \left(\tilde{\theta}^\top(\tau)\phi(\tau)\right)^2 d\tau$$

$$\geq \frac{1}{m_0}\left\{\frac{1}{2}\int_t^{t+T}(\tilde{\theta}^\top(t)\phi(\tau))^2 d\tau \right. \quad (4.8.35)$$

$$\left. - \int_t^{t+T}\left((\tilde{\theta}(\tau)-\tilde{\theta}(t))^\top\phi(\tau)\right)^2 d\tau\right\}$$

where $m_0 = \sup_{t \geq 0} m^2(t)$ is a constant. Because ϕ is PE, i.e.,

$$\int_t^{t+T_0} \phi(\tau)\phi^\top(\tau)d\tau \geq \alpha_0 T_0 I$$

for some T_0 and $\alpha_0 > 0$, we have

$$\int_t^{t+T_0}\left(\tilde{\theta}^\top(t)\phi(\tau)\right)^2 d\tau \geq \alpha_0 T_0 \tilde{\theta}^\top(t)\tilde{\theta}(t)$$

$$\geq 2\alpha_0 T_0 \lambda_{max}^{-1}(\Gamma^{-1})V(t) \quad (4.8.36)$$

$$= 2\alpha_0 T_0 \lambda_{min}(\Gamma)V(t)$$

On the other hand, we can write

$$\tilde{\theta}(\tau) - \tilde{\theta}(t) = \int_t^\tau \dot{\tilde{\theta}}(\sigma)d\sigma = -\int_t^\tau \Gamma \frac{\tilde{\theta}^\top(\sigma)\phi(\sigma)}{m^2(\sigma)}\phi(\sigma)d\sigma$$

and

$$(\tilde{\theta}(\tau)-\tilde{\theta}(t))^\top\phi(\tau) = -\int_t^\tau \left(\frac{\tilde{\theta}^\top(\sigma)\phi(\sigma)\Gamma\phi(\sigma)}{m^2}\right)^\top \phi(\tau)d\sigma$$

$$= -\int_t^\tau \frac{\tilde{\theta}^\top(\sigma)\phi(\sigma)}{m(\sigma)}\frac{\phi^\top(\tau)\Gamma\phi(\sigma)}{m(\sigma)}d\sigma \quad (4.8.37)$$

Noting that $m \geq 1$, it follows from (4.8.37) and the Schwartz inequality that

$$\int_t^{t+T_0}\left((\tilde{\theta}(\tau)-\tilde{\theta}(t))^\top\phi(\tau)\right)^2 d\tau$$

$$\leq \int_t^{t+T_0}\left(\int_t^\tau \left(\frac{\phi^\top(\tau)\Gamma\phi(\sigma)}{m(\sigma)}\right)^2 d\sigma \int_t^\tau \left(\frac{\tilde{\theta}^\top(\sigma)\phi(\sigma)}{m(\sigma)}\right)^2 d\sigma\right) d\tau$$

$$\leq \beta^4 \lambda_{max}^2(\Gamma)\int_t^{t+T_0}(\tau-t)\int_t^\tau \left(\frac{\tilde{\theta}^\top(\sigma)\phi(\sigma)}{m(\sigma)}\right)^2 d\sigma d\tau$$

4.8. PARAMETER CONVERGENCE PROOFS

where $\beta = \sup_{\tau \geq 0} |\phi(\tau)|$. Changing the sequence of integration, we have

$$\int_t^{t+T_0} \left((\tilde{\theta}(\tau) - \tilde{\theta}(t))^\top \phi(\tau) \right)^2 d\tau$$

$$\leq \beta^4 \lambda_{max}^2(\Gamma) \int_t^{t+T_0} \left(\frac{\tilde{\theta}^\top(\sigma)\phi(\sigma)}{m(\sigma)} \right)^2 \int_\sigma^{t+T_0} (\tau - t) d\tau d\sigma$$

$$\leq \beta^4 \lambda_{max}^2(\Gamma) \int_t^{t+T_0} \frac{(\tilde{\theta}^\top(\sigma)\phi(\sigma))^2}{m^2(\sigma)} \left\{ \frac{T_0^2 - (\sigma - t)^2}{2} \right\} d\sigma$$

$$\leq \frac{\beta^4 \lambda_{max}^2(\Gamma) T_0^2}{2} \int_t^{t+T_0} \frac{(\tilde{\theta}^\top(\sigma)\phi(\sigma))^2}{m^2(\sigma)} d\sigma \qquad (4.8.38)$$

Using (4.8.36) and (4.8.38) in (4.8.35) with $T = T_0$ we have

$$\int_t^{t+T_0} \frac{(\tilde{\theta}^\top(\tau)\phi(\tau))^2}{m^2(\tau)} d\tau \geq \frac{\alpha_0 T_0 \lambda_{min}(\Gamma)}{m_0} V(t)$$

$$- \frac{\beta^4 T_0^2 \lambda_{max}^2(\Gamma)}{2m_0} \int_t^{t+T_0} \frac{(\tilde{\theta}^\top(\sigma)\phi(\sigma))^2}{m^2(\sigma)} d\sigma$$

which implies

$$\int_t^{t+T_0} \frac{(\tilde{\theta}^\top(\tau)\phi(\tau))^2}{m^2(\tau)} d\tau \geq \frac{1}{1 + \frac{\beta^4 T_0^2 \lambda_{max}^2(\Gamma)}{2m_0}} \frac{\alpha_0 T_0 \lambda_{min}(\Gamma)}{m_0} V(t)$$

$$\triangleq \gamma_1 V(t) \qquad (4.8.39)$$

where $\gamma_1 = \frac{2\alpha_0 T_0 \lambda_{min}(\Gamma)}{2m_0 + \beta^4 T_0^2 \lambda_{max}^2(\Gamma)}$. Using (4.8.39) in (4.8.34) with $T = T_0$, it follows that

$$V(t + T_0) \leq V(t) - \gamma_1 V(t) = \gamma V(t) \qquad (4.8.40)$$

where $\gamma = 1 - \gamma_1$. Because $\gamma_1 > 0$ and $V(t + T_0) \geq 0$, we have $0 < \gamma < 1$. Because (4.8.40) holds for all $t \geq 0$, we can take $t = (n-1)T_0$ and use (4.8.40) successively to obtain

$$V(t) \leq V(nT_0) \leq \gamma^n V(0), \qquad \forall t \geq nT_0, \ n = 0, 1, \ldots$$

Hence, $V(t) \to 0$ as $t \to \infty$ exponentially fast which implies that $\tilde{\theta}(t) \to 0$ as $t \to \infty$ exponentially fast. □

4.8.4 Proof of Theorem 4.3.3 (iv)

In proving Theorem 4.3.3 (i) to (iii), we have shown that (see equation (4.3.64))

$$\dot{V}(t) = -\tilde{\theta}^\top(t) R(t) \tilde{\theta}(t)$$

where $V = \frac{\tilde{\theta}^\top \Gamma^{-1} \tilde{\theta}}{2}$. From equation (4.3.60), we have

$$R(t) = \int_0^t e^{-\beta(t-\tau)} \frac{\phi(\tau)\phi^\top(\tau)}{m^2(\tau)} d\tau$$

Because ϕ is PE and m is bounded, we have

$$\begin{aligned} R(t) &= \int_{t-T_0}^t e^{-\beta(t-\tau)} \frac{\phi(\tau)\phi^\top(\tau)}{m^2(\tau)} d\tau + \int_0^{t-T_0} e^{-\beta(t-\tau)} \frac{\phi(\tau)\phi^\top(\tau)}{m^2(\tau)} d\tau \\ &\geq \alpha_0' e^{-\beta T_0} \int_{t-T_0}^t \phi(\tau)\phi^\top(\tau) d\tau \\ &\geq \beta_1 e^{-\beta T_0} I \end{aligned}$$

for any $t \geq T_0$, where $\beta_1 = \alpha_0 \alpha_0' T_0$, $\alpha_0' = \sup_t \frac{1}{m^2(t)}$ and $\alpha_0, T_0 > 0$ are constants given by (4.3.40) in the definition of PE. Therefore,

$$\dot{V} \leq -\beta_1 e^{-\beta T_0} \tilde{\theta}^\top \tilde{\theta} \leq -2\beta_1 \lambda_{min}(\Gamma) e^{-\beta T_0} V$$

for $t \geq T_0$, which implies that $V(t)$ satisfies

$$V(t) \leq e^{-\alpha(t-T_0)} V(T_0), \quad t \geq T_0$$

where $\alpha = 2\beta_1 e^{-\beta T_0} \lambda_{min}(\Gamma)$. Thus, $V(t) \to 0$ as $t \to \infty$ exponentially fast with a rate equal to α. Using $\sqrt{2\lambda_{min}(\Gamma) V} \leq |\tilde{\theta}| \leq \sqrt{2\lambda_{max}(\Gamma) V}$, we have that

$$|\tilde{\theta}(t)| \leq \sqrt{2\lambda_{max}(\Gamma) V(T_0)} e^{-\frac{\alpha}{2}(t-T_0)} \leq \sqrt{\frac{\lambda_{max}(\Gamma)}{\lambda_{min}(\Gamma)}} |\tilde{\theta}(T_0)| e^{-\frac{\alpha}{2}(t-T_0)}, \quad t \geq T_0$$

Thus, $\theta(t) \to \theta^*$ exponentially fast with a rate of $\frac{\alpha}{2}$ as $t \to \infty$. Furthermore, for $\Gamma = \gamma I$, $\alpha = 2\beta_1 \gamma e^{-\beta T_0}$ and the rate of convergence ($\alpha/2$) can be made large by increasing the value of the adaptive gain. □

4.8.5 Proof of Theorem 4.3.4 (iv)

In proving Theorem 4.3.4 (i) to (iii), we have shown that $\tilde{\theta}(t)$ satisfies the following equation

$$\tilde{\theta}(t) = P(t) P_0^{-1} \tilde{\theta}(0)$$

We now show that $P(t) \to 0$ as $t \to \infty$ when ϕ satisfies the PE assumption.
Because P^{-1} satisfies

$$\frac{d}{dt} P^{-1} = \frac{\phi \phi^\top}{m^2}$$

4.8. PARAMETER CONVERGENCE PROOFS

using the condition that ϕ is PE, i.e., $\int_t^{t+T_0} \phi(\tau)\phi^\top(\tau)d\tau \geq \alpha_0 T_0 I$ for some constant $\alpha_0, T_0 > 0$, it follows that

$$P^{-1}(t) - P^{-1}(0) = \int_0^t \frac{\phi\phi^\top}{m^2}d\tau \geq n_0 \frac{\alpha_0 T_0}{\bar{m}} I \geq \left(\frac{t}{T_0} - 1\right)\frac{\alpha_0 T_0 I}{\bar{m}}$$

where $\bar{m} = \sup_t\{m^2(t)\}$ and n_0 is the largest integer that satisfies $n_0 \leq \frac{t}{T_0}$, i.e., $n_0 \geq \frac{t}{T_0} - 1$. Therefore,

$$\begin{aligned} P^{-1}(t) &\geq P^{-1}(0) + \left(\frac{t}{T_0} - 1\right)\frac{\alpha_0 T_0}{\bar{m}}I \\ &\geq \left(\frac{t}{T_0} - 1\right)\frac{\alpha_0 T_0}{\bar{m}}I, \quad \forall t > T_0 \end{aligned}$$

which, in turn, implies that

$$P(t) \leq \left(\left(\frac{t}{T_0} - 1\right)\alpha_0 T_0\right)^{-1} \bar{m} I, \quad \forall t > T_0 \tag{4.8.41}$$

Because $P(t) \geq 0$ for all $t \geq 0$ and the right-hand side of (4.8.41) goes to zero as $t \to \infty$, we can conclude that $P(t) \to 0$ as $t \to \infty$. Hence, $\tilde{\theta}(t) = P(t)P_0^{-1}\tilde{\theta}(0) \to 0$ as $t \to \infty$. □

4.8.6 Proof of Corollary 4.3.2

Let us denote $\Gamma = P^{-1}(t)$, then from (4.3.78) we have

$$\dot{\Gamma} = -\beta\Gamma + \frac{\phi\phi^\top}{m^2}, \quad \Gamma(0) = \Gamma^\top(0) = \Gamma_0 = P_0^{-1}$$

or

$$\Gamma(t) = e^{-\beta t}\Gamma_0 + \int_0^t e^{-\beta(t-\tau)}\frac{\phi(\tau)\phi^\top(\tau)}{m^2}d\tau$$

Using the condition that $\phi(t)$ is PE and $m \in \mathcal{L}_\infty$, we can show that for all $t \geq T_0$

$$\begin{aligned} \Gamma(t) &\geq \int_0^t e^{-\beta(t-\tau)}\frac{\phi\phi^\top}{m^2}d\tau \\ &= \int_{t-T_0}^t e^{-\beta(t-\tau)}\frac{\phi(\tau)\phi^\top(\tau)}{m^2}d\tau + \int_0^{t-T_0} e^{-\beta(t-\tau)}\frac{\phi(\tau)\phi^\top(\tau)}{m^2}d\tau \\ &\geq e^{-\beta T_0}\frac{\alpha_0 T_0}{\bar{m}}I \end{aligned} \tag{4.8.42}$$

where $\bar{m} = \sup_t m^2(t)$. For $t \leq T_0$, we have

$$\Gamma(t) \geq e^{-\beta t}\Gamma_0 \geq e^{-\beta T_0}\Gamma_0 \geq \lambda_{min}(\Gamma_0)e^{-\beta T_0}I \tag{4.8.43}$$

Conditions (4.8.42), (4.8.43) imply that

$$\Gamma(t) \geq \gamma_1 I \qquad (4.8.44)$$

for all $t \geq 0$ where $\gamma_1 = \min\{\frac{\alpha_0 T_0}{\beta_1}, \lambda_{min}(\Gamma_0)\}e^{-\beta T_0}$.

On the other hand, using the boundedness of ϕ, we can establish that for some constant $\beta_2 > 0$

$$\begin{aligned}\Gamma(t) &\leq \Gamma_0 + \beta_2 \int_0^t e^{-\beta(t-\tau)}d\tau I \\ &\leq \lambda_{max}(\Gamma_0)I + \frac{\beta_2}{\beta}I \leq \gamma_2 I\end{aligned} \qquad (4.8.45)$$

where $\gamma_2 = \lambda_{max}(\Gamma_0) + \frac{\beta_2}{\beta} > 0$.

Combining (4.8.44) and (4.8.45), we conclude

$$\gamma_1 I \leq \Gamma(t) \leq \gamma_2 I$$

for some $\gamma_1 > 0, \gamma_2 > 0$. Therefore,

$$\gamma_2^{-1} I \leq P(t) \leq \gamma_1^{-1} I$$

and consequently $P(t), P^{-1}(t) \in \mathcal{L}_\infty$. Because $P(t), P^{-1}(t) \in \mathcal{L}_\infty$, the exponential convergence of θ to θ^* can be proved using exactly the same procedure and arguments as in the proof of Theorem 4.3.2. □

4.8.7 Proof of Theorem 4.5.1(iii)

Consider the following differential equations which describe the behavior of the adaptive law (see (4.5.6) and (4.5.7)):

$$\begin{aligned}\dot{e} &= A_c e + B_c(-\rho^* \tilde{\theta}^\top \phi - \tilde{\rho}\xi - \epsilon n_s^2) \\ \dot{\tilde{\theta}} &= \Gamma \epsilon \phi \operatorname{sgn}(\rho^*) \\ \dot{\tilde{\rho}} &= \gamma \epsilon \xi \\ \epsilon &= C_c^\top e\end{aligned} \qquad (4.8.46)$$

Because $\xi, \rho \in \mathcal{L}_\infty$ and $\xi \in \mathcal{L}_2$, we can treat ξ, ρ as external input functions and write (4.8.46) as

$$\dot{x}_a = A_a x_a + B_a(-\tilde{\rho}\xi) \qquad (4.8.47)$$

where

$$x_a = \begin{bmatrix} e \\ \tilde{\theta} \end{bmatrix}, \quad A_a = \begin{bmatrix} A_c - n_s^2 B_c C_c^\top & -\rho^* B_c \phi^\top \\ \Gamma \operatorname{sgn}(\rho^*) \phi C_c^\top & 0 \end{bmatrix}, \quad B_a = \begin{bmatrix} B_c \\ 0 \end{bmatrix}$$

4.8. PARAMETER CONVERGENCE PROOFS

In proving Corollary 4.3.1, we have shown that when ϕ is PE and $\phi, \dot{\phi} \in \mathcal{L}_\infty$, the system $\dot{x} = A_a x$ is e.s. Therefore, the state transition matrix $\Phi_a(t, t_0)$ of (4.8.47) satisfies

$$\|\Phi_a(t, 0)\| \leq \alpha_0 e^{-\gamma_0 t} \qquad (4.8.48)$$

for some constants $\alpha_0, \gamma_0 > 0$, which together with $-B_a \tilde{\rho} \xi \in \mathcal{L}_2$ imply that $x_a(t) \to 0$ as $t \to \infty$.

4.8.8 Proof of Theorem 4.6.1 (iii)

From the proof of Theorem 4.6.1 (i) to (ii), we have the inequality (see (4.6.9))

$$V(k+1) - V(k) \leq -(2 - T_s \lambda_m) \int_{t_k}^{t_{k+1}} \epsilon^2(\tau) m^2(\tau) d\tau \qquad (4.8.49)$$

Using inequality (4.8.49) consecutively, we have

$$\begin{aligned} V(k+n) - V(k) &\leq -(2 - T_s \lambda_m) \int_{t_k}^{t_{k+n}} \epsilon^2(\tau) m^2(\tau) d\tau \\ &= -(2 - T_s \lambda_m) \sum_{i=0}^{n-1} \int_{t_{k+i}}^{t_{k+i+1}} \epsilon^2(\tau) m^2(\tau) d\tau \end{aligned} \qquad (4.8.50)$$

for any integer n. We now write

$$\begin{aligned} \int_{t_{k+i}}^{t_{k+i+1}} \epsilon^2(\tau) m^2(\tau) d\tau &= \int_{t_{k+i}}^{t_{k+i+1}} \frac{\left(\tilde{\theta}_{k+i}^\top \phi(\tau)\right)^2}{m^2(\tau)} d\tau \\ &= \int_{t_{k+i}}^{t_{k+i+1}} \frac{\left(\tilde{\theta}_k^\top \phi(\tau) + (\tilde{\theta}_{k+i} - \tilde{\theta}_k)^\top \phi(\tau)\right)^2}{m^2(\tau)} d\tau \end{aligned}$$

Using the inequality $(x+y)^2 \geq \frac{1}{2} x^2 - y^2$, we write

$$\int_{t_{k+i}}^{t_{k+i+1}} \epsilon^2(\tau) m^2(\tau) d\tau \geq \frac{1}{2} \int_{t_{k+i}}^{t_{k+i+1}} \frac{\left(\tilde{\theta}_k^\top \phi(\tau)\right)^2}{m^2(\tau)} d\tau - \int_{t_{k+i}}^{t_{k+i+1}} \frac{\left((\tilde{\theta}_{k+i} - \tilde{\theta}_k)^\top \phi(\tau)\right)^2}{m^2(\tau)} d\tau \qquad (4.8.51)$$

Because $\phi(\tau)/m(\tau)$ is bounded, we denote $c = \sup \frac{|\phi(\tau)|^2}{m^2(\tau)}$ and have

$$\int_{t_{k+i}}^{t_{k+i+1}} \frac{\left((\tilde{\theta}_{k+i} - \tilde{\theta}_k)^\top \phi(\tau)\right)^2}{m^2(\tau)} d\tau \leq c T_s |\tilde{\theta}_{k+i} - \tilde{\theta}_k|^2$$

From the hybrid adaptive algorithm, we have

$$\tilde{\theta}_{k+i} - \tilde{\theta}_k = \int_{t_k}^{t_{k+i}} \epsilon(\tau) \phi(\tau) d\tau, \quad i = 1, 2, \ldots, n$$

therefore, using the Schwartz inequality and the boundedness of $|\phi(t)|/m(t)$,

$$|\tilde{\theta}_{k+i} - \tilde{\theta}_k|^2 \leq \left(\int_{t_k}^{t_{k+i}} |\epsilon(\tau)||\phi(\tau)|d\tau\right)^2 = \left(\int_{t_k}^{t_{k+i}} |\epsilon(\tau)|m(\tau)\frac{|\phi(\tau)|}{m(\tau)}d\tau\right)^2$$

$$\leq \int_{t_k}^{t_{k+i}} \epsilon^2(\tau)m^2(\tau)d\tau \int_{t_k}^{t_{k+i}} \frac{|\phi(\tau)|^2}{m^2(\tau)}d\tau$$

$$\leq ciT_s \int_{t_k}^{t_{k+i}} \epsilon^2(\tau)m^2(\tau)d\tau$$

$$\leq ciT_s \int_{t_k}^{t_{k+n}} \epsilon^2(\tau)m^2(\tau)d\tau \quad (4.8.52)$$

Using the expression (4.8.52) in (4.8.51), we have

$$\int_{t_{k+i}}^{t_{k+i+1}} \epsilon^2(\tau)m^2(\tau)d\tau \geq \frac{1}{2}\int_{t_{k+i}}^{t_{k+i+1}} \frac{\left(\tilde{\theta}_k^T \phi(\tau)\right)^2}{m^2(\tau)}d\tau - c^2 iT_s^2 \int_{t_k}^{t_{k+n}} \epsilon^2(\tau)m^2(\tau)d\tau$$

which leads to

$$\int_{t_k}^{t_{k+n}} \epsilon^2(\tau)m^2(\tau)d\tau = \sum_{i=0}^{n-1} \int_{t_{k+i}}^{t_{k+i+1}} \epsilon^2(\tau)m^2(\tau)d\tau$$

$$\geq \sum_{i=0}^{n-1} \left(\frac{1}{2}\int_{t_{k+i}}^{t_{k+i+1}} \frac{\left(\tilde{\theta}_k^T \phi(\tau)\right)^2}{m^2(\tau)}d\tau - c^2 iT_s^2 \int_{t_k}^{t_{k+n}} \epsilon^2(\tau)m^2(\tau)d\tau\right)$$

$$\geq \frac{1}{2}\tilde{\theta}_k^T \sum_{i=0}^{n-1} \int_{t_{k+i}}^{t_{k+i+1}} \frac{\phi(\tau)\phi^T(\tau)}{m^2(\tau)}d\tau \tilde{\theta}_k - \sum_{i=0}^{n-1} c^2 iT_s^2 \int_{t_k}^{t_{k+n}} \epsilon^2(\tau)m^2(\tau)d\tau$$

$$= \tilde{\theta}_k^T \int_{t_k}^{t_{k+n}} \frac{\phi(\tau)\phi^T(\tau)}{2m^2(\tau)}d\tau \tilde{\theta}_k - \frac{n(n-1)}{2}c^2 T_s^2 \int_{t_k}^{t_{k+n}} \epsilon^2(\tau)m^2(\tau)d\tau$$

or equivalently

$$\int_{t_k}^{t_{k+n}} \epsilon^2(\tau)m^2(\tau)d\tau \geq \frac{1}{2(1+n(n-1)c^2 T_s^2/2)}\tilde{\theta}_k^T \int_{t_k}^{t_{k+n}} \frac{\phi(\tau)\phi^T(\tau)}{m^2(\tau)}d\tau \tilde{\theta}_k \quad (4.8.53)$$

Because ϕ is PE and $1 \leq m < \infty$, there exist constants α_1', α_2' and $T_0 > 0$ such that

$$\alpha_2' I \geq \int_t^{t+T_0} \frac{\phi(\tau)\phi^T(\tau)}{m^2}d\tau \geq \alpha_1' I$$

for any t. Therefore, for any integer k, n where n satisfies $nT_s \geq T_0$, we have

$$\tilde{\theta}_k^T \int_{t_k}^{t_{k+n}} \frac{\phi(\tau)\phi^T(\tau)}{m^2}d\tau \tilde{\theta}_k \geq \alpha_1' \tilde{\theta}_k^T \tilde{\theta}_k \geq \frac{V(k)}{\lambda_m}\alpha_1' \quad (4.8.54)$$

Using (4.8.53), (4.8.54) in (4.8.50), we obtain the following inequality:

$$V(k+n) - V(k) \leq -\frac{(2-T_s\lambda_m)\alpha_1'}{\lambda_m(2+n(n-1)c^2T_s^2)}V(k) \qquad (4.8.55)$$

hold for any integer n with $n \geq T_0/T_s$. Condition (4.8.55) is equivalent to

$$V(k+n) \leq \gamma V(k)$$

with

$$\gamma \triangleq 1 - \frac{(2-T_s\lambda_m)\alpha_1'}{\lambda_m(2+n(n-1)c^2T_s^2)} < 1$$

Therefore,

$$V(kn) = V((k-)n+n) \leq \gamma V((k-1)n) \leq \gamma^2 V((k-2)n) \leq \ldots \leq \gamma^k V(0)$$

or

$$|\tilde{\theta}_{kn}| \leq \sqrt{\frac{V(kn)}{\lambda_m}} \leq \sqrt{\frac{V(0)}{\lambda_m}}(\sqrt{\gamma})^k \qquad (4.8.56)$$

Because $0 < \gamma < 1$ and, therefore, $\sqrt{\gamma} < 1$, (4.8.56) implies that $|\tilde{\theta}_{kn}| \to 0$ exponentially fast as $t \to \infty$, which, together with the property of the hybrid adaptive algorithm (i.e., $|\tilde{\theta}_{k+1}| \leq |\tilde{\theta}_k|$), implies that θ_k converges to θ^* exponentially fast and the proof is complete.

4.9 Problems

4.1 Consider the differential equation

$$\dot{\tilde{\theta}} = -\gamma u^2 \tilde{\theta}$$

given by (4.2.7) where $\gamma > 0$. Show that a necessary and sufficient condition for $\tilde{\theta}(t)$ to converge to zero exponentially fast is that $u(t)$ satisfies (4.2.11), i.e.,

$$\int_t^{t+T_0} u^2(\tau)d\tau \geq \alpha_0 T_0$$

for all $t \geq 0$ and some constants $\alpha_0, T_0 > 0$. (Hint: Show that

$$e^{-\gamma \int_{t_1}^t u^2(\tau)d\tau} \leq \alpha e^{-\gamma_0(t-t_1)}$$

for all $t \geq t_1$ and some $\alpha, \gamma_0 > 0$ if and only if (4.2.11) is satisfied.)

4.2 Consider the second-order stable system

$$\dot{x} = \begin{bmatrix} a_{11} & a_{12} \\ a_{21} & 0 \end{bmatrix} x + \begin{bmatrix} b_1 \\ b_2 \end{bmatrix} u$$

where x, u are available for measurement, $u \in \mathcal{L}_\infty$ and $a_{11}, a_{12}, a_{21}, b_1, b_2$ are unknown parameters. Design an on-line estimator to estimate the unknown parameters. Simulate your scheme using $a_{11} = -0.25, a_{12} = 3, a_{21} = -5, b_1 = 1, b_2 = 2.2$ and $u = 10 sin 2t$. Repeat the simulation when $u = 10 sin 2t + 7 cos 3.6t$. Comment on your results.

4.3 Consider the nonlinear system

$$\dot{x} = a_1 f_1(x) + a_2 f_2(x) + b_1 g_1(x) u + b_2 g_2(x) u$$

where $u, x \in \mathcal{R}^1$, f_i, g_i are known nonlinear functions of x and a_i, b_i are unknown constant parameters. The system is such that $u \in \mathcal{L}_\infty$ implies $x \in \mathcal{L}_\infty$. If x, u can be measured at each time, design an estimation scheme for estimating the unknown parameters on-line.

4.4 Design and analyze an on-line estimation scheme for estimating θ^* in (4.3.26) when $L(s)$ is chosen so that $W(s)L(s)$ is biproper and SPR.

4.5 Design an on-line estimation scheme to estimate the coefficients of the numerator polynomial

$$Z(s) = b_{n-1} s^{n-1} + b_{n-2} s^{n-2} + \cdots + b_1 s + b_0$$

of the plant

$$y = \frac{Z(s)}{R(s)} u$$

when the coefficients of $R(s) = s^n + a_{n-1} s^{n-1} + \cdots + a_1 s + a_0$ are known. Repeat the same problem when $Z(s)$ is known and $R(s)$ is unknown.

4.6 Consider the cost function given by (4.3.51), i.e.

$$J(\theta) = \frac{(z - \theta^\top \phi)^2}{2m^2}$$

which we like to minimize w.r.t. θ. Derive the nonrecursive algorithm

$$\theta(t) = \frac{\phi z}{\phi^\top \phi}$$

provided $\phi^\top \phi \neq 0$ by solving the equation $\nabla J(\theta) = 0$.

4.7 Show that $\omega_0 = F\omega$, where $F \in \mathcal{R}^{m \times n}$ with $m \leq n$ is a constant matrix and $\omega \in \mathcal{L}_\infty$ is PE, is PE if and only if F is of full rank.

4.8 Show that if $\omega, \dot{\omega} \in \mathcal{L}_\infty$, ω is PE and either

(a) $e \in \mathcal{L}_2$ or

(b) $e \in \mathcal{L}_\infty$ and $e(t) \to 0$ as $t \to \infty$

is satisfied, then $\omega_0 = \omega + e$ is PE.

4.9 Consider the mass-spring-damper system shown in Figure 4.11.

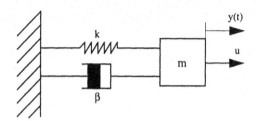

Figure 4.11 The mass-spring-damper system for Problem 4.9.

where β is the damping coefficient, k is the spring constant, u is the external force, and $y(t)$ is the displacement of the mass m resulting from the force u.

(a) Verify that the equations of the motion that describe the dynamic behavior of the system under small displacements are

$$m\ddot{y} + \beta \dot{y} + ky = u$$

(b) Design a gradient algorithm to estimate the constants m, β, k when y, u can be measured at each time t.

(c) Repeat (b) for a least squares algorithm.

(d) Simulate your algorithms in (b) and (c) on a digital computer by assuming that $m = 20$ kg, $\beta = 0.1$ kg/sec, $k = 5$ kg/sec^2 and inputs u of your choice.

(e) Repeat (d) when $m = 20$ kg for $0 \leq t \leq 20$ sec and $m = 20(2 - e^{-0.01(t-20)})$ kg for $t \geq 20$sec.

4.10 Consider the mass-spring-damper system shown in Figure 4.12.

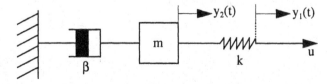

Figure 4.12 The mass-spring-damper system for Problem 4.10.

(a) Verify that the equations of motion are given by

$$k(y_1 - y_2) = u$$
$$k(y_1 - y_2) = m\ddot{y}_2 + \beta \dot{y}_2$$

(b) If y_1, y_2, u can be measured at each time t, design an on-line parameter estimator to estimate the constants k, m and β.

(c) We have the a priori knowledge that $0 \leq \beta \leq 1$, $k \geq 0.1$ and $m \geq 10$. Modify your estimator in (b) to take advantage of this a priori knowledge.

(d) Simulate your algorithm in (b) and (c) when $\beta = 0.2$ kg/sec, $m = 15$ kg, $k = 2$ kg/sec^2 and $u = 5 \sin 2t + 10.5$ kg $\cdot$ m/sec^2.

4.11 Consider the block diagram of a steer-by-wire system of an automobile shown in Figure 4.13.

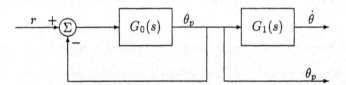

Figure 4.13 Block diagram of a steer-by-wire system for Problem 4.11.

where r is the steering command in degrees, θ_p is the pinion angle in degrees and $\dot{\theta}$ is the yaw rate in degree/sec. The transfer functions $G_0(s), G_1(s)$ are of the form

$$G_0(s) = \frac{k_0 \omega_0^2}{s^2 + 2\xi_0 \omega_0 s + \omega_0^2(1 - k_0)}$$

$$G_1(s) = \frac{k_1 \omega_1^2}{s^2 + 2\xi_1 \omega_1 s + \omega_1^2}$$

where $k_0, \omega_0, \xi_0, k_1, \omega_1, \xi_1$ are functions of the speed of the vehicle. Assuming that $r, \theta_p, \dot{\theta}$ can be measured at each time t, do the following:

(a) Design an on-line parameter estimator to estimate $k_i, \omega_i, \xi_i, i = 0, 1$ using the measurement of $\theta_p, \dot{\theta}, r$.

(b) Consider the values of the parameters shown in Table 4.6 at different speeds:

Table 4.6 Parameter values for the SBW system

Speed V	k_0	ω_0	ξ_0	k_1	ω_1	ξ_1
30 mph	0.81	19.75	0.31	0.064	14.0	0.365
60 mph	0.77	19.0	0.27	0.09	13.5	0.505

Assume that between speeds the parameters vary linearly. Use these values to simulate and test your algorithm in (a) when

(i) $r = 10\sin 0.2t + 8$ degrees and $V = 20$ mph.

(ii) $r = 5$ degrees and the vehicle speeds up from $V = 30$ mph to $V = 60$ mph in 40 second with constant acceleration and remains at 60 mph for 10 second.

4.12 Show that the hybrid adaptive law with projection presented in Section 4.6 guarantees the same properties as the hybrid adaptive law without projection.

4.13 Consider the equation of the motion of the mass-spring-damper system given in Problem 4.9, i.e.,
$$m\ddot{y} + \beta\dot{y} + ky = u$$
This system may be written in the form:
$$y = \rho^*(u - m\ddot{y} - \beta\dot{y})$$
where $\rho^* = \frac{1}{k}$ appears in a bilinear form with the other unknown parameters m, β. Use the adaptive law based on the bilinear parametric model to estimate ρ^*, m, β when u, y are the only signals available for measurement. Because $k > 0$, the sign of ρ^* may be assumed known. Simulate your adaptive law using the numerical values given in (d) and (e) of Problem 4.9.

Chapter 5

Parameter Identifiers and Adaptive Observers

5.1 Introduction

In Chapter 4, we developed a wide class of on-line parameter estimation schemes for estimating the unknown parameter vector θ^* that appears in certain general linear and bilinear parametric models. As shown in Chapter 2, these models are parameterizations of LTI plants, as well as of some special classes of nonlinear plants.

In this chapter we use the results of Chapter 4 to design *parameter identifiers* and *adaptive observers* for stable LTI plants. We define parameter identifiers as the on-line estimation schemes that guarantee convergence of the estimated parameters to the unknown parameter values. The design of such schemes includes the selection of the plant input so that a certain signal vector ϕ, which contains the I/O measurements, is PE. As shown in Chapter 4, the PE property of ϕ guarantees convergence of the estimated parameters to the unknown parameter values. A significant part of Section 5.2 is devoted to the characterization of the class of plant inputs that guarantee the PE property of ϕ. The rest of Section 5.2 is devoted to the design of parameter identifiers for plants with full and partial state measurements.

In Section 5.3 we consider the design of schemes that simultaneously estimate the plant state variables and parameters by processing the plant I/O measurements on-line. We refer to such schemes as *adaptive observers*.

The design of an adaptive observer is based on the combination of a state observer that could be used to estimate the state variables of a particular plant state-space representation with an on-line estimation scheme. The choice of the plant state-space representation is crucial for the design and stability analysis of the adaptive observer. We present several different plant state-space representations that we then use to design and analyze stable adaptive observers in Sections 5.3 to 5.5. In Section 5.6, we present all the lengthy proofs dealing with parameter convergence.

The stability properties of the adaptive observers developed are based on the assumption that the plant is stable and the plant input is bounded. This assumption is relaxed in Chapter 7, where adaptive observers are combined with control laws to form adaptive control schemes for unknown and possibly unstable plants.

5.2 Parameter Identifiers

Consider the LTI plant represented by the vector differential equation

$$\begin{aligned} \dot{x} &= Ax + Bu, \quad x(0) = x_0 \\ y &= C^\top x \end{aligned} \quad (5.2.1)$$

where $x \in \mathcal{R}^n, u \in \mathcal{R}^q, y \in \mathcal{R}^m$ and $n \geq m$. The elements of the plant input u are piecewise continuous, uniformly bounded functions of time. The plant parameters $A \in \mathcal{R}^{n \times n}$, $B \in \mathcal{R}^{n \times q}$, $C \in \mathcal{R}^{n \times m}$ are unknown constant matrices, and A is a stable matrix. We assume that the plant (5.2.1) is completely controllable and completely observable i.e., (5.2.1) is a minimal state-space representation of the plant.

The objective of this section is to design on-line parameter estimation schemes that guarantee convergence of the estimated plant parameters to the true ones. We refer to this class of schemes as *parameter identifiers*. The plant parameters to be estimated are the constant matrices A, B, C in (5.2.1) or any other set of parameters in an equivalent plant parameterization. The design of a parameter identifier consists of two steps.

In the first step we design an adaptive law for estimating the parameters of a convenient parameterization of the plant (5.2.1) by following any one of the procedures developed in Chapter 4.

In the second step, we select the input u so that the adaptive law designed in the first step guarantees that the estimated parameters converge to the true ones.

As shown in Chapter 4, parameter convergence requires additional conditions on a signal vector ϕ that are independent of the type of the adaptive law employed. In general the signal vector ϕ is related to the plant input or external input command u through the equation

$$\phi = H(s)u \qquad (5.2.2)$$

where $H(s)$ is some proper transfer matrix with stable poles. The objective of the second step is to choose the input u so that ϕ is PE, which, as shown in Chapter 4, guarantees convergence of the estimated parameters to their true values.

Before we embark on the design of parameter identifiers, let us first characterize the class of input signals u that guarantee the PE property of the signal vector ϕ in (5.2.2). The relationship between ϕ, $H(s)$ and the plant equation (5.2.1) will be explained in the sections to follow.

5.2.1 Sufficiently Rich Signals

Let us consider the first order plant

$$\dot{y} = -ay + bu, \quad y(0) = y_0$$

or

$$y = \frac{b}{s+a} u \qquad (5.2.3)$$

where $a > 0$, b are unknown constants and $y, u \in \mathcal{L}_\infty$ are measured at each time t.

We would like to estimate a, b by properly processing the input/output data y, u. It is clear that for the estimation of a, b to be possible, the input/output data should contain sufficient information about a, b.

For example, for $u = 0$ the output

$$y(t) = e^{-at} y_0$$

carries no information about the parameter b. If, in addition, $y_0 = 0$ then $y(t) \equiv 0 \ \forall t \geq 0$ obviously carries no information about any one of the plant

5.2. PARAMETER IDENTIFIERS

parameters. If we now choose $u = $ constant $c_0 \neq 0$, then

$$y(t) = e^{-at}\left(y_0 - \frac{b}{a}c_0\right) + \frac{b}{a}c_0$$

carries sufficient information about a, b provided $y_0 \neq ac_0/b$. This information disappears exponentially fast, and at steady state

$$y \approx \frac{b}{a}c_0$$

carries information about the zero frequency gain $\frac{b}{a}$ of the plant only, which is not sufficient to determine a, b uniquely.

Let us now choose $u(t) = \sin \omega_0 t$ for some $\omega_0 > 0$. The steady-state response of the plant (5.2.3) is given by

$$y \approx A \sin(\omega_0 t + \varphi)$$

where

$$A = \frac{|b|}{|j\omega_0 + a|} = \frac{|b|}{\sqrt{\omega_0^2 + a^2}}, \quad \varphi = (\text{sgn}(b) - 1)90° - \tan^{-1}\frac{\omega_0}{a} \quad (5.2.4)$$

It is clear that measurements of the amplitude A and phase φ uniquely determine a, b by solving (5.2.4) for the unknown a, b.

The above example demonstrates that for the on-line estimation of a, b to be possible, the input signal has to be chosen so that $y(t)$ carries sufficient information about a, b. This conclusion is obviously independent of the method or scheme used to estimate a, b.

Let us now consider an on-line estimation scheme for the first-order plant (5.2.3). For simplicity we assume that $\dot{y}$ is measured and write (5.2.3) in the familiar form of the linear parametric model

$$z = \theta^{*\mathsf{T}}\phi$$

where $z = \dot{y}, \theta^* = [b, a]^{\mathsf{T}}, \phi = [u, -y]^{\mathsf{T}}$.

Following the results of Chapter 4, we consider the following gradient algorithm for estimating θ^* given in Table 4.2.

$$\begin{aligned}\dot{\theta} &= \Gamma\epsilon\phi, \quad \Gamma = \Gamma^{\mathsf{T}} > 0 \\ \epsilon &= z - \hat{z}, \quad \hat{z} = \theta^{\mathsf{T}}\phi\end{aligned} \quad (5.2.5)$$

where $\theta(t)$ is the estimate of θ^* at time t and the normalizing signal $m^2 = 1$ due to $\phi \in \mathcal{L}_\infty$.

As we established in Chapter 4, (5.2.5) guarantees that $\epsilon, \dot\theta \in \mathcal{L}_2 \cap \mathcal{L}_\infty$, $\theta \in \mathcal{L}_\infty$, and if $\dot u \in \mathcal{L}_\infty$ then $\epsilon(t), \dot\theta(t) \to 0$ as $t \to \infty$. If in addition ϕ is PE, i.e., it satisfies

$$\alpha_1 I \geq \frac{1}{T_0} \int_t^{t+T_0} \phi(\tau)\phi^\top(\tau) d\tau \geq \alpha_0 I, \quad \forall t \geq 0 \qquad (5.2.6)$$

for some $T_0, \alpha_0, \alpha_1 > 0$, then $\theta(t) \to \theta^*$ exponentially fast. Because the vector ϕ is given by

$$\phi = \begin{bmatrix} 1 \\ -\frac{b}{s+a} \end{bmatrix} u$$

the convergence of $\theta(t)$ to θ^* is guaranteed if we choose u so that ϕ is PE, i.e., it satisfies (5.2.6). Let us try the choices of input u considered earlier. For $u = c_0$ the vector ϕ at steady state is given by

$$\phi \approx \begin{bmatrix} c_0 \\ -\frac{bc_0}{a} \end{bmatrix}$$

which does not satisfy the right-hand-side inequality in (5.2.6) for any constant c_0. Hence, for $u = c_0$, ϕ cannot be PE and, therefore, $\theta(t)$ cannot be guaranteed to converge to θ^* exponentially fast. This is not surprising since as we showed earlier, for $u = c_0$, $y(t)$ does not carry sufficient information about a, b at steady state.

On the other hand, for $u = \sin t$ we have that at steady state

$$\phi \approx \begin{bmatrix} \sin t \\ -A \sin(t + \varphi) \end{bmatrix}$$

where $A = \frac{|b|}{\sqrt{1+a^2}}$, $\varphi = (\text{sgn}(b) - 1)90° - \tan^{-1}\frac{1}{a}$, which can be shown to satisfy (5.2.6). Therefore, for $u = \sin t$ the signal vector ϕ carries sufficient information about a and b, ϕ is PE and $\theta(t) \to \theta^*$ exponentially fast.

We say that $u = \sin t$ is *sufficiently rich* for identifying the plant (5.2.3), i.e., it contains a sufficient number of frequencies to excite all the modes of the plant. Because $u = c_0 \neq 0$ can excite only the zero frequency gain of the plant, it is not sufficiently rich for the plant (5.2.3).

5.2. PARAMETER IDENTIFIERS

Let us consider the second order plant

$$y = \frac{b_1 s + b_0}{s^2 + a_1 s + a_0} u = G(s) u \qquad (5.2.7)$$

where $a_1, a_0 > 0$ and $G(s)$ has no zero-pole cancellations. We can show that for $u = \sin \omega_0 t$, $y(t)$ at steady state does not carry sufficient information to be able to uniquely determine a_1, a_0, b_1, b_0. On the other hand, $u(t) = \sin \omega_0 t + \sin \omega_1 t$ where $\omega_0 \neq \omega_1$ leads to the steady-state response

$$y(t) = A_0 \sin(\omega_0 t + \varphi_0) + A_1 \sin(\omega_1 t + \varphi_1)$$

where $A_0 = |G(j\omega_0)|, \varphi_0 = \angle G(j\omega_0), A_1 = |G(j\omega_1)|, \varphi_1 = \angle G(j\omega_1)$. By measuring $A_0, A_1, \varphi_0, \varphi_1$ we can determine uniquely a_1, a_0, b_1, b_0 by solving four algebraic equations.

Because each frequency in u contributes two equations, we can argue that the number of frequencies that u should contain, in general, is proportional to the number of unknown plant parameters to be estimated.

We are now in a position to give the following definition of sufficiently rich signals.

Definition 5.2.1 *A signal $u : \mathcal{R}^+ \to \mathcal{R}$ is called **sufficiently rich** of order n if it consists of at least $\frac{n}{2}$ distinct frequencies.*

For example, the input

$$u = \sum_{i=1}^{m} A_i \sin \omega_i t \qquad (5.2.8)$$

where $m \geq n/2$, $A_i \neq 0$ are constants and $\omega_i \neq \omega_k$ for $i \neq k$ is sufficiently rich of order n.

A more general definition of sufficient richness that includes signals that are not necessarily equal to a sum of sinusoids is presented in [201] and is given below.

Definition 5.2.2 *A signal $u : \mathcal{R}^+ \to \mathcal{R}^n$ is said to be **stationary** if the following limit exists uniformly in t_0*

$$R_u(t) = \lim_{T \to \infty} \frac{1}{T} \int_{t_0}^{t_0+T} u(\tau) u^\top(t + \tau) d\tau$$

The matrix $R_u(t) \in \mathcal{R}^{n \times n}$ is called the *autocovariance* of u. $R_u(t)$ is a positive semidefinite matrix and its Fourier transform given by

$$S_u(\omega) = \int_{-\infty}^{\infty} e^{-j\omega\tau} R_u(\tau) d\tau$$

is referred to as the *spectral measure* of u. If u has a sinusoidal component at frequency ω_0 then u is said to have a *spectral line* at frequency ω_0 and $S_u(\omega)$ has a *point mass* (a delta function) at ω_0 and $-\omega_0$. Given $S_u(\omega)$, $R_u(t)$ can be calculated using the inverse Fourier transform, i.e.,

$$R_u(t) = \frac{1}{2\pi} \int_{-\infty}^{\infty} e^{j\omega t} S_u(\omega) d\omega$$

Furthermore, we have

$$\int_{-\infty}^{\infty} S_u(\omega) d\omega = 2\pi R_u(0)$$

For further details about the properties of $R_u(t)$, $S_u(\omega)$, the reader is referred to [186, 201].

Definition 5.2.3 *A stationary signal $u : \mathcal{R}^+ \to \mathcal{R}$ is called* **sufficiently rich of order** n, *if the support of the spectral measure $S_u(\omega)$ of u contains at least n points.*

Definition 5.2.3 covers a wider class of signals that includes those specified by Definition 5.2.1. For example, the input (5.2.8) has a spectral measure with $2m$ points of support, i.e., at $\omega_i, -\omega_i$ for $i = 1, 2, \ldots m$, where $m \geq n/2$, and is, therefore, sufficiently rich of order n.

Let us now consider the equation

$$\phi = H(s)u \qquad (5.2.9)$$

where $H(s)$ is a proper transfer matrix with stable poles and $\phi \in \mathcal{R}^n$. The PE property of ϕ is related to the sufficient richness of u by the following theorem given in [201].

Theorem 5.2.1 *Let $u : \mathcal{R}^+ \mapsto \mathcal{R}$ be stationary and assume that $H(j\omega_1)$, ..., $H(j\omega_n)$ are linearly independent on $\mathcal{C}^n$ for all $\omega_1, \omega_2, \ldots, \omega_n \in \mathcal{R}$, where $\omega_i \neq \omega_k$ for $i \neq k$. Then ϕ is PE if, and only if, u is sufficiently rich of order n.*

5.2. PARAMETER IDENTIFIERS

The proof of Theorem 5.2.1 is given in Section 5.6.

The notion of persistence of excitation of the vector ϕ and richness of the input u attracted the interest of several researchers in the 1960s and 1970s who gave various interpretations to the properties of PE and sufficiently rich signals. The reader is referred to [1, 171, 201, 209, 242] for further information on the subject.

Roughly speaking, if u has at least one distinct frequency component for each two unknown parameters, then it is sufficiently rich. For example, if the number of unknown parameters is n, then $m \geq \frac{n}{2}$ distinct frequencies in u are sufficient for u to qualify as being sufficiently rich of order n. Of course, these statements are valid provided $H(j\omega_1), \ldots, H(j\omega_n)$ with $\omega_i \neq \omega_k$ are linearly independent on $\mathcal{C}^n$ for all $\omega_i \in \mathcal{R}, i = 1, 2, \ldots, n$. The vectors $H(j\omega_i), i = 1, 2, \ldots, n$ may become linearly dependent at some frequencies in $\mathcal{R}$ under certain conditions such as the one illustrated by the following example where zeros of the plant are part of the internal model of u.

Example 5.2.1 Let us consider the following plant:

$$y = \frac{b_0(s^2+1)}{(s+2)^3} u = G(s)u$$

where b_0 is the only unknown parameter. Following the procedure of Chapter 2, we rewrite the plant in the form of the linear parametric model

$$y = \theta^* \phi$$

where $\theta^* = b_0$ is the unknown parameter and

$$\phi = H(s)u, \qquad H(s) = \frac{s^2+1}{(s+2)^3}$$

According to Theorem 5.2.1, we first need to check the linear independence of $H(j\omega_1), \ldots, H(j\omega_n)$. For $n = 1$ this condition becomes $H(j\omega) \neq 0, \forall \omega \in \mathcal{R}$. It is clear that for $\omega = 1, H(j) = 0$, and, therefore, ϕ may not be PE if we simply choose u to be sufficiently rich of order 1. That is, for $u = \sin t$, the steady-state values of ϕ, y are equal to zero and, therefore, carry no information about the unknown b_0. We should note, however, that for $u = \sin \omega t$ and any $\omega \neq 1, 0$, ϕ is PE. $\triangledown$

Remark 5.2.1 The above example demonstrates that the condition for the linear independence of the vectors $H(j\omega_i), i = 1, 2, \cdots, n$ on $\mathcal{C}^n$ is sufficient to guarantee that ϕ is PE when u is sufficiently rich of order n.

It also demonstrates that when the plant is partially known, the input u does not have to be sufficiently rich of order n where n is the order of the plant. In this case, the condition on u can be relaxed, depending on the number of the unknown parameters. For further details on the problem of prior information and persistent of excitation, the reader is referred to [35].

In the following sections we use Theorem 5.2.1 to design the input signal u for a wide class of parameter estimators developed in Chapter 4.

5.2.2 Parameter Identifiers with Full-State Measurements

Let us consider the plant

$$\dot{x} = Ax + Bu, \quad x(0) = x_0 \qquad (5.2.10)$$

where $C^\top = I$, i.e., the state $x \in \mathcal{R}^n$ is available for measurement and A, B are constant matrices with unknown elements that we like to identify. We assume that A is stable and $u \in \mathcal{L}_\infty$.

As shown in Chapter 4, the following two types of parameter estimators may be used to estimate A, B from the measurements of x, u.

Series-Parallel

$$\begin{aligned}
\dot{\hat{x}} &= A_m \hat{x} + (\hat{A} - A_m)x + \hat{B}u \qquad (5.2.11) \\
\dot{\hat{A}} &= \gamma_1 \epsilon_1 x^\top, \quad \dot{\hat{B}} = \gamma_2 \epsilon_1 u^\top
\end{aligned}$$

where A_m is a stable matrix chosen by the designer, $\epsilon_1 = x - \hat{x}$, $\gamma_1, \gamma_2 > 0$ are the scalar adaptive gains.

Parallel

$$\begin{aligned}
\dot{\hat{x}} &= \hat{A}\hat{x} + \hat{B}u \qquad (5.2.12) \\
\dot{\hat{A}} &= \gamma_1 \epsilon_1 \hat{x}^\top, \quad \dot{\hat{B}} = \gamma_2 \epsilon_1 u^\top
\end{aligned}$$

where $\epsilon_1 = x - \hat{x}$ and $\gamma_1, \gamma_2 > 0$ are the scalar adaptive gains.

As shown in Chapter 4, if A is a stable matrix and $u \in \mathcal{L}_\infty$ then $\hat{x}, \hat{A}, \hat{B} \in \mathcal{L}_\infty$; $\|\dot{\hat{A}}(t)\|, \|\dot{\hat{B}}(t)\|, \epsilon_1 \in \mathcal{L}_2 \cap \mathcal{L}_\infty$ and $\epsilon_1(t)$ and the elements of $\dot{\hat{A}}(t), \dot{\hat{B}}(t)$ converge to zero as $t \to \infty$.

5.2. PARAMETER IDENTIFIERS

For the estimators (5.2.11) and (5.2.12) to become parameter identifiers, the input signal u has to be chosen so that $\hat{A}(t), \hat{B}(t)$ converge to the unknown plant parameters A, B, respectively, as $t \to \infty$.

For simplicity let us first consider the case where u is a scalar input, i.e., $B \in \mathcal{R}^{n \times 1}$.

Theorem 5.2.2 *Let (A, B) be a controllable pair. If the input $u \in \mathcal{R}^1$ is sufficiently rich of order $n + 1$, then the estimates $\hat{A}, \hat{B}$ generated by (5.2.11) or (5.2.12) converge exponentially fast to the unknown plant parameters A, B, respectively.*

The proof of Theorem 5.2.2 is quite long and is presented in Section 5.6.

An example of a sufficiently rich input u for the estimators (5.2.11), (5.2.12) is the input

$$u = \sum_{i=1}^{m} A_i \sin \omega_i t$$

for some constants $A_i \neq 0$ and $\omega_i \neq \omega_k$ for $i \neq k$ and for some integer $m \geq \frac{n+1}{2}$.

Example 5.2.2 Consider the second-order plant

$$\dot{x} = Ax + Bu$$

where $x = [x_1, x_2]^T$, and the matrices A, B are unknown, and A is a stable matrix. Using (5.2.11) with $A_m = \begin{bmatrix} -1 & 0 \\ 0 & -1 \end{bmatrix}$, the series-parallel parameter identifier is given by

$$\dot{\hat{x}} = \begin{bmatrix} -1 & 0 \\ 0 & -1 \end{bmatrix}(\hat{x} - x) + \begin{bmatrix} \hat{a}_{11}(t) & \hat{a}_{12}(t) \\ \hat{a}_{21}(t) & \hat{a}_{22}(t) \end{bmatrix} x + \begin{bmatrix} \hat{b}_1(t) \\ \hat{b}_2(t) \end{bmatrix} u$$

where $\hat{x} = [\hat{x}_1, \hat{x}_2]^T$ and

$$\dot{\hat{a}}_{ik} = (x_i - \hat{x}_i)x_k, \quad i = 1, 2; \ k = 1, 2$$
$$\dot{\hat{b}}_i = (x_i - \hat{x}_i)u, \quad i = 1, 2$$

The input u is selected as

$$u = 5 \sin 2.5t + 6 \sin 6.1t$$

which has more frequencies than needed since it is sufficiently rich of order 4. An input with the least number of frequencies that is sufficiently rich for the plant considered is
$$u = c_0 + \sin\omega_0 t$$
for some $c_0 \neq 0$ and $\omega_0 \neq 0$. ▽

If u is a vector, i.e., $u \in \mathcal{R}^q, q > 1$ then the following theorem may be used to select u.

Theorem 5.2.3 *Let (A, B) be a controllable pair. If each element $u_i, i = 1, 2, \ldots q$ of u is sufficiently rich of order $n + 1$ and uncorrelated, i.e., each u_i contains different frequencies, then $\hat{A}(t), \hat{B}(t)$ converge to A, B, respectively, exponentially fast.*

The proof of Theorem 5.2.3 is similar to that for Theorem 5.2.2, and is given in Section 5.6.

The controllability of the pair (A, B) is critical for the results of Theorems 5.2.2 and 5.2.3 to hold. If (A, B) is not a controllable pair, then the elements of (A, B) that correspond to the uncontrollable part cannot be learned from the output response because the uncontrollable parts decay to zero exponentially fast and are not affected by the input u.

The complexity of the parameter identifier may be reduced if some of the elements of the matrices (A, B) are known. In this case, the order of the adaptive law can be reduced to be equal to the number of the unknown parameters. In addition, the input u may not have to be sufficiently rich of order $n + 1$. The details of the design and analysis of such schemes are left as exercises for the reader and are included in the problem section.

In the following section we extend the results of Theorem 5.2.2 to the case where only the output of the plant, rather than the full state, is available for measurement.

5.2.3 Parameter Identifiers with Partial-State Measurements

In this section we concentrate on the SISO plant

$$\begin{aligned} \dot{x} &= Ax + Bu, \quad x(0) = x_0 \\ y &= C^\top x \end{aligned} \quad (5.2.13)$$

5.2. PARAMETER IDENTIFIERS

where A is a stable matrix, and $y, u \in \mathcal{R}^1$ are the only signals available for measurement.

Equation (5.2.13) may be also written as

$$y = C^\top(sI - A)^{-1}Bu + C^\top(sI - A)^{-1}x_0 \qquad (5.2.14)$$

where, because of the stability of A, $\epsilon_t = \mathcal{L}^{-1}\{C^\top(sI - A)^{-1}\}x_0$ is an exponentially decaying to zero term. We would like to design an on-line parameter identifier to estimate the parameters A, B, C. The triple (A, B, C) contains $n^2 + 2n$ unknown parameters to be estimated using only input/output data. The I/O properties of the plant (5.2.14) at steady state (where $\epsilon_t = 0$), however, are uniquely determined by at most $2n$ parameters. These parameters correspond to the coefficients of the transfer function

$$\frac{y(s)}{u(s)} = C^\top(sI - A)^{-1}B = \frac{b_m s^m + b_{m-1} s^{m-1} + \ldots + b_0}{s^n + a_{n-1} s^{n-1} + \ldots + a_0} \qquad (5.2.15)$$

where $m \leq n - 1$.

Because there is an infinite number of triples (A, B, C) that give the same transfer function (5.2.15), the triple (A, B, C) associated with the specific or physical state space representation in (5.3.13) cannot be determined, in general, from input-output data. The best we can do in this case is to estimate the $n + m + 1 \leq 2n$ coefficients of the plant transfer function (5.2.15) and try to calculate the corresponding triple (A, B, C) using some a priori knowledge about the structure of (A, B, C). For example, when (A, B, C) is in a canonical form, (A, B, C) can be uniquely determined from the coefficients of the transfer function by using the results of Chapter 2. In this section, we concentrate on identifying the $n + m + 1$ coefficients of the transfer function of the plant (5.2.13) rather than the $n^2 + 2n$ parameters of the triple (A, B, C).

The first step in the design of the parameter identifier is to develop an adaptive law that generates estimates for the unknown plant parameter vector

$$\theta^* = [b_m, b_{m-1}, \ldots, b_0, a_{n-1}, a_{n-2}, \ldots, , a_0]^\top$$

that contains the coefficients of the plant transfer function. As we have shown in Section 2.4.1, the vector θ^* satisfies the plant parametric equation

$$z = \theta^{*\top}\phi + \epsilon_t \qquad (5.2.16)$$

where

$$z = \frac{s^n}{\Lambda(s)}y, \quad \phi = \left[\frac{\alpha_m^T(s)}{\Lambda(s)}u, -\frac{\alpha_{n-1}^T(s)}{\Lambda(s)}y\right]^T$$

$$\epsilon_t = \frac{C^T \operatorname{adj}(sI-A)}{\Lambda(s)}x_0, \quad \alpha_i(s) = [s^i, s^{i-1}, \ldots, 1]^T$$

and $\Lambda(s)$ is an arbitrary monic Hurwitz polynomial of degree n.

Using (5.2.16), we can select any one of the adaptive laws presented in Tables 4.2, 4.3, and 4.5 of Chapter 4 to estimate θ^*. As an example, consider the gradient algorithm

$$\begin{aligned} \dot{\theta} &= \Gamma \epsilon \phi \\ \epsilon &= z - \hat{z}, \quad \hat{z} = \theta^T \phi \end{aligned} \qquad (5.2.17)$$

where $\Gamma = \Gamma^T > 0$ is the adaptive gain matrix, and θ is the estimate of θ^* at time t. The normalizing signal m^2 in this case is chosen as $m^2 = 1$ due to $\phi \in \mathcal{L}_\infty$.

The signal vector ϕ and signal z may be generated by the state equation

$$\begin{aligned} \dot{\phi}_0 &= \Lambda_c \phi_0 + lu, \quad \phi_0(0) = 0 \\ \phi_1 &= P_0 \phi_0 \\ \dot{\phi}_2 &= \Lambda_c \phi_2 - ly, \quad \phi_2(0) = 0 \\ \phi &= [\phi_1^T, \phi_2^T]^T \\ z &= y + \lambda^T \phi_2 \end{aligned} \qquad (5.2.18)$$

where $\phi_0 \in \mathcal{R}^n, \phi_1 \in \mathcal{R}^{m+1}, \phi_2 \in \mathcal{R}^n$

$$\Lambda_c = \left[\begin{array}{c} -\lambda^T \\ \hline I_{n-1} \mid 0 \end{array}\right], \quad l = \begin{bmatrix} 1 \\ 0 \\ \vdots \\ 0 \end{bmatrix}$$

$$P_0 = \begin{bmatrix} O_{(m+1)\times(n-m-1)} \mid I_{m+1} \end{bmatrix} \in \mathcal{R}^{(m+1)\times n}, \quad \lambda = \begin{bmatrix} \lambda_{n-1} \\ \lambda_{n-2} \\ \vdots \\ \lambda_0 \end{bmatrix}$$

5.2. PARAMETER IDENTIFIERS

I_i is the identity matrix of dimension $i \times i$, $O_{i \times k}$ is a matrix of dimension i by k with all elements equal to zero, and $\det(sI - \Lambda_c) = \Lambda(s) = s^n + \lambda^\top \alpha_{n-1}(s)$. When $m = n - 1$, the matrix P_0 becomes the identity matrix of dimension $n \times n$.

The state equations (5.2.18) are developed by using the identity

$$(sI - \Lambda_c)^{-1} l = \frac{\alpha_{n-1}(s)}{\Lambda(s)}$$

established in Chapter 2.

The adaptive law (5.2.17) or any other one obtained from Tables 4.2 and 4.3 in Chapter 4 guarantees that $\epsilon, \theta, \hat{z} \in \mathcal{L}_\infty$; $\epsilon, \dot{\theta} \in \mathcal{L}_2 \cap \mathcal{L}_\infty$ and $\epsilon(t), \dot{\theta}(t) \to 0$ as $t \to \infty$ for any piecewise bounded input u. For these adaptive laws to become parameter identifiers, the input signal u has to be chosen to be sufficiently rich so that ϕ is PE, which in turn guarantees that the estimated parameters converge to the actual ones.

Theorem 5.2.4 *Assume that the plant transfer function in (5.2.15) has no zero-pole cancellations. If u is sufficiently rich of order $n + m + 1$, then the adaptive law (5.2.17) or any other adaptive law from Tables 4.2, 4.3, and 4.5 of Chapter 4, based on the plant parametric model (5.2.16), guarantees that the estimated parameter vector $\theta(t)$ converges to θ^*. With the exception of the pure least-squares algorithm in Table 4.3 where the convergence is asymptotic, all the other adaptive laws guarantee exponential convergence of θ to θ^*.*

Proof In Chapter 4, we have established that $\theta(t)$ converges to θ^* if the signal ϕ is PE. Therefore we are left to show that ϕ is PE if u is sufficiently rich of order $n + m + 1$ and the transfer function has no zero-pole cancellations.

From the definition of ϕ, we can write

$$\phi = H(s)u$$

and

$$\begin{aligned} H(s) &= \frac{1}{\Lambda(s)} \left[\alpha_m^\top(s), -\alpha_{n-1}^\top G(s) \right]^\top \\ &= \frac{1}{\Lambda(s)R(s)} \left[\alpha_m^\top(s)R(s), -\alpha_{n-1}^\top(s)Z(s) \right]^\top \end{aligned} \quad (5.2.19)$$

where $G(s) = \frac{Z(s)}{R(s)}$ is the transfer function of the plant. We first show by contradiction that $\{H(j\omega_1), H(j\omega_2), \ldots, H(j\omega_{n+m+1})\}$ are linearly independent in $\mathcal{C}^{n+m+1}$ for any $\omega_1, \ldots, \omega_{n+m+1} \in \mathcal{R}$ and $\omega_i \neq \omega_k, i, k = 1, \ldots, n+m+1$.

Let us assume that there exist $\omega_1, \ldots, \omega_{n+m+1}$ such that $H(j\omega_1), \ldots, H(j\omega_{n+m+1})$ are linearly dependent, that is, there exists a vector $h = [c_m, c_{m-1}, \ldots, c_0, d_{n-1}, d_{n-2}, \ldots, d_0]^T \in \mathcal{C}^{n+m+1}$ such that

$$\begin{bmatrix} H^T(j\omega_1) \\ H^T(j\omega_2) \\ \vdots \\ H^T(j\omega_{n+m+1}) \end{bmatrix} h = 0_{n+m+1} \qquad (5.2.20)$$

where 0_{n+m+1} is a zero vector of dimension $n+m+1$. Using the expression (5.2.19) for $H(s)$, (5.2.20) can be written as

$$\frac{1}{\Lambda(j\omega_i) R(j\omega_i)} [b(j\omega_i) R(j\omega_i) + a(j\omega_i) Z(j\omega_i)] = 0, \quad i = 1, 2, \ldots, n+m+1$$

where

$$b(s) = c_m s^m + c_{m-1} s^{m-1} + \cdots + c_0$$
$$a(s) = -d_{n-1} s^{n-1} - d_{n-2} s^{n-2} - \cdots - d_0$$

Now consider the following polynomial:

$$f(s) \triangleq a(s) Z(s) + b(s) R(s)$$

Because $f(s)$ has degree of at most $m + n$ and it vanishes at $s = j\omega_i, i = 1, \ldots, n+m+1$ points, it must be identically equal to zero, that is

$$a(s) Z(s) + b(s) R(s) = 0$$

or equivalently

$$G(s) = \frac{Z(s)}{R(s)} = -\frac{b(s)}{a(s)} \qquad (5.2.21)$$

for all s. However, because $a(s)$ has degree at most $n-1$, (5.2.21) implies that $G(s)$ has at least one zero-pole cancellation, which contradicts the assumption of no zero-pole cancellation in $G(s)$. Thus, we have proved that $H(j\omega_i), i = 1, \ldots, n+m+1$ are linearly independent for any $\omega_i \neq \omega_k, i, k = 1, \ldots, n+m+1$. It then follows from Theorem 5.2.1 and the assumption of u being sufficiently rich of order $n + m + 1$ that ϕ is PE and therefore θ converges to θ^*. As shown in Chapter 4, the convergence of θ to θ^* is exponential for all the adaptive laws of Tables 4.2, 4.3, and 4.5 with the exception of the pure least-squares where the convergence is asymptotic. $\square$

5.2. PARAMETER IDENTIFIERS

An interesting question to ask at this stage is what happens to the parameter estimates when u is sufficiently rich of order less than $n + m + 1$, i.e., when its spectrum $S_u(\omega)$ is concentrated on $k < n + m + 1$ points.

Let us try to answer this question by considering the plant parametric equation (5.2.16) with $\epsilon_t = 0$, i.e.,

$$z = \theta^{*\mathsf{T}} \phi = \theta^{*\mathsf{T}} H(s) u$$

where $H(s)$ is given by (5.2.19).

The estimated value of z is given by

$$\hat{z} = \theta^{\mathsf{T}} \phi = \theta^{\mathsf{T}} H(s) u$$

The adaptive laws of Tables 4.2, 4.3, and 4.5 of Chapter 4 based on the parametric equation (5.2.16) guarantee that $\hat{z}(t) \to z(t)$ as $t \to \infty$ for any bounded input u. If u contains frequencies at $\omega_1, \omega_2, \ldots, \omega_k$, then as $t \to \infty$, $\hat{z} = \theta^{\mathsf{T}} H(s) u$ becomes equal to $z = \theta^{*\mathsf{T}} H(s) u$ at these frequencies. Therefore, as $t \to \infty$, θ must satisfy the equation

$$Q(j\omega_1, \ldots, j\omega_k) \theta = Q(j\omega_1, \ldots, j\omega_k) \theta^*$$

where

$$Q(j\omega_1, \ldots, j\omega_k) = \begin{bmatrix} H^{\mathsf{T}}(j\omega_1) \\ \vdots \\ H^{\mathsf{T}}(j\omega_k) \end{bmatrix}$$

is a $k \times n + m + 1$ matrix with k linearly independent row vectors. Hence,

$$Q(j\omega_1, \ldots, j\omega_k)(\theta - \theta^*) = 0$$

which is satisfied for any θ for which $\theta - \theta^*$ is in the null space of Q. If $k < n + m + 1$ it follows that the null space of Q contains points for which $\theta \neq \theta^*$. This means that $\hat{z}$ can match z at k frequencies even when $\theta \neq \theta^*$.

The following Theorem presented in [201] gives a similar result followed by a more rigorous proof.

Theorem 5.2.5 (Partial Convergence) *Assume that the plant transfer function in (5.2.15) has no zero-pole cancellations. If u is stationary, then*

$$\lim_{t \to \infty} R_\phi(0) (\theta(t) - \theta^*) = 0$$

where $\theta(t)$ is generated by (5.2.17) or any other adaptive law from Tables 4.2, 4.3, and 4.5 of Chapter 4 based on the plant parametric model (5.2.16).

The proof of Theorem 5.2.5 is given in Section 5.6.

Theorem 5.2.5 does not imply that $\theta(t)$ converges to a constant let alone to θ^*. It does imply, however, that $\tilde{\theta}(t) = \theta(t) - \theta^*$ converges to the null space of the autocovariance $R_\phi(0)$ of ϕ, which depends on $H(s)$ and the spectrum of u. In fact it can be shown [201] that $R_\phi(0)$ is related to the spectrum of u via the equation

$$R_\phi(0) = \sum_{i=1}^{k} H(-j\omega_i) H^\top(j\omega_i) S_u(\omega_i)$$

which indicates the dependence of the null space of $R_\phi(0)$ on the spectrum of the input u.

Example 5.2.3 Consider the second order plant

$$y = \frac{b_0}{s^2 + a_1 s + a_0} u$$

where $a_1, a_0 > 0$ and $b_0 \neq 0$ are the unknown plant parameters. We first express the plant in the form of

$$z = \theta^{*\top} \phi$$

where $\theta^* = [b_0, a_1, a_0]^\top$, $z = \frac{s^2}{\Lambda(s)} y$, $\phi = \left[\frac{1}{\Lambda(s)} u, -\frac{[s,1]}{\Lambda(s)} y \right]^\top$ and choose $\Lambda(s) = (s+2)^2$. Let us choose the pure least-squares algorithm from Table 4.3, i.e.,

$$\dot{\theta} = P\epsilon\phi, \quad \theta(0) = \theta_0$$
$$\dot{P} = -P\phi\phi^\top P, \quad P(0) = p_0 I$$
$$\epsilon = z - \theta^\top \phi$$

where θ is the estimate of θ^* and select $p_0 = 50$. The signal vector $\phi = [\phi_1^\top, \phi_2^\top]^\top$ is generated by the state equations

$$\dot{\phi}_0 = \Lambda_c \phi_0 + lu$$
$$\phi_1 = [0\ 1]\phi_0$$
$$\dot{\phi}_2 = \Lambda_c \phi_2 - ly$$
$$z = y + [4\ 4]\phi_2$$

where $\Lambda_c = \begin{bmatrix} -4 & -4 \\ 1 & 0 \end{bmatrix}, l = \begin{bmatrix} 1 \\ 0 \end{bmatrix}$. The reader can demonstrate via computer simulations that: (i) for $u = 5\sin t + 12\sin 3t$, $\theta(t) \to \theta^*$ as $t \to \infty$; (ii) for $u = 12\sin 3t$, $\theta(t) \to \bar{\theta}$ as $t \to \infty$ where $\bar{\theta}$ is a constant vector that depends on the initial condition $\theta(0)$. $\triangledown$

Remark 5.2.2 As illustrated with Example 5.2.3, one can choose any one of the adaptive laws presented in Tables 4.2, 4.3, and 4.5 to form parameter identifiers. The complete proof of the stability properties of such parameter identifiers follows directly from the results of Chapter 4 and Theorem 5.2.4. The reader is asked to repeat some of the stability proofs of parameter identifiers with different adaptive laws in the problem section.

5.3 Adaptive Observers

Consider the LTI SISO plant

$$\begin{aligned} \dot{x} &= Ax + Bu, \quad x(0) = x_0 \\ y &= C^\top x \end{aligned} \quad (5.3.1)$$

where $x \in \mathcal{R}^n$. We assume that u is a piecewise continuous and bounded function of time, and A is a stable matrix. In addition we assume that the plant is completely controllable and completely observable.

The problem is to construct a scheme that estimates both the parameters of the plant, i.e., A, B, C as well as the state vector x using only I/O measurements. We refer to such a scheme as the *adaptive observer*.

A good starting point for choosing the structure of the adaptive observer is the state observer, known as the Luenberger observer, used to estimate the state vector x of the plant (5.3.1) when the parameters A, B, C are known.

5.3.1 The Luenberger Observer

If the initial state vector x_0 is known, the estimate $\hat{x}$ of x in (5.3.1) may be generated by the state observer

$$\dot{\hat{x}} = A\hat{x} + Bu, \quad \hat{x}(0) = x_0 \quad (5.3.2)$$

where $\hat{x} \in \mathcal{R}^n$. Equations (5.3.1) and (5.3.2) imply that $\hat{x}(t) = x(t), \forall t \geq 0$. When x_0 is unknown and A is a stable matrix, the following state observer may be used to generate the estimate $\hat{x}$ of x:

$$\dot{\hat{x}} = A\hat{x} + Bu, \quad \hat{x}(0) = \hat{x}_0 \quad (5.3.3)$$

In this case, the state observation error $\tilde{x} = x - \hat{x}$ satisfies the equation

$$\dot{\tilde{x}} = A\tilde{x}, \qquad \tilde{x}(0) = x_0 - \hat{x}_0$$

which implies that $\tilde{x}(t) = e^{At}\tilde{x}(0)$. Because A is a stable matrix $\tilde{x}(t) \to 0$, i.e., $\hat{x}(t) \to x(t)$ as $t \to \infty$ exponentially fast with a rate that depends on the location of the eigenvalues of A. The observers (5.3.2), (5.3.3) contain no feedback terms and are often referred to as *open-loop observers*.

When x_0 is unknown and A is not a stable matrix, or A is stable but the state observation error is required to converge to zero faster than the rate with which $\| e^{At} \|$ goes to zero, the following observer, known as the *Luenberger observer*, is used:

$$\begin{aligned} \dot{\hat{x}} &= A\hat{x} + Bu + K(y - \hat{y}), \quad \hat{x}(0) = \hat{x}_0 \\ \hat{y} &= C^T \hat{x} \end{aligned} \qquad (5.3.4)$$

where K is a matrix to be chosen by the designer. In contrast to (5.3.2) and (5.3.3), the Luenberger observer (5.3.4) has a feedback term that depends on the output observation error $\tilde{y} = y - \hat{y}$.

The state observation error $\tilde{x} = x - \hat{x}$ for (5.3.4) satisfies

$$\dot{\tilde{x}} = (A - KC^T)\tilde{x}, \qquad \tilde{x}(0) = x_0 - \hat{x}_0 \qquad (5.3.5)$$

Because (C, A) is an observable pair, we can choose K so that $A - KC^T$ is a stable matrix. In fact, the eigenvalues of $A - KC^T$, and, therefore, the rate of convergence of $\tilde{x}(t)$ to zero can be arbitrarily chosen by designing K appropriately [95]. Therefore, it follows from (5.3.5) that $\hat{x}(t) \to x(t)$ exponentially fast as $t \to \infty$, with a rate that depends on the matrix $A - KC^T$. This result is valid for any matrix A and any initial condition x_0 as long as (C, A) is an observable pair and A, C are known.

Example 5.3.1 Consider the plant described by

$$\begin{aligned} \dot{x} &= \begin{bmatrix} -4 & 1 \\ -4 & 0 \end{bmatrix} x + \begin{bmatrix} 1 \\ 3 \end{bmatrix} u \\ y &= [1, 0]x \end{aligned}$$

The Luenberger observer for estimating the state x is given by

$$\begin{aligned} \dot{\hat{x}} &= \begin{bmatrix} -4 & 1 \\ -4 & 0 \end{bmatrix} \hat{x} + \begin{bmatrix} 1 \\ 3 \end{bmatrix} u + \begin{bmatrix} k_1 \\ k_2 \end{bmatrix} (y - \hat{y}) \\ \hat{y} &= [1, 0]\hat{x} \end{aligned}$$

where $K = [k_1, k_2]^\top$ is chosen so that

$$A_0 = \begin{bmatrix} -4 & 1 \\ -4 & 0 \end{bmatrix} - \begin{bmatrix} k_1 \\ k_2 \end{bmatrix} [1 \ 0] = \begin{bmatrix} -4-k_1 & 1 \\ -4-k_2 & 0 \end{bmatrix}$$

is a stable matrix. Let us assume that $\hat{x}(t)$ is required to converge to $x(t)$ faster than e^{-5t}. This requirement is achieved by choosing k_1, k_2 so that the eigenvalues of A_0 are real and less than -5, i.e., we choose the desired eigenvalues of A_0 to be $\lambda_1 = -6, \lambda_2 = -8$ and design k_1, k_2 so that

$$\det(sI - A_0) = s^2 + (4+k_1)s + 4 + k_2 = (s+6)(s+8)$$

which gives

$$k_1 = 10, \quad k_2 = 44 \qquad \triangledown$$

5.3.2 The Adaptive Luenberger Observer

Let us now consider the problem where both the state x and parameters A, B, C are to be estimated on-line simultaneously using an adaptive observer.

A straightforward procedure for choosing the structure of the adaptive observer is to use the same equation as the Luenberger observer in (5.3.4), but replace the unknown parameters A, B, C with their estimates $\hat{A}, \hat{B}, \hat{C}$, respectively, generated by some adaptive law. The problem we face with this procedure is the inability to estimate uniquely the $n^2 + 2n$ parameters of A, B, C from input/output data. As explained in Section 5.2.3, the best we can do in this case is to estimate the $n+m+1 \leq 2n$ parameters of the plant transfer function and use them to calculate $\hat{A}, \hat{B}, \hat{C}$. These calculations, however, are not always possible because the mapping of the $2n$ estimated parameters of the transfer function to the $n^2 + 2n$ parameters of $\hat{A}, \hat{B}, \hat{C}$ is not unique unless (A, B, C) satisfies certain structural constraints. One such constraint is that (A, B, C) is in the observer form, i.e., the plant is represented as

$$\dot{x}_\alpha = \begin{bmatrix} & \vdots & I_{n-1} \\ -a_p & \vdots & \ldots \\ & \vdots & 0 \end{bmatrix} x_\alpha + b_p u \qquad (5.3.6)$$

$$y = [1 \ 0 \ \ldots \ 0] x_\alpha$$

where $a_p = [a_{n-1}, a_{n-2}, \ldots a_0]^T$ and $b_p = [b_{n-1}, b_{n-2}, \ldots b_0]^T$ are vectors of dimension n and $I_{n-1} \in \mathcal{R}^{(n-1)\times(n-1)}$ is the identity matrix. The elements of a_p and b_p are the coefficients of the denominator and numerator, respectively, of the transfer function

$$\frac{y(s)}{u(s)} = \frac{b_{n-1}s^{n-1} + b_{n-2}s^{n-2} + \ldots + b_0}{s^n + a_{n-1}s^{n-1} + \ldots a_0} \qquad (5.3.7)$$

and can be estimated on-line from input/output data by using the techniques described in Chapter 4.

Because both (5.3.1) and (5.3.6) represent the same plant, we can assume the plant representation (5.3.6) and estimate x_α instead of x. The disadvantage is that in a practical situation x may represent physical variables that are of interest, whereas x_α may be an artificial state variable.

The adaptive observer for estimating the state x_α of (5.3.6) is motivated from the structure of the Luenberger observer (5.3.4) and is given by

$$\dot{\hat{x}} = \hat{A}(t)\hat{x} + \hat{b}_p(t)u + K(t)(y - \hat{y}) \qquad (5.3.8)$$
$$\hat{y} = [1\ 0\ \ldots\ 0]\hat{x}$$

where $\hat{x}$ is the estimate of x_α,

$$\hat{A}(t) = \begin{bmatrix} & \vdots & I_{n-1} & \\ -\hat{a}_p(t) & \vdots & \ldots & \\ & \vdots & 0 & \end{bmatrix}, \quad K(t) = a^* - \hat{a}_p(t)$$

$a^* \in \mathcal{R}^n$ is chosen so that

$$A^* = \begin{bmatrix} & \vdots & I_{n-1} & \\ -a^* & \vdots & \ldots & \\ & \vdots & 0 & \end{bmatrix} \qquad (5.3.9)$$

is a stable matrix and $\hat{a}_p(t)$ and $\hat{b}_p(t)$ are the estimates of the vectors a_p and b_p, respectively, at time t.

A wide class of adaptive laws may be used to generate $\hat{a}_p(t)$ and $\hat{b}_p(t)$ on-line. As an example, we can start with (5.3.7) to obtain as in Section 2.4.1 the parametric model

$$z = \theta^{*T}\phi \qquad (5.3.10)$$

5.3. ADAPTIVE OBSERVERS

where

$$\phi = \left[\frac{\alpha_{n-1}^T(s)}{\Lambda(s)}u, -\frac{\alpha_{n-1}^T(s)}{\Lambda(s)}y\right]^T = [\phi_1^T, \phi_2^T]^T$$

$$z = \frac{s^n}{\Lambda(s)}y = y + \lambda^T \phi_2$$

$$\Lambda(s) = s^n + \lambda^T \alpha_{n-1}(s)$$

and

$$\theta^* = [b_{n-1}, b_{n-2}, \ldots, a_{n-1}, a_{n-2}, \ldots, a_0]^T$$

is the parameter vector to be estimated and $\Lambda(s)$ is a Hurwitz polynomial of degree n chosen by the designer. A state-space representation for ϕ and z may be obtained as in (5.2.18) by using the identity $(sI - \Lambda_c)^{-1}l \triangleq \frac{\alpha_{n-1}(s)}{\Lambda(s)}$ where (Λ_c, l) is in the controller canonical form and $\det(sI - \Lambda_c) = \Lambda(s)$.

In view of (5.3.10), we can choose any adaptive law from Tables 4.2, 4.3 and 4.5 of Chapter 4 to estimate θ^* and, therefore, a_p, b_p on-line. We can form a wide class of adaptive observers by combining (5.3.8) with any adaptive law from Tables 4.2, 4.3 and 4.5 of Chapter 4 that is based on the parametric plant model (5.3.10).

We illustrate the design of such adaptive observer by using the gradient algorithm of Table 4.2 (A) in Chapter 4 as the adaptive law. The main equations of the observer are summarized in Table 5.1.

The stability properties of the class of adaptive observers formed by combining the observer equation (5.3.8) with an adaptive law from Tables 4.2 and 4.3 of Chapter 4 are given by the following theorem.

Theorem 5.3.1 *An adaptive observer for the plant (5.3.6) formed by combining the observer equation (5.3.8) and any adaptive law based on the plant parametric model (5.3.10) obtained from Tables 4.2 and 4.3 of Chapter 4 guarantees that*

(i) *All signals are u.b.*
(ii) *The output observation error $\tilde{y} = y - \hat{y}$ converges to zero as $t \to \infty$.*
(iii) *If u is sufficiently rich of order $2n$, then the state observation error $\tilde{x} = x_\alpha - \hat{x}$ and parameter error $\tilde{\theta} \triangleq \theta - \theta^*$ converge to zero. The rate of convergence is exponential for all adaptive laws except for the pure least-squares where the convergence is asymptotic.*

Table 5.1 Adaptive observer with gradient algorithm

Plant	$\dot{x}_\alpha = \begin{bmatrix} & \vdots & I_{n-1} \\ -a_p & \vdots & \ldots \\ & \vdots & 0 \end{bmatrix} x_\alpha + b_p u, \quad x_\alpha \in \mathcal{R}^n$ $y = [1, 0 \ldots 0] x_\alpha$
Observer	$\dot{\hat{x}} = \begin{bmatrix} & \vdots & I_{n-1} \\ -\hat{a}_p(t) & \vdots & \ldots \\ & \vdots & 0 \end{bmatrix} \hat{x} + \hat{b}_p(t) u + (a^* - \hat{a}_p(t))(y - \hat{y})$ $\hat{y} = [1 \ 0 \ldots 0] \hat{x}$
Adaptive law	$\dot{\theta} = \Gamma \epsilon \phi$ $\theta = \left[\hat{b}_p^\top(t), \hat{a}_p^\top(t) \right]^\top, \quad \epsilon = \frac{z - \hat{z}}{m^2},$ $\hat{z} = \theta^\top \phi, \quad \Gamma = \Gamma^\top > 0$ $\phi = \left[\frac{\alpha_{n-1}^\top(s)}{\Lambda(s)} u, -\frac{\alpha_{n-1}^\top(s)}{\Lambda(s)} y \right]^\top$ $z = \frac{s^n}{\Lambda(s)} y$
Design variables	a^* is chosen so that A^* in (5.3.9) is stable; $m^2 = 1$ or $m^2 = 1 + \phi^\top \phi$; $\Lambda(s)$ is a monic Hurwitz polynomial of degree n.

Proof (i) The adaptive laws of Tables 4.2 and 4.3 of Chapter 4 guarantee that $\epsilon, \epsilon m, \dot{\theta} \in \mathcal{L}_2 \cap \mathcal{L}_\infty$ and $\theta \in \mathcal{L}_\infty$ independent of the boundedness of u, y. Because $u \in \mathcal{L}_\infty$ and the plant is stable, we have $x_\alpha, y, \phi, m \in \mathcal{L}_\infty$. Because of the boundedness of y, ϕ we can also establish that $\epsilon, \epsilon m, \dot{\theta} \to 0$ as $t \to \infty$ by showing that $\dot{\epsilon} \in \mathcal{L}_\infty$ (which follows from $\dot{\theta}, \dot{\phi} \in \mathcal{L}_\infty$), which, together with $\epsilon \in \mathcal{L}_2$, implies that $\epsilon \to 0$ as $t \to \infty$ (see Lemma 3.2.5). Because $m, \phi \in \mathcal{L}_\infty$, the convergence of $\epsilon m, \dot{\theta}$ to zero follows.

The proof of (i) is complete if we establish that $\hat{x} \in \mathcal{L}_\infty$. We rewrite the observer

5.3. ADAPTIVE OBSERVERS

equation (5.3.8) in the form

$$\dot{\hat{x}} = A^*\hat{x} + \hat{b}_p(t)u + (\hat{A}(t) - A^*)x_\alpha \qquad (5.3.11)$$

Because $\theta = \left[\hat{b}_p^\top(t), \hat{a}_p^\top(t)\right]^\top, u, x_\alpha \in \mathcal{L}_\infty$ and A^* is a stable matrix, it follows that $\hat{x} \in \mathcal{L}_\infty$. Hence, the proof of (i) is complete.

(ii) Let $\tilde{x} \triangleq x_\alpha - \hat{x}$ be the state observation error. It follows from (5.3.11), (5.3.6) that

$$\dot{\tilde{x}} = A^*\tilde{x} - \tilde{b}_p u + \tilde{a}_p y, \quad \tilde{x}(0) = x_\alpha(0) - \hat{x}(0) \qquad (5.3.12)$$

where $\tilde{b}_p \triangleq \hat{b}_p - b_p, \tilde{a}_p \triangleq \hat{a}_p - a_p$. From (5.3.12), we obtain

$$\tilde{y} = C^\top \tilde{x}(s) = C^\top(sI - A^*)^{-1}(-\tilde{b}_p u + \tilde{a}_p y) + \epsilon_t$$

where $\epsilon_t = \mathcal{L}^{-1}\left\{C^\top(sI - A^*)^{-1}\right\}\tilde{x}(0)$ is an exponentially decaying to zero term. Because (C, A) is in the observer canonical form, we have

$$C^\top(sI - A^*)^{-1} = \frac{\alpha_{n-1}^\top(s)}{\det(sI - A^*)} = \frac{[s^{n-1}, s^{n-2}, \ldots s, 1]}{\det(sI - A^*)}$$

Letting $\Lambda^*(s) \triangleq \det(sI - A^*)$, we have

$$\tilde{y}(s) = \frac{1}{\Lambda^*(s)} \sum_{i=1}^n s^{n-i} \left[-\tilde{b}_{n-i}u + \tilde{a}_{n-i}y\right] + \epsilon_t$$

where $\tilde{b}_i, \tilde{a}_i$ is the ith element of $\tilde{b}_p$ and $\tilde{a}_p$, respectively, which may be written as

$$\tilde{y}(s) = \frac{\Lambda(s)}{\Lambda^*(s)} \sum_{i=1}^n \frac{s^{n-i}}{\Lambda(s)} \left[-\tilde{b}_{n-i}u + \tilde{a}_{n-i}y\right] + \epsilon_t \qquad (5.3.13)$$

where $\Lambda(s)$ is the Hurwitz polynomial of degree n defined in (5.3.10). We now apply Lemma A.1 (see Appendix A) for each term under the summation in (5.3.13) to obtain

$$\tilde{y} = \frac{\Lambda(s)}{\Lambda^*(s)} \sum_{i=1}^n \left[-\tilde{b}_{n-i}\frac{s^{n-i}}{\Lambda(s)}u + \tilde{a}_{n-i}\frac{s^{n-i}}{\Lambda(s)}y \right.$$
$$\left. - W_{ci}(s)(W_{bi}(s)u)\dot{\tilde{b}}_{n-i} + W_{ci}(s)(W_{bi}(s)y)\dot{\tilde{a}}_{n-i}\right] + \epsilon_t \qquad (5.3.14)$$

where the elements of $W_{ci}(s), W_{bi}(s)$ are strictly proper transfer functions with the same poles as $\Lambda(s)$. Using the definition of ϕ and parameter error $\tilde{\theta} = \theta - \theta^*$, we rewrite (5.3.14) as

$$\tilde{y} = \frac{\Lambda(s)}{\Lambda^*(s)}\left\{-\tilde{\theta}^\top \phi + \sum_{i=1}^n W_{ci}(s)\left(-(W_{bi}(s)u)\dot{\tilde{b}}_{n-i} + (W_{bi}(s)y)\dot{\tilde{a}}_{n-i}\right)\right\} + \epsilon_t$$

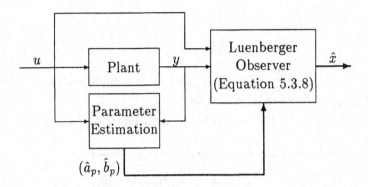

Figure 5.1 General structure of the adaptive Luenberger observer.

Because $\dot{\tilde{b}}_{n-i}, \dot{\tilde{a}}_{n-i} \in \mathcal{L}_2 \cap \mathcal{L}_\infty$ converge to zero as $t \to \infty$, $u, y \in \mathcal{L}_\infty$ and the elements of $W_{ci}(s), W_{bi}(s)$ are strictly proper stable transfer functions, it follows from Corollary 3.3.1 that all the terms under the summation are in $\mathcal{L}_2 \cap \mathcal{L}_\infty$ and converge to zero as $t \to \infty$. Furthermore, from $\epsilon m^2 = z - \hat{z} = -\tilde{\theta}^\top \phi, m \in \mathcal{L}_\infty, \epsilon \in \mathcal{L}_2 \cap \mathcal{L}_\infty$ and $\epsilon(t) \to 0$ as $t \to 0$, we have that $\tilde{\theta}^\top \phi \in \mathcal{L}_2 \cap \mathcal{L}_\infty$ converges to zero as $t \to \infty$. Hence, $\tilde{y}$ is the sum of an output of a proper stable transfer function whose input is in $\mathcal{L}_2$ and converges to zero as $t \to \infty$ and the exponentially decaying to zero term ϵ_t. Therefore, $\tilde{y}(t) \to 0$ as $t \to \infty$.

(iii) If ϕ is PE, then we can establish, using the results of Chapter 4, that $\tilde{a}_p(t), \tilde{b}_p(t)$ converge to zero. Hence, the input $-\tilde{b}_p u + \tilde{a}_p y$ converges to zero, which, together with the stability of A^*, implies that $\tilde{x}(t) \to 0$. With the exception of the pure least-squares, all the other adaptive laws guarantee that the convergence of $\tilde{b}_p, \tilde{a}_p$ to zero is exponential, which implies that $\tilde{x}$ also goes to zero exponentially fast .

The PE property of ϕ is established by using exactly the same steps as in the proof of Theorem 5.2.4. □

The general structure of the adaptive observer is shown in Figure 5.1.

The only a priori knowledge assumed about the plant (5.3.1) is that it is completely observable and completely controllable and its order n is known. The knowledge of n is used to choose the order of the observer, whereas the observability of (C, A) is used to guarantee the existence of the state space representation of the plant in the observer form that in turn enables us to design a stable adaptive observer. The controllability of (A, B) is not needed for stability, but it is used together with the observability of (C, A) to establish that ϕ is PE from the properties of the input u.

5.3. ADAPTIVE OBSERVERS

Theorem 5.3.1 shows that for the state x_α of the plant to be estimated exactly, the input has to be sufficiently rich of order $2n$, which implies that the adaptive law has to be a parameter identifier. Even with the knowledge of the parameters a_p, b_p and of the state x_α, however, it is not in general possible to calculate the original state of the plant x because of the usual nonunique mapping from the coefficients of the transfer function to the parameters of the state space representation.

Example 5.3.2 Let us consider the second order plant

$$\dot{x} = \begin{bmatrix} -a_1 & 1 \\ -a_0 & 0 \end{bmatrix} x + \begin{bmatrix} b_1 \\ b_0 \end{bmatrix} u$$
$$y = [1,0] x$$

where a_1, a_0, b_1, b_0 are the unknown parameters and u, y are the only signals available for measurement.

Using Table 5.1, the adaptive observer for estimating x, and the unknown parameters are described as follows: The observer equation is given by

$$\dot{\hat{x}} = \begin{bmatrix} -\hat{a}_1(t) & 1 \\ -\hat{a}_0(t) & 0 \end{bmatrix} \hat{x} + \begin{bmatrix} \hat{b}_1(t) \\ \hat{b}_0(t) \end{bmatrix} u + \begin{bmatrix} 9 - \hat{a}_1(t) \\ 20 - \hat{a}_0(t) \end{bmatrix} (y - \hat{y})$$
$$\hat{y} = [1,0] \hat{x}$$

where the constants $a_1^* = 9, a_0^* = 20$ are selected so that $A^* = \begin{bmatrix} -a_1^* & 1 \\ -a_0^* & 0 \end{bmatrix}$ has eigenvalues at $\lambda_1 = -5, \lambda_2 = -4$.

The adaptive law is designed by first selecting

$$\Lambda(s) = (s+2)(s+3) = s^2 + 5s + 6$$

and generating the information vector $\phi = [\phi_1^T, \phi_2^T]^T$

$$\dot{\phi}_1 = \begin{bmatrix} -5 & -6 \\ 1 & 0 \end{bmatrix} \phi_1 + \begin{bmatrix} 1 \\ 0 \end{bmatrix} u$$
$$\dot{\phi}_2 = \begin{bmatrix} -5 & -6 \\ 1 & 0 \end{bmatrix} \phi_2 + \begin{bmatrix} -1 \\ 0 \end{bmatrix} y$$

and the signals

$$z = y + [5,6]\phi_2$$
$$\hat{z} = [\hat{b}_1, \hat{b}_0] \phi_1 + [\hat{a}_1, \hat{a}_0] \phi_2$$
$$\epsilon = z - \hat{z}$$

The adaptive law is then given by

$$\begin{bmatrix} \dot{b}_1 \\ \dot{b}_2 \end{bmatrix} = \gamma_1 \epsilon \phi_1, \quad \begin{bmatrix} \dot{a}_1 \\ \dot{a}_2 \end{bmatrix} = \gamma_2 \epsilon \phi_2$$

The adaptive gains $\gamma_1, \gamma_2 > 0$ are usually chosen by trial and error using simulations in order to achieve a good rate of convergence. Small γ_1, γ_2 may result in slow convergent rate whereas large γ_1, γ_2 may make the differential equations "stiff" and difficult to solve numerically on a digital computer.

In order for the parameters to converge to their true values, the plant input u is chosen to be sufficiently rich of order 4. One possible choice for u is

$$u = A_1 \sin \omega_1 t + A_2 \sin \omega_2 t$$

for some constants $A_1, A_2 \neq 0$ and $\omega_1 \neq \omega_2$. $\triangledown$

5.3.3 Hybrid Adaptive Luenberger Observer

The adaptive law for the adaptive observer presented in Table 5.1 can be replaced with a hybrid one without changing the stability properties of the observer in any significant way. The hybrid adaptive law updates the parameter estimates only at specific instants of time t_k, where t_k is an unbounded monotonic sequence, and $t_{k+1} - t_k = T_s$ where T_s may be considered as the sampling period.

The hybrid adaptive observer is developed by replacing the continuous-time adaptive law with the hybrid one given in Table 4.5 as shown in Table 5.2. The combination of the discrete-time adaptive law with the continuous-time observer equation makes the overall system more difficult to analyze. The stability properties of the hybrid adaptive observer, however, are very similar to those of the continuous-time adaptive observer and are given by the following theorem.

Theorem 5.3.2 *The hybrid adaptive Luenberger observer presented in Table 5.2 guarantees that*

(i) *All signals are u.b.*

(ii) *The output observation error $\tilde{y} \triangleq y - \hat{y}$ converges to zero as $t \to \infty$.*

(iii) *If u is sufficiently rich of order $2n$, then the state observation error $\tilde{x} \triangleq x_\alpha - \hat{x}$ and parameter error $\tilde{\theta}_k \triangleq \theta_k - \theta^*$ converge to zero exponentially fast.*

5.3. ADAPTIVE OBSERVERS

Table 5.2 Hybrid adaptive Luenberger observer

Plant	$\dot{x}_\alpha = \begin{bmatrix} & \vdots & I_{n-1} & \\ -a_p & \vdots & \cdots & \\ & \vdots & 0 & \end{bmatrix} x_\alpha + b_p u, \quad x_\alpha \in \mathcal{R}^n$ $y = [1, 0 \ldots 0] u$
Observer	$\dot{\hat{x}} = \begin{bmatrix} & \vdots & I_{n-1} & \\ -\hat{a}_k & \vdots & \cdots & \\ & \vdots & 0 & \end{bmatrix} \hat{x} + \hat{b}_k u + (a^* - \hat{a}_k)(y - \hat{y})$ $\hat{y} = [1, 0, \ldots, 0] \hat{x}$ $\hat{x} \in \mathcal{R}^n, t \in [t_k, t_{k+1}), k = 0, 1, \ldots$
Hybrid adaptive law	$\theta_{k+1} = \theta_k + \Gamma \int_{t_k}^{t_{k+1}} \epsilon(\tau) \phi(\tau) d\tau$ $\epsilon = \frac{z - \hat{z}}{m^2}, \quad \hat{z}(t) = \theta_k^T \phi(t), \quad \forall t \in [t_k, t_{k+1})$ $m^2 = 1 + \alpha \phi^T \phi, \quad \alpha \geq 0$ $\phi = \left[\frac{\alpha_{n-1}^T(s)}{\Lambda(s)} u, -\frac{\alpha_{n-1}^T(s)}{\Lambda(s)} y \right]^T, z = \frac{s^n}{\Lambda(s)} y, \theta_k = \left[\hat{b}_k^T, \hat{a}_k^T \right]^T$
Design variable	a^* is chosen so that A^* in (5.3.9) is stable; $\Lambda(s)$ is monic Hurwitz of degree n; $T_s = t_{k+1} - t_k$ is the sampling period; $\Gamma = \Gamma^T > 0, T_s$ are chosen so that $2 - T_s \lambda_{max}(\Gamma) > \gamma_0$ for some $\gamma_0 > 0$

To prove Theorem 5.3.2 we need the following lemma:

Lemma 5.3.1 *Consider any piecewise constant function defined as*

$$f(t) = f_k \quad \forall t \in [kT_s, (k+1)T_s), \quad k = 0, 1, \ldots.$$

If the sequence $\{\Delta f_k\} \in \ell_2$ where $\Delta f_k \stackrel{\Delta}{=} f_{k+1} - f_k$, then there exists a continuous function $\bar{f}(t)$ such that $|f - \bar{f}| \in \mathcal{L}_2$ and $\dot{\bar{f}} \in \mathcal{L}_2$.

Proof The proof of Lemma 5.3.1 is constructive and rather simple. Consider the following linear interpolation:

$$\bar{f}(t) = f_k + \frac{f_{k+1} - f_k}{T_s}(t - kT_s) \quad \forall t \in [kT_s, (k+1)T_s)$$

It is obvious that $\bar{f}$ has the following properties: (i) it is continuous; (ii) $|\bar{f}(t) - f(t)| \leq |f_{k+1} - f_k|$; (iii) $\dot{\bar{f}}$ is piecewise continuous and $|\dot{\bar{f}}(t)| = \frac{1}{T_s}|f_{k+1} - f_k|$ $\forall t \in [kT_s, (k+1)T_s)$. Therefore, using the assumption $\Delta f_k \in \ell_2$ we have

$$\int_0^\infty |\bar{f}(t) - f(t)|^2 dt \leq T_s \sum_{k=0}^\infty |\Delta f_k|^2 < \infty$$

and

$$\int_0^\infty |\dot{\bar{f}}(t)|^2 dt \leq \sum_{k=0}^\infty |\Delta f_k|^2 < \infty$$

i.e., $\dot{\bar{f}} \in \mathcal{L}_2$. $\square$

Proof of Theorem 5.3.2

(i) We have shown in Chapter 4 that the hybrid adaptive law given in Table 4.5 guarantees that $\theta_k \in l_\infty$; $\epsilon, \epsilon m \in \mathcal{L}_\infty \cap \mathcal{L}_2$. Because $u \in \mathcal{L}_\infty$ and the plant is stable, we have $y, x_\alpha, \phi, m \in \mathcal{L}_\infty$. As we show in the proof of Theorem 5.3.1, we can write the observer equation in the form

$$\dot{\hat{x}} = A^* \hat{x} + \hat{b}_k u + (\hat{A}_k - A^*) x_\alpha, \quad \forall t \in [t_k, t_{k+1}) \qquad (5.3.15)$$

where A^* is a stable matrix and

$$\hat{A}_k \triangleq \begin{bmatrix} -\hat{a}_k & \begin{array}{c} I_{n-1} \\ \cdots \\ 0 \end{array} \end{bmatrix}$$

Because $\hat{a}_k, \hat{b}_k \in l_\infty$; $x_\alpha, u \in \mathcal{L}_\infty$, it follows from (5.3.15) that $\hat{x} \in \mathcal{L}_\infty$, and, therefore, all signals are u.b.

(ii) Following the same procedure as in the proof of Theorem 5.3.1, we can express $\tilde{y} = y - \hat{y}$ as

$$\tilde{y} = C^\top (sI - A^*)^{-1}(-\tilde{b}_k u + \tilde{a}_k y) + \epsilon_t \qquad (5.3.16)$$

where ϵ_t is an exponentially decaying to zero term.

From Lemma 5.3.1 and the properties of the hybrid adaptive law, i.e., $\tilde{\theta}_k \in l_\infty$ and $|\tilde{\theta}_{k+1} - \tilde{\theta}_k| \in \ell_2$, we conclude that there exists a continuous piecewise vector function $\tilde{\theta}$ such that $|\tilde{\theta}(t) - \tilde{\theta}_k(t)|$, $\dot{\tilde{\theta}} \in \mathcal{L}_\infty \bigcap \mathcal{L}_2$, where $\tilde{\theta}_k(t)$ is the piecewise constant function defined by $\tilde{\theta}_k(t) = \tilde{\theta}_k$, $\forall t \in [t_k, t_{k+1})$. Therefore, we can write (5.3.16) as

$$\tilde{y} = C^\top(sI - A^*)^{-1}(-\tilde{b}u + \tilde{a}y) + f(t) + \epsilon_t \qquad (5.3.17)$$

where

$$f(t) = C^\top(sI - A^*)^{-1}\left((\tilde{b} - \tilde{b}_k)u - (\tilde{a} - \tilde{a}_k)y\right)$$

Using Corollary 3.3.1, it follows from $u, y \in \mathcal{L}_\infty$ and $|\tilde{\theta} - \tilde{\theta}_k| \in \mathcal{L}_\infty \bigcap \mathcal{L}_2$ that $f \in \mathcal{L}_\infty \bigcap \mathcal{L}_2$ and $f(t) \to 0$ as $t \to \infty$.

Because now $\tilde{\theta}$ has the same properties as those used in the continuous adaptive Luenberger observer, we can follow exactly the same procedure as in the proof of Theorem 5.3.1 to shown that the first term in (5.3.17) converges to zero and, therefore, $\tilde{y}(t) \to 0$ as $t \to \infty$.

(iii) We have established in the proof of Theorem 5.2.4 that when u is sufficiently rich of order $2n$, ϕ is PE. Using the PE property of ϕ and Theorem 4.6.1 (iii) we have that $\theta_k \to \theta^*$ exponentially fast. $\square$

5.4 Adaptive Observer with Auxiliary Input

Another class of adaptive observers that attracted considerable interest [28, 123, 172] involves the use of auxiliary signals in the observer equation. An observer that belongs to this class is described by the equation

$$\begin{aligned}\dot{\hat{x}} &= \hat{A}(t)\hat{x} + \hat{b}_p(t)u + K(t)(y - \hat{y}) + v, \quad \hat{x}(0) = \hat{x}_0 \\ \hat{y} &= [1\ 0\ \ldots 0]\hat{x}\end{aligned} \qquad (5.4.1)$$

where $\hat{x}, \hat{A}(t), \hat{b}_p(t), K(t)$ are as defined in (5.3.8), and v is an auxiliary vector input to be designed. The motivation for introducing v is to be able to use the SPR-Lyapunov design approach to generate the adaptive laws for $\hat{A}(t), \hat{b}_p(t), K(t)$. This approach is different from the one taken in Section 5.3 where the adaptive law is developed independently without considering the observer structure.

The first step in the SPR-Lyapunov design approach is to obtain an error equation that relates the estimation or observation error with the parameter error as follows: By using the same steps as in the proof of Theorem 5.3.1, the state error $\tilde{x} = x_\alpha - \hat{x}$ satisfies

$$\dot{\tilde{x}} = A^*\tilde{x} + \tilde{a}_p y - \tilde{b}_p u - v, \quad \tilde{x}(0) = \tilde{x}_0$$

$$\tilde{y} = C^\top \tilde{x} \tag{5.4.2}$$

where $\tilde{a}_p, \tilde{b}_p$ are the parameter errors. Equation (5.4.2) is not in the familiar form studied in Chapter 4 that allows us to choose an appropriate Lyapunov-like function for designing an adaptive law and proving stability. The purpose of the signal vector v is to convert (5.4.2) into a form that is suitable for applying the Lyapunov approach studied in Chapter 4. The following Lemma establishes the existence of a vector v that converts the error equation (5.4.2) into one that is suitable for applying the SPR-Lyapunov design approach.

Lemma 5.4.1 *There exists a signal vector $v \in \mathcal{R}^n$, generated from measurements of known signals, for which the system (5.4.2) becomes*

$$\begin{aligned} \dot{e} &= A^* e + B_c(-\tilde{\theta}^\top \phi), \quad e(0) = \tilde{x}_0 \\ \tilde{y} &= C^\top e \end{aligned} \tag{5.4.3}$$

where $C^\top(sI - A^*)^{-1} B_c$ is SPR, $\phi \in \mathcal{R}^{2n}$ is a signal vector generated from input/output data, $\tilde{\theta} = [\tilde{b}_p^\top, \tilde{a}_p^\top]^\top$, and $e \in \mathcal{R}^n$ is a new state vector.

Proof Because (C, A^*) is in the observer form, we have

$$C^\top (sI - A^*)^{-1} = \frac{\alpha_{n-1}^\top(s)}{\Lambda^*(s)}$$

where $\alpha_{n-1}(s) = [s^{n-1}, s^{n-2}, \ldots, 1]^\top$, $\Lambda^*(s) = det(sI - A^*)$. Therefore, (5.4.2) can be expressed in the form

$$\begin{aligned} \tilde{y} &= C^\top(sI - A^*)^{-1} \left[\tilde{a}_p y - \tilde{b}_p u - v \right] + \epsilon_t \\ &= \frac{\Lambda(s)}{\Lambda^*(s)} \left\{ \sum_{i=1}^n \frac{s^{n-i}}{\Lambda(s)} \left(\tilde{a}_{n-i} y - \tilde{b}_{n-i} u \right) - \frac{\alpha_{n-1}^\top(s)}{\Lambda(s)} v \right\} + \epsilon_t \end{aligned} \tag{5.4.4}$$

where $\epsilon_t = \mathcal{L}^{-1}\left\{ C^\top (sI - A^*)^{-1} \right\} \tilde{x}_0$, $\Lambda(s) = s^{n-1} + \lambda^\top \alpha_{n-2}(s)$ is a Hurwitz polynomial, $\lambda^\top = [\lambda_{n-2}, \lambda_{n-3}, \ldots, \lambda_1, \lambda_0]$ is to be specified and $\tilde{a}_i, \tilde{b}_i$ is the ith element of $\tilde{a}_p, \tilde{b}_p$ respectively.

Applying the Swapping Lemma A.1 given in Appendix A to each term under the summation in (5.4.4), we obtain

$$\begin{aligned} \tilde{y} &= \frac{\Lambda(s)}{\Lambda^*(s)} \left\{ \sum_{i=1}^n \left(\tilde{a}_{n-i} \frac{s^{n-i}}{\Lambda(s)} y - \tilde{b}_{n-i} \frac{s^{n-i}}{\Lambda(s)} u \right) \right. \\ &\quad + \left. \sum_{i=1}^n \left\{ W_{ci}(s)(W_{bi}(s)y) \dot{\tilde{a}}_{n-i} - W_{ci}(s)(W_{bi}(s)u) \dot{\tilde{b}}_{n-i} \right\} - \frac{\alpha_{n-1}^\top(s)}{\Lambda(s)} v \right\} + \epsilon_t \end{aligned} \tag{5.4.5}$$

5.4. ADAPTIVE OBSERVER WITH AUXILIARY INPUT

We now use Lemma A.1 to obtain expressions for W_{ci}, W_{bi} in terms of the parameters of $\Lambda(s)$. Because W_{c1}, W_{b1} are the transfer functions resulting from swapping with the transfer function

$$\frac{s^{n-1}}{\Lambda(s)} = 1 - \frac{\alpha_{n-2}^T(s)\lambda}{\Lambda(s)} = 1 - C_0^T(sI - \Lambda_0)^{-1}\lambda = 1 + C_0^T(sI - \Lambda_0)^{-1}(-\lambda)$$

where $C_0 = [1, 0, \ldots, 0]^T \in \mathcal{R}^{n-1}$

$$\Lambda_0 = \begin{bmatrix} & \vdots & I_{n-2} & \\ -\lambda & \vdots & \cdots & \\ & \vdots & 0 & \end{bmatrix} \in \mathcal{R}^{(n-1)\times(n-1)}$$

it follows from Lemma A.1 that

$$W_{c1}(s) = -C_0^T(sI - \Lambda_0)^{-1} = \frac{-\alpha_{n-2}^T(s)}{\Lambda(s)}, \quad W_{b1}(s) = (sI - \Lambda_0)^{-1}(-\lambda) .$$

Similarly W_{ci}, W_{bi} $i = 2, 3, \ldots, n$ result from swapping with

$$\frac{s^{n-i}}{\Lambda(s)} = C_0^T(sI - \Lambda_0)^{-1}d_i$$

where $d_i = [0, \ldots, 0, 1, 0, \ldots, 0]^T \in \mathcal{R}^{n-1}$ has all its elements equal to zero except for the $(i-1)-th$ element that is equal to one. Therefore it follows from Lemma A.1 that

$$W_{ci}(s) = -C_0^T(sI - \Lambda_0)^{-1} = -\frac{\alpha_{n-2}^T(s)}{\Lambda(s)}, \quad i = 2, 3, \ldots, n$$

$$W_{bi}(s) = (sI - \Lambda_0)^{-1}d_i, \quad i = 2, 3, \ldots, n$$

If we define

$$\phi \triangleq \left[\frac{\alpha_{n-1}^T(s)}{\Lambda(s)}u, \; -\frac{\alpha_{n-1}^T(s)}{\Lambda(s)}y\right]^T$$

and use the expressions for $W_{ci}(s), W_{bi}(s)$, (5.4.5) becomes

$$\begin{aligned} \tilde{y} &= \frac{\Lambda(s)}{\Lambda^*(s)} \left\{ -\tilde{\theta}^T \phi - \frac{s^{n-1}}{\Lambda(s)} v_1 \right. \\ &\left. - \frac{\alpha_{n-2}^T(s)}{\Lambda(s)} \left[\sum_{i=1}^n \left\{ [(sI - \Lambda_0)^{-1}d_i y] \, \dot{\tilde{a}}_{n-i} - [(sI - \Lambda_0)^{-1}d_i u] \, \dot{\tilde{b}}_{n-i} \right\} + \bar{v} \right] \right\} + \epsilon_t \end{aligned}$$

where v is partitioned as $v = [v_1, \bar{v}^T]^T$ with $v_1 \in \mathcal{R}^1$ and $\bar{v} \in \mathcal{R}^{n-2}$ and $d_1 = -\lambda$.

Choosing $v_1 = 0$ and

$$\bar{v} = -\sum_{i=1}^{n}\left\{\left[(sI-\Lambda_0)^{-1}d_iy\right]\dot{\hat{a}}_{n-i} - \left[(sI-\Lambda_0)^{-1}d_iu\right]\dot{\hat{b}}_{n-i}\right\}$$

we obtain

$$\tilde{y} = \frac{\Lambda(s)}{\Lambda^*(s)}(-\tilde{\theta}^\top \phi) + \epsilon_t \qquad (5.4.6)$$

Because

$$\frac{\Lambda(s)}{\Lambda^*(s)} = \frac{s^{n-1} + \lambda^\top \alpha_{n-2}(s)}{\Lambda^*(s)} = \frac{\alpha_{n-1}^\top(s)b_\lambda}{\Lambda^*(s)} = C^\top(sI - A^*)^{-1}b_\lambda$$

where $b_\lambda = [1, \lambda^\top]^\top \in \mathcal{R}^n$ and $C = [1, 0, \ldots, 0]^\top \in \mathcal{R}^n$. For $B_c = b_\lambda$, (5.4.3) is a minimal state-space representation of (5.4.6). Because $\Lambda(s)$ is arbitrary, its coefficient vector λ can be chosen so that $\frac{\Lambda(s)}{\Lambda^*(s)}$ is SPR. The signal $\bar{v}$ is implementable because $\dot{\hat{a}}_i, \dot{\hat{b}}_i$ and u, y are available for measurement. $\square$

We can now use (5.4.3) instead of (5.4.2) to develop an adaptive law for generating $\theta = \left[\hat{b}_p^\top, \hat{a}_p^\top\right]^\top$. Using the results of Chapter 4, it follows that the adaptive law is given by

$$\dot{\theta} = \Gamma \epsilon \phi, \quad \epsilon = \tilde{y} = y - \hat{y} \qquad (5.4.7)$$

where $\Gamma = \Gamma^\top > 0$.

We summarize the main equations of the adaptive observer developed above in Table 5.3. The structure of the adaptive observer is shown in Figure 5.2.

Theorem 5.4.1 *The adaptive observer presented in Table 5.3 guarantees that for any bounded input signal u,*

(i) *all signals are u.b.*
(ii) $\tilde{y}(t) = y - \hat{y} \to 0$ *as* $t \to \infty$.
(iii) $\dot{\hat{a}}_p(t), \dot{\hat{b}}_p(t) \in \mathcal{L}_2 \cap \mathcal{L}_\infty$ *and converge to zero as* $t \to \infty$.

In addition, if u is sufficiently rich of order $2n$, then

(iv) $\tilde{x}(t) = x_\alpha(t) - \hat{x}(t)$, $\tilde{a}_p(t) = \hat{a}_p(t) - a_p$, $\tilde{b}_p(t) = \hat{b}_p(t) - b_p$ *converge to zero exponentially fast.*

Table 5.3 Adaptive observer with auxiliary input

Plant	$\dot{x}_\alpha = \begin{bmatrix} & I_{n-1} \\ -a_p & \cdots \\ & 0 \end{bmatrix} x_\alpha + b_p u, \quad y = [1\ 0\ldots 0]x_\alpha$ $a_p = [a_{n-1}, \ldots, a_0]^\top, \quad b_p = [b_{n-1}, \ldots, b_0]^\top$
Observer	$\dot{\hat{x}} = \begin{bmatrix} & I_{n-1} \\ -\hat{a}_p(t) & \cdots \\ & 0 \end{bmatrix}\hat{x} + \hat{b}_p(t)u + (a^* - \hat{a}_p(t))(y - \hat{y}) + v$ $\hat{y} = [1\ 0\ \ldots\ 0]\hat{x}$
Adaptive law	$\begin{bmatrix} \dot{\hat{b}}_p \\ \dot{\hat{a}}_p \end{bmatrix} = \Gamma \phi (y - \hat{y}), \quad \Gamma = \Gamma^\top > 0$ $\phi = \left[\frac{\alpha_{n-1}^\top(s)}{\Lambda(s)} u, -\frac{\alpha_{n-1}^\top(s)}{\Lambda(s)} y \right]^\top$
Auxiliary input	$v = \begin{bmatrix} 0 \\ \bar{v} \end{bmatrix}, \quad \bar{v} = \sum_{i=1}^n \left[-[W_i(s)y]\dot{\hat{a}}_{n-i} + [W_i(s)u]\dot{\hat{b}}_{n-i} \right]$ $W_i(s) = (sI - \Lambda_0)^{-1} d_i,$ $d_1 = -\lambda; \quad d_i^\top = [0\ldots 0, \underset{(i-1)}{1}, 0\ldots 0], \quad i = 2, \ldots, n$ $\Lambda_0 = \begin{bmatrix} & I_{n-2} \\ -\lambda & \cdots \\ & 0 \end{bmatrix}, \quad \lambda = [\lambda_{n-2}, \ldots, \lambda_0]^\top$ $\det(sI - \Lambda_0) = \Lambda(s) = s^{n-1} + \lambda^\top \alpha_{n-2}(s)$
Design variables	(i) a^* is chosen such that $A^* = \begin{bmatrix} & I_{n-1} \\ -a^* & \cdots \\ & 0 \end{bmatrix}$ is stable (ii) The vector λ is chosen such that Λ_0 is stable and $[1\ 0\ldots\ 0](sI - A^*)^{-1} \begin{bmatrix} 1 \\ \lambda \end{bmatrix}$ is SPR.

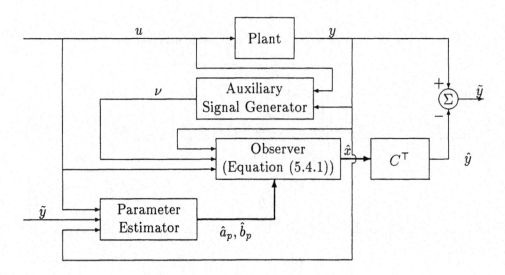

Figure 5.2 Structure of the adaptive observer with auxiliary input signal.

Proof The main equations that describe the stability properties of the adaptive observer are the error equation (5.4.3) that relates the parameter error with the state observation error and the adaptive law (5.4.7). Equations (5.4.3) and (5.4.7) are analyzed in Chapter 4 where it has been shown that $e, \tilde{y} \in \mathcal{L}_2 \cap \mathcal{L}_\infty$ and $\tilde{\theta} \in \mathcal{L}_\infty$ for any signal vector ϕ with piecewise continuous elements. Because $u, y \in \mathcal{L}_\infty$, it follows that $\phi \in \mathcal{L}_\infty$ and from (5.4.3) and (5.4.7) that $\dot{\theta}, \dot{e}, \dot{\tilde{y}} \in \mathcal{L}_\infty$ and $\dot{\theta} \in \mathcal{L}_2$. Using $e, \tilde{y} \in \mathcal{L}_2 \cap \mathcal{L}_\infty$ together with $\dot{e}, \dot{\tilde{y}} \in \mathcal{L}_\infty$, we have $\tilde{y}(t) \to 0$, $e(t) \to 0$ as $t \to \infty$, which implies that $\dot{\theta}(t) \to 0$ as $t \to \infty$. From $\dot{\theta} \in \mathcal{L}_2 \cap \mathcal{L}_\infty$, it follows that $v \in \mathcal{L}_2 \cap \mathcal{L}_\infty$ and $v(t) \to 0$ as $t \to \infty$. Hence, all inputs to the state equation (5.4.2) are in $\mathcal{L}_\infty$, which implies that $\tilde{x} \in \mathcal{L}_\infty$, i.e., $\hat{x} \in \mathcal{L}_\infty$.

In Chapter 4 we established that if $\phi, \dot{\phi} \in \mathcal{L}_\infty$ and ϕ is PE, then the error equations (5.4.3) and (5.4.7) guarantee that $\tilde{\theta}(t) \to 0$ as $t \to \infty$ exponentially fast. In our case $\phi, \dot{\phi} \in \mathcal{L}_\infty$ and therefore if u is chosen so that ϕ is PE then $\tilde{\theta}(t) \to 0$ as $t \to \infty$ exponentially fast, which implies that $e(t), \tilde{y}(t), \dot{\theta}(t), v(t), \tilde{a}_p(t)y, \tilde{b}_p(t)u$ and, therefore, $\tilde{x}(t)$ converge to zero exponentially fast. To explore the PE property of ϕ, we note that ϕ is related to u through the equation

$$\phi = \frac{1}{\Lambda(s)} \begin{bmatrix} \alpha_{n-1}(s) \\ -\alpha_{n-1}(s)C^\top(sI-A)^{-1}b_p \end{bmatrix} u$$

Because u is sufficiently rich of order $2n$, the PE property of ϕ can be established by following exactly the same steps as in the proof of Theorem 5.2.4. □

The auxiliary signal vector $\bar{v}$ can be generated directly from the signals

5.4. ADAPTIVE OBSERVER WITH AUXILIARY INPUT

ϕ and $\dot{\tilde{\theta}}$, i.e., the filters W_i for y, u do not have to be implemented. This simplification reduces the number of integrators required to generate ϕ, $\bar{v}$ considerably and it follows from the relationship

$$(sI - \Lambda_0)^{-1} d_i = Q_i \frac{\alpha_{n-1}(s)}{\Lambda(s)}, \quad i = 1, 2, \ldots, n$$

where $Q_i \in \mathcal{R}^{(n-1) \times (n-1)}$ are constant matrices whose elements depend on the coefficients of the numerator polynomials of $(sI - \Lambda_0)^{-1} d_i$ (see Problem 2.12).

As with the adaptive Luenberger observer, the adaptive observer with auxiliary input shown in Table 5.3 requires the input u to be sufficiently rich of order $2n$ in order to guarantee exact plant state observation. The only difference between the two observers is that the adaptive Luenberger observer may employ any one of the adaptive laws given in Tables 4.2, 4.3, and 4.5 whereas the one with the auxiliary input given in Table 5.3 relies on the SPR-Lyapunov design approach only. It can be shown, however, (see Problem 5.16) by modifying the proof of Lemma 5.4.1 that the observation error $\tilde{y}$ may be expressed in the form

$$\tilde{y} = -\tilde{\theta}^\top \phi \tag{5.4.8}$$

by properly selecting the auxiliary input v and ϕ. Equation (5.4.8) is in the form of the error equation that appears in the case of the linear parametric model $y = \theta^{*\top} \phi$ and allows the use of any one of the adaptive laws of Table 4.2, 4.3, and 4.5 leading to a wide class of adaptive observers with auxiliary input.

The following example illustrates the design of an adaptive observer with auxiliary input v and the generation of v from the signals $\phi, \dot{\theta}$.

Example 5.4.1 Let us consider the second order plant

$$\begin{aligned} \dot{x}_\alpha &= \begin{bmatrix} -a_1 & 1 \\ -a_0 & 0 \end{bmatrix} x_\alpha + \begin{bmatrix} b_1 \\ b_0 \end{bmatrix} u \\ y &= \begin{bmatrix} 1 & 0 \end{bmatrix} x_\alpha \end{aligned}$$

where $a_1 > 0, a_0 > 0, b_1, b_0$ are unknown constants and y, u are the only available signals for measurement.

We use Table 5.3 to develop the adaptive observer for estimating x_α and the unknown plant parameters. We start with the design variables. We choose $a^* = [4,4]^T$, $\lambda = \lambda_0, \Lambda(s) = s + \lambda_0$. Setting $\lambda_0 = 3$ we have

$$[1\ 0][sI - A^*]^{-1}\begin{bmatrix} 1 \\ 3 \end{bmatrix} = \frac{s+3}{(s+2)^2}$$

which is SPR.

The signal vector $\phi = \left[\frac{s}{s+3}u, \frac{1}{s+3}u, -\frac{s}{s+3}y, -\frac{1}{s+3}y\right]^T$ is realized as follows

$$\phi = [\phi_1, \phi_2, \phi_3, \phi_4]^T$$

where

$$\begin{aligned}
\phi_1 &= u - 3\bar{\phi}_1, & \dot{\bar{\phi}}_1 &= -3\bar{\phi}_1 + u \\
\phi_2 &= \bar{\phi}_1 \\
\phi_3 &= -y + 3\bar{\phi}_3, & \dot{\bar{\phi}}_3 &= -3\bar{\phi}_3 + y \\
\phi_4 &= -\bar{\phi}_3
\end{aligned}$$

with $\bar{\phi}_1(0) = 0, \bar{\phi}_3(0) = 0$.

For simplicity we choose the adaptive gain $\Gamma = \text{diag}\{10, 10, 10, 10\}$. The adaptive law is given by

$$\dot{\hat{b}}_1 = 10\phi_1(y - \hat{y}), \quad \dot{\hat{b}}_0 = 10\phi_2(y - \hat{y})$$
$$\dot{\hat{a}}_1 = 10\phi_3(y - \hat{y}), \quad \dot{\hat{a}}_0 = 10\phi_4(y - \hat{y})$$

and the signal vector $v = \begin{bmatrix} 0 \\ \bar{v} \end{bmatrix}$ by

$$\begin{aligned}
\bar{v} &= \left(\frac{3}{s+3}y\right)\dot{\hat{a}}_1 - \left(\frac{3}{s+3}u\right)\dot{\hat{b}}_1 - \left(\frac{1}{s+3}y\right)\dot{\hat{a}}_0 + \left(\frac{1}{s+3}u\right)\dot{\hat{b}}_0 \\
&= 3\dot{\hat{a}}_1\bar{\phi}_3 - 3\dot{\hat{b}}_1\bar{\phi}_1 - \dot{\hat{a}}_0\bar{\phi}_3 + \dot{\hat{b}}_0\bar{\phi}_1 \\
&= 10(y-\hat{y})(-3\phi_3\phi_4 - 3\phi_1\phi_2 + \phi_4^2 + \phi_2^2)
\end{aligned}$$

The observer equation becomes

$$\begin{aligned}
\dot{\hat{x}} &= \begin{bmatrix} -\hat{a}_1 & 1 \\ -\hat{a}_0 & 0 \end{bmatrix}\hat{x} + \begin{bmatrix} \hat{b}_1 \\ \hat{b}_0 \end{bmatrix}u + \begin{bmatrix} 4-\hat{a}_1 \\ 4-\hat{a}_0 \end{bmatrix}(y-\hat{y}) \\
&\quad + \begin{bmatrix} 0 \\ 10 \end{bmatrix}(y-\hat{y})[-3\phi_3\phi_4 - 3\phi_1\phi_2 + \phi_4^2 + \phi_2^2]
\end{aligned}$$

The input signal is chosen as $u = A_1\sin\omega_1 t + A_2\sin\omega_2 t$ for some $A_1, A_2 \neq 0$ and $\omega_1 \neq \omega_2$. $\triangledown$

5.5 Adaptive Observers for Nonminimal Plant Models

The adaptive observers presented in Sections 5.3 and 5.4 are suitable for estimating the states of a minimal state space realization of the plant that is expressed in the observer form. Simpler (in terms of the number of integrators required for implementation) adaptive observers may be constructed if the objective is to estimate the states of certain nonminimal state-space representations of the plant. Several such adaptive observers have been presented in the literature over the years [103, 108, 120, 123, 130, 172], in this section we present only those that are based on the two nonminimal plant representations developed in Chapter 2 and shown in Figures 2.2 and 2.3.

5.5.1 Adaptive Observer Based on Realization 1

Following the plant parameterization shown in Figure 2.2, the plant (5.3.1) is represented in the state space form

$$\begin{aligned}
\dot{\phi}_1 &= \Lambda_c \phi_1 + l u, \quad \phi_1(0) = 0 \\
\dot{\phi}_2 &= \Lambda_c \phi_2 - l y, \quad \phi_2(0) = 0 \\
\dot{\omega} &= \Lambda_c \omega, \quad \omega(0) = \omega_0 = B_0 x_0 \\
\eta_0 &= C_0^\top \omega \\
z &= y + \lambda^\top \phi_2 = \theta^{*\top} \phi + \eta_0 \\
y &= \theta^{*\top} \phi - \lambda^\top \phi_2 + \eta_0
\end{aligned} \quad (5.5.1)$$

where $\omega \in \mathcal{R}^n$, $\phi = [\phi_1^\top, \phi_2^\top]^\top$, $\phi_i \in \mathcal{R}^n$, $i = 1, 2$; $\Lambda_c \in \mathcal{R}^{n \times n}$ is a known stable matrix in the controller form; $l = [1, 0, \ldots, 0]^\top \in \mathcal{R}^n$ is a known vector such that $(sI - \Lambda_c)^{-1} l = \frac{\alpha_{n-1}(s)}{\Lambda(s)}$ and $\Lambda(s) = \det(sI - \Lambda_c) = s^n + \lambda^\top \alpha_{n-1}(s)$, $\lambda = [\lambda_{n-1}, \ldots, \lambda_0]^\top$; $\theta^* = [b_{n-1}, b_{n-2}, \ldots, b_0, a_{n-1}, a_{n-2}, \ldots, a_0]^\top \in \mathcal{R}^{2n}$ are the unknown parameters to be estimated; and $B_0 \in \mathcal{R}^{n \times n}$ is a constant matrix defined in Section 2.4.

The plant parameterization (5.5.1) is of order $3n$ and has $2n$ unknown parameters. The state ω and signal η_0 decay to zero exponentially fast with a rate that depends on Λ_c. Because Λ_c is arbitrary, it can be chosen so that η_0, ω go to zero faster than a certain given rate. Because $\phi_1(0) = \phi_2(0) = 0$, the states ϕ_1, ϕ_2 can be reproduced by the observer

$$\begin{aligned} \dot{\hat{\phi}}_1 &= \Lambda_c \hat{\phi}_1 + lu, & \hat{\phi}_1(0) &= 0 \\ \dot{\hat{\phi}}_2 &= \Lambda_c \hat{\phi}_2 - ly, & \hat{\phi}_2(0) &= 0 \end{aligned} \qquad (5.5.2)$$

which implies that $\hat{\phi}_i(t) = \phi_i(t), i = 1, 2 \ \forall t \geq 0$. The output of the observer is given by

$$\begin{aligned} z_0 &= \theta^{*\top} \phi \\ y_0 &= \theta^{*\top} \hat{\phi} - \lambda^\top \hat{\phi}_2 = \theta^{*\top} \phi - \lambda^\top \phi_2 \end{aligned} \qquad (5.5.3)$$

The state ω in (5.5.1) can not be reproduced exactly unless the initial condition x_0 and therefore ω_0 is known.

Equation (5.5.2) and (5.5.3) describe the nonminimal state observer for the plant (5.3.1) when θ^* is known. Because $\hat{\phi}(t) = \phi(t), \forall t \geq 0$, it follows that the output observation errors $e_z \triangleq z - z_0$, $e_0 \triangleq y - y_0$ satisfy

$$e_z = e_0 = \eta_0 = C_0^\top e^{\Lambda_c t} \omega_0$$

which implies that e_0, e_z decay to zero exponentially fast. The eigenvalues of Λ_c can be regarded as the eigenvalues of the observer and can be assigned arbitrarily through the design of Λ_c.

When θ^* is unknown, the observer equation (5.5.2) remains the same but (5.5.3) becomes

$$\begin{aligned} \hat{z} &= \hat{y} + \lambda^\top \hat{\phi}_2 \\ \hat{y} &= \theta^\top \hat{\phi} - \lambda^\top \hat{\phi}_2 \end{aligned} \qquad (5.5.4)$$

where $\theta(t)$ is the estimate of θ^* at time t and $\phi(t) = \hat{\phi}(t)$ is generated from (5.5.2). Because θ^* satisfies the parametric model

$$z = y + \lambda^\top \phi_2 = \theta^{*\top} \phi + \eta_0 \qquad (5.5.5)$$

where z, ϕ are available for measurement and η_0 is exponentially decaying to zero, the estimate $\theta(t)$ of θ^* may be generated using (5.5.5) and the results of Chapter 4. As shown in Chapter 4, the exponentially decaying to zero term η_0 does not affect the properties of the adaptive laws developed for $\eta_0 = 0$ in (5.5.5). Therefore, for design purposes, we can assume that

5.5. NONMINIMAL ADAPTIVE OBSERVER

$\eta_0 \equiv 0 \ \forall t \geq 0$ and select any one of the adaptive laws from Tables 4.2, 4.3, and 4.5 to generate $\theta(t)$. In the analysis, we include $\eta_0 \neq 0$ and verify that its presence does not affect the stability properties and steady state behavior of the adaptive observer. As an example, let us use Table 4.2 and choose the gradient algorithm

$$\dot{\theta} = \Gamma \epsilon \phi, \quad \epsilon = z - \hat{z} = y - \hat{y} \qquad (5.5.6)$$

where $\Gamma = \Gamma^\top > 0$ and $\phi = \hat{\phi}$.

Equations (5.5.2), (5.5.4) and (5.5.6) form the adaptive observer and are summarized in Table 5.4 and shown in Figure 5.3.

The stability properties of the adaptive observer in Table 5.4 are given by the following theorem.

Theorem 5.5.1 *The adaptive observer for the nonminimal plant representation (5.5.1) with the adaptive law based on the gradient algorithm or any other adaptive law from Tables 4.2, 4.3, and 4.5 that is based on the parametric model (5.5.5) guarantees that*

(i) $\hat{\phi}(t) = \phi(t) \ \forall t \geq 0$.

(ii) *All signals are u.b.*

(iii) *The output observation error $\epsilon(t) = y(t) - \hat{y}(t)$ converges to zero as $t \to \infty$.*

(iv) *If u is sufficiently rich of order $2n$, then $\theta(t)$ converges to θ^*. The convergence of $\theta(t)$ to θ^* is exponential for all the adaptive laws of Tables 4.2, 4.3, and 4.5 with the exception of the pure least squares where convergence is asymptotic.*

Proof (i) This proof follows directly from (5.5.1) and (5.5.2).

(ii) Because $\hat{\phi} = \phi$, the following properties of the adaptive laws can be established using the results of Chapter 4: (a) $\epsilon, \theta \in \mathcal{L}_\infty$; (b) $\epsilon, \dot{\theta} \in \mathcal{L}_\infty \cap \mathcal{L}_2$ for the continuous time adaptive laws and $|\theta_{k+1} - \theta_k| \in \ell_2$ for the hybrid one. From $\epsilon, \theta, \dot{\theta} \in \mathcal{L}_\infty$, $u \in \mathcal{L}_\infty$, the stability of Λ_c and the stability of the plant, we have that all signals are u.b.

(iii) For the continuous-time adaptive laws, we have that

$$\epsilon = z - \hat{z} = y + \lambda^\top \phi_2 - (\hat{y} + \lambda^\top \phi_2) = y - \hat{y}$$

We can verify that $\dot{\epsilon} \in \mathcal{L}_\infty$ which together with $\epsilon \in \mathcal{L}_2$ imply that $\epsilon(t) \to 0$ as $t \to \infty$.

Table 5.4 Adaptive observer (Realization 1)

Plant	$\dot{\phi}_1 = \Lambda_c \phi_1 + lu, \quad \phi_1(0) = 0$ $\dot{\phi}_2 = \Lambda_c \phi_2 - ly, \quad \phi_2(0) = 0$ $\dot{\omega} = \Lambda_c \omega, \quad \omega(0) = \omega_0$ $\eta_0 = C_0^T \omega$ $z = y + \lambda^T \phi_2 = \theta^{*T} \phi + \eta_0$ $y = \theta^{*T} \phi - \lambda^T \phi_2 + \eta_0$
Observer	$\dot{\hat{\phi}}_1 = \Lambda_c \hat{\phi}_1 + lu, \quad \hat{\phi}_1(0) = 0$ $\dot{\hat{\phi}}_2 = \Lambda_c \hat{\phi}_2 - ly, \quad \hat{\phi}_2(0) = 0$ $\hat{z} = \theta^T \hat{\phi}$ $\hat{y} = \hat{z} - \lambda^T \hat{\phi}_2$
Adaptive law	$\dot{\theta} = \Gamma \epsilon \hat{\phi}, \quad \Gamma = \Gamma^T > 0$ $\epsilon = z - \hat{z}, \quad \hat{z} = \theta^T \hat{\phi}$
Design variables	Λ_c is a stable matrix; (Λ_c, l) is in the controller form; $\Lambda(s) = \det(sI - \Lambda_c) = s^n + \lambda^T \alpha_{n-1}(s)$

For the hybrid adaptive law, we express ϵ as

$$\epsilon = \theta^{*T} \phi - \theta_k^T \phi = -\tilde{\theta}_k^T \phi, \quad \forall t \in [t_k, t_{k+1})$$

Because the hybrid adaptive law guarantees that (a) $\tilde{\theta}_k \in l_\infty$, (b) $|\tilde{\theta}_{k+1} - \tilde{\theta}_k| \in l_2$ and $|\tilde{\theta}_{k+1} - \tilde{\theta}_k| \to 0$ as $k \to \infty$, we can construct a continuous, piecewise linear function $\tilde{\theta}(t)$ from $\tilde{\theta}_k$ using linear interpolation that satisfies: (a) $|\tilde{\theta} - \tilde{\theta}_k|, \dot{\tilde{\theta}} \in \mathcal{L}_\infty \cap \mathcal{L}_2$ and (b) $|\tilde{\theta}(t) - \tilde{\theta}_k(t)| \to 0$ as $t \to \infty$. Therefore, we can write

$$\epsilon = -\tilde{\theta}_k^T \phi = -\tilde{\theta}^T \phi + (\tilde{\theta} - \tilde{\theta}_k)^T \phi \quad (5.5.7)$$

From $\epsilon \in \mathcal{L}_2, |\tilde{\theta} - \tilde{\theta}_k| \in \mathcal{L}_2$ and $\phi \in \mathcal{L}_\infty$, we have $\tilde{\theta}^T \phi \in \mathcal{L}_2$. Because $\frac{d}{dt}(\tilde{\theta}^T \phi) = \dot{\tilde{\theta}}^T \phi + \tilde{\theta}^T \dot{\phi}$ and $\tilde{\theta}, \dot{\tilde{\theta}}, \phi, \dot{\phi} \in \mathcal{L}_\infty$, we conclude that $\frac{d}{dt}(\tilde{\theta}^T \phi) \in \mathcal{L}_\infty$, which, together

5.5. NONMINIMAL ADAPTIVE OBSERVER

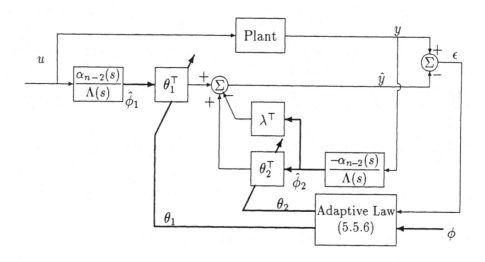

Figure 5.3 Adaptive observer using nonminimal Realization 1.

with $\tilde{\theta}^T \phi \in \mathcal{L}_2$, implies that $\tilde{\theta}^T \phi \to 0$ as $t \to \infty$. Therefore, we have established that $(\tilde{\theta} - \tilde{\theta}_k)^T \phi \to 0$ and $\tilde{\theta}^T \phi \to 0$ as $t \to \infty$. Thus, using (5.5.7) for ϵ, we have that $\epsilon(t) \to 0$ as $t \to \infty$.

(iv) The proof of this part follows directly from the properties of the adaptive laws developed in Chapter 4 and Theorem 5.2.4 by noticing that $\phi = H(s)u$ where $H(s)$ is the same as in the proof of Theorem 5.2.4. $\square$

The reader can verify that the adaptive observer of Table 5.4 is the state-space representation of the parameter identifier given by (5.2.17), (5.2.18) in Section 5.2.3. This same identifier is shared by the adaptive Luenberger observer indicating that the state ϕ of the nonminimal state space representation of the plant is also part of the state of the adaptive Luenberger observer. In addition to ϕ, the adaptive Luenberger observer estimates the state of a minimal state space representation of the plant expressed in the observer canonical form at the expense of implementing an additional nth order state equation referred to as the observer equation in Table 5.1. The parameter identifiers of Section 5.2 can, therefore, be viewed as adaptive observers based on a nonminimal state representation of the plant.

5.5.2 Adaptive Observer Based on Realization 2

An alternative nonminimal state-space realization for the plant (5.3.1), developed in Chapter 2 and shown in Figure 2.5, is described by the state equations

$$\begin{aligned}
\dot{\bar{x}}_1 &= -\lambda_0 \bar{x}_1 + \bar{\theta}^{*T}\phi, & \bar{x}_1(0) &= 0 \\
\dot{\phi}_1 &= \Lambda_c \phi_1 + lu, & \phi_1(0) &= 0 \\
\dot{\phi}_2 &= \Lambda_c \phi_2 - ly, & \phi_2(0) &= 0 \\
\phi &= [u, \phi_1^T, y, \phi_2^T]^T \\
\dot{\omega} &= \Lambda_c \omega, & \omega(0) &= \omega_0 \\
\eta_0 &= C_0^T \omega \\
y &= \bar{x}_1 + \eta_0
\end{aligned} \quad (5.5.8)$$

where $\bar{x}_1 \in \mathcal{R}$; $\phi_i \in \mathcal{R}^{n-1}, i = 1, 2$; $\bar{\theta}^* \in \mathcal{R}^{2n}$ is a vector of linear combinations of the unknown plant parameters $a = [a_{n-1}, a_{n-2}, \ldots, a_0]^T$, $b = [b_{n-1}, b_{n-2}, \ldots, b_0]^T$ in the plant transfer function (5.3.7) as shown in Section 2.4; $\lambda_0 > 0$ is a known scalar; $\Lambda_c \in \mathcal{R}^{(n-1)\times(n-1)}$ is a known stable matrix; and $(sI - \Lambda_c)^{-1}l = \frac{\alpha_{n-2}(s)}{\Lambda^*(s)}$, where $\Lambda^*(s) = s^{n-1} + q_{n-2}s^{n-2} + \ldots + q_0$ is a known Hurwitz polynomial. When θ^* is known, the states $\bar{x}_1, \phi_1, \phi_2$ can be generated exactly by the observer

$$\begin{aligned}
\dot{x}_{o1} &= -\lambda_0 x_{o1} + \bar{\theta}^{*T}\hat{\phi}, & x_{o1}(0) &= 0 \\
\dot{\hat{\phi}}_1 &= \Lambda_c \hat{\phi}_1 + lu, & \hat{\phi}_1(0) &= 0 \\
\dot{\hat{\phi}}_2 &= \Lambda_c \hat{\phi}_2 - ly, & \hat{\phi}_2(0) &= 0 \\
y_0 &= x_{o1}
\end{aligned} \quad (5.5.9)$$

where $x_{o1}, \hat{\phi}_i$ are the estimates of $\bar{x}_1, \phi$ respectively. As in Section 5.5.1, no attempt is made to generate an estimate of the state ω, because $\omega(t) \to 0$ as $t \to \infty$ exponentially fast. The state observer (5.5.9) guarantees that $x_{o1}(t) = \bar{x}_1(t), \hat{\phi}_1(t) = \phi_1(t), \hat{\phi}_2(t) = \phi_2(t) \ \forall t \geq 0$. The observation error $e_0 \triangleq y - y_0$ satisfies

$$e_0 = \eta_0 = C_0^T e^{\Lambda_c t} \omega_0$$

i.e., $e_0(t) \to 0$ as $t \to \infty$ with an exponential rate that depends on the matrix Λ_c that is chosen by the designer.

5.5. NONMINIMAL ADAPTIVE OBSERVER

When $\bar{\theta}^*$ is unknown, (5.5.9) motivates the adaptive observer

$$\begin{aligned}
\dot{\hat{x}}_1 &= -\lambda_0 \hat{x}_1 + \bar{\theta}^T \hat{\phi}, & \hat{x}_0(0) &= 0 \\
\dot{\hat{\phi}}_1 &= \Lambda_c \hat{\phi}_1 + lu, & \hat{\phi}_1(0) &= 0 \\
\dot{\hat{\phi}}_2 &= \Lambda_c \hat{\phi}_2 - ly, & \hat{\phi}_2(0) &= 0 \\
\hat{y} &= \hat{x}_1
\end{aligned} \quad (5.5.10)$$

where $\bar{\theta}(t)$ is the estimate of $\bar{\theta}^*$ to be generated by an adaptive law and $\hat{x}_1$ is the estimate of $\bar{x}_1$. The adaptive law for $\bar{\theta}$ is developed using the SPR-Lyapunov design approach as follows:

We define the observation error $\tilde{y} = y - \hat{y} = \tilde{x}_1 + \eta_0$, where $\tilde{x}_1 = \bar{x}_1 - \hat{x}_1$, and use it to develop the error equation

$$\dot{\tilde{y}} = -\lambda_0 \tilde{y} - \tilde{\theta}^T \phi + C_1^T \omega \quad (5.5.11)$$

where $C_1^T = \lambda_0 C_0^T + C_0^T \Lambda_c$ and $\tilde{\theta} = \bar{\theta} - \bar{\theta}^*$ by using (5.5.8) and (5.5.9), and the fact that $\phi = \hat{\phi}$. Except for the exponentially decaying to zero term $C_1^T \omega$, equation (5.5.11) is in the appropriate form for applying the SPR-Lyapunov design approach.

In the analysis below, we take care of the exponentially decaying term $C_1^T \omega$ by choosing the Lyapunov function candidate

$$V = \frac{\tilde{y}^2}{2} + \frac{\tilde{\theta}^T \Gamma^{-1} \tilde{\theta}}{2} + \beta \omega^T P_c \omega$$

where $\Gamma = \Gamma^T > 0$; $P_c = P_c^T > 0$ satisfies the Lyapunov equation

$$P_c \Lambda_c + \Lambda_c^T P_c = -I$$

and $\beta > 0$ is a scalar to be selected. The time derivative $\dot{V}$ of V along the solution of (5.5.11) is given by

$$\dot{V} = -\lambda_0 \tilde{y}^2 - \tilde{y} \tilde{\theta}^T \phi + \tilde{y} C_1^T \omega + \tilde{\theta}^T \Gamma^{-1} \dot{\tilde{\theta}} - \beta \omega^T \omega \quad (5.5.12)$$

As described earlier, if the adaptive law is chosen as

$$\dot{\tilde{\theta}} = \dot{\bar{\theta}} = \Gamma \tilde{y} \phi \quad (5.5.13)$$

(5.5.12) becomes

$$\dot{V} = -\lambda_0 \tilde{y}^2 - \beta \omega^T \omega + \tilde{y} C_1^T \omega$$

If β is chosen as $\beta > \frac{|c_1|^2}{2\lambda_0}$, then it can be shown by completing the squares that

$$\dot V \leq -\frac{\lambda_0}{2}\tilde y^2 - \frac{\beta}{2}\omega^T\omega$$

which implies that $V, \theta, \tilde y \in \mathcal{L}_\infty$ and $\tilde y \in \mathcal{L}_2$. Because from (5.5.11) $\dot{\tilde y} \in \mathcal{L}_\infty$ it follows that $\tilde y \to 0$ as $t \to \infty$, i.e., $\hat y(t) \to y(t)$ and $\hat x_1(t) \to \bar x_1(t)$ as $t \to \infty$. Hence, the overall state of the observer $\hat X = \left[\hat x_1, \hat\phi_1^T, \hat\phi_2^T\right]^T$ converges to the overall plant state $X = \left[\bar x_1, \phi_1^T, \phi_2^T\right]^T$ as $t \to \infty$. In addition $\dot{\bar\theta} \in \mathcal{L}_2 \cap \mathcal{L}_\infty$ and $\lim_{t\to\infty} \dot{\bar\theta}(t) = 0$.

The convergence of $\bar\theta(t)$ to $\bar\theta^*$ depends on the properties of the input u. We can show by following exactly the same steps as in the previous sections that if u is sufficiently rich of order $2n$, then $\hat\phi = \phi$ is PE, which, together with $\phi, \dot\phi \in \mathcal{L}_\infty$, implies that $\bar\theta(t)$ converges to $\bar\theta^*$ exponentially fast. For $\dot\phi \in \mathcal{L}_\infty$, we require, however, that $\dot u \in \mathcal{L}_\infty$.

The main equations of the adaptive observer are summarized in Table 5.5 and the block diagram of the observer is shown in Figure 5.4 where $\bar\theta = [\bar\theta_1^T, \bar\theta_2^T]^T$ is partitioned into $\bar\theta_1 \in \mathcal{R}^n, \bar\theta_2 \in \mathcal{R}^n$.

Example 5.5.1 Consider the LTI plant

$$y = \frac{b_1 s + b_0}{s^2 + a_1 s + a_0} u \qquad (5.5.14)$$

where $a_1, a_0 > 0$ and b_1, b_0 are the unknown parameters. We first obtain the plant representation 2 by following the results and approach presented in Chapter 2.

We choose $\Lambda(s) = (s + \lambda_0)(s + \lambda)$ for some $\lambda_0, \lambda > 0$. It follows from (5.5.14) that

$$\frac{s^2}{\Lambda(s)} y = [b_1, b_0]\begin{bmatrix} s \\ 1 \end{bmatrix}\frac{1}{\Lambda(s)} u - [a_1, a_0]\begin{bmatrix} s \\ 1 \end{bmatrix}\frac{1}{\Lambda(s)} y$$

Because $\frac{s^2}{\Lambda(s)} = 1 - \frac{(\lambda_0+\lambda)s + \lambda_0\lambda}{\Lambda(s)}$ we have

$$y = \theta_1^{*T}\frac{\alpha_1(s)}{\Lambda(s)} u - \theta_2^{*T}\frac{\alpha_1(s)}{\Lambda(s)} y + \bar\lambda^T\frac{\alpha_1(s)}{\Lambda(s)} y \qquad (5.5.15)$$

where $\theta_1^* = [b_1, b_0]^T, \theta_2^* = [a_1, a_0]^T, \bar\lambda = [\lambda_0 + \lambda, \lambda_0\lambda]^T$ and $\alpha_1(s) = [s, 1]^T$. Because $\Lambda(s) = (s + \lambda_0)(s + \lambda)$, equation (5.5.15) implies that

$$y = \frac{1}{s+\lambda_0}\left[\theta_1^{*T}\frac{\alpha_1(s)}{s+\lambda} u - \theta_2^{*T}\frac{\alpha_1(s)}{s+\lambda} y + \bar\lambda^T\frac{\alpha_1(s)}{s+\lambda} y\right]$$

5.5. NONMINIMAL ADAPTIVE OBSERVER

Table 5.5 Adaptive observer (Realization 2)

Plant	$\dot{\bar{x}}_1 = -\lambda_0 \bar{x}_1 + \bar{\theta}^{*T}\phi, \quad \bar{x}_1(0) = 0$ $\dot{\phi}_1 = \Lambda_c \phi_1 + lu, \quad \phi_1(0) = 0$ $\dot{\phi}_2 = \Lambda_c \phi_2 - ly, \quad \phi_2(0) = 0$ $\dot{\omega} = \Lambda_c \omega, \quad \omega(0) = \omega_0$ $\eta_0 = C_0^T \omega$ $y = \bar{x}_1 + \eta_0$ where $\phi = [u, \phi_1^T, y, \phi_2^T]^T$ $\phi_i \in \mathcal{R}^{n-1}, \quad i=1,2; \; \bar{x}_1 \in \mathcal{R}^1$
Observer	$\dot{\hat{x}}_1 = -\lambda_0 \hat{x}_1 + \hat{\theta}^T \hat{\phi}, \quad \hat{x}_1(0) = 0$ $\dot{\hat{\phi}}_1 = \Lambda_c \hat{\phi}_1 + lu, \quad \hat{\phi}_1(0) = 0$ $\dot{\hat{\phi}}_2 = \Lambda_c \hat{\phi}_2 - ly, \quad \hat{\phi}_2(0) = 0$ $\hat{y} = \hat{x}_1$ where $\hat{\phi} = [u, \hat{\phi}_1^T, y, \hat{\phi}_2^T]^T$ $\hat{\phi}_i \in \mathcal{R}^{n-1}, i=1,2, \; \hat{x}_1 \in \mathcal{R}^1$
Adaptive law	$\dot{\hat{\theta}} = \Gamma \tilde{y} \hat{\phi}, \quad \tilde{y} = y - \hat{y}$
Design variables	$\Gamma = \Gamma^T > 0; \; \Lambda_c \in \mathcal{R}^{(n-1)\times(n-1)}$ is any stable matrix, and $\lambda_0 > 0$ is any scalar

Substituting for

$$\frac{\alpha_1(s)}{s+\lambda} = \frac{1}{s+\lambda}\begin{bmatrix} s \\ 1 \end{bmatrix} = \begin{bmatrix} 1 \\ 0 \end{bmatrix} + \frac{1}{s+\lambda}\begin{bmatrix} -\lambda \\ 1 \end{bmatrix}$$

we obtain

$$y = \frac{1}{s+\lambda_0}\left[b_1 u + (b_0 - \lambda b_1)\frac{1}{s+\lambda}u - a_1 y - (a_0 - \lambda a_1)\frac{1}{s+\lambda}y \right.$$
$$\left. + (\lambda_0 + \lambda)y - \lambda^2 \frac{1}{s+\lambda}y\right]$$

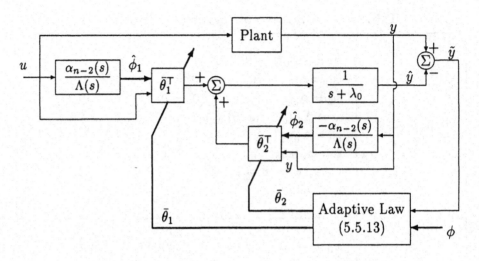

Figure 5.4 Adaptive observer using nonminimal Realization 2.

which implies that

$$\dot{\bar{x}}_1 = -\lambda_0 \bar{x}_1 + \bar{\theta}^{*T} \phi, \quad \bar{x}_1(0) = 0$$
$$\dot{\phi}_1 = -\lambda \phi_1 + u, \quad \phi_1(0) = 0$$
$$\dot{\phi}_2 = -\lambda \phi_2 - y, \quad \phi_2(0) = 0$$
$$y = \bar{x}_1$$

where $\phi = [u, \phi_1, y, \phi_2]^T$, $\bar{\theta}^* = [b_1, b_0 - \lambda b_1, \lambda_0 + \lambda - a_1, a_0 - \lambda a_1 + \lambda^2]^T$. Using Table 5.5, the adaptive observer for estimating $\bar{x}_1, \phi_1, \phi_2$ and θ^* is given by

$$\dot{\hat{x}}_1 = -\lambda_0 \hat{x}_1 + \bar{\theta}^T \hat{\phi}, \quad \hat{x}_1(0) = 0$$
$$\dot{\hat{\phi}}_1 = -\lambda \hat{\phi}_1 + u, \quad \hat{\phi}_1(0) = 0$$
$$\dot{\hat{\phi}}_2 = -\lambda \hat{\phi}_2 - y, \quad \hat{\phi}_2(0) = 0$$
$$\hat{y} = \hat{x}_1$$
$$\dot{\bar{\theta}} = \Gamma \hat{\phi}(y - \hat{y})$$

where $\hat{\phi} = [u, \hat{\phi}_1, y, \hat{\phi}_2]^T$ and $\Gamma = \Gamma^T > 0$. If in addition to $\bar{\theta}^*$, we like to estimate $\theta^* = [b_1, b_0, a_1, a_0]^T$, we use the relationships

$$\hat{b}_1 = \bar{\theta}_1$$
$$\hat{b}_0 = \bar{\theta}_2 + \lambda \bar{\theta}_1$$
$$\hat{a}_1 = -\bar{\theta}_3 + \lambda_0 + \lambda$$
$$\hat{a}_0 = \bar{\theta}_4 - \lambda \bar{\theta}_3 + \lambda \lambda_0$$

where $\bar{\theta}_i, i = 1, 2, 3, 4$ are the elements of $\bar{\theta}$ and $\hat{b}_i, \hat{a}_i, i = 1, 2$ are the estimates of $b_i, a_i, i = 0, 1$, respectively.

For parameter convergence we choose

$$u = 6\sin 2.6t + 8\sin 4.2t$$

which is sufficiently rich of order 4. $\square$

5.6 Parameter Convergence Proofs

In this section we present all the lengthy proofs of theorems dealing with convergence of the estimated parameters.

5.6.1 Useful Lemmas

The following lemmas are used in the proofs of several theorems to follow:

Lemma 5.6.1 *If the autocovariance of a function* $x : \mathcal{R}^+ \mapsto \mathcal{R}^n$ *defined as*

$$R_x(t) \triangleq \lim_{T \to \infty} \frac{1}{T} \int_{t_0}^{t_0+T} x(\tau) x^\top (t+\tau) d\tau \qquad (5.6.1)$$

exists and is uniform with respect to t_0*, then* x *is PE if and only if* $R_x(0)$ *is positive definite.*

Proof

If: The definition of the autocovariance $R_x(0)$ implies that there exists a $T_0 > 0$ such that

$$\frac{1}{2} R_x(0) \leq \frac{1}{T_0} \int_{t_0}^{t_0+T_0} x(\tau) x^\top(\tau) d\tau \leq \frac{3}{2} R_x(0), \quad \forall t \geq 0$$

If $R_x(0)$ is positive definite, there exist $\alpha_1, \alpha_2 > 0$ such that $\alpha_1 I \leq R_x(0) \leq \alpha_2 I$. Therefore,

$$\frac{1}{2}\alpha_1 I \leq \frac{1}{T_0} \int_{t_0}^{t_0+T_0} x(\tau) x^\top(\tau) d\tau \leq \frac{3}{2}\alpha_2 I$$

for all $t_0 \geq 0$ and thus x is PE.

Only if: If x is PE, then there exist constants $\alpha_0, T_1 > 0$ such that

$$\int_t^{t+T_1} x(\tau) x^\top(\tau) d\tau \geq \alpha_0 T_1 I$$

for all $t \geq 0$. For any $T > T_1$, we can write

$$\int_{t_0}^{t_0+T} x(\tau)x^\top(\tau)d\tau = \sum_{i=0}^{k-1} \int_{t_0+iT_1}^{t_0+(i+1)T_1} x(\tau)x^\top(\tau)d\tau + \int_{t_0+kT_1}^{t_0+T} x(\tau)x^\top(\tau)d\tau$$
$$\geq k\alpha_0 T_1 I$$

where k is the largest integer that satisfies $k \leq T/T_1$, i.e., $kT_1 \leq T < (k+1)T_1$. Therefore, we have

$$\frac{1}{T}\int_t^{t+T} x(\tau)x^\top(\tau)d\tau \geq \frac{kT_1}{T}\alpha_0 I$$

For $k \geq 2$, we have $\frac{kT_1}{T} = \frac{(k+1)T_1}{T} - \frac{T_1}{T} \geq 1 - \frac{T_1}{T} \geq \frac{1}{2}$, thus,

$$\frac{1}{T}\int_{t_0}^{t_0+T} x(\tau)x^\top(\tau)d\tau \geq \frac{\alpha_0}{2}I$$

and

$$R_x(0) = \lim_{T\to\infty} \frac{1}{T}\int_{t_0}^{t_0+T} x(\tau)x^\top(\tau)d\tau \geq \frac{\alpha_0}{2}I$$

which implies that $R_x(0)$ is positive definite. $\square$

Lemma 5.6.2 *Consider the system*

$$y = H(s)u$$

where $H(s)$ is a strictly proper transfer function matrix of dimension $m \times n$ with stable poles and real impulse response $h(t)$. If u is stationary, with autocovariance $R_u(t)$, then y is stationary, with autocovariance

$$R_y(t) = \int_{-\infty}^{\infty}\int_{-\infty}^{\infty} h(\tau_1)R_u(t+\tau_1-\tau_2)h^\top(\tau_2)d\tau_1 d\tau_2$$

and spectral distribution

$$S_y(\omega) = H(-j\omega)S_u(\omega)H^\top(j\omega)$$

Proof See [201].

Lemma 5.6.3 *Consider the system described by*

$$\begin{bmatrix} \dot{x}_1 \\ \dot{x}_2 \end{bmatrix} = \begin{bmatrix} A & -F^\top(t) \\ P_1 F(t) P_2 & 0 \end{bmatrix}\begin{bmatrix} x_1 \\ x_2 \end{bmatrix} \quad (5.6.2)$$

5.6. PARAMETER CONVERGENCE PROOF

where $x_1 \in \mathcal{R}^{n_1}, x_2 \in \mathcal{R}^{rn_1}$ for some integer $r, n_1 \geq 1$, A, P_1, P_2 are constant matrices and $F(t)$ is of the form

$$F(t) = \begin{bmatrix} z_1 I_{n_1} \\ z_2 I_{n_1} \\ \vdots \\ z_r I_{n_1} \end{bmatrix} \in \mathcal{R}^{rn_1 \times n_1}$$

where $z_i, i = 1, 2, \ldots, r$ are the elements of the vector $z \in \mathcal{R}^r$. Suppose that z is PE and there exists a matrix $P_0 > 0$ such that

$$\dot{P_0} + A_0^\top P_0 + P_0 A_0 + C_0 C_0^\top \leq 0 \qquad (5.6.3)$$

where

$$A_0 = \begin{bmatrix} A & -F^\top(t) \\ P_1 F(t) P_2 & 0 \end{bmatrix}, \quad C_0^\top = [I_{n_1}, 0]$$

Then the equilibrium $x_{1e} = 0, x_{2e} = 0$ of (5.6.2) is e.s. in the large.

Proof Consider the system (5.6.2) that we express as

$$\begin{aligned} \dot{x} &= A_0(t) x \\ y &= C_0^\top x = x_1 \end{aligned} \qquad (5.6.4)$$

where $x = [x_1^\top, x_2^\top]^\top$. We first show that (C_0, A_0) is UCO by establishing that $(C_0, A_0 + K C_0^\top)$ is UCO for some $K \in \mathcal{L}_\infty$ which according to Lemma 4.8.1 implies that (C_0, A_0) is UCO. We choose

$$K = \begin{bmatrix} -\gamma I_{n_1} - A \\ -P_1 F(t) P_2 \end{bmatrix}$$

for some $\gamma > 0$ and consider the following system associated with $(C_0, A_0 + K C_0^\top)$:

$$\begin{bmatrix} \dot{Y}_1 \\ \dot{Y}_2 \end{bmatrix} = \begin{bmatrix} -\gamma I_{n_1} & -F^\top(t) \\ 0 & 0 \end{bmatrix} \begin{bmatrix} Y_1 \\ Y_2 \end{bmatrix}$$
$$y_1 = [I_{n_1}\ 0] \begin{bmatrix} Y_1 \\ Y_2 \end{bmatrix} \qquad (5.6.5)$$

According to Lemma 4.8.4, the system (5.6.5) is UCO if

$$F_f(t) = \frac{1}{s + \gamma} F(t)$$

satisfies

$$\alpha I_{rn_1} \leq \frac{1}{T} \int_t^{t+T} F_f(\tau) F_f^\top(\tau) d\tau \leq \beta I_{rn_1}, \quad \forall t \geq 0 \qquad (5.6.6)$$

for some constants $\alpha, \beta, T > 0$. We prove (5.6.6) by first showing that $F(t)$ satisfies

$$\alpha' I_{rn_1} \le \frac{1}{T} \int_t^{t+T} F(\tau) F^\top(\tau) d\tau \le \beta' I_{rn_1}, \quad \forall t \ge 0$$

for some constants α', β' as follows: Using a linear transformation, we can express $F(t)$ as

$$F(t) = F_0 Z(t)$$

where $F_0 \in \mathcal{R}^{rn_1 \times rn_1}$ is a constant matrix of full rank, $Z \in \mathcal{R}^{rn_1 \times n_1}$ is a block diagonal matrix defined as $Z \triangleq \text{diag}\{\underbrace{z, z, \ldots, z}_{n_1}\}$, i.e.,

$$Z = \begin{bmatrix} z_1 & 0 & \cdots & 0 \\ z_2 & 0 & \cdots & 0 \\ \vdots & \vdots & & \vdots \\ z_{n_1} & 0 & \cdots & 0 \\ 0 & z_1 & & 0 \\ \vdots & \vdots & & \vdots \\ 0 & z_{n_1} & & 0 \\ \vdots & \vdots & & \vdots \\ 0 & 0 & \cdots & z_1 \\ \vdots & \vdots & & \vdots \\ 0 & 0 & \cdots & z_{n_1} \end{bmatrix}$$

Therefore, $FF^\top = F_0 Z Z^\top F_0^\top$ and

$$ZZ^\top = \text{diag}\{\underbrace{zz^\top, zz^\top, \ldots, zz^\top}_{n_1}\}$$

Because z is PE, we have

$$\alpha_1 I_r \le \frac{1}{T} \int_t^{t+T} zz^\top d\tau \le \alpha_2 I_r, \quad \forall t \ge 0$$

for some $\alpha_1, \alpha_2, T > 0$. Therefore,

$$\alpha_1 I_{rn_1} \le \frac{1}{T} \int_t^{t+T} Z(\tau) Z^\top(\tau) d\tau \le \alpha_2 I_{rn_1}, \quad \forall t \ge 0$$

which implies that

$$\alpha_1 F_0 F_0^\top \le \frac{1}{T} \int_t^{t+T} F(\tau) F^\top(\tau) d\tau \le \alpha_2 F_0 F_0^\top, \quad \forall t \ge 0$$

5.6. PARAMETER CONVERGENCE PROOF

Because F_0 is of full rank, we have

$$\beta_1 I_{rn_1} \leq F_0 F_0^T \leq \beta_2 I_{rn_1}$$

for some constants $\beta_1, \beta_2 > 0$. Hence,

$$\alpha' I_{rn_1} \leq \frac{1}{T} \int_t^{t+T} F(\tau) F^T(\tau) d\tau \leq \beta' I_{rn_1}, \quad \forall t \geq 0 \tag{5.6.7}$$

where $\beta' = \alpha_2 \beta_2, \alpha' = \alpha_1 \beta_1$.

Following the same arguments used in proving Lemma 4.8.3 (iv), one can show (see Problem 5.18) that (5.6.7) implies (5.6.6).

Because all the conditions in Lemma 4.8.4 are satisfied, we conclude, by applying Lemma 4.8.4 that (5.6.5) is UCO, which in turn implies that (5.6.2) is UCO. Therefore, it follows directly from Theorem 3.4.8 and (5.6.3) that the equilibrium $x_{1e} = 0, x_{2e} = 0$ of (5.6.2) is e.s. in the large. □

5.6.2 Proof of Theorem 5.2.1

According to Lemma 5.6.1, Theorem 5.2.1 can be proved if we establish that $R_\phi(0)$ is positive definite if and only if u is sufficiently rich of order n.

If: We will show the result by contradiction. Because u is stationary and $R_\phi(0)$ is uniform with respect to t, we take $t = 0$ and obtain [186]

$$R_\phi(0) = \lim_{T \to \infty} \frac{1}{T} \int_0^T \phi(\tau) \phi^T(\tau) d\tau = \frac{1}{2\pi} \int_{-\infty}^\infty S_\phi(\omega) d\omega \tag{5.6.8}$$

where $S_\phi(\omega)$ is the spectral distribution of ϕ. From Lemma 5.6.2, we have

$$S_\phi(\omega) = H(-j\omega) S_u(\omega) H^T(j\omega) \tag{5.6.9}$$

Using the condition that u is sufficiently rich of order n, i.e., u has spectral lines at n points, we can express $S_u(\omega)$ as

$$S_u(\omega) = \sum_{i=1}^n f_u(\omega_i) \delta(\omega - \omega_i) \tag{5.6.10}$$

where $f_u(\omega_i) > 0$. Using (5.6.9) and (5.6.10) in (5.6.8), we obtain

$$R_\phi(0) = \frac{1}{2\pi} \sum_{i=1}^n f_u(\omega_i) H(-j\omega_i) H^T(j\omega_i)$$

Suppose that $R_\phi(0)$ is not positive definite, then there exists $x \in \mathcal{R}^n$ with $x \neq 0$ such that

$$x^T R_\phi(0) x = \sum_{i=1}^n f_u(\omega_i) x^T H(-j\omega_i) H^T(j\omega_i) x = 0 \tag{5.6.11}$$

Because $f_u(\omega_i) > 0$ and each term under the summation is nonnegative, (5.6.11) can be true only if:

$$x^\top H(-j\omega_i) H^\top(j\omega_i) x = 0, \quad i = 1, 2, \ldots, n$$

or equivalently

$$x^\top H(-j\omega_i) = 0, \quad i = 1, 2, \ldots, n \tag{5.6.12}$$

However, (5.6.12) implies that $\{H(j\omega_1), H(j\omega_2), \ldots, H(j\omega_n)\}$ are linearly dependent, which contradicts with the condition that $H(j\omega_1), \ldots H(j\omega_n)$ are linearly independent for all $\omega_1, \ldots, \omega_n$. Hence, $R_\phi(0)$ is positive definite.

Only if: We also prove this by contradiction. Assume that $R_\phi(0)$ is positive definite but u is sufficiently rich of order $r < n$, then we can express $R_\phi(0)$ as

$$R_\phi(0) = \frac{1}{2\pi} \sum_{i=1}^{r} f_u(\omega_i) H(-j\omega_i) H^\top(-j\omega_i)$$

where $f_u(\omega_i) > 0$. Note that the right hand side is the sum of $r - dyads$, and the rank of $R_\phi(0)$ can be at most $r < n$, which contradicts with the assumption that $R_\phi(0)$ is positive definite. □

5.6.3 Proof of Theorem 5.2.2

We first consider the series-parallel scheme (5.2.11). From (5.2.10), (5.2.11) we obtain the error equations

$$\begin{aligned} \dot{\epsilon}_1 &= A_m \epsilon_1 - \tilde{B} u - \tilde{A} x \\ \dot{\tilde{A}} &= \gamma \epsilon_1 x^\top, \quad \dot{\tilde{B}} = \gamma \epsilon_1 u \end{aligned} \tag{5.6.13}$$

where $\tilde{A} \triangleq \hat{A} - A$, $\tilde{B} \triangleq \hat{B} - B$ and $\tilde{B} \in \mathcal{R}^{n \times 1}$, $\tilde{A} \in A^{n \times n}$. For simplicity, let us take $\gamma_1 = \gamma_2 = \gamma$. The parameter error $\tilde{A}$ is in the matrix form, which we rewrite in the familiar vector form to apply the stability theorems of Chapter 3 directly. Defining the vector

$$\tilde{\theta} \triangleq [\tilde{a}_1^\top, \tilde{a}_2^\top, \ldots, \tilde{a}_n^\top, \tilde{B}^\top]^\top \in \mathcal{R}^{n(n+1)}$$

where $\tilde{a}_i$ is the ith column of $\tilde{A}$, we can write

$$\tilde{A} x + \tilde{B} u = [\tilde{a}_1, \tilde{a}_2, \ldots, \tilde{a}_n] \begin{bmatrix} x_1 \\ x_2 \\ \vdots \\ x_n \end{bmatrix} + \tilde{B} u = F^\top(t) \tilde{\theta}$$

5.6. PARAMETER CONVERGENCE PROOF

where $F^T(t) \triangleq [x_1 I_n, x_2 I_n, \ldots, x_n I_n, u I_n] \in \mathcal{R}^{n \times n(n+1)}$. Because $\dot{\tilde{a}}_i = \gamma \epsilon_1 x_i$, $\dot{\tilde{B}} = \gamma \epsilon_1 u$, the matrix differential equations for $\tilde{A}, \tilde{B}$ can be rewritten as

$$\dot{\tilde{\theta}} = \gamma F(t) \epsilon_1$$

Therefore, (5.6.13) is equivalent to

$$\begin{cases} \dot{\epsilon}_1 = A_m \epsilon_1 - F^T(t) \tilde{\theta} \\ \dot{\tilde{\theta}} = \gamma F(t) \epsilon_1 \end{cases} \quad (5.6.14)$$

which is in the form of (5.6.2). To apply Lemma 5.6.3, we need to verify that all the conditions stated in the lemma are satisfied by (5.6.14).

We first prove that there exists a constant matrix $P_0 > 0$ such that

$$A_0^T P_0 + P_0 A_0 = -C_0 C_0^T$$

where

$$A_0 = \begin{bmatrix} A & -F^T(t) \\ \gamma F(t) & 0 \end{bmatrix} \in \mathcal{R}^{(n+n(n+1)) \times (n+n(n+1))}, C_0^T = [I_n, 0] \in \mathcal{R}^{n+n(n+1)}$$

In Section 4.2.3, we have shown that the time derivative of the Lyapunov function

$$V = \epsilon_1^T P \epsilon_1 + \text{tr}\left\{ \frac{\tilde{A}^T P \tilde{A}}{\gamma_1} \right\} + \text{tr}\left\{ \frac{\tilde{B}^T P \tilde{B}}{\gamma_2} \right\}$$

(with $\gamma_1 = \gamma_2 = \gamma$) satisfies

$$\dot{V} = -\epsilon_1^T \epsilon_1 \quad (5.6.15)$$

where P satisfies $A_m^T P + P A_m = -I_n$. Note that

$$\text{tr}\left\{ \tilde{A}^T P \tilde{A} \right\} = \sum_{i=1}^{n} \tilde{a}_i^T P \tilde{a}_i$$

$$\text{tr}\left\{ \tilde{B}^T P \tilde{B} \right\} = \tilde{B}^T P \tilde{B}$$

where the second equality is true because $\tilde{B} \in \mathcal{R}^{n \times 1}$. We can write

$$V = \epsilon_1^T P \epsilon_1 + \frac{1}{\gamma_1} \sum_{i=1}^{n} \tilde{a}_i^T P \tilde{a}_i + \frac{1}{\gamma_2} \tilde{B}^T P \tilde{B} = x^T P_0 x$$

where $x \triangleq [\epsilon_1^T, \tilde{\theta}^T]^T$ and P_0 is a block diagonal matrix defined as

$$P_0 \triangleq \text{diag}\{P, \underbrace{\gamma_1^{-1} P, \ldots, \gamma_1^{-1} P}_{n-\text{times}}, \gamma_2^{-1} P\} \in \mathcal{R}^{(n+n(n+1)) \times (n+n(n+1))}$$

Hence, (5.6.15) implies that

$$\dot{V} = x^\top (P_0 A_0 + A_0^\top P_0) x = -x^\top C_0 C_0^\top x$$

or equivalently

$$\dot{P}_0 + P_0 A_0 + A_0^\top P_0 = -C_0 C_0^\top, \quad \dot{P}_0 = 0 \qquad (5.6.16)$$

Next, we show that $z \triangleq [x_1, x_2, \ldots, x_n, u]^\top$ is PE. We write

$$z = H(s)u, \quad H(s) = \begin{bmatrix} (sI - A)^{-1} B \\ 1 \end{bmatrix}$$

If we can show that $H(j\omega_1), H(j\omega_2), \ldots, H(j\omega_{n+1})$ are linearly independent for any $\omega_1, \omega_2, \ldots, \omega_{n+1}$, then it follows immediately from Theorem 5.2.1 that z is PE if and only if u is sufficiently rich of order $n+1$.

Let $a(s) = \det(sI - A) = s^n + a_{n-1} s^{n-1} + \ldots + a_1 s + a_0$. We can verify using matrix manipulations that the matrix $(sI - A)^{-1}$ can be expressed as

$$(sI - A)^{-1} = \frac{1}{a(s)} \{ I s^{n-1} + (A + a_{n-1} I) s^{n-2} + (A^2 + a_{n-1} A + a_{n-2} I) s^{n-3}$$
$$+ \ldots + (A^{n-1} + a_{n-1} A^{n-2} + \ldots + a_1 I) \} \qquad (5.6.17)$$

Defining

$$b_1 = B, \ b_2 = (A + a_{n-1} I) B, \ b_3 = (A^2 + a_{n-1} A + a_{n-2} I) B, \ldots, b_n = (A^{n-1} + \ldots + a_1 I) B$$

$$L = [b_1, b_2, \ldots, b_n] \in \mathcal{R}^{n \times n}$$

and then using (5.6.17), $H(s)$ can be conveniently expressed as

$$H(s) = \begin{bmatrix} L \begin{bmatrix} s^{n-1} \\ s^{n-2} \\ \vdots \\ s \\ 1 \end{bmatrix} \\ a(s) \end{bmatrix} \frac{1}{a(s)} \qquad (5.6.18)$$

To explore the linear dependency of $H(j\omega_i)$, we define $\bar{H} \triangleq [H(j\omega_1), H(j\omega_2), \ldots, H(j\omega_{n+1})]$. Using the expression (5.6.18) for $H(s)$, we have

$$\bar{H} = \begin{bmatrix} L & 0_{n \times 1} \\ 0_{1 \times n} & 1 \end{bmatrix} \begin{bmatrix} (j\omega_1)^{n-1} & (j\omega_2)^{n-1} & \cdots & (j\omega_{n+1})^{n-1} \\ (j\omega_1)^{n-2} & (j\omega_2)^{n-2} & \cdots & (j\omega_{n+1})^{n-2} \\ \vdots & \vdots & & \vdots \\ j\omega_1 & j\omega_2 & \cdots & j\omega_{n+1} \\ 1 & 1 & \cdots & 1 \\ a(j\omega_1) & a(j\omega_2) & \cdots & a(j\omega_{n+1}) \end{bmatrix}$$

5.6. PARAMETER CONVERGENCE PROOF

$$\times \begin{bmatrix} \frac{1}{a(j\omega_1)} & 0 & \cdots & 0 \\ 0 & \frac{1}{a(j\omega_2)} & \cdots & 0 \\ \vdots & & \ddots & 0 \\ 0 & \cdots & 0 & \frac{1}{a(j\omega_{n+1})} \end{bmatrix}$$

From the assumption that (A, B) is controllable, we conclude that the matrix L is of full rank. Thus the matrix $\bar{H}$ has rank of $n+1$ if and only if the matrix V_1 defined as

$$V_1 \triangleq \begin{bmatrix} (j\omega_1)^{n-1} & (j\omega_2)^{n-1} & \cdots & (j\omega_{n+1})^{n-1} \\ (j\omega_1)^{n-2} & (j\omega_2)^{n-2} & \cdots & (j\omega_{n+1})^{n-2} \\ \vdots & \vdots & & \vdots \\ j\omega_1 & j\omega_2 & \cdots & j\omega_{n+1} \\ 1 & 1 & \cdots & 1 \\ a(j\omega_1) & a(j\omega_2) & \cdots & a(j\omega_{n+1}) \end{bmatrix}$$

has rank of $n+1$. Using linear transformations (row operations), we can show that V_1 is equivalent to the following Vandermonde matrix [62]:

$$V = \begin{bmatrix} (j\omega_1)^n & (j\omega_2)^n & \cdots & (j\omega_{n+1})^n \\ (j\omega_1)^{n-1} & (j\omega_2)^{n-1} & \cdots & (j\omega_{n+1})^{n-1} \\ (j\omega_1)^{n-2} & (j\omega_2)^{n-2} & \cdots & (j\omega_{n+1})^{n-2} \\ \vdots & \vdots & & \vdots \\ j\omega_1 & j\omega_2 & \cdots & j\omega_{n+1} \\ 1 & 1 & \cdots & 1 \end{bmatrix}$$

Because

$$\det(V) = \prod_{1 \leq i < k \leq n+1} (j\omega_i - j\omega_k)$$

V is of full rank for any ω_i with $\omega_i \neq \omega_k$ $i, k = 1, \ldots n+1$. This leads to the conclusion that V_1 and, therefore, $\bar{H}$ have rank of $n+1$, which implies that $H(j\omega_1), H(j\omega_2), \ldots, H(j\omega_{n+1})$ are linearly independent for any $\omega_1, \omega_2, \ldots \omega_{n+1}$. It then follows immediately from Theorem 5.2.1 that z is PE.

Because we have shown that all the conditions of Lemma 5.6.3 are satisfied by (5.6.14), we can conclude that $x_e = 0$ of (5.6.14) is e.s. in the large, i.e., $\epsilon_1, \tilde{\theta} \to 0$ exponentially fast as $t \to \infty$. Thus, $\hat{A} \to A, \hat{B} \to B$ exponentially fast as $t \to \infty$, and the proof of Theorem 5.2.2 for the series-parallel scheme is complete.

For the parallel scheme, we only need to establish that $\hat{z} \triangleq [\hat{x}^T, u]^T$ is PE. The rest of the proof follows by using exactly the same arguments and procedure as in the case of the series-parallel scheme.

Because x is the state of the plant and it is independent of the identification scheme, it follows from the previous analysis that z is PE under the conditions given in Theorem 5.2.2, i.e., (A, B) controllable and u sufficiently rich of order $n+1$. From the definition of ϵ_1, we have

$$\hat{x} = x - \epsilon_1, \quad \hat{z} = z - \begin{bmatrix} 0_{n \times 1} \\ 1 \end{bmatrix} \epsilon_1$$

thus the PE property of $\hat{z}$ follows immediately from Lemma 4.8.3 by using $\epsilon_1 \in \mathcal{L}_2$ and z being PE. $\square$

5.6.4 Proof of Theorem 5.2.3

We consider the proof for the series-parallel scheme. The proof for the parallel scheme follows by using the same arguments used in Section 5.6.3 in the proof of Theorem 5.2.2 for the parallel scheme.

Following the same procedure used in proving Theorem 5.2.2, we can write the differential equations

$$\begin{aligned}
\dot{\epsilon}_1 &= A_m \epsilon_1 + (A - \hat{A})x + (B - \hat{B})u \\
\dot{\hat{A}} &= \gamma \epsilon_1 x^\top \\
\dot{\hat{B}} &= \gamma \epsilon_1 u^\top
\end{aligned}$$

in the vector form as

$$\begin{aligned}
\dot{\epsilon}_1 &= A_m \epsilon_1 - F^\top(t)\tilde{\theta} \\
\dot{\tilde{\theta}} &= F(t)\epsilon_1
\end{aligned} \quad (5.6.19)$$

where $\tilde{\theta} \triangleq [\tilde{a}_1^\top, \tilde{a}_2^\top, \ldots, \tilde{a}^\top, \tilde{b}_1^\top, \tilde{b}_2^\top, \ldots, \tilde{b}_q^\top]^\top$ and $\tilde{a}_i, \tilde{b}_i$ denotes the ith column of $\tilde{A}, \tilde{B}$, respectively, $F^\top(t) \triangleq [x_1 I_n, x_2 I_n, \ldots, x_n I_n, u_1 I_n, u_2 I_n, \ldots, u_q I_n]$. Following exactly the same arguments used in the proof for Theorem 5.2.2, we complete the proof by showing (i) there exists a matrix $P_0 > 0$ such that $A_0^\top P_0 + P_0 A_0 = -C_0 C_0^\top$, where A_0, C_0 are defined the same way as in Section 5.6.3 and (ii) $z \triangleq [x_1, x_2, \ldots, x_n, u_1, u_2, \ldots, u_q]^\top$ is PE.

The proof for (i) is the same as that in Section 5.6.3. We prove (ii) by showing that the autocovariance of z, $R_z(0)$ is positive definite as follows: We express z as

$$z = \begin{bmatrix} (sI - A)^{-1} B \\ I_q \end{bmatrix} u = \sum_{i=1}^{q} \begin{bmatrix} (sI - A)^{-1} b_i \\ e_i \end{bmatrix} u_i$$

where $I_q \in \mathcal{R}^{q \times q}$ is the identity matrix and b_i, e_i denote the ith column of B, I_q, respectively. Assuming that $u_i, i = 1, \ldots, q$ are stationary and uncorrelated, the

5.6. PARAMETER CONVERGENCE PROOF

autocovariance of z can be calculated as

$$R_z(0) = \frac{1}{2\pi} \sum_{i=1}^{q} \int_{-\infty}^{\infty} H_i(-j\omega) S_{u_i}(\omega) H_i^T(j\omega)$$

where

$$H_i(s) = \begin{bmatrix} (sI - A)^{-1} b_i \\ e_i \end{bmatrix}$$

and $S_{u_i}(\omega)$ is the spectral distribution of u_i. Using the assumption that u_i is sufficiently rich of order $n+1$, we have

$$S_{u_i}(\omega) = \sum_{k=1}^{n+1} f_{u_i}(\omega_{ik}) \delta(\omega - \omega_{ik})$$

and

$$\frac{1}{2\pi} \int_{-\infty}^{\infty} H_i(-j\omega) S_{u_i}(\omega) H_i^T(j\omega) = \frac{1}{2\pi} \sum_{k=1}^{n+1} f_{u_i}(\omega_{ik}) H_i(-j\omega_{ik}) H_i^T(j\omega_{ik})$$

where $f_{u_i}(\omega_{ik}) > 0$. Therefore,

$$R_z(0) = \frac{1}{2\pi} \sum_{i=1}^{q} \sum_{k=1}^{n+1} f_{u_i}(\omega_{ik}) H_i(-j\omega_{ik}) H_i^T(j\omega_{ik}) \qquad (5.6.20)$$

Let us now consider the solution of the quadratic equation

$$x^T R_z(0) x = 0, \quad x \in \mathcal{R}^{n+q}$$

Because each term under the summation of the right-hand side of (5.6.20) is semi-positive definite, $x^T R_z(0) x = 0$ is true if and only if

$$f_{u_i}(\omega_{ik}) x^T H_i(-j\omega_{ik}) H_i^T(j\omega_{ik}) x = 0, \quad \begin{array}{l} i = 1, 2, \ldots q \\ k = 1, 2, \ldots n+1 \end{array}$$

or equivalently

$$H_i^T(j\omega_{ik}) x = 0, \quad \begin{array}{l} i = 1, 2, \ldots q \\ k = 1, 2, \ldots n+1 \end{array} \qquad (5.6.21)$$

Because $H_i(s) = \frac{1}{a(s)} \begin{bmatrix} \text{adj}(aI - A) b_i \\ a(s) e_i \end{bmatrix}$ where $a(s) = det(sI - A)$, (5.6.21) is equivalent to:

$$\bar{H}_i^T(j\omega_{ik}) x = 0, \quad \bar{H}_i(s) \triangleq a(s) H_i(s), \quad \begin{array}{l} i = 1, 2, \ldots q \\ k = 1, 2, \ldots n+1 \end{array} \qquad (5.6.22)$$

Noting that each element in $\bar{H}_i$ is a polynomial of order at most equal to n, we find that $g_i(s) \triangleq \bar{H}_i^T(s)x$ is a polynomial of order at most equal to n. Therefore, (5.6.22) implies that the polynomial $g_i(s)$ vanishes at $n+1$ points, which, in turn, implies that $g_i(s) \equiv 0$ for all $s \in \mathcal{C}$. Thus, we have

$$\bar{H}_i^T(s)x \equiv 0, \quad i = 1, 2, \ldots q \tag{5.6.23}$$

for all s. Equation (5.6.23) can be written in the matrix form

$$\left[\begin{array}{c} \operatorname{adj}(sI - A)B \\ a(s)I_q \end{array}\right]^T x \equiv \underline{0}_q \tag{5.6.24}$$

where $\underline{0}_q \in \mathcal{R}^q$ is a column vector with all elements equal to zero. Let $X = [x_1, \ldots, x_n] \in \mathcal{R}^n$, $Y = [x_{n+1}, \ldots, x_{n+q}] \in \mathcal{R}^q$, i.e., $x = [X^T, Y^T]^T$. Then (5.6.24) can be expressed as

$$(\operatorname{adj}(sI - A)B)^T X + a(s)Y = \underline{0}_q \tag{5.6.25}$$

Consider the following expressions for $\operatorname{adj}(sI - A)B$ and $a(s)$:

$$\begin{aligned} \operatorname{adj}(sI - A)B &= Bs^{n-1} + (AB + a_{n-1}B)s^{n-2} + (A^2B + a_{n-1}AB + a_{n-2}B)s^{n-3} \\ &\quad + \ldots + (A^{n-1}B + a_{n-1}A^{n-2}B + \ldots + a_1 B) \\ a(s) &= s^n + a_{n-1}s^{n-1} + \ldots + a_1 s + a_0 \end{aligned}$$

and equating the coefficients of s^i on both sides of equation (5.6.25), we find that X, Y must satisfy the following algebraic equations:

$$\begin{cases} Y = \underline{0}_q \\ B^T X + a_{n-1} Y = \underline{0}_q \\ (AB + a_{n-1}B)^T X + a_{n-2} Y = \underline{0}_q \\ \quad \vdots \\ (A^{n-1} + a_{n-1}A^{n-2}B + \ldots + a_1 B)^T X + a_0 Y = \underline{0}_q \end{cases}$$

or equivalently:

$$Y = \underline{0}_q \quad \text{and} \quad (B, AB, \ldots, A^{n-1}B)^T X = \underline{0}_{nq} \tag{5.6.26}$$

where $\underline{0}_{nq}$ is a zero column-vector of dimension nq. Because (A, B) is controllable, the matrix $(B, AB, \ldots A^{n-1}B)$ is of full rank; therefore, (5.6.26) is true if and only if $X = 0, Y = 0$, i.e. $x = 0$.

Thus, we have proved that $x^T R_z(0) x = 0$ if and only if $x = 0$, which implies that $R_z(0)$ is positive definite. Then it follows from Lemma 5.6.1 that z is PE.

Using exactly the same arguments as used in proving Theorem 5.2.2, we conclude from z being PE that $\epsilon_1(t), \tilde{\theta}(t)$ converge to zero exponentially fast as $t \to \infty$. From the definition of $\tilde{\theta}$ we have that $\hat{A}(t) \to A, \hat{B}(t) \to B$ exponentially fast as $t \to \infty$ and the proof is complete. □

5.6. PARAMETER CONVERGENCE PROOF

5.6.5 Proof of Theorem 5.2.5

Let us define
$$\bar{\epsilon}(t) = \tilde{\theta}^\top(t) R_\phi(0) \tilde{\theta}(t)$$
where $\tilde{\theta}(t) = \theta(t) - \theta^*$. We will show that $\bar{\epsilon}(t) \to 0$ as $t \to \infty$, i.e., for any given $\epsilon' > 0$, there exists a $t_1 \geq 0$ such that for all $t \geq t_1$, $\bar{\epsilon}(t) < \epsilon'$.

We express $\bar{\epsilon}$ as

$$\begin{aligned}
\bar{\epsilon}(t) &= \tilde{\theta}^\top(t) \frac{1}{T} \int_{t_1}^{t_1+T} \phi(\tau)\phi^\top(\tau) d\tau \, \tilde{\theta}(t) \\
&\quad + \tilde{\theta}^\top(t) \left(R_\phi(0) - \frac{1}{T} \int_{t_1}^{t_1+T} \phi(\tau)\phi^\top(\tau) d\tau \right) \tilde{\theta}(t) \\
&= \frac{1}{T} \int_{t_1}^{t_1+T} (\tilde{\theta}^\top(\tau)\phi(\tau))^2 d\tau + \frac{1}{T} \int_{t_1}^{t_1+T} \left\{ (\tilde{\theta}^\top(t)\phi(\tau))^2 - (\tilde{\theta}^\top(\tau)\phi(\tau))^2 \right\} d\tau \\
&\quad + \tilde{\theta}^\top(t) \left(R_\phi(0) - \frac{1}{T} \int_{t_1}^{t_1+T} \phi(\tau)\phi^\top(\tau) d\tau \right) \tilde{\theta}(t) \\
&\triangleq \bar{\epsilon}_1(t) + \bar{\epsilon}_2(t) + \bar{\epsilon}_3(t)
\end{aligned}$$

where t_1, T are arbitrary at this point and will be specified later. We evaluate each term on the right-hand side of the above equation separately.

Because $\epsilon(t) = \tilde{\theta}^\top(t)\phi(t) \to 0$ as $t \to \infty$, there exists a $t_1' \geq 0$ such that $|\epsilon(t)| < \sqrt{\frac{\epsilon'}{3}}$ for $t \geq t_1'$. Choosing $t_1 \geq t_1'$, we have

$$|\bar{\epsilon}_1(t)| = \frac{1}{T} \int_{t_1}^{t_1+T} \epsilon^2(\tau) d\tau < \frac{\epsilon'}{3} \tag{5.6.27}$$

For $\bar{\epsilon}_2(t)$, we have

$$\begin{aligned}
\bar{\epsilon}_2(t) &= \frac{1}{T} \int_{t_1}^{t_1+T} (\tilde{\theta}(t) - \tilde{\theta}(\tau))^\top \phi(\tau)\phi^\top(\tau)(\tilde{\theta}(t) + \tilde{\theta}(\tau)) d\tau \\
&= \frac{1}{T} \int_{t_1}^{t_1+T} \left(-\int_t^\tau \dot{\tilde{\theta}}(\sigma) d\sigma \right)^\top \phi(\tau)\phi^\top(\tau)(\tilde{\theta}(t) + \tilde{\theta}(\tau)) d\tau
\end{aligned}$$

Using the property of the adaptive law that $\dot{\tilde{\theta}} \to 0$ as $t \to \infty$, for any given T, ϵ' we can find a $t_2' \geq 0$ such that for all $t \geq t_2'$, $|\dot{\tilde{\theta}}(t)| \leq \frac{2\epsilon'}{3KT}$, where $K = \sup_{t,\tau} |\phi(\tau)\phi^\top(\tau)(\tilde{\theta}(t) + \tilde{\theta}(\tau))|$ is a finite constant because of the fact that both $\tilde{\theta}$ and ϕ are uniformly bounded signals. Choosing $t_1 \geq t_2'$, we have

$$|\bar{\epsilon}_2(t)| \leq \frac{1}{T} \int_{t_1}^{t_1+T} \frac{2\epsilon'(\tau-t)}{3KT} |\phi(\tau)\phi^\top(\tau)(\tilde{\theta}(t) + \tilde{\theta}(\tau))| d\tau < \frac{\epsilon'}{3} \tag{5.6.28}$$

for all $t \geq t_2'$.

Because the autocovariance $R_\phi(0)$ exists, there exists $T_0 \geq 0$ such that for all $T \geq T_0$,

$$\left\| R_\phi(0) - \frac{1}{T} \int_{t_1}^{t_1+T} \phi(\tau)\phi^\top(\tau) d\tau \right\| < \frac{\epsilon'}{3K'}$$

where $K' = \sup_t |\tilde{\theta}(t)|^2$ is a finite constant. Therefore,

$$|\bar{\epsilon}_3| < \frac{\epsilon'}{3} \qquad (5.6.29)$$

Combining (5.6.27) to (5.6.29), we have that for $t \geq \max\{t_1', t_2'\} \triangleq t_1$, $T \geq T_0$,

$$\bar{\epsilon}(t) = \tilde{\theta}^\top(t) R_\phi(0) \tilde{\theta}(t) < \epsilon'$$

which, in turn, implies that $\bar{\epsilon}$; therefore, $R_\phi(0)\tilde{\theta}$ converges to zero as $t \to \infty$. □

5.7 Problems

5.1 Show that

$$\phi = \begin{bmatrix} A\sin(t+\varphi) \\ \sin(t) \end{bmatrix}$$

where A, φ are nonzero constants is PE in $\mathcal{R}^2$.

5.2 Consider the following plant

$$y = \frac{b_1 s}{(s+1)^2} u$$

where b_1 is the only unknown parameter. Is $u = c_0$ (constant) $\neq 0$ sufficiently rich for identifying b_1? Explain. Design a parameter identifier to identify b_1 from the measurements of u, y. Simulate your scheme on a digital computer for $b_1 = 5$.

5.3 Consider the plant

$$y = \frac{b_2 s^2 + b_0}{(s+2)^3} u$$

where b_2, b_0 are the only unknown parameters. Is $u = \sin t$ sufficiently rich for identifying b_2, b_0 where b_2, b_0 can be any number in $\mathcal{R}$. Explain. Design a parameter identifier to identify b_2, b_0. Simulate your scheme for a) $b_2 = 1, b_0 = 1$ and b) $b_2 = 3, b_0 = 5$.

5.7. PROBLEMS

5.4 Consider the second order stable plant

$$\dot{x} = \begin{bmatrix} a_1 & 1 \\ a_0 & 0 \end{bmatrix} x + \begin{bmatrix} b_1 \\ 1 \end{bmatrix} u$$

where a_1, a_0, b_1 are the only unknown parameters and $u \in \mathcal{L}_\infty$.

(a) Design a parameter identifier to estimate the unknown parameters.

(b) Choose u with the least number of frequencies that guarantees parameter convergence.

(c) Simulate your scheme for $a_1 = -2, a_0 = -5, b_1 = 8$.

5.5 Simulate the series-parallel identifier in Example 5.2.2. Repeat Example 5.2.2 by designing and simulating a parallel identifier. In simulations use numerical values for A, B of your choice.

5.6 Perform the simulations requested in (i) and (ii) of Example 5.2.3 when $b_0 = -2, a_1 = 2.8, a_0 = 5.6$. Comment on your results.

5.7 Repeat Problem 5.6 when the pure least-squares algorithm in Example 5.2.3 is replaced with the least-squares with covariance resetting algorithm.

5.8 Repeat Problem 5.6 when the pure least-squares algorithm in Example 5.2.2 is replaced with the

(a) Integral algorithm
(b) Hybrid adaptive law.

5.9 Design an adaptive Luenberger observer for the plant

$$\dot{x} = \begin{bmatrix} -a_1 & 1 \\ -a_0 & 0 \end{bmatrix} x + \begin{bmatrix} b_1 \\ 1 \end{bmatrix} u$$
$$y = [1, 0]x$$

where $a_1, a_0 > 0$ and $b_1 \neq 0$ are the only unknown parameters using the following adaptive laws for on-line estimation:

(a) Integral algorithm
(b) Pure least-squares
(c) Hybrid adaptive law

In each case present the complete stability proof. Simulate the adaptive observers with inputs u of your choice. For simulation purposes assume that the unknown parameters have the following values: $a_1 = 2.5, a_0 = 3.6, b_1 = 4$.

5.10 Repeat Problem 5.9 by designing an adaptive observer with auxiliary input.

5.11 Consider the LTI plant

$$y = \frac{b_0}{s^2 + a_1 s + a_2} u$$

where $b_0 \neq 0$, and $a_1, a_2 > 0$. Represent the plant in the following forms:
(a) Observable form
(b) Nonminimal Realization 1
(c) Nonminimal Realization 2

5.12 Design an adaptive observer using an integral adaptive law for the plant of Problem 5.11 represented in the observer form.

5.13 Repeat Problem 5.12 for the same plant in the nonminimal Realization 1.

5.14 Design an adaptive observer for the plant of Problem 5.11 expressed in the nonminimal Realization 2.

5.15 Consider the following plant:
$$y = W_0(s)G(s)u$$
where $W_0(s)$ is a known proper transfer function with stable poles and $G(s)$ is a strictly proper transfer function of order n with stable poles but unknown coefficients.

(a) Design a parameter identifier to identify the coefficients of $G(s)$.

(b) Design an adaptive observer to estimate the states of a minimal realization of the plant.

5.16 Prove that there exists a signal vector $v \in \mathcal{R}^n$ available for measurement for which the system given by (5.4.2) becomes
$$\tilde{y} = -\tilde{\theta}^\top \phi$$
where $\phi = H(s) \begin{bmatrix} u \\ y \end{bmatrix} \in \mathcal{R}^{2n}$ and $H(s)$ is a known transfer matrix.

5.17 Use the result of Problem 5.16 to develop an adaptive observer with auxiliary input that employs a least-squares algorithm as a parameter estimator.

5.18 Let $F(t) : \mathcal{R} \mapsto \mathcal{R}^{n \times m}$ and $F, \dot{F} \in \mathcal{L}_\infty$. If there exist positive constants k_1, k_2, T_0 such that
$$k_1 I_n \leq \frac{1}{T_0} \int_t^{t+T_0} F(\tau) F^\top(\tau) d\tau \leq k_2 I_n$$
for any $t \geq 0$, where I_n is the identity matrix of dimension n, show that if
$$F_f = \frac{b}{s+a} F$$
with $a > 0$, $b \neq 0$, then there exist positive constants k_1', k_2', T_0' such that
$$k_1' I_n \leq \frac{1}{T_0'} \int_t^{t+T_0'} F_f(\tau) F_f^\top(\tau) d\tau \leq k_2' I_n$$
for any $t \geq 0$.

Chapter 6

Model Reference Adaptive Control

6.1 Introduction

Model reference adaptive control (MRAC) is one of the main approaches to adaptive control. The basic structure of a MRAC scheme is shown in Figure 6.1. The reference model is chosen to generate the desired trajectory, y_m, that the plant output y_p has to follow. The tracking error $e_1 \triangleq y_p - y_m$ represents the deviation of the plant output from the desired trajectory. The closed-loop plant is made up of an ordinary feedback control law that contains the plant and a controller $C(\theta)$ and an adjustment mechanism that generates the controller parameter estimates $\theta(t)$ on-line.

The purpose of this chapter is to design the controller and parameter adjustment mechanism so that all signals in the closed-loop plant are bounded and the plant output y_p tracks y_m as close as possible.

MRAC schemes can be characterized as *direct* or *indirect* and with *normalized* or *unnormalized* adaptive laws. In direct MRAC, the parameter vector θ of the controller $C(\theta)$ is updated directly by an adaptive law, whereas in indirect MRAC θ is calculated at each time t by solving a certain algebraic equation that relates θ with the on-line estimates of the plant parameters. In both direct and indirect MRAC with normalized adaptive laws, the form of $C(\theta)$, motivated from the known parameter case, is kept unchanged. The controller $C(\theta)$ is combined with an adaptive law (or an adaptive law and

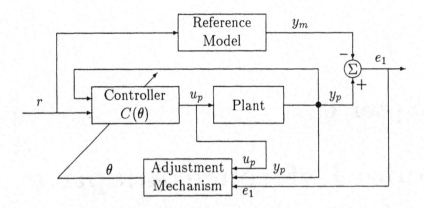

Figure 6.1 General structure of MRAC scheme.

an algebraic equation in the indirect case) that is developed independently by following the techniques of Chapter 4. This design procedure allows the use of a wide class of adaptive laws that includes gradient, least-squares and those based on the SPR-Lyapunov design approach. On the other hand, in the case of MRAC schemes with unnormalized adaptive laws, $C(\theta)$ is modified to lead to an error equation whose form allows the use of the SPR-Lyapunov design approach for generating the adaptive law. In this case, the design of $C(\theta)$ and adaptive law is more complicated in both the direct and indirect case, but the analysis is much simpler and follows from a consideration of a single Lyapunov-like function.

The chapter is organized as follows: In Section 6.2, we use several examples to illustrate the design and analysis of a class of simple direct MRAC schemes with unnormalized adaptive laws. These examples are used to motivate the more general and complicated designs treated in the rest of the chapter. In Section 6.3 we define the model reference control (MRC) problem for SISO plants and solve it for the case of known plant parameters. The control law developed in this section is used in the rest of the chapter to form MRAC schemes in the unknown parameter case.

The design of direct MRAC schemes with unnormalized adaptive laws is treated in Section 6.4 for plants with relative degree $n^* = 1, 2, 3$. The case of $n^* > 3$ follows by using the same techniques as in the case of $n^* = 3$ and is omitted because of the complexity of the control law that increases with n^*. In Section 6.5 we consider the design and analysis of a wide class of direct

6.2. SIMPLE DIRECT MRAC SCHEMES

MRAC schemes with normalized adaptive laws for plants with arbitrary but known relative degree.

The design of indirect MRAC with unnormalized and normalized adaptive laws is considered in Section 6.6. In Section 6.7, we briefly summarize some efforts and alternative approaches to relax some of the basic assumptions used in MRAC that include the minimum phase, known relative degree and upper bound on the order of the plant. In Section 6.8, we present all the long and more complicated proofs of theorems and lemmas.

6.2 Simple Direct MRAC Schemes

In this section, we use several examples to illustrate the design and analysis of some simple direct MRAC schemes with unnormalized adaptive laws. We concentrate on the SPR-Lyapunov approach for designing the adaptive laws. This approach dominated the literature of adaptive control for continuous-time plants with relative degree $n^* = 1$ because of the simplicity of design and stability analysis [48, 85, 172, 201].

6.2.1 Scalar Example: Adaptive Regulation

Consider the following scalar plant:

$$\dot{x} = ax + u, \quad x(0) = x_0 \tag{6.2.1}$$

where a is a constant but unknown. The control objective is to determine a bounded function $u = f(t, x)$ such that the state $x(t)$ is bounded and converges to zero as $t \to \infty$ for any given initial condition x_0. Let $-a_m$ be the desired closed-loop pole where $a_m > 0$ is chosen by the designer.

Control Law If the plant parameter a is known, the control law

$$u = -k^* x \tag{6.2.2}$$

with $k^* = a + a_m$ could be used to meet the control objective, i.e., with (6.2.2), the closed-loop plant is

$$\dot{x} = -a_m x$$

whose equilibrium $x_e = 0$ is e.s. in the large.

Because a is unknown, k^* cannot be calculated and, therefore, (6.2.2) cannot be implemented. A possible procedure to follow in the unknown parameter case is to use the same control law as given in (6.2.2) but with k^* replaced by its estimate $k(t)$, i.e., we use

$$u = -k(t)x \qquad (6.2.3)$$

and search for an adaptive law to update $k(t)$ continuously with time.

Adaptive Law The adaptive law for generating $k(t)$ is developed by viewing the problem as an on-line identification problem for k^*. This is accomplished by first obtaining an appropriate parameterization for the plant (6.2.1) in terms of the unknown k^* and then using a similar approach as in Chapter 4 to estimate k^* on-line. We illustrate this procedure below.

We add and subtract the desired control input $-k^*x$ in the plant equation to obtain

$$\dot{x} = ax - k^*x + k^*x + u.$$

Because $a - k^* = -a_m$ we have

$$\dot{x} = -a_m x + k^*x + u$$

or

$$x = \frac{1}{s + a_m}(u + k^*x) \qquad (6.2.4)$$

Equation (6.2.4) is a parameterization of the plant equation (6.2.1) in terms of the unknown controller parameter k^*. Because x, u are measured and $a_m > 0$ is known, a wide class of adaptive laws may be generated by simply using Tables 4.1 to 4.3 of Chapter 4. It turns out that the adaptive laws developed for (6.2.4) using the SPR-Lyapunov design approach without normalization simplify the stability analysis of the resulting closed-loop adaptive control scheme considerably. Therefore, as a starting point, we concentrate on the simple case and deal with the more general case that involves a wide class of adaptive laws in later sections.

Because $\frac{1}{s+a_m}$ is SPR we can proceed with the SPR-Lyapunov design approach of Chapter 4 and generate the estimate $\hat{x}$ of x as

$$\hat{x} = \frac{1}{s + a_m}[kx + u] = \frac{1}{s + a_m}(0) \qquad (6.2.5)$$

6.2. SIMPLE DIRECT MRAC SCHEMES

where the last equality is obtained by substituting the control law $u = -kx$. If we now choose $\hat{x}(0) = 0$, we have $\hat{x}(t) \equiv 0, \forall t \geq 0$, which implies that the estimation error ϵ_1 defined as $\epsilon_1 = x - \hat{x}$ is equal to the regulation error, i.e., $\epsilon_1 = x$, so that (6.2.5) does not have to be implemented to generate $\hat{x}$. Substituting for the control $u = -k(t)x$ in (6.2.4), we obtain the error equation that relates the parameter error $\tilde{k} = k - k^*$ with the estimation error $\epsilon_1 = x$, i.e.,

$$\dot{\epsilon}_1 = -a_m \epsilon_1 - \tilde{k}x, \qquad \epsilon_1 = x \tag{6.2.6}$$

or

$$\epsilon_1 = \frac{1}{s + a_m}\left(-\tilde{k}x\right)$$

As demonstrated in Chapter 4, the error equation (6.2.6) is in a convenient form for choosing an appropriate Lyapunov function to design the adaptive law for $k(t)$. We assume that the adaptive law is of the form

$$\dot{\tilde{k}} = \dot{k} = f_1(\epsilon_1, x, u) \tag{6.2.7}$$

where f_1 is some function to be selected, and propose

$$V(\epsilon_1, \tilde{k}) = \frac{\epsilon_1^2}{2} + \frac{\tilde{k}^2}{2\gamma} \tag{6.2.8}$$

for some $\gamma > 0$ as a potential Lyapunov function for the system (6.2.6), (6.2.7). The time derivative of V along the trajectory of (6.2.6), (6.2.7) is given by

$$\dot{V} = -a_m \epsilon_1^2 - \tilde{k}\epsilon_1 x + \frac{\tilde{k} f_1}{\gamma} \tag{6.2.9}$$

Choosing $f_1 = \gamma \epsilon_1 x$, i.e.,

$$\dot{k} = \gamma \epsilon_1 x = \gamma x^2, \quad k(0) = k_0 \tag{6.2.10}$$

we have

$$\dot{V} = -a_m \epsilon_1^2 \leq 0 \tag{6.2.11}$$

Analysis Because V is a positive definite function and $\dot{V} \leq 0$, we have $V \in \mathcal{L}_\infty$, which implies that $\epsilon_1, \tilde{k} \in \mathcal{L}_\infty$. Because $\epsilon_1 = x$, we also have that $x \in \mathcal{L}_\infty$ and therefore all signals in the closed-loop plant are bounded.

Furthermore, $\epsilon_1 = x \in \mathcal{L}_2$ and $\dot\epsilon_1 = \dot x \in \mathcal{L}_\infty$ (which follows from (6.2.6)) imply, according to Lemma 3.2.5, that $\epsilon_1(t) = x(t) \to 0$ as $t \to \infty$. From $x(t) \to 0$ and the boundedness of k, we establish that $\dot k(t) \to 0, u(t) \to 0$ as $t \to \infty$.

We have shown that the combination of the control law (6.2.3) with the adaptive law (6.2.10) meets the control objective in the sense that it forces the plant state to converge to zero while guaranteeing signal boundedness.

It is worth mentioning that as in the simple parameter identification examples considered in Chapter 4, we cannot establish that $k(t)$ converges to k^*, i.e., that the pole of the closed-loop plant converges to the desired one given by $-a_m$. The lack of parameter convergence is less crucial in adaptive control than in parameter identification because in most cases, the control objective can be achieved without requiring the parameters to converge to their true values.

The simplicity of this scalar example allows us to solve for $\epsilon_1 = x$ explicitly, and study the properties of $k(t), x(t)$ as they evolve with time. We can verify that

$$\begin{aligned}\epsilon_1(t) &= \frac{2ce^{-ct}}{c + k_0 - a + (c - k_0 + a)e^{-2ct}}\epsilon_1(0), \quad \epsilon_1 = x \\ k(t) &= a + \frac{c\left[(c + k_0 - a)e^{2ct} - (c - k_0 + a)\right]}{(c + k_0 - a)e^{2ct} + (c - k_0 + a)}\end{aligned} \quad (6.2.12)$$

where $c^2 = \gamma x_0^2 + (k_0 - a)^2$, satisfy the differential equations (6.2.6) and (6.2.10) of the closed-loop plant. Equation (6.2.12) can be used to investigate the effects of initial conditions and adaptive gain γ on the transient and asymptotic behavior of $x(t), k(t)$. We have $\lim_{t\to\infty} k(t) = a + c$ if $c > 0$ and $\lim_{t\to\infty} k(t) = a - c$ if $c < 0$, i.e.,

$$\lim_{t\to\infty} k(t) = k_\infty = a + \sqrt{\gamma x_0^2 + (k_0 - a)^2} \quad (6.2.13)$$

Therefore, for $x_0 \neq 0$, $k(t)$ converges to a stabilizing gain whose value depends on γ and the initial condition x_0, k_0. It is clear from (6.2.13) that the value of k_∞ is independent of whether k_0 is a destabilizing gain, i.e., $0 < k_0 < a$, or a stabilizing one, i.e., $k_0 > a$, as long as $(k_0 - a)^2$ is the same. The use of different k_0, however, will affect the transient behavior as it is obvious from (6.2.12). In the limit as $t \to \infty$, the closed-loop pole converges

6.2. SIMPLE DIRECT MRAC SCHEMES

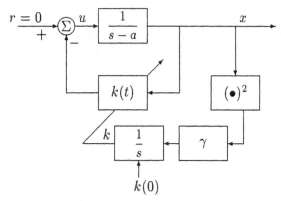

Figure 6.2 Block diagram for implementing the adaptive controller (6.2.14).

to $-(k_\infty - a)$, which may be different from $-a_m$. Because the control objective is to achieve signal boundedness and regulation of the state $x(t)$ to zero, the convergence of $k(t)$ to k^* is not crucial.

Implementation The adaptive control scheme developed and analyzed above is given by the following equations:

$$u = -k(t)x, \quad \dot{k} = \gamma \epsilon_1 x = \gamma x^2, \quad k(0) = k_0 \qquad (6.2.14)$$

where x is the measured state of the plant. A block diagram for implementing (6.2.14) is shown in Figure 6.2.

The design parameters in (6.2.14) are the initial parameter k_0 and the adaptive gain $\gamma > 0$. For signal boundedness and asymptotic regulation of x to zero, our analysis allows k_0, γ to be arbitrary. It is clear, however, from (6.2.12) that their values affect the transient performance of the closed-loop plant as well as the steady-state value of the closed-loop pole. For a given $k_0, x_0 \neq 0$, large γ leads to a larger value of c in (6.2.12) and, therefore, to a faster convergence of $x(t)$ to zero. Large γ, however, may make the differential equation for k "stiff" (i.e., $\dot{k}$ large) that will require a very small step size or sampling period to implement it on a digital computer. Small sampling periods make the adaptive scheme more sensitive to measurement noise and modeling errors.

Remark 6.2.1 In the proceeding example, we have not used any reference model to describe the desired properties of the closed-loop system. A reasonable choice for the reference model would be

$$\dot{x}_m = -a_m x_m, \quad x_m(0) = x_{m0} \qquad (6.2.15)$$

which, by following exactly the same procedure, would lead to the adaptive control scheme

$$u = -k(t)x, \quad \dot{k} = \gamma e_1 x$$

where $e_1 = x - x_m$. If $x_{m0} \neq x_0$, the use of (6.2.15) will affect the transient behavior of the tracking error but will have no effect on the asymptotic properties of the closed-loop scheme because x_m converges to zero exponentially fast.

6.2.2 Scalar Example: Adaptive Tracking

Consider the following first order plant:

$$\dot{x} = ax + bu \qquad (6.2.16)$$

where a, b are unknown parameters but the sign of b is known. The control objective is to choose an appropriate control law u such that all signals in the closed-loop plant are bounded and x tracks the state x_m of the reference model given by

$$\dot{x}_m = -a_m x_m + b_m r$$

i.e.,

$$x_m = \frac{b_m}{s + a_m} r \qquad (6.2.17)$$

for any bounded piecewise continuous signal $r(t)$, where $a_m > 0$, b_m are known and $x_m(t), r(t)$ are measured at each time t. It is assumed that a_m, b_m and r are chosen so that x_m represents the desired state response of the plant.

Control Law For x to track x_m for any reference input signal $r(t)$, the control law should be chosen so that the closed-loop plant transfer function

6.2. SIMPLE DIRECT MRAC SCHEMES

from the input r to output x is equal to that of the reference model. We propose the control law

$$u = -k^*x + l^*r \tag{6.2.18}$$

where k^*, l^* are calculated so that

$$\frac{x(s)}{r(s)} = \frac{bl^*}{s-a+bk^*} = \frac{b_m}{s+a_m} = \frac{x_m(s)}{r(s)} \tag{6.2.19}$$

Equation (6.2.19) is satisfied if we choose

$$l^* = \frac{b_m}{b}, \quad k^* = \frac{a_m + a}{b} \tag{6.2.20}$$

provided of course that $b \neq 0$, i.e., the plant (6.2.16) is controllable. The control law (6.2.18), (6.2.20) guarantees that the transfer function of the closed-loop plant, i.e., $\frac{x(s)}{r(s)}$ is equal to that of the reference model. Such a transfer function matching guarantees that $x(t) = x_m(t), \forall t \geq 0$ when $x(0) = x_m(0)$ or $|x(t) - x_m(t)| \to 0$ exponentially fast when $x(0) \neq x_m(0)$, for any bounded reference signal $r(t)$.

When the plant parameters a, b are unknown, (6.2.18) cannot be implemented. Therefore, instead of (6.2.18), we propose the control law

$$u = -k(t)x + l(t)r \tag{6.2.21}$$

where $k(t), l(t)$ is the estimate of k^*, l^*, respectively, at time t, and search for an adaptive law to generate $k(t), l(t)$ on-line.

Adaptive Law As in Example 6.2.1, we can view the problem as an on-line identification problem of the unknown constants k^*, l^*. We start with the plant equation (6.2.16) which we express in terms of k^*, l^* by adding and subtracting the desired input term $-bk^*x + bl^*r$ to obtain

$$\dot{x} = -a_m x + b_m r + b(k^*x - l^*r + u)$$

i.e.,

$$x = \frac{b_m}{s+a_m} r + \frac{b}{s+a_m}(k^*x - l^*r + u) \tag{6.2.22}$$

Because $x_m = \frac{b_m}{s+a_m} r$ is a known bounded signal, we express (6.2.22) in terms of the tracking error defined as $e \triangleq x - x_m$, i.e.,

$$e = \frac{b}{s+a_m}(k^*x - l^*r + u) \tag{6.2.23}$$

Because b is unknown, equation (6.2.23) is in the form of the bilinear parametric model considered in Chapter 4, and may be used to choose an adaptive law directly from Table 4.4 of Chapter 4.

Following the procedure of Chapter 4, the estimate $\hat{e}$ of e is generated as

$$\hat{e} = \frac{1}{s + a_m} \hat{b} \left(kx - lr + u \right) = \frac{1}{s + a_m} (0) \qquad (6.2.24)$$

where the last identity is obtained by substituting for the control law

$$u = -k(t)x + l(t)r$$

Equation (6.2.24) implies that the estimation error, defined as $\epsilon_1 \triangleq e - \hat{e}$, can be simply taken to be the tracking error, i.e., $\epsilon_1 = e$, and, therefore, there is no need to generate $\hat{e}$. Furthermore, since $\hat{e}$ is not generated, the estimate $\hat{b}$ of b is not required.

Substituting $u = -k(t)x + l(t)r$ in (6.2.23) and defining the parameter errors $\tilde{k} \triangleq k - k^*, \tilde{l} \triangleq l - l^*$, we have

$$\epsilon_1 = e = \frac{b}{s + a_m} \left(-\tilde{k}x + \tilde{l}r \right)$$

or

$$\dot{\epsilon}_1 = -a_m \epsilon_1 + b \left(-\tilde{k}x + \tilde{l}r \right), \quad \epsilon_1 = e = x - x_m \qquad (6.2.25)$$

As shown in Chapter 4, the development of the differential equation (6.2.25) relating the estimation error with the parameter error is a significant step in deriving the adaptive laws for updating $k(t), l(t)$. We assume that the structure of the adaptive law is given by

$$\dot{k} = f_1(\epsilon_1, x, r, u), \quad \dot{l} = f_2(\epsilon_1, x, r, u) \qquad (6.2.26)$$

where the functions f_1, f_2 are to be designed.

As shown in Example 6.2.1, however, the use of the SPR-Lyapunov approach without normalization allows us to design an adaptive law for k, l and analyze the stability properties of the closed-loop system using a single Lyapunov function. For this reason, we proceed with the SPR-Lyapunov approach without normalization and postpone the use of other approaches that are based on the use of the normalized estimation error for later sections.

6.2. SIMPLE DIRECT MRAC SCHEMES

Consider the function

$$V\left(\epsilon_1, \tilde{k}, \tilde{l}\right) = \frac{\epsilon_1^2}{2} + \frac{\tilde{k}^2}{2\gamma_1}|b| + \frac{\tilde{l}^2}{2\gamma_2}|b| \qquad (6.2.27)$$

where $\gamma_1, \gamma_2 > 0$ as a Lyapunov candidate for the system (6.2.25), (6.2.26). The time derivative $\dot{V}$ along any trajectory of (6.2.25), (6.2.26) is given by

$$\dot{V} = -a_m \epsilon_1^2 - b\tilde{k}\epsilon_1 x + b\tilde{l}\epsilon_1 r + \frac{|b|\tilde{k}}{\gamma_1}f_1 + \frac{|b|\tilde{l}}{\gamma_2}f_2 \qquad (6.2.28)$$

Because $|b| = b\mathrm{sgn}(b)$, the indefinite terms in (6.2.28) disappear if we choose $f_1 = \gamma_1 \epsilon_1 x \mathrm{sgn}(b)$, $f_2 = -\gamma_2 \epsilon_1 r\ \mathrm{sgn}(b)$. Therefore, for the adaptive law

$$\dot{k} = \gamma_1 \epsilon_1 x\ \mathrm{sgn}(b), \quad \dot{l} = -\gamma_2 \epsilon_1 r\ \mathrm{sgn}(b) \qquad (6.2.29)$$

we have

$$\dot{V} = -a_m \epsilon_1^2 \qquad (6.2.30)$$

Analysis Treating $x_m(t), r(t)$ in (6.2.25) as bounded functions of time, it follows from (6.2.27), (6.2.30) that V is a Lyapunov function for the third-order differential equation (6.2.25) and (6.2.29) where x_m is treated as a bounded function of time, and the equilibrium $\epsilon_{1e} = e_e = 0$, $\tilde{k}_e = 0, \tilde{l}_e = 0$ is u.s. Furthermore, $\epsilon_1, \tilde{k}, \tilde{l} \in \mathcal{L}_\infty$ and $\epsilon_1 \in \mathcal{L}_2$. Because $\epsilon_1 = e = x - x_m$ and $x_m \in \mathcal{L}_\infty$, we also have $x \in \mathcal{L}_\infty$ and $u \in \mathcal{L}_\infty$; therefore, all signals in the closed-loop are bounded. Now from (6.2.25) we have $\dot{\epsilon}_1 \in \mathcal{L}_\infty$, which, together with $\epsilon_1 \in \mathcal{L}_2$, implies that $\epsilon_1(t) = e(t) \to 0$ as $t \to \infty$.

We have established that the control law (6.2.21) together with the adaptive law (6.2.29) guarantee boundedness for all signals in the closed-loop system. In addition, the plant state $x(t)$ tracks the state of the reference model x_m asymptotically with time for any reference input signal r, which is bounded and piecewise continuous. These results do not imply that $k(t) \to k^*$ and $l(t) \to l^*$ as $t \to \infty$, i.e., the transfer function of the closed-loop plant may not approach that of the reference model as $t \to \infty$. To achieve such a result, the reference input r has to be sufficiently rich of order 2. For example, $r(t) = \sin \omega t$ for some $\omega \neq 0$ guarantees the exponential convergence of $x(t)$ to $x_m(t)$ and of $k(t), l(t)$ to k^*, l^*, respectively. In general, a sufficiently rich reference input $r(t)$ is not desirable especially

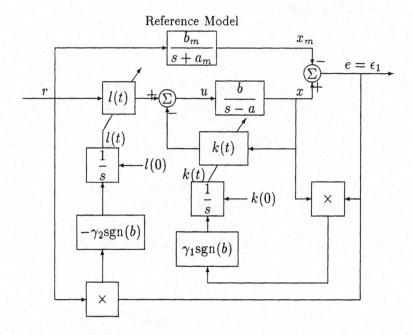

Figure 6.3 Block diagram for implementing the adaptive law (6.2.21) and (6.2.29).

in cases where the control objective involves tracking of signals that are not rich in frequencies. Parameter convergence and the conditions the reference input r has to satisfy are discussed later on in this chapter.

Implementation The MRAC control law (6.2.21), (6.2.29) can be implemented as shown in Figure 6.3. The adaptive gains γ_1, γ_2 are designed by following similar considerations as in the previous examples. The initial conditions $l(0), k(0)$ are chosen to be any a priori guess of the unknown parameters l^*, k^*, respectively. Small initial parameter error usually leads to better transient behavior. As we mentioned before, the reference model and input r are designed so that x_m describes the desired trajectory to be followed by the plant state.

Remark 6.2.2 The assumption that the sign of b is known may be relaxed by using the techniques of Chapter 4 summarized in Table 4.4 and is left as an exercise for the reader (see Problem 6.3).

6.2. SIMPLE DIRECT MRAC SCHEMES

6.2.3 Vector Case: Full-State Measurement

Let us now consider the nth order plant

$$\dot{x} = Ax + Bu, \quad x \in \mathcal{R}^n \tag{6.2.31}$$

where $A \in \mathcal{R}^{n \times n}, B \in \mathcal{R}^{n \times q}$ are unknown constant matrices and (A, B) is controllable. The control objective is to choose the input vector $u \in \mathcal{R}^q$ such that all signals in the closed-loop plant are bounded and the plant state x follows the state $x_m \in \mathcal{R}^n$ of a reference model specified by the LTI system

$$\dot{x}_m = A_m x_m + B_m r \tag{6.2.32}$$

where $A_m \in \mathcal{R}^{n \times n}$ is a stable matrix, $B_m \in \mathcal{R}^{n \times q}$, and $r \in \mathcal{R}^q$ is a bounded reference input vector. The reference model and input r are chosen so that $x_m(t)$ represents a desired trajectory that x has to follow.

Control Law If the matrices A, B were known, we could apply the control law

$$u = -K^* x + L^* r \tag{6.2.33}$$

and obtain the closed-loop plant

$$\dot{x} = (A - BK^*)x + BL^* r \tag{6.2.34}$$

Hence, if $K^* \in \mathcal{R}^{q \times n}$ and $L^* \in \mathcal{R}^{q \times q}$ are chosen to satisfy the algebraic equations

$$A - BK^* = A_m, \quad BL^* = B_m \tag{6.2.35}$$

then the transfer matrix of the closed-loop plant is the same as that of the reference model and $x(t) \to x_m(t)$ exponentially fast for any bounded reference input signal $r(t)$. We should note that given the matrices A, B, A_m, B_m, no K^*, L^* may exist to satisfy the matching condition (6.2.35) indicating that the control law (6.2.33) may not have enough structural flexibility to meet the control objective. In some cases, if the structure of A, B is known, A_m, B_m may be designed so that (6.2.35) has a solution for K^*, L^*.

Let us assume that K^*, L^* in (6.2.35) exist, i.e., that there is sufficient structural flexibility to meet the control objective, and propose the control law

$$u = -K(t)x + L(t)r \tag{6.2.36}$$

where $K(t), L(t)$ are the estimates of K^*, L^*, respectively, to be generated by an appropriate adaptive law.

Adaptive Law By adding and subtracting the desired input term, namely, $-B(K^*x - L^*r)$ in the plant equation and using (6.2.35), we obtain

$$\dot{x} = A_m x + B_m r + B(K^*x - L^*r + u) \qquad (6.2.37)$$

which is the extension of the scalar equation (6.2.22) in Example 6.2.2 to the vector case. Following the same procedure as in Section 6.2.2, we can show that the tracking error $e = x - x_m$ and parameter error $\tilde{K} \triangleq K - K^*$, $\tilde{L} \triangleq L - L^*$ satisfy the equation

$$\dot{e} = A_m e + B(-\tilde{K}x + \tilde{L}r) \qquad (6.2.38)$$

which also depends on the unknown matrix B. In the scalar case we manage to get away with the unknown B by assuming that its sign is known. An extension of the scalar assumption of Section 6.2.2 to the vector case is as follows; Let us assume that L^* is either positive definite or negative definite and $\Gamma^{-1} = L^* \text{sgn}(l)$, where $l = 1$ if L^* is positive definite and $l = -1$ if L^* is negative definite. Then $B = B_m L^{*-1}$ and (6.2.38) becomes

$$\dot{e} = A_m e + B_m L^{*-1}(-\tilde{K}x + \tilde{L}r)$$

We propose the following Lyapunov function candidate

$$V(e, \tilde{K}, \tilde{L}) = e^\top P e + \text{tr}[\tilde{K}^\top \Gamma \tilde{K} + \tilde{L}^\top \Gamma \tilde{L}]$$

where $P = P^\top > 0$ satisfies the Lyapunov equation

$$PA_m + A_m^\top P = -Q$$

for some $Q = Q^\top > 0$. Then,

$$\dot{V} = -e^\top Q e + 2e^\top P B_m L^{*-1}(-\tilde{K}x + \tilde{L}r) + 2\text{tr}[\tilde{K}^\top \Gamma \dot{\tilde{K}} + \tilde{L}^\top \Gamma \dot{\tilde{L}}]$$

Now

$$e^\top P B_m L^{*-1} \tilde{K} x = \text{tr}[x^\top \tilde{K}^\top \Gamma B_m^\top P e] \text{sgn}(l) = \text{tr}[\tilde{K}^\top \Gamma B_m^\top P e x^\top] \text{sgn}(l)$$

6.2. SIMPLE DIRECT MRAC SCHEMES

and
$$e^\top P B_m L^{*-1} \tilde{L} r = \text{tr}[\tilde{L}^\top \Gamma B_m^\top P e r^\top]\text{sgn}(l)$$

Therefore, for

$$\dot{\tilde{K}} = \dot{K} = B_m^\top P e x^\top \text{sgn}(l), \quad \dot{\tilde{L}} = \dot{L} = -B_m^\top P e r^\top \text{sgn}(l) \quad (6.2.39)$$

we have
$$\dot{V} = -e^\top Q e$$

Analysis From the properties of $V, \dot{V}$, we establish as in the scalar case that $K(t), L(t), e(t)$ are bounded and that $e(t) \to 0$ as $t \to \infty$.

Implementation The adaptive control scheme developed is given by (6.2.36) and (6.2.39). The matrix $B_m P$ acts as an adaptive gain matrix, where P is obtained by solving the Lyapunov equation $P A_m + A_m^\top P = -Q$ for some arbitrary $Q = Q^\top > 0$. Different choices of Q will not affect boundedness and the asymptotic behavior of the scheme, but they will affect the transient response. The assumption that the unknown L^* in the matching equation $BL^* = B_m$ is either positive or negative definite imposes an additional restriction on the structure and elements of B, B_m. Because B is unknown this assumption may not be realistic in some applications.

The case where B is completely unknown is treated in [172] using the adaptive control law

$$u = -L(t)K(t)x + L(t)r$$
$$\dot{K} = B_m^\top P e x^\top, \quad \dot{L} = -L B_m^\top e u^\top L \quad (6.2.40)$$

The result established, however, is only local which indicates that for stability $K(0), L(0)$ have to be chosen close to the equilibrium $K_e = K^*, L_e = L^*$ of (6.2.40). Furthermore, K^*, L^* are required to satisfy the matching equations $A - BL^*K^* = A_m, BL^* = B_m$.

6.2.4 Nonlinear Plant

The procedure of Sections 6.2.1 to 6.2.3 can be extended to some special classes of nonlinear plants as demonstrated briefly by using the following nonlinear example

$$\dot{x} = af(x) + bg(x)u \quad (6.2.41)$$

where a, b are unknown scalars, $f(x), g(x)$ are known functions with $g(x) > c > 0$ $\forall x \in \mathcal{R}^1$ and some constant $c > 0$. The sgn(b) is known and $f(x)$ is bounded for bounded x. It is desired that x tracks the state x_m of the reference model given by

$$\dot{x}_m = -a_m x_m + b_m r$$

for any bounded reference input signal r.

Control Law If a, b were known, the control law

$$u = \frac{1}{g(x)} [k_1^* f(x) + k_2^* x + l^* r] \qquad (6.2.42)$$

with

$$k_1^* = -\frac{a}{b}, \quad k_2^* = -\frac{a_m}{b}, \quad l^* = \frac{b_m}{b}$$

could meet the control objective exactly. For the case of a, b unknown, we propose a control law of the same form as (6.2.42) but with adjustable gains, i.e., we use

$$u = \frac{1}{g(x)} [k_1(t) f(x) + k_2(t) x + l(t) r] \qquad (6.2.43)$$

where k_1, k_2, l, are the estimates of the unknown controller gains k_1^*, k_2^*, l^* respectively to be generated by an adaptive law.

Adaptive Law As in the previous examples, we first rewrite the plant equation in terms of the unknown controller gains k_1^*, k_2^*, l^*, i.e., substituting for $a = -bk_1^*$ and adding and subtracting the term $b\left(k_1^* f(x) + k_2^* x + l^* r\right)$ in (6.2.41) and using the equation $bl^* = b_m$, $bk_2^* = -a_m$, we obtain

$$\dot{x} = -a_m x + b_m r + b\left[-k_1^* f(x) - k_2^* x - l^* r + g(x) u\right]$$

If we let $e \triangleq x - x_m$, $\tilde{k}_1 \triangleq k_1 - k_1^*$, $\tilde{k}_2 \triangleq k_2 - k_2^*$, $\tilde{l} \triangleq l - l^*$ to be the tracking and parameter errors, we can show as before that the tracking error satisfies the differential equation

$$\dot{e} = -a_m e + b\left(\tilde{k}_1 f(x) + \tilde{k}_2 x + \tilde{l} r\right), \quad \epsilon_1 = e$$

6.2. SIMPLE DIRECT MRAC SCHEMES

which we can use as in Section 6.2.2 to develop the adaptive laws

$$\begin{aligned}
\dot{k}_1 &= -\gamma_1 e f(x)\,\text{sgn}(b) \\
\dot{k}_2 &= -\gamma_2 e x\,\text{sgn}(b) \\
\dot{l} &= -\gamma_3 e r\,\text{sgn}(b)
\end{aligned} \qquad (6.2.44)$$

where $\gamma_i > 0, i = 1, 2, 3$ are the adaptive gains.

Analysis We can establish that all signals in the closed-loop plant (6.2.41), (6.2.43), and (6.2.44) are bounded and that $|e(t)| = |x(t) - x_m(t)| \to 0$ as $t \to \infty$ by using the Lyapunov function

$$V(e, \tilde{k}_1, \tilde{k}_2, \tilde{l}) = \frac{e^2}{2} + \frac{\tilde{k}_1^2}{2\gamma_1}|b| + \frac{\tilde{k}_2^2}{2\gamma_2}|b| + \frac{\tilde{l}^2}{2\gamma_3}|b|$$

in a similar way as in Section 6.2.2. The choice of the control law to cancel the nonlinearities and force the plant to behave as an LTI system is quite obvious for the case of the plant (6.2.41). Similar techniques may be used to deal with some more complicated nonlinear problems where the choice of the control law in the known and unknown parameter case is less obvious [105].

Remark 6.2.3 The simple adaptive control schemes presented in this section have the following characteristics:

(i) The adaptive laws are developed using the SPR-Lyapunov design approach and are driven by the estimation error rather than the normalized estimation error. The estimation error is equal to the regulation or tracking error that is to be driven to zero as a part of the control objective.

(ii) The design of the adaptive law and the stability analysis of the closed-loop adaptive scheme is accomplished by using a single Lyapunov function.

(iii) The full state vector is available for measurement.

Another approach is to use the procedure of Chapter 4 and develop adaptive laws based on the SPR-Lyapunov method that are driven by normalized estimation errors. Such schemes, however, are not as easy to analyze as the schemes with unnormalized adaptive laws developed in this section. The reason is that the normalized estimation error is not simply related to the regulation or tracking error and additional stability arguments are needed to complete the analysis of the respective adaptive control scheme.

The distinction between adaptive schemes with normalized and unnormalized adaptive laws is made clear in this chapter by analyzing them in separate sections. Some of the advantages and disadvantages of normalized and unnormalized adaptive laws are discussed in the sections to follow.

The assumption of full state measurement in the above examples is relaxed in the following sections where we formulate and solve the general MRAC problem.

6.3 MRC for SISO Plants

In Section 6.2 we used several examples to illustrate the design and analysis of MRAC schemes for plants whose state vector is available for measurement. The design of MRAC schemes for plants whose output rather than the full state is available for measurement follows a similar procedure as that used in Section 6.2. This design procedure is based on combining a control law whose form is the same as the one we would use in the known parameter case with an adaptive law that provides on-line estimates for the controller parameters.

In the general case, the design of the control law is not as straightforward as it appears to be in the case of the examples of Section 6.2. Because of this reason, we use this section to formulate the MRC problem for a general class of LTI SISO plants and solve it for the case where the plant parameters are known exactly. The significance of the existence of a control law that solves the MRC problem is twofold: First it demonstrates that given a set of assumptions about the plant and reference model, there is enough structural flexibility to meet the control objective; second, it provides the form of the control law that is to be combined with an adaptive law to form MRAC schemes in the case of unknown plant parameters to be treated in the sections to follow.

6.3. MRC FOR SISO PLANTS

6.3.1 Problem Statement

Consider the SISO, LTI plant described by the vector differential equation

$$\begin{aligned} \dot{x}_p &= A_p x_p + B_p u_p, \quad x_p(0) = x_0 \\ y_p &= C_p^T x_p \end{aligned} \quad (6.3.1)$$

where $x_p \in \mathcal{R}^n; y_p, u_p \in \mathcal{R}^1$ and A_p, B_p, C_p have the appropriate dimensions. The transfer function of the plant is given by

$$y_p = G_p(s) u_p \quad (6.3.2)$$

with $G_p(s)$ expressed in the form

$$G_p(s) = k_p \frac{Z_p(s)}{R_p(s)} \quad (6.3.3)$$

where Z_p, R_p are monic polynomials and k_p is a constant referred to as the *high frequency gain*.

The reference model, selected by the designer to describe the desired characteristics of the plant, is described by the differential equation

$$\begin{aligned} \dot{x}_m &= A_m x_m + B_m r, \quad x_m(0) = x_{m0} \\ y_m &= C_m^T x_m \end{aligned} \quad (6.3.4)$$

where $x_m \in \mathcal{R}^{p_m}$ for some integer $p_m; y_m, r \in \mathcal{R}^1$ and r is the reference input which is assumed to be a uniformly bounded and piecewise continuous function of time. The transfer function of the reference model given by

$$y_m = W_m(s) r$$

is expressed in the same form as (6.3.3), i.e.,

$$W_m(s) = k_m \frac{Z_m(s)}{R_m(s)} \quad (6.3.5)$$

where $Z_m(s), R_m(s)$ are monic polynomials and k_m is a constant.

The MRC objective is to determine the plant input u_p so that all signals are bounded and the plant output y_p tracks the reference model output y_m as close as possible for *any* given reference input $r(t)$ of the class defined above. We refer to the problem of finding the desired u_p to meet the control objective as the *MRC problem*.

In order to meet the MRC objective with a control law that is implementable, i.e., a control law that is free of differentiators and uses only measurable signals, we assume that the plant and reference model satisfy the following assumptions:

Plant Assumptions

    P1. $Z_p(s)$ is a monic Hurwitz polynomial of degree m_p

    P2. An upper bound n of the degree n_p of $R_p(s)$

    P3. the relative degree $n^* = n_p - m_p$ of $G_p(s)$, and

    P4. the sign of the high frequency gain k_p are known

Reference Model Assumptions:

    M1. $Z_m(s), R_m(s)$ are monic Hurwitz polynomials of degree q_m, p_m, respectively, where $p_m \leq n$.

    M2. The relative degree $n_m^* = p_m - q_m$ of $W_m(s)$ is the same as that of $G_p(s)$, i.e., $n_m^* = n^*$.

Remark 6.3.1 Assumption **P1** requires the plant transfer function $G_p(s)$ to be minimum phase. We make no assumptions, however, about the location of the poles of $G_p(s)$, i.e., the plant is allowed to have unstable poles. We allow the plant to be uncontrollable or unobservable, i.e., we allow common zeros and poles in the plant transfer function. Because, by assumption **P1**, all the plant zeros are in $\mathcal{C}^-$, any zero-pole cancellation can only occur in $\mathcal{C}^-$, which implies that the plant (6.3.1) is both stabilizable and detectable.

The minimum phase assumption (**P1**) is a consequence of the control objective which is met by designing an MRC control law that cancels the zeros of the plant and replaces them with those of the reference model in an effort to force the closed-loop plant transfer function from r to y_p to be equal to $W_m(s)$. For stability, such cancellations should occur in $\mathcal{C}^-$ which implies that $Z_p(s)$ should satisfy assumption **P1**. As we will show in Section 6.7, assumptions **P3**, **P4** can be relaxed at the expense of more complex control laws.

6.3.2 MRC Schemes: Known Plant Parameters

In addition to assumptions **P1** to **P4** and **M1**, **M2**, let us also assume that the plant parameters, i.e., the coefficients of $G_p(s)$ are known exactly. Because the plant is LTI and known, the design of the MRC scheme is achieved using linear system theory.

The MRC objective is met if u_p is chosen so that the closed-loop transfer function from r to y_p has stable poles and is equal to $W_m(s)$, the transfer function of the reference model. Such a transfer function matching guarantees that for any reference input signal $r(t)$, the plant output y_p converges to y_m exponentially fast.

A trivial choice for u_p is the cascade open-loop control law

$$u_p = C(s)r, \quad C(s) = \frac{k_m}{k_p}\frac{Z_m(s)}{R_m(s)}\frac{R_p(s)}{Z_p(s)} \tag{6.3.6}$$

which leads to the closed-loop transfer function

$$\frac{y_p}{r} = \frac{k_m}{k_p}\frac{Z_m}{R_m}\frac{R_p}{Z_p}\frac{k_p Z_p}{R_p} = W_m(s) \tag{6.3.7}$$

This control law, however, is feasible only when $R_p(s)$ is Hurwitz. Otherwise, (6.3.7) may involve zero-pole cancellations outside $\mathcal{C}^-$, which will lead to unbounded internal states associated with non-zero initial conditions [95]. In addition, (6.3.6) suffers from the usual drawbacks of open loop control such as deterioration of performance due to small parameter changes and inexact zero-pole cancellations.

Instead of (6.3.6), let us consider the feedback control law

$$u_p = \theta_1^{*T}\frac{\alpha(s)}{\Lambda(s)}u_p + \theta_2^{*T}\frac{\alpha(s)}{\Lambda(s)}y_p + \theta_3^* y_p + c_0^* r \tag{6.3.8}$$

shown in Figure 6.4 where

$$\alpha(s) \stackrel{\Delta}{=} \alpha_{n-2}(s) = [s^{n-2}, s^{n-3}, \ldots, s, 1]^T \quad \text{for } n \geq 2$$
$$\alpha(s) \stackrel{\Delta}{=} 0 \quad \text{for } n = 1$$

$c_0^*, \theta_3^* \in \mathcal{R}^1; \theta_1^*, \theta_2^* \in \mathcal{R}^{n-1}$ are constant parameters to be designed and $\Lambda(s)$ is an arbitrary monic Hurwitz polynomial of degree $n-1$ that contains $Z_m(s)$ as a factor, i.e.,

$$\Lambda(s) = \Lambda_0(s) Z_m(s)$$

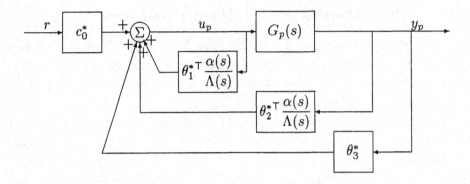

Figure 6.4 Structure of the MRC scheme (6.3.8).

which implies that $\Lambda_0(s)$ is monic, Hurwitz and of degree $n_0 = n - 1 - q_m$. The controller parameter vector

$$\theta^* = \left[\theta_1^{*T}, \theta_2^{*T}, \theta_3^*, c_0^*\right]^T \in \mathcal{R}^{2n}$$

is to be chosen so that the transfer function from r to y_p is equal to $W_m(s)$.

The I/O properties of the closed-loop plant shown in Figure 6.4 are described by the transfer function equation

$$y_p = G_c(s) r \qquad (6.3.9)$$

where

$$G_c(s) = \frac{c_0^* k_p Z_p \Lambda^2}{\Lambda \left[\left(\Lambda - \theta_1^{*T} \alpha(s) \right) R_p - k_p Z_p \left(\theta_2^{*T} \alpha(s) + \theta_3^* \Lambda \right) \right]} \qquad (6.3.10)$$

We can now meet the control objective if we select the controller parameters $\theta_1^*, \theta_2^*, \theta_3^*, c_0^*$ so that the closed-loop poles are stable and the closed-loop transfer function $G_c(s) = W_m(s)$, i.e.,

$$\frac{c_0^* k_p Z_p \Lambda^2}{\Lambda \left[\left(\Lambda - \theta_1^{*T} \alpha \right) R_p - k_p Z_p \left(\theta_2^{*T} \alpha + \theta_3^* \Lambda \right) \right]} = k_m \frac{Z_m}{R_m} \qquad (6.3.11)$$

is satisfied for all $s \in \mathcal{C}$. Because the degree of the denominator of $G_c(s)$ is $n_p + 2n - 2$ and that of $R_m(s)$ is $p_m \leq n$, for the matching equation (6.3.11) to hold, an additional $n_p + 2n - 2 - p_m$ zero-pole cancellations must occur in $G_c(s)$. Now because $Z_p(s)$ is Hurwitz by assumption and $\Lambda(s) = \Lambda_0(s) Z_m(s)$

6.3. MRC FOR SISO PLANTS

is designed to be Hurwitz, it follows that all the zeros of $G_c(s)$ are stable and therefore any zero-pole cancellation can only occur in $\mathcal{C}^-$. Choosing

$$c_0^* = \frac{k_m}{k_p} \tag{6.3.12}$$

and using $\Lambda(s) = \Lambda_0(s) Z_m(s)$ the matching equation (6.3.11) becomes

$$\left(\Lambda - \theta_1^{*T}\alpha\right) R_p - k_p Z_p \left(\theta_2^{*T}\alpha + \theta_3^*\Lambda\right) = Z_p \Lambda_0 R_m \tag{6.3.13}$$

or

$$\theta_1^{*T}\alpha(s) R_p(s) + k_p \left(\theta_2^{*T}\alpha(s) + \theta_3^*\Lambda(s)\right) Z_p(s) = \Lambda(s) R_p(s) - Z_p(s) \Lambda_0(s) R_m(s) \tag{6.3.14}$$

Equating the coefficients of the powers of s on both sides of (6.3.14), we can express (6.3.14) in terms of the algebraic equation

$$S\bar{\theta}^* = p \tag{6.3.15}$$

where $\bar{\theta}^* = \left[\theta_1^{*T}, \theta_2^{*T}, \theta_3^*\right]^T$, S is an $(n + n_p - 1) \times (2n - 1)$ matrix that depends on the coefficients of $R_p, k_p Z_p$ and Λ, and p is an $n + n_p - 1$ vector with the coefficients of $\Lambda R_p - Z_p \Lambda_0 R_m$. The existence of $\bar{\theta}^*$ to satisfy (6.3.15) and, therefore, (6.3.14) will very much depend on the properties of the matrix S. For example, if $n > n_p$, more than one $\bar{\theta}^*$ will satisfy (6.3.15), whereas if $n = n_p$ and S is nonsingular, (6.3.15) will have only one solution.

Remark 6.3.2 For the design of the control input (6.3.8), we assume that $n \geq n_p$. Because the plant is known exactly, there is no need to assume an upper bound for the degree of the plant, i.e., because n_p is known n can be set equal to n_p. We use $n \geq n_p$ on purpose in order to use the result in the unknown plant parameter case treated in Sections 6.4 and 6.5, where only the upper bound n for n_p is known.

Remark 6.3.3 Instead of using (6.3.15), one can solve (6.3.13) for $\theta_1^*, \theta_2^*, \theta_3^*$ as follows: Dividing both sides of (6.3.13) by $R_p(s)$, we obtain

$$\Lambda - \theta_1^{*T}\alpha - k_p \frac{Z_p}{R_p}(\theta_2^{*T}\alpha + \theta_3^*\Lambda) = Z_p \left(Q + k_p \frac{\Delta^*}{R_p}\right)$$

where $Q(s)$ (of degree $n - 1 - m_p$) is the quotient and $k_p\Delta^*$ (of degree at most $n_p - 1$) is the remainder of $\Lambda_0 R_m/R_p$, respectively. Then the solution for θ_i^*, $i = 1, 2, 3$ can be found by inspection, i.e.,

$$\theta_1^{*T}\alpha(s) = \Lambda(s) - Z_p(s)Q(s) \qquad (6.3.16)$$

$$\theta_2^{*T}\alpha(s) + \theta_3^*\Lambda(s) = \frac{Q(s)R_p(s) - \Lambda_0(s)R_m(s)}{k_p} \qquad (6.3.17)$$

where the equality in the second equation is obtained by substituting for $\Delta^*(s)$ using the identity $\frac{\Lambda_0 R_m}{R_p} = Q + \frac{k_p\Delta^*}{R_p}$. The parameters θ_i^*, $i = 1, 2, 3$ can now be obtained directly by equating the coefficients of the powers of s on both sides of (6.3.16), (6.3.17).

Equations (6.3.16) and (6.3.17) indicate that in general the controller parameters θ_i^*, $i = 1, 2, 3$ are nonlinear functions of the coefficients of the plant polynomials $Z_p(s), R_p(s)$ due to the dependence of $Q(s)$ on the coefficients of $R_p(s)$. When $n = n_p$ and $n^* = 1$, however, $Q(s) = 1$ and the θ_i^*'s are linear functions of the coefficients of $Z_p(s), R_p(s)$.

Lemma 6.3.1 *Let the degrees of $R_p, Z_p, \Lambda, \Lambda_0$ and R_m be as specified in (6.3.8). Then*
(i) *The solution $\bar{\theta}^*$ of (6.3.14) or (6.3.15) always exists.*
(ii) *In addition if R_p, Z_p are coprime and $n = n_p$, then the solution $\bar{\theta}^*$ is unique.*

Proof Let $R_p = \bar{R}_p(s)h(s)$ and $Z_p(s) = \bar{Z}_p(s)h(s)$ and $\bar{R}_p(s), \bar{Z}_p(s)$ be coprime, where $h(s)$ is a monic polynomial of degree r_0 (with $0 \leq r_0 \leq m_p$). Because $Z_p(s)$ is Hurwitz, it follows that $h(s)$ is also Hurwitz. If R_p, Z_p are coprime, $h(s) = 1$, i.e., $r_0 = 0$. If R_p, Z_p are not coprime, $r_0 \geq 1$ and $h(s)$ is their common factor. We can now write (6.3.14) as

$$\theta_1^{*T}\alpha\bar{R}_p + k_p(\theta_2^{*T}\alpha + \theta_3^*\Lambda)\bar{Z}_p = \Lambda\bar{R}_p - \bar{Z}_p\Lambda_0 R_m \qquad (6.3.18)$$

by canceling $h(s)$ from both sides of (6.3.14). Because $h(s)$ is Hurwitz, the cancellation occurs in $\mathcal{C}^-$. Equation (6.3.18) leads to $n_p + n - r_0 - 2$ algebraic equations with $2n - 1$ unknowns. It can be shown that the degree of $\Lambda\bar{R}_p - \bar{Z}_p\Lambda_0 R_m$ is $n_p + n - r_0 - 2$ because of the cancellation of the term $s^{n_p + n - r_0 - 1}$. Because $\bar{R}_p, \bar{Z}_p$ are coprime, it follows from Theorem 2.3.1 that there exists unique polynomials $a_0(s), b_0(s)$ of degree $n - 2, n_p - r_0 - 1$ respectively such that

$$a_0(s)\bar{R}_p(s) + b_0(s)\bar{Z}_p(s) = \Lambda(s)\bar{R}_p(s) - \bar{Z}_p(s)\Lambda_0(s)R_m(s) \qquad (6.3.19)$$

6.3. MRC FOR SISO PLANTS

is satisfied for $n \geq 2$. It now follows by inspection that

$$\theta_1^{*T}\alpha(s) = f(s)\bar{Z}_p(s) + a_0(s) \tag{6.3.20}$$

and

$$k_p(\theta_2^{*T}\alpha(s) + \theta_3^*\Lambda(s)) = -f(s)\bar{R}_p(s) + b_0(s) \tag{6.3.21}$$

satisfy (6.3.18), where $f(s)$ is any given polynomial of degree $n_f = n - n_p + r_0 - 1$. Hence, the solution $\theta_1^*, \theta_2^*, \theta_3^*$ of (6.3.18) can be obtained as follows: We first solve (6.3.19) for $a_0(s), b_0(s)$. We then choose an arbitrary polynomial $f(s)$ of degree $n_f = n - n_p + r_0 - 1$ and calculate $\theta_1^*, \theta_2^*, \theta_3^*$ from (6.3.20), (6.3.21) by equating coefficients of the powers of s. Because $f(s)$ is arbitrary, the solution $\theta_1^*, \theta_2^*, \theta_3^*$ is not unique. If, however, $n = n_p$ and $r_0 = 0$, i.e., R_p, Z_p are coprime, then $f(s) = 0$ and $\theta_1^{*T}\alpha(s) = a_0(s)$, $k_p(\theta_2^{*T}\alpha(s) + \theta_3^*\Lambda(s)) = b_0(s)$ which implies that the solution $\theta_1^*, \theta_2^*, \theta_3^*$ is unique due to the uniqueness of $a_0(s), b_0(s)$. If $n = n_p = 1$, then $\alpha(s) = 0$, $\Lambda(s) = 1$, $\theta_1^* = \theta_2^* = 0$ and θ_3^* given by (6.3.18) is unique. □

Remark 6.3.4 It is clear from (6.3.12), (6.3.13) that the control law (6.3.8) places the poles of the closed-loop plant at the roots of the polynomial $Z_p(s)\Lambda_0(s)R_m(s)$ and changes the high frequency gain from k_p to k_m by using the feedforward gain c_0^*. Therefore, the MRC scheme can be viewed as a special case of a general pole placement scheme where the desired closed-loop characteristic equation is given by

$$Z_p(s)\Lambda_0(s)R_m(s) = 0$$

The transfer function matching (6.3.11) is achieved by canceling the zeros of the plant, i.e., $Z_p(s)$, and replacing them by those of the reference model, i.e., by designing $\Lambda = \Lambda_0 Z_m$. Such a cancellation is made possible by assuming that $Z_p(s)$ is Hurwitz and by designing Λ_0, Z_m to have stable zeros.

We have shown that the control law (6.3.8) guarantees that the closed-loop transfer function $G_c(s)$ of the plant from r to y_p has all its poles in $\mathcal{C}^-$ and in addition, $G_c(s) = W_m(s)$. In our analysis we assumed zero initial conditions for the plant, reference model and filters. The transfer function matching, i.e., $G_c(s) = W_m(s)$, together with zero initial conditions guarantee that $y_p(t) = y_m(t), \forall t \geq 0$ and for any reference input $r(t)$ that is bounded and piecewise continuous. The assumption of zero initial conditions is common in most I/O control design approaches for LTI systems and is valid provided that any zero-pole cancellation in the closed-loop plant transfer function occurs in $\mathcal{C}^-$. Otherwise nonzero initial conditions may lead to unbounded internal states that correspond to zero-pole cancellations in $\mathcal{C}^+$.

In our design we make sure that all cancellations in $G_c(s)$ occur in $\mathcal{C}^-$ by assuming stable zeros for the plant transfer function and by using stable filters in the control law. Nonzero initial conditions, however, will affect the transient response of $y_p(t)$. As a result we can no longer guarantee that $y_p(t) = y_m(t)$ $\forall t \geq 0$ but instead that $y_p(t) \to y_m(t)$ exponentially fast with a rate that depends on the closed-loop dynamics. We analyze the effect of initial conditions by using state space representations for the plant, reference model, and controller as follows: We begin with the following state-space realization of the control law (6.3.8):

$$
\begin{aligned}
\dot{\omega}_1 &= F\omega_1 + g u_p, & \omega_1(0) &= 0 \\
\dot{\omega}_2 &= F\omega_2 + g y_p, & \omega_2(0) &= 0 \\
u_p &= \theta^{*T}\omega
\end{aligned}
\qquad (6.3.22)
$$

where $\omega_1, \omega_2 \in \mathcal{R}^{n-1}$,

$$\theta^* = \left[\theta_1^{*T}, \theta_2^{*T}, \theta_3^*, c_0^*\right]^T, \quad \omega = [\omega_1^T, \omega_2^T, y_p, r]^T$$

$$
F = \begin{bmatrix}
-\lambda_{n-2} & -\lambda_{n-3} & -\lambda_{n-4} & \cdots & -\lambda_0 \\
1 & 0 & 0 & \cdots & 0 \\
0 & 1 & 0 & \cdots & 0 \\
\vdots & \vdots & \ddots & \ddots & \vdots \\
0 & 0 & \cdots & 1 & 0
\end{bmatrix}, \quad g = \begin{bmatrix} 1 \\ 0 \\ \vdots \\ 0 \end{bmatrix} \qquad (6.3.23)
$$

λ_i are the coefficients of

$$\Lambda(s) = s^{n-1} + \lambda_{n-2}s^{n-2} + \ldots + \lambda_1 s + \lambda_0 = \det(sI - F)$$

and (F, g) is the state space realization of $\frac{\alpha(s)}{\Lambda(s)}$, i.e., $(sI - F)^{-1}g = \frac{\alpha(s)}{\Lambda(s)}$. The block diagram of the closed-loop plant with the control law (6.3.22) is shown in Figure 6.5.

6.3. MRC FOR SISO PLANTS

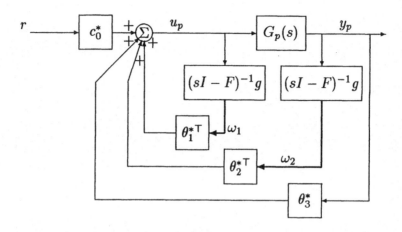

Figure 6.5 Block diagram of the MRC scheme (6.3.22).

We obtain the state-space representation of the overall closed-loop plant by augmenting the state x_p of the plant (6.3.1) with the states ω_1, ω_2 of the controller (6.3.22), i.e.,

$$\dot{Y}_c = A_c Y_c + B_c c_0^* r, \quad Y_c(0) = Y_0$$
$$y_p = C_c^T Y_c \qquad (6.3.24)$$

where

$$Y_c = \left[x_p^T, \omega_1^T, \omega_2^T\right]^T \in \mathcal{R}^{n_p + 2n - 2}$$

$$A_c = \begin{bmatrix} A_p + B_p \theta_3^* C_p^T & B_p \theta_1^{*T} & B_p \theta_2^{*T} \\ g\theta_3^* C_p^T & F + g\theta_1^{*T} & g\theta_2^{*T} \\ gC_p^T & 0 & F \end{bmatrix}, \quad B_c = \begin{bmatrix} B_p \\ g \\ 0 \end{bmatrix} \qquad (6.3.25)$$

$$C_c^T = \left[C_p^T, 0, 0\right]$$

and Y_0 is the vector with initial conditions. We have already established that the transfer function from r to y_p is given by

$$\frac{y_p(s)}{r(s)} = \frac{c_0^* k_p Z_p \Lambda^2}{\Lambda\left[(\Lambda - \theta_1^{*T}\alpha) R_p - k_p Z_p \left(\theta_2^{*T}\alpha + \theta_3^*\Lambda\right)\right]} = W_m(s)$$

which implies that

$$C_c^T (sI - A_c)^{-1} B_c c_0^* = \frac{c_0^* k_p Z_p \Lambda^2}{\Lambda\left[(\Lambda - \theta_1^{*T}\alpha) R_p - k_p Z_p \left(\theta_2^{*T}\alpha + \theta_3^*\Lambda\right)\right]} = W_m(s)$$

and, therefore,

$$\det(sI - A_c) = \Lambda\left[\left(\Lambda - \theta_1^{*T}\alpha\right)R_p - k_p Z_p\left(\theta_2^{*T}\alpha + \theta_3^*\Lambda\right)\right] = \Lambda Z_p \Lambda_0 R_m$$

where the last equality is obtained by using the matching equation (6.3.13). It is clear that the eigenvalues of A_c are equal to the roots of the polynomials Λ, Z_p and R_m; therefore, A_c is a stable matrix. The stability of A_c and the boundedness of r imply that the state vector Y_c in (6.3.24) is bounded.

Since $C_c^T(sI - A_c)^{-1}B_c c_0^* = W_m(s)$, the reference model may be realized by the triple $(A_c, B_c c_0^*, C_c)$ and described by the nonminimal state space representation

$$\begin{aligned} \dot{Y}_m &= A_c Y_m + B_c c_0^* r, \quad Y_m(0) = Y_{m0} \\ y_m &= C_c^T Y_m \end{aligned} \quad (6.3.26)$$

where $Y_m \in \mathcal{R}^{n_p+2n-2}$.

Letting $e = Y_c - Y_m$ to be the state error and $e_1 = y_p - y_m$ the output tracking error, it follows from (6.3.24) and (6.3.26) that

$$\dot{e} = A_c e, \quad e_1 = C_c^T e$$

i.e., the tracking error e_1 satisfies

$$e_1 = C_c^T e^{A_c t}(Y_c(0) - Y_m(0))$$

Because A_c is a stable matrix, $e_1(t)$ converges exponentially to zero. The rate of convergence depends on the location of the eigenvalues of A_c, which are equal to the roots of $\Lambda(s)\Lambda_0(s)R_m(s)Z_p(s) = 0$. We can affect the rate of convergence by designing $\Lambda(s)\Lambda_0(s)R_m(s)$ to have fast zeros, but we are limited by the dependence of A_c on the zeros of $Z_p(s)$, which are fixed by the given plant.

Example 6.3.1 Let us consider the second order plant

$$y_p = \frac{-2(s+5)}{s^2 - 2s + 1}u_p = \frac{-2(s+5)}{(s-1)^2}u_p$$

and the reference model

$$y_m = \frac{3}{s+3}r$$

6.3. MRC FOR SISO PLANTS

The order of the plant is $n_p = 2$. Its relative degree $n^* = 1$ is equal to that of the reference model.

We choose the polynomial $\Lambda(s)$ as

$$\Lambda(s) = s + 1 = \Lambda_0(s)$$

and the control input

$$u_p = \theta_1^* \frac{1}{s+1} u_p + \theta_2^* \frac{1}{s+1} y_p + \theta_3^* y_p + c_0^* r$$

which gives the closed-loop transfer function

$$\frac{y_p}{r} = \frac{-2c_0^*(s+5)(s+1)}{(s+1-\theta_1^*)(s-1)^2 + 2(s+5)(\theta_2^* + \theta_3^*(s+1))} = G_c(s)$$

Forcing $G_c(s) = 3/(s+3)$, we have $c_0^* = -3/2$ and the matching equation (6.3.14) becomes

$$\theta_1^*(s-1)^2 - 2(\theta_2^* + \theta_3^*(s+1))(s+5) = (s+1)(s-1)^2 - (s+5)(s+1)(s+3)$$

i.e.,

$$(\theta_1^* - 2\theta_3^*)s^2 + (-2\theta_1^* - 2\theta_2^* - 12\theta_3^*)s + \theta_1^* - 10(\theta_2^* + \theta_3^*) = -10s^2 - 24s - 14$$

Equating the powers of s we have

$$\theta_1^* - 2\theta_3^* = -10$$
$$\theta_1^* + \theta_2^* + 6\theta_3^* = 12$$
$$\theta_1^* - 10\theta_2^* - 10\theta_3^* = -14$$

i.e.,

$$\begin{bmatrix} 1 & 0 & -2 \\ 1 & 1 & 6 \\ 1 & -10 & -10 \end{bmatrix} \begin{bmatrix} \theta_1^* \\ \theta_2^* \\ \theta_3^* \end{bmatrix} = \begin{bmatrix} -10 \\ 12 \\ -14 \end{bmatrix}$$

which gives

$$\begin{bmatrix} \theta_1^* \\ \theta_2^* \\ \theta_3^* \end{bmatrix} = \begin{bmatrix} -4 \\ -2 \\ 3 \end{bmatrix}$$

The control input is therefore given by

$$u_p = -4 \frac{1}{s+1} u_p - 2 \frac{1}{s+1} y_p + 3 y_p - 1.5 r$$

and is implemented as follows:

$$\dot{\omega}_1 = -\omega_1 + u_p, \quad \dot{\omega}_2 = -\omega_2 + y_p$$
$$u_p = -4\omega_1 - 2\omega_2 + 3y_p - 1.5r$$

▽

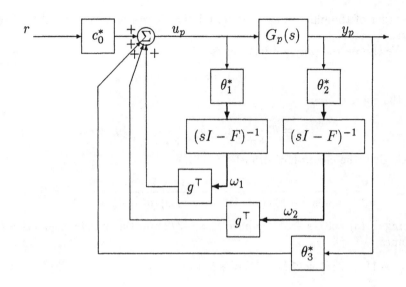

Figure 6.6 Implementation of the MRC scheme (6.3.28).

Remark 6.3.5 Because θ^* is a constant vector, the control law (6.3.8) may be also written as

$$u_p = \frac{\alpha^\top(s)}{\Lambda(s)}(\theta_1^* u_p) + \frac{\alpha^\top(s)}{\Lambda(s)}(\theta_2^* y_p) + \theta_3^* y_p + c_0^* r \qquad (6.3.27)$$

and implemented as

$$\begin{aligned}
\dot{\omega}_1 &= F^\top \omega_1 + \theta_1^* u_p \\
\dot{\omega}_2 &= F^\top \omega_2 + \theta_2^* y_p \\
u_p &= g^\top \omega_1 + g^\top \omega_2 + \theta_3^* y_p + c_0^* r
\end{aligned} \qquad (6.3.28)$$

where $\omega_1, \omega_2 \in \mathcal{R}^{n-1}$; F, g are as defined in (6.3.23), i.e., $g^\top(sI - F^\top)^{-1} = g^\top((sI - F)^{-1})^\top = \frac{\alpha^\top(s)}{\Lambda(s)}$. The block diagram for implementing (6.3.28) is shown in Figure 6.6.

Remark 6.3.6 The structure of the feedback control law (6.3.8) is not unique. For example, instead of the control law (6.3.8) we can also use

$$u_p = \theta_1^{*\top} \frac{\alpha(s)}{\Lambda_1(s)} u_p + \theta_2^{*\top} \frac{\alpha(s)}{\Lambda_1(s)} y_p + c_0^* r \qquad (6.3.29)$$

where $\Lambda_1(s) = \Lambda_0(s)Z_m(s)$ has degree n, $\alpha(s) = \alpha_{n-1}(s) = [s^{n-1}, s^{n-2}, \ldots, s, 1]^T$ and $\theta_1^*, \theta_2^* \in \mathcal{R}^n$. The overall desired controller parameter $\theta^* = \left[\theta_1^{*T}, \theta_2^{*T}, c_0^*\right]^T \in \mathcal{R}^{2n+1}$ has one dimension more than that in (6.3.8). The analysis of (6.3.29) is very similar to that of (6.3.8) and is left as an exercise (see Problem 6.7) for the reader.

Remark 6.3.7 We can express the control law (6.3.8) in the general feedback form shown in Figure 6.7 with a feedforward block

$$C(s) = \frac{c_0^* \Lambda(s)}{\Lambda(s) - \theta_1^{*T}\alpha(s)}$$

and a feedback block

$$F(s) = -\frac{\theta_2^{*T}\alpha(s) + \theta_3^*\Lambda(s)}{\Lambda(s)c_0^*}$$

where $c_0^*, \theta_1^*, \theta_2^*, \theta_3^*$ are chosen to satisfy the matching equation (6.3.12), (6.3.13). The general structure of Figure 6.7 allows us to analyze and study properties such as robustness, disturbance rejection, etc., of the MRC scheme using well established results from linear system theory [57, 95].

We have shown that the MRC law (6.3.8), whose parameters $\theta_i^*, i = 1, 2, 3$ and c_0^* are calculated using the matching equations (6.3.12), (6.3.13) meets the MRC objective. The solution of the matching equations for θ_i^* and c_0^* requires the knowledge of the coefficients of the plant polynomials $k_p Z_p(s)$, $R_p(s)$. In the following sections we combine the MRC law (6.3.8) with an adaptive law that generates estimates for θ_i^*, c_0^* on-line to deal with the case of unknown plant parameters.

6.4 Direct MRAC with Unnormalized Adaptive Laws

The main characteristics of the simple MRAC schemes developed in Section 6.2 are

(i) The adaptive laws are driven by the estimation error which, due to the special form of the control law, is equal to the regulation or tracking

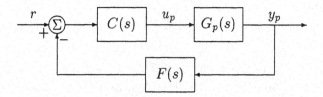

Figure 6.7 MRC in the general feedback block diagram where $C(s) = \frac{c_0^* \Lambda(s)}{\Lambda(s) - \theta_1^{*T} \alpha(s)}$, $F(s) = -\frac{\theta_2^{*T} \alpha(s) + \theta_3^* \Lambda(s)}{c_0^* \Lambda(s)}$.

error. They are derived using the SPR-Lyapunov design approach without the use of normalization

(ii) A simple Lyapunov function is used to design the adaptive law and establish boundedness for all signals in the closed-loop plant.

The extension of the results of Sections 6.2, 6.3 to the SISO plant (6.3.1) with unknown parameters became an active area of research in the 70's. In 1974, Monopoli [150] introduced the concept of the augmented error, a form of an estimation error without normalization, that he used to develop stable MRAC schemes for plants with relative degree 1 and 2, but not for plants with higher relative degree. Following the work of Monopoli, Feuer and Morse [54] designed and analyzed MRAC schemes with unnormalized adaptive laws that are applicable to plants with known relative degree of arbitrary positive value. This generalization came at the expense of additional complexity in the structure of MRAC schemes for plants with relative degree higher than 2. The complexity of the Feuer and Morse MRAC schemes relative to the ones using adaptive laws with normalized estimation errors, introduced during the same period [48, 72, 174], was responsible for their lack of popularity within the adaptive control research community. As a result, it was not until the early 1990s that the MRAC schemes designed for plants with relative degree higher than 2 and employing unnormalized adaptive laws were revived again and shown to offer some advantages over the MRAC with normalization when applied to certain classes of nonlinear plants [98, 116, 117, 118, 162].

In this section, we follow an approach very similar to that of Feuer and Morse and extend the results of Section 6.2 to the general case of higher

6.4. DIRECT MRAC WITH UNNORMALIZED ADAPTIVE LAWS

order SISO plants. The complexity of the schemes increases with the relative degree n^* of the plant. The simplest cases are the ones where $n^* = 1$ and 2. Because of their simplicity, they are still quite popular in the literature of continuous-time MRAC and are presented in separate sections.

6.4.1 Relative Degree $n^* = 1$

Let us assume that the relative degree of the plant

$$y_p = G_p(s)u_p = k_p \frac{Z_p(s)}{R_p(s)} u_p \qquad (6.4.1)$$

is $n^* = 1$. The reference model

$$y_m = W_m(s)r$$

is chosen to have the same relative degree and both $G_p(s), W_m(s)$ satisfy assumptions P1 to P4, and M1 and M2, respectively. In addition $W_m(s)$ is designed to be SPR.

The design of the MRAC law to meet the control objective defined in Section 6.3.1 proceeds as follows:

We have shown in Section 6.3.2 that the control law

$$\begin{aligned}
\dot{\omega}_1 &= F\omega_1 + g u_p, \quad \omega_1(0) = 0 \\
\dot{\omega}_2 &= F\omega_2 + g y_p, \quad \omega_2(0) = 0 \\
u_p &= \theta^{*T}\omega
\end{aligned} \qquad (6.4.2)$$

where $\omega = \left[\omega_1^T, \omega_2^T, y_p, r\right]^T$, and $\theta^* = \left[\theta_1^{*T}, \theta_2^{*T}, \theta_3^*, c_0^*\right]^T$ calculated from the matching equation (6.3.12) and (6.3.13) meets the MRC objective defined in Section 6.3.1. Because the parameters of the plant are unknown, the desired controller parameter vector θ^* cannot be calculated from the matching equation and therefore (6.4.2) cannot be implemented. A reasonable approach to follow in the unknown plant parameter case is to replace (6.4.2) with the control law

$$\begin{aligned}
\dot{\omega}_1 &= F\omega_1 + g u_p, \quad \omega_1(0) = 0 \\
\dot{\omega}_2 &= F\omega_2 + g y_p, \quad \omega_2(0) = 0 \\
u_p &= \theta^T\omega
\end{aligned} \qquad (6.4.3)$$

where $\theta(t)$ is the estimate of θ^* at time t to be generated by an appropriate adaptive law. We derive such an adaptive law by following a similar procedure as in the case of the examples of Section 6.2. We first obtain a composite state space representation of the plant and controller, i.e.,

$$\begin{aligned}\dot{Y}_c &= A_0 Y_c + B_c u_p, \quad Y_c(0) = Y_0 \\ y_p &= C_c^T Y_c \\ u_p &= \theta^T \omega\end{aligned}$$

where $Y_c = \left[x_p^T, \omega_1^T, \omega_2^T\right]^T$.

$$A_0 = \begin{bmatrix} A_p & 0 & 0 \\ 0 & F & 0 \\ gC_p^T & 0 & F \end{bmatrix}, \quad B_c = \begin{bmatrix} B_p \\ g \\ 0 \end{bmatrix}$$

$$C_c^T = \begin{bmatrix} C_p^T, 0, 0 \end{bmatrix}$$

and then add and subtract the desired input $B_c \theta^{*T} \omega$ to obtain

$$\dot{Y}_c = A_0 Y_c + B_c \theta^{*T} \omega + B_c \left(u_p - \theta^{*T} \omega\right)$$

If we now absorb the term $B_c \theta^{*T} \omega$ into the homogeneous part of the above equation, we end up with the representation

$$\begin{aligned}\dot{Y}_c &= A_c Y_c + B_c c_0^* r + B_c \left(u_p - \theta^{*T} \omega\right), \quad Y_c(0) = Y_0 \\ y_p &= C_c^T Y_c \end{aligned} \qquad (6.4.4)$$

where A_c is as defined in (6.3.25). Equation (6.4.4) is the same as the closed-loop equation (6.3.24) in the known parameter case except for the additional input term $B_c(u_p - \theta^{*T}\omega)$ that depends on the choice of the input u_p. It serves as the parameterization of the plant equation in terms of the desired controller parameter vector θ^*. Let $e = Y_c - Y_m$ and $e_1 = y_p - y_m$ where Y_m is the state of the nonminimal representation of the reference model given by (6.3.26), we obtain the error equation

$$\begin{aligned}\dot{e} &= A_c e + B_c(u_p - \theta^{*T}\omega), \quad e(0) = e_0 \\ e_1 &= C_c^T e \end{aligned} \qquad (6.4.5)$$

6.4. DIRECT MRAC WITH UNNORMALIZED ADAPTIVE LAWS

Because
$$C_c^T(sI - A_c)^{-1} B_c c_0^* = W_m(s)$$

we have
$$e_1 = W_m(s)\rho^* \left(u_p - \theta^{*T}\omega\right) \quad (6.4.6)$$

where $\rho^* = \frac{1}{c_0^*}$, which is in the form of the bilinear parametric model analyzed in Chapter 4. We can now use (6.4.6) to generate a wide class of adaptive laws for estimating θ^* by using the results of Chapter 4. We should note that (6.4.5) and (6.4.6) hold for any relative degree and will also be used in later sections.

The estimate $\hat{e}_1(t)$ of $e_1(t)$ based on $\theta(t)$, the estimate of θ^* at time t, is given by
$$\hat{e}_1 = W_m(s)\rho \left(u_p - \theta^T\omega\right) \quad (6.4.7)$$

where ρ is the estimate of ρ^*. Because the control input is given by
$$u_p = \theta^T(t)\omega$$

it follows that $\hat{e}_1 = W_m(s)[0]$; therefore, the estimation error ϵ_1 defined in Chapter 4 as $\epsilon_1 = e_1 - \hat{e}_1$ may be taken to be equal to e_1, i.e., $\epsilon_1 = e_1$. Consequently, (6.4.7) is not needed and the estimate ρ of ρ^* does not have to be generated. Substituting for the control law in (6.4.5), we obtain the error equation

$$\begin{aligned}\dot{e} &= A_c e + \bar{B}_c \rho^* \tilde{\theta}^T \omega, \quad e(0) = e_0 \\ e_1 &= C_c^T e \end{aligned} \quad (6.4.8)$$

where
$$\bar{B}_c = B_c c_0^*$$

or
$$e_1 = W_m(s)\rho^* \tilde{\theta}^T \omega$$

which relates the parameter error $\tilde{\theta} \triangleq \theta(t) - \theta^*$ with the tracking error e_1. Because $W_m(s) = C_c^T(sI - A_c)^{-1} B_c c_0^*$ is SPR and A_c is stable, equation (6.4.8) is in the appropriate form for applying the SPR-Lyapunov design approach.

We therefore proceed by proposing the Lyapunov-like function

$$V\left(\tilde{\theta}, e\right) = \frac{e^\top P_c e}{2} + \frac{\tilde{\theta}^\top \Gamma^{-1} \tilde{\theta}}{2}|\rho^*| \tag{6.4.9}$$

where $\Gamma = \Gamma^\top > 0$ and $P_c = P_c^\top > 0$ satisfies the algebraic equations

$$P_c A_c + A_c^\top P_c = -qq^\top - \nu_c L_c$$
$$P_c \bar{B}_c = C_c$$

where q is a vector, $L_c = L_c^\top > 0$ and $\nu_c > 0$ is a small constant, that are implied by the MKY lemma. The time derivative $\dot{V}$ of V along the solution of (6.4.8) is given by

$$\dot{V} = -\frac{e^\top qq^\top e}{2} - \frac{\nu_c}{2} e^\top L_c e + e^\top P_c \bar{B}_c \rho^* \tilde{\theta}^\top \omega + \tilde{\theta}^\top \Gamma^{-1} \dot{\tilde{\theta}} |\rho^*|$$

Because $e^\top P_c \bar{B}_c = e_1$ and $\rho^* = |\rho^*|\mathrm{sgn}(\rho^*)$, we can make $\dot{V} \leq 0$ by choosing

$$\dot{\tilde{\theta}} = \dot{\theta} = -\Gamma e_1 \omega \, \mathrm{sgn}(\rho^*) \tag{6.4.10}$$

which leads to

$$\dot{V} = -\frac{e^\top qq^\top e}{2} - \frac{\nu_c}{2} e^\top L_c e \tag{6.4.11}$$

Equations (6.4.9) and (6.4.11) imply that V and, therefore, $e, \tilde{\theta} \in \mathcal{L}_\infty$.

Because $e = Y_c - Y_m$ and $Y_m \in \mathcal{L}_\infty$, we have $Y_c \in \mathcal{L}_\infty$, which implies that $y_p, \omega_1, \omega_2 \in \mathcal{L}_\infty$. Because $u_p = \theta^\top \omega$ and $\theta, \omega \in \mathcal{L}_\infty$ we also have $u_p \in \mathcal{L}_\infty$. Therefore all the signals in the closed-loop plant are bounded. It remains to show that the tracking error $e_1 = y_p - y_m$ goes to zero as $t \to \infty$.

From (6.4.9) and (6.4.11) we establish that e and therefore $e_1 \in \mathcal{L}_2$. Furthermore, using $\theta, \omega, e \in \mathcal{L}_\infty$ in (6.4.8) we have that $\dot{e}, \dot{e}_1 \in \mathcal{L}_\infty$. Hence, $e_1, \dot{e}_1 \in \mathcal{L}_\infty$ and $e_1 \in \mathcal{L}_2$, which, by Lemma 3.2.5, imply that $e_1(t) \to 0$ as $t \to \infty$.

We summarize the main equations of the MRAC scheme in Table 6.1.

The stability properties of the MRAC scheme of Table 6.1 are given by the following theorem.

Theorem 6.4.1 *The MRAC scheme summarized in Table 6.1 guarantees that:*

6.4. DIRECT MRAC WITH UNNORMALIZED ADAPTIVE LAWS

Table 6.1 MRAC scheme: $n^* = 1$

Plant	$y_p = k_p \frac{Z_p(s)}{R_p(s)} u_p, \quad n^* = 1$
Reference model	$y_m = W_m(s)r, \quad W_m(s) = k_m \frac{Z_m(s)}{R_m(s)}$
Control law	$\dot{\omega}_1 = F\omega_1 + gu_p, \quad \omega_1(0) = 0$ $\dot{\omega}_2 = F\omega_2 + gy_p, \quad \omega_2(0) = 0$ $u_p = \theta^\top \omega$ $\omega = [\omega_1^\top, \omega_2^\top, y_p, r]^\top, \omega_1 \in \mathcal{R}^{n-1}, \omega_2 \in \mathcal{R}^{n-1}$
Adaptive law	$\dot{\theta} = -\Gamma e_1 \omega \, \text{sgn}(\rho^*)$ $e_1 = y_p - y_m, \quad \text{sgn}(\rho^*) = \text{sgn}(k_p/k_m)$
Assumptions	Z_p, R_p and $W_m(s)$ satisfy assumptions P1 to P4, and M1 and M2, respectively; $W_m(s)$ is SPR; $(sI - F)^{-1}g = \frac{\alpha(s)}{\Lambda(s)}, \alpha(s) = [s^{n-2}, s^{n-3}, \ldots s, 1]^\top$, where $\Lambda = \Lambda_0 Z_m$ is Hurwitz, and $\Lambda_0(s)$ is of degree $n-1-q_m$, q_m is the degree of $Z_m(s)$; $\Gamma = \Gamma^\top > 0$ is arbitrary

(i) *All signals in the closed-loop plant are bounded and the tracking error e_1 converges to zero asymptotically with time for any reference input $r \in \mathcal{L}_\infty$.*

(ii) *If r is sufficiently rich of order $2n$, $\dot{r} \in \mathcal{L}_\infty$ and $Z_p(s), R_p(s)$ are relatively coprime, then the parameter error $|\tilde{\theta}| = |\theta - \theta^*|$ and the tracking error e_1 converge to zero exponentially fast.*

Proof (i) This part has already been completed above.

(ii) Equations (6.4.8) and (6.4.10) have the same form as (4.3.30) and (4.3.35) with $n_s^2 = 0$ in Chapter 4 whose convergence properties are established by Corollary

4.3.1. Therefore, by using the same steps as in the proof of Corollary 4.3.1 we can establish that if $\omega, \dot{\omega} \in \mathcal{L}_\infty$ and ω is PE then $\tilde{\theta}(t) \to 0$ exponentially fast. If $\dot{r} \in \mathcal{L}_\infty$ then it follows from the results of part (i) that $\dot{\omega} \in \mathcal{L}_\infty$. For the proof to be complete, it remains to show that ω is PE.

We express ω as

$$\omega = \begin{bmatrix} (sI - F)^{-1} g G_p^{-1}(s) y_p \\ (sI - F)^{-1} g y_p \\ y_p \\ r \end{bmatrix} \tag{6.4.12}$$

Because $y_p = y_m + e_1 = W_m(s)r + e_1$ we have

$$\omega = \omega_m + \bar{\omega} \tag{6.4.13}$$

where

$$\omega_m = H(s) r, \quad \bar{\omega} = H_0(s) e_1$$

and

$$H(s) = \begin{bmatrix} (sI - F)^{-1} g G_p^{-1}(s) W_m(s) \\ (sI - F)^{-1} g W_m(s) \\ W_m(s) \\ 1 \end{bmatrix}, \quad H_0(s) = \begin{bmatrix} (sI - F)^{-1} g G_p^{-1}(s) \\ (sI - F)^{-1} g \\ 1 \\ 0 \end{bmatrix}$$

The vector $\bar{\omega}$ is the output of a proper transfer matrix whose poles are stable and whose input $e_1 \in \mathcal{L}_2 \cap \mathcal{L}_\infty$ and goes to zero as $t \to \infty$. Hence, from Corollary 3.3.1 we have $\bar{\omega} \in \mathcal{L}_2 \cap \mathcal{L}_\infty$ and $|\bar{\omega}(t)| \to 0$ as $t \to \infty$. It then follows from Lemma 4.8.3 that ω is PE if ω_m is PE.

It remains to show that ω_m is PE when r is sufficiently rich of order $2n$.

Because r is sufficiently rich of order $2n$, according to Theorem 5.2.1, we can show that ω_m is PE by proving that $H(j\omega_1), H(j\omega_2), \ldots, H(j\omega_{2n})$ are linearly independent on $\mathcal{C}^{2n}$ for any $\omega_1, \omega_2, \ldots, \omega_{2n} \in \mathcal{R}$ with $\omega_i \neq \omega_j$ for $i \neq j$.

From the definition of $H(s)$, we can write

$$H(s) = \frac{1}{k_p Z_p(s) \Lambda(s) R_m(s)} \begin{bmatrix} \alpha(s) R_p(s) k_m Z_m(s) \\ \alpha(s) k_p Z_p(s) k_m Z_m(s) \\ \Lambda(s) k_p Z_p(s) k_m Z_m(s) \\ \Lambda(s) k_p Z_p(s) R_m(s) \end{bmatrix} \triangleq \frac{1}{k_p Z_p(s) \Lambda(s) R_m(s)} H_1(s) \tag{6.4.14}$$

Because all the elements of $H_1(s)$ are polynomials of s with order less than or equal to that of $\Lambda(s) Z_p(s) R_m(s)$, we can write

$$H_1(s) = \bar{H} \begin{bmatrix} s^l \\ s^{l-1} \\ \vdots \\ 1 \end{bmatrix} \tag{6.4.15}$$

6.4. DIRECT MRAC WITH UNNORMALIZED ADAPTIVE LAWS

where $l \triangleq 2n - 1 + q_m$ is the order of the polynomial $\Lambda(s)Z_p(s)R_m(s)$, q_m is the degree of $Z_m(s)$ and $\bar{H} \in \mathcal{R}^{2n \times (l+1)}$ is a constant matrix.

We now prove by contradiction that $\bar{H}$ in (6.4.15) is of full rank, i.e., $rank(\bar{H}) = 2n$. Suppose $rank(\bar{H}) < 2n$, i.e., there exists a constant vector $C \in \mathcal{R}^{2n}$ with $C \neq 0$ such that
$$C^\top \bar{H} = 0$$
or equivalently
$$C^\top H_1(s) = 0 \qquad (6.4.16)$$
for all $s \in \mathcal{C}$. Let $C = [C_1^\top, C_2^\top, c_3, c_4]^\top$, where $C_1, C_2 \in \mathcal{R}^{n-1}, c_3, c_4 \in \mathcal{R}^1$, then (6.4.16) can be written as

$$\begin{aligned}
C_1^\top \alpha(s) R_p(s) k_m Z_m(s) &+ C_2^\top \alpha(s) k_p Z_p(s) k_m Z_m(s) \\
+ c_3 \Lambda(s) k_p Z_p(s) k_m Z_m(s) &+ c_4 \Lambda(s) R_m(s) k_p Z_p(s) = 0
\end{aligned} \qquad (6.4.17)$$

Because the leading coefficient of the polynomial on the left hand side is c_4, for (6.4.17) to hold, it is necessary that $c_4 = 0$. Therefore,

$$[C_1^\top \alpha(s) R_p(s) + C_2^\top \alpha(s) k_p Z_p(s) + c_3 \Lambda(s) k_p Z_p(s)] k_m Z_m(s) = 0$$

or equivalently
$$C_1^\top \alpha(s) R_p(s) + C_2^\top \alpha(s) k_p Z_p(s) + c_3 \Lambda(s) k_p Z_p(s) = 0 \qquad (6.4.18)$$

Equation (6.4.18) implies that
$$k_p \frac{Z_p(s)}{R_p(s)} = -\frac{C_1^\top \alpha(s)}{c_3 \Lambda(s) + C_2^\top \alpha(s)} \qquad (6.4.19)$$

Noting that $c_3 \Lambda(s) + C_2^\top \alpha(s)$ is of order at most equal to $n - 1$, (6.4.19) contradicts our assumption that $Z_p(s), R_p(s)$ are coprime. Therefore $\bar{H}$ must be of full rank.

Now consider the $2n \times 2n$ matrix $L(\omega_1, \ldots, \omega_{2n}) \triangleq [H(j\omega_1), \ldots, H(j\omega_{2n})]$. Using (6.4.14) and (6.4.15), we can express $L(\omega_1, \omega_2, \ldots, \omega_{2n})$ as

$$L(\omega_1, \ldots, \omega_{2n}) = \bar{H} \begin{bmatrix} (j\omega_1)^l & (j\omega_2)^l & \cdots & (j\omega_{2n})^l \\ (j\omega_1)^{l-1} & (j\omega_2)^{l-1} & \cdots & (j\omega_{2n})^{l-1} \\ \vdots & \vdots & & \vdots \\ 1 & 1 & \cdots & 1 \end{bmatrix}$$

$$\times \begin{bmatrix} \frac{1}{D(j\omega_1)} & 0 & \cdots & 0 \\ 0 & \frac{1}{D(j\omega_2)} & \cdots & 0 \\ 0 & & \ddots & 0 \\ 0 & 0 & \cdots & \frac{1}{D(j\omega_{2n})} \end{bmatrix} \qquad (6.4.20)$$

where $D(s) = k_p Z_p(s) \Lambda(s) R_m(s)$. Note that the matrix in the middle of the right-hand side of (6.4.20) is a submatrix of the Vandermonte matrix, which is always nonsingular for $\omega_i \neq \omega_k, i \neq k; i, k = 1, \ldots, 2n$. We, therefore, conclude from (6.4.20) that $L(\omega_1, \ldots, \omega_{2n})$ is of full rank which implies that $H(j\omega_1), \ldots, H(j\omega_{2n})$ are linearly independent on $\mathcal{C}^{2n}$ and the proof is complete. □

Example 6.4.1 Let us consider the second order plant

$$y_p = \frac{k_p(s + b_0)}{(s^2 + a_1 s + a_0)} u_p$$

where $k_p > 0, b_0 > 0$ and k_p, b_0, a_1, a_0 are unknown constants. The desired performance of the plant is specified by the reference model

$$y_m = \frac{1}{s+1} r$$

Using Table 6.1, the control law is designed as

$$\begin{aligned} \dot{\omega}_1 &= -2\omega_1 + u_p, \quad \omega_1(0) = 0 \\ \dot{\omega}_2 &= -2\omega_2 + y_p, \quad \omega_2(0) = 0 \\ u_p &= \theta_1 \omega_1 + \theta_2 \omega_2 + \theta_3 y_p + c_0 r \end{aligned}$$

by choosing $F = -2, g = 1$ and $\Lambda(s) = s + 2$. The adaptive law is given by

$$\dot{\theta} = -\Gamma e_1 \omega, \quad \theta(0) = \theta_0$$

where $e_1 = y_p - y_m, \theta = [\theta_1, \theta_2, \theta_3, c_0]^T$ and $\omega = [\omega_1, \omega_2, y_p, r]^T$. We can choose $\Gamma = \text{diag}\{\gamma_i\}$ for some $\gamma_i > 0$ and obtain the decoupled adaptive law

$$\dot{\theta}_i = -\gamma_i e_1 \omega_i, \quad i = 1, \ldots, 4$$

where $\theta_4 = c_0, \omega_3 = y_p, \omega_4 = r$; or we can choose Γ to be any positive definite matrix.

For parameter convergence, we choose r to be sufficiently rich of order 4. As an example, we select $r = A_1 \sin \omega_1 t + A_2 \sin \omega_2 t$ for some nonzero constants $A_1, A_2, \omega_1, \omega_2$ with $\omega_1 \neq \omega_2$. We should emphasize that we may not always have the luxury to choose r to be sufficiently rich. For example, if the control objective requires r =constant in order for y_p to follow a constant set point at steady state, then the use of a sufficiently rich input r of order 4 will destroy the desired tracking properties of the closed-loop plant.

The simulation results for the MRAC scheme for the plant with $b_0 = 3$, $a_1 = 3$, $a_0 = -10$, $k_p = 1$ are shown in Figures 6.8 and 6.9. The initial value of the

6.4. DIRECT MRAC WITH UNNORMALIZED ADAPTIVE LAWS

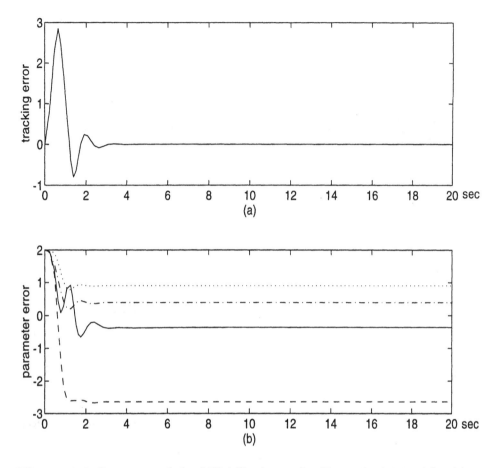

Figure 6.8 Response of the MRAC scheme for Example 6.4.1 with $r(t) =$ unit step function.

parameters are chosen as $\theta(0) = [3, -10, 2, 3]^T$. Figure 6.8(a, b) shows the response of the tracking error e_1 and estimated parameter error $\tilde{\theta}_i$ for $\gamma_i = 1$ and $r =$ unit step. Figure 6.9 shows the simulation results for $\Gamma = \text{diag}\{2, 6, 6, 2\}$ and $r = 0.5\sin 0.7t + 2\cos 5.9t$. From Figure 6.9 (b), we note that the estimated parameters converge to $\theta^* = [1, -12, 0, 1]^T$ due to the use of a sufficiently rich input. ▽

Remark 6.4.1 The error equation (6.4.8) takes into account the initial conditions of the plant states. Therefore the results of Theorem 6.4.1 hold

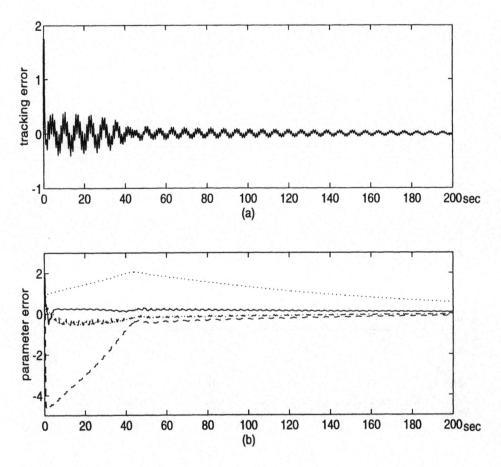

Figure 6.9 Response of the MRAC scheme for Example 6.4.1 with $r(t) = 0.5 \sin 0.7t + 2 \cos 5.9t$.

for any finite initial condition for the states of the plant and filters. In the analysis, we implicitly assumed that the nonlinear differential equations (6.4.8) and (6.4.10) with initial conditions $e(0) = e_0$, $\tilde{\theta}(0) = \tilde{\theta}_0$ possess a unique solution. For our analysis to be valid, the solution $\tilde{\theta}(t), e(t)$ has to exist for all $t \in [0, \infty)$. The existence and uniqueness of solutions of adaptive control systems is addressed in [191].

Remark 6.4.2 The proof of Theorem 6.4.1 part (i) may be performed by using a minimal state-space representation for the equation

$$e_1 = W_m(s)\rho^* \tilde{\theta}^\top \omega \qquad (6.4.21)$$

6.4. DIRECT MRAC WITH UNNORMALIZED ADAPTIVE LAWS

rather than the nonminimal state representation (6.4.8) to develop the adaptive law $\dot{\theta} = -\Gamma e_1 \omega$. In this case we establish that $e_1, \theta \in \mathcal{L}_\infty$ and $e_1 \in \mathcal{L}_2$ by using the LKY (instead of the MKY Lemma) and the properties of a Lyapunov-like function. The boundedness of e_1 implies that $y_p \in \mathcal{L}_\infty$.

The boundedness of ω and the rest of the signals requires the following additional arguments: We write ω as

$$\omega = \begin{bmatrix} (sI - F)^{-1} g G_p^{-1}(s) y_p \\ (sI - F)^{-1} g y_p \\ y_p \\ r \end{bmatrix}$$

Because $y_p \in \mathcal{L}_\infty$ and $(sI - F)^{-1} g G_p^{-1}(s)$, $(sI - F)^{-1} g$ are proper (note that the relative degree of $G_p(s)$ is 1) with stable poles, we have $\omega \in \mathcal{L}_\infty$. From $u_p = \theta^T \omega$ and $\theta, \omega \in \mathcal{L}_\infty$, it follows that $u_p \in \mathcal{L}_\infty$. The proof of $e_1(t) \to 0$ as $t \to \infty$ follows by applying Lemma 3.2.5 and using the properties $e_1 \in \mathcal{L}_2$, $\dot{e}_1 = s W_m(s) \rho^* \tilde{\theta}^T \omega \in \mathcal{L}_\infty$.

Remark 6.4.3 The effect of initial conditions may be accounted for by considering

$$e_1 = W_m(s) \rho^* \tilde{\theta}^T \omega + C_c^T (sI - A_c)^{-1} e(0)$$

instead of (6.4.21). Because the term that depends on $e(0)$ is exponentially decaying to zero, it does not affect the stability results. This can be shown by modifying the Lyapunov-like function to accommodate the exponentially decaying to zero term (see Problem 6.19).

6.4.2 Relative Degree $n^* = 2$

Let us consider again the parameterization of the plant in terms of θ^*, developed in the previous section, i.e.,

$$\begin{aligned} \dot{e} &= A_c e + B_c \left(u_p - \theta^{*T} \omega \right) \\ e_1 &= C_c^T e \end{aligned} \quad (6.4.22)$$

or
$$e_1 = W_m(s)\rho^* \left(u_p - \theta^{*T}\omega\right) \tag{6.4.23}$$

In the relative degree $n^* = 1$ case, we are able to design $W_m(s)$ to be SPR which together with the control law $u_p = \theta^T\omega$ enables us to obtain an error equation that is suitable for applying the SPR-Lyapunov design method.

With $n^* = 2$, $W_m(s)$ can no longer be designed to be SPR and therefore the procedure of Section 6.4.1 fails to apply here.

Instead, let us follow the techniques of Chapter 4 and use the identity $(s+p_0)(s+p_0)^{-1} = 1$ for some $p_0 > 0$ to rewrite (6.4.22), (6.4.23) as

$$\begin{aligned} \dot{e} &= A_c e + \bar{B}_c (s+p_0) \rho^* \left(u_f - \theta^{*T}\phi\right), \quad e(0) = e_0 \\ e_1 &= C_c^T e \end{aligned} \tag{6.4.24}$$

i.e.,
$$e_1 = W_m(s)(s+p_0)\rho^* \left(u_f - \theta^{*T}\phi\right) \tag{6.4.25}$$

where $\bar{B}_c = B_c c_0^*$,
$$u_f = \frac{1}{s+p_0} u_p, \quad \phi = \frac{1}{s+p_0}\omega$$

and $W_m(s), p_0 > 0$ are chosen so that $W_m(s)(s+p_0)$ is SPR.

We use ρ, θ, the estimate of ρ^*, θ^*, respectively, to generate the estimate $\hat{e}_1$ of e_1 as
$$\hat{e}_1 = W_m(s)(s+p_0)\rho(u_f - \theta^T\phi)$$

If we follow the same procedure as in Section 6.4.2, then the next step is to choose u_p so that $\hat{e}_1 = W_m(s)(s+p_0)[0]$, $\epsilon_1 = e_1$ and (6.4.24) is in the form of the error equation (6.4.8) where the tracking error e_1 is related to the parameter error $\tilde{\theta}$ through an SPR transfer function. The control law $u_p = \theta^T\omega$ (used in the case of $n^* = 1$) motivated from the known parameter case cannot transform (6.4.23) into the error equation we are looking for. Instead if we choose u_p so that
$$u_f = \theta^T \phi \tag{6.4.26}$$

we have $\hat{e}_1 = W_m(s)(s+p_0)[0]$ and by substituting (6.4.26) in (6.4.24), we obtain the error equation

$$\begin{aligned} \dot{e} &= A_c e + \bar{B}_c (s+p_0) \rho^* \tilde{\theta}^T \phi, \quad e(0) = e_0 \\ e_1 &= C_c^T e \end{aligned} \tag{6.4.27}$$

6.4. DIRECT MRAC WITH UNNORMALIZED ADAPTIVE LAWS

or in the transfer function form

$$e_1 = W_m(s)(s+p_0)\rho^*\tilde{\theta}^\top \phi$$

which can be transformed into the desired form by using the transformation

$$\bar{e} = e - \bar{B}_c \rho^* \tilde{\theta}^\top \phi \qquad (6.4.28)$$

i.e.,

$$\begin{aligned} \dot{\bar{e}} &= A_c \bar{e} + B_1 \rho^* \tilde{\theta}^\top \phi, \quad \bar{e}(0) = \bar{e}_0 \\ e_1 &= C_c^\top \bar{e} \end{aligned} \qquad (6.4.29)$$

where $B_1 = A_c \bar{B}_c + \bar{B}_c p_0$ and $C_c^\top \bar{B}_c = C_p^\top B_p c_0^* = 0$ due to $n^* = 2$. With (6.4.29), we can proceed as in the case of $n^* = 1$ and develop an adaptive law for θ. Let us first examine whether we can choose u_p to satisfy equation (6.4.26). We have

$$u_p = (s+p_0)u_f = (s+p_0)\theta^\top \phi$$

which implies that

$$u_p = \theta^\top \omega + \dot{\theta}^\top \phi \qquad (6.4.30)$$

Because $\dot{\theta}$ is made available by the adaptive law, the control law (6.4.30) can be implemented without the use of differentiators. Let us now go back to the error equation (6.4.29). Because

$$C_c^\top (sI - A_c)^{-1} B_1 = C_c^\top (sI - A_c)^{-1} \bar{B}_c (s+p_0) = W_m(s)(s+p_0)$$

is SPR, (6.4.29) is of the same form as (6.4.8) and the adaptive law for generating θ is designed by considering

$$V(\tilde{\theta}, \bar{e}) = \frac{\bar{e}^\top P_c \bar{e}}{2} + \frac{\tilde{\theta}^\top \Gamma^{-1} \tilde{\theta}}{2}|\rho^*|$$

where $P_c = P_c^\top > 0$ satisfies the MKY Lemma. As in the case of $n^* = 1$, for

$$\dot{\tilde{\theta}} = \dot{\theta} = -\Gamma e_1 \phi \operatorname{sgn}(\rho^*) \qquad (6.4.31)$$

the time derivative $\dot{V}$ of V along the solution of (6.4.29), (6.4.31) is given by

$$\dot{V} = -\frac{\bar{e}^\top q q^\top \bar{e}}{2} - \frac{\nu_c}{2}\bar{e}^\top L_c \bar{e} \le 0$$

which implies that $\bar{e}, \tilde{\theta}, e_1 \in \mathcal{L}_\infty$ and $\bar{e}, e_1 \in \mathcal{L}_2$. Because $e_1 = y_p - y_m$, we also have $y_p \in \mathcal{L}_\infty$. The signal vector ϕ is expressed as

$$\phi = \frac{1}{s+p_0} \begin{bmatrix} (sI-F)^{-1}gG_p^{-1}(s)y_p \\ (sI-F)^{-1}gy_p \\ y_p \\ r \end{bmatrix} \quad (6.4.32)$$

by using $u_p = G_p^{-1}(s)y_p$. We can observe that each element of ϕ is the output of a proper stable transfer function whose input is y_p or r. Because $y_p, r \in \mathcal{L}_\infty$ we have $\phi \in \mathcal{L}_\infty$. Now $\bar{e}, \theta, \phi \in \mathcal{L}_\infty$ imply (from (6.4.28)) that e and, therefore, $Y_c \in \mathcal{L}_\infty$. Because $\omega, \phi, e_1 \in \mathcal{L}_\infty$ we have $\dot{\theta} \in \mathcal{L}_\infty$ and $u_p \in \mathcal{L}_\infty$ and therefore all signals in the closed-loop plant are bounded. From (6.4.29) we also have that $\dot{\bar{e}} \in \mathcal{L}_\infty$, i.e., $\dot{e}_1 \in \mathcal{L}_\infty$, which, together with $e_1 \in \mathcal{L}_\infty \cap \mathcal{L}_2$, implies that $e_1(t) \to 0$ as $t \to \infty$.

We present the main equations of the overall MRAC scheme in Table 6.2 and summarize its stability properties by the following theorem.

Theorem 6.4.2 *The MRAC scheme of Table 6.2 guarantees that*

(i) *All signals in the closed-loop plant are bounded and the tracking error e_1 converges to zero asymptotically.*

(ii) *If R_p, Z_p are coprime and r is sufficiently rich of order $2n$, then the parameter error $|\tilde{\theta}| = |\theta - \theta^*|$ and the tracking error e_1 converge to zero exponentially fast.*

Proof (i) This part has been completed above.

(ii) Consider the error equations (6.4.29), (6.4.31) which have the same form as equations (4.3.10), (4.3.35) with $n_s^2 = 0$ in Chapter 4. Using Corollary 4.3.1 we have that if $\phi, \dot{\phi} \in \mathcal{L}_\infty$ and ϕ is PE, then the adaptive law (6.4.31) guarantees that $|\tilde{\theta}|$ converges to zero exponentially fast. We have already established that $y_p, \phi \in \mathcal{L}_\infty$. It follows from (6.4.32) and the fact that $\dot{e}_1$ and therefore $\dot{y}_p \in \mathcal{L}_\infty$ that $\dot{\phi} \in L_\infty$. Hence it remains to show that ϕ is PE.

As in the case of $n^* = 1$ we write ϕ as

$$\phi = \phi_m + \bar{\phi}$$

6.4. DIRECT MRAC WITH UNNORMALIZED ADAPTIVE LAWS

Table 6.2 MRAC scheme: $n^* = 2$

Plant	$y_p = k_p \frac{Z_p(s)}{R_p(s)} u_p, \quad n^* = 2$
Reference model	$y_m = W_m(s)r, \quad W_m(s) = k_m \frac{Z_m(s)}{R_m(s)}$
Control law	$\dot{\omega}_1 = F\omega_1 + g u_p, \quad \omega_1(0) = 0$ $\dot{\omega}_2 = F\omega_2 + g y_p, \quad \omega_2(0) = 0$ $\dot{\phi} = -p_0 \phi + \omega, \quad \phi(0) = 0$ $u_p = \theta^T \omega + \dot{\theta}^T \phi = \theta^T \omega - \phi^T \Gamma \phi e_1 \operatorname{sgn}(k_p/k_m)$ $\omega = [\omega_1^T, \omega_2^T, y_p, r]^T, \quad \omega_1 \in \mathcal{R}^{n-1}, \quad \omega_2 \in \mathcal{R}^{n-1}$
Adaptive law	$\dot{\theta} = -\Gamma e_1 \phi \operatorname{sgn}(k_p/k_m), \quad e_1 = y_p - y_m$
Assumptions	$Z_p(s)$ is Hurwitz; $W_m(s)(s+p_0)$ is strictly proper and SPR; F, g, Γ are as defined in Table 6.1; plant and reference model satisfy assumptions P1 to P4, and M1 and M2, respectively

where

$$\phi_m = \frac{1}{s+p_0} \begin{bmatrix} (sI - F)^{-1} g G_p^{-1}(s) W_m(s) \\ (sI - F)^{-1} g W_m(s) \\ W_m(s) \\ 1 \end{bmatrix} r$$

and

$$\bar{\phi} = \frac{1}{s+p_0} \begin{bmatrix} (sI - F)^{-1} g G_p^{-1}(s) \\ (sI - F)^{-1} g \\ 1 \\ 0 \end{bmatrix} e_1$$

Because $e_1 \in \mathcal{L}_2 \cap \mathcal{L}_\infty$ and $e_1 \to 0$ as $t \to \infty$ it follows (see Corollary 3.3.1) that $\bar{\phi} \in \mathcal{L}_2 \cap \mathcal{L}_\infty$ and $|\bar{\phi}| \to 0$ as $t \to \infty$.

Proceeding as in the proof of Theorem 6.4.1, we establish that ϕ_m is PE and use Lemma 4.8.3 to show that ϕ is also PE which implies, using the results of Chapter 4, that $|\tilde{\theta}| \to 0$ exponentially fast. Using (6.4.29) and the exponential convergence of $|\tilde{\theta}|$ to zero we obtain that e_1 converges to zero exponentially fast. □

Example 6.4.2 Let us consider the second order plant

$$y_p = \frac{k_p}{(s^2 + a_1 s + a_0)} u_p$$

where $k_p > 0$, and a_1, a_0 are constants. The reference model is chosen as

$$y_m = \frac{5}{(s+5)^2} r$$

Using Table 6.2 the control law is designed as

$$\dot{\omega}_1 = -2\omega_1 + u_p$$
$$\dot{\omega}_2 = -2\omega_2 + y_p$$
$$\dot{\phi} = -\phi + \omega$$
$$u_p = \theta^T \omega - \phi^T \Gamma \phi e_1$$

where $\omega = [\omega_1, \omega_2, y_p, r]^T$, $e_1 = y_p - y_m$, $p_0 = 1$, $\Lambda(s) = s + 2$ and $\frac{5(s+1)}{(s+5)^2}$ is SPR. The adaptive law is given by

$$\dot{\theta} = -\Gamma e_1 \phi$$

where $\Gamma = \Gamma^T > 0$ is any positive definite matrix and $\theta = [\theta_1, \theta_2, \theta_3, c_0]^T$.

For parameter convergence, the input u_p is chosen as

$$u_p = A_1 \sin \omega_1 t + A_2 \sin \omega_2 t$$

for some $A_1, A_2 \neq 0$ and $\omega_1 \neq \omega_2$.

Figures 6.10 and 6.11 show some simulation results of the MRAC scheme for the plant with $a_1 = 3, a_0 = -10, k_p = 1$. We start with an initial parameter vector $\theta(0) = [3, 18, -8, 3]^T$ that leads to an initially destabilizing controller. The tracking error and estimated parameter error response is shown in Figure 6.10 for $\Gamma = \mathrm{diag}\{2, 4, 0.8, 1\}$, and $r =$ unit step. Due to the initial destabilizing controller, the transient response is poor. The adaptive mechanism alters the unstable behavior of the initial controller and eventually drives the tracking error to zero. In Figure 6.11 we show the response of the same system when $r = 3\sin 4.9t + 0.5 \cos 0.7t$ is a sufficiently rich input of order 4. Because of the use of a sufficiently rich signal, $\theta(t)$ converges to $\theta^* = [1, 12, -10, 1]^T$. ▽

6.4. DIRECT MRAC WITH UNNORMALIZED ADAPTIVE LAWS

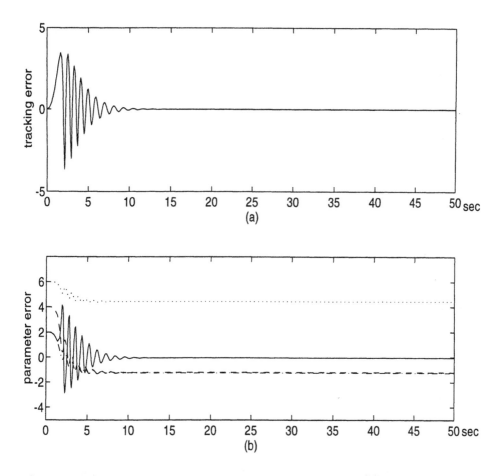

Figure 6.10 Response of the MRAC scheme for Example 6.4.2 with $r(t) =$ unit step function.

Remark 6.4.4 The control law (6.4.30) is a modification of the certainty equivalence control law $u_p = \theta^T \omega$ and is motivated from stability considerations. The additional term $\dot{\theta}^T \phi = -\phi^T \Gamma \phi e_1 \, \text{sgn}(\rho^*)$ is a nonlinear one that disappears asymptotically with time, i.e., $u_p = \theta^T \omega + \dot{\theta}^T \phi$ converges to the certainty equivalence control law as $t \to \infty$. The number and complexity of the additional terms in the certainty equivalence control law increase with the relative degree n^* as we demonstrate in the next section.

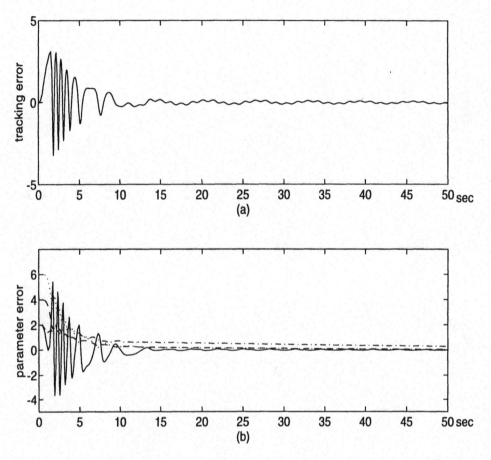

Figure 6.11 Response of the MRAC scheme for Example 6.4.2 with $r(t) = 3\sin 4.9t + 0.5\sin 0.7t$.

Remark 6.4.5 The proof of Theorem 6.4.2 may be accomplished by using a minimal state space realization for the error equation

$$e_1 = W_m(s)(s+p_0)\rho^*\tilde{\theta}^\top \phi$$

The details of such an approach are left as an exercise for the reader.

6.4.3 Relative Degree $n^* = 3$

As in the case of $n^* = 2$, the transfer function $W_m(s)$ of the reference model cannot be chosen to be SPR because according to assumption (M2),

6.4. DIRECT MRAC WITH UNNORMALIZED ADAPTIVE LAWS

$W_m(s)$ should have the same relative degree as the plant transfer function. Therefore, the choice of $u_p = \theta^\top \omega$ in the error equation

$$e_1 = W_m(s)\rho^*(u_p - \theta^{*\top}\omega) \qquad (6.4.33)$$

will not lead to the desired error equation where the tracking error is related to the parameter error through an SPR transfer function. As in the case of $n^* = 2$, let us rewrite (6.4.33) in a form that involves an SPR transfer function by using the techniques of Chapter 4, i.e., we express (6.4.33) as

$$e_1 = W_m(s)(s+p_0)(s+p_1)\rho^*\left(u_f - \theta^{*\top}\phi\right) \qquad (6.4.34)$$

where

$$u_f = \frac{1}{(s+p_0)(s+p_1)}u_p, \quad \phi = \frac{1}{(s+p_0)(s+p_1)}\omega$$

and $W_m(s), p_0, p_1$ are chosen so that $\bar{W}_m(s) \triangleq W_m(s)(s+p_0)(s+p_1)$ is SPR, which is now possible because the relative degree of $\bar{W}_m(s)$ is 1. For simplicity and without loss of generality let us choose

$$W_m(s) = \frac{1}{(s+p_0)(s+p_1)(s+q_0)}$$

for some $q_0 > 0$ so that

$$e_1 = \frac{1}{s+q_0}\rho^*(u_f - \theta^{*\top}\phi) \qquad (6.4.35)$$

The estimate of $\hat{e}_1$ of e_1 based on the estimates ρ, θ is given by

$$\hat{e}_1 = \frac{1}{s+q_0}\rho(u_f - \theta^\top \phi) \qquad (6.4.36)$$

If we proceed as in the case of $n^* = 2$ we would attempt to choose

$$u_f = \theta^\top \phi \qquad (6.4.37)$$

to make $\hat{e}_1 = \frac{1}{s+q_0}[0]$ and obtain the error equation

$$e_1 = \frac{1}{s+q_0}\rho^*\tilde{\theta}^\top \phi \qquad (6.4.38)$$

The adaptive law
$$\dot{\theta} = -\Gamma e_1 \phi \operatorname{sgn}(\rho^*) \qquad (6.4.39)$$
will then follow by using the standard procedure. Equation (6.4.37), however, implies the use of the control input
$$u_p = (s+p_0)(s+p_1)u_f = (s+p_0)(s+p_1)\theta^\top \phi \qquad (6.4.40)$$
which involves $\ddot{\theta}$, that is not available for measurement. Consequently the control law (6.4.40) cannot be implemented and the choice of $u_f = \theta^\top \phi$ is not feasible.

The difficulty of not being able to extend the results for $n^* = 1, 2$ to $n^* \geq 3$ became the major obstacle in advancing research in adaptive control during the 1970s. By the end of the 1970s and early 1980s, however, this difficulty was circumvented and several successful MRAC schemes were proposed using different approaches. Efforts to extend the procedure of $n^* = 1, 2$ to $n^* \geq 3$ continued during the early 1990s and led to new designs for MRAC. One such design proposed by Morse [164] employs the same control law as in (6.4.40) but the adaptive law for θ is modified in such a way that $\ddot{\theta}$ becomes an available signal. This modification, achieved at the expense of a higher-order adaptive law, led to a MRAC scheme that guarantees signal boundedness and convergence of the tracking error to zero.

Another successful MRAC design that has it roots in the paper of Feuer and Morse [54] is proposed in [162] for a third order plant with known high frequency gain. In this design, the adaptive law is kept unchanged but the control law is chosen as
$$u_p = \theta^\top \omega + u_a$$
where u_a is designed based on stability considerations. Below we present and analyze a very similar design as in [162].

We start by rewriting (6.4.35), (6.4.36) as
$$e_1 = \frac{1}{s+q_0}\rho^*(\tilde{\theta}^\top \phi + r_0), \quad \hat{e}_1 = \frac{1}{s+q_0}\rho r_0 \qquad (6.4.41)$$
where $r_0 = u_f - \theta^\top \phi$ and $\tilde{\theta} = \theta - \theta^*$.

Because r_0 cannot be forced to be equal to zero by setting $u_f = \theta^\top \phi$, we will focus on choosing u_p so that r_0 goes to zero as $t \to \infty$. In this case,

6.4. DIRECT MRAC WITH UNNORMALIZED ADAPTIVE LAWS

the estimation error $\epsilon_1 = e_1 - \hat{e}_1$ is not equal to e_1 because $\hat{e}_1 \neq 0$ owing to $\tilde{\rho}r_0 \neq 0$. However, it satisfies the error equation

$$\epsilon_1 = e_1 - \hat{e}_1 = \frac{1}{s+q_0}(\rho^*\tilde{\theta}^T\phi - \tilde{\rho}r_0) \qquad (6.4.42)$$

that leads to the adaptive law

$$\dot{\theta} = -\Gamma\epsilon_1\phi\,\text{sgn}(\rho^*), \quad \dot{\rho} = \gamma\epsilon_1 r_0 \qquad (6.4.43)$$

where $\Gamma = \Gamma^T$ and $\gamma > 0$ by considering the Lyapunov-like function

$$V = \frac{\epsilon_1^2}{2} + \frac{\tilde{\theta}^T\Gamma^{-1}\tilde{\theta}}{2}|\rho^*| + \frac{\tilde{\rho}^2}{2\gamma}$$

We now need to choose u_a in $u_p = \theta^T\omega + u_a$ to establish stability for the system (6.4.41) to (6.4.43). Let us now express r_0 as

$$r_0 = u_f - \theta^T\phi = \frac{1}{s+p_0}\left[u_1 - \dot{\theta}^T\phi - \theta^T\phi_1\right]$$

where

$$u_1 = \frac{1}{s+p_1}u_p, \quad \phi_1 = (s+p_0)\phi = \frac{1}{s+p_1}\omega$$

i.e.,

$$\dot{r}_0 = -p_0 r_0 + u_1 - \dot{\theta}^T\phi - \theta^T\phi_1 \qquad (6.4.44)$$

Substituting for $\dot{\theta}$, we obtain

$$\dot{r}_0 = -p_0 r_0 + u_1 + \phi^T\Gamma\phi\epsilon_1\,\text{sgn}(\rho^*) - \theta^T\phi_1 \qquad (6.4.45)$$

If we now choose $u_1 = -\phi^T\Gamma\phi\epsilon_1\text{sgn}(\rho^*) + \theta^T\phi_1$ then $\dot{r}_0 = -p_0 r_0$ and r_0 converges to zero exponentially fast. This choice of u_1, however, leads to a control input u_p that is not implementable since $u_p = (s+p_1)u_1$ will involve the first derivative of u_1 and, therefore, the derivative of e_1 that is not available for measurement. Therefore, the term $\phi^T\Gamma\phi\epsilon_1\,\text{sgn}(\rho^*)$ in (6.4.45) cannot be eliminated by u_1. Its effect, however, may be counteracted by introducing what is called a "nonlinear damping" term in u_1 [99]. That is, we choose

$$u_1 = \theta^T\phi_1 - \alpha_0\left(\phi^T\Gamma\phi\right)^2 r_0 \qquad (6.4.46)$$

where $\alpha_0 > 0$ is a design constant, and obtain

$$\dot{r}_0 = -\left[p_0 + \alpha_0\left(\phi^\top \Gamma \phi\right)^2\right] r_0 + \phi^\top \Gamma \phi \epsilon_1 \mathrm{sgn}(\rho^*)$$

The purpose of the nonlinear term $(\phi^\top \Gamma \phi)^2$ is to "damp out" the possible destabilizing effect of the nonlinear term $\phi^\top \Gamma \phi \epsilon_1$ as we show in the analysis to follow. Using (6.4.46), the control input $u_p = (s+p_1)u_1$ is given by

$$u_p = \theta^\top \omega + \dot{\theta}^\top \phi_1 - (s+p_1)\alpha_0(\phi^\top \Gamma \phi)^2 r_0 \tag{6.4.47}$$

If we now perform the differentiation in (6.4.47) and substitute for the derivative of r_0 we obtain

$$\begin{aligned} u_p &= \theta^\top \omega + \dot{\theta}^\top \phi_1 - 4\alpha_0 \phi^\top \Gamma \phi \left(\phi^\top \Gamma \dot{\phi}\right) r_0 - \alpha_0 (p_1 - p_0)\left(\phi^\top \Gamma \phi\right)^2 r_0 \\ &\quad + \alpha_0^2 \left(\phi^\top \Gamma \phi\right)^4 r_0 - \alpha_0 \left(\phi^\top \Gamma \phi\right)^3 \epsilon_1 \mathrm{sgn}(\rho^*) \end{aligned} \tag{6.4.48}$$

where $\dot{\phi}$ is generated from

$$\dot{\phi} = \frac{s}{(s+p_0)(s+p_1)}\omega$$

which demonstrates that u_p can be implemented without the use of differentiators.

We summarize the main equations of the MRAC scheme in Table 6.3.

The stability properties of the proposed MRAC scheme listed in Table 6.3 are summarized as follows.

Theorem 6.4.3 *The MRAC scheme of Table 6.3 guarantees that*
(i) *All signals in the closed-loop plant are bounded and $r_0(t)$, $e_1(t) \to 0$ as $t \to \infty$.*
(ii) *If k_p is known, r is sufficiently rich of order $2n$ and Z_p, R_p are coprime, then the parameter error $|\tilde{\theta}| = |\theta - \theta^*|$ and tracking error e_1 converge to zero exponentially fast.*
(iii) *If r is sufficiently rich of order $2n$ and Z_p, R_p are coprime, then $|\tilde{\theta}|$ and e_1 converge to zero asymptotically (not necessarily exponentially fast).*
(iv) *The estimate ρ converges to a constant $\bar{\rho}$ asymptotically independent of the richness of r.*

6.4. DIRECT MRAC WITH UNNORMALIZED ADAPTIVE LAWS

Table 6.3 MRAC scheme: $n^* = 3$

Plant	$y_p = k_p \dfrac{Z_p(s)}{R_p(s)} u_p, \quad n^* = 3$
Reference model	$y_m = W_m(s) r$
Control law	$\dot{\omega}_1 = F\omega_1 + g u_p, \quad \omega_1(0) = 0$ $\dot{\omega}_2 = F\omega_2 + g y_p, \quad \omega_2(0) = 0$ $\dot{r}_0 = -(p_0 + \alpha_0(\phi^T \Gamma \phi)^2) r_0 + \phi^T \Gamma \phi \epsilon_1 \mathrm{sgn}(\rho^*)$ $u_p = \theta^T \omega + u_a$ $u_a = \dot{\theta}^T \phi_1 - \alpha_0(p_1 - p_0)(\phi^T \Gamma \phi)^2 r_0 - 4\alpha_0 \phi^T \Gamma \phi (\phi^T \Gamma \dot{\phi}) r_0$ $\quad + \alpha_0^2 (\phi^T \Gamma \phi)^4 r_0 - \alpha_0 (\phi^T \Gamma \phi)^3 \epsilon_1 \mathrm{sgn}(\rho^*)$
Adaptive law	$\dot{\theta} = -\Gamma \epsilon_1 \phi \, \mathrm{sgn}(\rho^*), \quad \dot{\rho} = \gamma \epsilon_1 r_0$ $\epsilon_1 = e_1 - \hat{e}_1, \quad \hat{e}_1 = \dfrac{1}{s+q_0} \rho r_0$ $\phi = \dfrac{1}{(s+p_0)(s+p_1)} \omega, \quad \omega = \left[\omega_1^T, \omega_2^T, y_p, r \right]^T$ $\phi_1 = \dfrac{1}{s+p_1} \omega, \quad e_1 = y_p - y_m$
Design variables	$\Gamma = \Gamma^T > 0, \gamma > 0, \alpha_0 > 0$ are arbitrary design constants; $W_m(s)(s+p_0)(s+p_1)$ is strictly proper and SPR; F, g are as in the case of $n^* = 1$; $Z_p(s), R_p(s)$ and $W_m(s)$ satisfy assumptions P1 to P4, M1 and M2, respectively; $\mathrm{sgn}(\rho^*) = \mathrm{sgn}(k_p/k_m)$

Proof (i) The equations that describe the stability properties of the closed-loop plant are

$$\begin{aligned}
\dot{\epsilon}_1 &= -q_0 \epsilon_1 + \rho^* \tilde{\theta}^T \phi - \tilde{\rho} r_0 \\
\dot{r}_0 &= -(p_0 + \alpha_0(\phi^T \Gamma \phi)^2) r_0 + \phi^T \Gamma \phi \epsilon_1 \, \mathrm{sgn}(\rho^*) \quad (6.4.49) \\
\dot{\tilde{\theta}} &= -\Gamma \epsilon_1 \phi \, \mathrm{sgn}(\rho^*), \quad \dot{\tilde{\rho}} = \gamma \epsilon_1 r_0
\end{aligned}$$

We propose the Lyapunov-like function

$$V = \frac{\epsilon_1^2}{2} + |\rho^*|\tilde{\theta}^\top \frac{\Gamma^{-1}}{2}\tilde{\theta} + \frac{\tilde{\rho}^2}{2\gamma} + \gamma_0 \frac{r_0^2}{2}$$

where $\gamma_0 > 0$ is a constant to be selected. The time derivative of V along the trajectories of (6.4.49) is given by

$$\begin{aligned}\dot{V} &= -q_0\epsilon_1^2 - \gamma_0 p_0 r_0^2 - \gamma_0\alpha_0 r_0^2(\phi^\top\Gamma\phi)^2 + \gamma_0\epsilon_1 r_0\phi^\top\Gamma\phi\,\text{sgn}(\rho^*) \\ &\leq -q_0\epsilon_1^2 - \gamma_0 p_0 r_0^2 - \gamma_0\alpha_0 r_0^2(\phi^\top\Gamma\phi)^2 + \gamma_0|\epsilon_1|\,|r_0|\phi^\top\Gamma\phi\end{aligned}$$

By completing the squares we obtain

$$\begin{aligned}\dot{V} &\leq -q_0\frac{\epsilon_1^2}{2} - \frac{q_0}{2}\left[|\epsilon_1| - \gamma_0\frac{|r_0|\phi^\top\Gamma\phi}{q_0}\right]^2 + \gamma_0^2\frac{r_0^2(\phi^\top\Gamma\phi)^2}{2q_0} - \gamma_0 p_0 r_0^2 - \gamma_0\alpha_0 r_0^2(\phi^\top\Gamma\phi)^2 \\ &\leq -q_0\frac{\epsilon_1^2}{2} - \gamma_0 p_0 r_0^2 - \left[\alpha_0 - \frac{\gamma_0}{2q_0}\right]\gamma_0 r_0^2(\phi^\top\Gamma\phi)^2\end{aligned}$$

Because $\gamma_0 > 0$ is arbitrary, used for analysis only, for any given α_0 and $q_0 > 0$, we can choose it as $\gamma_0 = 2\alpha_0 q_0$ leading to

$$\dot{V} \leq -q_0\frac{\epsilon_1^2}{2} - \gamma_0 p_0 r_0^2 \leq 0$$

Hence, $\epsilon_1, r_0, \rho, \theta \in \mathcal{L}_\infty$ and $\epsilon_1, r_0 \in \mathcal{L}_2$. Because $\rho, r_0 \in \mathcal{L}_\infty$, it follows from (6.4.41) that $\hat{e}_1 \in \mathcal{L}_\infty$ which implies that $e_1 = \epsilon_1 + \hat{e}_1 \in \mathcal{L}_\infty$. Hence, $y_p \in \mathcal{L}_\infty$, which, together with

$$\phi = \frac{1}{(s+p_0)(s+p_1)}\begin{bmatrix}(sI-F)^{-1}gG_p^{-1}(s)y_p \\ (sI-F)^{-1}gy_p \\ y_p \\ r\end{bmatrix} \quad (6.4.50)$$

implies that $\phi \in \mathcal{L}_\infty$. Using $\epsilon_1, \phi \in \mathcal{L}_\infty$ and $\epsilon_1 \in \mathcal{L}_2$ in (6.4.49) we have $\dot{\theta}, \dot{\rho} \in \mathcal{L}_\infty \cap \mathcal{L}_2$. From the error equation (6.4.49) it follows that $\dot{\epsilon}_1$ and, therefore, $\dot{y}_p \in \mathcal{L}_\infty$ which imply that $\dot{\phi}$ and $\phi_1 \in \mathcal{L}_\infty$. The second derivative $\ddot{e}_1$ can be shown to be bounded by using $\dot{\theta}, \dot{\phi}, \dot{r}_0 \in \mathcal{L}_\infty$ in (6.4.41). Because $\ddot{e}_1 = \ddot{y}_p - s^2 W_m(s)r$ and $s^2 W_m(s)$ is proper, it follows that $\ddot{y}_p \in \mathcal{L}_\infty$, which, together with (6.4.50), implies that $\ddot{\phi} \in \mathcal{L}_\infty$. Because $\omega = (s+p_0)(s+p_1)\phi$, we have that $\omega \in \mathcal{L}_\infty$ and therefore u_p and all signals are bounded. Because $r_0 \in \mathcal{L}_\infty \cap \mathcal{L}_2$ and $\rho \in \mathcal{L}_\infty$, it follows from (6.4.41) that $\hat{e}_1 \in \mathcal{L}_\infty \cap \mathcal{L}_2$, which, together with $\epsilon_1 \in \mathcal{L}_\infty \cap \mathcal{L}_2$, implies that $e_1 \in \mathcal{L}_\infty \cap \mathcal{L}_2$. Because $\dot{e}_1 \in \mathcal{L}_\infty$ and $e_1 \in \mathcal{L}_2$, it follows that $e_1(t) \to 0$ as $t \to \infty$ and the proof of (i) is complete. From $r_0 \in \mathcal{L}_2, \dot{r}_0 \in \mathcal{L}_\infty$ we also have $r_0(t) \to 0$ as $t \to \infty$.

6.4. DIRECT MRAC WITH UNNORMALIZED ADAPTIVE LAWS

(ii) First, we show that ϕ is PE if r is sufficiently rich of order $2n$. Using the expression (6.4.50) for ϕ and substituting $y_p = W_m r + e_1$, we can write

$$\phi = \phi_m + \bar{\phi}$$

where

$$\phi_m = \frac{1}{(s+p_0)(s+p_1)} \begin{bmatrix} (sI - F)^{-1} g G_p^{-1} W_m \\ (sI - F)^{-1} g W_m \\ W_m \\ 1 \end{bmatrix} r$$

and

$$\bar{\phi} = \frac{1}{(s+p_0)(s+p_1)} \begin{bmatrix} (sI - F)^{-1} g G_p^{-1} \\ (sI - F)^{-1} g \\ 1 \\ 0 \end{bmatrix} e_1$$

Using the same arguments as in the proof of Theorem 6.4.2, we can establish that ϕ_m is PE provided r is sufficiently rich of order $2n$ and Z_p, R_p are coprime. Then the PE property of ϕ follows immediately from Lemma 4.8.3 and $e_1 \in \mathcal{L}_2$.

If k_p is known, then $\tilde{\rho} = 0$ and (6.4.49) is reduced to

$$\begin{aligned} \dot{\epsilon}_1 &= -q_0 \epsilon_1 + \rho^* \tilde{\theta}^T \phi \\ \dot{\tilde{\theta}} &= -\Gamma \epsilon_1 \phi \operatorname{sgn}(\rho^*) \end{aligned} \quad (6.4.51)$$

We can use the same steps as in the proof of Corollary 4.3.1 in Chapter 4 to show that the equilibrium $\epsilon_{1e} = 0, \tilde{\theta}_e = 0$ of (6.4.51) is e.s. provided ϕ is PE and $\phi, \dot{\phi} \in \mathcal{L}_\infty$. Since we have established in (i) that $\phi, \dot{\phi} \in \mathcal{L}_\infty$, (ii) follows.

(iii) When k_p is unknown, we have

$$\begin{aligned} \dot{\epsilon}_1 &= -q_0 \epsilon_1 + \rho^* \tilde{\theta}^T \phi - \tilde{\rho} r_0 \\ \dot{\tilde{\theta}} &= -\Gamma \epsilon_1 \phi \operatorname{sgn}(\rho^*) \end{aligned} \quad (6.4.52)$$

We consider (6.4.52) as a linear-time-varying system with $\epsilon_1, \tilde{\theta}$ as states and $\tilde{\rho} r_0$ as the external input. As shown in (ii), when $\tilde{\rho} r_0 = 0$ and ϕ is PE, the homogeneous part of (6.4.52) is e.s. We have shown in (i) that $\tilde{\rho} \in \mathcal{L}_\infty, r_0 \in \mathcal{L}_\infty \cap \mathcal{L}_2$ and $r_0(t) \to 0$ as $t \to \infty$. Therefore, it follows (by extending the results of Corollary 3.3.1) that $\epsilon_1, \tilde{\theta} \in \mathcal{L}_\infty \cap \mathcal{L}_2$ that $\epsilon_1(t), \tilde{\theta}(t) \to 0$ as $t \to \infty$.

(iv) Because $\epsilon_1, r_0 \in \mathcal{L}_2$ we have

$$\begin{aligned} \int_0^t |\dot{\tilde{\rho}}| d\tau &\leq \gamma \int_0^t |\epsilon_1| |r_0| d\tau \\ &\leq \gamma \left(\int_0^\infty \epsilon_1^2 d\tau\right)^{\frac{1}{2}} \left(\int_0^\infty r_0^2 d\tau\right)^{\frac{1}{2}} < \infty \end{aligned}$$

which for $t \to \infty$ implies that $\dot\rho = \dot{\tilde\rho} \in L_1$ and therefore $\rho, \tilde\rho$ converge to a constant as $t \to \infty$. $\square$

Example 6.4.3 Let us consider the third order plant

$$y_p = \frac{k_p}{s^3 + a_2 s^2 + a_1 s + a_0} u_p$$

where k_p, a_0, a_1, a_2 are unknown constants, and the sign of k_p is assumed to be known. The control objective is to choose u_p to stabilize the plant and force the output y_p to track the output y_m of the reference model given by

$$y_m = \frac{1}{(s+2)^3} r$$

Because $n^* = 3$, the MRAC scheme in Table 6.3 is considered. We choose $p_1 = p_0 = 2$ so that $W_m(s)(s+p_1)(s+p_0) = \frac{1}{s+2}$ is SPR. The signals w, ϕ, ϕ_1 are generated as

$$\dot\omega_1 = \begin{bmatrix} -10 & -25 \\ 1 & 0 \end{bmatrix} \omega_1 + \begin{bmatrix} 1 \\ 0 \end{bmatrix} u_p, \quad \omega_1(0) = \begin{bmatrix} 0 \\ 0 \end{bmatrix}$$

$$\dot\omega_2 = \begin{bmatrix} -10 & -25 \\ 1 & 0 \end{bmatrix} \omega_2 + \begin{bmatrix} 1 \\ 0 \end{bmatrix} y_p, \quad \omega_2(0) = \begin{bmatrix} 0 \\ 0 \end{bmatrix}$$

$$\omega = [\omega_1^T, \omega_2^T, y_p, r]^T$$

$$\phi_1 = \frac{1}{s+2}\omega, \quad \phi = \frac{1}{s+2}\phi_1$$

by choosing $\Lambda(s) = (s+5)^2$. Then, according to Table 6.3, the adaptive control law that achieves the control objective is given by

$$u_p = \theta^T \omega - \epsilon_1 \mathrm{sgn}(k_p) \phi^T \Gamma \phi_1 - 4\alpha_0 \phi^T \Gamma \phi (\phi^T \Gamma \dot\phi) r_0 + \alpha_0^2 (\phi^T \Gamma \phi)^4 r_0 - \alpha_0 (\phi^T \Gamma \phi)^3 \epsilon_1 \mathrm{sgn}(k_p)$$

$$\dot\theta = -\Gamma \epsilon_1 \phi \mathrm{sgn}(k_p), \quad \dot\rho = \gamma \epsilon_1 r_0$$

$$\epsilon_1 = e_1 - \hat e_1, \quad e_1 = y_p - y_m, \quad \hat e_1 = \frac{1}{s+2} \rho r_0$$

and r_0 is generated by the equation

$$\dot r_0 = -(2 + \alpha_0 (\phi^T \Gamma \phi)^2) r_0 + \phi^T \Gamma \phi \epsilon_1 \mathrm{sgn}(k_p)$$

The simulation results of the MRAC scheme for a unit step reference input are shown in Figure 6.12. The plant used for simulations is an unstable one with transfer function $G_p(s) = \frac{1}{s^3 + 6s^2 + 3s - 10}$. The initial conditions for the controller parameters are: $\theta_0 = [1.2, -9, 31, 160, -50, 9]^T$, and $\rho(0) = 0.2$. The design parameters used for simulations are: $\Gamma = 50 I, \gamma = 50, \alpha_0 = 0.01$. $\triangledown$

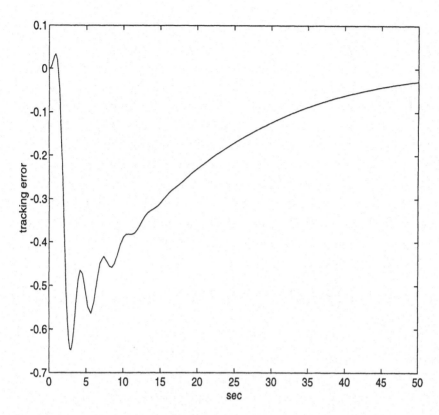

Figure 6.12 Response of the MRAC scheme for Example 6.4.3 with $r(t)=$ unit step function.

Remark 6.4.6 The effect of initial conditions can be taken into account by using
$$e_1 = W_m(s)\rho^*(u_p - \theta^{*\top}\omega) + C_c^\top(sI - A_c)^{-1}e(0)$$
instead of (6.4.33). The proof can be easily modified to take care of the exponentially decaying to zero term that is due to $e(0) \neq 0$ (see Problem 6.19).

Similarly, the results presented here are valid only if the existence and uniqueness of solutions of (6.4.49) can be established. For further discussion and details on the existence and uniqueness of solutions of the class of differential equations that arise in adaptive systems, the reader is referred to [191].

Remark 6.4.7 The procedure for $n^* = 3$ may be extended to the case of $n^* > 3$ by following similar steps. The complexity of the control input u_p, however, increases considerably with n^* to the point that it defeats any simplicity we may gain from analysis by using a single Lyapunov-like function to establish stability. In addition to complexity the highly nonlinear terms in the control law may lead to a "high bandwidth" control input that may have adverse effects on robustness with respect to modeling errors. We will address some of these robustness issues in Chapter 8. On the other hand, the idea of unnormalized adaptive laws together with the nonlinear modification of the certainty equivalence control laws were found to be helpful in solving the adaptive control problem for a class of nonlinear plants [98, 99, 105].

6.5 Direct MRAC with Normalized Adaptive Laws

In this section we present and analyze a class of MRAC schemes that dominated the literature of adaptive control due to the simplicity of their design as well as their robustness properties in the presence of modeling errors. Their design is based on the certainty equivalence approach that combines a control law, motivated from the known parameter case, with an adaptive law generated using the techniques of Chapter 4. The adaptive law is driven by the normalized estimation error and is based on an appropriate parameterization of the plant that involves the unknown desired controller parameters. While the design of normalized MRAC schemes follows directly from the results of Section 6.3 and Chapter 4, their analysis is more complicated than that of the unnormalized MRAC schemes presented in Section 6.4 for the case of $n^* = 1, 2$. However, their analysis, once understood, carries over to all relative degrees of the plant without additional complications.

6.5.1 Example: Adaptive Regulation

Let us consider the scalar plant

$$\dot{x} = ax + u, \quad x(0) = x_0 \qquad (6.5.1)$$

6.5. DIRECT MRAC WITH NORMALIZED ADAPTIVE LAWS

where a is an unknown constant and $-a_m$ is the desired closed-loop pole for some $a_m > 0$.

The desired control law

$$u = -k^* x, \quad k^* = a + a_m$$

that could be used to meet the control objective when a is known, is replaced with

$$u = -k(t)x \qquad (6.5.2)$$

where $k(t)$ is to be updated by an appropriate adaptive law. In Section 6.2.1 we updated $k(t)$ using an unnormalized adaptive law driven by the estimation error, which was shown to be equal to the regulation error x. In this section, we use normalized adaptive laws to update $k(t)$. These are adaptive laws driven by the normalized estimation error which is not directly related to the regulation error x. As a result, the stability analysis of the closed-loop adaptive system is more complicated.

As shown in Section 6.2.1, by adding and subtracting the term $-k^* x$ in the plant equation (6.5.1) and using $k^* = a + a_m$ to eliminate the unknown a, we can obtain the parametric plant model

$$\dot{x} = -a_m x + k^* x + u$$

whose transfer function form is

$$x = \frac{1}{s + a_m}(k^* x + u) \qquad (6.5.3)$$

If we now put (6.5.3) into the form of the general parametric model $z = W(s)\theta^{*\top}\psi$ considered in Chapter 4, we can simply pick any adaptive law for estimating k^* on-line from Tables 4.1 to 4.3 of Chapter 4. Therefore, let us rewrite (6.5.3) as

$$z = \frac{1}{s + a_m} k^* x \qquad (6.5.4)$$

where $z = x - \frac{1}{s+a_m} u$ is available from measurement.

Using Table 4.1 of Chapter 4, the SPR-Lyapunov design approach gives the adaptive law

$$\dot{k} = \gamma \epsilon x$$

$$\epsilon = z - \hat{z} - \frac{1}{s + a_m}\epsilon n_s^2 \qquad (6.5.5)$$

$$\hat{z} = \frac{1}{s + a_m}kx, \quad n_s^2 = x^2$$

where $\gamma > 0$ is the adaptive gain and $L(s)$ in Table 4.1 is taken as $L(s) = 1$.

Rewriting (6.5.4) as $z = k^*\phi$, $\phi = \frac{1}{s+a_m}x$, we use Table 4.2(A) to obtain the gradient algorithm

$$\dot{k} = \gamma\epsilon\phi$$

$$\epsilon = \frac{z - \hat{z}}{m^2}, \quad m^2 = 1 + \phi^2 \qquad (6.5.6)$$

$$\phi = \frac{1}{s + a_m}x, \quad \hat{z} = k\phi, \quad \gamma > 0$$

and from Table 4.3(A), the least-squares algorithm

$$\dot{k} = p\epsilon\phi, \quad \dot{p} = -\frac{p^2\phi^2}{m^2}, \quad p(0) > 0$$

$$\epsilon = \frac{z - \hat{z}}{m^2}, \quad \hat{z} = k\phi, \quad m^2 = 1 + \phi^2, \quad \phi = \frac{1}{s + a_m}x \qquad (6.5.7)$$

The control law (6.5.2) with any one of the three adaptive laws (6.5.5) to (6.5.7) forms an adaptive control scheme.

We analyze the stability properties of each scheme when applied to the plant (6.5.1) as follows: We start by writing the closed-loop plant equation as

$$x = \frac{1}{s + a_m}(-\tilde{k}x) \qquad (6.5.8)$$

by substituting $u = -kx$ in (6.5.3). As shown in Chapter 4 all three adaptive laws guarantee that $\tilde{k} \in \mathcal{L}_\infty$ independent of the boundedness of x, u, which implies from (6.5.8) that x cannot grow or decay faster than an exponential. However, the boundedness of $\tilde{k}$ by itself does not imply that $x \in \mathcal{L}_\infty$, let alone $x(t) \to 0$ as $t \to \infty$. To analyze (6.5.8), we need to exploit the properties of $\tilde{k}x$ by using the properties of the specific adaptive law that generates $k(t)$.

Let us start with the adaptive law (6.5.5). As shown in Chapter 4 using the Lyapunov-like function

$$V = \frac{\epsilon^2}{2} + \frac{\tilde{k}^2}{2\gamma}$$

6.5. DIRECT MRAC WITH NORMALIZED ADAPTIVE LAWS

and its time derivative

$$\dot V = -a_m \epsilon^2 - \epsilon^2 n_s^2 \leq 0$$

the adaptive law (6.5.5) guarantees that $\epsilon, \tilde k \in \mathcal{L}_\infty$ and $\epsilon, \epsilon n_s, \dot{\tilde k} \in \mathcal{L}_2$ independent of the boundedness of x. The normalized estimation error ϵ is related to $\tilde k x$ through the equation

$$\epsilon = z - \hat z - \frac{1}{s + a_m}\epsilon n_s^2 = \frac{1}{s + a_m}(-\tilde k x - \epsilon n_s^2) \qquad (6.5.9)$$

where $n_s^2 = x^2$. Using (6.5.9) and $\epsilon n_s^2 = \epsilon n_s x$ in (6.5.8), we obtain

$$x = \epsilon + \frac{1}{s + a_m}\epsilon n_s^2 = \epsilon + \frac{1}{s + a_m}\epsilon n_s x \qquad (6.5.10)$$

Because $\epsilon \in \mathcal{L}_\infty \cap \mathcal{L}_2$ and $\epsilon n_s \in \mathcal{L}_2$ the boundedness of x is established by taking absolute values on each side of (6.5.10) and applying the B-G lemma. We leave this approach as an exercise for the reader.

A more elaborate but yet more systematic method that we will follow in the higher order case involves the use of the properties of the $\mathcal{L}_{2\delta}$ norm and the B-G Lemma. We present such a method below and use it to understand the higher-order case to be considered in the sections to follow.

Step 1. *Express the plant output y (or state x) and plant input u in terms of the parameter error $\tilde k$.* We have

$$x = \frac{1}{s + a_m}(-\tilde k x), \quad u = (s - a)x = \frac{(s - a)}{s + a_m}(-\tilde k x) \qquad (6.5.11)$$

The above integral equations may be expressed in the form of algebraic inequalities by using the properties of the $\mathcal{L}_{2\delta}$ norm $\|(\cdot)_t\|_{2\delta}$, which for simplicity we denote by $\|\cdot\|$.

We have

$$\|x\| \leq c\|\tilde k x\|, \quad \|u\| \leq c\|\tilde k x\| \qquad (6.5.12)$$

where $c \geq 0$ is a generic symbol used to denote any finite constant. Let us now define

$$m_f^2 \triangleq 1 + \|x\|^2 + \|u\|^2 \qquad (6.5.13)$$

The significance of the signal m_f is that it bounds $|x|, |\dot x|$ and $|u|$ from above provided $k \in \mathcal{L}_\infty$. Therefore if we establish that $m_f \in \mathcal{L}_\infty$ then the boundedness of all signals follows. The boundedness of $|x|/m_f, |\dot x|/m_f, |u|/m_f$ follows from $\tilde{k} \in \mathcal{L}_\infty$ and the properties of the $\mathcal{L}_{2\delta}$-norm given by Lemma 3.3.2, i.e., from (6.5.11) we have

$$\frac{|x(t)|}{m_f} \leq \left\|\frac{1}{s+a_m}\right\|_{2\delta} |\tilde{k}| \frac{\|x\|}{m_f} \leq c$$

and

$$\frac{|\dot{x}(t)|}{m_f} \leq a_m \frac{|x(t)|}{m_f} + |\tilde{k}| \frac{|x(t)|}{m_f} \leq c$$

Similarly,

$$\frac{|u(t)|}{m_f} \leq |k| \frac{|x|}{m_f} \leq c$$

Because of the normalizing properties of m_f, we refer to it as the *fictitious normalizing signal*.

It follows from (6.5.12), (6.5.13) that

$$m_f^2 \leq 1 + c\|\tilde{k}x\|^2 \qquad (6.5.14)$$

Step 2. *Use the Swapping Lemma and properties of the $\mathcal{L}_{2\delta}$ norm to upper bound $\|\tilde{k}x\|$ with terms that are guaranteed by the adaptive law to have finite $\mathcal{L}_2$ gains.* We use the Swapping Lemma A.2 given in Appendix A to write the identity

$$\tilde{k}x = \left(1 - \frac{\alpha_0}{s+\alpha_0}\right)\tilde{k}x + \frac{\alpha_0}{s+\alpha_0}\tilde{k}x = \frac{1}{s+\alpha_0}(\dot{\tilde{k}}x + \tilde{k}\dot{x}) + \frac{\alpha_0}{s+\alpha_0}\tilde{k}x$$

where $\alpha_0 > 0$ is an arbitrary constant. Since, from (6.5.11), $\tilde{k}x = -(s+a_m)x$, we have

$$\tilde{k}x = \frac{1}{s+\alpha_0}(\dot{\tilde{k}}x + \tilde{k}\dot{x}) - \alpha_0 \frac{(s+a_m)}{(s+\alpha_0)}x \qquad (6.5.15)$$

which imply that

$$\|\tilde{k}x\| \leq \left\|\frac{1}{s+\alpha_0}\right\|_{\infty\delta} (\|\dot{\tilde{k}}x\| + \|\tilde{k}\dot{x}\|) + \alpha_0 \left\|\frac{s+a_m}{s+\alpha_0}\right\|_{\infty\delta} \|x\|$$

For $\alpha_0 > 2a_m > \delta$, we have $\|\frac{1}{s+\alpha_0}\|_{\infty\delta} = \frac{2}{2\alpha_0-\delta} < \frac{2}{\alpha_0}$, therefore,

$$\|\tilde{k}x\| \leq \frac{2}{\alpha_0}(\|\dot{\tilde{k}}x\| + \|\tilde{k}\dot{x}\|) + \alpha_0 c\|x\|$$

6.5. DIRECT MRAC WITH NORMALIZED ADAPTIVE LAWS

where $c = \|\frac{s+a_m}{s+\alpha_0}\|_{\infty\delta}$. Since $\frac{x}{m_f}, \frac{\dot{x}}{m_f} \in \mathcal{L}_\infty$, it follows that

$$\|\tilde{k}x\| \leq \frac{c}{\alpha_0}(\|\dot{\tilde{k}}m_f\| + \|\tilde{k}m_f\|) + \alpha_0 c\|x\| \qquad (6.5.16)$$

Equation (6.5.16) is independent of the adaptive law used to update $k(t)$. The term $\frac{c}{\alpha_0}\|\dot{\tilde{k}}m_f\|$ in (6.5.16) is "small" because $\dot{\tilde{k}} \in \mathcal{L}_2$ (guaranteed by any one of the adaptive laws (6.5.5) - (6.5.7)), whereas the term $\frac{c}{\alpha_0}\|\tilde{k}m_f\|$ can be made small by choosing α_0 large but finite. Large α_0, however, may make $\alpha_0 c\|x\|$ large unless $\|x\|$ is also small in some sense. We establish the smallness of the regulation error x by exploiting its relationship with the normalized estimation error ϵ. This relationship depends on the specific adaptive law used. For example, for the adaptive law (6.5.5) that is based on the SPR-Lyapunov design approach, we have established that

$$x = \epsilon + \frac{1}{s+a_m}\epsilon n_s^2$$

which together with $|\epsilon n_s^2| \leq |\epsilon n_s|\frac{|x|}{m_f}m_f \leq c\epsilon n_s m_f$ imply that

$$\|x\| \leq \|\epsilon\| + c\|\epsilon n_s m_f\|$$

hence,

$$\|\tilde{k}x\| \leq \frac{c}{\alpha_0}(\|\dot{\tilde{k}}m_f\| + \|\tilde{k}m_f\|) + \alpha_0 c\|\epsilon\| + \alpha_0 c\|\epsilon n_s m_f\| \qquad (6.5.17)$$

Similarly, for the gradient or least-squares algorithms, we have

$$x = \epsilon m^2 + \frac{1}{s+a_m}\dot{k}\phi \qquad (6.5.18)$$

obtained by using the equation

$$\frac{1}{s+a_m}kx = k\phi - \frac{1}{s+a_m}\dot{k}\phi$$

that follows from Swapping Lemma A.1 together with the equation for ϵm^2 in (6.5.6). Equation (6.5.18) implies that

$$\|x\| \leq \|\epsilon\| + \|\epsilon n_s^2\| + c\|\dot{k}\phi\|$$

Because $n_s^2 = \phi^2$ and $\phi = \frac{1}{s+a_m}x$, we have $|\phi(t)| \leq c\|x\|$ which implies that $\frac{\phi}{m_f} \in \mathcal{L}_\infty$ and, therefore,

$$\|x\| \leq \|\epsilon\| + \|\epsilon n_s m_f\| + c\|\dot{k} m_f\|$$

Substituting for $\|x\|$ in (6.5.16), we obtain the same expression for $\|\tilde{k}x\|$ as in (6.5.17).

Step 3. *Use the B-G Lemma to establish boundedness.* From (6.5.14) and (6.5.17), we obtain

$$m_f^2 \leq 1 + \alpha_0^2 c + \frac{c}{\alpha_0^2}(\|\dot{\tilde{k}} m_f\|^2 + \|\tilde{k} m_f\|^2) + c\alpha_0^2 \|\epsilon n_s m_f\|^2 \qquad (6.5.19)$$

by using the fact that $\epsilon \in \mathcal{L}_\infty \cap \mathcal{L}_2$. We can express (6.5.19) as

$$m_f^2 \leq 1 + \alpha_0^2 c + \frac{c}{\alpha_0^2}\|m_f\|^2 + c\alpha_0^2 \|\tilde{g} m_f\|^2 \qquad (6.5.20)$$

where $\tilde{g}^2 \triangleq |\epsilon n_s|^2 + \frac{|\dot{\tilde{k}}|^2}{\alpha_0^4}$. Because the adaptive laws guarantee that $\epsilon n_s, \dot{\tilde{k}} \in \mathcal{L}_2$ it follows that $\tilde{g} \in \mathcal{L}_2$. Using the definition of the $\mathcal{L}_{2\delta}$ norm, inequality (6.5.20) may be rewritten as

$$m_f^2 \leq 1 + c\alpha_0^2 + c\int_0^t e^{-\delta(t-\tau)}\left(\alpha_0^2 \tilde{g}^2(\tau) + \frac{1}{\alpha_0^2}\right) m_f^2(\tau)d\tau$$

Applying the B-G Lemma III, we obtain

$$m_f^2 \leq (1 + c\alpha_0^2)e^{-\delta(t-\tau)}\Phi(t,t_0) + (1 + c\alpha_0^2)\delta \int_{t_0}^t e^{-\delta(t-\tau)}\Phi(t,\tau)d\tau$$

where

$$\Phi(t,\tau) = e^{\frac{c}{\alpha_0^2}(t-\tau)} e^{c\int_\tau^t \alpha_0^2 \tilde{g}^2(\sigma)d\sigma}$$

Choosing α_0 so that $\frac{c}{\alpha_0^2} \leq \frac{\delta}{2}$, $\alpha_0 > 2a_m$ and using $\tilde{g} \in \mathcal{L}_2$, it follows that $m_f \in \mathcal{L}_\infty$. Because m_f bounds $x, \dot{x}, u$ from above, it follows that all signals in the closed-loop adaptive system are bounded.

Step 4. *Establish convergence of the regulation error to zero.* For the adaptive law (6.5.5), it follows from (6.5.9), (6.5.10) that $x \in \mathcal{L}_2$ and from (6.5.8) that $\dot{x} \in \mathcal{L}_\infty$. Hence, using Lemma 3.2.5, we have $x(t) \to 0$ as $t \to \infty$. For the adaptive law (6.5.6) or (6.5.7) we have from (6.5.18) that $x \in \mathcal{L}_2$ and from (6.5.8) that $\dot{x} \in \mathcal{L}_\infty$, hence, $x(t) \to 0$ as $t \to \infty$.

6.5.2 Example: Adaptive Tracking

Let us consider the tracking problem defined in Section 6.2.2 for the first order plant

$$\dot{x} = ax + bu \qquad (6.5.21)$$

where a, b are unknown (with $b \neq 0$). The control law

$$u = -k^*x + l^*r \qquad (6.5.22)$$

where

$$k^* = \frac{a_m + a}{b}, \quad l^* = \frac{b_m}{b} \qquad (6.5.23)$$

guarantees that all signals in the closed-loop plant are bounded and the plant state x converges exponentially to the state x_m of the reference model

$$x_m = \frac{b_m}{s + a_m} r \qquad (6.5.24)$$

Because a, b are unknown, we replace (6.5.22) with

$$u = -k(t)x + l(t)r \qquad (6.5.25)$$

where $k(t), l(t)$ are the on-line estimates of k^*, l^*, respectively. We design the adaptive laws for updating $k(t)$, $l(t)$ by first developing appropriate parametric models for k^*, l^* of the form studied in Chapter 4. We then choose the adaptive laws from Tables 4.1 to 4.5 of Chapter 4 based on the parametric model satisfied by k^*, l^*.

As in Section 6.2.2, if we add and subtract the desired input $-bk^*x + bl^*r$ in the plant equation (6.5.21) and use (6.5.23) to eliminate the unknown a, we obtain

$$\dot{x} = -a_m x + b_m r + b(u + k^*x - l^*r)$$

which together with (6.5.24) and the definition of $e_1 = x - x_m$ give

$$e_1 = \frac{b}{s + a_m}(u + k^*x - l^*r) \qquad (6.5.26)$$

Equation (6.5.26) can also be rewritten as

$$e_1 = b(\theta^{*T}\phi + u_f) \qquad (6.5.27)$$

where $\theta^* = [k^*, l^*]$, $\phi = \frac{1}{s+a_m}[x, -r]^T$, $u_f = \frac{1}{s+a_m}u$. Both equations are in the form of the parametric models given in Table 4.4 of Chapter 4. We can use them to choose any adaptive law from Table 4.4. As an example, let us choose the gradient algorithm listed in Table 4.4(D) that does not require the knowledge of sign b. We have

$$\dot{k} = N(w)\gamma_1 \epsilon \phi_1$$
$$\dot{l} = N(w)\gamma_2 \epsilon \phi_2$$
$$\dot{\hat{b}} = N(w)\gamma \epsilon \xi$$
$$N(w) = w^2 \cos w, \quad w = w_0 + \frac{\hat{b}^2}{2\gamma}$$
$$\dot{w}_0 = \epsilon^2 m^2, \quad w_0(0) = 0$$
$$\epsilon = \frac{e_1 - \hat{e}_1}{m^2}, \quad \hat{e}_1 = N(w)\hat{b}\xi \qquad (6.5.28)$$
$$\xi = k\phi_1 + l\phi_2 + u_f, \quad u_f = \frac{1}{s+a_m}u$$
$$\phi_1 = \frac{1}{s+a_m}x, \quad \phi_2 = -\frac{1}{s+a_m}r$$
$$m^2 = 1 + n_s^2, \quad n_s^2 = \phi_1^2 + \phi_2^2 + u_f^2$$
$$\gamma_1, \gamma_2, \gamma > 0$$

As shown in Chapter 4, the above adaptive law guarantees that $k, l, w, w_0 \in \mathcal{L}_\infty$ and $\epsilon, \epsilon n_s, \dot{k}, \dot{l}, \dot{\hat{b}} \in \mathcal{L}_\infty \cap \mathcal{L}_2$ independent of the boundedness of u, e_1, ϕ.

Despite the complexity of the adaptive law (6.5.28), the stability analysis of the closed-loop adaptive system described by the equations (6.5.21), (6.5.25), (6.5.28) is not more complicated than that of any other adaptive law from Table 4.4. We carry out the stability proof by using the properties of the $\mathcal{L}_{2\delta}$-norm and B-G Lemma in a similar way as in Section 6.5.1.

Step 1. *Express the plant output x and input u in terms of the parameter errors $\tilde{k}, \tilde{l}$.* From (6.5.24), (6.5.25) and (6.5.26) we have

$$x = x_m - \frac{b}{s+a_m}\left(\tilde{k}x - \tilde{l}r\right) = \frac{1}{s+a_m}\left(b_m r + b\tilde{l}r - b\tilde{k}x\right) \qquad (6.5.29)$$

6.5. DIRECT MRAC WITH NORMALIZED ADAPTIVE LAWS

and from (6.5.21), (6.5.29)

$$u = \frac{(s-a)}{b}x = \frac{(s-a)}{b(s+a_m)}\left[b_m r + b\tilde{l}r - b\tilde{k}x\right] \quad (6.5.30)$$

For simplicity, let us denote $\|(\cdot)_t\|_{2\delta}$ by $\|\cdot\|$. Again for the sake of clarity and ease of exposition, let us also denote any positive finite constant whose actual value does not affect stability with the same symbol c. Using the properties of the $\mathcal{L}_{2\delta}$-norm in (6.5.29), (6.5.30) and the fact that $r, \tilde{l} \in \mathcal{L}_\infty$ we have

$$\|x\| \leq c + c\|\tilde{k}x\|, \quad \|u\| \leq c + c\|\tilde{k}x\|$$

for any $\delta \in [0, 2a_m)$, which imply that the fictitious normalizing signal defined as

$$m_f^2 \triangleq 1 + \|x\|^2 + \|u\|^2$$

satisfies

$$m_f^2 \leq c + c\|\tilde{k}x\|^2 \quad (6.5.31)$$

We verify, using the boundedness of $r, \tilde{l}, \tilde{k}$, that $\phi_1/m_f, \dot{x}/m_f, n_s/m_f \in \mathcal{L}_\infty$ as follows: From the definition of ϕ_1, we have $|\phi_1(t)| \leq c\|x\| \leq cm_f$. Similarly, from (6.5.29) and the boundedness of $r, \tilde{l}, \tilde{k}$, we have

$$|x(t)| \leq c + c\|x\| \leq c + cm_f$$

Because $\dot{x} = -a_m x + b_m r + b\tilde{l}r - b\tilde{k}x$, it follows that $|\dot{x}| \leq c + cm_f$. Next, let us consider the signal $n_s^2 = 1 + \phi_1^2 + \phi_2^2 + u_f^2$. Because $|u_f| \leq c\|u\| \leq cm_f$ and $\frac{\phi_1}{m_f}, \phi_2 \in \mathcal{L}_\infty$, it follows that $n_s \leq cm_f$.

Step 2. *Use the Swapping Lemma and properties of the $\mathcal{L}_{2\delta}$ norm to upper bound $\|\tilde{k}x\|$ with terms that are guaranteed by the adaptive law to have finite $\mathcal{L}_2$-gains.* We start with the identity

$$\tilde{k}x = \left(1 - \frac{\alpha_0}{s+\alpha_0}\right)\tilde{k}x + \frac{\alpha_0}{s+\alpha_0}\tilde{k}x = \frac{1}{s+\alpha_0}\left(\dot{\tilde{k}}x + \tilde{k}\dot{x}\right) + \frac{\alpha_0}{s+\alpha_0}\tilde{k}x \quad (6.5.32)$$

where $\alpha_0 > 0$ is an arbitrary constant. From (6.5.29) we also have that

$$\tilde{k}x = -\frac{(s+a_m)}{b}e_1 + \tilde{l}r$$

where $e_1 = x - x_m$, which we substitute in the second term of the right-hand side of (6.5.32) to obtain

$$\tilde{k}x = \frac{1}{s+\alpha_0}\left(\dot{\tilde{k}}x + \tilde{k}\dot{x}\right) - \frac{\alpha_0(s+a_m)}{b(s+\alpha_0)}e_1 + \frac{\alpha_0}{s+\alpha_0}\tilde{l}r$$

Because $\tilde{k}, \tilde{l}, r \in \mathcal{L}_\infty$ we have

$$\|\tilde{k}x\| \leq \frac{c}{\alpha_0}\|\dot{\tilde{k}}x\| + \frac{c}{\alpha_0}\|\dot{x}\| + \alpha_0 c\|e_1\| + c \qquad (6.5.33)$$

for any $0 < \delta < 2a_m < \alpha_0$.

As in Section 6.5.1, the gain of the first two terms on the right-hand side of (6.5.33) can be reduced by choosing α_0 large. So the only term that needs further examination is $\alpha_0 c\|e_1\|$. The tracking error e_1, however, is related to the normalized estimation error ϵ through the equation

$$e_1 = \epsilon m^2 + N(w)\hat{b}\xi = \epsilon + \epsilon n_s^2 + N(w)\hat{b}\xi$$

that follows from (6.5.28).

Because $\epsilon, \epsilon n_s \in \mathcal{L}_\infty \cap \mathcal{L}_2$ and $N(w)\hat{b} \in \mathcal{L}_\infty$, the signal we need to concentrate on is ξ which is given by

$$\xi = k\phi_1 + l\phi_2 + \frac{1}{s+a_m}u$$

We consider the equation

$$\frac{1}{s+a_m}u = \frac{1}{s+a_m}(-kx + lr) = -k\phi_1 - l\phi_2 + \frac{1}{s+a_m}\left(\dot{k}\phi_1 + \dot{l}r\right)$$

obtained by using the Swapping Lemma A.1 or the equation

$$(s+a_m)(k\phi_1 + l\phi_2) = kx - lr + (\dot{k}\phi_1 + \dot{l}\phi_2) = u + \dot{k}\phi_1 + \dot{l}\phi_2$$

Using any one of the above equations to substitute for $k_1\phi_1 + l\phi_2$ in the equation for ξ we obtain

$$\xi = \frac{1}{s+a_m}(\dot{k}\phi_1 + \dot{l}\phi_2)$$

hence

$$e_1 = \epsilon + \epsilon n_s^2 + N(w)\hat{b}\frac{1}{s+a_m}(\dot{k}\phi_1 + \dot{l}\phi_2) \qquad (6.5.34)$$

6.5. DIRECT MRAC WITH NORMALIZED ADAPTIVE LAWS

Because $\epsilon, N(w), \hat{b}, \dot{l}, \phi_2, r \in \mathcal{L}_\infty$ it follows from (6.5.34) that

$$\|e_1\| \le c + \|\epsilon n_s^2\| + c\|\dot{k}\phi_1\| \tag{6.5.35}$$

and, therefore, (6.5.33) and (6.5.35) imply that

$$\|\tilde{k}x\| \le c + c\alpha_0 + \frac{c}{\alpha_0}\|\dot{\tilde{k}}x\| + \frac{c}{\alpha_0}\|\dot{x}\| + \alpha_0 c\|\epsilon n_s^2\| + \alpha_0 c\|\dot{k}\phi_1\| \tag{6.5.36}$$

Step 3. *Use the B-G Lemma to establish boundedness.* Using (6.5.36) and the normalizing properties of m_f, we can write (6.5.31) in the form

$$\begin{aligned}m_f^2 &\le c + c\alpha_0^2 + \frac{c}{\alpha_0^2}\|\dot{\tilde{k}}m_f\|^2 + \frac{c}{\alpha_0^2}\|m_f\|^2 + \alpha_0^2 c\|\epsilon n_s m_f\|^2 + \alpha_0^2 c\|\dot{k}m_f\|^2 \\ &\le c + c\alpha_0^2 + c\alpha_0^2\|\tilde{g}m_f\|^2 + \frac{c}{\alpha_0^2}\|m_f\|^2 \end{aligned} \tag{6.5.37}$$

where $\tilde{g}^2 \triangleq \frac{1}{\alpha_0^4}|\dot{k}|^2 + |\epsilon^2 n_s^2| + |\dot{k}|$. Inequality (6.5.37) has exactly the same form and properties as inequality (6.5.20) in Section 6.5.1. Therefore the boundedness of m_f follows by applying the B-G Lemma and choosing $\alpha_0^2 \ge \max\{4a_m^2, \frac{2c}{\delta}\}$ as in the example of Section 6.5.1.

From $m_f \in \mathcal{L}_\infty$ we have $x, \dot{x}, n_s, \phi_1 \in \mathcal{L}_\infty$, which imply that u and all signals in the closed-loop plant are bounded.

Step 4. *Establish convergence of the tracking error to zero.* We show the convergence of the tracking error to zero by using (6.5.34). From $\epsilon, \epsilon n_s, \dot{k}, \dot{l} \in \mathcal{L}_2$ and $n_s, N(w)\hat{b}, \phi_1, \phi_2 \in \mathcal{L}_\infty$ we can establish, using (6.5.34), that $e_1 \in \mathcal{L}_2$ which together with $\dot{e}_1 = \dot{x} - \dot{x}_m \in \mathcal{L}_\infty$ imply (see Lemma 3.2.5) that $e_1(t) \to 0$ as $t \to \infty$.

6.5.3 MRAC for SISO Plants

In this section we extend the design approach and analysis used in the examples of Sections 6.5.1 and 6.5.2 to the general SISO plant (6.3.1). We consider the same control objective as in Section 6.3.1 where the plant (6.3.1) and reference model (6.3.4) satisfy assumptions P1 to P4, and M1 and M2, respectively.

The design of MRAC schemes for the plant (6.3.1) with unknown parameters is based on the certainty equivalence approach and is conceptually simple. With this approach, we develop a wide class of MRAC schemes by combining the MRC law (6.3.22), where θ^* is replaced by its estimate $\theta(t)$, with different adaptive laws for generating $\theta(t)$ on-line. We design the adaptive laws by first developing appropriate parametric models for θ^* which we then use to pick up the adaptive law of our choice from Tables 4.1 to 4.5 in Chapter 4.

Let us start with the control law

$$u_p = \theta_1^T(t)\frac{\alpha(s)}{\Lambda(s)}u_p + \theta_2^T(t)\frac{\alpha(s)}{\Lambda(s)}y_p + \theta_3(t)y_p + c_0(t)r \qquad (6.5.38)$$

whose state-space realization is given by

$$\begin{aligned} \dot\omega_1 &= F\omega_1 + gu_p, \quad \omega_1(0) = 0 \\ \dot\omega_2 &= F\omega_2 + gy_p, \quad \omega_2(0) = 0 \\ u_p &= \theta^T\omega \end{aligned} \qquad (6.5.39)$$

where $\theta = [\theta_1^T, \theta_2^T, \theta_3, c_0]^T$ and $\omega = [\omega_1^T, \omega_2^T, y_p, r]^T$, and search for an adaptive law to generate $\theta(t)$, the estimate of the desired parameter vector θ^*.

In Section 6.4.1 we develop the bilinear parametric model

$$e_1 = W_m(s)\rho^*[u_p - \theta^{*T}\omega] \qquad (6.5.40)$$

where $\rho^* = \frac{1}{c_0^*}, \theta^* = [\theta_1^{*T}, \theta_2^{*T}, \theta_3^*, c_0^*]^T$ by adding and subtracting the desired control input $\theta^{*T}\omega$ in the overall representation of the plant and controller states (see (6.4.6)). The same parametric model may be developed by using the matching equation (6.3.13) to substitute for the unknown plant polynomial $R_p(s)$ in the plant equation and by cancelling the Hurwitz polynomial $Z_p(s)$. The parametric model (6.5.40) holds for any relative degree of the plant transfer function.

A linear parametric model for θ^* may be developed from (6.5.40) as follows: Because $\rho^* = \frac{1}{c_0^*}$ and $\theta^{*T}\omega = \theta_0^{*T}\omega_0 + c_0^*r$ where $\theta_0^* = [\theta_1^{*T}, \theta_2^{*T}, \theta_3^*]^T$ and $\omega_0 = [\omega_1^T, \omega_2^T, y_p]^T$, we rewrite (6.5.40) as

$$W_m(s)u_p = c_0^*e_1 + W_m(s)\theta_0^{*T}\omega_0 + c_0^*W_m(s)r$$

Substituting for $e_1 = y_p - y_m$ and using $y_m = W_m(s)r$ we obtain

$$W_m(s)u_p = c_0^*y_p + W_m(s)\theta_0^{*T}\omega_0$$

6.5. DIRECT MRAC WITH NORMALIZED ADAPTIVE LAWS

which may be written as
$$z = \theta^{*\mathsf{T}} \phi_p \qquad (6.5.41)$$

where
$$\begin{aligned}
z &= W_m(s) u_p \\
\phi_p &= [W_m(s)\omega_1^{\mathsf{T}}, W_m(s)\omega_2^{\mathsf{T}}, W_m(s)y_p, y_p]^{\mathsf{T}} \\
\theta^* &= [\theta_1^{*\mathsf{T}}, \theta_2^{*\mathsf{T}}, \theta_3^*, c_0^*]^{\mathsf{T}}
\end{aligned}$$

In view of (6.5.40), (6.5.41) we can now develop a wide class of MRAC schemes by using Tables 4.1 to 4.5 to choose an adaptive law for θ based on the bilinear parametric model (6.5.40) or the linear one (6.5.41).

Before we do that, let us compare the two parametric models (6.5.40), (6.5.41). The adaptive laws based on (6.5.40) listed in Table 4.4 of Chapter 4 generate estimates for c_0^* as well as for $\rho^* = \frac{1}{c_0^*}$. In addition, some algorithms require the knowledge of the $\mathrm{sgn}(\rho^*)$ and of a lower bound for $|\rho^*|$. On the other hand the adaptive laws based on the linear model (6.5.41) generate estimates of c_0^* only, without any knowledge of the $\mathrm{sgn}(\rho^*)$ or lower bound for $|\rho^*|$. This suggests that (6.5.41) is a more appropriate parameterization of the plant than (6.5.40). It turns out, however, that in the stability analysis of the MRAC schemes whose adaptive laws are based on (6.5.41), $1/c_0(t)$ is required to be bounded. This can be guaranteed by modifying the adaptive laws for $c_0(t)$ using projection so that $|c_0(t)| \geq \underline{c}_0 > 0, \forall t \geq 0$ for some constant $\underline{c}_0 \leq |c_0^*| = |\frac{k_m}{k_p}|$. Such a projection algorithm requires the knowledge of the $\mathrm{sgn}(c_0^*) = \mathrm{sgn}\left(\frac{k_m}{k_p}\right)$ and the lower bound $\underline{c}_0$, which is calculated from the knowledge of an upper bound for $|k_p|$. Consequently, as far as a priori knowledge is concerned, (6.5.41) does not provide any special advantages over (6.5.40). In the following we modify all the adaptive laws that are based on (6.5.41) using the gradient projection method so that $|c_0(t)| \geq \underline{c}_0 > 0 \ \forall t \geq 0$.

The main equations of several MRAC schemes formed by combining (6.5.39) with an adaptive law from Tables 4.1 to 4.5 based on (6.5.40) or (6.5.41) are listed in Tables 6.4 to 6.7.

The basic block diagram of the MRAC schemes described in Table 6.4 is shown in Figure 6.13. An equivalent representation that is useful for analysis is shown in Figure 6.14.

Table 6.4 MRAC schemes

Plant	$y_p = k_p \frac{Z_p(s)}{R_p(s)} u_p$
Reference model	$y_m = k_m \frac{Z_m(s)}{R_m(s)} r = W_m(s) r$
Control law	$\dot{\omega}_1 = F\omega_1 + g u_p,\ \omega_1(0) = 0$ $\dot{\omega}_2 = F\omega_2 + g y_p,\ \omega_2(0) = 0$ $u_p = \theta^\top \omega$ $\theta = [\theta_1^\top, \theta_2^\top, \theta_3, c_0]^\top,\ \omega = [\omega_1^\top, \omega_2^\top, y_p, r]^\top$ $\omega_i \in \mathcal{R}^{n-1},\ i = 1, 2$
Adaptive law	Any adaptive law from Tables 6.5, 6.6
Assumptions	Plant and reference model satisfy assumptions P1 to P4 and M1, M2 respectively.
Design variables	F, g chosen so that $(sI - F)^{-1} g = \frac{\alpha(s)}{\Lambda(s)}$, where $\alpha(s) = [s^{n-2}, s^{n-3}, \ldots, s, 1]^\top$ for $n \geq 2$ and $\alpha(s) = 0$ for $n = 1$; $\Lambda(s) = s^{n-1} + \lambda_{n-2} s^{n-2} + \cdots + \lambda_0$ is Hurwitz.

Figure 6.14 is obtained by rewriting u_p as

$$u_p = \theta^{*\top} \omega + \tilde{\theta}^\top \omega$$

where $\tilde{\theta} \triangleq \theta - \theta^*$, and by using the results of Section 6.4.1, in particular, equation (6.4.6) to absorb the term $\theta^{*\top} \omega$.

For $\tilde{\theta} = \theta - \theta^* = 0$, the closed-loop MRAC scheme shown in Figure 6.14 reverts to the one in the known parameter case shown in Figure 6.5. For $\tilde{\theta} \neq 0$, the stability of the closed-loop MRAC scheme depends very much on the properties of the input $\frac{1}{c_0^*} \tilde{\theta}^\top \omega$, which, in turn, depend on the properties of the adaptive law that generates the trajectory $\tilde{\theta}(t) = \theta(t) - \theta^*$.

Table 6.5 Adaptive laws based on $e_1 = W_m(s)\rho^*(u_p - \theta^{*T}\omega)$

\multicolumn{2}{c}{A. Based on the SPR-Lyapunov approach}
Parametric model	$e_1 = W_m(s)L(s)[\rho^*(u_f - \theta^{*T}\phi)]$
Adaptive law	$\dot{\theta} = -\Gamma\epsilon\phi \, \text{sgn}(k_p/k_m)$ $\dot{\rho} = \gamma\epsilon\xi, \quad \epsilon = e_1 - \hat{e}_1 - W_m(s)L(s)(\epsilon n_s^2)$ $\hat{e}_1 = W_m(s)L(s)[\rho(u_f - \theta^T\phi)]$ $\xi = u_f - \theta^T\phi, \quad \phi = L^{-1}(s)\omega$ $u_f = L^{-1}(s)u_p, \quad n_s^2 = \phi^T\phi + u_f^2$
Assumptions	Sign (k_p) is known
Design variables	$\Gamma = \Gamma^T > 0, \gamma > 0$; $W_m(s)L(s)$ is proper and SPR; $L^{-1}(s)$ is proper and has stable poles
\multicolumn{2}{c}{B. Gradient algorithm with known $\text{sgn}(k_p)$}
Parametric model	$e_1 = \rho^*(u_f - \theta^{*T}\phi)$
Adaptive law	$\dot{\theta} = -\Gamma\epsilon\phi \, \text{sgn}(k_p/k_m)$ $\dot{\rho} = \gamma\epsilon\xi$ $\epsilon = \frac{e_1 - \hat{e}_1}{m^2}$ $\hat{e}_1 = \rho(u_f - \theta^T\phi)$ $\phi = W_m(s)\omega, \quad u_f = W_m(s)u_p$ $\xi = u_f - \theta^T\phi$ $m^2 = 1 + \phi^T\phi + u_f^2$
Design variables	$\Gamma = \Gamma^T > 0, \gamma > 0$

Table 6.5 (Continued)

C. Gradient algorithm with unknown sgn(k_p)	
Parametric model	$e_1 = \rho^*(u_f - \theta^{*T}\phi)$
Adaptive law	$\dot{\theta} = -N(x_0)\Gamma\epsilon\phi$ $\dot{\rho} = N(x_0)\gamma\epsilon\xi$ $N(x_0) = x_0^2 \cos x_0$ $x_0 = w_0 + \frac{\rho^2}{2\gamma}$, $\dot{w}_0 = \epsilon^2 m^2$, $w_0(0) = 0$ $\epsilon = \frac{e_1 - \hat{e}_1}{m^2}$, $\hat{e}_1 = N(x_0)\rho(u_f - \theta^T\phi)$ $\phi = W_m(s)\omega$, $u_f = W_m(s)u_p$ $\xi = u_f - \theta^T\phi$, $m^2 = 1 + n_s^2$, $n_s^2 = \phi^T\phi + u_f^2$
Design variables	$\Gamma = \Gamma^T > 0$, $\gamma > 0$

The following theorem gives the stability properties of the MRAC scheme shown in Figures 6.13 and 14 when the adaptive laws in Tables 6.5 and 6.6 are used to update $\theta(t)$ on-line.

Theorem 6.5.1 *The closed-loop MRAC scheme shown in Figure 6.13 and described in Table 6.4 with any adaptive law from Tables 6.5 and 6.6 has the following properties:*

(i) *All signals are uniformly bounded.*

(ii) *The tracking error $e_1 = y_p - y_m$ converges to zero as $t \to \infty$.*

(iii) *If the reference input signal r is sufficiently rich of order $2n$, $\dot{r} \in \mathcal{L}_\infty$ and R_p, Z_p are coprime, the tracking error e_1 and parameter error $\tilde{\theta} = \theta - \theta^*$ converge to zero for the adaptive law with known sgn(k_p). The convergence is asymptotic in the case of the adaptive law of Table 6.5(A, B) and exponential in the case of the adaptive law of Table 6.6.*

6.5. DIRECT MRAC WITH NORMALIZED ADAPTIVE LAWS

Table 6.6 Adaptive laws based on $W_m(s)u_p = \theta^{*T}\phi_p$

Constraint	$g(\theta) = \underline{c}_0 - c_0 \operatorname{sgn} c_0 \leq 0$ $\theta = [\theta_1^T, \theta_2^T, \theta_3, c_0]^T$
Projection operator	$P_r[\Gamma x] \triangleq \begin{cases} \Gamma x & \text{if } |c_0(t)| > \underline{c}_0 \text{ or} \\ & \text{if } |c_0(t)| = \underline{c}_0 \text{ and } (\Gamma x)^T \nabla g \leq 0 \\ \Gamma x - \Gamma \frac{\nabla g \nabla g^T}{\nabla g^T \Gamma \nabla g} \Gamma x & \text{otherwise} \end{cases}$
A. Gradient Algorithm	
Adaptive law	$\dot{\theta} = Pr[\Gamma \epsilon \phi_p]$
Design variable	$\Gamma = \Gamma^T > 0$
B. Integral gradient aAlgorithm	
Adaptive law	$\dot{\theta} = Pr[-\Gamma(R\theta + Q)]$ $\dot{R} = -\beta R + \frac{\phi_p \phi_p^T}{m^2}, \quad R(0) = 0$ $\dot{Q} = -\beta Q - \frac{\phi_p}{m^2} z, \quad Q(0) = 0$ $z = W_m(s) u_p$
Design variable	$\Gamma = \Gamma^T > 0, \beta > 0$
C. Least-squares with covariance resetting	
Adaptive law	$\dot{\theta} = Pr[P\epsilon\phi_p]$ $\dot{P} = \begin{cases} \frac{-P\phi_p\phi_p^T P}{m^2} & \text{if } |c_0(t)| > \underline{c}_0 \text{ or} \\ & \text{if } |c_0(t)| = \underline{c}_0 \text{ and } (P\epsilon\phi_p)^T \nabla g \leq 0 \\ 0 & \text{otherwise} \end{cases}$ $P(t_r^+) = P_0 = \rho_0 I$
Design variable	$P(0) = P^T(0) > 0$, t_r is the time for which $\lambda_{min}(P(t)) \leq \rho_1, \rho_1 > \rho_0 > 0$
Common signals and variables	
	$\epsilon = \frac{W_m(s)u_p - \hat{z}}{m^2}, \hat{z} = \theta^T \phi_p, m^2 = 1 + \phi_p^T \phi_p$ $\phi_p = [W_m(s)\omega_1^T, W_m(s)\omega_2^T, W_m(s)y_p, y_p]^T$ $|c_0(0)| \geq \underline{c}_0, 0 < \underline{c}_0 \leq |c_0^*|, \ \operatorname{sgn}(c_0(0)) = \operatorname{sgn}(k_p/k_m)$

Table 6.7 Hybrid MRAC

Parametric model	$W_m(s)u_p = \theta^{*\top}\phi_p$
Control law	$\dot{\omega}_1 = F\omega_1 + gu_p$ $\dot{\omega}_2 = F\omega_2 + gy_p$ $u_p(t) = \theta_k^\top \omega(t), \quad t \in (t_k, t_{k+1}]$
Hybrid adaptive law	$\theta_{k+1} = \bar{P}r\left\{\theta_k + \Gamma\int_{t_k}^{t_{k+1}}\epsilon(\tau)\phi_p(\tau)d\tau\right\}; k=0,1,2,\ldots$ $\epsilon = \frac{W_m(s)u_p - \hat{z}}{m^2}, \quad \hat{z}(t) = \theta_k^\top \phi_p, \quad t \in (t_k, t_{k+1}]$ $m^2 = 1 + \beta\phi_p^\top\phi_p$ where $\bar{P}r[\cdot]$ is the discrete-time projection $\bar{P}r\{x\} \triangleq \begin{cases} x & \text{if } \|x_0\| \geq \underline{c}_0 \\ x + \frac{\gamma_0}{\gamma_{00}}(\underline{c}_0 - x_0)\text{sgn}(c_0) & \text{otherwise} \end{cases}$ where x_0 is the last element of the vector x; γ_0 is the last column of the matrix Γ, and γ_{00} is the last element of γ_0
Design variables	$\beta > 1; t_k = kT_s, T_s > 0; \Gamma = \Gamma^\top > 0$ $2 - T_s\lambda_{max}(\Gamma) > \gamma$ for some $\gamma > 0$

Outline of Proof The proof follows the same procedure as that for the examples presented in Sections 6.5.1, 6.5.2. It is completed in five steps.

Step 1. *Express the plant input and output in terms of the adaptation error $\tilde{\theta}^\top \omega$.* Using Figure 6.14 we can verify that the transfer function between the input $r + \frac{1}{c_0^*}\tilde{\theta}^\top \omega$ and the plant output y_p is given by

$$y_p = G_c(s)\left(r + \frac{1}{c_0^*}\tilde{\theta}^\top \omega\right)$$

where

$$G_c(s) = \frac{c_0^* k_p Z_p}{(1 - \theta_1^{*\top}(sI - F)^{-1}g)R_p - k_p Z_p[\theta_2^{*\top}(sI - F)^{-1}g + \theta_3^*]}$$

Because of the matching equations (6.3.12) and (6.3.13), and the fact that $(sI - F)^{-1}g = \frac{\alpha(s)}{\Lambda(s)}$, we have, after cancellation of all the stable common zeros and poles

6.5. DIRECT MRAC WITH NORMALIZED ADAPTIVE LAWS

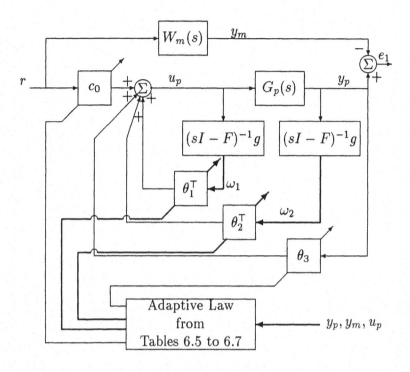

Figure 6.13 Block diagram of direct MRAC with normalized adaptive law.

in $G_c(s)$, that $G_c(s) = W_m(s)$. Therefore, the plant output may be written as

$$y_p = W_m(s)\left(r + \frac{1}{c_0^*}\tilde{\theta}^\top \omega\right) \quad (6.5.42)$$

Because $y_p = G_p(s)u_p$ and $G_p^{-1}(s)$ has stable poles, we have

$$u_p = G_p^{-1}(s)W_m(s)\left(r + \frac{1}{c_0^*}\tilde{\theta}^\top \omega\right) \quad (6.5.43)$$

where $G_p^{-1}(s)W_m(s)$ is biproper Because of Assumption M2.

We now define the fictitious normalizing signal m_f as

$$m_f^2 \triangleq 1 + \|u_p\|^2 + \|y_p\|^2 \quad (6.5.44)$$

where $\|\cdot\|$ denotes the $\mathcal{L}_{2\delta}$-norm for some $\delta > 0$. Using the properties of the $\mathcal{L}_{2\delta}$ norm it follows that

$$m_f \leq c + c\|\tilde{\theta}^\top \omega\| \quad (6.5.45)$$

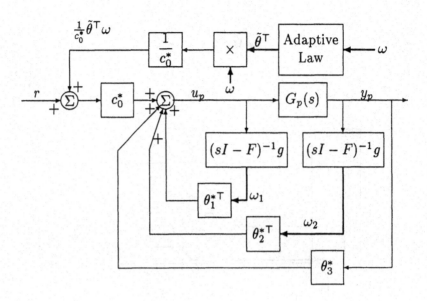

Figure 6.14 Equivalent representation of the MRAC scheme of Table 6.4.

where c is used to denote any finite constant and $\delta > 0$ is such that $W_m(s - \frac{\delta}{2}), G_p^{-1}(s - \frac{\delta}{2})$ have stable poles. Furthermore, for $\theta \in \mathcal{L}_\infty$ (guaranteed by the adaptive law), the signal m_f bounds most of the signals and their derivatives from above.

Step 2. *Use the Swapping Lemmas and properties of the $\mathcal{L}_{2\delta}$ norm to upper bound $\|\tilde{\theta}^T \omega\|$ with terms that are guaranteed by the adaptive law to have finite $\mathcal{L}_2$ gains.* This is the most complicated step and it involves the use of Swapping Lemmas A.1 and A.2 to obtain the inequality

$$\|\tilde{\theta}^T \omega\| \leq \frac{c}{\alpha_0} m_f + c\alpha_0^{n^*} \|\tilde{g} m_f\| \qquad (6.5.46)$$

where $\tilde{g}^2 \triangleq \epsilon^2 n_s^2 + |\dot{\theta}|^2 + \epsilon^2$ and $\tilde{g}$ is guaranteed by the adaptive law to belong to $\mathcal{L}_2$ and $\alpha_0 > 0$ is an arbitrary constant to be chosen.

Step 3. *Use the B-G Lemma to establish boundedness.* From (6.5.45) and (6.5.46), it follows that

$$m_f^2 \leq c + \frac{c}{\alpha_0^2} m_f^2 + c\alpha_0^{2n^*} \|\tilde{g} m_f\|^2 \qquad (6.5.47)$$

or

$$m_f^2 \leq c + c \int_0^t \alpha_0^{2n^*} \tilde{g}^2(\tau) m_f^2(\tau) d\tau$$

6.5. DIRECT MRAC WITH NORMALIZED ADAPTIVE LAWS

for any $\alpha_0 > \alpha_0^*$ and some $\alpha_0^* > 0$.

Applying the B-G Lemma and using $\tilde{g} \in \mathcal{L}_2$, the boundedness of m_f follows. Using $m_f \in \mathcal{L}_\infty$, we establish the boundedness of all the signals in the closed-loop plant.

Step 4. *Show that the tracking error converge to zero.* The convergence of e_1 to zero is established by showing that $e_1 \in \mathcal{L}_2$ and $\dot{e}_1 \in \mathcal{L}_\infty$ and using Lemma 3.2.5.

Step 5. *Establish that the parameter error converges to zero.* The convergence of the estimated parameters to their true values is established by first showing that the signal vector ϕ or ϕ_p, which drives the adaptive law under consideration, can be expressed as

$$\phi \text{ or } \phi_p = H(s)r + \bar{\phi}$$

where $H(s)$ is a stable transfer matrix and $\bar{\phi} \in \mathcal{L}_2$. If r is sufficiently rich of order $2n$ and Z_p, R_p are coprime then it follows from the results of Section 6.4 that $\phi_m = H(s)r$ is PE, which implies that ϕ or ϕ_p is PE. The PE property of ϕ_p or ϕ guarantees that $\tilde{\theta}$ and e_1 converge to zero as shown in Chapter 4.

A detailed proof of Theorem 6.5.1 is given in Section 6.8. $\square$

The MRAC scheme of Table 6.4 with any adaptive law from Table 6.5 guarantees that $\xi \in \mathcal{L}_2$ which together with $\epsilon \in \mathcal{L}_2$ implies that $\dot{\rho} \in \mathcal{L}_1$, i.e.,

$$\int_0^t |\dot{\rho}| d\tau \leq \gamma \left(\int_0^\infty \epsilon^2 d\tau \right)^{\frac{1}{2}} \left(\int_0^\infty \xi^2 d\tau \right)^{\frac{1}{2}} < \infty$$

which, in turn, implies that $\rho(t)$ converges to a constant as $t \to \infty$ independent of the richness of r.

The stability properties of the hybrid MRAC scheme of Table 6.7 are similar to the continuous-time MRAC schemes and are summarized by the following Theorem.

Theorem 6.5.2 *The closed-loop system obtained by applying the hybrid MRAC scheme of Table 6.7 to the plant given in Table 6.4 has the following properties:*

(i) *All signals are bounded.*

(ii) *The tracking error e_1 converges to zero as $t \to \infty$.*

(iii) *If r is sufficiently rich of order $2n$ and Z_p, R_p are coprime, then the parameter error $\tilde{\theta}_k = \theta_k - \theta^*$ converges exponentially fast to zero as k, $t \to \infty$.*

Proof First, we show that the projection algorithm used in Table 6.7 guarantees $|c_{ok}| \geq \underline{c}_0 \ \forall k \geq 0$ without affecting (i) to (iii) of the hybrid adaptive law established in Theorem 4.6.1. We rewrite the adaptive law of Table 6.7 as

$$\theta_{k+1} = \theta_{k+1}^p + \Delta p$$

$$\theta_{k+1}^p \triangleq \theta_k + \Gamma \int_{t_k}^{t_{k+1}} \epsilon(\tau)\phi_p(\tau) d\tau \quad (6.5.48)$$

$$\Delta p \triangleq \begin{cases} 0 & \text{if } |c_{0(k+1)}^p| \geq \underline{c}_0 \\ \frac{\gamma_0}{\gamma_{00}}(\underline{c}_0 - c_{0(k+1)}^p)\text{sgn}(c_0) & \text{otherwise} \end{cases}$$

where $c_{0(k+1)}^p$ is the last element of the vector θ_{k+1}^p. We can view θ_{k+1}^p, θ_{k+1} as the pre- and post-projection estimate of θ^* respectively. It is obvious, from the definition of γ_0, γ_{00} that $c_{0(k+1)} = \underline{c}_0 \text{sgn}(c_0)$ if $|c_{0(k+1)}^p| < \underline{c}_0$ and $c_{0(k+1)} = c_{0(k+1)}^p$ if $|c_{0(k+1)}^p| \geq \underline{c}_0$. Therefore, the constraint $c_{0k} \geq \underline{c}_0$ is satisfied for all k.

Now consider the same Lyapunov function used in proving Theorem 4.6.1, i.e.,

$$V(k) = \tilde{\theta}_k^T \Gamma^{-1} \tilde{\theta}_k$$

we have

$$V(k+1) = \tilde{\theta}_{k+1}^{p\,T} \Gamma^{-1} \tilde{\theta}_{k+1}^p + 2\tilde{\theta}_{k+1}^{p\,T} \Gamma^{-1} \Delta p + (\Delta p)^T \Gamma^{-1} \Delta p \quad (6.5.49)$$

In the proof of Theorem 4.6.1, we have shown that the first term in (6.5.49) satisfies

$$\tilde{\theta}_{k+1}^{p\,T} \Gamma^{-1} \tilde{\theta}_{k+1}^p \leq V(k) - (2 - T_s \lambda_m) \int_{t_k}^{t_{k+1}} \epsilon^2(\tau) m^2(\tau) d\tau \quad (6.5.50)$$

For simplicity, let us consider the case $\text{sgn}(c_0) = 1$. Exactly the same analysis can be carried out when $\text{sgn}(c_0) = -1$. Using $\Gamma^{-1}\Gamma = I$ and the definition of γ_0, γ_{00}, we have

$$\Gamma^{-1}\gamma_0 = \begin{bmatrix} 0 \\ 0 \\ \vdots \\ 0 \\ 1 \end{bmatrix}$$

Therefore, for $c_{0(k+1)}^p < \underline{c}_0$, the last two terms in (6.5.49) can be expressed as

$$2\tilde{\theta}_{k+1}^{p\,T}\Gamma^{-1}\Delta p + (\Delta p)^T \Gamma^{-1} \Delta p = \frac{1}{\gamma_{00}}(c_{0(k+1)}^p - c_0^*)(\underline{c}_0 - c_0^p) + \frac{1}{\gamma_{00}}(\underline{c}_0 - c_{0(k+1)}^p)^2$$

$$= \frac{1}{\gamma_{00}}(\underline{c}_0 - c_0^*)(\underline{c}_0 + c_{0(k+1)}^p - 2c_0^*) < 0$$

6.5. DIRECT MRAC WITH NORMALIZED ADAPTIVE LAWS

where the last inequality follows because $c_0^* > \underline{c}_0 > c_{0(k+1)}^p$. For $c_{0(k+1)}^p \geq \underline{c}_0$, we have $2\tilde{\theta}_{k+1}^{pT}\Gamma^{-1}\Delta p + (\Delta p)^T\Gamma^{-1}\Delta p = 0$. Hence,

$$2\tilde{\theta}_{k+1}^{pT}\Gamma^{-1}\Delta p + (\Delta p)^T\Gamma^{-1}\Delta p \leq 0, \quad \forall k \geq 0 \tag{6.5.51}$$

Combining (6.5.49)-(6.5.51), we have

$$\Delta V(k) \triangleq V(k+1) - V(k) \leq -(2 - T_s\lambda_m)\int_{t_k}^{t_{k+1}} \epsilon^2(\tau)m^2(\tau)d\tau$$

which is similar to (4.6.9) established for the hybrid adaptive law without projection. Therefore, the projection does not affect the ideal asymptotic properties of the hybrid adaptive law.

The rest of the stability proof is similar to that of Theorem 6.5.1 and is briefly outlined as follows: Noting that $\theta(t) = \theta_k, \forall t \in [kT_s, (k+1)T_s)$ is a piecewise constant function with discontinuities at $t = kT_s$, $k = 0, 1, \ldots$, and not differentiable, we make the following change in the proof to accommodate the discontinuity in θ_k. We write $\theta(t) = \bar{\theta}(t) + (\theta(t) - \bar{\theta}(t))$ where $\bar{\theta}(t)$ is obtained by linearly interpolating θ_k, θ_{k+1} on the interval $[kT_s, (k+1)T_s)$. Because $\theta_k \in \mathcal{L}_\infty, \Delta\theta_k \in \mathcal{L}_2$, we can show that $\bar{\theta}$ has the following properties: (i) $\bar{\theta}$ is continuous, (ii) $\dot{\bar{\theta}}(t) = \theta_{k+1} - \theta_k \forall t \in [kT_s, (k+1)T_s)$, and $\dot{\bar{\theta}} \in \mathcal{L}_\infty \cap \mathcal{L}_2$, (iii) $|\theta(t) - \bar{\theta}(t)| \leq |\theta_{k+1} - \theta_k|$ and $|\theta(t) - \bar{\theta}(t)| \in \mathcal{L}_\infty \cap \mathcal{L}_2$. Therefore, we can use $\bar{\theta}$ in the place of θ and the error resulting from this substitution is $(\theta - \bar{\theta})^T\omega$, which has an $\mathcal{L}_2$ gain because $|\theta - \bar{\theta}| \in \mathcal{L}_2$. The rest of the proof is then the same as that for the continuous scheme except that an additional term $(\theta - \bar{\theta})^T\omega$ appears in the equations. This term, however, doesnot affect the stability analysis since it has an $\mathcal{L}_2$ gain. $\square$

Remark 6.5.1 In the analysis of the MRAC schemes in this section we assume that the nonlinear differential equations describing the stability properties of the schemes possess a unique solution. This assumption is essential for the validity of our analysis. We can establish that these differential equations do possess a unique solution by using the results on existence and uniqueness of solutions of adaptive systems given in [191].

6.5.4 Effect of Initial Conditions

The analysis of the direct MRAC schemes with normalized adaptive laws presented in the previous sections is based on the assumption that the initial conditions for the plant and reference model are equal to zero. This assumption allows us to use transfer function representations and other I/O tools that help improve the clarity of presentation.

Nonzero initial conditions introduce exponentially decaying to zero terms in the parametric models (6.5.40) and (6.5.41) as follows:

$$
\begin{aligned}
e_1 &= W_m(s)\rho^*[u_p - \theta^{*\top}\omega] + \epsilon_t \\
z &= \theta^{*\top}\phi_p + \epsilon_t
\end{aligned}
\qquad (6.5.52)
$$

where $\epsilon_t = C_c^\top(sI - A_c)^{-1}e(0)$ and $A_c, e(0)$ are as defined in Section 6.4. Because A_c is a stable matrix, the properties of the adaptive laws based on (6.5.52) with $\epsilon_t = 0$ remain unchanged when $\epsilon_t \neq 0$ as established in Section 4.3.7. Similarly, the ϵ_t terms also appear in (6.5.42) and (6.5.43) as follows:

$$
\begin{aligned}
y_p &= W_m(s)(r + \frac{1}{c_0^*}\tilde{\theta}^\top\omega) + \epsilon_t \\
u_p &= G_p^{-1}(s)W_m(s)(r + \frac{1}{c_0^*}\tilde{\theta}^\top\omega) + \epsilon_t
\end{aligned}
\qquad (6.5.53)
$$

where in this case ϵ_t denotes exponentially decaying to zero terms because of nonzero initial conditions. The exponentially decaying term only contributes to the constant in the inequality

$$ m_f^2 \leq c + \epsilon_t + \frac{c}{\alpha_0^2}\|m_f\|^2 + \alpha_0^{2n^*}c\|\tilde{g}m_f\|^2 $$

where $\tilde{g} \in \mathcal{L}_2$. Applying the B-G Lemma to the above inequality we can establish, as in the case of $\epsilon_t = 0$, that $m_f \in \mathcal{L}_\infty$. Using $m_f \in \mathcal{L}_\infty$, the rest of the analysis follows using the same arguments as in the zero initial condition case.

6.6 Indirect MRAC

In the previous sections, we used the direct approach to develop stable MRAC schemes for controlling a wide class of plants with unknown parameters. The assumption on the plant and the special form of the controller enabled us to obtain appropriate parameterizations for the unknown controller vector θ^* that in turn allows us to develop adaptive laws for estimating the controller parameter vector $\theta(t)$ directly.

6.6. INDIRECT MRAC

An alternative way of controlling the same class of plants is to use the indirect approach, where the high frequency gain k_p and coefficients of the plant polynomials $Z_p(s), R_p(s)$ are estimated and the estimates are, in turn, used to determine the controller parameter vector $\theta(t)$ at each time t. The MRAC schemes based on this approach are referred to as *indirect MRAC schemes* since $\theta(t)$ is estimated indirectly using the plant parameter estimates.

The block diagram of an indirect MRAC scheme is shown in Figure 6.15. The coefficients of the plant polynomials $Z_p(s), R_p(s)$ and high frequency gain k_p are represented by the vector θ_p^*. The on-line estimate $\theta_p(t)$ of θ_p^*, generated by an adaptive law, is used to calculate the controller parameter vector $\theta(t)$ at each time t using the same mapping $f : \theta_p \mapsto \theta$ as the mapping $f : \theta_p^* \mapsto \theta^*$ defined by the matching equations (6.3.12), (6.3.16), (6.3.17). The adaptive law generating θ_p may share the same filtered values of u_p, y_p, i.e., ω_1, ω_2 as the control law leading to some further interconnections not shown in Figure 6.15.

In the following sections we develop a wide class of indirect MRAC schemes that are based on the same assumptions and have the same stability properties as their counterparts direct MRAC schemes developed in Sections 6.4 and 6.5.

6.6.1 Scalar Example

Let us consider the plant
$$\dot{x} = ax + bu \quad (6.6.1)$$

where a, b are unknown constants and $\text{sgn}(b)$ is known. It is desired to choose u such that all signals in the closed-loop plant are bounded and the plant state x tracks the state x_m of the reference model

$$\dot{x}_m = -a_m x_m + b_m r \quad (6.6.2)$$

where $a_m > 0$, b_m and the reference input signal r are chosen so that $x_m(t)$ represents the desired state response of the plant.

Control Law As in Section 6.2.2, if the plant parameters a, b were known, the control law
$$u = -k^* x + l^* r \quad (6.6.3)$$

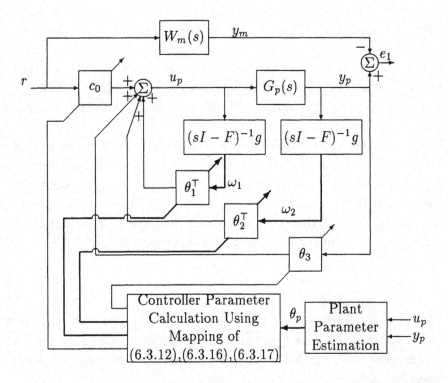

Figure 6.15 Block diagram of an indirect MRAC scheme.

with
$$k^* = \frac{a_m + a}{b}, \quad l^* = \frac{b_m}{b} \tag{6.6.4}$$

could be used to meet the control objective. In the unknown parameter case, we propose
$$u = -k(t)x + l(t)r \tag{6.6.5}$$

where $k(t), l(t)$ are the on-line estimates of k^*, l^* at time t, respectively. In direct adaptive control, $k(t), l(t)$ are generated directly by an adaptive law. In indirect adaptive control, we follow a different approach. We evaluate $k(t), l(t)$ by using the relationship (6.6.4) and the estimates $\hat{a}, \hat{b}$ of the unknown parameters a, b as follows:

$$k(t) = \frac{a_m + \hat{a}(t)}{\hat{b}(t)}, \quad l(t) = \frac{b_m}{\hat{b}(t)} \tag{6.6.6}$$

6.6. INDIRECT MRAC

where $\hat{a}, \hat{b}$ are generated by an adaptive law that we design.

Adaptive Law The adaptive law for generating $\hat{a}, \hat{b}$ is obtained by following the same procedure as in the identification examples of Chapter 4, i.e., we rewrite (6.6.1) as

$$x = \frac{1}{s+a_m}[(a+a_m)x + bu]$$

and generate $\hat{x}$, the estimate of x, from

$$\hat{x} = \frac{1}{s+a_m}[(\hat{a}+a_m)x + \hat{b}u] = x_m \qquad (6.6.7)$$

where the last equality is obtained by using (6.6.5), (6.6.6). As in Section 6.2.2, the estimation error $\epsilon_1 = x - x_m = e_1$ is the same as the tracking error and satisfies the differential equation

$$\dot{e}_1 = -a_m e_1 - \tilde{a}x - \tilde{b}u \qquad (6.6.8)$$

where

$$\tilde{a} \stackrel{\triangle}{=} \hat{a} - a, \quad \tilde{b} \stackrel{\triangle}{=} \hat{b} - b$$

are the parameter errors. Equation (6.6.8) motivates the choice of

$$V = \frac{1}{2}\left(e_1^2 + \frac{\tilde{a}^2}{\gamma_1} + \frac{\tilde{b}^2}{\gamma_2}\right) \qquad (6.6.9)$$

for some $\gamma_1, \gamma_2 > 0$, as a potential Lyapunov-like function candidate for (6.6.8). The time derivative of V along any trajectory of (6.6.8) is given by

$$\dot{V} = -a_m e_1^2 - \tilde{a}xe_1 - \tilde{b}ue_1 + \frac{\tilde{a}\dot{\tilde{a}}}{\gamma_1} + \frac{\tilde{b}\dot{\tilde{b}}}{\gamma_2} \qquad (6.6.10)$$

Hence, for

$$\dot{\tilde{a}} = \dot{\hat{a}} = \gamma_1 e_1 x, \quad \dot{\tilde{b}} = \dot{\hat{b}} = \gamma_2 e_1 u \qquad (6.6.11)$$

we have

$$\dot{V} = -a_m e_1^2 \leq 0$$

which implies that $e_1, \hat{a}, \hat{b} \in \mathcal{L}_\infty$ and that $e_1 \in \mathcal{L}_2$ by following the usual arguments. Furthermore, $x_m, e_1 \in \mathcal{L}_\infty$ imply that $x \in \mathcal{L}_\infty$. The boundedness of u, however, cannot be established unless we show that $k(t), l(t)$

are bounded. The boundedness of $\frac{1}{\hat{b}}$ and therefore of $k(t), l(t)$ cannot be guaranteed by the adaptive law (6.6.11) because (6.6.11) may generate estimates $\hat{b}(t)$ arbitrarily close or even equal to zero. The requirement that $\hat{b}(t)$ is bounded away from zero is a controllability condition for the estimated plant that the control law (6.6.5) is designed for. One method for avoiding $\hat{b}(t)$ going through zero is to modify the adaptive law for $\hat{b}(t)$ so that adaptation takes place in a closed subset of $\mathcal{R}^1$ which doesnot include the zero element. Such a modification is achieved by using the following a priori knowledge:

The sgn(b) and a lower bound $b_0 > 0$ for $|b|$ is known $\hspace{2em}$ (A2)

Applying the projection method with the constraint $\hat{b}\,\text{sgn}(b) \geq b_0$ to the adaptive law (6.6.11), we obtain

$$\dot{\hat{a}} = \gamma_1 e_1 x, \quad \dot{\hat{b}} = \begin{cases} \gamma_2 e_1 u & \text{if } |\hat{b}| > b_0 \text{ or} \\ & \text{if } |\hat{b}| = b_0 \text{ and } e_1 u\,\text{sgn}(b) \geq 0 \\ 0 & \text{otherwise} \end{cases} \quad (6.6.12)$$

where $\hat{b}(0)$ is chosen so that $\hat{b}(0)\text{sgn}(b) \geq b_0$.

Analysis It follows from (6.6.12) that if $\hat{b}(0)\text{sgn}(b) \geq b_0$, then whenever $\hat{b}(t)\text{sgn}(b) = |\hat{b}(t)|$ becomes equal to b_0 we have $\dot{\hat{b}}\hat{b} \geq 0$ which implies that $|\hat{b}(t)| \geq b_0$, $\forall t \geq 0$. Furthermore the time derivative of (6.6.9) along the trajectory of (6.6.8), (6.6.12) satisfies

$$\dot{V} = \begin{cases} -a_m e_1^2 & \text{if } |\hat{b}| > b_0 \text{ or } |\hat{b}| = b_0 \text{ and } e_1 u\,\text{sgn}(b) \geq 0 \\ -a_m e_1^2 - \tilde{b} e_1 u & \text{if } |\hat{b}| = b_0 \text{ and } e_1 u\,\text{sgn}(b) < 0 \end{cases}$$

Now for $|\hat{b}| = b_0$, we have $(\hat{b} - b)\text{sgn}(b) < 0$. Therefore, for $|\hat{b}| = b_0$ and $e_1 u\,\text{sgn}(b) < 0$, we have

$$\tilde{b} e_1 u = (\hat{b} - b) e_1 u = (\hat{b} - b)\text{sgn}(b)(e_1 u \text{sgn}(b)) > 0$$

which implies that

$$\dot{V} \leq -a_m e_1^2 \leq 0, \quad \forall t \geq 0$$

Therefore, the function V given by (6.6.9) is a Lyapunov function for the system (6.6.8), (6.6.12) since u, x in (6.6.8) can be expressed in terms of e_1 and x_m where $x_m(t)$ is treated as an arbitrary bounded function of time.

6.6. INDIRECT MRAC

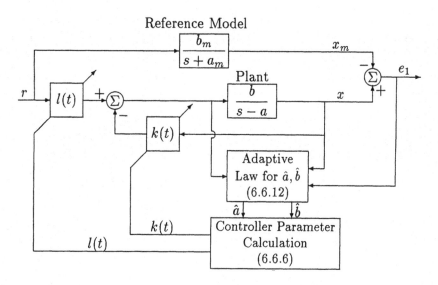

Figure 6.16 Block diagram for implementing the indirect MRAC scheme given by (6.6.5), (6.6.6), and (6.6.12).

Hence the equilibrium $e_{1e} = 0, \hat{a}_e = a, \hat{b}_e = b$ is u.s. and $e_1, \hat{b}, \hat{a} \in \mathcal{L}_\infty$. Using the usual arguments, we have $e_1 \in \mathcal{L}_2$ and $\dot{e}_1 \in \mathcal{L}_\infty$ which imply that $e_1(t) = x(t) - x_m(t) \to 0$ as $t \to \infty$ and therefore that $\dot{\hat{a}}(t), \dot{\hat{b}}(t) \to 0$ as $t \to \infty$.

As in the direct case it can be shown that if the reference input signal $r(t)$ is sufficiently rich of order 2 then $\tilde{b}, \tilde{a}$ and, therefore, $\tilde{k}, \tilde{l}$ converge to zero exponentially fast.

Implementation The proposed indirect MRAC scheme for (6.6.1) described by (6.6.5), (6.6.6), and (6.6.12) is implemented as shown in Figure 6.16.

6.6.2 Indirect MRAC with Unnormalized Adaptive Laws

As in the case of direct MRAC considered in Section 6.5, we are interested in extending the indirect MRAC scheme for the scalar plant of Section 6.6.1 to a higher order plant. The basic features of the scheme of Section 6.6.1 is that the adaptive law is driven by the tracking error and a single Lyapunov function is used to design the adaptive law and establish signal boundedness.

In this section, we extend the results of Section 6.6.1 to plants with relative degree $n^* = 1$. The same methodology is applicable to the case of $n^* \geq 2$ at the expense of additional algebraic manipulations. We assign these more complex cases as problems for the ambitious reader in the problem section.

Let us start by considering the same plant and control objective as in the direct MRAC scheme of Section 6.4.1 where the relative degree of the plant is assumed to be $n^* = 1$. We propose the same control law

$$\begin{aligned}
\dot{\omega}_1 &= F\omega_1 + gu_p, \quad \omega_1(0) = 0 \\
\dot{\omega}_2 &= F\omega_2 + gy_p, \quad \omega_2(0) = 0 \\
u_p &= \theta^T \omega
\end{aligned} \quad (6.6.13)$$

as in the direct MRAC case where $\theta(t)$ is calculated using the estimate of k_p and the estimates of the coefficients of the plant polynomials $Z_p(s), R_p(s)$, represented by the vector $\theta_p(t)$, at each time t. Our goal is to develop an adaptive law that generates the estimate $\theta_p(t)$ and specify the mapping from $\theta_p(t)$ to $\theta(t)$ that allows us to calculate $\theta(t)$ at each time t. We start with the mapping that relates the unknown vectors $\theta^* = \begin{bmatrix} \theta_1^{*T}, \theta_2^{*T}, \theta_3^*, c_0^* \end{bmatrix}^T$ and θ_p^* specified by the matching equations (6.3.12), (6.3.16), and (6.3.17) (with $Q(s) = 1$ due to $n^* = 1$), i.e.,

$$\begin{aligned}
c_0^* &= \frac{k_m}{k_p} \\
\theta_1^{*T} \alpha(s) &= \Lambda(s) - Z_p(s), \\
\theta_2^{*T} \alpha(s) + \theta_3^* \Lambda(s) &= \frac{R_p(s) - \Lambda_0(s) R_m(s)}{k_p}
\end{aligned} \quad (6.6.14)$$

To simplify (6.6.14) further, we express $Z_p(s), R_p(s), \Lambda(s), \Lambda_0(s)R_m(s)$ as

$$\begin{aligned}
Z_p(s) &= s^{n-1} + p_1^T \alpha_{n-2}(s) \\
R_p(s) &= s^n + a_{n-1} s^{n-1} + p_2^T \alpha_{n-2}(s) \\
\Lambda(s) &= s^{n-1} + \lambda^T \alpha_{n-2}(s) \\
\Lambda_0(s) R_m(s) &= s^n + r_{n-1} s^{n-1} + \nu^T \alpha_{n-2}(s)
\end{aligned}$$

where $p_1, p_2 \in \mathcal{R}^{n-1}$, a_{n-1} are the plant parameters, i.e., $\theta_p^* = [k_p, p_1^T, a_{n-1}, p_2^T]^T$; $\lambda, \nu \in \mathcal{R}^{n-1}$ and r_{n-1} are the coefficients of the known polynomials $\Lambda(s)$,

6.6. INDIRECT MRAC

$\Lambda_0(s)R_m(s)$ and $\alpha_{n-2}(s) = [s^{n-2}, s^{n-3}, \ldots, s, 1]^\top$, which we then substitute in (6.6.14) to obtain the equations

$$\begin{aligned}
c_0^* &= \frac{k_m}{k_p} \\
\theta_1^* &= \lambda - p_1 \\
\theta_2^* &= \frac{p_2 - a_{n-1}\lambda + r_{n-1}\lambda - \nu}{k_p} \\
\theta_3^* &= \frac{a_{n-1} - r_{n-1}}{k_p}
\end{aligned} \quad (6.6.15)$$

If we let $\hat{k}_p(t), \hat{p}_1(t), \hat{p}_2(t), \hat{a}_{n-1}(t)$ be the estimate of k_p, p_1, p_2, a_{n-1} respectively at each time t, then $\theta(t) = [\theta_1^\top, \theta_2^\top, \theta_3, c_0]^\top$ may be calculated as

$$\begin{aligned}
c_0(t) &= \frac{k_m}{\hat{k}_p(t)} \\
\theta_1(t) &= \lambda - \hat{p}_1(t) \\
\theta_2(t) &= \frac{\hat{p}_2(t) - \hat{a}_{n-1}(t)\lambda + r_{n-1}\lambda - \nu}{\hat{k}_p(t)} \\
\theta_3(t) &= \frac{\hat{a}_{n-1}(t) - r_{n-1}}{\hat{k}_p(t)}
\end{aligned} \quad (6.6.16)$$

provided $|\hat{k}_p(t)| \neq 0, \forall t \geq 0$.

The adaptive laws for generating $\hat{p}_1, \hat{p}_2, \hat{a}_{n-1}, \hat{k}_p$ on-line can be developed by using the techniques of Chapter 4. In this section we concentrate on adaptive laws that are driven by the tracking error e_1 rather than the normalized estimation error, and are developed using the SPR-Lyapunov design approach. We start with the parametric model given by equation (6.4.6), i.e.,

$$e_1 = W_m(s)\rho^*(u_p - \theta^{*\top}\omega) \quad (6.6.17)$$

where $\rho^* = \frac{1}{c_0^*} = \frac{k_p}{k_m}$. As in the direct case, we choose $W_m(s)$, the transfer function of the reference model, to be SPR with relative degree $n^* = 1$. The adaptive law for $\theta_p = [\hat{k}_p, \hat{p}_1^\top, \hat{a}_{n-1}, \hat{p}_2^\top]^\top$ is developed by first relating e_1 with the parameter error $\tilde{\theta}_p = \theta_p - \theta_p^*$ through the SPR transfer function $W_m(s)$ and then proceeding with the Lyapunov design approach as follows:

We rewrite (6.6.17) as

$$e_1 = W_m(s)\frac{1}{k_m}\left(k_p u_p - k_p \theta^{*\top}\omega - \hat{k}_p u_p + \hat{k}_p \theta^\top \omega\right) \tag{6.6.18}$$

where $-\hat{k}_p u_p + \hat{k}_p \theta^\top \omega = 0$ because of (6.6.13). If we now substitute for $k_p\theta^*, \hat{k}_p\theta$ from (6.6.15) and (6.6.16), respectively, in (6.6.18) we obtain

$$e_1 = W_m(s)\frac{1}{k_m}\left[\tilde{k}_p\left(\lambda^\top\omega_1 - u_p\right) + \tilde{p}_2^\top\omega_2 + \tilde{a}_{n-1}\left(y_p - \lambda^\top\omega_2\right) - \hat{k}_p\hat{p}_1^\top\omega_1 + k_p p_1^\top\omega_1\right]$$

where $\tilde{k}_p \triangleq \hat{k}_p - k_p, \tilde{p}_2 \triangleq \hat{p}_2 - p_2, \tilde{a}_{n-1} \triangleq \hat{a}_{n-1} - a_{n-1}$ are the parameter errors. Because $-\hat{k}_p \hat{p}_1^\top \omega_1 + k_p p_1^\top \omega_1 + k_p \hat{p}_1^\top \omega_1 - k_p \hat{p}_1^\top \omega_1 = -\tilde{k}_p \hat{p}_1^\top \omega_1 - k_p \tilde{p}_1^\top \omega_1$ we have

$$e_1 = W_m(s)\frac{1}{k_m}\left[\tilde{k}_p \xi_1 + \tilde{a}_{n-1}\xi_2 + \tilde{p}_2^\top\omega_2 - k_p \tilde{p}_1^\top\omega_1\right] \tag{6.6.19}$$

where

$$\xi_1 \triangleq \lambda^\top\omega_1 - u_p - \hat{p}_1^\top\omega_1, \quad \xi_2 \triangleq y_p - \lambda^\top\omega_2, \quad \tilde{p}_1 \triangleq \hat{p}_1 - p_1$$

A minimal state-space representation of (6.6.19) is given by

$$\begin{aligned}\dot{e} &= A_c e + B_c\left[\tilde{k}_p \xi_1 + \tilde{a}_{n-1}\xi_2 + \tilde{p}_2^\top\omega_2 - k_p \tilde{p}_1^\top\omega_1\right] \\ e_1 &= C_c^\top e\end{aligned} \tag{6.6.20}$$

where $C_c^\top(sI - A_c)^{-1}B_c = W_m(s)\frac{1}{k_m}$. Defining the Lyapunov-like function

$$V = \frac{e^\top P_c e}{2} + \frac{\tilde{k}_p^2}{2\gamma_p} + \frac{\tilde{a}_{n-1}^2}{2\gamma_1} + \frac{\tilde{p}_1^\top \Gamma_1^{-1}\tilde{p}_1}{2}|k_p| + \frac{\tilde{p}_2^\top \Gamma_2^{-1}\tilde{p}_2}{2}$$

where $P_c = P_c^\top > 0$ satisfies the algebraic equations of the LKY Lemma, $\gamma_1, \gamma_p > 0$ and $\Gamma_i = \Gamma_i^\top > 0$, $i = 1, 2$, it follows that by choosing the adaptive laws

$$\begin{aligned}\dot{\hat{a}}_{n-1} &= -\gamma_1 e_1 \xi_2 \\ \dot{\hat{p}}_1 &= \Gamma_1 e_1 \omega_1 \operatorname{sgn}(k_p) \\ \dot{\hat{p}}_2 &= -\Gamma_2 e_1 \omega_2 \\ \dot{\hat{k}}_p &= \begin{cases} -\gamma_p e_1 \xi_1 & \text{if } |\hat{k}_p| > k_0 \\ & \text{or if } |\hat{k}_p| = k_0 \text{ and } e_1\xi_1 \operatorname{sgn}(k_p) \leq 0 \\ 0 & \text{otherwise}\end{cases}\end{aligned} \tag{6.6.21}$$

6.6. INDIRECT MRAC

where $\hat{k}_p(0)\,\mathrm{sgn}(k_p) \geq k_0 > 0$ and k_0 is a known lower bound for $|k_p|$, we have

$$\dot{V} = \begin{cases} -e^\top \frac{qq^\top}{2} e - \nu_c e^\top \frac{L_c}{2} e & \text{if } |\hat{k}_p| > 0 \text{ or if } |\hat{k}_p| = k_0 \text{ and} \\ & e_1 \xi_1 \,\mathrm{sgn}(k_p) \leq 0 \\ -e^\top \frac{qq^\top}{2} e - \nu_c e^\top \frac{L_c}{2} e + e_1 \xi_1 \tilde{k}_p & \text{if } |\hat{k}_p| = k_0 \text{ and } e_1 \xi_1 \,\mathrm{sgn}(k_p) > 0 \end{cases}$$

where the scalar $\nu_c > 0$, matrix $L_c = L_c^\top > 0$ and vector q are defined in the LKY Lemma.

Because for $e_1 \xi_1 \,\mathrm{sgn}(k_p) > 0$ and $|\hat{k}_p| = k_0$ we have $(\hat{k}_p - k_p)\,\mathrm{sgn}(k_p) < 0$ and $e_1 \xi_1 \tilde{k}_p < 0$, it follows that

$$\dot{V} \leq -\nu_c \frac{e^\top L_c e}{2}$$

which implies that $e_1, e, \hat{a}_{n-1}, \hat{p}_1, \hat{p}_2, \hat{k}_p \in \mathcal{L}_\infty$ and $e, e_1 \in \mathcal{L}_2$. As in Section 6.4.1, $e_1 \in \mathcal{L}_\infty$ implies that $y_p, \omega_1, \omega_2 \in \mathcal{L}_\infty$, which, together with $\theta \in \mathcal{L}_\infty$ (guaranteed by (6.6.16) and the boundedness of $\theta_p, \frac{1}{\hat{k}_p}$), implies that $u_p \in \mathcal{L}_\infty$. Therefore, all signals in the closed-loop system are bounded. The convergence of e_1 to zero follows from $e_1 \in \mathcal{L}_2$ and $\dot{e}_1 \in \mathcal{L}_\infty$ guaranteed by (6.6.18).

We summarize the stability properties of above scheme, whose main equations are listed in Table 6.8, by the following theorem.

Theorem 6.6.1 *The indirect MRAC scheme shown in Table 6.8 guarantees that all signals are u.b., and the tracking error e_1 converges to zero as $t \to \infty$.*

We should note that the properties of Theorem 6.6.1 are established under the assumption that the calculation of $\theta(t)$ is performed instantaneously. This assumption is quite reasonable if we consider implementation with a fast computer. However, we can relax this assumption by using a hybrid adaptive law to update θ_p at discrete instants of time thus providing sufficient time for calculating $\theta(t)$. The details of such a hybrid scheme are left as an exercise for the reader.

As in the case of the direct MRAC schemes with unnormalized adaptive laws, the complexity of the indirect MRAC without normalization increases with the relative degree n^* of the plant. The details of such schemes for the case of $n^* = 2$ and higher are given as exercises in the problem section.

Table 6.8 Indirect MRAC scheme with unnormalized adaptive law for $n^* = 1$

Plant	$y_p = k_p \frac{Z_p(s)}{R_p(s)} u_p, \quad n^* = 1$
Reference model	$y_m = W_m(s)r, \quad W_m(s) = k_m \frac{Z_m(s)}{R_m(s)}$
Control law	$\dot{\omega}_1 = F\omega_1 + gu_p$ $\dot{\omega}_2 = F\omega_2 + gy_p$ $u_p = \theta^\top \omega$ $\theta = \begin{bmatrix} \theta_1^\top, \theta_2^\top, \theta_3, c_0 \end{bmatrix}; \omega = [\omega_1^\top, \omega_2^\top, y_p, r]^\top$
Adaptive law	$\dot{\hat{k}}_p = \begin{cases} -\gamma_p e_1 \xi_1 & \text{if } |\hat{k}_p| > k_0 \text{ or} \\ & \text{if } |\hat{k}_p| = k_0 \text{ and } e_1\xi_1\text{sgn}(k_p) \leq 0 \\ 0 & \text{otherwise} \end{cases}$ $\dot{\hat{a}}_{n-1} = -\gamma_1 e_1 \xi_2$ $\dot{\hat{p}}_1 = \Gamma_1 e_1 \omega_1 \, \text{sgn}(k_p)$ $\dot{\hat{p}}_2 = -\Gamma_2 e_1 \omega_2$ $e_1 = y_p - y_m$ $\xi_1 = \lambda^\top \omega_1 - u_p - \hat{p}_1^\top \omega_1; \quad \xi_2 = y_p - \lambda^\top \omega_2$
Calculation of $\theta(t)$	$c_0(t) = k_m/\hat{k}_p(t)$ $\theta_1(t) = \lambda - \hat{p}_1(t)$ $\theta_2(t) = (\hat{p}_2(t) - \hat{a}_{n-1}(t)\lambda + r_{n-1}\lambda - \nu)/\hat{k}_p(t)$ $\theta_3(t) = (\hat{a}_{n-1}(t) - r_{n-1})/\hat{k}_p(t)$
Design variables	k_0 : lower bound for $|k_p| \geq k_0 > 0$; $\lambda \in \mathcal{R}^{n-1}$: coefficient vector of $\Lambda(s) - s^{n-1}$; $r_{n-1} \in \mathcal{R}^1$: coefficient of s^{n-1} in $\Lambda_0(s)R_m(s)$; $\nu \in \mathcal{R}^{n-1}$: coefficient vector of $\Lambda_0(s)R_m(s) - s^n - r_{n-1}s^{n-1}$; $\Lambda(s), \Lambda_0(s)$ as defined in Section 6.4.1

6.6. INDIRECT MRAC

Another interesting class of indirect MRAC schemes with unnormalized adaptive laws is developed in [116, 117, 118] using a systematic recursive procedure, called backstepping. The procedure is based on a specific state space representation of the plant and leads to control and adaptive laws that are highly nonlinear.

6.6.3 Indirect MRAC with Normalized Adaptive Law

As in the direct MRAC case, the design of indirect MRAC with normalized adaptive laws is conceptually simple. The simplicity arises from the fact that the control and adaptive laws are designed independently and are combined using the certainty equivalence approach. As mentioned earlier, in indirect MRAC the adaptive law is designed to provide on-line estimates of the high frequency gain k_p and of the coefficients of the plant polynomials $Z_p(s), R_p(s)$ by processing the plant input and output measurements. These estimates are used to compute the controller parameters at each time t by using the relationships defined by the matching equations (6.3.12), (6.3.16), (6.3.17).

The adaptive law is developed by first expressing the plant in the form of a linear parametric model as shown in Chapters 2, 4, and 5. Starting with the plant equation (6.3.2) that we express in the form

$$y_p = \frac{b_m s^m + b_{m-1} s^{m-1} + \cdots + b_0}{s^n + a_{n-1} s^{n-1} + \cdots + a_0} u_p$$

where $b_m = k_p$ is the high frequency gain, and using the results of Section 2.4.1 we obtain the following plant parametric model:

$$z = \theta_p^{*\top} \phi \qquad (6.6.22)$$

where

$$z = \frac{s^n}{\Lambda_p(s)} y_p, \quad \phi = \left[\frac{\alpha_{n-1}^\top(s)}{\Lambda_p(s)} u_p, -\frac{\alpha_{n-1}^\top(s)}{\Lambda_p(s)} y_p \right]^\top$$

$$\theta_p^* = [\underbrace{0, \ldots, 0}_{n-m-1}, b_m, \cdots, b_0, a_{n-1}, \ldots, a_0]^\top$$

and $\Lambda_p(s) = s^n + \lambda_p^\top \alpha_{n-1}(s)$ with $\lambda_p = [\lambda_{n-1}, \ldots, \lambda_0]^\top$ is a Hurwitz polynomial. Since in this case m is known, the first $n - m - 1$ elements of θ_p^* are known to be equal to zero.

The parametric model (6.6.22) may be used to generate a wide class of adaptive laws by using Tables 4.2 to 4.5 from Chapter 4. Using the estimate $\theta_p(t)$ of θ_p^*, the MRAC law may be formed as follows:

The controller parameter vectors $\theta_1(t), \theta_2(t), \theta_3(t), c_0(t)$ in the control law

$$u_p = \theta_1^T \frac{\alpha(s)}{\Lambda(s)} u_p + \theta_2^T \frac{\alpha(s)}{\Lambda(s)} y_p + \theta_3 y_p + c_0 r = \theta^T \omega \quad (6.6.23)$$

where $\omega = \left[\frac{\alpha^T(s)}{\Lambda(s)} u_p, \frac{\alpha^T(s)}{\Lambda(s)} y_p, y_p, r \right]^T$, $\alpha(s) = \alpha_{n-2}(s)$ and $\theta = [\theta_1^T, \theta_2^T, \theta_3, c_0]^T$ is calculated using the mapping $\theta(t) = f(\theta_p(t))$. The mapping $f(\cdot)$ is obtained by using the matching equations (6.3.12), (6.3.16), (6.3.17), i.e.,

$$c_0^* = \frac{k_m}{k_p}$$

$$\theta_1^{*T} \alpha(s) = \Lambda(s) - Z_p(s) Q(s) \quad (6.6.24)$$

$$\theta_2^{*T} \alpha(s) + \theta_3^* \Lambda(s) = \frac{Q(s) R_p(s) - \Lambda_0(s) R_m(s)}{k_p}$$

where $Q(s)$ is the quotient of $\frac{\Lambda_0 R_m}{R_p}$ and $\Lambda(s) = \Lambda_0(s) Z_m(s)$. That is, if $\hat{R}_p(s,t), \hat{\bar{Z}}_p(s,t)$ are the estimated values of the polynomials $R_p(s), \bar{Z}_p(s) \triangleq k_p Z_p(s)$ respectively at each time t, then $c_0, \theta_1, \theta_2, \theta_3$ are obtained as solutions to the following polynomial equations:

$$c_0 = \frac{k_m}{\hat{k}_p}$$

$$\theta_1^T \alpha(s) = \Lambda(s) - \frac{1}{\hat{k}_p} \hat{\bar{Z}}_p(s,t) \cdot \hat{Q}(s,t) \quad (6.6.25)$$

$$\theta_2^T \alpha(s) + \theta_3 \Lambda(s) = \frac{1}{\hat{k}_p} [\hat{Q}(s,t) \cdot \hat{R}_p(s,t) - \Lambda_0(s) R_m(s)]$$

provided $\hat{k}_p \neq 0$, where $\hat{Q}(s,t)$ is the quotient of $\frac{\Lambda_0(s) R_m(s)}{\hat{R}_p(s,t)}$. Here $A(s,t) \cdot B(s,t)$ denotes the frozen time product of two operators $A(s,t), B(s,t)$.

The polynomials $\hat{R}_p(s,t), \hat{\bar{Z}}_p(s,t)$ are evaluated from the estimate

$$\theta_p = [\underbrace{0, \ldots, 0}_{n-m-1}, \hat{b}_m, \ldots, \hat{b}_0, \hat{a}_{n-1}, \ldots, \hat{a}_0]^T$$

of θ_p^*, i.e.,

$$\hat{R}_p(s,t) = s^n + \hat{a}_{n-1} s^{n-1} + \cdots + \hat{a}_0$$

6.6. INDIRECT MRAC

$$\hat{\bar{Z}}_p(s,t) = \hat{b}_m s^m + \hat{b}_{m-1} s^{m-1} + \cdots + \hat{b}_0$$

$$\hat{k}_p = \hat{b}_m$$

As in Section 6.6.2, the estimate $\hat{b}_m = \hat{k}_p$ should be constrained from going through zero by using projection.

The equations of the indirect MRAC scheme are described by (6.6.23), (6.6.25) where θ is generated by any adaptive law from Tables 4.2 to 4.5 based on the parametric model (6.6.22). Table 6.9 summarizes the main equations of an indirect MRAC scheme with the gradient algorithm as the adaptive law. Its stability properties are summarized by the following theorem:

Theorem 6.6.2 *The indirect MRAC scheme summarized in Table 6.9 guarantees that all signals are u.b., and the tracking error $e_1 = y_p - y_m$ converges to zero as $t \to \infty$.*

The proof of Theorem 6.6.2 is more complex than that for a direct MRAC scheme due to the nonlinear transformation $\theta_p \mapsto \theta$. The details of the proof are given in Section 6.8. The number of filters required to generate the signals ω, ϕ in Table 6.9 may be reduced from $4n - 2$ to $2n$ by selecting $\Lambda_p(s) = (s + \lambda_0)\Lambda(s)$ for some $\lambda_0 > 0$ and sharing common signals in the control and adaptive law.

Remark 6.6.1 Instead of the gradient algorithm, a least-squares or a hybrid adaptive law may be used in Table 6.9. The hybrid adaptive law will simplify the computations considerably since the controller parameter vector θ will be calculated only at discrete points of time rather than continuously.

The indirect MRAC scheme has certain advantages over the corresponding direct scheme. First, the order of the adaptive law in the indirect case is $n + m + 1$ compared to $2n$ in the direct case. Second, the indirect scheme allows us to utilize any apriori information about the plant parameters to initialize the parameter estimates or even reduce the order of the adaptive law further as indicated by the following example.

Example 6.6.1 Consider the third order plant

$$y_p = \frac{1}{s^2(s+a)} u_p \qquad (6.6.26)$$

Table 6.9 Indirect MRAC scheme with normalized adaptive law

Plant	$y_p = G_p(s)u_p, \quad G_p(s) = \bar{Z}_p(s)/R_p(s), \quad \bar{Z}_p(s) = k_p Z_p(s)$
Reference model	$y_m = k_m \dfrac{Z_m(s)}{R_m(s)} r$
Control law	$u_p = \theta^\top \omega$ $\omega = [\omega_1^\top, \omega_2^\top, y_p, r]^\top, \quad \theta = [\theta_1^\top, \theta_2^\top, \theta_3, c_0]^\top$ $\omega_1 = \dfrac{\alpha_{n-2}(s)}{\Lambda(s)} u_p, \quad \omega_2 = \dfrac{\alpha_{n-2}(s)}{\Lambda(s)} y_p$
Adaptive law	$\dot{\hat{p}}_1 = \Gamma_1 \bar{\phi}_1 \epsilon, \quad \dot{\hat{p}}_2 = \Gamma_2 \phi_2 \epsilon$ $\dot{\hat{k}}_p = \begin{cases} \gamma_m \epsilon \phi_{1m} & \text{if } |\hat{k}_p| > k_0 \text{ or} \\ & \text{if } |\hat{k}_p| = k_0 \text{ and } \phi_{1m} \epsilon \operatorname{sgn}(k_p) \geq 0 \\ 0 & \text{otherwise} \end{cases}$ $\hat{k}_p(0)\operatorname{sgn}(k_p) \geq k_0 > 0$ $\epsilon = (z - \hat{z})/m^2, \quad z = y_p + \lambda_p^\top \phi_2$ $\hat{z} = \theta_p^\top \phi, \quad m^2 = 1 + \phi^\top \phi, \quad \phi = [\phi_1^\top, \phi_2^\top]^\top$ $\theta_p = [\underbrace{0, \ldots, 0}_{n-m-1}, \hat{k}_p, \hat{p}_1^\top, \hat{p}_2^\top]^\top$ $\phi_1 = \dfrac{\alpha_{n-1}(s)}{\Lambda_p(s)} u_p, \quad \phi_2 = -\dfrac{\alpha_{n-1}(s)}{\Lambda_p(s)} y_p$ $\phi_1 = [\phi_0^\top, \bar{\phi}_1^\top]^\top, \quad \phi_0 \in \mathcal{R}^{n-m}, \bar{\phi}_1 \in \mathcal{R}^m$ $\phi_{1m} \in \mathcal{R}^1$ is the last element of ϕ_0 $\hat{p}_1 = [\hat{b}_{m-1}, \ldots, \hat{b}_0]^\top, \hat{p}_2 = [\hat{a}_{n-1}, \ldots, \hat{a}_0]^\top$ $\hat{\bar{Z}}_p(s,t) = \hat{k}_p s^m + \hat{p}_1^\top \alpha_{m-1}(s), \hat{R}_p(s,t) = s^n + \hat{p}_2^\top \alpha_{n-1}(s)$
Calculation of θ	$c_0(t) = \dfrac{k_m}{\hat{k}_p(t)}$ $\theta_1^\top(t) \alpha_{n-2}(s) = \Lambda(s) - \dfrac{1}{\hat{k}_p(t)} \hat{\bar{Z}}_p(s,t) \cdot \hat{Q}(s,t)$ $\theta_2^\top(t)\alpha_{n-2}(s) + \theta_3(t)\Lambda(s)$ $\quad = \dfrac{1}{\hat{k}_p}(\hat{Q}(s,t) \cdot \hat{R}_p(s,t) - \Lambda_0(s) R_m(s))$ $\hat{Q}(s,t) = $ quotient of $\Lambda_0(s)R_m(s)/\hat{R}_p(s,t)$
Design variables	k_0: lower bound for $|\hat{k}_p| \geq k_0 > 0$; $\Lambda_p(s)$: monic Hurwitz of degree n; For simplicity, $\Lambda_p(s) = (s + \lambda_0)\Lambda(s), \lambda_0 > 0$; $\Lambda(s) = \Lambda_0(s) Z_m(s); \Gamma_1 = \Gamma_1^\top > 0, \Gamma_2 = \Gamma_2^\top > 0; \Gamma_1 \in \mathcal{R}^{m \times m}$, $\Gamma_2 \in \mathcal{R}^{n \times n}; \lambda_p \in \mathcal{R}^n$ is the coefficient vector of $\Lambda_p(s) - s^n$

6.6. INDIRECT MRAC

where a is the only unknown parameter. The output y_p is required to track the output of y_m of the reference model

$$y_m = \frac{1}{(s+2)^3} r$$

The control law is given by

$$u_p = \theta_{11} \frac{s}{(s+\lambda_1)^2} u_p + \theta_{12} \frac{1}{(s+\lambda_1)^2} u_p + \theta_{21} \frac{s}{(s+\lambda_1)^2} y_p + \theta_{22} \frac{1}{(s+\lambda_1)^2} y_p + \theta_3 y_p + c_0 r$$

where $\theta = [\theta_{11}, \theta_{12}, \theta_{21}, \theta_{22}, \theta_3, c_0]^T \in \mathcal{R}^6$. In direct MRAC, θ is generated by a sixth-order adaptive law. In indirect MRAC, θ is calculated from the adaptive law as follows:

Using Table 6.9, we have

$$\theta_p = [0, 0, 1, \hat{a}, 0, 0]^T$$

$$\dot{\hat{a}} = \gamma_a \phi_a \epsilon$$

$$\epsilon = \frac{z - \hat{z}}{1 + \phi^T \phi}, \quad \hat{z} = \theta_p^T \phi, \quad z = y_p + \lambda_p^T \phi_2$$

$$\phi = [\phi_1^T, \phi_2^T]^T, \quad \phi_1 = \frac{[s^2, s, 1]^T}{(s+\lambda_1)^3} u_p, \quad \phi_2 = -\frac{[s^2, s, 1]^T}{(s+\lambda_1)^3} y_p$$

where $\Lambda_p(s)$ is chosen as $\Lambda_p(s) = (s+\lambda_1)^3$, $\lambda_p = [3\lambda_1, 3\lambda_1^2, \lambda_1^3]^T$.

$$\phi_a = [0, 0, 0, 1, 0, 0,]\phi = -\frac{s^2}{(s+\lambda_1)^3} y_p$$

and $\gamma_a > 0$ is a constant. The controller parameter vector is calculated as

$$c_0 = 1, \quad \theta_1^T \begin{bmatrix} s \\ 1 \end{bmatrix} = (s+\lambda_1)^2 - \hat{Q}(s,t)$$

$$\theta_2^T \begin{bmatrix} s \\ 1 \end{bmatrix} + \theta_3(s+\lambda_1)^2 = \hat{Q}(s,t) \cdot [s^3 + \hat{a}s^2] - (s+\lambda_1)^2(s+2)^3$$

where $\hat{Q}(s,t)$ is the quotient of $\frac{(s+\lambda_1)^2(s+2)^3}{s^3+\hat{a}s^2}$. ▽

The example demonstrates that for the plant (6.6.26), the indirect scheme requires a first order adaptive law whereas the direct scheme requires a sixth-order one.

6.7 Relaxation of Assumptions in MRAC

The stability properties of the MRAC schemes of the previous sections are based on assumptions P1 to P4, given in Section 6.3.1. While these assumptions are shown to be sufficient for the MRAC schemes to meet the control objective, it has often been argued whether they are also necessary. We have already shown in the previous sections that assumption P4 can be completely relaxed at the expense of a more complex adaptive law, therefore P4 is no longer necessary for meeting the MRC objective. In this section we summarize some of the attempts to relax assumptions P1 to P3 during the 1980s and early 1990s.

6.7.1 Assumption P1: Minimum Phase

This assumption is a consequence of the control objective in the known parameter case that requires the closed-loop plant transfer function to be equal to that of the reference model. Since this objective can only be achieved by cancelling the zeros of the plant and replacing them by those of the reference model, $Z_p(s)$ has to be Hurwitz, otherwise zero-pole cancellations in $\mathcal{C}^+$ will take place and lead to some unbounded state variables within the closed-loop plant. The assumption of minimum phase in MRAC has often been considered as one of the limitations of adaptive control in general, rather than a consequence of the MRC objective, and caused some confusion to researchers outside the adaptive control community. One of the reasons for such confusion is that the closed-loop MRAC scheme is a nonlinear dynamic system and zero-pole cancellations no longer make much sense. In this case, the minimum phase assumption manifests itself as a condition for proving that the plant input is bounded by using the boundedness of other signals in the closed-loop.

For the MRC objective and the structures of the MRAC schemes presented in the previous sections, the minimum phase assumption seems to be not only sufficient, but also necessary for stability. If, however, we modify the MRC objective not to include cancellations of unstable plant zeros, then it seems reasonable to expect to be able to relax assumption P1. For example, if we can restrict ourselves to changing only the poles of the plant and tracking a restricted class of signals whose internal model is known, then we may be able to allow plants with unstable zeros. The details of such designs

6.7. RELAXATION OF ASSUMPTIONS IN MRAC

that fall in the category of general pole placement are given in Chapter 7. It has often been argued that if we assume that the unstable zeros of the plant are known, we can include them to be part of the zeros of the reference model and design the MRC or MRAC scheme in a way that allows only the cancellation of the stable zeros of the plant. Although such a design seems to be straightforward, the analysis of the resulting MRAC scheme requires the incorporation of an adaptive law with projection. The projection in turn requires the knowledge of a convex set in the parameter space where the estimation is to be constrained. The development of such a convex set in the higher order case is quite awkward, if possible. The details of the design and analysis of MRAC for plants with known unstable zeros for discrete-time plants are given in [88, 194].

The minimum phase assumption is one of the main drawbacks of MRAC for the simple reason that the corresponding discrete-time plant of a sampled minimum phase continuous-time plant is often nonminimum phase [14].

6.7.2 Assumption P2: Upper Bound for the Plant Order

The knowledge of an upper bound n for the plant order is used to determine the order of the MRC law. This assumption can be completely relaxed if the MRC objective is modified. For example, it has been shown in [102, 159, 160] that the control objective of regulating the output of a plant of unknown order to zero can be achieved by using simple adaptive controllers that are based on high-gain feedback, provided the plant is minimum phase and the plant relative degree n^* is known. The principle behind some of these high-gain stabilizing controllers can be explained for a minimum phase plant with $n^* = 1$ and arbitrary order as follows: Consider the following minimum phase plant with relative degree $n^* = 1$:

$$y_p = \frac{k_p Z_p(s)}{R_p(s)} u_p$$

From root locus arguments, it is clear that the input

$$u_p = -\theta y_p \text{sgn}(k_p)$$

with sufficiently large gain θ will force the closed loop characteristic equation

$$R_p(s) + |k_p|\theta Z_p(s) = 0$$

to have roots in $Re\,[s] < 0$. Having this result in mind, it can be shown that the adaptive control law

$$u_p = -\theta y_p \text{sgn}\,(k_p), \qquad \dot{\theta} = y_p^2$$

can stabilize any minimum phase plant with $n^* = 1$ and of arbitrary order. As n^* increases, the structure and analysis of the high gain adaptive controllers becomes more complicated [159, 160].

Another class of adaptive controllers for regulation that attracted considerable interest in the research community is based on search methods and discontinuous adjustments of the controller gains [58, 137, 147, 148]. Of particular theoretical interest is the controller proposed in [137] referred to as *universal controller* that is based on the rather weak assumption that only the order n_c of a stabilizing linear controller needs to be known for the stabilization and regulation of the output of the unknown plant to zero. The universal controller is based on an automated dense search throughout the set of all possible n_c-order linear controllers until it passes through a subset of stabilizing controllers in a way that ensures asymptotic regulation and the termination of the search. One of the drawbacks of the universal controller is the possible presence of large overshoots as pointed out in [58] which limits its practicality.

An interesting MRAC scheme that is also based on high gain feedback and discontinuous adjustment of controller gains is given in [148]. In this case, the MRC objective is modified to allow possible nonzero tracking errors that can be forced to be less than a prespecified (arbitrarily small) constant after an (arbitrarily short) prespecified period of time, with an (arbitrarily small) prespecified upper bound on the amount of overshoot. The only assumption made about the unknown plant is that it is minimum phase.

The adaptive controllers of [58, 137, 147, 148] where the controller gains are switched from one constant value to another over intervals of times based on some cost criteria and search methods are referred to in [163] as *nonidentifier-based* adaptive controllers to distinguish them from the class of *identifier-based* ones that are studied in this book.

6.7.3 Assumption P3: Known Relative Degree n^*

The knowledge of the relative degree n^* of the plant is used in the MRAC schemes of the previous sections in order to develop control laws that are free

6.7. RELAXATION OF ASSUMPTIONS IN MRAC

of differentiators. This assumption may be relaxed at the expense of additional complexity in the control and adaptive laws. For the identifier-based schemes, several approaches have been proposed that require the knowledge of an upper bound n_u^* for n^* [157, 163, 217]. In the approach of [163], n_u^* parameterized controllers $C_i, i = 1, 2, \ldots, n_u^*$ are constructed in a way that C_i can meet the MRC objective for a reference model M_i of relative degree i when the unknown plant has a relative degree i. A switching logic with hysteresis is then designed that switches from one controller to another based on some error criteria. It is established in [165] that switching stops in finite time and the MRC objective is met exactly.

In another approach given in [217], the knowledge of an upper bound n_u^* and lower bound n_l^* of n^* are used to construct a feedforward dynamic term that replaces $c_0 r$ in the standard MRC law, i.e., u_p is chosen as

$$u_p = \theta_1^T \frac{\alpha(s)}{\Lambda(s)} + \theta_2^T \frac{\alpha(s)}{\Lambda(s)} + \theta_3 y_p + \theta_4^T b_1(s) n_1(s) W_m(s) r$$

where $b_1(s) = \begin{bmatrix} 1, s, \ldots, s^{\bar{n}^*} \end{bmatrix}^T$, $\bar{n}^* = n_u^* - n_l^*$, $\theta_4^* \in \mathcal{R}^{\bar{n}^*+1}$ and $n_1(s)$ is an arbitrary monic Hurwitz polynomial of degree n_l^*. The relative degree of the transfer function $W_m(s)$ of the reference model is chosen to be equal to n_u^*. It can be shown that for some constant vectors $\theta_1^*, \theta_2^*, \theta_3^*, \theta_4^*$ the MRC objective is achieved exactly, provided θ_4^* is chosen so that $k_p \theta_4^{*T} b_1(s)$ is a monic Hurwitz polynomial of degree $n^* - n_l^* \leq \bar{n}^*$, which implies that the last $\bar{n}^* - n^* - n_l^*$ elements of θ^* are equal to zero.

The on-line estimate θ_i of $\theta_i^*, i = 1, \ldots, 4$ is generated by an adaptive law designed by following the procedure of the previous sections. The adaptive law for θ_4, however, is modified using projection so that θ_4 is constrained to be inside a convex set $\mathcal{C}$ which guarantees that $k_p \theta_4^T b_1(s)$ is a monic Hurwitz polynomial at each time t. The development of such set is trivial when the uncertainty in the relative degree, i.e., $\bar{n}^*$ is less or equal to 2. For uncertainties greater than 2, the calculation of $\mathcal{C}$, however, is quite involved.

6.7.4 Tunability

The concept of tunability introduced by Morse in [161] is a convenient tool for analyzing both identifier and nonidentifier-based adaptive controllers and for discussing the various questions that arise in conjunction with assumptions

P1 to P4 in MRAC. Most of the adaptive control schemes may be represented by the equations

$$\dot{x} = A(\theta)x + B(\theta)r$$
$$\epsilon_1 = C(\theta)x \qquad (6.7.1)$$

where $\theta : \mathcal{R}^+ \mapsto \mathcal{R}^n$ is the estimated parameter vector and ϵ_1 is the estimation or tuning error.

Definition 6.7.1 *[161, 163] The system (6.7.1) is said to be tunable on a subset $\mathcal{S} \subset \mathcal{R}^n$ if for each $\theta \in \mathcal{S}$ and each bounded input r, every possible trajectory of the system for which $\epsilon_1(t) = 0, t \in [0, \infty)$ is bounded.*

Lemma 6.7.1 *The system (6.7.1) is tunable on $\mathcal{S}$ if and only if $\{C(\theta), A(\theta)\}$ is detectable for each $\theta \in \mathcal{S}$.*

If $\{C(\theta), A(\theta)\}$ is not detectable, then it follows that the state x may grow unbounded even when adaptation is successful in driving ϵ_1 to zero by adjusting θ. One scheme that may exhibit such a behavior is a MRAC of the type considered in previous sections that is designed for a nonminimum-phase plant. In this case, it can be established that the corresponding system of equations is not detectable and therefore not tunable.

The concept of tunability may be used to analyze the stability properties of MRAC schemes by following a different approach than those we discussed in the previous sections. The details of this approach are given in [161, 163] where it is used to analyze a wide class of adaptive control algorithms. The analysis is based on deriving (6.7.1) and establishing that $\{C(\theta), A(\theta)\}$ is detectable, which implies tunability. Detectability guarantees the existence of a matrix $H(\theta)$ such that for each fixed $\theta \in \mathcal{S}$ the matrix $A_c(\theta) = A(\theta) - H(\theta)C(\theta)$ is stable. Therefore, (6.7.1) may be written as

$$\dot{x} = [A(\theta) - H(\theta)C(\theta)]x + B(\theta)r + H(\theta)\epsilon_1 \qquad (6.7.2)$$

by using the so called output injection. Now from the properties of the adaptive law that guarantees $\dot{\theta}, \frac{\epsilon_1}{m} \in \mathcal{L}_2$ where $m = 1 + (C_0^\top x)^2$ for some vector C_0, we can establish that the homogeneous part of (6.7.2) is u.a.s., which, together with the B-G Lemma, guarantees that $x \in \mathcal{L}_\infty$. The boundedness of x can then be used in a similar manner as in the previous sections to establish the boundedness of all signals in the closed loop and the convergence of the tracking error to zero.

6.8 Stability Proofs of MRAC Schemes

6.8.1 Normalizing Properties of Signal m_f

In Section 6.5, we have defined the fictitious normalizing signal m_f

$$m_f^2 \triangleq 1 + \|u_p\|^2 + \|y_p\|^2 \tag{6.8.1}$$

where $\|\cdot\|$ denotes the $\mathcal{L}_{2\delta}$-norm, and used its normalizing properties to establish stability of the closed-loop MRAC for the adaptive tracking and regulation examples. We now extend the results to the general SISO MRAC scheme as follows:

Lemma 6.8.1 *Consider the plant equation (6.3.2) and control law (6.5.39). There exists a $\delta > 0$ such that:*
(i) $\omega_1/m_f, \omega_2/m_f \in \mathcal{L}_\infty$.
(ii) *If $\theta \in \mathcal{L}_\infty$, then $u_p/m_f, y_p/m_f, \omega/m_f, W(s)\omega/m_f \in \mathcal{L}_\infty$ where $W(s)$ is a proper transfer function with stable poles.*
(iii) *If $\dot{r}, \dot{\theta} \in \mathcal{L}_\infty$, then in addition to (i) and (ii), we have $\|\dot{y}_p\|/m_f, \|\dot{\omega}\|/m_f \in \mathcal{L}_\infty$.*

Proof (i) Because

$$\omega_1 = \frac{\alpha(s)}{\Lambda(s)} u_p, \quad \omega_2 = \frac{\alpha(s)}{\Lambda(s)} y_p$$

and each element of $\frac{\alpha(s)}{\Lambda(s)}$ has relative degree greater or equal to 1, it follows from Lemma 3.3.2 and the definition of m_f that $\omega_1/m_f, \omega_2/m_f \in \mathcal{L}_\infty$.

(ii) We can apply Lemma 3.3.2 to equation (6.5.42), i.e.,

$$y_p = W_m(s)\left(r + \frac{1}{c_0^*}\tilde{\theta}^\top \omega\right)$$

to obtain (using $\theta \in \mathcal{L}_\infty$) that

$$|y_p(t)| \leq c + c\|\tilde{\theta}^\top \omega\| \leq c + c\|\omega\|$$

where $c \geq 0$ denotes any finite constant and $\|\cdot\|$ the $\mathcal{L}_{2\delta}$ norm. On the other hand, we have

$$\|\omega\| \leq \|\omega_1\| + \|\omega_2\| + \|y_p\| + \|r\| \leq c\|u_p\| + c\|y_p\| + c \leq c m_f + c$$

therefore, $\frac{y_p}{m_f} \in \mathcal{L}_\infty$. Because $\omega = [\omega_1^\top, \omega_2^\top, y_p, r]^\top$ and ω_1/m_f, ω_2/m_f, y_p/m_f, $r/m_f \in \mathcal{L}_\infty$, it follows that $\omega/m_f \in \mathcal{L}_\infty$. From $u_p = \theta^\top \omega$ and $\omega/m_f, \theta \in \mathcal{L}_\infty$ we conclude that $u_p/m_f \in \mathcal{L}_\infty$. Consider $\phi \triangleq W(s)\omega$. We have $|\phi| \leq c\|\omega\|$ for some $\delta > 0$ such that $W(s - \frac{\delta}{2})$ is stable. Hence, $\frac{\phi}{m_f} \in \mathcal{L}_\infty$ due to $\|\omega\|/m_f \in \mathcal{L}_\infty$.

(iii) Note that $\dot{y}_p = sW_m(s)[r + \frac{1}{c_0^*}\tilde{\theta}^\top \omega]$ and $sW_m(s)$ is a proper stable transfer function, therefore from Lemma 3.3.2 we have

$$\|\dot{y}_p\| \leq c + c\|\tilde{\theta}^\top \omega\| \leq c + cm_f$$

i.e., $\|\dot{y}_p\|/m_f \in \mathcal{L}_\infty$. Similarly, because $\|\dot{\omega}\| \leq \|\dot{\omega}_1\| + \|\dot{\omega}_2\| + \|\dot{y}_p\| + \|\dot{r}\|$, applying Lemma 3.3.2 we have $\|\dot{\omega}_1\| \leq c\|u_p\|$, $\|\dot{\omega}_2\| \leq c\|y_p\|$ which together with $\frac{\|\dot{y}_p\|}{m_f}, \dot{r} \in \mathcal{L}_\infty$ imply that $\|\dot{\omega}\|/m_f \in \mathcal{L}_\infty$ and the lemma is proved. In (i) to (iii), $\delta > 0$ is chosen so that $\frac{1}{\Lambda(s)}, W_m(s), W(s)$ are analytic in $\text{Re}[s] \geq \delta/2$. $\square$

6.8.2 Proof of Theorem 6.5.1: Direct MRAC

In this section we present all the details of the proof in the five steps outlined in Section 6.5.3.

Step 1. *Express the plant input and output in terms of the adaptation error $\tilde{\theta}^\top \omega$.* From Figure 6.14, we can verify that the transfer function between the input $r + \frac{1}{c_0^*}\tilde{\theta}^\top \omega$ and the plant output y_p is given by

$$y_p = G_c(s)\left(r + \frac{1}{c_0^*}\tilde{\theta}^\top \omega\right)$$

where

$$G_c(s) = \frac{c_0^* k_p Z_p}{(1 - \theta_1^{*\top}(sI - F)^{-1}g)R_p - k_p Z_p[\theta_2^{*\top}(sI - F)^{-1}g + \theta_3^*]}$$

Using (6.3.12), (6.3.13), and $(sI - F)^{-1}g = \frac{\alpha(s)}{\Lambda(s)}$, we have, after cancellation of all the stable common zeros and poles in $G_c(s)$, that $G_c(s) = W_m(s)$. Therefore, the plant output may be written as

$$y_p = W_m(s)\left(r + \frac{1}{c_0^*}\tilde{\theta}^\top \omega\right) \tag{6.8.2}$$

Because $y_p = G_p(s)u_p$ and $G_p^{-1}(s)$ has stable poles, we have

$$u_p = G_p^{-1}(s)W_m(s)\left(r + \frac{1}{c_0^*}\tilde{\theta}^\top \omega\right) \tag{6.8.3}$$

where $G_p^{-1}(s)W_m(s)$ is stable (due to assumption P1) and biproper (due to assumption M2). For simplicity, let us denote the $\mathcal{L}_{2\delta}$-norm $\|(\cdot)_t\|_{2\delta}$ for some $\delta > 0$ by $\|\cdot\|$. Using the fictitious normalizing signal $m_f^2 \triangleq 1 + \|u_p\|^2 + \|y_p\|^2$, it follows from (6.8.2), (6.8.3) and Lemma 3.3.2 that

$$m_f^2 \leq c + c\|\tilde{\theta}^\top \omega\|^2 \tag{6.8.4}$$

6.8. STABILITY PROOFS OF MRAC SCHEMES

holds for some $\delta > 0$ such that $W_m(s-\delta/2)G_p^{-1}(s-\delta/2)$ is a stable transfer function, where $c \geq 0$ in (6.8.4) and in the remainder of this section denotes any finite constant. The normalizing properties of m_f have been established in Section 6.8.1, i.e., all signals in the closed-loop adaptive system and some of their derivatives are bounded from above by m_f provided $\theta \in \mathcal{L}_\infty$.

Step 2. *Use the swapping lemmas and properties of the $\mathcal{L}_{2\delta}$ norm to bound $\|\tilde{\theta}^\top \omega\|$ from above with terms that are guaranteed by the adaptive law to have finite $\mathcal{L}_2$ gains.* Using the Swapping Lemma A.2 in Appendix A, we can express $\tilde{\theta}^\top \omega$ as

$$\tilde{\theta}^\top \omega = F_1(s, \alpha_0)\left(\dot{\tilde{\theta}}^\top \omega + \tilde{\theta}^\top \dot{\omega}\right) + F(s, \alpha_0)\left(\tilde{\theta}^\top \omega\right) \qquad (6.8.5)$$

where $F(s, \alpha_0) = \frac{\alpha_0^{n^*}}{(s+\alpha_0)^{n^*}}$, $F_1(s, \alpha_0) = \frac{1-F(s,\alpha_0)}{s}$, $\alpha_0 > 0$ is an arbitrary constant and n^* is the relative degree of $W_m(s)$. On the other hand, using Swapping Lemma A.1, we can write

$$\tilde{\theta}^\top \omega = W^{-1}(s)\left(\tilde{\theta}^\top W(s)\omega + W_c(s)((W_b(s)\omega^\top)\dot{\tilde{\theta}})\right) \qquad (6.8.6)$$

where $W(s)$ is a strictly proper transfer function with poles and zeros in $\mathcal{C}^-$ that we will specify later. Using (6.8.6) in (6.8.5) we obtain

$$\tilde{\theta}^\top \omega = F_1[\dot{\tilde{\theta}}^\top \omega + \tilde{\theta}^\top \dot{\omega}] + FW^{-1}[\tilde{\theta}^\top W(s)\omega + W_c((W_b\omega^\top)\dot{\tilde{\theta}})] \qquad (6.8.7)$$

where $F(s)W^{-1}(s)$ can be made proper by choosing $W(s)$ appropriately.

We obtain a bound for $\|\tilde{\theta}^\top \omega\|$ by considering each adaptive law separately.

Adaptive Law of Table 6.5. We express the normalized estimation error as

$$\epsilon = W_m L(\rho^* \tilde{\theta}^\top \phi - \tilde{\rho}\xi - \epsilon n_s^2)$$

i.e.,

$$\tilde{\theta}^\top \phi = \tilde{\theta}^\top L^{-1}\omega = \frac{1}{\rho^*}(W_m^{-1}L^{-1}\epsilon + \tilde{\rho}\xi + \epsilon n_s^2) \qquad (6.8.8)$$

Choosing $W(s) = L^{-1}(s)$ and substituting for $\tilde{\theta}^\top W(s)\omega = \tilde{\theta}^\top L^{-1}(s)\omega$ from (6.8.8) into (6.8.7), we obtain

$$\tilde{\theta}^\top \omega = F_1(\dot{\tilde{\theta}}^\top \omega + \tilde{\theta}^\top \dot{\omega}) + \frac{1}{\rho^*}FW_m^{-1}\epsilon + FL\left[\frac{\tilde{\rho}}{\rho^*}\xi + \frac{\epsilon n_s^2}{\rho^*} + W_c(W_b\omega^\top)\dot{\tilde{\theta}}\right]$$

Now from the definition of ξ and using Swapping Lemma A.1 with $W(s) = L^{-1}(s)$, we have $\xi = u_f - \theta^\top \phi = L^{-1}u_p - \theta^\top \phi$, i.e.,

$$\xi = -\theta^\top \phi + L^{-1}\theta^\top \omega = W_c[(W_b\omega^\top)\dot{\theta}]$$

Therefore,

$$\tilde{\theta}^T \omega = F_1[\dot{\tilde{\theta}}^T \omega + \tilde{\theta}^T \dot{\omega}] + \frac{1}{\rho^*} F W_m^{-1} \epsilon + \frac{1}{\rho^*} F L[\rho W_c(W_b \omega^T)\dot{\theta} + \epsilon n_s^2] \qquad (6.8.9)$$

From the definition of $F(s)$, $F_1(s)$ and Swapping Lemma A.2, it follows that for $\alpha_0 > \delta$

$$\|F_1(s)\|_{\infty\delta} \leq \frac{c}{\alpha_0}, \quad \|F(s)W_m^{-1}(s)\|_{\infty\delta} \leq c\alpha_0^{n^*}$$

Therefore,

$$\|\tilde{\theta}^T \omega\| \leq \frac{c}{\alpha_0}(\|\dot{\tilde{\theta}}^T \omega\| + \|\tilde{\theta}^T \dot{\omega}\|) + c\alpha_0^{n^*}(\|\epsilon\| + \|\bar{\omega}\| + \|\epsilon n_s^2\|) \qquad (6.8.10)$$

where $\bar{\omega} = \rho W_c(s)[Q_b(t)\dot{\theta}], Q_b = W_b(s)\omega^T$. Using Lemma 3.3.2, we can show that $Q_b/m_f \in \mathcal{L}_\infty$ and, therefore,

$$\|\bar{\omega}\| \leq c\|\dot{\tilde{\theta}} m_f\| \qquad (6.8.11)$$

Using the normalizing properties of m_f established in Lemma 6.8.1, we have

$$\|\dot{\tilde{\theta}}^T \omega\| \leq c\|\dot{\theta} m_f\|, \quad \|\tilde{\theta}^T \dot{\omega}\| \leq c\|\dot{\omega}\| \leq cm_f, \quad \|\epsilon n_s^2\| \leq c\|\epsilon n_s m_f\|$$

It, therefore, follows from (6.8.10), (6.8.11) that

$$\|\tilde{\theta}^T \omega\| \leq \frac{c}{\alpha_0}\|\dot{\theta} m_f\| + \frac{c}{\alpha_0} m_f + c\alpha_0^{n^*}(\|\epsilon\| + \|\dot{\theta} m_f\| + \|\epsilon n_s m_f\|) \qquad (6.8.12)$$

which we express as

$$\|\tilde{\theta}^T \omega\| \leq c\alpha_0^{n^*}\|\tilde{g} m_f\| + \frac{c}{\alpha_0} m_f \qquad (6.8.13)$$

where $\tilde{g}^2 = \epsilon^2 n_s^2 + \epsilon^2 + |\dot{\theta}|^2$, by taking $\alpha_0 > 1$ so that $\frac{1}{\alpha_0} \leq \alpha_0^{n^*}$ and using $\|\epsilon\| \leq \|\epsilon m_f\|$ due to $m_f \geq 1$. Since $\epsilon n_s, \epsilon, \dot{\theta} \in \mathcal{L}_2$, it follows that $\tilde{g} \in \mathcal{L}_2$.

Adaptive Laws of Table 6.6. The normalized estimation error is given by

$$\epsilon = \frac{W_m(s)u_p - \hat{z}}{1 + n_s^2} = -\frac{\tilde{\theta}^T \phi_p}{1 + n_s^2}, \quad n_s^2 = \phi_p^T \phi_p$$

where $\phi_p = [W_m(s)\omega_1^T, W_m(s)\omega_2^T, W_m(s)y_p, y_p]^T$. Let

$$\omega_p = [\omega_1^T, \omega_2^T, y_p, W_m^{-1} y_p]^T$$

then $\phi_p = W_m(s)\omega_p$. To relate $\tilde{\theta}^T \omega$ with ϵ, we write

$$\tilde{\theta}^T \omega = \tilde{\theta}_0^T \omega_0 + \tilde{c}_0 r$$

6.8. STABILITY PROOFS OF MRAC SCHEMES

where $\omega_0 = [\omega_1^T, \omega_2^T, y_p]^T$ and $\tilde{\theta}_0 = [\tilde{\theta}_1^T, \tilde{\theta}_2^T, \tilde{\theta}_3]^T$. From

$$y_p = W_m(r + \frac{1}{c_0^*}\tilde{\theta}^T\omega)$$

we have $r = W_m^{-1}y_p - \frac{1}{c_0^*}\tilde{\theta}^T\omega$, therefore,

$$\tilde{\theta}^T\omega = \tilde{\theta}_0^T\omega_0 + \tilde{c}_0 W_m^{-1}[y_p] - \frac{\tilde{c}_0}{c_0^*}\tilde{\theta}^T\omega = \tilde{\theta}^T\omega_p - \frac{\tilde{c}_0}{c_0^*}\tilde{\theta}^T\omega$$

i.e.,

$$\tilde{\theta}^T\omega = \frac{c_0^*}{c_0}\tilde{\theta}^T\omega_p \tag{6.8.14}$$

Using $\phi_p = W_m(s)\omega_p$ and applying Swapping Lemma A.1 we have

$$W_m(s)\tilde{\theta}^T\omega_p = \tilde{\theta}^T\phi_p + W_c(W_b\omega_p^T)\dot{\tilde{\theta}}$$

From Swapping Lemma A.2, we have

$$\tilde{\theta}^T\omega_p = F_1(\dot{\tilde{\theta}}^T\omega_p + \tilde{\theta}^T\dot{\omega}_p) + F(\tilde{\theta}^T\omega_p)$$

where $F_1(s), F(s)$ satisfy $\|F_1(s)\|_{\infty\delta} \leq \frac{c}{\alpha_0}$, $\|F(s)W_m^{-1}(s)\|_{\infty\delta} \leq c\alpha_0^{n^*}$ for $\alpha_0 > \delta$. Using the above two inequalities, we obtain

$$\tilde{\theta}^T\omega_p = F_1(\dot{\tilde{\theta}}^T\omega_p + \tilde{\theta}^T\dot{\omega}_p) + FW_m^{-1}[\tilde{\theta}^T\phi_p + W_c(W_b\omega_p^T)\dot{\tilde{\theta}}] \tag{6.8.15}$$

Using (6.8.14) we obtain

$$\dot{\tilde{\theta}}^T\omega_p + \tilde{\theta}^T\dot{\omega}_p = \frac{\dot{c}_0}{c_0^*}\tilde{\theta}^T\omega + \frac{c_0}{c_0^*}\dot{\tilde{\theta}}^T\omega + \frac{c_0}{c_0^*}\tilde{\theta}^T\dot{\omega}$$

which we use together with $\tilde{\theta}^T\phi_p = -\epsilon - \epsilon n_s^2$ to express (6.8.15) as

$$\tilde{\theta}^T\omega_p = F_1\left(\frac{\dot{c}_0}{c_0^*}\tilde{\theta}^T\omega + \frac{c_0}{c_0^*}\dot{\tilde{\theta}}^T\omega + \frac{c_0}{c_0^*}\tilde{\theta}^T\dot{\omega}\right) + FW_m^{-1}[-\epsilon - \epsilon n_s^2 + W_c(W_b\omega_p^T)\dot{\tilde{\theta}}] \tag{6.8.16}$$

Due to the boundedness of $\frac{1}{c_0}$, it follows from (6.8.14) that

$$\|\tilde{\theta}^T\omega\| \leq c\|\tilde{\theta}^T\omega_p\| \tag{6.8.17}$$

Using the same arguments as we used in establishing (6.8.13) for the adaptive laws of Table 6.5, we use (6.8.16), (6.8.17) and the normalizing properties of m_f to obtain

$$\|\tilde{\theta}^T\omega\| \leq c\|\tilde{\theta}^T\omega_p\| \leq c\alpha_0^{n^*}\|\tilde{g}m_f\| + \frac{c}{\alpha_0}m_f \tag{6.8.18}$$

where $\tilde{g} \in \mathcal{L}_2$.

Step 3. *Use the B-G Lemma to establish signal boundedness.* Using (6.8.13) or (6.8.18) in $m_f^2 \leq c + c\|\tilde{\theta}^\top \omega\|^2$, we have

$$m_f^2 \leq c + c\alpha_0^{2n^*}\|\tilde{g}m_f\|^2 + \frac{c}{\alpha_0^2}m_f^2 \qquad (6.8.19)$$

which for large α_0 implies

$$m_f^2 \leq c + c\alpha_0^{2n^*}\|\tilde{g}m_f\|^2$$

for some other constant $c \geq 0$. The boundedness of m_f follows from $\tilde{g} \in \mathcal{L}_2$ and the B-G Lemma.

Because $m_f, \theta \in \mathcal{L}_\infty$, it follows from Lemma 6.8.1 that $u_p, y_p, \omega \in \mathcal{L}_\infty$ and therefore all signals in the closed-loop plant are bounded.

Step 4. *Establish convergence of the tracking error to zero.* Let us now consider the equation for the tracking error that relates e_1 with signals that are guaranteed by the adaptive law to be in $\mathcal{L}_2$. For the adaptive law of Table 6.5(A), we have

$$e_1 = \epsilon + W_m L \epsilon n_s^2 + W_m L \rho \xi$$

Because $\xi = L^{-1}\theta^\top \omega - \theta^\top \phi = W_c(W_b \omega^\top)\dot{\theta}$, $\epsilon, \epsilon n_s, \dot{\theta} \in \mathcal{L}_2$ and $\omega, n_s \in \mathcal{L}_\infty$, it follows that $e_1 \in \mathcal{L}_2$. In addition we can establish, using (6.8.2), that $\dot{e}_1 \in \mathcal{L}_\infty$ and therefore from Lemma 3.2.5, we conclude that $e_1(t) \to 0$ as $t \to \infty$.

For the adaptive laws of Table 6.5(B,C) the proof follows by using $L(s) = W_m^{-1}(s)$ and following exactly the same arguments as above. For the adaptive laws of Table 6.6 we have

$$\epsilon m^2 = W_m \theta^\top \omega - \theta^\top \phi_p$$

Applying Swapping Lemma A.1, we have

$$W_m \theta^\top \omega = \theta^\top W_m \omega + W_c(W_b \omega^\top)\dot{\theta}$$

i.e.,

$$\epsilon m^2 = \theta^\top [W_m \omega - \phi_p] + W_c(W_b \omega^\top)\dot{\theta}$$

Now, $W_m \omega - \phi_p = [0, \ldots, 0, -e_1]^\top$, hence,

$$c_0 e_1 = -\epsilon m^2 + W_c(W_b \omega^\top)\dot{\theta}$$

Because $\frac{1}{c_0}, m, \omega$ are bounded and $\epsilon m, \dot{\theta} \in \mathcal{L}_2$, it follows that $e_1 \in \mathcal{L}_2$. As before we can use the boundedness of the signals to show that $\dot{e}_1 \in \mathcal{L}_\infty$ which together with $e_1 \in \mathcal{L}_2$ imply that $e_1(t) \to 0$ as $t \to \infty$.

6.8. STABILITY PROOFS OF MRAC SCHEMES

Step 5. *Establish parameter convergence.* First, we show that ϕ, ϕ_p are PE if r is sufficiently rich of order $2n$. From the definition of ϕ, we can write

$$\phi = H(s) \begin{bmatrix} (sI - F)^{-1} g u_p \\ (sI - F)^{-1} g y_p \\ y_p \\ r \end{bmatrix}$$

where $H(s) = W_m(s)$ if the adaptive law of Table 6.5(B) is used and $H(s) = L^{-1}(s)$ if that of Table 6.5(A) is used. Using $u_p = G_p^{-1}(s) y_p$ and $y_p = W_m(s) r + e_1$, we have

$$\phi = \phi_m + \bar{\phi}$$

where

$$\phi_m = H(s) \begin{bmatrix} (sI - F)^{-1} g G_p^{-1}(s) W_m(s) \\ (sI - F)^{-1} g W_m(s) \\ W_m(s) \\ 1 \end{bmatrix} r, \quad \bar{\phi} = H(s) \begin{bmatrix} (sI - F)^{-1} g G_p^{-1}(s) \\ (sI - F)^{-1} g \\ 1 \\ 0 \end{bmatrix} e_1$$

Because $e_1 \in \mathcal{L}_2$, it follows from the properties of the PE signals that ϕ is PE if and only if ϕ_m is PE. In the proof of Theorem 6.4.1 and 6.4.2, we have proved that

$$\phi_0 \triangleq \begin{bmatrix} (sI - F)^{-1} g G_p^{-1}(s) W_m(s) \\ (sI - F)^{-1} g W_m(s) \\ W_m(s) \\ 1 \end{bmatrix} r$$

is PE provided r is sufficiently rich of order $2n$. Because $H(s)$ is stable and minimum phase and $\phi_0 \in \mathcal{L}_\infty$ owing to $\dot{r} \in \mathcal{L}_\infty$, it follows from Lemma 4.8.3 (iv) that $\phi_m = H(s) \phi_0$ is PE.

From the definition of ϕ_p, we have

$$\phi_p = W_m(s) \begin{bmatrix} (sI-F)^{-1} g G_p^{-1}(s) W_m(s) \\ (sI - F)^{-1} g W_m(s) \\ W_m(s) \\ 1 \end{bmatrix} r + \begin{bmatrix} W_m(s)(sI-F)^{-1} g G_p^{-1}(s) \\ W_m(s)(sI - F)^{-1} g \\ W_m(s) \\ 1 \end{bmatrix} e_1$$

Because ϕ_p has the same form as ϕ, the PE property of ϕ_p follows by using the same arguments.

We establish the convergence of the parameter error and tracking error to zero as follows: First, let us consider the adaptive laws of Table 6.5. For the adaptive law based on the SPR-Lyapunov approach (Table 6.5(A)), we have

$$\begin{align} \epsilon &= W_m L(\rho^* \tilde{\theta}^\top \phi - \tilde{\rho} \xi - \epsilon n_s^2) \\ \dot{\tilde{\theta}} &= -\Gamma \epsilon \phi \operatorname{sgn}(k_p/k_m) \end{align} \tag{6.8.20}$$

where $C_c^T(sI - A_c)^{-1}B_c = W_m(s)L(s)$ is SPR. The stability properties of (6.8.20) are established by Theorem 4.5.1 in Chapter 4. According to Theorem 4.5.1 (iii), $\tilde{\theta}(t) \to 0$ as $t \to \infty$ provided $\phi, \dot{\phi} \in \mathcal{L}_\infty$, ϕ is PE and $\xi \in \mathcal{L}_2$. Because $\xi = W_c(W_b\omega^T)\dot{\theta}$ and $\omega \in \mathcal{L}_\infty$, $\dot{\theta} \in \mathcal{L}_2$, we have $\xi \in \mathcal{L}_2$. From $u_p, y_p \in \mathcal{L}_\infty$ and the expression for ϕ, we can establish that $\phi, \dot{\phi} \in \mathcal{L}_\infty$. Because ϕ is shown to be PE, the convergence of $\tilde{\theta}(t)$ to zero follows. From $\tilde{\theta}(t) \to 0$ as $t \to \infty$ and $\tilde{\rho}\xi, \epsilon n_s^2 \in \mathcal{L}_2$ and the stability of A_c, we have that $e(t) \to 0$ as $t \to \infty$. In fact, the convergence of $e(t)$ to zero follows from the properties of the Lyapunov-like function used to analyze (6.8.20) in Theorem 4.5.1 without requiring ϕ to be PE.

The proof for the adaptive law of Table 6.5(B) follows from the above arguments by replacing $L^{-1}(s)$ with $W_m(s)$ and using the results of Theorem 4.5.2 (iii).

For the adaptive laws of Table 6.6, we have established in Chapter 4 that without projection, $\tilde{\theta}(t) \to 0$ as $t \to \infty$ exponentially fast provided ϕ_p is PE. Since projection can only make the derivative $\dot{V}$ of the Lyapunov-like function V, used to analyze the stability properties of the adaptive law, more negative the exponential convergence of $\tilde{\theta}$ to zero can be established by following exactly the same steps as in the case of no projection.

6.8.3 Proof of Theorem 6.6.2: Indirect MRAC

We follow the same steps as in proving stability for the direct MRAC scheme:

Step 1. *Express y_p, u_p in terms of $\tilde{\theta}^T \omega$.* Because the control law for the indirect MRAC is the same as the direct MRAC scheme, equations (6.8.2), (6.8.3) still hold, i.e., we have

$$y_p = W_m(s)\left(r + \frac{1}{c_0^*}\tilde{\theta}^T \omega\right), \quad u_p = G_p^{-1}(s)W_m(s)\left(r + \frac{1}{c_0^*}\tilde{\theta}^T \omega\right)$$

As in the direct case, we define the fictitious normalizing signal

$$m_f^2 \triangleq 1 + \|u_p\|^2 + \|y_p\|^2$$

where $\|\cdot\|$ denotes the $\mathcal{L}_{2\delta}$-norm for some $\delta > 0$. Using the same arguments as in the proof of Theorem 6.5.1, we have

$$m_f^2 \leq c + c\|\tilde{\theta}^T \omega\|^2$$

Step 2. *We upper bound $\|\tilde{\theta}^T \omega\|$ with terms that are guaranteed by the adaptive laws to have $\mathcal{L}_2$ gains.* From (6.6.25), we have

$$\theta_1^T \alpha_{n-2}(s) = \Lambda(s) - \frac{1}{k_p}\hat{\bar{Z}}_p(s,t) \cdot \hat{Q}(s,t) \qquad (6.8.21)$$

6.8. STABILITY PROOFS OF MRAC SCHEMES

$$\theta_2^T \alpha_{n-2}(s) + \theta_3 \Lambda(s) = \frac{1}{\hat{k}_p}[\hat{Q}(s,t) \cdot \hat{R}_p(s,t) - \Lambda_0(s)R_m(s)] \qquad (6.8.22)$$

Consider the above polynomial equations as operator equations. We apply (6.8.21) to the signal $\frac{W_m(s)}{\Lambda(s)} u_p$, and (6.8.22) to $\frac{W_m(s)}{\Lambda(s)} y_p$ to obtain

$$\theta_1^T W_m(s)\omega_1 = W_m(s)u_p - \frac{1}{\hat{k}_p}\hat{\bar{Z}}_p(s,t) \cdot \hat{Q}(s,t)\frac{W_m(s)}{\Lambda(s)}u_p$$

$$\theta_2^T W_m(s)\omega_2 + \theta_3 W_m(s)y_p = \frac{1}{\hat{k}_p}[\hat{Q}(s,t) \cdot \hat{R}_p(s,t) - \Lambda_0(s)R_m(s)]\frac{W_m(s)}{\Lambda(s)}y_p$$

Combining these two equations, we have

$$\theta_0^T W_m(s)\omega_0 = W_m(s)u_p - \frac{1}{\hat{k}_p}\hat{\bar{Z}}_p(s,t) \cdot \hat{Q}(s,t)\frac{W_m(s)}{\Lambda(s)}u_p$$

$$+ \frac{1}{\hat{k}_p}[\hat{Q}(s,t) \cdot \hat{R}_p(s,t) - \Lambda_0(s)R_m(s)]\frac{W_m(s)}{\Lambda(s)}y_p \qquad (6.8.23)$$

where $\theta_0 = [\theta_1^T, \theta_2^T, \theta_3]^T$, $\omega_0 = [\omega_1^T, \omega_2^T, y_p]^T$. Repeating the same algebraic manipulation, but replacing θ, θ_p by θ^*, θ_p^* in the polynomial equations, we have

$$\theta_0^{*T} W_m(s)\omega_0 = W_m(s)u_p - \frac{\bar{Z}_p(s)Q(s)}{k_p}\frac{W_m(s)}{\Lambda(s)}u_p$$

$$+ \frac{Q(s)R_p(s) - \Lambda_0(s)R_m(s)}{k_p}\frac{W_m(s)}{\Lambda(s)}y_p \qquad (6.8.24)$$

where $Q(s)$ is the quotient of $\Lambda_0(s)R_m(s)/R_p(s)$ whose order is $n^* - 1$ and n^* is the relative degree of $G_p(s)$. Subtracting (6.8.24) from (6.8.23), we have

$$\tilde{\theta}_0^T W_m(s)\omega_0 = \left\{-\frac{1}{\hat{k}_p}\hat{\bar{Z}}_p(s,t) \cdot \hat{Q}(s,t)\frac{W_m(s)}{\Lambda(s)}u_p + \frac{1}{\hat{k}_p}\hat{Q}(s,t) \cdot \hat{R}_p(s,t)\frac{W_m(s)}{\Lambda(s)}y_p\right\}$$

$$- \left\{\frac{1}{\hat{k}_p}\frac{\Lambda_0(s)R_m(s)}{\Lambda(s)}W_m(s)y_p - \frac{1}{k_p}\frac{\Lambda_0(s)R_m(s)}{\Lambda(s)}W_m(s)y_p\right\}$$

$$+ \left\{\frac{\bar{Z}_p(s)Q(s)}{k_p\Lambda(s)}W_m(s)u_p - \frac{R_p(s)Q(s)}{k_p\Lambda(s)}W_m(s)y_p\right\}$$

$$\stackrel{\triangle}{=} e_f - e_{1f} + e_{2f} \qquad (6.8.25)$$

where $\tilde{\theta}_0 \stackrel{\triangle}{=} \theta_0 - \theta_0^*$ and

$$e_f \stackrel{\triangle}{=} -\frac{1}{\hat{k}_p}\hat{\bar{Z}}_p(s,t) \cdot \hat{Q}(s,t)\frac{W_m(s)}{\Lambda(s)}u_p + \frac{1}{\hat{k}_p}\hat{Q}(s,t) \cdot \hat{R}_p(s,t)\frac{W_m(s)}{\Lambda(s)}y_p$$

e_{1f}, e_{2f} are defined as the terms in the second and third brackets of (6.8.25), respectively. Because $c_0 = \frac{k_m}{k_p}, c_0^* = \frac{k_m}{k_p}$, $\Lambda(s) = \Lambda_0(s)Z_m(s)$ and $\bar{Z}_p(s)u_p = R_p(s)y_p$, we have

$$e_{1f} = (c_0 - c_0^*)y_p, \quad e_{2f} = 0 \qquad (6.8.26)$$

Using (6.8.26) in (6.8.25) and defining $\omega_p \triangleq [\omega_0^T, W_m^{-1}(s)y_p]^T$ we can write

$$\tilde{\theta}^T W_m(s)\omega_p = e_f \qquad (6.8.27)$$

Because $\tilde{\theta}^T \omega = \frac{c_0^*}{c_0}\tilde{\theta}^T \omega_p$, proved in Section 6.8.2 (see equation (6.8.14)), we use Swapping Lemma A.2 to write

$$\tilde{\theta}^T \omega = \frac{c_0^*}{c_0}\left(F_1(s,\alpha_0)(\dot{\tilde{\theta}}^T \omega_p + \tilde{\theta}^T \dot{\omega}_p) + F(s,\alpha_0)\tilde{\theta}^T \omega_p\right) \qquad (6.8.28)$$

where $F(s, \alpha_0)$ and $F_1(s, \alpha_0)$ are as defined in Section 6.8.2 and satisfy

$$\|F_1(s,\alpha_0)\|_{\infty\delta} \leq \frac{c}{\alpha_0}, \quad \|F(s,\alpha_0)W_m^{-1}(s)\|_{\infty\delta} \leq c\alpha_0^{n^*}$$

for any $\alpha_0 > \delta > 0$. Applying Swapping Lemma A.1 to $W_m(s)\tilde{\theta}^T \omega_p$ and using (6.8.27), we obtain

$$\begin{aligned}\tilde{\theta}^T \omega_p &= W_m^{-1}(s)\left(\tilde{\theta}^T W_m(s)\omega_p + W_c(s)(W_b(s)\omega_p^T)\theta\right) \\ &= W_m^{-1}(s)\left(e_f + W_c(s)(W_b(s)\omega_p^T)\theta\right)\end{aligned} \qquad (6.8.29)$$

Substituting (6.8.29) in (6.8.28), we have

$$\tilde{\theta}^T \omega = \frac{c_0^*}{c_0}\left(F_1(s,\alpha_0)(\dot{\tilde{\theta}}^T \omega_p + \tilde{\theta}^T \dot{\omega}_p) + F(s,\alpha_0)W_m^{-1}(s)\left(e_f + W_c(s)(W_b(s)\omega_p^T)\theta\right)\right)$$

Because c_0 is bounded from below, i.e., $|c_0| > \underline{c}_0 > 0$, it follows from Lemma 3.3.2 that

$$\|\tilde{\theta}^T \omega\| \leq \frac{c}{\alpha_0}\left(\|\dot{\tilde{\theta}}^T \omega_p\| + \|\tilde{\theta}^T \dot{\omega}_p\|\right) + c\alpha_0^{n^*}(\|e_f\| + \|\bar{\omega}\dot{\tilde{\theta}}\|)\right) \qquad (6.8.30)$$

where $\bar{\omega} = W_b(\omega_p)^T$. From $\tilde{\theta} \in \mathcal{L}_\infty$ and the normalizing properties of m_f, we have

$$\|\tilde{\theta}^T \omega\| \leq \frac{c}{\alpha_0}\|\dot{\theta}m_f\| + \frac{c}{\alpha_0}m_f + c\alpha_0^{n^*}(\|e_f\| + \|\dot{\theta}m_f\|) \qquad (6.8.31)$$

We now concentrate on the term $\|e_f\|$ in (6.8.31). Let us denote

$$\hat{Q}(s,t) \triangleq q^T\alpha_{n^*-1}(s), \quad \hat{\bar{Z}}_p(s,t) = \hat{b}_p^T\alpha_m(s), \quad \hat{R}_p(s,t) = s^n + \hat{a}_p^T\alpha_{n-1}(s)$$

6.8. STABILITY PROOFS OF MRAC SCHEMES

where $\hat{b}_p = [\hat{k}_p, \hat{p}_1^\top]^\top$, $\hat{a}_p = \hat{p}_2$ and $\hat{p}_1, \hat{p}_2$ are defined in Table 6.9. Treating s as the differentiation operator, using $\Lambda_p(s) = \Lambda(s)(s + \lambda_0)$ and Swapping Lemma A.3 (i), we have

$$\frac{1}{\hat{k}_p}\tilde{\hat{Z}}_p(s,t) \cdot \hat{Q}(s,t) \frac{W_m(s)}{\Lambda(s)} u_p = \frac{1}{\hat{k}_p} \hat{Q}(s,t) \cdot \tilde{\hat{Z}}_p(s,t) \frac{W_m(s)(s+\lambda_0)}{\Lambda_p(s)} u_p$$

$$= \frac{1}{\hat{k}_p}\left\{ \hat{Q}(s,t)\left(\tilde{\hat{Z}}_p(s,t)\frac{W_m(s)(s+\lambda_0)}{\Lambda_p(s)}u_p\right)\right. \quad (6.8.32)$$

$$\left. -q^\top D_{n^*-2}(s)\left[\alpha_{n^*-2}(s)\left(\alpha_m^\top(s)\frac{W_m(s)(s+\lambda_0)}{\Lambda_p(s)}u_p\right)\right]\dot{\hat{b}}_p\right\}$$

(by taking $A(s,t) = \hat{Q}(s,t), B(s,t) = \tilde{\hat{Z}}_p(s,t), f = \frac{W_m(s)}{\Lambda(s)}u_p$) and

$$\frac{1}{\hat{k}_p}\hat{Q}(s,t)\cdot\tilde{\hat{R}}_p(s,t)\frac{W_m(s)}{\Lambda(s)}y_p$$

$$= \frac{1}{\hat{k}_p}\left\{\hat{Q}(s,t)\left(\tilde{\hat{R}}_p(s,t)\frac{W_m(s)(s+\lambda_0)}{\Lambda_p(s)}y_p\right)\right. \quad (6.8.33)$$

$$\left. -q^\top D_{n^*-2}(s)\left[\alpha_{n^*-2}(s)\left(\alpha_{n-1}^\top(s)\frac{W_m(s)(s+\lambda_0)}{\Lambda_p(s)}y_p\right)\right]\dot{\hat{a}}_p\right\}$$

(by taking $A(s,t) = \hat{Q}(s,t), B(s,t) = \tilde{\hat{R}}_p(s,t), f = \frac{W_m(s)}{\Lambda(s)}y_p$). Using (6.8.32), (6.8.33) in the expression for e_f and noting that

$$\hat{R}_p(s,t)\frac{W_m(s)(s+\lambda_0)}{\Lambda_p(s)}y_p - \hat{Z}_p(s,t)\frac{W_m(s)(s+\lambda_0)}{\Lambda_p(s)}u_p = -\tilde{\theta}_p^\top W_m(s)(s+\lambda_0)\phi$$

we have

$$e_f = \frac{1}{\hat{k}_p}\left\{-\tilde{\theta}_p^\top W_m(s)(s+\lambda_0)\phi + e_f'\right\} \quad (6.8.34)$$

where θ_p, ϕ are as defined in Table 6.9 and

$$e_f' \triangleq q^\top D_{n^*-2}(s)\left(\alpha_{n^*-2}(s)\left[\alpha_m^\top(s)\frac{W_m(s)(s+\lambda_0)}{\Lambda_p(s)}u_p\right.\right.$$

$$\left.\left. -\alpha_{n-1}^\top(s)\frac{W_m(s)(s+\lambda_0)}{\Lambda_p(s)}y_p\right]\begin{bmatrix}\dot{\hat{b}}_p\\ \dot{\hat{a}}_p\end{bmatrix}\right) \quad (6.8.35)$$

Using the normalizing properties of m_f and the fact that $y_p = W_m(s)(r + \frac{1}{c_0^*}\tilde{\theta}^\top\omega)$ and $\theta \in \mathcal{L}_\infty$, we have

$$\|e_f'\| \leq c\|\dot{\theta}_p m_f\|$$

Applying Swapping Lemma A.1 and using $\tilde{\theta}_p^\top \phi = -\epsilon m^2$, we can write

$$\tilde{\theta}_p^\top W_m(s)(s+\lambda_0)\phi = -W_m(s)(s+\lambda_0)\epsilon m^2 - W_c[W_b\phi^\top]\dot{\tilde{\theta}}_p \quad (6.8.36)$$

428   CHAPTER 6. MODEL REFERENCE ADAPTIVE CONTROL

Again using the normalizing properties of m_f, it follows from (6.8.34), (6.8.36) that

$$\|e_f\| \leq c(\|\epsilon m m_f\| + \|\dot{\theta}_p m_f\|) \tag{6.8.37}$$

for some constant c. Combining (6.8.31), (6.8.37), we have

$$\|\tilde{\theta}^T \omega\| \leq \frac{c}{\alpha_0} m_f + \frac{c}{\alpha_0}\|\dot{\theta} m_f\| + c\alpha_0^{n^*}\|\dot{\theta} m_f\| + c\alpha_0^{n^*}(\|\epsilon m m_f\| + \|\dot{\theta}_p m_f\|) \tag{6.8.38}$$

From (6.6.25), we can establish, using $\theta_p \in \mathcal{L}_\infty$, $\frac{1}{k_p} \in \mathcal{L}_\infty$, $\dot{\theta}_p \in \mathcal{L}_2$, that $\theta \in \mathcal{L}_\infty$ and $\dot{\theta} \in \mathcal{L}_2$. Hence,

$$\|\tilde{\theta}^T \omega\| \leq \frac{c}{\alpha_0} m_f + c\|\tilde{g} m_f\|$$

where $\tilde{g}^2 = \frac{1}{\alpha_0^2}|\dot{\theta}|^2 + \alpha_0^{2n^*}(|\dot{\theta}|^2 + \epsilon^2 m^2 + |\dot{\theta}_p|^2)$ and $\tilde{g} \in \mathcal{L}_2$.

Step 3. *Use the B-G Lemma to establish boundedness.* This step is identical to Step 3 in the proof of stability for the direct MRAC scheme and is omitted.

Step 4. *Convergence of the tracking error to zero.* From

$$y_p = W_m(s)(r + \frac{1}{c_0^*}\tilde{\theta}^T \omega)$$

and $\tilde{\theta}^T \omega = \frac{c_0^*}{c_0}\tilde{\theta}^T \omega_p$, we have

$$\begin{aligned} e_1 &= \frac{1}{c_0^*}W_m(s)\tilde{\theta}^T\omega = W_m(s)\frac{1}{c_0}\tilde{\theta}^T\omega_p \\ &= \frac{1}{c_0}W_m(s)\tilde{\theta}^T\omega_p - W_c(s)\left((W_b(s)\tilde{\theta}^T\omega_p)\frac{c_0}{c_0^2}\right) \end{aligned} \tag{6.8.39}$$

where the last equality is obtained by applying the Swapping Lemma A.1. Substituting $\tilde{\theta}_p^T \omega$ from (6.8.29) in (6.8.39), we have

$$e_1 = \frac{1}{c_0}\left(e_f + W_c(s)(W_b(s)\omega_p^T)\theta\right) - W_c(s)\left((W_b(s)\tilde{\theta}^T\omega_p)\frac{c_0}{c_0^2}\right)$$

Note from (6.8.35) and $\dot{\tilde{a}}_p, \dot{\tilde{b}}_p \in \mathcal{L}_2$, $u_p, y_p \in \mathcal{L}_\infty$ that $e'_f \in \mathcal{L}_2$. From $\epsilon m, \dot{\theta}_p \in \mathcal{L}_2$ and the boundedness of $\frac{1}{k_p}, m$ and ϕ, it also follows from (6.8.34), (6.8.37) that $e_f \in \mathcal{L}_2$. From $e_f, \dot{\theta}, \dot{c}_0 \in \mathcal{L}_2$, it follows that $e_1 \in \mathcal{L}_2$ which together with $\dot{e}_1 \in \mathcal{L}_\infty$ imply that $e_1(t) \to 0$ as $t \to \infty$. □

6.9 Problems

6.1 Consider the first order plant
$$y = \frac{b}{s-1}u$$
where $b > 0$ is the only unknown parameter. Design and analyze a direct MRAC scheme that can stabilize the plant and force y to follow the output y_m of the reference model
$$y_m = \frac{2}{s+2}r$$
for any bounded and continuous reference signal r.

6.2 The dynamics of a throttle to speed subsystem of a vehicle may be represented by the first-order system
$$V = \frac{b}{s+a}\theta + d$$
where V is the vehicle speed, θ is the throttle angle and d is a constant load disturbance. The parameters $b > 0, a$ are unknown constants whose values depend on the operating state of the vehicle that is defined by the gear state, steady-state velocity, drag, etc. We would like to design a cruise control system by choosing the throttle angle θ so that V follows a desired velocity V_m generated by the reference model
$$V_m = \frac{0.5}{s+0.5}V_s$$
where V_s is the desired velocity set by the driver.

(a) Assume that a, b, and d are known exactly. Design an MRC law that meets the control objective.

(b) Design and analyze a direct MRAC scheme to be used in the case of $a, b,$ and d (with $b > 0$) being unknown.

(c) Simulate your scheme in (b) by assuming $V_s = 35$ and using the following values for $a, b,$ and d: (i) $a = 0.02, b = 1.3, d = 10$; (ii) $a = 0.02(2 + \sin 0.01t), b = 1.3, d = 10\sin 0.02t$.

6.3 Consider the adaptive tracking problem discussed in Section 6.2.2, i.e., the output y of the plant
$$\dot{x} = ax + bu$$
$$y = x$$
is required to track the output y_m of the reference model
$$\dot{x}_m = -a_m x_m + b_m r, \quad y_m = x_m$$
where a and b are unknown constants with $b \neq 0$ and sign(b) unknown. Design an adaptive controller to meet the control objective.

6.4 Consider the following plant
$$\dot{x} = -x + bu$$
where $b \neq 0$ is unknown. Design an adaptive controller that will force x to track x_m, the state of the reference model
$$\dot{x}_m = -x_m + r$$
for any bounded reference input r.

6.5 Consider the following SISO plant
$$y_p = k_p \frac{Z_p(s)}{R_p(s)} u_p$$
where $k_p, Z_p(s)$, and $R_p(s)$ are known. Design an MRC scheme for y_p to track y_m generated by the reference model
$$y_m = k_m \frac{Z_m(s)}{R_m(s)} r$$
where $R_m(s)$ has the same degree as $R_p(s)$. Examine stability when $Z_p(s)$ is Hurwitz and when it is not. Comment on your results.

6.6 Consider the third order plant
$$y_p = \frac{k_p(s + b_0)}{s^3 + a_2 s^2 + a_1 s + a_0} u_p$$
where $a_i, i = 0, 1, 2; b_0, k_p$ are constants and $b_0 > 0$. The transfer function of the reference model is given by
$$y_m = \frac{1}{(s+1)^2} r$$

(a) Assuming that a_i, b_0, and k_p are known, design an MRC law that guarantees closed-loop stability and $y_p \to y_m$ as $t \to \infty$ for any bounded reference input r.

(b) Repeat (a) when a_i, b_0, and k_p are unknown and $k_p > 0$.

(c) If in (b) $a_2 = 0, a_1 = 0, a_0 = 1$ are known but k_p, b_0 are unknown, indicate the simplification that results in the control law.

6.7 Show that the MRC law given by (6.3.29) in Remark 6.3.6 meets the MRC objective for the plant given by (6.3.1).

6.8 Show that the MRC law given by (6.3.27) or (6.3.28) in Remark 6.3.5 meets the MRC objective for the plant (6.3.1) for any given nonzero initial conditions.

6.9. PROBLEMS

6.9 Repeat the proof of Theorem 6.4.2 by using a minimal state space representation of the error system $e_1 = W_m(s)(s+p_0)\rho^* \tilde{\theta}^T \phi$ as explained in Remark 6.4.5.

6.10 Consider the third-order plant

$$y_p = \frac{1}{s^3 + a_2 s^2 + a_1 s + a_0} u_p$$

where $a_i, i = 0, 1, 2$ are unknown constants and the reference model

$$y_m = \frac{2}{(s+1)(s+2)(s+2.5)} r$$

(a) Design a direct MRAC law with unnormalized adaptive law so that all signals in the closed-loop plant are bounded and $y_p(t) \to y_m(t)$ as $t \to \infty$ for any bounded reference input r.

(b) Simulate your scheme by assuming the following values for the plant parameters

$$a_0 = 1, \ a_1 = -1.5, \ a_2 = 0$$

and examine the effect of your choice of r on parameter convergence.

6.11 Design and analyze a direct MRAC with normalized adaptive law for the plant

$$y_p = \frac{b}{s+a} u_p$$

where $b > 0.5, a$ are unknown constants. The reference model is given by

$$y_m = \frac{3}{s+3} r$$

(a) Design a direct MRAC scheme based on the gradient algorithm

(b) Repeat (a) for a least-squares algorithm

(c) Repeat (a) using the SPR-Lyapunov design approach with normalization

(d) Simulate your design in (c) with and without normalization. For simulations use $b = 1.2$, $a = -1$ and r a signal of your choice.

6.12 Repeat Problem 6.11(a), for a hybrid MRAC scheme. Simulate your scheme using the values of $b = 1.2, a = -1$ and r a signal of your choice.

6.13 Consider the plant

$$y_p = \frac{(s+b_0)}{(s+a)^2} u_p$$

where $b_0 > 0.2$ and a are unknown constants. The reference model is given by

$$y_m = \frac{1}{s+1} r$$

(a) Design a direct MRAC scheme with unnormalized adaptive law

(b) Repeat (a) with a normalized adaptive law

(c) Design an indirect MRAC scheme with an unnormalized adaptive law

(d) Repeat (c) with a normalized adaptive law

(e) Simulate one direct and one indirect MRAC scheme of your choice from (a) to (d) and compare their performance when $b_0 = 2$, $a = -5$, a nd $r = 2 + sin 0.8t$. Comment.

6.14 Consider the control law (6.3.27) for the plant (6.3.1) where $\theta_i^*, i = 1, 2, 3, c_0^*$ are the desired controller parameters. Design and analyze a direct MRAC scheme based on this control law.

6.15 Consider the control law given by equation (6.3.29) in Remark 6.3.6 designed for the plant (6.3.1). Design and analyze a direct MRAC scheme based on this control law.

6.16 Consider the SISO plant

$$y_p = \frac{k_p Z_p(s)}{R_p(s)} u_p$$

where $Z_p(s)$, $R_p(s)$ are monic, $Z_p(s)$ is Hurwitz and the relative degree $n^* = 1$. The order n of R_p is unknown. Show that the adaptive control law

$$u_p = -\theta y_p \operatorname{sgn}(k_p), \quad \dot{\theta} = y_p^2$$

guarantees signal boundedness and convergence of y_p to zero for any finite n. (Hint: The plant may be represented as

$$\dot{x}_1 = A_{11} x_1 + A_{12} y_p, \quad x_1 \in \mathcal{R}^{n-1}$$
$$\dot{y}_p = A_{21} x_1 + a_0 y_p + k_p u_p$$

where A_{11} is stable.)

6.17 Following the same procedure used in Section 6.6.2, derive an indirect MRAC scheme using unnormalized adaptive laws for a plant with $n^* = 2$.

6.18 Let $\omega \in \mathcal{L}_\infty$ be PE and $e \in \mathcal{S}(\mu) \cap \mathcal{L}_\infty$ where $\mu \geq 0$. Let $\omega_\mu = \omega + e$. Show that there exists a $\mu^* > 0$ such that for any $\mu \in [0, \mu^*)$, ω_μ is PE.

6.19 Consider the MRAC problem of Section 6.4.1. It has been shown (see Remark 6.4.3) that the nonzero initial condition appears in the error equation as

$$e_1 = W_m(s)\rho^*(u_p - \theta^{*T}\omega) + C_c^T (sI - A_c)^{-1} e(0)$$

6.9. PROBLEMS

Show that the same stability results as in the case of $e(0) = 0$ can be established when $e(0) \neq 0$ by using the new Lyapunov-like function

$$V = \frac{e^\top P_c e}{2} + \frac{\tilde{\theta}^\top \Gamma^{-1} \tilde{\theta}}{2}|\rho^*| + \beta e_0^\top P_0 e_0$$

where e_0 is the zero-input response, i.e.,

$$\dot{e}_0 = A_c e_0, \qquad e_0(0) = e(0)$$

P_0 satisfies $A_c^\top P_0 + P_0 A_c = -I$ and $\beta > 0$ is an arbitrary positive constant.

Chapter 7

Adaptive Pole Placement Control

7.1 Introduction

In Chapter 6 we considered the design of a wide class of MRAC schemes for LTI plants with stable zeros. The assumption that the plant is minimum phase, i.e., it has stable zeros, is rather restrictive in many applications. For example, the approximation of time delays often encountered in chemical and other industrial processes leads to plant models with unstable zeros. As we discussed in Chapter 6, the minimum phase assumption is a consequence of the MRC objective that requires cancellation of the plant zeros in an effort to make the closed-loop plant transfer function equal to that of the reference model. The same assumption is also used to express the desired controller parameters in the form of a linear or bilinear parametric model, and is, therefore, crucial for parameter estimation and the stability of the overall adaptive control scheme.

Another class of control schemes that is popular in the known parameter case are those that change the poles of the plant and do not involve plant zero-pole cancellations. These schemes are referred to as *pole placement* schemes and are applicable to both minimum and nonminimum phase LTI plants. The combination of a pole placement control law with a parameter estimator or an adaptive law leads to an *adaptive pole placement control* (APPC) scheme that can be used to control a wide class of LTI plants with

7.1. INTRODUCTION

unknown parameters.

As in the MRAC case, the APPC schemes may be divided into two classes: The *indirect APPC* schemes where the adaptive law generates on-line estimates of the coefficients of the plant transfer function that are then used to calculate the parameters of the pole placement control law by solving a certain algebraic equation; and the *direct APPC* where the parameters of the pole placement control law are generated directly by an adaptive law without any intermediate calculations that involve estimates of the plant parameters.

The direct APPC schemes are restricted to scalar plants and to special classes of plants where the desired parameters of the pole placement controller can be expressed in the form of the linear or bilinear parametric models. Efforts to develop direct APPC schemes for a general class of LTI plants led to APPC schemes where both the controller and plant parameters are estimated on-line simultaneously [49, 112], leading to a rather complex adaptive control scheme.

The indirect APPC schemes, on the other hand, are easy to design and are applicable to a wide class of LTI plants that are not required to be minimum phase or stable. The main drawback of indirect APPC is the possible loss of stabilizability of the estimated plant based on which the calculation of the controller parameters is performed. This drawback can be eliminated by modifying the indirect APPC schemes at the expense of adding more complexity. Because of its flexibility in choosing the controller design methodology (state feedback, compensator design, linear quadratic, etc.) and adaptive law (least squares, gradient, or SPR-Lyapunov type), indirect APPC is the most general class of adaptive control schemes. This class also includes indirect MRAC as a special case where some of the poles of the plant are assigned to be equal to the zeros of the plant to facilitate the required zero-pole cancellation for transfer function matching. Indirect APPC schemes have also been known as self-tuning regulators in the literature of adaptive control to distinguish them from direct MRAC schemes.

The chapter is organized as follows: In Section 7.2 we use several examples to illustrate the design and analysis of APPC. These examples are used to motivate the more complicated designs in the general case treated in the rest of the chapter. In Section 7.3, we define the pole placement control (PPC) problem for a general class of SISO, LTI plants and solve it for the case of known plant parameters using several different control laws. The significance of Section 7.3 is that it provides several pole placement control laws to be used together with adaptive laws to form APPC schemes. The design and analysis of indirect APPC schemes for a general class of SISO, LTI plants is presented in Section 7.4. Section 7.5 is devoted to the design and analysis of hybrid APPC schemes where the estimated plant parameters are updated at distinct points in time. The problem of stabilizability of the estimated plant model at each instant of time is treated in Section 7.6. A simple example is first used to illustrate the possible loss of stabilizability and a modified indirect APPC scheme is then proposed and analyzed. The modified scheme is guaranteed to meet the control objective and is therefore not affected by the possible loss of stabilizability during parameter estimation. Section 7.7 is devoted to stability proofs of the various theorems given in previous sections.

7.2 Simple APPC Schemes

In this section we use several examples to illustrate the design and analysis of simple APPC schemes. The important features and characteristics of these schemes are used to motivate and understand the more complicated ones to be introduced in the sections to follow.

7.2.1 Scalar Example: Adaptive Regulation

Consider the scalar plant

$$\dot{y} = ay + bu \tag{7.2.1}$$

where a and b are unknown constants, and the sign of b is known. The control objective is to choose u so that the closed-loop pole is placed at $-a_m$, where $a_m > 0$ is a given constant, y and u are bounded, and $y(t)$ converges to zero as $t \to \infty$.

If a and b were known and $b \neq 0$ then the control law

$$u = -ky + r \tag{7.2.2}$$

$$k = \frac{a + a_m}{b} \tag{7.2.3}$$

7.2. SIMPLE APPC SCHEMES

where r is a reference input, would lead to the closed-loop plant

$$\dot{y} = -a_m y + br \qquad (7.2.4)$$

i.e., the control law described by (7.2.2) and (7.2.3) changes the pole of the plant from a to $-a_m$ but preserves the zero structure. This is in contrast to MRC, where the zeros of the plant are canceled and replaced with new ones. It is clear from (7.2.4) that the pole placement law (7.2.2) and (7.2.3) with $r=0$ meets the control objective exactly.

Let us now consider the case where a and b are unknown. As in the MRAC case, we use the certainty equivalence approach to form adaptive pole placement control schemes as follows: We use the same control law as in (7.2.2) but replace the unknown controller parameter k with its on-line estimate $\hat{k}$. The estimate $\hat{k}$ may be generated in two different ways: The direct one where $\hat{k}$ is generated by an adaptive law and the indirect one where $\hat{k}$ is calculated from

$$\hat{k} = \frac{\hat{a} + a_m}{\hat{b}} \qquad (7.2.5)$$

provided $\hat{b} \neq 0$ where $\hat{a}$ and $\hat{b}$ are the on-line estimates of a and b, respectively. We consider each design approach separately.

Direct Adaptive Regulation In this case the time-varying gain $\hat{k}$ in the control law

$$u = -\hat{k}y + r, \qquad r = 0 \qquad (7.2.6)$$

is updated directly by an adaptive law. The adaptive law is developed as follows: We add and subtract the desired control input, $u^* = -ky + r$ with $k = \frac{a+a_m}{b}$ in the plant equation, i.e.,

$$\dot{y} = ay + bu^* - bu^* + bu = -a_m y - b(\hat{k} - k)y + br$$

to obtain, for $r = 0$, the error equation

$$\dot{y} = -a_m y - b\tilde{k}y \qquad (7.2.7)$$

where $\tilde{k} \triangleq \hat{k} - k$, that relates the parameter error term $b\tilde{k}y$ and regulation error y through the SPR transfer function $\frac{1}{s+a_m}$. As shown in Chapter 4,

(7.2.7) is in the appropriate form for applying the Lyapunov design approach. Choosing
$$V = \frac{y^2}{2} + \frac{\tilde{k}^2 |b|}{2\gamma}$$
it follows that for
$$\dot{\tilde{k}} = \gamma y^2 \operatorname{sgn}(b) \tag{7.2.8}$$
we have
$$\dot{V} = -a_m y^2 \leq 0$$
which implies that $y, \tilde{k}, u \in \mathcal{L}_\infty$ and $y \in \mathcal{L}_2$. From (7.2.7) and $y, \tilde{k} \in \mathcal{L}_\infty$ we have $\dot{y} \in \mathcal{L}_\infty$; therefore, using Lemma 3.2.5 we obtain $y(t) \to 0$ as $t \to \infty$.

In summary, the direct APPC scheme in (7.2.6) and (7.2.8) guarantees signal boundedness and regulation of the plant state $y(t)$ to zero. The scheme, however, does not guarantee that the closed-loop pole of the plant is placed at $-a_m$ even asymptotically with time. To achieve such a pole placement result, we need to show that $\hat{k} \to \frac{a+a_m}{b}$ as $t \to \infty$. For parameter convergence, however, y is required to be PE which is in conflict with the objective of regulating y to zero. The conflict between parameter identification and regulation or control is well known in adaptive control and cannot be avoided in general.

Indirect Adaptive Regulation   In this approach, the gain $\hat{k}(t)$ in the control law
$$u = -\hat{k}(t)y + r, \qquad r = 0 \tag{7.2.9}$$
is calculated by using the algebraic equation
$$\hat{k} = \frac{\hat{a} + a_m}{\hat{b}} \tag{7.2.10}$$
with $\hat{b} \neq 0$ and the on-line estimates $\hat{a}$ and $\hat{b}$ of the plant parameters a and b, respectively. The adaptive law for generating $\hat{a}$ and $\hat{b}$ is constructed by using the techniques of Chapter 4 as follows:

We construct the series-parallel model
$$\dot{y}_m = -a_m(y_m - y) + \hat{a}y + \hat{b}u \tag{7.2.11}$$

7.2. SIMPLE APPC SCHEMES

then subtract (7.2.11) from the plant equation (7.2.1) to obtain the error equation

$$\dot{e} = -a_m e - \tilde{a}y - \tilde{b}u \qquad (7.2.12)$$

where $e \triangleq y - y_m$, $\tilde{a} \triangleq \hat{a} - a$, $\tilde{b} \triangleq \hat{b} - b$. Using the Lyapunov-like function

$$V = \frac{e^2}{2} + \frac{\tilde{a}^2}{2\gamma_1} + \frac{\tilde{b}^2}{2\gamma_2}$$

for some $\gamma_1, \gamma_2 > 0$ and choosing

$$\dot{\hat{a}} = \gamma_1 ey, \quad \dot{\hat{b}} = \gamma_2 eu \qquad (7.2.13)$$

we have

$$\dot{V} = -a_m e^2 \leq 0$$

which implies that $e, \hat{a}, \hat{b} \in \mathcal{L}_\infty$ and $e \in \mathcal{L}_2$. These properties, however, do not guarantee that $\hat{b}(t) \neq 0 \ \forall t \geq 0$, a condition that is required for $\hat{k}$, given by (7.2.10), to be finite. In fact, for $\hat{k}$ to be uniformly bounded, we should have $|\hat{b}(t)| \geq b_0 > 0 \ \forall t \geq 0$ and some constant b_0. Because such a condition cannot be guaranteed by the adaptive law, we modify (7.2.13) by assuming that $|b| \geq b_0 > 0$ where b_0 and $\text{sgn}(b)$ are known a priori, and use the projection techniques of Chapter 4 to obtain

$$\dot{\hat{a}} = \gamma_1 ey \qquad (7.2.14)$$
$$\dot{\hat{b}} = \begin{cases} \gamma_2 eu & \text{if } |\hat{b}| > b_0 \text{ or if } |\hat{b}| = b_0 \text{ and } \text{sgn}(b)eu \geq 0 \\ 0 & \text{otherwise} \end{cases}$$

where $\hat{b}(0)$ is chosen so that $\hat{b}(0)\text{sgn}(b) \geq b_0$. The modified adaptive law guarantees that $|\hat{b}(t)| \geq b_0 \ \forall t \geq 0$. Furthermore, the time derivative $\dot{V}$ of V along the solution of (7.2.12) and (7.2.14) satisfies

$$\dot{V} = \begin{cases} -a_m e^2 & \text{if } |\hat{b}| > b_0 \text{ or if } |\hat{b}| = b_0 \text{ and } \text{sgn}(b)eu \geq 0 \\ -a_m e^2 - \tilde{b}eu & \text{if } |\hat{b}| = b_0 \text{ and } \text{sgn}(b)eu < 0 \end{cases}$$

Now for $|\hat{b}| = b_0$ and $\text{sgn}(b)eu < 0$, since $|b| \geq b_0$, we have

$$\tilde{b}eu = \hat{b}eu - beu = (|\hat{b}| - |b|)eu\,\text{sgn}(b) = (b_0 - |b|)eu\,\text{sgn}(b) \geq 0$$

therefore,

$$\dot{V} \leq -a_m e^2$$

Hence, $e, \tilde{a}, \tilde{b} \in \mathcal{L}_\infty; e \in \mathcal{L}_2$ and $|\hat{b}(t)| \geq b_0 \; \forall t \geq 0$, which implies that $\tilde{k} \in \mathcal{L}_\infty$. Substituting for the control law (7.2.9) and (7.2.10) in (7.2.11), we obtain $\dot{y}_m = -a_m y_m$, which implies that $y_m \in \mathcal{L}_\infty, y_m(t) \to 0$ as $t \to \infty$ and, therefore, $y, u \in \mathcal{L}_\infty$. From (7.2.12), we have $\dot{e} \in \mathcal{L}_\infty$, which, together with $e \in \mathcal{L}_2$, implies that $e(t) = y(t) - y_m(t) \to 0$ as $t \to \infty$. Therefore, it follows that $y(t) = e(t) + y_m(t) \to 0$ as $t \to \infty$.

The indirect adaptive pole placement scheme given by (7.2.9), (7.2.10), and (7.2.14) has, therefore, the same stability properties as the direct one. It has also several differences. The main difference is that the gain $\hat{k}$ is updated indirectly by solving an algebraic time varying equation at each time t. According to (7.2.10), the control law (7.2.9) is designed to meet the pole placement objective for the estimated plant at each time t rather than the actual plant. Therefore, for such a design to be possible, the estimated plant has to be controllable or at least stabilizable at each time t, which implies that $|\hat{b}(t)| \neq 0 \; \forall t \geq 0$. In addition, for uniform signal boundedness, $\hat{b}$ should satisfy $|\hat{b}(t)| \geq b_0 > 0$ where b_0 is a lower bound for $|b|$ that is known a priori.

The fact that only the sign of b is needed in the direct case, whereas the knowledge of a lower bound b_0 is also needed in the indirect case motivated the authors of [46] to come up with a modified indirect scheme presented in the next section where only the sign of b is needed.

7.2.2 Modified Indirect Adaptive Regulation

In the indirect case, the form of the matching equation

$$a - bk = -a_m$$

is used to calculate $\hat{k}$ from the estimates of $\hat{a}, \hat{b}$ by selecting $\hat{k}$ to satisfy

$$\hat{a} - \hat{b}\hat{k} = -a_m$$

as indicated by (7.2.10).

If instead of calculating $k(t)$ we update it using an adaptive law driven by the error

$$\varepsilon_c = \hat{a}(t) - \hat{b}(t)\hat{k}(t) + a_m$$

then we end up with a modified scheme that does not require the knowledge of a lower bound for $|b|$ as shown below. The error ε_c is motivated from the

7.2. SIMPLE APPC SCHEMES

fact that as $\varepsilon_c \to 0$ the estimated closed-loop pole $\hat{a}(t) - \hat{b}(t)\hat{k}(t)$ converges to $-a_m$, the desired pole. The adaptive law for $\hat{k}$ is developed by expressing ε_c in terms of $\tilde{k}, \tilde{a}, \tilde{b}$ by using $a_m = bk - a$ and adding and subtracting the term $b\hat{k}$, i.e.,

$$\varepsilon_c = \hat{a} - \hat{b}\hat{k} + bk - a + b\hat{k} - b\hat{k} = \tilde{a} - \tilde{b}\hat{k} - b\tilde{k}$$

The adaptive law for $\hat{k}$ is then obtained by using the gradient method to minimize $\frac{\varepsilon_c^2}{2}$ w.r.t. $\tilde{k}$, i.e.,

$$\dot{\tilde{k}} = -\gamma \frac{1}{2}\frac{\partial \varepsilon_c^2}{\partial \tilde{k}} = \gamma b \varepsilon_c$$

where $\gamma > 0$ is an arbitrary constant. Because $b = |b|\mathrm{sgn}(b)$ and γ is arbitrary, it follows that

$$\dot{\tilde{k}} = \dot{\hat{k}} = \gamma_0 \varepsilon_c \mathrm{sgn}(b)$$

for some other arbitrary constant $\gamma_0 > 0$. To assure stability of the overall scheme, the adaptive laws for $\tilde{a}, \tilde{b}$ in (7.2.13) are modified as shown below:

$$\dot{\tilde{a}} = \dot{\hat{a}} = \gamma_1 ey - \gamma_1 \frac{1}{2}\frac{\partial \varepsilon_c^2}{\partial \tilde{a}} = \gamma_1 ey - \gamma_1 \varepsilon_c$$
$$\dot{\tilde{b}} = \dot{\hat{b}} = \gamma_2 eu - \gamma_2 \frac{1}{2}\frac{\partial \varepsilon_c^2}{\partial \tilde{b}} = \gamma_2 eu + \gamma_2 \hat{k}\varepsilon_c$$

The overall modified indirect scheme is summarized as follows:

$$u = -\hat{k}y$$
$$\dot{\hat{k}} = \gamma_0 \varepsilon_c \mathrm{sgn}(b)$$
$$\dot{\hat{a}} = \gamma_1 ey - \gamma_1 \varepsilon_c$$
$$\dot{\hat{b}} = \gamma_2 eu + \gamma_2 \hat{k}\varepsilon_c$$
$$\varepsilon_c = \hat{a} - \hat{b}\hat{k} + a_m, \quad e = y - y_m$$
$$\dot{y}_m = -a_m(y_m - y) + \hat{a}y + \hat{b}u$$

Stability Analysis Let us choose the Lyapunov-like function

$$V = \frac{e^2}{2} + \frac{\tilde{a}^2}{2\gamma_1} + \frac{\tilde{b}^2}{2\gamma_2} + \frac{\tilde{k}^2|b|}{2\gamma_0}$$

where $e = y - y_m$ satisfies the error equation
$$\dot{e} = -a_m e - \tilde{a}y - \tilde{b}u \qquad (7.2.12)$$
The time-derivative $\dot{V}$ of V along the trajectories of the overall system is given by
$$\dot{V} = -a_m e^2 - \varepsilon_c(\tilde{a} - \tilde{b}\hat{k} - b\tilde{k}) = -a_m e^2 - \varepsilon_c^2 \leq 0$$
which implies that $e, \tilde{a}, \tilde{b}, \tilde{k} \in \mathcal{L}_\infty$; $e, \varepsilon_c \in \mathcal{L}_2$. The boundedness of y_m and, therefore, of y, u is established as follows: The series-parallel model equation (7.2.11) can be rewritten as
$$\dot{y}_m = -a_m(y_m - y) + (\hat{a} - \hat{b}\hat{k})y = -(a_m - \varepsilon_c)y_m + \varepsilon_c e$$
Because $e, \varepsilon_c \in \mathcal{L}_\infty \bigcap \mathcal{L}_2$, it follows that $y_m \in \mathcal{L}_\infty$ and, therefore, $y, u \in \mathcal{L}_\infty$. From $\dot{\varepsilon}_c, \dot{e} \in \mathcal{L}_\infty$ and $e, \varepsilon_c \in \mathcal{L}_2$, we have $e(t), \varepsilon_c(t) \to 0$ as $t \to \infty$, which imply that $y_m(t) \to 0$ as $t \to \infty$. Therefore, $y(t) \to 0$ as $t \to \infty$.

The modified indirect scheme demonstrates that we can achieve the same stability result as in the direct case by using the same a priori knowledge about the plant, namely, the knowledge of the sign of b. For the modified scheme, the controller parameter $\hat{k}$ is adjusted dynamically by using the error ε_c between the closed-loop pole of the estimated plant, i.e., $\hat{a} - \hat{b}\hat{k}$ and the desired pole $-a_m$. The use of ε_c introduces an additional gradient term to the adaptive law for $\hat{a}, \hat{b}$ and increases the complexity of the overall scheme. In [46] it was shown that this method can be extended to higher order plants with stable zeros.

7.2.3 Scalar Example: Adaptive Tracking

Let us consider the same plant (7.2.1) as in Section 7.2.1, i.e.,
$$\dot{y} = ay + bu$$
The control objective is modified to include tracking and is stated as follows: Choose the plant input u so that the closed-loop pole is at $-a_m$; $u, y \in \mathcal{L}_\infty$ and $y(t)$ tracks the reference signal $y_m(t) = c, \forall t \geq 0$, where $c \neq 0$ is a finite constant.

Let us first consider the case where a and b are known exactly. It follows from (7.2.1) that the tracking error $e = y - c$ satisfies
$$\dot{e} = ae + ac + bu \qquad (7.2.15)$$

7.2. SIMPLE APPC SCHEMES

Because a, b, and c are known, we can choose

$$u = -k_1 e - k_2 \qquad (7.2.16)$$

where

$$k_1 = \frac{a + a_m}{b}, \quad k_2 = \frac{ac}{b}$$

(provided $b \neq 0$) to obtain

$$\dot{e} = -a_m e \qquad (7.2.17)$$

It is clear from (7.2.17) that $e(t) = y(t) - y_m \to 0$ exponentially fast.

Let us now consider the design of an APPC scheme to meet the control objective when a and b are constant but unknown. The certainty equivalence approach suggests the use of the same control law as in (7.2.16) but with k_1 and k_2 replaced with their on-line estimates $\hat{k}_1(t)$ and $\hat{k}_2(t)$, respectively, i.e.,

$$u = -\hat{k}_1(t) e - \hat{k}_2(t) \qquad (7.2.18)$$

As in Section 7.2.1, the updating of $\hat{k}_1$ and $\hat{k}_2$ may be done directly, or indirectly via calculation using the on-line estimates $\hat{a}$ and $\hat{b}$ of the plant parameters. We consider each case separately.

Direct Adaptive Tracking In this approach we develop an adaptive law that updates $\hat{k}_1$ and $\hat{k}_2$ in (7.2.18) directly without any intermediate calculation. By adding and subtracting the desired input $u^* = -k_1 e - k_2$ in the error equation (7.2.15), we have

$$\begin{aligned}\dot{e} &= ae + ac + b(-k_1 e - k_2) + b(k_1 e + k_2) + bu \\ &= -a_m e + b(u + k_1 e + k_2)\end{aligned}$$

which together with (7.2.18) imply that

$$\dot{e} = -a_m e - b(\tilde{k}_1 e + \tilde{k}_2) \qquad (7.2.19)$$

where $\tilde{k}_1 \triangleq \hat{k}_1 - k_1, \tilde{k}_2 \triangleq \hat{k}_2 - k_2$. As in Chapter 4, equation (7.2.19) motivates the Lyapunov-like function

$$V = \frac{e^2}{2} + \frac{\tilde{k}_1^2 |b|}{2\gamma_1} + \frac{\tilde{k}_2^2 |b|}{2\gamma_2} \qquad (7.2.20)$$

whose time-derivative $\dot{V}$ along the trajectory of (7.2.19) is forced to satisfy

$$\dot{V} = -a_m e^2 \qquad (7.2.21)$$

by choosing

$$\dot{\hat{k}}_1 = \gamma_1 e^2 \text{sgn}(b), \quad \dot{\hat{k}}_2 = \gamma_2 e\text{sgn}(b) \qquad (7.2.22)$$

From (7.2.20) and (7.2.21) we have that $e, \hat{k}_1, \hat{k}_2 \in \mathcal{L}_\infty$ and $e \in \mathcal{L}_2$, which, in turn, imply that $y, u \in \mathcal{L}_\infty$ and $e(t) \to 0$ as $t \to \infty$ by following the usual arguments as in Section 7.2.1.

The APPC scheme (7.2.18), (7.2.22) may be written as

$$\begin{aligned} u &= -\hat{k}_1 e - \gamma_2 \text{sgn}(b) \int_0^t e(\tau) d\tau \\ \dot{\hat{k}}_1 &= \gamma_1 e^2 \text{sgn}(b) \end{aligned} \qquad (7.2.23)$$

indicating the proportional control action for stabilization and the integral action for rejecting the constant term ac in the error equation (7.2.15). We refer to (7.2.23) as the *direct adaptive proportional plus integral (PI) controller*. The same approach may be repeated when y_m is a known bounded signal with known $\dot{y}_m \in \mathcal{L}_\infty$. The reader may verify that in this case, the adaptive control scheme

$$\begin{aligned} u &= -\hat{k}_1 e - \hat{k}_2 y_m - \hat{k}_3 \dot{y}_m \\ \dot{\hat{k}}_1 &= \gamma_1 e^2 \text{sgn}(b) \\ \dot{\hat{k}}_2 &= \gamma_2 e y_m \text{sgn}(b), \quad \dot{\hat{k}}_3 = \gamma_3 e \dot{y}_m \text{sgn}(b) \end{aligned} \qquad (7.2.24)$$

where $e = y - y_m$, guarantees that all signals in the closed-loop plant in (7.2.1) and (7.2.24) are bounded, and $y(t) \to y_m(t)$ as $t \to \infty$.

Indirect Adaptive Tracking In this approach, we use the same control law as in the direct case, i.e.,

$$u = -\hat{k}_1 e - \hat{k}_2 \qquad (7.2.25)$$

but with $\hat{k}_1, \hat{k}_2$ calculated using the equations

$$\hat{k}_1 = \frac{\hat{a} + a_m}{\hat{b}}, \quad \hat{k}_2 = \frac{\hat{a}c}{\hat{b}} \qquad (7.2.26)$$

7.2. SIMPLE APPC SCHEMES

provided $\hat{b} \neq 0$, where $\hat{a}$ and $\hat{b}$ are the on-line estimates of the plant parameters a and b, respectively.

We generate $\hat{a}$ and $\hat{b}$ using an adaptive law as follows: We start with the series-parallel model

$$\dot{e}_m = -a_m(e_m - e) + \hat{a}(e+c) + \hat{b}u$$

based on the tracking error equation (7.2.15) and define the error $e_0 \triangleq e - e_m$, which satisfies the equation

$$\dot{e}_0 = -a_m e_0 - \tilde{a}(e+c) - \tilde{b}u \qquad (7.2.27)$$

The following adaptive laws can now be derived by using the same approach as in Section 7.2.1:

$$\dot{\hat{a}} = \gamma_1 e_0(e+c) = \gamma_1 e_0 y \qquad (7.2.28)$$

$$\dot{\hat{b}} = \begin{cases} \gamma_2 e_0 u & \text{if } |\hat{b}| > b_0 \text{ or if } |\hat{b}| = b_0 \text{ and } \operatorname{sgn}(b) e_0 u \geq 0 \\ 0 & \text{otherwise} \end{cases}$$

where $\hat{b}(0)$ satisfies $\hat{b}(0)\operatorname{sgn}(b) \geq b_0$. The reader may verify that the time derivative of the Lyapunov-like function

$$V = \frac{e_0^2}{2} + \frac{\tilde{a}^2}{2\gamma_1} + \frac{\tilde{b}^2}{2\gamma_2}$$

along the trajectories of (7.2.27), (7.2.28) satisfies

$$\dot{V} \leq -a_m e_0^2$$

which implies $e_0, \tilde{a}, \tilde{b} \in \mathcal{L}_\infty$ and $e_0 \in \mathcal{L}_2$. Because of (7.2.25) and (7.2.26), we have $\dot{e}_m = -a_m e_m$, and, thus, $e_m \in \mathcal{L}_\infty$ and $e_m(t) \to 0$ as $t \to \infty$. Therefore, we conclude that e, u and $\dot{e}_0 \in \mathcal{L}_\infty$. From $e_0 \in \mathcal{L}_2, \dot{e}_0 \in \mathcal{L}_\infty$ we have $e_0(t) \to 0$ as $t \to \infty$, which implies that $e(t) \to 0$ as $t \to \infty$. As in Section 7.2.2, the assumption that a lower bound b_0 for b is known can be relaxed by modifying the indirect scheme (7.2.25) to (7.2.28). In this case both $\hat{k}_1$ and $\hat{k}_2$ are adjusted dynamically rather than calculated from (7.2.26).

Example 7.2.1 Adaptive Cruise Control Most of today's automobiles are equipped with the so-called cruise control system. The cruise control system is responsible for maintaining a certain vehicle speed by automatically controlling the throttle angle. The mass of air and fuel that goes into the combustion cylinders and generates the engine torque is proportional to the throttle angle. The driver sets the desired speed V_d manually by speeding to V_d and then switching on the cruise control system. The driver can also set the speed V_d by using the "accelerator" button to accelerate from a lower speed to V_d through the use of the cruise control system. Similarly, if the driver interrupts a previously set speed by using the brake, the "resume" button may be used to allow the cruise control system to accelerate to the preset desired speed. Because of the changes in the dynamics of vehicles that are due to mechanical wear, loads, aerodynamic drag, etc., the use of adaptive control seems to be attractive for this application.

The linearized model of the throttle system with throttle angle θ as the input and speed V_s as the output is of the form

$$\dot{V}_s = -aV_s + b\theta + d \qquad (7.2.29)$$

where a and b are constant parameters that depend on the speed of the vehicle, aerodynamic drag and on the type of the vehicle and its condition. The variable d represents load disturbances resulting from uphill situations, drag effects, number of people in the vehicle, road condition, etc. The uncertainty in the values of $a, b,$ and d makes adaptive control suitable for this application.

Equation (7.2.29) is of the same form as equation (7.2.15); therefore, we can derive a direct adaptive PI control scheme by following the same procedure, i.e., the adaptive cruise control system is described by the following equations:

$$\begin{aligned} \theta &= -\hat{k}_1 \bar{V}_s - \hat{k}_2 \\ \dot{\hat{k}}_1 &= \gamma_1 \bar{V}_s^2 \\ \dot{\hat{k}}_2 &= \gamma_2 \bar{V}_s \end{aligned} \qquad (7.2.30)$$

where $\bar{V}_s = V_s - V_d$ and V_d is the desired speed set by the driver and $\gamma_1, \gamma_2 > 0$ are the adaptive gains. In (7.2.30), we use the a priori knowledge of $\text{sgn}(b) > 0$ which is available from experimental data.

The direct adaptive cruise control scheme given by (7.2.30) is tested on an actual vehicle [92]. The actual response for a particular test is shown in Figure 7.1.

The design of an indirect adaptive cruise control scheme follows in a similar manner by modifying the approach of Section 7.2.3 and is left as an exercise for the reader. ∇

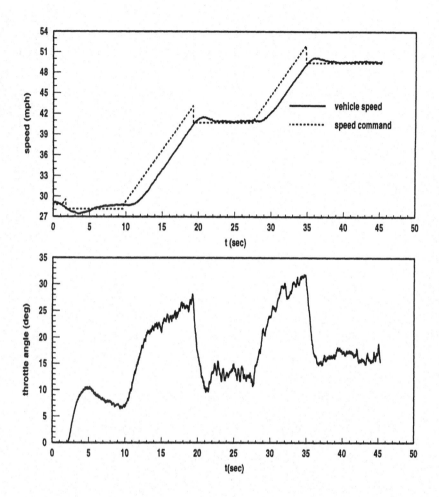

Figure 7.1 Response of the adaptive cruise control system.

7.3 PPC: Known Plant Parameters

As we demonstrated in Sections 6.6 and 7.2, an indirect adaptive control scheme consists of three parts: the adaptive law that provides on-line estimates for the plant parameters; the mapping between the plant and controller parameters that is used to calculate the controller parameters from the on-line estimates of the plant parameters; and the control law.

The form of the control law and the mapping between plant parameter estimates and controller parameters are the same as those used in the known plant parameter case. The purpose of this section is to develop several control laws that can meet the pole placement control objective when the plant parameters are known exactly. The form of these control laws as well as the mapping between the controller and plant parameters will be used in Section 7.4 to form indirect APPC schemes for plants with unknown parameters.

7.3.1 Problem Statement

Consider the SISO LTI plant

$$y_p = G_p(s)u_p, \quad G_p(s) = \frac{Z_p(s)}{R_p(s)} \qquad (7.3.1)$$

where $G_p(s)$ is proper and $R_p(s)$ is a monic polynomial. The control objective is to choose the plant input u_p so that the closed-loop poles are assigned to those of a given monic Hurwitz polynomial $A^*(s)$. The polynomial $A^*(s)$, referred to as the desired closed-loop characteristic polynomial, is chosen based on the closed-loop performance requirements. To meet the control objective, we make the following assumptions about the plant:

P1. $R_p(s)$ is a monic polynomial whose degree n is known.

P2. $Z_p(s), R_p(s)$ are coprime and degree$(Z_p) < n$.

Assumptions (P1) and (P2) allow Z_p, R_p to be non-Hurwitz in contrast to the MRC case where Z_p is required to be Hurwitz. If, however, Z_p is Hurwitz, the MRC problem is a special case of the general pole placement problem defined above with $A^*(s)$ restricted to have Z_p as a factor. We will explain the connection between the MRC and the PPC problems in Section 7.3.2.

In general, by assigning the closed-loop poles to those of $A^*(s)$, we can guarantee closed-loop stability and convergence of the plant output y_p to zero provided there is no external input. We can also extend the PPC objective to include tracking, where y_p is required to follow a certain class of reference signals y_m, by using the internal model principle discussed in Chapter 3 as follows: The reference signal $y_m \in \mathcal{L}_\infty$ is assumed to satisfy

$$Q_m(s)y_m = 0 \qquad (7.3.2)$$

7.3. PPC: KNOWN PLANT PARAMETERS

where $Q_m(s)$, the internal model of y_m, is a known monic polynomial of degree q with nonrepeated roots on the $j\omega$-axis and satisfies

P3. $Q_m(s), Z_p(s)$ are coprime.

For example, if y_p is required to track the reference signal $y_m = 2 + sin(2t)$, then $Q_m(s) = s(s^2 + 4)$ and, therefore, according to P3, $Z_p(s)$ should not have s or $s^2 + 4$ as a factor.

The effect of $Q_m(s)$ on the tracking error $e_1 = y_p - y_m$ is explained in Chapter 3 for a general feedback system and it is analyzed again in the sections to follow.

In addition to assumptions P1 to P3, let us also assume that the coefficients of $Z_p(s), R_p(s)$, i.e., the plant parameters are known exactly and propose several control laws that meet the control objective. The knowledge of the plant parameters is relaxed in Section 7.4.

7.3.2 Polynomial Approach

We consider the control law

$$Q_m(s)L(s)u_p = -P(s)y_p + M(s)y_m \qquad (7.3.3)$$

where $P(s), L(s), M(s)$ are polynomials (with $L(s)$ monic) of degree $q+n-1, n-1, q+n-1$, respectively, to be found and $Q_m(s)$ satisfies (7.3.2) and assumption P3.

Applying (7.3.3) to the plant (7.3.1), we obtain the closed-loop plant equation

$$y_p = \frac{Z_p M}{LQ_m R_p + PZ_p} y_m \qquad (7.3.4)$$

whose characteristic equation

$$LQ_m R_p + PZ_p = 0 \qquad (7.3.5)$$

has order $2n + q - 1$. The objective now is to choose P, L such that

$$LQ_m R_p + PZ_p = A^* \qquad (7.3.6)$$

is satisfied for a given monic Hurwitz polynomial $A^*(s)$ of degree $2n+q-1$. Because assumptions P2 and P3 guarantee that $Q_m R_p, Z_p$ are coprime, it follows from Theorem 2.3.1 that L, P satisfying (7.3.6) exist and are unique. The solution for the coefficients of $L(s), P(s)$ of equation (7.3.6) may be obtained by solving the algebraic equation

$$S_l \beta_l = \alpha_l^* \qquad (7.3.7)$$

where S_l is the Sylvester matrix of $Q_m R_p, Z_p$ of dimension $2(n+q) \times 2(n+q)$

$$\beta_l = [l_q^T, p^T]^T, \alpha_l^* = [\underbrace{0,\ldots,0}_{q}, 1, \alpha^{*T}]^T$$

$$l_q = [\underbrace{0,\ldots,0}_{q}, 1, l^T]^T \in \mathcal{R}^{n+q}$$

$$l = [l_{n-2}, l_{n-3}, \ldots, l_1, l_0]^T \in \mathcal{R}^{n-1}$$

$$p = [p_{n+q-1}, p_{n+q-2}, \ldots, p_1, p_0]^T \in \mathcal{R}^{n+q}$$

$$\alpha^* = [a^*_{2n+q-2}, a^*_{2n+q-3}, \ldots, a^*_1, a^*_0]^T \in \mathcal{R}^{2n+q-1}$$

l_i, p_i, a_i^* are the coefficients of

$$L(s) = s^{n-1} + l_{n-2}s^{n-2} + \cdots + l_1 s + l_0 = s^{n-1} + l^T \alpha_{n-2}(s)$$

$$P(s) = p_{n+q-1}s^{n+q-1} + p_{n+q-2}s^{n+q-2} + \cdots + p_1 s + p_0 = p^T \alpha_{n+q-1}(s)$$

$$A^*(s) = s^{2n+q-1} + a^*_{2n+q-2}s^{2n+q-2} + \cdots + a^*_1 s + a^*_0 = s^{2n+q-1} + \alpha^{*T} \alpha_{2n+q-2}(s)$$

The coprimeness of $Q_m R_p, Z_p$ guarantees that S_l is nonsingular; therefore, the coefficients of $L(s), P(s)$ may be computed from the equation

$$\beta_l = S_l^{-1} \alpha_l^*$$

Using (7.3.6), the closed-loop plant is described by

$$y_p = \frac{Z_p M}{A^*} y_m \qquad (7.3.8)$$

Similarly, from the plant equation in (7.3.1) and the control law in (7.3.3) and (7.3.6), we obtain

$$u_p = \frac{R_p M}{A^*} y_m \qquad (7.3.9)$$

7.3. PPC: KNOWN PLANT PARAMETERS

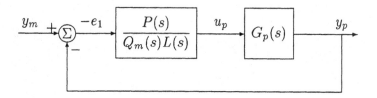

Figure 7.2 Block diagram of pole placement control.

Because $y_m \in \mathcal{L}_\infty$ and $\frac{Z_p M}{A^*}, \frac{R_p M}{A^*}$ are proper with stable poles, it follows that $y_p, u_p \in \mathcal{L}_\infty$ for any polynomial $M(s)$ of degree $n + q - 1$. Therefore, the pole placement objective is achieved by the control law (7.3.3) without having to put any additional restrictions on $M(s), Q_m(s)$. When $y_m = 0$, (7.3.8), (7.3.9) imply that y_p, u_p converge to zero exponentially fast.

When $y_m \neq 0$, the tracking error $e_1 = y_p - y_m$ is given by

$$e_1 = \frac{Z_p M - A^*}{A^*} y_m = \frac{Z_p}{A^*}(M - P)y_m - \frac{LR_p}{A^*} Q_m y_m \qquad (7.3.10)$$

For zero tracking error, (7.3.10) suggests the choice of $M(s) = P(s)$ to null the first term in (7.3.10). The second term in (7.3.10) is nulled by using $Q_m y_m = 0$. Therefore, for $M(s) = P(s)$, we have

$$e_1 = \frac{Z_p}{A^*}[0] - \frac{LR_p}{A^*}[0]$$

Because $\frac{Z_p}{A^*}, \frac{LR_p}{A^*}$ are proper with stable poles, it follows that e_1 converges exponentially to zero. Therefore, the pole placement and tracking objective are achieved by using the control law

$$Q_m L u_p = -P(y_p - y_m) \qquad (7.3.11)$$

which is implemented as shown in Figure 7.2 using $n + q - 1$ integrators to realize $C(s) = \frac{P(s)}{Q_m(s)L(s)}$. Because $L(s)$ is not necessarily Hurwitz, the realization of (7.3.11) with $n+q-1$ integrators may have a transfer function, namely $C(s)$, with poles outside $\mathcal{C}^-$. An alternative realization of (7.3.11) is obtained by rewriting (7.3.11) as

$$u_p = \frac{\Lambda - LQ_m}{\Lambda} u_p - \frac{P}{\Lambda}(y_p - y_m) \qquad (7.3.12)$$

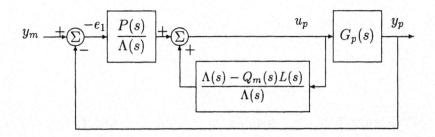

Figure 7.3 An alternative realization of the pole placement control.

where Λ is any monic Hurwitz polynomial of degree $n+q-1$. The control law (7.3.12) is implemented as shown in Figure 7.3 using $2(n+q-1)$ integrators to realize the proper stable transfer functions $\frac{\Lambda - LQ_m}{\Lambda}, \frac{P}{\Lambda}$. We summarize the main equations of the control law in Table 7.1.

Remark 7.3.1 The MRC law of Section 6.3.2 shown in Figure 6.1 is a special case of the general PPC law (7.3.3), (7.3.6). We can obtain the MRC law of Section 6.3.2 by choosing

$$Q_m = 1, \quad A^* = Z_p \Lambda_0 R_m, \quad M(s) = \frac{R_m \Lambda_0}{k_p}$$

$$L(s) = \Lambda(s) - \theta_1^{*T} \alpha_{n-2}(s), \quad P(s) = -(\theta_2^{*T} \alpha_{n-2}(s) + \theta_3^* \Lambda(s))$$

$$\Lambda = \Lambda_0 Z_m, \quad y_m = k_m \frac{Z_m}{R_m} r$$

where Z_p, Λ_0, R_m are Hurwitz and $\Lambda_0, R_m, k_p, \theta_1^*, \theta_2^*, \theta_3^*, r$ are as defined in Section 6.3.2.

Example 7.3.1 Consider the plant

$$y_p = \frac{b}{s+a} u_p$$

where a and b are known constants. The control objective is to choose u_p such that the poles of the closed-loop plant are placed at the roots of $A^*(s) = (s+1)^2$ and y_p tracks the constant reference signal $y_m = 1$. Clearly the internal model of y_m is $Q_m(s) = s$, i.e., $q = 1$. Because $n = 1$, the polynomials L, P, Λ are of the form

$$L(s) = 1, \quad P(s) = p_1 s + p_0, \quad \Lambda = s + \lambda_0$$

7.3. PPC: KNOWN PLANT PARAMETERS

Table 7.1 PPC law: polynomial approach

Plant	$y_p = \frac{Z_p(s)}{R_p(s)} u_p$
Reference input	$Q_m(s) y_m = 0$
Calculation	Solve for $L(s) = s^{n-1} + l^T \alpha_{n-2}(s)$ and $P(s) = p^T \alpha_{n+q-1}(s)$ the polynomial equation $L(s)Q_m(s)R_p(s) + P(s)Z_p(s) = A^*(s)$ or solve for β_l the algebraic equation $S_l \beta_l = \alpha_l^*$, where S_l is the Sylvester matrix of $R_p Q_m, Z_p$ $\beta_l = [l_q^T, p^T]^T \in \mathcal{R}^{2(n+q)}$ $l_q = [\underbrace{0,\ldots,0}_{q}, 1, l^T]^T \in \mathcal{R}^{n+q}$ $A^*(s) = s^{2n+q-1} + \alpha^{*T} \alpha_{2n+q-2}(s)$ $\alpha_l^* = [\underbrace{0,\ldots,0}_{q}, 1, \alpha^{*T}]^T \in \mathcal{R}^{2(n+q)}$
Control law	$u_p = \frac{\Lambda - LQ_m}{\Lambda} u_p - \frac{P}{\Lambda} e_1$ $e_1 = y_p - y_m$
Design variables	$A^*(s)$ is monic Hurwitz; $Q_m(s)$ is a monic polynomial of degree q with nonrepeated roots on $j\omega$ axis; $\Lambda(s)$ is a monic Hurwitz polynomial of degree $n + q - 1$

where $\lambda_0 > 0$ is arbitrary and p_0, p_1 are calculated by solving

$$s(s+a) + (p_1 s + p_0)b = (s+1)^2 \tag{7.3.13}$$

Equating the coefficients of the powers of s in (7.3.13), we obtain

$$p_1 = \frac{2-a}{b}, \quad p_0 = \frac{1}{b}$$

Equation (7.3.13) may be also written in the form of the algebraic equation (7.3.7), i.e., the Sylvester matrix of $s(s+a), b$ is given by

$$S_l = \begin{bmatrix} 1 & 0 & 0 & 0 \\ a & 1 & 0 & 0 \\ 0 & a & b & 0 \\ 0 & 0 & 0 & b \end{bmatrix}$$

and

$$\beta_l = \begin{bmatrix} 0 \\ 1 \\ p_1 \\ p_0 \end{bmatrix}, \quad \alpha_l^* = \begin{bmatrix} 0 \\ 1 \\ 2 \\ 1 \end{bmatrix}$$

Therefore, the PPC law is given by

$$\begin{aligned} u_p &= \frac{(s+\lambda_0 - s)}{s+\lambda_0} u_p - \left[\frac{2-a}{b}s + \frac{1}{b}\right] \frac{1}{s+\lambda_0} e_1 \\ &= \frac{\lambda_0}{s+\lambda_0} u_p - \frac{(2-a)s+1}{b(s+\lambda_0)} e_1 \end{aligned}$$

where $e_1 = y_p - y_m = y_p - 1$. A state-space realization of the control law is given by

$$\begin{aligned} \dot{\phi}_1 &= -\lambda_0 \phi_1 + u_p \\ \dot{\phi}_2 &= -\lambda_0 \phi_2 + e_1 \\ u_p &= \lambda_0 \phi_1 - \frac{1 - 2\lambda_0 + a\lambda_0}{b} \phi_2 - \frac{2-a}{b} e_1 \end{aligned}$$

$\triangledown$

7.3.3 State-Variable Approach

An alternative approach for deriving a PPC law is to use a state observer and state-variable feedback.

We start by considering the expression

$$e_1 = \frac{Z_p(s)}{R_p(s)} u_p - y_m \qquad (7.3.14)$$

for the tracking error. Filtering each side of (7.3.14) with $\frac{Q_m(s)}{Q_1(s)}$, where $Q_1(s)$ is an arbitrary monic Hurwitz polynomial of degree q, and using $Q_m(s)y_m = 0$ we obtain

$$e_1 = \frac{Z_p Q_1}{R_p Q_m} \bar{u}_p \qquad (7.3.15)$$

7.3. PPC: KNOWN PLANT PARAMETERS

where

$$\bar{u}_p = \frac{Q_m}{Q_1} u_p \qquad (7.3.16)$$

With (7.3.15), we have converted the tracking problem into the regulation problem of choosing $\bar{u}_p$ to regulate e_1 to zero.

Let (A, B, C) be a state space realization of (7.3.15) in the observer canonical form, i.e.,

$$\begin{aligned} \dot{e} &= Ae + B\bar{u}_p \\ e_1 &= C^\top e \end{aligned} \qquad (7.3.17)$$

where

$$A = \left[\begin{array}{c|c} & I_{n+q-1} \\ -\theta_1^* & --- \\ & 0 \end{array}\right], \quad B = \theta_2^*, \quad C = [1, 0, \ldots, 0]^\top \qquad (7.3.18)$$

$\theta_1^*, \theta_2^* \in \mathcal{R}^{n+q}$ are the coefficient vectors of the polynomials $R_p(s)Q_m(s) - s^{n+q}$ and $Z_p(s)Q_1(s)$, respectively. Because $R_p Q_m, Z_p$ are coprime, any possible zero-pole cancellation in (7.3.15) between $Q_1(s)$ and $R_p(s)Q_m(s)$ will occur in $\mathcal{C}^-$ due to $Q_1(s)$ being Hurwitz, which implies that (A, B) is always stabilizable.

We consider the feedback law

$$\bar{u}_p = -K_c \hat{e}, \quad u_p = \frac{Q_1}{Q_m} \bar{u}_p \qquad (7.3.19)$$

where $\hat{e}$ is the state of the full-order Luenberger observer

$$\dot{\hat{e}} = A\hat{e} + B\bar{u}_p - K_o(C^\top \hat{e} - e_1) \qquad (7.3.20)$$

and K_c and K_o are calculated from

$$\det(sI - A + BK_c) = A_c^*(s) \qquad (7.3.21)$$

$$\det(sI - A + K_o C^\top) = A_o^*(s) \qquad (7.3.22)$$

where A_c^* and A_o^* are given monic Hurwitz polynomials of degree $n + q$. The roots of $A_c^*(s) = 0$ represent the desired pole locations of the transfer function of the closed-loop plant whereas the roots of $A_o^*(s)$ are equal to the poles of the observer dynamics. As in every observer design, the roots of $A_o^*(s) = 0$ are chosen to be faster than those of $A_c^*(s) = 0$ in order to reduce the effect of the observer dynamics on the transient response of the tracking error e_1. The existence of K_c in (7.3.21) follows from the controllability of (A, B). If (A, B) is stabilizable but not controllable because of the common factors between $Q_1(s), R_p(s)$, the solution for K_c in (7.3.21) still exists provided $A_c^*(s)$ is chosen to contain the common factors of Q_1, R_p. Because $A_c^*(s)$ is chosen based on the desired closed-loop performance requirements and $Q_1(s)$ is an arbitrary monic Hurwitz polynomial of degree q, we can choose $Q_1(s)$ to be a factor of $A_c^*(s)$ and, therefore, guarantee the existence of K_c in (7.3.21) even when (A, B) is not controllable. The existence of K_o in (7.3.22) follows from the observability of (C, A). Because of the special form of (7.3.17) and (7.3.18), the solution of (7.3.22) is given by $K_o = \alpha_0^* - \theta_1^*$ where α_0^* is the coefficient vector of $A_o^*(s)$.

Theorem 7.3.1 *The PPC law (7.3.19) to (7.3.22) guarantees that all signals in the closed-loop plant are bounded and e_1 converges to zero exponentially fast.*

Proof We define the observation error $e_o \triangleq e - \hat{e}$. Subtracting (7.3.20) from (7.3.17) we have
$$\dot{e}_o = (A - K_o C^\top) e_o \tag{7.3.23}$$
Using (7.3.19) in (7.3.20) we obtain
$$\dot{\hat{e}} = (A - B K_c)\hat{e} + K_o C^\top e_o \tag{7.3.24}$$
Because $A - K_o C^\top, A - B K_c$ are stable, the equilibrium $e_{oe} = 0, \hat{e}_e = 0$ of (7.3.23), (7.3.24) is e.s. in the large. Therefore $\hat{e}, e_o \in \mathcal{L}_\infty$ and $\hat{e}(t), e_o(t) \to 0$ as $t \to \infty$. From $e_o = e - \hat{e}$ and $\bar{u}_p = -K_c \hat{e}$, it follows that $e, \bar{u}_p, e_1 \in \mathcal{L}_\infty$ and $e(t), \bar{u}_p(t), e_1(t) \to 0$ as $t \to \infty$. The boundedness of y_p follows from that of e_1 and y_m. We prove that $u_p \in \mathcal{L}_\infty$ as follows: Because of Assumption P3 and the stability of $Q_1(s)$, the polynomials $Z_p Q_1, R_p Q_m$ have no common unstable zeros. Therefore there exists polynomials X, Y of degree $n + q - 1$ with X monic that satisfy the Diophantine equation
$$R_p Q_m X + Z_p Q_1 Y = A^* \tag{7.3.25}$$
where A^* is a monic Hurwitz polynomial of degree $2(n + q) - 1$ that contains the common zeros of $Q_1(s), R_p(s) Q_m(s)$. Dividing each side of (7.3.25) by A^* and using it to filter u_p, we obtain the equation
$$\frac{R_p Q_m X}{A^*} u_p + \frac{Q_1 Y Z_p}{A^*} u_p = u_p$$

7.3. PPC: KNOWN PLANT PARAMETERS

Because $Q_m u_p = Q_1 \bar{u}_p$ and $Z_p u_p = R_p y_p$, we have

$$u_p = \frac{R_p X Q_1}{A^*} \bar{u}_p + \frac{Q_1 Y R_p}{A^*} y_p$$

Because the transfer functions operating on $\bar{u}_p, y_p$ are proper with poles in $\mathcal{C}^-$, then $\bar{u}_p, y_p \in \mathcal{L}_\infty$ imply that $u_p \in \mathcal{L}_\infty$ and the proof is complete. □

The main equations of the state variable feedback law are summarized in Table 7.2.

Example 7.3.2 Let us consider the same plant and control objective as in Example 7.3.1, i.e.,

Plant $\quad y_p = \frac{b}{s+a} u_p$

Control Objective Choose u_p such that the closed-loop poles are placed at the roots of $A^*(s) = (s+1)^2$ and y_p tracks $y_m = 1$.

Clearly, the internal model of y_m is $Q_m(s) = s$ and the tracking error $e_1 = y_p - y_m$ satisfies

$$e_1 = \frac{b}{s+a} u_p - y_m$$

Filtering each side of the above equation with $\frac{s}{s+1}$, i.e., using $Q_1(s) = s+1$, we obtain

$$e_1 = \frac{b(s+1)}{(s+a)s} \bar{u}_p \quad (7.3.26)$$

where $\bar{u}_p = \frac{s}{s+1} u_p$. The state-space realization of (7.3.26) is given by

$$\dot{e} = \begin{bmatrix} -a & 1 \\ 0 & 0 \end{bmatrix} e + \begin{bmatrix} 1 \\ 1 \end{bmatrix} b\bar{u}_p$$

$$e_1 = [1 \ 0]e$$

The control law is then chosen as follows:

Observer $\quad \dot{\hat{e}} = \begin{bmatrix} -a & 1 \\ 0 & 0 \end{bmatrix} \hat{e} + \begin{bmatrix} 1 \\ 1 \end{bmatrix} b\bar{u}_p - K_0([1 \ 0]\hat{e} - e_1)$

Control Law $\quad \bar{u}_p = -K_c \hat{e}, \quad u_p = \frac{(s+1)}{s} \bar{u}_p = \bar{u}_p + \int_0^t \bar{u}_p(\tau) d\tau$

where $K_o = [k_{o_1}, k_{o_2}]^T$, $K_c = [k_{c_1}, k_{c_2}]$ are calculated using (7.3.21) and (7.3.22). We select the closed-loop polynomial $A_c^*(s) = (s+1)^2$ and the observer polynomial $A_o^*(s) = (s+5)^2$ and solve

$$\det(sI - A + BK_c) = (s+1)^2, \quad \det(sI - A + K_o C^T) = (s+5)^2$$

Table 7.2 State-space pole placement control law

Plant	$y_p = \frac{Z_p(s)}{R_p(s)} u_p$
Reference input	$Q_m(s) y_m = 0$
Observer	$\dot{\hat{e}} = A\hat{e} + B\bar{u}_p - K_o(C^\top \hat{e} - e_1)$ $A = \begin{bmatrix} -\theta_1^* & \vline & I_{n+q-1} \\ & \vline & ---- \\ & \vline & 0 \end{bmatrix},\ B = \theta_2^*$ $C^\top = [1, 0, \ldots, 0],\ e_1 = y_p - y_m$ where $\theta_1^*, \theta_2^* \in \mathcal{R}^{n+q}$ are the coefficient vectors of $R_p(s)Q_m(s) - s^{n+q}$ and $Z_p(s)Q_1(s)$, respectively
Calculation	$K_o = \alpha_0^* - \theta_1^*$ where α_0^* is the coefficient vector of $A_o^*(s) - s^{n+q}$; K_c is solved from $\det(sI - A + BK_c) = A_c^*(s)$
Control law	$\bar{u}_p = -K_c \hat{e},\quad u_p = \frac{Q_1(s)}{Q_m(s)} \bar{u}_p$
Design variables	$A_o^*(s), A_c^*(s)$ are monic Hurwitz polynomials of degree $n+q$; $Q_1(s)$ is a monic Hurwitz polynomial of degree q; $A_c^*(s)$ contains $Q_1(s)$ as a factor; $Q_m(s)$ is a monic polynomial of degree q with nonrepeated roots on the $j\omega$ axis

where

$$A = \begin{bmatrix} -a & 1 \\ 0 & 0 \end{bmatrix},\ B = \begin{bmatrix} 1 \\ 1 \end{bmatrix} b,\ C^\top = [1, 0]$$

for K_o, K_c to obtain

$$K_c = \frac{1}{b}[1-a, 1],\quad K_o = [10-a, 25]^\top$$

7.3. PPC: KNOWN PLANT PARAMETERS

Note that the solution for K_c holds for any a and $b \neq 0$. For $a = 1$ and $b \neq 0$ the pair

$$\left\{ \begin{bmatrix} -a & 1 \\ 0 & 0 \end{bmatrix}, \begin{bmatrix} 1 \\ 1 \end{bmatrix} b \right\}$$

is uncontrollable due to a zero-pole cancellation in (7.3.26) at $s = -1$. Because, however, $A_c^*(s) = (s+1)^2$ contains $s+1$ as a factor the solution of K_c still exists. The reader can verify that for $A_c^*(s) = (s + 2)^2$ (i.e., $A_c^*(s)$ doesnot have $Q_1(s) = s + 1$ as a factor) and $a = 1$, no finite value of K_c exists to satisfy (7.3.21). ▽

7.3.4 Linear Quadratic Control

Another method for solving the PPC problem is using an optimization technique to design a control input that guarantees boundedness and regulation of the plant output or tracking error to zero by minimizing a certain cost function that reflects the performance of the closed-loop system. As we have shown in Section 7.3.3, the system under consideration is

$$\begin{aligned} \dot{e} &= Ae + B\bar{u}_p \\ e_1 &= C^\top e \end{aligned} \quad (7.3.27)$$

where $\bar{u}_p$ is to be chosen so that the closed-loop system has eigenvalues that are the same as the zeros of a given Hurwitz polynomial $A_c^*(s)$. If the state e is available for measurement, then the control input

$$\bar{u}_p = -K_c e$$

where K_c is chosen so that $A - BK_c$ is a stable matrix, leads to the closed-loop system

$$\dot{e} = (A - BK_c)e$$

whose equilibrium $e_e = 0$ is exponentially stable. The existence of such K_c is guaranteed provided (A, B) is controllable (or stabilizable). In Section 7.3.3, K_c is chosen so that $\det(sI - A + BK_c) = A_c^*(s)$ is a Hurwitz polynomial. This choice of K_c leads to a bounded input $\bar{u}_p$ that forces e, e_1 to converge to zero exponentially fast with a rate that depends on the location of the eigenvalues of $A - BK_c$, i.e., the zeros of $A_c^*(s)$. The rate of decay of e_1 to zero depends on how negative the real parts of the eigenvalues of $A - BK_c$ are. It can be shown [95] that the more negative these values are, the larger the value of K_c and, therefore, the higher the required signal energy in $\bar{u}_p$. In the limit, as the eigenvalues of $A - BK_c$ are forced to $-\infty$, the input $\bar{u}_p$ becomes a string of impulsive functions that restore $e(t)$ instantaneously to zero. These facts indicate that there exists a trade-off between the rate of decay of e_1, e to zero and the energy of the input $\bar{u}_p$. This trade-off motivates the following linear quadratic control problem where the control input $\bar{u}_p$ is chosen to minimize the quadratic cost

$$J = \int_0^\infty (e_1^2 + \lambda \bar{u}_p^2) dt$$

where $\lambda > 0$, a weighting coefficient to be designed, penalizes the level of the control input signal. The optimum control input $\bar{u}_p$ that minimizes J is [95]

$$\bar{u}_p = -K_c e, \quad K_c = \lambda^{-1} B^T P \qquad (7.3.28)$$

where $P = P^T > 0$ satisfies the algebraic equation

$$A^T P + PA - PB\lambda^{-1}B^T P + CC^T = 0 \qquad (7.3.29)$$

known as the *Riccati* equation.

Because (A, B) is stabilizable, owing to Assumption P3 and the fact that $Q_1(s)$ is Hurwitz, the existence and uniqueness of $P = P^T > 0$ satisfying (7.3.29) is guaranteed [95]. It is clear that as $\lambda \to 0$, a situation known as *low cost of control*, $\|K_c\| \to \infty$, which implies that $\bar{u}_p$ may become unbounded. On the other hand if $\lambda \to \infty$, a situation known as *high cost of control*, $\bar{u}_p \to 0$ if the open-loop system is stable. If the open loop is unstable, then $\bar{u}_p$ is the one that minimizes $\int_0^\infty \bar{u}_p^2 dt$ among all stabilizing control laws. In this case, the real part of the eigenvalues of $A - BK_c$ may not be negative enough indicating that the tracking or regulation error may not go to zero fast enough. With $\lambda > 0$ and finite, however, (7.3.28), (7.3.29) guarantee that $A - BK_c$ is a stable matrix, e, e_1 converge to zero exponentially fast, and $\bar{u}_p$ is bounded. The location of the eigenvalues of $A - BK_c$ depends on the particular choice of λ. For a given $\lambda > 0$, the polynomial

$$f(s) \triangleq R_p(s)Q_m(s)R_p(-s)Q_m(-s) + \lambda^{-1} Z_p(s)Q_1(s)Z_p(-s)Q_1(-s)$$

7.3. PPC: KNOWN PLANT PARAMETERS

is an even function of s and $f(s) = 0$ has a total of $2(n+q)$ roots with $n+q$ of them in $\mathcal{C}^-$ and the other $n+q$ in $\mathcal{C}^+$. It can be shown that the poles corresponding to the closed-loop LQ control are the same as the roots of $f(s) = 0$ that are located in $\mathcal{C}^-$ [6]. On the other hand, however, given a desired polynomial $A_c^*(s)$, there may not exist a λ so that $\det(sI - A + B\lambda^{-1}B^\top P) = A_c^*(s)$. Hence, the LQ control solution provides us with a procedure for designing a stabilizing control input for the system (7.3.27). It doesnot guarantee, however, that the closed-loop system has the same eigenvalues as the roots of the desired polynomial $A_c^*(s)$. The significance of the LQ solution also relies on the fact that the resulting closed-loop has good sensitivity properties.

As in Section 7.3.3, the state e of (7.3.28) may not be available for measurement. Therefore, instead of (7.3.28), we use

$$\bar{u}_p = -K_c \hat{e}, \quad K_c = \lambda^{-1}B^\top P \qquad (7.3.30)$$

where $\hat{e}$ is the state of the observer equation

$$\dot{\hat{e}} = A\hat{e} + B\bar{u}_p - K_o(C^\top \hat{e} - e_1) \qquad (7.3.31)$$

and $K_o = \alpha_0^* - \theta_1^*$ as in (7.3.20). As in Section 7.3.3, the control input is given by

$$u_p = \frac{Q_1(s)}{Q_m(s)}\bar{u}_p \qquad (7.3.32)$$

Theorem 7.3.2 *The LQ control law (7.3.30) to (7.3.32) guarantees that all signals in the closed-loop plant are bounded and $e_1(t) \to 0$ as $t \to \infty$ exponentially fast.*

Proof As in Section 7.3.3, we consider the system described by the error equations

$$\begin{aligned}\dot{e}_o &= (A - K_o C^\top)e_o \\ \dot{\hat{e}} &= (A - BK_c)\hat{e} + K_o C^\top e_o\end{aligned} \qquad (7.3.33)$$

where $K_c = \lambda^{-1}B^\top P$ and K_o is chosen to assign the eigenvalues of $A - K_o C^\top$ to the zeros of a given Hurwitz polynomial $A_o^*(s)$. Therefore, the equilibrium $e_{oe} = 0$, $\hat{e}_e = 0$ is e.s. in the large if and only if the matrix $A - BK_c$ is stable. We establish the stability of $A - BK_c$ by considering the system

$$\dot{\bar{e}} = (A - BK_c)\bar{e} = (A - \lambda^{-1}BB^\top P)\bar{e} \qquad (7.3.34)$$

and proving that the equilibrium $\bar{e}_e = 0$ is e.s. in the large. We choose the Lyapunov function

$$V(\bar{e}) = \bar{e}^\top P \bar{e}$$

where $P = P^\top > 0$ satisfies the Riccati equation (7.3.29). Then $\dot{V}$ along the trajectories of (7.3.34) is given by

$$\dot{V} = -\bar{e}^\top CC^\top \bar{e} - \lambda^{-1} \bar{e}^\top PBB^\top P\bar{e} = -(C^\top \bar{e})^2 - \frac{1}{\lambda}(B^\top P\bar{e})^2 \leq 0$$

which implies that $\bar{e}_e = 0$ is stable, $\bar{e} \in \mathcal{L}_\infty$ and $C^\top \bar{e}, B^\top P\bar{e} \in \mathcal{L}_2$. We now rewrite (7.3.34) as

$$\dot{\bar{e}} = (A - K_o C^\top)\bar{e} + K_o C^\top \bar{e} - \frac{1}{\lambda} BB^\top P\bar{e}$$

by using *output injection*, i.e., adding and subtracting the term $K_o C^\top \bar{e}$. Because $A - K_o C^\top$ is stable and $C^\top \bar{e}, B^\top P\bar{e} \in \mathcal{L}_\infty \bigcap \mathcal{L}_2$, it follows from Corollary 3.3.1 that $\bar{e} \in \mathcal{L}_\infty \bigcap \mathcal{L}_2$ and $\bar{e} \to 0$ as $t \to \infty$. Using the results of Section 3.4.5 it follows that the equilibrium $\bar{e}_e = 0$ is e.s. in the large which implies that

$$A - BK_c = A - \lambda^{-1} BB^\top P$$

is a stable matrix. The rest of the proof is the same as that of Theorem 7.3.1 and is omitted. $\square$

The main equations of the LQ control law are summarized in Table 7.3.

Example 7.3.3 Let us consider the scalar plant

$$\begin{aligned} \dot{x} &= -ax + bu_p \\ y_p &= x \end{aligned}$$

where a and b are known constants and $b \neq 0$. The control objective is to choose u_p to stabilize the plant and regulate y_p, x to zero. In this case $Q_m(s) = Q_1(s) = 1$ and no observer is needed because the state x is available for measurement. The control law that minimizes

$$J = \int_0^\infty (y_p^2 + \lambda u_p^2) dt$$

is given by

$$u_p = -\frac{1}{\lambda} bp y_p$$

7.3. PPC: KNOWN PLANT PARAMETERS

Table 7.3 LQ control law

Plant	$y_p = \frac{Z_p(s)}{R_p(s)} u_p$
Reference input	$Q_m(s) y_m = 0$
Observer	As in Table 7.2
Calculation	Solve for $P = P^\top > 0$ the equation $A^\top P + PA - PB\lambda^{-1}B^\top P + CC^\top = 0$ A, B, C as defined in Table 7.2
Control law	$\bar{u}_p = -\lambda^{-1} B^\top P \hat{e}, \quad u_p = \frac{Q_1(s)}{Q_m(s)} \bar{u}_p$
Design variables	$\lambda > 0$ penalizes the control effort; $Q_1(s), Q_m(s)$ as in Table 7.2

where $p > 0$ is a solution of the scalar Riccati equation

$$-2ap - \frac{p^2 b^2}{\lambda} + 1 = 0$$

The two possible solutions of the above quadratic equation are

$$p_1 = \frac{-\lambda a + \sqrt{\lambda^2 a^2 + b^2 \lambda}}{b^2}, \quad p_2 = \frac{-\lambda a - \sqrt{\lambda^2 a^2 + b^2 \lambda}}{b^2}$$

It is clear that $p_1 > 0$ and $p_2 < 0$; therefore, the solution we are looking for is $p = p_1 > 0$. Hence, the control input is given by

$$u_p = -\left(\frac{-a}{b} + \frac{\sqrt{a^2 + \frac{b^2}{\lambda}}}{b} \right) y_p$$

that leads to the closed-loop plant

$$\dot{x} = -\sqrt{a^2 + \frac{b^2}{\lambda}}\,x$$

which implies that for any finite $\lambda > 0$, we have $x \in \mathcal{L}_\infty$ and $x(t) \to 0$ as $t \to \infty$ exponentially fast. It is clear that for $\lambda \to 0$ the closed-loop eigenvalue goes to $-\infty$. For $\lambda \to \infty$ the closed-loop eigenvalue goes to $-|a|$, which implies that if the open-loop system is stable then the eigenvalue remains unchanged and if the open-loop system is unstable, the eigenvalue of the closed-loop system is flipped to the left half plane reflecting the minimum effort that is required to stabilize the unstable system. The reader may verify that for the control law chosen above, the cost function J becomes

$$J = \lambda \frac{\sqrt{a^2 + \frac{b^2}{\lambda}} - a}{b^2} x^2(0)$$

It is clear that if $a > 0$, i.e., the plant is open-loop stable, the cost J is less than when $a < 0$, i.e., the plant is open-loop unstable. More details about the LQ problem may be found in several books [16, 95, 122]. ▽

Example 7.3.4 Let us consider the same plant as in Examples 7.3.1, i.e.,

Plant $\quad\quad\quad\quad y_p = \dfrac{b}{s+a} u_p$

Control Objective Choose u_p so that the closed-loop poles are stable and y_p tracks the reference signal $y_m = 1$.

Tracking Error Equations The problem is converted to a regulation problem by considering the tracking error equation

$$e_1 = \frac{b(s+1)}{(s+a)s}\bar{u}_p, \quad \bar{u}_p = \frac{s}{s+1} u_p$$

where $e_1 = y_p - y_m$ generated as shown in Example 7.3.2. The state-space representation of the tracking error equation is given by

$$\dot{e} = \begin{bmatrix} -a & 1 \\ 0 & 0 \end{bmatrix} e + \begin{bmatrix} 1 \\ 1 \end{bmatrix} b\bar{u}_p$$

$$e_1 = [1, 0]e$$

Observer The observer equation is the same as in Example 7.3.2, i.e.,

$$\dot{\hat{e}} = \begin{bmatrix} -a & 1 \\ 0 & 0 \end{bmatrix} \hat{e} + \begin{bmatrix} 1 \\ 1 \end{bmatrix} b\bar{u}_p - K_o([1\ 0]\hat{e} - e_1)$$

7.3. PPC: KNOWN PLANT PARAMETERS

where $K_o = [10-a, 25]^T$ is chosen so that the observer poles are equal to the roots of $A_o^*(s) = (s+5)^2$.

Control law The control law, according to (7.3.30) to (7.3.32), is given by

$$\bar{u}_p = -\lambda^{-1}[b,b]P\hat{e}, \quad u_p = \frac{s+1}{s}\bar{u}_p$$

where P satisfies the Riccati equation

$$\begin{bmatrix} -a & 1 \\ 0 & 0 \end{bmatrix}^T P + P\begin{bmatrix} -a & 1 \\ 0 & 0 \end{bmatrix} - P\begin{bmatrix} b \\ b \end{bmatrix}\lambda^{-1}[b\ b]P + \begin{bmatrix} 1 & 0 \\ 0 & 0 \end{bmatrix} = 0 \quad (7.3.35)$$

where $\lambda > 0$ is a design parameter to be chosen. For $\lambda = 0.1, a = -0.5, b = 2$, the positive definite solution of (7.3.35) is

$$P = \begin{bmatrix} 0.1585 & 0.0117 \\ 0.0117 & 0.0125 \end{bmatrix}$$

which leads to the control law

$$\bar{u}_p = -[3.4037\ \ 0.4829]\hat{e}, \quad u_p = \frac{s+1}{s}\bar{u}_p$$

This control law shifts the open-loop eigenvalues from $\lambda_1 = 0.5, \lambda_2 = 0$ to $\lambda_1 = -1.01, \lambda_2 = -6.263$.

For $\lambda = 1, a = -0.5, b = 2$ we have

$$P = \begin{bmatrix} 0.5097 & 0.1046 \\ 0.1046 & 0.1241 \end{bmatrix}$$

leading to

$$\bar{u}_p = -[1.2287\ \ 0.4574]\hat{e}$$

and the closed-loop eigenvalues $\lambda_1 = -1.69, \lambda_2 = -1.19$.

Let us consider the case where $a = 1, b = 1$. For these values of a, b the pair

$$\left\{\begin{bmatrix} -1 & 1 \\ 0 & 0 \end{bmatrix}, \begin{bmatrix} 1 \\ 1 \end{bmatrix}\right\}$$

is uncontrollable but stabilizable and the open-loop plant has eigenvalues at $\lambda_1 = 0$, $\lambda_2 = -1$. The part of the system that corresponds to $\lambda_2 = -1$ is uncontrollable. In this case, $\lambda = 0.1$ gives

$$P = \begin{bmatrix} 0.2114 & 0.0289 \\ 0.0289 & 0.0471 \end{bmatrix}$$

and $\bar{u}_p = -[2.402\ \ 0.760]\hat{e}$, which leads to a closed-loop plant with eigenvalues at -1.0 and -3.162. As expected, the uncontrollable dynamics that correspond to $\lambda_2 = -1$ remained unchanged. $\triangledown$

7.4 Indirect APPC Schemes

Let us consider the plant given by (7.3.1), i.e.,

$$y_p = G_p(s)u_p, \quad G_p(s) = \frac{Z_p(s)}{R_p(s)}$$

where $R_p(s), Z_p(s)$ satisfy Assumptions P1 and P2. The control objective is to choose u_p so that the closed-loop poles are assigned to the roots of the characteristic equation $A^*(s) = 0$, where $A^*(s)$ is a given monic Hurwitz polynomial, and y_p is forced to follow the reference signal $y_m \in \mathcal{L}_\infty$ whose internal model $Q_m(s)$, i.e.,

$$Q_m(s)y_m = 0$$

is known and satisfies assumption P3.

In Section 7.3, we assume that the plant parameters (i.e., the coefficients of $Z_p(s), R_p(s)$) are known exactly and propose several control laws that meet the control objective. In this section, we assume that $Z_p(s), R_p(s)$ satisfy Assumptions P1 to P3 but their coefficients are unknown constants; and use the certainty equivalence approach to design several indirect APPC schemes to meet the control objective. As mentioned earlier, with this approach we combine the PPC laws developed in Section 7.3 for the known parameter case with adaptive laws that generate on-line estimates for the unknown plant parameters. The adaptive laws are developed by first expressing (7.3.1) in the form of the parametric models considered in Chapter 4, where the coefficients of Z_p, R_p appear in a linear form, and then using Tables 4.1 to 4.5 to pick up the adaptive law of our choice. We illustrate the design of adaptive laws for the plant (7.3.1) in the following section.

7.4.1 Parametric Model and Adaptive Laws

We consider the plant equation

$$R_p(s)y_p = Z_p(s)u_p$$

where $R_p(s) = s^n + a_{n-1}s^{n-1} + \cdots + a_1 s + a_0$, $Z_p(s) = b_{n-1}s^{n-1} + \cdots + b_1 s + b_0$, which may be expressed in the form

$$[s^n + \theta_a^{*T}\alpha_{n-1}(s)]y_p = \theta_b^{*T}\alpha_{n-1}(s)u_p \tag{7.4.1}$$

7.4. INDIRECT APPC SCHEMES

where $\alpha_{n-1}(s) = [s^{n-1}, \ldots, s, 1]^T$ and $\theta_a^* = [a_{n-1}, \ldots, a_0]^T$, $\theta_b^* = [b_{n-1}, \ldots, b_0]^T$ are the unknown parameter vectors. Filtering each side of (7.4.1) with $\frac{1}{\Lambda_p(s)}$, where $\Lambda_p(s) = s^n + \lambda_{n-1} s^{n-1} + \cdots \lambda_0$ is a Hurwitz polynomial, we obtain

$$z = \theta_p^{*T} \phi \qquad (7.4.2)$$

where

$$z = \frac{s^n}{\Lambda_p(s)} y_p, \quad \theta_p^* = [\theta_b^{*T}, \theta_a^{*T}]^T, \quad \phi = \left[\frac{\alpha_{n-1}^T(s)}{\Lambda_p(s)} u_p, -\frac{\alpha_{n-1}^T(s)}{\Lambda_p(s)} y_p \right]^T$$

Equation (7.4.2) is in the form of the linear parametric model studied in Chapter 4, thus leading to a wide class of adaptive laws that can be picked up from Tables 4.1 to 4.5 for estimating θ_p^*.

Instead of (7.4.1), we can also write

$$y_p = (\Lambda_p - R_p) \frac{1}{\Lambda_p} y_p + Z_p \frac{1}{\Lambda_p} u_p$$

that leads to the linear parametric model

$$y_p = \theta_\lambda^{*T} \phi \qquad (7.4.3)$$

where $\theta_\lambda^* = \left[\theta_b^{*T}, (\theta_a^* - \lambda_p)^T \right]^T$ and $\lambda_p = [\lambda_{n-1}, \lambda_{n-2}, \ldots, \lambda_0]^T$ is the coefficient vector of $\Lambda_p(s) - s^n$. Equation (7.4.3) can also be used to generate a wide class of adaptive laws using the results of Chapter 4.

The plant parameterizations in (7.4.2) and (7.4.3) assume that the plant is strictly proper with known order n but unknown relative degree $n^* \geq 1$. The number of the plant zeros, i.e., the degree of $Z_p(s)$, however, is unknown. In order to allow for the uncertainty in the number of zeros, we parameterize $Z_p(s)$ to have degree $n - 1$ where the coefficients of s^i for $i = m+1, m+2, \ldots, n-1$ are equal to zero and m is the degree of $Z_p(s)$. If $m < n - 1$ is known, then the dimension of the unknown vector θ_p^* is reduced to $n + m + 1$.

The adaptive laws for estimating on-line the vector θ_p^* or θ_λ^* in (7.4.2), (7.4.3) have already been developed in Chapter 4 and are presented in Tables 4.1 to 4.5. In the following sections, we use (7.4.2) or (7.4.3) to pick up adaptive laws from Tables 4.1 to 4.5 of and combine them with the PPC laws of Section 7.3 to form APPC schemes.

7.4.2 APPC Scheme: The Polynomial Approach

Let us first illustrate the design and analysis of an APPC scheme based on the PPC scheme of Section 7.3.2 using a first order plant model. Then we consider the general case that is applicable to an nth-order plant.

Example 7.4.1 Consider the same plant as in example 7.3.1, i.e.,

$$y_p = \frac{b}{s+a} u_p \qquad (7.4.4)$$

where a and b are unknown constants and u_p is to be chosen so that the poles of the closed-loop plant are placed at the roots of $A^*(s) = (s+1)^2 = 0$ and y_p tracks the constant reference signal $y_m = 1 \; \forall t \geq 0$.

Let us start by designing each block of the APPC scheme, i.e., the adaptive law for estimating the plant parameters a and b; the mapping from the estimates of a, b to the controller parameters; and the control law.

Adaptive Law We start with the following parametric model for (7.4.4)

$$z = \theta_p^{*\top} \phi$$

where

$$z = \frac{s}{s+\lambda} y_p, \quad \phi = \frac{1}{s+\lambda} \begin{bmatrix} u_p \\ -y_p \end{bmatrix}, \quad \theta_p^* = \begin{bmatrix} b \\ a \end{bmatrix} \qquad (7.4.5)$$

and $\lambda > 0$ is an arbitrary design constant. Using Tables 4.1 to 4.5 of Chapter 4, we can generate a number of adaptive laws for estimating θ_p^*. For this example, let us choose the gradient algorithm of Table 4.2

$$\dot{\theta}_p = \Gamma \varepsilon \phi \qquad (7.4.6)$$

$$\varepsilon = \frac{z - \theta_p^\top \phi}{m^2}, \quad m^2 = 1 + \phi^\top \phi$$

where $\Gamma = \Gamma^\top > 0$, $\theta_p = [\hat{b}, \hat{a}]^\top$ and $\hat{a}(t), \hat{b}(t)$ is the estimate of a and b respectively.

Calculation of Controller Parameters As shown in Section 7.3.2, the control law

$$u_p = \frac{\Lambda - LQ_m}{\Lambda} u_p - \frac{P}{\Lambda} e_1 \qquad (7.4.7)$$

can be used to achieve the control objective, where $\Lambda(s) = s + \lambda_0$, $L(s) = 1$, $Q_m(s) = s$, $P(s) = p_1 s + p_0$, $e_1 \triangleq y_p - y_m$ and the coefficients p_1, p_0 of $P(s)$ satisfy the Diophantine equation

$$s(s+a) + (p_1 s + p_0)b = (s+1)^2 \qquad (7.4.8)$$

7.4. INDIRECT APPC SCHEMES

or equivalently the algebraic equation

$$\begin{bmatrix} 1 & 0 & 0 & 0 \\ a & 1 & 0 & 0 \\ 0 & a & b & 0 \\ 0 & 0 & 0 & b \end{bmatrix} \begin{bmatrix} 0 \\ 1 \\ p_1 \\ p_0 \end{bmatrix} = \begin{bmatrix} 0 \\ 1 \\ 2 \\ 1 \end{bmatrix} \qquad (7.4.9)$$

whose solution is

$$p_1 = \frac{2-a}{b}, \quad p_0 = \frac{1}{b}$$

Because a and b are unknown, the certainty equivalence approach suggests the use of the same control law but with the controller polynomial $P(s) = p_1 s + p_0$ calculated by using the estimates $\hat{a}(t)$ and $\hat{b}(t)$ of a and b at each time t as if they were the true parameters, i.e., $P(s,t) = \hat{p}_1(t)s + \hat{p}_0(t)$ is generated by solving the polynomial equation

$$s \cdot (s + \hat{a}) + (\hat{p}_1 s + \hat{p}_0) \cdot \hat{b} = (s+1)^2 \qquad (7.4.10)$$

for $\hat{p}_1$ and $\hat{p}_0$ by treating $\hat{a}(t)$ and $\hat{b}(t)$ as frozen parameters at each time t, or by solving the algebraic time varying equation

$$\begin{bmatrix} 1 & 0 & 0 & 0 \\ \hat{a}(t) & 1 & 0 & 0 \\ 0 & \hat{a}(t) & \hat{b}(t) & 0 \\ 0 & 0 & 0 & \hat{b}(t) \end{bmatrix} \begin{bmatrix} 0 \\ 1 \\ \hat{p}_1(t) \\ \hat{p}_0(t) \end{bmatrix} = \begin{bmatrix} 0 \\ 1 \\ 2 \\ 1 \end{bmatrix} \qquad (7.4.11)$$

for $\hat{p}_0$ and $\hat{p}_1$. The solution of (7.4.10), where $\hat{a}(t)$ and $\hat{b}(t)$ are treated as constants at each time t, is referred to as *pointwise* to distinguish it from solutions that may be obtained with s treated as a differential operator, and $\hat{a}(t)$ and $\hat{b}(t)$ treated as differentiable functions of time.

The Diophantine equation (7.4.10) or algebraic equation (7.4.11) has a unique solution provided that $(s + \hat{a}), \hat{b}$ are coprime, i.e., provided $\hat{b} \neq 0$. The solution is given by

$$\hat{p}_1(t) = \frac{2 - \hat{a}}{\hat{b}}, \quad \hat{p}_0(t) = \frac{1}{\hat{b}}$$

In fact for $\hat{a}, \hat{b} \in \mathcal{L}_\infty$ to imply that $\hat{p}_0, \hat{p}_1 \in \mathcal{L}_\infty$, $(s+\hat{a}), \hat{b}$ have to be strongly coprime, i.e., $|\hat{b}| \geq b_0$ for some constant $b_0 > 0$. For this simple example, the adaptive law (7.4.6) can be modified to guarantee that $|\hat{b}(t)| \geq b_0 > 0 \; \forall t \geq 0$ provided $\text{sgn}(b)$ and a lower bound b_0 of $|b|$ are known as shown in Section 7.2 and previous chapters. For clarity of presentation, let us assume that the adaptive law (7.4.6) is modified using projection (as in Section 7.2) to guarantee $|\hat{b}(t)| \geq b_0 \; \forall t \geq 0$ and proceed with the rest of the design and analysis.

Control Law The estimates $\hat{p}_0(t), \hat{p}_1(t)$ are used in place of the unknown p_0, p_1 to form the control law

$$u_p = \frac{\lambda_0}{s + \lambda_0} u_p - \left(\hat{p}_1(t) \frac{s}{s + \lambda_0} + \hat{p}_0(t) \frac{1}{s + \lambda_0} \right) (y_p - y_m) \qquad (7.4.12)$$

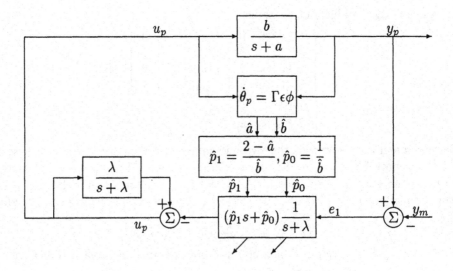

Figure 7.4 Block diagram of APPC for a first-order plant.

where $\lambda_0 > 0$ is an arbitrary design constant. For simplicity of implementation, λ_0 may be taken to be equal to the design parameter λ used in the adaptive law (7.4.6), so that the same signals can be shared by the adaptive and control law.

Implementation The block diagram of the APPC scheme with $\lambda = \lambda_0$ for the first order plant (7.4.4) is shown in Figure 7.4.

For $\lambda = \lambda_0$, the APPC scheme may be realized by the following equations:

Filters

$$\begin{aligned}
\dot{\phi}_1 &= -\lambda \phi_1 + u_p, & \phi_1(0) &= 0 \\
\dot{\phi}_2 &= -\lambda \phi_2 - y_p, & \phi_2(0) &= 0 \\
\dot{\phi}_m &= -\lambda \phi_m + y_m, & \phi_m(0) &= 0 \\
z &= \lambda \phi_2 + y_p = -\dot{\phi}_2
\end{aligned}$$

Adaptive Law

$$\dot{\hat{b}} = \begin{cases} \gamma_1 \epsilon \phi_1 & \text{if } |\hat{b}| > b_0 \text{ or if } |\hat{b}| = b_0 \text{ and } \epsilon \phi_1 \operatorname{sgn}\hat{b} \geq 0 \\ 0 & \text{otherwise} \end{cases}$$

$$\dot{\hat{a}} = \gamma_2 \epsilon \phi_2$$

$$\epsilon = \frac{z - \hat{b}\phi_1 - \hat{a}\phi_2}{m^2}, \quad m^2 = 1 + \phi_1^2 + \phi_2^2$$

7.4. INDIRECT APPC SCHEMES

Control Law

$$u_p = \lambda\phi_1 - (\hat{p}_1(t)\lambda - \hat{p}_0(t))(\phi_2 + \phi_m) - \hat{p}_1(t)(y_p - y_m)$$
$$\hat{p}_1 = \frac{2-\hat{a}}{\hat{b}}, \quad \hat{p}_0 = \frac{1}{\hat{b}}$$

where $\gamma_1, \gamma_2, \lambda > 0$ are design constants and $\hat{b}(0)\text{sgn}(b) \geq b_0$.

Analysis The stability analysis of the indirect APPC scheme is carried out in the following four steps:

Step 1. *Manipulate the estimation error and control law equations to express y_p, u_p in terms of the estimation error ϵ.* We start with the expression for the normalized estimation error

$$\epsilon m^2 = z - \hat{b}\phi_1 - \hat{a}\phi_2 = -\phi_2 - \hat{b}\phi_1 - \hat{a}\phi_2$$

which implies that
$$\dot{\phi}_2 = -\hat{b}\phi_1 - \hat{a}\phi_2 - \epsilon m^2 \tag{7.4.13}$$

From the control law, we have

$$u_p = \lambda\phi_1 + \hat{p}_1(\dot{\phi}_2 + \dot{\phi}_m) + \hat{p}_0\phi_2 + \hat{p}_0\phi_m$$

Because $u_p - \lambda\phi_1 = \dot{\phi}_1$, it follows that

$$\dot{\phi}_1 = \hat{p}_1\dot{\phi}_2 + \hat{p}_0\phi_2 + \bar{y}_m$$

where $\bar{y}_m \triangleq \hat{p}_1\dot{\phi}_m + \hat{p}_0\phi_m$. Substituting for $\dot{\phi}_2$ from (7.4.13) we obtain

$$\dot{\phi}_1 = -\hat{p}_1\hat{b}\phi_1 - (\hat{p}_1\hat{a} - \hat{p}_0)\phi_2 - \hat{p}_1\epsilon m^2 + \bar{y}_m \tag{7.4.14}$$

Equations (7.4.13), (7.4.14) form the following state space representation for the APPC scheme:

$$\dot{x} = A(t)x + b_1(t)\epsilon m^2 + b_2\bar{y}_m$$
$$\begin{bmatrix} u_p \\ -y_p \end{bmatrix} = \dot{x} + \lambda x = (A(t) + \lambda I)x + b_1(t)\epsilon m^2 + b_2\bar{y}_m \tag{7.4.15}$$

where

$$x = \begin{bmatrix} \phi_1 \\ \phi_2 \end{bmatrix}, \quad A(t) = \begin{bmatrix} -\hat{p}_1\hat{b} & \hat{p}_0 - \hat{p}_1\hat{a} \\ -\hat{b} & -\hat{a} \end{bmatrix}, \quad b_1(t) = \begin{bmatrix} -\hat{p}_1 \\ -1 \end{bmatrix}, \quad b_2 = \begin{bmatrix} 1 \\ 0 \end{bmatrix}$$

$m^2 = 1 + x^T x$ and $\bar{y}_m \in \mathcal{L}_\infty$.

Step 2. *Show that the homogeneous part of (7.4.15) is e.s.* For each fixed t, $\det(sI - A(t)) = (s+\hat{a})s + \hat{b}(\hat{p}_1 s + \hat{p}_0) = (s+1)^2$, i.e., $\lambda(A(t)) = -1, \forall t \geq 0$. As shown in Chapter 4, the adaptive law guarantees that $\epsilon, \hat{a}, \hat{b} \in \mathcal{L}_\infty$; $\epsilon, \epsilon m, \dot{\hat{a}}, \dot{\hat{b}} \in \mathcal{L}_\infty \cap \mathcal{L}_2$. From $\hat{p}_1 = \frac{2-\hat{a}}{\hat{b}}, \hat{p}_0 = \frac{1}{\hat{b}}$ and $\hat{b}^{-1} \in \mathcal{L}_\infty$ (because of projection), it follows that $\hat{p}_1, \hat{p}_0 \in \mathcal{L}_\infty$ and $\dot{\hat{p}}_1, \dot{\hat{p}}_0 \in \mathcal{L}_\infty \cap \mathcal{L}_2$. Hence, $\|\dot{A}(t)\| \in \mathcal{L}_\infty \cap \mathcal{L}_2$ which together with $\lambda(A(t)) = -1 \ \forall t \geq 0$ and Theorem 3.4.11 imply that the state transition matrix $\Phi(t, \tau)$ associated with $A(t)$ satisfies

$$\|\Phi(t,\tau)\| \leq k_1 e^{-k_2(t-\tau)}, \quad \forall t \geq \tau \geq 0$$

for some constants $k_1, k_2 > 0$.

Step 3. *Use the properties of the $\mathcal{L}_{2\delta}$ norm and B-G Lemma to establish boundedness.* For simplicity, let us now denote $\|(\cdot)_t\|_{2\delta}$ for some $\delta > 0$ with $\|\cdot\|$. Applying Lemma 3.3.3 to (7.4.15) we obtain

$$\|x\| \leq c\|\epsilon m^2\| + c, \quad |x(t)| \leq c\|\epsilon m^2\| + c \qquad (7.4.16)$$

for any $\delta \in [0, \delta_1)$ where $\delta_1 > 0$ is any constant less than $2k_2$, and some finite constants $c \geq 0$.

As in the MRAC case, we define the fictitious normalizing signal

$$m_f^2 \triangleq 1 + \|u_p\|^2 + \|y_p\|^2$$

From (7.4.15) we have $\|u_p\| + \|y_p\| \leq c\|x\| + c\|\epsilon m^2\| + c$, which, together with (7.4.16), implies that

$$m_f^2 \leq c\|\epsilon m^2\|^2 + c$$

Because $|\phi_1| \leq c\|u_p\|, |\phi_2| \leq c\|y_p\|$ for $\delta \in [0, 2\lambda)$, it follows that $m = \sqrt{1 + \phi^\top \phi} \leq cm_f$ and, therefore,

$$m_f^2 \leq c\|\tilde{g}m_f\|^2 + c$$

where $\tilde{g} \triangleq \epsilon m \in \mathcal{L}_2$ because of the properties of the adaptive law, or

$$m_f^2 \leq c \int_0^t e^{-\delta(t-\tau)} \tilde{g}^2(\tau) m_f^2(\tau) d\tau + c$$

where $0 < \delta \leq \delta^*$ and $\delta^* = \min[2\lambda, \delta_1], \delta_1 \in (0, 2k_2)$. Applying the B-G Lemma, we can establish that $m_f \in \mathcal{L}_\infty$. Because $m \leq cm_f$, we have m and therefore $\phi_1, \phi_2, x, \dot{x}, u_p, y_p \in \mathcal{L}_\infty$.

Step 4. *Establish tracking error convergence.* We consider the estimation error equation

$$\epsilon m^2 = -\dot{\phi}_2 - \hat{a}\phi_2 - \hat{b}\phi_1$$

7.4. INDIRECT APPC SCHEMES

or
$$\epsilon m^2 = (s + \hat{a})\frac{1}{s+\lambda}y_p - \hat{b}\frac{1}{s+\lambda}u_p \qquad (7.4.17)$$

Operating on each side of (7.4.17) with $s \triangleq \frac{d}{dt}$, we obtain

$$s(\epsilon m^2) = s(s+\hat{a})\frac{1}{s+\lambda}y_p - \hat{b}\frac{s}{s+\lambda}u_p - \dot{\hat{b}}\frac{1}{s+\lambda}u_p \qquad (7.4.18)$$

by using the property $s(xy) = \dot{x}y + x\dot{y}$. For $\lambda = \lambda_0$, it follows from the control law (7.4.12) that

$$\frac{s}{s+\lambda}u_p = -(\hat{p}_1 s + \hat{p}_0)\frac{1}{s+\lambda}e_1$$

which we substitute in (7.4.18) to obtain

$$s(\epsilon m^2) = s(s+\hat{a})\frac{1}{s+\lambda}y_p + \hat{b}(\hat{p}_1 s + \hat{p}_0)\frac{1}{s+\lambda}e_1 - \dot{\hat{b}}\frac{1}{s+\lambda}u_p \qquad (7.4.19)$$

Now, because $s(s+\hat{a})\frac{1}{s+\lambda}y_p = (s+\hat{a})\frac{s}{s+\lambda}y_p + \dot{\hat{a}}\frac{1}{s+\lambda}y_p$ and $se_1 = sy_p - sy_m = sy_p$ (note that $sy_m = 0$), we have

$$s(s+\hat{a})\frac{1}{s+\lambda}y_p = (s+\hat{a})\frac{s}{s+\lambda}e_1 + \dot{\hat{a}}\frac{1}{s+\lambda}y_p$$

which we substitute in (7.4.19) to obtain

$$s(\epsilon m^2) = \left[(s+\hat{a})s + \hat{b}(\hat{p}_1 s + \hat{p}_0)\right]\frac{1}{s+\lambda}e_1 + \dot{\hat{a}}\frac{1}{s+\lambda}y_p - \dot{\hat{b}}\frac{1}{s+\lambda}u_p$$

Using (7.4.10) we have $(s+\hat{a})s + \hat{b}(\hat{p}_1 s + \hat{p}_0) = (s+1)^2$ and therefore

$$s(\epsilon m^2) = \frac{(s+1)^2}{s+\lambda}e_1 + \dot{\hat{a}}\frac{1}{s+\lambda}y_p - \dot{\hat{b}}\frac{1}{s+\lambda}u_p$$

or
$$e_1 = \frac{s(s+\lambda)}{(s+1)^2}\epsilon m^2 - \frac{s+\lambda}{(s+1)^2}\dot{\hat{a}}\frac{1}{s+\lambda}y_p + \frac{s+\lambda}{(s+1)^2}\dot{\hat{b}}\frac{1}{s+\lambda}u_p \qquad (7.4.20)$$

Because $u_p, y_p, m, \epsilon \in \mathcal{L}_\infty$ and $\dot{\hat{a}}, \dot{\hat{b}}, \epsilon m \in \mathcal{L}_\infty \cap \mathcal{L}_2$, it follows from Corollary 3.3.1 that $e_1 \in \mathcal{L}_\infty \cap \mathcal{L}_2$. Hence, if we show that $\dot{e}_1 \in \mathcal{L}_\infty$, then by Lemma 3.2.5 we can conclude that $e_1 \to 0$ as $t \to \infty$. Since $\dot{e}_1 = \dot{y}_p = ay_p + bu_p \in \mathcal{L}_\infty$, it follows that $e_1 \to 0$ as $t \to \infty$.

We can continue our analysis and establish that $\epsilon, \dot{\hat{a}}, \dot{\hat{b}}, \dot{\hat{p}}_0, \dot{\hat{p}}_1 \to 0$ as $t \to \infty$. There is no guarantee, however, that $\hat{p}_0, \hat{p}_1, \hat{a}, \hat{b}$ will converge to the actual values p_0, p_1, a, b respectively unless the reference signal y_m is sufficiently rich of order 2, which is not the case for the example under consideration.

As we indicated earlier the calculation of the controller parameters $\hat{p}_0(t), \hat{p}_1(t)$ at each time is possible provided the estimated plant polynomials $(s+\hat{a}(t)), \hat{b}(t)$ are strongly coprime, i.e., provided $|\hat{b}(t)| \geq b_0 > 0 \ \forall t \geq 0$. This condition implies that at each time t, the estimated plant is strongly controllable. This is not surprising because the control law is calculated at each time t to meet the control objective for the estimated plant. As we will show in Section 7.6, the adaptive law without projection cannot guarantee that $|\hat{b}(t)| \geq b_0 > 0 \ \forall t \geq 0$. Projection requires the knowledge of b_0 and $\text{sgn}(b)$ and constrains $\hat{b}(t)$ to be in the region $|\hat{b}(t)| \geq b_0$ where controllability is always satisfied. In the higher-order case, the problem of controllability or stabilizability of the estimated plant is more difficult as we demonstrate below for a general nth-order plant. $\triangledown$

General Case

Let us now consider the nth-order plant

$$y_p = \frac{Z_p(s)}{R_p(s)} u_p$$

where $Z_p(s), R_p(s)$ satisfy assumptions P1, P2, and P3 with the same control objective as in Section 7.3.1, except that in this case the coefficients of Z_p, R_p are unknown. The APPC scheme that meets the control objective for the unknown plant is formed by combining the control law (7.3.12), summarized in Table 7.1, with an adaptive law based on the parametric model (7.4.2) or (7.4.3). The adaptive law generates on-line estimates θ_a, θ_b of the coefficient vectors, θ_a^* of $R_p(s) = s^n + \theta_a^{*T}\alpha_{n-1}(s)$ and θ_b^* of $Z_p(s) = \theta_b^{*T}\alpha_{n-1}(s)$ respectively, to form the estimated plant polynomials

$$\hat{R}_p(s,t) = s^n + \theta_a^T \alpha_{n-1}(s), \quad \hat{Z}_p(s,t) = \theta_b^T \alpha_{n-1}(s)$$

The estimated plant polynomials are used to compute the estimated controller polynomials $\hat{L}(s,t), \hat{P}(s,t)$ by solving the Diophantine equation

$$\hat{L}Q_m \cdot \hat{R}_p + \hat{P} \cdot \hat{Z}_p = A^* \qquad (7.4.21)$$

for $\hat{L}, \hat{P}$ pointwise in time or the algebraic equation

$$\hat{S}_l \hat{\beta}_l = \alpha_l^* \qquad (7.4.22)$$

7.4. INDIRECT APPC SCHEMES

for $\hat{\beta}_l$, where $\hat{S}_l$ is the Sylvester matrix of $\hat{R}_p Q_m, \hat{Z}_p$; $\hat{\beta}_l$ contains the coefficients of $\hat{L}, \hat{P}$; and α_l^* contains the coefficients of $A^*(s)$ as shown in Table 7.1. The control law in the unknown parameter case is then formed as

$$u_p = (\Lambda - \hat{L}Q_m)\frac{1}{\Lambda}u_p - \hat{P}\frac{1}{\Lambda}(y_p - y_m) \qquad (7.4.23)$$

Because different adaptive laws may be picked up from Tables 4.2 to 4.5, a wide class of APPC schemes may be developed. As an example, we present in Table 7.4 the main equations of an APPC scheme that is based on the gradient algorithm of Table 4.2.

The implementation of the APPC scheme of Table 7.4 requires that the solution of the polynomial equation (7.4.21) for $\hat{L}, \hat{P}$ or of the algebraic equation (7.4.22) for $\hat{\beta}_l$ exists at each time. The existence of this solution is guaranteed provided that $\hat{R}_p(s,t)Q_m(s), \hat{Z}_p(s,t)$ are coprime at each time t, i.e., the Sylvester matrix $\hat{S}_l(t)$ is nonsingular at each time t. In fact for the coefficient vectors l, p of the polynomials $\hat{L}, \hat{P}$ to be uniformly bounded for bounded plant parameter estimates θ_p, the polynomials $\hat{R}_p(s,t)Q_m(s), \hat{Z}_p(s,t)$ have to be strongly coprime which implies that their Sylvester matrix should satisfy

$$|\det(S_l(t))| \geq \nu_0 > 0$$

for some constant ν_0 at each time t. Such a strong condition cannot be guaranteed by the adaptive law without any additional modifications, giving rise to the so called *"stabilizability"* or *"admissibility"* problem to be discussed in Section 7.6. As in the scalar case, the stabilizability problem arises from the fact that the control law is chosen to stabilize the estimated plant (characterized by $\hat{Z}_p(s,t), \hat{R}_p(s,t)$) at each time. For such a control law to exist, the estimated plant has to satisfy the usual observability, controllability conditions which in this case translate into the equivalent condition of $\hat{R}_p(s,t)Q_m(s), \hat{R}_p(s,t)$ being coprime. The stabilizability problem is one of the main drawbacks of indirect APPC schemes in general and it is discussed in Section 7.6. In the meantime let us assume that the estimated plant is stabilizable, i.e., $\hat{R}_p Q_m, \hat{Z}_p$ are strongly coprime $\forall t \geq 0$ and proceed with the analysis of the APPC scheme presented in Table 7.4.

Theorem 7.4.1 *Assume that the estimated plant polynomials $\hat{R}_p Q_m, \hat{Z}_p$ are strongly coprime at each time t. Then all the signals in the closed-loop*

Table 7.4 APPC scheme: polynomial approach.

Plant	$y_p = \frac{Z_p(s)}{R_p(s)} u_p$ $Z_p(s) = \theta_b^{*\top} \alpha_{n-1}(s)$ $R_p(s) = s^n + \theta_a^{*\top} \alpha_{n-1}(s)$ $\alpha_{n-1}(s) = [s^{n-1}, s^{n-2}, \ldots, s, 1]^\top, \theta_p^* = [\theta_b^{*\top}, \theta_a^{*\top}]^\top$
Reference signal	$Q_m(s) y_m = 0$
Adaptive law	Gradient algorithm from Table 4.2 $\dot{\theta}_p = \Gamma \epsilon \phi, \quad \Gamma = \Gamma^\top > 0$ $\epsilon = (z - \theta_p^\top \phi)/m^2, \quad m^2 = 1 + \phi^\top \phi$ $\phi = [\frac{\alpha_{n-1}^\top(s)}{\Lambda_p(s)} u_p, -\frac{\alpha_{n-1}^\top(s)}{\Lambda_p(s)} y_p]^\top$ $z = \frac{s^n}{\Lambda_p(s)} y_p, \quad \theta_p = [\theta_b^\top, \theta_a^\top]^\top$ $\hat{Z}_p(s,t) = \theta_b^\top \alpha_{n-1}(s), \quad \hat{R}_p(s,t) = s^n + \theta_a^\top \alpha_{n-1}(s)$
Calculation	Solve for $\hat{L}(s,t) = s^{n-1} + l^\top \alpha_{n-2}(s)$, $\hat{P}(s,t) = p^\top \alpha_{n+q-1}(s)$ the polynomial equation: $\hat{L}(s,t) \cdot Q_m(s) \cdot \hat{R}_p(s,t) + \hat{P}(s,t) \cdot \hat{Z}_p(s,t) = A^*(s)$ or solve for $\hat{\beta}_l$ the algebraic equation $\hat{S}_l \hat{\beta}_l = \alpha_l^*$ where $\hat{S}_l$ is the Sylverster matrix of $\hat{R}_p Q_m, \hat{Z}_p$ $\hat{\beta}_l = [l_q^\top, p^\top]^\top \in \mathcal{R}^{2(n+q)}, l_q = [\underbrace{0, \ldots, 0}_{q}, 1, l^\top]^\top \in \mathcal{R}^{n+q}$ $A^*(s) = s^{2n+q-1} + \alpha^{*\top} \alpha_{2n+q-2}(s)$ $\alpha_l^* = [\underbrace{0, \ldots, 0}_{q}, 1, \alpha^{*\top}]^\top \in \mathcal{R}^{2(n+q)}$
Control law	$u_p = (\Lambda - \hat{L} Q_m) \frac{1}{\Lambda} u_p - \hat{P} \frac{1}{\Lambda}(y_p - y_m)$
Design variables	$A^*(s)$ monic Hurwitz; $\Lambda(s)$ monic Hurwitz of degree $n + q - 1$; for simplicity, $\Lambda(s) = \Lambda_p(s)\Lambda_q(s)$, where $\Lambda_p(s), \Lambda_q(s)$ are monic Hurwitz of degree $n, q-1$, respectively

7.4. INDIRECT APPC SCHEMES

APPC scheme of Table 7.4 are u.b. and the tracking error converges to zero asymptotically with time. The same result holds if we replace the gradient algorithm in Table 7.4 with any other adaptive law from Tables 4.2 and 4.3.

Outline of Proof: The proof is completed in the following four steps as in Example 7.4.1:

Step 1. *Manipulate the estimation error and control law equations to express the plant input u_p and output y_p in terms of the estimation error.* This step leads to the following equations:

$$\begin{aligned} \dot{x} &= A(t)x + b_1(t)\epsilon m^2 + b_2 \bar{y}_m \\ u_p &= C_1^\top x + d_1 \epsilon m^2 + d_2 \bar{y}_m \\ y_p &= C_2^\top x + d_3 \epsilon m^2 + d_4 \bar{y}_m \end{aligned} \quad (7.4.24)$$

where $\bar{y}_m \in \mathcal{L}_\infty$; $A(t), b_1(t)$ are u.b. because of the boundedness of the estimated plant and controller parameters (which is guaranteed by the adaptive law and the stabilizability assumption); b_2 is a constant vector; C_1 and C_2 are vectors whose elements are u.b.; and d_1 to d_4 are u.b. scalars.

Step 2. *Establish the e.s. of the homogeneous part of (7.4.24).* The matrix $A(t)$ has stable eigenvalues at each frozen time t that are equal to the roots of $A^*(s) = 0$. In addition $\dot{\theta}_p, \dot{l}, \dot{p} \in \mathcal{L}_2$ (guaranteed by the adaptive law and the stabilizability assumption), imply that $\|\dot{A}(t)\| \in \mathcal{L}_2$. Therefore, using Theorem 3.4.11, we conclude that the homogeneous part of (7.4.24) is u.a.s.

Step 3. *Use the properties of the $\mathcal{L}_{2\delta}$ norm and B-G Lemma to establish boundedness.* Let $m_f^2 \triangleq 1 + \|u_p\|^2 + \|y_p\|^2$ where $\|\cdot\|$ denotes the $\mathcal{L}_{2\delta}$ norm. Using the results established in Steps 1 and 2 and the normalizing properties of m_f, we show that

$$m_f^2 \leq c\|\epsilon m m_f\|^2 + c \quad (7.4.25)$$

which implies that

$$m_f^2 \leq c \int_0^t e^{-\delta(t-\tau)} \epsilon^2 m^2 m_f^2 d\tau + c$$

Because $\epsilon m \in \mathcal{L}_2$, the boundedness of m_f follows by applying the B-G lemma. Using the boundedness of m_f, we can establish the boundedness of all signals in the closed-loop plant.

Step 4. *Establish that the tracking error e_1 converges to zero.* The convergence of e_1 to zero follows by using the control and estimation error equations to express e_1 as the output of proper stable LTI systems whose inputs are in $\mathcal{L}_2 \cap \mathcal{L}_\infty$.

The details of the proof of Theorem 7.4.1 are given in Section 7.7. □

7.4.3 APPC Schemes: State-Variable Approach

As in Section 7.4.2, let us start with a scalar example to illustrate the design and analysis of an APPC scheme formed by combining the control law of Section 7.3.3 developed for the case of known plant parameters with an adaptive law.

Example 7.4.2 We consider the same plant as in Example 7.3.2, i.e.,

$$y_p = \frac{b}{s+a} u_p \tag{7.4.26}$$

where a and b are unknown constants with $b \neq 0$ and u_p is to be chosen so that the poles of the closed-loop plant are placed at the roots of $A^*(s) = (s+1)^2 = 0$ and y_p tracks the reference signal $y_m = 1$. As we have shown in Example 7.3.2, if a, b are known, the following control law can be used to meet the control objective:

$$\dot{\hat{e}} = \begin{bmatrix} -a & 1 \\ 0 & 0 \end{bmatrix} \hat{e} + \begin{bmatrix} 1 \\ 1 \end{bmatrix} b\bar{u}_p - K_o([1\ 0]\hat{e} - e_1)$$

$$\bar{u}_p = -K_c \hat{e}, \quad u_p = \frac{s+1}{s} \bar{u}_p \tag{7.4.27}$$

where K_o, K_c are calculated by solving the equations

$$\det(sI - A + BK_c) = (s+1)^2$$

$$\det(sI - A + K_o C^T) = (s+5)^2$$

where

$$A = \begin{bmatrix} -a & 1 \\ 0 & 0 \end{bmatrix}, \quad B = \begin{bmatrix} 1 \\ 1 \end{bmatrix} b, \quad C^T = [1, 0]$$

i.e.,

$$K_c = \frac{1}{b}[1-a, 1], \quad K_o = [10-a, 25]^T \tag{7.4.28}$$

The APPC scheme for the plant (7.4.26) with unknown a and b may be formed by replacing the unknown parameters a and b in (7.4.27) and (7.4.28) with their on-line estimates $\hat{a}$ and $\hat{b}$ generated by an adaptive law as follows:

Adaptive Law The adaptive law uses the measurements of the plant input u_p and output y_p to generate $\hat{a}, \hat{b}$. It is therefore independent of the choice of the control law and the same adaptive law as the one used in Example 7.4.1 can be employed, i.e.,

$$\dot{\theta}_p = \Pr\{\Gamma \epsilon \phi\}, \quad \Gamma = \Gamma^T > 0$$

7.4. INDIRECT APPC SCHEMES

$$\epsilon = \frac{z - \theta_p^T \phi}{m^2}, \quad m^2 = 1 + \phi^T \phi, \quad z = \frac{s}{s+\lambda} y_p \qquad (7.4.29)$$

$$\theta_p = \begin{bmatrix} \hat{b} \\ \hat{a} \end{bmatrix}, \quad \phi = \frac{1}{s+\lambda} \begin{bmatrix} u_p \\ -y_p \end{bmatrix}$$

where $\lambda > 0$ is a design constant and $\Pr\{\cdot\}$ is the projection operator as defined in Example 7.4.1 that guarantees $\hat{b}(t)| \geq b_0 > 0 \ \forall t \geq 0$.

State Observer

$$\dot{\hat{e}} = \hat{A}(t)\hat{e} + \hat{B}(t)\bar{u}_p - \hat{K}_o(C^T \hat{e} - e_1) \qquad (7.4.30)$$

where

$$\hat{A} = \begin{bmatrix} -\hat{a} & 1 \\ 0 & 0 \end{bmatrix}, \quad \hat{B} = \begin{bmatrix} 1 \\ 1 \end{bmatrix} \hat{b}, \quad C^T = [1, 0]$$

Calculation of Controller Parameters Calculate $\hat{K}_c, \hat{K}_o$ by solving

$$\det(sI - \hat{A} + \hat{B}\hat{K}_c) = A_c^*(s) = (s+1)^2, \quad \det(sI - \hat{A} + \hat{K}_o C^T) = (s+5)^2$$

for each frozen time t which gives

$$\hat{K}_c = \frac{1}{\hat{b}}[1 - \hat{a}, 1], \quad \hat{K}_o = [10 - \hat{a}, 25]^T \qquad (7.4.31)$$

Control Law:

$$\bar{u}_p = -\hat{K}_c \hat{e}, \quad u_p = \frac{s+1}{s} \bar{u}_p \qquad (7.4.32)$$

The solution for the controller parameter vector $\hat{K}_c$ exists for any monic Hurwitz polynomial $A_c^*(s)$ of degree 2 provided $(\hat{A}, \hat{B})$ is stabilizable and $A_c^*(s)$ contains the uncontrollable eigenvalues of $\hat{A}$ as roots. For the example considered, $(\hat{A}, \hat{B})$ loses its controllability when $\hat{b} = 0$. It also loses its controllability when $\hat{b} \neq 0$ and $\hat{a} = 1$. In this last case, however, $(\hat{A}, \hat{B})$, even though uncontrollable, is stabilizable and the uncontrollable eigenvalue is at $s = -1$, which is a zero of $A_c^*(s) = (s+1)^2$. Therefore, as it is also clear from (7.4.31), $\hat{K}_c$ exists for all $\hat{a}, \hat{b}$ provided $\hat{b} \neq 0$. Because the projection operator in (7.4.29) guarantees as in Example 7.4.1 that $|\hat{b}(t)| \geq b_0 > 0 \ \forall t \geq 0$, the existence and boundedness of $\hat{K}_c$ follows from $\hat{a}, \hat{b} \in \mathcal{L}_\infty$.

Analysis

Step 1. *Develop the state error equations for the closed-loop APPC scheme.* The state error equations for the closed-loop APPC scheme include the tracking error equation and the observer equation. The tracking error equation

$$e_1 = \frac{b(s+1)}{s(s+a)} \bar{u}_p, \quad \bar{u}_p = \frac{s}{s+1} u_p$$

CHAPTER 7. ADAPTIVE POLE PLACEMENT CONTROL

is expressed in the state space form

$$\dot{e} = \begin{bmatrix} -a & 1 \\ 0 & 0 \end{bmatrix} e + \begin{bmatrix} 1 \\ 1 \end{bmatrix} b\bar{u}_p \qquad (7.4.33)$$

$$e_1 = C^T e = [1, 0]e$$

Let $e_o = e - \hat{e}$ be the observation error. Then from (7.4.30) and (7.4.33) we obtain the state equations

$$\dot{\hat{e}} = A_c(t)\hat{e} + \begin{bmatrix} -\hat{a} + 10 \\ 25 \end{bmatrix} C^T e_o$$

$$\dot{e}_o = \begin{bmatrix} -10 & 1 \\ -25 & 0 \end{bmatrix} e_o + \begin{bmatrix} 1 \\ 0 \end{bmatrix} \tilde{a} e_1 - \begin{bmatrix} 1 \\ 1 \end{bmatrix} \tilde{b}\bar{u}_p$$

$$\bar{u}_p = -\hat{K}_c \hat{e} \qquad (7.4.34)$$

where $A_c(t) \triangleq \hat{A}(t) - \hat{B}\hat{K}_c, \tilde{a} \triangleq \hat{a} - a, \tilde{b} \triangleq \hat{b} - b$. The plant output is related to $e_0, \hat{e}$ as follows:

$$y_p = e_1 + y_m = C^T(e_0 + \hat{e}) + y_m \qquad (7.4.35)$$

The relationship between u_p and $e_0, \hat{e}$ may be developed as follows:

The coprimeness of $b, s(s+a)$ implies the existence of polynomials $X(s), Y(s)$ of degree 1 and with $X(s)$ monic such that

$$s(s+a)X(s) + b(s+1)Y(s) = A^*(s) \qquad (7.4.36)$$

where $A^*(s) = (s+1)a^*(s)$ and $a^*(s)$ is any monic polynomial of degree 2. Choosing $a^*(s) = (s+1)^2$, we obtain $X(s) = s+1$ and $Y(s) = \frac{(2-a)s+1}{b}$. Equation (7.4.36) may be written as

$$\frac{s(s+a)}{(s+1)^2} + \frac{(2-a)s+1}{(s+1)^2} = 1 \qquad (7.4.37)$$

which implies that

$$u_p = \frac{s(s+a)}{(s+1)^2} u_p + \frac{(2-a)s+1}{(s+1)^2} u_p$$

Using $u_p = \frac{s+a}{b} y_p, su_p = (s+1)\bar{u}_p$ and $\bar{u}_p = -\hat{K}_c \hat{e}$, we have

$$u_p = -\frac{s+a}{s+1}\hat{K}_c\hat{e} + \frac{(2-a)s+1}{(s+1)^2}\frac{(s+a)}{b} y_p \qquad (7.4.38)$$

Equations (7.4.34), (7.4.35), and (7.4.38) describe the stability properties of the closed-loop APPC scheme.

Step 2. *Establish the e.s. of the homogeneous part of (7.4.34).* The homogeneous part of (7.4.34) is considered to be the part with the input $\tilde{a}e_1, \tilde{b}\bar{u}_p$ set equal

7.4. INDIRECT APPC SCHEMES

to zero. The e.s. of the homogeneous part of (7.4.34) can be established by showing that $A_c(t)$ is a u.a.s matrix and $A_o = \begin{bmatrix} -10 & 1 \\ -25 & 0 \end{bmatrix}$ is a stable matrix. Because A_o is stable by design, it remains to show that $A_c(t) = \hat{A}(t) - \hat{B}(t)\hat{K}_c(t)$ is u.a.s.

The projection used in the adaptive law guarantees as in Example 7.4.1 that $|\hat{b}(t)| \geq b_0 > 0, \forall t \geq 0$. Because $\hat{b}, \hat{a} \in \mathcal{L}_\infty$ it follows that the elements of $\hat{K}_c, A_c$ are u.b. Furthermore, at each time t, $\lambda(A_c(t)) = -1, -1$, i.e., $A_c(t)$ is a stable matrix at each frozen time t. From (7.4.31) and $\dot{\hat{a}}, \dot{\hat{b}} \in \mathcal{L}_2$, we have $|\dot{\hat{K}}_c| \in \mathcal{L}_2$. Hence, $\|\dot{A}_c(t)\| \in \mathcal{L}_2$ and the u.a.s of $A_c(t)$ follows by applying Theorem 3.4.11.

In the rest of the proof, we exploit the u.a.s of the homogeneous part of (7.4.34) and the relationships of the inputs $\tilde{a}e_1$, $\tilde{b}\bar{u}_p$ with the properties of the adaptive law in an effort to first establish signal boundedness and then convergence of the tracking error to zero. In the analysis we employ the $\mathcal{L}_{2\delta}$ norm $\|(\cdot)_t\|_{2\delta}$ which for clarity of presentation we denote by $\|\cdot\|$.

Step 3. *Use the properties of the $\mathcal{L}_{2\delta}$ norm and B-G Lemma to establish boundedness.* Applying Lemmas 3.3.2 and 3.3.3 to (7.4.34), (7.4.35), and (7.4.38), we obtain

$$\begin{aligned} \|\hat{e}\| &\leq c\|C^\top e_o\| \\ \|y_p\| &\leq c\|C^\top e_o\| + c\|\hat{e}\| + c \\ \|u_p\| &\leq c\|\hat{e}\| + c\|y_p\| + c \end{aligned}$$

for some $\delta > 0$ where $c \geq 0$ denotes any finite constant, which imply that

$$\begin{aligned} \|y_p\| &\leq c\|C^\top e_o\| + c \\ \|u_p\| &\leq c\|C^\top e_o\| + c \end{aligned}$$

Therefore, the fictitious normalizing signal

$$m_f^2 \triangleq 1 + \|y_p\|^2 + \|u_p\|^2 \leq c\|C^\top e_o\|^2 + c \qquad (7.4.39)$$

We now need to find an upper bound for $\|C^\top e_o\|$, which is a function of the $\mathcal{L}_2$ signals $\epsilon m, \dot{\hat{a}}, \dot{\hat{b}}$. From equation (7.4.34) we write

$$C^\top e_o = \frac{s}{(s+5)^2}\tilde{a}e_1 - \frac{s+1}{(s+5)^2}\tilde{b}\frac{s}{s+1}u_p$$

Applying the Swapping Lemma A.1 (see Appendix A) to the above equation and using the fact that $se_1 = sy_p$ we obtain

$$C^\top e_o = \tilde{a}\frac{s}{(s+5)^2}y_p - \tilde{b}\frac{s}{(s+5)^2}u_p + W_{c1}(W_{b1}e_1)\dot{\hat{a}} - W_{c2}(W_{b2}u_p)\dot{\hat{b}} \qquad (7.4.40)$$

where the elements of $W_{ci}(s), W_{bi}(s)$ are strictly proper transfer functions with poles at $-1, -5$. To relate the first two terms on the right-hand side of (7.4.40) with ϵm^2, we use (7.4.29) and $z = \theta_p^{*T} \phi$ to write

$$\epsilon m^2 = z - \theta_p^T \phi = -\tilde{\theta}_p^T \phi = \tilde{a}\frac{1}{s+\lambda}y_p - \tilde{b}\frac{1}{s+\lambda}u_p \qquad (7.4.41)$$

We filter both sides of (7.4.41) with $\frac{s(s+\lambda)}{(s+5)^2}$ and then apply the Swapping Lemma A.1 to obtain

$$\frac{s(s+\lambda)}{(s+5)^2}\epsilon m^2 = \tilde{a}\frac{s}{(s+5)^2}y_p - \tilde{b}\frac{s}{(s+5)^2}u_p + W_c\left\{(W_b y_p)\dot{\tilde{a}} - (W_b u_p)\dot{\tilde{b}}\right\} \qquad (7.4.42)$$

where the elements of $W_c(s), W_b(s)$ are strictly proper transfer functions with poles at -5. Using (7.4.42) in (7.4.40), we have that

$$C^T e_o = \frac{s(s+\lambda)}{(s+5)^2}\epsilon m^2 + G(s, \dot{\tilde{a}}, \dot{\tilde{b}}) \qquad (7.4.43)$$

where

$$G(s, \dot{\tilde{a}}, \dot{\tilde{b}}) = W_{c1}(W_{b1}e_1)\dot{\tilde{a}} - W_{c2}(W_{b2}u_p)\dot{\tilde{b}} - W_c\left\{(W_b y_p)\dot{\tilde{a}} - (W_b u_p)\dot{\tilde{b}}\right\}$$

Using Lemma 3.3.2 we can establish that $\phi/m_f, m/m_f, W_{b1}e_1/m_f, W_{b2}u_p/m_f, W_b y_p/m_f, W_b u_p/m_f \in \mathcal{L}_\infty$. Using the same lemma we have from (7.4.43) that

$$\|C^T e_o\| \leq c\|\epsilon m m_f\| + c\|m_f \dot{\tilde{a}}\| + c\|m_f \dot{\tilde{b}}\| \qquad (7.4.44)$$

Combining (7.4.39), (7.4.44), we have

$$m_f^2 \leq c\|gm_f\|^2 + c \qquad (7.4.45)$$

or

$$m_f^2 \leq c\int_0^t e^{-\delta(t-\tau)}g^2(\tau)m_f^2(\tau)d\tau + c$$

where $g^2(\tau) = \epsilon^2 m^2 + \dot{\tilde{a}}^2 + \dot{\tilde{b}}^2$. Because $\epsilon m, \dot{\tilde{a}}, \dot{\tilde{b}} \in \mathcal{L}_2$, it follows that $g \in \mathcal{L}_2$ and therefore by applying B-G Lemma to (7.4.45), we obtain $m_f \in \mathcal{L}_\infty$. The boundedness of m_f implies that $\phi, m, W_{b1}e_1, W_{b2}u_p, W_b y_p, W_b u_p \in \mathcal{L}_\infty$. Because $\dot{\tilde{a}}, \dot{\tilde{b}}, \epsilon m^2 \in \mathcal{L}_\infty \bigcap \mathcal{L}_2$, it follows that $C^T e_o \in \mathcal{L}_\infty \bigcap \mathcal{L}_2$ by applying Corollary 3.3.1 to (7.4.43). Now by using $C^T e_o \in \mathcal{L}_\infty \bigcap \mathcal{L}_2$ in (7.4.34), it follows from the stability of $A_c(t)$ and Lemma 3.3.3 or Corollary 3.3.1 that $\hat{e} \in \mathcal{L}_\infty \bigcap \mathcal{L}_2$ and $\hat{e}(t) \to 0$ as $t \to \infty$. Hence, from (7.4.35), we have $y_p \in \mathcal{L}_\infty$ and $e_1 \in \mathcal{L}_\infty \bigcap \mathcal{L}_2$ and from (7.4.38) that $u_p \in \mathcal{L}_\infty$.

7.4. INDIRECT APPC SCHEMES

Step 4. *Convergence of the tracking error to zero.* Because $\dot{e}_1 \in \mathcal{L}_\infty$ (due to $\dot{e}_1 = \dot{y}_p$ and $\dot{y}_p \in \mathcal{L}_\infty$) and $e_1 \in \mathcal{L}_\infty \cap \mathcal{L}_2$, it follows from Lemma 3.2.5 that $e_1(t) \to 0$ as $t \to \infty$.

We can continue the analysis and establish that $\epsilon, \epsilon m \to 0$ as $t \to \infty$, which implies that $\dot{\theta}_p, \dot{\hat{K}}_c \to 0$ as $t \to \infty$.

The convergence of θ_p to $\theta_p^* = [b, a]^\top$ and of $\hat{K}_c$ to $K_c = \frac{1}{b}[1-a, 1]$ cannot be guaranteed, however, unless the signal vector ϕ is PE. For ϕ to be PE, the reference signal y_m has to be sufficiently rich of order 2. Because $y_m = 1$ is sufficiently rich of order 1, $\phi \in \mathcal{R}^2$ is not PE. ▽

General Case

Let us now extend the results of Example 7.4.2 to the nth-order plant (7.3.1). We design an APPC scheme for the plant (7.3.1) by combining the state feedback control law of Section 7.3.3 summarized in Table 7.2 with any appropriate adaptive law from Tables 4.2 to 4.5 based on the plant parametric model (7.4.2) or (7.4.3).

The adaptive law generates the on-line estimates $\hat{R}_p(s,t), \hat{Z}_p(s,t)$ of the unknown plant polynomials $R_p(s), Z_p(s)$, respectively. These estimates are used to generate the estimates $\hat{A}$ and $\hat{B}$ of the unknown matrices A and B, respectively, that are used to calculate the controller parameters and form the observer equation.

Without loss of generality, we concentrate on parametric model (7.4.2) and select the gradient algorithm given in Table 4.2. The APPC scheme formed is summarized in Table 7.5.

The algebraic equation for calculating the controller parameter vector $\hat{K}_c$ in Table 7.5 has a finite solution for $\hat{K}_c(t)$ at each time t provided the pair $(\hat{A}, \hat{B})$ is controllable which is equivalent to $\hat{Z}_p(s,t)Q_1(s), \hat{R}_p(s,t)Q_m(s)$ being coprime at each time t. In fact, for $\hat{K}_c(t)$ to be uniformly bounded, $(\hat{A}, \hat{B})$ has to be strongly controllable, i.e., the absolute value of the determinant of the Sylvester matrix of $\hat{Z}_p(s,t)Q_1(s), \hat{R}_p(s,t)Q_m(s)$ has to be greater than some constant $\nu_0 > 0$ for all $t \geq 0$. This strong controllability condition may be relaxed by choosing $Q_1(s)$ to be a factor of the desired closed-loop Hurwitz polynomial $A_c^*(s)$ as indicated in Table 7.5. By doing so, we allow $Q_1(s), \hat{R}_p(s,t)$ to have common factors without affecting the solvability of the algebraic equation for $\hat{K}_c$, because such common factors are

Table 7.5 APPC scheme: state feedback law

Plant	$y_p = \frac{Z_p(s)}{R_p(s)} u_p$ $Z_p(s) = \theta_b^{*T} \alpha_{n-1}(s), \quad R_p(s) = s^n + \theta_a^{*T} \alpha_{n-1}(s)$ $\alpha_{n-1}(s) = [s^{n-1}, s^{n-2}, \ldots s, 1]^T$
Reference signal	$Q_m(s) y_m = 0$
Adaptive law	Gradient algorithm based on $z = \theta_p^{*T} \phi$ $\dot{\theta}_p = \Gamma \epsilon \phi, \quad \Gamma = \Gamma^T > 0$ $\epsilon = \frac{z - \theta_p^T \phi}{m^2}, \quad m^2 = 1 + \phi^T \phi$ $\phi = \left[\frac{\alpha_{n-1}^T(s)}{\Lambda_p(s)} u_p, -\frac{\alpha_{n-1}^T(s)}{\Lambda_p(s)} y_p \right]^T, \quad z = \frac{s^n}{\Lambda_p(s)} y_p$ $\theta_p = \left[\theta_b^T, \theta_a^T \right]^T$ $\hat{Z}_p(s,t) = \theta_b^T(t) \alpha_{n-1}(s), \hat{R}_p(s,t) = s^n + \theta_a^T(t) \alpha_{n-1}(s)$
State observer	$\dot{\hat{e}} = \hat{A}\hat{e} + \hat{B}\bar{u}_p - \hat{K}_o(t)(C^T \hat{e} - e_1), \quad \hat{e} \in \mathcal{R}^{n+q}$ $\hat{A} = \begin{bmatrix} & I_{n+q-1} \\ -\theta_1(t) & \hline \\ & 0 \end{bmatrix}, \hat{B} = \theta_2(t), C^T = [1, 0, \ldots, 0]$ $\theta_1 \in \mathcal{R}^{n+q}$ is the coefficient vector of $\hat{R}_p Q_m - s^{n+q}$ $\theta_2 \in \mathcal{R}^{n+q}$ is the coefficient vector of $\hat{Z}_p Q_1$ $\hat{K}_o = \alpha^* - \theta_1$, and α^* is the coefficient vector of $A_o^*(s) - s^{n+q}$
Calculation of controller parameters	Solve for $\hat{K}_c$ pointwise in time the equation $\det(sI - \hat{A} + \hat{B}\hat{K}_c) = A_c^*(s)$
Control law	$\bar{u}_p = -\hat{K}_c(t)\hat{e}, \quad u_p = \frac{Q_1}{Q_m} \bar{u}_p$
Design variables	$Q_m(s)$ monic of degree q with nonrepeated roots on the $j\omega$-axis; $Q_1(s)$ monic Hurwitz of degree q; $A_o^*(s)$ monic Hurwitz of degree $n+q$; $A_c^*(s)$ monic Hurwitz of degree $n+q$ and with $Q_1(s)$ as a factor; $\Lambda_p(s)$ monic Hurwitz of degree n

7.4. INDIRECT APPC SCHEMES

also included in $A_c^*(s)$. Therefore the condition that guarantees the existence and uniform boundedness of $\hat{K}_c$ is that $\hat{Z}_p(s,t)$, $\hat{R}_p(s,t)Q_m(s)$ are strongly coprime at each time t. As we mentioned earlier, such a condition cannot be guaranteed by any one of the adaptive laws developed in Chapter 4 without additional modifications, thus giving rise to the so-called stabilizability or admissibility problem to be discussed in Section 7.6. In this section, we assume that the polynomials $\hat{Z}_p(s,t)$, $\hat{R}_p(s,t)Q_m(s)$ are strongly coprime at each time t and proceed with the analysis of the APPC scheme of Table 7.5. We relax this assumption in Section 7.6 where we modify the APPC schemes to handle the possible loss of stabilizability of the estimated plant.

Theorem 7.4.2 *Assume that the polynomials $\hat{Z}_p$, $\hat{R}_p Q_m$ are strongly coprime at each time t. Then all the signals in the closed-loop APPC scheme of Table 7.5 are uniformly bounded and the tracking error e_1 converges to zero asymptotically with time. The same result holds if we replace the gradient algorithm with any other adaptive law from Tables 4.2 to 4.4 that is based on the plant parametric model (7.4.2) or (7.4.3).*

Outline of Proof
Step 1. Develop the state error equations for the closed-loop APPC scheme, i.e.,

$$\begin{aligned}
\dot{\hat{e}} &= A_c(t)\hat{e} + \hat{K}_o C^\top e_o \\
\dot{e}_o &= A_o e_o + \tilde{\theta}_1 e_1 - \tilde{\theta}_2 \bar{u}_p \\
y_p &= C^\top e_o + C^\top \hat{e} + y_m \\
u_p &= W_1(s)\hat{K}_c(t)\hat{e} + W_2(s) y_p \\
\bar{u}_p &= -\hat{K}_c \hat{e}
\end{aligned} \qquad (7.4.46)$$

where $e_o \triangleq e - \hat{e}$ is the observation error, A_o is a constant stable matrix, $W_1(s)$ and $W_2(s)$ are strictly proper transfer functions with stable poles, and $A_c(t) = \hat{A} - \hat{B}\hat{K}_c$.

Step 2. *Establish e.s. for the homogeneous part of (7.4.46).* The gain $\hat{K}_c$ is chosen so that the eigenvalues of $A_c(t)$ at each time t are equal to the roots of the Hurwitz polynomial $A_c^*(s)$. Because $\hat{A}, \hat{B} \in \mathcal{L}_\infty$ (guaranteed by the adaptive law) and $\hat{Z}_p, \hat{R}_p Q_m$ are strongly coprime (by assumption), we conclude that $(\hat{A}, \hat{B})$ is stabilizable in a strong sense and $\hat{K}_c, A_c \in \mathcal{L}_\infty$. Using $\dot{\theta}_a, \dot{\theta}_b \in \mathcal{L}_2$, guaranteed by the adaptive law, we have $\dot{\hat{K}}_c, \dot{A}_c \in \mathcal{L}_2$. Therefore, applying Theorem 3.4.11, we have that $A_c(t)$ is a u.a.s. matrix. Because A_o is a constant stable matrix, the e.s. of the homogeneous part of (7.4.46) follows.

Step 3. *Use the properties of the $\mathcal{L}_{2\delta}$ norm and the B-G Lemma to establish signal boundedness.* We use the properties of the $\mathcal{L}_{2\delta}$ norm and equation (7.4.46) to establish the inequality

$$m_f^2 \leq c\|gm_f\|^2 + c$$

where $g^2 = \epsilon^2 m^2 + |\dot{\theta}_a|^2 + |\dot{\theta}_b|^2$ and $m_f^2 \triangleq 1 + \|u_p\|^2 + \|y_p\|^2$ is the fictitious normalizing signal. Because $g \in \mathcal{L}_2$, it follows that $m_f \in \mathcal{L}_\infty$ by applying the B-G Lemma. Using $m_f \in \mathcal{L}_\infty$, we establish the boundedness of all signals in the closed-loop plant.

Step 4. *Establish the convergence of the tracking error e_1 to zero.* This is done by following the same procedure as in Example 7.4.2. □

The details of the proof of Theorem 7.4.2 are given in Section 7.7.

7.4.4 Adaptive Linear Quadratic Control (ALQC)

The linear quadratic (LQ) controller developed in Section 7.3.4 can be made adaptive and used to meet the control objective when the plant parameters are unknown. This is achieved by combining the LQ control law (7.3.28) to (7.3.32) with an adaptive law based on the plant parametric model (7.4.2) or (7.4.3).

We demonstrate the design and analysis of ALQ controllers using the following examples:

Example 7.4.3 We consider the same plant and control objective as in Example 7.3.3, given by

$$\begin{aligned} \dot{x} &= -ax + bu_p \\ y_p &= x \end{aligned} \qquad (7.4.47)$$

where the plant input u_p is to be chosen to stabilize the plant and regulate y_p to zero. In contrast to Example 7.3.3, the parameters a and b are unknown constants. The control law $u_p = -\frac{1}{\lambda} b p y_p$ in Example 7.3.3 is modified by replacing the unknown plant parameters a, b with their on-line estimates $\hat{a}$ and $\hat{b}$ generated by the same adaptive law used in Example 7.4.2, as follows:

Adaptive Law

$$\dot{\theta}_p = \Gamma \epsilon \phi, \quad \Gamma = \Gamma^\top > 0$$

$$\epsilon = \frac{z - \theta_p^\top \phi}{m^2}, \quad m^2 = 1 + \phi^\top \phi, \quad z = \frac{s}{s + \lambda_0} y_p$$

7.4. INDIRECT APPC SCHEMES

$$\theta_p = \begin{bmatrix} \hat{b} \\ \hat{a} \end{bmatrix}, \quad \phi = \frac{1}{s+\lambda_0} \begin{bmatrix} u_p \\ -y_p \end{bmatrix}$$

where $\lambda_0 > 0$ is a design constant.

Control Law

$$u_p = -\frac{1}{\lambda}\hat{b}(t)p(t)y_p \tag{7.4.48}$$

Riccati Equation Solve the equation

$$-2\hat{a}(t)p(t) - \frac{p^2(t)\hat{b}^2(t)}{\lambda} + 1 = 0$$

at each time t for $p(t) > 0$, i.e.,

$$p(t) = \frac{-\lambda\hat{a} + \sqrt{\lambda^2\hat{a}^2 + \hat{b}^2\lambda}}{\hat{b}^2} > 0 \tag{7.4.49}$$

As in the previous examples, for the solution $p(t)$ in (7.4.49) to be finite, the estimate $\hat{b}$ should not cross zero. In fact, for $p(t)$ to be uniformly bounded, $\hat{b}(t)$ should satisfy $|\hat{b}(t)| \geq b_0 > 0, \forall t \geq 0$ for some constant b_0 that satisfies $|b| \geq b_0$. Using the knowledge of b_0 and $\text{sgn}(b)$, the adaptive law for $\hat{b}$ can be modified as before to guarantee $|\hat{b}(t)| \geq b_0, \forall t \geq 0$ and at the same time retain the properties that $\theta_p \in \mathcal{L}_\infty$ and $\epsilon, \epsilon m, \dot{\theta}_p \in \mathcal{L}_2 \cap \mathcal{L}_\infty$. The condition $|\hat{b}(t)| \geq b_0$ implies that the estimated plant, characterized by the parameters $\hat{a}$, $\hat{b}$, is strongly controllable at each time t, a condition required for the solution $p(t) > 0$ of the Riccati equation to exist and be uniformly bounded.

Analysis For this first-order regulation problem, the analysis is relatively simple and can be accomplished in the following four steps:

Step 1. *Develop the closed-loop error equation.* The closed-loop plant can be written as

$$\dot{x} = -(\hat{a} + \frac{\hat{b}^2 p}{\lambda})x + \tilde{a}x - \tilde{b}u_p \tag{7.4.50}$$

by adding and subtracting $\hat{a}x - \hat{b}u_p$ and using $u_p = -\hat{b}px/\lambda$. The inputs $\tilde{a}x, \tilde{b}u_p$ are due to the parameter errors $\tilde{a} \triangleq \hat{a} - a, \tilde{b} \triangleq \hat{b} - b$.

Step 2. *Establish the e.s. of the homogeneous part of (7.4.50).* The eigenvalue of the homogeneous part of (7.4.50) is

$$-(\hat{a} + \frac{\hat{b}^2 p}{\lambda})$$

which is guaranteed to be negative by the choice of $p(t)$ given by (7.4.49), i.e.,

$$-(\hat{a} + \frac{\hat{b}^2 p}{\lambda}) = -\sqrt{\hat{a}^2 + \frac{\hat{b}^2}{\lambda}} \leq -\frac{b_0}{\sqrt{\lambda}} < 0$$

provided, of course, the adaptive law is modified by using projection to guarantee $|\hat{b}(t)| \geq b_0, \forall t \geq 0$. Hence, the homogeneous part of (7.4.50) is e.s.

Step 3. *Use the properties of the $\mathcal{L}_{2\delta}$ norm and B-G Lemma to establish boundedness.* The properties of the input $\tilde{a}x - \tilde{b}u_p$ in (7.4.50) depend on the properties of the adaptive law that generates $\tilde{a}$ and $\tilde{b}$. The first task in this step is to establish the smallness of the input $\tilde{a}x - \tilde{b}u_p$ by relating it with the signals $\dot{\tilde{a}}, \dot{\tilde{b}}$, and εm that are guaranteed by the adaptive law to be in $\mathcal{L}_2$.

We start with the estimation error equation

$$\varepsilon m^2 = z - \theta_p^T \phi = -\tilde{\theta}_p^T \phi = \tilde{a}\frac{1}{s+\lambda_0}x - \tilde{b}\frac{1}{s+\lambda_0}u_p \quad (7.4.51)$$

Operating with $(s + \lambda_0)$ on both sides of (7.4.51) and using the property of differentiation, i.e., $sxy = x\dot{y} + \dot{x}y$ where $s \triangleq \frac{d}{dt}$ is treated as the differential operator, we obtain

$$(s+\lambda_0)\varepsilon m^2 = \tilde{a}x - \tilde{b}u_p + \dot{\tilde{a}}\frac{1}{s+\lambda_0}x - \dot{\tilde{b}}\frac{1}{s+\lambda_0}u_p \quad (7.4.52)$$

Therefore,

$$\tilde{a}x - \tilde{b}u_p = (s+\lambda_0)\varepsilon m^2 - \dot{\tilde{a}}\frac{1}{s+\lambda_0}x + \dot{\tilde{b}}\frac{1}{s+\lambda_0}u_p$$

which we substitute in (7.4.50) to obtain

$$\dot{x} = -(\hat{a} + \frac{\hat{b}^2 p}{\lambda})x + (s+\lambda_0)\varepsilon m^2 - \dot{\tilde{a}}\frac{1}{s+\lambda_0}x + \dot{\tilde{b}}\frac{1}{s+\lambda_0}u_p \quad (7.4.53)$$

If we define $\bar{e} \triangleq x - \varepsilon m^2$, (7.4.53) becomes

$$\dot{\bar{e}} = -(\hat{a} + \frac{\hat{b}^2 p}{\lambda})\bar{e} + (\lambda_0 - \hat{a} - \frac{\hat{b}^2 p}{\lambda})\varepsilon m^2 - \dot{\tilde{a}}\frac{1}{s+\lambda_0}x + \dot{\tilde{b}}\frac{1}{s+\lambda_0}u_p \quad (7.4.54)$$
$$x = \bar{e} + \varepsilon m^2$$

Equation (7.4.54) has a homogeneous part that is e.s. and an input that is small in some sense because of $\varepsilon m, \dot{\tilde{a}}, \dot{\tilde{b}} \in \mathcal{L}_2$.

Let us now use the properties of the $\mathcal{L}_{2\delta}$ norm, which for simplicity is denoted by $\|\cdot\|$ to analyze (7.4.54). The fictitious normalizing signal m_f satisfies

$$m_f^2 \triangleq 1 + \|y_p\|^2 + \|u_p\|^2 \leq 1 + c\|x\|^2 \quad (7.4.55)$$

7.4. INDIRECT APPC SCHEMES

for some $\delta > 0$ because of the control law chosen and the fact that $\hat{b}, p \in \mathcal{L}_\infty$. Because $x = \bar{e} + \varepsilon m^2$, we have $\|x\| \leq \|\bar{e}\| + \|\varepsilon m^2\|$, which we use in (7.4.55) to obtain

$$m_f^2 \leq 1 + c\|\bar{e}\|^2 + c\|\varepsilon m^2\|^2 \tag{7.4.56}$$

From (7.4.54), we have

$$\|\bar{e}\|^2 \leq c\|\varepsilon m^2\|^2 + c\|\dot{\tilde{a}}\bar{x}\|^2 + c\|\dot{\tilde{b}}\bar{u}_p\|^2 \tag{7.4.57}$$

where $\bar{x} = \frac{1}{s+\lambda_0} x$, $\bar{u}_p = \frac{1}{s+\lambda_0} u_p$. Using the properties of the $\mathcal{L}_{2\delta}$ norm, it can be shown that m_f bounds from above $m, \bar{x}, \bar{u}_p$ and therefore it follows from (7.4.56), (7.4.57) that

$$m_f^2 \leq 1 + c\|\varepsilon m m_f\|^2 + c\|\dot{\tilde{a}} m_f\|^2 + c\|\dot{\tilde{b}} m_f\|^2 \tag{7.4.58}$$

which implies that

$$m_f^2 \leq 1 + c \int_0^t e^{-\delta(t-\tau)} g^2(\tau) m_f^2(\tau) d\tau$$

where $g^2 \triangleq \varepsilon^2 m^2 + \dot{\tilde{a}}^2 + \dot{\tilde{b}}^2$. Since $\varepsilon m, \dot{\tilde{a}}, \dot{\tilde{b}} \in \mathcal{L}_2$ imply that $g \in \mathcal{L}_2$, the boundedness of m_f follows by applying the B-G Lemma.

Now $m_f \in \mathcal{L}_\infty$ implies that $m, \bar{x}, \bar{u}_p$ and, therefore, $\phi \in \mathcal{L}_\infty$. Using (7.4.54), and the fact that $\dot{\tilde{a}}, \dot{\tilde{b}}, \varepsilon m^2, \bar{x} = (1/(s+\lambda_0))x, \bar{u}_p = (1/(s+\lambda_0))u_p \in \mathcal{L}_\infty$, we have $\bar{e} \in \mathcal{L}_\infty$, which implies that $x = y_p \in \mathcal{L}_\infty$, and therefore u_p and all signals in the closed loop plant are bounded.

Step 4. *Establish that* $x = y_p \to 0$ *as* $t \to \infty$. We proceed as follows: Using (7.4.54), we establish that $\bar{e} \in \mathcal{L}_2$, which together with $\varepsilon m^2 \in \mathcal{L}_2$ imply that $x = \bar{e} + \varepsilon m^2 \in \mathcal{L}_2$. Because (7.4.53) implies that $\dot{x} \in \mathcal{L}_\infty$, it follows from Lemma 3.2.5 that $x(t) \to 0$ as $t \to \infty$.

The analysis of the ALQ controller presented above is simplified by the fact that the full state is available for measurement; therefore, no state observer is necessary. Furthermore, the u.a.s. of the homogeneous part of (7.4.50) is established by simply showing that the time–varying scalar $\hat{a} + \hat{b}^2 p/\lambda \geq b_0/\sqrt{\lambda} > 0$, $\forall t \geq 0$, i.e., that the closed–loop eigenvalue is stable at each time t, which in the scalar case implies u.a.s. $\triangledown$

In the following example, we consider the tracking problem for the same scalar plant given by (7.4.47). In this case the analysis requires some additional arguments due to the use of a state observer and higher dimensionality.

Example 7.4.4 Let us consider the same plant as in Example 7.4.3 but with the following control objective: Choose u_p to stabilize the plant and force y_p to follow the constant reference signal $y_m(t) = 1$. This is the same control problem we solved in Example 7.3.4 under the assumption that the plant parameters a and b are known exactly. The control law when a, b are known given in Example 7.3.4 is summarized below:

State Observer $\quad \dot{\hat{e}} = A\hat{e} + B\bar{u}_p - K_o(C^T\hat{e} - e_1), \quad e_1 = y_p - y_m$

Control Law $\quad \bar{u}_p = -\lambda^{-1}[b, b]P\hat{e}, \quad u_p = \dfrac{s+1}{s}\bar{u}_p$

Riccati Equation $\quad A^T P + PA - PBB^T P\lambda^{-1} + CC^T = 0$

where $A = \begin{bmatrix} -a & 1 \\ 0 & 0 \end{bmatrix}, \; B = \begin{bmatrix} 1 \\ 1 \end{bmatrix}b, \; C^T = [1, 0], \; K_o = [10-a, 25]^T$.

In this example, we assume that a and b are unknown constants and use the certainty equivalence approach to replace the unknown a, b with their estimates $\hat{a}, \hat{b}$ generated by an adaptive law as follows:

State Observer
$$\dot{\hat{e}} = \hat{A}(t)\hat{e} + \hat{B}\bar{u}_p - \hat{K}_o(t)([1\ 0]\hat{e} - e_1)$$

$$\hat{A} = \begin{bmatrix} -\hat{a} & 1 \\ 0 & 0 \end{bmatrix}, \; \hat{B} = \hat{b}\begin{bmatrix} 1 \\ 1 \end{bmatrix}, \; \hat{K}_o = \begin{bmatrix} 10-\hat{a} \\ 25 \end{bmatrix} \quad (7.4.59)$$

Control Law
$$\bar{u}_p = -\frac{\hat{b}}{\lambda}[1\ 1]P\hat{e}, \quad u_p = \frac{s+1}{s}\bar{u}_p \quad (7.4.60)$$

Riccati Equation Solve for $P(t) = P^T(t) > 0$ at each time t the equation

$$\hat{A}^T P + P\hat{A} - P\frac{\hat{B}\hat{B}^T}{\lambda}P + CC^T = 0, \quad C^T = [1,\ 0] \quad (7.4.61)$$

The estimates $\hat{a}(t)$ and $\hat{b}(t)$ are generated by the same adaptive law as in Example 7.4.3.

For the solution $P = P^T > 0$ of (7.4.61) to exist, the pair $(\hat{A}, \hat{B})$ has to be stabilizable. Because $(\hat{A}, \hat{B}, C)$ is the realization of $\frac{\hat{b}(s+1)}{(s+\hat{a})s}$, the stabilizability of $(\hat{A}, \hat{B})$ is guaranteed provided $\hat{b} \neq 0$ (note that for $\hat{a} = 1$, the pair $(\hat{A}, \hat{B})$ is no longer controllable but it is still stabilizable). In fact for $P(t)$ to be uniformly bounded, we require $|\hat{b}(t)| \geq b_0 > 0$, for some constant b_0, which is a lower bound for $|b|$. As in the previous examples, the adaptive law for $\hat{b}$ can be modified to guarantee $|\hat{b}(t)| \geq b_0, \forall t \geq 0$ by assuming that b_0 and $\mathrm{sgn}(b)$ are known a priori.

Analysis The analysis is very similar to that given in Example 7.4.2. The tracking error equation is given by

$$\dot{e} = \begin{bmatrix} -a & 1 \\ 0 & 0 \end{bmatrix}e + \begin{bmatrix} 1 \\ 1 \end{bmatrix}b\bar{u}_p, \quad e_1 = [1, 0]e$$

7.4. INDIRECT APPC SCHEMES

If we define $e_o = e - \hat{e}$ to be the observation error and use the control law (7.4.60) in (7.4.59), we obtain the same error equation as in Example 7.4.2, i.e.,

$$\dot{\hat{e}} = A_c(t)\hat{e} + \begin{bmatrix} -\hat{a} + 10 \\ 25 \end{bmatrix} C^\top e_o$$

$$\dot{e}_o = \begin{bmatrix} -10 & 1 \\ -25 & 0 \end{bmatrix} e_o + \begin{bmatrix} 1 \\ 0 \end{bmatrix} \tilde{a} e_1 - \begin{bmatrix} 1 \\ 1 \end{bmatrix} \tilde{b} \bar{u}_p$$

where $A_c(t) = \hat{A}(t) - \hat{B}\hat{B}^\top P/\lambda$ and y_p, u_p are related to e_o, $\hat{e}$ through the equations

$$y_p = C^\top e_0 + C^\top \hat{e} + y_m$$

$$u_p = -\frac{s+a}{s+1} \frac{\hat{B}^\top P}{\lambda} \hat{e} + \frac{(2-a)s+1}{(s+1)^2} \frac{s+a}{b} y_p$$

If we establish the u.a.s. of $A_c(t)$, then the rest of the analysis is exactly the same as that for Example 7.4.2.

Using the results of Section 7.3.4, we can establish that the matrix $A_c(t)$ at each frozen time t has all its eigenvalues in the open left half s-plane. Furthermore,

$$\|\dot{A}_c(t)\| \leq \|\dot{\hat{A}}(t)\| + \frac{2\|\dot{\hat{B}}(t)\| \|\hat{B}(t)\| \|P(t)\|}{\lambda} + \frac{\|\hat{B}(t)\|^2 \|\dot{P}(t)\|}{\lambda}$$

where $\|\dot{\hat{A}}(t)\|$, $\|\dot{\hat{B}}(t)\| \in \mathcal{L}_2$ due to $\dot{\hat{a}}, \dot{\hat{b}} \in \mathcal{L}_2$ guaranteed by the adaptive law. By taking the first-order derivative on each side of (7.4.61), $\dot{P}$ can be shown to satisfy

$$\dot{P} A_c + A_c^\top \dot{P} = -Q \qquad (7.4.62)$$

where

$$Q(t) = \dot{\hat{A}}^\top P + P \dot{\hat{A}} - P \frac{\dot{\hat{B}} \hat{B}^\top}{\lambda} P - P \frac{\hat{B} \dot{\hat{B}}^\top}{\lambda} P$$

For a given $\hat{A}$, $\hat{B}$, and P, (7.4.62) is a Lyapunov equation and its solution $\dot{P}$ exists and is continuous with respect to Q, i.e., $\|\dot{P}(t)\| \leq c\|Q(t)\|$ for some constant $c \geq 0$. Because of $\|\dot{\hat{A}}(t)\|, \|\dot{\hat{B}}(t)\| \in \mathcal{L}_2$ and $\hat{A}, \hat{B}, P \in \mathcal{L}_\infty$, we have $\|\dot{P}(t)\| \in \mathcal{L}_2$ and, thus, $\dot{A}_c(t) \in \mathcal{L}_2$. Because $A_c(t)$ is a stable matrix at each frozen time t, we can apply Theorem 3.4.11 to conclude that A_c is u.a.s. The rest of the analysis follows by using exactly the same steps as in the analysis of Example 7.4.2 and is, therefore, omitted. ▽

General Case

Following the same procedure as in Examples 7.4.3 and 7.4.4, we can design a wide class of ALQ control schemes for the nth-order plant (7.3.1) by combining the LQ control law of Section 7.3.4 with adaptive laws based on the parametric model (7.4.2) or (7.4.3) from Tables 4.2 to 4.5.

Table 7.6 gives such an ALQ scheme based on a gradient algorithm for the nth-order plant (7.3.1).

As with the previous APPC schemes, the ALQ scheme depends on the solvability of the algebraic Riccati equation. The Riccati equation is solved for each time t by using the on-line estimates $\hat{A}, \hat{B}$ of the plant parameters. For the solution $P(t) = P^{\mathsf{T}}(t) > 0$ to exist, the pair $(\hat{A}, \hat{B})$ has to be stabilizable at each time t. This implies that the polynomials $\hat{R}_p(s,t)Q_m(s)$ and $\hat{Z}_p(s,t)Q_1(s)$ should not have any common unstable zeros at each frozen time t. Because $Q_1(s)$ is Hurwitz, a sufficient condition for $(\hat{A}, \hat{B})$ to be stabilizable is that the polynomials $\hat{R}_p(s,t)Q_m(s)$, $\hat{Z}_p(s,t)$ are coprime at each time t. For $P(t)$ to be uniformly bounded, however, we will require $\hat{R}_p(s,t)Q_m(s)$, $\hat{Z}_p(s,t)$ to be strongly coprime at each time t.

In contrast to the simple examples considered, the modification of the adaptive law to guarantee the strong coprimeness of $\hat{R}_p Q_m$, $\hat{Z}_p$ without the use of additional a priori information about the unknown plant is not clear. This problem known as the stabilizability problem in indirect APPC is addressed in Section 7.6. In the meantime, let us assume that the stabilizability of the estimated plant is guaranteed and proceed with the following theorem that states the stability properties of the ALQ control scheme given in Table 7.6.

Theorem 7.4.3 *Assume that the polynomials $\hat{R}_p(s,t)Q_m(s)$, $\hat{Z}_p(s,t)$ are strongly coprime at each time t. Then the ALQ control scheme of Table 7.6 guarantees that all signals in the closed-loop plant are bounded and the tracking error e_1 converges to zero as $t \to \infty$. The same result holds if we replace the gradient algorithm in Table 7.6 with any other adaptive law from Tables 4.2 to 4.4 based on the plant parametric model (7.4.2) or (7.4.3).*

Proof The proof is almost identical to that of Theorem 7.4.2, except for some minor details. The same error equations as in the proof of Theorem 7.4.2 that relate $\hat{e}$ and the observation error $e_o = e - \hat{e}$ with the plant input and output also

7.4. INDIRECT APPC SCHEMES

Table 7.6 Adaptive linear quadratic control scheme

Plant	$y_p = \frac{Z_p(s)}{R_p(s)} u_p$ $Z_p(s) = \theta_b^{*T} \alpha_{n-1}(s)$, $R_p(s) = s^n + \theta_a^{*T} \alpha_{n-1}(s)$ $\alpha_{n-1}(s) = [s^{n-1}, s^{n-2}, \ldots, s, 1]^T$
Reference signal	$Q_m(s) y_m = 0$
Adaptive law	Same gradient algorithm as in Table 7.5 to generate $\hat{Z}_p(s,t) = \theta_b^T(t) \alpha_{n-1}(s)$, $\hat{R}_p(s,t) = s^n + \theta_a^T(t) \alpha_{n-1}(s)$
State observer	$\dot{\hat{e}} = \hat{A}(t)\hat{e} + \hat{B}\bar{u}_p - \hat{K}_o(t)(C^T \hat{e} - e_1)$ $\hat{A}(t) = \left[\begin{array}{c\|c} & I_{n+q-1} \\ -\theta_1 & ---- \\ & 0 \end{array} \right]$, $\hat{B}(t) = \theta_2(t)$ $\hat{K}_o(t) = a^* - \theta_1$, $C = [1, 0, \ldots, 0]^T \in \mathcal{R}^{n+q}$ θ_1 is the coefficient vector of $\hat{R}_p(s,t) Q_m(s) - s^{n+q}$ θ_2 is the coefficient vector of $\hat{Z}_p(s,t) Q_1(s)$ a^* is the coefficient vector of $A_o^*(s) - s^{n+q}$ $\theta_1, \theta_2, a^* \in \mathcal{R}^{n+q}$
Riccati equation	Solve for $P(t) = P^T(t) > 0$ the equation $\hat{A}^T P + P\hat{A} - \frac{1}{\lambda} P \hat{B} \hat{B}^T P + CC^T = 0$
Control law	$\bar{u}_p = -\frac{1}{\lambda} \hat{B}^T P \hat{e}$, $u_p = \frac{Q_1(s)}{Q_m(s)} \bar{u}_p$
Design variables	$Q_m(s)$ is a monic polynomial of degree q with nonrepeated roots on the $j\omega$ axis; $A_o^*(s)$ is a monic Hurwitz polynomial of degree $n+q$ with relatively fast zero; $\lambda > 0$ as in Table 7.3; $Q_1(s)$ is a monic Hurwitz polynomial of degree q.

hold here, i.e.,

$$\dot{\hat{e}} = A_c(t)\hat{e} + \hat{K}_o C^\top e_o$$
$$\dot{e}_o = A_o e_o + \tilde{\theta}_1 e_1 + \tilde{\theta}_2 \hat{K}_c \hat{e}$$
$$y_p = C^\top e_o + C^\top \hat{e} + y_m$$
$$u_p = W_1(s)\hat{K}_c(t)\hat{e} + W_2(s) y_p$$

The only difference is that in $A_c = \hat{A} - \hat{B}\hat{K}_c$, we have $\hat{K}_c = \hat{B}^\top(t) P(t)/\lambda$. If we establish that $\hat{K}_c \in \mathcal{L}_\infty$, and A_c is u.a.s., then the rest of the proof is identical to that of Theorem 7.4.2.

The strong coprimeness assumption about $\hat{R}_p Q_m$, $\hat{Z}_p$ guarantees that the solution $P(t) = P^\top(t) > 0$ of the Riccati equation exists at each time t and $P \in \mathcal{L}_\infty$. This, together with the boundedness of the plant parameter estimates, guarantee that $\hat{B}$, and therefore $\hat{K}_c \in \mathcal{L}_\infty$. Furthermore, using the results of Section 7.3.4, we can establish that $A_c(t)$ is a stable matrix at each frozen time t. As in Example 7.4.4, we have

$$\|\dot{A}_c(t)\| \leq \|\dot{\hat{A}}(t)\| + \frac{2\|\dot{\hat{B}}(t)\|\,\|\hat{B}(t)\|\,\|P(t)\|}{\lambda} + \frac{\|\hat{B}(t)\|^2 \|\dot{P}(t)\|}{\lambda}$$

and

$$\dot{P} A_c + A_c^\top \dot{P} = -Q$$

where

$$Q = \dot{\hat{A}}^\top P + P\dot{\hat{A}} - \frac{P\dot{\hat{B}}\hat{B}^\top P}{\lambda} - \frac{P\hat{B}\dot{\hat{B}}^\top P}{\lambda}$$

which, as shown earlier, imply that $\|\dot{P}(t)\|$ and, thus, $\|\dot{A}_c(t)\| \in \mathcal{L}_2$.

The pointwise stability of A_c together with $\|\dot{A}_c(t)\| \in \mathcal{L}_2$ imply, by Theorem 3.4.11, that A_c is a u.a.s. matrix. The rest of the proof is completed by following exactly the same steps as in the proof of Theorem 7.4.2. □

7.5 Hybrid APPC Schemes

The stability properties of the APPC schemes presented in Section 7.4 are based on the assumption that the algebraic equations used to calculate the controller parameters are solved continuously and instantaneously. In practice, even with high speed computers and advanced software tools, a short time interval is always required to complete the calculations at a given time t. The robustness and stability properties of the APPC schemes of Section 7.4

7.5. HYBRID APPC SCHEMES

Table 7.7 Hybrid adaptive law

Plant	$y_p = \frac{Z_p(s)}{R_p(s)} u_p$ $Z_p(s) = \theta_b^{*\top} \alpha_{n-1}(s),\ R_p(s) = s^n + \theta_a^{*\top} \alpha_{n-1}(s)$ $\alpha_{n-1}(s) = [s^{n-1}, s^{n-2}, \ldots, s, 1]^\top$
Adaptive law	$\theta_{p(k+1)} = \theta_{pk} + \Gamma \int_{t_k}^{t_{k+1}} \epsilon(\tau)\phi(\tau)d\tau,\ k = 0, 1, \ldots$ $\phi = \left[\frac{\alpha_{n-1}^\top(s)}{\Lambda_p(s)} u_p, -\frac{\alpha_{n-1}^\top(s)}{\Lambda_p(s)} y_p\right]^\top,\ z = \frac{s^n}{\Lambda_p(s)} y_p$ $\epsilon = \frac{z - \theta_{pk}^\top \phi}{m^2},\ m^2 = 1 + \phi^\top \phi,\ \forall t \in [t_k, t_{k+1})$ $\theta_{pk} = [\theta_{bk}^\top, \theta_{ak}^\top]^\top$ $\hat{R}_p(s, t_k) = s^n + \theta_{a(k-1)}^\top \alpha_{n-1}(s)$ $\hat{Z}_p(s, t_k) = \theta_{b(k-1)}^\top \alpha_{n-1}(s)$
Design variables	$T_s = t_{k+1} - t_k > T_m;\ 2 - T_s \lambda_{\max}(\Gamma) > \gamma$, for some $\gamma > 0$; $\Lambda_p(s)$ is monic and Hurwitz with degree n

with respect to such computational real time delays can be considerably improved by using a hybrid adaptive law for parameter estimation. The sampling rate of the hybrid adaptive law may be chosen appropriately to allow for the computations of the control law to be completed within the sampling interval.

Let T_m be the maximum time for performing the computations required to calculate the control law. Then the sampling period T_s of the hybrid adaptive law may be chosen as $T_s = t_{k+1} - t_k > T_m$ where $\{t_k : k = 1, 2, \ldots\}$ is a time sequence. Table 7.7 presents a hybrid adaptive law based on parametric model (7.4.2). It can be used to replace the continuous-time

Table 7.8 Hybrid APPC scheme: polynomial approach

Plant	$y_p = \dfrac{Z_p(s)}{R_p(s)} u_p$
Reference signal	$Q_m(s) y_m = 0$
Adaptive law	Hybrid adaptive law of Table 7.7
Algebraic equation	Solve for $\hat{L}(s, t_k) = s^{n-1} + l^\top(t_k)\alpha_{n-2}(s)$ $\hat{P}(s, t_k) = p^\top(t_k)\alpha_{n+q-1}(s)$ from equation $\hat{L}(s,t_k)Q_m(s)\hat{R}_p(s,t_k) + \hat{P}(s,t_k)\hat{Z}_p(s,t_k) = A^*(s)$
Control law	$u_p = \dfrac{\Lambda(s) - \hat{L}(s,t_k)Q_m(s)}{\Lambda(s)} u_p - \dfrac{\hat{P}(s,t_k)}{\Lambda(s)}(y_p - y_m)$
Design variables	A^* monic Hurwitz of degree $2n + q - 1$; $Q_m(s)$ monic of degree q with nonrepeated roots on the $j\omega$ axis; $\Lambda(s) = \Lambda_p(s)\Lambda_q(s)$; $\Lambda_p(s), \Lambda_q(s)$ monic and Hurwitz with degree n, $q-1$, respectively

adaptive laws of the APPC schemes discussed in Section 7.4 as shown in Tables 7.8 to 7.10.

The controller parameters in the hybrid adaptive control schemes of Tables 7.8 to 7.10 are updated at discrete times by solving certain algebraic equations. As in the continuous-time case, the solution of these equations exist provided the estimated polynomials $\hat{R}_p(s, t_k)Q_m(s)$, $\hat{Z}_p(s, t_k)$ are strongly coprime at each time t_k.

The following theorem summarizes the stability properties of the hybrid APPC schemes presented in Tables 7.8 to 7.10.

Theorem 7.5.1 *Assume that the polynomials $\hat{R}_p(s, t_k)Q_m(s)$, $\hat{Z}_p(s, t_k)$ are strongly coprime at each time $t = t_k$. Then the hybrid APPC schemes given*

7.5. HYBRID APPC SCHEMES

Table 7.9 Hybrid APPC scheme: state variable approach

Plant	$y_p = \dfrac{Z_p(s)}{R_p(s)} u_p$
Reference signal	$Q_m(s) y_m = 0$
Adaptive law	Hybrid adaptive law of Table 7.7.
State observer	$\dot{\hat{e}} = \hat{A}_{k-1}\hat{e} + \hat{B}_{k-1}\bar{u}_p - \hat{K}_{o(k-1)}[C^\top \hat{e} - e_1]$ $\hat{A}_{k-1} = \begin{bmatrix} -\theta_{1(k-1)} & \begin{array}{c} I_{n+q-1} \\ \hline 0 \end{array} \end{bmatrix}$ $\hat{B}_{k-1} = \theta_{2(k-1)},\ C^\top = [1, 0, \ldots, 0]$ $\hat{K}_{o(k-1)} = a^* - \theta_{1(k-1)}$ $\theta_{1(k-1)},\ \theta_{2(k-1)}$ are the coefficient vectors of $\hat{R}_p(s, t_k) Q_m(s) - s^{n+q},\ \hat{Z}_p(s, t_k) Q_1(s)$, respectively, a^* is the coefficient vector of $A_o^*(s) - s^{n+q}$
Algebraic equation	Solve for $\hat{K}_{c(k-1)}$ the equation $\det[sI - \hat{A}_{k-1} + \hat{B}_{k-1}\hat{K}_{c(k-1)}] = A_c^*(s)$
Control law	$\bar{u}_p = -\hat{K}_{c(k-1)}\hat{e},\quad u_p = \dfrac{Q_1(s)}{Q_m(s)} \bar{u}_p$
Design variables	Choose $Q_m, Q_1, A_o^*, A_c^*, Q_1$ as in Table 7.5

in Tables 7.8 to 7.10 guarantee signal boundedness and convergence of the tracking error to zero asymptotically with time.

The proof of Theorem 7.5.1 is similar to that of the theorems in Section 7.4, with minor modifications that take into account the discontinuities in the parameters and is given in Section 7.7.

Table 7.10 Hybrid adaptive LQ control scheme

Plant	$y_p = \dfrac{Z_p(s)}{R_p(s)} u_p$
Reference signal	$Q_m(s) y_m = 0$
Adaptive law	Hybrid adaptive law of Table 7.7.
State observer	$\dot{\hat{e}} = \hat{A}_{k-1}\hat{e} + \hat{B}_{k-1}\bar{u}_p - \hat{K}_{o(k-1)}[C^\mathsf{T}\hat{e} - e_1]$ $\hat{K}_{o(k-1)}, \hat{A}_{k-1}, \hat{B}_{k-1}, C$ as in Table 7.9
Riccati equation	Solve for $P_{k-1} = P_{k-1}^\mathsf{T} > 0$ the equation $\hat{A}_{k-1}^\mathsf{T} P_{k-1} + P_{k-1}\hat{A}_{k-1} - \frac{1}{\lambda}P_{k-1}\hat{B}_{k-1}\hat{B}_{k-1}^\mathsf{T} P_{k-1} + CC^\mathsf{T} = 0$
Control law	$\bar{u}_p = -\dfrac{1}{\lambda}\hat{B}_{k-1}^\mathsf{T} P_{k-1}\hat{e}, \quad u_p = \dfrac{Q_1(s)}{Q_m(s)}\bar{u}_p$
Design variables	Choose $\lambda, Q_1(s), Q_m(s)$ as in Table 7.6

The major advantage of the hybrid adaptive control schemes described in Tables 7.7 to 7.10 over their continuous counterparts is the smaller computational effort required during implementation. Another possible advantage is better robustness properties in the presence of measurement noise, since the hybrid scheme does not respond instantaneously to changes in the system, which may be caused by measurement noise.

7.6 Stabilizability Issues and Modified APPC

The main drawbacks of the APPC schemes of Sections 7.4 and 7.5 is that the adaptive law cannot guarantee that the estimated plant parameters or polynomials satisfy the appropriate controllability or stabilizability condition at each time t, which is required to calculate the controller parameter

7.6. STABILIZABILITY ISSUES AND MODIFIED APPC

vector θ_c. Loss of stabilizability or controllability may lead to computational problems and instability.

In this section we concentrate on this problem of the APPC schemes and propose ways to avoid it. We call the estimated plant parameter vector θ_p at time t *stabilizable* if the corresponding algebraic equation is solvable for the controller parameters. Because we are dealing with time-varying estimates, uniformity with respect to time is guaranteed by requiring the *level of stabilizability* to be greater than some constant $\varepsilon^* > 0$. For example, the level of stabilizability can be defined as the absolute value of the determinant of the Sylvester matrix of the estimated plant polynomials.

We start with a simple example that demonstrates the loss of stabilizability that leads to instability.

7.6.1 Loss of Stabilizability: A Simple Example

Let us consider the first order plant

$$\dot{y} = y + bu \tag{7.6.1}$$

where $b \neq 0$ is an unknown constant. The control objective is to choose u such that $y, u \in \mathcal{L}_\infty$, and $y(t) \to 0$ as $t \to \infty$.

If b were known then the control law

$$u = -\frac{2}{b}y \tag{7.6.2}$$

would meet the control objective exactly. When b is unknown, a natural approach to follow is to use the certainty equivalence control (CEC) law

$$u_c = -\frac{2}{\hat{b}}y \tag{7.6.3}$$

where $\hat{b}(t)$ is the estimate of b at time t, generated on-line by an appropriate adaptive law.

Let us consider the following two adaptive laws:

(i) Gradient

$$\dot{\hat{b}} = \gamma \phi \varepsilon, \quad \hat{b}(0) = \hat{b}_0 \neq 0 \tag{7.6.4}$$

where $\gamma > 0$ is the constant adaptive gain.

(ii) Pure Least-Squares

$$\dot{\hat{b}} = P\phi\varepsilon, \quad \hat{b}(0) = \hat{b}_0 \neq 0$$

$$\dot{P} = -P^2 \frac{\phi^2}{1 + \beta_0 \phi^2}, \quad P(0) = p_0 > 0 \qquad (7.6.5)$$

where $P, \phi \in \mathcal{R}^1$,

$$\varepsilon = \frac{z - \hat{b}\phi}{1 + \beta_0 \phi^2}, \quad y_f = \frac{1}{s+1} y, \quad \phi = \frac{1}{s+1} u$$

$$z = \dot{y}_f - y_f \qquad (7.6.6)$$

It can be shown that for $\beta_0 > 0$, the control law (7.6.3) with $\hat{b}$ generated by (7.6.4) or (7.6.5) meets the control objective provided that $\hat{b}(t) \neq 0 \ \forall t \geq 0$.

Let us now examine whether (7.6.4) or (7.6.5) can satisfy the condition $\hat{b}(t) \neq 0, \ \forall t \geq 0$.

From (7.6.1) and (7.6.6), we obtain

$$\varepsilon = -\frac{\tilde{b}\phi}{1 + \beta_0 \phi^2} \qquad (7.6.7)$$

where $\tilde{b} \triangleq \hat{b} - b$ is the parameter error. Using (7.6.7) in (7.6.4), we have

$$\dot{\hat{b}} = -\gamma \frac{\phi^2}{1 + \beta_0 \phi^2} (\hat{b} - b), \quad \hat{b}(0) = \hat{b}_0 \qquad (7.6.8)$$

Similarly, (7.6.5) can be rewritten as

$$\dot{\hat{b}} = -P \frac{\phi^2}{1 + \beta_0 \phi^2} (\hat{b} - b), \quad \hat{b}(0) = \hat{b}_0$$

$$P(t) = \frac{p_0}{1 + p_0 \int_0^t \frac{\phi^2}{1+\beta_0\phi^2} d\tau}, \quad p_0 > 0 \qquad (7.6.9)$$

It is clear from (7.6.8) and (7.6.9) that for $\hat{b}(0) = b$, $\dot{\hat{b}}(t) = 0$ and $\hat{b}(t) = b$, $\forall t \geq 0$; therefore, the control objective can be met exactly with such an initial condition for $\hat{b}$.

If $\phi(t) = 0$ over a nonzero finite time interval, we will have $\dot{\hat{b}} = 0$, $u = y = 0$, which is an equilibrium state (not necessarily stable though) and the control objective is again met.

7.6. STABILIZABILITY ISSUES AND MODIFIED APPC

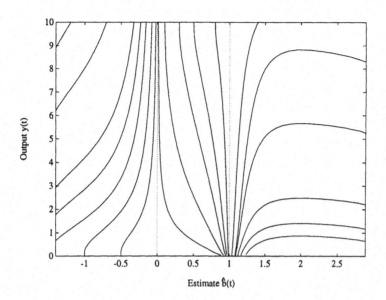

Figure 7.5 Output $y(t)$ versus estimate $\hat{b}(t)$ for different initial conditions $y(0)$ and $\hat{b}(0)$ using the CEC $u_c = -2y/\hat{b}$.

For analysis purposes, let us assume that $b > 0$ (unknown to the designer). For $\phi \neq 0$, both (7.6.8), (7.6.9) imply that

$$\operatorname{sgn}(\dot{\hat{b}}) = -\operatorname{sgn}(\hat{b}(t) - b)$$

and, therefore, for $b > 0$ we have

$$\dot{\hat{b}}(t) > 0 \text{ if } \hat{b}(0) < b \quad \text{and} \quad \dot{\hat{b}}(t) < 0 \text{ if } \hat{b}(0) > b$$

Hence, for $\hat{b}(0) < 0 < b$, $\hat{b}(t)$ is monotonically increasing and crosses zero leading to an unbounded control u_c.

Figure 7.5 shows the plots of $y(t)$ vs $\hat{b}(t)$ for different initial conditions $\hat{b}(0)$, $y(0)$, demonstrating that for $\hat{b}(0) < 0 < b$, $\hat{b}(t)$ crosses zero leading to unbounded closed-loop signals. The value of $b = 1$ is used for this simulation.

The above example demonstrates that the CEC law (7.6.3) with (7.6.4) or (7.6.5) as adaptive laws for generating $\hat{b}$ is not guaranteed to meet the control objective. If the sign of b and a lower bound for $|b|$ are known, then the adaptive laws (7.6.4), (7.6.5) can be modified using projection to constrain $\hat{b}(t)$ from changing sign. This projection approach works for this simple example but its extension to the higher order case is awkward due to the lack of any procedure for constructing the appropriate convex parameter sets for projecting the estimated parameters.

7.6.2 Modified APPC Schemes

The stabilizability problem has attracted considerable interest in the adaptive control community and several solutions have been proposed. We list the most important ones below with a brief explanation regarding their advantages and drawbacks.

(a) Stabilizability is assumed. In this case, no modifications are introduced and stabilizability is assumed to hold for all $t \geq 0$. Even though there is no theoretical justification for such an assumption to hold, it has been often argued that in most simulation studies, no stabilizability problems usually arise. The example presented above illustrates that no stabilizability problem would arise if the initial condition of $\hat{b}(0)$ happens to be in the region $\hat{b}(0) > b$. In the higher order case, loss of stabilizability occurs at certain isolated manifolds in the parameter space when visited by the estimated parameters. Therefore, one can easily argue that the loss of stabilizability is not a frequent phenomenon that occurs in the implementation of APPC schemes.

(b) Parameter projection methods [73, 109, 111]. In this approach, the adaptive laws used to estimate the plant parameters on-line are modified using the gradient projection method. The parameter estimates are constrained to lie inside a convex subset $\mathcal{C}_0$ of the parameter space that is assumed to have the following properties:

(i) The unknown plant parameter vector $\theta_p^* \in \mathcal{C}_0$.

(ii) Every member θ_p of $\mathcal{C}_0$ has a corresponding level of stabilizability greater than ϵ^* for some known constant $\epsilon^* > 0$.

We have already demonstrated this approach for the scalar plant

$$y_p = \frac{b}{s+a} u_p$$

7.6. STABILIZABILITY ISSUES AND MODIFIED APPC

In this case, the estimated polynomials are $s + \hat{a}, \hat{b}$ which, for the APPC schemes of Section 7.4 to be stable, are required to be strongly coprime. This implies that $\hat{b}$ should satisfy $|\hat{b}(t)| \geq b_0$ for some $b_0 > 0$ for all $t \geq 0$. The subset $\mathcal{C}_0$ in this case is defined as

$$\mathcal{C}_0 = \left\{ \hat{b} \in \mathcal{R}^1 | \; \hat{b} \operatorname{sgn}(b) \geq b_0 \right\}$$

where the unknown b is assumed to belong to $\mathcal{C}_0$, i.e., $|b| \geq b_0$. As shown in Examples 7.4.1 and 7.4.2, we guaranteed that $|\hat{b}(t)| \geq b_0$ by using projection to constrain $\hat{b}(t)$ to be inside $\mathcal{C}_0$ $\forall t \geq 0$. This modification requires that b_0 and the $sgn(b)$ are known *a priori*.

Let us now consider the general case of Sections 7.4 and 7.5 where the estimated polynomials $\hat{R}_p(s,t)Q_m(s), \hat{Z}_p(s,t)$ are required to be strongly coprime. This condition implies that the Sylvester matrix $\mathcal{S}_e(\theta_p)$ of $\hat{R}_p(s,t)$ $Q_m(s), \hat{Z}_p(s,t)$ satisfies

$$|\det \mathcal{S}_e(\theta_p)| \geq \epsilon^*$$

where $\theta_p \in \mathcal{R}^{2n}$ is the vector containing the coefficients of $\hat{R}_p(s,t) - s^n$ and $\hat{Z}_p(s,t)$, and $\epsilon^* > 0$ is a constant. If $\epsilon^* > 0$ is chosen so that

$$|\det \mathcal{S}_e(\theta_p^*)| \geq \epsilon^* > 0$$

where $\theta_p^* \in \mathcal{R}^{2n}$ is the corresponding vector with the coefficients of the unknown polynomials $R_p(s), Z_p(s)$, then the subset $\mathcal{C}_0$ may be defined as

$$\mathcal{C}_0 = \text{convex subset of } \mathcal{D} \in \mathcal{R}^{2n} \text{ that contains } \theta_p^*$$

where

$$\mathcal{D} = \left\{ \theta_p \in \mathcal{R}^{2n} | \; |\det \mathcal{S}_e(\theta_p)| \geq \epsilon^* > 0 \right\}$$

Given such a convex set $\mathcal{C}_0$, the stabilizability of the estimated parameters at each time t is ensured by incorporating a projection algorithm in the adaptive law to guarantee that the estimates are in $\mathcal{C}_0, \forall t \geq 0$. The projection is based on the gradient projection method and does not alter the usual properties of the adaptive law that are used in the stability analysis of the overall scheme.

This approach is simple but relies on the rather strong assumption that the set $\mathcal{C}_0$ is known. No procedure has been proposed for constructing such a set $\mathcal{C}_0$ for a general class of plants.

An extension of this approach has been proposed in [146]. It is assumed that a finite number of convex subsets $\mathcal{C}_1, \ldots, \mathcal{C}_p$ are known such that

(i) $\theta_p^* \in \cup_{i=1}^p \mathcal{C}_i$ and the stabilizability degree of the corresponding plant is greater than some known $\varepsilon^* > 0$.

(ii) For every $\theta_p \in \cup_{i=1}^p \mathcal{C}_i$ the corresponding plant model is stabilizable with a stabilizability degree greater than ε^*.

In this case, p adaptive laws with a projection, one for each subset $\mathcal{C}_i$, are used in parallel. A suitable performance index is used to select the adaptive law at each time t whose parameter estimates are to be used to calculate the controller parameters. The price paid in this case is the use of p parallel adaptive laws with projection instead of one. As in the case of a single convex subset, there is no effective procedure for constructing $\mathcal{C}_i$, $i = 1, 2, \ldots, p$, with properties (i) and (ii) in general. The assumption, however, that $\theta_p^* \in \cup_{i=1}^p \mathcal{C}_i$ is weaker.

(c) *Correction Approach* [40]. In this approach, a subset $\mathcal{D}$ in the parameter space is known with the
following properties:

(i) $\theta_p^* \in \mathcal{D}$ and the stabilizability degree of the plant is greater than some known constant $\varepsilon^* > 0$.

(ii) For every $\theta_p \in \mathcal{D}$, the corresponding plant model is stabilizable with a degree greater than ε^*.

Two least-squares estimators with estimates $\hat{\theta}_p, \bar{\theta}_p$ of θ_p^* are run in parallel. The controller parameters are calculated from $\bar{\theta}_p$ as long as $\bar{\theta}_p \in \mathcal{D}$. When $\bar{\theta}_p \notin \mathcal{D}$, $\bar{\theta}_p$ is reinitialized as follows:

$$\bar{\theta}_p = \hat{\theta}_p + P^{1/2}\gamma$$

where P is the covariance matrix for the least-squares estimator of θ_p^*, and γ is a vector chosen so that $\bar{\theta}_p \in \mathcal{D}$. The search for the appropriate γ can be systematic or random.

The drawbacks of this approach are (1) added complexity due to the two parallel estimators, and (2) the search procedure for γ can be tedious and time-consuming. The advantages of this approach, when compared with the projection one, is that the subset $\mathcal{D}$ does not have to be convex. The

7.6. STABILIZABILITY ISSUES AND MODIFIED APPC

importance of this advantage, however, is not clear since no procedure is given as to how to construct $\mathcal{D}$ to satisfy conditions (i), (ii) above.

(d) *Persistent excitation approach* [17, 49]. In this approach, the reference input signal or an external signal is chosen to be sufficiently rich in frequencies so that the signal information vector is PE over an interval. The PE property guarantees that the parameter estimate $\hat{\theta}_p$ of θ_p^* converges exponentially to θ_p^* (provided the covariance matrix in the case of least squares is prevented from becoming singular). Using this PE property, and by assuming that a lower bound $\varepsilon^* > 0$ for the stabilizability degree of the plant is known, the following modification is used: When the stabilizability degree of the estimated plant is greater than ε^*, the controller parameters are computed using $\hat{\theta}_p$; otherwise the controller parameters are frozen to their previous value. Since $\hat{\theta}_p$ converges to θ_p^*, the stabilizability degree of the estimated plant is guaranteed to be greater than ε^* asymptotically with time.

The main drawback of this approach is that the reference signal or external signal has to be on all the time, in addition to being sufficiently rich, which implies that accurate regulation or tracking of signals that are not rich is not possible. Thus the stabilizability problem is overcome at the expense of destroying the desired tracking or regulation properties of the adaptive scheme. Another less serious drawback is that a lower bound $\varepsilon^* > 0$ for the stabilizability degree of the unknown plant is assumed to be known a priori.

An interesting method related to PE is proposed in [114] for the stabilization of unknown plants. In this case the PE property of the signal information vector over an interval is generated by a "rich" nonlinear feedback term that disappears asymptotically with time. The scheme of [114] guarantees exact regulation of the plant output to zero. In contrast to other PE methods [17, 49], both the plant and the controller parameters are estimated on-line leading to a higher order adaptive law.

(e) *The cyclic switching strategy* [166]. In this approach the control input is switched between the CEC law and a finite member of specially constructed controllers with fixed parameters. The fixed controllers have the property that at least one of them makes the resulting closed-loop plant observable through a certain error signal. The switching logic is based on a cyclic switching rule. The proposed scheme does not rely on persistent excitation and does not require the knowledge of a lower bound for the level of stabilizability. One can argue, however, that the concept of PE to help cross the points in the parameter space where stabilizability is weak or lost is implicitly used by the scheme because the switching between different controllers which are not necessarily stabilizing may cause considerable excitation over intervals of time.

Some of the drawbacks of the cyclic switching approach are the complexity and the possible bad transient of the plant or tracking error response during the initial stages of adaptation when switching is active.

(f) *Switched-excitation approach* [63]. This approach is based on the use of an open loop rich excitation signal that is switched on whenever the calculation of the CEC law is not possible due to the loss of stabilizability of the estimated plant. It differs from the PE approach described in (d) in that the switching between the rich external input and the CEC law terminates in finite time after which the CEC law is on and no stabilizability issues arise again. We demonstrate this method in subsection 7.6.3.

Similar methods as above have been proposed for APPC schemes for discrete-time plants [5, 33, 64, 65, 128, 129, 189, 190].

7.6.3 Switched-Excitation Approach

Let us consider the same plant (7.6.1) and control objective as in subsection 7.6.1. Instead of the control law (7.6.3), let us propose the following switching control law

$$u = \begin{cases} -2y/\hat{b} & \text{if } t \in [0, t_1) \cup (t_k + j_k\tau, t_{k+1}) \\ c & \text{if } t \in [t_k, t_k + j_k\tau] \end{cases} \quad (7.6.10)$$

where $c \neq 0$ is a constant.

$$\dot{\hat{b}} = P\phi\varepsilon, \quad \hat{b}(0) = \hat{b}_0 \neq 0 \quad (7.6.11)$$

$$\dot{P} = -\frac{P^2\phi^2}{1 + \beta_0\phi^2}, \quad P(0) = P(t_k) = P(t_k + j\tau) = p_0 I > 0, \quad j = 1, 2, \ldots, j_k$$

$$\beta_0 = \begin{cases} 1 & \text{if } t \in [0, t_1) \cup (t_k + j_k\tau, t_{k+1}) \\ 0 & \text{if } t \in [t_k, t_k + j_k\tau] \end{cases}$$

where ε, ϕ are as defined in (7.6.6), and

7.6. STABILIZABILITY ISSUES AND MODIFIED APPC

- $k = 1, 2, \ldots$

- t_1 is the first time instant for which

$$|\hat{b}(t_1)| = \nu(1)$$

where

$$\nu(k) = \frac{|\hat{b}_0|}{2k} = \frac{\nu(1)}{k}, \quad k = 1, 2, \ldots$$

- t_k is the first time instant after $t = t_{k-1} + j_{k-1}\tau$ for which

$$|\hat{b}(t_k)| = \nu(k + \sum_{i=1}^{k} j_i) = \frac{\nu(1)}{k + \sum_{i=1}^{k} j_i}$$

where $j_k = 1, 2, \ldots$ is the smallest integer for which

$$|\hat{b}(t_k + j_k\tau)| > \nu(k + \sum_{i=1}^{k} j_i)$$

and $\tau > 0$ is a design constant.

Even though the description of the above scheme appears complicated, the intuition behind it is very simple. We start with an initial guess $\hat{b}_0 \neq 0$ for $\hat{b}$ and apply $u = u_c = -2y/\hat{b}$. If $|\hat{b}(t)|$ reaches the chosen threshold $\nu(1) = |\hat{b}_0|/2$, say at $t = t_1$, u switches from $u = u_c$ to $u = u_r = c \neq 0$, where u_r is a rich signal for the plant considered. The signal u_r is applied for an interval τ, where τ is a design constant, and $|\hat{b}(t_1 + \tau)|$ is compared with the new threshold $\nu(2) = \nu(1)/2$. If $|\hat{b}(t_1+\tau)| > \nu(2)$, then we switch back to $u = u_c$ at $t = t_1 + \tau$. We continue with $u = u_c$ unless $|\hat{b}(t_2)| = \nu(3) = \nu(1)/3$ for some finite t_2 in which case we switch back to $u = u_r$. If $|\hat{b}(t_1+\tau)| \leq \nu(2)$, we continue with $u = u_r$ until $t = t_1 + 2\tau$ and check for the condition

$$|\hat{b}(t_1 + 2\tau)| > \nu(3) = \frac{\nu(1)}{3} \tag{7.6.12}$$

If $\hat{b}(t_1 + 2\tau)$ satisfies (7.6.12), we switch to $u = u_c$ and repeat the same procedure by reducing the threshold $\nu(k)$ at each step. If $\hat{b}(t_1 + 2\tau)$ does not satisfy (7.6.12) then we continue with $u = u_r$ for another interval τ and check for

$$|\hat{b}(t_1 + 3\tau)| > \nu(4) = \frac{\nu(1)}{4}$$

and repeat the same procedure. We show that the sequences t_k, $\nu(k)$ converge in a finite number of steps, and therefore $|\hat{b}(t)| > \nu^* > 0$ for some constant ν^* and $u = u_c$ for all t greater than some finite time t^*.

In the sequel, for the sake of simplicity and without loss of generality, we take $c = 1, p_0 \tau = 1$ and adjust the initial condition $\phi(t_k)$ for the filter $\phi = \frac{1}{s+1} u$ to be equal to $\phi(t_k) = 1$ so that $\phi(t) = 1$ for $t \in [t_k, t_k + j_k \tau]$ and $\forall k \geq 1$.

Let us start with a "wrong" initial guess for $\hat{b}(0)$, i.e., assume that $b > 0$ and take $\hat{b}(0) < 0$. Because

$$\dot{\hat{b}} = -P\phi^2(\hat{b} - b)/(1 + \beta_0 \phi^2)$$

and $P(t) > 0$, for any finite time t, we have that $\dot{\hat{b}} \geq 0$ for $\hat{b}(t) < b$ where $\dot{\hat{b}}(t) = 0$ if and only if $\phi(t) = 0$. Because $\phi(t) = 1$ $\forall t \geq 0$, $\hat{b}(t)$, starting from $\hat{b}(0) < 0$, is monotonically increasing. As $\hat{b}(t)$ increases, approaching zero, it satisfies, at some time $t = t_1$,

$$\hat{b}(t_1) = -\nu(1) = -|\hat{b}_0|/2$$

and, therefore, signals the switching of the control law from $u = u_c$ to $u = u_r$, i.e., for $t \geq t_1$, we have

$$u = u_r = 1$$

$$\dot{P} = -P^2 \phi^2, \quad P(t_1) = p_0 \qquad (7.6.13)$$

$$\dot{\hat{b}} = -P\phi^2(\hat{b} - b) \qquad (7.6.14)$$

The solutions of (7.6.13) and (7.6.14) are given by

$$P(t) = \frac{p_0}{1 + p_0(t - t_1)}, \quad t \geq t_1$$

$$\hat{b}(t) = b + \frac{1}{1 + p_0(t - t_1)}(\hat{b}(t_1) - b), \quad t \geq t_1$$

We now need to monitor $\hat{b}(t)$ at $t = t_1 + j_1 \tau, j_1 = 1, 2, \ldots$ until

$$|\hat{b}(t_1 + j_1 \tau)| = \left| b - \frac{\nu(1) + b}{1 + j_1} \right| > \frac{\nu(1)}{1 + j_1} \quad \text{(because } p_0 \tau = 1\text{)} \qquad (7.6.15)$$

7.6. STABILIZABILITY ISSUES AND MODIFIED APPC

is satisfied for some j_1 and switch to $u = u_c$ at $t = t_1 + j_1\tau$. We have

$$|\hat{b}(t_1 + j_1\tau)| = \frac{bj_1 - \nu(1)}{1 + j_1} \qquad (7.6.16)$$

Let j_1^* be the smallest integer for which $bj_1^* > \nu(1)$. Then, $\forall j_1 \geq j_1^*$, condition (7.6.15) is the same as

$$|\hat{b}(t_1 + j_1\tau)| = \hat{b}(t_1 + j_1\tau) = \frac{bj_1 - \nu(1)}{1 + j_1} > \frac{\nu(1)}{1 + j_1} \qquad (7.6.17)$$

Hence, for $j_1 = 2j_1^*$, (7.6.17) is satisfied, i.e., by applying the rich signal $u = u_r = 1$ for $2j_1^*$ intervals of length τ, $\hat{b}(t)$ passes through zero and exceeds the value of $\nu^* = \nu(1)/(1 + 2j_1^*) > 0$. Because $\dot{\hat{b}}(t) \geq 0$, we have $\hat{b}(t) > \nu^*$, $\forall t \geq t_1 + 2j_1^*\tau$ and therefore $u = u_c = -2y/\hat{b}$, without any further switching.

Figure 7.6 illustrates typical time responses of y, u and $\hat{b}$ when the switched-excitation approach is applied to the first order plant given by (7.6.1). The simulations are performed with $b = 1$ and $\hat{b}(0) = -1.5$. At $t = t_1 \approx 0.2s$, $\hat{b}(t_1) = \nu(1) = \hat{b}(0)/2 = -0.75$, the input $u = u_c = -2y/\hat{b}$ is switched to $u = u_r = 1$ for a period $\tau = 0.25s$. Because at time $t = t_1 + \tau$, $\hat{b}(t_1 + \tau)$ is less than $\nu(2) = \nu(1)/2$, $u = u_r = 1$ is applied for another period τ. Finally, at time $t = t_1 + 2\tau$, $\hat{b}(t) > \nu(3) = \nu(1)/3$, therefore u is switched back to $u = u_c$. Because $\hat{b}(t) > \nu(1)/3$, $\forall t \geq 0.7s$, no further switching occurs and the exact regulation of y to zero is achieved.

General Case

The above example may be extended to the general plant (7.3.1) and control objectives defined in the previous subsections as follows.

Let u_c denote the certainly equivalence control law based on any one of the approaches presented in Section 7.4. Let

$$C_d(\theta_p) \triangleq \det|\mathcal{S}_e(\theta_p)|$$

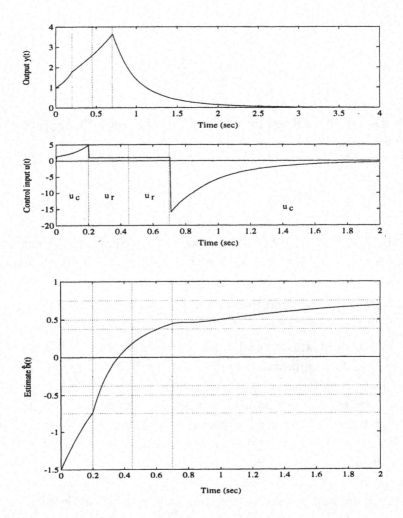

Figure 7.6 The time response of the output $y(t)$, control input $u(t)$ and estimate $\hat{b}(t)$ for the switched-excitation approach.

where $\mathcal{S}_e(\theta_p)$ denotes the Sylvester matrix of the estimated polynomials $\hat{R}_p(s,t)\ Q_m(s)$, $\hat{Z}_p(s,t)$ and $\theta_p(t)$ is the vector with the plant parameter estimates.

Following the scalar example, we propose the modified control law

$$u = \begin{cases} u_c(t) & \text{if } t \in [0, t_1) \cup (t_k + j_k\tau, t_{k+1}) \\ u_r(t) & \text{if } t \in [t_k, t_k + j_k\tau] \end{cases} \quad (7.6.18)$$

7.6. STABILIZABILITY ISSUES AND MODIFIED APPC

Adaptive Law

$$\dot{\theta}_p = P\varepsilon\phi, \quad \theta_p(0) = \theta_0 \qquad (7.6.19)$$

$$\dot{P} = -\frac{P\phi\phi^T P}{1 + \beta_0 \phi^T \phi}, \quad P(0) = P_0 = P_0^T > 0$$

where

$$\varepsilon = (z - \theta_p^T \phi)/(1 + \beta_0 \phi^T \phi)$$

and θ_0 is chosen so that $C_d(\theta_0) > 0$. Furthermore,

$$P(0) = P(t_k) = P(t_k + j\tau) = k_0^{-1} I, \ j = 1, 2, \ldots, j_k \qquad (7.6.20)$$

where $k_0 = $ constant > 0, and

$$\beta_0 = \begin{cases} 1 & \text{if } t \in [0, t_1) \cup (t_k + j_k \tau, t_{k+1}) \\ 0 & \text{if } t \in [t_k, t_k + j_k \tau] \end{cases} \qquad (7.6.21)$$

where
- $k = 1, 2, \ldots$
- t_1 is the first time instant for which $C_d(\theta_p(t_1)) = \nu(1) > 0$, where

$$\nu(k) = \frac{C_d(\theta_0)}{2k} = \frac{\nu(1)}{k}, \ k = 1, 2, \ldots \qquad (7.6.22)$$

- t_k ($k \geq 2$) is the first time instant after $t = t_{k-1} + j_{k-1}\tau$ for which

$$C_d(\theta_p(t_k)) = \nu\left(k + \sum_{i=1}^{k-1} j_i\right) = \frac{\nu(1)}{(k + \sum_{i=1}^{k-1} j_i)} \qquad (7.6.23)$$

and $j_k = 1, 2, \ldots$ is the smallest integer for which

$$C_d(\theta_p(t_k + j_k \tau)) > \nu\left(k + \sum_{i=1}^{k} j_i\right) \qquad (7.6.24)$$

where $\tau > 0$ is a design constant.
- $u_c(t)$ is the certainty equivalence control given in Section 7.4.
- $u_r(t)$ is any bounded stationary signal which is sufficiently rich of order $2n$. For example, one can choose

$$u_r(t) = \sum_{i=1}^{n} A_i \sin\omega_i t$$

where $A_i \neq 0$, $i = 1, \ldots, n$, and $\omega_i \neq \omega_j$ for $i \neq j$.

From (7.6.18), we see that in the time intervals $(t_k + j_k \tau, t_{k+1})$, the stabilizability degree $C_d(\theta_p(t))$ is above the threshold $\nu(k + \sum_{i=1}^{k} j_i)$ and the adaptive control system includes a normalized least-squares estimator and a pole placement controller. In the time intervals $[t_{k+1}, t_{k+1} + j_{k+1}\tau]$, the control input is equal to an external exciting input $u_r(t)$ and the parameter vector estimate $\theta_p(t)$ is generated by an unnormalized least-squares estimator.

The switching (at time $t = t_k$) from the pole placement control u_c to the external rich signal u_r occurs when the stabilizability degree $C_d(\theta_p(t))$ of the estimated model reaches the threshold $\nu(k+\sum_{i=1}^{k} j_i)$. We keep applying $u = u_r$ during successive time intervals of fixed length τ, until time $t = t_k + j_k\tau$ for which the condition $C_d(\theta_p(t_k + j_k\tau)) > \nu(k + \sum_{i=1}^{k} j_i)$ is satisfied and u switches back to the CEC law. The idea behind this approach is that when the estimated model is stabilizable, the control objective is pole placement and closed-loop stabilization; but when the estimation starts to deteriorate, the control priority becomes the "improvement" of the quality of estimation, so that the estimated parameters can cross the hypersurfaces that contain the points where $C_d(\theta_p)$ is close to zero.

The following theorem establishes the stability properties of the proposed adaptive pole placement scheme.

Theorem 7.6.1 *All the signals in the closed-loop (7.3.1) and (7.6.18) to (7.6.24) are bounded and the tracking error converges to zero as $t \to \infty$. Furthermore, there exist finite constants $\nu^*, T^* > 0$ such that for $t \geq T^*$, we have $C_d(\theta_p(t)) \geq \nu^*$ and $u(t) = u_c$.*

The proof of Theorem 7.6.1 is rather long and can be found in [63].

The design of the switching logic in the above modified controllers is based on a simple and intuitive idea that when the quality of parameter estimation is "poor," the objective changes from pole placement to parameter identification. Parameter identification is aided by an external open-loop sufficiently rich signal that is kept on until the quality of the parameter estimates is acceptable for control design purposes. One of the advantages of the switched-excitation algorithm is that it is intuitive and easy to implement. It may suffer, however, from the same drawbacks as other switching algorithms, that is, the transient performance may be poor during switching.

7.7. STABILITY PROOFS

The adaptive control scheme (7.6.18) to (7.6.24) may be simplified when a lower bound $\nu^* > 0$ for $C_d(\theta_p^*)$ is known. In this case, $\nu(k) = \nu^* \; \forall k$. In the proposed scheme, the sequence $\nu(k)$ converges to ν^* and therefore the lower bound for $C_d(\theta_p^*)$ is also identified. The idea behind the identification of ν^* is due to [189] where very similar to the switched-excitation approach methods are used to solve the stabilizability problem of APPC for discrete-time plants.

7.7 Stability Proofs

In this section we present all the long proofs of the theorems of the previous subsections. In most cases, these proofs follow directly from those already presented for the simple examples and are repeated for the sake of completeness.

7.7.1 Proof of Theorem 7.4.1

Step 1. *Let us start by establishing the expressions (7.4.24).* We rewrite the control law (7.4.23) and the normalized estimation error as

$$\hat{L}Q_m \frac{1}{\Lambda} u_p = -\hat{P}\frac{1}{\Lambda}(y_p - y_m) \tag{7.7.1}$$

$$\epsilon m^2 = z - \theta_p^\top \phi = \hat{R}_p \frac{1}{\Lambda_p} y_p - \hat{Z}_p \frac{1}{\Lambda_p} u_p \tag{7.7.2}$$

where $\Lambda(s), \Lambda_p(s)$ are monic, Hurwitz polynomials of degree $n+q-1$, n, respectively, $\hat{R}_p = s^n + \theta_a^\top \alpha_{n-1}(s)$, $\hat{Z}_p = \theta_b^\top \alpha_{n-1}(s)$. From Table 7.4, we have $\Lambda(s) = \Lambda_p(s)\Lambda_q(s)$, where $\Lambda_q(s)$ is a monic Hurwitz polynomial of degree $q-1$. This choice of Λ simplifies the proof. We should point out that the same analysis can also be carried out with Λ, Λ_p being Hurwitz but otherwise arbitrary, at the expense of some additional algebra.

Let us define

$$u_f \triangleq \frac{1}{\Lambda} u_p, \quad y_f \triangleq \frac{1}{\Lambda} y_p$$

and write (7.7.1), (7.7.2) as

$$\hat{P} y_f + \hat{L} Q_m u_f = y_{m1}, \quad \hat{R}_p \Lambda_q y_f - \hat{Z}_p \Lambda_q u_f = \epsilon m^2 \tag{7.7.3}$$

where $y_{m1} \triangleq \hat{P}\frac{1}{\Lambda} y_m \in \mathcal{L}_\infty$. By expressing the polynomials $\hat{R}_p(s)\Lambda_q(s)$, $\hat{Z}_p(s)\Lambda_q(s)$, $\hat{P}(s), \hat{L}(s)Q_m(s)$ as

$$\hat{R}_p(s)\Lambda_q(s) = s^{n+q-1} + \bar{\theta}_1^\top \alpha_{n+q-2}(s), \quad \hat{Z}_p(s)\Lambda_q(s) = \bar{\theta}_2^\top \alpha_{n+q-2}(s)$$

$$\hat{P}(s) = p_0 s^{n+q-1} + \bar{p}^T \alpha_{n+q-2}(s), \quad \hat{L}(s)Q_m(s) = s^{n+q-1} + \bar{l}^T \alpha_{n+q-2}(s)$$

we can rewrite (7.7.3) in the form of

$$\begin{aligned}
y_f^{(n+q-1)} &= -\bar{\theta}_1^T \alpha_{n+q-2}(s) y_f + \bar{\theta}_2^T \alpha_{n+q-2}(s) u_f + \epsilon m^2 \\
u_f^{(n+q-1)} &= (p_0 \bar{\theta}_1 - \bar{p})^T \alpha_{n+q-2}(s) y_f - (p_0 \bar{\theta}_2 + \bar{l})^T \alpha_{n+q-2}(s) u_f \\
&\quad - p_0 \epsilon m^2 + y_{m1}
\end{aligned} \qquad (7.7.4)$$

where the second equation is obtained from (7.7.3) by substituting for $y_f^{(n+q-1)}$. Defining the state $x \triangleq \left[y_f^{(n+q-2)}, \ldots, \dot{y}_f, y_f, u_f^{(n+q-2)}, \ldots, \dot{u}_f, u_f \right]^T$, we obtain

$$\dot{x} = A(t)x + b_1(t)\epsilon m^2 + b_2 y_{m1} \qquad (7.7.5)$$

where

$$A(t) = \begin{bmatrix} -\bar{\theta}_1^T & | & \bar{\theta}_2^T \\ \hline & 0 & & \\ I_{n+q-2} & \vdots & | & O_{(n+q-2)\times(n+q-1)} \\ & 0 & & \\ \hline p_0 \bar{\theta}_1^T - \bar{p}^T & | & -p_0 \bar{\theta}_2^T - \bar{l}^T \\ \hline & & & 0 \\ O_{(n+q-2)\times(n+q-1)} & | & I_{n+q-2} & \vdots \\ & & & 0 \end{bmatrix}, \quad b_1(t) = \begin{bmatrix} 1 \\ \vdots \\ 0 \\ -p_0 \\ 0 \\ \vdots \\ 0 \end{bmatrix}, \quad b_2 = \begin{bmatrix} 0 \\ \vdots \\ 0 \\ 1 \\ 0 \\ \vdots \\ 0 \end{bmatrix}$$

$O_{(n+q-2)\times(n+q-1)}$ is an $(n+q-2)$ by $(n+q-1)$ matrix with all elements equal to zero. Now, because $u_p = \Lambda u_f = u_f^{(n+q-1)} + \lambda^T \alpha_{n+q-2}(s) u_f$, $y_p = \Lambda y_f = y_f^{(n+q-1)} + \lambda^T \alpha_{n+q-2}(s) y_f$ where λ is the coefficient vector of $\Lambda(s) - s^{n+q-1}$, we have

$$\begin{aligned}
u_p &= \underbrace{[0,\ldots,0,}_{n+q-1} \underbrace{1,0,\ldots,0]}_{n+q-1} \dot{x} + [0,\ldots,0, \underbrace{\lambda^T]}_{n+q-1} x \\
y_p &= \underbrace{[1,0,\ldots,0,}_{n+q-1} \underbrace{0,0,\ldots,0]}_{n+q-1} \dot{x} + [\lambda^T, \underbrace{0,\ldots,0]}_{n+q-1} x
\end{aligned} \qquad (7.7.6)$$

Step 2. *Establish the e.s. of the homogeneous part of (7.7.5)*. Because $\hat{P}, \hat{L}$ satisfy the Diophantine equation (7.4.21), we can show that for each frozen time[1] t,

$$\det(sI - A(t)) = \hat{R}_p \Lambda_q \cdot \hat{L}Q_m + \hat{P} \cdot \bar{Z}_p \Lambda_q = A^* \Lambda_q \qquad (7.7.7)$$

---

[1] $X \cdot Y$ denotes the algebraic product of two polynomials that may have time-varying coefficients.

7.7. STABILITY PROOFS

i.e., $A(t)$ is a stable matrix for each frozen time t. One way to verify (7.7.7) is to consider (7.7.3) with the coefficients of $\hat{R}_p, \hat{Z}_p, \hat{P}, \hat{L}$ frozen at each time t. It follows from (7.7.3) that

$$y_f = \frac{1}{\hat{R}_p \Lambda_q \cdot \hat{L} Q_m + \hat{P} \cdot \hat{Z}_p \Lambda_q}(\hat{L} Q_m \epsilon m^2 + \hat{Z}_p \Lambda_q y_{m1})$$

whose state space realization is given by (7.7.5). Because

$$\hat{R}_p \Lambda_q \cdot \hat{L} Q_m + \hat{P} \cdot \hat{Z}_p \Lambda_q = A^* \Lambda_q$$

(7.7.7) follows.

We now need to show that $\|\dot{A}(t)\| \in \mathcal{L}_2, \|A(t)\| \in \mathcal{L}_\infty$ from the properties $\theta_p \in \mathcal{L}_\infty$ and $\dot{\theta}_p \in \mathcal{L}_2$ which are guaranteed by the adaptive law of Table 7.4. Using the assumption that the polynomials $\hat{R}_p Q_m$ and $\hat{Z}_p$ are strongly coprime at each time t, we conclude that the Sylvester matrix $\hat{S}_l$ (defined in Table 7.4) is uniformly nonsingular, i.e., $|\det(\hat{S}_l)| > \nu_0$ for some $\nu_0 > 0$, and thus $\theta_p \in \mathcal{L}_\infty$ implies that $S_l, S_l^{-1} \in \mathcal{L}_\infty$. Therefore, the solution $\hat{\beta}_l$ of the algebraic equation $\hat{S}_l \hat{\beta}_l = \alpha_l^*$ which can be expressed as

$$\hat{\beta}_l = \hat{S}_l^{-1} \alpha_l^*$$

is u.b. On the other hand, because $\theta_p \in \mathcal{L}_\infty$ and $\dot{\theta}_p \in \mathcal{L}_2$, it follows from the definition of the Sylvester matrix that $\|\dot{\hat{S}}_l(t)\| \in \mathcal{L}_2$. Noting that

$$\dot{\hat{\beta}}_l = -\hat{S}_l^{-1} \dot{\hat{S}}_l \hat{S}_l^{-1} \alpha_l^*$$

we have $\dot{\hat{\beta}}_l \in \mathcal{L}_2$ which is implied by $\hat{S}_l, \hat{S}_l^{-1} \in \mathcal{L}_\infty$ and $\|\dot{\hat{S}}_l(t)\| \in \mathcal{L}_2$.

Because the vectors $\bar{\theta}_1, \bar{\theta}_2, \bar{p}, \bar{l}$ are linear combinations of $\theta_p, \hat{\beta}_l$ and all elements in $A(t)$ are uniformly bounded, we have

$$\|\dot{A}(t)\| \leq c(|\dot{\hat{\beta}}_l(t)| + |\dot{\theta}_p(t)|)$$

which implies that $\|\dot{A}(t)\| \in \mathcal{L}_2$. Using Theorem 3.4.11, it follows that the homogeneous part of (7.7.5) is e.s.

Step 3. *Use the properties of the $\mathcal{L}_{2\delta}$ norm and B-G Lemma to establish boundedness.* As before, for clarity of presentation, we denote the $\mathcal{L}_{2\delta}$ norm as $\|\cdot\|$, then from Lemma 3.3.3 and (7.7.5) we have

$$\|x\| \leq c\|\epsilon m^2\| + c, \quad |x(t)| \leq c\|\epsilon m^2\| + c \qquad (7.7.8)$$

for some $\delta > 0$. Defining $m_f^2 \triangleq 1 + \|y_p\|^2 + \|u_p\|^2$, it follows from (7.7.5), (7.7.6), (7.7.8) and Lemma 3.3.2 that $\phi, m \leq m_f$ and

$$m_f^2 \leq c + c\|x\|^2 + c\|\epsilon m^2\|^2 \leq c\|\epsilon m m_f\|^2 + c \qquad (7.7.9)$$

i.e.,
$$m_f^2 \leq c + c \int_0^t e^{-\delta(t-\tau)} \epsilon^2(\tau) m^2(\tau) m_f^2(\tau) d\tau$$

Because $\epsilon m \in \mathcal{L}_2$, guaranteed by the adaptive law, the boundedness of m_f can be established by applying the B-G Lemma. The boundedness of the rest of the signals follows from $m_f \in \mathcal{L}_\infty$ and the properties of the $\mathcal{L}_{2\delta}$ norm that is used to show that m_f bounds most of the signals from above.

Step 4. *Establish that the tracking error converges to zero.* The tracking properties of the APPC scheme are established by manipulating (7.7.1) and (7.7.2) as follows: From the normalized estimation error equation we have

$$\epsilon m^2 = \hat{R}_p \frac{1}{\Lambda_p} y_p - \hat{Z}_p \frac{1}{\Lambda_p} u_p$$

Filtering each side of the above equation with $\bar{L} Q_m \frac{1}{\Lambda_q}$ where $\bar{L}(s,t) \triangleq s^{n-1} + \alpha_{n-2}^\top(s) l$ and $l_c = [1, l^\top]^\top$ is the coefficient vector of $\hat{L}(s,t)$, it follows that

$$\bar{L} Q_m \frac{1}{\Lambda_q} (\epsilon m^2) = \bar{L} Q_m \frac{1}{\Lambda_q} \left(\hat{R}_p \frac{1}{\Lambda_p} y_p - \hat{Z}_p \frac{1}{\Lambda_p} u_p \right) \qquad (7.7.10)$$

Noting that $\Lambda = \Lambda_q \Lambda_p$, and applying the Swapping Lemma A.1, we obtain the following equations:

$$\frac{Q_m}{\Lambda_q} \hat{R}_p \frac{1}{\Lambda_p} y_p = \hat{R}_p \frac{Q_m}{\Lambda} y_p + r_1, \quad \frac{Q_m}{\Lambda_q} \left(\hat{Z}_p \frac{1}{\Lambda_p} u_p \right) = \hat{Z}_p \frac{Q_m}{\Lambda} u_p + r_2 \qquad (7.7.11)$$

where

$$r_1 \triangleq W_{c1}(s) \left(\left(W_{b1}(s) \frac{\alpha_{n-1}^\top(s)}{\Lambda_p} y_p \right) \dot{\theta}_a \right), \quad r_2 \triangleq W_{c1}(s) \left(\left(W_{b1}(s) \frac{\alpha_{n-1}^\top(s)}{\Lambda_p} u_p \right) \dot{\theta}_b \right)$$

and W_{c1}, W_{b1} are as defined in Swapping Lemma A.1 with $W = \frac{Q_m}{\Lambda_q}$. Because $u_p, y_p \in \mathcal{L}_\infty$ and $\dot{\theta}_a, \dot{\theta}_b \in \mathcal{L}_2$, it follows that $r_1, r_2 \in \mathcal{L}_2$. Using (7.7.11) in (7.7.10), we have

$$\bar{L} Q_m \frac{1}{\Lambda_q} (\epsilon m^2) = \bar{L} \left(\hat{R}_p \frac{Q_m}{\Lambda} y_p - \hat{Z}_p \frac{Q_m}{\Lambda} u_p + r_1 - r_2 \right) \qquad (7.7.12)$$

Noting that $Q_m y_p = Q_m (e_1 + y_m) = Q_m e_1$, we can write (7.7.12) as

$$\bar{L} Q_m \frac{1}{\Lambda_q} (\epsilon m^2) = \bar{L} \left(\hat{R}_p \frac{Q_m}{\Lambda} e_1 + r_1 - r_2 \right) - \bar{L} \left(\hat{Z}_p \frac{Q_m}{\Lambda} u_p \right) \qquad (7.7.13)$$

7.7. STABILITY PROOFS

Applying the Swapping Lemma A.4 (i) to the second term in the right-hand side of (7.7.13), we have

$$\bar{L}\left(\hat{Z}_p \frac{Q_m}{\Lambda} u_p\right) = \bar{Z}_p\left(\hat{L}(s,t)\frac{Q_m}{\Lambda} u_p\right) + r_3 \qquad (7.7.14)$$

(by taking $\frac{Q_m}{\Lambda} u_p = f$ in Lemma A.4 (i)), where $\bar{Z}_p = \alpha_{n-1}^\top(s)\theta_b$ and

$$r_3 = \alpha_{n-2}^\top F(l,\theta_b)\alpha_{n-2}(s)\frac{Q_m}{\Lambda(s)} u_p$$

where $\|F(l,\theta_b)\| \leq c_1|\dot{l}| + c_2|\dot{\theta}_b|$ for some constants $c_1, c_2 > 0$. Because

$$\hat{L}(s,t)Q_m(s) = \hat{L}(s,t) \cdot Q_m(s)$$

we use the control law (7.7.1) to write

$$\hat{L}(s,t)\frac{Q_m(s)}{\Lambda(s)} u_p = \hat{L}(s,t) \cdot \frac{Q_m(s)}{\Lambda(s)} u_p = -\hat{P}\frac{1}{\Lambda(s)} e_1 \qquad (7.7.15)$$

Substituting (7.7.15) in (7.7.14) and then in (7.7.13), we obtain

$$\bar{L}Q_m \frac{1}{\Lambda_q}(\epsilon m^2) = \bar{L}\left(\hat{R}_p \frac{Q_m}{\Lambda(s)} e_1\right) + \bar{Z}_p\left(\hat{P}\frac{1}{\Lambda(s)} e_1\right) + \bar{L}(r_1 - r_2) - r_3 \qquad (7.7.16)$$

According to Swapping Lemma A.4 (ii) (with $f = \frac{1}{\Lambda}e_1$ and $\Lambda_0(s) = 1$), we can write

$$\bar{L}\left(\hat{R}_p \frac{Q_m}{\Lambda(s)} e_1\right) = \overline{\hat{L}(s,t) \cdot \hat{R}_p(s,t)Q_m(s)}\frac{1}{\Lambda(s)} e_1 + r_4$$
$$\bar{Z}_p\left(\hat{P}\frac{1}{\Lambda(s)} e_1\right) = \overline{\hat{Z}_p(s,t) \cdot \hat{P}(s,t)}\frac{1}{\Lambda(s)} e_1 + r_5 \qquad (7.7.17)$$

where

$$r_4 \triangleq \alpha_{n-1}^\top(s)G(s,e_1,l,\theta_a), \quad r_5 \triangleq \alpha_{n-1}^\top(s)G'(s,e_1,\theta_b,p) \qquad (7.7.18)$$

and $G(s,e_1,l,\theta_a), G'(s,e_1,\theta_b,p)$ are defined in the Swapping Lemma A.4 (ii). Because $e_1 \in \mathcal{L}_\infty$ and $\dot{l},\dot{p},\dot{\theta}_a,\dot{\theta}_b \in \mathcal{L}_2$, it follows from the definition of G, G' that $G, G' \in \mathcal{L}_2$. Using (7.7.17) in (7.7.16) we have

$$\bar{L}Q_m \frac{1}{\Lambda_q}(\epsilon m^2) = \left(\overline{\hat{L}(s,t) \cdot \hat{R}_p(s,t)Q_m(s)} + \overline{\hat{Z}_p \cdot \hat{P}(s,t)}\right)\frac{1}{\Lambda(s)} e_1 + \bar{L}(r_1-r_2)-r_3+r_4+r_5 \qquad (7.7.19)$$

In the above equations, we use $\bar{X}(s,t)$ to denote the swapped polynomial of $X(s,t)$, i.e.,

$$X(s,t) \triangleq p_x^\top(t)\alpha_n(s), \quad \bar{X}(s,t) \triangleq \alpha_n^\top(s)p_x(t)$$

Because
$$\hat{L}(s,t) \cdot \hat{R}_p(s,t)Q_m(s) + \hat{Z}_p(s,t) \cdot \hat{P}(s,t) = A^*(s)$$
we have
$$\overline{\hat{L}(s,t) \cdot \hat{R}_p(s,t)Q_m(s) + \hat{Z}_p(s,t) \cdot \hat{P}(s,t)} = \overline{A^*(s)} = A^*(s)$$
where the second equality holds because the coefficients of $A^*(s)$ are constant. Therefore, (7.7.19) can be written as
$$A^*(s)\frac{1}{\Lambda(s)}e_1 = v$$
i.e.,
$$e_1 = \frac{\Lambda(s)}{A^*(s)}v \qquad (7.7.20)$$
where
$$v \triangleq \bar{L}\left(Q_m\frac{1}{\Lambda_q(s)}(\epsilon m^2) - r_1 + r_2\right) + r_3 - r_4 - r_5$$
Because $\bar{L}(s) = \alpha_{n-1}^\mathsf{T}(s)l_c$, $l_c = [1, l^\mathsf{T}]^\mathsf{T}$ we have
$$v = \alpha_{n-1}^\mathsf{T}(s)l_c[\frac{Q_m}{\Lambda_q}(\epsilon m^2) - r_1 + r_2] + r_3 - r_4 - r_5$$
Therefore, it follows from (7.7.20) that
$$e_1 = \frac{\Lambda(s)}{A^*(s)}\alpha_{n-1}^\mathsf{T}(s)l_c\frac{Q_m(s)}{\Lambda_q(s)}(\epsilon m^2) + v_0 \qquad (7.7.21)$$
where
$$v_0 = \frac{\Lambda(s)}{A^*(s)}[\alpha_{n-1}^\mathsf{T}(s)l_c(r_2 - r_1) + \alpha_{n-2}^\mathsf{T}(s)v_1 - \alpha_{n-1}^\mathsf{T}(s)v_2]$$
$$v_1 = F(l, \theta_b)\alpha_{n-2}(s)\frac{Q_m(s)}{\Lambda(s)}u_p$$
$$v_2 = G(s, e_1, l, \theta_a) + G'_{\!*}(s, e_1, p, \theta_b)$$
Because $\frac{\Lambda(s)\alpha_{n-1}^\mathsf{T}(s)}{A^*(s)}$, $\frac{\alpha_{n-2}(s)Q_m(s)}{\Lambda(s)}$ are strictly proper and stable, $l_c \in \mathcal{L}_\infty$, and $r_1, r_2, v_1, v_2 \in \mathcal{L}_2$, it follows from Corollary 3.3.1 that $v_0 \in \mathcal{L}_2$ and $v_0(t) \to 0$ as $t \to \infty$.

Applying the Swapping Lemma A.1 to the first term on the right side of (7.7.21) we have
$$e_1 = l_c^\mathsf{T}\frac{\alpha_{n-1}(s)\Lambda(s)Q_m(s)}{A^*(s)\Lambda_q(s)}(\epsilon m^2) + W_c(s)(W_b(s)(\epsilon m^2))\dot{l}_c + v_0$$

7.7. STABILITY PROOFS

where $W_c(s), W_b(s)$ have strictly proper stable elements. Because $\frac{\alpha_{n-1}Q_m\Lambda}{A^*\Lambda_q}$ is proper, $l_c, \epsilon m \in \mathcal{L}_\infty \cap \mathcal{L}_2$ and $v_0 \in \mathcal{L}_\infty \cap \mathcal{L}_2$, it follows that $e_1 \in \mathcal{L}_\infty \cap \mathcal{L}_2$.

The plant equation $y_p = \frac{Z_p}{R_p}u_p$ assumes a minimal state space representation of the form

$$\dot{Y} = AY + Bu_p$$

$$y_p = C^\top Y$$

where (C, A) is observable due to the coprimeness of Z_p, R_p. Using output injection we have

$$\dot{Y} = (A - KC^\top)Y + Bu_p + Ky_p$$

where K is chosen so that $A_{co} \triangleq A - KC^\top$ is a stable matrix. Because $y_p, u_p \in \mathcal{L}_\infty$ and A_{co} is stable, we have $Y \in \mathcal{L}_\infty$, which implies that $\dot{Y}, \dot{y}_p \in \mathcal{L}_\infty$. Therefore, $\dot{e}_1 = \dot{y}_p - \dot{y}_m \in \mathcal{L}_\infty$, which, together with $e_1 \in \mathcal{L}_2$ and Lemma 3.2.5, implies that $e_1(t) \to 0$ as $t \to \infty$.

7.7.2 Proof of Theorem 7.4.2

Step 1. *Develop the closed-loop state error equations.* We start by representing the tracking error equation

$$e_1 = \frac{Z_p Q_1}{R_p Q_m}\bar{u}_p$$

in the following state-space form

$$\dot{e} = \begin{bmatrix} -\theta_1^* & I_{n+q-1} \\ & ---- \\ & 0 \end{bmatrix} e + \theta_2^* \bar{u}_p, \quad e \in \mathcal{R}^{n+q} \quad (7.7.22)$$

$$e_1 = C^\top e$$

where $C^\top = [1, 0, \ldots, 0] \in \mathcal{R}^{n+q}$ and θ_1^*, θ_2^* are the coefficient vectors of $R_p Q_m - s^{n+q}, Z_p Q_1$, respectively.

Let $e_o \triangleq e - \hat{e}$ be the state observation error. Then from the equation for $\hat{e}$ in Table 7.5 and (7.7.22), we have

$$\begin{aligned} \dot{\hat{e}} &= A_c(t)\hat{e} + \hat{K}_o C^\top e_o \\ \dot{e}_o &= A_o e_o + \tilde{\theta}_1 e_1 - \tilde{\theta}_2 \bar{u}_p \end{aligned} \quad (7.7.23)$$

where

$$A_o \triangleq \begin{bmatrix} -a^* & I_{n+q-1} \\ & ---- \\ & 0 \end{bmatrix}$$

is a stable matrix; $A_c(t) \triangleq \hat{A} - \hat{B}\hat{K}_c$ and $\tilde{\theta}_1 \triangleq \theta_1 - \theta_1^*$, $\tilde{\theta}_2 \triangleq \theta_2 - \theta_2^*$. The plant input and output satisfy

$$y_p = C^\top e_o + C^\top \hat{e} + y_m$$
$$u_p = \frac{R_p X Q_1}{A^*}\bar{u}_p + \frac{Q_1 Y R_p}{A^*}y_p \qquad (7.7.24)$$

where $X(s), Y(s)$ are polynomials of degree $n+q-1$ that satisfy (7.3.25) and $A^*(s)$ is a Hurwitz polynomial of degree $2(n+q)-1$. Equation (7.7.24) for u_p is established in the proof of Theorem 7.3.1 and is used here without proof.

Step 2. *Establish e.s. for the homogeneous part of (7.7.23).* Let us first examine the stability properties of $A_c(t)$ in (7.7.23). For each frozen time t, we have

$$\det(sI - A_c) = \det(sI - \hat{A} + \hat{B}\hat{K}_c) = A_c^*(s) \qquad (7.7.25)$$

i.e., $A_c(t)$ is stable at each frozen time t. If $\hat{Z}_p(s,t)Q_1(s)$, $\hat{R}_p(s,t)Q_m(s)$ are strongly coprime, i.e., $(\hat{A}, \hat{B})$ is strongly controllable at each time t, then the controller gains $\hat{K}_c$ may be calculated at each time t using Ackermann's formula [95], i.e.,

$$\hat{K}_c = [0, 0, \ldots, 0, 1]\mathcal{G}_c^{-1} A_c^*(\hat{A})$$

where

$$\mathcal{G}_c \triangleq [\hat{B}, \hat{A}\hat{B}, \ldots, \hat{A}^{n+q-1}\hat{B}]$$

is the controllability matrix of the pair $(\hat{A}, \hat{B})$. Because $(\hat{A}, \hat{B})$ is assumed to be strongly controllable and $\hat{A}, \hat{B} \in \mathcal{L}_\infty$ due to $\theta_p \in \mathcal{L}_\infty$, we have $\hat{K}_c \in \mathcal{L}_\infty$. Now,

$$\dot{\hat{K}}_c = [0, 0, \ldots, 0, 1]\left\{-\mathcal{G}_c^{-1}\dot{\mathcal{G}}_c \mathcal{G}_c^{-1} A_c^*(\hat{A}) + \mathcal{G}_c^{-1}\frac{d}{dt}A_c^*(\hat{A})\right\}$$

Because $\theta_p \in \mathcal{L}_\infty$ and $\dot{\theta}_p \in \mathcal{L}_2$, it follows that $\|\dot{\hat{K}}_c(t)\| \in \mathcal{L}_2$, which, in turn, implies that $\|\dot{A}_c(t)\| \in \mathcal{L}_2$. From A_c being pointwise stable and $\|\dot{A}_c(t)\| \in \mathcal{L}_2$, we have that $A_c(t)$ is a u.a.s matrix by applying Theorem 3.4.11. If $\hat{Z}_p(s,t)Q_1(s)$, $\hat{R}_p(s,t)Q_m(s)$ are not strongly coprime but $\hat{Z}_p(s,t), \hat{R}_p(s,t)Q_m(s)$ are, the boundedness of $\hat{K}_c$ and $\|\dot{\hat{K}}_c(t)\| \in \mathcal{L}_2$ can still be established by decomposing $(\hat{A}, \hat{B})$ into the strongly controllable and the stable uncontrollable or weakly controllable parts and using the results in [95] to obtain an expression for $\hat{K}_c$. Because A_o is a stable matrix the homogeneous part of (7.7.23) is e.s.

Step 3. *Use the properties of the $\mathcal{L}_{2\delta}$ norm and the B-G Lemma to establish boundedness.* As in Example 7.4.2., we apply Lemmas 3.3.3, 3.3.2 to (7.7.23) and (7.7.24), respectively, to obtain

$$\begin{aligned}\|\hat{e}\| &\leq c\|C^\top e_o\| \\ \|y_p\| &\leq c\|C^\top e_o\| + c\|\hat{e}\| + c \leq c\|C^\top e_o\| + c \\ \|u_p\| &\leq c\|\hat{e}\| + c\|y_p\| \leq c\|C^\top e_o\| + c\end{aligned} \qquad (7.7.26)$$

7.7. STABILITY PROOFS

where $\|\cdot\|$ denotes the $\mathcal{L}_{2\delta}$ norm for some $\delta > 0$.

We relate the term $C^\top e_o$ with the estimation error by using (7.7.23) to express $C^\top e_o$ as

$$C^\top e_o = C^\top (sI - A_o)^{-1}(\tilde{\theta}_1 e_1 - \tilde{\theta}_2 \bar{u}_p) \qquad (7.7.27)$$

Noting that (C, A_o) is in the observer canonical form, i.e., $C^\top (sI - A_o)^{-1} = \frac{\alpha_{n+q-1}(s)}{A_o^*(s)}$, we have

$$C^\top e_o = \sum_{i=0}^{\bar{n}} \frac{s^{\bar{n}-i}}{A_o^*}(\tilde{\theta}_{1i} e_1 - \tilde{\theta}_{2i} \bar{u}_p), \quad \bar{n} = n + q - 1$$

where $\tilde{\theta}_i = [\tilde{\theta}_{i1}, \tilde{\theta}_{i2}, \ldots, \tilde{\theta}_{i\bar{n}}]^\top, i = 1, 2$. Applying Swapping Lemma A.1 to each term under the summation, we have

$$\frac{s^{\bar{n}-i}}{A_o^*(s)} \tilde{\theta}_{1i} e_1 = \frac{\Lambda_p(s) Q_1(s)}{A_o^*(s)} \left(\tilde{\theta}_{1i} \frac{s^{\bar{n}-i}}{\Lambda_p(s) Q_1(s)} e_1 + W_{ci}(s)(W_{bi}(s) e_1) \dot{\tilde{\theta}}_{1i} \right)$$

and

$$\frac{s^{\bar{n}-i}}{A_o^*(s)} \tilde{\theta}_{2i} \bar{u}_p = \frac{\Lambda_p(s) Q_1(s)}{A_o^*(s)} \left(\tilde{\theta}_{2i} \frac{s^{\bar{n}-i}}{\Lambda_p(s) Q_1(s)} \bar{u}_p + W_{ci}(s)(W_{bi}(s) \bar{u}_p) \dot{\tilde{\theta}}_{2i} \right)$$

where $W_{ci}, W_{bi}, i = 0, \ldots, n+q-1$ are transfer matrices defined in Lemma A.1 with $W(s) = \frac{s^{\bar{n}-i}}{\Lambda_p(s) Q_1(s)}$. Therefore, $C^\top e_o$ can be expressed as

$$\begin{aligned}
C^\top e_o &= \frac{\Lambda_p(s) Q_1(s)}{A_o^*(s)} \sum_{i=0}^{\bar{n}} \left(\tilde{\theta}_{1i} \frac{s^{\bar{n}-i}}{\Lambda_p(s) Q_1(s)} e_1 - \tilde{\theta}_{2i} \frac{s^{\bar{n}-i}}{\Lambda_p(s) Q_1(s)} \bar{u}_p \right) + r_1 \\
&= \frac{\Lambda_p(s) Q_1(s)}{A_o^*(s)} \left(\tilde{\theta}_1^\top \frac{\alpha_{n+q-1}(s)}{\Lambda_p(s) Q_1(s)} e_1 - \tilde{\theta}_2^\top \frac{\alpha_{n+q-1}(s)}{\Lambda_p(s) Q_1(s)} \bar{u}_p \right) + r_1 \quad (7.7.28)
\end{aligned}$$

where

$$r_1 \triangleq \frac{\Lambda_p(s) Q_1(s)}{A_o^*(s)} \sum_{i=0}^{\bar{n}} W_{ci}(s)[(W_{bi}(s) e_1) \dot{\tilde{\theta}}_{1i} - (W_{bi}(s) \bar{u}_p) \dot{\tilde{\theta}}_{2i}]$$

From the definition of $\tilde{\theta}_1$, we have

$$\begin{aligned}
\tilde{\theta}_1^\top \alpha_{n+q-1}(s) &= \theta_1^\top \alpha_{n+q-1}(s) - \theta_1^{*\top} \alpha_{n+q-1}(s) \\
&= \hat{R}_p(s,t) Q_m(s) - s^{n+q} - R_p(s) Q_m(s) + s^{n+q} \\
&= (\hat{R}_p(s,t) - R_p(s)) Q_m(s) = \tilde{\theta}_a^\top \alpha_{n-1}(s) Q_m(s) \quad (7.7.29)
\end{aligned}$$

where $\tilde{\theta}_a \triangleq \theta_a - \theta_a^*$ is the parameter error. Similarly,

$$\tilde{\theta}_2^\top \alpha_{n+q-1}(s) = \tilde{\theta}_b^\top \alpha_{n-1}(s) Q_1(s) \qquad (7.7.30)$$

where $\tilde{\theta}_b \triangleq \theta_b - \theta_b^*$. Using (7.7.29) and (7.7.30) in (7.7.28), we obtain

$$\begin{aligned} C^T e_o &= \frac{\Lambda_p(s)Q_1(s)}{A_o^*(s)} \left(\tilde{\theta}_a^T \alpha_{n-1}(s) \frac{Q_m(s)}{Q_1(s)} \frac{1}{\Lambda_p(s)} e_1 - \tilde{\theta}_b^T \alpha_{n-1}(s) \frac{1}{\Lambda_p(s)} \bar{u}_p \right) + r_1 \\ &= \frac{\Lambda_p(s)Q_1(s)}{A_o^*(s)} \left(\tilde{\theta}_a^T \alpha_{n-1}(s) \frac{Q_m(s)}{Q_1(s)\Lambda_p(s)} y_p - \tilde{\theta}_b^T \alpha_{n-1}(s) \frac{Q_m(s)}{\Lambda_p(s)Q_1(s)} u_p \right) \\ &\quad + r_1 \end{aligned} \qquad (7.7.31)$$

where the second equality is obtained using

$$Q_m(s)e_1 = Q_m(s)y_p, \quad Q_1(s)\bar{u}_p = Q_m(s)u_p$$

Noting that

$$\alpha_{n-1}(s)\frac{1}{\Lambda_p(s)} u_p = \phi_1, \quad \alpha_{n-1}(s)\frac{1}{\Lambda_p(s)} y_p = -\phi_2$$

we use Swapping Lemma A.1 to obtain the following equalities:

$$\frac{Q_m(s)}{Q_1(s)} \tilde{\theta}_a^T \phi_2 = -\tilde{\theta}_a^T \alpha_{n-1}(s) \frac{Q_m(s)}{Q_1(s)\Lambda_p(s)} y_p + W_{cq} \left(W_{bq}(s)\phi_2^T \right) \dot{\tilde{\theta}}_a$$

$$\frac{Q_m(s)}{Q_1(s)} \tilde{\theta}_b^T \phi_1 = \tilde{\theta}_b^T \alpha_{n-1}(s) \frac{Q_m(s)}{Q_1(s)\Lambda_p(s)} u_p + W_{cq} \left(W_{bq}(s)\phi_1^T \right) \dot{\tilde{\theta}}_b$$

where W_{cq}, W_{bq} are as defined in Swapping Lemma A.1 with $W(s) = \frac{Q_m(s)}{Q_1(s)}$. Using the above equalities in (7.7.31) we obtain

$$C^T e_o = -\frac{\Lambda_p(s)Q_m(s)}{A_o^*(s)} \tilde{\theta}_p^T \phi + r_2 \qquad (7.7.32)$$

where

$$r_2 \triangleq r_1 + \frac{\Lambda_p(s)Q_1(s)}{A_o^*(s)} \left(W_{cq}(s) \left(W_{bq}(s)\phi_1^T \right) \dot{\tilde{\theta}}_b + W_{cq}(s) \left(W_{bq}(s)\phi_2^T \right) \dot{\tilde{\theta}}_a \right)$$

From Table 7.5, the normalized estimation error satisfies the equation

$$\epsilon m^2 = -\tilde{\theta}_p^T \phi$$

which can be used in (7.7.32) to yield

$$C^T e_o = \frac{\Lambda_p(s)Q_m(s)}{A_o^*(s)} \epsilon m^2 + r_2 \qquad (7.7.33)$$

From the definition of $m_f^2 \triangleq 1 + \|u_p\|^2 + \|y_p\|^2$ and Lemma 3.3.2, we can show that m_f is a normalizing signal in the sense that $\phi/m_f, m/m_f \in \mathcal{L}_\infty$ for some

7.7. STABILITY PROOFS

$\delta > 0$. From the expression of r_1, r_2 and the normalizing properties of m_f, we use Lemma 3.3.2 in (7.7.33) and obtain

$$\|C^T e_o\| \leq c\|\epsilon m m_f\| + c\|\dot{\theta}_p m_f\| \qquad (7.7.34)$$

Using (7.7.34) in (7.7.26) and in the definition of m_f, we have the following inequality:

$$m_f^2 \leq c\|\epsilon m m_f\|^2 + \|\dot{\theta}_p m_f\|^2 + c$$

or

$$m_f^2 \leq c\|g m_f\|^2 + c$$

where $g^2 \triangleq \epsilon^2 m^2 + |\dot{\theta}_p|^2$ and $g \in \mathcal{L}_2$, to which we can apply the B-G Lemma to show that $m_f \in \mathcal{L}_\infty$. From $m_f \in \mathcal{L}_\infty$ and the properties of the $\mathcal{L}_{2\delta}$ norm we can establish boundedness for the rest of the signals.

Step 4. *Convergence of the tracking error to zero.* The convergence of the tracking error to zero can be proved using the following arguments: Because all signals are bounded, we can establish that $\frac{d}{dt}(\epsilon m^2) \in \mathcal{L}_\infty$, which, together with $\epsilon m^2 \in \mathcal{L}_2$, implies that $\epsilon(t) m^2(t) \to 0$, and, therefore, $\dot{\theta}_p(t) \to 0$ as $t \to \infty$. From the expressions of r_1, r_2 we can conclude, using Corollary 3.3.1, that $r_2 \in \mathcal{L}_2$ and $r_2(t) \to 0$ as $t \to \infty$. From (7.7.33), i.e.,

$$C^T e_o = \frac{\Lambda_p(s) Q_m(s)}{A_o^*(s)} \epsilon m^2 + r_2$$

and $\epsilon m^2, r_2 \in \mathcal{L}_2$ and $r_2 \to 0$ as $t \to \infty$, it follows from Corollary 3.3.1 that $|C^T e_o| \in \mathcal{L}_2$ and $|C^T e_o(t)| \to 0$ as $t \to \infty$. Because, from (7.7.23), we have

$$\dot{\hat{e}} = A_c(t)\hat{e} + K_o C^T e_o$$

it follows from the u.a.s property of $A_c(t)$ and the fact that $|C^T e_o| \in \mathcal{L}_2$, $|C^T e_o| \to 0$ as $t \to \infty$ that $\hat{e} \to 0$ as $t \to \infty$. From $e_1 \triangleq y_p - y_m = C^T e_o + C^T \hat{e}$ (see (7.7.24)), we conclude that $e_1(t) \to 0$ as $t \to \infty$ and the proof is complete.

7.7.3 Proof of Theorem 7.5.1

We present the stability proof for the hybrid scheme of Table 7.9. The proof of stability for the schemes in Tables 7.8 and 7.10 follows using very similar tools and arguments and is omitted.

First, we establish the properties of the hybrid adaptive law using the following lemma.

Lemma 7.7.1 *Let $\theta_{1k}, \theta_{2k}, \hat{K}_{ok}, \hat{K}_{ck}$ be as defined in Table 7.9 and θ_{pk} as defined in Table 7.7. The hybrid adaptive law of Table 7.7 guarantees the following:*

(i) $\theta_{pk} \in \ell_\infty$ and $\Delta\theta_{pk} \in \ell_2, \varepsilon m \in \mathcal{L}_2$

(ii) $\theta_{1k}, \theta_{2k}, \hat{K}_{ok} \in \ell_\infty$ and $\Delta\theta_{1k}, \Delta\theta_{2k} \in \ell_2$

where $\Delta x_k \triangleq x_{k+1} - x_k; \ k = 0, 1, 2, \ldots$

(iii) If $\hat{R}_p(s, t_k)Q_m(s), \hat{Z}_p(s, t_k)$ are strongly coprime for each $k = 0, 1, \ldots$, then $\hat{K}_{ck} \in \ell_\infty$.

Proof The proof for (i) is given in Section 4.6.

By definition, θ_{1k}, θ_{2k} are the coefficient vectors of $\hat{R}_p(s, t_k)Q_m(s) - s^{n+q}$, $\hat{Z}_p(s, t_k)Q_1(s)$ respectively, where $\hat{R}_p(s, t_k) = s^n + \theta_{ak}^\top \alpha_{n-1}(s)$, $\hat{Z}_p(s, t_k) = \theta_{bk}^\top \alpha_{n-1}(s)$. We can write

$$\theta_{1k} = F_1 \theta_{ak}, \quad \theta_{2k} = F_2 \theta_{bk}$$

where $F_1, F_2 \in \mathcal{R}^{(n+q) \times n}$ are constant matrices which depend only on the coefficients of $Q_1(s), Q_m(s)$, respectively. Therefore, the properties $\theta_{1k}, \theta_{2k} \in \ell_\infty$ and $\Delta\theta_{1k}, \Delta\theta_{2k} \in \ell_2$ follow directly from (i). Because $\hat{K}_{ok} = a^* - \theta_{1k}$, we have $\hat{K}_{ok} \in \ell_\infty$ and the proof of (ii) is complete.

Part (iii) is a direct result of linear system theory. $\square$

Lemma 7.5.1 indicates that the hybrid adaptive law given in Table 7.7 has essentially the same properties as its continuous counterpart. We use Lemma 7.7.1 to prove Theorem 7.5.1 for the adaptive law given in Table 7.9.

As in the continuous time case, the proof is completed in four steps as follows:

Step 1. *Develop the state error equation for the closed-loop APPC scheme.* Because the controller is unchanged for the hybrid adaptive scheme, we follow exactly the same steps as in the proof of Theorem 7.4.2 to obtain the error equations

$$\begin{aligned} \dot{\hat{e}} &= A_{ck}(t)\hat{e} + \hat{K}_{ok}(t)C^\top e_o \\ \dot{e}_o &= A_o e_o + \tilde{\theta}_{1k}(t)e_1 - \tilde{\theta}_{2k}\bar{u}_p \\ y_p &= C^\top e_o + C^\top \hat{e} + y_m \quad (7.7.35) \\ u_p &= W_1(s)\hat{K}_{ck}(t)\hat{e} + W_2(s)y_p \\ \bar{u}_p &= -\hat{K}_{ck}(t)\hat{e} \end{aligned}$$

where $A_{ck}, \hat{K}_{ok}, \hat{K}_{ck}, \tilde{\theta}_{1k}, \tilde{\theta}_{2k}$ are as defined in Section 7.4.3 with $\theta_b(t), \theta_a(t)$ replaced by their discrete versions θ_{bk}, θ_{ak} respectively, i.e., $\hat{K}_{ok} = \alpha^* - \theta_{1k}$ and $\hat{K}_{ck}$ is solved from $det(sI - \hat{A}_k + \hat{B}_k\hat{K}_{ck}) = A_c^*(s)$; A_o is a constant stable matrix, $A_{ck} = \hat{A}_k - \hat{B}_k\hat{K}_{ck}$ and $W_1(s), W_2(s)$ are proper stable transfer functions. Furthermore, $A_{ck}(t), \hat{K}_{ck}(t), \hat{K}_{ok}(t), \tilde{\theta}_{1k}(t), \tilde{\theta}_{2k}(t)$ are piecewise constant functions defined as $f_k(t) \triangleq f_k, \forall t \in [kT_s, (k+1)T_s)$.

7.7. STABILITY PROOFS

Step 2. *Establish e.s. for the homogeneous part of (7.7.35).* Consider the system

$$\dot{z} = \bar{A}_k z, \quad z(t_0) = z_0 \tag{7.7.36}$$

where

$$\bar{A}_k(t) = \begin{bmatrix} A_{ck}(t) & \hat{K}_{ok}(t)C^\top \\ 0 & A_0 \end{bmatrix}, \quad t \in [kT_s, (k+1)T_s)$$

Because $\Delta\theta_{1k}, \Delta\theta_{2k} \in \ell_2$ and $(\hat{A}_k, \hat{B}_k)$ is strongly controllable, one can verify that $\Delta\hat{K}_{ok}, \Delta\hat{K}_{ck} \in \ell_2$ and thus $\Delta A_{ck}, \Delta\bar{A}_k \in \ell_2$. Therefore, for any given small number $\mu > 0$, we can find an integer N_μ such that[2]

$$\|\bar{A}_k - \bar{A}_{N_\mu}\| < \mu, \quad \forall k \geq N_\mu$$

We write (7.7.36) as

$$\dot{z} = \bar{A}_{N_\mu} z + (\bar{A}_k - \bar{A}_{N_\mu}) z, \quad \forall t \geq N_\mu T_s \tag{7.7.37}$$

Because $\bar{A}_{N_\mu}$ is a constant matrix that is stable and $\|\bar{A}_k - \bar{A}_{N_\mu}\| < \mu, \forall k \geq N_\mu$, we can fix μ to be small enough so that the matrix $\bar{A}_k = \bar{A}_{N_\mu} + (\bar{A}_k - \bar{A}_{N_\mu})$ is stable which implies that

$$|z(t)| \leq c_1 e^{-\alpha(t - N_\mu T_s)} |z(N_\mu T_s)|, \quad \forall t \geq N_\mu T_s$$

for some constants $c_1, \alpha > 0$. Because

$$z(N_\mu T_s) = e^{\bar{A}_{N_\mu} T_s} e^{\bar{A}_{N_\mu - 1} T_s} \cdots e^{\bar{A}_1 T_s} e^{\bar{A}_0 T_s} z_0$$

and N_μ is a fixed integer, we have

$$|z(N_\mu T_s)| \leq c_2 e^{-\alpha(N_\mu T_s - t_0)} |z_0|$$

for some constant $c_2 > 0$. Therefore,

$$|z(t)| \leq c_1 c_2 e^{-\alpha(t - t_0)} |z_0|, \quad \forall t \geq N_\mu T_s \tag{7.7.38}$$

where c_1, c_2, α are independent of t_0, z_0.

On the other hand, for $t \in [t_0, N_\mu T_s)$, we have

$$z(t) = e^{\bar{A}_i (t - iT_s)} e^{\bar{A}_{i-1} T_s} \cdots e^{\bar{A}_1 T_s} e^{\bar{A}_0 T_s} z_0 \tag{7.7.39}$$

---

[2] If a sequence $\{f_k\}$ satisfies $\Delta f_k \in \ell_2$, then $\lim_{n \to \infty} \sum_{i=n}^{\infty} \|\Delta f_i\| = 0$, which implies that for every given $\mu > 0$, there exists an integer N_μ that depends on μ such that $\sum_{i=N_\mu}^{k} \|\Delta f_i\| < \mu, \forall k \geq N_\mu$. Therefore, $\|f_k - f_{N_\mu}\| = \left\|\sum_{i=N_\mu}^{k} \Delta f_i\right\| \leq \sum_{i=N_\mu}^{k} \|\Delta f_i\| < \mu$.

where i is the largest integer that satisfies $i \leq \frac{t}{T_s}$ and $i \leq N_\mu$. Because N_μ is a fixed integer, it follows from (7.7.39) that

$$|z(t)| \leq c_3'|z_0| \leq c_3 e^{-\alpha(t-t_0)}|z_0|, \quad \forall t \in [t_0, N_\mu T_s) \qquad (7.7.40)$$

for some constant c_3' and $c_3 \triangleq c_3' e^{\alpha N_\mu T_s}$, which is independent of t_0, z_0. Combining (7.7.38) and (7.7.40), we have

$$|z(t)| \leq c e^{-\alpha(t-t_0)}|z(t_0)|, \quad \forall t \geq t_0$$

for some constants $c, \alpha > 0$ which are independent of t_0, z_0 and therefore (7.7.36), the homogeneous part of (7.7.35), is e.s.

Step 3. *Use the properties of the $\mathcal{L}_{2\delta}$ norm and B-G Lemma to establish signal boundedness.* Following exactly the same procedure as in the proof of Theorem 7.4.2 presented in Section 7.7.2 for Step 3 of the continuous APPC, we can derive

$$\|m_f\|^2 \leq c + c\|C^\top e_o\|^2$$

where $\|\cdot\|$ denotes the $\mathcal{L}_{2\delta}$ norm, and

$$C^\top e_o = C^\top (sI - A_o)^{-1}(\tilde{\theta}_{1k} e_1 - \tilde{\theta}_{2k} \bar{u}_p) \qquad (7.7.41)$$

Because $\Delta \tilde{\theta}_{ak}, \Delta \tilde{\theta}_{bk} \in l_2$, according to Lemma 5.3.1, there exist vectors $\bar{\theta}_a(t), \bar{\theta}_b(t)$ whose elements are continuous functions of time such that $|\bar{\theta}_{ak}(t) - \bar{\theta}_a(t)| \in \mathcal{L}_2$, $|\bar{\theta}_{bk}(t) - \bar{\theta}_b(t)| \in \mathcal{L}_2$ and $\dot{\bar{\theta}}_a(t), \dot{\bar{\theta}}_b(t) \in \mathcal{L}_2$. Let $\bar{\theta}_1, \bar{\theta}_2$ be the vectors calculated from Table 7.9 by replacing θ_{ak}, θ_{bk} with $\bar{\theta}_a, \bar{\theta}_b$ respectively. As we have shown in the proof of Theorem 7.4.2, $\bar{\theta}_1, \bar{\theta}_2$ depend linearly on $\bar{\theta}_a, \bar{\theta}_b$. Therefore, given the fact that $\dot{\bar{\theta}}_a, \dot{\bar{\theta}}_b, |\bar{\theta}_a - \bar{\theta}_{ak}|, |\bar{\theta}_b - \bar{\theta}_{bk}| \in \mathcal{L}_2$, we have $\dot{\bar{\theta}}_1, \dot{\bar{\theta}}_2, |\bar{\theta}_1 - \bar{\theta}_{1k}|, |\bar{\theta}_2 - \bar{\theta}_{2k}| \in \mathcal{L}_2$. Thus, we can write (7.7.41) as

$$C^\top e_o = C^\top (sI - A_o)^{-1}(\bar{\theta}_1 e_1 - \bar{\theta}_2 \bar{u}_p + \bar{e}) \qquad (7.7.42)$$

where $\bar{e} \triangleq (\tilde{\theta}_{1k}(t) - \bar{\theta}_1(t))e_1 - (\tilde{\theta}_{2k}(t) - \bar{\theta}_2(t))\bar{u}_p$. Now, $\bar{\theta}_1, \bar{\theta}_2$ have exactly the same properties as $\tilde{\theta}_1, \tilde{\theta}_2$ in equation (7.7.27). Therefore, we can follow exactly the same procedure given in subsection 7.7.2 (from equation (7.7.27) to (7.7.33)) with the following minor changes:

- Replace $\tilde{\theta}_1, \tilde{\theta}_2$ by $\bar{\theta}_1, \bar{\theta}_2$
- Replace $\tilde{\theta}_a, \tilde{\theta}_b$ by $\bar{\theta}_a, \bar{\theta}_b$
- Add an additional term $C^\top (sI - A_o)^{-1} \bar{e}$ in all equations for $C^\top e_o$
- In the final stage of Step 3, to relate $\bar{\theta}_a^\top \phi_2 - \bar{\theta}_b^\top \phi_1$ with ϵm^2, we write

$$\bar{\theta}_a^\top \phi_2 - \bar{\theta}_b^\top \phi_1 = \tilde{\theta}_{ak}^\top \phi_2 - \tilde{\theta}_{bk}^\top \phi_1 + \bar{e}_2$$

where $\bar{e}_2 \triangleq (\bar{\theta}_a - \tilde{\theta}_{ak})^\top \phi_2 - (\bar{\theta}_b - \tilde{\theta}_{bk})^\top \phi_1$, i.e.,

$$\epsilon m^2 = \tilde{\theta}_{ak}^\top \phi_2 - \tilde{\theta}_{bk}^\top \phi_1 = \bar{\theta}_a^\top \phi_2 - \bar{\theta}_b^\top \phi_1 - \bar{e}_2$$

7.8. PROBLEMS

At the end of this step, we derive the inequality

$$m_f^2 \leq c\|\tilde{g}m_f\|^2 + c$$

where $\tilde{g}^2 \triangleq \epsilon^2 m^2 + |\dot{\bar{\theta}}_p|^2 + |\tilde{\theta}_{pk} - \bar{\theta}_p|^2$, $\bar{\theta}_p = [\bar{\theta}_a^T, \bar{\theta}_b^T]^T$, and $\tilde{g} \in \mathcal{L}_2$. The rest of the proof is then identical to that of Theorem 7.4.2.

Step 4. *Establish tracking error converges to zero.* This is completed by following the same procedure as in the continuous case and is omitted here. □

7.8 Problems

7.1 Consider the regulation problem for the first order plant

$$\dot{x} = ax + bu$$
$$y = x$$

(a) Assume a and b are known and $b \neq 0$. Design a controller using pole placement such that the closed-loop pole is located at -5.

(b) Repeat (a) using LQ control. Determine the value of λ in the cost function so that the closed-loop system has a pole at -5.

(c) Repeat (a) when a is known but b is unknown.

(d) Repeat (a) when b is known but a is unknown.

7.2 For the LQ control problem, the closed-loop poles can be determined from $G(s) = Z_p(s)/R_p(s)$ (the open-loop transfer function) and $\lambda > 0$ (the weighting of the plant input in the quadratic cost function) as follows: Define $F(s) \triangleq R_p(s)R_p(-s) + \lambda^{-1}Z_p(s)Z_p(-s)$. Because $F(s) = F(-s)$, it can be factorized as

$$F(s) = (s + p_1)(s + p_2)\cdots(s + p_n)(-s + p_1)(-s + p_2)\cdots(-s + p_n)$$

where $p_i > 0$. Then, the closed-loop poles of the LQ control are equal to $p_1, p_2, \ldots, p_n$.

(a) Using this property, give a procedure for designing an ALQC without having to solve a Riccati equation.

(b) What are the advantages and disadvantages of this procedure compared with that of the standard ALQC described in Section 7.4.3?

7.3 Consider the speed control system given in Problem 6.2, where a, b, d are assumed unknown.

(a) Design an APPC law to achieve the following performance specifications:
  (i) The time constant of the closed-loop system is less than 2 sec.
  (ii) The steady state error is zero for any constant disturbance d

(b) Design an ALQC law such that

$$J = \int_0^\infty (y^2 + \lambda u^2) dt$$

is minimized. Simulate the closed-loop ALQC system with $a = 0.02, b = 1.3, d = 0.5$ for different values of λ. Comment on your results.

7.4 Consider the following system:

$$y = \frac{\omega_n^2}{s^2 + 2\xi\omega_n s + \omega_n^2} u$$

where the parameter ω_n (the natural frequency) is known, but the damping ratio ξ is unknown. The performance specifications for the closed-loop system are given in terms of the unit step response as follows: (a) the peak overshoot is less than 5% and (b) the settling time is less than 2 sec.

(a) Design an estimation scheme to estimate ξ when ω_n is known.

(b) Design an indirect APPC and analyze the stability properties of the closed-loop system.

7.5 Consider the plant

$$y = \frac{s+b}{s(s+a)} u$$

(a) Design an adaptive law to generate $\hat{a}$ and $\hat{b}$, the estimate of a and b, respectively, on-line.

(b) Design an APPC scheme to stabilize the plant and regulate y to zero.

(c) Discuss the stabilizability condition $\hat{a}$ and $\hat{b}$ have to satisfy at each time t.

(d) What additional assumptions you need to impose on the parameters a and b so that the adaptive algorithm can be modified to guarantee the stabilizability condition? Use these assumptions to propose a modified APPC scheme.

7.6 Repeat Problem 7.3 using a hybrid APPC scheme.

7.7 Solve the MRAC problem given by Problem 6.10 in Chapter 6 using Remark 7.3.1 and an APPC scheme.

7.8 Use Remark 7.3.1 to verify that the MRC law of Section 6.3.2 shown in Figure 6.1 is a special case of the general PPC law given by equations (7.3.3), (7.3.6).

7.8. PROBLEMS

7.9 Establish the stability properties of the hybrid ALQ control scheme given in Table 7.10.

7.10 Establish the stability properties of the hybrid APPC scheme of Table 7.8.

7.11 For $n = 2, q = 1$, show that $A(t)$ defined in (7.7.5) satisfies $det(sI - A(t)) = A^*\Lambda_q$, where A^*, Λ_q are defined in Table 7.4.

Chapter 8

Robust Adaptive Laws

8.1 Introduction

The adaptive laws and control schemes developed and analyzed in Chapters 4 to 7 are based on a plant model that is free of noise, disturbances and unmodeled dynamics. These schemes are to be implemented on actual plants that most likely deviate from the plant models on which their design is based. An actual plant may be infinite dimensional, nonlinear and its measured input and output may be corrupted by noise and external disturbances. The effect of the discrepancies between the plant model and the actual plant on the stability and performance of the schemes of Chapters 4 to 7 may not be known until these schemes are implemented on the actual plant. In this chapter, we take an intermediate step, and examine the stability and robustness of the schemes of Chapters 4 to 7 when applied to more complex plant models that include a class of uncertainties and external disturbances that are likely to be present in the actual plant.

The question of how well an adaptive scheme of the class developed in Chapters 4 to 7 performs in the presence of plant model uncertainties and bounded disturbances was raised in the late 1970s. It was shown, using simple examples, that an adaptive scheme designed for a disturbance free plant model may go unstable in the presence of small disturbances [48]. These examples demonstrated that the adaptive schemes of Chapters 4 to 7 are not robust with respect to external disturbances. This nonrobust behavior of adaptive schemes became a controversial issue in the early 1980s when more examples of instabilities were published demonstrating lack of robustness in the presence of unmodeled dynamics or bounded disturbances [85, 197]. This motivated many researchers to study the mechanisms of instabilities and find ways to counteract them. By the mid-1980s, several new designs and modifications were proposed and analyzed leading to a body of work known as *robust adaptive control*.

The purpose of this chapter is to analyze various instability mechanisms that may arise when the schemes of Chapters 4 to 7 are applied to plant models with uncertainties and propose ways to counteract them.

We start with Section 8.2 where we characterize various plant model uncertainties to be used for testing the stability and robustness of the schemes of Chapters 4 to 7. In Section 8.3, we analyze several instability mechanisms exhibited by the schemes of Chapters 4 to 7, in the presence of external disturbances or unmodeled dynamics. The understanding of these instability phenomena helps the reader understand the various modifications presented in the rest of the chapter. In Section 8.4, we use several techniques to modify the adaptive schemes of Section 8.3 and establish robustness with respect to bounded disturbances and unmodeled dynamics. The examples presented in Sections 8.3 and 8.4 demonstrate that the cause of the nonrobust behavior of the adaptive schemes is the adaptive law that makes the closed-loop plant nonlinear and time varying. The remaining sections are devoted to the development of adaptive laws that are robust with respect to a wide class of plant model uncertainties. We refer to them as *robust adaptive laws*. These robust adaptive laws are combined with control laws to generate robust adaptive control schemes in Chapter 9.

8.2 Plant Uncertainties and Robust Control

The first task of a control engineer in designing a control system is to obtain a mathematical model that describes the actual plant to be controlled. The actual plant, however, may be too complex and its dynamics may not be completely understood. Developing a mathematical model that describes accurately the physical behavior of the plant over an operating range is a challenging task. Even if a detailed mathematical model of the plant is available, such a model may be of high order leading to a complex controller whose implementation may be costly and whose operation may not be well understood. This makes the modeling task even more challenging because the mathematical model of the plant is required to describe accurately the plant as well as be simple enough from the control design point of view. While a simple model leads to a simpler control design, such a design must possess a sufficient degree of robustness with respect to the unmodeled plant characteristics. To study and improve the robustness properties of control designs, we need a characterization of the types of plant uncertainties that are likely to be encountered in practice. Once the plant uncertainties are characterized in some mathematical form, they can be used to analyze the stability and performance properties of controllers designed using simplified plant models but applied to plants with uncertainties.

8.2.1 Unstructured Uncertainties

Let us start with an example of the frequency response of a stable plant. Such a response can be obtained in the form of a Bode diagram by exciting the plant with a sinusoidal input at various frequencies and measuring its steady state output response. A typical frequency response of an actual stable plant with an output y may have the form shown in Figure 8.1.

It is clear that the data obtained for $\omega \geq \omega_m$ are unreliable because at high frequencies the measurements are corrupted by noise, unmodeled high frequency dynamics, etc. For frequencies below ω_m, the data are accurate enough to be used for approximating the plant by a finite-order model. An approximate model for the plant, whose frequency response is shown in Figure 8.1, is a second-order transfer function $G_0(s)$ with one stable zero and two poles, which disregards the phenomena beyond, say $\omega \geq \omega_m$. The modeling error resulting from inaccuracies in the zero-pole locations and high frequency phenomena can be characterized by an upper bound in the frequency domain.

Now let us use the above example to motivate the following relationships between the actual transfer function of the plant denoted by $G(s)$ and the transfer function of the nominal or modeled part of the plant denoted by $G_0(s)$.

Definition 8.2.1 (Additive Perturbations) *Suppose that $G(s)$ and $G_0(s)$ are related by*

$$G(s) = G_0(s) + \Delta_a(s) \qquad (8.2.1)$$

8.2. PLANT UNCERTAINTIES AND ROBUST CONTROL

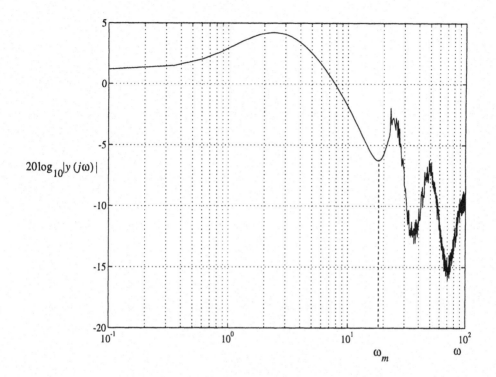

Figure 8.1 An example of a frequency response of a stable plant.

where $\Delta_a(s)$ is stable. Then $\Delta_a(s)$ is called an additive plant perturbation or uncertainty.

The structure of $\Delta_a(s)$ is usually unknown but $\Delta_a(s)$ is assumed to satisfy an upper bound in the frequency domain, i.e.,

$$|\Delta_a(j\omega)| \leq \delta_a(\omega) \quad \forall \omega \qquad (8.2.2)$$

for some known function $\delta_a(\omega)$. In view of (8.2.1) and (8.2.2) defines a family of plants described by

$$\Pi_a = \{G \mid |G(j\omega) - G_0(j\omega)| \leq \delta_a(\omega)\} \qquad (8.2.3)$$

The upper bound $\delta_a(\omega)$ of $\Delta_a(j\omega)$ may be obtained from frequency response experiments. In robust control [231], $G_0(s)$ is known exactly and the uncertainties of the zeros and poles of $G(s)$ are included in $\Delta_a(s)$. In adaptive control, the parameters of $G_0(s)$ are unknown and therefore zero-pole inaccuracies do not have to be included in $\Delta_a(s)$. Because the main topic of the book is adaptive control, we adopt Definition 8.2.1, which requires $\Delta_a(s)$ to be stable.

Definition 8.2.2 (Multiplicative Perturbations) *Let $G(s), G_0(s)$ be related by*

$$G(s) = G_0(s)(1 + \Delta_m(s)) \tag{8.2.4}$$

where $\Delta_m(s)$ is stable. Then $\Delta_m(s)$ is called a multiplicative plant perturbation or uncertainty.

In the case of multiplicative plant perturbations, $\Delta_m(s)$ may be constrained to satisfy an upper bound in the frequency domain, i.e.,

$$|\Delta_m(j\omega)| \leq \delta_m(\omega) \tag{8.2.5}$$

for some known $\delta_m(\omega)$ which may be generated from frequency response experiments. Equations (8.2.4) and (8.2.5) describe a family of plants given by

$$\Pi_m = \left\{ G \,\middle|\, \frac{|G(j\omega) - G_0(j\omega)|}{|G_0(j\omega)|} \leq \delta_m(\omega) \right\} \tag{8.2.6}$$

For the same reason as in the additive perturbation case, we adopt Definition 8.2.2 which requires $\Delta_m(s)$ to be stable instead of the usual definition in robust control where $\Delta_m(s)$ is allowed to be unstable for a certain family of plants.

Definition 8.2.3 (Stable Factor Perturbations) *Let $G(s)$, $G_0(s)$ have the following coprime factorizations [231]:*

$$G(s) = \frac{N_0(s) + \Delta_1(s)}{D_0(s) + \Delta_2(s)}, \quad G_0(s) = \frac{N_0(s)}{D_0(s)} \tag{8.2.7}$$

where N_0 and D_0 are proper stable rational transfer functions that are coprime,[1] and $\Delta_1(s)$ and $\Delta_2(s)$ are stable. Then $\Delta_1(s)$ and $\Delta_2(s)$ are called stable factor plant perturbations.

---

[1]Two proper transfer functions $P(s), Q(s)$ are coprime if and only if they have no finite common zeros in the closed right half s-plane and at least one of them has relative degree zero [231].

8.2. PLANT UNCERTAINTIES AND ROBUST CONTROL

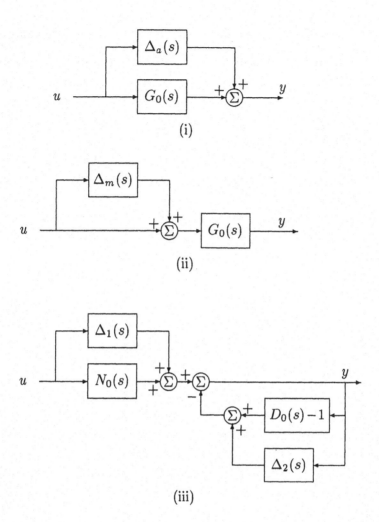

Figure 8.2 Block diagram representations of plant models with (i) additive, (ii) multiplicative, and (iii) stable factor perturbations.

Figure 8.2 shows a block diagram representation of the three types of plant model uncertainties.

The perturbations $\Delta_a(s), \Delta_m(s), \Delta_1(s),$ and $\Delta_2(s)$ defined above with no additional restrictions are usually referred to as *unstructured plant model uncertainties*.

8.2.2 Structured Uncertainties: Singular Perturbations

In many applications, the plant perturbations may have a special form because they may originate from variations of physical parameters or arise because of a deliberate reduction in the complexity of a higher order mathematical model of the plant. Such perturbations are usually referred to as *structured plant model perturbations*.

The knowledge of the structure of plant model uncertainties can be exploited in many control problems to achieve better performance and obtain less conservative results.

An important class of structured plant model perturbations that describe a wide class of plant dynamic uncertainties, such as fast sensor and actuator dynamics, is given by *singular perturbation models* [106].

For a SISO, LTI plant, the following singular perturbation model in the state space form

$$\begin{aligned} \dot{x} &= A_{11}x + A_{12}z + B_1 u, \quad x \in \mathcal{R}^n \\ \mu \dot{z} &= A_{21}x + A_{22}z + B_2 u, \quad z \in \mathcal{R}^m \\ y &= C_1^T x + C_2^T z \end{aligned} \quad (8.2.8)$$

can be used to describe the *slow (or dominant)* and *fast (or parasitic)* phenomena of the plant. The scalar μ represents all the small parameters such as small time constants, small masses, etc., to be neglected. In most applications, the representation (8.2.8) with a single parameter μ can be achieved by proper scaling as shown in [106]. All the matrices in (8.2.8) are assumed to be constant and independent of μ. As explained in [106], this assumption is for convenience only and leads to a minor loss of generality.

The two time scale property of (8.2.8) is evident if we use the change of variables

$$z_f = z + L(\mu)x \quad (8.2.9)$$

where $L(\mu)$ is required to satisfy the algebraic equation

$$A_{21} - A_{22}L + \mu L A_{11} - \mu L A_{12} L = 0 \quad (8.2.10)$$

to transform (8.2.8) into

$$\begin{aligned} \dot{x} &= A_s x + A_{12} z_f + B_1 u \\ \mu \dot{z}_f &= A_f z_f + B_s u \\ y &= C_s^T x + C_2^T z_f \end{aligned} \quad (8.2.11)$$

8.2. PLANT UNCERTAINTIES AND ROBUST CONTROL

where $A_s = A_{11} - A_{12}L$, $A_f = A_{22} + \mu L A_{12}$, $B_s = B_2 + \mu L B_1$, $C_s^T = C_1^T - C_2^T L$. As shown in [106], if A_{22} is nonsingular, then for all $\mu \in [0, \mu^*)$ and some $\mu^* > 0$, a solution of the form

$$L = A_{22}^{-1} A_{21} + O(\mu)$$

satisfying (8.2.10) exists. It is clear that for $u = 0$, i.e.,

$$\begin{aligned} \dot{x} &= A_s x + A_{12} z_f \\ \mu \dot{z}_f &= A_f z_f \end{aligned} \qquad (8.2.12)$$

the eigenvalues of (8.2.12) are equal to those of A_s and A_f/μ, which, for small μ and for A_f nonsingular, are of $O(1)$ and $O(1/\mu)$, respectively[2]. The smaller the value of μ, the wider the distance between the eigenvalues of A_s and A_f/μ, and the greater the separation of time scales. It is clear that if A_f is stable then the smaller the value of μ is, the faster the state variable z_f goes to zero. In the limit as $\mu \to 0$, z_f converges instantaneously, i.e., infinitely fast to zero. Thus for small μ, the effect of stable fast dynamics is reduced considerably after a very short time. Therefore, when A_{22} is stable (which for small μ implies that A_f is stable), a reasonable approximation of (8.2.8) is obtained by setting $\mu = 0$, solving for z from the second equation of (8.2.8) and substituting for its value in the first equation of (8.2.8), i.e.,

$$\begin{aligned} \dot{x}_0 &= A_0 x_0 + B_0 u, \quad x_0 \in \mathcal{R}^n \\ y_0 &= C_0^T x_0 + D_0 u \end{aligned} \qquad (8.2.13)$$

where $A_0 = A_{11} - A_{12} A_{22}^{-1} A_{21}$, $B_0 = B_1 - A_{12} A_{22}^{-1} B_2$, $C_0^T = C_1^T - C_2^T A_{22}^{-1} A_{21}$ and $D_0 = -C_2^T A_{22}^{-1} B_2$. With μ set to zero, the dimension of the state space of (8.2.8) reduces from $n + m$ to n because the differential equation for z in (8.2.8) degenerates into the algebraic equation

$$0 = A_{21} x_0 + A_{22} z_0 + B_2 u$$

i.e.,

$$z_0 = -A_{22}^{-1}(A_{21} x_0 + B_2 u) \qquad (8.2.14)$$

---

[2] We say that a function $f(x)$ is $O(|x|)$ in $\mathcal{D} \subset \mathcal{R}^n$ if there exists a finite constant $c > 0$ such that $|f(x)| \leq c|x|$ for all $x \in \mathcal{D}$ where $x = 0 \in \mathcal{D}$.

where the subscript 0 is used to indicate that the variables belong to the system with $\mu = 0$. The transfer function

$$G_0(s) = C_0^\mathsf{T}(sI - A_0)^{-1}B_0 + D_0 \qquad (8.2.15)$$

represents the nominal or slow or dominant part of the plant.

We should emphasize that even though the transfer function $G(s)$ from u to y of the full-order plant given by (8.2.8) is strictly proper, the nominal part $G_0(s)$ may be biproper because $D_0 = -C_2^\mathsf{T} A_{22}^{-1} B_2$ may not be equal to zero. The situation where the throughput $D_0 = -C_2^\mathsf{T} A_{22}^{-1} B_2$ induced by the fast dynamics is nonzero is referred to as *strongly observable parasitics* [106]. As discussed in [85, 101], if $G_0(s)$ is assumed to be equal to $C_0^\mathsf{T}(sI - A_0)^{-1}B_0$ instead of (8.2.15), the control design based on $G_0(s)$ and applied to the full-order plant with $\mu \geq 0$ may lead to instability. One way to eliminate the effect of strongly controllable and strongly observable parasitics is to augment (8.2.8) with a low pass filter as follows: We pass y through the filter $\frac{f_1}{s+f_0}$ for some $f_1, f_0 > 0$, i.e.,

$$\dot{y}_f = -f_0 y_f + f_1 y \qquad (8.2.16)$$

and augment (8.2.8) with (8.2.16) to obtain the system of order $(n+1+m)$

$$\begin{aligned} \dot{x}_a &= \hat{A}_{11} x_a + \hat{A}_{12} z + \hat{B}_1 u \\ \mu \dot{z} &= \hat{A}_{21} x_a + A_{22} z + B_2 u \\ \hat{y} &= \hat{C}_1^\mathsf{T} x_a \end{aligned} \qquad (8.2.17)$$

where $x_a = [y_f, x^\mathsf{T}]^\mathsf{T}$ and $\hat{A}_{11}, \hat{A}_{12}, \hat{A}_{21}, \hat{B}_1, \hat{C}_1$ are appropriately defined. The nominal transfer function of (8.2.17) is now

$$\hat{G}_0(s) = \hat{C}_0^\mathsf{T}(sI - \hat{A}_0)^{-1}\hat{B}_0$$

which is strictly proper.

Another convenient representation of (8.2.8) is obtained by using the change of variables

$$\eta = z_f + A_f^{-1} B_s u \qquad (8.2.18)$$

i.e., the new state η represents the difference between the state z_f and the "quasi steady" state response of (8.2.11) due to $\mu \neq 0$ obtained by approxi-

8.2. PLANT UNCERTAINTIES AND ROBUST CONTROL

mating $\mu \dot{z}_f \approx 0$. Using (8.2.18), we obtain

$$\begin{aligned} \dot{x} &= A_s x + A_{12}\eta + \bar{B}_s u \\ \mu \dot{\eta} &= A_f \eta + \mu A_f^{-1} B_s \dot{u} \\ y &= C_s^T x + C_2^T \eta + D_s u \end{aligned} \quad (8.2.19)$$

where $\bar{B}_s = B_1 - A_{12} A_f^{-1} B_s$, $D_s = -C_2^T A_f^{-1} B_s$, provided u is differentiable. Because for $|\dot{u}| = O(1)$, the slow component of η is of $O(\mu)$, i.e., at steady state $|\eta| = O(\mu)$, the state η is referred to as the *parasitic state*. It is clear that for $|\dot{u}| = O(1)$ the effect of η on x at steady state is negligible for small μ whereas for $|\dot{u}| \geq O(1/\mu)$, $|\eta|$ is of $O(1)$ at steady state, and its effect on the slow state x may be significant. The effect of $\dot{u}$ and η on x is examined in later sections in the context of robustness of adaptive systems.

8.2.3 Examples of Uncertainty Representations

We illustrate various types of plant uncertainties by the following examples:

Example 8.2.1 Consider the following equations describing the dynamics of a DC motor

$$J\dot{\omega} = k_1 i$$

$$L\frac{di}{dt} = -k_2 \omega - Ri + v$$

where i, v, R and L are the armature current, voltage, resistance, and inductance, respectively; J is the moment of inertia; ω is the angular speed; $k_1 i$ and $k_2 \omega$ are the torque and back e.m.f., respectively.

Defining $x = \omega, z = i$ we have

$$\dot{x} = b_0 z, \quad \mu \dot{z} = -\alpha_2 x - \alpha_1 z + v$$
$$y = x$$

where $b_0 = k_1/J, \alpha_2 = k_2, \alpha_1 = R$ and $\mu = L$, which is in the form of the singular perturbation model (8.2.8). The transfer function between the input v and the output y is given by

$$\frac{y(s)}{v(s)} = \frac{b_0}{\mu s^2 + \alpha_1 s + \alpha_0} = G(s, \mu s)$$

where $\alpha_0 = b_0 \alpha_2$. In most DC motors, the inductance $L = \mu$ is small and can be neglected leading to the reduced order or nominal plant transfer function

$$G_0(s) = \frac{b_0}{\alpha_1 s + \alpha_0}$$

Using $G_0(s)$ as the nominal transfer function, we can express $G(s, \mu s)$ as

$$G(s, \mu s) = G_0(s) + \Delta_a(s, \mu s), \quad \Delta_a(s, \mu s) = -\mu \frac{b_0 s^2}{(\mu s^2 + \alpha_1 s + \alpha_0)(\alpha_1 s + \alpha_0)}$$

where $\Delta_a(s, \mu s)$ is strictly proper and stable since $\mu, \alpha_1, \alpha_0 > 0$ or as

$$G(s, \mu s) = G_0(s)(1 + \Delta_m(s, \mu s)), \quad \Delta_m(s, \mu s) = -\mu \frac{s^2}{\mu s^2 + \alpha_1 s + \alpha_0}$$

where $\Delta_m(s, \mu s)$ is proper and stable.

Let us now use Definition 8.2.3 and express $G(s, \mu s)$ in the form of stable factor perturbations. We write $G(s, \mu s)$ as

$$G(s, \mu s) = \frac{\frac{b_0}{(s+\lambda)} + \Delta_1(s, \mu s)}{\frac{\alpha_1 s + \alpha_0}{(s+\lambda)} + \Delta_2(s, \mu s)}$$

where

$$\Delta_1(s, \mu s) = -\mu \frac{b_0 s}{(\mu s + \alpha_1)(s + \lambda)}, \quad \Delta_2(s, \mu s) = -\mu \frac{\alpha_0 s}{(\mu s + \alpha_1)(s + \lambda)}$$

and $\lambda > 0$ is an arbitrary constant. $\triangledown$

Example 8.2.2 Consider a system with the transfer function

$$G(s) = e^{-\tau s} \frac{1}{s^2}$$

where $\tau > 0$ is a small constant. As a first approximation, we can set $\tau = 0$ and obtain the reduced order or nominal plant transfer function

$$G_0(s) = \frac{1}{s^2}$$

leading to

$$G(s) = G_0(s)(1 + \Delta_m(s))$$

with $\Delta_m(s) = e^{-\tau s} - 1$. Using Definition 8.2.3, we can express $G(s)$ as

$$G(s) = \frac{N_0(s) + \Delta_1(s)}{D_0(s) + \Delta_2(s)}$$

where $N_0(s) = \frac{1}{(s+\lambda)^2}, D_0(s) = \frac{s^2}{(s+\lambda)^2}, \Delta_1(s) = \frac{e^{-\tau s}-1}{(s+\lambda)^2}$, and $\Delta_2(s) = 0$ where $\lambda > 0$ is an arbitrary constant.

It is clear that for small τ, $\Delta_m(s), \Delta_1(s)$ are approximately equal to zero at low frequencies. $\triangledown$

8.2. PLANT UNCERTAINTIES AND ROBUST CONTROL

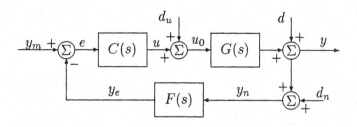

Figure 8.3 General feedback system.

8.2.4 Robust Control

The ultimate goal of any control design is to meet the performance requirements when implemented on the actual plant. In order to meet such a goal, the controller has to be designed to be insensitive, i.e., robust with respect to the class of plant uncertainties that are likely to be encountered in real life. In other words, the robust controller should guarantee closed-loop stability and acceptable performance not only for the nominal plant model but also for a family of plants, which, most likely, include the actual plant.

Let us consider the feedback system of Figure 8.3 where C, F are designed to stabilize the nominal part of the plant model whose transfer function is $G_0(s)$. The transfer function of the actual plant is $G(s)$ and d_u, d, d_n, y_m are external bounded inputs as explained in Section 3.6. The difference between $G(s)$ and $G_0(s)$ is the plant uncertainty that can be any one of the forms described in the previous section. Thus $G(s)$ may represent a family of plants with the same nominal transfer function $G_0(s)$ and plant uncertainty characterized by some upper bound in the frequency domain. We say that *the controller (C, F) is robust with respect to the plant uncertainties in $G(s)$ if, in addition to $G_0(s)$, it also stabilizes $G(s)$*. The property of C, F to stabilize $G(s)$ is referred to as *robust stability*.

The following theorem defines the class of plant uncertainties for which the controller C, F guarantees robust stability.

Theorem 8.2.1 *Let us consider the feedback system of Figure 8.3 where*

$$\begin{aligned} (i) \quad G(s) &= G_0(s) + \Delta_a(s) \\ (ii) \quad G(s) &= G_0(s)(1 + \Delta_m(s)) \end{aligned}$$

$$\text{(iii)} \quad G(s) = \frac{N_0(s) + \Delta_1(s)}{D_0(s) + \Delta_2(s)}, \quad G_0(s) = \frac{N_0(s)}{D_0(s)}$$

where $\Delta_a(s), \Delta_m(s), \Delta_1(s), \Delta_2(s), N_0(s)$, and $D_0(s)$ are as defined in Section 8.2.1 and assume that C, F are designed to internally stabilize the feedback system when $G(s) = G_0(s)$. Then the feedback system with $G(s)$ given by (i), (ii), (iii) is internally stable provided conditions

$$\text{(i)} \quad \left\| \frac{C(s)F(s)}{1 + C(s)F(s)G_0(s)} \right\|_\infty \delta_a(\omega) < 1 \qquad (8.2.20)$$

$$\text{(ii)} \quad \left\| \frac{C(s)F(s)G_0(s)}{1 + C(s)F(s)G_0(s)} \right\|_\infty \delta_m(\omega) < 1 \qquad (8.2.21)$$

$$\text{(iii)} \quad \left\| \frac{\Delta_2(s) + C(s)F(s)\Delta_1(s)}{D_0(s) + C(s)F(s)N_0(s)} \right\|_\infty < 1 \qquad (8.2.22)$$

are satisfied for all ω, respectively.

Proof As in Sections 3.6.1 and 3.6.2, the feedback system is internally stable if and only if each element of the transfer matrix

$$H(s) = \frac{1}{1 + FCG} \begin{bmatrix} 1 & -FG & -F & -F \\ C & 1 & -FC & -FC \\ CG & G & 1 & -FCG \\ CG & G & 1 & 1 \end{bmatrix}$$

has stable poles. Because C, F are designed such that each element of $H(s)$ with $G = G_0$ has stable poles and $\Delta_a, \Delta_m, \Delta_1$, and Δ_2 are assumed to have stable poles, the only instability which may arise in the feedback system is from any unstable root of

$$1 + F(s)C(s)G(s) = 0 \qquad (8.2.23)$$

(i) Let us consider $G = G_0 + \Delta_a$, then (8.2.23) can be written as

$$1 + \frac{FC\Delta_a}{1 + FCG_0} = 0 \qquad (8.2.24)$$

Because $\frac{FC}{1+FCG_0}, \Delta_a$ have stable poles, it follows from the Nyquist criterion that the roots of (8.2.24) are in the open left-half s-plane if the Nyquist plot of $\frac{FC\Delta_a}{1+FCG_0}$ does not encircle the $(-1, j0)$ point in the complex $s = \sigma + j\omega$ plane. This condition is satisfied if

$$\left| \frac{C(j\omega)F(j\omega)\delta_a(\omega)}{1 + C(j\omega)F(j\omega)G_0(j\omega)} \right| < 1, \quad \forall \omega \in \mathcal{R}$$

8.2. PLANT UNCERTAINTIES AND ROBUST CONTROL

The above condition implies that the Nyquist plot of $\frac{FC\Delta_a}{1+FCG_0}$ does not encircle the $(-1, j0)$ point for any $\Delta_a(s)$ that satisfies $|\Delta_a(j\omega)| \leq \delta_a(\omega)$. Hence, the feedback system is internally stable for any $G(s)$ in the family

$$\Pi_a = \{G \mid |G(j\omega) - G_0(j\omega)| \leq \delta_a(\omega)\}$$

(ii) If $G = G_0(1 + \Delta_m)$ equation (8.2.23) may be written as

$$1 + \frac{FCG_0}{1+FCG_0}\Delta_m = 0$$

and the proof follows as in (i).

(iii) If $G = \frac{N_0 + \Delta_1}{D_0 + \Delta_2}$, equation (8.2.23) may be written as

$$1 + \frac{1}{(D_0 + FCN_0)}[\Delta_2 + FC\Delta_1] = 0$$

Because $\frac{1}{D_0+FCN_0}, \frac{FC}{D_0+FCN_0}, \Delta_2, \Delta_1$ have stable poles, the result follows by applying the Nyquist criterion as in (i). □

We should emphasize that conditions (8.2.20) to (8.2.22) are not only sufficient for stability but also necessary in the sense that if they are violated, then within the family of plants considered, there exists a plant G for which the feedback system with compensators F, C is unstable.

Conditions (8.2.20) to (8.2.22) are referred to as conditions for *robust stability*. They may be used to choose F, C such that in addition to achieving internal stability for the nominal plant, they also guarantee robust stability with respect to a class of plant uncertainties.

As in Section 3.6.3, let us consider the performance of the feedback system of Figure 8.3 with respect to the external inputs in the presence of dynamic plant uncertainties. We concentrate on the case where $F(s) = 1$ and $d_n = d_u = 0$ and the plant transfer function is given by

$$G(s) = G_0(s)(1 + \Delta_m(s)) \qquad (8.2.25)$$

By performance, we mean that the plant output y is as close to the reference input y_m as possible for all plants G in the family Π_m despite the presence of the external inputs y_m, d. From Figure 8.3, we have

$$y = \frac{CG}{1+CG}y_m + \frac{1}{1+CG}d$$

or
$$e = y_m - y = \frac{1}{1+CG}(y_m - d) \qquad (8.2.26)$$

Because $1 + CG = 1 + CG_0 + CG_0\Delta_m$, (8.2.26) can be expressed as

$$e = \frac{S_0}{1+T_0\Delta_m}(y_m - d) \qquad (8.2.27)$$

where $S_0 = \frac{1}{1+CG_0}, T_0 = \frac{CG_0}{1+CG_0}$ are the sensitivity and complementary sensitivity functions for the nominal plant, respectively.

For robust stability we require

$$\|T_0(s)\|_\infty \delta_m(\omega) = \left\|\frac{C(s)G_0(s)}{1+C(s)G_0(s)}\right\|_\infty \delta_m(\omega) < 1$$

which suggests that the loop gain $L_0 = CG_0$ should be much less than 1 whenever $\delta_m(\omega)$ exceeds 1. For good tracking performance, i.e., for small error e, however, (8.2.27) implies that $S_0 = \frac{1}{1+CG_0}$ should be small, which, in turn, implies that the loop gain L_0 should be large. This is the classical trade-off between nominal performance and robust stability that is well known in the area of robust control. A good compromise may be found when $\delta_m(\omega) < 1$ at low frequencies and $v = y_m - d$ is small at high frequencies. In this case the loop gain $L_0 = CG_0$ can be shaped, through the choice of C, to be large at low frequencies and small at high frequencies.

The trade-off design of the compensator C to achieve robust stability and performance can be formulated as an optimal control problem where the $\mathcal{L}_2$ norm of e, i.e.,

$$\|e\|_2^2 = \int_0^\infty e^2 dt$$

is minimized with respect to C for the "worst" plant within the family Π_m and the "worst" input v within the family of the inputs considered. This "min-max" approach is used in the so-called H_2 and H_∞ optimal control designs that are quite popular in the literature of robust control [45, 152].

8.3 Instability Phenomena in Adaptive Systems

In Chapters 4 to 7, we developed a wide class of adaptive schemes that meet certain control objectives successfully for a wide class of LTI plants whose

8.3. INSTABILITY PHENOMENA IN ADAPTIVE SYSTEMS

parameters are unknown. Some of the basic assumptions made in designing and analyzing these schemes are the following:

- The plant is free of noise disturbances, unmodeled dynamics and unknown nonlinearities.

- The unknown parameters are constant for all time.

Because in applications most of these assumptions will be violated, it is of interest from the practical point of view to examine the stability properties of these schemes when applied to the actual plant with disturbances, unmodeled dynamics, etc.

In this section we show that some of the simple adaptive schemes analyzed in the previous chapters can be driven unstable by bounded disturbances and unmodeled dynamics. We study several instability mechanisms whose understanding helps find ways to counteract them by redesigning the adaptive schemes of the previous chapters.

8.3.1 Parameter Drift

Let us consider the same plant as in the example of Section 4.2.1, where the plant output is now corrupted by some unknown bounded disturbance $d(t)$, i.e.,

$$y = \theta^* u + d$$

The adaptive law for estimating θ^* derived in Section 4.2.1 for $d(t) = 0$ $\forall t \geq 0$ is given by

$$\dot{\theta} = \gamma \epsilon_1 u, \quad \epsilon_1 = y - \theta u \qquad (8.3.1)$$

where $\gamma > 0$ and $\theta(t)$ is the on-line estimate of θ^*. We have shown that for $d(t) = 0$ and $u, \dot{u} \in \mathcal{L}_\infty$, (8.3.1) guarantees that $\theta, \epsilon_1 \in \mathcal{L}_\infty$ and $\epsilon_1(t) \to 0$ as $t \to \infty$. Let us now analyze (8.3.1) when $d(t) \neq 0$. Defining $\tilde{\theta} \triangleq \theta - \theta^*$, we have

$$\epsilon_1 = -\tilde{\theta} u + d$$

and

$$\dot{\tilde{\theta}} = -\gamma u^2 \tilde{\theta} + \gamma d u \qquad (8.3.2)$$

We analyze (8.3.2) by considering the function

$$V(\tilde{\theta}) = \frac{\tilde{\theta}^2}{2\gamma} \quad (8.3.3)$$

Along the trajectory of (8.3.2), we have

$$\dot{V} = -\tilde{\theta}^2 u^2 + d\tilde{\theta}u = -\frac{\tilde{\theta}^2 u^2}{2} - \frac{1}{2}(\tilde{\theta}u - d)^2 + \frac{d^2}{2} \quad (8.3.4)$$

For the class of inputs considered, i.e., $u \in \mathcal{L}_\infty$ we cannot conclude that $\tilde{\theta}$ is bounded from considerations of (8.3.3), (8.3.4), i.e., we cannot find a constant $V_0 > 0$ such that for $V > V_0$, $\dot{V} \leq 0$. In fact, for $\theta^* = 2, \gamma = 1$, $u = (1+t)^{-\frac{1}{2}} \in \mathcal{L}_\infty$ and

$$d(t) = (1+t)^{-\frac{1}{4}}\left(\frac{5}{4} - 2(1+t)^{-\frac{1}{4}}\right) \to 0 \quad as \quad t \to \infty$$

we have

$$y(t) = \frac{5}{4}(1+t)^{-\frac{1}{4}} \to 0 \quad as \quad t \to \infty$$

$$\epsilon_1(t) = \frac{1}{4}(1+t)^{-\frac{1}{4}} \to 0 \quad as \quad t \to \infty$$

and

$$\theta(t) = (1+t)^{\frac{1}{4}} \to \infty \quad as \quad t \to \infty$$

i.e., the estimated parameter $\theta(t)$ drifts to infinity with time. We refer to this instability phenomenon as *parameter drift*.

It can be easily verified that for $u(t) = (1+t)^{-\frac{1}{2}}$ the equilibrium $\tilde{\theta}_e = 0$ of the homogeneous part of (8.3.2) is u.s. and a.s., but not u.a.s. Therefore, it is not surprising that we are able to find a bounded input γdu that leads to an unbounded solution $\tilde{\theta}(t) = \theta(t) - 2$. If, however, we restrict $u(t)$ to be PE with level $\alpha_0 > 0$, say, by choosing $u^2 = \alpha_0$, then it can be shown (see Problem 8.4) that the equilibrium $\tilde{\theta}_e = 0$ of the homogeneous part of (8.3.2) is u.a.s. (also e.s.) and that

$$\lim_{t \to \infty} \sup_{\tau \geq t} |\tilde{\theta}(\tau)| \leq \frac{d_0}{\alpha_0}, \quad d_0 = \sup_t |d(t)u(t)|$$

i.e., $\tilde{\theta}(t)$ converges exponentially to the residual set

$$D_\theta = \left\{\tilde{\theta} \,\middle|\, |\tilde{\theta}| \leq \frac{d_0}{\alpha_0}\right\}$$

8.3. INSTABILITY PHENOMENA IN ADAPTIVE SYSTEMS

indicating that the parameter error at steady state is of the order of the disturbance level.

Unfortunately, we cannot always choose u to be PE especially in the case of adaptive control where u is no longer an external signal but is generated from feedback.

Another case of parameter drift can be demonstrated by applying the adaptive control scheme

$$u = -kx, \quad \dot{k} = \gamma x^2 \qquad (8.3.5)$$

developed in Section 6.2.1 for the ideal plant $\dot{x} = ax + u$ to the plant

$$\dot{x} = ax + u + d \qquad (8.3.6)$$

where $d(t)$ is an unknown bounded input disturbance. It can be verified that for $k(0) = 5, x(0) = 1, a = 1, \gamma = 1$ and

$$d(t) = (1+t)^{-\frac{1}{5}} \left(5 - (1+t)^{-\frac{1}{5}} - 0.4(1+t)^{-\frac{6}{5}} \right) \to 0 \quad as \ t \to \infty$$

the solution of (8.3.5), (8.3.6) is given by

$$x(t) = (1+t)^{-\frac{2}{5}} \to 0 \quad as \ t \to \infty$$

and

$$k(t) = 5(1+t)^{\frac{1}{5}} \to \infty \quad as \ t \to \infty$$

We should note that since $k(0) = 5 > 1$ and $k(t) \geq k(0) \ \forall t \geq 0$ parameter drift can be stopped at any time t by switching off adaptation, i.e., setting $\gamma = 0$ in (8.3.5), and still have $x, u \in \mathcal{L}_\infty$ and $x(t) \to 0$ as $t \to \infty$. For example, if $\gamma = 0$ for $t \geq t_1 > 0$, then $k(t) = \bar{k} \ \forall t \geq t_1$, where $\bar{k} \geq 5$ is a stabilizing gain, which guarantees that $u, x \in \mathcal{L}_\infty$ and $x(t) \to 0$ as $t \to \infty$.

One explanation of the parameter drift phenomenon may be obtained by solving for the "quasi" steady state response of (8.3.6) with $u = -kx$, i.e.,

$$x_s \approx \frac{d}{k-a}$$

Clearly, for a given a and d, the only way for x_s to go to zero is for $k \to \infty$. That is, in an effort to eliminate the effect of the input disturbance d and send x to zero, the adaptive control scheme creates a high gain feedback. This high gain feedback may lead to unbounded plant states when in addition to bounded disturbances, there are dynamic plant uncertainties, as explained next.

8.3.2 High-Gain Instability

Consider the following plant:

$$\frac{y}{u} = \frac{1-\mu s}{(s-a)(1+\mu s)} = \frac{1}{s-a}\left[1 - \frac{2\mu s}{1+\mu s}\right] \qquad (8.3.7)$$

where μ is a small positive number which may be due to a small time constant in the plant.

A reasonable approximation of the second-order plant may be obtained by approximating μ with zero leading to the first-order plant model

$$\frac{\bar{y}}{u} = \frac{1}{s-a} \qquad (8.3.8)$$

where $\bar{y}$ denotes the output y when $\mu = 0$. The plant equation (8.3.7) is in the form of $y = G_0(1 + \Delta_m)u$ where $G_0(s) = \frac{1}{s-a}$ and $\Delta_m(s) = -\frac{2\mu s}{1+\mu s}$. It may be also expressed in the singular perturbation state-space form discussed in Section 8.2, i.e.,

$$\begin{aligned} \dot{x} &= ax + z - u \\ \mu \dot{z} &= -z + 2u \\ y &= x \end{aligned} \qquad (8.3.9)$$

Setting $\mu = 0$ in (8.3.9) we obtain the state-space representation of (8.3.8), i.e.,

$$\begin{aligned} \dot{\bar{x}} &= a\bar{x} + u \\ \bar{y} &= \bar{x} \end{aligned} \qquad (8.3.10)$$

where $\bar{x}$ denotes the state x when $\mu = 0$. Let us now design an adaptive controller for the simplified plant (8.3.10) and use it to control the actual second order plant where $\mu > 0$. As we have shown in Section 6.2.1, the adaptive law

$$u = -k\bar{x}, \quad \dot{k} = \gamma \epsilon_1 \bar{x}, \quad \epsilon_1 = \bar{x}$$

can stabilize (8.3.10) and regulate $\bar{y} = \bar{x}$ to zero. Replacing $\bar{y}$ with the actual output of the plant $y = x$, we have

$$u = -kx, \quad \dot{k} = \gamma \epsilon_1 x, \quad \epsilon_1 = x \qquad (8.3.11)$$

which, when applied to (8.3.9), gives us the closed-loop plant

$$\begin{aligned} \dot{x} &= (a+k)x + z \\ \mu \dot{z} &= -z - 2kx \end{aligned} \qquad (8.3.12)$$

8.3. INSTABILITY PHENOMENA IN ADAPTIVE SYSTEMS

whose equilibrium $x_e = 0, z_e = 0$ with $k = k^*$ is a.s. if and only if the eigenvalues of (8.3.12) with $k = k^*$ are in the left-half s-plane, which implies that

$$\frac{1}{\mu} - a > k^* > a$$

Because $\dot{k} \geq 0$, it is clear that

$$k(0) > \frac{1}{\mu} - a \Rightarrow k(t) > \frac{1}{\mu} - a, \quad \forall t \geq 0$$

that is, from such a $k(0)$ the equilibrium of (8.3.12) cannot be reached. Moreover, with $k > \frac{1}{\mu} - a$ the linear feedback loop is unstable even when the adaptive loop is disconnected, i.e., $\gamma = 0$.

We refer to this form of instability as *high-gain instability*. The adaptive control law (8.3.11) can generate a high gain feedback which excites the unmodeled dynamics and leads to instability and unbounded solutions. This type of instability is well known in robust control with no adaptation [106] and can be avoided by keeping the controller gains small (leading to a small loop gain) so that the closed-loop bandwidth is away from the frequency range where the unmodeled dynamics are dominant.

8.3.3 Instability Resulting from Fast Adaptation

Let us consider the second-order plant

$$\begin{aligned} \dot{x} &= -x + bz - u \\ \mu \dot{z} &= -z + 2u \\ y &= x \end{aligned} \quad (8.3.13)$$

where $b > 1/2$ is an unknown constant and $\mu > 0$ is a small number. For $u = 0$, the equilibrium $x_e = 0, z_e = 0$ is e.s. for all $\mu \geq 0$. The objective here, however, is not regulation but tracking. That is, the output $y = x$ is required to track the output x_m of the reference model

$$\dot{x}_m = -x_m + r \quad (8.3.14)$$

where r is a bounded piecewise continuous reference input signal. For $\mu = 0$, the reduced-order plant is

$$\begin{aligned} \dot{\bar{x}} &= -\bar{x} + (2b - 1)u \\ \bar{y} &= \bar{x} \end{aligned} \quad (8.3.15)$$

which has a pole at $s = -1$, identical to that of the reference model, and an unknown gain $2b - 1 > 0$. The adaptive control law

$$u = lr, \quad \dot{l} = -\gamma \epsilon_1 r, \quad \epsilon_1 = \bar{x} - x_m \quad (8.3.16)$$

guarantees that $\bar{x}(t), l(t)$ are bounded and $|\bar{x}(t) - x_m(t)| \to 0$ as $t \to \infty$ for any bounded reference input r. If we now apply (8.3.16) with $\bar{x}$ replaced by x to the actual plant (8.3.13) with $\mu > 0$, the closed-loop plant is

$$\begin{aligned} \dot{x} &= -x + bz - lr \\ \mu \dot{z} &= -z + 2lr \\ \dot{l} &= -\gamma r(x - x_m) \end{aligned} \quad (8.3.17)$$

When $\gamma = 0$ that is when l is fixed and finite, the signals $x(t)$ and $z(t)$ are bounded. Thus, no instability can occur in (8.3.17) for $\gamma = 0$.

Let us assume that $r =$ constant. Then, (8.3.17) is an LTI system of the form

$$\dot{Y} = AY + B\gamma r x_m \quad (8.3.18)$$

where $Y = [x, z, l]^T, B = [0, 0, 1]^T$,

$$A = \begin{bmatrix} -1 & b & -r \\ 0 & -1/\mu & 2r/\mu \\ -\gamma r & 0 & 0 \end{bmatrix}$$

and $\gamma r x_m$ is treated as a bounded input. The characteristic equation of (8.3.18) is then given by

$$\det(sI - A) = s^3 + (1 + \frac{1}{\mu})s^2 + (\frac{1}{\mu} - \gamma r^2)s + \frac{\gamma}{\mu} r^2 (2b - 1) \quad (8.3.19)$$

Using the Routh-Hurwitz criterion, we have that for

$$\gamma r^2 > \frac{1}{\mu} \frac{(1 + \mu)}{(2b + \mu)} \quad (8.3.20)$$

two of the roots of (8.3.19) are in the open right-half s-plane. Hence given any $\mu > 0$, if γ, r satisfy (8.3.20), the solutions of (8.3.18) are unbounded in the sense that $|Y(t)| \to \infty$ as $t \to \infty$ for almost all initial conditions. Large γr^2 increases the speed of adaptation, i.e., $\dot{l}$, which in turn excites the unmodeled dynamics and leads to instability. The effect of $\dot{l}$ on the unmodeled dynamics

8.3. INSTABILITY PHENOMENA IN ADAPTIVE SYSTEMS

can be seen more clearly by defining a new state variable $\eta = z - 2lr$, called the parasitic state [85, 106], to rewrite (8.3.17) as

$$\begin{aligned} \dot{x} &= -x + (2b-1)lr + b\eta \\ \mu\dot{\eta} &= -\eta + 2\mu\gamma r^2(x - x_m) - 2\mu l\dot{r} \\ \dot{l} &= -\gamma r(x - x_m) \end{aligned} \quad (8.3.21)$$

where for $r =$ constant, $\dot{r} \equiv 0$. Clearly, for a given μ, large γr^2 may lead to a fast adaptation and large parasitic state η which acts as a disturbance in the dominant part of the plant leading to false adaptation and unbounded solutions. For stability and bounded solutions, γr^2 should be kept small, i.e., the speed of adaptation should be slow relative to the speed of the parasitics characterized by $\frac{1}{\mu}$.

8.3.4 High-Frequency Instability

Let us now assume that in the adaptive control scheme described by (8.3.21), both γ and $|r|$ are kept small, i.e., adaptation is slow. We consider the case where $r = r_0 \sin\omega t$ and r_0 is small but the frequency ω can be large. At lower to middle frequencies and for small $\gamma r^2, \mu$, we can approximate $\mu\dot{\eta} \approx 0$ and solve (8.3.21) for η, i.e., $\eta \approx 2\mu\gamma r^2(x - x_m) - 2\mu l\dot{r}$, which we substitute into the first equation in (8.3.21) to obtain

$$\begin{aligned} \dot{x} &= -(1 - 2\mu b\gamma r^2)x + (gr - 2b\mu\dot{r})l - 2\mu b\gamma r^2 x_m \\ \dot{l} &= -\gamma r(x - x_m) \end{aligned} \quad (8.3.22)$$

where $g = 2b - 1$. Because γr is small we can approximate $l(t)$ with a constant and solve for the sinusoidal steady-state response of x from the first equation of (8.3.22) where the small $\mu\gamma r^2$ terms are neglected, i.e.,

$$x_{ss} = \frac{r_0}{1+\omega^2}[(g - 2b\mu\omega^2)\sin\omega t - \omega(g + 2b\mu)\cos\omega t]l$$

Now substituting for $x = x_{ss}$ in the second equation of (8.3.22) we obtain

$$\dot{l}_s = -\frac{\gamma r_0^2 \sin\omega t}{1+\omega^2}\left[(g - 2b\mu\omega^2)\sin\omega t - \omega(g + 2b\mu)\cos\omega t\right]l_s + \gamma x_m r_0 \sin\omega t$$

i.e.,

$$\dot{l}_s = \alpha(t)l_s + \gamma x_m r_0 \sin\omega t \quad (8.3.23)$$

$$\alpha(t) \triangleq -\frac{\gamma r_0^2 \sin \omega t}{1+\omega^2}\left[(g - 2b\mu\omega^2)\sin\omega t - \omega(g + 2b\mu)\cos\omega t\right]$$

where the subscript 's' indicates that (8.3.23) is approximately valid for slow adaptation, that is, for $\gamma r_0^2(1+\omega^2)^{-1/2}$ sufficiently small. The slow adaptation is unstable if the integral of the periodic coefficient $\alpha(t)$ over the period is positive, that is, when $\mu\omega^2 > g/(2b)$, or $\omega^2 > g/(2b\mu)$ [3].

The above approximate instability analysis indicates that if the reference input signal $r(t)$ has frequencies in the parasitic range, (i.e., of the order of $\frac{1}{\mu}$ and higher), these frequencies may excite the unmodeled dynamics, cause the signal-to-noise ratio to be small and, therefore, lead to the wrong adjustment of the parameters and eventually to instability. We should note again that by setting $\gamma = 0$, i.e., switching adaptation off, instability ceases.

A more detailed analysis of slow adaptation and high-frequency instability using averaging is given in [3, 201].

8.3.5 Effect of Parameter Variations

Adaptive control was originally motivated as a technique for compensating large variations in the plant parameters with respect to time. It is, therefore, important to examine whether the adaptive schemes developed in Chapters 4 to 7 can meet this challenge.

We consider the linear time varying scalar plant

$$\dot{x} = a(t)x + b(t)u \qquad (8.3.24)$$

where $a(t)$ and $b(t)$ are time varying parameters that are bounded and have bounded first order derivatives for all t and $b(t) \geq b_0 > 0$. We further assume that $|\dot{a}(t)|, |\dot{b}(t)|$ are small, i.e., the plant parameters vary slowly with time. The objective is to find u such that x tracks the output of the LTI reference model

$$\dot{x}_m = -x_m + r$$

for all reference input signals r that are bounded and piecewise continuous. If $a(t), b(t)$ were known for all $t \geq 0$, then the following control law

$$u = -k^*(t)x + l^*(t)r$$

where

$$k^*(t) = \frac{1+a(t)}{b(t)}, \quad l^*(t) = \frac{1}{b(t)}$$

8.3. INSTABILITY PHENOMENA IN ADAPTIVE SYSTEMS

would achieve the control objective exactly. Since $a(t)$ and $b(t)$ are unknown, we try
$$u = -k(t)x + l(t)r$$
together with an adaptive law for adjusting $k(t)$ and $l(t)$. Let us use the same adaptive law as the one that we would use if $a(t)$ and $b(t)$ were constants, i.e., following the approach of Section 6.2.2, we have

$$\dot{k} = \gamma_1 \epsilon_1 x, \quad \dot{l} = -\gamma_2 \epsilon_1 r, \quad \gamma_1, \gamma_2 > 0 \qquad (8.3.25)$$

where $\epsilon_1 = e = x - x_m$ satisfies

$$\dot{\epsilon}_1 = -\epsilon_1 + b(t)(-\tilde{k}x + \tilde{l}r) \qquad (8.3.26)$$

where $\tilde{k}(t) \triangleq k(t) - k^*(t), \tilde{l}(t) \triangleq l(t) - l^*(t)$. Because $k^*(t), l^*(t)$ are time varying, $\dot{\tilde{k}} = \dot{k} - \dot{k}^*, \dot{\tilde{l}} = \dot{l} - \dot{l}^*$ and therefore (8.3.25), rewritten in terms of the parameter error, becomes

$$\dot{\tilde{k}} = \gamma_1 \epsilon_1 x - \dot{k}^*, \quad \dot{\tilde{l}} = -\gamma_2 \epsilon_1 r - \dot{l}^* \qquad (8.3.27)$$

Hence, the effect of the time variations is the appearance of the disturbance terms $\dot{k}^*, \dot{l}^*$, which are bounded because $\dot{a}(t), \dot{b}(t)$ are assumed to be bounded. Let us now choose the same Lyapunov-like function as in the LTI case for analyzing the stability properties of (8.3.26), (8.3.27), i.e.,

$$V(\epsilon_1, \tilde{k}, \tilde{l}) = \frac{\epsilon_1^2}{2} + b(t)\frac{\tilde{k}^2}{2\gamma_1} + b(t)\frac{\tilde{l}^2}{2\gamma_2} \qquad (8.3.28)$$

where $b(t) \geq b_0 > 0$. Then along the trajectory of (8.3.26), (8.3.27) we have

$$\begin{aligned}
\dot{V} &= -\epsilon_1^2 - \epsilon_1 b \tilde{k} x + \epsilon_1 b \tilde{l} r + \dot{b}\frac{\tilde{k}^2}{2\gamma_1} + b\tilde{k}\epsilon_1 x - \frac{b\tilde{k}\dot{k}^*}{\gamma_1} \\
&\quad + \dot{b}\frac{\tilde{l}^2}{2\gamma_2} - b\tilde{l}\epsilon_1 r - b\frac{\tilde{l}\dot{l}^*}{\gamma_2} \\
&= -\epsilon_1^2 + \frac{\dot{b}}{2}\left(\frac{\tilde{k}^2}{\gamma_1} + \frac{\tilde{l}^2}{\gamma_2}\right) - b\left(\frac{\tilde{k}\dot{k}^*}{\gamma_1} + \frac{\tilde{l}\dot{l}^*}{\gamma_2}\right) \qquad (8.3.29)
\end{aligned}$$

From (8.3.28) and (8.3.29) we can say nothing about the boundedness of $\epsilon_1, \tilde{k}, \tilde{l}$ unless $\dot{k}^*, \dot{l}^*$ are either zero or decaying to zero exponentially fast.

Even if $\dot{k}^*(t), \dot{l}^*(t)$ are sufficiently small, i.e., $a(t), b(t)$ vary sufficiently slowly with time, the boundedness of $\epsilon_1, \tilde{k}, \tilde{l}$ can not be assured from the properties of $V, \dot{V}$ given by (8.3.28), (8.3.29) no matter how small $\dot{a}, \dot{b}$ are.

It has been established in [9] by using a different Lyapunov-like function and additional arguments that the above adaptive control scheme guarantees signal boundedness for any bounded $a(t), b(t)$ with arbitrary but finite speed of variation. This approach, however, has been restricted to a special class of plants and has not been extended to the general case. Furthermore, the analysis does not address the issue of tracking error performance that is affected by the parameter variations.

The above example demonstrates that the empirical observation that an adaptive controller designed for an LTI plant will also perform well when the plant parameters are slowly varying with time, held for a number of years in the area of adaptive control as a justification for dealing with LTI plants, may not be valid. In Chapter 9, we briefly address the adaptive control problem of linear time varying plants and show how the adaptive and control laws designed for LTI plants can be modified to meet the control objective in the case of linear plants with time varying parameters. For more details on the design and analysis of adaptive controllers for linear time varying plants, the reader is referred to [226].

8.4 Modifications for Robustness: Simple Examples

The instability examples presented in Section 8.3 demonstrate that the adaptive schemes designed in Chapters 4 to 7 for ideal plants, i.e., plants with no modeling errors may easily go unstable in the presence of disturbances or unmodeled dynamics. The lack of robustness is primarily due to the adaptive law which is nonlinear in general and therefore more susceptible to modeling error effects.

The lack of robustness of adaptive schemes in the presence of bounded disturbances was demonstrated as early as 1979 [48] and became a hot issue in the early 1980s when several adaptive control examples are used to show instability in the presence of unmodeled dynamics and bounded disturbances [86, 197]. It was clear from these examples that new approaches and adaptive

8.4. MODIFICATIONS FOR ROBUSTNESS: SIMPLE EXAMPLES

laws were needed to assure boundedness of all signals in the presence of plant uncertainties. These activities led to a new body of work referred to as *robust adaptive control*.

The purpose of this section is to introduce and analyze some of the techniques used in the 1980s to modify the adaptive laws designed for ideal plants so that they can retain their stability properties in the presence of "reasonable" classes of modeling errors.

We start with the simple plant

$$y = \theta^* u + \eta \tag{8.4.1}$$

where η is the unknown modeling error term and y, u are the plant output and input that are available for measurement. Our objective is to design an adaptive law for estimating θ^* that is *robust* with respect to the modeling error term η. By robust we mean the properties of the adaptive law for $\eta \neq 0$ are "close" (within the order of the modeling error) to those with $\eta = 0$.

Let us start with the adaptive law

$$\dot{\theta} = \gamma \epsilon_1 u, \quad \epsilon_1 = y - \theta u \tag{8.4.2}$$

developed in Section 4.2.1 for the plant (8.4.1) with $\eta = 0$ and $u \in \mathcal{L}_\infty$, and shown to guarantee the following two properties:

$$\text{(i)} \quad \epsilon_1, \theta, \dot{\theta} \in \mathcal{L}_\infty, \quad \text{(ii)} \quad \epsilon_1, \dot{\theta} \in \mathcal{L}_2 \tag{8.4.3}$$

If in addition $\dot{u} \in \mathcal{L}_\infty$, the adaptive law also guarantees that $\epsilon_1, \dot{\theta} \to 0$ as $t \to \infty$. When $\eta = 0$ but $u \notin \mathcal{L}_\infty$, instead of (8.4.2) we use the normalized adaptive law

$$\dot{\theta} = \gamma \epsilon u, \quad \epsilon = \frac{y - \theta u}{m^2} \tag{8.4.4}$$

where m is the normalizing signal that has the property of $\frac{u}{m} \in \mathcal{L}_\infty$. A typical choice for m in this case is $m^2 = 1 + u^2$. The normalized adaptive law (8.4.4) guarantees that

$$\text{(i)} \quad \epsilon, \epsilon m, \theta, \dot{\theta} \in \mathcal{L}_\infty, \quad \text{(ii)} \quad \epsilon, \epsilon m, \dot{\theta} \in \mathcal{L}_2 \tag{8.4.5}$$

We refer to properties (i) and (ii) given by (8.4.3) or (8.4.5) and established for $\eta \equiv 0$ as the *ideal properties* of the adaptive laws. When $\eta \neq 0$, the adaptive law (8.4.2) or (8.4.4) can no longer guarantee properties (i) and (ii) given by (8.4.3) or (8.4.5). As we have shown in Section 8.3.1, we can easily find a bounded modeling error term η, such as a bounded output disturbance, that can cause the parameter estimate $\theta(t)$ to drift to infinity.

In the following sections, we introduce and analyze several techniques that can be used to modify the adaptive laws in (8.4.2) and (8.4.4). The objective of these modifications is to guarantee that the properties of the modified adaptive laws are as close as possible to the ideal properties (i) and (ii) given by (8.4.3) and (8.4.5) despite the presence of the modeling error term $\eta \neq 0$.

We first consider the case where η is due to a bounded output disturbance and $u \in \mathcal{L}_\infty$. We then extend the results to the case where $u \in \mathcal{L}_\infty$ and η is due to a dynamic plant uncertainty in addition to a bounded output disturbance.

8.4.1 Leakage

The idea behind *leakage* is to modify the adaptive law so that the time derivative of the Lyapunov function used to analyze the adaptive scheme becomes negative in the space of the parameter estimates when these parameters exceed certain bounds. We demonstrate the use of leakage by considering the adaptive law (8.4.2)

$$\dot{\theta} = \gamma \epsilon_1 u, \quad \epsilon_1 = y - \theta u$$

with $u \in \mathcal{L}_\infty$ for the plant (8.4.1), which we rewrite in terms of the parameter error $\tilde{\theta} \triangleq \theta - \theta^*$ as

$$\dot{\tilde{\theta}} = \gamma \epsilon_1 u, \quad \epsilon_1 = -\tilde{\theta} u + \eta \tag{8.4.6}$$

The modeling error term η, which in this case is assumed to be bounded, affects the estimation error which in turn affects the evolution of $\tilde{\theta}(t)$. Let us consider the Lyapunov function

$$V(\tilde{\theta}) = \frac{\tilde{\theta}^2}{2\gamma} \tag{8.4.7}$$

that is used to establish properties (i) and (ii), given by (8.4.3), when $\eta = 0$. The time derivative of V along the solution of (8.4.6) is given by

$$\dot{V} = \tilde{\theta} \epsilon_1 u = -\epsilon_1^2 + \epsilon_1 \eta \leq -|\epsilon_1|(|\epsilon_1| - d_0) \tag{8.4.8}$$

8.4. MODIFICATIONS FOR ROBUSTNESS: SIMPLE EXAMPLES

where $d_0 > 0$ is an upper bound for the error term η. From (8.4.7) and (8.4.8), we cannot conclude anything about the boundedness of $\tilde{\theta}$ because the situation where $|\epsilon_1| < d_0$, i.e., $\dot{V} \geq 0$ and $\tilde{\theta}(t) \to \infty$ as $t \to \infty$ cannot be excluded by the properties of V and $\dot{V}$ for all input signals $u \in \mathcal{L}_\infty$.

One way to avoid this situation and establish boundedness in the presence of η is to modify the adaptive law as

$$\dot{\theta} = \gamma \epsilon_1 u - \gamma w \theta, \quad \epsilon_1 = y - \theta u \tag{8.4.9}$$

where the term $w\theta$ with $w > 0$ converts the "pure" integral action of the adaptive law given by (8.4.6) to a "leaky" integration and is therefore referred to as the *leakage* modification. The design variable $w(t) \geq 0$ is to be chosen so that for $V \geq V_0 > 0$ and some V_0, which may depend on d_0, the time derivative $\dot{V} \leq 0$. Such a property of $V, \dot{V}$ will allow us to apply Theorem 3.4.3 and establish boundedness for V and, therefore, for θ. The stability properties of the adaptive law (8.4.9) are described by the differential equation

$$\dot{\tilde{\theta}} = \gamma \epsilon_1 u - \gamma w \theta, \quad \epsilon_1 = -\tilde{\theta} u + \eta \tag{8.4.10}$$

Let us analyze (8.4.10) for various choices of the leakage term $w(t)$.

(a) σ-*modification* [85, 86]. The simplest choice for $w(t)$ is

$$w(t) = \sigma > 0, \quad \forall t \geq 0$$

where σ is a small constant and is referred to as the *fixed* σ-*modification*. The adaptive law becomes

$$\dot{\theta} = \gamma \epsilon_1 u - \gamma \sigma \theta$$

and in terms of the parameter error

$$\dot{\tilde{\theta}} = \gamma \epsilon_1 u - \gamma \sigma \theta, \quad \epsilon = -\tilde{\theta} u + \eta \tag{8.4.11}$$

The time derivative of $V = \frac{\tilde{\theta}^2}{2\gamma}$ along any trajectory of (8.4.11) is given by

$$\dot{V} = -\epsilon_1^2 + \epsilon_1 \eta - \sigma \tilde{\theta} \theta \leq -\epsilon_1^2 + |\epsilon_1| d_0 - \sigma \tilde{\theta} \theta \tag{8.4.12}$$

Using completion of squares, we write

$$-\epsilon_1^2 + |\epsilon_1|d_0 \le -\frac{\epsilon_1^2}{2} - \frac{1}{2}[\epsilon_1 - d_0]^2 + \frac{d_0^2}{2} \le -\frac{\epsilon_1^2}{2} + \frac{d_0^2}{2}$$

and

$$-\sigma\tilde{\theta}\theta = -\sigma\tilde{\theta}(\tilde{\theta} + \theta^*) \le -\sigma\tilde{\theta}^2 + \sigma|\tilde{\theta}||\theta^*| \le -\frac{\sigma\tilde{\theta}^2}{2} + \frac{\sigma|\theta^*|^2}{2} \quad (8.4.13)$$

Therefore,

$$\dot{V} \le -\frac{\epsilon_1^2}{2} - \frac{\sigma\tilde{\theta}^2}{2} + \frac{d_0^2}{2} + \frac{\sigma|\theta^*|^2}{2}$$

Adding and subtracting the term αV for some $\alpha > 0$, we obtain

$$\dot{V} \le -\alpha V - \frac{\epsilon_1^2}{2} - \left(\sigma - \frac{\alpha}{\gamma}\right)\frac{\tilde{\theta}^2}{2} + \frac{d_0^2}{2} + \frac{\sigma|\theta^*|^2}{2}$$

If we choose $0 < \alpha \le \sigma\gamma$, we have

$$\dot{V} \le -\alpha V + \frac{d_0^2}{2} + \frac{\sigma|\theta^*|^2}{2} \quad (8.4.14)$$

which implies that for $V \ge V_0 = \frac{1}{2\alpha}(d_0^2 + \sigma|\theta^*|^2)$, $\dot{V} \le 0$. Therefore, using Theorem 3.4.3, we have that $\tilde{\theta}, \theta \in \mathcal{L}_\infty$ which, together with $u \in \mathcal{L}_\infty$, imply that $\epsilon_1, \dot{\theta} \in \mathcal{L}_\infty$. Hence, with the σ-modification, we managed to extend property (i) given by (8.4.3) for the ideal case to the case where $\eta \ne 0$ provided $\eta \in \mathcal{L}_\infty$. In addition, we can establish by integrating (8.4.14) that

$$V(\tilde{\theta}(t)) = \frac{\tilde{\theta}^2}{2\gamma} \le e^{-\alpha t}\frac{\tilde{\theta}^2(0)}{2\gamma} + \frac{1}{2\alpha}(d_0^2 + \sigma|\theta^*|^2)$$

which implies that $\tilde{\theta}$ converges exponentially to the residual set

$$D_\sigma = \left\{\tilde{\theta} \in \mathcal{R} \,\middle|\, \tilde{\theta}^2 \le \frac{\gamma}{\alpha}(d_0^2 + \sigma|\theta^*|^2)\right\}$$

Let us now examine whether we can extend the ideal property (ii) ϵ_1, $\dot{\theta} \in \mathcal{L}_2$ given by (8.4.3) for the case of $\eta = 0$ to the case where $\eta \ne 0$ but $\eta \in \mathcal{L}_\infty$.

We consider the following expression for $\dot{V}$:

$$\dot{V} \le -\frac{\epsilon_1^2}{2} - \sigma\frac{\theta^2}{2} + \frac{d_0^2}{2} + \sigma\frac{|\theta^*|^2}{2} \quad (8.4.15)$$

obtained by using the inequality

$$-\sigma\tilde{\theta}\theta \leq -\sigma\frac{\theta^2}{2} + \sigma\frac{|\theta^*|^2}{2}$$

instead of (8.4.13). Integrating (8.4.15) we obtain

$$\int_t^{t+T} \epsilon_1^2 d\tau + \int_t^{t+T} \sigma\theta^2 d\tau \leq (d_0^2+\sigma|\theta^*|^2)T + 2[V(t)-V(t+T)] \leq c_0(d_0^2+\sigma)T + c_1 \quad (8.4.16)$$

$\forall t \geq 0$ and $T > 0$ where $c_0 = max[1, |\theta^*|^2]$, $c_1 = 2\sup_t[V(t) - V(t+T)]$. Expression (8.4.16) implies that $\epsilon_1, \sqrt{\sigma}\theta$ are $(d_0^2+\sigma)$-small in the mean square sense (m.s.s.), i.e., $\epsilon_1, \sqrt{\sigma}\theta \in \mathcal{S}(d_0^2 + \sigma)$. Because

$$|\dot{\theta}|^2 \leq 2\gamma^2\epsilon_1^2 u^2 + 2\gamma^2\sigma^2\theta^2$$

it follows that $\dot{\theta}$ is also $(d_0^2 + \sigma)$-small in the m.s.s., i.e., $\dot{\theta} \in \mathcal{S}(d_0^2 + \sigma)$.

Therefore, the ideal property (ii) extends to

$$\text{(ii)}' \quad \epsilon_1, \dot{\theta} \in \mathcal{S}(d_0^2 + \sigma) \quad (8.4.17)$$

The $\mathcal{L}_2$ property of $\epsilon_1, \dot{\theta}$ can therefore no longer be guaranteed in the presence of the nonzero modeling error term η. The $\mathcal{L}_2$ property of ϵ_1 was used in Section 4.2.1 together with $\dot{u} \in \mathcal{L}_\infty$ to establish that $\epsilon_1(t) \to 0$ as $t \to \infty$. It is clear from (8.4.17) and the analysis above that even when the disturbance η disappears, i.e., $\eta = 0$, the adaptive law with the σ-modification given by (8.4.11) cannot guarantee that $\epsilon_1, \dot{\theta} \in \mathcal{L}_2$ or that $\epsilon_1(t) \to 0$ as $t \to \infty$ unless $\sigma = 0$. Thus robustness is achieved at the expense of destroying the ideal property given by (ii) in (8.4.3) and having possible nonzero estimation errors at steady state. This drawback of the σ-modification motivated what is called the *switching σ-modification* described next.

(b) *Switching-σ* [91]. Because the purpose of the σ-modification is to avoid parameter drift, it does not need to be active when the estimated parameters are within some acceptable bounds. Therefore a more reasonable modification would be

$$w(t) = \sigma_s, \quad \sigma_s = \begin{cases} 0 & \text{if } |\theta| < M_0 \\ \sigma_0 & \text{if } |\theta| \geq M_0 \end{cases}$$

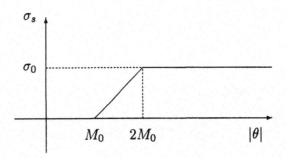

Figure 8.4 Continuous switching σ-modification.

where $M_0 > 0, \sigma_0 > 0$ are design constants and M_0 is chosen to be large enough so that $M_0 > |\theta^*|$. With the above choice of σ_s, however, the adaptive law (8.4.9) is a discontinuous one and may not guarantee the existence of the solution $\theta(t)$. It may also cause oscillations on the switching surface $|\theta| = M_0$ during implementation. The discontinuous switching σ-modification can be replaced with a continuous one such as

$$w(t) = \sigma_s, \quad \sigma_s = \begin{cases} 0 & \text{if } |\theta(t)| < M_0 \\ \sigma_0 \left(\frac{|\theta(t)|}{M_0} - 1 \right) & \text{if } M_0 \leq |\theta(t)| \leq 2M_0 \\ \sigma_0 & \text{if } |\theta(t)| > 2M_0 \end{cases} \quad (8.4.18)$$

shown in Figure 8.4 where the design constants M_0, σ_0 are as defined in the discontinuous switching σ-modification. In this case the adaptive law is given by

$$\dot{\theta} = \gamma \epsilon_1 u - \gamma \sigma_s \theta, \quad \epsilon_1 = y - \theta u$$

and in terms of the parameter error

$$\dot{\tilde{\theta}} = \gamma \epsilon_1 u - \gamma \sigma_s \theta, \quad \epsilon_1 = -\tilde{\theta} u + \eta \quad (8.4.19)$$

The time derivative of $V(\tilde{\theta}) = \frac{\tilde{\theta}^2}{2\gamma}$ along the solution of (8.4.18), (8.4.19) is given by

$$\dot{V} = -\epsilon_1^2 + \epsilon_1 \eta - \sigma_s \tilde{\theta} \theta \leq -\frac{\epsilon_1^2}{2} - \sigma_s \tilde{\theta} \theta + \frac{d_0^2}{2} \quad (8.4.20)$$

Now

$$\sigma_s \tilde{\theta} \theta = \sigma_s (\theta - \theta^*) \theta \geq \sigma_s |\theta|^2 - \sigma_s |\theta| |\theta^*|$$
$$\geq \sigma_s |\theta| (|\theta| - M_0 + M_0 - |\theta^*|)$$

8.4. MODIFICATIONS FOR ROBUSTNESS: SIMPLE EXAMPLES

i.e.,
$$\sigma_s \tilde{\theta}\theta \geq \sigma_s|\theta|(|\theta| - M_0) + \sigma_s|\theta|(M_0 - |\theta^*|) \geq 0 \qquad (8.4.21)$$
where the last inequality follows from the fact that $\sigma_s \geq 0$, $\sigma_s(|\theta| - M_0) \geq 0$ and $M_0 > |\theta^*|$. Hence, $-\sigma_s\tilde{\theta}\theta \leq 0$. Therefore, in contrast to the fixed σ-modification, the switching σ can only make $\dot{V}$ more negative. We can also verify that
$$-\sigma_s\tilde{\theta}\theta \leq -\sigma_0\tilde{\theta}\theta + 2\sigma_0 M_0^2 \qquad (8.4.22)$$
which we can use in (8.4.20) to obtain
$$\dot{V} \leq -\frac{\epsilon_1^2}{2} - \sigma_0\tilde{\theta}\theta + 2\sigma_0 M_0^2 + \frac{d_0^2}{2} \qquad (8.4.23)$$

The boundedness of V and, therefore, of θ follows by using the same procedure as in the fixed σ case to manipulate the term $\sigma_0\tilde{\theta}\theta$ in (8.4.23) and express $\dot{V}$ as
$$\dot{V} \leq -\alpha V + 2\sigma_0 M_0^2 + \frac{d_0^2}{2} + \sigma_0\frac{|\theta^*|^2}{2}$$
where $0 \leq \alpha \leq \sigma_0\gamma$ which implies that $\tilde{\theta}$ converges exponentially to the residual set
$$D_s = \left\{\tilde{\theta} \,\middle|\, |\tilde{\theta}|^2 \leq \frac{\gamma}{\alpha}(d_0^2 + \sigma_0|\theta^*|^2 + 4\sigma_0 M_0^2)\right\}$$
The size of D_s is larger than that of D_σ in the fixed σ-modification case because of the additional term $4\sigma_0 M_0^2$. The boundedness of $\tilde{\theta}$ implies that $\theta, \dot{\theta}, \epsilon_1 \in \mathcal{L}_\infty$ and, therefore, property (i) of the unmodified adaptive law in the ideal case is preserved by the switching σ-modification when $\eta \neq 0$ but $\eta \in \mathcal{L}_\infty$.

As in the fixed σ-modification case, the switching σ cannot guarantee the $\mathcal{L}_2$ property of $\epsilon_1, \dot{\theta}$ in general when $\eta \neq 0$. The bound for ϵ_1 in m.s.s. follows by integrating (8.4.20) to obtain
$$2\int_t^{t+T}\sigma_s\tilde{\theta}\theta d\tau + \int_t^{t+T}\epsilon_1^2 d\tau \leq d_0^2 T + c_1$$
where $c_1 = 2\sup_{t\geq 0}[V(t) - V(t+T)], \forall t \geq 0, T \geq 0$. Because $\sigma_s\tilde{\theta}\theta \geq 0$ it follows that $\sqrt{\sigma_s\tilde{\theta}\theta}, \epsilon_1 \in \mathcal{S}(d_0^2)$. From (8.4.21) we have $\sigma_s\tilde{\theta}\theta \geq \sigma_s|\theta|(M_0 - |\theta^*|)$ and, therefore,
$$\sigma_s^2|\theta|^2 \leq \frac{(\sigma_s\tilde{\theta}\theta)^2}{(M_0 - |\theta^*|)^2} \leq c\sigma_s\tilde{\theta}\theta$$

for some constant c that depends on the bound for $\sigma_0|\theta|$, which implies that $\sigma_s|\theta| \in \mathcal{S}(d_0^2)$. Because

$$|\dot{\theta}|^2 \leq 2\gamma^2 \epsilon_1^2 u^2 + 2\gamma^2 \sigma_s^2 |\theta|^2$$

it follows that $|\dot{\theta}| \in \mathcal{S}(d_0^2)$. Hence, the adaptive law with the switching-σ modification given by (8.4.19) guarantees that

$$\text{(ii)}' \quad \epsilon_1, \dot{\theta} \in \mathcal{S}(d_0^2)$$

In contrast to the fixed σ-modification, the switching σ preserves the ideal properties of the adaptive law, i.e., when η disappears ($\eta = d_0 = 0$), equation (8.4.20) implies (because $-\sigma_s \tilde{\theta}\theta \leq 0$) that $\epsilon_1 \in \mathcal{L}_2$ and $\sqrt{\sigma_s \tilde{\theta}\theta} \in \mathcal{L}_2$, which, in turn, imply that $\dot{\theta} \in \mathcal{L}_2$. In this case if $\dot{u} \in \mathcal{L}_\infty$, we can also establish as in the the ideal case that $\epsilon_1(t), \dot{\theta}(t) \to 0$ as $t \to \infty$, i.e., the switching σ does not destroy any of the ideal properties of the unmodified adaptive law. The only drawback of the switching σ when compared with the fixed σ is that it requires the knowledge of an upper bound M_0 for $|\theta^*|$. If M_0 in (8.4.18) happens not to be an upper bound for $|\theta^*|$, then the adaptive law (8.4.18) has the same properties and drawbacks as the fixed σ-modification (see Problem 8.7).

(c) ϵ_1-*modification* [172]. Another attempt to eliminate the main drawback of the fixed σ-modification led to the following modification:

$$w(t) = |\epsilon_1|\nu_0$$

where $\nu_0 > 0$ is a design constant. The adaptive law becomes

$$\dot{\theta} = \gamma\epsilon_1 u - \gamma|\epsilon_1|\nu_0\theta, \quad \epsilon_1 = y - \theta u$$

and in terms of the parameter error

$$\dot{\tilde{\theta}} = \gamma\epsilon_1 u - \gamma|\epsilon_1|\nu_0\theta, \quad \epsilon_1 = -\tilde{\theta}u + \eta \qquad (8.4.24)$$

The logic behind this choice of w is that because in the ideal case ϵ_1 is guaranteed to converge to zero (when $\dot{u} \in \mathcal{L}_\infty$), then the leakage term $w(t)\theta$ will go to zero with ϵ_1 when $\eta = 0$; therefore, the ideal properties of the adaptive law (8.4.24) when $\eta = 0$ will not be affected by the leakage.

8.4. MODIFICATIONS FOR ROBUSTNESS: SIMPLE EXAMPLES

The time derivative of $V(\tilde{\theta}) = \frac{\tilde{\theta}^2}{2\gamma}$ along the solution of (8.4.24) is given by

$$\dot{V} = -\epsilon_1^2 + \epsilon_1\eta - |\epsilon_1|\nu_0\tilde{\theta}\theta \leq -|\epsilon_1|\left(|\epsilon_1| + \nu_0\frac{\tilde{\theta}^2}{2} - \nu_0\frac{|\theta^*|^2}{2} - d_0\right) \quad (8.4.25)$$

where the inequality is obtained by using $\nu_0\tilde{\theta}\theta \leq -\nu_0\frac{\tilde{\theta}^2}{2} + \nu_0\frac{|\theta^*|^2}{2}$. It is clear that for $|\epsilon_1| + \nu_0\frac{\tilde{\theta}^2}{2} \geq \nu_0\frac{|\theta^*|^2}{2} + d_0$, i.e., for $V \geq V_0 \triangleq \frac{1}{\gamma\nu_0}(\nu_0\frac{|\theta^*|^2}{2} + d_0)$ we have $\dot{V} \leq 0$, which implies that V and, therefore, $\tilde{\theta}, \theta \in \mathcal{L}_\infty$. Because $\epsilon_1 = -\tilde{\theta}u + \eta$ and $u \in \mathcal{L}_\infty$ we also have that $\epsilon_1 \in \mathcal{L}_\infty$, which, in turn, implies that $\dot{\theta} \in \mathcal{L}_\infty$. Hence, property (i) is also guaranteed by the ϵ_1-modification despite the presence of $\eta \neq 0$.

Let us now examine the $\mathcal{L}_2$ properties of $\epsilon_1, \dot{\theta}$ guaranteed by the unmodified adaptive law ($w(t) \equiv 0$) when $\eta = 0$. We rewrite (8.4.25) as

$$\dot{V} \leq -\frac{\epsilon_1^2}{2} + \frac{d_0^2}{2} - |\epsilon_1|\nu_0\tilde{\theta}^2 - |\epsilon_1|\nu_0\theta^*\tilde{\theta} \leq -\frac{\epsilon_1^2}{2} - \frac{|\epsilon_1|\nu_0\tilde{\theta}^2}{2} + |\epsilon_1|\nu_0\frac{|\theta^*|^2}{2} + \frac{d_0^2}{2}$$

by using the inequality $-a^2 \pm ab \leq -\frac{a^2}{2} + \frac{b^2}{2}$. If we repeat the use of the same inequality we obtain

$$\dot{V} \leq -\frac{\epsilon_1^2}{4} + \frac{d_0^2}{2} + \nu_0^2\frac{|\theta^*|^4}{4} \quad (8.4.26)$$

Integrating on both sides of (8.4.26), we establish that $\epsilon_1 \in \mathcal{S}(d_0^2 + \nu_0^2)$. Because $u, \theta \in \mathcal{L}_\infty$, it follows that $|\dot{\theta}| \leq c|\epsilon_1|$ for some constant $c \geq 0$ and therefore $\dot{\theta} \in \mathcal{S}(d_0^2 + \nu_0^2)$. Hence, the adaptive law with the ϵ_1-modification guarantees that

$$\text{(ii)}' \quad \epsilon_1, \dot{\theta} \in \mathcal{S}(d_0^2 + \nu_0^2)$$

which implies that $\epsilon_1, \dot{\theta}$ are of order of d_0, ν_0 in m.s.s.

It is clear from the above analysis that in the absence of the disturbance i.e., $\eta = 0$, $\dot{V}$ cannot be shown to be negative definite or semidefinite unless $\nu_0 = 0$. The term $|\epsilon_1|\nu_0\tilde{\theta}\theta$ in (8.4.25) may make $\dot{V}$ positive even when $\eta = 0$ and therefore the ideal properties of the unmodified adaptive law cannot be guaranteed by the adaptive law with the ϵ_1-modification when $\eta = 0$ unless $\nu_0 = 0$. This indicates that the initial rationale for developing the ϵ_1-modification is not valid. It is shown in [172], however, that if u is PE, then $\epsilon_1(t)$ and therefore $w(t) = \nu_0|\epsilon_1(t)|$ do converge to zero as $t \to \infty$ when $\eta(t) \equiv 0, \forall t \geq 0$. Therefore the ideal properties of the adaptive law can be recovered with the ϵ_1-modification provided u is PE.

Remark 8.4.1

(i) Comparing the three choices for the leakage term $w(t)$, it is clear that the fixed σ- and ϵ_1-modification require no a priori information about the plant, whereas the switching-σ requires the design constant M_0 to be larger than the unknown $|\theta^*|$. In contrast to the fixed σ- and ϵ_1-modifications, however, the switching σ achieves robustness without having to destroy the ideal properties of the adaptive scheme. Such ideal properties are also possible for the ϵ_1-modification under a PE condition[172].

(ii) The leakage $-w\theta$ with $w(t) \geq 0$ introduces a term in the adaptive law that has the tendency to drive θ towards $\theta = 0$ when the other term (i.e., $\gamma\epsilon_1 u$ in the case of (8.4.9)) is small. If $\theta^* \neq 0$, the leakage term may drive θ towards zero and possibly further away from the desired θ^*. If an a priori estimate $\hat{\theta}^*$ of θ^* is available the leakage term $-w\theta$ may be replaced with the *shifted leakage* $-w(\theta - \hat{\theta}^*)$, which shifts the tendency of θ from zero to $\hat{\theta}^*$, a point that may be closer to θ^*. The analysis of the adaptive laws with the shifted leakage is very similar to that of $-w\theta$ and is left as an exercise for the reader.

(iii) One of the main drawbacks of the leakage modifications is that the estimation error ϵ_1 and $\dot{\theta}$ are only guaranteed to be of the order of the disturbance and, with the exception of the switching σ-modification, of the order of the size of the leakage design parameter, in m.s.s. This means that at steady state, we cannot guarantee that ϵ_1 is of the order of the modeling error. The m.s.s. bound of ϵ_1 may allow ϵ_1 to exhibit "bursting," i.e., ϵ_1 may assume values higher than the order of the modeling error for some finite intervals of time. One way to avoid bursting is to use PE signals or a *dead-zone* modification as it will be explained later on in this chapter

8.4. MODIFICATIONS FOR ROBUSTNESS: SIMPLE EXAMPLES

(iv) The leakage modification may be also derived by modifying the cost function $J(\theta) = \frac{\epsilon_1^2}{2} = \frac{(y-\theta u)^2}{2}$, used in the ideal case, to

$$J(\theta) = \frac{(y-\theta u)^2}{2} + w\frac{\theta^2}{2} \qquad (8.4.27)$$

Using the gradient method, we now obtain

$$\dot{\theta} = -\gamma \nabla J = \gamma \epsilon_1 u - \gamma w \theta$$

which is the same as (8.4.9). The modified cost now penalizes θ in addition to ϵ_1 which explains why for certain choices of $w(t)$ the drifting of θ to infinity due to the presence of modeling errors is counteracted.

8.4.2 Parameter Projection

An effective method for eliminating parameter drift and keeping the parameter estimates within some a priori defined bounds is to use the gradient projection method to constrain the parameter estimates to lie inside a *bounded convex set* in the parameter space. Let us illustrate the use of projection for the adaptive law

$$\dot{\theta} = \gamma \epsilon_1 u, \quad \epsilon_1 = y - \theta u$$

We like to constrain θ to lie inside the convex bounded set

$$g(\theta) \triangleq \left\{ \theta \,\middle|\, \theta^2 \leq M_0^2 \right\}$$

where $M_0 \geq |\theta^*|$. Applying the gradient projection method, we obtain

$$\dot{\tilde{\theta}} = \dot{\theta} = \begin{cases} \gamma \epsilon_1 u & \text{if } |\theta| < M_0 \\ & \text{or if } |\theta| = M_0 \text{ and } \theta \epsilon_1 u \leq 0 \\ 0 & \text{if } |\theta| = M_0 \text{ and } \theta \epsilon_1 u > 0 \end{cases} \qquad (8.4.28)$$

$$\epsilon_1 = y - \theta u = -\tilde{\theta} u + \eta$$

which for $|\theta(0)| \leq M_0$ guarantees that $|\theta(t)| \leq M_0, \forall t \geq 0$. Let us now analyze the above adaptive law by considering the Lyapunov function

$$V = \frac{\tilde{\theta}^2}{2\gamma}$$

whose time derivative $\dot{V}$ along (8.4.28) is given by

$$\dot{V} = \begin{cases} -\epsilon_1^2 + \epsilon_1 \eta & \text{if } |\theta| < M_0 \\ & \text{or if } |\theta| = M_0 \text{ and } \theta \epsilon_1 u \le 0 \\ 0 & \text{if } |\theta| = M_0 \text{ and } \theta \epsilon_1 u > 0 \end{cases} \quad (8.4.29)$$

Let us consider the case when $\dot{V} = 0$, $|\theta| = M_0$ and $\theta \epsilon_1 u > 0$. Using the expression $\epsilon_1 = -\tilde{\theta} u + \eta$, we write $\dot{V} = 0 = -\epsilon_1^2 + \epsilon_1 \eta - \tilde{\theta} \epsilon_1 u$. The last term in the expression of $\dot{V}$ can be written as

$$\tilde{\theta} \epsilon_1 u = (\theta - \theta^*) \epsilon_1 u = M_0 \text{sgn}(\theta) \epsilon_1 u - \theta^* \epsilon_1 u$$

Therefore, for $\theta \epsilon_1 u > 0$ and $|\theta| = M_0$ we have

$$\tilde{\theta} \epsilon_1 u = M_0 |\epsilon_1 u| - \theta^* \epsilon_1 u \ge M_0 |\epsilon_1 u| - |\theta^*||\epsilon_1 u| \ge 0$$

where the last inequality is obtained by using the assumption that $M_0 \ge |\theta^*|$, which implies that for $|\theta| = M_0$ and $\theta \epsilon_1 u > 0$ we have $\tilde{\theta} \epsilon_1 u \ge 0$ and that $\dot{V} = 0 \le -\epsilon_1^2 + \epsilon_1 \eta$. Therefore, (8.4.29) implies that

$$\dot{V} \le -\epsilon_1^2 + \epsilon_1 \eta \le -\frac{\epsilon_1^2}{2} + \frac{d_0^2}{2}, \quad \forall t \ge 0 \quad (8.4.30)$$

A bound for ϵ_1 in m.s.s. may be obtained by integrating both sides of (8.4.30) to get

$$\int_t^{t+T} \epsilon_1^2 d\tau \le d_0^2 T + 2 \{V(t) - V(t+T)\}$$

$\forall t \ge 0$ and any $T \ge 0$. Because $V \in \mathcal{L}_\infty$, it follows that $\epsilon_1 \in \mathcal{S}(d_0^2)$.

The projection algorithm has very similar properties and the same advantages and disadvantages as the switching σ-modification. For this reason, the switching σ-modification has often been referred to as "soft projection." It is soft in the sense that it allows $|\theta(t)|$ to exceed the bound M_0 but it does not allow $|\theta(t)|$ to depart from M_0 too much.

8.4.3 Dead Zone

Let us consider the adaptive error equation (8.4.6) i.e.,

$$\dot{\theta} = \gamma \epsilon_1 u, \quad \epsilon_1 = -\tilde{\theta} u + \eta \quad (8.4.31)$$

8.4. MODIFICATIONS FOR ROBUSTNESS: SIMPLE EXAMPLES

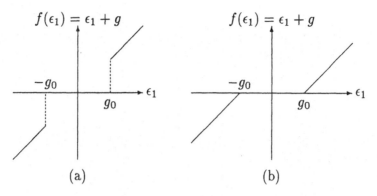

Figure 8.5 Dead zone functions: (a) discontinuous (b) continuous.

Because $\sup_t |\eta(t)| \leq d_0$, it follows that if $|\epsilon_1| \gg d_0$ then the signal $\tilde{\theta}u$ is dominant in ϵ_1. If, however, $|\epsilon_1| < d_0$ then η may be dominant in ϵ_1. Therefore small ϵ_1 relative to d_0 indicates the possible presence of a large noise (because of η) to signal ($\tilde{\theta}u$) ratio, whereas large ϵ_1 relative to d_0 indicates a small noise to signal ratio. Thus, it seems reasonable to update the parameter estimate θ only when the signal $\tilde{\theta}u$ in ϵ_1 is large relative to the disturbance η and switch-off adaptation when ϵ_1 is small relative to the size of η. This method of adaptation is referred to as *dead zone* because of the presence of a zone or interval where θ is constant, i.e., no updating occurs. The use of a dead zone is illustrated by modifying the adaptive law $\dot{\theta} = \gamma \epsilon_1 u$ used when $\eta = 0$ to

$$\dot{\theta} = \gamma u(\epsilon_1 + g), \quad g = \begin{cases} 0 & \text{if } |\epsilon_1| \geq g_0 \\ -\epsilon_1 & \text{if } |\epsilon_1| < g_0 \end{cases} \quad (8.4.32)$$

$$\epsilon_1 = y - \theta u$$

where $g_0 > 0$ is a known strict upper bound for $|\eta(t)|$, i.e., $g_0 > d_0 \geq \sup_t |\eta(t)|$. The function $f(\epsilon_1) = \epsilon_1 + g$ is known as the *dead zone function* and is shown in Figure 8.5 (a).

It follows from (8.4.32) that for small ϵ_1, i.e., $|\epsilon_1| < g_0$, we have $\dot{\theta} = 0$ and no adaptation takes place, whereas for "large" ϵ_1, i.e., $|\epsilon_1| \geq g_0$, we adapt the same way as if there was no disturbance, i.e., $\dot{\theta} = \gamma \epsilon_1 u$. Because of the dead zone, $\dot{\theta}$ in (8.4.32) is a discontinuous function which may give rise to problems related to existence and uniqueness of solutions as well as to computational problems at the switching surface [191]. One way to avoid these problems is to use the continuous dead zone shown in Figure 8.5(b).

Using the continuous dead-zone function from Figure 8.5 (b), the adaptive law (8.4.32) becomes

$$\dot{\theta} = \gamma u(\epsilon_1 + g), \quad g = \begin{cases} g_0 & \text{if } \epsilon_1 < -g_0 \\ -g_0 & \text{if } \epsilon_1 > g_0 \\ -\epsilon_1 & \text{if } |\epsilon_1| \leq g_0 \end{cases} \quad (8.4.33)$$

which together with

$$\epsilon_1 = -\tilde{\theta}u + \eta$$

describe the stability properties of the modified adaptive law. We analyze (8.4.33) by considering the Lyapunov function

$$V = \frac{\tilde{\theta}^2}{2\gamma}$$

whose time derivative $\dot{V}$ along the trajectory of (8.4.33) is given by

$$\dot{V} = \tilde{\theta}u(\epsilon_1 + g) = -(\epsilon_1 - \eta)(\epsilon_1 + g) \quad \cdot \quad (8.4.34)$$

Now,

$$(\epsilon_1 - \eta)(\epsilon_1 + g) = \begin{cases} (\epsilon_1 - \eta)(\epsilon_1 + g_0) > 0 & \text{if } \epsilon_1 < -g_0 < -|\eta| \\ (\epsilon_1 - \eta)(\epsilon_1 - g_0) > 0 & \text{if } \epsilon_1 > g_0 > |\eta| \\ 0 & \text{if } |\epsilon_1| \leq g_0 \end{cases}$$

i.e., $(\epsilon_1 - \eta)(\epsilon_1 + g) \geq 0 \; \forall t \geq 0$ and, therefore,

$$\dot{V} = -(\epsilon_1 - \eta)(\epsilon_1 + g) \leq 0$$

which implies that $V, \tilde{\theta}, \theta \in \mathcal{L}_\infty$ and that $\sqrt{(\epsilon_1 - \eta)(\epsilon_1 + g)} \in \mathcal{L}_2$. The boundedness of θ, u implies that $\epsilon_1, \dot{\theta} \in \mathcal{L}_\infty$. Hence, the adaptive law with the dead zone guarantees the property $\epsilon_1, \dot{\theta} \in \mathcal{L}_\infty$ in the presence of nonzero $\eta \in \mathcal{L}_\infty$. Let us now examine the $\mathcal{L}_2$ properties of $\epsilon_1, \dot{\theta}$ that are guaranteed by the unmodified adaptive law when $\eta = 0$. We can verify that

$$(\epsilon_1 + g)^2 \leq (\epsilon_1 - \eta)(\epsilon_1 + g)$$

for each choice of g given by (8.4.33). Since $\sqrt{(\epsilon_1 - \eta)(\epsilon_1 + g)} \in \mathcal{L}_2$ it follows that $(\epsilon_1 + g) \in \mathcal{L}_2$ which implies that $\dot{\theta} \in \mathcal{L}_2$. Hence, the dead zone preserves the $\mathcal{L}_2$ property of $\dot{\tilde{\theta}}$ despite the presence of the bounded disturbance η. Let us now examine the properties of ϵ_1 by rewriting (8.4.34) as

$$\begin{aligned}\dot{V} &= -\epsilon_1^2 - \epsilon_1 g + \epsilon_1 \eta + \eta g \\ &\leq -\epsilon_1^2 + |\epsilon_1| g_0 + |\epsilon_1| d_0 + d_0 g_0\end{aligned}$$

By completing the squares, we obtain

$$\dot{V} \leq -\frac{\epsilon_1^2}{2} + \frac{(d_0 + g_0)^2}{2} + d_0 g_0$$

which implies that $\epsilon_1 \in \mathcal{S}(d_0^2 + g_0^2)$. Therefore, our analysis cannot guarantee that ϵ_1 is in $\mathcal{L}_2$ but instead that ϵ_1 is of the order of the disturbance and design parameter g_0 in m.s.s. Furthermore, even when $\eta = 0$, i.e., $d_0 = 0$ we cannot establish that ϵ_1 is in $\mathcal{L}_2$ unless we set $g_0 = 0$.

A bound for $|\epsilon_1|$ at steady state may be established when u, η are uniformly continuous functions of time. The uniform continuity of u, η, together with $\tilde{\theta} \in \mathcal{L}_\infty$, implies that $\epsilon_1 = -\tilde{\theta} u + \eta$ is uniformly continuous, which, in turn, implies that $\epsilon_1 + g$ is a uniformly continuous function of time. Because $\epsilon_1 + g \in \mathcal{L}_2$, it follows from Barbălat's Lemma that $|\epsilon_1 + g| \to 0$ as $t \to \infty$ which by the expression for g in (8.4.33) implies that

$$\lim_{t \to \infty} \sup_{\tau \geq t} |\epsilon_1(\tau)| \leq g_0 \qquad (8.4.35)$$

The above bound indicates that at steady state the estimation error ϵ_1 is of the order of the design parameter g_0 which is an upper bound for $|\eta(t)|$. Hence if g_0 is designed properly, phenomena such as bursting (i.e., large errors relative to the level of the disturbance at steady state and over short intervals of time) that may arise in the case of the leakage modification and projection will not occur in the case of dead zone.

Another important property of the adaptive law with dead-zone is that it guarantees parameter convergence, something that cannot be guaranteed in general by the unmodified adaptive laws when $\eta = 0$ unless the input u is PE. We may establish this property by using two different methods as follows:

The first method relies on the simplicity of the example and is not applicable to the higher order case. It uses the property of $V = \frac{\tilde{\theta}^2}{2\gamma}$ and $\dot{V} \leq 0$ to conclude that $V(t)$ and, therefore, $\tilde{\theta}(t)$ have a limit as $t \to \infty$.

The second method is applicable to the higher order case as well and proceeds as follows:

We have
$$\dot{V} = -(\epsilon_1 - \eta)(\epsilon_1 + g) = -|\epsilon_1 - \eta||\epsilon_1 + g| \qquad (8.4.36)$$
where the second equality holds due to the choice of g. Because $|\epsilon_1 + g| = 0$ if $|\epsilon_1| \leq g_0$, it follows from (8.4.36) that
$$\dot{V} = -|\epsilon_1 - \eta||\epsilon_1 + g| \leq -|g_0 - |\eta|||\epsilon_1 + g| \qquad (8.4.37)$$

Integrating on both sides of (8.4.37), we obtain
$$\inf_t |g_0 - |\eta(t)|| \int_0^\infty |\epsilon_1 + g| d\tau$$
$$\leq \int_0^\infty |\epsilon_1 - \eta||\epsilon_1 + g| d\tau = V(0) - V_\infty$$

Because $g_0 > \sup_t |\eta(t)|$, it follows that $\epsilon_1 + g \in \mathcal{L}_1$. We have $|\dot{\tilde{\theta}}| \leq \gamma |u||\epsilon_1 + g|$ which implies that $\dot{\tilde{\theta}} \in \mathcal{L}_1$ which in turn implies that
$$\lim_{t \to \infty} \int_0^t \dot{\tilde{\theta}} d\tau = \lim_{t \to \infty} \tilde{\theta}(t) - \tilde{\theta}(0)$$

exists and, therefore, $\theta(t)$ converges to some limit $\bar{\theta}$.

The parameter convergence property of the adaptive law with dead zone is very helpful in analyzing and understanding adaptive controllers incorporating such adaptive law. The reason is that the initially nonlinear system that represents the closed-loop adaptive control scheme with dead zone converges to an LTI one as $t \to \infty$.

One of the main drawbacks of the dead zone is the assumption that an upper bound for the modeling error is known a priori. As we point out in later sections, this assumption becomes more restrictive in the higher order case. If the bound of the disturbance is under estimated, then the properties of the adaptive law established above can no longer be guaranteed.

Another drawback of the dead zone is that in the absence of the disturbance, i.e., $\eta(t) = 0$, we cannot establish that $\epsilon_1 \in \mathcal{L}_2$ and/or that $\epsilon_1(t) \to 0$ as $t \to \infty$ (when $\dot{u} \in \mathcal{L}_\infty$) unless we remove the dead-zone, i.e., we set $g_0 = 0$. Therefore robustness is achieved at the expense of destroying some of the ideal properties of the adaptive law.

8.4.4 Dynamic Normalization

In the previous sections, we design several modified adaptive laws to estimate the parameter θ^* in the parametric model

$$y = \theta^* u + \eta \qquad (8.4.38)$$

when $u \in \mathcal{L}_\infty$ and η is an unknown bounded signal that may arise because of a bounded output disturbance etc.

Let us now extend the results of the previous sections to the case where u is not necessarily bounded and η is either bounded or is bounded from above by $|u|$ i.e., $|\eta| \leq c_1|u| + c_2$ for some constants $c_1, c_2 \geq 0$. In this case a normalizing signal such as $m^2 = 1 + u^2$ may be used to rewrite (8.4.38) as

$$\bar{y} = \theta^* \bar{u} + \bar{\eta}$$

where $\bar{x} \triangleq \frac{x}{m}$ denotes the normalized value of x. Since with $m^2 = 1 + u^2$ we have $\bar{y}, \bar{u}, \bar{\eta} \in \mathcal{L}_\infty$, the same procedure as in the previous sections may be used to develop adaptive laws that are robust with respect to the bounded modeling error term $\bar{\eta}$. The design details and analysis of these adaptive laws is left as an exercise for the reader.

Our objective in this section is to go a step further and consider the case where η in (8.4.38) is not necessarily bounded from above by $|u|$ but it is related to u through some transfer function. Such a case may arise in the on-line estimation problem of the constant θ^* in the plant equation

$$y = \theta^*(1 + \Delta_m(s))u + d \qquad (8.4.39)$$

where $\Delta_m(s)$ is an unknown multiplicative perturbation that is strictly proper and stable and d is an unknown bounded disturbance. Equation (8.4.39) may be rewritten in the form of (8.4.38), i.e.,

$$y = \theta^* u + \eta, \quad \eta = \theta^* \Delta_m(s) u + d \qquad (8.4.40)$$

To apply the procedure of the previous sections to (8.4.40), we need to find a normalizing signal m that allow us to rewrite (8.4.40) as

$$\bar{y} = \theta^* \bar{u} + \bar{\eta} \qquad (8.4.41)$$

where $\bar{y} \triangleq \frac{y}{m}, \bar{u} \triangleq \frac{u}{m}, \bar{\eta} \triangleq \frac{\eta}{m}$ and $\bar{y}, \bar{u}, \bar{\eta} \in \mathcal{L}_\infty$. Because of $\Delta_m(s)$ in (8.4.40), the normalizing signal given by $m^2 = 1 + u^2$ is not guaranteed to bound η from above. A new choice for m is found by using the properties of the $\mathcal{L}_{2\delta}$ norm presented in Chapter 3 to obtain the inequality

$$|\eta(t)| \leq |\theta^*| \|\Delta_m(s)\|_{2\delta} \|u_t\|_{2\delta} + |d(t)| \qquad (8.4.42)$$

that holds for any finite $\delta \geq 0$ provided $\Delta_m(s)$ is strictly proper and analytic in $\text{Re}[s] \geq -\frac{\delta}{2}$. If we now assume that in addition to being strictly proper, $\Delta_m(s)$ satisfies the following assumption:

A1. $\Delta_m(s)$ is analytic in $\text{Re}[s] \geq -\frac{\delta_0}{2}$ for some known $\delta_0 > 0$

then we can rewrite (8.4.42) as

$$|\eta(t)| \leq \mu_0 \|u_t\|_{2\delta_0} + d_0 \qquad (8.4.43)$$

where $\mu_0 = |\theta^*| \|\Delta_m(s)\|_{2\delta_0}$ and d_0 is an upper bound for $|d(t)|$. Because μ_0, d_0 are constants, inequality (8.4.43) motivates the normalizing signal m given by

$$m^2 = 1 + u^2 + \|u_t\|_{2\delta_0}^2 \qquad (8.4.44)$$

that bounds both u, η from above. It may be generated by the equations

$$\begin{aligned} \dot{m}_s &= -\delta_0 m_s + u^2, \quad m_s(0) = 0 \\ n_s^2 &= m_s, \quad m^2 = 1 + u^2 + n_s^2 \end{aligned} \qquad (8.4.45)$$

We refer to $m_s = n_s^2$ as the *dynamic normalizing signal* in order to distinguish it from the *static* one given by $m^2 = 1 + u^2$.

Because m bounds u, η in (8.4.40), we can generate a wide class of adaptive laws for the now bounded modeling error term $\bar{\eta} = \frac{\eta}{m}$ by following the procedure of the previous sections. By considering the normalized equation (8.4.41) with m given by (8.4.45) and using the results of Section 8.3.1, we obtain

$$\dot{\theta} = \gamma \bar{\epsilon}_1 \bar{u} - \gamma w \theta \qquad (8.4.46)$$

where

$$\bar{\epsilon}_1 = \bar{y} - \hat{\bar{y}} = \frac{y - \hat{y}}{m}, \quad \hat{y} = \theta u$$

and w is the leakage term to be chosen.

8.4. MODIFICATIONS FOR ROBUSTNESS: SIMPLE EXAMPLES

As in the ideal case considered in Chapter 4, we can express (8.4.46) in terms of the normalized estimation error

$$\epsilon = \frac{y - \hat{y}}{m^2}, \quad \hat{y} = \theta u$$

to obtain

$$\dot{\theta} = \gamma \epsilon u - \gamma w \theta \qquad (8.4.47)$$

or we can develop (8.4.47) directly by using the gradient method to minimize

$$J(\theta) = \frac{\epsilon^2 m^2}{2} + \frac{w\theta^2}{2} = \frac{(y - \theta u)^2}{2m^2} + \frac{w\theta^2}{2}$$

with respect to θ.

Let us now summarize the main equations of the adaptive law (8.4.47) for the parametric model $y = \theta^* u + \eta$ given by (8.4.40). We have

$$\begin{aligned}
\dot{\theta} &= \gamma \epsilon u - \gamma w \theta, \quad \epsilon = \frac{y - \hat{y}}{m^2}, \quad \hat{y} = \theta u \\
m^2 &= 1 + u^2 + n_s^2, \quad m_s = n_s^2 \\
\dot{m}_s &= -\delta_0 m_s + u^2, \quad m_s(0) = 0
\end{aligned} \qquad (8.4.48)$$

The analysis of (8.4.48) for the various choices of the leakage term $w(t)$ is very similar to that presented in Section 8.4.1. As a demonstration, we analyze (8.4.48) for the fixed σ-modification, i.e., $w(t) = \sigma$. From (8.4.40), (8.4.48) we obtain the error equations

$$\epsilon = \frac{-\tilde{\theta}u + \eta}{m^2}, \quad \dot{\tilde{\theta}} = \gamma \epsilon u - \gamma \sigma \theta \qquad (8.4.49)$$

where $\tilde{\theta} \triangleq \theta - \theta^*$. We consider the Lyapunov function

$$V = \frac{\tilde{\theta}^2}{2\gamma}$$

whose time derivative $\dot{V}$ along the solution of (8.4.49) is given by

$$\dot{V} = \epsilon \tilde{\theta} u - \sigma \tilde{\theta} \theta = -\epsilon^2 m^2 + \epsilon \eta - \sigma \tilde{\theta} \theta$$

By completing the squares, we obtain

$$\dot{V} \leq -\frac{\epsilon^2 m^2}{2} - \frac{\sigma \tilde{\theta}^2}{2} + \frac{\eta^2}{2m^2} + \frac{\sigma |\theta^*|^2}{2}$$

Because $\frac{\eta}{m} \in \mathcal{L}_\infty$ it follows that $V, \tilde{\theta} \in \mathcal{L}_\infty$ and $\epsilon m \in \mathcal{S}(\frac{\eta^2}{m^2} + \sigma)$. Because $\dot{\tilde{\theta}} = \gamma \epsilon m \frac{u}{m} - \gamma \sigma \theta$ and $\frac{u}{m} \in \mathcal{L}_\infty$, we can establish as in Section 8.4.1 that $\dot{\tilde{\theta}} \in \mathcal{S}(\frac{\eta^2}{m^2} + \sigma)$. The boundedness of $\tilde{\theta}, \frac{u}{m}, \frac{\eta}{m}$ also implies by (8.4.49) that $\epsilon, \epsilon m, \dot{\theta} \in \mathcal{L}_\infty$. The properties of (8.4.48) with $w = \sigma$ are therefore given by

- (i) $\epsilon, \epsilon m, \theta, \dot{\theta} \in \mathcal{L}_\infty$

- (ii) $\epsilon, \epsilon m, \dot{\theta} \in \mathcal{S}(\frac{\eta^2}{m^2} + \sigma)$

In a similar manner, other choices for $w(t)$ and modifications, such as dead zone and projections, may be used together with the normalizing signal (8.4.45) to design adaptive laws that are robust with respect to the dynamic uncertainty η.

The use of dynamic normalization was first introduced in [48] to handle the effects of bounded disturbances which could also be handled by static normalization. The use of dynamic normalization in dealing with dynamic uncertainties in adaptive control was first pointed out in [193] where it was used to develop some of the first global results in robust adaptive control for discrete-time plants. The continuous-time version of the dynamic normalization was first used in [91, 113] to design robust MRAC schemes for plants with additive and multiplicative plant uncertainties. Following the work of [91, 113, 193], a wide class of robust adaptive laws incorporating dynamic normalizations are developed for both continuous- and discrete-time plants.

Remark 8.4.2 The leakage, projection, and dead-zone modifications are not necessary for signal boundedness when u in the parametric model (8.4.1) is bounded and PE and $\eta \in \mathcal{L}_\infty$, as indicated in Section 8.3.1. The PE property guarantees exponential stability in the absence of modeling errors which in turn guarantees bounded states in the presence of bounded modeling error inputs provided the modeling error term doesnot destroy the PE property of the input. In this case the steady state bounds for the parameter and estimation error are of the order of the modeling error, which implies that phenomena, such as bursting, where the estimation error assumes values larger than the order of the modeling error at steady state, are not present. The use of PE signals to improve robustness and performance is discussed in subsequent sections. When u, η are not necessarily bounded, the above remarks hold provided $\frac{u}{m} \in \mathcal{L}_\infty$ is PE.

8.5 Robust Adaptive Laws

In Chapter 4 we developed a wide class of adaptive laws for estimating on-line a constant parameter vector θ^* in certain plant parametric models. The vector θ^* could contain the unknown coefficients of the plant transfer function or the coefficients of various other plant parameterizations. The adaptive laws of Chapter 4 are combined with control laws in Chapters 6, 7 to form adaptive control schemes that are shown to meet the control objectives for the plant model under consideration.

A crucial assumption made in the design and analysis of the adaptive schemes of Chapters 4 to 7 is that the plant model is an ideal one, i.e., it is free of disturbances and modeling errors and the plant parameters are constant for all time.

The simple examples of Section 8.3 demonstrate that the stability properties of the adaptive schemes developed in Chapters 4 to 7 can no longer be guaranteed in the presence of bounded disturbances, unmodeled plant dynamics and parameter variations. The main cause of possible instabilities in the presence of modeling errors is the adaptive law that makes the overall adaptive scheme time varying and nonlinear. As demonstrated in Section 8.4 using a simple example, the destabilizing effects of bounded disturbances and of a class of dynamic uncertainties may be counteracted by using simple modifications that involve leakage, dead-zone, projection, and dynamic normalization. In this section we extend the results of Section 8.4 to a general class of parametric models with modeling errors that may arise in the on-line parameter estimation problem of a wide class of plants.

We start with the following section where we develop several parametric models with modeling errors that are used in subsequent sections to develop adaptive laws that are robust with respect to uncertainties. We refer to such adaptive laws as *robust adaptive laws*.

8.5.1 Parametric Models with Modeling Error

Linear Parametric Models

Let us start with the plant

$$y = G_0(s)(1 + \Delta_m(s))u + d \qquad (8.5.1)$$

where

$$G_0(s) = \frac{Z(s)}{R(s)} \qquad (8.5.2)$$

represents the dominant or modeled part of the plant transfer function with $Z(s) = b_{n-1}s^{n-1} + b_{n-2}s^{n-2} + \ldots + b_1 s + b_0$ and $R(s) = s^n + a_{n-1}s^{n-1} + \ldots + a_1 s + a_0$; $\Delta_m(s)$ is a multiplicative perturbation and d is a bounded disturbance.

We would like to express (8.5.1) in the form where the coefficients of $Z(s), R(s)$ lumped in the vector

$$\theta^* = [b_{n-1}, b_{n-2}, \ldots, b_0, a_{n-1}, a_{n-2}, \ldots, a_0]^\top$$

are separated from signals as done in the ideal case, where $\Delta_m = 0, d = 0$, presented in Section 2.4.1. From (8.5.1) we have

$$Ry = Zu + Z\Delta_m u + Rd \qquad (8.5.3)$$

As in Section 2.4.1, to avoid the presence of derivatives of signals in the parametric model, we filter each side of (8.5.3) with $\frac{1}{\Lambda(s)}$, where $\Lambda(s)$ is a monic Hurwitz polynomial of degree n, to obtain

$$\frac{R}{\Lambda}y = \frac{Z}{\Lambda}u + \eta_m$$

where

$$\eta_m = \frac{Z\Delta_m}{\Lambda}u + \frac{R}{\Lambda}d$$

is the modeling error term because of Δ_m, d. If instead of (8.5.1), we consider a plant with an additive perturbation, i.e.,

$$y = G_0(s)u + \Delta_a(s)u + d$$

we obtain

$$\frac{R}{\Lambda}y = \frac{Z}{\Lambda}u + \eta_a$$

8.5. ROBUST ADAPTIVE LAWS

where
$$\eta_a = \frac{R}{\Lambda}\Delta_a u + \frac{R}{\Lambda}d$$

Similarly, if we consider a plant with stable factor perturbations, i.e.,
$$y = \frac{N_0(s) + \Delta_1(s)}{D_0(s) + \Delta_2(s)}u + d$$

where
$$N_0(s) = \frac{Z(s)}{\Lambda(s)}, \quad D_0(s) = \frac{R(s)}{\Lambda(s)}$$

and N_0, D_0 are proper stable transfer functions that are coprime, we obtain
$$\frac{R}{\Lambda}y = \frac{Z}{\Lambda}u + \eta_s$$

where
$$\eta_s = \Delta_1 u - \Delta_2 y + (\frac{R}{\Lambda} + \Delta_2)d$$

Therefore, without loss of generality we can consider the plant parameterization
$$\frac{R}{\Lambda}y = \frac{Z}{\Lambda}u + \eta \tag{8.5.4}$$

where
$$\eta = \Delta_y(s)y + \Delta_u(s)u + d_1 \tag{8.5.5}$$

is the modeling error with Δ_y, Δ_u being stable transfer functions and d_1 being a bounded disturbance and proceed as in the ideal case to obtain a parametric model that involves θ^*. If we define

$$z \triangleq \frac{s^n}{\Lambda(s)}y, \quad \phi \triangleq \left[\frac{\alpha_{n-1}^T(s)}{\Lambda(s)}u, -\frac{\alpha_{n-1}^T(s)}{\Lambda(s)}y\right]^T$$

where $\alpha_i(s) \triangleq [s^i, s^{i-1}, \ldots, s, 1]^T$, as in Section 2.4.1, we can rewrite (8.5.4) in the form
$$z = \theta^{*T}\phi + \eta \tag{8.5.6}$$

or in the form
$$y = \theta_\lambda^{*T}\phi + \eta \tag{8.5.7}$$

where $\theta_\lambda^{*T} = [\theta_1^{*T}, \theta_2^{*T} - \lambda^T]^T$, $\theta_1^* = [b_{n-1}, \ldots, b_0]^T$, and $\theta_2^* = [a_{n-1}, \ldots, a_0]^T$; $\lambda = [\lambda_{n-1}, \ldots, \lambda_0]^T$ is the coefficient vector of $\Lambda(s) - s^n = \lambda_{n-1}s^{n-1} + \cdots + \lambda_0$.

The effect of possible nonzero initial conditions in the overall plant state-space representation may be also included in (8.5.6), (8.5.7) by following the same procedure as in the ideal case. It can be shown that the initial conditions appear as an exponentially decaying to zero term η_0, i.e.,

$$z = \theta^{*T}\phi + \eta + \eta_0$$
$$y = \theta_\lambda^{*T}\phi + \eta + \eta_0 \qquad (8.5.8)$$

where η_0 is the output of the system

$$\dot{\omega} = \Lambda_c \omega, \quad \omega(0) = \omega_0$$
$$\eta_0 = C_0^T \omega$$

Λ_c is a stable matrix whose eigenvalues are equal to the poles of $\Delta_m(s)$ or $\Delta_a(s)$, or $\Delta_1(s)$ and $\Delta_2(s)$ and $\Lambda(s)$, and the degree of ω is equal to the order of the overall plant.

The parametric models given by (8.5.6) to (8.5.8) correspond to Parameterization 1 in Section 2.4.1. Parameterization 2 developed in the same section for the plant with $\eta = 0$ may be easily extended to the case of $\eta \neq 0$ to obtain

$$y = W_m(s)\theta_\lambda^{*T}\psi + \eta$$
$$z = W_m(s)\theta^{*T}\psi + \eta \qquad (8.5.9)$$

where $W_m(s)$ is a stable transfer function with relative degree 1 and $\psi = W_m^{-1}(s)\phi$.

Parametric models (8.5.6), (8.5.7) and (8.5.9) may be used to estimate on-line the parameter vector θ^* associated with the dominant part of the plant characterized by the transfer function $G_0(s)$. The only signals available for measurements in these parametric models are the plant output y and the signals ϕ, ψ, and z that can be generated by filtering the plant input u and output y as in Section 2.4.1. The modeling error term η due to the unmodeled dynamics and bounded disturbance is unknown, and is to be treated as an unknown disturbance term that is not necessarily bounded. If, however, Δ_y, Δ_u are proper then $d_1, y, u \in \mathcal{L}_\infty$ will imply that $\eta \in \mathcal{L}_\infty$. The properness of Δ_y, Δ_u may be established by assuming that $G_0(s)$ and the overall plant transfer function are proper. In fact Δ_y, Δ_u can be made strictly proper by filtering each side of (8.5.4) with a first order stable filter without affecting the form of the parametric models (8.5.6) to (8.5.9).

8.5. ROBUST ADAPTIVE LAWS

Bilinear Parametric Models

Let us now extend the bilinear parametric model of Section 2.4.2 developed for the ideal plant to the plant that includes a multiplicative perturbation and a bounded disturbance, i.e., consider the plant

$$y = G_0(s)(1 + \Delta_m(s))u + d \qquad (8.5.10)$$

where

$$G_0(s) = k_p \frac{Z_p(s)}{R_p(s)}$$

k_p is a constant, $R_p(s)$ is monic and of degree n, $Z_p(s)$ is monic Hurwitz of degree $m < n$ and k_p, Z_p, R_p satisfy the Diophantine equation

$$R_p Q + k_p Z_p P = Z_p A \qquad (8.5.11)$$

where

$$Q(s) = s^{n-1} + q^\top \alpha_{n-2}(s), \quad P(s) = p^\top \alpha_{n-1}(s)$$

and $A(s)$ is a monic Hurwitz polynomial of degree $2n - 1$. As in Section 2.4.2, our objective is to obtain a parameterization of the plant in terms of the coefficient vectors q, p of Q, P, respectively, by using (8.5.11) to substitute for $Z_p(s), R_p(s)$ in (8.5.10). This parameterization problem appears in direct MRAC where q, p are the controller parameters that we like to estimate on-line. Following the procedure of Section 2.4.2, we express (8.5.10) in the form

$$R_p y = k_p Z_p u + k_p Z_p \Delta_m u + R_p d$$

which implies that

$$Q R_p y = k_p Z_p Q u + k_p Z_p Q \Delta_m u + Q R_p d \qquad (8.5.12)$$

From (8.5.11) we have $Q R_p = Z_p A - k_p Z_p P$, which we use in (8.5.12) to obtain

$$Z_p(A - k_p P)y = k_p Z_p Q u + k_p Z_p Q \Delta_m u + Q R_p d$$

Because $Z_p(s), A(s)$ are Hurwitz we can filter both sides of the above equation with $1/(Z_p A)$ and rearrange the terms to obtain

$$y = k_p \frac{P}{A} y + k_p \frac{Q}{A} u + k_p \frac{Q \Delta_m}{A} u + \frac{Q R_p}{A Z_p} d$$

Substituting for $P(s) = p^\top \alpha_{n-1}(s)$, $Q(s) = s^{n-1} + q^\top \alpha_{n-2}(s)$, we obtain

$$y = \frac{\Lambda(s)}{A(s)} k_p \left[p^\top \frac{\alpha_{n-1}(s)}{\Lambda(s)} y + q^\top \frac{\alpha_{n-2}(s)}{\Lambda(s)} u + \frac{s^{n-1}}{\Lambda(s)} u \right] + \eta \qquad (8.5.13)$$

where $\Lambda(s)$ is a Hurwitz polynomial of degree n_λ that satisfies $2n - 1 \geq n_\lambda \geq n - 1$ and

$$\eta = \frac{k_p Q \Delta_m}{A} u + \frac{Q R_p}{A Z_p} d \qquad (8.5.14)$$

is the modeling error resulting from Δ_m, d. We can verify that $\frac{Q\Delta_m}{A}, \frac{QR_p}{AZ_p}$ are strictly proper and biproper respectively with stable poles provided the overall plant transfer function is strictly proper. From (8.5.13), we obtain the bilinear parametric model

$$y = W(s)\rho^*(\theta^{*\top}\phi + z_0) + \eta \qquad (8.5.15)$$

where $\rho^* = k_p$,

$$W(s) = \frac{\Lambda(s)}{A(s)}$$

is a proper transfer function with stable poles and zeros and

$$\theta^* = [q^\top, p^\top]^\top, \quad \phi = \left[\frac{\alpha_{n-2}^\top(s)}{\Lambda(s)} u, \frac{\alpha_{n-1}^\top(s)}{\Lambda(s)} y \right]^\top, \quad z_0 = \frac{s^{n-1}}{\Lambda(s)} u$$

If instead of (8.5.10), we use the plant representation with an additive plant perturbation, we obtain (8.5.15) with

$$\eta = \frac{A - k_p P}{A}(\Delta_a u + d)$$

and in the case of a plant with stable factor perturbations Δ_1, Δ_2 we obtain (8.5.15) with

$$\eta = \frac{\Lambda Q}{A Z_p}(\Delta_1 u - \Delta_2 y) + \frac{\Lambda Q}{A Z_p}\left(\frac{R_p}{\Lambda} + \Delta_2\right) d$$

where Λ is now restricted to have degree n. If we assume that the overall plant transfer function is strictly proper for the various plant representations with perturbations, then a general form for the modeling error term is

$$\eta = \Delta_u u + \Delta_y y + d_1 \qquad (8.5.16)$$

8.5. ROBUST ADAPTIVE LAWS

where $\Delta_u(s), \Delta_y(s)$ are strictly proper with stable poles and d_1 is a bounded disturbance term.

Example 8.5.1 Consider the following plant

$$y = e^{-\tau s} \frac{b_1 s + b_0}{s^2 + a_1 s + a_0} \frac{1 - \mu s}{1 + \mu s} u$$

where $\mu > 0, \tau > 0$ are small constants. An approximate model of the plant is obtained by setting μ, τ to zero, i.e.,

$$y = G_0(s)u = \frac{b_1 s + b_0}{s^2 + a_1 s + a_0} u$$

Treating $G_0(s)$ as the nominal or modeled part of the plant, we can express the overall plant in terms of $G_0(s)$ and a multiplicative plant uncertainty, i.e.,

$$y = \frac{b_1 s + b_0}{s^2 + a_1 s + a_0}(1 + \Delta_m(s))u$$

where

$$\Delta_m(s) = \frac{(e^{-\tau s} - 1) - \mu s(1 + e^{-\tau s})}{1 + \mu s}$$

Let us now obtain a parametric model for the plant in terms of the parameter vector $\theta^* = [b_1, b_0, a_1, a_0]^T$. We have

$$(s^2 + a_1 s + a_0)y = (b_1 s + b_0)u + (b_1 s + b_0)\Delta_m(s)u$$

i.e.,

$$s^2 y = (b_1 s + b_0)u - (a_1 s + a_0)y + (b_1 s + b_0)\Delta_m(s)u$$

Filtering each side with $\frac{1}{(s+2)^2}$ we obtain

$$z = \theta^{*T}\phi + \eta$$

where

$$z = \frac{s^2}{(s+2)^2}y, \quad \phi = \left[\frac{s}{(s+2)^2}u, \frac{1}{(s+2)^2}u, -\frac{s}{(s+2)^2}y, -\frac{1}{(s+2)^2}y\right]^T$$

$$\eta = \frac{(b_1 s + b_0)}{(s+2)^2} \frac{[(e^{-\tau s} - 1) - \mu s(1 + e^{-\tau s})]}{(1 + \mu s)} u \qquad \triangledown$$

In the following sections, we use the linear and bilinear parametric models to design and analyze adaptive laws for estimating the parameter vector θ^* (and ρ^*) by treating η as an unknown modeling error term of the form given by (8.5.16). To avoid repetitions, we will consider the general parametric models

$$z = W(s)\theta^{*T}\psi + \eta \tag{8.5.17}$$

$$z = W(s)\rho^*[\theta^{*T}\psi + z_0] + \eta \tag{8.5.18}$$

$$\eta = \Delta_u(s)u + \Delta_y(s)y + d_1 \tag{8.5.19}$$

where $W(s)$ is a known proper transfer function with stable poles, z, ψ, and z_0 are signals that can be measured and Δ_u, Δ_y are strictly proper stable transfer functions, and $d_1 \in \mathcal{L}_\infty$.

We will show that the adaptive laws of Chapter 4 developed for the parametric models (8.5.17), (8.5.18) with $\eta = 0$ can be modified to handle the case where $\eta \neq 0$ is of the form (8.5.19).

8.5.2 SPR-Lyapunov Design Approach with Leakage

Let us consider the general parametric model (8.5.17) which we rewrite as

$$z = W(s)(\theta^{*T}\psi + W^{-1}(s)\eta)$$

and express it in the SPR form

$$z = W(s)L(s)(\theta^{*T}\phi + \eta_s) \tag{8.5.20}$$

where $\phi = L^{-1}(s)\psi$, $\eta_s = L^{-1}(s)W^{-1}(s)\eta$ and $L(s)$ is chosen so that $W(s)L(s)$ is a proper SPR transfer function and $L^{-1}(s)$ is proper and stable. The procedure for designing an adaptive law for estimating on-line the constant vector θ^* in the presence of the unknown modeling error term

$$\eta_s = L^{-1}(s)W^{-1}(s)(\Delta_u(s)u + \Delta_y(s)y + d_1) \tag{8.5.21}$$

is very similar to that in the ideal case ($\eta_s = 0$) presented in Chapter 4. As in the ideal case, the predicted value $\hat{z}$ of z based on the estimate θ of θ^* and the normalized estimation error ϵ are generated as

$$\hat{z} = W(s)L(s)\theta^T\phi, \quad \epsilon = z - \hat{z} - W(s)L(s)\epsilon n_s^2 \tag{8.5.22}$$

where $m^2 = 1 + n_s^2$ is the normalizing signal to be designed. From (8.5.20) and (8.5.22), we obtain the error equation

$$\epsilon = WL[-\tilde{\theta}^T\phi - \epsilon n_s^2 + \eta_s] \tag{8.5.23}$$

8.5. ROBUST ADAPTIVE LAWS

where $\tilde{\theta} \triangleq \theta - \theta^*$. Without loss of generality, we assume that WL is strictly proper and therefore (8.5.23) may assume the following minimal state representation

$$\dot{e} = A_c e + B_c[-\tilde{\theta}^T \phi - \epsilon n_s^2 + \eta_s]$$
$$\epsilon = C_c^T e.$$

Because $C_c^T(sI - A_c)^{-1} B_c = W(s)L(s)$ is SPR, the triple (A_c, B_c, C_c) satisfies the following equations given by the LKY Lemma

$$P_c A_c + A_c^T P_c = -qq^T - \nu_c L_c$$
$$P_c B_c = C_c$$

where $P_c = P_c^T > 0$, q is a vector, $\nu_c > 0$ is a scalar and $L_c = L_c^T > 0$. The adaptive law is now developed by choosing the Lyapunov-like function

$$V(e, \tilde{\theta}) = \frac{e^T P_c e}{2} + \frac{\tilde{\theta}^T \Gamma^{-1} \tilde{\theta}}{2}$$

where $\Gamma = \Gamma^T > 0$ is arbitrary, and designing $\dot{\tilde{\theta}} = \dot{\theta}$ so that the properties of $\dot{V}$ allow us to conclude boundedness for V and, therefore, $e, \tilde{\theta}$. We have

$$\dot{V} = -\frac{1}{2} e^T qq^T e - \frac{\nu_c}{2} e^T L_c e - \tilde{\theta}^T \phi \epsilon - \epsilon^2 n_s^2 + \epsilon \eta_s + \tilde{\theta}^T \Gamma^{-1} \dot{\theta} \quad (8.5.24)$$

If $\eta_s = 0$, we would proceed as in Chapter 4 and choose $\dot{\theta} = \Gamma \epsilon \phi$ in order to eliminate the indefinite term $-\tilde{\theta}^T \phi \epsilon$. Because $\eta_s \neq 0$ and η_s is not necessarily bounded we also have to deal with the indefinite and possibly unbounded term $\epsilon \eta_s$ in addition to the term $-\tilde{\theta}^T \phi \epsilon$. Our objective is to design the normalizing signal $m^2 = 1 + n_s^2$ to take care of the term $\epsilon \eta_s$.

For any constant $\alpha \in (0,1)$ to be specified later, we have

$$\begin{aligned}
-\epsilon^2 n_s^2 + \epsilon \eta_s &= -\epsilon^2 n_s^2 + \epsilon \eta_s - \alpha \epsilon^2 + \alpha \epsilon^2 \\
&= -(1-\alpha)\epsilon^2 n_s^2 + \alpha \epsilon^2 - \alpha \epsilon^2 m^2 + \epsilon \eta_s \\
&= -(1-\alpha)\epsilon^2 n_s^2 + \alpha \epsilon^2 - \alpha \left(\epsilon m - \frac{\eta_s}{2\alpha m}\right)^2 + \frac{\eta_s^2}{4\alpha m^2} \\
&\leq -(1-\alpha)\epsilon^2 n_s^2 + \alpha \epsilon^2 + \frac{\eta_s^2}{4\alpha m^2}
\end{aligned}$$

which together with $|\epsilon| \leq |C_c||e|$ and

$$-\frac{\nu_c}{4} e^T L_c e \leq -\frac{\nu_c \lambda_{min}}{4}(L_c)|e|^2 \leq -\frac{\nu_c}{4} \frac{\lambda_{min}(L_c)}{|C_c|^2} \epsilon^2$$

imply that

$$\dot{V} \le -\lambda_0|e|^2 - (\beta_0 - \alpha)\epsilon^2 - (1-\alpha)\epsilon^2 n_s^2 + \frac{\eta_s^2}{4\alpha m^2} - \tilde{\theta}^T \phi\epsilon + \tilde{\theta}^T \Gamma^{-1}\dot{\tilde{\theta}}$$

where $\lambda_0 = \frac{\nu_c}{4}\lambda_{min}(L_c)$, $\beta_0 = \lambda_0/|C_c|^2$. Choosing $\alpha = min(\frac{1}{2}, \beta_0)$, we obtain

$$\dot{V} \le -\lambda_0|e|^2 - \frac{1}{2}\epsilon^2 n_s^2 + \frac{\eta_s^2}{4\alpha m^2} - \tilde{\theta}^T \phi\epsilon + \tilde{\theta}^T \Gamma^{-1}\dot{\tilde{\theta}}$$

If we now design the normalizing signal $m^2 = 1 + n_s^2$ so that $\eta_s/m \in \mathcal{L}_\infty$ (in addition to $\phi/m \in \mathcal{L}_\infty$ required in the ideal case), then the positive term $\eta_s^2/4\alpha m^2$ is guaranteed to be bounded from above by a finite constant and can, therefore, be treated as a bounded disturbance.

Design of the Normalizing Signal

The design of m is achieved by considering the modeling error term

$$\eta_s = W^{-1}(s)L^{-1}(s)[\Delta_u(s)u + \Delta_y(s)y + d_1]$$

and the following assumptions:

(A1) $W^{-1}(s)L^{-1}(s)$ is analytic in $\text{Re}[s] \ge -\frac{\delta_0}{2}$ for some known $\delta_0 > 0$

(A2) Δ_u, Δ_y are strictly proper and analytic in $\text{Re}[s] \ge -\frac{\delta_0}{2}$

Because WL is strictly proper and SPR, it has relative degree 1 which means that $W^{-1}L^{-1}\Delta_u, W^{-1}L^{-1}\Delta_y$ are proper and at most biproper transfer functions. Hence η_s may be expressed as

$$\eta_s = c_1 u + c_2 y + \Delta_1(s)u + \Delta_2(s)y + d_2$$

where $c_1 + \Delta_1(s) = W^{-1}(s)L^{-1}(s)\Delta_u(s)$, $c_2 + \Delta_2(s) = W^{-1}(s)L^{-1}(s)\Delta_y(s)$, $d_2 = W^{-1}(s)L^{-1}(s)d_1$ and Δ_1, Δ_2 are strictly proper. Using the properties of the $\mathcal{L}_{2\delta}$-norm from Chapter 3, we have

$$|\eta_s(t)| \le c(|u(t)| + |y(t)|) + \|\Delta_1(s)\|_{2\delta_0}\|u_t\|_{2\delta_0}$$
$$+ \|\Delta_2(s)\|_{2\delta_0}\|y_t\|_{2\delta_0} + |d_2(t)|$$

8.5. ROBUST ADAPTIVE LAWS

Because $\|\Delta_1(s)\|_{2\delta_0}, \|\Delta_2(s)\|_{2\delta_0}$ are finite and $d_2 \in \mathcal{L}_\infty$, it follows that the choice

$$m^2 = 1 + n_s^2, \quad n_s^2 = m_s + \phi^\top \phi + u^2 + y^2$$
$$\dot{m}_s = -\delta_0 m_s + u^2 + y^2, \quad m_s(0) = 0 \tag{8.5.25}$$

will guarantee that $\eta_s/m \in \mathcal{L}_\infty$ and $\phi/m \in \mathcal{L}_\infty$.

With the above choice for m, the term $\eta_s^2/(4\alpha m^2) \in \mathcal{L}_\infty$ which motivates the adaptive law

$$\dot{\theta} = \Gamma \epsilon \phi - w(t)\Gamma \theta \tag{8.5.26}$$

where $\Gamma = \Gamma^\top > 0$ is the adaptive gain and $w(t) \geq 0$ is a scalar signal to be designed so that $\dot{V} \leq 0$ whenever $V \geq V_0$ for some $V_0 \geq 0$. In view of (8.5.26), we have

$$\dot{V} \leq -\lambda_0 |e|^2 - \frac{1}{2}\epsilon^2 n_s^2 - w\tilde{\theta}^\top \theta + \frac{\eta_s^2}{4\alpha m^2} \tag{8.5.27}$$

which indicates that if w is chosen to make $\dot{V}$ negative whenever $|\tilde{\theta}|$ exceeds a certain bound that depends on $\eta_s^2/(4\alpha m^2)$, then we can establish the existence of $V_0 \geq 0$ such that $\dot{V} \leq 0$ whenever $V \geq V_0$.

The following theorems describe the stability properties of the adaptive law (8.5.26) for different choices of $w(t)$.

Theorem 8.5.1 (Fixed σ Modification) *Let*

$$w(t) = \sigma > 0, \quad \forall t \geq 0$$

where σ is a small constant. The adaptive law (8.5.26) guarantees that

(i) $\theta, \epsilon \in \mathcal{L}_\infty$

(ii) $\epsilon, \epsilon n_s, \dot{\theta} \in \mathcal{S}(\sigma + \eta_s^2/m^2)$

(iii) *In addition to properties (i), (ii), if $n_s, \phi, \dot{\phi} \in \mathcal{L}_\infty$, and ϕ is PE with level $\alpha_0 > 0$ that is independent of η_s, then the parameter error $\tilde{\theta} = \theta - \theta^*$ converges exponentially to the residual set*

$$D_\sigma = \left\{ \tilde{\theta} \,\middle|\, |\tilde{\theta}| \leq c(\sigma + \bar{\eta}) \right\}$$

where $c \in \mathcal{R}^+$ depends on α_0 and $\bar{\eta} = \sup_t |\eta_s|$.

Proof We have

$$-\sigma \tilde{\theta}^T \theta \leq -\sigma(\tilde{\theta}^T \tilde{\theta} - |\tilde{\theta}||\theta^*|) \leq -\frac{\sigma}{2}|\tilde{\theta}|^2 + \frac{\sigma}{2}|\theta^*|^2$$

and

$$-\sigma \tilde{\theta}^T \theta \leq -\frac{\sigma}{2}|\theta|^2 + \frac{\sigma}{2}|\theta^*|^2$$

Hence, for $w = \sigma$, (8.5.27) becomes

$$\dot{V} \leq -\lambda_0 |e|^2 - \frac{\epsilon^2 n_s^2}{2} - \frac{\sigma}{2}|\tilde{\theta}|^2 + c_0 \frac{\eta_s^2}{m^2} + \sigma \frac{|\theta^*|^2}{2} \qquad (8.5.28)$$

where $c_0 = \frac{1}{4\alpha}$ or

$$\dot{V} \leq -\beta V + c_0 \bar{\eta}_m + \sigma \frac{|\theta^*|^2}{2} + \beta \left(\frac{e^T P_c e}{2} + \frac{\tilde{\theta}^T \Gamma^{-1} \tilde{\theta}}{2} \right) - \lambda_0 |e|^2 - \frac{\sigma}{2}|\tilde{\theta}|^2 \qquad (8.5.29)$$

where $\bar{\eta}_m \triangleq \sup_t \frac{n_s^2}{m^2}$ and β is an arbitrary constant to be chosen.

Because $e^T P_c e + \tilde{\theta}^T \Gamma^{-1} \tilde{\theta} \leq |e|^2 \lambda_{max}(P_c) + |\tilde{\theta}|^2 \lambda_{max}(\Gamma^{-1})$, it follows that for

$$\beta = \min \left[\frac{2\lambda_0}{\lambda_{max}(P_c)}, \frac{\sigma}{\lambda_{max}(\Gamma^{-1})} \right]$$

(8.5.29) becomes

$$\dot{V} \leq -\beta V + c_0 \bar{\eta}_m + \sigma \frac{|\theta^*|^2}{2}$$

Hence, for $V \geq V_0 = \frac{2c_0 \bar{\eta}_m + \sigma |\theta^*|^2}{2\beta}$, $\dot{V} \leq 0$ which implies that $V \in \mathcal{L}_\infty$ and therefore $e, \epsilon, \tilde{\theta}, \theta \in \mathcal{L}_\infty$.

We can also use $-\sigma \tilde{\theta}^T \theta \leq -\frac{\sigma}{2}|\theta|^2 + \frac{\sigma}{2}|\theta^*|^2$ and rewrite (8.5.27) as

$$\dot{V} \leq -\lambda_0 |e|^2 - \frac{\epsilon^2 n_s^2}{2} - \frac{\sigma}{2}|\theta|^2 + c_0 \frac{\eta_s^2}{m^2} + \frac{\sigma}{2}|\theta^*|^2 \qquad (8.5.30)$$

Integrating both sides of (8.5.30), we obtain

$$\int_{t_0}^{t} \left(\lambda_0 |e|^2 + \frac{\epsilon^2 n_s^2}{2} + \frac{\sigma}{2}|\theta|^2 \right) d\tau \leq \int_{t_0}^{t} \left(c_0 \frac{\eta_s^2}{m^2} + \sigma \frac{|\theta^*|^2}{2} \right) d\tau + V(t_0) - V(t)$$

$\forall t \geq t_0$ and any $t_0 \geq 0$. Because $V \in \mathcal{L}_\infty$ and $\epsilon = C_c^T e$, it follows that $e, \epsilon, \epsilon n_s, \sqrt{\sigma}|\theta| \in \mathcal{S}(\sigma + \eta_s^2/m^2)$. Using the property $\frac{\phi}{m} \in \mathcal{L}_\infty$, it follows from (8.5.26) with $w = \sigma$ that

$$|\dot{\theta}|^2 \leq c(|\epsilon m|^2 + \sigma|\theta|^2)$$

for some constant $c > 0$ and, therefore, $\epsilon, \epsilon n_s, \sqrt{\sigma}|\theta| \in \mathcal{S}(\sigma + \eta_s^2/m^2)$ and $\theta \in \mathcal{L}_\infty$ imply that ϵm and $\dot{\theta} \in \mathcal{S}(\sigma + \eta_s^2/m^2)$.

8.5. ROBUST ADAPTIVE LAWS

To establish property (iii), we write (8.5.23), (8.5.26) as

$$\begin{aligned} \dot{e} &= A_c e + B_c(-\tilde{\theta}^\top \phi - \epsilon n_s^2) + B_c \eta_s \\ \dot{\tilde{\theta}} &= \Gamma \epsilon \phi - \sigma \Gamma \theta \\ \epsilon &= C_c^\top e \end{aligned} \qquad (8.5.31)$$

Consider (8.5.31) as a linear time-varying system with $\eta_s, \sigma\theta$ as inputs and $e, \tilde{\theta}$ as states. In Section 4.8.2, we have shown that the homogeneous part of (8.5.31) with $\eta_s \equiv 0, \sigma \equiv 0$ is e.s. provided that $n_s, \phi, \dot{\phi} \in \mathcal{L}_\infty$ and ϕ is PE. Therefore, the state transition matrix of the homogeneous part of (8.5.31) satisfies

$$\|\Phi(t, t_0)\| \le \beta_1 e^{-\beta_2(t-t_0)} \qquad (8.5.32)$$

for any $t, t_0 > 0$ and some constants $\beta_1, \beta_2 > 0$. Equations (8.5.31) and (8.5.32) imply that $\tilde{\theta}$ satisfies the inequality

$$|\tilde{\theta}| \le \beta_0 e^{-\beta_2 t} + \beta_1' \int_0^t e^{-\beta_2(t-\tau)}(|\eta_s| + \sigma|\theta|)d\tau \qquad (8.5.33)$$

for some constants $\beta_0, \beta_1' > 0$, provided that the PE property of ϕ is not destroyed by $\eta_s \ne 0$. Because $\theta \in \mathcal{L}_\infty$, which is established in (i), and $\phi \in \mathcal{L}_\infty$, it follows from (8.5.33) that

$$|\tilde{\theta}| \le \beta_0 e^{-\beta_2 t} + \frac{\beta_1'}{\beta_2}(\bar{\eta} + \sigma \sup_t |\theta|)$$

where $\bar{\eta} = \sup_t |\eta_s(t)|$. Hence, (iii) is proved by setting $c = \frac{\beta_1'}{\beta_2} max\{1, \sup_t |\theta|\}$. □

Theorem 8.5.2 (Switching σ) *Let*

$$w(t) = \sigma_s, \quad \sigma_s = \begin{cases} 0 & if\ |\theta| \le M_0 \\ \left(\frac{|\theta|}{M_0} - 1\right)^{q_0} \sigma_0 & if\ M_0 < |\theta| \le 2M_0 \\ \sigma_0 & if\ |\theta| > 2M_0 \end{cases}$$

where $q_0 \ge 1$ is any finite integer and σ_0, M_0 are design constants with $M_0 > |\theta^*|$ and $\sigma_0 > 0$. Then the adaptive law (8.5.26) guarantees that

(i) $\theta, \epsilon \in \mathcal{L}_\infty$

(ii) $\epsilon, \epsilon n_s, \dot{\theta} \in \mathcal{S}(\eta_s^2/m^2)$

(iii) *In the absence of modeling errors, i.e., when $\eta_s = 0$, property (ii) can be replaced with* (ii') $\epsilon, \epsilon n_s, \dot{\tilde{\theta}} \in \mathcal{L}_2$.

(iv) *In addition, if $n_s, \phi, \dot{\phi} \in \mathcal{L}_\infty$ and ϕ is PE with level $\alpha_0 > 0$ that is independent of η_s, then*

(a) *$\tilde{\theta}$ converges exponentially to the residual set*

$$D_s = \left\{ \tilde{\theta} \,\middle|\, |\tilde{\theta}| \leq c(\sigma_0 + \bar{\eta}) \right\}$$

where $c \in \mathcal{R}^+$ and $\bar{\eta} = \sup_t |\eta_s|$

(b) *There exists a constant $\bar{\eta}^* > 0$ such that for $\bar{\eta} < \bar{\eta}^*$, the parameter error $\tilde{\theta}$ converges exponentially fast to the residual set*

$$\bar{D}_s = \left\{ \tilde{\theta} \,\middle|\, |\tilde{\theta}| \leq c\bar{\eta} \right\}$$

Proof We have

$$\sigma_s \tilde{\theta}^\top \theta = \sigma_s(|\theta|^2 - \theta^{*\top}\theta) \geq \sigma_s |\theta|(|\theta| - M_0 + M_0 - |\theta^*|)$$

Because $\sigma_s(|\theta| - M_0) \geq 0$ and $M_0 > |\theta^*|$, it follows that

$$\sigma_s \tilde{\theta}^\top \theta \geq \sigma_s |\theta|(|\theta| - M_0) + \sigma_s |\theta|(M_0 - |\theta^*|) \geq \sigma_s |\theta|(M_0 - |\theta^*|) \geq 0$$

i.e.,

$$\sigma_s |\theta| \leq \sigma_s \frac{\tilde{\theta}^\top \theta}{M_0 - |\theta^*|} \qquad (8.5.34)$$

The inequality (8.5.27) for $\dot{V}$ with $w = \sigma_s$ can be written as

$$\dot{V} \leq -\lambda_0 |e|^2 - \frac{\epsilon^2 n_s^2}{2} - \sigma_s \tilde{\theta}^\top \theta + c_0 \frac{\eta_s^2}{m^2} \qquad (8.5.35)$$

Because for $|\theta| = |\tilde{\theta} + \theta^*| > 2M_0$, the term $-\sigma_s \tilde{\theta}^\top \theta = -\sigma_0 \tilde{\theta}^\top \theta \leq -\frac{\sigma_0}{2}|\tilde{\theta}|^2 + \frac{\sigma_0}{2}|\theta^*|^2$ behaves as the equivalent fixed-σ term, we can follow the same procedure as in the proof of Theorem 8.5.1 to show the existence of a constant $V_0 > 0$ for which $\dot{V} \leq 0$ whenever $V \geq V_0$ and conclude that $V, e, \epsilon, \theta, \tilde{\theta} \in \mathcal{L}_\infty$.

Integrating both sides of (8.5.35) from t_0 to t, we obtain that $e, \epsilon, \epsilon n_s, \epsilon m, \sqrt{\sigma_s \tilde{\theta}^\top \theta} \in \mathcal{S}(\eta_s^2/m^2)$. From (8.5.34), it follows that

$$\sigma_s^2 |\theta|^2 \leq c_2 \sigma_s \tilde{\theta}^\top \theta$$

for some constant $c_2 > 0$ that depends on the bound for $\sigma_0 |\theta|$, and, therefore,

$$|\dot{\theta}|^2 \leq c(|\epsilon m|^2 + \sigma_s \tilde{\theta}^\top \theta), \quad \text{for some} \ c \in \mathcal{R}^+$$

8.5. ROBUST ADAPTIVE LAWS

Because $\epsilon m, \sqrt{\sigma_s \tilde{\theta}^T \theta} \in \mathcal{S}(\eta_s^2/m^2)$, it follows that $\dot{\theta} \in \mathcal{S}(\eta_s^2/m^2)$.

The proof for part (iii) follows from (8.5.35) by setting $\eta_s = 0$, using $-\sigma_s \tilde{\theta}^T \theta \leq 0$ and repeating the above calculations for $\eta_s = 0$.

The proof of (iv)(a) is almost identical to that of Theorem 8.5.1 (iii) and is omitted.

To prove (iv)(b), we follow the same arguments used in the proof of Theorem 8.5.1 (iii) to obtain the inequality

$$|\tilde{\theta}| \leq \beta_0 e^{-\beta_2 t} + \beta_1' \int_0^t e^{-\beta_2(t-\tau)}(|\eta_s| + \sigma_s |\theta|)d\tau$$

$$\leq \beta_0 e^{-\beta_2 t} + \frac{\beta_1'}{\beta_2}\bar{\eta} + \beta_1' \int_0^t e^{-\beta_2(t-\tau)} \sigma_s |\theta| d\tau \quad (8.5.36)$$

From (8.5.34), we have

$$\sigma_s |\theta| \leq \frac{1}{M - |\theta^*|} \sigma_s \tilde{\theta}^T \theta \leq \frac{1}{M - |\theta^*|} \sigma_s |\theta| \, |\tilde{\theta}| \quad (8.5.37)$$

Therefore, using (8.5.37) in (8.5.36), we have

$$|\tilde{\theta}| \leq \beta_0 e^{-\beta_2 t} + \frac{\beta_1'}{\beta_2}\bar{\eta} + \beta_1'' \int_0^t e^{-\beta_2(t-\tau)} \sigma_s |\theta| \, |\tilde{\theta}| d\tau \quad (8.5.38)$$

where $\beta_1'' = \frac{\beta_1'}{M_0 - |\theta^*|}$. Applying B-G Lemma III to (8.5.38), it follows that

$$|\tilde{\theta}| \leq (\beta_0 + \frac{\beta_1'}{\beta_2}\bar{\eta})e^{-\beta_2(t-t_0)} e^{\beta_1'' \int_{t_0}^t \sigma_s |\theta| ds} + \beta_1' \bar{\eta} \int_{t_0}^t e^{-\beta_2(t-\tau)} e^{\beta'' \int_t^\tau \sigma_s |\theta| ds} d\tau \quad (8.5.39)$$

Note from (8.5.34), (8.5.35) that $\sqrt{\sigma_s |\theta|} \in \mathcal{S}(\eta_s^2/m^2)$, i.e.,

$$\int_{t_0}^t \sigma_s |\theta| d\tau \leq c_1 \bar{\eta}^2 (t - t_0) + c_0$$

$\forall t \geq t_0 \geq 0$ and some constants c_0, c_1. Therefore,

$$|\tilde{\theta}| \leq \bar{\beta}_1 e^{-\bar{\alpha}(t-t_0)} + \bar{\beta}_2 \bar{\eta} \int_{t_0}^t e^{-\bar{\alpha}(t-\tau)} d\tau \quad (8.5.40)$$

where $\bar{\alpha} = \beta_2 - \beta_2'' c_1 \bar{\eta}^2$ and $\bar{\beta}_1, \bar{\beta}_2 \geq 0$ are some constants that depend on c_0 and the constants in (8.5.39). Hence, for any $\bar{\eta} \in [0, \bar{\eta}^*)$, where $\bar{\eta}^* = \sqrt{\frac{\beta_2}{\beta_1'' c_1}}$ we have $\bar{\alpha} > 0$ and (8.5.40) implies that

$$|\tilde{\theta}| \leq \frac{\bar{\beta}_2}{\bar{\alpha}}\bar{\eta} + ce^{-\bar{\alpha}(t-t_0)}$$

for some constant c and for all $t \geq t_0 \geq 0$. Therefore the proof for (iv) is complete.
□

Theorem 8.5.3 (ϵ-Modification) *Let*

$$w(t) = |\epsilon m|\nu_0$$

where $\nu_0 > 0$ is a design constant. Then the adaptive law (8.5.26) with $w(t) = |\epsilon m|\nu_0$ guarantees that

(i) $\theta, \epsilon \in \mathcal{L}_\infty$

(ii) $\epsilon, \epsilon n_s, \dot{\theta} \in \mathcal{S}(\nu_0 + \eta_s^2/m^2)$

(iii) *In addition, if $n_s, \phi, \dot{\phi} \in \mathcal{L}_\infty$ and ϕ is PE with level $\alpha_0 > 0$ that is independent of η_s, then $\tilde{\theta}$ converges exponentially to the residual set*

$$D_\epsilon \triangleq \left\{ \tilde{\theta} \, \big| \, |\tilde{\theta}| \leq c(\nu_0 + \bar{\eta}) \right\}$$

where $c \in \mathcal{R}^+$ and $\bar{\eta} = \sup_t |\eta_s|$.

Proof Letting $w(t) = |\epsilon m|\nu_0$ and using (8.5.26) in the equation for $\dot{V}$ given by (8.5.24), we obtain

$$\dot{V} \leq -\nu_c \frac{e^\top L_c e}{2} - \epsilon^2 n_s^2 + |\epsilon m|\frac{|\eta_s|}{m} - |\epsilon m|\nu_0 \tilde{\theta}^\top \theta$$

Because $-\tilde{\theta}^\top \theta \leq -\frac{|\tilde{\theta}|^2}{2} + \frac{|\theta^*|^2}{2}$, it follows that

$$\dot{V} \leq -2\lambda_0 |e|^2 - \epsilon^2 n_s^2 - |\epsilon m|\left(\nu_0 \frac{|\tilde{\theta}|^2}{2} - \frac{|\eta_s|}{m} - \frac{|\theta^*|^2}{2}\nu_0\right) \quad (8.5.41)$$

where $\lambda_0 = \frac{\nu_c \lambda_{min}(L_c)}{4}$, which implies that for large $|e|$ or large $|\tilde{\theta}|$, $\dot{V} \leq 0$. Hence, by following a similar approach as in the proof of Theorem 8.5.1, we can show the existence of a constant $V_0 > 0$ such that for $V > V_0$, $\dot{V} \leq 0$, therefore, $V, e, \epsilon, \theta, \tilde{\theta} \in \mathcal{L}_\infty$.

Because $|e| \geq \frac{|\epsilon|}{|C_c|}$, we can write (8.5.41) as

$$\dot{V} \leq -\lambda_0 |e|^2 - \beta_0 \epsilon^2 - \epsilon^2 n_s^2 + \alpha_0 \epsilon^2 (1 + n_s^2) - \alpha_0 \epsilon^2 m^2 - |\epsilon m|\nu_0 \frac{|\tilde{\theta}|^2}{2}$$
$$+ |\epsilon m|\frac{|\eta_s|}{m} + |\epsilon m|\frac{|\theta^*|^2}{2}\nu_0$$

by adding and subtracting the term $\alpha_0 \epsilon^2 m^2 = \alpha_0 \epsilon^2 (1 + n_s^2)$, where $\beta_0 = \frac{\lambda_0}{|C_c|^2}$ and $\alpha_0 > 0$ is an arbitrary constant. Setting $\alpha_0 = min(1, \beta_0)$, we have

$$\dot{V} \leq -\lambda_0 |e|^2 - \alpha_0 \epsilon^2 m^2 + |\epsilon m|\frac{|\eta_s|}{m} + |\epsilon m|\frac{|\theta^*|^2}{2}\nu_0$$

8.5. ROBUST ADAPTIVE LAWS

By completing the squares and using the same approach as in the proof of Theorem 8.5.1, we can establish that $\epsilon, \epsilon n_s, \epsilon m \in \mathcal{S}(\nu_0 + \eta_s^2/m^2)$, which, together with $|\dot\theta| \leq \|\Gamma\| |\epsilon m| \frac{|\phi|}{m} + \nu_0 \|\Gamma\| |\epsilon m| |\theta| \leq c|\epsilon m|$ for some $c \in \mathcal{R}^+$, implies that $\dot\theta \in \mathcal{S}(\nu_0 + \eta_s^2/m^2)$.

The proof of (iii) is very similar to the proof of Theorem 8.5.1 (iii), which can be completed by treating the terms due to η_s and the ϵ-modification as bounded inputs to an e.s. linear time-varying system. □

Remark 8.5.1 The normalizing signal m given by (8.5.25) involves the dynamic term m_s and the signals ϕ, u, y. Under some conditions, the signals ϕ and/or u, y do not need to be included in the normalizing signal. These conditions are explained as follows:

(i) If $\phi = H(s)[u, y]^T$ where $H(s)$ has strictly proper elements that are analytic in $\text{Re}[s] \geq -\delta_0/2$, then $\frac{\phi}{1+m_s} \in \mathcal{L}_\infty$ and therefore the term $\phi^T\phi$ in the expression for n_s^2 can be dropped.

(ii) If $W(s)L(s)$ is chosen to be biproper, then $W^{-1}L^{-1}\Delta_u, W^{-1}L^{-1}\Delta_y$ are strictly proper and the terms u^2, y^2 in the expression for n_s^2 can be dropped.

(iii) The parametric model equation (8.5.20) can be filtered on both sides by a first order filter $\frac{f_0}{s+f_0}$ where $f_0 > \frac{\delta_0}{2}$ to obtain

$$z_f = W(s)L(s)(\theta^{*T}\phi_f + \eta_f)$$

where $x_f \triangleq \frac{f_0}{s+f_0} x$ denotes the filtered output of the signal x. In this case

$$\eta_f = L^{-1}(s)W^{-1}(s)\frac{f_0}{s+f_0}[\Delta_u(s)u + \Delta_y(s)y + d_1]$$

is bounded from above by m^2 given by

$$\begin{aligned} m^2 &= 1 + n_s^2, \quad n_s^2 = m_s, \\ \dot m_s &= -\delta_0 m_s + u^2 + y^2, \quad m_s(0) = 0 \end{aligned}$$

The choice of m is therefore dependent on the expression for the modeling error term η in the parametric model and the properties of the signal vector ϕ.

Remark 8.5.2 The assumption that the level $\alpha_0 > 0$ of PE of ϕ is independent of η_s is used to guarantee that the modeling error term η_s does not destroy or weaken the PE property of ϕ. Therefore, the constant β_2 in the bound for the transition matrix given by (8.5.32) guaranteed to be greater than zero for $\eta_s \equiv 0$ is not affected by $\eta_s \neq 0$.

8.5.3 Gradient Algorithms with Leakage

As in the ideal case presented in Chapter 4, the linear parametric model with modeling error can be rewritten in the form

$$z = \theta^{*T}\phi + \eta, \quad \eta = \Delta_u(s)u + \Delta_y(s)y + d_1 \tag{8.5.42}$$

where $\phi = W(s)\psi$. The estimate $\hat{z}$ of z and the normalized estimation error are constructed as

$$\hat{z} = \theta^T \phi, \quad \epsilon = \frac{z - \hat{z}}{m^2} = \frac{z - \theta^T \phi}{m^2} \tag{8.5.43}$$

where θ is the estimate of θ^* and $m^2 = 1 + n_s^2$ and n_s is the normalizing signal to be designed. For analysis purposes, we express (8.5.43) in terms of the parameter error $\tilde{\theta} = \theta - \theta^*$, i.e.,

$$\epsilon = \frac{-\tilde{\theta}^T \phi + \eta}{m^2} \tag{8.5.44}$$

If we now design m so that

$$\frac{\phi}{m}, \frac{\eta}{m} \in \mathcal{L}_\infty \tag{A1}$$

then the signal ϵm is a reasonable measure of the parameter error $\tilde{\theta}$ since for any piecewise continuous signal ϕ and η, large ϵm implies large $\tilde{\theta}$.

Design of the Normalizing Signal

Assume that $\Delta_u(s), \Delta_y(s)$ are strictly proper and analytic in $\text{Re}[s] \geq -\delta_0/2$. Then, according to Lemma 3.3.2, we have

$$|\eta(t)| \leq \|\Delta_u(s)\|_{2\delta_0}\|u_t\|_{2\delta_0} + \|\Delta_y(s)\|_{2\delta_0}\|y_t\|_{2\delta_0} + \|d_1\|_{2\delta_0}$$

which motivates the following normalizing signal

$$\begin{aligned} m^2 &= 1 + n_s^2, \quad n_s^2 = m_s + \phi^T\phi \\ \dot{m}_s &= -\delta_0 m_s + u^2 + y^2, \quad m_s(0) = 0 \end{aligned} \tag{8.5.45}$$

8.5. ROBUST ADAPTIVE LAWS

that satisfies assumption (A1). If $\phi = H(s)[u,y]^\top$ where all the elements of $H(s)$ are strictly proper and analytic in $Re[s] \geq -\frac{\delta_0}{2}$, then $\frac{\phi}{1+m_s} \in \mathcal{L}_\infty$ and the term $\phi^\top \phi$ can be dropped from the expression of n_s^2 without violating condition (A1).

If instead of being strictly proper, Δ_u, Δ_y are biproper, the expression for n_s^2 should be changed to

$$n_s^2 = 1 + u^2 + y^2 + \phi^\top \phi + m_s$$

to satisfy (A1).

The adaptive laws for estimating θ^* can now be designed by choosing appropriate cost functions that we minimize w.r.t. θ using the gradient method. We start with the instantaneous cost function

$$J(\theta, t) = \frac{\epsilon^2 m^2}{2} + \frac{w(t)}{2}\theta^\top \theta = \frac{(z-\theta^\top \phi)^2}{2m^2} + \frac{w(t)}{2}\theta^\top \theta$$

where $w(t) \geq 0$ is a design function that acts as a weighting coefficient. Applying the gradient method, we obtain

$$\dot{\theta} = -\Gamma \nabla J = \Gamma \epsilon \phi - w\Gamma \theta \qquad (8.5.46)$$
$$\epsilon m^2 = z - \theta^\top \phi = -\tilde{\theta}^\top \phi + \eta$$

where $\Gamma = \Gamma^\top > 0$ is the adaptive gain. The adaptive law (8.5.46) has the same form as (8.5.26) except that ϵ and ϕ are defined differently. The weighting coefficient $w(t)$ in the cost appears as a leakage term in the adaptive law in exactly the same way as with (8.5.26).

Instead of the instantaneous cost, we can also use the integral cost with a forgetting factor $\beta > 0$ given by

$$J(\theta, t) = \frac{1}{2}\int_0^t e^{-\beta(t-\tau)}\frac{[z(\tau) - \theta^\top(t)\phi(\tau)]^2}{m^2(\tau)}d\tau + \frac{1}{2}w(t)\theta^\top(t)\theta(t)$$

where $w(t) \geq 0$ is a design weighting function. As in Chapter 4, the application of the gradient method yields

$$\dot{\theta} = -\Gamma \nabla J = \Gamma \int_0^t e^{-\beta(t-\tau)}\frac{[z(\tau) - \theta^\top(t)\phi(\tau)]}{m^2(\tau)}\phi(\tau)d\tau - \Gamma w\theta$$

which can be implemented as

$$\dot{\theta} = -\Gamma(R\theta + Q) - \Gamma w\theta$$
$$\dot{R} = -\beta R + \frac{\phi\phi^T}{m^2}, \quad R(0) = 0 \qquad (8.5.47)$$
$$\dot{Q} = -\beta Q - \frac{z\phi}{m^2}, \quad Q(0) = 0$$

where $R \in \mathcal{R}^{n \times n}, Q \in \mathcal{R}^{n \times 1}$ and n is the dimension of the vector ϕ.

The properties of (8.5.46), (8.5.47) for the various choices of $w(t)$ are given by the following Theorems:

Theorem 8.5.4 *The adaptive law (8.5.46) or (8.5.47) guarantees the following properties:*

(A) *For the fixed-σ modification, i.e., $w(t) = \sigma > 0$, we have*

   (i) $\epsilon, \epsilon n_s, \theta, \dot{\theta} \in \mathcal{L}_\infty$
  (ii) $\epsilon, \epsilon n_s, \dot{\theta} \in \mathcal{S}(\sigma + \eta^2/m^2)$
 (iii) *If $n_s, \phi \in \mathcal{L}_\infty$ and ϕ is PE with level $\alpha_0 > 0$ independent of η, then $\tilde{\theta}$ converges exponentially to the residual set*

$$D_\sigma = \left\{ \tilde{\theta} \,\middle|\, |\tilde{\theta}| \leq c(\sigma + \bar{\eta}) \right\}$$

where $\bar{\eta} = \sup_t \frac{|\eta(t)|}{m(t)}$ and $c \geq 0$ is some constant.

(B) *For the switching-σ modification, i.e., for $w(t) = \sigma_s$ where σ_s is as defined in Theorem 8.5.2, we have*

   (i) $\epsilon, \epsilon n_s, \theta, \dot{\theta} \in \mathcal{L}_\infty$
  (ii) $\epsilon, \epsilon n_s, \dot{\theta} \in \mathcal{S}(\eta^2/m^2)$
 (iii) *In the absence of modeling error, i.e., for $\eta = 0$, the properties of (8.5.46), (8.5.47) are the same as those of the respective unmodified adaptive laws (i.e., with $w = 0$)*

8.5. ROBUST ADAPTIVE LAWS

(iv) If $n_s, \phi \in \mathcal{L}_\infty$ and ϕ is PE with level $\alpha_0 > 0$ independent of η, then

(a) $\tilde{\theta}$ converges exponentially fast to the residual set
$$D_s = \left\{ \tilde{\theta} \,\middle|\, |\tilde{\theta}| \leq c(\sigma_0 + \bar{\eta}) \right\}$$
where $c \geq 0$ is some constant.

(b) There exists an $\bar{\eta}^* > 0$ such that if $\bar{\eta} < \bar{\eta}^*$, then $\tilde{\theta}$ converges exponentially to the residual set $\bar{D}_s = \left\{ \tilde{\theta} \,\middle|\, |\tilde{\theta}| \leq c\bar{\eta} \right\}$.

Proof We first consider the gradient algorithm (8.5.46) based on the instantaneous cost function. The proof of (i), (ii) in (A), (B) can be completed by using the Lyapunov-like function $V(\tilde{\theta}) = \frac{\tilde{\theta}^\top \Gamma^{-1} \tilde{\theta}}{2}$ and following the same procedure as in the proof of Theorems 8.5.1, 8.5.2.

To show that the parameter error converges exponentially to a residual set for persistently exciting ϕ, we write (8.5.46) as

$$\dot{\tilde{\theta}} = -\Gamma \frac{\phi \phi^\top}{m^2} \tilde{\theta} + \frac{\Gamma \eta \phi}{m^2} - w(t)\Gamma\theta \qquad (8.5.48)$$

In Section 4.8.3 (proof of Theorem 4.3.2 (iii)), we have shown that the homogeneous part of (8.5.48) with $\eta \equiv 0$ and $w \equiv 0$ is e.s., i.e., the state transition matrix of the homogeneous part of (8.5.48) satisfies

$$\|\Phi(t, t_0)\| \leq \beta_1 e^{-\beta_2 (t - t_0)}$$

for some positive constants β_1, β_2. Viewing (8.5.48) as a linear time-varying system with inputs $\frac{\eta}{m}, w\theta$, and noting that $\frac{\phi}{m} \in \mathcal{L}_\infty$ and the PE level α_0 of ϕ is independent of η, we have

$$|\tilde{\theta}| \leq \beta_0 e^{-\beta_2 t} + \beta_1' \int_0^t e^{-\beta_2 (t-\tau)} \left(\frac{|\eta|}{m} + |w(\tau)\theta| \right) d\tau \qquad (8.5.49)$$

for some constant $\beta_1' > 0$. For the fixed σ-modification, i.e., $w(t) \equiv \sigma$, (A) (iii) follows immediately from (8.5.49) and $\theta \in \mathcal{L}_\infty$. For the switching σ-modification, we write

$$|\tilde{\theta}| \leq \beta_0 e^{-\beta_2 t} + \frac{\beta_1' \bar{\eta}}{\beta_2} + \beta_1' \int_0^t e^{-\beta_2 (t-\tau)} |w(\tau)\theta| d\tau \qquad (8.5.50)$$

To obtain a tighter bound for $\tilde{\theta}$, we use the property of σ_s to write (see equation (8.5.34))

$$\sigma_s |\theta| \leq \sigma_s \frac{\tilde{\theta}^\top \theta}{M_0 - |\theta^*|} \leq \frac{1}{M_0 - |\theta^*|} \sigma_s |\theta| \, |\tilde{\theta}| \qquad (8.5.51)$$

Using (8.5.51) in (8.5.50), we have

$$|\tilde{\theta}| \leq \beta_0 e^{-\beta_2 t} + \frac{\beta_1' \bar{\eta}}{\beta_2} + \beta_1'' \int_0^t e^{-\beta_2(t-\tau)} \sigma_s |\theta| \, |\tilde{\theta}| d\tau \qquad (8.5.52)$$

with $\beta_1'' = \frac{\beta_1'}{M_0 - |\theta^*|}$. Applying B-G Lemma III, we obtain

$$|\tilde{\theta}| \leq \left(\beta_0 + \frac{\beta_1'}{\beta_2}\bar{\eta}\right) e^{-\beta_2(t-t_0)} e^{\beta_1'' \int_{t_0}^t \sigma_s |\theta| ds} + \beta_1' \bar{\eta} \int_{t_0}^t e^{-\beta_2(t-\tau)} e^{\beta_1'' \int_\tau^t \sigma_s |\theta| ds} d\tau \qquad (8.5.53)$$

Now consider the Lyapunov function $V(\tilde{\theta}) \triangleq \frac{\tilde{\theta}^\top \Gamma^{-1} \tilde{\theta}}{2}$. Along the solution of the adaptive law given by (8.5.46) with $w = \sigma_s$, we have

$$\dot{V} = -\frac{(\tilde{\theta}^\top \phi)^2}{m^2} + \frac{\eta}{m} \frac{\tilde{\theta}^\top \phi}{m} - \sigma_s \theta^\top \tilde{\theta} \leq c\bar{\eta} - \sigma_s \theta^\top \tilde{\theta} \qquad (8.5.54)$$

for some constant $c \geq 0$, where the second inequality is obtained by using the properties that $\tilde{\theta}, \frac{\phi}{m} \in \mathcal{L}_\infty$. Equations (8.5.51) and (8.5.54) imply that

$$\sigma_s |\theta| \leq \frac{c\bar{\eta} - \dot{V}}{M_0 - |\theta^*|}$$

which, together with the boundedness of V, leads to

$$\int_{t_0}^t \sigma_s |\theta| d\tau \leq c_1 \bar{\eta}(t - t_0) + c_0 \qquad (8.5.55)$$

for all $t \geq t_0 \geq 0$ and for some constants c_1, c_0. Using (8.5.55) in (8.5.53), we have

$$|\tilde{\theta}| \leq \bar{\beta}_0 e^{-\bar{\alpha}(t-t_0)} + \bar{\beta}_1 \bar{\eta} \int_{t_0}^t e^{-\bar{\alpha}(t-\tau)} d\tau \qquad (8.5.56)$$

where $\bar{\alpha} = \beta_2 - \beta_1'' c_1 \bar{\eta}$. Therefore for $\bar{\eta}^* = \frac{\beta_2}{\beta_1'' c_1}$ and $\bar{\eta} < \bar{\eta}^*$, we have $\bar{\alpha} > 0$ and

$$|\tilde{\theta}| \leq \frac{\bar{\beta}_1 \bar{\eta}}{\bar{\alpha}} + \epsilon_t$$

where ϵ_t is an exponentially decaying to zero term and the proof of (B)(iv) is complete.

The proof for the integral adaptive law (8.5.47) is different and is presented below. From (8.5.47), we can verify that

$$Q(t) = -R(t)\theta^* - \int_0^t e^{-\beta(t-\tau)} \frac{\phi(\tau)}{m^2(\tau)} \eta(\tau) d\tau$$

8.5. ROBUST ADAPTIVE LAWS

and therefore

$$\dot{\theta} = \dot{\tilde{\theta}} = -\Gamma R\tilde{\theta} + \Gamma \int_0^t e^{-\beta(t-\tau)} \frac{\phi(\tau)\eta(\tau)}{m^2(\tau)} d\tau - \Gamma w\theta \qquad (8.5.57)$$

Let us now choose the same $V(\tilde{\theta})$ as in the ideal case, i.e.,

$$V(\tilde{\theta}) = \frac{\tilde{\theta}^\top \Gamma^{-1} \tilde{\theta}}{2}$$

Then along the solution of (8.5.57), we have

$$\begin{aligned}
\dot{V} &= -\tilde{\theta}^\top R\tilde{\theta} + \tilde{\theta}^\top(t) \int_0^t e^{-\beta(t-\tau)} \frac{\phi(\tau)\eta(\tau)}{m^2(\tau)} d\tau - w\tilde{\theta}^\top \theta \\
&\leq -\tilde{\theta}^\top R\tilde{\theta} + \left(\int_0^t e^{-\beta(t-\tau)} \frac{(\tilde{\theta}^\top(t)\phi(\tau))^2}{m^2(\tau)} \right)^{\frac{1}{2}} \left(\int_0^t e^{-\beta(t-\tau)} \frac{\eta^2(\tau)}{m^2(\tau)} d\tau \right)^{\frac{1}{2}} \\
&\quad - w\tilde{\theta}^\top \theta
\end{aligned} \qquad (8.5.58)$$

where the inequality is obtained by applying the Schwartz inequality. Using

$$(\tilde{\theta}^\top(t)\phi(\tau))^2 = \tilde{\theta}^\top(t)\phi(\tau)\phi^\top(\tau)\tilde{\theta}(t)$$

and the definition of the norm $\|\cdot\|_{2\delta}$, we have

$$\dot{V} \leq -\tilde{\theta}^\top R\tilde{\theta} + (\tilde{\theta}^\top R\tilde{\theta})^{\frac{1}{2}} \left\| (\frac{\eta}{m})_t \right\|_{2\beta} - w\tilde{\theta}^\top \theta$$

Using the inequality $-a^2 + ab \leq -\frac{a^2}{2} + \frac{b^2}{2}$, we obtain

$$\dot{V} \leq -\frac{\tilde{\theta}^\top R\tilde{\theta}}{2} + \frac{1}{2} \left(\left\| (\frac{\eta}{m})_t \right\|_{2\beta} \right)^2 - w\tilde{\theta}^\top \theta \qquad (8.5.59)$$

Let us now consider the following two choices for w:
(A) *Fixed σ* For $w = \sigma$, we have $-\sigma\tilde{\theta}^\top \theta \leq -\frac{\sigma}{2}|\tilde{\theta}|^2 + \frac{\sigma}{2}|\theta^*|^2$ and therefore

$$\dot{V} \leq -\frac{\tilde{\theta}^\top R\tilde{\theta}}{2} - \frac{\sigma}{2}|\tilde{\theta}|^2 + \frac{\sigma}{2}|\theta^*|^2 + \frac{1}{2}\left(\left\| (\frac{\eta}{m})_t \right\|_{2\beta} \right)^2 \qquad (8.5.60)$$

Because $R = R^\top \geq 0$, $\sigma > 0$ and $\|(\frac{\eta}{m})_t\|_{2\beta} \leq c_m$ for some constant $c_m \geq 0$, it follows that $\dot{V} \leq 0$ whenever $V(\tilde{\theta}) \geq V_0$ for some $V_0 \geq 0$ that depends on $\bar{\eta}, \sigma$ and $|\theta^*|^2$. Hence, $V, \tilde{\theta} \in \mathcal{L}_\infty$ which from (8.5.44) implies that $\epsilon, \epsilon m, \epsilon n_s \in \mathcal{L}_\infty$. Because $R, Q \in \mathcal{L}_\infty$ and $\theta \in \mathcal{L}_\infty$, we also have that $\dot{\theta} \in \mathcal{L}_\infty$.

Integrating on both sides of (8.5.60), we obtain

$$\frac{1}{2}\int_{t_0}^{t}(\tilde{\theta}^\top R\tilde{\theta} + \sigma|\tilde{\theta}|^2)d\tau \leq V(t_0) - V(t) + \frac{1}{2}\int_{t_0}^{t}\sigma|\theta^*|^2 d\tau$$
$$+ \frac{1}{2}\int_{t_0}^{t}\int_{0}^{\tau} e^{-\beta(\tau-s)} \frac{\eta^2(s)}{m^2(s)} ds d\tau$$

By interchanging the order of the double integration, i.e., using the identity

$$\int_{t_0}^{t}\int_{0}^{\tau} f(\tau)g(s)ds d\tau = \int_{t_0}^{t}\int_{s}^{t} f(\tau)g(s)d\tau ds + \int_{0}^{t_0}\int_{t_0}^{t} f(\tau)g(s)d\tau ds$$

we establish that

$$\int_{t_0}^{t}\int_{0}^{\tau} e^{-\beta(\tau-s)} \frac{\eta^2(s)}{m^2(s)} ds d\tau \leq \frac{1}{\beta}\int_{t_0}^{t} \frac{\eta^2}{m^2} d\tau + \frac{c}{\beta^2} \qquad (8.5.61)$$

where $c = \sup_t \frac{\eta^2(t)}{m^2(t)}$, which together with $V \in \mathcal{L}_\infty$ imply that

$$R^{\frac{1}{2}}\tilde{\theta}, \sqrt{\sigma}|\tilde{\theta}| \in \mathcal{S}(\sigma + \eta^2/m^2)$$

It can also be shown using the inequality $-\sigma\tilde{\theta}^\top\theta \leq -\frac{\sigma}{2}|\tilde{\theta}|^2 + \frac{\sigma}{2}|\theta^*|^2$ that $\sqrt{\sigma}|\theta| \in \mathcal{S}(\sigma + \eta^2/m^2)$. To examine the properties of $\epsilon, \epsilon n_s$ we proceed as follows: We have

$$\frac{d}{dt}\tilde{\theta}^\top R\tilde{\theta} = 2\tilde{\theta}^\top R\dot{\tilde{\theta}} + \tilde{\theta}^\top \dot{R}\tilde{\theta}$$
$$= -2\tilde{\theta}^\top R\Gamma R\tilde{\theta} + 2\tilde{\theta}^\top R\Gamma \int_0^t e^{-\beta(t-\tau)} \frac{\phi(\tau)\eta(\tau)}{m^2(\tau)} d\tau$$
$$-2\sigma\tilde{\theta}^\top R\Gamma\theta - \beta\tilde{\theta}^\top R\tilde{\theta} + \frac{(\tilde{\theta}^\top\phi)^2}{m^2}$$

where the second equality is obtained by using (8.5.57) with $w = \sigma$ and the expression for $\dot{R}$ given by (8.5.47). Because $\epsilon = \frac{-\tilde{\theta}^\top\phi+\eta}{m^2}$, it follows that

$$\frac{d}{dt}\tilde{\theta}^\top R\tilde{\theta} = -2\tilde{\theta}^\top R\Gamma R\tilde{\theta} + 2\tilde{\theta}^\top R\Gamma \int_0^t e^{-\beta(t-\tau)} \frac{\phi(\tau)\eta(\tau)}{m^2(\tau)} d\tau$$
$$-2\sigma\tilde{\theta}^\top R\Gamma\theta - \beta\tilde{\theta}^\top R\tilde{\theta} + (\epsilon m - \frac{\eta}{m})^2$$

Using $(\epsilon m - \frac{\eta}{m})^2 \geq \frac{\epsilon^2 m^2}{2} - \frac{\eta^2}{m^2}$, we obtain the inequality

$$\frac{\epsilon^2 m^2}{2} \leq \frac{d}{dt}\tilde{\theta}^\top R\tilde{\theta} + 2\tilde{\theta}^\top R\Gamma R\tilde{\theta} - 2\tilde{\theta}^\top R\Gamma \int_0^t e^{-\beta(t-\tau)} \frac{\phi(\tau)\eta(\tau)}{m^2(\tau)} d\tau$$
$$+ 2\sigma\tilde{\theta}^\top R\Gamma\theta + \beta\tilde{\theta}^\top R\tilde{\theta} + \frac{\eta^2}{m^2} \qquad (8.5.62)$$

8.5. ROBUST ADAPTIVE LAWS

Noting that

$$2\left|\tilde{\theta}^T R\Gamma \int_0^t e^{-\beta(t-\tau)}\frac{\phi(\tau)\eta(\tau)}{m^2(\tau)}d\tau\right|$$
$$\leq |\tilde{\theta}^T R\Gamma|^2 + \int_0^t e^{-\beta(t-\tau)}\frac{|\phi(\tau)|^2}{m^2(\tau)}d\tau \int_0^t e^{-\beta(t-\tau)}\frac{\eta^2(\tau)}{m^2(\tau)}d\tau$$

where the last inequality is obtained by using $2ab \leq a^2 + b^2$ and the Schwartz inequality, it follows from (8.5.62) that

$$\frac{\epsilon^2 m^2}{2} \leq \frac{d}{dt}\tilde{\theta}^T R\tilde{\theta} + 2\tilde{\theta}^T R\Gamma R\tilde{\theta} + 2\sigma\tilde{\theta}^T R\Gamma\theta + \beta\tilde{\theta}^T R\tilde{\theta}$$
$$+ |\tilde{\theta}^T R\Gamma)|^2 + \int_0^t e^{-\beta(t-\tau)}\frac{|\phi(\tau)|^2}{m^2(\tau)}d\tau \int_0^t e^{-\beta(t-\tau)}\frac{\eta^2(\tau)}{m^2(\tau)}d\tau + \frac{\eta^2}{m^2}$$

Because $R^{\frac{1}{2}}\tilde{\theta}, \sqrt{\sigma}|\theta| \in \mathcal{S}(\sigma + \eta^2/m^2)$ and $\frac{\phi}{m}, \theta, R \in \mathcal{L}_\infty$, it follows from the above inequality that $\epsilon, \epsilon n_s, \epsilon m \in \mathcal{S}(\sigma + \eta^2/m^2)$. Now from (8.5.57) with $w = \sigma$ we have

$$|\dot{\theta}| \leq \|\Gamma R^{\frac{1}{2}}\|\|R^{\frac{1}{2}}\tilde{\theta}\| + \sigma\|\Gamma\|\|\theta\|$$
$$+\|\Gamma\|\left(\int_0^t e^{-\beta(t-\tau)}\frac{|\phi(\tau)|^2}{m^2(\tau)}d\tau\right)^{\frac{1}{2}}\left(\int_0^t e^{-\beta(t-\tau)}\frac{\eta^2(\tau)}{m^2(\tau)}d\tau\right)^{\frac{1}{2}}$$

which can be used to show that $\dot{\theta} \in \mathcal{S}(\sigma + \eta^2/m^2)$ by performing similar manipulations as in (8.5.61) and using $\sqrt{\sigma}|\theta| \in \mathcal{S}(\sigma + \eta^2/m^2)$.

To establish the parameter error convergence properties, we write (8.5.57) with $w = \sigma$ as

$$\dot{\tilde{\theta}} = -\Gamma R(t)\tilde{\theta} + \Gamma \int_0^t e^{-\beta(t-\tau)}\frac{\phi\eta}{m^2}d\tau - \Gamma\sigma\theta \quad (8.5.63)$$

In Section 4.8.4, we have shown that the homogeneous part of (8.5.63), i.e., $\dot{\tilde{\theta}} = -\Gamma R(t)\tilde{\theta}$ is e.s. provided that ϕ is PE. Noting that $\left|\int_0^t e^{-\beta(t-\tau)}\frac{\phi\eta}{m^2}d\tau\right| \leq c\bar{\eta}$ for some constant $c \geq 0$, the rest of the proof is completed by following the same procedure as in the proof of Theorem 8.5.1 (iii).

(B) *Switching σ* For $w(t) = \sigma_s$ we have, as shown in the proof of Theorem 8.5.2, that

$$\sigma_s\tilde{\theta}^T\theta \geq \sigma_s|\theta|(M_0 - |\theta^*|) \geq 0, \quad i.e., \quad \sigma_s|\theta| \leq \sigma_s\frac{\tilde{\theta}^T\theta}{M_0 - |\theta^*|} \quad (8.5.64)$$

and for $|\theta| = |\tilde{\theta} + \theta^*| > 2M_0$ we have

$$-\sigma_s\tilde{\theta}^T\theta \leq -\frac{\sigma_0}{2}|\tilde{\theta}|^2 + \frac{\sigma_0}{2}|\theta^*|^2$$

Furthermore, the inequality (8.5.59) with $w(t) = \sigma_s$ can be written as

$$\dot{V} \leq -\frac{\tilde{\theta}^\top R \tilde{\theta}}{2} + \frac{1}{2}\left(\left\|(\frac{\eta}{m})_t\right\|_{2\beta}\right)^2 - \sigma_s \tilde{\theta}^\top \theta \qquad (8.5.65)$$

Following the same approach as the one used in the proof of part (A), we can show that $\dot{V} < 0$ for $V > V_0$ and for some $V_0 > 0$, i.e., $V \in \mathcal{L}_\infty$ which implies that $\epsilon, \theta, \epsilon n_s \in \mathcal{L}_\infty$. Integrating on both sides of (8.5.65), we obtain that $R^{\frac{1}{2}}\tilde{\theta}, \sqrt{\sigma_s \tilde{\theta}^\top \theta} \in \mathcal{S}(\eta^2/m^2)$. Proceeding as in the proof of part (A) and making use of (8.5.64), we show that $\epsilon, \epsilon n_s \in \mathcal{S}(\eta^2/m^2)$. Because

$$\sigma_s^2|\theta|^2 \leq c_1 \sigma_s \tilde{\theta}^\top \theta$$

where $c_1 \in \mathcal{R}^+$ depends on the bound for $\sigma_0|\theta|$, we can follow the same procedure as in part (A) to show that $\dot{\theta} \in \mathcal{S}(\eta^2/m^2)$. Hence, the proof for (i), (ii) of part (B) is complete.

The proof of part (B) (iii) follows directly by setting $\eta = 0$ and repeating the same steps.

The proof of part (B) (iv) is completed by using similar arguments as in the case of part (A) (iii) as follows: From (8.5.57), we have

$$\dot{\tilde{\theta}} = -\Gamma R(t)\tilde{\theta} + \Gamma \int_0^t e^{-\beta(t-\tau)} \frac{\phi\eta}{m^2} d\tau - \Gamma w \theta, \quad w = \sigma_s \qquad (8.5.66)$$

In Section 4.8.4, we have proved that the homogeneous part of (8.5.66) is e.s. if $\phi \in \mathcal{L}_\infty$ and ϕ is PE. Therefore, we can treat (8.5.66) as an exponentially stable linear time-varying system with inputs $\sigma_s \theta, \int_0^t e^{-\beta(t-\tau)} \frac{\phi\eta}{m^2} d\tau$. Because $\frac{|\eta|}{m} \leq \bar{\eta}$ and $\frac{\phi}{m} \in \mathcal{L}_\infty$, we have

$$\left|\int_0^t e^{-\beta(t-\tau)} \frac{\phi\eta}{m^2} d\tau\right| \leq c\bar{\eta}$$

for some constant $c \geq 0$. Therefore, the rest of the parameter convergence proof can be completed by following exactly the same steps as in the proof of part (A) (iii). $\square$

ϵ-Modification

Another choice for $w(t)$ in the adaptive law (8.5.46) is

$$w(t) = |\epsilon m|\nu_0$$

where $\nu_0 > 0$ is a design constant. For the adaptive law (8.5.47), however, the ϵ-modification takes a different form and is given by

$$w(t) = \nu_0 \|\epsilon(t, \cdot)m(\cdot)_t\|_{2\beta}$$

8.5. ROBUST ADAPTIVE LAWS

where

$$\epsilon(t,\tau) \triangleq \frac{z(\tau) - \theta^\top(t)\phi(\tau)}{m^2(\tau)}, \quad t \geq \tau$$

and

$$\|\epsilon(t,\cdot)m(\cdot)\|_{2\beta}^2 = \int_0^t e^{-\beta(t-\tau)}\epsilon^2(t,\tau)m^2(\tau)d\tau$$

We can verify that this choice of $w(t)$ may be implemented as

$$w(t) = (r_0 + 2\theta^\top Q + \theta^\top R\theta)^{\frac{1}{2}}\nu_0$$

$$\dot{r}_0 = -\beta r_0 + \frac{z^2}{m^2}, \quad r_0(0) = 0 \tag{8.5.67}$$

The stability properties of the adaptive laws with the ϵ-modification are given by the following theorem.

Theorem 8.5.5 *Consider the adaptive law (8.5.46) with $w = |\epsilon m|\nu_0$ and the adaptive law (8.5.47) with $w(t) = \nu_0 \|\epsilon(t,\cdot)m(\cdot)\|_{2\beta}$. Both adaptive laws guarantee that*

(i) $\epsilon, \epsilon n_s, \theta, \dot{\theta} \in \mathcal{L}_\infty$.
(ii) $\epsilon, \epsilon n_s, \dot{\theta} \in \mathcal{S}(\nu_0 + \eta^2/m^2)$.
(iii) *If $n_s, \phi \in \mathcal{L}_\infty$ and ϕ is PE with level $\alpha_0 > 0$ that is independent of η, then $\tilde{\theta}$ converges exponentially fast to the residual set*

$$D_\epsilon = \left\{\tilde{\theta} \,\Big|\, |\tilde{\theta}| \leq c(\nu_0 + \bar{\eta})\right\}$$

where $c \geq 0$ and $\bar{\eta} = \sup_t \frac{|\eta|}{m}$.

Proof The proof of (i) and (ii) for the adaptive law (8.5.46) with $w(t) = \nu_0|\epsilon m|$ follows directly from that of Theorem 8.5.3 by using the Lyapunov-like function $V = \frac{\tilde{\theta}^\top \Gamma^{-1}\tilde{\theta}}{2}$. The time derivative $\dot{V}$ of $V = \frac{\tilde{\theta}^\top \Gamma^{-1}\tilde{\theta}}{2}$ along the solution of (8.5.47) with $w(t) = \nu_0\|\epsilon(t,\cdot)m(\cdot)\|_{2\beta}$ becomes

$$\dot{V} = -\tilde{\theta}^\top[R\theta + Q] - w\tilde{\theta}^\top \theta$$
$$= -\tilde{\theta}^\top(t)\left[\int_0^t e^{-\beta(t-\tau)}\frac{\phi(\tau)[\theta^\top(t)\phi(\tau) - z(\tau)]}{m^2(\tau)}d\tau\right]$$
$$\quad - \|\epsilon(t,\cdot)m(\cdot)\|_{2\beta}\nu_0\tilde{\theta}^\top\theta$$

Using $\epsilon(t,\tau)m^2(\tau) = z(\tau) - \theta^\top(t)\phi(\tau) = -\tilde{\theta}^\top(t)\phi(\tau) + \eta(\tau)$ we have

$$\begin{aligned}
\dot{V} &= -\int_0^t e^{-\beta(t-\tau)}\epsilon(t,\tau)[\epsilon(t,\tau)m^2(\tau) - \eta(\tau)]d\tau - \|\epsilon(t,\cdot)m(\cdot)\|_{2\beta}\nu_0\tilde{\theta}^\top\theta \\
&\leq -\left(\|\epsilon(t,\cdot)m(\cdot)\|_{2\beta}\right)^2 + \|\epsilon(t,\cdot)m(\cdot)\|_{2\beta}\left\|\left(\frac{\eta}{m}\right)_t\right\|_{2\beta} - \|\epsilon(t,\cdot)m(\cdot)\|_{2\beta}\nu_0\tilde{\theta}^\top\theta \\
&= -\|\epsilon(t,\cdot)m(\cdot)\|_{2\beta}\left[\|\epsilon(t,\cdot)m(\cdot)\|_{2\beta} - \left\|\left(\frac{\eta}{m}\right)_t\right\|_{2\beta} + \nu_0\tilde{\theta}^\top\theta\right]
\end{aligned}$$

where the second term in the inequality is obtained by using the Schwartz inequality. Using $-\tilde{\theta}^\top\theta \leq -\frac{|\tilde{\theta}|^2}{2} + \frac{|\theta^*|^2}{2}$ and the same arguments as in the proof of Theorem 8.5.3, we establish that $V, \epsilon, \epsilon n_s, \theta, \dot{\theta} \in \mathcal{L}_\infty$. From (8.5.59) and $w(t) = \nu_0\|\epsilon(t,\cdot)m(\cdot)\|_{2\beta}$, we have

$$\begin{aligned}
\dot{V} &\leq -\frac{\tilde{\theta}^\top R\tilde{\theta}}{2} + \frac{1}{2}\left(\left\|\left(\frac{\eta}{m}\right)_t\right\|_{2\beta}\right)^2 - \|(\epsilon(t,\cdot)m(\cdot))_t\|_{2\beta}\nu_0\tilde{\theta}^\top\theta \\
&\leq -\frac{\tilde{\theta}^\top R\tilde{\theta}}{2} + \frac{1}{2}\left(\left\|\left(\frac{\eta}{m}\right)_t\right\|_{2\beta}\right)^2 - \|(\epsilon(t,\cdot)m(\cdot))_t\|_{2\beta}\nu_0\left(\frac{|\tilde{\theta}|^2}{2} - \frac{|\theta^*|^2}{2}\right)
\end{aligned}$$

Integrating on both sides of the above inequality and using the boundedness of V, ϵm and (8.5.61) we can establish that $R^{\frac{1}{2}}\tilde{\theta}, \sqrt{\nu_0}\left(\|(\epsilon(t,\tau)m(\tau))_t\|_{2\beta}\right)^{\frac{1}{2}}|\tilde{\theta}| \in \mathcal{S}(\nu_0 + \eta^2/m^2)$. Following the same steps as in the proof of Theorem 8.5.3, we can conclude that $\epsilon, \epsilon n_s, \dot{\theta} \in \mathcal{S}(\nu_0 + \eta^2/m^2)$ and the proof of (i), (ii) is complete.

Because $\epsilon m, \|\epsilon(t,\cdot)m(\cdot)_t\|_{2\beta} \in \mathcal{L}_\infty$, the proof of part (iii) can be completed by repeating the same arguments as in the proof of (A) (iii) and (B) (iv) of Theorem 8.5.4 for the σ-modification. $\square$

8.5.4 Least-Squares with Leakage

The least-squares algorithms with leakage follow directly from Chapter 4 by considering the cost function

$$\begin{aligned}
J(\theta,t) &= \int_0^t e^{-\beta(t-\tau)}\left\{\frac{(z(\tau) - \theta^\top(t)\phi(\tau))^2}{2m^2(\tau)} + w(\tau)\theta^\top(t)\theta(\tau)\right\}d\tau \\
&\quad + \frac{1}{2}e^{-\beta t}(\theta - \theta_0)^\top Q_0(\theta - \theta_0)
\end{aligned} \qquad (8.5.68)$$

where $\beta \geq 0$ is the forgetting factor, $w(t) \geq 0$ is a design weighting function and $m(t)$ is as designed in Section 8.5.3. Following the same procedure as in the ideal case of Chapter 4, we obtain

$$\dot{\theta} = P(\epsilon\phi - w\theta) \qquad (8.5.69)$$

8.5. ROBUST ADAPTIVE LAWS

$$\dot{P} = \beta P - P\frac{\phi\phi^\top}{m^2}P, \quad P(0) = P_0 = Q_0^{-1} \quad (8.5.70)$$

where $P_0 = P_0^\top > 0$ and (8.5.70) can be modified when $\beta = 0$ by using covariance resetting, i.e.,

$$\dot{P} = -P\frac{\phi\phi^\top}{m^2}P, \quad P(t_r^+) = P_0 = \rho_0 I \quad (8.5.71)$$

where t_r is the time for which $\lambda_{min}(P(t)) \leq \rho_1$ for some $\rho_0 > \rho_1 > 0$. When $\beta > 0$, (8.5.70) may be modified as

$$\dot{P} = \begin{cases} \beta P - P\frac{\phi\phi^\top}{m^2}P & \text{if } \|P(t)\| \leq R_0 \\ 0 & \text{otherwise} \end{cases} \quad (8.5.72)$$

with $\|P_0\| \leq R_0$ for some scalar $R_0 > 0$.

As we have shown in Chapter 4, both modifications for P guarantee that $P, P^{-1} \in \mathcal{L}_\infty$ and therefore the stability properties of (8.5.69) with (8.5.71) or (8.5.72) follow directly from those of the gradient algorithm (8.5.46) given by Theorems 8.5.4, 8.5.5. These properties can be established for the various choices of the leakage term $w(t)$ by considering the Lyapunov-like function $V(\tilde{\theta}) = \frac{\tilde{\theta}^\top P^{-1}\tilde{\theta}}{2}$ where P is given by (8.5.71) or (8.5.72) and by following the same procedure as in the proofs of Theorem 8.5.4 to 8.5.5. The details of these proofs are left as an exercise for the reader.

8.5.5 Projection

The two crucial techniques that we used in Sections 8.5.2 to 8.5.4 to develop robust adaptive laws are the dynamic normalization m and leakage. The normalization guarantees that the normalized modeling error term η/m is bounded and therefore acts as a bounded input disturbance in the adaptive law. Since a bounded disturbance may cause parameter drift, the leakage modification is used to guarantee bounded parameter estimates. Another effective way to guarantee bounded parameter estimates is to use projection to constrain the parameter estimates to lie inside some known convex bounded set in the parameter space that contains the unknown θ^*. Adaptive laws with projection have already been introduced and analyzed in Chapter 4. In this section, we illustrate the use of projection for a gradient algorithm that is used to estimate θ^* in the parametric model (8.5.42):

$$z = \theta^{*\top}\phi + \eta, \quad \eta = \Delta_u(s)u + \Delta_y(s)y + d_1$$

To avoid parameter drift in θ, the estimate of θ^*, we constrain θ to lie inside a convex bounded set that contains θ^*. As an example, consider the set

$$\mathcal{P} = \left\{ \theta \mid g(\theta) = \theta^\top \theta - M_0^2 \leq 0 \right\}$$

where M_0 is chosen so that $M_0 \geq |\theta^*|$. The adaptive law for θ is obtained by using the gradient projection method to minimize

$$J = \frac{(z - \theta^\top \phi)^2}{2m^2}$$

subject to $\theta \in \mathcal{P}$, where m is designed as in the previous sections to guarantee that $\phi/m, \eta/m \in \mathcal{L}_\infty$.

Following the results of Chapter 4, we obtain

$$\dot{\theta} = \begin{cases} \Gamma \epsilon \phi & \text{if } \theta^\top \theta < M_0^2 \\ & \text{or if } \theta^\top \theta = M_0^2 \text{ and } (\Gamma \epsilon \phi)^\top \theta \leq 0 \\ (I - \frac{\Gamma \theta \theta^\top}{\theta^\top \Gamma \theta}) \Gamma \epsilon \phi & \text{otherwise} \end{cases} \qquad (8.5.73)$$

where $\theta(0)$ is chosen so that $\theta^\top(0)\theta(0) \leq M_0^2$ and $\epsilon = \frac{z - \theta^\top \phi}{m^2}$, $\Gamma = \Gamma^\top > 0$.

The stability properties of (8.5.73) for estimating θ^* in (8.5.42) in the presence of the modeling error term η are given by the following Theorem.

Theorem 8.5.6 *The gradient algorithm with projection described by the equation (8.5.73) and designed for the parametric model (8.5.42) guarantees that*

(i) $\epsilon, \epsilon n_s, \theta, \dot{\theta} \in \mathcal{L}_\infty$

(ii) $\epsilon, \epsilon n_s, \dot{\theta} \in \mathcal{S}(\eta^2/m^2)$

(iii) *If* $\eta = 0$ *then* $\epsilon, \epsilon n_s, \dot{\theta} \in \mathcal{L}_2$

(iv) *If* $n_s, \phi \in \mathcal{L}_\infty$ *and* ϕ *is PE with level* $\alpha_0 > 0$ *that is independent of* η, *then*

   (a) $\tilde{\theta}$ *converges exponentially to the residual set*

$$D_p = \left\{ \tilde{\theta} \mid |\tilde{\theta}| \leq c(f_0 + \bar{\eta}) \right\}$$

   where $\bar{\eta} = \sup_t \frac{|\eta|}{m}$, $c \geq 0$ *is a constant and* $f_0 \geq 0$ *is a design constant.*

8.5. ROBUST ADAPTIVE LAWS

(b) *There exists an $\bar{\eta}^* > 0$ such that if $\bar{\eta} < \bar{\eta}^*$, then $\tilde{\theta}$ converges exponentially fast to the residual set*

$$D_p = \left\{ \tilde{\theta} \,\big|\, |\tilde{\theta}| \leq c\bar{\eta} \right\}$$

for some constant $c \geq 0$.

Proof As established in Chapter 4, the projection guarantees that $|\theta(t)| \leq M_0$, $\forall t \geq 0$ provided $|\theta(0)| \leq M_0$. Let us choose the Lyapunov-like function

$$V = \frac{\tilde{\theta}^\top \Gamma^{-1} \tilde{\theta}}{2}$$

Along the trajectory of (8.5.73) we have

$$\dot{V} = \begin{cases} -\epsilon^2 m^2 + \epsilon\eta & \text{if } \theta^\top \theta < M_0^2 \\ & \text{or if } \theta^\top \theta = M_0^2 \text{ and } (\Gamma\epsilon\phi)^\top \theta \leq 0 \\ -\epsilon^2 m^2 + \epsilon\eta - \frac{\tilde{\theta}^\top \theta}{\theta^\top \Gamma \theta} \theta^\top \Gamma\epsilon\phi & \text{if } \theta^\top \theta = M_0^2 \text{ and } (\Gamma\epsilon\phi)^\top \theta > 0 \end{cases}$$

For $\theta^\top \theta = M_0^2$ and $(\Gamma\epsilon\phi)^\top \theta = \theta^\top \Gamma\epsilon\phi > 0$, we have $\operatorname{sgn}\left\{\frac{\tilde{\theta}^\top \theta \theta^\top \Gamma\epsilon\phi}{\theta^\top \Gamma \theta}\right\} = \operatorname{sgn}\{\tilde{\theta}^\top \theta\}$. For $\theta^\top \theta = M_0^2$, we have

$$\tilde{\theta}^\top \theta = \theta^\top \theta - \theta^{*\top}\theta \geq M_0^2 - |\theta^*||\theta| = M_0(M_0 - |\theta^*|) \geq 0$$

where the last inequality is obtained using the assumption that $M_0 \geq |\theta^*|$. Therefore, it follows that $\frac{\tilde{\theta}^\top \theta \theta^\top \Gamma\epsilon\phi}{\theta^\top \Gamma \theta} \geq 0$ when $\theta^\top \theta = M_0^2$ and $(\Gamma\epsilon\phi)^\top \theta = \theta^\top \Gamma\epsilon\phi > 0$. Hence, the term due to projection can only make $\dot{V}$ more negative and, therefore,

$$\dot{V} = -\epsilon^2 m^2 + \epsilon\eta \leq -\frac{\epsilon^2 m^2}{2} + \frac{\eta^2}{m^2}$$

Because V is bounded due to $\theta \in \mathcal{L}_\infty$ which is guaranteed by the projection, it follows that $\epsilon m \in \mathcal{S}(\eta^2/m^2)$ which implies that $\epsilon, \epsilon n_s \in \mathcal{S}(\eta^2/m^2)$. From $\tilde{\theta} \in \mathcal{L}_\infty$ and $\phi/m, \eta/m \in \mathcal{L}_\infty$, we have $\epsilon, \epsilon n_s \in \mathcal{L}_\infty$. Now for $\theta^\top \theta = M_0^2$ we have $\frac{\|\Gamma\theta\theta^\top\|}{\theta^\top \Gamma \theta} \leq c$ for some constant $c \geq 0$ which implies that

$$|\dot{\theta}| \leq c|\epsilon\phi| \leq c|\epsilon m|$$

Hence, $\dot{\theta} \in \mathcal{S}(\eta^2/m^2)$ and the proof of (i) and (ii) is complete. The proof of part (iii) follows by setting $\eta = 0$, and it has already been established in Chapter 4.

The proof for parameter error convergence is completed as follows: Define the function

$$f \triangleq \begin{cases} \frac{\theta^\top \Gamma\epsilon\phi}{\theta^\top \Gamma\theta} & \text{if } \theta^\top \theta = M_0^2 \text{ and } (\Gamma\epsilon\phi)^\top \theta > 0 \\ 0 & \text{otherwise} \end{cases} \tag{8.5.74}$$

It is clear from the analysis above that $f(t) \geq 0 \ \forall t \geq 0$. Then, (8.5.73) may be written as

$$\dot{\theta} = \Gamma \epsilon \phi - \Gamma f \theta. \tag{8.5.75}$$

We can establish that $f\tilde{\theta}^T \theta$ has very similar properties as $\sigma_s \tilde{\theta}^T \theta$, i.e.,

$$f|\tilde{\theta}||\theta| \geq f\tilde{\theta}^T \theta \geq f|\theta|(M_0 - |\theta^*|), \quad f \geq 0$$

and $|f(t)| \leq f_0 \ \forall t \geq 0$ for some constant $f_0 \geq 0$. Therefore the proof of (iv) (a), (b) can be completed by following exactly the same procedure as in the proof of Theorem 8.5.4 (B) illustrated by equations (8.5.48) to (8.5.56). □

Similar results may be obtained for the SPR-Lyapunov type adaptive laws and least-squares by using projection to constrain the estimated parameters to remain inside a bounded convex set, as shown in Chapter 4.

8.5.6 Dead Zone

Let us consider the normalized estimation error

$$\epsilon = \frac{z - \theta^T \phi}{m^2} = \frac{-\tilde{\theta}^T \phi + \eta}{m^2} \tag{8.5.76}$$

for the parametric model

$$z = \theta^{*T}\phi + \eta, \quad \eta = \Delta_u(s)u + \Delta_y(s)y + d_1$$

The signal ϵ is used to "drive" the adaptive law in the case of the gradient and least-squares algorithms. It is a measure of the parameter error $\tilde{\theta}$, which is present in the signal $\tilde{\theta}^T\phi$, and of the modeling error η. When $\eta = 0$ and $\tilde{\theta} = 0$ we have $\epsilon = 0$ and no adaptation takes place. Because $\frac{\eta}{m}, \frac{\phi}{m} \in \mathcal{L}_\infty$, large ϵm implies that $\frac{\tilde{\theta}^T\phi}{m}$ is large which in turn implies that $\tilde{\theta}$ is large. In this case, the effect of the modeling error η is small and the parameter estimates driven by ϵ move in a direction which reduces $\tilde{\theta}$. When ϵm is small, however, the effect of η may be more dominant than that of the signal $\tilde{\theta}^T\phi$ and the parameter estimates may be driven in a direction dictated mostly by η. The principal idea behind the dead zone is to monitor the size of the estimation error and adapt only when the estimation error is large relative to the modeling error η, as shown below:

8.5. ROBUST ADAPTIVE LAWS

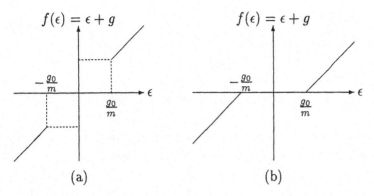

Figure 8.6 Normalized dead zone functions: (a) discontinuous; (b) continuous.

We first consider the gradient algorithm for the linear parametric model (8.5.42). We consider the same cost function as in the ideal case, i.e.,

$$J(\theta, t) = \frac{\epsilon^2 m^2}{2}$$

and write

$$\dot{\theta} = \begin{cases} -\Gamma \nabla J(\theta) & \text{if } |\epsilon m| > g_0 > \frac{|\eta|}{m} \\ 0 & \text{otherwise} \end{cases} \quad (8.5.77)$$

In other words we move in the direction of the steepest descent only when the estimation error is large relative to the modeling error, i.e., when $|\epsilon m| > g_0$, and switch adaptation off when ϵm is small, i.e., $|\epsilon m| \leq g_0$. In view of (8.5.77) we have

$$\dot{\theta} = \Gamma \phi (\epsilon + g), \quad g = \begin{cases} 0 & \text{if } |\epsilon m| > g_0 \\ -\epsilon & \text{if } |\epsilon m| \leq g_0 \end{cases} \quad (8.5.78)$$

To avoid any implementation problems which may arise due to the discontinuity in (8.5.78), the dead zone function is made continuous as follows:

$$\dot{\theta} = \Gamma \phi (\epsilon + g), \quad g = \begin{cases} \frac{g_0}{m} & \text{if } \epsilon m < -g_0 \\ -\frac{g_0}{m} & \text{if } \epsilon m > g_0 \\ -\epsilon & \text{if } |\epsilon m| \leq g_0 \end{cases} \quad (8.5.79)$$

The continuous and discontinuous dead zone functions are shown in Figure 8.6(a, b). Because the size of the dead zone depends on m, this dead zone function is often referred to as the *variable* or *relative dead zone*. Similarly the least-squares algorithm with the dead zone becomes

$$\dot{\theta} = P\phi(\epsilon + g) \tag{8.5.80}$$

where ϵ, g are as defined in (8.5.79) and P is given by either (8.5.71) or (8.5.72).

The dead zone modification can also be incorporated in the integral adaptive law (8.5.47). The principal idea behind the dead zone remains the same as before, i.e., shut off adaptation when the normalized estimation error is small relative to the modeling error. However, for the integral adaptive law, the shut-off process is no longer based on a pointwise-in-time comparison of the normalized estimation error and the a priori bound on the normalized modeling error. Instead, the decision to shut off adaptation is based on the comparison of the $\mathcal{L}_{2\delta}$ norms of certain signals as shown below:

We consider the same integral cost

$$J(\theta, t) = \frac{1}{2} \int_0^t e^{-\beta(t-\tau)} \frac{[z(\tau) - \theta^\top(t)\phi(\tau)]^2}{m^2(\tau)} d\tau$$

as in the ideal case and write

$$\dot{\theta} = \begin{cases} -\Gamma \nabla J(\theta) & \text{if } \|\epsilon(t,\cdot)m(\cdot)\|_{2\beta} > g_0 \geq \sup_t \|(\frac{\eta}{m})_t\|_{2\beta} + \nu \\ 0 & \text{otherwise} \end{cases} \tag{8.5.81}$$

where $\beta > 0$ is the forgetting factor, $\nu > 0$ is a small design constant,

$$\epsilon(t, \tau) \triangleq \frac{z(\tau) - \theta^\top(t)\phi(\tau)}{m^2(\tau)}$$

and $\|\epsilon(t,\cdot)m(\cdot)\|_{2\beta} \triangleq [\int_0^t e^{-\beta(t-\tau)} \epsilon^2(t,\tau) m^2(\tau) d\tau]^{1/2}$ is implemented as

$$\begin{aligned} \|\epsilon(t,\cdot)m(\cdot)\|_{2\beta} &= (r_0 + 2\theta^\top Q + \theta^\top R\theta)^{1/2} \\ \dot{r}_0 &= -\beta r_0 + \frac{z^2}{m^2}, \quad r_0(0) = 0 \end{aligned} \tag{8.5.82}$$

where Q, R are defined in the integral adaptive law given by (8.5.47).

In view of (8.5.81) we have

$$\begin{aligned} \dot{\theta} &= -\Gamma(R\theta + Q - g) \\ \dot{R} &= -\beta R + \frac{\phi \phi^\top}{m^2}, \quad R(0) = 0 \\ \dot{Q} &= -\beta Q - \frac{z\phi}{m^2}, \quad Q(0) = 0 \end{aligned} \tag{8.5.83}$$

8.5. ROBUST ADAPTIVE LAWS

where

$$g = \begin{cases} 0 & \text{if } \|\epsilon(t,\cdot)m(\cdot)\|_{2\beta} > g_0 \\ (R\theta + Q) & \text{otherwise} \end{cases}$$

To avoid any implementation problems which may arise due to the discontinuity in g, the dead zone function is made continuous as follows:

$$g = \begin{cases} 0 & \text{if } \|\epsilon(t,\cdot)m(\cdot)\|_{2\beta} > 2g_0 \\ (R\theta + Q)\left(2 - \frac{\|\epsilon(t,\cdot)m(\cdot)\|_{2\beta}}{g_0}\right) & \text{if } g_0 < \|\epsilon(t,\cdot)m(\cdot)\|_{2\beta} \leq 2g_0 \\ (R\theta + Q) & \text{if } \|\epsilon(t,\cdot)m(\cdot)\|_{2\beta} \leq g_0 \end{cases} \quad (8.5.84)$$

The following theorem summarizes the stability properties of the adaptive laws developed above.

Theorem 8.5.7 *The adaptive laws (8.5.79) and (8.5.80) with P given by (8.5.71) or (8.5.72) and the integral adaptive law (8.5.83) with g given by (8.5.84) guarantee the following properties:*

(i) $\epsilon, \epsilon n_s, \theta, \dot\theta \in \mathcal{L}_\infty$.
(ii) $\epsilon, \epsilon n_s, \dot\theta \in \mathcal{S}(g_0 + \eta^2/m^2)$.
(iii) $\dot\theta \in \mathcal{L}_2 \cap \mathcal{L}_1$.
(iv) $\lim_{t\to\infty} \theta(t) = \bar\theta$ *where* $\bar\theta$ *is a constant vector.*
(v) *If* $n_s, \phi \in \mathcal{L}_\infty$ *and* ϕ *is PE with level* $\alpha_0 > 0$ *independent of* η, *then* $\tilde\theta(t)$ *converges exponentially to the residual set*

$$D_d = \left\{\tilde\theta \in \mathcal{R}^n \,\Big|\, |\tilde\theta| \leq c(g_0 + \bar\eta)\right\}$$

where $\bar\eta = \sup_t \frac{|\eta(t)|}{m(t)}$ *and* $c \geq 0$ *is a constant.*

Proof Adaptive Law (8.5.79) We consider the function

$$V(\tilde\theta) = \frac{\tilde\theta^\top \Gamma^{-1} \tilde\theta}{2}$$

whose time derivative $\dot V$ along the solution of (8.5.79) where $\epsilon m^2 = -\tilde\theta^\top \phi + \eta$ is given by

$$\dot V = \tilde\theta^\top \phi(\epsilon + g) = -(\epsilon m^2 - \eta)(\epsilon + g) \quad (8.5.85)$$

Now

$$(\epsilon m^2 - \eta)(\epsilon + g) = \begin{cases} (\epsilon m + g_0)^2 - (g_0 + \frac{\eta}{m})(\epsilon m + g_0) > 0 & \text{if } \epsilon m < -g_0 \\ (\epsilon m - g_0)^2 + (g_0 - \frac{\eta}{m})(\epsilon m - g_0) > 0 & \text{if } \epsilon m > g_0 \\ 0 & \text{if } |\epsilon m| \leq g_0 \end{cases} \quad (8.5.86)$$

Hence, $(\epsilon m^2 - \eta)(\epsilon + g) \geq 0, \forall t \geq 0$ and $\dot{V} \leq 0$, which implies that $V, \theta \in \mathcal{L}_\infty$ and $\sqrt{(\epsilon m^2 - \eta)(\epsilon + g)} \in \mathcal{L}_2$. Furthermore, $\theta \in \mathcal{L}_\infty$ implies that $\tilde{\theta}, \epsilon, \epsilon n_s \in \mathcal{L}_\infty$. From (8.5.79) we have

$$\dot{\theta}^\top \dot{\theta} = \frac{\phi^\top \Gamma \Gamma \phi}{m^2}(\epsilon + g)^2 m^2 \tag{8.5.87}$$

However,

$$(\epsilon + g)^2 m^2 = (\epsilon m + gm)^2 = \begin{cases} (\epsilon m + g_0)^2 & \text{if } \epsilon m < -g_0 \\ (\epsilon m - g_0)^2 & \text{if } \epsilon m > g_0 \\ 0 & \text{if } |\epsilon m| \leq g_0 \end{cases}$$

which, together with (8.5.86), implies that

$$0 \leq (\epsilon + g)^2 m^2 \leq (\epsilon m^2 - \eta)(\epsilon + g), \quad i.e., (\epsilon + g)m \in \mathcal{L}_2$$

Hence, from (8.5.87) and $\frac{\phi}{m} \in \mathcal{L}_\infty$ we have that $\dot{\theta} \in \mathcal{L}_2$. Equation (8.5.85) can also be written as

$$\dot{V} \leq -\epsilon^2 m^2 + |\epsilon m|\frac{|\eta|}{m} + |\epsilon m|g_0 + \frac{|\eta|}{m}g_0$$

by using $|g| \leq \frac{g_0}{m}$. Then by completing the squares we have

$$\dot{V} \leq -\frac{\epsilon^2 m^2}{2} + \frac{|\eta|^2}{m^2} + g_0^2 + \frac{|\eta|}{m}g_0$$
$$\leq -\frac{\epsilon^2 m^2}{2} + \frac{3|\eta|^2}{2m^2} + \frac{3}{2}g_0^2$$

which, together with $V \in \mathcal{L}_\infty$, implies that $\epsilon m \in \mathcal{S}(g_0^2 + \frac{\eta^2}{m^2})$. Because $m^2 = 1 + n_s^2$ we have $\epsilon, \epsilon n_s \in \mathcal{S}(g_0^2 + \frac{\eta^2}{m^2})$. Because $|g| \leq \frac{g_0}{m} \leq c_1 g_0$, we can show that $|\dot{\theta}| \leq (|\epsilon m| + g_0)$, which implies that $\dot{\theta} \in \mathcal{S}(g_0^2 + \frac{\eta^2}{m^2})$ due to $\epsilon m \in \mathcal{S}(g_0^2 + \frac{\eta^2}{m^2})$. Because g_0 is a constant, we can absorb it in one of the constants in the definition of m.s.s. and write $\epsilon m, \dot{\theta} \in \mathcal{S}(g_0 + \frac{\eta^2}{m^2})$ to preserve compatibility with the other modifications.

To show that $\lim_{t \to \infty} \theta = \bar{\theta}$, we use (8.5.85) and the fact that $(\epsilon m^2 - \eta)(\epsilon + g) \geq 0$ to obtain

$$\dot{V} = -(\epsilon m^2 - \eta)(\epsilon + g) = -\left|\epsilon m - \frac{\eta}{m}\right||\epsilon + g|m$$
$$\leq -\left||\epsilon m| - \frac{|\eta|}{m}\right||\epsilon + g|m$$
$$\leq -\left|g_0 - \frac{|\eta|}{m}\right||\epsilon + g|m \quad (\text{because } |\epsilon + g| = 0 \text{ if } |\epsilon m| \leq g_0)$$

Because $\frac{\eta}{m} \in \mathcal{L}_\infty$ and $g_0 > \frac{|\eta|}{m}$ we integrate both sides of the above inequality and use the fact that $V \in \mathcal{L}_\infty$ to obtain that $(\epsilon + g)m \in \mathcal{L}_1$. Then from the adaptive

8.5. ROBUST ADAPTIVE LAWS

law (8.5.79) and the fact that $\frac{\phi}{m} \in \mathcal{L}_\infty$, it follows that $|\dot{\tilde{\theta}}| \in \mathcal{L}_1$, which, in turn, implies that the $\lim_{t\to\infty} \int_0^t \dot{\theta}d\tau$ exists and, therefore, θ converges to $\bar{\theta}$.

To show that $\tilde{\theta}$ converges exponentially to the residual set D_d for persistently exciting ϕ, we follow the same steps as in the case of the fixed σ-modification or the ϵ_1-modification. We express (8.5.79) in terms of the parameter error

$$\dot{\tilde{\theta}} = -\Gamma\frac{\phi\phi^\top}{m^2}\tilde{\theta} + \Gamma\frac{\phi\eta}{m^2} + \Gamma\phi g \tag{8.5.88}$$

where g satisfies $|g| \leq \frac{g_0}{m} \leq c_1 g_0$ for some constant $c_1 \geq 0$. Since the homogeneous part of (8.5.88) is e.s. when ϕ is PE, a property that is established in Section 4.8.3, the exponential convergence of $\tilde{\theta}$ to the residual set D_d can be established by repeating the same steps as in the proof of Theorem 8.5.4 (iii).

Adaptive Law (8.5.80) The proof for the adaptive law (8.5.80) is very similar to that of (8.5.79) presented above and is omitted.

Adaptive Law (8.5.83) with g Given by (8.5.84) This adaptive law may be rewritten as

$$\dot{\theta} = \sigma_e \Gamma \int_0^t e^{-\beta(t-\tau)}\epsilon(t,\tau)\phi(\tau)d\tau \tag{8.5.89}$$

where

$$\sigma_e = \begin{cases} 1 & \text{if } \|\epsilon(t,\cdot)m(\cdot)\|_{2\beta} > 2g_0 \\ \frac{\|\epsilon(t,\cdot)m(\cdot)\|_{2\beta}}{g_0} - 1 & \text{if } g_0 < \|\epsilon(t,\cdot)m(\cdot)\|_{2\beta} \leq 2g_0 \\ 0 & \text{if } \|\epsilon(t,\cdot)m(\cdot)\|_{2\beta} \leq g_0 \end{cases} \tag{8.5.90}$$

Once again we consider the positive definite function $V(\tilde{\theta}) = \frac{\tilde{\theta}^\top \Gamma^{-1}\tilde{\theta}}{2}$ whose time derivative along the solution of (8.5.83) is given by

$$\dot{V} = \sigma_e \tilde{\theta}^\top(t) \int_0^t e^{-\beta(t-\tau)}\epsilon(t,\tau)\phi(\tau)d\tau$$

Using $\tilde{\theta}^\top(t)\phi(\tau) = -\epsilon(t,\tau)m^2(\tau) + \eta(\tau)$, we obtain

$$\dot{V} = -\sigma_e \int_0^t e^{-\beta(t-\tau)}\epsilon(t,\tau)m(\tau)\left[\epsilon(t,\tau)m(\tau) - \frac{\eta(\tau)}{m(\tau)}\right]d\tau \tag{8.5.91}$$

Therefore, by using the Schwartz inequality, we have

$$\dot{V} \leq -\sigma_e \|\epsilon(t,\cdot)m(\cdot)\|_{2\beta}\left[\|\epsilon(t,\cdot)m(\cdot)\|_{2\beta} - \|(\frac{\eta}{m})_t\|_{2\beta}\right]$$

From the definition of σ_e and the fact that $g_0 \geq \sup_t \|(\frac{\eta}{m})_t\|_{2\beta} + \nu$, it follows that $\dot{V} \leq 0$ which implies that $V, \theta, \epsilon \in \mathcal{L}_\infty$ and $\lim_{t\to\infty} = V_\infty$ exists. Also $\theta \in \mathcal{L}_\infty$ implies that $\epsilon n_s \in \mathcal{L}_\infty$. Using $g_0 \geq \sup_t \|(\frac{\eta}{m})_t\|_{2\beta} + \nu$ in (8.5.91), we obtain

$$\dot{V} \leq -\nu\sigma_e\|\epsilon(t,\cdot)m(\cdot)\|_{2\beta}$$

Now integrating on both sides of the above inequality and using $V \in \mathcal{L}_\infty$, we have

$$\sigma_e \|\epsilon(t,\cdot)m(\cdot)\|_{2\beta} \in \mathcal{L}_1 \qquad (8.5.92)$$

Using (8.5.89) and the boundedness property of $\frac{\phi}{m}$, we can establish that $|\dot\theta| \in \mathcal{L}_1$. Moreover, from (8.5.83) with g given by (8.5.84) we also have that $\dot\theta$ is uniformly continuous which together with $\dot\theta \in \mathcal{L}_1$ imply that $\lim_{t\to\infty} \dot\theta(t) = 0$. Furthermore, using the same arguments as in the proof for the adaptive law (8.5.79), we can conclude from $|\dot\theta| \in \mathcal{L}_1$ that $\lim_{t\to\infty} \theta = \bar\theta$ for some constant vector $\bar\theta$.

It remains to show that $\epsilon, \epsilon n_s \in \mathcal{S}(g_0 + \frac{\eta^2}{m^2})$. This can be done as follows: Instead of (8.5.91), we use the following expression for $\dot V$:

$$\begin{aligned}
\dot V &= \tilde\theta^\top(t) \left[\int_0^t e^{-\beta(t-\tau)} \epsilon(t,\tau)\phi(\tau)d\tau\right] + \tilde\theta^\top g \\
&= -\tilde\theta^\top R \tilde\theta + \tilde\theta^\top(t) \int_0^t e^{-\beta(t-\tau)} \frac{\phi(\tau)\eta(\tau)}{m^2(\tau)} d\tau + \tilde\theta^\top g
\end{aligned}$$

obtained by using (8.5.83) instead of (8.5.89), which has the same form as equation (8.5.58) in the proof of Theorem 8.5.4. Hence, by following the same calculations that led to (8.5.59), we obtain

$$\dot V \leq -\frac{\tilde\theta^\top R \tilde\theta}{2} + \frac{1}{2}\left(\left\|\left(\frac{\eta}{m}\right)_t\right\|_{2\beta}\right)^2 + \tilde\theta^\top g \qquad (8.5.93)$$

Using the definition of $\epsilon(t,\tau)$, we obtain

$$\begin{aligned}
|R\theta + Q| &= \left|\int_0^t e^{-\beta(t-\tau)} \frac{\phi(\tau)(\theta^\top(\tau)\phi(\tau) - z(\tau))}{m^2} d\tau\right| \\
&= \left|\int_0^t e^{-\beta(t-\tau)} \phi(\tau)\epsilon(t,\tau) d\tau\right| \\
&\leq \left[\int_0^t e^{-\beta(t-\tau)} \frac{|\phi(\tau)|^2}{m^2(\tau)} d\tau\right]^{\frac{1}{2}} \|\epsilon(t,\cdot)m(\cdot)_t\|_{2\beta}
\end{aligned}$$

where the last inequality is obtained by using the Schwartz inequality. Because $\frac{\phi}{m} \in \mathcal{L}_\infty$, we have

$$|R\theta + Q| \leq c_2 \|\epsilon(t,\cdot)m(\cdot)_t\|_{2\beta} \qquad (8.5.94)$$

for some constant $c_2 \geq 0$. Using the definition of g given in (8.5.84) and (8.5.94), we have

$$|g| \leq c g_0$$

for some constant $c \geq 0$ and therefore it follows from (8.5.93) that

$$\dot V \leq -\frac{\tilde\theta^\top R \tilde\theta}{2} + \frac{1}{2}\left(\left\|\left(\frac{\eta}{m}\right)_t\right\|_{2\beta}\right)^2 + c g_0$$

8.5. ROBUST ADAPTIVE LAWS

where $c \geq 0$ depends on the bound for $|\tilde{\theta}|$.

The above inequality has the same form as (8.5.59) and we can duplicate the steps illustrated by equations (8.5.59) to (8.5.63) in the proof of Theorem 8.5.4 (A) to first show that $R^{\frac{1}{2}}\tilde{\theta} \in \mathcal{S}(g_0 + \eta^2/m^2)$ and then that $\epsilon, \epsilon n_s, \dot{\theta} \in \mathcal{S}(g_0 + \eta^2/m^2)$.

The proof for part (v) proceeds as follows: We write

$$\dot{\tilde{\theta}} = -\Gamma R\tilde{\theta} + \Gamma \int_0^t e^{-\beta(t-\tau)} \frac{\phi\eta}{m^2} d\tau + \Gamma g \qquad (8.5.95)$$

Because the equilibrium $\tilde{\theta}_e = 0$ of the homogeneous part of (8.5.95), i.e., $\dot{\tilde{\theta}} = -\Gamma R\tilde{\theta}$, is e.s. for persistently exciting ϕ, a property that has been established in Section 4.8.4, we can show as in the proof of Theorem 8.5.4 A(iii) that

$$|\tilde{\theta}| \leq \beta_0 e^{-\beta_2 t} + \beta_1 \int_0^t e^{-\beta_2(t-\tau)} |d(\tau)| d\tau$$

where $d \triangleq \int_0^t e^{-\beta(t-\tau)} \frac{\phi\eta}{m^2} d\tau + g$, β_0, β_1 are nonnegative constants and $\beta_2 > 0$. Because $\frac{\phi}{m} \in \mathcal{L}_\infty$ and $\frac{|\eta|}{m} \leq \bar{\eta}$, we have $|d| \leq c_1 \bar{\eta} + |g|$. Because $|g| \leq cg_0$, we have

$$|\tilde{\theta}| \leq c(\bar{\eta} + g_0) + \epsilon_t$$

for some constant $c \geq 0$, where ϵ_t is an exponentially decaying to zero term which completes the proof of part (v). □

8.5.7 Bilinear Parametric Model

In this section, we consider the bilinear parametric model

$$\begin{aligned} z &= W(s)\rho^*(\theta^{*T}\psi + z_0) + \eta \\ \eta &= \Delta_u(s)u + \Delta_y(s)y + d_1 \end{aligned} \qquad (8.5.96)$$

where z, ψ, z_0 are measurable signals, $W(s)$ is proper and stable, ρ^*, θ^* are the unknown parameters to be estimated on-line and η is the modeling error term due to a bounded disturbance d_1 and unmodeled dynamics $\Delta_u(s)u, \Delta_y(s)y$. The perturbations $\Delta_u(s), \Delta_y(s)$ are assumed without loss of generality to be strictly proper and analytic in $\text{Re}[s] \geq -\delta_0/2$ for some known $\delta_0 > 0$.

The techniques of the previous sections and the procedure of Chapter 4 may be used to develop robust adaptive laws for (8.5.96) in a straightforward manner.

As an example, we illustrate the extension of the gradient algorithm to the model (8.5.96). We express (8.5.96) as

$$z = \rho^*(\theta^{*\mathsf{T}}\phi + z_1) + \eta$$

where $\phi = W(s)\psi, z_1 = W(s)z_0$. Then the gradient algorithm (based on the instantaneous cost) is given by

$$\begin{aligned}\dot{\theta} &= \Gamma\epsilon\phi\,\mathrm{sgn}(\rho^*) - w_1\Gamma\theta \\ \dot{\rho} &= \gamma\epsilon\xi - w_2\gamma\rho \\ \epsilon &= \frac{z-\hat{z}}{m^2} = \frac{z - \rho(\theta^{\mathsf{T}}\phi + z_1)}{m^2} \\ \xi &= \theta^{\mathsf{T}}\phi + z_1\end{aligned} \qquad (8.5.97)$$

where θ, ρ is the estimate of θ^*, ρ^* at time t; $m > 0$ is chosen so that $\frac{\phi}{m}, \frac{z_1}{m}, \frac{\eta}{m} \in \mathcal{L}_\infty$, w_1, w_2 are leakage modifications and $\Gamma = \Gamma^{\mathsf{T}} > 0, \gamma > 0$ are the adaptive gains. Using Lemma 3.3.2, the normalizing signal m may be chosen as

$$m^2 = 1 + \phi^{\mathsf{T}}\phi + z_1^2 + n_s^2, \quad n_s^2 = m_s$$
$$\dot{m}_s = -\delta_0 m_s + u^2 + y^2, \quad m_s(0) = 0$$

If Δ_u, Δ_y are biproper, then m^2 may be modified to include u, y, i.e.,

$$m^2 = 1 + \phi^{\mathsf{T}}\phi + z_1^2 + n_s^2 + u^2 + y^2$$

If $\phi = H(s)[u, y]^{\mathsf{T}}, z_1 = h(s)[u, y]^{\mathsf{T}}$, where $H(s), h(s)$ are strictly proper and analytic in $\mathrm{Re}[s] \geq -\delta_0/2$, then the terms $\phi^{\mathsf{T}}\phi, z_1^2$ may be dropped from the expression of m^2.

The leakage terms w_1, w_2 may be chosen as in Section 8.5.2 to 8.5.4. For example, if we use the switching σ-modification we have

$$w_1 = \sigma_{1s}, \quad w_2 = \sigma_{2s} \qquad (8.5.98)$$

where

$$\sigma_{is} = \begin{cases} 0 & \text{if } |x_i| \leq M_i \\ \sigma_0(\frac{|x_i|}{M_i} - 1) & \text{if } M_i < |x_i| \leq 2M_i \\ \sigma_0 & \text{if } |x_i| > 2M_i \end{cases}$$

with $i = 1, 2$ and $|x_1| = |\theta|, |x_2| = |\rho|$, and $\sigma_0 > 0$ is a design constant. The properties of (8.5.97) and (8.5.98) can be established in a similar manner as in Section 8.5.3 and are summarized below.

8.5. ROBUST ADAPTIVE LAWS

Theorem 8.5.8 *The gradient algorithm (8.5.97) and (8.5.98) for the bilinear parametric model (8.5.96) guarantees the following properties.*

(i) $\epsilon, \epsilon n_s, \rho, \theta, \dot\rho, \dot\theta \in \mathcal{L}_\infty$

(ii) $\epsilon, \epsilon n_s, \dot\rho, \dot\theta \in \mathcal{S}(\eta^2/m^2)$

The proof of Theorem 8.5.8 is very similar to those for the linear parametric model, and can be completed by exploring the properties of the Lyapunov function $V = |\rho^*|\frac{\tilde\theta^\top \Gamma \tilde\theta}{2} + \frac{\tilde\rho^2}{2\gamma}$.

As we discussed in Chapter 4, another way of handling the bilinear case is to express (8.5.96) in the linear form as follows

$$z = W(s)(\bar\theta^{*\top}\bar\psi) + \eta \qquad (8.5.99)$$

where $\bar\theta^* = [\bar\theta_1^*, \bar\theta_2^{*\top}]^\top$, $\bar\theta_1^* = \rho^*$, $\bar\theta_2^* = \rho^*\theta^*$ and $\bar\psi = [z_0, \psi^\top]^\top$. Then using the procedure of Sections 8.5.2 to 8.5.4, we can develop a wide class of robust adaptive laws for estimating $\bar\theta^*$. From the estimate $\bar\theta$ of $\bar\theta^*$, we calculate the estimate θ, ρ of θ^* and ρ^* respectively as follows:

$$\rho = \bar\theta_1, \quad \theta = \frac{\bar\theta_2}{\bar\theta_1} \qquad (8.5.100)$$

where $\bar\theta_1$ is the estimate of $\bar\theta_1^* = \rho^*$ and $\bar\theta_2$ is the estimate of $\bar\theta_2^* = \rho^*\theta^*$. In order to avoid the possibility of division by zero in (8.5.100), we use the gradient projection method to constrain the estimate $\bar\theta_1$ to be in the set

$$\mathcal{C} = \left\{\bar\theta \in \mathcal{R}^{n+1} \,\middle|\, g(\bar\theta) = \rho_0 - \bar\theta_1 \mathrm{sgn}\rho^* \leq 0\right\} \qquad (8.5.101)$$

where $\rho_0 > 0$ is a lower bound for $|\rho^*|$. We illustrate this method for the gradient algorithm developed for the parametric model $z = \bar\theta^{*\top}\bar\phi + \eta$, $\bar\phi = W(s)\bar\psi = [z_1, \phi^\top]^\top$ given by

$$\begin{aligned}\dot{\bar\theta} &= \Gamma\epsilon\bar\phi - w\Gamma\bar\theta \\ \epsilon &= \frac{z - \hat z}{m^2} = \frac{z - \bar\theta^\top\bar\phi}{m^2}\end{aligned} \qquad (8.5.102)$$

where $m > 0$ is designed so that $\frac{\bar\phi}{m}, \frac{\eta}{m} \in \mathcal{L}_\infty$. We now apply the gradient projection method in order to constrain $\bar\theta(t) \in \mathcal{C}\ \forall t \geq t_0$, i.e., instead of

(8.5.102) we use

$$\dot{\bar{\theta}} = \begin{cases} \Gamma(\epsilon\bar{\phi} - w\bar{\theta}) & \text{if } \bar{\theta} \in \mathcal{C}_0 \\ & \text{or if } \theta \in \delta(\mathcal{C}) \text{ and } [\Gamma(\epsilon\bar{\phi} - w\bar{\theta})]^T \nabla g \leq 0 \\ \left(I - \Gamma \frac{\nabla g \nabla g^T}{\nabla g^T \Gamma \nabla g}\right) \Gamma(\epsilon\bar{\phi} - w\bar{\theta}) & \text{otherwise.} \end{cases}$$
(8.5.103)

where $\mathcal{C}_0, \delta(\mathcal{C})$ denote the interior and boundary of $\mathcal{C}$ respectively.

Because $\nabla g = [-\text{sgn}(\rho^*), 0, \ldots, 0]^T$, (8.5.103) can be simplified by choosing $\Gamma = diag(\gamma_0, \Gamma_0)$ where $\gamma_0 > 0$ and $\Gamma_0 = \Gamma_0^T > 0$, i.e.,

$$\dot{\bar{\theta}}_1 = \begin{cases} \gamma_0 \epsilon z_1 - \gamma_0 w \bar{\theta}_1 & \text{if } \bar{\theta}_1 \text{sgn}(\rho^*) > \rho_0 \\ & \text{or if } \bar{\theta}_1 \text{sgn}(\rho^*) = \rho_0 \text{ and } (\epsilon z_1 - w\bar{\theta}_1)\text{sgn}(\rho^*) \geq 0 \\ 0 & \text{otherwise} \end{cases}$$

$$\dot{\bar{\theta}}_2 = \Gamma_0(\epsilon\phi - w\bar{\theta}_2)$$
(8.5.104)

The adaptive law (8.5.104) guarantees the same properties as the adaptive law (8.5.97), (8.5.98) described by Theorem 8.5.8.

8.5.8 Hybrid Adaptive Laws

The adaptive laws developed in Sections 8.5.2 to 8.5.7 update the estimate $\theta(t)$ of the unknown parameter vector θ^* continuously with time, i.e., at each time t we have a new estimate. For computational and robustness reasons, it may be desirable to update the estimates only at specific instants of time t_k where $\{t_k\}$ is an unbounded monotonically increasing sequence in $\mathcal{R}^+$. Let $t_k = kT_s$ where $T_s = t_{k+1} - t_k$ is the "sampling" period and $k \in \mathcal{N}^+$ and consider the design of an adaptive law that generates the estimate of the unknown θ^* at the discrete instances of time $t = 0, T_s, 2T_s, \ldots$.

We can develop such an adaptive law for the gradient algorithms of Section 8.5.3. For example, let us consider the adaptive law

$$\dot{\theta} = \Gamma\epsilon\phi - w\Gamma\theta$$
$$\epsilon = \frac{z - \hat{z}}{m^2} = \frac{z - \theta^T\phi}{m^2}$$
(8.5.105)

where z is the output of the linear parametric model

$$z = \theta^{*T}\phi + \eta$$

8.5. ROBUST ADAPTIVE LAWS

and m is designed so that $\frac{\phi}{m}, \frac{\eta}{m} \in \mathcal{L}_\infty$. Integrating (8.5.105) from $t_k = kT_s$ to $t_{k+1} = (k+1)T_s$ we have

$$\theta_{k+1} = \theta_k + \Gamma \int_{t_k}^{t_{k+1}} (\epsilon(\tau)\phi(\tau) - w(\tau)\theta_k)d\tau \qquad (8.5.106)$$

where $\theta_k \triangleq \theta(t_k)$. Equation (8.5.106) generates a sequence of estimates, i.e., $\theta_0 = \theta(0), \theta_1 = \theta(T_s), \theta_2 = \theta(2T_s), \ldots, \theta_k = \theta(kT_s)$ of θ^*. Although ϵ, ϕ may vary with time continuously, $\theta(t) = \theta_k = $ constant for $t \in [t_k, t_{k+1})$. In (8.5.106) the estimate $\hat{z}$ of z and ϵ are generated by using θ_k, i.e.,

$$\hat{z}(t) = \theta_k^\top \phi(t), \quad \epsilon = \frac{z - \hat{z}}{m^2}, \quad t \in [t_k, t_{k+1}) \qquad (8.5.107)$$

We shall refer to (8.5.106) and (8.5.107) as the *robust hybrid adaptive law*.

The leakage term $w(t)$ chosen as in Section 8.5.3 has to satisfy some additional conditions for the hybrid adaptive law to guarantee similar properties as its continuous counterpart. These conditions arise from analysis and depend on the specific choice for $w(t)$. We present these conditions and properties of (8.5.106) for the switching σ-modification

$$w(t) = \sigma_s, \quad \sigma_s = \begin{cases} 0 & \text{if } |\theta_k| < M_0 \\ \sigma_0 & \text{if } |\theta_k| \geq M_0 \end{cases} \qquad (8.5.108)$$

where $\sigma_0 > 0$ and $M_0 \geq 2|\theta^*|$. Because of the discrete-time nature of (8.5.106), σ_s can now be discontinuous as given by (8.5.108).

The following theorem establishes the stability properties of (8.5.106), (8.5.108):

Theorem 8.5.9 *Let m, σ_0, T_s, Γ be chosen so that*

(a) $\frac{\eta}{m} \in \mathcal{L}_\infty, \frac{\phi^\top \phi}{m^2} \leq 1$

(b) $2T_s \lambda_m < 1, 2\sigma_0 \lambda_m T_s < 1$

where $\lambda_m = \lambda_{max}(\Gamma)$. Then the hybrid adaptive law (8.5.106), (8.5.108) guarantees that

(i) $\epsilon, \epsilon n_s \in \mathcal{L}_\infty, \theta_k \in \ell_\infty$

(ii) $\epsilon, \epsilon m \in \mathcal{S}(\frac{\eta^2}{m^2}), \Delta\theta_k \in \mathcal{D}(\frac{\eta^2}{m^2})$ where $\Delta\theta_k = \theta_{k+1} - \theta_k$ and

$$\mathcal{D}(y) \triangleq \left\{ \{x_k\} \Big| \sum_{k=k_0}^{k_0+N} x_k^T x_k \leq c_0 \int_{t_{k_0}}^{t_{k_0}+NT_s} y(\tau) d\tau + c_1 \right\}$$

for some $c_0, c_1 \in \mathcal{R}^+$ and any $k_0, N \in \mathcal{N}^+$.

(iii) If $n_s, \phi \in \mathcal{L}_\infty$ and ϕ is PE with a level of excitation $\alpha_0 > 0$ that is independent of η, then

(a) $\tilde{\theta}_k = \theta_k - \theta^*$ converges exponentially to the residual set

$$\mathcal{D}_0 = \left\{ \tilde{\theta} \Big| |\tilde{\theta}| \leq c(\sigma_0 + \bar{\eta}) \right\}$$

where $c \geq 0$, $\bar{\eta} = \sup_t |\frac{\eta}{m}|$.

(b) There exists a constant $\bar{\eta}^* > 0$ such that if $\bar{\eta} < \bar{\eta}^*$, then $\tilde{\theta}_k$ converges exponentially fast to the residual set

$$\mathcal{D}_\theta = \left\{ \tilde{\theta} \Big| |\tilde{\theta}| \leq c\bar{\eta} \right\}$$

Proof Consider the function

$$V(k) = \tilde{\theta}_k^T \Gamma^{-1} \tilde{\theta}_k \qquad (8.5.109)$$

Using $\tilde{\theta}_{k+1} = \tilde{\theta}_k + \Delta\theta_k$ in (8.5.109), where $\Delta\theta_k = \Gamma \int_{t_k}^{t_{k+1}} (\epsilon(\tau)\phi(\tau) - w(\tau)\theta_k) d\tau$, we can write

$$\begin{aligned} V(k+1) &= V(k) + 2\tilde{\theta}_k^T \Gamma^{-1} \Delta\theta_k + \Delta\theta_k^T \Gamma^{-1} \Delta\theta_k \\ &= V(k) + 2\tilde{\theta}_k^T \int_{t_k}^{t_{k+1}} (\epsilon(\tau)\phi(\tau) - w(\tau)\theta_k) d\tau \qquad (8.5.110) \\ &\quad + \int_{t_k}^{t_{k+1}} (\epsilon(\tau)\phi(\tau) - w(\tau)\theta_k)^T d\tau \Gamma \int_{t_k}^{t_{k+1}} (\epsilon(\tau)\phi(\tau) - w(\tau)\theta_k) d\tau \end{aligned}$$

Because $\tilde{\theta}_k^T \phi = -\epsilon m^2 + \eta$, we have

$$\tilde{\theta}_k^T \int_{t_k}^{t_{k+1}} \epsilon(\tau)\phi(\tau) d\tau = \int_{t_k}^{t_{k+1}} (-\epsilon^2 m^2 + \epsilon\eta) d\tau \leq -\int_{t_k}^{t_{k+1}} \frac{\epsilon^2 m^2}{2} d\tau + \int_{t_k}^{t_{k+1}} \frac{\eta^2}{2m^2} d\tau$$
$$(8.5.111)$$

8.5. ROBUST ADAPTIVE LAWS

where the last inequality is obtained by using the inequality $-a^2 + ab \leq -\frac{a^2}{2} + \frac{b^2}{2}$. Now consider the last term in (8.5.110), since

$$\int_{t_k}^{t_{k+1}} (\epsilon(\tau)\phi(\tau) - w(\tau)\theta_k)^T d\tau \Gamma \int_{t_k}^{t_{k+1}} (\epsilon(\tau)\phi(\tau) - w(\tau)\theta_k) d\tau$$

$$\leq \lambda_m \left| \int_{t_k}^{t_{k+1}} (\epsilon(\tau)\phi(\tau) - w(\tau)\theta_k) d\tau \right|^2$$

where $\lambda_m = \lambda_{max}(\Gamma^{-1})$, it follows from the inequality $(a+b)^2 \leq 2a^2 + 2b^2$ that

$$\int_{t_k}^{t_{k+1}} (\epsilon(\tau)\phi(\tau) - w(\tau)\theta_k)^T d\tau \Gamma \int_{t_k}^{t_{k+1}} (\epsilon(\tau)\phi(\tau) - w(\tau)\theta_k) d\tau$$

$$\leq 2\lambda_m \left| \int_{t_k}^{t_{k+1}} \epsilon(\tau) m(\tau) \frac{\phi(\tau)}{m(\tau)} d\tau \right|^2 + 2\lambda_m \left| \int_{t_k}^{t_{k+1}} w(\tau)\theta_k d\tau \right|^2$$

$$\leq 2\lambda_m \int_{t_k}^{t_{k+1}} \epsilon^2(\tau) m^2(\tau) d\tau \int_{t_k}^{t_{k+1}} \frac{|\phi|^2}{m^2} d\tau + 2\lambda_m \sigma_s^2 T_s^2 |\theta_k|^2$$

$$\leq 2\lambda_m T_s \int_{t_k}^{t_{k+1}} \epsilon^2(\tau) m^2(\tau) d\tau + 2\lambda_m \sigma_s^2 T_s^2 |\theta_k|^2 \qquad (8.5.112)$$

In obtaining (8.5.112), we have used the Schwartz inequality and the assumption $\frac{|\phi|}{m} \leq 1$. Using (8.5.111), (8.5.112) in (8.5.110), we have

$$V(k+1) \leq V(k) - (1 - 2\lambda_m T_s) \int_{t_k}^{t_{k+1}} \epsilon^2(\tau) m^2(\tau) d\tau + \int_{t_k}^{t_{k+1}} \frac{|\eta|^2}{m^2} d\tau$$

$$- 2\sigma_s T_s \tilde{\theta}_k^T \theta_k + 2\lambda_m \sigma_s^2 T_s^2 |\theta_k|^2$$

$$\leq V(k) - (1 - 2\lambda_m T_s) \int_{t_k}^{t_{k+1}} \epsilon^2(\tau) m^2(\tau) d\tau + \bar{\eta}^2 T_s$$

$$- 2\sigma_s T_s (\tilde{\theta}_k^T \theta_k - \lambda_m \sigma_0 T_s |\theta_k|^2)$$

$$\leq V(k) - (1 - 2\lambda_m T_s) \int_{t_k}^{t_{k+1}} \epsilon^2(\tau) m^2(\tau) d\tau + \bar{\eta}^2 T_s$$

$$- 2\sigma_s T_s \left((\frac{1}{2} - \lambda_m \sigma_0 T_s) |\theta_k|^2 - \frac{|\theta^*|^2}{2} \right) \qquad (8.5.113)$$

where the last inequality is obtained by using $\tilde{\theta}_k^T \theta_k \geq \frac{|\theta_k|^2}{2} - \frac{|\theta^*|^2}{2}$. Therefore, for $1 - 2\sigma_0 \lambda_m T_s > 0$ and $1 - 2\lambda_m T_s > 0$, it follows from (8.5.113) that $V(k+1) \leq V(k)$ whenever

$$|\theta_k|^2 \geq max \left\{ M_0^2, \frac{\bar{\eta}^2 + \sigma_0 |\theta^*|^2}{\sigma_0(1 - 2\lambda_m \sigma_0 T_s)} \right\}$$

Thus, we can conclude that $V(k)$ and $\theta_k \in \ell_\infty$. The boundedness of ϵ, ϵm follows immediately from the definition of ϵ and the normalizing properties of m, i.e., $\frac{\phi}{m}, \frac{\eta}{m} \in \mathcal{L}_\infty$.

To establish (ii), we use the inequality

$$\sigma_s \tilde{\theta}_k^T \theta_k = \frac{\sigma_s}{2}\theta_k^T\theta_k + (\frac{\sigma_s}{2}\theta_k^T\theta_k - \sigma_s\theta_k^T\theta^*)$$

$$\geq \frac{\sigma_s}{2}|\theta_k|^2 + \frac{\sigma_s}{2}|\theta_k|(|\theta_k| - 2|\theta^*|)$$

Because $\sigma_s = 0$ for $|\theta_k| \leq M_0$ and $\sigma_s > 0$ for $|\theta_k| \geq M_0 \geq 2|\theta^*|$, we have $\sigma_s|\theta_k|(|\theta_k|-2|\theta^*|) \geq 0 \ \forall k \geq 0$, therefore,

$$\sigma_s \tilde{\theta}_k^T \theta_k \geq \frac{\sigma_s}{2}|\theta_k|^2$$

which together with $2\sigma_0 \lambda_m T_s < 1$ imply that

$$\sigma_s(\tilde{\theta}_k^T \theta_k - \lambda_m \sigma_0 T_s |\theta_k|^2) \geq c_\sigma \sigma_s |\theta_k|^2 \quad (8.5.114)$$

where $c_\sigma \triangleq \frac{1}{2} - \lambda_m \sigma_0 T_s$. From (8.5.114) and the second inequality of (8.5.113), we have

$$(1 - 2\lambda_m T_s)\int_{t_k}^{t_{k+1}} \epsilon^2(\tau)m^2(\tau)d\tau + c_\sigma \sigma_s |\theta_k|^2 \leq V(k) - V(k+1) + \int_{t_k}^{t_{k+1}} \frac{\eta^2}{m^2}d\tau$$
$$(8.5.115)$$

which implies that $\epsilon m \in \mathcal{S}(\frac{\eta^2}{m^2})$ and $\sqrt{\sigma_s}\theta_k \in \mathcal{D}(\frac{\eta^2}{m^2})$. Because $|\epsilon| \leq |\epsilon m|$ (because $m \geq 1, \forall t \geq 0$), we have $\epsilon \in \mathcal{S}(\frac{\eta^2}{m^2})$.

Note from (8.5.112) that $\Delta \theta_k$ satisfies

$$\sum_{k=k_0}^{k_0+N}(\Delta\theta_k)^T\Delta\theta_k \leq 2\lambda_m T_s \sum_{k=k_0}^{k_0+N}\int_{t_k}^{t_{k+1}}\epsilon^2(\tau)m^2(\tau)d\tau + 2\lambda_m T_s^2 \sum_{k=k_0}^{k_0+N}\sigma_s^2|\theta_k|^2$$

$$\leq 2\lambda_m T_s \int_{t_{k_0}}^{t_{k_0+N}}\epsilon^2(\tau)m^2(\tau)d\tau + 2\lambda_m T_s^2 \sum_{k=k_0}^{k_0+N}\sigma_s^2|\theta_k|^2 \quad (8.5.116)$$

Using the properties that $\epsilon m \in \mathcal{S}(\frac{\eta^2}{m^2}), \sigma_s \theta_k \in \mathcal{D}(\frac{\eta^2}{m^2})$, we have

$$\sum_{k=k_0}^{k_0+N}(\Delta\theta_k)^T\Delta\theta_k \leq c_1 + c_2 \int_{t_{k_0}}^{t_{k_0+N}}\frac{\eta^2}{m^2}d\tau$$

for some constant $c_1, c_2 > 0$. Thus, we conclude that $\Delta\theta_k \in \mathcal{D}(\frac{\eta^2}{m^2})$.

Following the same arguments we used in Sections 8.5.2 to 8.5.6 to prove parameter convergence, we can establish (iii) as follows: We have

$$\tilde{\theta}_{k+1} = \tilde{\theta}_k - \Gamma \int_{t_k}^{t_{k+1}} \frac{\phi(\tau)\phi^T(\tau)}{m^2(\tau)}d\tau \tilde{\theta}_k + \Gamma \int_{t_k}^{t_{k+1}}\frac{\phi(\tau)\eta(\tau)}{m^2(\tau)}d\tau - \Gamma \sigma_s \theta_k T_s$$

8.5. ROBUST ADAPTIVE LAWS

which we express in the form

$$\tilde{\theta}_{k+1} = A(k)\tilde{\theta}_k + B\nu(k) \qquad (8.5.117)$$

where $A(k) = I - \Gamma \int_{t_k}^{t_{k+1}} \frac{\phi(\tau)\phi^\top(\tau)}{m^2(\tau)} d\tau$, $B = \Gamma$, $\nu(k) = \int_{t_k}^{t_{k+1}} \frac{\phi(\tau)\eta(\tau)}{m^2(\tau)} d\tau - \sigma_s \theta_k T_s$. We can establish parameter error convergence using the e.s. property of the homogeneous part of (8.5.117) when ϕ is PE. In Chapter 4, we have shown that ϕ being PE implies that the equilibrium $\tilde{\theta}_e = 0$ of $\tilde{\theta}_{k+1} = A(k)\tilde{\theta}_k$ is e.s., i.e., the solution $\tilde{\theta}_k$ of (8.5.117) satisfies

$$|\tilde{\theta}_k| \le \beta_0 \gamma^k + \beta_1 \sum_{i=0}^{k} \gamma^{k-i} |\nu_i| \qquad (8.5.118)$$

for some $0 < \gamma < 1$ and $\beta_0, \beta_1 > 0$. Since $|\nu_i| \le c(\bar{\eta} + \sigma_0)$ for some constant $c \ge 0$, the proof of (iii) (a) follows from (8.5.118). From the definition of ν_k, we have

$$|\nu_k| \le c_0 \bar{\eta} + c_1 |\sigma_s \theta_k| \le c_0 \bar{\eta} + c_1' |\sigma_s \theta_k| \, |\tilde{\theta}_k| \qquad (8.5.119)$$

for some constants c_0, c_1, c_1', where the second inequality is obtained by using $\sigma_s |\theta_k| \le \sigma_s \frac{\tilde{\theta}_k^\top \theta_k}{M_0 - |\theta^*|} \le \frac{1}{M_0 - |\theta^*|} |\sigma_s \theta_k| \, |\tilde{\theta}_k|$. Using (8.5.119) in (8.5.118), we have

$$|\tilde{\theta}_k| \le \beta_0 \gamma^k + \beta_1' \bar{\eta} + \beta_2 \sum_{i=0}^{k-1} \gamma^{k-i} |\sigma_s \theta_i| |\tilde{\theta}_i| \qquad (8.5.120)$$

where $\beta_1' = \frac{c_0 \beta_1}{1-\gamma}, \beta_2 = \beta_1 c_1'$.

To proceed with the parameter convergence analysis, we need the following discrete version of the B-G Lemma.

Lemma 8.5.1 *Let $x(k), f(k), g(k)$ be real valued sequences defined for $k = 0, 1, 2, \ldots,$, and $f(k), g(k), x(k) \ge 0\ \forall k$. If $x(k)$ satisfies*

$$x(k) \le f(k) + \sum_{i=0}^{k-1} g(i) x(i), \quad k = 0, 1, 2, \ldots$$

then,

$$x(k) \le f(k) + \sum_{i=0}^{k-1} \left[\prod_{j=i+1}^{k-1} (1 + g(j)) \right] g(i) f(i), \quad k = 0, 1, 2, \ldots$$

where $\prod_{j=i+1}^{k-1}(1 + g(j))$ is set equal to 1 when $i = k-1$.

The proof of Lemma 8.5.1 can be found in [42].

Let us continue the proof of Theorem 8.5.7 by using Lemma 8.5.1 to obtain from (8.5.120) that

$$|\tilde{\theta}_k| \leq \beta_0 \gamma^k + \beta_1' \bar{\eta} + \sum_{i=0}^{k-1} \gamma^k \left\{ \prod_{j=i+1}^{k-1} (1 + \beta_2 \sigma_s |\theta_j|) \right\} \beta_2 |\sigma_s \theta_i| (\beta_0 + \beta_1' \bar{\eta} \gamma^{-i}) \quad (8.5.121)$$

Using the inequality $\prod_{i=1}^{n} x_i \leq (\sum_{i=1}^{n} x_i/n)^n$ which holds for any integer n and positive x_i, we have

$$\prod_{j=i+1}^{k-1} (1 + \beta_2 \sigma_s |\theta_j|) \leq \left(\frac{\sum_{j=i+1}^{k-1} (1 + \beta_2 \sigma_s |\theta_j|)}{k-i-1} \right)^{k-i-1} = \left(1 + \frac{\sum_{j=i+1}^{k-1} \beta_2 \sigma_s |\theta_j|}{k-i-1} \right)^{k-i-1} \quad (8.5.122)$$

Because $\sigma_s |\theta_k| \leq \frac{\sigma_s |\tilde{\theta}_k| |\theta_k|}{M_0 - |\theta^*|}$, it follows from (8.5.115) that

$$\sigma_s |\theta_k| \leq c \left\{ V(k) - V(k+1) + \int_{t_k}^{t_{k+1}} \frac{\eta^2}{m^2} d\tau \right\}$$

for some constant c, which implies that

$$\sum_{j=i+1}^{k-1} \beta_2 \sigma_s |\theta_j| \leq c_1 \bar{\eta}^2 (k-i-1) + c_0$$

for some constant $c_1, c_0 > 0$. Therefore, it follows from (8.5.122) that

$$\prod_{j=i+1}^{k-1} (1 + \beta_2 \sigma_s |\theta_j|) \leq \left(1 + \frac{c_1 \bar{\eta}^2 (k-i-1) + c_0}{k-i-1} \right)^{k-i-1}$$

$$\leq \left(1 + c_1 \bar{\eta}^2 + \frac{c_0}{k-i-1} \right)^{k-i-1} \quad (8.5.123)$$

Because for $x > 0$, the inequality $(1 + \frac{x}{n})^n \leq e^x$ holds for any integer $n > 0$, it follows from (8.5.123) that

$$\prod_{j=i+1}^{k-1} (1 + \beta_2 \sigma_s |\theta_j|) \leq (1 + c_1 \bar{\eta}^2)^{k-i-1} e^{\frac{c_0}{1+c_1 \bar{\eta}^2}} \quad (8.5.124)$$

Using (8.5.124) in (8.5.121), we have

$$|\tilde{\theta}_k| \leq \beta_0 \gamma^k + \beta_1' \bar{\eta} + \sum_{i=0}^{k-1} \gamma^k (1 + c_1 \bar{\eta}^2)^{k-i-1} e^{\frac{c_0}{1+c_1 \bar{\eta}^2}} \beta_2 \sigma_s |\theta_i| (\beta_0 + \beta_1' \bar{\eta} \gamma^{-i}) \quad (8.5.125)$$

8.6. SUMMARY OF ROBUST ADAPTIVE LAWS

Because $\beta_2 \sigma_s |\theta_i| \in \mathcal{L}_\infty$ and $\gamma < 1$, the inequality (8.5.125) leads to

$$|\tilde{\theta}_k| \leq \beta_0 \gamma^k + \beta_1' \bar{\eta} + \beta_3 (\sqrt{\gamma})^k \sum_{i=0}^{k-1} \{\sqrt{\gamma}(1 + c_1 \bar{\eta}^2)\}^{k-i-1} + \beta_4 \bar{\eta} \sum_{i=0}^{k-1} [\gamma(1 + c_1 \bar{\eta}^2)]^{k-i}$$

for some constants $\beta_3, \beta_4 > 0$. If $\bar{\eta}$ is chosen to satisfy $\sqrt{\gamma}(1 + c_1 \bar{\eta}^2) < 1$, i.e., for $\bar{\eta}^{*2} = \frac{1-\sqrt{\gamma}}{c_1 \sqrt{\gamma}} > 0$ and $\bar{\eta}^2 < \bar{\eta}^{*2}$, we obtain

$$|\tilde{\theta}_k| \leq \beta_0 \gamma^k + \beta_1 \bar{\eta} + \beta_3' (\sqrt{\gamma})^k + \beta_4' \bar{\eta}$$

for some constants $\beta_3', \beta_4' > 0$. Therefore, $\tilde{\theta}_k$ converges exponentially to a residual set whose size is proportional to $\bar{\eta}$. □

8.5.9 Effect of Initial Conditions

The effect of initial conditions of plants such as the one described by the transfer function form in (8.5.1) is the presence of an additive exponentially decaying to zero term η_0 in the plant parametric model as indicated by equation (8.5.8). The term η_0 is given by

$$\dot{\omega} = \Lambda_c \omega, \quad \omega(0) = \omega_0$$
$$\eta_0 = C_c^\top \omega$$

where Λ_c is a stable matrix and ω_0 contains the initial conditions of the overall plant. As in the ideal case presented in Section 4.3.7, the term η_0 can be taken into account by modifying the Lyapunov-like functions $V(\tilde{\theta})$ used in the previous sections to

$$V_m(\tilde{\theta}) = V(\tilde{\theta}) + \omega^\top P_0 \omega$$

where $P_0 = P_0^\top > 0$ satisfies the Lyapunov equation $P_0 \Lambda_c + \Lambda_c^\top P_0 = -\gamma_0 I$ for some $\gamma_0 > 0$ to be chosen. As in Section 4.3.7, it can be shown that the properties of the robust adaptive laws presented in the previous sections are not affected by the initial conditions.

8.6 Summary of Robust Adaptive Laws

A robust adaptive law is constructed in a straight forward manner by modifying the adaptive laws listed in Tables 4.1 to 4.5 with the two crucial modifications studied in the previous sections. These are: the normalizing signal which is now chosen to bound the modeling error term η from above in addition to bounding the signal vector ϕ, and the leakage or dead zone or projection that changes the "pure" integral action of the adaptive law.

We have illustrated the procedure for modifying the adaptive laws of Tables 4.1 to 4.5 for the parametric model

$$z = W(s)[\theta^{*T}\psi + W^{-1}(s)\eta], \quad \eta = \Delta_u(s)u + \Delta_y(s)y + d_1 \qquad (8.6.1)$$

which we can also rewrite as

$$z = \theta^{*T}\phi + \eta \qquad (8.6.2)$$

where $\phi = W(s)\psi$. Without loss of generality, let us assume the following:

S1. $W(s)$ is a known proper transfer function and $W(s), W^{-1}(s)$ are analytic in $\text{Re}[s] \geq -\delta_0/2$ for some known $\delta_0 > 0$.

S2. The perturbations $\Delta_u(s), \Delta_y(s)$ are strictly proper and analytic in $\text{Re}[s] \geq -\delta_0/2$ and $\delta_0 > 0$ is known.

S3. $\psi = H_0(s)[u,y]^T$ where $H_0(s)$ is proper and analytic in $\text{Re}[s] \geq -\delta_0/2$.

For the LTI plant models considered in this book, assumptions S1 and S3 are satisfied by choosing the various design polynomials appropriately. The strict properness of Δ_u, Δ_y in assumption S2 follows from that of the overall plant transfer function.

Let us first discuss the choice of the normalizing signal for parametric model (8.6.1) to be used with the SPR-Lyapunov design approach. We rewrite (8.6.1) as

$$z_f = W(s)L(s)[\theta^{*T}\phi_f + \eta_f] \qquad (8.6.3)$$

with

$$z_f = \frac{h_0}{s + h_0}z, \quad \phi_f = \frac{h_0}{s + h_0}L^{-1}(s)\psi$$

$$\eta_f = L^{-1}(s)W^{-1}(s)\frac{h_0}{s + h_0}(\Delta_u(s)u + \Delta_y(s)y + d_1)$$

by filtering each side of (8.6.1) with the filter $\frac{h_0}{s+h_0}$. We note that $L(s)$ is chosen so that WL is strictly proper and SPR. In addition, $L^{-1}(s), L(s)$ are analytic in $\text{Re}[s] \geq -\delta_0/2$ and $h_0 > \delta_0/2$. Using Lemma 3.3.2, we have from assumptions S1 to S3 that the normalizing signal m generated by

$$m^2 = 1 + n_s^2, \quad n_s^2 = m_s$$
$$\dot{m}_s = -\delta_0 m_s + u^2 + y^2, \quad m_s(0) = 0 \qquad (8.6.4)$$

bounds both ϕ_f, η_f from above. The same normalizing signal bounds η, ϕ in the parametric model (8.6.2).

In Tables 8.1 to 8.4, we summarize several robust adaptive laws based on parametric models (8.6.1) and (8.6.2), which are developed by modifying the adaptive laws of Chapter 4. Additional robust adaptive laws based on least-squares and the integral adaptive law can be developed by following exactly the same procedure.

For the bilinear parametric model

$$z = \rho^*(\theta^{*\top}\phi + z_1) + \eta, \quad \eta = \Delta_u(s)u + \Delta_y(s)y + d \qquad (8.6.5)$$

the procedure is the same. Tables 8.5 to 8.7 present the robust adaptive laws with leakage, projection and dead zone for the parametric model (8.6.5) when the sign of ρ^* is known.

8.7 Problems

8.1 Consider the following system

$$y = e^{-\tau s} \frac{200}{(s-p)(s+100)} u$$

where $0 < \tau \ll 1$ and $p \approx 1$. Choose the dominant part of the plant and express the unmodeled part as a (i) multiplicative; (ii) additive; and (iii) stable factor perturbation.

Design an output feedback that regulates y to zero when $\tau = 0.02$.

8.2 Express the system in Problem 8.1 in terms of the general singular perturbation model presented in Section 8.2.2.

8.3 Establish the stability properties of the switching-σ modification for the example given in Section 8.4.1 when $0 < M_0 \leq |\theta^*|$.

Table 8.1 Robust adaptive law based on the SPR-Lyapunov method

Parametric model	$z = W(s)\theta^{*T}\phi + \eta,\ \eta = \Delta_u(s)u + \Delta_y(s)y + d$ $\phi = H_0(s)[u, y]^T$
Filtered parametric model	$z_f = W(s)L(s)(\theta^{*T}\phi_f + \eta_f),\ \phi_f = \frac{h_0}{s+h_0}W(s)L^{-1}(s)\phi$ $z_f = \frac{h_0}{s+h_0}z,\ \eta_f = W^{-1}(s)L^{-1}(s)\frac{h_0}{s+h_0}\eta$
Adaptive law	$\dot{\theta} = \Gamma\epsilon\phi - \Gamma w\theta$ $\epsilon = z - \hat{z} - W(s)L(s)\epsilon n_s^2$
Normalizing signal	$m^2 = 1 + n_s^2,\ n_s^2 = m_s$ $\dot{m}_s = -\delta_0 m_s + u^2 + y^2,\ m_s(0) = 0$
Leakage (a) Fixed σ	$w = \sigma$
(b) Switching σ	$w = \sigma_s = \begin{cases} 0 & \text{if } |\theta| \leq M_0 \\ \left(\frac{|\theta|}{M_0} - 1\right)\sigma_0 & \text{if } M_0 < |\theta| \leq 2M_0 \\ \sigma_0 & \text{if } |\theta| > 2M_0 \end{cases}$
(c) ϵ-modification	$w = |\epsilon m|\nu_0$
Assumptions	(i) Δ_u, Δ_y strictly proper and analytic in $Re[s] \geq -\delta_0/2$ for some known $\delta_0 > 0$; (ii) $W(s), H_0(s)$ are known and proper and $W(s), W^{-1}(s), H_0(s)$ are analytic in $Re[s] \geq -\delta_0/2$
Design variables	$\Gamma = \Gamma^T > 0;\ \sigma_0 > 0;\ \nu_0 > 0;\ h_0 > \delta_0/2;\ L^{-1}(s)$ is proper and $L(s), L^{-1}(s)$ are analytic in $Re[s] \geq -\delta_0/2;\ W(s)L(s)$ is proper and SPR
Properties	(i) $\epsilon, \epsilon m, \theta, \dot{\theta} \in \mathcal{L}_\infty$; (ii) $\epsilon, \epsilon m, \dot{\theta} \in \mathcal{S}(f_0 + \eta^2/m^2)$, where $f_0 = \sigma$ for $w = \sigma$, $f_0 = 0$ for $w = \sigma_s$, and $f_0 = \nu_0$ for $w = |\epsilon m|\nu_0$

8.7. PROBLEMS

Table 8.2 Robust adaptive law with leakage based on the gradient method

Parametric model	$z = \theta^{*T}\phi + \eta, \quad \eta = \Delta_u(s)u + \Delta_y(s)y + d$
Adaptive law	$\dot{\theta} = \Gamma\epsilon\phi - \Gamma w\theta$ $\epsilon = \dfrac{z - \theta^T\phi}{m^2}$
Normalizing signal	As in Table 8.1
Leakage w	As in Table 8.1
Assumptions	(i) Δ_u, Δ_y strictly proper and analytic in $Re[s] \geq -\delta_0/2$ for some known $\delta_0 > 0$; (ii) $\phi = H(s)[u,y]^T$, where $H(s)$ is strictly proper and analytic in $Re[s] \geq -\delta_0/2$.
Design variables	$\Gamma = \Gamma^T > 0$; The constants in w are as in Table 8.1.
Properties	(i) $\epsilon, \epsilon m, \theta, \dot{\theta} \in \mathcal{L}_\infty$; (ii) $\epsilon, \epsilon m, \dot{\theta} \in \mathcal{S}(\eta^2/m^2 + f_0)$, where f_0 is as defined in Table 8.1.

8.4 Consider the following singular perturbation model

$$\dot{x} = A_{11}x + A_{12}z + b_1 u$$

$$\mu\dot{z} = A_{22}z + b_2 u$$

where $x \in \mathcal{R}^n, z \in \mathcal{R}^m, y \in \mathcal{R}^1$ and A_{22} is a stable matrix. The scalar parameter μ is a small positive constant, i.e., $0 < \mu \ll 1$.

(a) Obtain an nth-order approximation of the above system.

(b) Use the transformation $\eta = z + A_{22}^{-1}b_2 u$ to transform the system into one with states x, η.

(c) Show that if $u = -Kx$ stabilizes the reduced-order system, then there exists a $\mu^* > 0$ such that $u = -Kx$ also stabilizes the full-order system for any $\mu \in [0, \mu^*)$.

Table 8.3 Robust adaptive law with projection based on the gradient method

Parametric model	$z = \theta^{*T}\phi + \eta, \quad \eta = \Delta_u(s)u + \Delta_y(s)y + d$
Adaptive law	$\dot{\theta} = \begin{cases} \Gamma\epsilon\phi & \text{if } |\theta| < M_0 \\ & \text{or if } |\theta| = M_0 \text{ and } (\Gamma\epsilon\phi)^T\theta \leq 0 \\ \left(I - \frac{\Gamma\theta\theta^T}{\theta^T\Gamma\theta}\right)\Gamma\epsilon\phi & \text{otherwise} \end{cases}$ $\epsilon = \dfrac{z - \theta^T\phi}{m^2}$
Normalizing signal	As in Table 8.1
Assumptions	As in Table 8.1
Design variables	$|\theta(0)| \leq M_0; \quad M_0 \geq |\theta^*|; \quad \Gamma = \Gamma^T > 0$
Properties	(i) $\epsilon, \epsilon m, \theta, \dot{\theta} \in \mathcal{L}_\infty$; (ii) $\epsilon, \epsilon m, \dot{\theta} \in \mathcal{S}(\eta^2/m^2)$

8.5 Establish the stability properties of the shifted leakage modification $-w(\theta - \theta^*)$ for the three choices of w and example given in Section 8.4.1.

8.6 Repeat the results of Section 8.4.1 when u is piecewise continuous but not necessarily bounded.

8.7 Repeat the results of Section 8.4.2 when u is piecewise continuous but not necessarily bounded.

8.8 Repeat the results of Section 8.4.3 when u is piecewise continuous but not necessarily bounded.

8.9 Simulate the adaptive control scheme given by equation (8.3.16) in Section 8.3.3 for the plant given by (8.3.13) when $b = 2$. Demonstrate the effect of large γ and large constant reference input r as well as the effect of switching-off adaptation by setting $\gamma = 0$ when instability is detected.

8.7. PROBLEMS

Table 8.4 Robust adaptive law with dead zone based on the gradient Method

Parametric model	$z = \theta^{*T}\phi + \eta, \quad \eta = \Delta_u(s)u + \Delta_y(s)y + d$
Adaptive law	$\dot{\theta} = \Gamma\phi(\epsilon + g)$ $g = \begin{cases} \frac{g_0}{m} & \text{if } \epsilon m < -g_0 \\ -\frac{g_0}{m} & \text{if } \epsilon m > g_0 \\ -\epsilon & \text{if } |\epsilon m| \leq g_0 \end{cases}$ $\epsilon = \dfrac{z - \theta^T\phi}{m^2}$
Normalizing signal	As in Table 8.1
Assumptions	As in Table 8.1
Design variables	$g_0 > \frac{|\eta|}{m}; \quad \Gamma = \Gamma^T > 0$
Properties	(i) $\epsilon, \epsilon m, \theta, \dot{\theta} \in \mathcal{L}_\infty$; (ii) $\epsilon, \epsilon m, \dot{\theta} \in \mathcal{S}(\eta^2/m^2 + g_0)$; (iii) $\lim_{t \to \infty} \theta(t) = \bar{\theta}$

8.10 Consider the closed-loop adaptive control scheme of Section 8.3.2, i.e.,

$$\dot{x} = ax + z - u$$
$$\mu\dot{z} = -z + 2u$$
$$y = x$$
$$u = -kx, \quad \dot{k} = \gamma x^2$$

Show that there exists a region of attraction $\mathcal{D}$ whose size is of $O(\frac{1}{\mu^\alpha})$ for some $\alpha > 0$ such that for $x(0), z(0), k(0) \in \mathcal{D}$ we have $x(t), z(t) \to 0$ and $k(t) \longrightarrow k(\infty)$ as $t \to \infty$. (Hint: use the transformation $\eta = z - 2u$ and choose $V = \frac{x^2}{2} + \frac{(k-k^*)^2}{2\gamma} + \mu\frac{(x+\eta)^2}{2}$ where $k^* > a$.)

Table 8.5 Robust adaptive law with leakage for the bilinear model

Parametric model	$z = \rho^*(\theta^{*T}\phi + z_1) + \eta, \quad \eta = \Delta_u(s)u + \Delta_y(s)y + d$
Adaptive law	$\dot{\theta} = \Gamma\epsilon\phi\,\text{sgn}(\rho^*) - w_1\Gamma\theta$ $\dot{\rho} = \gamma\epsilon\xi - w_2\gamma\rho$ $\epsilon = \dfrac{z - \rho\xi}{m^2}, \quad \xi = \theta^T\phi + z_1$
Normalizing signal	As in Table 8.1.
Leakage w_i $i = 1, 2$	As in Table 8.1
Assumptions	(i) Δ_u, Δ_y are as in Table 8.1; (ii) $\phi = H(s)[u, y]^T$, $z_1 = h_1(s)u + h_2(s)y$, where $H(s), h_1(s), h_2(s)$ are strictly proper and analytic in $Re[s] \geq -\delta_0/2$
Design variables	$\Gamma = \Gamma^T > 0, \ \gamma > 0$; the constants in w_i are as defined in Table 8.1
Proporties	(i) $\epsilon, \epsilon m, \rho, \theta, \dot{\rho}, \dot{\theta} \in \mathcal{L}_\infty$; (ii) $\epsilon, \epsilon m, \dot{\rho}, \dot{\theta} \in \mathcal{S}(\eta^2/m^2 + f_0)$, where f_0 is as defined in Table 8.1

8.11 Perform simulations to compare the properties of the various choices of leakage given in Section 8.4.1 using an example of your choice.

8.12 Consider the system
$$y = \theta^* u + \eta$$
$$\eta = \Delta(s)u$$

where y, u are available for measurement, θ^* is the unknown constant to be estimated and η is a modeling error signal with $\Delta(s)$ being proper and analytic in $Re[s] \geq -0.5$. The input u is piecewise continuous. Design an adaptive law with a switching-σ to estimate θ^*.

8.7. PROBLEMS

Table 8.6 Robust adaptive law with projection for the bilinear model

Parametric model	Same as in Table 8.5
Adaptive law	$\dot{\theta}_i = \begin{cases} \Gamma_i \epsilon \phi_i & \text{if } |\theta_i| < M_i \\ & \text{or if } |\theta_i| = M_i \text{ and } (\Gamma_i \epsilon \phi_i)^\top \theta_i \leq 0 \\ \left(I - \frac{\Gamma_i \theta_i \theta_i^\top}{\theta_i^\top \Gamma_i \theta_i}\right) \Gamma_i \epsilon \phi_i & \text{otherwise} \end{cases}$ $i = 1, 2$ with $\theta_1 = \theta, \theta_2 = \rho, \phi_1 = \phi \operatorname{sgn}(\rho^*), \phi_2 = \xi, \Gamma_1 = \Gamma, \Gamma_2 = \gamma$ $\epsilon = \frac{z - \rho \xi}{m^2}, \quad \xi = \theta^\top \phi + z_1$
Assumptions	As in Table 8.5
Normalizing signal	As in Table 8.1
Design variables	$|\theta(0)| \leq M_1, |\rho(0)| \leq M_2; \Gamma = \Gamma^\top > 0, \gamma > 0$
Properties	(i) $\epsilon, \epsilon m, \rho, \theta, \dot{\rho}, \dot{\theta} \in \mathcal{L}_\infty$; (ii) $\epsilon, \epsilon m, \dot{\rho}, \dot{\theta} \in \mathcal{S}(\eta^2/m^2)$

8.13. The linearized dynamics of a throttle angle θ to vehicle speed V subsystem are given by the 3rd order system

$$V = \frac{bp_1 p_2}{(s+a)(s+p_1)(s+p_2)} \theta + d$$

where $p_1, p_2 > 20, 1 \geq a > 0$ and d is a load disturbance.

(a) Obtain a parametric model for the parameters of the dominant part of the system.

(b) Design a robust adaptive law for estimating on-line these parameters.

(c) Simulate your estimation scheme when $a = 0.1, b = 1, p_1 = 50, p_2 = 100$ and $d = 0.02 \sin 5t$.

Table 8.7 Robust adaptive law with dead zone for the bilinear model

Parametric model	Same as in Table 8.5
Adaptive law	$\dot{\theta} = \Gamma\phi(\epsilon + g)\mathrm{sgn}(\rho^*)$ $\dot{\rho} = \gamma\xi(\epsilon + g)$ $g = \begin{cases} \frac{g_0}{m} & \text{if } \epsilon m < -g_0 \\ -\frac{g_0}{m} & \text{if } \epsilon m > g_0 \\ -\epsilon & \text{if } |\epsilon m| \leq g_0 \end{cases}$ $\epsilon = \frac{z - \rho\xi}{m^2}, \quad \xi = \theta^\top \phi + z_1$
Assumptions	As in Table 8.5
Normalizing signal	Same as in Table 8.1
Design variables	$g_0 > \frac{|\eta|}{m}; \quad \Gamma = \Gamma^\top > 0, \gamma > 0$
Properties	(i) $\epsilon, \epsilon m, \rho, \theta, \dot{\rho}, \dot{\theta} \in \mathcal{L}_\infty$; (ii) $\epsilon, \epsilon m, \dot{\rho}, \dot{\theta} \in \mathcal{S}(\eta^2/m^2 + g_0)$; (iii) $\lim_{t\to\infty} \theta(t) = \bar{\theta}$

8.14 Consider the parameter error differential equation

$$\dot{\tilde{\theta}} = -\gamma u^2 \tilde{\theta} + \gamma d u$$

that arises in the estimation problem of Section 8.3.1 in the presence of a bounded disturbance d.

(a) Show that for $d = 0$ and $u = \frac{1}{(1+t)^{\frac{1}{2}}}$, the equilibrium $\tilde{\theta}_e = 0$ is u.s and a.s but not u.a.s. Verify that for

$$d(t) = (1+t)^{-\frac{1}{4}}\left(\frac{5}{4} - 2(1+t)^{-\frac{1}{4}}\right)$$

8.7. PROBLEMS

$u = (1+t)^{-\frac{1}{2}}$ and $\gamma = 1$ we have $y \to 0$ as $t \to \infty$ and $\tilde{\theta}(t) \to \infty$ as $t \to \infty$.

(b) Repeat the same stability analysis for $u = u_0$ where $u_0 \neq 0$ is a constant, and show that for $d = 0$, the equilibrium $\tilde{\theta}_e = 0$ is u.a.s. Verify that $\tilde{\theta}(t)$ is bounded for any bounded d and obtain an upper bound for $|\tilde{\theta}(t)|$.

8.15 Repeat Problem 8.12 for an adaptive law with (i) dead zone; (ii) projection.

8.16 Consider the dynamic uncertainty

$$\eta = \Delta_u(s)u + \Delta_y(s)y$$

where Δ_u, Δ_y are proper transfer functions analytic in $\text{Re}[s] \geq -\frac{\delta_0}{2}$ for some known $\delta_0 > 0$.

(a) Design a normalizing signal m that guarantees $\eta/m \in \mathcal{L}_\infty$ when

   (i) Δ_u, Δ_y are biproper.

   (ii) Δ_u, Δ_y are strictly proper.

In each case specify the upper bound for $\frac{|\eta|}{m}$.

(b) Calculate the bound for $|\eta|/m$ when

   (i) $\Delta_u(s) = \dfrac{e^{-\tau s} - 1}{s + 2}, \Delta_y(s) = \mu \dfrac{s^2}{(s+1)^2}$

   (ii) $\Delta_u(s) = \dfrac{\mu s}{\mu s + 2}, \Delta_y(s) = \dfrac{\mu s}{(\mu s + 1)^2}$

where $0 < \mu \ll 1$ and $0 < \tau \ll 1$.

8.17 Consider the system

$$y = \frac{e^{-\tau s} b}{(s+a)(\mu s + 1)} u$$

where $0 < \tau \ll 1, 0 < \mu \ll 1$ and a, b are unknown constants. Obtain a parametric model for $\theta^* = [b, a]^\top$ by assuming $\tau \approx 0, \mu \approx 0$. Show the effect of the neglected dynamics on the parametric model.

Chapter 9

Robust Adaptive Control Schemes

9.1 Introduction

As we have shown in Chapter 8, the adaptive control schemes of Chapters 4 to 7 may go unstable in the presence of small disturbances or unmodeled dynamics. Because such modeling error effects will exist in any implementation, the nonrobust behavior of the schemes of Chapters 4 to 7 limits their applicability.

The purpose of this chapter is to redesign the adaptive schemes of the previous chapters and establish their robustness properties with respect to a wide class of bounded disturbances and unmodeled dynamics that are likely to be present in most practical applications.

We start with the parameter identifiers and adaptive observers of Chapter 5 and show that their robustness properties can be guaranteed by designing the plant input to be dominantly rich. A dominantly rich input is sufficiently rich for the simplified plant model, but it maintains its richness outside the high frequency range of the unmodeled dynamics. Furthermore, its amplitude is higher than the level of any bounded disturbance that may be present in the plant. As we will show in Section 9.2, a dominantly rich input guarantees exponential convergence of the estimated parameter errors to residual sets whose size is of the order of the modeling error.

While the robustness of the parameter identifiers and adaptive observers of Chapter 5 can be established by simply redesigning the plant input without having to modify the adaptive laws, this is not the case with the adaptive control schemes of Chapters 6 and 7 where the plant input is no longer a design variable. For the adaptive control schemes presented in Chapters 6 and 7, robustness is established by simply replacing the adaptive laws with robust adaptive laws developed in Chapter 8.

In Section 9.3, we use the robust adaptive laws of Chapter 8 to modify the MRAC schemes of Chapter 6 and establish their robustness with respect to bounded disturbances and unmodeled dynamics. In the case of the MRAC schemes with unnormalized adaptive laws, semiglobal boundedness results are established. The use of a dynamic normalizing signal in the case of MRAC with normalized adaptive laws enables us to establish global boundedness results and mean-square tracking error bounds. These bounds are further improved by modifying the MRAC schemes using an additional feedback term in the control input. By choosing a certain scalar design parameter τ in the control law, the modified MRAC schemes are shown to guarantee arbitrarily small $\mathcal{L}_\infty$ bounds for the steady state tracking error despite the presence of input disturbances. In the presence of unmodeled dynamics, the choice of τ is limited by the trade-off between nominal performance and robust stability.

The robustification of the APPC schemes of Chapter 7 is presented in Section 9.5. It is achieved by replacing the adaptive laws used in Chapter 7 with the robust ones developed in Chapter 8.

9.2 Robust Identifiers and Adaptive Observers

The parameter identifiers and adaptive observers of Chapter 5 are designed for the SISO plant model

$$y = G_0(s)u \qquad (9.2.1)$$

where $G_0(s)$ is strictly proper with stable poles and of known order n. In this section we apply the schemes of Chapter 5 to a more realistic plant model described by

$$y = G_0(s)(1 + \Delta_m(s))(u + d_u) \qquad (9.2.2)$$

where $G(s) = G_0(s)(1 + \Delta_m(s))$ is strictly proper of unknown degree; $\Delta_m(s)$ is an unknown multiplicative perturbation with stable poles and d_u is a bounded disturbance. Our objective is to estimate the coefficients of $G_0(s)$ and the states of a minimal or nonminimal state representation that corresponds to $G_0(s)$, despite the presence of $\Delta_m(s)$, d_u. This problem is, therefore, similar to the one we would face in an actual application, i.e., (9.2.1) represents the plant model on which our adaptive observer design is based and (9.2.2) the plant to which the observer will be applied.

Most of the effects of $\Delta_m(s), d_u$ on the robustness and performance of the schemes presented in Chapter 5 that are designed based on (9.2.1) may be illustrated and understood using the following simple examples.

Example 9.2.1 Effect of bounded disturbance. Let us consider the simple plant model

$$y = \theta^* u + d$$

where d is an external bounded disturbance, i.e., $|d(t)| \leq d_0, \forall t \geq 0, u \in \mathcal{L}_\infty$ and θ^* is the unknown constant to be identified. The adaptive law based on the parametric model with $d = 0$ is

$$\dot{\theta} = \gamma \epsilon_1 u, \quad \epsilon_1 = y - \theta u \qquad (9.2.3)$$

which for $d = 0$ guarantees that $\epsilon_1, \theta \in \mathcal{L}_\infty$ and $\epsilon_1, \dot{\theta} \in \mathcal{L}_2$. If, in addition, u is PE, then $\theta(t) \to \theta^*$ as $t \to \infty$ exponentially fast. When $d \neq 0$, the error equation that describes the stability properties of (9.2.3) is

$$\dot{\tilde{\theta}} = -\gamma u^2 \tilde{\theta} + \gamma u d \qquad (9.2.4)$$

where $\tilde{\theta} = \theta - \theta^*$. As shown in Section 8.3, if u is PE, then the homogeneous part of (9.2.4) is exponentially stable, and, therefore, the bounded input $\gamma u d$ implies bounded $\tilde{\theta}$. When u is not PE, the homogeneous part of (9.2.4) is only u.s. and therefore a bounded input does not guarantee bounded $\tilde{\theta}$. In fact, as shown in Section 8.3, we can easily find an input u that is not PE, and a bounded disturbance d that will cause $\tilde{\theta}$ to drift to infinity. One way to counteract parameter drift and establish boundedness for $\tilde{\theta}$ is to modify (9.2.3) using the techniques of Chapter 8. In this section, we are concerned with the parameter identification of stable plants which, for accurate parameter estimates, requires u to be PE independent of whether we have disturbances or not. Because the persistent excitation of u guarantees exponential convergence, we can establish robustness without modifying (9.2.3). Let us, therefore, proceed with the analysis of (9.2.4) by assuming that u is PE with some level $\alpha_0 > 0$, i.e., u satisfies

$$\int_t^{t+T} u^2 d\tau \geq \alpha_0 T, \quad \forall t \geq 0, \text{ for some } T > 0$$

9.2. ROBUST ADAPTIVE OBSERVERS

Then from (9.2.4), we obtain

$$|\tilde\theta(t)| \leq k_1 e^{-\gamma\alpha_0' t}|\tilde\theta(0)| + \frac{k_1}{\alpha_0'}(1 - e^{-\gamma\alpha_0' t})\sup_{\tau\leq t}|u(\tau)d(\tau)|$$

for some constants $k_1, \alpha_0' > 0$, where α_0' depends on α_0. Therefore, we have

$$\lim_{t\to\infty}\sup_{\tau\geq t}|\tilde\theta(\tau)| \leq \frac{k_1}{\alpha_0'}\lim_{t\to\infty}\sup_{\tau\geq t}|u(\tau)d(\tau)| = \frac{k_1}{\alpha_0'}\sup_\tau|u(\tau)d(\tau)| \quad (9.2.5)$$

The bound (9.2.5) indicates that the parameter identification error at steady state is of the order of the disturbance, i.e., as $d \to 0$ the parameter error also reduces to zero. For this simple example, it is clear that if we choose $u = u_0$, where u_0 is a constant different from zero, then $\alpha_0' = u_0^2, k_1 = 1$; therefore, the bound for $|\tilde\theta|$ is $\sup_t|d(t)|/u_0$. Thus the larger the u_0 is, the smaller the parameter error. Large u_0 relative to $|d|$ implies large signal-to-noise ratio and therefore better accuracy of identification.

Example 9.2.2 (Unmodeled Dynamics) Let us consider the plant

$$y = \theta^*(1 + \Delta_m(s))u$$

where $\Delta_m(s)$ is a proper perturbation with stable poles, and use the adaptive law (9.2.3) that is designed for $\Delta_m(s) = 0$ to identify θ^* in the presence of $\Delta_m(s)$. The parameter error equation in this case is given by

$$\dot{\tilde\theta} = -\gamma u^2 \tilde\theta + \gamma u \eta, \quad \eta = \theta^* \Delta_m(s) u \quad (9.2.6)$$

Because u is bounded and $\Delta_m(s)$ is stable, it follows that $\eta \in \mathcal{L}_\infty$ and therefore the effect of $\Delta_m(s)$ is to introduce the bounded disturbance term η in the adaptive law. Hence, if u is PE with level $\alpha_0 > 0$, we have, as in the previous example, that

$$\lim_{t\to\infty}\sup_{\tau\geq t}|\tilde\theta(\tau)| \leq \frac{k_1}{\alpha_0'}\sup_t|u(t)\eta(t)|$$

The question now is how to choose u so that the above bound for $|\tilde\theta|$ is as small as possible. The answer to this question is not as straightforward as in Example 9.2.1 because η is also a function of u. The bound for $|\tilde\theta|$ depends on the choice of u and the properties of $\Delta_m(s)$. For example, for constant $u = u_0 \neq 0$, we have $\alpha_0' = u_0^2, k_1 = 1$ and $\eta = \Delta_m(s)u_0$, i.e., $\lim_{t\to\infty}|\eta(t)| = |\Delta_m(0)||u_0|$, and, therefore,

$$\lim_{t\to\infty}\sup_{\tau\geq t}|\tilde\theta(\tau)| \leq |\Delta_m(0)|$$

If the plant is modeled properly, $\Delta_m(s)$ represents a perturbation that is small (for $s = j\omega$) in the low-frequency range, which is usually the range of interest. Therefore, for $u = u_0$, we should have $|\Delta_m(0)|$ small if not zero leading to the above bound, which is independent of u_0. Another choice of a PE input is $u = \cos\omega_0 t$ for some $\omega_0 \neq 0$. For this choice of u, because

$$e^{-\gamma \int_0^t \cos^2 \omega_0 \tau d\tau} = e^{-\frac{\gamma}{2}(t + \frac{\sin 2\omega_0 t}{2\omega_0})} = e^{-\frac{\gamma}{4}(t + \frac{\sin 2\omega_0 t}{\omega_0})} e^{-\frac{\gamma}{4}t} \leq e^{-\frac{\gamma}{4}t}$$

(where we used the inequality $t + \frac{\sin 2\omega_0 t}{\omega_0} \geq 0, \forall t \geq 0$) and $\sup_t |\eta(t)| \leq |\Delta_m(j\omega_0)|$, we have

$$\limsup_{t \to \infty} |\tilde{\theta}(\tau)| \leq 4|\Delta_m(j\omega_0)|$$

This bound indicates that for small parameter error, ω_0 should be chosen so that $|\Delta_m(j\omega_0)|$ is as small as possible. If $\Delta_m(s)$ is due to high frequency unmodeled dynamics, $|\Delta_m(j\omega_0)|$ is small provided ω_0 is a low frequency. As an example, consider

$$\Delta_m(s) = \frac{\mu s}{1 + \mu s}$$

where $\mu > 0$ is a small constant. It is clear that for low frequencies $|\Delta_m(j\omega)| = O(\mu)$ and $|\Delta_m(j\omega)| \to 1$ as $\omega \to \infty$. Because

$$|\Delta_m(j\omega_0)| = \frac{|\mu\omega_0|}{\sqrt{1 + \mu^2\omega_0^2}}$$

it follows that for $\omega_0 = \frac{1}{\mu}$, we have $\Delta_m(j\omega_0)| = \frac{1}{\sqrt{2}}$ whereas for $\omega_0 < \frac{1}{\mu}$ we have $|\Delta_m(j\omega_0)| = O(\mu)$. Therefore for accurate parameter identification, the input signal should be chosen to be PE but the PE property should be achieved with frequencies that do not excite the unmodeled dynamics. For the above example of $\Delta_m(s)$, $u = u_0$ does not excite $\Delta_m(s)$ at all, i.e., $\Delta_m(0) = 0$, whereas for $u = \sin\omega_0 t$ with $\omega_0 \ll \frac{1}{\mu}$ the excitation of $\Delta_m(s)$ is small leading to an $O(\mu)$ steady state error for $|\tilde{\theta}|$. $\triangledown$

In the following section we define the class of excitation signals that guarantee PE but with frequencies outside the range of the high frequency unmodeled dynamics.

9.2.1 Dominantly Rich Signals

Let us consider the following plant:

$$y = G_0(s)u + \Delta_a(s)u \tag{9.2.7}$$

9.2. ROBUST ADAPTIVE OBSERVERS

where $G_0(s), \Delta_a(s)$ are proper and stable, $\Delta_a(s)$ is an additive perturbation of the modeled part $G_0(s)$. We like to identify the coefficients of $G_0(s)$ by exciting the plant with the input u and processing the I/O data.

Because $\Delta_a(s)u$ is treated as a disturbance, the input u should be chosen so that at each frequency ω_i contained in u, we have $|G_0(j\omega_i)| \gg |\Delta_a(j\omega_i)|$. Furthermore, u should be rich enough to excite the modeled part of the plant that corresponds to $G_0(s)$ so that y contains sufficient information about the coefficients of $G_0(s)$. For such a choice of u to be possible, the spectrums of $G_0(s)$ and $\Delta_a(s)$ should be separated, or $|G_0(j\omega)| \gg |\Delta_a(j\omega)|$ at all frequencies. If $G_0(s)$ is chosen properly, then $|\Delta_a(j\omega)|$ should be small relative to $|G_0(j\omega)|$ in the frequency range of interest. Because we are usually interested in the system response at low frequencies, we would assume that $|G_0(j\omega)| \gg |\Delta_a(j\omega)|$ in the low-frequency range for our analysis. But at high frequencies, we may have $|G_0(j\omega)|$ being of the same order or smaller than $|\Delta_a(j\omega)|$. The input signal u should therefore be designed to be sufficiently rich for the modeled part of the plant but its richness should be achieved in the low-frequency range for which $|G_0(j\omega)| \gg |\Delta_a(j\omega)|$. An input signal with these two properties is called *dominantly rich* [90] because it excites the modeled or dominant part of the plant much more than the unmodeled one.

Example 9.2.3 Consider the plant

$$y = \frac{b}{s+a} \frac{1-2\mu s}{1+\mu s} u \tag{9.2.8}$$

where $\mu > 0$ is a small constant and $a > 0$, b are unknown constants. Because μ is small, we choose $G_0(s) = \frac{b}{s+a}$ as the dominant part and rewrite the plant as

$$y = \frac{b}{s+a}\left(1 - \frac{3\mu s}{1+\mu s}\right)u = \frac{b}{s+a}u + \Delta_a(\mu s, s)u$$

where

$$\Delta_a(\mu s, s) = -\frac{3b\mu s}{(1+\mu s)(s+a)}$$

is treated as an additive unmodeled perturbation. Because

$$|G_0(j\omega_0)|^2 = \frac{b^2}{\omega_0^2 + a^2}, \quad |\Delta_a(j\mu\omega_0, j\omega_0)|^2 = \frac{9b^2\mu^2\omega_0^2}{(\omega_0^2 + a^2)(1+\mu^2\omega_0^2)}$$

it follows that $|G_0(j\omega_0)| \gg |\Delta_a(j\mu\omega_0, j\omega_0)|$ provided that

$$|\omega_0| \ll \frac{1}{\sqrt{8\mu}}$$

Therefore, the input $u = sin\omega_0 t$ qualifies as a dominantly rich input of order 2 for the plant (9.2.8) provided $0 < \omega_0 < O(1/\mu)$ and μ is small. If $\omega_0 = \frac{1}{\mu}$, then

$$|G_0(j\omega_0)|^2 = \frac{\mu^2 b^2}{\mu^2 a^2 + 1} = O(\mu^2)$$

and

$$|\Delta_a(j\mu\omega_0, j\omega_0)|^2 = \frac{9b^2\mu^2}{2(\mu^2 a^2 + 1)} > |G_0(j\omega_0)|^2$$

Hence, for $\omega_0 \geq O(\frac{1}{\mu})$, the input $u = \sin(\omega_0 t)$ is not dominantly rich because it excites the unmodeled dynamics and leads to a small signal to noise ratio.

Let us now examine the effect of the dominantly rich input $u = sin(\omega_0 t)$, with $0 \leq \omega_0 < O(1/\mu)$, on the performance of an identifier designed to estimate the parameters a, b of the plant (9.2.8). Equation (9.2.8) can be expressed as

$$z = \theta^{*T}\phi + \eta \qquad (9.2.9)$$

where

$$z = \frac{s}{s+\lambda}y, \quad \phi = \left[\frac{1}{s+\lambda}u, -\frac{1}{s+\lambda}y\right]^T, \quad \theta^* = [b, a]^T$$

$\lambda > 0$ is a design parameter and

$$\eta = -\frac{3b\mu s}{(s+\lambda)(1+\mu s)}u$$

is the modeling error term, which is bounded because $u \in \mathcal{L}_\infty$ and $\lambda, \mu > 0$. The gradient algorithm for estimating θ^* when $\eta = 0$ is given by

$$\begin{aligned}\dot{\theta} &= \Gamma\epsilon\phi \\ \epsilon &= z - \hat{z}, \quad \hat{z} = \theta^T\phi\end{aligned} \qquad (9.2.10)$$

The error equation that describes the stability properties of (9.2.10) when applied to (9.2.9) with $\eta \neq 0$ is given by

$$\dot{\tilde{\theta}} = -\Gamma\phi\phi^T\tilde{\theta} + \Gamma\phi\eta \qquad (9.2.11)$$

Because $\phi \in \mathcal{L}_\infty$, it follows from the results of Chapter 4 that if ϕ is PE with level $\alpha_0 > 0$, then the homogeneous part of (9.2.11) is exponentially stable which implies that

$$\lim_{t\to\infty}\sup_{\tau \geq t}|\tilde{\theta}(\tau)| \leq \frac{c}{\beta_0}\sup_t|\phi(t)\eta(t)| \qquad (9.2.12)$$

9.2. ROBUST ADAPTIVE OBSERVERS

where $c = \|\Gamma\|$ and β_0 depends on Γ, α_0 in a way that as $\alpha_0 \to 0$, $\beta_0 \to 0$ (see proof of Theorem 4.3.2 in Section 4.8). Therefore, for a small error bound, the input u should be chosen to guarantee that ϕ is PE with level $\alpha_0 > 0$ as high as possible (without increasing $\sup_t |\phi(t)|$) and with $\sup_t |\eta(t)|$ as small as possible.

To see how the unmodeled dynamics affect the dominant richness properties of the excitation signal, let us consider the input signal $u = \sin\omega_0 t$ where $\omega_0 = O(\frac{1}{\mu})$. As discussed earlier, this signal u is *not* dominantly rich because it excites the high-frequency unmodeled dynamics. The loss of the dominant richness property in this case can also be seen by exploring the level of excitation of the signal ϕ. If we assume zero initial conditions, then from the definition of ϕ, we can express ϕ as

$$\phi = \begin{bmatrix} A_1(\omega_0) \sin(\omega_0 t + \varphi_1) \\ A_2(\omega_0) \sin(\omega_0 t + \varphi_2) \end{bmatrix}$$

where

$$A_1(\omega_0) = \frac{1}{\sqrt{\omega_0^2 + \lambda^2}}, \quad A_2(\omega_0) = \frac{|b|\sqrt{1+4\mu^2\omega_0^2}}{\sqrt{(\omega_0^2+\lambda^2)(\omega_0^2+a^2)(1+\mu^2\omega_0^2)}}$$

and φ_1, φ_2 also depend on ω_0. Thus, we can calculate

$$\int_t^{t+\frac{2\pi}{\omega_0}} \phi(\tau)\phi^\top(\tau)d\tau = \frac{\pi}{\omega_0} \begin{bmatrix} A_1^2 & A_1 A_2 \cos(\varphi_1 - \varphi_2) \\ A_1 A_2 \cos(\varphi_1 - \varphi_2) & A_2^2 \end{bmatrix} \quad (9.2.13)$$

For any $t \geq 0$ and $T \geq \frac{2\pi}{\omega_0}$, it follows from (9.2.13) that

$$\frac{1}{T}\int_t^{t+T} \phi(\tau)\phi^\top(\tau)d\tau = \frac{1}{T}\left\{\sum_{i=0}^{n_T-1} \int_{t+i\frac{2\pi}{\omega_0}}^{t+(i+1)\frac{2\pi}{\omega_0}} \phi\phi^\top d\tau + \int_{t+n_T}^{t+T} \phi\phi^\top d\tau\right\}$$

$$\leq \frac{1}{T}\left(\frac{n_T \pi}{\omega_0} + \frac{\pi}{\omega_0}\right)\begin{bmatrix} A_1^2 & A_1 A_2 \cos(\varphi_1 - \varphi_2) \\ A_1 A_2 \cos(\varphi_1 - \varphi_2) & A_2^2 \end{bmatrix}$$

where n_T is the largest integer that satisfies $n_T \leq \frac{T\omega_0}{2\pi}$. By noting that $\frac{n_T \pi}{T\omega_0} \leq \frac{1}{2}$ and $\frac{\pi}{T\omega_0} \leq \frac{1}{2}$, it follows that

$$\frac{1}{T}\int_t^{t+T} \phi(\tau)\phi^\top(\tau)d\tau \leq \begin{bmatrix} A_1^2 & A_1 A_2 \cos(\varphi_1 - \varphi_2) \\ A_1 A_2 \cos(\varphi_1 - \varphi_2) & A_2^2 \end{bmatrix}$$

$$\leq \alpha \begin{bmatrix} A_1^2 & 0 \\ 0 & A_2^2 \end{bmatrix}$$

for any constant $\alpha \geq 2$. Taking $\omega_0 = \frac{c}{\mu}$ for some constant $c > 0$ (i.e., $\omega_0 = O(\frac{1}{\mu})$), we have

$$A_1 = \frac{\mu}{\sqrt{c^2 + \lambda^2 \mu^2}}, \quad A_2 = \frac{|b|\sqrt{1+4c^2\mu^2}}{\sqrt{(c^2+\lambda^2\mu^2)(c^2+a^2\mu^2)(1+c^2)}}$$

which implies that

$$\frac{1}{T}\int_t^{t+T} \phi(\tau)\phi^T(\tau)d\tau \leq O(\mu^2)I \qquad (9.2.14)$$

for all $\mu \in [0,\mu^*]$ where $\mu^* > 0$ is any finite constant. That is, the level of PE of ϕ is of $O(\mu^2)$. As $\mu \to 0$ and $\omega_0 \to \infty$, we have $\frac{1}{T}\int_t^{t+T}\phi(\tau)\phi^T(\tau)d\tau \to 0$ and the level of PE vanishes to zero. Consequently, the upper bound for $\lim_{t\to\infty}\sup_{\tau\geq t}|\tilde{\theta}(\tau)|$ given by (9.2.12) will approach infinity which indicates that a quality estimate for θ^* cannot be guaranteed in general if the nondominantly rich signal $u = sin\omega_0 t$ with $\omega_0 = O(\frac{1}{\mu})$ is used to excite the plant.

On the other hand if $u = sin\omega_0 t$ is dominantly rich, i.e., $0 < |\omega_0| < O(\frac{1}{\mu})$, then we can show that the level of PE of ϕ is strictly greater than $O(\mu)$ and the bound for $\sup_t |\eta(t)|$ is of $O(\mu)$ (see Problem 9.1). $\triangledown$

Example 9.2.3 indicates that if the reference input signal is dominantly rich, i.e., it is rich for the dominant part of the plant but with frequencies away from the parasitic range, then the signal vector ϕ in the adaptive law is PE with a high level of excitation relative to the modeling error. This high level of excitation guarantees that the PE property of $\phi(t)$ cannot be destroyed by the unmodeled dynamics. Furthermore, it is sufficient for the parameter error to converge exponentially fast (in the case of the gradient algorithm (9.2.10)) to a small residual set.

Let us now consider the more general stable plant

$$y = G_0(s)u + \mu\Delta_a(\mu s, s)u + d \qquad (9.2.15)$$

where $G_0(s)$ corresponds to the dominant part of the plant and $\mu\Delta_a(\mu s, s)$ is an additive perturbation with the property $\lim_{\mu\to 0}\mu\Delta_a(\mu s, s) = 0$ for any given s and d is a bounded disturbance. The variable $\mu > 0$ is the small singular perturbation parameter that can be used as a measure of the separation of the spectrums of the dominant dynamics and the unmodeled high frequency ones.

Definition 9.2.1 (Dominant Richness) *A stationary signal $u : \mathcal{R}^+ \mapsto \mathcal{R}$ is called dominantly rich of order n for the plant (9.2.15) if (i) it consists of distinct frequencies $\omega_1, \omega_2, \ldots, \omega_N$ where $N \geq \frac{n}{2}$; (ii) ω_i satisfies*

$$|\omega_i| < O\left(\frac{1}{\mu}\right), \quad |\omega_i - \omega_k| > O(\mu), \quad \text{for } i \neq k; \quad i,k = 1,\ldots,N$$

9.2. ROBUST ADAPTIVE OBSERVERS

and (iii) the amplitude of the spectral line at ω_i defined as

$$f_u(\omega_i) \triangleq \lim_{T \to \infty} \frac{1}{T} \int_t^{t+T} u(\tau) e^{-j\omega_i \tau} d\tau \quad \forall t \geq 0$$

satisfies

$$|f_u(\omega_i)| > O(\mu) + O(d), \quad i = 1, 2, \ldots, N$$

For example, the input $u = A_1 sin\omega_1 t + A_2 sin\omega_2 t$ is dominantly rich of order 4 for the plant (9.2.15) with a second order $G_0(s)$ and $d = 0$ provided $|\omega_1 - \omega_2| > O(\mu); |\omega_1|, |\omega_2| < O(\frac{1}{\mu})$ and $|A_1|, |A_2| > O(\mu)$.

9.2.2 Robust Parameter Identifiers

Let us consider the plant

$$y = G_0(s)(1 + \mu \Delta_m(\mu s, s))u = G_0(s)u + \mu \Delta_a(\mu s, s)u \qquad (9.2.16)$$

where

$$G_0(s) = \frac{b_m s^m + b_{m-1} s^{m-1} + \cdots + b_0}{s^n + a_{n-1} s^{n-1} + \cdots + a_0} = \frac{Z_p(s)}{R_p(s)}, \Delta_a(\mu s, s) = G_0(s) \Delta_m(\mu s, s)$$

and $G_0(s)$ is the transfer function of the modeled or dominant part of the plant, $\Delta_m(\mu s, s)$ is a multiplicative perturbation that is due to unmodeled high frequency dynamics as well as other small perturbations; $\mu > 0$ and $\lim_{\mu \to 0} \mu \Delta_a(\mu s, s) = 0$ for any given s. The overall transfer function is assumed to be strictly proper with stable poles. Our objective is to identify the coefficients of $G_0(s)$ despite the presence of the modeling error term $\mu \Delta_a(\mu s, s) u$.

As in the ideal case, (9.2.16) can be expressed in the linear parametric form

$$z = \theta^{*T} \phi + \mu \eta \qquad (9.2.17)$$

where $\theta^* = [b_m, \ldots, b_0, a_{n-1}, \ldots, a_0]^T$,

$$z = \frac{s^n}{\Lambda(s)} y, \quad \phi = \left[\frac{\alpha_m^T(s)}{\Lambda(s)} u, -\frac{\alpha_{n-1}^T(s)}{\Lambda(s)} y \right]^T, \quad \eta = \frac{Z_p(s)}{\Lambda(s)} \Delta_m(\mu s, s) u$$

$\alpha_i(s) = [s^i, s^{i-1}, \ldots, 1]^T$ and $\Lambda(s)$ is a Hurwitz polynomial of degree n. As shown in Chapter 8, the adaptive law

$$\dot{\theta} = \Gamma \epsilon \phi, \quad \epsilon = z - \hat{z}, \quad \hat{z} = \theta^T \phi \qquad (9.2.18)$$

developed using the gradient method to identify θ^* on-line may not guarantee the boundedness of θ for a general class of bounded inputs u. If, however, we choose u to be dominantly rich of order $n+m+1$ for the plant (9.2.16), then the following theorem establishes that the parameter error $\theta(t) - \theta^*$ converges exponentially fast to a residual set whose size is of $O(\mu)$.

Theorem 9.2.1 *Assume that the nominal transfer function $G_0(s)$ has no zero-pole cancellation. If u is dominantly rich of order $n+m+1$, then there exists a $\mu^* > 0$ such that for $\mu \in [0, \mu^*)$, the adaptive law (9.2.18) or any stable adaptive law from Tables 4.2, 4.3, and 4.5, with the exception of the pure least-squares algorithm, based on the parametric model (9.2.17) with $\eta = 0$ guarantees that $\epsilon, \theta \in \mathcal{L}_\infty$ and θ converges exponentially fast to the residual set*

$$\mathcal{R}_e = \{\theta \mid |\theta - \theta^*| \leq c\mu\}$$

where c is a constant independent of μ.

To prove Theorem 9.2.1 we need to use the following Lemma that gives conditions that guarantee the PE property of ϕ for $\mu \neq 0$. We express the signal vector ϕ as

$$\phi = H_0(s)u + \mu H_1(\mu s, s)u \qquad (9.2.19)$$

where

$$H_0(s) = \frac{1}{\Lambda(s)} \left[\alpha_m^\top(s), -\alpha_{n-1}^\top(s)G_0(s) \right]^\top$$

$$H_1(\mu s, s) = \frac{1}{\Lambda(s)} \left[0, \ldots, 0, -\alpha_{n-1}^\top(s)G_0(s)\Delta_m(\mu s, s) \right]^\top$$

Lemma 9.2.1 *Let $u : \mathcal{R}^+ \mapsto \mathcal{R}$ be stationary and $H_0(s), H_1(\mu s, s)$ satisfy the following assumptions:*

(a) *The vectors $H_0(j\omega_1), H_0(j\omega_2), \ldots, H_0(j\omega_{\bar{n}})$ are linearly independent on $\mathcal{C}^{\bar{n}}$ for all possible $\omega_1, \omega_2, \ldots, \omega_{\bar{n}} \in \mathcal{R}$, where $\bar{n} \triangleq n+m+1$ and $\omega_i \neq \omega_k$ for $i \neq k$.*

(b) *For any set $\{\omega_1, \omega_2, \ldots, \omega_{\bar{n}}\}$ satisfying $|\omega_i - \omega_k| > O(\mu)$ for $i \neq k$ and $|\omega_i| < O(\frac{1}{\mu})$, we have $|\det(\bar{H})| > O(\mu)$ where $\bar{H} \triangleq [H_0(j\omega_1), H_0(j\omega_2), \ldots, H_0(j\omega_{\bar{n}})]$.*

9.2. ROBUST ADAPTIVE OBSERVERS

(c) $|H_1(j\mu\omega, j\omega)| \leq c$ *for some constant c independent of μ and for all $\omega \in \mathcal{R}$.*

Then there exists a $\mu^* > 0$ such that for $\mu \in [0, \mu^*)$, ϕ is PE of order $\bar{n}$ with level of excitation $\alpha_1 > O(\mu)$ provided that the input signal u is dominantly rich of order $\bar{n}$ for the plant (9.2.16).

Proof of Lemma 9.2.1: Let us define

$$\phi_0 = H_0(s)u, \quad \phi_1 = H_1(\mu s, s)u$$

Because ϕ_0 is the signal vector for the ideal case, i.e., $H_0(s)$ does not depend on μ, u being sufficiently rich of order $\bar{n}$ together with the assumed properties (a), (b) of $H_0(s)$ imply, according to Theorem 5.2.1, that ϕ_0 is PE with level $\alpha_0 > 0$ and α_0 is independent of μ, i.e.,

$$\frac{1}{T}\int_t^{t+T} \phi_0(\tau)\phi_0^\top(\tau)d\tau \geq \alpha_0 I \qquad (9.2.20)$$

$\forall t \geq 0$ and some $T > 0$. On the other hand, because $H_1(\mu s, s)$ is stable and $|H_1(j\mu\omega, j\omega)| \leq c$ for all $\omega \in \mathcal{R}$, we have $\phi_1 \in \mathcal{L}_\infty$ and

$$\frac{1}{T}\int_t^{t+T} \phi_1(\tau)\phi_1^\top(\tau)d\tau \leq \beta I \qquad (9.2.21)$$

for some constant β which is independent of μ. Note that

$$\frac{1}{T}\int_t^{t+T} \phi(\tau)\phi^\top(\tau)d\tau = \frac{1}{T}\int_t^{t+T} (\phi_0(\tau) + \mu\phi_1(\tau))(\phi_0^\top(\tau) + \mu\phi_1^\top(\tau))d\tau$$

$$\geq \frac{1}{T}\left\{\int_t^{t+T} \frac{\phi_0(\tau)\phi_0^\top(\tau)}{2}d\tau - \mu^2\int_t^{t+T}\phi_1(\tau)\phi_1^\top(\tau)d\tau\right\}$$

where the second inequality is obtained by using $(x+y)(x+y)^\top \geq \frac{xx^\top}{2} - yy^\top$, we obtain that

$$\frac{1}{T}\int_t^{t+T}\phi(\tau)\phi^\top(\tau)d\tau \geq \frac{\alpha_0}{2}I - \mu^2\beta I$$

which implies that ϕ has a level of PE $\alpha_1 = \frac{\alpha_0}{4}$, say, for $\mu \in [0, \mu^*)$ where $\mu^* \triangleq \sqrt{\frac{\alpha_0}{4\beta}}$.
□

Lemma 9.2.1 indicates that if u is dominantly rich, then the PE level of ϕ_0, the signal vector associated with the dominant part of the plant, is much higher than that of $\mu\phi_1$, the signal vector that is due to the unmodeled part of the plant, provided of course that μ is relatively small. The smaller the parameter μ is, the bigger the separation of the spectrums of the dominant and unmodeled high frequency dynamics.

Let us now use Lemma 9.2.1 to prove Theorem 9.2.1.

Proof of Theorem 9.2.1 The error equation that describes the stability properties of the parameter identifier is given by

$$\dot{\tilde{\theta}} = -\Gamma\phi\phi^T\tilde{\theta} + \mu\Gamma\phi\eta \qquad (9.2.22)$$

where $\tilde{\theta} \triangleq \theta - \theta^*$ is the parameter error. Let us first assume that all the conditions of Lemma 9.2.1 are satisfied so that for a dominantly rich input and for all $\mu \in [0, \mu^*)$, the signal vector ϕ is PE with level $\alpha_1 > O(\mu)$. Using the results of Chapter 4 we can show that the homogeneous part of (9.2.22) is u.a.s., i.e., there exists constants $\alpha > 0, \beta > 0$ independent of μ such that the transition matrix $\Phi(t, t_0)$ of the homogeneous part of (9.2.22) satisfies

$$\|\Phi(t, t_0)\| \leq \beta e^{-\alpha(t-t_0)} \qquad (9.2.23)$$

Therefore, it follows from (9.2.22), (9.2.23) that

$$|\tilde{\theta}(t)| \leq ce^{-\alpha t} + \mu c, \quad \forall \mu \in [0, \mu^*)$$

where $c \geq 0$ is a finite constant independent of μ, which implies that $\theta, \epsilon \in \mathcal{L}_\infty$ and $\theta(t)$ converges to the residual set $\mathcal{R}_e$ exponentially fast.

Let us now verify that all the conditions of Lemma 9.2.1 assumed above are satisfied. These conditions are

(a) $H_0(j\omega_1), \ldots, H_0(j\omega_{\bar{n}})$ are linearly independent for all possible $\omega_1, \omega_2, \ldots, \omega_{\bar{n}}$ where $\omega_i \neq \omega_k$, $i, k = 1, \ldots, \bar{n}$; $\bar{n} = n + m + 1$

(b) For any set $\{\omega_1, \omega_2, \ldots, \omega_{\bar{n}}\}$ where $|\omega_i - \omega_k| > O(\mu), i \neq k$ and $|\omega_i| < O(\frac{1}{\mu})$, we have

$$|\det\{[H_0(j\omega_1), H_0(j\omega_2), \ldots, H_0(j\omega_{\bar{n}})]\}| > O(\mu)$$

(c) $|H_1(\mu j\omega, j\omega)| \leq c$ for all $\omega \in \mathcal{R}$

It has been shown in the proof of Theorem 5.2.4 that the coprimeness of the numerator and denominator polynomials of $G_0(s)$ implies the linear independence of $H_0(j\omega_1), H_0(j\omega_2), \ldots, H_0(j\omega_{\bar{n}})$ for any $\omega_1, \omega_2, \ldots, \omega_{\bar{n}}$ with $\omega_i \neq \omega_k$ and thus (a) is verified. From the definition of $H_1(\mu s, s)$ and the assumption of $G_0(s)\Delta_m(\mu s, s)$ being proper, we have $|H_1(\mu j\omega, j\omega)| \leq c$ for some constant c, which verifies (c).

9.2. ROBUST ADAPTIVE OBSERVERS

To establish condition (b), we proceed as follows: From the definition of $H_0(s)$, we can write

$$H_0(s) = Q_0 \frac{1}{\Lambda(s)R_p(s)} \begin{bmatrix} s^{\bar{n}-1} \\ s^{\bar{n}-2} \\ \vdots \\ s \\ 1 \end{bmatrix}$$

where $Q_0 \in \mathcal{R}^{\bar{n} \times \bar{n}}$ is a constant matrix. Furthermore, Q_0 is nonsingular, otherwise, we can conclude that $H_0(j\omega_1), H_0(j\omega_2), \ldots, H_0(j\omega_{\bar{n}})$ are linearly dependent that contradicts with (a) which we have already proven to be true. Therefore,

$$\begin{aligned}&[H_0(j\omega_1),\ldots,H_0(j\omega_{\bar{n}})] \\ &= Q_0 \begin{bmatrix} (j\omega_1)^{\bar{n}-1} & (j\omega_2)^{\bar{n}-1} & \cdots & (j\omega_{\bar{n}})^{\bar{n}-1} \\ (j\omega_1)^{\bar{n}-2} & (j\omega_2)^{\bar{n}-2} & \cdots & (j\omega_{\bar{n}})^{\bar{n}-2} \\ \vdots & \vdots & \vdots & \\ j\omega_1 & j\omega_2 & \cdots & j\omega_{\bar{n}} \\ 1 & 1 & \cdots & 1 \end{bmatrix} diag\left\{\frac{1}{\Lambda(j\omega_i)R_p(j\omega_i)}\right\} \end{aligned} \quad (9.2.24)$$

Noting that the middle factor matrix on the right hand side of (9.2.24) is a Vandermonde matrix [62], we have

$$\det\{[H_0(j\omega_1),\ldots,H_0(j\omega_{\bar{n}})]\} = \det(Q_0) \prod_{i=1}^{\bar{n}} \frac{1}{\Lambda(j\omega_i)R_p(j\omega_i)} \prod_{1 \leq i < k \leq \bar{n}} (j\omega_i - j\omega_k)$$

and, therefore, (b) follows immediately from the assumption that $|\omega_i - \omega_k| > O(\mu)$.
□

Theorem 9.2.1 indicates that if the input u is dominantly rich of order $n + m + 1$ and there is sufficient separation between the spectrums of the dominant and unmodeled high frequency dynamics, i.e., μ is small, then the parameter error bound at steady state is small. The condition of dominant richness is also necessary for the parameter error to converge exponentially fast to a small residual set in the sense that if u is sufficiently rich but not dominantly rich then we can find an example for which the signal vector ϕ loses its PE property no matter how fast the unmodeled dynamics are. In the case of the pure least-squares algorithm where the matrix Γ in (9.2.22) is replaced with P generated from $\dot{P} = -P\frac{\phi\phi^\top}{m^2}P$, we cannot establish the u.a.s. of the homogeneous part of (9.2.22) even when ϕ is PE. As a result, we are unable to establish (9.2.23) and therefore the convergence of θ to a residual set even when u is dominantly rich.

9.2.3 Robust Adaptive Observers

In this section we examine the stability properties of the adaptive observers developed in Chapter 5 for the plant model

$$y = G_0(s)u \qquad (9.2.25)$$

when applied to the plant

$$y = G_0(s)(1 + \mu\Delta_m(\mu s, s))u \qquad (9.2.26)$$

with multiplicative perturbations. As in Section 9.2.2, the overall plant transfer function in (9.2.26) is assumed to be strictly proper with stable poles and $\lim_{\mu \to 0} \mu G_0(s)\Delta_m(\mu s, s) = 0$ for any given s. A minimal state-space representation of the dominant part of the plant associated with $G_0(s)$ is given by

$$\dot{x}_\alpha = \left[\begin{array}{c|c} -a_p & \begin{array}{c} I_{n-1} \\ \hline 0 \end{array} \end{array}\right] x_\alpha + b_p u, \quad x_\alpha \in \mathcal{R}^n$$

$$y = [1, 0, \ldots, 0]x_\alpha + \mu\eta$$

$$\eta = G_0(s)\Delta_m(\mu s, s)u \qquad (9.2.27)$$

Because the plant is stable and $u \in \mathcal{L}_\infty$, the effect of the plant perturbation appears as an output bounded disturbance.

Let us consider the adaptive Luenberger observer designed and analyzed in Section 5.3, i.e.,

$$\dot{\hat{x}} = \hat{A}(t)\hat{x} + \hat{b}_p(t)u + K(t)(y - \hat{y}), \quad \hat{x} \in \mathcal{R}^n$$

$$\hat{y} = [1, 0, \ldots, 0]\hat{x} \qquad (9.2.28)$$

where $\hat{x}$ is the estimate of x_α,

$$\hat{A}(t) = \left[\begin{array}{c|c} -\hat{a}_p & \begin{array}{c} I_{n-1} \\ \hline 0 \end{array} \end{array}\right], \quad K(t) = a^* - \hat{a}_p$$

$\hat{a}_p, \hat{b}_p$ are the estimates of a_p, b_p, respectively, and $a^* \in \mathcal{R}^n$ is a constant vector, and such that

$$A^* = \left[\begin{array}{c|c} -a^* & \begin{array}{c} I_{n-1} \\ \hline 0 \end{array} \end{array}\right]$$

9.2. ROBUST ADAPTIVE OBSERVERS

is a stable matrix. It follows that the observation error $\tilde{x} \triangleq x_\alpha - \hat{x}$ satisfies

$$\dot{\tilde{x}} = A^*\tilde{x} - \tilde{b}_p u + \tilde{a}_p y + \mu(a_p - a^*)\eta \qquad (9.2.29)$$

where A^* is a stable matrix; $\tilde{a}_p \triangleq \hat{a}_p - a_p, \tilde{b}_p \triangleq \hat{b}_p - b_p$ are the parameter errors. The parameter vectors a_p, b_p contain the coefficients of the denominator and numerator of $G_0(s)$ respectively and can be estimated using the adaptive law described by equation (9.2.18) presented in Section 9.2.2 or any other adaptive law from Tables 4.2, 4.3. The stability properties of the adaptive Luenberger observer described by (9.2.28) and (9.2.18) or any adaptive law from Tables 4.2 and 4.3 based on the parametric model (9.2.17) are given by the following theorem:

Theorem 9.2.2 *Assume that the input u is dominantly rich of order $2n$ for the plant (9.2.26) and $G_0(s)$ has no zero-pole cancellation. The adaptive Luenberger observer consisting of equation (9.2.28) with (9.2.18) or any adaptive law from Tables 4.2 and 4.3 (with the exception of the pure least-squares algorithm) based on the parametric model (9.2.17) with $\mu = 0$ applied to the plant (9.2.26) with $\mu \neq 0$ has the following properties: There exists a $\mu^* > 0$ such that for all $\mu \in [0, \mu^*)$*

(i) *All signals are u.b.*

(ii) *The state observation error $\tilde{x}$ and parameter error $\tilde{\theta}$ converge exponentially fast to the residual set*

$$\mathcal{R}_e = \left\{ \tilde{x}, \tilde{\theta} \Big| \; |\tilde{x}| + |\tilde{\theta}| \leq c\mu \right\}$$

where $c \geq 0$ is a constant independent of μ.

Proof: The proof follows directly from the results of Section 9.2.2 where we have established that a dominantly rich input guarantees signal boundedness and exponential convergence of the parameter estimates to a residual set where the parameter error $\tilde{\theta}$ satisfies $|\tilde{\theta}| \leq c\mu$ provided $\mu \in [0, \mu^*)$ for some $\mu^* > 0$. Because $|\tilde{b}_p|, |\tilde{a}_p|$ converge exponentially to residual sets whose size is of $O(\mu)$ and u, y, η are bounded for any $\mu \in [0, \mu^*)$, the convergence of $\tilde{x}, \tilde{\theta}$ to the set $\mathcal{R}_e$ follows directly from the stability of the matrix A^* and the error equation (9.2.29). $\square$

Theorem 9.2.2 shows that for stable plants, the combination of a state observer with an adaptive law from Tables 4.2 and 4.3 developed for modeling error and disturbance free parametric models leads to a robust adaptive observer provided that the input signal u is restricted to be dominantly rich. If u is not dominantly rich, then the results of Theorem 9.2.2 are no longer valid. If, however, instead of the adaptive laws of Tables 4.2 and 4.3, we use robust adaptive laws from Tables 8.1 to 8.4, we can then establish signal boundedness even when u is not dominantly rich. Dominant richness, however, is still needed to establish the convergence of the parameter and state observation errors to residual sets whose size is of the order of the modeling error. As in the case of the identifiers, we are unable to establish the convergence results of Theorem 9.2.1 for the pure least-squares algorithm.

9.3 Robust MRAC

In Chapter 6, we designed and analyzed a wide class of MRAC schemes for a plant that is assumed to be finite dimensional, LTI, whose input and output could be measured exactly and whose transfer function $G_0(s)$ satisfies assumptions P1 to P4 given in Section 6.3.1. We have shown that under these ideal assumptions, an adaptive controller can be designed to guarantee signal boundedness and convergence of the tracking error to zero.

As we explained in Chapter 8, in practice no plant can be modeled exactly by an LTI finite-dimensional system. Furthermore, the measurement of signals such as the plant input and output are usually contaminated with noise. Consequently, it is important from the practical viewpoint to examine whether the MRAC schemes of Chapter 6 can perform well in a realistic environment where $G_0(s)$ is not the actual transfer function of the plant but instead an approximation of the overall plant transfer function, and the plant input and output are affected by unknown disturbances. In the presence of modeling errors, exact model-plant transfer function matching is no longer possible in general and therefore the control objective of zero tracking error at steady state for any reference input signal may not be achievable. The best one can hope for, in the nonideal situation in general, is signal boundedness and small tracking errors that are of the order of the modeling error at steady state.

In this section we consider MRAC schemes that are designed for a simpli-

9.3. ROBUST MRAC

fied model of the plant but are applied to a higher-order plant. For analysis purposes, the neglected perturbations in the plant model can be characterized as unstructured uncertainties of the type considered in Chapter 8. In our analysis we concentrate on multiplicative type of perturbations. The same analysis can be carried out for additive and stable factor perturbations and is left as an exercise for the reader.

We assume that the plant is of the form

$$y_p = G_0(s)(1 + \Delta_m(s))(u_p + d_u) \tag{9.3.1}$$

where $G_0(s)$ is the modeled part of the plant, $\Delta_m(s)$ is an unknown multiplicative perturbation with stable poles and d_u is a bounded input disturbance. We assume that the overall plant transfer function and G_0 are strictly proper. This implies that $G_0 \Delta_m$ is also strictly proper. We design the MRAC scheme assuming that the plant model is of the form

$$y_p = G_0(s) u_p, \quad G_0(s) = k_p \frac{Z_p(s)}{R_p(s)} \tag{9.3.2}$$

where $G_0(s)$ satisfies assumptions P1 to P4 in Section 6.3.1, but we implement it on the plant (9.3.1). The effect of the dynamic uncertainty $\Delta_m(s)$ and disturbance d_u on the stability and performance of the MRAC scheme is analyzed in the next sections.

We first treat the case where the parameters of $G_0(s)$ are known exactly. In this case no adaptation is needed and therefore the overall closed-loop plant can be studied using linear system theory.

9.3.1 MRC: Known Plant Parameters

Let us consider the MRC law

$$u_p = \theta_1^{*\top} \frac{\alpha(s)}{\Lambda(s)} u_p + \theta_2^{*\top} \frac{\alpha(s)}{\Lambda(s)} y_p + \theta_3^* y_p + c_0^* r \tag{9.3.3}$$

developed in Section 6.3.2 and shown to meet the MRC objective for the plant model (9.3.2). Let us now apply (9.3.3) to the actual plant (9.3.1) and analyze its properties with respect to the multiplicative uncertainty $\Delta_m(s)$ and input disturbance d_u.

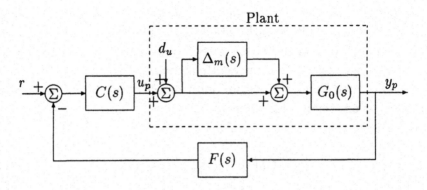

Figure 9.1 Closed-loop MRC scheme.

Following the analysis of Section 6.3.2 (see Figure 6.4), the closed-loop plant is represented as shown in Figure 9.1 where

$$C(s) = \frac{\Lambda(s)c_0^*}{\Lambda(s) - \theta_1^{*T}\alpha(s)}, \quad F(s) = -\frac{\theta_2^{*T}\alpha(s) + \theta_3^*\Lambda(s)}{c_0^*\Lambda(s)}$$

The closed-loop plant is in the form of the general feedback system in Figure 3.2 analyzed in Chapter 3. Therefore using equation (3.6.1) we have

$$\begin{bmatrix} u_p \\ y_p \end{bmatrix} = \begin{bmatrix} \dfrac{C}{1+FCG} & \dfrac{-FCG}{1+FCG} \\ \dfrac{CG}{1+FCG} & \dfrac{G}{1+FCG} \end{bmatrix} \begin{bmatrix} r \\ d_u \end{bmatrix} \quad (9.3.4)$$

where $G = G_0(1 + \Delta_m)$ is the overall transfer function which is assumed to be strictly proper. The stability properties of (9.3.4) are given by the following Theorem:

Theorem 9.3.1 *Assume that $\theta^* = [\theta_1^{*T}, \theta_2^{*T}, \theta_3^*, c_0^*]^T$ is chosen so that for $\Delta_m(s) = 0, d_u = 0$, the closed-loop plant is stable and the matching equation*

$$\frac{CG_0}{1+FCG_0} = W_m$$

where $W_m(s) = k_m \frac{Z_m(s)}{R_m(s)}$ is the transfer function of the reference model, is satisfied, i.e., the MRC law (9.3.3) meets the MRC objective for the simplified plant model (9.3.2). If

$$\left\| \frac{\theta_2^{*T}\alpha(s) + \theta_3^*\Lambda(s)}{\Lambda(s)} \frac{k_p}{k_m} W_m(s)\Delta_m(s) \right\|_\infty < 1 \quad (9.3.5)$$

9.3. ROBUST MRAC

then the closed-loop plant is internally stable. Furthermore there exists a constant $\delta^* > 0$ such that for any $\delta \in [0, \delta^*]$ the tracking error $e_1 = y_p - y_m$ satisfies

$$\limsup_{t \to \infty \atop \tau \geq t} |e_1(\tau)| \leq \Delta r_0 + c d_0 \quad (9.3.6)$$

where r_0, d_0 are upper bounds for $|r(t)|, |d_u(t)|$, $c \geq 0$ is a finite constant

$$\Delta \triangleq \left\| \frac{W_m(s)(\Lambda(s) - C_1^*(s))R_p(s)}{Z_p(s)[k_m\Lambda(s) - k_p W_m(s) D_1^*(s) \Delta_m(s)]} W_m(s)\Delta_m(s) \right\|_{2\delta}$$

with $C_1^*(s) = \theta_1^{*\mathsf{T}} \alpha(s), D_1^*(s) = \theta_2^{*\mathsf{T}} \alpha(s) + \theta_3^* \Lambda(s)$ and $y_m = W_m(s)r$ is the output of the reference model.

Proof A necessary and sufficient condition for $r, d_u \in \mathcal{L}_\infty$ to imply that $y_p, u_p \in \mathcal{L}_\infty$ is that the poles of each element of the transfer matrix in (9.3.4) are in $\mathcal{C}^-$. It can be shown using the expressions of F, C and G that the characteristic equation of each element of the transfer matrix in (9.3.4) is given by

$$(\Lambda - C_1^*)R_p - D_1^* k_p Z_p (1 + \Delta_m) = 0 \quad (9.3.7)$$

Using the matching equation (6.3.13), i.e., $(\Lambda - C_1^*)R_p - k_p D_1^* Z_p = Z_p \Lambda_0 R_m$, the characteristic equation (9.3.7) becomes

$$Z_p(\Lambda_0 R_m - k_p D_1^* \Delta_m) = 0$$

where Λ_0 is a factor of $\Lambda = \Lambda_0 Z_m$. Because Z_p is Hurwitz, we examine the roots of

$$\Lambda_0 R_m - k_p D_1^* \Delta_m = 0$$

which can be also written as

$$1 - \frac{D_1^* k_p}{\Lambda_0 R_m} \Delta_m = 1 - \frac{k_p D_1^* W_m \Delta_m}{k_m \Lambda} = 0$$

Because the poles of $\frac{D_1^* W_m \Delta_m}{\Lambda}$ are in $\mathcal{C}^-$, it follows from the Nyquist criterion that for all Δ_m satisfying

$$\left\| \frac{k_p}{k_m} \frac{D_1^*(s) W_m(s) \Delta_m(s)}{\Lambda(s)} \right\|_\infty < 1 \quad (9.3.8)$$

the roots of (9.3.7) are in $\mathcal{C}^-$. Hence, (9.3.8), (9.3.4) imply that for $r, d_u \in \mathcal{L}_\infty$, we have $u_p, y_p \in \mathcal{L}_\infty$.

Because $F(s) = -\frac{D_1^*(s)}{c_0^* \Lambda(s)}$ has stable poles and $r, d_u, u_p, y_p \in \mathcal{L}_\infty$, it follows from Figure 9.1 that all signals in the closed-loop scheme are bounded.

The tracking error $e_1 = y_p - y_m$ is given by

$$e_1 = \frac{W_m \Delta_m}{1 + FCG_0(1 + \Delta_m)} r + \frac{G_0(1 + \Delta_m)}{1 + FCG_0(1 + \Delta_m)} d_u \qquad (9.3.9)$$

Substituting for F, C, G_0 and using the matching equation, we have

$$e_1 = \frac{W_m(\Lambda - C_1^*)R_p}{Z_p(k_m\Lambda - k_p W_m D_1^* \Delta_m)} W_m \Delta_m r + \frac{k_p(\Lambda - C_1^*)W_m(1 + \Delta_m)}{(k_m\Lambda - k_p W_m D_1^* \Delta_m)} d_u \qquad (9.3.10)$$

Due to $G(s), G_0(s)$ being strictly proper and the fact that $W_m(s)$ has the same relative degree as $G_0(s)$, it follows that $W_m(s)\Delta_m(s)$ is strictly proper, which implies that the transfer function $e_1(s)/r(s)$ is strictly proper and $e_1(s)/d_u(s)$ is proper. Furthermore, both $e_1(s)/r(s), e_1(s)/d_u(s)$ have stable poles (due to (9.3.5)), which implies that there exists a constant $\delta^* > 0$ such that $\frac{e_1(s)}{r(s)}, \frac{e_1(s)}{d_u(s)}$ are analytic in $\text{Re}[s] \geq -\frac{\delta^*}{2}$. Using the properties of the $\mathcal{L}_{2\delta}$ norm, Lemma 3.3.2, (9.3.10) and the fact that $r, d_u \in \mathcal{L}_\infty$, the bound for the tracking error given by (9.3.6) follows. $\square$

Remark 9.3.1 The expression for the tracking error given by (9.3.9) suggests that by increasing the loop gain FCG_0, we may be able to improve the tracking performance. Because F, C, G_0 are constrained by the matching equation, any changes in the loop gain must be performed under the constraint of the matching equation (6.3.13). One can also select $\Lambda(s)$ and $W_m(s)$ in an effort to minimize the H_∞ norm in the $\mathcal{L}_{2e}$ bound for e_1, i.e.,

$$\|e_{1t}\|_2 \leq \left\| \frac{W_m \Delta_m}{1 + FCG_0(1 + \Delta_m)} \right\|_\infty \|r_t\|_2 + \left\| \frac{G_0(1 + \Delta_m)}{1 + FCG_0(1 + \Delta_m)} \right\|_\infty \|d_{ut}\|_2$$

under the constraint of the matching equation (6.3.13). Such a constrained minimization problem is beyond the scope of this book.

Example 9.3.1 Let us consider the plant

$$y = \frac{1}{s+a}(1 + \Delta_m(s))u = G(s)u \qquad (9.3.11)$$

where $G(s)$ is the overall strictly proper transfer function of the plant and Δ_m is a multiplicative perturbation. The control objective is to choose u such that all

9.3. ROBUST MRAC

signals in the closed-loop plant are bounded and y tracks as close as possible the output y_m of the reference model

$$y_m = \frac{b_m}{s + a_m} r$$

where $a_m, b_m > 0$. The plant (9.3.11) can be modeled as

$$y = \frac{1}{s + a} u \qquad (9.3.12)$$

The MRC law based on (9.3.12) given by

$$u = \theta^* y + b_m r$$

where $\theta^* = a - a_m$ meets the control objective for the plant model (9.3.12) exactly. Let us now implement the same control law on the actual plant (9.3.11). The closed-loop plant is given by

$$y = \frac{b_m(1 + \Delta_m(s))}{s + a_m - \theta^* \Delta_m(s)} r$$

whose characteristic equation is

$$s + a_m - \theta^* \Delta_m(s) = 0$$

or

$$1 - \frac{\theta^* \Delta_m(s)}{s + a_m} = 0$$

Because $\frac{\theta^* \Delta_m(s)}{s + a_m}$ is strictly proper with stable poles, it follows from the Nyquist criterion that a sufficient condition for the closed-loop system to be stable is that $\Delta_m(s)$ satisfies

$$\left\| \frac{\theta^* \Delta_m(s)}{s + a_m} \right\|_\infty = \left\| \frac{(a - a_m)\Delta_m(s)}{s + a_m} \right\|_\infty < 1 \qquad (9.3.13)$$

The tracking error $e_1 = y - y_m$ satisfies

$$e_1 = \frac{b_m}{(s + a_m)} \frac{(s + a_m + \theta^*)}{(s + a_m - \theta^* \Delta_m(s))} \Delta_m(s) r$$

Because $r \in \mathcal{L}_\infty$ and the transfer function $e_1(s)/r(s)$ has stable poles for all Δ_m satisfying (9.3.13), we have that

$$\lim_{t \to \infty} \sup_{\tau \geq t} |e_1(\tau)| \leq \Delta r_0$$

where $|r(t)| \leq r_0$ and

$$\Delta = \left\| \frac{b_m}{s + a_m} \frac{(s + a_m + \theta^*)}{(s + a_m - \theta^* \Delta_m(s))} \Delta_m(s) \right\|_{2\delta}$$

Furthermore,

$$\|e_{1t}\|_2 \leq \left\| \frac{b_m \Delta_m(s)(1 + \frac{\theta^*}{(s+a_m)})}{(s+a_m)(1 - \frac{\theta^* \Delta_m(s)}{(s+a_m)})} \right\|_\infty \|r_t\|_2$$

Therefore, the smaller the term $\|\frac{\Delta_m(s)}{s+a_m}\|_\infty$ is, the better the stability margin and tracking performance will be.

Let us consider the case where

$$\Delta_m(s) = -\frac{2\mu s}{1+\mu s}$$

and $\mu > 0$ is small, that arises from the parameterization of the nonminimum phase plant

$$y = \frac{1-\mu s}{(s+a)(1+\mu s)} u = \frac{1}{s+a}\left(1 - \frac{2\mu s}{1+\mu s}\right) u \qquad (9.3.14)$$

For this $\Delta_m(s)$, condition (9.3.13) becomes

$$\left\| \frac{(a-a_m)}{(s+a_m)} \frac{2\mu s}{(1+\mu s)} \right\|_\infty < 1$$

which is satisfied provided

$$\mu < \frac{1}{2|a-a_m|}$$

Similarly, it can be shown that $\Delta = O(\mu)$, i.e., the faster the unmodeled pole and zero in (9.3.14) are, the better the tracking performance. As $\mu \to 0$, $\Delta \to 0$ and therefore $\lim_{t\to\infty} \sup_{\tau \geq t} |e_1(\tau)| \to 0$. $\triangledown$

9.3.2 Direct MRAC with Unnormalized Adaptive Laws

The adaptive control schemes of Chapter 6 designed for the simplified plant model (9.3.2) can no longer be guaranteed to be stable when applied to the actual plant (9.3.1) with $\Delta_m(s) \neq 0$ or $d_u \neq 0$. We have already demonstrated various types of instability that may arise due to the presence of modeling errors using simple examples in Section 8.3. The main cause of instability is the adaptive law that makes the overall closed-loop plant nonlinear and more susceptible to modeling error effects. We have already proposed various types of modifications for changing the adaptive laws to counteract instabilities and improve performance in Chapter 8. In this section, we use some of these modification techniques to improve the robustness properties

9.3. ROBUST MRAC

of the direct MRAC schemes with unnormalized adaptive laws developed in Section 6.4.

We start by considering the same plant as in Example 9.3.1, i.e.,

$$y = \frac{1}{s+a}(1+\Delta_m(s))u \qquad (9.3.15)$$

where a is unknown and u is to be chosen so that the closed-loop plant is stable and y tracks the output y_m of the reference model

$$y_m = \frac{b_m}{s+a_m}r$$

as close as possible. The control law based on the modeled part of the plant obtained by neglecting $\Delta_m(s)$ is given by

$$u = \theta y + b_m r \qquad (9.3.16)$$

where θ is to be generated by an adaptive law. Letting $\tilde{\theta} = \theta - \theta^*$ where $\theta^* = a - a_m$ is the unknown controller parameter (see Example 9.3.1), we write the control law as

$$u = \theta^* y + b_m r + \tilde{\theta} y$$

and use it in (9.3.15) to obtain the closed-loop plant

$$y = W(s)(\tilde{\theta}y + b_m r)$$

where

$$W(s) = \frac{1+\Delta_m(s)}{s+a-\theta^*-\theta^*\Delta_m(s)} = \frac{1+\Delta_m(s)}{s+a_m-\theta^*\Delta_m(s)}$$

The condition for $W(s)$ to have stable poles is the same as in the nonadaptive case (i.e., $\tilde{\theta} = 0$) and is given by

$$\left\|\frac{\theta^*\Delta_m(s)}{s+a_m}\right\|_\infty < 1 \qquad (9.3.17)$$

Defining the tracking error $e_1 = y - y_m$, we obtain the error equation

$$e_1 = W(s)\tilde{\theta}y + \left(W(s) - \frac{1}{s+a_m}\right)b_m r$$

or
$$e_1 = W(s)\tilde{\theta}y + \frac{\Delta_m(s)(s+a)}{(s+a_m)(s+a_m-\theta^*\Delta_m(s))}b_m r \qquad (9.3.18)$$

The ideal error equation obtained by setting $\Delta_m(s) \equiv 0$ is
$$e_1 = \frac{1}{s+a_m}\tilde{\theta}y$$

which, based on the results of Chapter 4, suggests the adaptive law
$$\dot{\theta} = \dot{\tilde{\theta}} = -\gamma e_1 y \qquad (9.3.19)$$

due to the SPR property of $\frac{1}{s+a_m}$. As we showed in Chapter 4, the adaptive control law (9.3.16), (9.3.19) meets the control objective for the plant (9.3.15) provided $\Delta_m(s) \equiv 0$.

The presence of $\Delta_m(s)$ introduces an external unknown input that depends on Δ_m, r and acts as a bounded disturbance in the error equation. Furthermore $\Delta_m(s)$ changes the SPR transfer function that relates e_1 to $\tilde{\theta}y$ from $\frac{1}{s+a_m}$ to $W(s)$ as shown by (9.3.18). As we showed in Chapter 8, adaptive laws such as (9.3.19) may lead to instability when applied to (9.3.15) due to the presence of the disturbance term that appears in the error equation. One way to counteract the effect of the disturbance introduced by $\Delta_m(s) \neq 0$ and $r \neq 0$ is to modify (9.3.19) using leakage, dead zone, projection etc. as shown in Chapter 8. For this example, let us modify (9.3.19) using the fixed σ-modification, i.e.,
$$\dot{\theta} = -\gamma e_1 y - \gamma\sigma\theta \qquad (9.3.20)$$

where $\sigma > 0$ is small.

Even with this modification, however, we will have difficulty establishing global signal boundedness unless $W(s)$ is SPR. Because $W(s)$ depends on $\Delta_m(s)$ which is unknown and belongs to the class defined by (9.3.17), the SPR property of $W(s)$ cannot be guaranteed unless $\Delta_m(s) \equiv 0$. Below we treat the following cases that have been considered in the literature of robust adaptive control.

Case I: $W(s)$ Is SPR–Global Boundedness

Assume that $\Delta_m(s)$ is such that $W(s)$ is guaranteed to be SPR. The error equation (9.3.18) may be expressed as
$$e_1 = W(s)(\tilde{\theta}y + d) \qquad (9.3.21)$$

9.3. ROBUST MRAC

where $d = \left(1 - \frac{W^{-1}(s)}{s+a_m}\right) b_m r$ is guaranteed to be bounded due to $r \in \mathcal{L}_\infty$ and the stability of $W^{-1}(s)$ which is implied by the SPR property of $W(s)$. The state-space representation of (9.3.21) given by

$$\dot{e} = A_c e + B_c(\tilde{\theta} y + d) \qquad (9.3.22)$$
$$e_1 = C_c^\top e$$

where $C_c^\top (sI - A_c)^{-1} B_c = W(s)$ motivates the Lyapunov-like function

$$V = \frac{e^\top P_c e}{2} + \frac{\tilde{\theta}^2}{2\gamma}$$

with $P_c = P_c^\top > 0$ satisfies the equations in the LKY Lemma. The time derivative of V along the solution of (9.3.20) and (9.3.22) is given by

$$\dot{V} = -e^\top q q^\top e - \nu_c e^\top L_c e + e_1 d - \sigma \tilde{\theta} \theta$$

where $\nu_c > 0$, $L_c = L_c^\top > 0$ and q are defined by the LKY Lemma. We have

$$\dot{V} \leq -\nu_c \lambda_c |e|^2 + c|e||d| - \sigma|\tilde{\theta}|^2 + \sigma|\tilde{\theta}||\theta^*|$$

where $c = \|C^\top\|$, $\lambda_c = \lambda_{min}(L_c)$. Completing the squares and adding and subtracting αV, we obtain

$$\dot{V} \leq -\alpha V + \frac{c^2 |d|^2}{2\nu_c \lambda_c} + \frac{\sigma |\theta^*|^2}{2} \qquad (9.3.23)$$

where $\alpha = min\{\frac{\nu_c \lambda_c}{\lambda_p}, \sigma\gamma\}$ and $\lambda_p = \lambda_{max}(P_c)$, which implies that V and, therefore, $e, \theta, e_1 \in \mathcal{L}_\infty$ and that e_1 converges to a residual set whose size is of the order of the disturbance term $|d|$ and the design parameter σ. If, instead of the fixed-σ, we use the switching σ or the projection, we can verify that as $\Delta_m(s) \to 0$, i.e., $d \to 0$, the tracking error e_1 reduces to zero too.

Considerable efforts have been made in the literature of robust adaptive control to establish that the unmodified adaptive law (9.3.19) can be used to establish stability in the case where y is PE [85, 86]. It has been shown [170] that if $W(s)$ is SPR and y is PE with level $\alpha_0 \geq \gamma_0 \nu_0 + \gamma_1$ where ν_0 is an upper bound for $|d(t)|$ and γ_0, γ_1 are some positive constants, then all signals in the closed-loop system (9.3.18), (9.3.19) are bounded. Because α_0, ν_0 are proportional to the amplitude of r, the only way to generate the high level

α_0 of PE relative to ν_0 is through the proper selection of the spectrum of r. If $\Delta_m(s)$ is due to fast unmodeled dynamics, then the richness of r should be achieved in the low frequency range, i.e., r should be dominantly rich. Intuitively the spectrum of r should be chosen so that $|W(j\omega)| \gg |W(j\omega) - \frac{1}{j\omega+a_m}|$ for all ω in the frequency range of interest.

An example of $\Delta_m(s)$ that guarantees $W(s)$ to be SPR is $\Delta_m(s) = \frac{\mu s}{1+\mu s}$ where $\mu > 0$ is small enough to guarantee that $\Delta_m(s)$ satisfies (9.3.17).

Case II: $W(s)$ Is Not SPR–Semiglobal Stability

When $W(s)$ is not SPR, the error equation (9.3.18) may not be the appropriate one for analysis. Instead, we express the closed-loop plant (9.3.15), (9.3.16) as

$$y = \frac{1}{s+a}(\theta^* y + b_m r + \tilde{\theta} y) + \frac{1}{s+a}\Delta_m(s)u$$

and obtain the error equation

$$e_1 = \frac{1}{s+a_m}(\tilde{\theta} y + \Delta_m(s)u)$$

or

$$\begin{aligned} \dot{e}_1 &= -a_m e_1 + \tilde{\theta} y + \eta \\ \eta &= \Delta_m(s)u = \Delta_m(s)(\theta y + b_m r) \end{aligned} \quad (9.3.24)$$

Let us analyze (9.3.24) with the modified adaptive law (9.3.20)

$$\dot{\theta} = -\gamma e_1 y - \gamma \sigma \theta$$

The input η to the error equation (9.3.24) cannot be guaranteed to be bounded unless u is bounded. But because u is one of the signals whose boundedness is to be established, the equation $\eta = \Delta_m(s)u$ has to be analyzed together with the error equation and adaptive law. For this reason, we need to express $\eta = \Delta_m(s)u$ in the state space form. This requires to assume some structural information about $\Delta_m(s)$. In general $\Delta_m(s)$ is assumed to be proper and small in some sense. $\Delta_m(s)$ may be small at all frequencies, i.e., $\Delta_m(s) = \mu\Delta_1(s)$ where $\mu > 0$ is a small constant whose upper bound is to be established and $\Delta_1(s)$ is a proper stable transfer function. $\Delta_m(s)$ could also be small at low frequencies and large at high frequencies. This is the class of $\Delta_m(s)$ often encountered in applications where $\Delta_m(s)$ contains

9.3. ROBUST MRAC

all the fast unmodeled dynamics which are usually outside the frequency range of interest. A typical example is

$$\Delta_m(s) = -\frac{2\mu s}{1+\mu s} \qquad (9.3.25)$$

which makes (9.3.15) a second-order nonminimum-phase plant and $W(s)$ in (9.3.18) nonminimum-phase and, therefore, non-SPR. To analyze the stability of the closed-loop plant with $\Delta_m(s)$ specified in (9.3.25), we express the error equation (9.3.24) and the adaptive law (9.3.20) in the state-space form

$$\begin{aligned} \dot{e}_1 &= -a_m e_1 + \tilde{\theta} y + \eta \\ \mu \dot{\eta} &= -\eta - 2\mu \dot{u} \\ \dot{\tilde{\theta}} &= -\gamma e_1 y - \gamma \sigma \theta \end{aligned} \qquad (9.3.26)$$

Note that in this representation $\mu = 0 \Rightarrow \eta = 0$, i.e., the state η is a good measure of the effect of the unmodeled dynamics whose size is characterized by the value of μ. The stability properties of (9.3.26) are analyzed by considering the following positive definite function:

$$V(e_1, \eta, \tilde{\theta}) = \frac{e_1^2}{2} + \frac{\tilde{\theta}^2}{2\gamma} + \frac{\mu}{2}(\eta + e_1)^2 \qquad (9.3.27)$$

For each $\mu > 0$ and some constants $c_0 > 0, \alpha > 0$, the inequality

$$V(e_1, \eta, \tilde{\theta}) \le c_0 \mu^{-2\alpha}$$

defines a closed sphere $\mathcal{L}(\mu, \alpha, c_0)$ in $\mathcal{R}^3$ space. The time derivative of V along the trajectories of (9.3.26) is

$$\begin{aligned} \dot{V} &= -a_m e_1^2 - \sigma \tilde{\theta} \theta - \eta^2 + \mu[\dot{e}_1 - 2\dot{u}](\eta + e_1) \\ &\le -a_m e_1^2 - \frac{\sigma \tilde{\theta}^2}{2} - \eta^2 + \frac{\sigma \theta^{*2}}{2} + \mu c [e_1^4 + |e_1|^3 + e_1^2 + e_1^2|\tilde{\theta}| + e_1^2 \tilde{\theta}^2 + |e_1|\tilde{\theta}^2 \\ &\quad + |e_1||\tilde{\theta}| + |e_1||\tilde{\theta}||\eta| + |e_1|^3|\eta| + |e_1| + |\eta| + \eta^2 + |e_1||\eta| \\ &\quad + |\eta||\tilde{\theta}| + |\eta|\tilde{\theta}^2 + e_1^2|\eta| + |e_1|\tilde{\theta}^2||\eta| + |\tilde{\theta}|\eta^2 + |e_1||\dot{r}| + |\eta||\dot{r}|] \end{aligned} \qquad (9.3.28)$$

for some constant $c \ge 0$, where the last inequality is obtained by substituting for $\dot{e}_1$, $\dot{u} = \dot{\theta} y + \theta \dot{y} + b_m \dot{r}$ and taking bounds. Using the inequality $2\alpha\beta \le$

$\alpha^2 + \beta^2$, the multiplicative terms in (9.3.28) can be manipulated so that after some tedious calculations (9.3.28) is rewritten as

$$\dot{V} \leq -\frac{a_m e_1^2}{2} - \frac{\sigma \tilde{\theta}^2}{4} - \frac{\eta^2}{2} - \eta^2 \left[\frac{1}{2} - \mu c(|\tilde{\theta}| + 1)\right]$$
$$- e_1^2 \left[\frac{a_m}{2} - \mu c(e_1^2 + |e_1| + |\tilde{\theta}| + \tilde{\theta}^2 + |\eta| + |e_1||\eta| + 1)\right]$$
$$- \tilde{\theta}^2 \left[\frac{\sigma}{4} - \mu c(|e_1| + 1 + |e_1||\eta|)\right] + \mu c |\dot{r}|^2 + \mu c + \frac{\sigma |\theta^*|^2}{2} \quad (9.3.29)$$

Inside $\mathcal{L}(\mu, \alpha, c_0)$, $|e_1|, |\tilde{\theta}|$ can grow up to $O(\mu^{-\alpha})$ and $|\eta|$ can grow up to $O(\mu^{-1/2-\alpha})$. Hence, there exist positive constants k_1, k_2, k_3 such that inside $\mathcal{L}(\mu, \alpha, c_0)$, we have

$$|e_1| < k_1 \mu^{-\alpha}, \quad |\tilde{\theta}| < k_2 \mu^{-\alpha}, \quad |\eta| < k_3 \mu^{-1/2-\alpha}$$

For all $e_1, \eta, \tilde{\theta}$ inside $\mathcal{L}(\mu, \alpha, c_0)$, (9.3.29) can be simplified to

$$\dot{V} \leq -\frac{a_m}{4} e_1^2 - \frac{\eta^2}{2} - \frac{\sigma}{4} \tilde{\theta}^2 - \eta^2 \left(\frac{1}{2} - \beta_2 \mu^{1-\alpha}\right)$$
$$- e_1^2 \left(\frac{a_m}{2} - \mu^{1/2 - 2\alpha} \beta_1\right) - \tilde{\theta}^2 \left(\frac{\sigma}{4} - \beta_3 \mu^{1-\alpha}\right) \quad (9.3.30)$$
$$+ \mu c |\dot{r}|^2 + \mu c + \frac{\sigma \theta^{*2}}{2}$$

for some positive constants $\beta_1, \beta_2, \beta_3$. If we now fix $\sigma > 0$ then for $0 < \alpha < 1/4$, there exists a $\mu^* > 0$ such that for each $\mu \in (0, \mu^*]$

$$\frac{a_m}{2} \geq \mu^{1/2 - 2\alpha} \beta_1, \quad \frac{1}{2} \geq \beta_2 \mu^{1-\alpha}, \quad \frac{\sigma}{4} > \beta_3 \mu^{1-\alpha}$$

Hence, for each $\mu \in (0, \mu^*]$ and $e_1, \eta, \tilde{\theta}$ inside $\mathcal{L}(\mu, \alpha, c_0)$, we have

$$\dot{V} < -\frac{a_m}{2} e_1^2 - \frac{\eta^2}{2} - \frac{\sigma}{4} \tilde{\theta}^2 + \frac{\sigma \theta^{*2}}{2} + \mu c |\dot{r}|^2 + \mu c \quad (9.3.31)$$

On the other hand, we can see from the definition of V that

$$V(e_1, \eta, \tilde{\theta}) = \frac{e_1^2}{2} + \frac{\tilde{\theta}^2}{2\gamma} + \frac{\mu}{2}(\eta + e_1)^2 \leq c_4 \left(\frac{a_m}{2} e_1^2 + \frac{\eta^2}{2} + \frac{\sigma}{4} \tilde{\theta}^2\right)$$

9.3. ROBUST MRAC

where $c_4 = max\{\frac{1+2\mu}{a_m}, \frac{2}{\gamma\sigma}, 2\mu\}$. Thus, for any $0 < \beta \leq 1/c_4$, we have

$$\dot{V} < -\beta V + \frac{\sigma\theta^{*2}}{2} + \mu c|\dot{r}|^2 + \mu c$$

Because $r, \dot{r}$ are uniformly bounded, we define the set

$$D_0(\mu) \triangleq \left\{ e_1, \tilde{\theta}, \eta \middle| V(e_1, \eta, \tilde{\theta}) < \frac{1}{\beta}\left[\frac{\sigma\theta^{*2}}{2} + \mu c|\dot{r}|^2 + \mu c\right] \right\}$$

which for fixed α and σ and for sufficiently small μ is inside $\mathcal{L}(\mu, \alpha, c_0)$. Outside $D_0(\mu)$ and inside $\mathcal{L}(\mu, \alpha, c_0)$, $\dot{V} < 0$ and, therefore, $V(e_1, \eta, \tilde{\theta})$ decreases. Hence there exist positive constants c_1, c_2 such that the set

$$D(\mu) \triangleq \left\{ e_1, \eta, \tilde{\theta} \middle| |e_1| + |\tilde{\theta}| < c_1\mu^{-\alpha}, |\eta| < c_2\mu^{-\alpha-1/2} \right\}$$

is inside $\mathcal{L}(\mu, \alpha, c_0)$ and any solution $e_1(t), \eta(t), \tilde{\theta}(t)$ which starts in $D(\mu)$ remains inside $\mathcal{L}(\mu, \alpha, c_0)$. Furthermore, every solution of (9.3.26) that starts from $D(\mu)\backslash D_0(\mu)$ will enter $D_0(\mu)$ at $t = t_1$ for some finite time t_1 and remain in $D_0(\mu)$ thereafter. Similarly, any solution starting at $t = 0$ from $D_0(\mu)$ will remain in $D_0(\mu)$ for all $t \geq 0$.

The stability results obtained in this example are semiglobal in the sense that as $\mu \to 0$ the size of $D(\mu)$ becomes the whole space.

The above analysis demonstrates that the unnormalized adaptive law with σ-modification guarantees a semiglobal boundedness result in the presence of fast unmodeled dynamics. The speed of unmodeled dynamics is characterized by the positive scalar $\mu > 0$ in such a way that, as $\mu \to 0$, the unmodeled dynamics become infinitely fast and reduce to their steady-state value almost instantaneously. The region of attraction from which every solution is bounded and converges to a small residual set becomes the whole space as $\mu \to 0$.

A similar stability result is established for the general higher order case in [85, 86] using the fixed σ-modification. Other modifications such as dead zone, projection and various other choices of leakage can be used to establish a similar semiglobal boundedness result in the presence of unmodeled dynamics.

Case III: $W(s)$ Is Not SPR–Method of Averaging: Local Stability

The error system (9.3.18), (9.3.19) is a special case of the more general error system

$$e_1 = W(s)\tilde{\theta}^\top \omega + d \qquad (9.3.32)$$

$$\dot{\tilde{\theta}} = -\gamma e_1 \omega \qquad (9.3.33)$$

that includes the case of higher order plants. In (9.3.32), $W(s)$ is strictly proper and stable and d is a bounded disturbance. The signal vector $\omega \in \mathcal{R}^{2n}$ is a vector with the filtered values of the plant input and output and the reference input r. For the example considered above, $\omega = y$. In the error system (9.3.32), (9.3.33), the adaptive law is not modified.

With the method of averaging, we assume that the adaptive gain $\gamma > 0$ is sufficiently small (slow adaptation) so that for e_1, ω bounded, $\tilde{\theta}$ varies slowly with time. This allows us to approximate (9.3.32) with

$$e_1 \approx \tilde{\theta}^\top W(s)\omega + d$$

which we use in (9.3.33) to obtain

$$\dot{\tilde{\theta}} = -\gamma \omega (W(s)\omega)^\top \tilde{\theta} - \gamma \omega d \qquad (9.3.34)$$

Let us consider the homogeneous part of (9.3.34)

$$\dot{\tilde{\theta}} = -\gamma R(t)\tilde{\theta} \qquad (9.3.35)$$

where $R(t) = \omega(t)(W(s)\omega(t))^\top$. Even though $R(t)$ may vary rapidly with time, the variations of $\tilde{\theta}$ are slow due to small γ. This fact allows us to obtain stability conditions for (9.3.35) in terms of sample averages

$$\bar{R}_i(T_i) \triangleq \frac{1}{T_i} \int_{t_{i-1}}^{t_i} R(t)\,dt, \quad t_i = t_{i-1} + T_i$$

over finite sample intervals $0 < T_i < \infty$. We consider the case where $R(t) = R(t+T)$ is bounded and periodic with period $T > 0$. By taking $T_i = T$, the value of $\bar{R} = \bar{R}_i(T)$ is independent of the interval. It has been established in [3] that there exists a constant $\gamma^* > 0$ such that (9.3.35) is exponentially stable for $\gamma \in (0, \gamma^*)$ if and only if

$$\min_i \operatorname{Re}\lambda_i(\bar{R}) > 0, \quad i = 1, 2, \ldots, 2n \qquad (9.3.36)$$

9.3. ROBUST MRAC

where $\bar{R} = \frac{1}{T}\int_0^T R(t)dt$. Condition (9.3.36) is equivalent to the existence of a matrix $P = P^\top > 0$ such that

$$P\bar{R} + \bar{R}^\top P > 0$$

When the above inequality is satisfied for $P = I$, namely,

$$\bar{R} + \bar{R}^\top > 0 \tag{9.3.37}$$

its meaning can be explained in terms of the frequency content of ω by letting

$$\omega = \sum_{i=-\infty}^{\infty} Q_i e^{j\omega_i t}, \quad \omega_i = \frac{2\pi i}{T}$$

where Q_{-i} is the conjugate of Q_i. Because

$$W(s)\omega = \sum_{i=-\infty}^{\infty} Q_i W(j\omega_i) e^{j\omega_i t}$$

it follows that

$$\bar{R} = \frac{1}{T}\int_0^T \omega(t)(W(s)\omega(t))^\top dt = \sum_{i=-\infty}^{\infty} W(j\omega_i) Q_i Q_{-i}^\top \tag{9.3.38}$$

For $\bar{R}$ to be nonsingular, it is necessary that

$$\sum_{i=-\infty}^{\infty} Q_i Q_{-i}^\top > 0$$

which can be shown to imply that ω must be PE. Applying the condition (9.3.37) to (9.3.38) we obtain a signal dependent "average SPR" condition

$$\sum_{i=-\infty}^{\infty} \text{Re}\{W(j\omega_i)\} \text{Re}\{Q_i Q_{-i}^\top\} > 0 \tag{9.3.39}$$

Clearly, (9.3.39) is satisfied if ω is PE and $W(s)$ is SPR. Condition (9.3.39) allows $\text{Re}W(j\omega_i) < 0$ for some ω_i provided that in the sum of (9.3.39), the terms with $\text{Re}W(j\omega_i) > 0$ dominate.

The exponential stability of (9.3.35) implies that $\tilde{\theta}$ in (9.3.34) is bounded and converges exponentially to a residual set whose size is proportional to the bounds for ω and d. Because ω consists of the filtered values of the plant input and output, it cannot be shown to be bounded or PE unless certain assumptions are made about the stability of the plant, the initial conditions $\bar{e}_1(0), \tilde{\theta}(0)$ and disturbance d. Let ω_m represent ω when $\tilde{\theta} = 0, e_1 = 0$. The signal ω_m, often referred to as the *"tuned solution,"* can be made to satisfy (9.3.39) by the proper choice of reference signal r. If we express $\omega = \omega_m + e$ where e is the state of a state space representation of (9.3.32) with the same order as ω and $e_1 = C^\top e$, then ω is bounded, PE and satisfies (9.3.39) provided e is sufficiently small. This can be achieved by assuming that the initial conditions $e(0), \tilde{\theta}(0)$ and disturbance d are sufficiently small and the plant is stable. Therefore, the above stability result is local. Even though the above results are based on several restrictive assumptions, they are very valuable in understanding the stability and instability mechanisms of adaptive schemes in the presence of modeling errors. For further details on averaging and the extension of the above results to a wider class of adaptive schemes, the reader is referred to [3].

9.3.3 Direct MRAC with Normalized Adaptive Laws

The MRAC schemes with normalized adaptive laws discussed in Section 6.5 have the same robustness problems as those with the unnormalized adaptive laws, i.e., they can no longer guarantee boundedness of signals in the presence of unmodeled dynamics and bounded disturbances. For MRAC with normalized adaptive laws, however, the robustification can be achieved by simply using the certainty equivalence principle to combine the MRC law with any one of the robust adaptive laws presented in Section 8.5. The design procedure is the same as that in the ideal case, that is, we use the same control law as in the known parameter case but replace the unknown controller parameters with their on-line estimates generated by a robust adaptive law. The stability analysis of the resulting closed-loop adaptive scheme is also similar to the ideal case, with a few minor changes that incorporate the small-in-the-mean, instead of the $\mathcal{L}_2$, properties of $\dot{\theta}$ and ϵ into the analysis.

In this section, we first use an example to illustrate the design and stability analysis of a robust MRAC scheme with a normalized adaptive law. We then extend the results to a general SISO plant with unmodeled dynamics and bounded disturbances.

9.3. ROBUST MRAC

Example 9.3.2 Consider the following SISO plant

$$y = \frac{1}{s-a}(1 + \Delta_m(s))u \qquad (9.3.40)$$

with a strictly proper transfer function, where a is unknown and $\Delta_m(s)$ is a multiplicative plant uncertainty. Let us consider the following adaptive control law

$$u = -\theta y \qquad (9.3.41)$$

$$\begin{aligned}
\dot{\theta} &= \gamma\epsilon\phi, \quad \gamma > 0 \\
\epsilon &= \frac{z - \hat{z}}{m^2}, \quad m^2 = 1 + \phi^2 \\
\phi &= \frac{1}{s + a_m}y, \quad \hat{z} = \theta\phi, \quad z = y - \frac{1}{s + a_m}u
\end{aligned} \qquad (9.3.42)$$

where $-a_m$ is the desired closed loop pole and θ is the estimate of $\theta^* = a + a_m$.

As we have shown in Chapter 8, when (9.3.41), (9.3.42) designed for the plant model

$$y = \frac{1}{s-a}u$$

is applied to the plant (9.3.40) with $\Delta_m(s) \neq 0$, the plant uncertainty $\Delta_m(s)$ introduces a disturbance term in the adaptive law that may easily cause θ to drift to infinity and certain signals to become unbounded no matter how small $\Delta_m(s)$ is. The adaptive control law (9.3.41), (9.3.42) is, therefore, not robust with respect to the plant uncertainty $\Delta_m(s)$. This adaptive control scheme, however, can be made robust if we replace the adaptive law (9.3.42) with a robust one developed by following the procedure of Chapter 8 as follows:

We first express the desired controller parameter $\theta^* = a + a_m$ in the form of a linear parametric model by rewriting (9.3.40) as

$$z = \theta^*\phi + \eta$$

where z, ϕ are as defined in (9.3.42) and

$$\eta = \frac{1}{s + a_m}\Delta_m(s)u$$

is the modeling error term. If we now assume that a bound for the stability margin of the poles of $\Delta_m(s)$ is known, that is, $\Delta_m(s)$ is analytic in $Re[s] \geq -\frac{\delta_0}{2}$ for some known constant $\delta_0 > 0$, then we can verify that the signal m generated as

$$m^2 = 1 + m_s, \quad \dot{m}_s = -\delta_0 m_s + u^2 + y^2, \quad m_s(0) = 0, \quad \delta_0 < 2a_m \qquad (9.3.43)$$

guarantees that $\eta/m, \phi/m \in \mathcal{L}_\infty$ and therefore qualifies to be used as a normalizing signal. Hence, we can combine normalization with any modification, such as leakage, dead zone, or projection, to form a robust adaptive law. Let us consider the switching-σ modification, i.e.,

$$\dot{\theta} = \gamma\epsilon\phi - \sigma_s\gamma\theta$$
$$\epsilon = \frac{z - \theta\phi}{m^2} \qquad (9.3.44)$$

where σ_s is as defined in Chapter 8. According to Theorem 8.5.4, the robust adaptive law given by (9.3.43), (9.3.44) guarantees $\epsilon, \epsilon m, \theta, \dot{\theta} \in \mathcal{L}_\infty$ and $\epsilon, \epsilon m, \dot{\theta} \in \mathcal{S}\left(\frac{\eta^2}{m^2}\right)$. Because from the properties of the $\mathcal{L}_{2\delta}$ norm we have

$$|\eta(t)| \leq \Delta_2 \|u_t\|_{2\delta_0}, \quad \Delta_2 \overset{\triangle}{=} \left\|\frac{1}{s + a_m}\Delta_m(s)\right\|_{2\delta_0}$$

and $m^2 = 1 + \|u_t\|^2_{2\delta_0} + \|y_t\|^2_{2\delta_0}$, it follows that $\frac{|\eta|}{m} \leq \Delta_2$. Therefore, $\epsilon, \epsilon m, \dot{\theta} \in \mathcal{S}(\Delta_2^2)$.

We analyze the stability properties of the MRAC scheme described by (9.3.41), (9.3.43), and (9.3.44) when applied to the plant (9.3.40) with $\Delta_m(s) \neq 0$ as follows:

As in the ideal case considered in Section 6.5.1 for the same example but with $\Delta_m(s) \equiv 0$, we start by writing the closed-loop plant equation as

$$y = \frac{1}{s + a_m}(-\tilde{\theta}y + \Delta_m(s)u) = -\frac{1}{s + a_m}\tilde{\theta}y + \eta, \quad u = -\theta y \qquad (9.3.45)$$

Using the Swapping Lemma A.1 and noting that $\epsilon m^2 = -\tilde{\theta}\phi + \eta$, we obtain from (9.3.45) that

$$y = -\tilde{\theta}\phi + \frac{1}{s + a_m}(\dot{\tilde{\theta}}\phi) + \eta$$
$$= \epsilon m^2 + \frac{1}{s + a_m}\dot{\tilde{\theta}}\phi \qquad (9.3.46)$$

Using the properties of the $\mathcal{L}_{2\delta}$ norm $\|(\cdot)_t\|_{2\delta}$, which for simplicity we denote by $\|\cdot\|$, it follows from (9.3.46) that

$$\|y\| \leq \|\epsilon m^2\| + \|\frac{1}{s + a_m}\|_{\infty\delta}\|\dot{\tilde{\theta}}\phi\|$$

for any $0 < \delta \leq \delta_0$. Because $u = -\theta y$ and $\theta \in \mathcal{L}_\infty$, it follows that

$$\|u\| \leq c\|y\| \leq c\|\epsilon m^2\| + c\|\dot{\tilde{\theta}}\phi\|$$

where $c \geq 0$ is used to denote any finite constant. Therefore, the fictitious normalizing signal

$$m_f^2 \overset{\triangle}{=} 1 + \|u\|^2 + \|y\|^2 \leq 1 + c\|\epsilon m^2\|^2 + c\|\dot{\tilde{\theta}}\phi\|^2 \qquad (9.3.47)$$

9.3. ROBUST MRAC

We can establish as in Chapter 6 that m_f guarantees that $m/m_f, \phi/m_f, \eta/m_f \in \mathcal{L}_\infty$ by using the properties of the $\mathcal{L}_{2\delta}$ norm, which implies that (9.3.47) can be written as

$$m_f^2 \leq 1 + c\|\epsilon m m_f\|^2 + c\|\dot{\tilde{\theta}} m_f\|^2 \leq 1 + c\|\tilde{g} m_f\|^2$$

where $\tilde{g}^2 \triangleq \epsilon^2 m^2 + \dot{\tilde{\theta}}^2$ or

$$m_f^2 \leq 1 + c \int_0^t e^{-\delta(t-\tau)} \tilde{g}^2(\tau) m_f^2(\tau) d\tau \qquad (9.3.48)$$

Applying the B-G Lemma III to (9.3.48), we obtain

$$m_f^2(t) \leq \Phi(t,0) + \delta \int_0^t \Phi(t,\tau) d\tau$$

where

$$\Phi(t,\tau) = e^{-\delta(t-\tau)} e^{c \int_\tau^t \tilde{g}^2(s) ds}$$

Because the robust adaptive law guarantees that $\epsilon m, \dot{\tilde{\theta}} \in \mathcal{S}(\Delta_2^2)$, we have $\tilde{g} \in \mathcal{S}(\Delta_2^2)$ and

$$\Phi(t,\tau) \leq e^{-(\delta - c\Delta_2^2)(t-\tau)}$$

Hence, for

$$c\Delta_2^2 < \delta$$

$\Phi(t,\tau)$ is bounded from above by a decaying to zero exponential, which implies that $m_f \in \mathcal{L}_\infty$. Because of the normalizing properties of m_f, we have $\phi, y, u \in \mathcal{L}_\infty$ and all signals are bounded. The condition $c\Delta_2^2 < \delta$ implies that the multiplicative plant uncertainty $\Delta_m(s)$ should satisfy

$$\left\|\frac{\Delta_m(s)}{s+a_m}\right\|_{2\delta_0}^2 < \frac{\delta}{c}$$

where c can be calculated by keeping track of all the constants. It can be shown that the constant c depends on $\|\frac{1}{s+a_m}\|_{\infty\delta}$ and the upper bound for $|\theta(t)|$.

Because $0 < \delta \leq \delta_0$ is arbitrary, we can choose it to be equal to δ_0. The bound for $\sup_t |\theta(t)|$ can be calculated from the Lyapunov-like function used to analyze the adaptive law. Such a bound, however, may be conservative. If, instead of the switching σ, we use projection, then the bound for $|\theta(t)|$ is known a priori and the calculation of the constant c is easier.

The effect of the unmodeled dynamics on the regulation error y is analyzed as follows: From (9.3.46) and $m, \phi \in \mathcal{L}_\infty$, we have

$$\int_t^{t+T} y^2 d\tau \leq \int_t^{t+T} \epsilon^2 m^2 d\tau + c \int_t^{t+T} \dot{\tilde{\theta}}^2 d\tau, \quad \forall t \geq 0$$

for some constant $c \geq 0$ and any $T > 0$. Then using the m.s.s. property of $\epsilon m, \dot{\theta}$, we have

$$\frac{1}{T}\int_t^{t+T} y^2 d\tau \leq c\Delta_2^2 + \frac{c}{T}$$

therefore $y \in \mathcal{S}(\Delta_2^2)$, i.e., the regulation error is of the order of the modeling error in m.s.s.

The m.s.s. bound for y^2 does not imply that at steady state, y^2 is of the order of the modeling error characterized by Δ_2. A phenomenon known as bursting, where y^2 assumes large values over short intervals of time, cannot be excluded by the m.s.s. bound. The phenomenon of bursting is discussed in further detail in Section 9.4.

The stability and robustness analysis for the MRAC scheme presented above is rather straightforward due to the simplicity of the plant. It cannot be directly extended to the general case without additional steps. A more elaborate but yet more systematic method that extends to the general case is presented below by the following steps:

Step1. *Express the plant input u and output y in terms of the parameter error $\tilde{\theta}$*. We have

$$\begin{aligned} y &= -\frac{1}{s+a_m}\tilde{\theta}y + \eta \\ u &= (s-a)y - \Delta_m(s)u = -\frac{s-a}{s+a_m}\tilde{\theta}y + \eta_u \end{aligned} \quad (9.3.49)$$

where $\eta_u = -\frac{(a+a_m)}{s+a_m}\Delta_m(s)u$. Using the $\mathcal{L}_{2\delta}$ norm for some $\delta \in (0, 2a_m)$ we have

$$\|y\| \leq c\|\tilde{\theta}y\| + \|\eta\|, \quad \|u\| \leq c\|\tilde{\theta}y\| + \|\eta_u\|$$

and, therefore, the fictitious normalizing signal m_f satisfies

$$m_f^2 = 1 + \|u\|^2 + \|y\|^2 \leq 1 + c(\|\tilde{\theta}y\|^2 + \|\eta\|^2 + \|\eta_u\|^2) \quad (9.3.50)$$

Step 2. *Use the swapping lemmas and properties of the $\mathcal{L}_{2\delta}$ norm to upper bound $\|\tilde{\theta}y\|$ with terms that are guaranteed by the adaptive law to have small in m.s.s. gains.* We use the Swapping Lemma A.2 given in Appendix A to write the identity

$$\tilde{\theta}y = (1 - \frac{\alpha_0}{s+\alpha_0})\tilde{\theta}y + \frac{\alpha_0}{s+\alpha_0}\tilde{\theta}y = \frac{1}{s+\alpha_0}(\dot{\tilde{\theta}}y + \tilde{\theta}\dot{y}) + \frac{\alpha_0}{s+\alpha_0}\tilde{\theta}y \quad (9.3.51)$$

where $\alpha_0 > 0$ is an arbitrary constant. Now from the equation for y in (9.3.49) we obtain

$$\tilde{\theta}y = (s+a_m)(\eta - y)$$

9.3. ROBUST MRAC

which we substitute in the last term in (9.3.51) to obtain

$$\tilde{\theta}y = \frac{1}{s+\alpha_0}(\dot{\tilde{\theta}}y + \tilde{\theta}\dot{y}) + \alpha_0 \frac{(s+a_m)}{s+\alpha_0}(\eta - y)$$

Therefore, by choosing δ, α_0 to satisfy $\alpha_0 > a_m > \frac{\delta}{2} > 0$, we obtain

$$\|\tilde{\theta}y\| \leq \left\|\frac{1}{s+\alpha_0}\right\|_{\infty\delta}(\|\dot{\tilde{\theta}}y\| + \|\tilde{\theta}\dot{y}\|) + \alpha_0 \left\|\frac{s+a_m}{s+\alpha_0}\right\|_{\infty\delta}(\|\eta\| + \|y\|)$$

Hence,

$$\|\tilde{\theta}y\| \leq \frac{2}{\alpha_0}(\|\dot{\tilde{\theta}}y\| + \|\tilde{\theta}\dot{y}\|) + \alpha_0 c(\|\eta\| + \|y\|)$$

where $c = \|\frac{s+a_m}{s+\alpha_0}\|_{\infty\delta}$. Using (9.3.46), it follows that

$$\|y\| \leq \|\epsilon m^2\| + c\|\dot{\tilde{\theta}}\phi\|$$

therefore,

$$\|\tilde{\theta}y\| \leq \frac{2}{\alpha_0}(\|\dot{\tilde{\theta}}y\| + \|\tilde{\theta}\dot{y}\|) + \alpha_0 c(\|\epsilon m^2\| + \|\dot{\tilde{\theta}}\phi\| + \|\eta\|) \qquad (9.3.52)$$

The gain of the first term in the right-hand side of (9.3.52) can be made small by choosing large α_0. The m.s.s. gain of the second and third terms is guaranteed by the adaptive law to be of the order of the modeling error denoted by the bound Δ_2, i.e., $\epsilon m, \dot{\tilde{\theta}} \in \mathcal{S}(\Delta_2^2)$. The last term has also a gain which is of the order of the modeling error. This implies that the gain of $\|\tilde{\theta}y\|$ is small provided Δ_2 is small and α_0 is chosen to be large.

Step 3. *Use the B-G Lemma to establish boundedness.* The normalizing properties of m_f and $\theta \in \mathcal{L}_\infty$ guarantee that $y/m_f, \phi/m_f, m/m_f \in \mathcal{L}_\infty$. Because

$$\|\dot{y}\| \leq |a|\|y\| + \|\Delta_m(s)\|_{\infty\delta}\|u\| + \|u\|$$

it follows that $\|\dot{y}\|/m_f \in \mathcal{L}_\infty$. Due to the fact that $\Delta_m(s)$ is proper and analytic in $Re[s] \geq -\delta_0/2$, $\|\Delta_m(s)\|_{\infty\delta}$ is a finite number provided $0 < \delta \leq \delta_0$. Furthermore, $\|\eta\|/m_f \leq \Delta_\infty$ where

$$\Delta_\infty = \left\|\frac{\Delta_m(s)}{s+a_m}\right\|_{\infty\delta} \qquad (9.3.53)$$

and, therefore, (9.3.52) may be written in the form

$$\|\tilde{\theta}y\| \leq \frac{c}{\alpha_0}(\|\dot{\tilde{\theta}}m_f\| + m_f) + \alpha_0 c(\|\epsilon m m_f\| + \|\dot{\tilde{\theta}}m_f\| + \Delta_\infty m_f) \qquad (9.3.54)$$

Using (9.3.54) and $\|\eta\|/m_f \leq c\Delta_\infty, \|\eta_u\|/m_f \leq c\Delta_\infty$ in (9.3.50), we obtain

$$m_f^2 \leq 1 + c\left(\frac{1}{\alpha_0^2} + \alpha_0^2 \Delta_\infty^2\right) m_f^2 + c\|\tilde{g}m_f\|^2$$

where $\tilde{g}^2 = \frac{|\dot{\tilde{\theta}}|^2}{\alpha_0^2} + \alpha_0^2|\epsilon m|^2 + \alpha_0^2|\dot{\tilde{\theta}}|^2$ and $\alpha_0 \geq 1$. For $c\left(\frac{1}{\alpha_0^2} + \alpha_0^2\Delta_\infty^2\right) < 1$, we have

$$m_f^2 \leq c + c\|\tilde{g}m_f\|^2 = c + c\int_0^t e^{-\delta(t-\tau)}\tilde{g}^2(\tau)m_f^2(\tau)d\tau$$

Applying the B-G Lemma III, we obtain

$$m_f^2 \leq ce^{-\delta t}e^{c\int_0^t \tilde{g}^2(\tau)d\tau} + c\delta\int_0^t e^{-\delta(t-s)}e^{c\int_s^t \tilde{g}^2(\tau)d\tau}ds$$

Because $\dot{\tilde{\theta}}, \epsilon m \in \mathcal{S}(\Delta_2^2)$, it follows that

$$c\int_s^t \tilde{g}^2(\tau)d\tau \leq c\Delta_2^2\left(\frac{1}{\alpha_0^2} + \alpha_0^2\right)(t-s) + c$$

Hence, for

$$c\Delta_2^2\left(\frac{1}{\alpha_0^2} + \alpha_0^2\right) < \delta \qquad (9.3.55)$$

we have

$$e^{-\delta t}e^{c\int_0^t \tilde{g}^2(\tau)d\tau} \leq e^{-\bar{\alpha}t}$$

where $\bar{\alpha} = \delta - c\Delta_2^2\left(\frac{1}{\alpha_0^2} + \alpha_0^2\right)$ which implies that m_f is bounded. The boundedness of m_f implies that all the other signals are bounded too. The constant δ in (9.3.55) may be replaced by δ_0 since no restriction on δ is imposed except that $\delta \in (0, \delta_0]$. The constant $c > 0$ may be determined by following the calculations in each of the steps and is left as an exercise for the reader.

Step 4. *Obtain a bound for the regulation error y.* The regulation error, i.e., y, is expressed in terms of signals that are guaranteed by the adaptive law to be of the order of the modeling error in m.s.s. This is achieved by using the Swapping Lemma A.1 for the error equation (9.3.45) and the equation $\epsilon m^2 = -\tilde{\theta}\phi + \eta$ to obtain (9.3.46), which as shown before implies that $y \in \mathcal{S}(\Delta_2^2)$. That is, the regulation error is of the order of the modeling error in m.s.s.

The conditions that $\Delta_m(s)$ has to satisfy for robust stability are summarized as follows:

$$c\left(\frac{1}{\alpha_0^2} + \alpha_0^2\Delta_\infty^2\right) < 1, \quad c\Delta_2^2\left(\frac{1}{\alpha_0^2} + \alpha_0^2\right) < \delta_0$$

where

$$\Delta_\infty = \left\|\frac{\Delta_m(s)}{s + a_m}\right\|_{\infty\delta_0}, \quad \Delta_2 = \left\|\frac{\Delta_m(s)}{s + a_m}\right\|_{2\delta_0}$$

The constant $\delta_0 > 0$ is such that $\Delta_m(s)$ is analytic in $\text{Re}[s] \geq -\delta_0/2$ and c denotes finite constants that can be calculated. The constant $\alpha_0 > \max\{1, \delta_0/2\}$ is arbitrary and can be chosen to satisfy the above inequalities for small Δ_2, Δ_∞.

9.3. ROBUST MRAC

Let us now simulate the above robust MRAC scheme summarized by the equations

$$u = -\theta y$$
$$\dot{\theta} = \gamma\epsilon\phi - \sigma_s\gamma\theta, \quad \epsilon = \frac{z - \theta\phi}{m^2}$$
$$\phi = \frac{1}{s + a_m}y, \quad z = y - \frac{1}{s + a_m}u$$
$$m^2 = 1 + m_s, \quad \dot{m}_s = -\delta_0 m_s + u^2 + y^2, \quad m_s(0) = 0$$

where σ_s is the switching σ, and applied to the plant

$$y = \frac{1}{s - a}(1 + \Delta_m(s))u$$

where for simulation purposes we assume that $a = 1$ and $\Delta_m(s) = -\frac{2\mu s}{1+\mu s}$ with $\mu \geq 0$. It is clear that for $\mu > 0$ the plant is nonminimum phase. Figure 9.2 shows the response of $y(t)$ for different values of μ that characterize the size of the perturbation $\Delta_m(s)$. For small μ, we have boundedness and good regulation performance. As μ increases, stability deteriorates and for $\mu = 0.35$ the plant becomes unstable. ▽

General Case

Let us now consider the SISO plant given by

$$y_p = G_0(s)(1 + \Delta_m(s))(u_p + d_u) \tag{9.3.56}$$

where

$$G_0(s) = k_p \frac{Z_p(s)}{R_p(s)} \tag{9.3.57}$$

is the transfer function of the modeled part of the plant. The high frequency gain k_p and the polynomials $Z_p(s), R_p(s)$ satisfy assumptions P1 to P4 given in Section 6.3 and the overall transfer function of the plant is strictly proper. The multiplicative uncertainty $\Delta_m(s)$ satisfies the following assumptions:

S1. $\Delta_m(s)$ is analytic in $\text{Re}[s] \geq -\delta_0/2$ for some known $\delta_0 > 0$.

S2. There exists a strictly proper transfer function $W(s)$ analytic in $\text{Re}[s] \geq -\delta_0/2$ and such that $W(s)\Delta_m(s)$ is strictly proper.

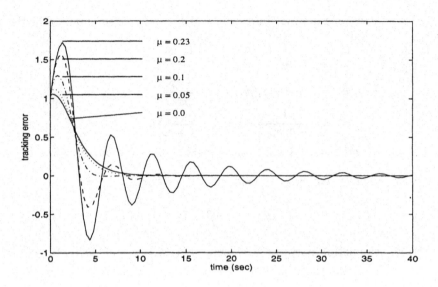

Figure 9.2 Simulation results of the MRAC scheme of Example 9.3.2 for different μ.

Assumptions S1 and S2 imply that Δ_∞, Δ_2 defined as

$$\Delta_\infty \triangleq \|W(s)\Delta_m(s)\|_{\infty\delta_0}, \quad \Delta_2 \triangleq \|W(s)\Delta_m(s)\|_{2\delta_0}$$

are finite constants. We should note that the strict properness of the overall plant transfer function and of $G_0(s)$ imply that $G_0(s)\Delta_m(s)$ is a strictly proper transfer function.

The control objective is to choose u_p and specify the bounds for Δ_∞, Δ_2 so that all signals in the closed-loop plant are bounded and the output y_p tracks, as close as possible, the output of the reference model y_m given by

$$y_m = W_m(s)r = k_m \frac{Z_m(s)}{R_m(s)} r$$

for any bounded reference signal $r(t)$. The transfer function $W_m(s)$ of the reference model satisfies assumptions M1 and M2 given in Section 6.3.

The design of the control input u_p is based on the plant model with $\Delta_m(s) \equiv 0$ and $d_u \equiv 0$. The control objective, however, has to be achieved for the plant with $\Delta_m(s) \neq 0$ and $d_u \neq 0$.

9.3. ROBUST MRAC

We start with the control law developed in Section 6.5.3 for the plant model with $\Delta_m(s) \equiv 0$, $d_u \equiv 0$, i.e.,

$$u_p = \theta^\top \omega \tag{9.3.58}$$

where $\theta = [\theta_1^\top, \theta_2^\top, \theta_3, c_0]^\top$, $\omega = [\omega_1^\top, \omega_2^\top, y_p, r]^\top$. The parameter vector θ is to be generated on-line by an adaptive law. The signal vectors ω_1, ω_2 are generated, as in Section 6.5.3, by filtering the plant input u_p and output y_p. The control law (9.3.58) will be robust with respect to the plant uncertainties $\Delta_m(s), d_u$ if we use robust adaptive laws from Chapter 8, instead of the adaptive laws used in Section 6.5.3, to update the controller parameters.

The derivation of the robust adaptive laws is achieved by first developing the appropriate parametric models for the desired controller parameter vector θ^* and then choosing the appropriate robust adaptive law by employing the results of Chapter 8 as follows:

We write the plant in the form

$$R_p y_p = k_p Z_p (1 + \Delta_m)(u_p + d_u) \tag{9.3.59}$$

and then use the matching equation

$$(\Lambda - \theta_1^{*\top}\alpha)R_p - k_p(\theta_2^{*\top}\alpha + \Lambda\theta_3^*)Z_p = Z_p \Lambda_0 R_m \tag{9.3.60}$$

where $\alpha = \alpha_{n-2}(s) = [s^{n-2}, \cdots, s, 1]^\top$, (developed in Section 6.3 and given by Equation (6.3.12)) satisfied by the desired parameter vector

$$\theta^* = [\theta_1^{*\top}, \theta_2^{*\top}, \theta_3^*, c_0^*]^\top$$

to eliminate the unknown polynomials $R_p(s)$, $Z_p(s)$ from the plant equation (9.3.59). From (9.3.59) we have

$$(\Lambda - \theta_1^{*\top}\alpha)R_p y_p = (\Lambda - \theta_1^{*\top}\alpha)k_p Z_p (1 + \Delta_m)(u_p + d_u)$$

which together with (9.3.60) imply that

$$Z_p(k_p(\theta_2^{*\top}\alpha + \Lambda\theta_3^*) + \Lambda_0 R_m)y_p = (\Lambda - \theta_1^{*\top}\alpha)k_p Z_p (1 + \Delta_m)(u_p + d_u)$$

Filtering each side with the stable filter $\frac{1}{\Lambda Z_p}$ and rearranging the terms, we obtain

$$k_p\left(\theta_2^{*\top}\frac{\alpha}{\Lambda} + \theta_3^*\right)y_p + \frac{R_m}{Z_m}y_p = k_p u_p - k_p \theta_1^{*\top}\frac{\alpha}{\Lambda}u_p$$

$$+ k_p \frac{\Lambda - \theta_1^{*\top}\alpha}{\Lambda}(\Delta_m(u_p + d_u) + d_u)$$

or
$$\left(\theta_1^{*T}\frac{\alpha}{\Lambda}u_p+\theta_2^{*T}\frac{\alpha}{\Lambda}y_p+\theta_3^*y_p-u_p\right)=-\frac{k_m}{k_p}W_m^{-1}y_p+\frac{\Lambda-\theta_1^{*T}\alpha}{\Lambda}(\Delta_m(u_p+d_u)+d_u)$$

Because $c_0^* = \frac{k_m}{k_p}$ it follows that

$$W_m\left(\theta_1^{*T}\frac{\alpha}{\Lambda}u_p+\theta_2^{*T}\frac{\alpha}{\Lambda}y_p+\theta_3^*y_p-u_p\right)=-c_0^*y_p+W_m(s)\eta_0 \qquad (9.3.61)$$

where
$$\eta_0 = \frac{\Lambda-\theta_1^{*T}\alpha}{\Lambda}(\Delta_m(u_p+d_u)+d_u)$$

is the modeling error term due to the unknown Δ_m, d_u.

As in the ideal case, (9.3.61) can be written as

$$W_m(s)u_p = \theta^{*T}\phi_p - \eta \qquad (9.3.62)$$

where

$$\theta^* = [\theta_1^{*T}, \theta_2^{*T}, \theta_3^*, c_0^*]^T, \quad \phi_p = [W_m\frac{\alpha^T}{\Lambda}u_p, W_m\frac{\alpha^T}{\Lambda}y_p, W_m y_p, y_p]^T$$

$$\eta = W_m(s)\eta_0 = \frac{\Lambda-\theta_1^{*T}\alpha}{\Lambda}W_m(s)[(\Delta_m(u_p+d_u)+d_u)]$$

Equation (9.3.62) is in the form of the linear parametric model (8.5.6) considered in Chapter 8.

Another convenient representation of (9.3.61) is obtained by adding the term $c_0^*y_m = c_0^*W_m r$ on each side of (9.3.61) to obtain

$$W_m(\theta_1^{*T}\frac{\alpha}{\Lambda}u_p+\theta_2^{*T}\frac{\alpha}{\Lambda}y_p+\theta_3^{*T}y_p+c_0^*r-u_p) = -c_0^*y_p+c_0^*y_m+W_m(s)\eta_0$$

or
$$W_m(\theta^{*T}\omega - u_p) = -c_0^*e_1 + W_m(s)\eta_0$$

leading to
$$e_1 = W_m(s)\rho^*(u_p - \theta^{*T}\omega + \eta_0) \qquad (9.3.63)$$

where $e_1 = y_p - y_m$,

$$\rho^* = \frac{1}{c_0^*}, \quad \omega = \left[\frac{\alpha^T}{\Lambda}u_p, \frac{\alpha^T}{\Lambda}y_p, y_p, r\right]^T$$

9.3. ROBUST MRAC

which is in the form of the bilinear parametric model (8.5.15) considered in Chapter 8.

Using (9.3.62) or (9.3.63), a wide class of robust MRAC schemes can be developed by simply picking up a robust adaptive law based on (9.3.62) or (9.3.63) from Chapter 8 and use it to update $\theta(t)$ in the control law (9.3.58). Table 9.1 gives a summary of MRAC schemes whose robust adaptive laws are based on (9.3.62) or (9.3.63) and are presented in Tables 9.2 to 9.4.

The robust adaptive laws of Tables 9.2 to 9.4 are based on the SPR-Lyapunov design approach and the gradient method with an instantaneous cost. Additional MRAC schemes may be constructed in exactly the same way using the least-squares method or the gradient method with an integral cost. The properties of this additional class of MRAC schemes are very similar to those presented in Tables 9.2 to 9.4 and can be established using exactly the same tools and procedure.

The following theorem summarizes the stability properties of the MRAC scheme of Table 9.1 with robust adaptive laws given in Tables 9.2 to 9.4.

Theorem 9.3.2 *Consider the MRAC schemes of Table 9.1 designed for the plant model $y_p = G_0(s)u_p$ but applied to the plant (9.3.56) with nonzero plant uncertainties $\Delta_m(s)$ and d_u. If*

$$c\left(\frac{1}{\alpha_0^2} + \alpha_0^{2k}\Delta_\infty^2\right) < 1, \quad c\left(\frac{1}{\alpha_0^2} + \alpha_0^{2k}\right)(f_0 + \Delta_i^2) \leq \frac{\delta}{2} \qquad (9.3.64)$$

where

- $\Delta_i = \Delta_{02}$ and $k = n^* + 1$ for the adaptive law of Table 9.2

- $\Delta_i = \Delta_2$ and $k = n^*$ for the adaptive laws of Tables 9.3, 9.4

- $\Delta_\infty = \|W(s)\Delta_m(s)\|_{\infty\delta_0}$

- $\Delta_{02} = \left\|\frac{\Lambda(s) - \theta_1^{*T}\alpha(s)}{\Lambda(s)}L^{-1}(s)\frac{h_0}{s+h_0}\Delta_m(s)\right\|_{2\delta_0}$

- $\Delta_2 = \left\|\frac{\Lambda(s) - \theta_1^{*T}\alpha(s)}{\Lambda(s)}W_m(s)\Delta_m(s)\right\|_{2\delta_0}$

- $\delta \in (0, \delta_0)$ is such that $G_0^{-1}(s)$ is analytic in $\text{Re}[s] \geq -\delta/2$

- $\alpha_0 > \max\{1, \delta_0/2\}$ is an arbitrary constant

- $h_0 > \delta_0/2$ is an arbitrary constant

- $c \geq 0$ denotes finite constants that can be calculated and

    (a) $f_0 = \sigma$ in the case of fixed σ-modification

    (b) $f_0 = \nu_0$ in the case of ϵ-modification

    (c) $f_0 = g_0$ in the case of dead zone modification

    (d) $f_0 = 0$ in the case of switching σ-modification and projection

Then all the signals in the closed-loop plant are bounded and the tracking error e_1 satisfies

$$\frac{1}{T}\int_t^{t+T} e_1^2 d\tau \leq c(\Delta^2 + d_0^2 + f_0) + \frac{c}{T}, \quad \forall t \geq 0$$

and for any $T > 0$, where d_0 is an upper bound for $|d_u|$ and $\Delta^2 = 1/\alpha_0^2 + \Delta_\infty^2 + \Delta_2^2 + \Delta_{02}^2$ for the scheme of Table 9.1 with the adaptive law given in Table 9.2 and $\Delta^2 = \Delta_2^2$ for the scheme of Table 9.1 with adaptive laws given in Tables 9.3, 9.4.

If, in addition, the reference signal r is dominantly rich of order $2n$ and Z_p, R_p are coprime, then the parameter error $\tilde{\theta}$ and tracking error e_1 converge to the residual set

$$\mathcal{S} = \left\{ \tilde{\theta} \in \mathcal{R}^{2n}, e_1 \in \mathcal{R} \,\middle|\, |\tilde{\theta}| + |e_1| \leq c(f_0 + \Delta + d_0) \right\}$$

where f_0, Δ are as defined above. The convergence to the residual set $\mathcal{S}$ is exponential in the case of the scheme of Table 9.1 with the adaptive law given in Table 9.4.

Outline of Proof The main tools and Lemmas as well as the steps for the proof of Theorem 9.3.2 are very similar to those in the proof of Theorem 6.5.1 for the ideal case. A summary of the main steps of the proof are given below. The details are presented in Section 9.8.

Step 1. *Express the plant input and output in terms of the parameter error term $\tilde{\theta}^\top \omega$.* In the presence of unmodeled dynamics and bounded disturbances, the closed-loop MRAC scheme can be represented by the block

9.3. ROBUST MRAC

Table 9.1 Robust MRAC Schemes

Actual plant	$y_p = k_p \frac{Z_p(s)}{R_p(s)}(1 + \Delta_m(s))(u_p + d_u)$
Plant model	$y_p = k_p \frac{Z_p(s)}{R_p(s)} u_p$
Reference model	$y_m = W_m(s)r = k_m \frac{Z_m(s)}{R_m(s)} r$
Control law	$\dot{\omega}_1 = F\omega_1 + gu_p$ $\dot{\omega}_2 = F\omega_2 + gy_p$ $u_p = \theta^\top \omega$ $\theta = [\theta_1^\top, \theta_2^\top, \theta_3, c_0]^\top, \omega = [\omega_1^\top, \omega_2^\top, y_p, r]^\top$ $\omega_i \in \mathcal{R}^{n-1}, \quad i = 1, 2$
Adaptive law	Any robust adaptive law from Tables 9.2 to 9.4
Assumptions	(i) Plant model and reference model satisfy assumptions P1 to P4, and M1 and M2 given in Section 6.3.1 for the ideal case; (ii) $\Delta_m(s)$ is analytic in $\text{Re}[s] \geq -\delta_0/2$ for some known $\delta_0 > 0$; (iii) overall plant transfer function is strictly proper
Design variables	W_m, W_m^{-1} and $\frac{1}{\Lambda(s)}$, where $\Lambda(s) = \det(sI - F)$ are designed to be analytic in $\text{Re}[s] \geq -\delta_0/2$

diagram shown in Figure 9.3. From Figure 9.3 and the matching equation (6.3.11), we can derive

$$y_p = W_m\left(r + \frac{1}{c_0^*}\tilde{\theta}^\top \omega\right) + \eta_y$$

$$u_p = G_0^{-1} W_m\left(r + \frac{1}{c_0^*}\tilde{\theta}^\top \omega\right) + \eta_u \qquad (9.3.65)$$

Table 9.2 Robust adaptive laws based on the SPR-Lyapunov approach and bilinear model (9.3.63)

Parametric model (9.3.63)	$e_1 = W_m(s)\rho^*(u_p - \theta^{*T}\omega + \eta_0)$ $\eta_0 = \frac{\Lambda - \theta_1^{*T}\alpha}{\Lambda}[\Delta_m(u_p + d_u) + d_u]$
Filtered parametric model	$e_f = W_m(s)L(s)\rho^*(u_f - \theta^{*T}\phi + \eta_f)$ $e_f = \frac{h_0}{s+h_0}e_1$ $\eta_f = L_0(s)\eta_0, u_f = L_0(s)u_p, \phi = L_0(s)\omega$ $L_0(s) = L^{-1}(s)\frac{h_0}{s+h_0}, h_0 > \delta_0/2$ $L(s)$ is chosen so that $W_m L$ is proper and SPR $L(s), L^{-1}(s)$ are analytic in $\text{Re}(s) \geq -\frac{\delta_0}{2}$
Normalized estimation error	$\epsilon = e_f - \hat{e}_f - W_m L \epsilon n_s^2, \hat{e}_f = W_m L \rho(u_f - \theta^T \phi)$ $n_s^2 = m_s, \dot{m}_s = -\delta_0 m_s + u_p^2 + y_p^2, m_s(0) = 0$
Robust adaptive laws with (a) Leakage	$\dot{\theta} = \Gamma\epsilon\phi\text{sgn}(\rho^*) - w_1\Gamma\theta, \dot{\rho} = \gamma\epsilon\xi - w_2\gamma\rho$ $\xi = u_f - \theta^T\phi$ where w_1, w_2 are as defined in Table 8.1 and $\Gamma = \Gamma^T > 0, \gamma > 0$
(b) Dead zone	$\dot{\theta} = \Gamma\phi(\epsilon + g)\text{sgn}(\rho^*), \dot{\rho} = \gamma\xi(\epsilon + g)$ where g is as defined in Table 8.7
(c) Projection	$\dot{\theta} = \Pr[\Gamma\phi\epsilon\text{sgn}(\rho^*)], \dot{\rho} = \Pr[\gamma\epsilon\xi]$ where the operator $\Pr[\cdot]$ is as defined in Table 8.6
Properties	(i) $\epsilon, \epsilon n_s, \theta, \rho \in \mathcal{L}_\infty$, (ii) $\epsilon, \epsilon n_s, \dot{\theta}, \dot{\rho} \in \mathcal{S}(f_0 + \frac{\eta_f^2}{m^2})$, where $m^2 = 1 + n_s^2$ and f_0 is a design parameter, depending on the choice of w_1, w_2 and g in (a), (b), and $f_0 = 0$ for (c)

9.3. ROBUST MRAC

Table 9.3 Robust adaptive laws based on the gradient method and bilinear model (9.3.63)

Parametric model (9.3.63)	$e_1 = W_m(s)\rho^*(u_p - \theta^{*T}\omega + \eta_0)$ $\eta_0 = \frac{\Lambda - \theta_1^{*T}\alpha}{\Lambda}[\Delta_m(u_p + d_u) + d_u]$
Parametric model rewritten	$e_1 = \rho^*(u_f - \theta^{*T}\phi + \eta)$, $u_f = W_m u_p$, $\phi = W_m \omega$ $\eta = \frac{\Lambda - \theta_1^{*T}\alpha}{\Lambda} W_m[\Delta_m(u_p + d_u) + d_u]$
Normalized estimation error	$\epsilon = \frac{e_1 - \rho\xi}{m^2}$, $\xi = u_f - \theta^T\phi$, $m^2 = 1 + n_s^2$ $n_s^2 = m_s$, $\dot{m}_s = -\delta_0 m_s + u_p^2 + y_p^2$, $m_s(0) = 0$
Robust adaptive laws	Same expressions as in Table 9.2 but with ϵ, ϕ, ξ, u_f as given above
Properties	(i) $\epsilon, \epsilon n_s, \theta, \rho, \dot\theta, \dot\rho \in \mathcal{L}_\infty$; (ii) $\epsilon, \epsilon n_s, \dot\theta, \dot\rho \in \mathcal{S}(f_0 + \frac{\eta_f^2}{m^2})$, where f_0 is as defined in Table 9.2

where

$$\eta_y = \frac{\Lambda - C_1^*}{c_0^*\Lambda} W_m[\Delta_m(u_p + d_u) + d_u], \quad \eta_u = \frac{D_1^*}{c_0^*\Lambda} W_m[\Delta_m(u_p + d_u) + d_u]$$

where $C_1^*(s) = \theta_1^{*T}\alpha(s)$, $D_1^* = \theta_2^{*T}\alpha(s) + \theta_3^*\Lambda(s)$. Using the properties of the $\mathcal{L}_{2\delta}$ norm $\|(\cdot)_t\|_{2\delta}$, which for simplicity we denote as $\|(\cdot)\|$, and the stability of W_m, G_0^{-1}, it follows that there exists $\delta \in (0, \delta_0]$ such that

$$\|y_p\| \leq c + c\|\tilde\theta^T\omega\| + \|\eta_y\|$$
$$\|u_p\| \leq c + c\|\tilde\theta^T\omega\| + \|\eta_u\|$$

The constant $\delta > 0$ is such that $G_0^{-1}(s)$ is analytic in $\text{Re}[s] \geq -\delta/2$. Therefore, the fictitious normalizing signal $m_f^2 \triangleq 1 + \|u_p\|^2 + \|y_p\|^2$ satisfies

$$m_f^2 \leq c + c\|\tilde\theta^T\omega\|^2 + c\|\eta_y\|^2 + c\|\eta_u\|^2 \qquad (9.3.66)$$

Table 9.4 Robust adaptive laws based on the linear model (9.3.62)

Parametric model	$z = \theta^{*T}\phi_p - \eta,\ \eta = \frac{\Lambda - \theta_1^{*T}\alpha}{\Lambda} W_m[\Delta_m(u_p + d_u) + d_u]$ $z = W_m u_p,\ \phi_p = [W_m\frac{\alpha^T}{\Lambda}u_p, W_m\frac{\alpha^T}{\Lambda}y_p, W_m y_p, y_p]^T$
Normalized estimation error	$\epsilon = \frac{z-\hat{z}}{m^2},\ \hat{z} = \theta^T\phi_p,\ \theta = [\theta_1^T, \theta_2^T, \theta_3, c_0]^T$ $m^2 = 1 + m_s,\ \dot{m}_s = -\delta_0 m_s + u_p^2 + y_p^2,\ m_s(0) = 0$
Constraint	$B(\theta) = \underline{c}_0 - c_0 sgn(\frac{k_p}{k_m}) \leq 0$ for some $\underline{c}_0 > 0$ satisfying $0 < \underline{c}_0 \leq |c_0^*|$
Projection operator	$\Pr[f] = \begin{cases} f & \text{if } |c_0| > \underline{c}_0 \\ & \text{or if } |c_0| = \underline{c}_0 \text{ and } f^T \nabla B \leq 0 \\ f - \frac{\Gamma \nabla B (\nabla B)^T}{(\nabla B)^T \Gamma \nabla B} f & \text{otherwise} \end{cases}$ $\nabla B = -[0,\ldots,0,1]^T sgn(k_p/k_m)$ $\Gamma = \Gamma^T > 0$
Robust adaptive law	$\dot{\theta} = \Pr[f],\ \|c_0(0)\| \geq \underline{c}_0$
(a) Leakage	$f = \Gamma\epsilon\phi_p - w\Gamma\theta,\ w$ as given in Table 8.1
(b) Dead zone	$f = \Gamma\phi_p(\epsilon + g),\ g$ as given in Table 8.4
(c) Projection	$\dot{\theta} = \begin{cases} \Pr[\Gamma\epsilon\phi_p] & \text{if } |\theta| < M_0 \\ & \text{or } |\theta| = M_0 \text{ and } (\Pr[\Gamma\epsilon\phi_p])^T\theta \leq 0 \\ (I - \frac{\Gamma\theta\theta^T}{\theta^T\Gamma\theta})\Pr[\Gamma\epsilon\phi_p] & \text{otherwise} \end{cases}$ where $M_0 \geq \|\theta^*\|, \|\theta(0)\| \leq M_0$ and $\Pr[\cdot]$ is the projection operator defined above
Properties	(i) $\epsilon, \epsilon n_s, \theta, \dot{\theta} \in \mathcal{L}_\infty$; (ii) $\epsilon, \epsilon n_s, \dot{\theta} \in \mathcal{S}(f_0 + \frac{\eta^2}{m^2})$, where f_0 is as defined in Table 9.2

9.3. ROBUST MRAC

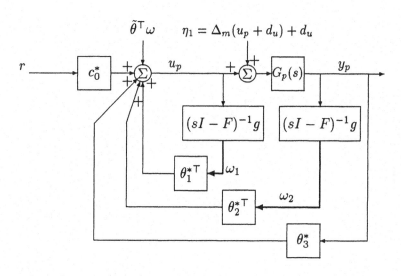

Figure 9.3 The closed-loop MRAC scheme in the presence of unmodeled dynamics and bounded disturbances.

Step 2. *Use the swapping lemmas and properties of the $\mathcal{L}_{2\delta}$ norm to bound $\|\tilde{\theta}^T\omega\|$ from above with terms that are guaranteed by the robust adaptive laws to have small in m.s.s. gains.* In this step we use the Swapping Lemma A.1 and A.2 and the properties of the $\mathcal{L}_{2\delta}$-norm to obtain the expression

$$\|\tilde{\theta}^T\omega\| \leq c\|gm_f\| + c(\frac{1}{\alpha_0} + \alpha_0^k \Delta_\infty)m_f + cd_0 \qquad (9.3.67)$$

where $\alpha_0 > \max\{1, \delta_0/2\}$ is arbitrary and $g \in \mathcal{S}(f_0 + \Delta_i^2 + \frac{d_0^2}{m^2})$ with $\Delta_i = \Delta_{02}, k = n^* + 1$ in the case of the adaptive law of Table 9.2, $\Delta_i = \Delta_2, k = n^*$ in the case of the adaptive laws of Tables 9.3 and 9.4, and d_0 is an upper bound for $|d_u|$.

Step 3. *Use the B-G Lemma to establish boundedness.* Using (9.3.67) in (9.3.66) we obtain

$$m_f^2 \leq c + c\|gm_f\|^2 + c\left(\frac{1}{\alpha_0^2} + \alpha_0^{2k}\Delta_\infty^2\right)m_f^2 + cd_0^2$$

We choose α_0 large enough so that for small Δ_∞

$$c\left(\frac{1}{\alpha_0^2} + \alpha_0^{2k}\Delta_\infty^2\right) < 1$$

We then have $m_f^2 \leq c + c\|gm_f\|^2$ for some constants $c \geq 0$, which implies that
$$m_f^2 \leq c + c\int_0^t e^{-\delta(t-\tau)}g^2(\tau)m_f^2(\tau)d\tau$$

Applying the B-G Lemma III, we obtain
$$m_f^2 \leq ce^{-\delta t}e^{c\int_0^t g^2(\tau)d\tau} + c\delta\int_0^t e^{-\delta(t-s)}e^{c\int_s^t g^2(\tau)d\tau}ds$$

Because
$$\int_s^t g^2(\tau)d\tau \leq c(f_0 + \Delta_i^2)(t-s) + c\int_s^t \frac{d_0^2}{m^2}d\tau + c$$

$\forall t \geq s \geq 0$, it follows that for $c(f_0 + \Delta_i^2) < \delta/2$ we have
$$m_f^2 \leq ce^{-\frac{\delta}{2}t}e^{c\int_0^t (d_0^2/m^2)d\tau} + c\delta\int_0^t e^{-\frac{\delta}{2}(t-s)}e^{c\int_s^t (d_0^2/m^2)d\tau}ds$$

The boundedness of m_f follows directly if we establish that $c\frac{d_0^2}{m^2(t)} < \delta/2$, $\forall t \geq 0$. This condition is satisfied if we design the signal n_s^2 as $n_s^2 = \beta_0 + m_s$ where β_0 is a constant chosen large enough to guarantee $c\frac{d_0^2}{m^2(t)} < c\frac{d_0^2}{\beta_0} < \delta/2, \forall t \geq 0$. This approach guarantees that the normalizing signal $m^2 = 1 + n_s^2$ is much larger than the level of the disturbance all the time. Such a large normalizing signal may slow down the speed of adaptation and, in fact, improve robustness. It may, however, have an adverse effect on transient performance.

The boundedness of all signals can be established, without having to modify n_s^2, by using the properties of the $\mathcal{L}_{2\delta}$ norm over an arbitrary interval of time. Considering the arbitrary interval $[t_1, t)$ for any $t_1 \geq 0$ we can establish by following a similar procedure as in Steps 1 and 2 the inequality

$$\begin{aligned} m^2(t) &\leq c(1 + m^2(t_1))e^{-\frac{\delta}{2}(t-t_1)}e^{c\int_{t_1}^t \frac{d_0^2}{m^2}d\tau} \\ &+ c\delta\int_{t_1}^t e^{-\frac{\delta}{2}(t-s)}e^{c\int_s^t \frac{d_0^2}{m(\tau)^2}d\tau}ds, \forall t \geq t_1 \geq 0 \end{aligned} \quad (9.3.68)$$

We assume that $m^2(t)$ goes unbounded. Then for any given large number $\bar{\alpha} > 0$ there exists constants $t_2 > t_1 > 0$ such that $m^2(t_1) = \bar{\alpha}$, $m^2(t_2) > f_1(\bar{\alpha})$, where $f_1(\bar{\alpha})$ is any static function satisfying $f_1(\bar{\alpha}) > \bar{\alpha}$. Using the fact that m^2 cannot grow or decay faster than an exponential, we can choose f_1

9.3. ROBUST MRAC

properly so that $m^2(t) \geq \bar{\alpha}$ $\forall t \in [t_1, t_2]$ for some $t_1 \geq \bar{\alpha}$ where $t_2 - t_1 > \bar{\alpha}$. Choosing $\bar{\alpha}$ large enough so that $d_0^2/\bar{\alpha} < \delta/2$, it follows from (9.3.68) that

$$m^2(t_2) \leq c(1+\bar{\alpha})e^{-\frac{\delta}{2}\bar{\alpha}}e^{cd_0^2} + c$$

We can now choose $\bar{\alpha}$ large enough so that $m^2(t_2) < \bar{\alpha}$ which contradicts with the hypothesis that $m^2(t_2) > \bar{\alpha}$ and therefore $m \in \mathcal{L}_\infty$. Because m bounds u_p, y_p, ω from above, we conclude that all signals are bounded.

Step 4. *Establish bounds for the tracking error e_1.* Bounds for e_1 in m.s.s. are established by relating e_1 with the signals that are guaranteed by the adaptive law to be of the order of the modeling error in m.s.s.

Step 5. *Establish convergence of estimated parameter and tracking error to residual sets.* Parameter convergence is established by expressing the parameter and tracking error equations as a linear system whose homogeneous part is e.s. and whose input is bounded.

The details of the algebra and calculations involved in Steps 1 to 5 are presented in Section 9.8.

Remark 9.3.2 *Effects of initial conditions.* The results of Theorem 9.3.2 are established using a transfer function representation for the plant. Because the transfer function is defined for zero initial conditions the results of Theorem 9.3.2 are valid provided the initial conditions of the state space plant representation are equal to zero. For nonzero initial conditions the same steps as in the proof of Theorem 9.3.2 can be followed to establish that

$$m_f^2(t) \leq c + cp_0 + cp_0 e^{-\delta t} + c\int_0^t e^{-\delta(t-\tau)}g^2(\tau)m_f^2(\tau)d\tau$$

where $p_0 \geq 0$ depends on the initial conditions. Applying the B-G Lemma III, we obtain

$$m_f^2(t) \leq (c + cp_0)e^{-\delta t}e^{c\int_0^t g^2(s)ds} + \delta(c + cp_0)\int_0^t e^{-\delta(t-\tau)}e^{c\int_\tau^t g^2(s)ds}d\tau$$

where $g \in \mathcal{S}(f_0 + \Delta_i^2 + d_0^2/m^2)$ and f_0, Δ_i are as defined in Theorem 9.3.2. Therefore the robustness bounds, obtained for zero initial conditions, will not be affected by the non-zero initial conditions. The bounds for m_f and tracking error e_1, however, will be affected by the size of the initial conditions.

Remark 9.3.3 *Robustness without dynamic normalization.* The results of Theorem 9.3.2 are based on the use of a dynamic normalizing signal $m_s = n_s^2$ so that $m = \sqrt{1 + n_s^2}$ bounds both the signal vector ϕ and modeling error term η from above. The question is whether the signal m_s is necessary for the results of Theorem 9.3.2 to hold. In [168, 240], it was shown that if m is chosen as $m^2 = 1 + \phi^\top \phi$, i.e., the same normalization used in the ideal case, then the projection modification alone is sufficient to obtain the same qualitative results as those of Theorem 9.3.2. The proof of these results is based on arguments over intervals of time, an approach that was also used in some of the original results on robustness with respect to bounded disturbances [48]. The extension of these results to modifications other than projection is not yet clear.

Remark 9.3.4 *Calculation of robustness bounds.* The calculation of the constants $c, \delta, \Delta_i, \Delta_\infty$ is tedious but possible as shown in [221, 227]. These constants depend on the properties of various transfer functions, namely their $H_{\infty\delta}, H_{2\delta}$ bounds and stability margins, and on the bounds for the estimated parameters. The bounds for the estimated parameters can be calculated from the Lyapunov-like functions which are used to analyze the adaptive laws. In the case of projection, the bounds for the estimated parameters are known a priori. Because the constants $c, \Delta_i, \Delta_\infty, \delta$ depend on unknown transfer functions and parameters such as $G_0(s), G_0^{-1}(s), \theta^*$, the conditions for robust stability are quite difficult to check for a given plant. The importance of the robustness bounds is therefore more qualitative than quantitative, and this is one of the reasons we did not explicitly specify every constant in the expression of the bounds.

Remark 9.3.5 *Existence and uniqueness of solutions.* Equations (9.3.65) together with the adaptive laws for generating $\theta = \tilde{\theta} + \theta^*$ used to establish the results of Theorem 9.3.2 are nonlinear time varying equations. The proof of Theorem 9.3.2 is based on the implicit assumption that these equations have a unique solution $\forall t \in [0, \infty)$. Without

9.3. ROBUST MRAC

this assumption, most of the stability arguments used in the proof of Theorem 9.3.2 are not valid. The problem of existence and uniqueness of solutions for a class of nonlinear equations, including those of Theorem 9.3.2 has been addressed in [191]. It is shown that the stability properties of a wide class of adaptive schemes do possess unique solutions provided the adaptive law contains no discontinuous modifications, such as switching σ and dead zone with discontinuities. An exception is the projection which makes the adaptive law discontinuous but does not affect the existence and uniqueness of solutions.

The condition for robust stability given by (9.3.64) also indicates that the design parameter f_0 has to satisfy certain bounds. In the case of switching σ and projection, $f_0 = 0$, and therefore (9.3.64) doesnot impose any restriction on the design parameters of these modifications. For modifications, such as the ϵ-modification, fixed σ, and dead zone, the design parameters have to be chosen small enough to satisfy (9.3.64). Because (9.3.64) depends on unknown constants, the design of f_0 can only be achieved by trial and error.

9.3.4 Robust Indirect MRAC

The indirect MRAC schemes developed and analyzed in Chapter 6 suffer from the same nonrobust problems the direct schemes do. Their robustification is achieved by using, as in the case of direct MRAC, the robust adaptive laws developed in Chapter 8 for on-line parameter estimation.

In the case of indirect MRAC with unnormalized adaptive laws, robustification leads to semiglobal stability in the presence of unmodeled high frequency dynamics. The analysis is the same as in the case of direct MRAC with unnormalized adaptive laws and is left as an exercise for the reader. The failure to establish global results in the case of MRAC with robust but unnormalized adaptive laws is due to the lack of an appropriate normalizing signal that could be used to bound from above the effect of dynamic uncertainties.

In the case of indirect MRAC with normalized adaptive laws, global stability is possible in the presence of a wide class of unmodeled dynamics by using robust adaptive laws with dynamic normalization as has been done in the case of direct MRAC in Section 9.3.3.

We illustrate the robustification of an indirect MRAC with normalized adaptive laws using the following example:

Example 9.3.3 Consider the MRAC problem for the following plant:

$$y_p = \frac{b}{s-a}(1 + \Delta_m(s))u_p \qquad (9.3.69)$$

where Δ_m is a proper transfer function and analytic in $\text{Re}[s] \geq -\delta_0/2$ for some known $\delta_0 > 0$, and a and b are unknown constants. The reference model is given by

$$y_m = \frac{b_m}{s+a_m}r, \qquad a_m > 0$$

If we assume $\Delta_m(s) = 0$, the following simple indirect MRAC scheme can be used to meet the control objective:

$$u_p = -k(t)y_p + l(t)r \qquad (9.3.70)$$
$$k(t) = \frac{a_m + \hat{a}}{\hat{b}}, \quad l = \frac{b_m}{\hat{b}}$$

$$\dot{\hat{a}} = \gamma_2 \epsilon \phi_2, \quad \dot{\hat{b}} = \begin{cases} \gamma_1 \epsilon \phi_1, & \text{if } |\hat{b}| > b_0 \text{ or} \\ & \text{if } |\hat{b}| = b_0 \text{ and } \epsilon\phi_1 \text{sgn}(b) \geq 0 \\ 0 & \text{otherwise} \end{cases} \qquad (9.3.71)$$

where

$$m^2 = 1 + \phi_1^2 + \phi_2^2, \quad \phi_1 = \frac{1}{s+\lambda}u_p; \quad \phi_2 = \frac{1}{s+\lambda}y_p$$

$$\epsilon = \frac{z - \hat{z}}{m^2}, \quad z = \frac{s}{s+\lambda}y_p, \quad \hat{z} = \hat{b}\phi_1 + \hat{a}\phi_2$$

$\hat{b}(0) \geq b_0$, b_0 is a known lower bound for $|b|$ and $\lambda > 0$ is a design constant. When (9.3.70) and (9.3.71) are applied to the plant (9.3.69), the ideal properties, such as stability and asymptotic tracking, can no longer be guaranteed when $\Delta_m(s) \neq 0$. As in the case of the direct MRAC, the boundedness of the signals can be lost due to the presence of unmodeled dynamics.

The indirect MRAC scheme described by (9.3.70) and (9.3.71) can be made robust by using the techniques developed in Chapter 8. For example, instead of the adaptive law (9.3.71), we use the robust adaptive law

$$\dot{\hat{a}} = \gamma_2 \epsilon \phi_2 - \sigma_s \gamma_1 \hat{a}$$
$$\dot{\hat{b}} = \begin{cases} \gamma_1 \epsilon \phi_1 - \gamma_1 \sigma_s \hat{b}, & \text{if } |\hat{b}| > b_0 \text{ or} \\ & \text{if } |\hat{b}| = b_0 \text{ and } (\epsilon\phi_1 - \sigma_s\hat{b})\text{sgn}(b) \geq 0 \\ 0 & \text{otherwise} \end{cases} \qquad (9.3.72)$$

9.3. ROBUST MRAC

where

$$\epsilon = \frac{z - \hat{z}}{m^2}, \quad \hat{z} = \hat{b}\phi_1 + \hat{a}\phi_2$$

$$\phi_1 = \frac{1}{s+\lambda}u_p, \quad \phi_2 = \frac{1}{s+\lambda}y_p, \quad z = \frac{s}{s+\lambda}y_p$$

$$m^2 = 1 + n_s^2, \quad n_s^2 = m_s, \quad \dot{m}_s = -\delta_0 m_s + u_p^2 + y_p^2, \quad m_s(0) = 0$$

The design constants λ and a_m are chosen as $\lambda > \delta_0/2, a_m > \delta_0/2$, σ_s is the switching σ-modification as defined in Chapter 8, b_0 is a lower bound satisfying $0 < b_0 \leq |b|$. The above robust adaptive law is developed using the parametric model

$$z = \theta^{*T}\phi + \eta$$

for the plant (9.3.69) where $\theta^* = [b, a]^T$,

$$\phi = \left[\frac{1}{s+\lambda}u_p, \frac{1}{s+\lambda}y_p\right]^T = [\phi_1, \phi_2]^T, \quad \eta = \frac{b\Delta_m(s)}{s+\lambda}u_p$$

As we have shown in Chapter 8, the adaptive law (9.3.72) guarantees that

(i) $\epsilon, \epsilon m, \hat{a}, \hat{b}, \dot{\hat{a}}, \dot{\hat{b}} \in \mathcal{L}_\infty$

(ii) $\epsilon, \epsilon n_s, \dot{\hat{a}}, \dot{\hat{b}} \in \mathcal{S}(\Delta_2^2)$, where $\Delta_2 = \|\frac{\Delta_m(s)}{s+\lambda}\|_{2\delta_0}$ $\qquad \triangledown$

Let us now apply the MRAC scheme given by the control law (9.3.70) and robust adaptive law (9.3.72) to the plant (9.3.69).

Theorem 9.3.3 *The closed-loop indirect MRAC scheme given by (9.3.69), (9.3.70), and (9.3.72) has the following properties: If $r, \dot{r} \in \mathcal{L}_\infty$ and the plant uncertainty $\Delta_m(s)$ satisfies the inequalities*

$$\frac{c}{\alpha_0^2} + c\Delta_\infty^2 + c\alpha_0^2\Delta_\lambda^2 < 1, \quad c\Delta_2^2 \leq \frac{\delta_0}{2}$$

where

$$\Delta_\infty = \|W_m(s)\Delta_m(s)\|_{\infty\delta_0}, \quad \Delta_2 = \left\|\frac{\Delta_m(s)}{s+\lambda}\right\|_{2\delta_0}, \Delta_\lambda = \left\|\frac{\Delta_m(s)}{s+\lambda}\right\|_{\infty\delta_0}$$

where $\alpha_0 > \delta_0$ is an arbitrary constant and $c \geq 0$ denotes any finite constant, then all signals are bounded and the tracking error e_1 satisfies

$$\int_t^{t+T} e_1^2 d\tau \leq c\Delta_2^2 + c$$

for all $t \geq 0$ and any $T > 0$.

Proof As in the direct case, the proof may be completed by using the following steps:

Step 1. *Express y_p, u_p in terms of the plant parameter error $\tilde{a} \triangleq \hat{a}-a, \tilde{b} \triangleq \hat{b}-b$.* First we write the control law as

$$u_p = -ky_p + lr = -k^* y_p + l^* r - \tilde{k} y_p + \tilde{l} r \qquad (9.3.73)$$

where $k^* = (a_m + a)/b, l^* = b_m/b$. Because $\tilde{k} = \tilde{a}/b - k\tilde{b}/b$ and $\tilde{l} = -l\tilde{b}/b$, (9.3.73) can also be written as

$$u_p = -k^* y_p + l^* r - \frac{1}{b}(\tilde{a} y_p + \tilde{b} u_p) \qquad (9.3.74)$$

by using the identity $-ky_p + lr = u_p$. Using (9.3.74) in (9.3.69), we obtain

$$y_p = W_m(s)\left(r - \frac{1}{b_m}\tilde{\theta}^T \omega_p + \frac{1}{l^*}\Delta_m(s)u_p\right) \qquad (9.3.75)$$

where $\omega_p = [u_p, y_p]^T$ and $\tilde{\theta} = [\tilde{b}, \tilde{a}]^T$.

Equation (9.3.74) may be also written in the compact form

$$u_p = -k^* y_p + l^* r - \frac{1}{b}\tilde{\theta}^T \omega_p \qquad (9.3.76)$$

Using the properties of the $\mathcal{L}_{2\delta}$ norm which for simplicity we denote by $\|(\cdot)\|$, we have

$$\|y_p\| \leq c + c\|\tilde{\theta}^T \omega_p\| + \frac{1}{|l^*|}\|W_m(s)\Delta_m(s)\|_{\infty\delta}\|u_p\|$$

$$\|u_p\| \leq c + c\|y_p\| + c\|\tilde{\theta}^T \omega_p\|$$

which imply that

$$\|u_p\|, \|y_p\| \leq c + c\|\tilde{\theta}^T \omega_p\| + c\Delta_\infty \|u_p\|$$

where $\Delta_\infty \triangleq \|W_m(s)\Delta_m(s)\|_{\infty\delta}$ for some $\delta \in (0, \delta_0]$. The fictitious normalizing signal $m_f^2 \triangleq 1 + \|u_p\|^2 + \|y_p\|^2$ satisfies

$$m_f^2 \leq c + c\|\tilde{\theta}^T \omega_p\|^2 + c\Delta_\infty^2 m_f^2 \qquad (9.3.77)$$

Step 2. *Obtain an upper bound for $\|\tilde{\theta}^T \omega_p\|$ in terms of signals that are guaranteed by the adaptive law to have small in m.s.s. gains.* We use the Swapping Lemma A.2 to express $\tilde{\theta}^T \omega_p$ as

$$\tilde{\theta}^T \omega_p = \left(1 - \frac{\alpha_0}{s + \alpha_0}\right)\tilde{\theta}^T \omega_p + \frac{\alpha_0}{s + \alpha_0}\tilde{\theta}^T \omega_p$$

$$= \frac{1}{s + \alpha_0}(\dot{\tilde{\theta}}^T \omega_p + \tilde{\theta}^T \dot{\omega}_p) + \frac{\alpha_0(s + \lambda)}{s + \alpha_0}\frac{1}{s + \lambda}\tilde{\theta}^T \omega_p \qquad (9.3.78)$$

9.3. ROBUST MRAC

where $\alpha_0 > \delta_0$ is arbitrary. From Swapping Lemma A.1, we have

$$\frac{1}{s+\lambda}\tilde{\theta}^\top \omega_p = \tilde{\theta}^\top \phi - \frac{1}{s+\lambda}\phi^\top \dot{\tilde{\theta}} \qquad (9.3.79)$$

where $\phi = \frac{1}{s+\lambda}\omega_p$. From the estimation error equation

$$\epsilon m^2 = z - \hat{z} = -\tilde{\theta}^\top \phi + \eta$$

we have

$$\tilde{\theta}^\top \phi = -\epsilon m^2 + \eta$$

and, therefore, (9.3.79) may be expressed as

$$\frac{1}{s+\lambda}\tilde{\theta}^\top \omega_p = -\epsilon m^2 + \eta - \frac{1}{s+\lambda}\phi^\top \dot{\tilde{\theta}} \qquad (9.3.80)$$

Substituting (9.3.80) in (9.3.78), we obtain

$$\tilde{\theta}^\top \omega_p = \frac{1}{s+\alpha_0}[\dot{\tilde{\theta}}^\top \omega_p + \tilde{\theta}^\top \dot{\omega}_p] + \frac{\alpha_0(s+\lambda)}{s+\alpha_0}\left(-\epsilon m^2 + \eta - \frac{1}{s+\lambda}\phi^\top \dot{\tilde{\theta}}\right) \qquad (9.3.81)$$

The signal m_f bounds from above $|\omega_p|, \|\dot{\omega}_p\|, m, \frac{1}{s+\lambda}\omega_p$. This can be shown as follows: Because the adaptive law guarantees that $\tilde{\theta} \in \mathcal{L}_\infty$, it follows from (9.3.75) that

$$|y_p(t)| \leq c + c\|\omega_p\| + c\Delta_1\|u_p\| \leq c + c\|u_p\| + c\|y_p\|$$

where $\Delta_1 = \|W_m(s)\Delta_m(s)\|_{2\delta}$, and, therefore, $y_p/m_f \in \mathcal{L}_\infty$. From $u_p = -ky_p + lr$ and $k, l \in \mathcal{L}_\infty$, it follows that $u_p/m_f \in \mathcal{L}_\infty$ and, therefore, $\omega_p/m_f \in \mathcal{L}_\infty$. Using (9.3.75) we have

$$\|\dot{y}_p\| \leq \|sW_m(s)\|_{\infty\delta}(c + c\|\omega_p\|) + c\|sW_m(s)\Delta_m(s)\|_{\infty\delta}\|u_p\|$$
$$\leq c + c\|\omega_p\| + c\|u_p\| \leq c + cm_f$$

Therefore, $\|\dot{y}_p\|/m_f \in \mathcal{L}_\infty$. Now $\dot{u}_p = -\dot{k}y_p + k\dot{y}_p + \dot{l}r + l\dot{r}$. Because $\dot{k}, \dot{l} \in \mathcal{L}_\infty$ and by assumption $r, \dot{r} \in \mathcal{L}_\infty$, it follows from $|y_p|/m_f \in \mathcal{L}_\infty$ and $\|\dot{y}_p\|/m_f \in \mathcal{L}_\infty$ that $\|\dot{u}_p\|/m_f \in \mathcal{L}_\infty$. From $\|\dot{\omega}_p\| \leq \|\dot{u}_p\| + \|\dot{y}_p\|$, it follows that $\|\dot{\omega}_p\|/m_f \in \mathcal{L}_\infty$. The boundedness of $m/m_f, \phi/m_f$ follows in a similar manner by using the properties of the $\mathcal{L}_{2\delta}$ norm.

Using the normalizing properties of m_f and the properties of the $\mathcal{L}_{2\delta}$ norm, we obtain from (9.3.81) the following inequality:

$$\|\tilde{\theta}^\top \omega_p\| \leq \frac{c}{\alpha_0}(\|\dot{\tilde{\theta}}m_f\| + m_f) + \alpha_0(\|\epsilon m m_f\| + \Delta_\lambda m_f + \|\dot{\tilde{\theta}}m_f\|) \qquad (9.3.82)$$

where for the $\|\eta\|$ term we used the inequality

$$\|\eta\| \leq \Delta_\lambda \|u_p\| \leq c\Delta_\lambda m_f$$

where $\Delta_\lambda = \|\frac{\Delta_m(s)}{s+\lambda}\|_\infty \delta$. The "smallness" of $\|\tilde{\theta}^\top \omega_p\|$ follows from $\dot{\tilde{\theta}}, \epsilon m \in \mathcal{S}(\Delta_2^2)$ and by choosing α_0 large.

Step 3. *Use the B-G Lemma to establish boundedness.* Using (9.3.82) in (9.3.77), we obtain

$$m_f^2 \leq c + c\|\tilde{g}m_f\|^2 + \frac{c}{\alpha_0^2}m_f^2 + c\alpha_0^2 \Delta_\lambda^2 m_f^2 + c\Delta_\infty^2 m_f^2$$

where $\tilde{g}^2 = (\frac{1}{\alpha_0^2} + \alpha_0^2)|\dot{\tilde{\theta}}|^2 + \alpha_0^2|\epsilon m|^2$, i.e., $\tilde{g} \in \mathcal{S}(\Delta_2^2)$. Therefore, for

$$\frac{c}{\alpha_0^2} + c\alpha_0^2 \Delta_\lambda^2 + c\Delta_\infty^2 < 1$$

we have

$$m_f^2 \leq c + c\|\tilde{g}m_f\|^2$$

or

$$m_f^2(t) \leq c + c\int_0^t e^{-\delta(t-\tau)} \tilde{g}^2(\tau) m_f^2(\tau) d\tau$$

Applying the B-G Lemma III, we obtain

$$m_f^2(t) \leq ce^{-\delta t} e^{c\int_0^t \tilde{g}^2(\tau)d\tau} + c\delta \int_0^t e^{-\delta(t-s)} e^{c\int_s^t \tilde{g}^2(\tau)d\tau} ds$$

Because $\tilde{g} \in \mathcal{S}(\Delta_2^2)$, i.e., $\int_s^t \tilde{g}^2(\tau)d\tau \leq c\Delta_2^2(t-s) + c$, it follows that for $c\Delta_2^2 < \delta$, $m_f^2 \in \mathcal{L}_\infty$ which implies that all signals are bounded.

Step 4. *Error bounds for the tracking error.* Using (9.3.75), the tracking error $e_1 = y_p - y_m$ is given by

$$e_1 = -\frac{1}{s+a_m}\tilde{\theta}^\top \omega_p + \frac{s+\lambda}{s+a_m}\eta = -\frac{s+\lambda}{s+a_m}\left[\frac{1}{s+\lambda}\tilde{\theta}^\top \omega_p - \eta\right]$$

Using (9.3.80), we have

$$e_1 = \frac{s+\lambda}{s+a_m}\left(\epsilon m^2 + \frac{1}{s+\lambda}\phi^\top \dot{\tilde{\theta}}\right) \qquad (9.3.83)$$

Because $\epsilon m, \dot{\tilde{\theta}} \in \mathcal{S}(\Delta_2^2)$ and $\phi, m \in \mathcal{L}_\infty$, it follows from (9.3.83) and Corollary 3.3.3 that $e_1 \in \mathcal{S}(\Delta_2^2)$, i.e.,

$$\int_t^{t+T} e_1^2 d\tau \leq c\Delta_2^2 T + c$$

$\forall t \geq 0$ and any $T \geq 0$. □

The extension of this example to the general case follows from the material presented in Section 9.3.3 for the direct case and that in Chapter 6 for indirect MRAC and is left as an exercise for the reader.

9.4 Performance Improvement of MRAC

In Chapter 6 we have established that under certain assumptions on the plant and reference model, we can design MRAC schemes that guarantee signal boundedness and asymptotic convergence of the tracking error to zero. These results, however, provide little information about the rate of convergence and the behavior of the tracking error during the initial stages of adaptation. Of course if the reference signal is sufficiently rich we have exponential convergence and therefore more information about the asymptotic and transient behavior of the scheme can be inferred. Because in most situations we are not able to use sufficiently rich reference inputs without violating the tracking objective, the transient and asymptotic properties of the MRAC schemes in the absence of rich signals are very crucial.

The robustness modifications, introduced in Chapter 8 and used in the previous sections for robustness improvement of MRAC, provide no guarantees of transient and asymptotic performance improvement. For example, in the absence of dominantly rich input signals, the robust MRAC schemes with normalized adaptive laws guarantee signal boundedness for any finite initial conditions, and a tracking error that is of the order of the modeling error in m.s.s. Because smallness in m.s.s. does not imply smallness pointwise in time, the possibility of having tracking errors that are much larger than the order of the modeling error over short time intervals at steady state cannot be excluded. A phenomenon known as "bursting," where the tracking error, after reaching a steady-state behavior, bursts into oscillations of large amplitude over short intervals of time, have often been observed in simulations. Bursting cannot be excluded by the m.s.s. bounds obtained in the previous sections unless the reference signal is dominantly rich and/or an adaptive law with a dead zone is employed. Bursting is one of the most annoying phenomena in adaptive control and can take place even in simulations of some of the ideal MRAC schemes of Chapter 6. The cause of bursting in this case could be the computational error which acts as a small bounded disturbance. There is a significant number of research results on bursting and other undesirable phenomena, mainly for discrete-time plants [2, 68, 81, 136, 203, 239]. We use the following example to explain one of the main mechanisms of bursting.

Example 9.4.1 (Bursting) Let us consider the following MRAC scheme:

Plant $\quad \dot{x} = ax + bu + d$

Reference Model $\quad \dot{x}_m = -x_m + r, \quad x_m(0) = 0$

Adaptive Controller

$$u = \theta^\top \omega, \quad \theta = [\theta_0, c_0]^\top, \quad \omega = [x, r]^\top$$

$$\dot{\theta} = Pr[-e_1 \omega \operatorname{sgn}(b)], \quad e_1 = x - x_m \qquad (9.4.1)$$

The projection operator in (9.4.1) constrains θ to lie inside a bounded set, where $|\theta| \leq M_0$ and $M_0 > 0$ is large enough to satisfy $|\theta^*| \leq M_0$ where $\theta^* = \left[-\frac{a+1}{b}, \frac{1}{b}\right]^\top$ is the desired controller parameter vector. The input d in the plant equation is an arbitrary unknown bounded disturbance. Let us assume that $|d(t)| \leq d_0, \forall t \geq 0$ for some $d_0 > 0$.

If we use the analysis of the previous sections, we can establish that all signals are bounded and the tracking error $e_1 \in \mathcal{S}(d_0^2)$, i.e.,

$$\int_{t_0}^{t} e_1^2 d\tau \leq d_0^2(t - t_0) + k_0, \quad \forall t \geq t_0 \geq 0 \qquad (9.4.2)$$

where k_0 depends on initial conditions. Furthermore, if r is dominantly rich, then $e_1, \tilde{\theta} = \theta - \theta^*$ converge exponentially to the residual set

$$\mathcal{S}_0 = \left\{ e_1, \tilde{\theta} \,\middle|\, |e_1| + |\tilde{\theta}| \leq cd_0 \right\}$$

Let us consider the tracking error equation

$$\dot{e}_1 = -e_1 + b\tilde{\theta}^\top \omega + d \qquad (9.4.3)$$

and choose the following disturbance

$$d = h(t)\operatorname{sat}\{-b(\bar{\theta} - \theta^*)^\top \omega\}$$

where $h(t)$ is a square wave of period 100π sec and amplitude 1,

$$\operatorname{sat}\{x\} \triangleq \begin{cases} x & \text{if } |x| < d_0 \\ d_0 & \text{if } x \geq d_0 \\ -d_0 & \text{if } x \leq -d_0 \end{cases}$$

and $\bar{\theta}$ is an arbitrary constant vector such that $|\bar{\theta}| < M_0$. It is clear that $|d(t)| < d_0$ $\forall t \geq 0$. Let us consider the case where r is sufficiently rich but not dominantly rich, i.e., $r(t)$ has at least one frequency but $|r| \ll d_0$. Consider a time interval $[t_1, t_1 + T_1]$ over which $|e_1(t)| \leq d_0$. Such an interval not only exists but is also large due to the uniform continuity of $e_1(t)$ and the inequality (9.4.2). Because $|e_1| \leq d_0$

9.4. PERFORMANCE IMPROVEMENT OF MRAC

and $0 < |r| \ll d_0$, we could have $|b(\bar{\theta} - \theta^*)^\top \omega| < d_0$ for some values of $|\bar{\theta}| < M_0$ over a large interval $[t_2, t_2 + T_2] \subset [t_1, t_1 + T]$. Therefore, for $t \in [t_2, t_2 + T_2]$ we have $d = -h(t)b(\bar{\theta} - \theta^*)^\top \omega$ and equation (9.4.3) becomes

$$\dot{e}_1 = -e_1 + b(\theta - \bar{\theta})^\top \bar{\omega} \quad \forall t \in [t_2, t_2 + T_2] \tag{9.4.4}$$

where $\bar{\omega} = h(t)\omega$. Because r is sufficiently rich, we can establish that $\bar{\omega}$ is PE and therefore θ converges exponentially towards $\bar{\theta}$. If T_2 is large, then θ will get very close to $\bar{\theta}$ as $t \to t_2 + T_2$. If we now choose $\bar{\theta}$ to be a destabilizing gain or a gain that causes a large mismatch between the closed-loop plant and the reference model, then as $\theta \to \bar{\theta}$ the tracking error will start increasing, exceeding the bound of d_0. In this case d will reach the saturation bound d_0 and equation (9.4.4) will no longer hold. Since (9.4.2) does not allow large intervals of time over which $|e_1| > d_0$, we will soon have $|e_1| \leq d_0$ and the same phenomenon will be repeated again.

The simulation results of the above scheme for $a = 1$, $b = 1$, $r = 0.1\sin 0.01t$, $d_0 = 0.5$, $M_0 = 10$ are given in Figures 9.4 and 9.5. In Figure 9.4, a stabilizing $\bar{\theta} = [-3, 4]^\top$ is used and therefore no bursting occurred. The result with $\bar{\theta} = [0, 4]^\top$, where $\bar{\theta}$ corresponds to a destabilizing controller parameter, is shown in Figure 9.5. The tracking error e_1 and parameter $\theta_1(t)$, the first element of θ, are plotted as functions of time. Note that in both cases, the controller parameter $\theta_1(t)$ converges to $\bar{\theta}_1$, i.e., to -3 (a stabilizing gain) in Figure 9.4, and to 0 (a destabilizing gain) in Figure 9.5 over the period where $d = -h(t)b(\bar{\theta} - \theta^*)^\top \omega$. The value of $\bar{\theta}_2 = 4$ is larger than $\theta_2^* = 1$ and is responsible for some of the nonzero values of e_1 at steady state shown in Figures 9.4 and 9.5. ▽

Bursting is not the only phenomenon of bad behavior of robust MRAC. Other phenomena such as chaos, bifurcation and large transient oscillations could also be present without violating the boundedness results and m.s.s. bounds developed in the previous sections [68, 81, 136, 239].

One way to eliminate most of the undesirable phenomena in MRAC is to use reference input signals that are dominantly rich. These signals guarantee a high level of excitation relative to the level of the modeling error, that in turn guarantees exponential convergence of the tracking and parameter error to residual sets whose size is of the order of the modeling error. The use of dominantly rich reference input signals is not always possible especially in the case of regulation or tracking of signals that are not rich. Therefore, by forcing the reference signal to be dominantly rich, we eliminate bursting and other undesirable phenomena at the expense of destroying the tracking properties of the scheme in the case where the desired reference signal is not

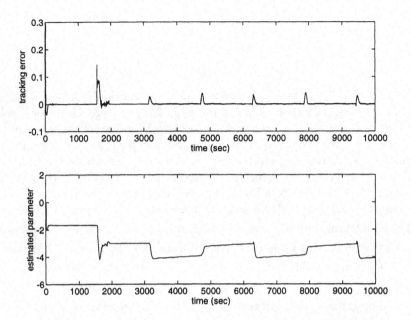

Figure 9.4 Simulation results for Example 9.4.1: No bursting because of the stabilizing $\bar{\theta} = [-3, 4]^\mathsf{T}$.

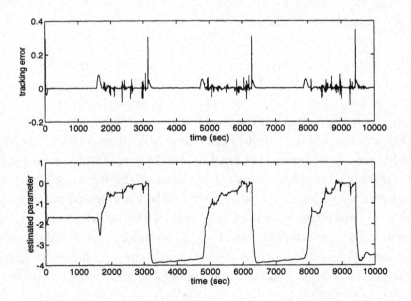

Figure 9.5 Simulation results for Example 9.4.1: Bursting because of the destabilizing $\bar{\theta} = [0, 4]^\mathsf{T}$.

9.4. PERFORMANCE IMPROVEMENT OF MRAC

rich. Another suggested method for eliminating bursting is to use adaptive laws with dead zones. Such adaptive laws guarantee convergence of the estimated parameters to constant values despite the presence of modeling error provided of course the size of the dead zone is higher than the level of the modeling error. The use of dead zones, however, does not guarantee good transient performance or zero tracking error in the absence of modeling errors.

In an effort to improve the transient and steady-state performance of MRAC, a high gain scheme was proposed in [148], whose gains are switched from one value to another based on a certain rule, that guarantees arbitrarily good transient and steady state tracking performance. The scheme does not employ any of the on-line parameter estimators developed in this book. The improvement in performance is achieved by modifying the MRC objective to one of "approximate tracking." As a result, non-zero tracking errors remained at steady state. Eventhough the robustness properties of the scheme of [148] are not analyzed, the high-gain nature of the scheme is expected to introduce significant trade-offs between stability and performance in the presence of unmodeled dynamics.

In the following sections, we propose several modified MRAC schemes that guarantee reduction of the size of bursts and an improved steady-state tracking error performance.

9.4.1 Modified MRAC with Unnormalized Adaptive Laws

The MRAC schemes of Chapter 6 and of the previous sections are designed using the certainty equivalence approach to combine a control law, that works in the case of known parameters, with an adaptive law that provides on-line parameter estimates to the controller. The design of the control law does not take into account the fact that the parameters are unknown, but blindly considers the parameter estimates provided by the adaptive laws to be the true parameters. In this section we take a slightly different approach. We modify the control law design to one that takes into account the fact that the plant parameters are not exactly known and reduces the effect of the parametric uncertainty on stability and performance as much as possible. This control law, which is robust with respect to parametric uncertainty, can then be combined with an adaptive law to enhance stability and performance. We illustrate this design methodology using the same plant and control objective as in Example 9.4.1, i.e.,

Plant $\qquad \dot{x} = ax + bu + d$
Reference Model $\qquad \dot{x}_m = -x_m + r, \quad x_m(0) = 0$

Let us choose a control law that employs no adaptation and meets the control objective of stability and tracking as close as possible even though the plant parameters a and b are unknown.

We consider the control law

$$u = \bar{\theta}_0 x + \bar{c}_0 r + u_\alpha \qquad (9.4.5)$$

where $\bar{\theta}_0$, $\bar{c}_0$ are constants that depend on some nominal known values of a, b if available, and u_α is an auxiliary input to be chosen. With the input (9.4.5), the closed-loop plant becomes

$$\dot{x} = -x + b\left(\bar{\theta}_0 + \frac{a+1}{b}\right)x + b\bar{c}_0 r + bu_\alpha + d \qquad (9.4.6)$$

and the tracking error equation is given by

$$\dot{e}_1 = -e_1 + b\tilde{\bar{\theta}}_0 e_1 + b\tilde{\bar{\theta}}_0 x_m + b\tilde{\bar{c}}_0 r + bu_\alpha + d \qquad (9.4.7)$$

where $\tilde{\bar{\theta}}_0 = \bar{\theta}_0 + \frac{a+1}{b}$, $\tilde{\bar{c}}_0 = \bar{c}_0 - \frac{1}{b}$ are the constant parameter errors. Let us now choose

$$u_\alpha = -\frac{s+1}{\tau s}\operatorname{sgn}(b) e_1 \qquad (9.4.8)$$

where $\tau > 0$ is a small design constant. The closed-loop error equation becomes

$$e_1 = \frac{\tau s}{\tau s^2 + (\tau + |b| - \tau b\tilde{\bar{\theta}}_0)s + |b|}[b\tilde{\bar{\theta}}_0 x_m + b\tilde{\bar{c}}_0 r + d] \qquad (9.4.9)$$

If we now choose τ to satisfy

$$0 < \tau < \frac{1}{|\tilde{\bar{\theta}}_0|} \qquad (9.4.10)$$

the closed-loop tracking error transfer function is stable which implies that $e_1, \frac{1}{s}e_1 \in \mathcal{L}_\infty$ and therefore all signals are bounded.

9.4. PERFORMANCE IMPROVEMENT OF MRAC

Another expression for the tracking error obtained using $x = e_1 + x_m$ and (9.4.7) is

$$e_1 = \frac{\tau s}{(s+1)(\tau s + |b|)}[b\tilde{\theta}_0 x + b\tilde{c}_0 r + d] \quad (9.4.11)$$

or for $|b| \neq \tau$ we have

$$e_1 = \frac{\tau}{|b| - \tau}\left[\frac{|b|}{\tau s + |b|} - \frac{1}{s+1}\right](w + d) \quad (9.4.12)$$

where $w \triangleq b(\tilde{\theta}_0 x + \tilde{c}_0 r)$ is due to the parametric uncertainty. Because $x \in \mathcal{L}_\infty$, for any given $\tau \in (0, 1/|\tilde{\theta}_0|)$, we can establish that there exists a constant $w_0 \geq 0$ independent of τ such that $\sup_t |w(t)| \leq w_0$. It, therefore, follows from (9.4.12) that

$$|e_1(t)| \leq \frac{\tau}{|b| - \tau}\left(\frac{|b|}{\tau}e^{-\frac{|b|}{\tau}t} - e^{-t}\right)|e_1(0)| + \frac{2\tau}{|b| - \tau}(w_0 + d_0) \quad (9.4.13)$$

where d_0 is the upper bound for $|d(t)| \leq d_0, \forall t \geq 0$. It is, therefore, clear that if we use the modified control law

$$u = \bar{\theta}_0 x + \bar{c}_0 r - \frac{s+1}{\tau s}\text{sgn}(b)e_1$$

with

$$0 < \tau < \min\left[|b|, \frac{1}{|\tilde{\theta}_0|}\right] \quad (9.4.14)$$

then the tracking error will converge exponentially fast to the residual set

$$\mathcal{S}_e = \left\{e_1 \,\Big|\, |e_1| \leq \frac{2\tau}{|b| - \tau}(w_0 + d_0)\right\} \quad (9.4.15)$$

whose size reduces to zero as $\tau \to 0$.

The significance of the above control law is that no matter how we choose the finite gains $\bar{\theta}_0, \bar{c}_0$, there always exist a range of nonzero design parameter values τ for stability. Of course the further $\bar{\theta}_0$ is away from the desired θ_0^* where $\theta_0^* = -\frac{a+1}{b}$, the smaller the set of values of τ for stability as indicated by (9.4.14). Even though the tracking performance of this modified control law can be arbitrarily improved by making τ arbitrarily small, we cannot guarantee that the tracking error converges to zero as $t \to \infty$ even when the disturbance $d = 0$. This is because the parameter error $\tilde{\theta}_0, \tilde{c}_0$

is nonzero and acts as a disturbance in the tracking error equation. The asymptotic convergence of the tracking error to zero in the case of $d = 0$ can be reestablished if, instead of using the constant gains $\bar{\theta}_0, \bar{c}_0$ in (9.4.5), we use an adaptive law to generate on-line estimates for the controller gains. Therefore, instead of combining (9.4.5) with $u_\alpha = 0$ with an adaptive law as we did in Chapter 6, we combine the modified control law

$$u = \theta_0(t)x + c_0(t)r - \frac{s+1}{\tau s}\text{sgn}(b)e_1 \qquad (9.4.16)$$

with an adaptive law that generates $\theta_0(t), c_0(t)$, the estimates of θ_0^*, c_0^* respectively. With (9.4.16) the tracking error equation may be written as

$$\dot{e}_1 = -e_1 + b\tilde{\theta}^T\omega - |b|\frac{s+1}{\tau s}e_1 + d \qquad (9.4.17)$$

where $\tilde{\theta} = \theta - \theta^*, \theta = [\theta_0, c_0]^T, \theta^* = [\theta_0^*, c_0^*]^T, \theta_0^* = -\frac{a+1}{b}, c_0^* = \frac{1}{b}, \omega = [x, r]^T$ or

$$\dot{e}_1 = -\left(1 + \frac{|b|}{\tau}\right)e_1 + b\tilde{\theta}^T\omega - \frac{|b|}{\tau}e_2 + d$$
$$\dot{e}_2 = e_1 \qquad (9.4.18)$$

We develop the adaptive laws for θ_0, c_0 by considering the following Lyapunov-like function

$$V = \frac{(e_1 + e_2)^2}{2} + l_0\frac{e_2^2}{2} + \frac{\tilde{\theta}^T\tilde{\theta}}{2}|b| \qquad (9.4.19)$$

where $l_0 > 0$ is an arbitrary constant to be selected.

The time-derivative $\dot{V}$ along the solution of (9.4.18) may be written as

$$\dot{V} = -\left(1 + \frac{|b|}{\tau}\right)e_1^2 - \frac{|b|}{\tau}e_2^2 + (e_1 + e_2)d + e_1 e_2\left(l_0 - \frac{2|b|}{\tau} - 1\right)$$
$$+ (e_1 + e_2)b\tilde{\theta}^T\omega + |b|\tilde{\theta}^T Pr[-(e_1 + e_2)\omega\text{sgn}(b)] \qquad (9.4.20)$$

by choosing the adaptive law

$$\dot{\theta} = Pr[-(e_1 + e_2)\omega\text{sgn}(b)] \qquad (9.4.21)$$

$$= \begin{cases} -(e_1 + e_2)\omega\text{sgn}(b) & \text{if } |\theta| < M_0 \text{ or} \\ & \text{if } |\theta| = M_0 \text{ and } \theta^T(e_1 + e_2)\omega\text{sgn}(b) \geq 0 \\ -\left(I - \frac{\theta\theta^T}{\theta^T\theta}\right)(e_1 + e_2)\omega\text{sgn}(b) & \text{otherwise} \end{cases}$$

9.4. PERFORMANCE IMPROVEMENT OF MRAC

where $|\theta(0)| \leq M_0$ and $M_0 \geq |\theta^*|$. Selecting $l_0 = \frac{2|b|}{\tau} + 1$, completing the squares and using the properties of projection, we can establish that

$$\dot{V} \leq -\frac{|b|}{2\tau}e_1^2 - \frac{|b|}{2\tau}e_2^2 + \frac{\tau d^2}{|b|} \qquad (9.4.22)$$

which implies that $\theta, e_1, e_2 \in \mathcal{L}_\infty$ and $e_1, e_2 \in \mathcal{S}(\tau d^2)$, i.e., the m.s.s. bound for e_1, e_2 can be made arbitrarily small by choosing an arbitrarily small τ. In addition, $x, \omega \in \mathcal{L}_\infty$. Furthermore if $d = 0$, we can establish that $e_1, e_2 \to 0$ as $t \to \infty$.

Let us now obtain an $\mathcal{L}_\infty$-bound for e_1. From (9.4.17) we have

$$e_1 = \frac{\tau s}{\tau s^2 + (\tau + |b|)s + |b|}(w + d) \qquad (9.4.23)$$

where $w = b\tilde{\theta}^\top \omega$. Because we have shown above that $w \in \mathcal{L}_\infty$ for any $\tau > 0$, we can treat w as a bounded disturbance term. As in the non-adaptive case we express e_1 as

$$e_1 = \frac{\tau}{|b| - \tau}\left[\frac{|b|}{\tau s + |b|} - \frac{1}{s+1}\right](w+d) \qquad (9.4.24)$$

for $|b| \neq \tau$, which implies that

$$|e_1(t)| \leq \frac{\tau}{|b| - \tau}\left|\frac{|b|}{\tau}e^{-\frac{|b|}{\tau}t} - e^{-t}\right||e_1(0)| + \frac{2\tau}{|b| - \tau}(w_0 + d_0) \qquad (9.4.25)$$

where w_0 and d_0 are the upper bounds for w and d respectively. It is therefore clear that the modified adaptive control scheme guarantees that the tracking error converges exponentially to the residual set

$$\mathcal{S}_e = \left\{e_1 \,\middle|\, |e_1| \leq \frac{2\tau}{|b| - \tau}(w_0 + d_0)\right\} \qquad (9.4.26)$$

whose size can be made arbitrarily small by choosing an arbitrarily small τ. The residual set is qualitatively the same as in the non-adaptive case. The difference is in the value of w_0, the bound for the parametric uncertainty term $b\tilde{\theta}^\top \omega$. One can argue that w_0 should be smaller in the adaptive case than in the non-adaptive case due to the learning capability of the adaptive scheme. Another significant difference is that in the absence of the disturbance, i.e.,

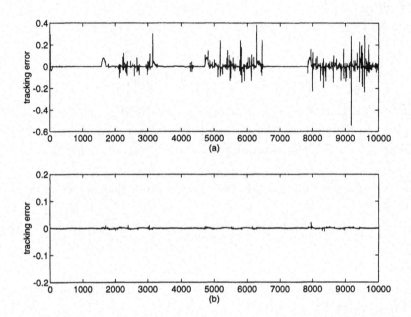

Figure 9.6 Simulation for (a) MRAC without modification; (b) modified MRAC with $\tau = 0.5$.

$d = 0$, the modified adaptive scheme guarantees that $e_1 \to 0$ as $t \to \infty$ for any $\tau > 0$.

The significance of the modified adaptive control scheme is that it guarantees m.s.s. and $\mathcal{L}_\infty$ bounds for the tracking error that can be made small by choosing a small design parameter τ. This means that by the proper choice of τ, we can significantly reduce bursting and improve the tracking error performance of the adaptive control scheme.

We demonstrate the effectiveness of the modified scheme by simulating it with the same disturbance as in Example 9.4.1 The simulation results with $\tau = 0.5$ (a moderate value for the design parameter) are shown in Figures 9.6. It can be seen that the bursting is suppressed by the additional compensation term.

The methodology used in the above example can be extended to the general case of MRAC with unnormalized adaptive laws.

9.4.2 Modified MRAC with Normalized Adaptive Laws

The method used in the previous section can be extended to MRAC with normalized adaptive laws. In this section we briefly describe two modifications that can be used to improve the performance of MRAC with normalized adaptive laws [36, 39, 211, 212].

We consider the general plant

$$y_p = k_p \frac{Z_p(s)}{R_p(s)}(u_p + d_u) \tag{9.4.27}$$

where k_p, Z_p, and R_p satisfy assumptions P1 to P4 given in Section 6.3 and d_u is a bounded input disturbance. For simplicity we assume that k_p is known. The coefficients of Z_p, R_p are completely unknown. The reference model is given by

$$y_m = W_m(s)r = k_m \frac{Z_m(s)}{R_m(s)} r \tag{9.4.28}$$

where $W_m(s)$ satisfies assumptions M1 and M2 given in Section 6.3.

We consider the control law

$$u_p = \theta_0^T(t)\omega_0 + c_0^* r + u_a \tag{9.4.29}$$

where $\theta_0 = [\theta_1^T, \theta_2^T, \theta_3^T]^T$, $\omega_0 = [\omega_1^T, \omega_2^T, y_p]^T$, $c_0^* = k_m/k_p$, $\omega_1 = \frac{\alpha(s)}{\Lambda(s)} u_p$, $\omega_2 = \frac{\alpha(s)}{\Lambda(s)} y_p$, $\alpha(s) = [s^{n-2}, \cdots, s, 1]^T$ and $\Lambda = \Lambda_0 Z_m$ is a Hurwitz polynomial of degree $n-1$.

The auxiliary input u_a is to be chosen for performance improvement. The parameter vector $\theta_0(t)$ may be generated by any one of the robust adaptive laws of Chapter 8. As an example let us use the gradient algorithm with projection based on the parametric plant model $z = \theta_0^{*T} \phi_0 - d_\eta$, developed using equation (9.3.62), where $z = W_m(s) u_p - c_0^* y_p$, $d_\eta = \frac{\Lambda - \theta_1^{*T} \alpha}{\Lambda} W_m d_u$, i.e.,

$$\begin{aligned}
\dot{\theta}_0 &= \Pr[\Gamma \epsilon \phi_0] \\
\epsilon &= \frac{z - \theta_0^T \phi_0}{m^2} \\
m^2 &= 1 + n_s^2, \quad n_s^2 = m_s \\
\dot{m}_s &= -\delta_0 m_s + u_p^2 + y_p^2, \quad m_s(0) = 0 \\
\phi_0 &= W_m(s) \omega_0
\end{aligned} \tag{9.4.30}$$

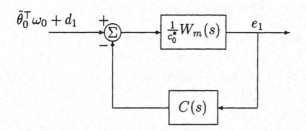

Figure 9.7 A closed-loop representation of the modified MRAC scheme.

where the projection operator $\Pr[\cdot]$ constrains θ_0 to satisfy $|\theta_0(t)| \leq M_0\ \forall t \geq 0$ for some $M_0 \geq |\theta_0^*|$, $\Gamma = \Gamma^\top > 0$ and $\delta_0 > 0$ is chosen so that $W_m(s)$ and the filters for generating ω_0 are analytic in $Re[s] \geq -\delta_0/2$.

With (9.4.29) the tracking error equation may be written as

$$e_1 = \frac{1}{c_0^*} W_m(s)[\tilde{\theta}_0^\top \omega_0 + u_a + d_1] \tag{9.4.31}$$

where $d_1 = \frac{\Lambda - \theta^{*\top}\alpha}{\Lambda} d_u$, $\tilde{\theta}_0 = \theta_0 - \theta_0^*$ by following the same procedure as the one used to develop equation (9.3.65). The idea now is to choose u_a so that all signals in the closed-loop plant remain bounded as in the case of $u_a = 0$ and the effect of $\tilde{\theta}_0^\top \omega_0 + d_1$ on e_1 is reduced as much as possible. We present two different methods that lead to two different choices of u_a achieving similar results.

Method 1. This method is developed in [36, 212] and is described as follows:

We choose u_a as

$$u_a = -C(s)e_1 \tag{9.4.32}$$

where $C(s)$ is a proper transfer function to be designed. With (9.4.32) the tracking error system may be represented by the Figure 9.7.

The set of all stabilizing compensators $C(s)$ for this system is given by

$$C(s) = \frac{Q(s)}{1 - \frac{1}{c_0^*} W_m(s)Q(s)} \tag{9.4.33}$$

where $Q(s)$ ranges over the set of all stable rational transfer functions [36].

9.4. PERFORMANCE IMPROVEMENT OF MRAC

With this choice of $C(s)$, the tracking error equation becomes

$$e_1 = \left[1 - \frac{1}{c_0^*}W_m(s)Q(s)\right]\frac{1}{c_0^*}W_m(s)[\tilde{\theta}_0^\top \omega_0 + d_1]$$

which implies that

$$\|e_{1t}\|_\infty \leq \|h_m\|_1 (\|(\tilde{\theta}_0^\top \omega_0)_t\|_\infty + \|d_{1t}\|_\infty)$$

where h_m is the impulse response of the system with transfer function $\frac{1}{c_0^*}W_m(s)(1 - \frac{1}{c_0^*}W_m(s)Q(s))$. The problem now reduces to choosing a proper stable rational function $Q(s)$ to minimize the $\mathcal{L}_1$ norm of h_m, i.e., to make $\|h_m\|_1$ as small as possible.

Lemma 9.4.1 *The transfer function*

$$Q(s) = \frac{c_0^* W_m^{-1}(s)}{(\tau s + 1)^{n^*}} \qquad (9.4.34)$$

where n^* is the relative degree of the plant and $\tau > 0$ is a design constant guarantees that

$$\|h_m\|_1 \to 0 \quad as \quad \tau \to 0$$

Proof We have

$$\left[1 - \frac{1}{c_0^*}W_m(s)Q(s)\right]\frac{W_m(s)}{c_0^*} = \frac{(\tau s + 1)^{n^*} - 1}{(\tau s + 1)^{n^*}} \frac{W_m(s)}{c_0^*}$$

$$= \tau s \frac{W_m(s)}{c_0^*}\left[\frac{1}{\tau s + 1} + \frac{1}{(\tau s + 1)^2} + \cdots + \frac{1}{(\tau s + 1)^{n^*}}\right]$$

Hence,

$$\|h_m\|_1 \leq \tau \|h_0\|_1 [\|h_1\|_1 + \|h_2\|_1 + \cdots \|h_{n^*}\|_1]$$

where $h_0(t)$ is the impulse response of $sW_m(s)/c_0^*$ and $h_i(t)$ the impulse response of $1/(\tau s + 1)^i, i = 1, 2, \ldots, n^*$. It is easy to verify that $\|h_0\|_1, \|h_i\|_1$ are bounded from above by a constant $c > 0$ independent of τ. Therefore $\|h_m\|_1 \leq \tau c$ and the proof is complete. □

Using (9.4.32), (9.4.33) and (9.4.34), the control law (9.4.29) becomes

$$u_p = \theta_0^\top(t)\omega_0 + c_0^* r - C(s)e_1 \qquad (9.4.35)$$

$$C(s) = \frac{c_0^* W_m^{-1}(s)}{(\tau s + 1)^{n^*} - 1}$$

which together with the adaptive law (9.4.30) and $c_0^* = k_m/k_p$ describe the modified MRAC scheme.

The control law (9.4.35) poses no implementation problems because $C(s)$ is proper, and e_1 is measured.

Theorem 9.4.1 *The MRAC scheme described by (9.4.30) and (9.4.35) with $\tau \in (0, 1/\delta_0)$ guarantees the following properties when applied to the plant (9.4.27):*

(i) *All signals are bounded.*

(ii) *The tracking error satisfies*

$$\lim_{t \to \infty} \sup_{\tau_0 \geq t} |e_1(\tau_0)| \leq \tau(c + d_0)$$

where d_0 is an upper bound for the disturbance d_u and $c \geq 0$ is a constant independent of τ.

(iii) *When $d_u = 0$, $e_1(t) \to 0$ as $t \to \infty$.*

The proof of Theorem 9.4.1 follows from the proofs of the standard robust MRAC schemes and is left as an exercise for the reader.

Theorem 9.4.1 indicates that the steady state value of e_1 can be made arbitrarily small by choosing a small design parameter τ. Small τ implies the presence of a high gain equal to $1/\tau$ in the control law (9.4.35). Such a high gain is expected to have adverse effects on robust stability demonstrating the classical trade-off between robust stability and nominal performance that is present in every robust control scheme in the nonadaptive case. In the presence of unmodeled dynamics it is expected that τ has to meet a lower bound for preservation of robust stability.

Method 2. This method is developed and analyzed in [39, 211] and is described as follows:

We consider the tracking error equation (9.4.31)

$$e_1 = \frac{1}{c_0^*} W_m(s)(\tilde{\theta}_0^T \omega_0 + u_a + d_1)$$

9.4. PERFORMANCE IMPROVEMENT OF MRAC

Using Swapping Lemma A.1 we have

$$W_m(s)\tilde{\theta}_0^\top \omega_0 = \tilde{\theta}_0^\top \phi_0 + W_c(s)(W_b(s)\omega_0^\top)\dot{\tilde{\theta}}_0$$

where the elements of W_c, W_b are strictly proper transfer functions with the same poles as $W_m(s)$. From the expression of the normalized estimation error in (9.4.30), we have

$$\epsilon m^2 = -\tilde{\theta}_0^\top \phi_0 - d_\eta$$

where $d_\eta = W_m(s)d_1$. Therefore, $\epsilon m^2 = -\tilde{\theta}_0^\top \phi_0 - W_m(s)d_1$ leading to the tracking error equation

$$e_1 = \frac{1}{c_0^*}\left[-\epsilon m^2 + W_c(W_b \omega_0^\top)\dot{\tilde{\theta}}_0 + W_m u_a\right]$$

Because $\epsilon, W_c, W_b, \omega_0, \dot{\tilde{\theta}}_0$ are known, the input u_a is to be chosen to reduce the effect of $\tilde{\theta}_0, \dot{\tilde{\theta}}_0$ on e_1.

Let us choose

$$u_a = -Q(s)[-\epsilon m^2 + W_c(s)(W_b(s)\omega_0^\top)\dot{\tilde{\theta}}_0] = -Q(s)W_m(s)(\tilde{\theta}_0^\top \omega_0 + d_1) \quad (9.4.36)$$

where $Q(s)$ is a proper stable transfer function to be designed. With this choice of u_a, we have

$$e_1 = \frac{1}{c_0^*}[(1 - W_m(s)Q(s))W_m(s)(\tilde{\theta}_0^\top \omega_0 + d_1)]$$

which implies that

$$\|e_{1t}\|_\infty \leq \|h_m\|_1(\|(\tilde{\theta}_0^\top \omega_0)_t\|_\infty + \|d_{1t}\|_\infty)$$

where $h_m(t)$ is the impulse response of $(1 - W_m(s)Q(s))W_m(s)$.

As in Method 1 if we choose $Q(s)$ as

$$Q(s) = \frac{W_m^{-1}(s)}{(\tau s + 1)^{n^*}} \quad (9.4.37)$$

we can establish that $\|h_m\|_1 \to 0$ as $\tau \to 0$.

With (9.4.37) the modified control law (9.4.29), (9.4.36) becomes

$$u_p = \theta_0^\top \omega_0 + c_0^* r - \frac{W_m^{-1}(s)}{(\tau s + 1)^{n^*}}(-\epsilon m^2 + W_c(s)(W_b(s)\omega_0^\top)\dot{\tilde{\theta}}_0) \qquad (9.4.38)$$

where W_c, W_b can be calculated from the Swapping Lemma A.1 and $\dot{\tilde{\theta}}_0 = \dot{\theta}_0$ is available from the adaptive law.

Theorem 9.4.2 *The modified MRAC scheme described by (9.4.30), (9.4.38) with $\tau \in (0, 1/\delta_0)$ guarantees the following properties when applied to the plant (9.4.27):*

(i) *All signals are bounded.*

(ii) *The tracking error e_1 satisfies*

$$\limsup_{t \to \infty} \sup_{\tau_0 \geq t} |e_1(\tau_0)| \leq \tau(c + d_0)$$

where d_0 is an upper bound for the disturbance d_u and $c \geq 0$ is a finite constant independent of τ.

(iii) *When $d_u = 0$ we have $|e_1(t)| \to 0$ as $t \to \infty$.*

The proof of Theorem 9.4.2 follows from the proofs of the robust MRAC schemes presented in Section 9.3 and is given in [39].

In [39], the robustness properties of the modified MRAC scheme are analyzed by applying it to the plant

$$y_p = k_p \frac{Z(s)}{R(s)}(1 + \mu \Delta_m(s))u_p$$

where $\mu \Delta_m(s)$ is a multiplicative perturbation and $\mu > 0$. It is established that for $\tau \in (\tau_{min}, \frac{1}{\delta_0})$ where $0 < \tau_{min} < \frac{1}{\delta_0}$, there exists a $\mu^*(\tau_{min}) > 0$ such that all signals are bounded and

$$\limsup_{t \to \infty} \sup_{\tau_0 \geq t} |e_1(\tau_0)| \leq \tau c + \mu c$$

where $c \geq 0$ is a constant independent of τ, μ. The function $\mu^*(\tau_{min})$ is such that as $\tau_{min} \to 0, \mu^*(\tau_{min}) \to 0$ demonstrating that for a given size of unmodeled dynamics characterized by the value of μ^*, τ cannot be made arbitrarily small.

Remark 9.4.1 The modified MRAC schemes proposed above are based on the assumption that the high frequency gain k_p is known. The case of unknown k_p is not as straightforward. It is analyzed in [37] under the assumption that a lower and an upper bound for k_p is known a priori.

Remark 9.4.2 The performance of MRAC that includes transient as well as steady-state behavior is a challenging problem especially in the presence of modeling errors. The effect of the various design parameters, such as adaptive gains and filters, on the performance and robustness of MRAC is not easy to quantify and is unknown in general. Tuning of some of the design parameters for improved performance is found to be essential even in computer simulations, let alone real-time implementations, especially for high order plants. For additional results on the performance of MRAC, the reader is referred to [37, 119, 148, 182, 211, 227, 241].

9.5 Robust APPC Schemes

In Chapter 7, we designed and analyzed a wide class of APPC schemes for a plant that is assumed to be finite dimensional, LTI, free of any noise and external disturbances and whose transfer function satisfies assumptions P1 to P3.

In this section, we consider APPC schemes that are designed for a simplified plant model but are applied to a higher-order plant with unmodeled dynamics and bounded disturbances. In particular, we consider the higher order plant model which we refer to it as the actual plant

$$y_p = G_0(s)[1 + \Delta_m(s)](u_p + d_u) \qquad (9.5.1)$$

where $G_0(s)$ satisfies P1 to P3 given in Chapter 7, $\Delta_m(s)$ is an unknown multiplicative uncertainty, d_u is a bounded input disturbance and the overall plant transfer function $G(s) = G_0(s)(1+\Delta_m(s))$ is strictly proper. We design APPC schemes for the lower-order plant model

$$y_p = G_0(s)u_p \qquad (9.5.2)$$

but apply and analyze them for the higher order plant model (9.5.1). The effect of perturbation Δ_m and disturbance d_u on the stability and performance of the APPC schemes is investigated in the following sections.

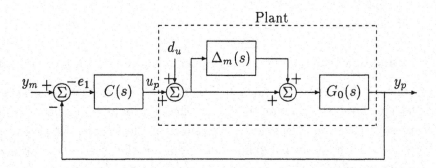

Figure 9.8 Closed-loop PPC schemes with unmodeled dynamics and bounded disturbances.

We first consider the nonadaptive case where $G_0(s)$ is known exactly.

9.5.1 PPC: Known Parameters

Let us consider the control laws of Section 7.3.3 that are designed for the simplified plant model (9.5.2) with known plant parameters and apply them to the higher order plant (9.5.1). The block diagram of the closed-loop plant is shown in Figure 9.8 where $C(s)$ is the transfer function of the controller. The expression for $C(s)$ for each of the PPC laws developed in Chapter 7 is given as follows.

For the control law in (7.3.6) and (7.3.11) of Section 7.3.2 which is based on the polynomial approach, i.e.,

$$Q_m L u_p = P(y_m - y_p), \quad L Q_m R_p + P Z_p = A^* \tag{9.5.3}$$

where $Q_m(s)$ is the internal model of the reference signal y_m, i.e., $Q_m(s) y_m = 0$, we have

$$C(s) = \frac{P(s)}{Q_m(s) L(s)} \tag{9.5.4}$$

The control law (7.3.19), (7.3.20) of Section 7.3.3 based on the state-variable approach, i.e.,

$$\begin{aligned} \dot{\hat{e}} &= A\hat{e} + B\bar{u}_p - K_o(C^\top \hat{e} - e_1) \\ \bar{u}_p &= -K_c \hat{e}, \quad u_p = \frac{Q_1}{Q_m} \bar{u}_p \end{aligned} \tag{9.5.5}$$

9.5. ROBUST APPC SCHEMES

where K_c satisfies (7.3.21) and K_o satisfies (7.3.22), can be put in the form of Figure 9.8 as follows:

We have
$$\hat{e}(s) = (sI - A + K_oC^\mathsf{T})^{-1}(B\bar{u}_p + K_oe_1)$$

and
$$\bar{u}_p = -K_c(sI - A + K_oC^\mathsf{T})^{-1}(B\bar{u}_p + K_oe_1)$$

i.e.,
$$\bar{u}_p = -\frac{K_c(sI - A + K_oC^\mathsf{T})^{-1}K_o}{1 + K_c(sI - A + K_oC^\mathsf{T})^{-1}B}e_1$$

and, therefore,
$$C(s) = \frac{K_c(sI - A + K_oC^\mathsf{T})^{-1}K_o}{(1 + K_c(sI - A + K_oC^\mathsf{T})^{-1}B)}\frac{Q_1(s)}{Q_m(s)} \qquad (9.5.6)$$

For the LQ control of Section 7.3.4, the same control law (9.5.5) is used, but K_c is calculated by solving the algebraic equation

$$A^\mathsf{T}P + PA - PB\lambda^{-1}B^\mathsf{T}P + CC^\mathsf{T} = O, \quad K_c = \lambda^{-1}B^\mathsf{T}P \qquad (9.5.7)$$

Therefore, the expression for the transfer function $C(s)$ is exactly the same as (9.5.6) except that (9.5.7) should be used to calculate K_c.

We express the closed-loop PPC plant into the general feedback form discussed in Section 3.6 to obtain

$$\begin{bmatrix} u_p \\ y_p \end{bmatrix} = \begin{bmatrix} \frac{C}{1+CG} & \frac{-CG}{1+CG} \\ \frac{CG}{1+CG} & \frac{G}{1+CG} \end{bmatrix} \begin{bmatrix} y_m \\ d_u \end{bmatrix} \qquad (9.5.8)$$

where $G = G_0(1 + \Delta_m)$ is the overall transfer function and C is different for different pole placement schemes. The stability properties of (9.5.8) are given by the following theorem:

Theorem 9.5.1 *The closed-loop plant described by (9.5.8) is internally stable provided*

$$\|T_0(s)\Delta_m(s)\|_\infty < 1$$

where $T_0(s) = \frac{CG_0}{1+CG_0}$ is the complementary sensitivity function of the nominal plant. Furthermore, the tracking error e_1 converges exponentially to the residual set

$$\mathcal{D}_e = \{e_1 \,|\, |e_1| \leq cd_0\} \qquad (9.5.9)$$

where d_0 is an upper bound for $|d_u|$ and $c > 0$ is a constant.

Proof The proof of the first part of Theorem 9.5.1 follows immediately from equation (9.5.8) and the small gain theorem by expressing the characteristic equation of (9.5.8) as

$$1 + \frac{CG_0}{1+CG_0}\Delta_m = 0$$

To establish (9.5.9), we use (9.5.8) to write

$$e_1 = \left[\frac{CG}{1+CG} - 1\right]y_m + \frac{G}{1+CG}d_u = -\frac{1}{1+CG}y_m + \frac{G}{1+CG}d_u$$

It follows from (9.5.4), (9.5.6) that the controller $C(s)$ is of the form $C(s) = \frac{C_0(s)}{Q_m(s)}$ for some $C_0(s)$. Therefore

$$e_1 = -\frac{Q_m}{Q_m + C_0 G}y_m + \frac{GQ_m}{Q_m + C_0 G}d_u$$

where $G = G_0(1 + \Delta_m)$. Since $Q_m y_m = 0$ and the closed-loop plant is internally stable due to $\|T_0(s)\Delta_m(s)\|_\infty < 1$, we have

$$e_1 = \frac{(1+\Delta_m)G_0 Q_m}{Q_m + C_0 G_0 + \Delta_m C_0 G_0}d_u + \epsilon_t \qquad (9.5.10)$$

where ϵ_t is an exponentially decaying to zero term. Therefore, (9.5.9) is implied by (9.5.10) and the internal stability of the closed-loop plant. $\square$

It should be pointed out that the tracking error at steady state is not affected by y_m despite the presence of the unmodeled dynamics. That is, if $d_u \equiv 0$ and $\Delta_m \neq 0$, we still have $e_1(t) \to 0$ as $t \to \infty$ provided the closed-loop plant is internally stable. This is due to the incorporation of the internal model of y_m in the control law. If $Q_m(s)$ contains the internal model of d_u as a factor, i.e., $Q_m(s) = Q_d(s)\bar{Q}_m(s)$ where $Q_d(s)d_u = 0$ and $\bar{Q}_m(s)y_m = 0$, then it follows from (9.5.10) that $e_1 = \epsilon_t$, i.e., the tracking error converges to zero exponentially despite the presence of the input disturbance. The internal model of d_u can be constructed if we know the frequencies of d_u. For example, if d_u is a slowly varying signal of unknown magnitude we could choose $Q_d(s) = s$.

The robustness and performance properties of the PPC schemes given by Theorem 9.5.1 are based on the assumption that the parameters of the modeled part of the plant, i.e., the coefficients of $G_0(s)$ are known exactly. When

9.5. ROBUST APPC SCHEMES

the coefficients of $G_0(s)$ are unknown, the PPC laws (9.5.3) and (9.5.5) are combined with adaptive laws that provide on-line estimates for the unknown parameters leading to a wide class of APPC schemes. The design of these APPC schemes so that their robustness and performance properties are as close as possible to those described by Theorem 9.5.1 for the known parameter case is a challenging problem in robust adaptive control and is treated in the following sections.

9.5.2 Robust Adaptive Laws for APPC Schemes

The adaptive control schemes of Chapter 7 can meet the control objective for the ideal plant (9.5.2) but not for the actual plant (9.5.1) where $\Delta_m(s) \neq 0, d_u \neq 0$. The presence of Δ_m and/or d_u may easily lead to various types of instability. As in the case of MRAC, instabilities can be counteracted and robustness properties improved if instead of the adaptive laws used in Chapter 7, we use robust adaptive laws to update or estimate the unknown parameters.

The robust adaptive laws to be combined with PPC laws developed for the known parameter case are generated by first expressing the unknown parameters of the modeled part of the plant in the form of the parametric models considered in Chapter 8 and then applying the results of Chapter 8 directly.

We start by writing the plant equation (9.5.1) as

$$R_p y_p = Z_p (1 + \Delta_m)(u_p + d_u) \quad (9.5.11)$$

where $Z_p = \theta_b^{*\top} \alpha_{n-1}(s)$, $R_p = s^n + \theta_a^{*\top} \alpha_{n-1}(s)$; θ_b^*, θ_a^* are the coefficient vectors of Z_p, R_p respectively and $\alpha_{n-1}(s) = [s^{n-1}, s^{n-2}, \ldots, s, 1]^\top$. Filtering each side of (9.5.11) with $\frac{1}{\Lambda_p(s)}$, where $\Lambda_p(s) = s^n + \lambda_1 s^{n-1} + \cdots + \lambda_n$ is Hurwitz, we obtain

$$z = \theta_p^{*\top} \phi + \eta \quad (9.5.12)$$

where

$$z = \frac{s^n}{\Lambda_p(s)} y_p, \quad \theta_p^* = [\theta_b^{*\top}, \theta_a^{*\top}]^\top$$

$$\phi = \left[\frac{\alpha_{n-1}^\top(s)}{\Lambda_p} u_p, -\frac{\alpha_{n-1}^\top(s)}{\Lambda_p} y_p \right]^\top, \quad \eta = \frac{Z_p}{\Lambda_p}[\Delta_m u_p + (1 + \Delta_m) d_u]$$

Equation (9.5.12) is in the form of the linear parametric model considered in Chapter 8, and therefore it can be used to generate a wide class of robust adaptive laws of the gradient and least-squares type. As we have shown in Chapter 8, one of the main ingredients of a robust adaptive law is the normalizing signal m that needs to be chosen so that $\frac{|\phi|}{m}, \frac{\eta}{m} \in \mathcal{L}_\infty$. We apply Lemma 3.3.2 and write

$$|\eta(t)| \leq \left\|\frac{Z_p(s)}{\Lambda_p(s)}\Delta_m(s)\right\|_{2\delta} \|(u_p)_t\|_{2\delta} + \left\|\frac{Z_p(s)(1+\Delta_m(s))}{\Lambda_p(s)}\right\|_{2\delta} \frac{d_0}{\sqrt{\delta}} + \epsilon_t \quad (9.5.13)$$

for some $\delta > 0$, where ϵ_t is an exponentially decaying to zero term and $d_0 = \sup_t |d_u(t)|$. Similarly,

$$|\phi(t)| \leq \sum_{i=1}^{n} \left\|\frac{s^{n-i}}{\Lambda_p(s)}\right\|_{2\delta} (\|(u_p)_t\|_{2\delta} + \|(y_p)_t\|_{2\delta}) \quad (9.5.14)$$

The above $H_{2\delta}$ norms exist provided $1/\Lambda_p(s)$ and $\Delta_m(s)$ are analytic in $\text{Re}[s] \geq -\delta/2$ and $\frac{Z_p \Delta_m}{\Lambda_p}$ is strictly proper. Because the overall plant transfer function $G(s)$ and $G_0(s)$ are assumed to be strictly proper, it follows that $G_0 \Delta_m$ and therefore $\frac{Z_p \Delta_m}{\Lambda_p}$ are strictly proper. Let us now assume that $\Delta_m(s)$ is analytic in $\text{Re}[s] \geq -\delta_0/2$ for some known $\delta_0 > 0$. If we design $\Lambda_p(s)$ to have roots in the region $\text{Re}[s] < -\delta_0/2$ then it follows from (9.5.13), (9.5.14) by setting $\delta = \delta_0$ that the normalizing signal m given by

$$m^2 = 1 + \|u_{pt}\|_{2\delta_0}^2 + \|y_{pt}\|_{2\delta_0}^2$$

bounds η, ϕ from above. The signal m may be generated from the equations

$$\begin{aligned} m^2 &= 1 + n_s^2, \quad n_s^2 = m_s \\ \dot{m}_s &= -\delta_0 m_s + u_p^2 + y_p^2, \quad m_s(0) = 0 \end{aligned} \quad (9.5.15)$$

Using (9.5.15) and the parametric model (9.5.12), a wide class of robust adaptive laws may be generated by employing the results of Chapter 8 or by simply using Tables 8.2 to 8.4.

As an example let us consider a robust adaptive law based on the gradient algorithm and switching σ-modification to generate on-line estimates of θ^* in (9.5.12). We have from Table 8.2 that

$$\dot{\theta}_p = \Gamma \epsilon \phi - \sigma_s \Gamma \theta_p$$

9.5. ROBUST APPC SCHEMES

$$\epsilon = \frac{z - \theta_p^T \phi}{m^2}, \quad m^2 = 1 + n_s^2, \quad n_s^2 = m_s$$

$$z = \frac{s^n}{\Lambda_p(s)} y_p \qquad (9.5.16)$$

$$\dot{m}_s = -\delta_0 m_s + u_p^2 + y_p^2, \quad m_s(0) = 0$$

$$\sigma_s = \begin{cases} 0 & \text{if } |\theta_p| \leq M_0 \\ \sigma_0(\frac{|\theta_p|}{M_0} - 1) & \text{if } M_0 < |\theta_p| \leq 2M_0 \\ \sigma_0 & \text{if } |\theta_p| > M_0 \end{cases}$$

where θ_p is the estimate of θ_p^*, $M_0 > |\theta_p^*|$, $\sigma_0 > 0$ and $\Gamma = \Gamma^T > 0$. As established in Chapter 8, the above adaptive law guarantees that (i) $\epsilon, \epsilon n_s, \theta_p, \dot{\theta}_p \in \mathcal{L}_\infty$, (ii) $\epsilon, \epsilon n_s, \dot{\theta}_p \in \mathcal{S}(\eta^2/m^2)$ independent of the boundedness of ϕ, z, m.

The adaptive law (9.5.16) or any other robust adaptive law based on parametric model (9.5.12) can be combined with the PPC laws (9.5.3), (9.5.5) and (9.5.7) to generate a wide class of robust APPC schemes as demonstrated in the following sections.

9.5.3 Robust APPC: Polynomial Approach

Let us combine the PPC law (9.5.3) with a robust adaptive law by replacing the unknown polynomials R_p, Z_p of the modeled part of the plant with their on-line estimates $\hat{R}_p(s,t), \hat{Z}_p(s,t)$ generated by the adaptive law. The resulting robust APPC scheme is summarized in Table 9.5.

We like to examine the properties of the APPC schemes designed for (9.5.2) but applied to the actual plant (9.5.1) with $\Delta_m(s) \neq 0$ and $d_u \neq 0$.

As in the MRAC case, we first start with a simple example and then generalize it to the plant (9.5.1).

Example 9.5.1 Let us consider the plant

$$y_p = \frac{b}{s + a}(1 + \Delta_m(s))u_p \qquad (9.5.17)$$

where a, b are unknown constants and $\Delta_m(s)$ is a multiplicative plant uncertainty. The input u_p has to be chosen so that the poles of the closed-loop modeled part of the plant (i.e., with $\Delta_m(s) \equiv 0$) are placed at the roots of $A^*(s) = (s+1)^2$ and y_p tracks the constant reference signal $y_m = 1$ as close as possible. The same control problem has been considered and solved in Example 7.4.1 for the case where $\Delta_m(s) \equiv 0$.

Table 9.5 Robust APPC scheme: polynomial approach

Actual plant	$y_p = \frac{Z_p}{R_p}(1+\Delta_m)(u_p + d_u)$
Plant model	$y_p = \frac{Z_p(s)}{R_p(s)} u_p, Z_p(s) = \theta_b^{*\top}\alpha(s), R_p(s) = s^n + \theta_a^{*\top}\alpha(s)$ $\theta_p^* = [\theta_b^{*\top}, \theta_a^{*\top}]^\top, \alpha(s) = \alpha_{n-1}(s)$
Reference signal	$Q_m(s) y_m = 0$
Assumptions	(i) Modeled part of the plant $y_p = \frac{Z_p}{R_p} u_p$ and $Q_m(s)$ satisfy assumptions P1 to P3 given in Section 7.3.1; (ii) $\Delta_m(s)$ is analytic in $\mathrm{Re}[s] \geq -\delta_0/2$ for some known $\delta_0 > 0$
Robust adaptive law	Use any robust adaptive law from Tables 8.2 to 8.4 based on the parametric model $z = \theta_p^{*\top}\phi + \eta$ with $z = \frac{s^n}{\Lambda_p(s)} y_p, \phi = [\frac{\alpha^\top(s)}{\Lambda_p(s)} u_p, -\frac{\alpha^\top(s)}{\Lambda_p(s)} y_p]^\top$ $\eta = \frac{Z_p}{\Lambda_p}[\Delta_m(u_p + d_u) + d_u]$
Calculation of controller parameters	Solve for $\hat{L}(s,t) = s^{n-1} + l^\top(t)\alpha_{n-2}(s)$ $\hat{P}(s,t) = p^\top(t)\alpha_{n+q-1}(s)$ from the equation $\hat{L}(s,t) \cdot Q_m(s) \cdot \hat{R}_p(s,t) + \hat{P}(s,t) \cdot \hat{Z}_p(s,t) = A^*(s)$, where $\hat{Z}_p(s,t) = \theta_b^\top(t)\alpha(s), \hat{R}_p(s,t) = s^n + \theta_a^\top(t)\alpha(s)$
Control law	$u_p = (\Lambda - \hat{L} Q_m)\frac{1}{\Lambda} u_p - \hat{P}\frac{1}{\Lambda}(y_p - y_m)$
Design variables	$\Lambda(s) = \Lambda_p(s)\Lambda_q(s)$, Λ_p, Λ_q are monic and Hurwitz of degree n and $q-1$ respectively and with roots in $\mathrm{Re}[s] < -\delta_0/2$; $A^*(s)$ is monic Hurwitz of degree $2n+q-1$ with roots in $\mathrm{Re}[s] < -\delta_0/2$; $Q_m(s)$ is monic of degree q with nonrepeated roots on the $j\omega$-axis

9.5. ROBUST APPC SCHEMES

We design each block of the robust APPC for the above plant as follows:

Robust Adaptive Law The parametric model for the plant is

$$z = \theta_p^{*\top}\phi + \eta$$

where

$$z = \frac{s}{s+\lambda}y_p, \quad \theta_p^* = [b, a]^\top$$

$$\phi = \frac{1}{s+\lambda}[u_p, -y_p]^\top, \quad \eta = \frac{b}{s+\lambda}\Delta_m(s)u_p$$

We assume that $\Delta_m(s)$ is analytic in $\text{Re}[s] \geq -\delta_0/2$ for some known $\delta_0 > 0$ and design $\lambda > \delta_0/2$. Using the results of Chapter 8, we develop the following robust adaptive law:

$$\dot{\theta}_p = \Pr\{\Gamma(\epsilon\phi - \sigma_s\theta_p)\}, \quad \Gamma = \Gamma^\top > 0 \quad (9.5.18)$$

$$\epsilon = \frac{z - \theta_p^\top\phi}{m^2}, \quad m^2 = 1 + n_s^2, \quad n_s^2 = m_s$$

$$\dot{m}_s = -\delta_0 m_s + |u_p|^2 + |y_p|^2, \quad m_s(0) = 0$$

$$\sigma_s = \begin{cases} 0 & \text{if } |\theta_p| < M_0 \\ \left(\frac{|\theta_p|}{M_0} - 1\right)\sigma_0 & \text{if } M_0 \leq |\theta_p| < 2M_0 \\ \sigma_0 & \text{if } |\theta_p| \geq 2M_0 \end{cases}$$

where $\theta_p = [\hat{b}, \hat{a}]^\top$; $\Pr\{\cdot\}$ is the projection operator which keeps the estimate $|\hat{b}(t)| \geq b_0 > 0 \ \forall t \geq 0$ as defined in Example 7.4.1 where b_0 is a known lower bound for $|b|$; and $M_0 > |\theta^*|, \sigma_0 > 0$ are design constants.

Control Law The control law is given by

$$u_p = \frac{\lambda}{s+\lambda}u_p - (\hat{p}_1 s + \hat{p}_0)\frac{1}{s+\lambda}(y_p - y_m) \quad (9.5.19)$$

where $\hat{p}_1, \hat{p}_0$ satisfy

$$s \cdot (s + \hat{a}) + (\hat{p}_1 s + \hat{p}_0) \cdot \hat{b} = (s+1)^2$$

leading to

$$\hat{p}_1 = \frac{2 - \hat{a}}{\hat{b}}, \quad \hat{p}_0 = \frac{1}{\hat{b}}$$

As shown in Chapter 8, the robust adaptive law (9.5.18) guarantees that

(i) $\epsilon, \epsilon m, \theta \in \mathcal{L}_\infty$, (ii) $\epsilon, \epsilon m, \dot{\theta} \in \mathcal{S}(\eta^2/m^2)$

Because of the projection that guarantees $|\hat{b}(t)| \geq b_0 > 0 \ \forall t \geq 0$, we also have that $\hat{p}_0, \hat{p}_1 \in \mathcal{L}_\infty$ and $\dot{\hat{p}}_0, \dot{\hat{p}}_1 \in \mathcal{S}(\eta^2/m^2)$. Because

$$\frac{|\eta|}{m} \leq \left\| \frac{b}{s+\lambda} \Delta_m(s) \right\|_{2\delta_0} \triangleq \Delta_2 \qquad (9.5.20)$$

we have $\dot{\hat{p}}_0, \dot{\hat{p}}_1, \epsilon, \epsilon m, \dot{\theta} \in \mathcal{S}(\Delta_2^2)$.

We use the properties of the adaptive law to analyze the stability properties of the robust APPC scheme (9.5.18) and (9.5.19) when applied to the plant (9.5.17) with $\Delta_m(s) \neq 0$. The steps involved in the analysis are the same as in the ideal case presented in Chapter 7 and are elaborated below:

Step 1. *Express u_p, y_p in terms of the estimation error.* As in Example 7.4.1, we define the states

$$\phi_1 = \frac{1}{s+\lambda} u_p, \quad \phi_2 = -\frac{1}{s+\lambda} y_p, \quad \phi_m = \frac{1}{s+\lambda} y_m$$

From the adaptive law, we have

$$\epsilon m^2 = \frac{s}{s+\lambda} y_p - \theta_p^\top \frac{1}{s+\lambda} [u_p, -y_p]^\top = -\dot{\phi}_2 - \theta_p^\top [\phi_1, \phi_2]^\top$$

i.e.,

$$\dot{\phi}_2 = -\hat{b}\phi_1 - \hat{a}\phi_2 - \epsilon m^2 \qquad (9.5.21)$$

From the control law, we have

$$u_p = \lambda\phi_1 + \hat{p}_1\dot{\phi}_2 + \hat{p}_0\phi_2 + \hat{p}_0\phi_m + \hat{p}_1\dot{\phi}_m$$

Because $u_p = \dot{\phi}_1 + \lambda\phi_1$, it follows from above and the equation (9.5.21) that

$$\dot{\phi}_1 = -\hat{p}_1\hat{b}\phi_1 - (\hat{p}_1\hat{a} - \hat{p}_0)\phi_2 - \hat{p}_1\epsilon m^2 + \bar{y}_m \qquad (9.5.22)$$

where $\bar{y}_m \triangleq \hat{p}_1\dot{\phi}_m + \hat{p}_0\phi_m \in \mathcal{L}_\infty$ due to $\hat{p}_0, \hat{p}_1 \in \mathcal{L}_\infty$. Combining (9.5.21), (9.5.22), we obtain exactly the same equation (7.4.15) as in the ideal case, i.e.,

$$\begin{aligned} \dot{x} &= A(t)x + b_1(t)\epsilon m^2 + b_2\bar{y}_m \\ u_p &= \dot{x}_1 + \lambda x_1, \quad y_p = -\dot{x}_2 - \lambda x_2 \end{aligned} \qquad (9.5.23)$$

where $x = [x_1, x_2]^\top \triangleq [\phi_1, \phi_2]^\top$,

$$A(t) = \begin{bmatrix} -\hat{p}_1\hat{b} & -(\hat{p}_1\hat{a} - \hat{p}_0) \\ -\hat{b} & -\hat{a} \end{bmatrix}, \quad b_1(t) = \begin{bmatrix} -\hat{p}_1 \\ -1 \end{bmatrix}, \quad b_2 = \begin{bmatrix} 1 \\ 0 \end{bmatrix}$$

Note that in deriving (9.5.23), we only used equation $\epsilon m^2 = z - \theta_p^\top \phi$ and the control law (9.5.19). In both equations, η or $\Delta_m(s)$ do not appear explicitly.

9.5. ROBUST APPC SCHEMES

Equation (9.5.23) is exactly the same as (7.4.15) and for each fixed t, we have $det(sI - A(t)) = (s + 1)^2$. However, in contrast to the ideal case where $\epsilon m, \|\dot{A}(t)\| \in \mathcal{L}_2 \cap \mathcal{L}_\infty$, here we can only establish that $\epsilon m, \|\dot{A}(t)\| \in \mathcal{S}(\Delta_2^2) \cap \mathcal{L}_\infty$, which follows from the fact that $\epsilon m, \dot{\theta} \in \mathcal{S}(\Delta_2^2) \cap \mathcal{L}_\infty, \theta \in \mathcal{L}_\infty$ and $|\hat{b}(t)| \geq b_0$ guaranteed by the adaptive law.

Step 2. *Show that the homogeneous part of (9.5.23) is e.s.* For each fixed t, $det(sI - A(t)) = (s + \hat{a})s + \hat{b}(\hat{p}_1 s + \hat{p}_0) = (s + 1)^2$, i.e., $\lambda(A(t)) = -1, \forall t \geq 0$. From $\dot{\hat{a}}, \dot{\hat{b}}, \dot{\hat{p}}_1, \dot{\hat{p}}_0 \in \mathcal{S}(\Delta_2^2) \cap \mathcal{L}_\infty$, we have $\|\dot{A}(t)\| \in \mathcal{S}(\Delta_2^2) \cap \mathcal{L}_\infty$. Applying Theorem 3.4.11(b), it follows that for $\Delta_2 < \Delta^*$ for some $\Delta^* > 0$, the matrix $A(t)$ is u.a.s. which implies that the transition matrix $\Phi(t, \tau)$ of the homogeneous part of (9.5.23) satisfies

$$\|\Phi(t,\tau)\| \leq k_1 e^{-k_2(t-\tau)}, \quad \forall t \geq \tau \geq 0$$

for some constants $k_1, k_2 > 0$.

Step 3. *Use the B-G lemma to establish signal boundedness.* Proceeding the same way as in the ideal case and applying Lemma 3.3.3 to (9.5.23), we obtain

$$|x(t)|, \|x\| \leq c\|\epsilon m^2\| + c$$

where $\|(\cdot)\|$ denotes the $\mathcal{L}_{2\delta}$ norm $\|(\cdot)_t\|_{2\delta}$ for any $0 < \delta < \min[\delta_0, 2k_2]$ and $c \geq 0$ denotes any finite constant. From (9.5.23), we also have

$$\|u_p\|, \|y_p\| \leq c\|\epsilon m^2\| + c$$

Therefore, the fictitious normalizing signal $m_f^2 \triangleq 1 + \|u_p\|^2 + \|y_p\|^2$ satisfies

$$m_f^2 \leq c\|\epsilon m^2\|^2 + c \quad .$$

The signal m_f bounds m, x, u_p, y_p from above. This property of m_f is established by using Lemma 3.3.2 to first show that $\phi_1/m_f, \phi_2/m_f$ and therefore $x/m_f \in \mathcal{L}_\infty$. From $\delta < \delta_0$ we also have that $m/m_f \in \mathcal{L}_\infty$. Because the elements of $A(t)$ are bounded (guaranteed by the adaptive law and the coprimeness of the estimated polynomials) and $\epsilon m \in \mathcal{L}_\infty$, it follows from (9.5.23) that $\dot{x}/m_f \in \mathcal{L}_\infty$ and therefore $u_p/m_f, y_p/m_f \in \mathcal{L}_\infty$. The signals $\phi_1, \phi_2, x, u_p, y_p$ can also be shown to be bounded from above by m due to $\lambda > \delta_0/2$. We can therefore write

$$m_f^2 \leq c\|\epsilon m m_f\|^2 + c = c + c\int_0^t e^{-\delta(t-\tau)} \tilde{g}^2(\tau) m_f^2(\tau) d\tau$$

where $\tilde{g} = \epsilon m \in \mathcal{S}(\Delta_2^2)$. Applying B-G Lemma III, we obtain

$$m_f^2 \leq c e^{-\delta t} e^{c\int_0^t \tilde{g}^2(\tau)d\tau} + c\delta \int_0^t e^{-\delta(t-s)} e^{c\int_s^t \tilde{g}^2(\tau)d\tau} ds$$

Because $c \int_s^t \tilde{g}^2(\tau)d\tau \le c\Delta_2^2(t-s) + c$, it follows that for $c\Delta_2^2 < \delta$, we have $m_f \in \mathcal{L}_\infty$. From $m_f \in \mathcal{L}_\infty$, we have $x, u_p, y_p, m \in \mathcal{L}_\infty$ and therefore all signals are bounded.

The condition on Δ_2 for signal boundedness is summarized as follows:

$$c\Delta_2 < \min[\sqrt{\delta}, c\Delta^*], \quad 0 < \delta < \min[\delta_0, 2k_2]$$

where as indicated before $\Delta^* > 0$ is the bound for Δ_2 for $A(t)$ to be u.a.s.

Step 4. *Establish tracking error bounds.* As in the ideal case considered in Example 7.4.1 (Step 4), we can establish by following exactly the same procedure that

$$e_1 = \frac{s(s+\lambda)}{(s+1)^2}\epsilon m^2 - \frac{s+\lambda}{(s+1)^2}(\dot{\tilde{a}}\frac{1}{s+\lambda}y_p - \dot{\tilde{b}}\frac{1}{s+\lambda}u_p) \tag{9.5.24}$$

This equation is exactly the same as in the ideal case except that $\epsilon m, \dot{\tilde{a}}, \dot{\tilde{b}} \in \mathcal{S}(\Delta_2^2)$ instead of being in $\mathcal{L}_2$. Because $y_p, u_p, m \in \mathcal{L}_\infty$, it follows that

$$\epsilon m^2, \quad (\dot{\tilde{a}}\frac{1}{s+\lambda}y_p - \dot{\tilde{b}}\frac{1}{s+\lambda}u_p) \in \mathcal{S}(\Delta_2^2)$$

Therefore by writing $\frac{s(s+\lambda)}{(s+1)^2} = 1 + \frac{(\lambda-2)s-1}{(s+1)^2}$, using $\epsilon m^2 \in \mathcal{S}(\Delta_2^2)$ and applying Corollary 3.3.3 to (9.5.24), we obtain

$$\int_0^t e_1^2 d\tau \le c\Delta_2^2 t + c \tag{9.5.25}$$

which implies that the mean value of e_1^2 is of the order of the modeling error characterized by Δ_2.

Let us now simulate the APPC scheme (9.5.18), (9.5.19) applied to the plant given by (9.5.17). For simulation purposes, we use $a = -1, b = 1$ and $\Delta_m(s) = \frac{-2\mu s}{1+\mu s}$. The plant output response y_p versus t is shown in Figure 9.9 for different values of μ that indicate the size of $\Delta_m(s)$. As μ increases from 0, the response of y_p deteriorates and for $\mu = 0.28$, the closed loop becomes unstable. $\triangledown$

Remark 9.5.1 The calculation of the maximum size of unmodeled dynamics characterized by Δ_2 for robust stability is tedious and involves several conservative steps. The most complicated step is the calculation of Δ^* using the proof of Theorem 3.4.11 and the rate of decay of the state transition matrix of $A(t)$. In addition, these calculations involve the knowledge or bounds of the unknown parameters. The robustness results obtained are therefore more qualitative than quantitative.

9.5. ROBUST APPC SCHEMES

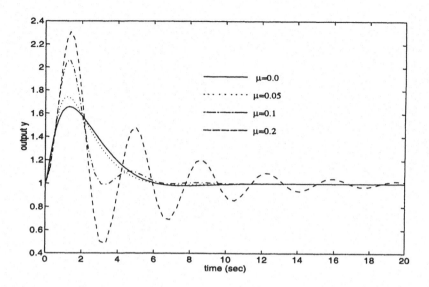

Figure 9.9 Plant output response for the APPC scheme of Example 9.5.1 for different values of μ.

Remark 9.5.2 In the above example the use of projection guarantees that $|b(t)| \geq b_0 > 0 \ \forall t \geq 0$ where b_0 is a known lower bound for $|b|$ and therefore stabilizability of the estimated plant is assured. As we showed in Chapter 7, the problem of stabilizability becomes more difficult to handle in the higher order case since no procedure is yet available for the development of convex parameter sets where stabilizability is guaranteed.

Let us now extend the results of Example 9.5.1 to higher order plants. We consider the APPC scheme of Table 9.5 that is designed based on the plant model (9.5.2) and applied to the actual plant (9.5.1).

Theorem 9.5.2 *Assume that the estimated polynomials $\hat{R}_p(s,t), \hat{Z}_p(s,t)$ of the plant model are such that $\hat{R}_p Q_m, \hat{Z}_p$ are strongly coprime at each time t. There exists a $\delta^* > 0$ such that if*

$$c(f_0 + \Delta_2^2) < \delta^*, \quad \text{where } \Delta_2 \triangleq \left\| \frac{Z_p(s)}{\Lambda_p(s)} \Delta_m(s) \right\|_{2\delta_0}$$

then the APPC schemes of Table 9.5 guarantee that all signals are bounded and the tracking error e_1 satisfies

$$\int_0^t e_1^2 d\tau \leq c(\Delta_2^2 + d_0^2 + f_0)t + c, \quad \forall t \geq 0$$

where $f_0 = 0$ in the case of switching-σ and projection and $f_0 > 0$ in the case of fixed-σ ($f_0 = \sigma$), dead zone ($f_0 = g_0$) and ϵ-modification ($f_0 = \nu_0$) and d_0 is an upper bound for $|d_u|$.

The proof of Theorem 9.5.2 for $d_u = 0$ follows directly from the analysis of Example 9.5.1 and the proof for the ideal case in Chapter 7. When $d_u \neq 0$ the proof involves a contradiction argument similar to that in the MRAC case. The details of the proof are presented in Section 9.9.1.

Remark 9.5.3 As discussed in Chapter 7, the assumption that the estimated time varying polynomials $\hat{R}_p Q_m, \hat{Z}_p$ are strongly coprime cannot be guaranteed by the adaptive law. The modifications discussed in Chapter 7 could be used to relax this assumption without changing the qualitative nature of the results of Theorem 9.5.2.

9.5.4 Robust APPC: State Feedback Law

A robust APPC scheme based on a state feedback law can be formed by combining the PPC law (9.5.5) with a robust adaptive law as shown in Table 9.6. The design of the APPC scheme is based on the plant model (9.5.2) but is applied to the actual plant (9.5.1). We first demonstrate the design and analysis of the robust APPC scheme using the following example.

Example 9.5.2 Let us consider the same plant as in Example 9.5.1,

$$y_p = \frac{b}{s+a}(1 + \Delta_m(s))u_p \qquad (9.5.26)$$

which for control design purposes is modeled as

$$y_p = \frac{b}{s+a} u_p$$

where a, b are unknown and $\Delta_m(s)$ is a multiplicative plant uncertainty that is analytic in $\text{Re}[s] \geq -\delta_0/2$ for some known $\delta_0 > 0$. The control objective is to

9.5. ROBUST APPC SCHEMES

Table 9.6 Robust APPC scheme: state feedback law

Actual plant	$y_p = \frac{Z_p}{R_p}(1 + \Delta_m)(u_p + d_u)$
Plant model	$y_p = \frac{Z_p}{R_p} u_p$
Reference signal	$Q_m(s) y_m = 0$
Assumptions	Same as in Table 9.5
Robust adaptive law	Same as in Table 9.5; it generates the estimates $\hat{Z}_p(s,t), \hat{R}_p(s,t)$
State observer	$\dot{\hat{e}} = \hat{A}\hat{e} + \hat{B}\bar{u}_p - \hat{K}_o(t)(C^T \hat{e} - e_1), \quad \hat{e} \in \mathcal{R}^{n+q}$ $\hat{A} = \begin{bmatrix} & I_{n+q-1} \\ -\theta_1(t) & ---- \\ & 0 \end{bmatrix}, \hat{B} = \theta_2(t), C^T = [1, 0, \ldots, 0]$ $\hat{R}_p Q_m = s^{n+q} + \theta_1^T(t)\alpha_{n+q-1}(s)$ $\hat{Z}_p Q_1 = \theta_2^T(t)\alpha_{n+q-1}(s)$ $\hat{K}_o = a^* - \theta_1, A_o^*(s) = s^{n+q} + a^{*T}\alpha_{n+q-1}(s)$ $e_1 = y_p - y_m$
Calculation of controller parameters	Solve $\hat{K}_c(t)$ from $\det(sI - \hat{A} + \hat{B}\hat{K}_c) = A_c^*(s)$ at each time t
Control law	$\bar{u}_p = -\hat{K}_c(t)\hat{e}, \quad u_p = \frac{Q_1}{Q_m}\bar{u}_p$
Design variables	$Q_1(s), A_o^*(s), A_c^*(s)$ monic of degree $q, n+q, n+q$, respectively, with roots in $\text{Re}[s] < -\delta_0/2$; $A_c^*(s)$ has $Q_1(s)$ as a factor; $Q_m(s)$ as in Table 9.5.

stabilize the actual plant and force y_p to track the reference signal $y_m = 1$ as close as possible. As in Example 9.5.1, the parametric model for the actual plant is given by

$$z = \theta_p^{*T}\phi + \eta \qquad (9.5.27)$$

where $z = \frac{s}{s+\lambda}y_p, \phi = \frac{1}{s+\lambda}[u_p, -y_p]^T, \theta_p^* = [b, a]^T, \eta = \frac{b}{s+\lambda}\Delta_m u$ and $\lambda > 0$ is chosen to satisfy $\lambda > \delta_0/2$. A robust adaptive law based on (9.5.27) is

$$\dot{\theta}_p = \Pr[\Gamma(\epsilon\phi - \sigma_s\theta_p)] \qquad (9.5.28)$$

where $\theta_p = [\hat{b}, \hat{a}]^T$; $\Pr(\cdot)$ is the projection operator that guarantees that $|\hat{b}(t)| \geq b_0 > 0 \ \forall t \geq 0$ where b_0 is a known lower bound for $|b|$. The other signals used in (9.5.28) are defined as

$$\epsilon = \frac{z - \theta_p^T\phi}{m^2}, \quad m^2 = 1 + n_s^2, \quad n_s^2 = m_s$$
$$\dot{m}_s = -\delta_0 m_s + u_p^2 + y_p^2, \quad m_s(0) = 0 \qquad (9.5.29)$$

and σ_s is the switching σ-modification. Because $y_m = 1$ we have $Q_m(s) = s$ and select $Q_1(s) = s + 1$, i.e., we assume that $\delta_0 < 2$. Choosing $\hat{K}_o = [10, 25]^T - [\hat{a}, 0]^T$, the state-observer is given by

$$\dot{\hat{e}} = \begin{bmatrix} -\hat{a} & 1 \\ 0 & 0 \end{bmatrix}\hat{e} + \begin{bmatrix} 1 \\ 1 \end{bmatrix}\hat{b}\bar{u}_p - \begin{bmatrix} -\hat{a} + 10 \\ 25 \end{bmatrix}([1, 0]\hat{e} - e_1) \qquad (9.5.30)$$

where the poles of the observer, i.e., the eigenvalues of $\hat{A} - \hat{K}_0 C$, are chosen as the roots of $A_o(s) = s^2 + 10s + 25 = (s + 5)^2$. The closed-loop poles chosen as the roots of $A_c^*(s) = (s + 1)^2$ are used to calculate the controller parameter gain $\hat{K}_c(t) = [\hat{k}_1, \hat{k}_2]$ using

$$\det(sI - \hat{A} + \hat{B}\hat{K}_c) = (s + 1)^2$$

which gives

$$\hat{k}_1 = \frac{1 - \hat{a}}{\hat{b}}, \quad \hat{k}_2 = \frac{1}{\hat{b}}$$

The control law is given by

$$\bar{u}_p = -[\hat{k}_1, \hat{k}_2]\hat{e}, \quad u_p = \frac{s+1}{s}\bar{u}_p \qquad (9.5.31)$$

The closed-loop plant is described by equations (9.5.26) to (9.5.31) and analyzed by using the following steps.

Step 1. *Develop the state error equations for the closed-loop plant.* We start with the plant equation

$$(s + a)y_p = b(1 + \Delta_m(s))u_p$$

9.5. ROBUST APPC SCHEMES

which we rewrite as
$$(s+a)sy_p = b(1+\Delta_m(s))su_p$$

Using $sy_p = se_1$ and filtering each side with $\frac{1}{Q_1(s)} = \frac{1}{s+1}$, we obtain

$$(s+a)\frac{s}{s+1}e_1 = b(1+\Delta_m)\bar{u}_p, \quad \bar{u}_p = \frac{s}{s+1}u_p$$

which implies that
$$e_1 = \frac{b(s+1)}{s(s+a)}(1+\Delta_m)\bar{u}_p \quad (9.5.32)$$

We consider the following state representation of (9.5.32)

$$\dot{e} = \begin{bmatrix} -a & 1 \\ 0 & 0 \end{bmatrix} e + \begin{bmatrix} 1 \\ 1 \end{bmatrix} b\bar{u}_p + \begin{bmatrix} \lambda - a + 1 \\ \lambda \end{bmatrix} \eta$$

$$e_1 = [1,0]e + \eta, \quad \eta = \frac{b}{s+\lambda}\Delta_m(s)u_p \quad (9.5.33)$$

Let $e_o = e - \hat{e}$ be the observation error. It follows from (9.5.30), (9.5.33) that e_o satisfies

$$\dot{e}_o = \begin{bmatrix} -10 & 1 \\ -25 & 0 \end{bmatrix} e_o + \begin{bmatrix} 1 \\ 0 \end{bmatrix} \tilde{a}e_1 - \begin{bmatrix} 1 \\ 1 \end{bmatrix} \tilde{b}\bar{u}_p + \begin{bmatrix} \lambda - 9 \\ \lambda - 25 \end{bmatrix} \eta \quad (9.5.34)$$

The plant output is related to $\hat{e}, e_o, e_1$ as follows:

$$y_p = C^T e_o + C^T \hat{e} + \eta + y_m, \quad e_1 = y_p - y_m \quad (9.5.35)$$

where $C^T = [1,0]$. A relationship between u_p and $\hat{e}, e_o, \eta$ that involves stable transfer functions is developed by using the identity

$$\frac{s(s+a)}{(s+1)^2} + \frac{(2-a)s+1}{(s+1)^2} = 1 \quad (9.5.36)$$

developed in Section 7.4.3 (see (7.4.37)) under the assumption that $b, s(s+a)$ are coprime, i.e., $b \neq 0$. From (9.5.36), we have

$$u_p = \frac{s(s+a)}{(s+1)^2}u_p + \frac{(2-a)s+1}{(s+1)^2}u_p$$

Using $su_p = (s+1)\bar{u}_p = -(s+1)\hat{K}_c\hat{e}$ and $u_p = \frac{s+a}{b}y_p - \Delta_m u_p$ in the above equation, we obtain

$$u_p = -\frac{s+a}{s+1}[\hat{k}_1, \hat{k}_2]\hat{e} + \frac{[(2-a)s+1](s+a)}{b(s+1)^2}y_p - \frac{[(2-a)s+1](s+\lambda)}{b(s+1)^2}\eta \quad (9.5.37)$$

Substituting for $\bar{u}_p = -\hat{K}_c(t)\hat{e}$ in (9.5.30) and using $C^\top \hat{e} - e_1 = -C^\top e_o - \eta$, $C^\top = [1, 0]$, we obtain

$$\dot{\hat{e}} = A_c(t)\hat{e} + \begin{bmatrix} -\hat{a} + 10 \\ 25 \end{bmatrix} (C^\top e_o + \eta) \qquad (9.5.38)$$

where

$$A_c(t) = \begin{bmatrix} -\hat{a} & 1 \\ 0 & 0 \end{bmatrix} - \hat{b} \begin{bmatrix} 1 \\ 1 \end{bmatrix} [\hat{k}_1, \hat{k}_2] = \begin{bmatrix} -1 & 0 \\ -(1-\hat{a}) & -1 \end{bmatrix} \qquad (9.5.39)$$

Equations (9.5.34) to (9.5.39) are the error equations to be analyzed in the steps to follow.

Step 2. *Establish the e.s. of the homogeneous parts of (9.5.34) and (9.5.38).* The homogeneous part of (9.5.34), i.e.,

$$\dot{Y}_0 = AY_0, \quad A = \begin{bmatrix} -10 & 1 \\ -25 & 0 \end{bmatrix}$$

is e.s. because $\det(sI - A) = (s+5)^2$. The homogeneous part of (9.5.38) is $\dot{Y} = A_c(t)Y$ where $\det(sI - A_c(t)) = (s+1)^2$ at each time t. Hence, $\lambda(A_c(t)) = -1$ $\forall t \geq 0$. The adaptive law guarantees that $|\hat{b}(t)| \geq b_0 > 0$ $\forall t \geq 0$ and $\hat{a}, \hat{b}, \dot{\hat{a}}, \dot{\hat{b}}, \hat{k}_i, \dot{\hat{k}}_i \in \mathcal{L}_\infty$ and $\dot{\hat{a}}, \dot{\hat{b}}, \dot{\hat{k}}_i \in \mathcal{S}(\eta^2/m^2)$. Because $\frac{|\eta|}{m} \leq \left\| b\frac{\Delta_m(s)}{s+\lambda} \right\|_{2\delta_0} \triangleq \Delta_2$, it follows that $\|\dot{A}_c(t)\| \in \mathcal{S}(\Delta_2^2)$. Applying Theorem 3.4.11(b), we have that $A_c(t)$ is e.s. for $\Delta_2 < \Delta^*$ and some $\Delta^* > 0$.

Step 3. *Use the properties of the $\mathcal{L}_{2\delta}$ norm and B-G lemma to establish signal boundedness.* Let us denote $\|(\cdot)_t\|_{2\delta}$ for some $\delta \in (0, \delta_0)$ with $\|(\cdot)\|$. Using Lemma 3.3.2 and the properties of the $\mathcal{L}_{2\delta}$-norm, we have from (9.5.35), (9.5.37) that for $\delta < 2$

$$\|y_p\| \leq \|C^\top e_o\| + c\|\hat{e}\| + \|\eta\| + c$$

$$\|u_p\| \leq c(\|\hat{e}\| + \|y_p\| + \|\eta\|)$$

Combining the above inequalities, we establish that the fictitious normalizing signal $m_f^2 \triangleq 1 + \|u_p\|^2 + \|y_p\|^2$ satisfies

$$m_f^2 \leq c + c\|C^\top e_o\|^2 + c\|\hat{e}\|^2 + c\|\eta\|^2$$

If we now apply Lemma 3.3.3 to (9.5.38), we obtain

$$\|\hat{e}\| \leq c(\|C^\top e_o\| + \|\eta\|)$$

and, therefore,

$$m_f^2 \leq c + c\|C^\top e_o\|^2 + c\|\eta\|^2 \qquad (9.5.40)$$

9.5. ROBUST APPC SCHEMES

To evaluate the term $\|C^\top e_o\|$ in (9.5.40), we use (9.5.34) to write

$$C^\top e_o = \frac{s}{(s+5)^2}\tilde{a}e_1 - \frac{s+1}{(s+5)^2}\tilde{b}\frac{s}{s+1}u_p + \frac{(\lambda-9)s+\lambda-25}{(s+5)^2}\eta$$

Applying the Swapping Lemma A.1 and using the fact that $se_1 = sy_p$, we have

$$\begin{aligned} C^\top e_o &= \tilde{a}\frac{s}{(s+5)^2}y_p - \tilde{b}\frac{s}{(s+5)^2}u_p \\ &\quad + W_{c1}(W_{b1}e_1)\dot{\tilde{a}} - W_{c2}(W_{b2}u_p)\dot{\tilde{b}} + \frac{(\lambda-9)s+(\lambda-25)}{(s+5)^2}\eta \end{aligned} \quad (9.5.41)$$

where $W_{ci}(s), W_{bi}(s)$ are strictly proper transfer functions with poles at -5. The first two terms on the right side of (9.5.41) can be related to the normalized estimation error. Using $z = \theta^{*\top}\phi + \eta$ we express equation (9.5.29) as

$$\begin{aligned} \epsilon m^2 &= \theta^{*\top}\phi + \eta - \theta_p^\top\phi \\ &= \tilde{a}\frac{1}{s+\lambda}y_p - \tilde{b}\frac{1}{s+\lambda}u_p + \eta \end{aligned} \quad (9.5.42)$$

We now filter each side of (9.5.42) with $\frac{s(s+\lambda)}{(s+5)^2}$ and apply the Swapping Lemma A.1 to obtain

$$\begin{aligned} \frac{s(s+\lambda)}{(s+5)^2}\epsilon m^2 &= \tilde{a}\frac{s}{(s+5)^2}y_p - \tilde{b}\frac{s}{(s+5)^2}u_p \\ &\quad + W_c\left\{(W_b y_p)\dot{\tilde{a}} - (W_b u_p)\dot{\tilde{b}}\right\} + \frac{s(s+\lambda)}{(s+5)^2}\eta \end{aligned} \quad (9.5.43)$$

where W_c, W_b are strictly proper transfer functions with poles at -5. Using (9.5.43) in (9.5.41) we obtain

$$\begin{aligned} C^\top e_o &= \frac{s(s+\lambda)}{(s+5)^2}\epsilon m^2 - W_c\left\{(W_b y_p)\dot{\tilde{a}} - (W_b u_p)\dot{\tilde{b}}\right\} \\ &\quad + W_{c1}(W_{b1}e_1)\dot{\tilde{a}} - W_{c2}(W_{b2}u_p)\dot{\tilde{b}} - \frac{s^2+9s+25-\lambda}{(s+5)^2}\eta \end{aligned}$$

Using the fact that $W(s)y_p, W(s)u_p$ are bounded from above by m_f for any strictly proper $W(s)$ that is analytic in $Re[s] \geq -\frac{\delta}{2}$ and $m \leq m_f$, we apply Lemma 3.3.2 to obtain

$$\|C^\top e_o\| \leq c\|\epsilon m m_f\| + c\|\dot{\tilde{a}}m_f\| + c\|\dot{\tilde{b}}m_f\| + c\|\eta\| \quad (9.5.44)$$

which together with (9.5.40) imply that

$$m_f^2 \leq c + c\|\tilde{g}m_f\|^2 + c\|\eta\|^2$$

where $\tilde{g}^2 = |\dot{\tilde{a}}|^2 + |\dot{\tilde{b}}|^2 + |\epsilon m|^2$. Since

$$\frac{\|\eta\|}{m_f} \leq \left\|\frac{b\Delta_m(s)}{s+\lambda}\right\|_{\infty\delta} \triangleq \Delta_\infty$$

we have

$$m_f^2 \leq c + c\|\tilde{g}m_f\|^2 + c\Delta_\infty^2 m_f^2$$

Therefore, for

$$c\Delta_\infty^2 < 1$$

we have

$$m_f^2 \leq c + c\|\tilde{g}m_f\|^2 = c + c\int_0^t e^{-\delta(t-\tau)}\tilde{g}^2(\tau)m_f^2(\tau)d\tau$$

Applying B-G Lemma III we obtain

$$m_f^2 \leq ce^{-\delta t}e^{c\int_0^t \tilde{g}^2(\tau)d\tau} + c\delta\int_0^t e^{-\delta(t-s)}e^{c\int_s^t \tilde{g}^2(\tau)d\tau}ds$$

Because $c\int_s^t \tilde{g}^2(\tau)d\tau \leq c\Delta_2^2(t-s) + c$, it follows that for

$$c\Delta_2^2 < \delta$$

we have $m_f \in \mathcal{L}_\infty$. The boundedness of all the signals is established as follows: Because $m \leq m_f$, $\|\eta\| \leq \Delta_\infty m_f$, $|\eta| < \Delta_2 m$, we have $m, \|\eta\|, \eta \in \mathcal{L}_\infty$. Applying Lemma 3.3.3 to (9.5.38) and (9.5.34), we obtain

$$\begin{aligned}\|\hat{e}\|, |\hat{e}(t)| &\leq c\|C^T e_o\| + c\|\eta\| \\ \|e_o\|, |e_o(t)| &\leq c\|\hat{e}\| + c\|\eta\| + c\|e_1\| \\ \|e_1\| &\leq \|C^T e_0\| + c\|\hat{e}\| + \|\eta\|\end{aligned} \quad (9.5.45)$$

From $\epsilon, m, m_f, \|\eta\|, \dot{\tilde{a}}, \dot{\tilde{b}} \in \mathcal{L}_\infty$, it follows from (9.5.44) that $\|C^T e_o\| \in \mathcal{L}_\infty$ and therefore using the above inequalities, we have $\|\hat{e}\|, |\hat{e}|, \|e_o\|, |e_o|, \|e_1\| \in \mathcal{L}_\infty$. From (9.5.35), (9.5.37) and $\eta \in \mathcal{L}_\infty$, it follows that $y_p, u_p \in \mathcal{L}_\infty$ which together with $e = e_o + \hat{e}$ and $e_o, \hat{e} \in \mathcal{L}_\infty$, we can conclude that all signals are bounded.

The conditions on $\Delta_m(s)$ for robust stability are summarized as follows:

$$c\Delta_2 < \min[c\Delta^*, \sqrt{\delta}], \quad c\Delta_\infty^2 < 1$$

$$\Delta_2 = \left\|\frac{b\Delta_m(s)}{s+\lambda}\right\|_{2\delta_0}, \quad \Delta_\infty = \left\|\frac{b\Delta_m(s)}{s+\lambda}\right\|_{\infty\delta}$$

where Δ^* is a bound for the stability of $A_c(t)$ and $0 < \delta \leq \delta_0$ is a measure of the stability margin of $A_c(t)$ and $c > 0$ represents finite constants that do not affect the qualitative nature of the above bounds.

9.5. ROBUST APPC SCHEMES

Step 4. *Establish tracking error bounds.* We have, from (9.5.35) that
$$e_1 = C^\top e_o + C^\top \hat{e} + \eta$$
Because (9.5.44), (9.5.45) hold for $\delta = 0$, it follows that
$$\|e_{1t}\|_2^2 \triangleq \int_0^t e_1^2(\tau)d\tau \leq \|(C^\top e_o)_t\|_2^2 + c\|\hat{e}_t\|_2^2 + \|\eta_t\|_2^2$$
$$\leq c\|(gm_f)_t\|_2^2 + c\|\eta_t\|_2^2$$
Because $m_f \in \mathcal{L}_\infty$, $\frac{|\eta|}{m} \leq \Delta_2$ and $\|g_t\|_2^2 = \int_0^t g^2(\tau)d\tau \leq \Delta_2^2 t + c$, we have
$$\int_0^t e_1^2 d\tau \leq c\Delta_2^2 t + c$$
and the stability analysis is complete. ▽

Let us now extend the results of Example 9.5.2 to the general scheme presented in Table 9.6.

Theorem 9.5.3 *Assume that $\hat{R}_p(s,t)Q_m(s)$, $\hat{Z}_p(s,t)$ are strongly coprime and consider the APPC scheme of Table 9.6 that is designed for the plant model but applied to the actual plant with $\Delta_m, d_u \neq 0$. There exists constants $\delta^*, \Delta_\infty^* > 0$ such that for all $\Delta_m(s)$ satisfying*
$$\Delta_2^2 + f_0 \leq \delta^*, \quad \Delta_\infty \leq \Delta_\infty^*$$
where
$$\Delta_\infty \triangleq \left\|\frac{Z_p(s)}{\Lambda(s)}\Delta_m(s)\right\|_{\infty\delta}, \quad \Delta_2 \triangleq \left\|\frac{Z_p(s)}{\Lambda(s)}\Delta_m(s)\right\|_{2\delta_0}$$
and $\Lambda(s)$ is an arbitrary polynomial of degree n with roots in $Re[s] < -\frac{\delta_0}{2}$, all signals in the closed-loop plant are bounded and the tracking error satisfies
$$\int_0^t e_1^2(\tau)d\tau \leq c(\Delta_2^2 + d_0^2 + f_0)t + c$$
where $f_0 = 0$ in the case of switching σ and projection and $f_0 > 0$ in the case of fixedσ ($f_0 = \sigma$), ϵ-modification ($f_0 = \nu_0$), and dead zone ($f_0 = g_0$).

The proof for the input disturbance free ($d_u = 0$) case or the case where $\frac{d_u}{m}$ is sufficiently small follows from the proof in the ideal case and in Example 9.5.2. The proof for the general case where $\frac{d_u}{m}$ is not necessarily small involves some additional arguments and the use of the $\mathcal{L}_{2\delta}$ norm over an arbitrary time interval and is presented in Section 9.9.2.

9.5.5 Robust LQ Adaptive Control

Robust adaptive LQ control (ALQC) schemes can be formed by combining the LQ control law (9.5.5), where K_c is calculated from the Riccati Equation (9.5.7), with robust adaptive laws as shown in Table 9.7.

The ALQC is exactly the same as the one considered in Section 9.5.4 and shown in Table 9.6 with the exception that the controller gain matrix $\hat{K}_c$ is calculated using the algebraic Riccati equation at each time t instead of the pole placement equation of Table 9.6.

The stability properties of the ALQC scheme of Table 9.7 applied to the actual plant are the same as those of APPC scheme based on state feedback and are summarized by the following theorem.

Theorem 9.5.4 *Assume that $\hat{R}_p(s,t)Q_m(s)$, $\hat{Z}_p(s,t)$ are strongly coprime. There exists constants $\delta^*, \Delta_\infty^* > 0$ such that for all $\Delta_m(s)$ satisfying*

$$\Delta_2^2 + f_0 \leq \delta^*, \quad \Delta_\infty \leq \Delta_\infty^*$$

where

$$\Delta_\infty \triangleq \left\| \frac{Z_p(s)}{\Lambda(s)} \Delta_m(s) \right\|_{\infty\delta}, \quad \Delta_2 \triangleq \left\| \frac{Z_p(s)}{\Lambda(s)} \Delta_m(s) \right\|_{2\delta_0}$$

all signals are bounded and the tracking error e_1 satisfies

$$\int_0^t e_1^2(\tau)d\tau \leq c(\Delta_2^2 + d_0^2 + f_0)t + c$$

where $f_0, \Lambda(s)$ are as defined in Theorem 9.5.3.

Proof The proof is almost identical to that of Theorem 9.5.3. The only difference is that the feedback gain $\hat{K}_c$ is calculated using a different equation. Therefore, if we can show that the feedback gain calculated from the Riccati equation guarantees that $\|\dot{\hat{A}}_c(t)\| \in \mathcal{S}(\frac{\eta^2}{m^2} + f_0)$, then the rest of the proof can be completed by following exactly the same steps as in the proof of Theorem 9.5.3. The proof for $\|\dot{\hat{A}}_c(t)\| \in \mathcal{S}(\frac{\eta^2}{m^2} + f_0)$ is established by using the same steps as in the proof of Theorem 7.4.3 to show that $\|\dot{\hat{A}}_c(t)\| \leq c\|\dot{\hat{A}}(t)\| + c\|\dot{\hat{B}}(t)\|$. Because $\|\dot{\hat{A}}(t)\|, \|\dot{\hat{B}}(t)\| \leq c|\dot{\theta}_p|$ and $\dot{\theta}_p \in \mathcal{S}(\frac{\eta^2}{m^2} + f_0)$ we have $\|\dot{\hat{A}}_c(t)\| \in \mathcal{S}(\frac{\eta^2}{m^2} + f_0)$. $\square$

9.5. ROBUST APPC SCHEMES

Table 9.7 Robust adaptive linear quadratic control scheme

Actual plant	$y_p = \frac{Z_p(s)}{R_p(s)}(1+\Delta_m)(u_p + d_u)$
Plant model	$y_p = \frac{Z_p(s)}{R_p(s)} u_p$
Reference signal	$Q_m(s) y_m = 0$
Assumptions	Same as in Table 9.5
Robust adaptive law	Same as in Table 9.5 to generate $\hat{Z}_p(s,t), \hat{R}_p(s,t)$ and $\hat{A}, \hat{B}$
State observer	Same as in Table 9.6
Calculation of controller parameters	$\hat{K}_c = \lambda^{-1} \hat{B}^\top P$ $\hat{A}^\top P + P \hat{A} - \frac{1}{\lambda} P \hat{B} \hat{B}^\top P + CC^\top = 0$ $C^\top = [1, 0, \ldots, 0], P = P^\top > 0$
Control law	$\bar{u}_p = -\hat{K}_c(t)\hat{e}, \quad u_p = \frac{Q_1}{Q_m} \bar{u}_p$
Design variables	$\lambda > 0$ and Q_1, Q_m as in Table 9.6

Example 9.5.3 Consider the same plant as in Example 9.5.2, i.e.,

$$y_p = \frac{b}{s+a}(1+\Delta_m(s))u_p$$

and the same control objective that requires the output y_p to track the constant reference signal $y_m = 1$. A robust ALQC scheme can be constructed using Table 9.7 as follows:

Adaptive Law

$$\dot{\theta}_p = \Pr[\Gamma(\epsilon\phi - \sigma_s \theta_p)]$$

$$\epsilon = \frac{z - \theta_p^T \phi}{m^2}, \quad m^2 = 1 + n_s^2, \quad n_s^2 = m_s$$
$$\dot{m}_s = -\delta_0 m_s + u_p^2 + y_p^2, \quad m_s(0) = 0$$
$$z = \frac{s}{s+\lambda} y_p, \quad \phi = \frac{1}{s+\lambda}[u_p, -y_p]^T$$
$$\theta_p = [\hat{b}, \hat{a}]^T$$

The projection operator $\Pr[\cdot]$ constraints $\hat{b}$ to satisfy $|\hat{b}(t)| \geq b_0 \; \forall t \geq 0$ where b_0 is a known lower bound for $|b|$ and σ_s is the switching σ-modification.

State Observer
$$\dot{\hat{e}} = \hat{A}\hat{e} + \hat{B}\bar{u}_p - \hat{K}_o([1\; 0]\hat{e} - e_1)$$

where
$$\hat{A} = \begin{bmatrix} -\hat{a} & 1 \\ 0 & 0 \end{bmatrix}, \quad \hat{B} = \hat{b}\begin{bmatrix} 1 \\ 1 \end{bmatrix}, \quad \hat{K}_o = \begin{bmatrix} 10 - \hat{a} \\ 25 \end{bmatrix}$$

Controller Parameters $\hat{K}_c = \lambda^{-1}\hat{B}^T P$ where $P = P^T > 0$ is solved pointwise in time using
$$\hat{A}^T P + P\hat{A} - P\hat{B}\lambda^{-1}\hat{B}^T P + CC^T = O, \quad C^T = [1\; 0]$$

Control Law $\quad \bar{u}_p = -\hat{K}_c(t)\hat{e}, \quad u_p = \dfrac{s+1}{s}\bar{u}_p \hfill \triangledown$

9.6 Adaptive Control of LTV Plants

One of the main reasons for considering adaptive control in applications is to compensate for large variations in the plant parameters. One can argue that if the plant model is LTI with unknown parameters a sufficient number of tests can be performed off-line to calculate these parameters with sufficient accuracy and therefore, there is no need for adaptive control when the plant model is LTI. One can also go further and argue that if some nominal values of the plant model parameters are known, robust control may be adequate as long as perturbations around these nominal values remain within certain bounds. And again in this case there is no need for adaptive control. In many applications, however, such as aerospace, process control, etc., LTI plant models may not be good approximations of the plant due to drastic changes in parameter values that may arise due to changes in operating

9.6. ADAPTIVE CONTROL OF LTV PLANTS

points, partial failure of certain components, wear and tear effects, etc. In such applications, linear time varying (LTV) plant models of the form

$$\begin{aligned} \dot{x} &= A(t)x + B(t)u \\ y &= C^\mathsf{T}(t)x + D(t)u \end{aligned} \qquad (9.6.1)$$

where A, B, C, and D consist of unknown time varying elements, may be necessary. Even though adaptive control was motivated for plants modeled by (9.6.1), most of the work on adaptive control until the mid 80's dealt with LTI plants. For some time, adaptive control for LTV plants whose parameters vary slowly with time was considered to be a trivial extension of that for LTI plants. This consideration was based on the intuitive argument that an adaptive system designed for an LTI plant should also work for a linear plant whose parameters vary slowly with time. This argument was later on proved to be invalid. In fact, attempts to apply adaptive control to simple LTV plants led to similar unsolved stability and robustness problems as in the case of LTI plants with modeling errors. No significant progress was made towards the design and analysis of adaptive controllers for LTV plants until the mid-1980s when some of the fundamental robustness problems of adaptive control for LTI plants were resolved.

In the early attempts [4, 74], the notion of the PE property of certain signals in the adaptive control loop was employed to guarantee the exponential stability of the unperturbed error system, which eventually led to the local stability of the closed-loop time-varying (TV) plant. Elsewhere, the restriction of the type of time variations of the plant parameters also led to the conclusion that an adaptive controller could be used in the respective environment. More specifically, in [31] the parameter variations were assumed to be perturbations of some nominal constant parameters, which are small in the norm and modeled as a martingale process with bounded covariance. A treatment of the parameter variations as small TV perturbations, in an $\mathcal{L}_2$-gain sense, was also considered in [69] for a restrictive class of LTI-nominal plants. Special models of parameter variations, such as exponential or $1/t$-decaying or finite number of jump discontinuities, were considered in [70, 138, 181]. The main characteristic of these early studies was that no modifications to the adaptive laws were necessary due to either the use of PE or the restriction of the parameter variations to a class that introduces no persistent modeling error effects in the adaptive control scheme.

The adaptive control problem for general LTV plants was initially treated as a robustness problem where the effects of slow parameter variations were treated as unmodeled perturbations [9, 66, 145, 222]. The same robust adaptive control techniques used for LTI plants in the presence of bounded disturbances and unmodeled dynamics were shown to work for LTV plants when their parameters are smooth and vary slowly with time or when they vary discontinuously, i.e., experience large jumps in their values but the discontinuities occur over large intervals of time [225].

The robust MRAC and APPC schemes presented in the previous sections can be shown to work well with LTV plants whose parameters belong to the class of smooth and slowly varying with respect to time or the class with infrequent jumps in their values. The difficulty in analyzing these schemes with LTV plants has to do with the representations of the plant model. In the LTI case the transfer function and related input/output results are used to design and analyze adaptive controllers. For an LTV plant, the notion of a transfer function and of poles and zeros is no longer applicable which makes it difficult to extend the results of the LTI case to the LTV one. This difficulty was circumvented in [223, 224, 225, 226] by using the notion of the polynomial differential operator and the polynomial integral operator to describe an LTV plant such as (9.6.1) in an input-output form that resembles a transfer function description.

The details of these mathematical preliminaries as well as the design and analysis of adaptive controllers for LTV plants of the form (9.6.1) are presented in a monograph [226] and in a series of papers [223, 224, 225]. Interested readers are referred to these papers for further information.

9.7 Adaptive Control for Multivariable Plants

The design of adaptive controllers for MIMO plant models is more complex than in the SISO case. In the MIMO case we are no longer dealing with a single transfer function but with a transfer matrix whose elements are transfer functions describing the coupling between inputs and outputs. As in the SISO case, the design of an adaptive controller for a MIMO plant can be accomplished by combining a control law, that meets the control objective when the plant parameters are known, with an adaptive law that generates estimates for the unknown parameters. The design of the control and adaptive law, however, requires the development of certain parameterizations of the plant model that are more complex than those in the SISO case.

In this section we briefly describe several approaches that can be used to design adaptive controllers for MIMO plants.

9.7.1 Decentralized Adaptive Control

Let us consider the MIMO plant model

$$y = H(s)u \qquad (9.7.1)$$

where $y \in \mathcal{R}^N, u \in \mathcal{R}^N$ and $H(s) \in \mathcal{C}^{N \times N}$ is the plant transfer matrix that is assumed to be proper. Equation (9.7.1) may be also expressed as

$$y_i = h_{ii}(s)u_i + \sum_{\substack{1 \le j \le N \\ j \ne i}} h_{ij}(s)u_j, \quad i = 1, 2, \ldots, N \qquad (9.7.2)$$

where $h_{ij}(s)$, the elements of $H(s)$, are transfer functions.

Another less obvious but more general decomposition of (9.7.1) is

$$y_i = h_{ii}(s)u_i + \sum_{\substack{1 \le j \le N \\ j \ne i}} \left(h_{ij}(s)u_j + q_{ij}(s)y_j \right), \quad i = 1, 2, \ldots, N \qquad (9.7.3)$$

for some different transfer functions $h_{ij}(s), q_{ij}(s)$.

If the MIMO plant model (9.7.3) is such that the interconnecting or coupling transfer functions $h_{ij}(s), q_{ij}(s)$ $(i \ne j)$ are stable and small in some sense, then they can be treated as modeling error terms in the control design. This means that instead of designing an adaptive controller for the MIMO plant (9.7.3), we can design N adaptive controllers for N SISO plant models of the form

$$y_i = h_{ii}(s)u_i, \quad i = 1, 2, \ldots, N \qquad (9.7.4)$$

If these adaptive controllers are designed based on robustness considerations, then the effect of the small unmodeled interconnections present in the MIMO plant will not destroy stability. This approach, known as *decentralized adaptive control*, has been pursued in [38, 59, 82, 185, 195].

The analysis of decentralized adaptive control designed for the plant model (9.7.4) but applied to (9.7.3) follows directly from that of robust adaptive control for plants with unmodeled dynamics considered in the previous sections and is left as an exercise for the reader.

9.7.2 The Command Generator Tracker Approach

The command generator tracker (CGT) theory was first proposed in [26] for the model following problem with known parameters. An adaptive control algorithm based on the CGT method was subsequently developed in [207, 208]. Extensions and further improvements of the adaptive CGT algorithms for finite [18, 100] as well as for infinite [233, 234] dimensional plant models followed the work of [207, 208].

The adaptive CGT algorithms are based on a plant/reference model structural matching and on an SPR condition. They are developed using the SPR-Lyapunov design approach.

The plant model under consideration in the CGT approach is in the state space form

$$\dot{x} = Ax + Bu, \quad x(0) = x_0$$
$$y = C^\top x \qquad (9.7.5)$$

where $x \in \mathcal{R}^n; y, u \in \mathcal{R}^m$; and A, B, and C are constant matrices of appropriate dimensions. The control objective is to choose u so that all signals in the closed-loop plant are bounded and the plant output y tracks the output y_m of the reference model described by

$$\dot{x}_m = A_m x_m + B_m r \quad x_m(0) = x_{m0}$$
$$y_m = C_m^\top x_m \qquad (9.7.6)$$

where $x_m \in \mathcal{R}^{n_m}, r \in \mathcal{R}^{p_m}$ and $y_m \in \mathcal{R}^m$.

The only requirement on the reference model at this point is that its output y_m has the same dimension as the plant output. The dimension of x_m can be much lower than that of x.

Let us first consider the case where the plant parameters A, B, and C are known and use the following assumption.

Assumption 1 (CGT matching condition) There exist matrices S_{11}^*, $S_{12}^*, S_{21}^*, S_{22}^*$ such that the desired plant input u^* that meets the control objective satisfies

$$\begin{bmatrix} x^* \\ u^* \end{bmatrix} = \begin{bmatrix} S_{11}^* & S_{12}^* \\ S_{21}^* & S_{22}^* \end{bmatrix} \begin{bmatrix} x_m \\ r \end{bmatrix} \qquad (9.7.7)$$

9.7. ADAPTIVE CONTROL OF MIMO PLANTS

where x^* is equal to x when $u = u^*$, i.e., x^* satisfies

$$\dot{x}^* = Ax^* + Bu^*$$

$$y^* = C^T x^* = y_m$$

Assumption 2 (Output stabilizability condition) There exists a matrix G^* such that $A + BG^*C^T$ is a stable matrix.

If assumptions 1, 2 are satisfied we can use the control law

$$u = G^*(y - y_m) + S_{21}^* x_m + S_{22}^* r \qquad (9.7.8)$$

that yields the closed-loop plant

$$\dot{x} = Ax + BG^* e_1 + BS_{21}^* x_m + BS_{22}^* r$$

where $e_1 = y - y_m$. Let $e = x - x^*$ be the state error between the actual and desired plant state. Note the e is not available for measurement and is used only for analysis. The error e satisfies the equation

$$\dot{e} = (A + BG^*C^T)e$$

which implies that e and therefore e_1 are bounded and converge to zero exponentially fast. If x_m, r are also bounded then we can conclude that all signals are bounded and the control law (9.7.8) meets the control objective exactly.

If A, B, and C are unknown, the matrices G^*, S_{21}^*, and S_{22}^* cannot be calculated, and, therefore, (9.7.8) cannot be implemented. In this case we use the control law

$$u = G(t)(y - y_m) + S_{21}(t)x_m + S_{22}(t)r \qquad (9.7.9)$$

where $G(t), S_{21}(t), S_{22}(t)$ are the estimates of G^*, S_{21}^*, S_{22}^* to be generated by an adaptive law. The adaptive law is developed using the SPR-Lyapunov design approach and employs the following assumption.

Assumption 3 (SPR condition) There exists a matrix G^* such that $C^T(sI - A - BG^*C^T)^{-1}B$ is an SPR transfer matrix.

Assumption 3 puts a stronger condition on the output feedback gain G^*, i.e., in addition to making $A + BG^*C$ stable it should also make the closed-loop plant transfer matrix SPR.

For the closed-loop plant (9.7.5), (9.7.9), the equation for $e = x - x^*$ becomes

$$\dot{e} = (A + BG^*C^T)e + B\tilde{G}e_1 + B\tilde{S}\omega$$
$$e_1 = C^T e \qquad (9.7.10)$$

where

$$\tilde{G}(t) = G(t) - G^*, \quad \tilde{S}(t) = S(t) - S^*$$
$$S(t) = [S_{21}(t), S_{22}(t)], \quad S^* = [S_{21}^*, S_{22}^*]$$
$$\omega = [x_m^T, r^T]^T$$

We propose the Lyapunov-like function

$$V = e^T P_c e + tr[\tilde{G}^T \Gamma_1^{-1} \tilde{G}] + tr[\tilde{S}^T \Gamma_2^{-1} \tilde{S}]$$

where $P_c = P_c^T > 0$ satisfies the equations of the matrix form of the LKY Lemma (see Lemma 3.5.4) because of assumption 3, and Γ_1, Γ_2 are symmetric positive definite matrices.

Choosing

$$\dot{\tilde{G}} = \dot{G} = -\Gamma_1 e_1 e_1^T$$
$$\dot{\tilde{S}} = \dot{S} = -\Gamma_2 e_1 \omega^T \qquad (9.7.11)$$

it follows as in the SISO case that

$$\dot{V} = -e^T Q Q^T e - \nu_c e^T L_c e$$

where $L_c = L_c^T > 0$ and $\nu_c > 0$ is a small constant and Q is a constant matrix. Using similar arguments as in the SISO case we can establish that e, G, S are bounded and $e \in \mathcal{L}_2$. If in addition x_m, r are bounded we can establish that all signals are bounded and e, e_1 converge to zero as $t \to \infty$. Therefore the adaptive control law (9.7.9), (9.7.11) meets the control objective.

Due to the restrictive nature of Assumptions 1 to 3, the CGT approach did not receive as much attention as other adaptive control methods. As

9.7. ADAPTIVE CONTROL OF MIMO PLANTS

shown in [233, 234], the CGT matching condition puts strong conditions on the plant and model that are difficult to satisfy when the reference signal r is allowed to be bounded but otherwise arbitrary. If r is restricted to be equal to a constant these conditions become weaker and easier to satisfy. The class of plants that becomes SPR by output feedback is also very restrictive. In an effort to expand this class of plants and therefore relax assumption 3, the plant is augmented with parallel dynamics [18, 100], i.e., the augmented plant is described by

$$y_a = [C^T(sI - A)^{-1}B + W_a(s)]u$$

where $W_a(s)$ represents the parallel dynamics. If $W_a(s)$ is chosen so that assumptions 1, 2, 3 are satisfied by the augmented plant, then we can establish signal boundedness and convergence of y_a to y_m. Because $y_a \neq y$ the convergence of the true tracking error to zero cannot be guaranteed.

One of the advantages of the adaptive control schemes based on the CGT approach is that they are simple to design and analyze.

The robustness of the adaptive CGT based schemes with respect to bounded disturbances and unmodeled dynamics can be established using the same modifications and analysis as in the case of MRAC with unnormalized adaptive laws and is left as an exercise for the reader.

9.7.3 Multivariable MRAC

In this section we discuss the extension of some of the MRAC schemes for SISO plants developed in Chapter 6 to the MIMO plant model

$$y = G(s)u \qquad (9.7.12)$$

where $y \in \mathcal{R}^N, u \in \mathcal{R}^N$ and $G(s)$ is an $N \times N$ transfer matrix. The reference model to be matched by the closed-loop plant is given by

$$y_m = W_m(s)r \qquad (9.7.13)$$

where $y_m, r \in \mathcal{R}^N$. Because $G(s)$ is not a scalar transfer function, the multivariable counterparts of high frequence gain, relative degree, zeros and order need to be developed. The following Lemma is used to define the counterpart of the high frequency gain and relative degree for MIMO plants.

Lemma 9.7.1 *For any $N \times N$ strictly proper rational full rank transfer matrix $G(s)$, there exists a (non-unique) lower triangular polynomial matrix $\xi_m(s)$, defined as the* modified left interactor *(MLI) matrix of $G(s)$, of the form*

$$\xi_m(s) = \begin{bmatrix} d_1(s) & 0 & \cdots & 0 & 0 \\ h_{21}(s) & d_2(s) & 0 & \cdots & 0 \\ \vdots & & & \ddots & 0 \\ h_{N1}(s) & \cdots & \cdots & h_{N(N-1)}(s) & d_N(s) \end{bmatrix}$$

where $h_{ij}(s), j = 1, \ldots, N-1, i = 2, \ldots, N$ are some polynomials and $d_i(s)$, $i = 1, \ldots, N$, are arbitrary monic Hurwitz polynomials of certain degree $l_i > 0$, such that $\lim_{s \to \infty} \xi_m(s)G(s) = K_p$, the high-frequency-gain matrix of $G(s)$, is finite and nonsingular.

For a proof of Lemma 9.7.1, see [216].

A similar concept to that of the left interactor matrix, which was introduced in [228] for a discrete-time plant is the *modified right interactor* (MRI) matrix used in [216] for the design of MRAC for MIMO plants.

To meet the control objective we make the following assumptions about the plant:

A1. $G(s)$ is strictly proper, has full rank and a known MLI matrix $\xi_m(s)$.

A2. All zeros of $G(s)$ are stable.

A3. An upper bound $\bar{\nu}_0$ on the observability index ν_0 of $G(s)$ is known [50, 51, 73, 172].

A4. A matrix S_p that satisfies $K_p S_p = (K_p S_p)^\top > 0$ is known.

Furthermore we assume that the transfer matrix $W_m(s)$ of the reference model is designed to satisfy the following assumptions:

M1. All poles and zeros of $W_m(s)$ are stable.

M2. The zero structure at infinity of $W_m(s)$ is the same as that of $G(s)$, i.e., $\lim_{s \to \infty} \xi_m(s) W_m(s)$ is finite and nonsingular. Without loss of generality we can choose $W_m(s) = \xi_m^{-1}(s)$.

9.7. ADAPTIVE CONTROL OF MIMO PLANTS

Assumptions A1 to A4, and M1 and M2 are the extensions of the assumptions P1 to P4, and M1 and M2 in the SISO case to the MIMO one. The knowledge of the interactor matrix is equivalent to the knowledge of the relative degree in the SISO case. Similarly the knowledge of S_p in A4 is equivalent to the knowledge of the sign of the high frequency gain.

The a priori knowledge about $G(s)$ which is necessary for $\xi_m(s)$ to be known is very crucial in adaptive control because the parameters of $G(s)$ are considered to be unknown. It can be verified that if $\xi_m(s)$ is diagonal and the relative degree of each element of $G(s)$ is known then $\xi_m(s)$ can be completely specified without any apriori knowledge about the parameters of $G(s)$. When $\xi_m(s)$ is not diagonal, it is shown in [172, 205] that a diagonal stable dynamic pre-compensator $W_p(s)$ can be found, using only the knowledge of the relative degree for each element of $G(s)$, such that for most cases $G(s)W_p(s)$ has a diagonal $\xi_m(s)$. In another approach shown in [216] one could check both the MLI and MRI matrices and choose the one that is diagonal, if any, before proceeding with the search for a compensator. The design of MRAC using the MRI is very similar to that using the MLI that is presented below.

We can design a MRAC scheme for the plant (9.7.12) by using the certainty equivalence approach as we did in the SISO case. We start by assuming that all the parameters of $G(s)$ are known and propose the control law

$$u = \theta_1^{*T}\omega_1 + \theta_2^{*T}\omega_2 + \theta_3^* r = \theta^{*T}\omega \qquad (9.7.14)$$

where $\theta^{*T} = [\theta_1^{*T}, \theta_2^{*T}, \theta_3^*]$, $\omega = [\omega_1^T, \omega_2^T, r^T]^T$.

$$\omega_1 = \frac{A(s)}{\Lambda(s)}u, \quad \omega_2 = \frac{A(s)}{\Lambda(s)}y$$

$$A(s) = [Is^{\bar{\nu}_0-1}, Is^{\bar{\nu}_0-2}, \ldots, Is, I]^T$$

$$\theta_1^* = [\theta_{11}^*, \ldots, \theta_{1\bar{\nu}_0}^*]^T, \theta_2^* = [\theta_{21}^*, \ldots, \theta_{2\bar{\nu}_0}^*]^T$$

$$\theta_3^*, \theta_{ij}^* \in \mathcal{R}^{N \times N}, \quad i = 1, 2, \quad j = 1, 2, \ldots, \bar{\nu}_0$$

and $\Lambda(s)$ is a monic Hurwitz polynomial of degree $\bar{\nu}_0$.

It can be shown [216] that the closed-loop plant transfer matrix from y to r is equal to $W_m(s)$ provided $\theta_3^* = K_p^{-1}$ and θ_1^*, θ_2^* are chosen to satisfy the matching equation

$$I - \theta_1^{*T}\frac{A(s)}{\Lambda(s)} - \theta_2^{*T}\frac{A(s)}{\Lambda(s)}G(s) = \theta_3^* W_m^{-1}(s)G(s) \quad (9.7.15)$$

The same approach as in the SISO case can be used to show that the control law (9.7.14) with $\theta_i^*, i = 1, 2, 3$ as chosen above guarantees that all closed-loop signals are bounded and the elements of the tracking error $e_1 = y - y_m$ converge to zero exponentially fast.

Because the parameters of $G(s)$ are unknown, instead of (9.7.14) we use the control law

$$u = \theta^T(t)\omega \quad (9.7.16)$$

where $\theta(t)$ is the on-line estimate of the matrix θ^* to be generated by an adaptive law. The adaptive law is developed by first obtaining an appropriate parametric model for θ^* and then using a similar procedure as in Chapter 4 to design the adaptive law.

From the plant and matching equations (9.7.12), (9.7.15) we obtain

$$u - \theta_1^{*T}\omega_1 - \theta_2^{*T}\omega_2 = \theta_3^* W_m^{-1}(s)y$$

$$\theta_3^{*-1}(u - \theta^{*T}\omega) = W_m^{-1}(s)(y - y_m) = \xi_m(s)e_1 \quad (9.7.17)$$

Let d_m be the maximum degree of $\xi_m(s)$ and choose a Hurwitz polynomial $f(s)$ of degree d_m. Filtering each side of (9.7.17) with $1/f(s)$ we obtain

$$z = \psi^*[\theta^{*T}\phi + z_0] \quad (9.7.18)$$

where

$$z = -\frac{\xi_m(s)}{f(s)}e_1, \quad \psi^* = \theta_3^{*-1} = K_p$$

$$\phi = \frac{1}{f(s)}\omega, \quad z_0 = -\frac{1}{f(s)}u$$

which is the multivariable version of the bilinear parametric model for SISO plants considered in Chapter 4.

Following the procedure of Chapter 4 we generate the estimated value

$$\hat{z} = \psi(t)[\theta^T(t)\phi + z_0]$$

9.7. ADAPTIVE CONTROL OF MIMO PLANTS

of z and the normalized estimation error

$$\epsilon = \frac{z - \hat{z}}{m^2}$$

We can verify that

$$\epsilon = -\frac{\tilde{\psi}\xi + \psi^* \tilde{\theta}^\top \phi}{m^2}$$

where $\xi = z_0 + \theta^\top \phi$ and m^2 could be chosen as $m^2 = 1 + |\xi|^2 + |\phi|^2$. The adaptive law for θ can now be developed by using the Lyapunov-like function

$$V = \frac{1}{2} tr[\tilde{\theta}^\top \Gamma_p \tilde{\theta}] + \frac{1}{2} tr[\tilde{\psi}^\top \Gamma^{-1} \tilde{\psi}]$$

where $\Gamma_p = K_p^\top S_p^{-1} = (S_p^{-1})^\top (K_p S_p)^\top S_p^{-1}$ and $\Gamma = \Gamma^\top > 0$. If we choose

$$\dot{\theta}^\top = -S_p \epsilon \phi^\top, \quad \dot{\psi} = -\Gamma \epsilon \xi^\top \qquad (9.7.19)$$

it follows that

$$\dot{V} = -\epsilon^\top \epsilon m^2 \leq 0$$

which implies that θ, ψ are bounded and $|\epsilon m| \in \mathcal{L}_2$.

The stability properties of the MRAC (9.7.12), (9.7.16), (9.7.19) are similar to those in the SISO case and can be established following exactly the same procedure as in the SISO case. The reader is referred to [216] for the stability proofs and properties of (9.7.12), (9.7.16), (9.7.19).

The multivariable MRAC scheme (9.7.12), (9.7.16), (9.7.19) can be made robust by choosing the normalizing signal as

$$m^2 = 1 + m_s, \quad \dot{m}_s = -\delta_0 m_s + |u|^2 + |y|^2, m_s(0) = 0 \qquad (9.7.20)$$

where $\delta_0 > 0$ is designed as in the SISO case and by modifying the adaptive laws (9.7.19) using exactly the same techniques as in Chapter 8 for the SISO case.

For further information on the design and analysis of adaptive controllers for MIMO plants, the reader is referred to [43, 50, 51, 73, 151, 154, 172, 201, 205, 213, 214, 216, 218, 228].

9.8 Stability Proofs of Robust MRAC Schemes

9.8.1 Properties of Fictitious Normalizing Signal

As in Chapter 6, the stability analysis of the robust MRAC schemes involves the use of the $\mathcal{L}_{2\delta}$ norm and its properties as well as the properties of the signal

$$m_f^2 \triangleq 1 + \|u_p\|^2 + \|y_p\|^2 \tag{9.8.1}$$

where $\|(\cdot)\|$ denotes the $\mathcal{L}_{2\delta}$ norm $\|(\cdot)_t\|_{2\delta}$. The signal m_f has the property of bounding from above almost all the signals in the closed-loop plant. In this, book, m_f is used for analysis only and is referred to as the fictitious normalizing signal, to be distinguished from the normalizing signal m used in the adaptive law. For the MRAC law presented in Table 9.1, the properties of the signal m_f are given by the following Lemma. These properties are independent of the adaptive law employed.

Lemma 9.8.1 *Consider the MRAC scheme of Table 9.1. For any given $\delta \in (0, \delta_0]$, the signal m_f given by (9.8.1) guarantees that:*

(i) $\frac{\|\omega\|}{m_f}, \frac{|\omega_i|}{m_f}, i = 1, 2$ and $\frac{n_s}{m_f} \in \mathcal{L}_\infty$.

(ii) *If $\theta \in \mathcal{L}_\infty$, then $\frac{u_p}{m_f}, \frac{y_p}{m_f}, \frac{\omega}{m_f}, \frac{W(s)\omega}{m_f}, \frac{\|u_p\|}{m_f}, \frac{\|\dot{y}_p\|}{m_f} \in \mathcal{L}_\infty$ where $W(s)$ is any proper transfer function that is analytic in $Re[s] \geq -\delta_0/2$.*

(iii) *If $\theta, \dot{r} \in \mathcal{L}_\infty$, then $\frac{\|\dot{\omega}\|}{m_f} \in \mathcal{L}_\infty$.*

(iv) *For $\delta = \delta_0$, (i) to (iii) are satisfied by replacing m_f with $m = \sqrt{1 + n_s^2}$, where $n_s^2 = m_s$ and $\dot{m}_s = -\delta_0 m_s + u_p^2 + y_p^2, m_s(0) = 0$.*

Proof (i) We have

$$\omega_1 = \frac{\alpha(s)}{\Lambda(s)} u_p, \quad \omega_2 = \frac{\alpha(s)}{\Lambda(s)} y_p$$

Because the elements of $\frac{\alpha(s)}{\Lambda(s)}$ are strictly proper and analytic in $Re[s] \geq -\delta_0/2$, it follows from Lemma 3.3.2 that $\frac{\|\omega_i\|}{m_f}, \frac{|\omega_i|}{m_f} \in \mathcal{L}_\infty, i = 1, 2$ for any $\delta \in (0, \delta_0]$. Now $\|\omega\| \leq \|\omega_1\| + \|\omega_2\| + \|y_p\| + c \leq c m_f + c$ and therefore $\frac{\|\omega\|}{m_f} \in \mathcal{L}_\infty$. Because $n_s^2 = m_s = \|u_{pt}\|_{2\delta_0}^2 + \|y_{pt}\|_{2\delta_0}^2$ and $\|(\cdot)_t\|_{2\delta_0} \leq \|(\cdot)_t\|_{2\delta}$ for any $\delta \in (0, \delta_0]$, it follows that $n_s/m_f \in \mathcal{L}_\infty$.

(ii) From equation (9.3.63) and $u_p = \theta^\top \omega$ we have

$$y_p = W_m(s)\rho^* \tilde{\theta}^\top \omega + \rho^* \eta + W_m(s) r$$

where $\eta = \frac{\Lambda - \theta_1^{*\top}\alpha}{\Lambda} W_m[\Delta_m(u_p + d_u) + d_u]$. Because $W_m, W_m \Delta_m$ are strictly proper and analytic in $Re[s] \geq -\delta_0/2$, and $\tilde{\theta} \in \mathcal{L}_\infty$, it follows that for any $\delta \in (0, \delta_0]$

$$|y_p(t)| \leq c\|\omega\| + c\|u_p\| + c \leq c m_f + c$$

i.e., $y_p/m_f \in \mathcal{L}_\infty$. Since $\omega = [\omega_1^T, \omega_2^T, y_p, r]^T$ and $r, \omega_i/m_f, i = 1, 2$ and $y_p/m_f \in \mathcal{L}_\infty$, it follows that $\omega/m_f \in \mathcal{L}_\infty$. From $u_p = \theta^T \omega$ and $\theta, \frac{\|\omega\|}{m_f}, \frac{|\omega_i|}{m_f} \in \mathcal{L}_\infty$, we have $\frac{\|u_p\|}{m_f}, \frac{u_p}{m_f} \in \mathcal{L}_\infty$. We have $|W(s)\omega| \leq \|W_0(s)\|_{2\delta}\|\omega\| + |d|\|\omega\| \leq cm_f$ where $W_0(s) + d = W(s)$ and, therefore, $\frac{W(s)\omega}{m_f} \in \mathcal{L}_\infty$. The signal $\dot{y}_p$ is given by

$$\dot{y}_p = sW_m(s)r + sW_m(s)\rho^* \tilde{\theta}^T \omega + \rho^* \dot{\eta}$$

Because $sW_m(s)$ is at most biproper we have

$$\|\dot{y}_p\| \leq c\|sW_m(s)\|_{\infty\delta}\|\omega\| + c\|\dot{\eta}\| + c$$

where

$$\begin{aligned}\|\dot{\eta}\| &\leq \left\|\frac{\Lambda - \theta_1^{*T}\alpha(s)}{\Lambda(s)} sW_m(s)\Delta_m(s)\right\|_{\infty\delta} \|u_p\| + cd_0 \\ &\leq cm_f + cd_0\end{aligned}$$

which together with $\frac{\|\omega\|}{m_f} \in \mathcal{L}_\infty$ imply that $\frac{\|\dot{y}_p\|}{m_f} \in \mathcal{L}_\infty$.

(iii) We have $\dot{\omega} = [\frac{s\alpha(s)}{\Lambda(s)}u_p, \frac{s\alpha(s)}{\Lambda(s)}y_p, \dot{y}_p, \dot{r}]^T$. Because the elements of $\frac{s\alpha(s)}{\Lambda(s)}$ are proper and analytic in $\text{Re}[s] \geq -\delta_0/2$ and $\dot{r}, \frac{\|\dot{y}_p\|}{m_f} \in \mathcal{L}_\infty$, it follows that $\frac{\|\dot{\omega}\|}{m_f} \in \mathcal{L}_\infty$.

(iv) For $\delta = \delta_0$ we have $m_f^2 = 1 + \|u_{pt}\|_{2\delta_0}^2 + \|y_{pt}\|_{2\delta_0}^2$. Because $m^2 = 1 + \|u_{pt}\|_{2\delta_0}^2 + \|y_{pt}\|_{2\delta_0}^2 = m_f^2$, the proof of (iv) follows. □

Lemma 9.8.2 *Consider the systems*

$$v = H(s)v_p \tag{9.8.2}$$

and

$$\begin{aligned}\dot{m}_s &= -\delta_0 m_s + |u_p|^2 + |y_p|^2, \quad m_s(0) = 0 \\ m^2 &= 1 + m_s\end{aligned}$$

where $\delta_0 > 0, \|v_{pt}\|_{2\delta_0} \leq cm(t) \ \forall t \geq 0$ and some constant $c > 0$ and $H(s)$ is a proper transfer function with stable poles.

Assume that either $\|v_t\|_{2\delta_0} \leq cm(t) \ \forall t \geq 0$ for some constant $c > 0$ or $H(s)$ is analytic in $\text{Re}[s] \geq -\delta_0/2$. Then there exists a constant $\delta > 0$ such that

(i) $\|v_{t,t_1}\| \leq ce^{-\frac{\delta}{2}(t-t_1)}m(t_1) + \|H(s)\|_{\infty\delta}\|v_{p_{t,t_1}}\|, \ \forall t \geq t_1 \geq 0$

(ii) *If $H(s)$ is strictly proper, then*

$$|v(t)| \leq ce^{-\frac{\delta}{2}(t-t_1)}m(t_1) + \|H(s)\|_{2\delta}\|v_{p_{t,t_1}}\|, \ \forall t \geq t_1 \geq 0$$

where $\|(\cdot)_{t,t_1}\|$ denotes the $\mathcal{L}_{2\delta}$ norm over the interval $[t_1, t]$

Proof Let us represent (9.8.2) in the minimal state space form

$$\dot{x} = Ax + Bv_p, \quad x(0) = 0$$
$$v = C^\top x + Dv_p \qquad (9.8.3)$$

Because A is a stable matrix, we have $\|e^{A(t-\tau)}\| \leq \lambda_0 e^{-\alpha_0(t-\tau)}$ for some $\lambda_0, \alpha_0 > 0$. Applying Lemma 3.3.5 to (9.8.3) we obtain

$$\|v_{t,t_1}\| \leq ce^{-\frac{\delta}{2}(t-t_1)}|x(t_1)| + \|H(s)\|_\infty \delta \|v_{p_t,t_1}\| \qquad (9.8.4)$$

and for $H(s)$ strictly proper

$$|v(t)| \leq ce^{-\frac{\delta}{2}(t-t_1)}|x(t_1)| + \|H(s)\|_{2\delta}\|v_{p_t,t_1}\| \qquad (9.8.5)$$

for any $0 < \delta < 2\alpha_0$ and for some constant $c \geq 0$. If $H(s)$ is analytic in $Re[s] \geq -\delta_0/2$ then $A - \frac{\delta_0}{2}I$ is a stable matrix and from Lemma 3.3.3 we have

$$|x(t)| \leq c\|v_{pt}\|_{2\delta_0} \leq cm(t) \quad \forall t \geq 0$$

Hence, $|x(t_1)| \leq cm(t_1)$, (i) and (ii) follow from (9.8.4) and (9.8.5), respectively.

When $H(s)$ is not analytic in $Re[s] \geq -\delta_0/2$ we use output injection to rewrite (9.8.3) as

$$\dot{x} = (A - KC^\top)x + Bv_p + KC^\top x$$

or

$$\dot{x} = A_c x + B_c v_p + Kv \qquad (9.8.6)$$

where $A_c = A - KC^\top$, $B_c = B - KD$ and K is chosen so that $A_c - \frac{\delta_0}{2}I$ is a stable matrix. The existence of such a K is guaranteed by the observability of (C, A)[95]. Applying Lemma 3.3.3 (i) to (9.8.6), we obtain

$$|x(t)| \leq c\|v_{pt}\|_{2\delta_0} + c\|v_t\|_{2\delta_0} \leq cm(t) \quad \forall t \geq 0 \qquad (9.8.7)$$

where the last inequality is established by using the assumption $\|v_t\|_{2\delta_0}, \|v_{pt}\|_{2\delta_0} \leq cm(t)$ of the Lemma. Hence, $\|x(t_1)\| \leq cm(t_1)$ and (i), (ii) follow directly from (9.8.4), (9.8.5). □

Instead of m_f given by (9.8.1), let us consider the signal

$$m_{f_1}^2(t) \triangleq e^{-\delta(t-t_1)}m^2(t_1) + \|u_{p_t,t_1}\|^2 + \|y_{p_t,t_1}\|^2, \quad t \geq t_1 \geq 0 \qquad (9.8.8)$$

The signal m_{f_1} has very similar normalizing properties as m_f as indicated by the following lemma:

Lemma 9.8.3 *The signal m_{f_1} given by (9.8.8) guarantees that*

9.8. STABILITY PROOFS OF ROBUST MRAC SCHEMES

(i) $\omega_i(t)/m_{f_1}, i = 1, 2; \|\omega_{t,t_1}\|/m_{f_1}$ and $n_s/m_{f_1} \in \mathcal{L}_\infty$

(ii) If $\theta \in \mathcal{L}_\infty$, then $y_p/m_{f_1}, u_p/m_{f_1}, \omega/m_{f_1}, W(s)\omega/m_{f_1} \in \mathcal{L}_\infty$ and $\|u_{p_{t,t_1}}\|/m_{f_1}$, $\|y_{p_{t,t_1}}\|/m_{f_1} \in \mathcal{L}_\infty$

(iii) If $\theta, \dot{r} \in \mathcal{L}_\infty$, then $\|\dot{\omega}_{t,t_1}\|/m_{f_1} \in \mathcal{L}_\infty$

where $\|(\cdot)_{t,t_1}\|$ denotes the $\mathcal{L}_{2\delta}$-norm defined over the interval $[t_1, t], t \geq t_1 \geq 0$; δ is any constant in the interval $(0, \delta_0]$ and $W(s)$ is a proper transfer function, which is analytic in $\operatorname{Re}[s] \geq -\delta_0/2$.

Proof (i) We have $\omega_1 = \frac{\alpha(s)}{\Lambda(s)} u_p, \omega_2 = \frac{\alpha(s)}{\Lambda(s)} y_p$. Since each element of $\frac{\alpha(s)}{\Lambda(s)}$ is strictly proper and analytic in $\operatorname{Re}[s] \geq -\frac{\delta_0}{2}$ and $\|u_{pt}\|_{2\delta_0}, \|y_{pt}\|_{2\delta_0} \leq cm$, it follows from Lemma 9.8.2 that

$$|\omega_1(t)|, \|\omega_{1_{t,t_1}}\| \leq ce^{-\frac{\delta}{2}(t-t_1)} m(t_1) + c\|u_{p_{t,t_1}}\| \leq cm_{f_1}$$

$$|\omega_2(t)|, \|\omega_{2_{t,t_1}}\| \leq ce^{-\frac{\delta}{2}(t-t_1)} m(t_1) + c\|y_{p_{t,t_1}}\| \leq cm_{f_1}$$

and, therefore, $|\omega_i(t)|, \|\omega_{i_{t,t_1}}\|, i = 1, 2$ are bounded from above by m_{f_1}. Because

$$\|\omega_{t,t_1}\|^2 \leq \|\omega_{1_{t,t_1}}\|^2 + \|\omega_{2_{t,t_1}}\|^2 + \|y_{p_{t,t_1}}\|^2 + c \leq cm_{f_1}^2$$

therefore, $\|\omega_{t,t_1}\|$ is bounded from above by m_{f_1}. We have $n_s^2 = m_s = m^2 - 1$ and

$$\begin{aligned} n_s^2 &= m_s(t) = e^{-\delta_0(t-t_1)}(m^2(t_1) - 1) + \|u_{p_{t,t_1}}\|_{2\delta_0}^2 + \|y_{p_{t,t_1}}\|_{2\delta_0}^2 \\ &\leq e^{-\delta(t-t_1)} m^2(t_1) + \|u_{p_{t,t_1}}\|^2 + \|y_{p_{t,t_1}}\|^2 = m_{f_1}^2 \end{aligned}$$

for any given $\delta \in (0, \delta_0]$, and the proof of (i) is complete.

(ii) We have

$$y_p = W_m(s)\rho^* \tilde{\theta}^\top \omega + \rho^* \eta + W_m(s) r$$

Because $W_m(s)$ is strictly proper and analytic in $Re[s] \geq -\delta_0/2$, and Lemma 9.8.1 together with $\tilde{\theta} \in \mathcal{L}_\infty$ imply that $\|\rho^* \tilde{\theta}^\top \omega_t\|_{2\delta_0} \leq cm(t) \; \forall t \geq 0$, it follows from Lemma 9.8.2 that

$$|y_p(t)|, \|y_{p_{t,t_1}}\| \leq ce^{-\frac{\delta}{2}(t-t_1)} m(t_1) + c\|\omega_{t,t_1}\| + c\|\eta_{t,t_1}\| + c, \quad \forall t \geq t_1 \geq 0$$

Now $\eta = \Delta(s) u_p + d_\eta$ where $\Delta(s)$ is strictly proper and analytic in $Re[s] \geq -\frac{\delta_0}{2}$, $d_\eta \in \mathcal{L}_\infty$ and $\|u_{p_t}\|_{2\delta_0} \leq m(t)$. Hence from Lemma 9.8.2 we have

$$\|\eta_{t,t_1}\| \leq ce^{-\frac{\delta}{2}(t-t_1)} m(t_1) + c\|u_{p_{t,t_1}}\| + c$$

Because $u_p = \theta^\top \omega$ and $\theta \in \mathcal{L}_\infty$, we have $\|u_{p_{t,t_1}}\| \leq c\|\omega_{t,t_1}\|$, and from part (i) we have $\|\omega_{t,t_1}\| \leq cm_{f_1}$. Therefore $\|u_{p_{t,t_1}}\| \leq cm_{f_1}, \|\eta_{t,t_1}\| \leq ce^{-\frac{\delta}{2}(t-t_1)} m(t_1) + cm_{f_1} + c$ and

$$|y_p(t)|, \|y_{p_{t,t_1}}\| \leq ce^{-\frac{\delta}{2}(t-t_1)} m(t_1) + cm_{f_1} + c \leq cm_{f_1}$$

In a similar manner, we show $\|\dot{y}_{p_t,t_1}\| \leq cm_{f_1}$. From $|\omega(t)| \leq |\omega_1(t)| + |\omega_2(t)| + |y_p(t)| + c$, it follows that $|\omega(t)| \leq cm_{f_1}$. Because $u_p = \theta^\top \omega$ and $\theta \in \mathcal{L}_\infty$, it follows directly that $|u_p(t)| \leq cm_{f_1}$.

Consider $v \triangleq W(s)\omega$ where $W(s)$ is proper and analytic in $Re[s] \geq -\delta_0/2$. Because from Lemma 9.8.1 $\|\omega_t\|_{2\delta_0} \leq cm(t)$, it follows from Lemma 9.8.2 that

$$|v(t)| \leq ce^{-\frac{\delta}{2}(t-t_1)}m(t_1) + c\|\omega_{t,t_1}\| \leq cm_{f_1}$$

(iii) We have $\dot{\omega} = [\dot{\omega}_1^\top, \dot{\omega}_2^\top, \dot{y}_p, \dot{r}]^\top$, where $\dot{\omega}_1 = \frac{s\alpha(s)}{\Lambda(s)}u_p, \dot{\omega}_2 = \frac{s\alpha(s)}{\Lambda(s)}y_p$. Because the elements of $\frac{s\alpha(s)}{\Lambda(s)}$ are proper, it follows from the results of (i), (ii) that $\|\dot{\omega}_{i_t,t_1}\| \leq cm_{f_1}, i = 1,2$ which together with $\dot{r} \in \mathcal{L}_\infty$ and $\|\dot{y}_{p_t,t_1}\| \leq cm_{f_1}$ imply (iii). $\square$

9.8.2 Proof of Theorem 9.3.2

We complete the proof of Theorem 9.3.2 in five steps outlined in Section 9.3.

Step 1. *Express the plant input and output in terms of the parameter error term $\tilde{\theta}^\top \omega$.* We use Figure 9.3 to express u_p, y_p in terms of the parameter error $\tilde{\theta}$ and modeling error input. We have

$$y_p = \frac{G_0 \Lambda c_0^*}{(\Lambda - C_1^*) - G_0 D_1^*}\left[r + \frac{1}{c_0^*}\tilde{\theta}^\top \omega + \frac{\Lambda - C_1^*}{c_0^* \Lambda}\eta_1\right]$$

$$\eta_1 = \Delta_m(s)(u_p + d_u) + d_u$$

$$u_p = \frac{\Lambda c_0^*}{(\Lambda - C_1^*) - G_0 D_1^*}\left[r + \frac{1}{c_0^*}\tilde{\theta}^\top \omega\right] + \frac{G_0 D_1^*}{(\Lambda - C_1^*) - G_0 D_1^*}\eta_1$$

where $C_1^*(s) = \theta_1^{*\top}\alpha(s), D_1^* = \theta_3^*\Lambda(s) + \theta_2^{*\top}\alpha(s)$. Using the matching equation

$$\frac{G_0 \Lambda c_0^*}{(\Lambda - C_1^*) - G_0 D_1^*} = W_m$$

we obtain

$$\begin{aligned} y_p &= W_m\left(r + \frac{1}{c_0^*}\tilde{\theta}^\top \omega\right) + \eta_y \\ u_p &= G_0^{-1}W_m\left(r + \frac{1}{c_0^*}\tilde{\theta}^\top \omega\right) + \eta_u \end{aligned} \quad (9.8.9)$$

where

$$\eta_u = \frac{\theta_3^* \Lambda + \theta_2^{*\top}\alpha}{c_0^* \Lambda}W_m\eta_1, \quad \eta_y = \frac{\Lambda - \theta_1^{*\top}\alpha}{c_0^* \Lambda}W_m\eta_1$$

9.8. STABILITY PROOFS OF ROBUST MRAC SCHEMES

Let us simplify the notation by denoting $\|(\cdot)_t\|_{2\delta}$ with $\|\cdot\|$. From (9.8.9) and the stability of $W_m, G_0^{-1}W_m$, we obtain

$$\|y_p\| \leq c + c\|\tilde{\theta}^T\omega\| + \|\eta_y\|, \quad \|u_p\| \leq c + c\|\tilde{\theta}^T\omega\| + \|\eta_u\| \tag{9.8.10}$$

for some $\delta > 0$ by applying Lemma 3.3.2. Using the expressions for η_u, η_y, we have

$$\|\eta_y\| \leq \left\|\frac{\Lambda(s) - \theta_1^{*T}\alpha(s)}{c_0^*\Lambda(s)}\right\|_{\infty\delta} \|W_m(s)\Delta_m(s)\|_{\infty\delta}\|u_p\| + cd_0$$

$$\|\eta_u\| \leq \left\|\frac{\theta_3^*\Lambda(s) + \theta_2^{*T}\alpha(s)}{c_0^*\Lambda(s)}\right\|_{\infty\delta} \|W_m(s)\Delta_m(s)\|_{\infty\delta}\|u_p\| + cd_0$$

where d_0 is the upper bound for d_u and $c \geq 0$ denotes any finite constant, which implies that

$$\|\eta_y\| \leq c\Delta_\infty m_f + cd_0, \quad \|\eta_u\| \leq c\Delta_\infty m_f + cd_0 \tag{9.8.11}$$

where $\Delta_\infty \triangleq \|W_m(s)\Delta_m(s)\|_{\infty\delta}$. From (9.8.10) and (9.8.11), it follows that the fictitious normalizing signal $m_f^2 \triangleq 1 + \|u_p\|^2 + \|y_p\|^2$ satisfies

$$m_f^2 \leq c + c\|\tilde{\theta}^T\omega\|^2 + c\Delta_\infty^2 m_f^2 \tag{9.8.12}$$

Step 2. *Use the swapping lemmas and properties of the $\mathcal{L}_{2\delta}$-norm to bound $\|\tilde{\theta}^T\omega\|$ from above.* Using the Swapping Lemma A.2 from Appendix A, we express $\tilde{\theta}^T\omega$ as

$$\tilde{\theta}^T\omega = F_1(s,\alpha_0)(\dot{\tilde{\theta}}^T\omega + \tilde{\theta}^T\dot{\omega}) + F(s,\alpha_0)\tilde{\theta}^T\omega \tag{9.8.13}$$

where $F(s,\alpha_0) = \frac{\alpha_0^k}{(s+\alpha_0)^k}$, $F_1(s,\alpha_0) = \frac{1-F(s,\alpha_0)}{s}$, α_0 is an arbitrary constant to be chosen, $k \geq n^*$ and n^* is the relative degree of $G_0(s)$ and $W_m(s)$. Using Swapping Lemma A.1, we have

$$\tilde{\theta}^T\omega = W^{-1}(s)\left[\tilde{\theta}^T W(s)\omega + W_c(W_b\omega^T)\dot{\tilde{\theta}}\right] \tag{9.8.14}$$

where $W(s)$ is any strictly proper transfer function with the property that $W(s)$, $W^{-1}(s)$ are analytic in $\text{Re}[s] \geq -\delta_0/2$. The transfer matrices $W_c(s)$, $W_b(s)$ are strictly proper and have the same poles as $W(s)$. Substituting for $\tilde{\theta}^T\omega$ given by (9.8.14) to the right hand side of (9.8.13), we obtain

$$\tilde{\theta}^T\omega = F_1[\dot{\tilde{\theta}}^T\omega + \tilde{\theta}^T\dot{\omega}] + FW^{-1}[\tilde{\theta}^T W\omega + W_c(W_b\omega^T)\dot{\tilde{\theta}}] \tag{9.8.15}$$

where FW^{-1} can be made proper by choosing $W(s)$ appropriately.

We now use equations (9.8.13) to (9.8.15) together with the properties of the robust adaptive laws to obtain an upper bound for $\|\tilde{\theta}^T\omega\|$. We consider each adaptive law of Tables 9.2 to 9.4 separately as follows:

Robust Adaptive Law of Table 9.2 We have

$$\epsilon = e_f - \hat{e}_f - W_m L \epsilon n_s^2 = W_m L[\rho^* \tilde{\theta}^T \phi - \tilde{\rho}\xi - \epsilon n_s^2 + \rho^* \eta_f]$$

therefore,

$$\tilde{\theta}^T \phi = \frac{1}{\rho^*}\left(W_m^{-1} L^{-1}\epsilon + \tilde{\rho}\xi + \epsilon n_s^2\right) - \eta_f \qquad (9.8.16)$$

where $\phi = L_0(s)\omega$, $L_0(s) = L^{-1}(s)\frac{h_0}{s+h_0}$. Choosing $W(s) = L_0(s)$ in (9.8.15) we obtain

$$\tilde{\theta}^T \omega = F_1[\dot{\tilde{\theta}}^T \omega + \tilde{\theta}^T \dot{\omega}] + FL_0^{-1}[\tilde{\theta}^T L_0 \omega + W_c(W_b \omega^T)\dot{\tilde{\theta}}]$$

Substituting for $\tilde{\theta}^T \phi = \tilde{\theta}^T L_0 \omega$ and $L_0(s) = L^{-1}(s)\frac{h_0}{s+h_0}$, and using (9.8.16), we obtain

$$\tilde{\theta}^T \omega = F_1[\dot{\tilde{\theta}}^T \omega + \tilde{\theta}^T \dot{\omega}]$$
$$+ FW_m^{-1}\frac{s+h_0}{h_0 \rho^*}\epsilon + FL_0^{-1}\left[\frac{\tilde{\rho}\xi + \epsilon n_s^2}{\rho^*} - \eta_f + W_c(W_b \omega^T)\dot{\tilde{\theta}}\right]$$

Substituting for $\xi = u_f - \theta^T \phi = L_0 \theta^T \omega - \theta^T L_0 \phi = W_c(W_b \omega^T)\dot{\theta}$, where the last equality is obtained by applying Swapping Lemma A.1 to $L_0(s)\theta^T \omega$, we obtain

$$\tilde{\theta}^T \omega = F_1[\dot{\tilde{\theta}}^T \omega + \tilde{\theta}^T \dot{\omega}] + FW_m^{-1}\frac{s+h_0}{h_0 \rho^*}\epsilon + FL_0^{-1}\left[\frac{\epsilon n_s^2}{\rho^*} - \eta_f + \frac{\rho}{\rho^*}W_c(W_b \omega^T)\dot{\tilde{\theta}}\right] \qquad (9.8.17)$$

Choosing $k = n^* + 1$ in the expression for $F(s, \alpha_0)$, it follows that $FW_m^{-1}(s+h_0)$ is biproper, $FL_0^{-1} = FL\frac{s+h_0}{h_0}$ is strictly proper and both are analytic in $Re[s] \geq -\delta_0/2$. Using the properties of F, F_1 given by Swapping Lemma A.2, and noting that $\|\frac{1}{(s+\alpha_0)^k}\frac{s+h_0}{h_0}W_m^{-1}(s)\|_{\infty\delta}$ and $\|\frac{1}{(s+\alpha_0)^k}\frac{s+h_0}{h_0}L(s)\|_{\infty\delta}$ are finite constants, we have

$$\|F_1(s, \alpha_0)\|_{\infty\delta} \leq \frac{c}{\alpha_0}, \quad \left\|F(s, \alpha_0)W_m^{-1}(s)\frac{s+h_0}{h_0}\right\|_{\infty\delta} \leq c\alpha_0^k$$

$$\|F(s, \alpha_0)L_0^{-1}(s)\|_{\infty\delta} = \left\|F(s, \alpha_0)L(s)\frac{s+h_0}{h_0}\right\|_{\infty\delta} \leq c\alpha_0^k \qquad (9.8.18)$$

for any $\delta \in (0, \delta_0]$.

We can use (9.8.17), (9.8.18) and the properties of the $\mathcal{L}_{2\delta}$ norm given by Lemma 3.3.2 to derive the inequality

$$\|\tilde{\theta}^T \omega\| \leq \frac{c}{\alpha_0}(\|\dot{\tilde{\theta}}^T \omega\| + \|\tilde{\theta}^T \dot{\omega}\|) + c\alpha_0^k\|\epsilon\| + c\alpha_0^k\|\epsilon n_s^2\|$$
$$+ c\alpha_0^k\|W_c(W_b\omega^T)\dot{\tilde{\theta}}\| + \|FL_0^{-1}\eta_f\| \qquad (9.8.19)$$

9.8. STABILITY PROOFS OF ROBUST MRAC SCHEMES

From Lemma 9.8.1 and $\theta \in \mathcal{L}_\infty$, we have

$$\frac{\omega}{m_f},\ \frac{\|\dot{\omega}\|}{m_f},\ \frac{W_b\omega^T}{m_f},\ \frac{n_s}{m_f} \in \mathcal{L}_\infty$$

$$\|W_c(W_b\omega^T)\dot{\theta}\| \le c\|\dot{\theta}m_f\|,\ \|\tilde{\theta}^T\dot{\omega}\| \le cm_f,\ \|\epsilon n_s^2\| \le \|\epsilon n_s m_f\|$$

Furthermore,

$$\|FL_0^{-1}\eta_f\| = \|F\eta_0\| = \|FW_m^{-1}\eta\| \le \|F(s)W_m^{-1}(s)\|_{\infty\delta}\|\eta\| \le c\alpha_0^k\|\eta\|$$

where

$$\eta = \frac{\Lambda(s) - \theta_1^{*T}\alpha(s)}{\Lambda(s)}W_m(s)[\Delta_m(s)(u_p + d_u) + d_u]$$

and

$$\|\eta\| \le \left\|\frac{\Lambda(s) - \theta_1^{*T}\alpha(s)}{\Lambda(s)}\right\|_{\infty\delta}\|W_m(s)\Delta_m(s)\|_{\infty\delta}\|u_p\| + cd_0$$

$$\le c\Delta_\infty m_f + cd_0$$

Hence,

$$\|FL_0^{-1}\eta_f\| \le c\alpha_0^k \Delta_\infty m_f + cd_0$$

and (9.8.19) may be rewritten as

$$\|\tilde{\theta}^T\omega\| \le \frac{c}{\alpha_0}\|\dot{\tilde{\theta}}m_f\| + \frac{c}{\alpha_0}m_f + c\alpha_0^k(\|\epsilon n_s m_f\| + \|\epsilon m_f\| + \|\dot{\tilde{\theta}}m_f\|) + c\alpha_0^k\Delta_\infty m_f + cd_0 \qquad (9.8.20)$$

where $c \ge 0$ denotes any finite constant and for ease of presentation, we use the inequality $\|\epsilon\| \le \|\epsilon m_f\|$ in order to simplify the calculations. We can also express (9.8.20) in the compact form

$$\|\tilde{\theta}^T\omega\| \le c\|\tilde{g}m_f\| + c\left(\frac{1}{\alpha_0} + \alpha_0^k\Delta_\infty\right)m_f + cd_0 \qquad (9.8.21)$$

where $\tilde{g}^2 = \frac{|\dot{\tilde{\theta}}|^2}{\alpha_0^2} + \alpha_0^{2k}(|\epsilon n_s|^2 + |\dot{\tilde{\theta}}|^2 + \epsilon^2)$. Since $\epsilon, \epsilon n_s, \dot{\tilde{\theta}} \in \mathcal{S}(f_0 + \frac{\eta_f^2}{m^2})$, it follows that $\tilde{g} \in \mathcal{S}(f_0 + \frac{\eta_f^2}{m^2})$. Because

$$\eta_f = \frac{\Lambda(s) - \theta_1^{*T}\alpha(s)}{\Lambda(s)}L^{-1}(s)\frac{h_0}{s+h_0}[\Delta_m(s)(u_p + d_u) + d_u]$$

we have

$$|\eta_f(t)| \le \Delta_{02}m(t) + cd_0$$

where

$$\Delta_{02} \triangleq \left\|\frac{\Lambda(s) - \theta_1^{*T}\alpha(s)}{\Lambda(s)}L^{-1}(s)\frac{h_0}{s+h_0}\Delta_m(s)\right\|_{2\delta_0} \qquad (9.8.22)$$

Hence, in (9.8.21), $g \in S(f_0 + \Delta_{02}^2 + \frac{d_0^2}{m^2})$.

Robust Adaptive Law of Table 9.3 This adaptive law follows directly from that of Table 9.2 by taking $L^{-1}(s) = W_m(s)$, $L_0(s) = W_m(s)$ and by replacing $\frac{h_0}{s+h_0}$ by unity. We can therefore go directly to equation (9.8.17) and obtain

$$\tilde{\theta}^T \omega = F_1(\dot{\tilde{\theta}}^T \omega + \tilde{\theta}^T \dot{\omega}) + \frac{FW_m^{-1}}{\rho^*}\epsilon + FW_m^{-1}\left[\frac{\epsilon n_s^2}{\rho^*} + \frac{\rho}{\rho^*}W_c(W_b\omega^T)\dot{\tilde{\theta}} - \eta\right]$$

where

$$\eta = \frac{\Lambda(s) - \theta_1^{*T}\alpha(s)}{\Lambda(s)} W_m(s)[\Delta_m(s)(u_p + d_u) + d_u]$$

The value of k in the expression for $F(s, \alpha_0)$ can be taken as $k = n^*$ (even though $k = n^* + 1$ will also work) leading to FW_m^{-1} being biproper.

Following the same procedure as in the case of the adaptive law of Table 9.2, we obtain

$$\|\tilde{\theta}^T \omega\| \leq \frac{c}{\alpha_0}(\|\dot{\tilde{\theta}}^T \omega\| + \|\tilde{\theta}^T \dot{\omega}\|) + c\alpha_0^k \|\epsilon\| + c\alpha_0^k \|\epsilon n_s^2\|$$
$$+ c\alpha_0^k \|W_c(W_b\omega^T)\dot{\tilde{\theta}}\| + \|FW_m^{-1}\eta\|$$

which may be rewritten in the form

$$\|\tilde{\theta}^T\omega\| \leq c\|\tilde{g}m_f\| + c\left(\frac{c}{\alpha_0} + \alpha_0^k \Delta_\infty\right)m_f + cd_0 \qquad (9.8.23)$$

where $\tilde{g}^2 = \frac{|\dot{\tilde{\theta}}|^2}{\alpha_0^2} + \alpha_0^{2k}(|\epsilon n_s|^2 + |\dot{\tilde{\theta}}|^2 + \epsilon^2)$. Because $\epsilon, \epsilon n_s, \dot{\tilde{\theta}} \in S(f_0 + \frac{\eta^2}{m^2})$, it follows that $\tilde{g} \in S(f_0 + \frac{\eta^2}{m^2})$. Because

$$|\eta(t)| \leq \Delta_2 m(t) + cd_0$$

where

$$\Delta_2 = \left\|\frac{\Lambda(s) - \theta_1^{*T}\alpha(s)}{\Lambda(s)} W_m(s)\Delta_m(s)\right\|_{2\delta_0} \qquad (9.8.24)$$

it follows that in (9.8.23) $\tilde{g} \in S(f_0 + \Delta_2^2 + \frac{d_0^2}{m^2})$.

Robust Adaptive Law of Table 9.4 We have $\epsilon m^2 = z - \hat{z} = -\tilde{\theta}^T\phi_p - \eta$, i.e., $\tilde{\theta}^T\phi_p = -\eta - \epsilon m^2$. We need to relate $\tilde{\theta}^T\phi_p$ with $\tilde{\theta}^T\omega$. Consider the identities

$$\tilde{\theta}^T\phi_p = \tilde{\theta}_0^T\phi_0 + \tilde{c}_0 y_p, \quad \tilde{\theta}^T\omega = \tilde{\theta}_0^T\omega_0 + \tilde{c}_0 r$$

where $\tilde{\theta}_0 = [\tilde{\theta}_1^T, \tilde{\theta}_2^T, \tilde{\theta}_3]^T$, $\omega_0 = [\omega_1^T, \omega_2^T, y_p]^T$ and $\phi_0 = W_m(s)\omega_0$. Using the above equations, we obtain

$$\tilde{\theta}_0^T\phi_0 = \tilde{\theta}^T\phi_p - \tilde{c}_0 y_p = -\epsilon m^2 - \tilde{c}_0 y_p - \eta \qquad (9.8.25)$$

9.8. STABILITY PROOFS OF ROBUST MRAC SCHEMES

Let us now use the Swapping Lemma A.1 to write

$$W_m(s)\tilde{\theta}^\top \omega = \tilde{\theta}_0^\top W_m(s)\omega_0 + \tilde{c}_0 y_m + W_c(W_b \omega^\top)\dot{\tilde{\theta}}$$

Because $\tilde{\theta}_0^\top \phi_0 = \tilde{\theta}_0^\top W_m(s)\omega_0$, it follows from above and (9.8.25) that

$$W_m(s)\tilde{\theta}^\top \omega = -\epsilon m^2 - \tilde{c}_0 y_p - \eta + \tilde{c}_0 y_m + W_c(W_b \omega^\top)\dot{\tilde{\theta}} \quad (9.8.26)$$

Substituting for

$$y_p = y_m + \frac{1}{c_0^*} W_m(s)\tilde{\theta}^\top \omega + \eta_y$$

in (9.8.26) and using $\eta = c_0^* \eta_y$, where

$$\eta_y = \frac{\Lambda(s) - \theta_1^{*\top}\alpha(s)}{c_0^* \Lambda(s)} W_m(s)[\Delta_m(s)(u_p + d_u) + d_u]$$

we obtain

$$W_m(s)\tilde{\theta}^\top \omega = -\epsilon m^2 - \frac{\tilde{c}_0}{c_0^*} W_m(s)\tilde{\theta}^\top \omega + W_c(W_b \omega^\top)\dot{\tilde{\theta}} - c_0 \eta_y$$

Because $1 + \frac{\tilde{c}_0}{c_0^*} = \frac{c_0}{c_0^*}$ and $\frac{1}{c_0} \in \mathcal{L}_\infty$, we have

$$W_m(s)\tilde{\theta}^\top \omega = \frac{c_0^*}{c_0}\left[-\epsilon m^2 + W_c(W_b \omega^\top)\dot{\tilde{\theta}} - c_0 \eta_y\right] \quad (9.8.27)$$

Rewriting (9.8.13) as

$$\tilde{\theta}^\top \omega = F_1(\dot{\tilde{\theta}}^\top \omega + \tilde{\theta}^\top \dot{\omega}) + FW_m^{-1}(W_m \tilde{\theta}^\top \omega)$$

and substituting for $W_m(s)\tilde{\theta}^\top \omega$ from (9.8.27), we obtain

$$\begin{aligned}\tilde{\theta}^\top \omega &= F_1(\dot{\tilde{\theta}}^\top \omega + \tilde{\theta}^\top \dot{\omega}) \\ &+ FW_m^{-1}\frac{c_0^*}{c_0}\left[-\epsilon m^2 + W_c(W_b \omega^\top)\dot{\tilde{\theta}} - c_0 \eta_y\right]\end{aligned} \quad (9.8.28)$$

Following the same approach as with the adaptive law of Table 9.3, we obtain

$$\|\tilde{\theta}^\top \omega\| \leq c\|\tilde{g}m_f\| + c\left(\frac{c}{\alpha_0} + \alpha_0^k \Delta_\infty\right) m_f + c d_0 \quad (9.8.29)$$

where $\tilde{g} \in \mathcal{S}(f_0 + \frac{\eta^2}{m^2})$ or $\tilde{g} \in \mathcal{S}(f_0 + \Delta_2^2 + \frac{d_0^2}{m^2})$ and $\tilde{g}^2 \triangleq \frac{|\dot{\theta}|^2}{\alpha_0^2} + \alpha_0^{2k}(|\epsilon n_s^2| + |\dot{\theta}|^2 + \epsilon^2)$.

Step 3. *Use the B-G Lemma to establish boundedness.* The bound for $\|\tilde{\theta}^\top \omega\|$ in (9.8.21), (9.8.23), and (9.8.29) has exactly the same form for all three adaptive laws given in Tables 9.2 to 9.4. Substituting for the bound of $\|\tilde{\theta}^T \omega\|$ in (9.8.12) we obtain

$$m_f^2 \leq c + c\|\tilde{g} m_f\|^2 + c\left(\frac{1}{\alpha_0^2} + \alpha_0^{2k}\Delta_\infty^2\right) m_f^2 + c d_0^2$$

For

$$c\left(\frac{1}{\alpha_0^2} + \alpha_0^{2k}\Delta_\infty^2\right) < 1 \qquad (9.8.30)$$

we have

$$m_f^2 \leq c_0 + c\|\tilde{g} m_f\|^2$$

where c_0 depends on d_0^2, which may be rewritten as

$$m_f^2 \leq c_0 + c\int_0^t e^{-\delta(t-\tau)} \tilde{g}^2(\tau) m_f^2(\tau) d\tau$$

Applying the B-G Lemma III, we obtain

$$m_f^2 \leq c e^{-\delta t} e^{c \int_0^t \tilde{g}^2(\tau) d\tau} + c_0 \delta \int_0^t e^{-\delta(t-s)} e^{c \int_s^t \tilde{g}^2(\tau) d\tau} ds$$

Because $\tilde{g} \in \mathcal{S}(f_0 + \Delta_i^2 + \frac{d_0^2}{m^2})$ where $\Delta_i = \Delta_{02}$ for the SPR-Lyapunov based adaptive law and $\Delta_i = \Delta_2$ for the adaptive laws of Tables 9.2 and 9.4, we obtain

$$m_f^2 \leq c e^{-\frac{\delta}{2} t} e^{c \int_0^t \frac{d_0^2}{m^2(\tau)} d\tau} + c_0 \delta \int_0^t e^{-\frac{\delta}{2}(t-s)} e^{c \int_s^t \frac{d_0^2}{m^2(\tau)} d\tau} ds \qquad (9.8.31)$$

provided

$$c(f_0 + \Delta_i^2) \leq \frac{\delta}{2} \qquad (9.8.32)$$

where c in (9.8.32) is proportional to $\frac{1}{\alpha_0^2} + \alpha_0^{2k}$ and can be calculated by keeping track of all the constants in each step. The constant c also depends on the $H_{\infty\delta}$ and $H_{2\delta}$ norms of the transfer functions involved and the upper bound for the estimated parameters.

To establish the boundedness of m_f, we have to show that $\frac{c d_0^2}{m^2(t)} < \frac{\delta}{2}$ for all $t \geq 0$ or for most of the time. The boundedness of m_f will follow directly if we modify the normalizing signal as $m^2 = 1 + n_s^2, n_s^2 = \beta_0 + m_s$ and choose the constant β_0 large enough so that

$$\frac{c d_0^2}{m^2(t)} \leq \frac{c d_0^2}{\beta_0} \leq \frac{\delta}{2}$$

$\forall t \geq 0$. This means that m is always larger than the level of the disturbance. Such a large m will slow down the speed of adaptation and may in fact improve robustness. A slow adaptation, however, may have an adverse effect on performance.

The boundedness of signals can be established, however, without having to modify the normalizing signal by using the properties of the $\mathcal{L}_{2\delta}$ norm defined over

9.8. STABILITY PROOFS OF ROBUST MRAC SCHEMES

an arbitrary interval $[t_1, t]$ given by Lemma 9.8.2, 9.8.3 and by repeating steps 1 to 3 as follows:

We apply Lemma 9.8.2 to (9.8.9) to obtain

$$\|u_{p_{t,t_1}}\|^2, \|y_{p_{t,t_1}}\|^2 \leq ce^{-\delta(t-t_1)}m^2(t_1) + c\|\tilde{\theta}^T \omega_{t,t_1}\|^2 + c\Delta_\infty^2 \|u_{p_{t,t_1}}\|^2 + cd_0^2$$

where we use the fact that $\|\tilde{\theta}^T \omega_t\|_{2\delta_0}, \|y_{pt}\|_{2\delta_0}, \|u_{pt}\|_{2\delta_0} \leq cm(t)$. Therefore, the fictitious signal $m_{f_1}^2 \triangleq e^{-\delta(t-t_1)}m^2(t_1) + \|u_{p_{t,t_1}}\|^2 + \|y_{p_{t,t_1}}\|^2$ satisfies

$$m_{f_1}^2 \leq ce^{-\delta(t-t_1)}m^2(t_1) + c\|\tilde{\theta}^T \omega_{t,t_1}\|^2 + c\Delta_\infty^2 m_{f_1}^2 + cd_0^2 \quad \forall t \geq t_1 \geq 0$$

Following the same procedure as in step 2 and using Lemma 9.8.2, 9.8.3 we obtain

$$\|\tilde{\theta}^T \omega\|^2 \leq ce^{-\delta(t-t_1)}m^2(t_1) + c\|(\tilde{g}m_{f_1})_{t,t_1}\|^2 + c\left(\frac{1}{\alpha_0^2} + \alpha_0^{2k}\right)\Delta_\infty^2 m_{f_1}^2 + cd_0^2$$

where $\tilde{g}$ is as defined before. Therefore,

$$m_{f_1}^2 \leq ce^{-\delta(t-t_1)}m^2(t_1) + c\|(\tilde{g}m_{f_1})_{t,t_1}\|^2 + c\left(\frac{1}{\alpha_0^2} + \alpha_0^{2k}\right)\Delta_\infty^2 m_{f_1}^2 + cd_0^2$$

Using (9.8.30), we obtain

$$m_{f_1}^2 \leq ce^{-\delta(t-t_1)}m^2(t_1) + c\|(\tilde{g}m_{f_1})_{t,t_1}\|^2 + cd_0^2, \quad \forall t \geq t_1$$

or

$$m_{f_1}^2(t) \leq c + ce^{-\delta(t-t_1)}m^2(t_1) + c\int_{t_1}^t e^{-\delta(t-\tau)}\tilde{g}^2(\tau)m_{f_1}^2(\tau)d\tau$$

Applying the B-G Lemma III, we obtain

$$m_{f_1}^2(t) \leq ce^{-\delta(t-t_1)}(1 + m^2(t_1))e^{c\int_{t_1}^t \tilde{g}^2(\tau)d\tau} + c\delta \int_{t_1}^t e^{-\delta(t-s)}e^{c\int_s^t \tilde{g}^2(\tau)d\tau}ds$$

$\forall t \geq t_1 \geq 0$. Because $\tilde{g} \in \mathcal{S}(f_0 + \Delta_i^2 + \frac{d_0^2}{m^2})$, it follows as before that for $cf_0 + c\Delta_i^2 \leq \delta/2$, we have

$$m^2(t) \leq m_{f_1}^2(t) \leq ce^{-\delta/2(t-t_1)}(1 + m^2(t_1))e^{c\int_{t_1}^t d_0^2/m^2 d\tau}$$
$$+ c\delta \int_{t_1}^t e^{-\delta/2(t-s)}e^{c\int_s^t d_0^2/m^2(\tau)d\tau}ds \qquad (9.8.33)$$

$\forall t \geq t_1$, where the inequality $m^2(t) \leq m_{f_1}^2$ follows from the definition of m_{f_1}. If we establish that $m \in \mathcal{L}_\infty$, then it will follow from Lemma 9.8.1 that all signals are bounded. The boundedness of m is established by contradiction as follows: Let us assume that $m^2(t)$ grows unbounded. Because $\theta \in \mathcal{L}_\infty$, it follows that

$$m^2(t) \leq e^{k_1(t-t_0)}m^2(t_0)$$

for some $k_1 > 0$, i.e., $m^2(t)$ cannot grow faster than an exponential. As $m^2(t)$ grows unbounded, we can find a $t_0 > \bar{\alpha} > 0$ and a $t_2 > t_0$ with $\bar{\alpha} > t_2 - t_0$ such that $m^2(t_2) > \bar{\alpha} e^{k_1 \bar{\alpha}}$ for some large constant $\bar{\alpha} > 0$. We have

$$\bar{\alpha} e^{k_1 \bar{\alpha}} < m^2(t_2) \le e^{k_1(t_2-t_0)} m^2(t_0)$$

which implies that
$$\ln m^2(t_0) > \ln \bar{\alpha} + k_1[\bar{\alpha} - (t_2 - t_0)]$$

Because $\bar{\alpha} > t_2 - t_0$, it follows that $\ln m^2(t_0) > \ln \bar{\alpha}$, i.e., $m^2(t_0) > \bar{\alpha}$ for $t_0 \in (t_2 - \bar{\alpha}, t_2)$.

Let $t_1 = \sup_{\tau \le t_2} \{arg(m^2(\tau) = \bar{\alpha})\}$. Since $m^2(t_0) > \bar{\alpha}$ for all $t_0 \in (t_2 - \bar{\alpha}, t_2)$, it follows that $t_1 \le t_2 - \bar{\alpha}$ and $m^2(t) \ge \bar{\alpha} \; \forall t \in [t_1, t_2)$ and $t_2 - t_1 \ge \bar{\alpha}$. We now consider (9.8.33) with t_1 as defined above and $t = t_2$. We have

$$m^2(t_2) \le c(1+\bar{\alpha})e^{-\beta(t_2-t_1)} + c\delta \int_{t_1}^{t_2} e^{-\beta(t-s)} ds$$

where $\beta = \frac{\delta}{2} - \frac{cd_0^2}{\bar{\alpha}}$. For large $\bar{\alpha}$, we have $\beta = \frac{\delta}{2} - \frac{cd_0^2}{\bar{\alpha}} > 0$ and

$$m^2(t_2) \le c(1+\bar{\alpha})e^{-\beta \bar{\alpha}} + \frac{c\delta}{\beta}$$

Hence, for sufficiently large $\bar{\alpha}$, we have $m^2(t_2) < c < \bar{\alpha}$, which contradicts the hypothesis that $m^2(t_2) > \bar{\alpha} e^{k_1 \bar{\alpha}} > \bar{\alpha}$. Therefore, $m \in \mathcal{L}_\infty$, which, together with Lemma 9.8.1, implies that all signals are bounded.

Step 4. *Establish bounds for the tracking error.* In this step, we establish bounds for the tracking error e_1 by relating it with signals that are guaranteed by the adaptive law to be of the order of the modeling error in m.s.s. We consider each adaptive law separately.

Robust Adaptive Law of Table 9.2 Consider the tracking error equation

$$e_1 = W_m(s)\rho^* \tilde{\theta}^T \omega + \eta_y$$

We have
$$|e_1(t)| \le \|W_m(s)\|_{2\delta} \|\tilde{\theta}^T \omega\| |\rho^*| + |\eta_y|$$

Therefore,
$$|e_1(t)| \le c\|\tilde{\theta}^T \omega\| + c|\eta_y|$$

Using (9.8.21) and $m_f \in \mathcal{L}_\infty$ in the above equation, we obtain

$$|e_1(t)|^2 \le c\|\tilde{g}\|^2 + c\left(\frac{1}{\alpha_0} + \alpha_0^k \Delta_\infty\right)^2 + cd_0^2 + c|\eta_y|^2$$

9.8. STABILITY PROOFS OF ROBUST MRAC SCHEMES

We can also establish that
$$|\eta_y(t)|^2 \leq c\Delta_2^2 + cd_0^2$$
Therefore,
$$|e_1(t)|^2 \leq c\|\tilde{g}\|^2 + c\left(\frac{1}{\alpha_0} + \alpha_0^k \Delta_\infty\right)^2 + cd_0^2 + c\Delta_2^2$$
where $\tilde{g} \in \mathcal{S}(f_0 + \Delta_{02}^2 + d_0^2)$. Using Corollary 3.3.3, we can establish that $\|\tilde{g}\| \in \mathcal{S}(f_0 + \Delta_{02}^2 + d_0^2)$ and, therefore,
$$\int_t^{t+T} |e_1|^2 d\tau \leq c(f_0 + \Delta^2 + d_0^2)T + c$$
where $\Delta = \frac{1}{\alpha_0^2} + \Delta_\infty^2 + \Delta_2^2 + \Delta_{02}^2$.

Robust Adaptive Law of Table 9.3 It follows from the equation of the estimation error that
$$e_1 = \rho\xi + \epsilon m^2$$
Because $\epsilon, \xi, \epsilon m \in \mathcal{S}(f_0 + \Delta_2^2 + d_0^2)$ and $\rho, m \in \mathcal{L}_\infty$, it follows that $e_1 \in \mathcal{S}(f_0 + \Delta_2^2 + d_0^2)$.

Robust Adaptive Law of Table 9.4 Substituting (9.8.27) in the tracking error equation
$$e_1 = \frac{1}{c_0^*} W_m(s)\tilde{\theta}^\top \omega + \eta_y$$
and using $\eta = c_0^* \eta_y$, we obtain
$$e_1 = \frac{1}{c_0}\left(-\epsilon m^2 + W_c(W_b \omega^\top)\dot{\tilde{\theta}}\right)$$

Using $\omega \in \mathcal{L}_\infty$, $\dot{\theta} \in \mathcal{S}(f_0 + \Delta_2^2 + d_0^2)$, we have that the signal $\xi = W_c(W_b\omega^\top)\dot{\tilde{\theta}} \in \mathcal{S}(f_0 + \Delta_2^2 + d_0^2)$. Since $\frac{1}{c_0}, m \in \mathcal{L}_\infty$ and $\epsilon m \in \mathcal{S}(f_0 + \Delta_2^2 + d_0^2)$, it follows that $e_1 \in \mathcal{S}(f_0 + \Delta_2^2 + d_0^2)$.

Step 5: *Establish parameter and tracking error convergence.* As in the ideal case, we first show that ϕ, ϕ_p is PE if r is dominantly rich. From the definition of ϕ, ϕ_p, we have

$$\phi = H(s)\begin{bmatrix} \frac{\alpha(s)}{\Lambda(s)} u_p \\ \frac{\alpha(s)}{\Lambda(s)} y_p \\ y_p \\ r \end{bmatrix}, \quad \phi_p = \begin{bmatrix} W_m(s)\frac{\alpha(s)}{\Lambda(s)} u_p \\ W_m(s)\frac{\alpha(s)}{\Lambda(s)} y_p \\ W_m(s) y_p \\ y_p \end{bmatrix}$$

where $H(s) = L_0(s)$ for the adaptive law of Table 9.2 and $H(s) = W_m(s)$ for the adaptive law of Table 9.3. Using $y_p = W_m(s)r + e_1$, $u_p = G_0^{-1}(s)(W_m(s)r + e_1) - \eta_1$, $\eta_1 = \Delta_m(s)(u_p + d_u) + d_u$, we can write
$$\phi = \phi_m + \phi_e, \quad \phi_p = \phi_{pm} + \phi_{pe}$$

where

$$\phi_m = H(s) \begin{bmatrix} \frac{\alpha(s)}{\Lambda(s)} G_0^{-1}(s) W_m(s) \\ \frac{\alpha(s)}{\Lambda(s)} W_m(s) \\ W_m(s) \\ 1 \end{bmatrix} r, \quad \phi_{pm} = W_m(s) \begin{bmatrix} \frac{\alpha(s)}{\Lambda(s)} G_0^{-1}(s) W_m(s) \\ \frac{\alpha(s)}{\Lambda(s)} W_m(s) \\ W_m(s) \\ 1 \end{bmatrix} r$$

$$\phi_e = H(s) \begin{bmatrix} \frac{\alpha(s)}{\Lambda(s)} G_0^{-1}(s) \\ \frac{\alpha(s)}{\Lambda(s)} \\ 1 \\ 0 \end{bmatrix} e_1 - H(s) \begin{bmatrix} \frac{\alpha(s)}{\Lambda(s)} \\ 0 \\ 0 \\ 0 \end{bmatrix} \eta_1$$

$$\phi_{pe} = \begin{bmatrix} W_m(s) \frac{\alpha(s)}{\Lambda(s)} G_0^{-1}(s) \\ W_m(s) \frac{\alpha(s)}{\Lambda(s)} \\ W_m(s) \\ 1 \end{bmatrix} e_1 - \begin{bmatrix} W_m(s) \frac{\alpha(s)}{\Lambda(s)} \\ 0 \\ 0 \\ 0 \end{bmatrix} \eta_1$$

Because we have established that $e_1 \in \mathcal{S}(\Delta^2 + d_0^2 + f_0)$ and $u_p \in \mathcal{L}_\infty$, we conclude that $\phi_e, \phi_{pe} \in \mathcal{S}(\Delta^2 + d_0^2 + f_0)$. In Chapter 8, we have proved that ϕ_m, ϕ_{pm} are PE with level $\alpha_0 > O(\Delta^2 + d_0^2)$ provided that r is dominantly rich and Z_p, R_p are coprime, i.e., there exist $T_0 > 0, T_{p0} > 0, \alpha_0 > 0, \alpha_{p0} > 0$ such that

$$\frac{1}{T_0} \int_t^{t+T_0} \phi_m(\tau) \phi_m^\top(\tau) d\tau \geq \alpha_0 I, \quad \frac{1}{T_{p0}} \int_t^{t+T_{p0}} \phi_{pm}(\tau) \phi_{pm}^\top(\tau) d\tau \geq \alpha_{p0} I$$

$\forall t \geq 0$. Note that

$$\frac{1}{nT_0} \int_t^{t+nT_0} \phi(\tau) \phi^\top(\tau) d\tau \geq \frac{1}{2nT_0} \int_t^{t+nT_0} \phi_m(\tau) \phi_m^\top(\tau) d\tau$$
$$- \frac{1}{nT_0} \int_t^{t+nT_0} \phi_e(\tau) \phi_e^\top(\tau) d\tau$$
$$\geq \frac{\alpha_0}{2} I - \left(c(\Delta^2 + d_0^2 + f_0) + \frac{c}{nT_0} \right) I$$

where n is any positive integer. If we choose n to satisfy $\frac{c}{nT_0} < \frac{\alpha_0}{8}$, then for $c(\Delta^2 + d_0^2 + f_0) < \frac{\alpha_0}{8}$, we have

$$\frac{1}{nT_0} \int_t^{t+nT_0} \phi(\tau) \phi^\top(\tau) d\tau \geq \frac{\alpha_0}{4} I$$

which implies that ϕ is PE. Similarly, we can establish the PE property for ϕ_p.

Using the results of Chapters 4 and 8, we can establish that when ϕ, or ϕ_p is PE, the robust adaptive laws guarantee that $\tilde{\theta}$ converges to a residual set whose size is of the order of the modeling error, i.e., $\tilde{\theta}$ satisfies

$$|\tilde{\theta}(t)| \leq c(f_0 + \Delta + d_0) + r_{\tilde{\theta}}(t) \qquad (9.8.34)$$

where $r_{\tilde{\theta}}(t) \to 0$ as $t \to \infty$. Furthermore, for the robust adaptive law of Table 9.4 based on the linear parametric model, we have $r_{\tilde{\theta}}(t) \to 0$ exponentially fast. Now

$$e_1(t) = \frac{1}{c_0^*} W_m(s) \tilde{\theta}^T \omega + \eta_y \qquad (9.8.35)$$

where $\eta_y = \frac{\Lambda - \theta_1^{*T} \alpha}{c_0^* \Lambda} W_m(d_u + \Delta_m(s)(u_p + d_u))$. Because $\omega \in \mathcal{L}_\infty$ and $|\eta_y| \leq c(\Delta_2 + d_0)$, we can conclude from (9.8.34) and (9.8.35) that

$$|e_1(t)| \leq c(f_0 + \Delta + d_0) + c r_e(t) \qquad (9.8.36)$$

where $r_e(t) \triangleq \int_0^t h_m(t-\tau) r_{\tilde{\theta}}(\tau) d\tau$ and $h_m(t) = \mathcal{L}^{-1}\{W_m(s)\}$. Therefore, we have $r_e(t) \to 0$ as $t \to \infty$. Furthermore, when $r_{\tilde{\theta}}(t)$ converges to zero exponentially fast (i.e., the adaptive law of Table 9.4 is used), $r_e(t) \to 0$ exponentially fast. Combining (9.8.34) and (9.8.36), the parameter error and tracking error convergence to $\mathcal{S}$ follows.

9.9 Stability Proofs of Robust APPC Schemes

9.9.1 Proof of Theorem 9.5.2

The proof is completed by following the same steps as in the ideal case and Example 9.5.1.

Step 1. *Express u_p, y_p in terms of the estimation error.* Following exactly the same steps as in the ideal case given in Section 7.7.1, we can show that the input u_p and output y_p satisfy the same equations as (7.4.24), that is

$$\begin{aligned} \dot{x} &= A(t)x + b_1(t)\epsilon m^2 + b_2 \bar{y}_m \\ y_p &= C_1^T x + d_1 \epsilon m^2 + d_2 \bar{y}_m \\ u_p &= C_2^T x + d_3 \epsilon m^2 + d_4 \bar{y}_m \end{aligned} \qquad (9.9.1)$$

where $x, A(t), b_1(t), b_2$ and $C_i, i = 1, 2, d_k, k = 1, 2, 3, 4$ are as defined in Section 7.7.1 and $\bar{y}_m = P \frac{1}{\Lambda(s)} y_m$. As illustrated in Example 9.5.1, the modeling error terms due to $\Delta_m(s), d_u$ do not appear explicitly in (9.9.1). Their effect, however, is manifested in $\epsilon, \epsilon m, \|\dot{A}(t)\|$ where instead of belonging to $\mathcal{L}_2$ as in the ideal case, they belong to $\mathcal{S}(\frac{\eta^2}{m^2} + f_0)$. Because $\frac{\eta^2}{m^2} \leq \Delta_2^2 + \frac{d_0^2}{m^2}$, we have $\epsilon, \epsilon m, \|\dot{A}(t)\| \in \mathcal{S}(\Delta_2^2 + \frac{d_0^2}{m^2} + f_0)$.

Step 2. *Establish the e.s. property of $A(t)$.* As in the ideal case, the APPC law guarantees that $\det(sI - A(t)) = A^*(s)$ for each time t where $A^*(s)$ is Hurwitz. If we apply Theorem 3.4.11 (b) to the homogeneous part of (9.9.1), we can establish that $A(t)$ is u.a.s which is equivalent to e.s. provided

$$c(f_0 + \Delta_2^2 + \frac{1}{T}\int_t^{t+T} \frac{d_0^2}{m^2} d\tau) < \mu^* \qquad (9.9.2)$$

$\forall t \geq 0$, any $T \geq 0$ and some $\mu^* > 0$ where $c \geq 0$ is a finite constant. Because $m^2 > 0$, condition (9.9.2) may not be satisfied for small Δ_2, f_0 unless d_0, the upper bound for the input disturbance, is zero or sufficiently small. As in the MRAC case, we can deal with the disturbance in two different ways. One way is to modify $m^2 = 1 + m_s$ to $m^2 = 1 + \beta_0 + m_s$ where β_0 is chosen to be large enough so that $c\frac{d_0^2}{m^2} \leq c\frac{d_0^2}{\beta_0} < \frac{\mu^*}{2}$, say, so that for

$$c(f_0 + \Delta_2^2) < \frac{\mu^*}{2}$$

condition (9.9.2) is always satisfied and $A(t)$ is e.s. The other way is to keep the same m^2 and establish that when m^2 grows large over an interval of time $I_1 = [t_1, t_2]$, say, the state transition matrix $\Phi(t, \tau)$ of $A(t)$ satisfies $\|\Phi(t,\tau)\| \leq k_1 e^{-k_2(t-\tau)}$ $\forall t \geq \tau$ and $t, \tau \in I_1$. This property of $A(t)$ (when m^2 grows large) is used in Step 3 to contradict the hypothesis that m^2 could grow unbounded and conclude boundedness.

Let us start by assuming that m^2 grows unbounded. Because all the elements of the state x are the outputs of strictly proper transfer functions with the same poles as the roots of $\Lambda(s)$ and inputs u_p, y_p (see Section 7.7.1) and the roots of $\Lambda(s)$ are located in $\text{Re}[s] < -\delta_0/2$, it follows from Lemma 3.3.2 that $\frac{x}{m} \in \mathcal{L}_\infty$. Because $\bar{y}_m, \epsilon m \in \mathcal{L}_\infty$, it follows from (9.9.1) that $\frac{y_p}{m}, \frac{u_p}{m} \in \mathcal{L}_\infty$. Because u_p^2, y_p^2 are bounded from above by m^2, it follows from the equation for m^2 that m^2 cannot grow faster than an exponential, i.e., $m^2(t) \leq e^{k_1(t-t_0)} m^2(t_0), \forall t \geq t_0 \geq 0$ for some $k > 0$. Because $m^2(t)$ is assumed to grow unbounded, we can find a $t_0 > \bar{\alpha} > 0$ for any arbitrary constant $\bar{\alpha} > t_2 - t_0$ such that $m^2(t_2) > \bar{\alpha} e^{k_1 \bar{\alpha}}$. We have

$$\bar{\alpha} e^{k_1 \bar{\alpha}} < m^2(t_2) \leq e^{k_1(t_2 - t_0)} m^2(t_0)$$

which implies that
$$\ln m^2(t_0) > \ln \bar{\alpha} + k_1[\bar{\alpha} - (t_2 - t_0)]$$

Because $\bar{\alpha} > t_2 - t_0$ and $t_0 \in (t_2 - \bar{\alpha}, t_2)$, it follows that

$$m^2(t_0) > \bar{\alpha}, \quad \forall t_0 \in (t_2 - \bar{\alpha}, t_2)$$

Let $t_1 = \sup_{\tau \leq t_2} \{arg(m^2(\tau) = \bar{\alpha})\}$. Then, $m^2(t_1) = \bar{\alpha}$ and $m^2(t) \geq \bar{\alpha}, \forall t \in [t_1, t_2)$ where $t_1 \leq t_2 - \bar{\alpha}$, i.e., $t_2 - t_1 \geq \bar{\alpha}$. Let us now consider the behavior of the homogeneous part of (9.9.1), i.e.,

$$\dot{Y} = A(t)Y \qquad (9.9.3)$$

over the interval $I_1 \triangleq [t_1, t_2]$ for which $m^2(t) \geq \bar{\alpha}$ and $t_2 - t_1 \geq \bar{\alpha}$ where $\bar{\alpha} > 0$ is an arbitrary constant. Because $\det(sI - A(t)) = A^*(s)$, i.e., $A(t)$ is a pointwise stable matrix, the Lyapunov equation

$$A^\top(t)P(t) + P(t)A(t) = -I \qquad (9.9.4)$$

9.9. STABILITY PROOFS OF ROBUST APPC SCHEMES

has the solution $P(t) = P^\top(t) > 0$ for each $t \in I_1$. If we consider the Lyapunov function
$$V(t) = Y^\top(t) P(t) Y(t)$$
then along the trajectory of (9.9.3), we have
$$\dot{V} = -Y^\top Y + Y^\top \dot{P} Y \leq -Y^\top Y + \|\dot{P}(t)\| Y^\top Y$$
As in the proof of Theorem 3.4.11, we can use (9.9.4) and the boundedness of P, A to establish that $\|\dot{P}(t)\| \leq c\|\dot{A}(t)\|$. Because $\lambda_1 Y^\top Y \leq V \leq \lambda_2 Y^\top Y$ for some $0 < \lambda_1 < \lambda_2$, it follows that
$$\dot{V} \leq -(\lambda_2^{-1} - c\lambda_1^{-1} \|\dot{A}(t)\|) V$$
i.e.,
$$V(t) \leq e^{-\int_\tau^t (\lambda_2^{-1} - c\lambda_1^{-1} \|\dot{A}(s)\|) ds} V(\tau)$$
$\forall\ t \geq \tau \geq 0$. For the interval $I_1 = [t_1, t_2)$, we have $m^2(t) \geq \bar{\alpha}$ and, therefore,
$$\int_\tau^t \|\dot{A}(\tau)\| d\tau \leq c(\Delta_2^2 + f_0 + \frac{d_0^2}{\bar{\alpha}})(t - \tau) + c$$
and therefore,
$$V(t) \leq e^{-\lambda_0 (t-\tau)} V(\tau), \quad \forall t, \tau \in [t_1, t_2) \tag{9.9.5}$$
and $t \geq \tau$ provided
$$c(f_0 + \Delta_2^2 + \frac{d_0^2}{\bar{\alpha}}) < \lambda_0 \tag{9.9.6}$$
where $\lambda_0 = \frac{\lambda_2^{-1}}{2}$. From (9.9.5) we have
$$\lambda_1 Y^\top(t) Y(t) \leq Y^\top(t) P Y(t) \leq e^{-\lambda_0 (t-\tau)} \lambda_2 Y^\top(\tau) Y(\tau)$$
which implies that
$$|Y(t)| \leq \sqrt{\frac{\lambda_2}{\lambda_1}} e^{-\lambda_0(t-\tau)} |Y(\tau)|, \quad \forall t, \tau \in [t_1, t_2)$$
which, in turn, implies that the transition matrix $\Phi(t, \tau)$ of (9.9.3) satisfies
$$\|\Phi(t, \tau)\| \leq \beta_0 e^{-\alpha_0 (t-\tau)}, \quad \forall t, \tau \in [t_1, t_2) \tag{9.9.7}$$
where $\beta_0 = \sqrt{\frac{\lambda_2}{\lambda_1}}, \alpha_0 = \frac{\lambda_0}{2}$. Condition (9.9.6) can now be satisfied by choosing $\bar{\alpha}$ large enough and by requiring Δ_2, f_0 to be smaller than some constant, i.e., $c(f_0 + \Delta_2^2) < \frac{\lambda_2^{-1}}{4}$, say. In the next step we use (9.9.7) and continue our argument over the interval I_1 in order to establish boundedness by contradiction.

Step 3. *Boundedness using the B-G Lemma and contradiction.* Let us apply Lemma 3.3.6 to (9.9.1) for $t \in [t_1, t_2)$. We have

$$\|x_{t,t_1}\| \leq ce^{-\delta/2(t-t_1)}|x(t_1)| + c\|(\epsilon m^2)_{t,t_1}\| + c$$

where $\|(\cdot)_{t,t_1}\|$ denotes the $\mathcal{L}_{2\delta}$ norm $\|(\cdot)_{t,t_1}\|_{2\delta}$ defined over the interval $[t_1, t)$, for any $0 < \delta < \delta_1 < 2\alpha_0$. Because $\frac{x}{m} \in \mathcal{L}_\infty$ it follows that

$$\|x_{t,t_1}\| \leq ce^{-\delta/2(t-t_1)}m(t_1) + c\|(\epsilon m^2)_{t,t_1}\| + c$$

Because $\|y_{pt,t_1}\|, \|u_{pt,t_1}\| \leq c\|x_{t,t_1}\| + c\|(\epsilon m^2)_{t,t_1}\| + c$, it follows that

$$\|y_{pt,t_1}\|, \|u_{pt,t_1}\| \leq ce^{-\delta/2(t-t_1)}m(t_1) + c\|(\epsilon m^2)_{t,t_1}\| + c$$

Now $m^2(t) = 1 + m_s(t)$ and

$$m_s(t) = e^{-\delta_0(t-t_1)}m_s(t_1) + \|y_{pt,t_1}\|_{2\delta_0}^2 + \|u_{pt,t_1}\|_{2\delta_0}^2$$

Because $\|(\cdot)_{t,t_1}\|_{2\delta_0} \leq \|(\cdot)_{t,t_1}\|$ for $\delta \leq \delta_0$, it follows that

$$m^2(t) = 1 + m_s(t) \leq 1 + e^{-\delta_0(t-t_1)}m^2(t_1) + \|y_{pt,t_1}\|^2 + \|u_{pt,t_1}\|^2 \quad \forall t \geq t_1$$

Substituting for the bound for $\|y_{pt,t_1}\|, \|u_{pt,t_1}\|$ we obtain

$$m^2(t) \leq ce^{-\delta(t-t_1)}m^2(t_1) + c\|(\epsilon m^2)_{t,t_1}\|^2 + c \quad \forall t \geq t_1 \geq 0$$

or

$$m^2(t) \leq c + ce^{-\delta(t-t_1)}m^2(t_1) + c\int_{t_1}^{t} e^{-\delta(t-\tau)}\epsilon^2 m^2 m^2(\tau)d\tau \quad (9.9.8)$$

Applying B-G Lemma III we obtain

$$m^2(t) \leq c(1 + m^2(t_1))e^{-\delta(t-t_1)}e^{c\int_{t_1}^{t}\epsilon^2 m^2 d\tau} + c\delta\int_{t_1}^{t} e^{-\delta(t-s)}e^{c\int_{s}^{t}\epsilon^2 m^2 d\tau}ds, \quad \forall t \geq t_1$$

For $t, s \in [t_1, t_2)$ we have

$$c\int_{s}^{t}\epsilon^2 m^2 d\tau \leq c\left(\Delta_2^2 + f_0 + \frac{d_0^2}{\bar{\alpha}}\right)(t-s) + c$$

By choosing $\bar{\alpha}$ large enough so that $c\frac{d_0^2}{\bar{\alpha}} < \frac{\delta}{4}$ and by requiring

$$c(\Delta_2^2 + f_0) < \frac{\delta}{4}$$

we have

$$\begin{aligned}
m^2(t_2) &\leq c(1 + m^2(t_1))e^{-\frac{\delta}{2}(t_2-t_1)} + c\delta\int_{t_1}^{t_2} e^{-\frac{\delta}{2}(t_2-s)}ds \\
&\leq c(1 + m^2(t_1))e^{-\frac{\delta}{2}(t_2-t_1)} + c
\end{aligned}$$

9.9. STABILITY PROOFS OF ROBUST APPC SCHEMES

Because $t_2 - t_1 \geq \bar{\alpha}, m^2(t_1) = \bar{\alpha}$ and $m^2(t_2) > \bar{\alpha}$, we have

$$\bar{\alpha} < m^2(t_2) \leq c(1+\bar{\alpha})e^{-\frac{\delta \bar{\alpha}}{2}} + c$$

Therefore, we can choose $\bar{\alpha}$ large enough so that $m^2(t_2) < \bar{\alpha}$ which contradicts the hypothesis that $m^2(t_2) > \bar{\alpha}$. Therefore, $m \in \mathcal{L}_\infty$ which implies that $x, u_p, y_p \in \mathcal{L}_\infty$.

The condition for robust stability is, therefore,

$$c(f_0 + \Delta_2^2) < \min\{\frac{\lambda_2^{-1}}{2}, \frac{\delta}{4}\} \triangleq \delta^*$$

for some finite constant $c > 0$.

Step 4. *Establish bounds for the tracking error.* A bound for the tracking error e_1 is obtained by expressing e_1 in terms of signals that are guaranteed by the adaptive law to be of the order of the modeling error in m.s.s. The tracking error equation has exactly the same form as in the ideal case in Section 7.7.1 and is given by

$$e_1 = \frac{\Lambda(s)s^{n-1}Q_m(s)}{A^*(s)}\epsilon m^2 + \frac{\Lambda(s)\alpha_{n-2}^\top(s)}{A^*(s)}v_0$$

(see equation (7.7.21)) where v_0 is the output of proper stable transfer functions whose inputs are elements of $\dot{\theta}_p$ multiplied by bounded signals. Because $\dot{\theta}_p, \epsilon m^2 \in \mathcal{S}(\frac{\eta^2}{m^2} + f_0)$ and $\frac{\eta^2}{m^2} \leq c(\Delta_2^2 + d_0^2)$, due to $m \in \mathcal{L}_\infty$, it follows from Corollary 3.3.3 that $e_1 \in \mathcal{S}(\Delta_2^2 + d_0^2 + f_0)$ and the proof is complete.

9.9.2 Proof of Theorem 9.5.3

We use the same steps as in the proof of Example 9.5.2.

Step 1. *Develop the state error equations for the closed-loop plant.* We start with the plant equation

$$R_p y_p = Z_p(1 + \Delta_m)(u_p + d_u)$$

Operating with $\frac{Q_m(s)}{Q_1(s)}$ on each side, we obtain

$$R_p \frac{Q_m}{Q_1} y_p = R_p \frac{Q_m}{Q_1} e_1 = Z_p \bar{u}_p + Z_p \frac{Q_m}{Q_1}[\Delta_m(u_p + d_u) + d_u]$$

i.e.,

$$e_1 = \frac{Z_p Q_1}{R_p Q_m}\bar{u}_p + \frac{Z_p Q_m}{R_p Q_m}[\Delta_m(u_p + d_u) + d_u]$$

Because $\frac{Z_p}{R_p}\Delta_m$ is strictly proper, we can find an arbitrary polynomial $\Lambda(s)$ whose roots are in $\text{Re}[s] < -\delta_0/2$ and has the same degree as R_p, i.e., n and express e_1 as

$$e_1 = \frac{Z_p Q_1}{R_p Q_m}\bar{u}_p + \frac{\Lambda Q_m}{R_p Q_m}\eta \quad (9.9.9)$$

where
$$\eta = \frac{Z_p}{\Lambda}[\Delta_m(u_p + d_u) + d_u]$$

We express (9.9.9) in the following canonical state-space form

$$\dot{e} = Ae + B\bar{u}_p + B_1\eta$$
$$e_1 = C^T e + d_1\eta \qquad (9.9.10)$$

where $C^T(sI - A)^{-1}B = \frac{Z_p Q_1}{R_p Q_m}, C^T(sI - A)^{-1}B_1 + d_1 = \frac{\Lambda Q_m}{R_p Q_m}$ with

$$A = \begin{bmatrix} -\theta_1^* & | & I_{n+q-1} \\ & | & ---- \\ & | & 0 \end{bmatrix}, \quad B = \theta_2^*, \quad C^T = [1, 0, \ldots, 0]$$

and θ_1^*, θ_2^* are defined as $R_p Q_m = s^{n+q} + \theta_1^{*T}\alpha_{n+q-1}(s), Z_p Q_1 = \theta_2^{*T}\alpha_{n+q-1}$.

It follows that $e_o = e - \hat{e}$ is the state observation error. From the equation of $\hat{e}$ in Table 9.6 and (9.9.10), and the control law $\bar{u}_p = -K_c(t)\hat{e}$, we obtain

$$\dot{\hat{e}} = A_c(t)\hat{e} + \hat{K}_o C^T e_o + \hat{K}_o d_1 \eta \qquad (9.9.11)$$

$$\dot{e}_o = A_o e_o + \tilde{\theta}_1 e_1 - \tilde{\theta}_2 \bar{u}_p + B_2 \eta \qquad (9.9.12)$$

where $B_2 = B_1 - \tilde{\theta}_1 d_1 - \hat{K}_o(t)d_1$ and

$$A_o = \begin{bmatrix} -a^* & | & I_{n+q-1} \\ & | & ---- \\ & | & 0 \end{bmatrix}$$

whose eigenvalues are equal to the roots of $A_o^*(s)$ and therefore are located in $\text{Re}[s] < -\delta_0/2$; $A_c(t) = \hat{A} - \hat{B}\hat{K}_c$ and $\tilde{\theta}_1 \triangleq \theta_1 - \theta_1^*, \tilde{\theta}_2 \triangleq \theta_2 - \theta_2^*$. The plant output satisfies

$$y_p = C^T e_o + C^T \hat{e} + y_m + d_1 \eta \qquad (9.9.13)$$

As shown in Section 7.3.3, the polynomials Z_p, Q_1, R_p, Q_m satisfy the equation (7.3.25), i.e.,

$$R_p Q_m X + Z_p Q_1 Y = A^* \qquad (9.9.14)$$

where X, Y have degree $n+q-1$, X is monic and A^* is an arbitrary monic Hurwitz polynomial of degree $2(n+q) - 1$ that contains the common zeros of $Q_1, R_p Q_m$. Without loss of generality, we require $A^*(s)$ to have all its roots in $\text{Re}[s] < -\delta_0/2$. From (9.9.14) and $Q_m u_p = Q_1 \bar{u}_p$, it follows that

$$u_p = \frac{R_p X Q_1}{A^*}\bar{u}_p + \frac{Q_1 Y Z_p}{A^*} u_p$$

9.9. STABILITY PROOFS OF ROBUST APPC SCHEMES

Because $R_p y_p = Z_p u_p + Z_p[\Delta_m(u_p + d_u) + d_u]$, we have

$$u_p = \frac{R_p X Q_1}{A^*}\bar{u}_p + \frac{Q_1 Y R_p}{A^*}y_p - \frac{Q_1 Y \Lambda}{A^*}\eta \tag{9.9.15}$$

by using the definition of Λ, η given earlier.

Equations (9.9.11) to (9.9.15) together with $\bar{u}_p = -K_c(t)\hat{e}$ describe the stability properties of the closed-loop APPC scheme of Table 9.6.

Step 2. *Establish stability of the homogeneous part of (9.9.11) and (9.9.12).* The stability of the homogeneous part of (9.9.12) is implied by the stability of the matrix A_o whose eigenvalues are the same as the roots of $A_o^*(s)$. Because $\det(sI - A_c(t)) = A_c^*(s)$, we have that $A_c(t)$ is pointwise stable. As we showed in the ideal case in Section 7.7.2, the assumption that $\hat{Z}_p, Q_1 \hat{R}_p$ are strongly coprime implies that $(\hat{A}, \hat{B})$ is a stabilizable pair [95] in the strong sense which can be used to show that $\hat{K}_c, \dot{\hat{K}}_c$ are bounded provided $\theta_1, \theta_2, \dot{\theta}_1, \dot{\theta}_2 \in \mathcal{L}_\infty$. Because the coefficients of $\hat{Z}_p(s,t), \hat{R}_p(s,t)$ generated by the adaptive law are bounded and their derivatives belong to $\in \mathcal{S}(\frac{\eta^2}{m^2} + f_0)$, it follows that $|\theta_i|, |\dot{\theta}_i|, |\hat{K}_c|, |\dot{\hat{K}}_c| \in \mathcal{L}_\infty$ and $|\dot{\theta}_i|, |\dot{\hat{K}}_c| \in \mathcal{S}(\frac{\eta^2}{m^2}+f_0)$ for $i=1,2$. Because $A_c = \hat{A} - \hat{B}\hat{K}_c$, it follows that $\|A_c(t)\|, \|\dot{A}_c(t)\| \in \mathcal{L}_\infty$ and $\|\dot{A}_c(t)\| \in \mathcal{S}(\eta^2/m^2 + f_0)$. Because $\frac{|\eta|^2}{m^2} \leq \Delta_2 + \int_0^t \frac{d_0^2}{m^2(\tau)}d\tau$ where d_0 is an upper bound for $|d_u|$, the stability of $A_c(t)$ cannot be established using Theorem 3.4.11 (b) unless $\frac{d_0}{m}$ is assumed to be small. The term d_0/m can be made small by modifying m to be greater than a large constant β_0, say, all the time as discussed in the proof of Theorem 9.5.2. In this proof we keep the same m as in Table 9.6 and consider the behavior of the state transition matrix $\Phi(t,\tau)$ of $A_c(t)$ over a particular finite interval where $m(t)$ has grown to be greater than an arbitrary constant $\bar{\alpha} > 0$. The properties of $\Phi(t,\tau)$ are used in Step 3 to contradict the hypothesis that m^2 can grow unbounded and conclude boundedness.

Let us start by showing that m cannot grow faster than an exponential due to the boundedness of the estimated parameters $\hat{A}, \hat{B}$. Because (C, A) in (9.9.10) is observable, it follows from Lemma 3.3.4 that

$$|e(t)|, \|e_t\|_{2\delta_0} \leq c\|\bar{u}_{pt}\|_{2\delta_0} + c\|\eta_t\|_{2\delta_0} + c\|y_{pt}\|_{2\delta_0} + c$$

Because $\bar{u}_p = \frac{Q_m}{Q_1}u_p$, it follows from Lemma 3.3.2 that $\|\bar{u}_{pt}\|_{2\delta_0} \leq cm$ which together with $\|\eta_t\|_{2\delta_0} \leq \Delta_2 m$ imply that $e/m \in \mathcal{L}_\infty$. Similarly, applying Lemma 3.3.3 to (9.9.12) and using $\theta_i \in \mathcal{L}_\infty$, we can establish that $e_o/m \in \mathcal{L}_\infty$ which implies that $\hat{e} = e - e_o$ is bounded from above by m. From (9.9.13), (9.9.15), it follows that $u_p/m, y_p/m \in \mathcal{L}_\infty$ which together with the equation for m^2 imply that $m^2(t) \leq e^{k_1(t-t_0)}m(t_0)$ $\forall t \geq t_0 \geq 0$ for some constant k_1.

We assume that $m^2(t)$ grows unbounded and proceed exactly the same way as in Step 2 of the proof of Theorem 9.5.2 in Section 9.9.1 to show that for

$$c(\Delta_2^2 + f_0) < \lambda$$

and some constant $\lambda > 0$ that depends on the pointwise stability of $A_c(t)$, we have

$$\|\Phi(t,\tau)\| \leq \lambda_0 e^{-\alpha_0(t-\tau)} \quad \forall t,\tau \in [t_1, t_2]$$

where $t_2 - t_1 > \bar{\alpha}$, $m^2(t_1) = \bar{\alpha}$, $m^2(t_2) \geq \bar{\alpha} e^{k_1 \bar{\alpha}} > \bar{\alpha}$, $m^2(t) \geq \bar{\alpha} \; \forall t \in [t_1, t_2]$ and $\bar{\alpha}$ is large enough so that $d_0/\bar{\alpha} < c\lambda$.

Step 3. *Boundedness using the B-G Lemma and contradiction.* Let us apply Lemma 3.3.6 to (9.9.11) for $t \in [t_1, t_2]$. We have

$$\|\hat{e}_{t,t_1}\| \leq c e^{-\frac{\delta}{2}(t-t_1)}|\hat{e}(t_1)| + c\|(C^\top e_o)_{t,t_1}\| + c\|\eta_{t,t_1}\|$$

where $\|(\cdot)_{t,t_1}\|$ denotes the $\mathcal{L}_{2\delta}$-norm $\|(\cdot)_{t,t_1}\|_{2\delta}$ for some $\delta > 0$. Because $\hat{e}/m \in \mathcal{L}_\infty$, we have

$$\|\hat{e}_{t,t_1}\| \leq c e^{-\frac{\delta}{2}(t-t_1)}m(t_1) + c\|(C^\top e_o)_{t,t_1}\| + c\|\eta_{t,t_1}\| \quad (9.9.16)$$

From (9.9.13), we have

$$\|y_{pt,t_1}\| \leq c\|(C^\top e_o)_{t,t_1}\| + c\|\hat{e}_{t,t_1}\| + c\|\eta_{t,t_1}\| + c \quad (9.9.17)$$

Applying Lemma 3.3.6 to (9.9.15) and noting that the states of any minimal state representation of the transfer functions in (9.9.15) are bounded from above by m due to the location of the roots of $A^*(s)$ in $\text{Re}[s] < -\delta_0/2$ and using $\bar{u}_p = -\bar{K}_c \hat{e}$, we obtain

$$\|u_{pt,t_1}\| \leq c\|\hat{e}_{t,t_1}\| + c\|y_{pt,t_1}\| + c\|\eta_{t,t_1}\| + c e^{-\frac{\delta}{2}(t-t_1)}m(t_1) \quad (9.9.18)$$

Combining (9.9.16), (9.9.17) and (9.9.18) we have

$$\begin{aligned}
m_{f_1}^2(t) &\triangleq e^{-\delta_0(t-t_1)}m^2(t_1) + \|u_{pt,t_1}\|^2 + \|y_{pt,t_1}\|^2 \quad (9.9.19) \\
&\leq c e^{-\delta(t-t_1)}m^2(t_1) + c\|(C^\top e_o)_{t,t_1}\|^2 + c\|\eta_{t,t_1}\| + c, \forall t \in [t_1, t_2]
\end{aligned}$$

Now from (9.9.12) we have

$$\begin{aligned}
C^\top e_o &= C^\top(sI - A_o)^{-1}(\tilde{\theta}_1 e_1 - \tilde{\theta}_2 \bar{u}_p) + C^\top(sI - A_o)^{-1}B_2 \eta \\
&= \frac{\alpha_{n+q-1}(s)}{A_o^*(s)}(\tilde{\theta}_1 e_1 - \tilde{\theta}_2 \bar{u}_p) + \frac{\alpha_{n+q-1}(s)}{A_o^*(s)}B_2 \eta \quad (9.9.20)
\end{aligned}$$

Following exactly the same procedure as in Step 3 of the proof of Theorem 7.4.2 in Section 7.7.2 in dealing with the first term of (9.9.20), we obtain

$$C^\top e_0 = \frac{\Lambda_p(s)Q_m(s)}{A_o^*(s)}\epsilon m^2 + r_2 + W(s)\eta \quad (9.9.21)$$

where r_2 consists of terms of the form $W_1(s)\bar{\omega}^\top \dot{\theta}_p$ with $\bar{\omega}/m \in \mathcal{L}_\infty$ and $W_1(s), W(s)$ being proper and analytic in $\text{Re}[s] \geq -\delta_0/2$. Applying Lemma 3.3.5 to (9.9.21)

9.9. STABILITY PROOFS OF ROBUST APPC SCHEMES

and noting that any state of a minimal state-space representation of the transfer functions in (9.9.21) is bounded from above by m, we obtain

$$\|(C^T e_0)_{t,t_1}\| \leq c e^{-\frac{\delta}{2}(t-t_1)} m(t_1) + c\|(\epsilon m^2)_{t,t_1}\| + c\|(\dot{\theta}_p m)_{t,t_1}\| + c\|\eta_{t,t_1}\| \quad (9.9.22)$$

Using (9.9.22) in (9.9.19), and noting that $\|\eta_{t,t_1}\| \leq \Delta_\infty m_{f_1}$, we obtain

$$m_{f_1}^2(t) \leq c e^{-\delta(t-t_1)}(1 + m^2(t_1)) + c\|(\tilde{g}m)_{t,t_1}\|^2 + c\Delta_\infty m_{f_1}(t), \quad \forall t \in [t_1, t_2)$$

where $\tilde{g}^2 = \epsilon^2 m^2 + |\dot{\theta}_p|^2$. Therefore, for

$$c\Delta_\infty^2 < 1 \quad (9.9.23)$$

we have $m^2(t) \leq m_{f_1}^2(t)$ and

$$m^2(t) \leq c e^{-\delta(t-t_1)}(1 + m^2(t_1)) + c \int_{t_1}^t \tilde{g}^2(\tau) m^2(\tau) d\tau, \quad \forall t \in [t_1, t_2) \quad (9.9.24)$$

Applying B-G Lemma III we obtain

$$m^2(t) \leq c(1 + m^2(t_1)) e^{-\delta(t-t_1)} e^{c \int_{t_1}^t \tilde{g}^2(\tau) d\tau} + c\delta \int_{t_1}^t e^{-\delta(t-s)} e^{c \int_s^t \tilde{g}^2(\tau) d\tau} ds$$

Because $m^2(t) \geq \bar{\alpha} \; \forall t \in [t_1, t_2)$, we have $c \int_s^t \tilde{g}^2(\tau) d\tau \leq c(\Delta_2^2 + f_0 + \frac{d_0^2}{\bar{\alpha}})(t-s) + c, \forall t, s \in [t_1, t_2)$. Choosing $\bar{\alpha}$ large enough so that $c\frac{d_0^2}{\bar{\alpha}} < \frac{\delta}{4}$, and restricting Δ_2, f_0 to satisfy

$$c(\Delta_2^2 + f_0) < \frac{\delta}{4} \quad (9.9.25)$$

we obtain

$$m^2(t_2) \leq c(1 + m^2(t_1)) e^{-\frac{\delta}{2}(t_2 - t_1)} + c$$

Hence,

$$\bar{\alpha} < m^2(t_2) \leq c(1 + \bar{\alpha}) e^{-\frac{\delta \bar{\alpha}}{2}} + c$$

which implies that for $\bar{\alpha}$ large, $m^2(t_2) < \bar{\alpha}$ which contradicts the hypothesis that $m^2(t_2) > \bar{\alpha}$. Hence, $m \in \mathcal{L}_\infty$ which implies that $e, \hat{e}, e_o, y_p, u_p \in \mathcal{L}_\infty$.

Using (9.9.23), (9.9.25) and $c(\Delta_2^2 + f_0) < \lambda$ in Step 2, we can define the constants $\Delta_\infty^*, \delta^* > 0$ stated in the theorem.

Step 4. *Tracking error bounds.* The tracking error equation is given by

$$e_1 = C^T e_o + C^T \hat{e} + d_1 \eta$$

We can verify that the $\mathcal{L}_{2e}$-norm of $\hat{e}$ and $C^T e_o$ satisfy

$$\|\hat{e}_t\|_{2e} \leq c\|(C^T e_o)_t\|_{2e} + c\|\eta_t\|_{2e}$$
$$\|(C^T e_o)_t\|_{2e} \leq c\|\epsilon m_t\|_{2e} + c\|\dot{\theta}_{pt}\|_{2e} + c\|\eta_t\|_{2e}$$

by following the same approach as in the previous steps and using $\delta = 0$, which imply that

$$\|e_{1t}\|_{2e}^2 \triangleq \int_0^t e_1^2 d\tau \leq c \int_0^t (\epsilon^2 m^2 + |\dot\theta_p|^2 + \eta^2)d\tau$$
$$\leq c(\Delta_2^2 + f_0 + d_0^2)t + c$$

and the proof is complete. $\square$

9.10 Problems

9.1 Show that if $u = sin\omega_0 t$ is dominantly rich, i.e., $0 < \omega_0 < O(1/\mu)$ in Example 9.2.3, then the signal vector ϕ is PE with level of excitation $\alpha_0 > O(\mu)$.

9.2 Consider the nonminimum-phase plant

$$y_p = G(s)u_p, \quad G(s) = \frac{1 - \mu s}{s^2 + as + b}$$

where $\mu > 0$ is a small number.

(a) Express the plant in the form of equation (9.3.1) so that the nominal part of the plant is minimum-phase and the unmodeled part is small for small μ.

(b) Design a robust MRAC scheme with a reference model chosen as

$$W_m(s) = \frac{1}{s^2 + 1.4s + 1}$$

based on the nominal part of the plant

(c) Simulate the closed-loop MRAC system with $a = -3, b = 2$ and $r =$step function for $\mu = 0, 0.01, 0.05, 0.1$, respectively. Comment on your simulation results.

9.3 In Problem 9.2, the relative degree of the overall plant transfer function is $n^* = 1$ when $\mu \neq 0$. Assume that the reference model is chosen to be $W_{m1}(s) = \frac{a_m}{s+a_m}$. Discuss the consequences of designing an MRAC scheme for the full order plant using $W_{m1}(s)$ as the reference model. Simulate the closed-loop scheme using the values given in part (c) of Problem 9.2.

9.4 For the MRAC problem given in Problem 9.2,

(a) Choose one reference input signal r that is sufficiently rich but not dominantly rich and one that is dominantly rich.

(b) Simulate the MRAC scheme developed in Problem 9.2 using the input signals designed in (a).

(c) Compare the simulation results with those obtained in Problem 9.2. Comment on your observations.

9.5 Consider the plant

$$y_p = \frac{1}{s(s+a)} u_p$$

where a is an unknown constant and

$$y_m = \frac{9}{(s+3)^2} r$$

is the reference model.

(a) Design a modified MRAC scheme using Method 1 given in Section 9.4.2

(b) Simulate the modified MRAC scheme for different values of the design parameter τ.

9.6 Consider Problem 9.2.

(a) Replace the standard robust MRAC scheme with a modified one from Section 9.4. Simulate the modified MRAC scheme using the same parameters as in Problem 9.2 (c) and $\tau = 0.1$.

(b) For a fixed μ (for example, $\mu = 0.01$), simulate the closed MRAC scheme for different τ.

(c) Comment on your results and observations.

9.7 Consider the speed control problem described in Problem 6.2 of Chapter 6. Suppose the system dynamics are described by

$$V = \frac{b}{s+a}(1 + \Delta_m(s))\theta + d$$

where d is a bounded disturbance and Δ_m represents the unmodeled dynamics, and the reference model

$$V_m = \frac{0.5}{s+0.5} V_s$$

is as described in Problem 6.2.

(a) Design a robust MRAC scheme with and without normalization

(b) Simulate the two schemes for $a = 0.02\sin 0.01t$, $b = 1.3$, $\Delta_m(s) = -\frac{2\mu s}{\mu s + 1}$ and $d = 0$ for $\mu = 0, 0.01, 0.2$. Comment on your simulation results.

9.8 Consider the plant
$$y = \frac{1}{s-a}u - \mu\Delta_a(s)u$$
where $\mu > 0$ is a small parameter, a is unknown and $\Delta_a(s)$ is a strictly proper stable unknown transfer function perturbation independent of μ. The control objective is to choose u so that all signals are bounded and y tracks, as close as possible, the output y_m of the reference model
$$y_m = \frac{1}{s+1}r$$
for any bounded reference input r as close as possible.

(a) Design a robust MRAC to meet the control objective.

(b) Develop bounds for robust stability.

(c) Develop a bound for the tracking error $e_1 = y - y_m$.

Repeat (a), (b), (c) for the plant
$$y = \frac{e^{-\tau s}}{s-a}u$$
where $\tau > 0$ is a small constant.

9.9 Consider the following expressions for the plant
$$y_p = G_0(s)u_p + \Delta_a(s)u_p \qquad (9.10.1)$$
$$y_p = \frac{N_0(s) + \Delta_1(s)}{D_0(s) + \Delta_2(s)}u_p \qquad (9.10.2)$$
where $\Delta_a, \Delta_1, \Delta_2$ are plant perturbations as defined in Section 8.2 of Chapter 8.

(a) Design a model reference controller based on the dominant part of the plant given by
$$y_p = G_0(s)u_p \qquad (9.10.3)$$
where $G_0(s) = \frac{N_0(s)}{D_0(s)} = k_p\frac{Z_p(s)}{R_p(s)}$ satisfies the MRAC Assumptions P1 to P4 and
$$y_m = W_m(s)r = k_m\frac{Z_m(s)}{R_m(s)}r$$
is the reference model that satisfies Assumptions M1 and M2 given in Section 6.3 of Chapter 6.

(b) Apply the MRC law designed using (9.10.3) to the full-order plants given by (9.10.1), (9.10.2) and obtain bounds for robust stability.

9.10. PROBLEMS

(c) Obtain bounds for the tracking error $e_1 = y_p - y_m$.

9.10 Consider the plant
$$y_p = G_0(s)u_p + \Delta_a(s)u_p$$
where
$$y_p = G_0(s)u_p$$
is the plant model and $\Delta_a(s)$ is an unknown additive perturbation. Consider the PPC laws given by (9.5.3), (9.5.5), and (9.5.7) designed based on the plant model but applied to the plant with $\Delta_a(s) \neq 0$. Obtain a bound for $\Delta_a(s)$ for robust stability.

9.11 Consider the plant
$$y_p = \frac{N_0 + \Delta_1}{D_0 + \Delta_2} u_p$$
where $G_0 = \frac{N_0}{D_0}$ is the modeled part and Δ_1, Δ_2 are stable factor perturbations. Apply the PPC laws of Problem 9.10 to the above plant and obtain bounds for robust stability.

9.12 Consider the robust APPC scheme of Example 9.5.1 given by (9.5.18), (9.5.19) designed for the plant model
$$y_p = \frac{b}{s+a} u_p$$
but applied to the following plants

(i) $\quad y_p = \dfrac{b}{s+a} u_p + \Delta_a(s) u_p$

(ii) $\quad y_p = \dfrac{\frac{b}{s+\lambda_0} + \Delta_1(s)}{\frac{s+a}{s+\lambda_0} + \Delta_2(s)} u_p$

where $\Delta_a(s)$ is an additive perturbation and $\Delta_1(s), \Delta_2(s)$ are stable factor perturbations and $\lambda_0 > 0$.

(a) Obtain bounds and conditions for $\Delta_a(s), \Delta_1(s), \Delta_2(s)$ for robust stability.

(b) Obtain a bound for the mean square value of the tracking error.

(c) Simulate the APPC scheme for the plant (i) when $b = 2(1+0.5\sin 0.01t)$, $a = -2\sin 0.002t$, $\Delta_a(s) = -\frac{\mu s}{(s+5)^2}$ for $\mu = 0, 0.1, 0.5, 1$.

9.13 Consider the robust APPC scheme based on state feedback of Example 9.5.2 designed for the plant model
$$y_p = \frac{b}{s+a} u_p$$

but applied to the following plants:

(i) $$y_p = \frac{b}{s+a}u_p + \Delta_a(s)u_p$$

(ii) $$y_p = \frac{\frac{b}{s+\lambda_0} + \Delta_1(s)}{\frac{s+a}{s+\lambda_0} + \Delta_2(s)}u_p$$

where $\Delta_a(s)$ is an additive perturbation and $\Delta_1(s), \Delta_2(s)$ are stable factor perturbations and $\lambda_0 > 0$.

(a) Obtain bounds and conditions for $\Delta_a(s), \Delta_1(s), \Delta_2(s)$ for robust stability.

(b) Obtain a bound for the mean square value of the tracking error.

(c) Simulate the APPC scheme with plant (ii) when

$$b = 1, \quad a = -2\sin 0.01t, \quad \lambda_0 = 2, \quad \Delta_1(s) = \frac{(e^{-\tau s} - 1)}{s+2}, \quad \Delta_2(s) = 0$$

for $\tau = 0, 0.1, 0.5, 1$.

9.14 Simulate the ALQC scheme of Example 9.5.3 for the plant

$$y_p = \frac{b}{s+a}(1 + \Delta_m(s))u_p$$

where $\Delta_m(s) = -\frac{2\mu s}{1+\mu s}$ and $\mu \geq 0$.

(a) For simulation purposes, assume $b = -1, a = -2(1 + 0.02\sin 0.1t)$. Consider the following values of μ: $\mu = 0, \mu = 0.05, \mu = 0.2, \mu = 0.5$.

(b) Repeat (a) with an adaptive law that employs a dead zone.

9.15 Consider the following MIMO plant

$$\begin{bmatrix} y_1 \\ y_2 \end{bmatrix} = \begin{bmatrix} h_{11}(s) & h_{12}(s) \\ 0 & h_{22}(s) \end{bmatrix}\begin{bmatrix} u_1 \\ u_2 \end{bmatrix}$$

where $h_{11} = \frac{b_1}{s+a_1}, h_{12} = \mu\Delta(s), h_{22} = \frac{b_2}{s+a_2}$ and $\Delta(s)$ is strictly proper and stable.

(a) Design a decentralized MRAC scheme so that y_1, y_2 tracks y_{m1}, y_{m2}, the outputs of the reference model

$$\begin{bmatrix} y_{m1} \\ y_{m2} \end{bmatrix} = \begin{bmatrix} \frac{1}{s+1} & 0 \\ 0 & \frac{2}{s+2} \end{bmatrix}\begin{bmatrix} r_1 \\ r_2 \end{bmatrix}$$

for any bounded reference input signal r_1, r_2 as close as possible.

(b) Calculate a bound for $\mu\Delta(s)$ for robust stability.

9.10. PROBLEMS

9.16 Design a MIMO MRAC scheme using the CGT approach for the plant

$$\dot{x} = Ax + Bu$$
$$y = C^\top x$$

where $x \in \mathcal{R}^2, u \in \mathcal{R}^2, y \in \mathcal{R}^2$ and (A, B, C) corresponds to a minimal state representation. The reference model is given by

$$\dot{x}_m = \begin{bmatrix} -2 & 0 \\ 0 & -2 \end{bmatrix} x_m + \begin{bmatrix} 1 \\ 1 \end{bmatrix} r$$
$$y_m = x_m$$

Simulate the MRAC scheme when

$$A = \begin{bmatrix} -3 & 1 \\ 0 & 0.2 \end{bmatrix}, \quad B = \begin{bmatrix} 1 & 0 \\ 0 & 2 \end{bmatrix}, \quad C = \begin{bmatrix} 1 & 0 \\ 0 & 1 \end{bmatrix}$$

Appendix

A Swapping Lemmas

The following lemmas are useful in establishing stability in most of the adaptive control schemes presented in this book:

Lemma A.1 (Swapping Lemma A.1) *Let $\tilde{\theta}, w : \mathcal{R}^+ \mapsto \mathcal{R}^n$ and $\tilde{\theta}$ be differentiable. Let $W(s)$ be a proper stable rational transfer function with a minimal realization (A, B, C, d), i.e.,*

$$W(s) = C^\top (sI - A)^{-1} B + d$$

Then

$$W(s)\tilde{\theta}^\top \omega = \tilde{\theta}^\top W(s)\omega + W_c(s)\left((W_b(s)\omega^\top)\dot{\tilde{\theta}}\right)$$

where

$$W_c(s) = -C^\top (sI - A)^{-1}, \quad W_b(s) = (sI - A)^{-1} B$$

Proof We have

$$\begin{aligned}
W(s)\tilde{\theta}^\top \omega &= W(s)\omega^\top \tilde{\theta} = d\tilde{\theta}^\top \omega + C^\top \int_0^t e^{A(t-\tau)} B\omega^\top \tilde{\theta} d\tau \\
&= d\tilde{\theta}^\top \omega + C^\top e^{At}\left(\left.\int_0^\tau e^{-A\sigma}B\omega^\top(\sigma)d\sigma\tilde{\theta}(\tau)\right|_{\tau=0}^{\tau=t}\right. \\
&\qquad \left. - \int_0^t \int_0^\tau e^{-A\sigma} B\omega^\top(\sigma)d\sigma\dot{\tilde{\theta}}(\tau)d\tau\right) \qquad (A.1) \\
&= \tilde{\theta}^\top \left[d\omega + C^\top \int_0^t e^{A(t-\sigma)} B\omega(\sigma)d\sigma \right] \\
&\qquad - C^\top \int_0^t e^{A(t-\tau)} \int_0^\tau e^{A(\tau-\sigma)} B\omega^\top(\sigma)d\sigma\dot{\tilde{\theta}}(\tau)d\tau
\end{aligned}$$

A. SWAPPING LEMMAS

Noting that

$$d\omega + C^\top \int_0^t e^{A(t-\sigma)} B\omega(\sigma)d\sigma = (d + C^\top(sI-A)^{-1}B)\omega = W(s)\omega,$$

$$\int_0^t e^{A(t-\sigma)} B\omega^\top(\sigma)d\sigma = (sI-A)^{-1}B\omega^\top$$

and

$$C^\top \int_0^t e^{A(t-\tau)} f(\tau)d\tau = C^\top(sI-A)^{-1}f$$

we can express (A.1) as

$$W(s)\tilde{\theta}^\top \omega = \tilde{\theta}^\top W(s)\omega - C^\top(sI-A)^{-1}\left\{\left((sI-A)^{-1}B\omega^\top\right)\dot{\tilde{\theta}}\right\} \qquad (A.2)$$

and the proof is complete. □

Other proofs of Lemma A.1 can be found in [201, 221].

Lemma A.2 (Swapping Lemma A.2) *Let $\tilde{\theta}, \omega : \mathcal{R}^+ \mapsto \mathcal{R}^n$ and $\tilde{\theta}, \omega$ be differentiable. Then*

$$\tilde{\theta}^\top \omega = F_1(s,\alpha_0)\left[\dot{\tilde{\theta}}^\top \omega + \tilde{\theta}^\top \dot{\omega}\right] + F(s,\alpha_0)[\tilde{\theta}^\top \omega] \qquad (A.3)$$

where $F(s,\alpha_0) \triangleq \frac{\alpha_0^k}{(s+\alpha_0)^k}, F_1(s,\alpha_0) \triangleq \frac{1-F(s,\alpha_0)}{s}, k \geq 1$ and $\alpha_0 > 0$ is an arbitrary constant. Furthermore, for $\alpha_0 > \delta$ where $\delta \geq 0$ is any given constant, $F_1(s,\alpha_0)$ satisfies

$$\|F_1(s,\alpha_0)\|_{\infty\delta} \leq \frac{c}{\alpha_0}$$

for a finite constant $c \in \mathcal{R}^+$ which is independent of α_0.

Proof Let us write

$$\tilde{\theta}^\top \omega = (1 - F(s,\alpha_0))\tilde{\theta}^\top \omega + F(s,\alpha_0)\tilde{\theta}^\top \omega$$

Note that

$$(s+\alpha_0)^k = \sum_{i=0}^k C_k^i s^i \alpha_0^{k-i} = \alpha_0^k + \sum_{i=1}^k C_k^i s^i \alpha_0^{k-i} = \alpha_0^k + s\sum_{i=1}^k C_k^i s^{i-1}\alpha_0^{k-i}$$

where $C_k^i \triangleq \frac{k!}{i!(k-i)!}$ ($0! \triangleq 1$), we have

$$1 - F(s, \alpha_0) = \frac{(s+\alpha_0)^k - \alpha_0^k}{(s+\alpha_0)^k} = s\frac{\sum_{i=1}^k C_k^i s^{i-1} \alpha_0^{k-i}}{(s+\alpha_0)^k}$$

Defining

$$F_1(s, \alpha_0) \triangleq \frac{\sum_{i=1}^k C_k^i s^{i-1} \alpha_0^{k-i}}{(s+\alpha_0)^k} = \frac{1 - F(s, \alpha_0)}{s}$$

we can write

$$\begin{aligned}
\tilde{\theta}^\top \omega &= F_1(s, \alpha_0) s(\tilde{\theta}^\top \omega) + F(s, \alpha_0)(\tilde{\theta}^\top \omega) \\
&= F_1(s, \alpha_0)(\dot{\tilde{\theta}}^\top \omega + \tilde{\theta}^\top \dot{\omega}) + F(s, \alpha_0)(\tilde{\theta}^\top \omega)
\end{aligned} \qquad (A.4)$$

To show that $\|F_1(s, \alpha_0)\|_{\infty\delta} \leq \frac{c}{\alpha_0}$ for some constant c, we write

$$F_1(s, \alpha_0) = \frac{\sum_{i=1}^k C_k^i s^{i-1} \alpha_0^{k-i}}{(s+\alpha_0)^k} = \frac{1}{s+\alpha_0} \sum_{i=1}^k C_k^i \frac{s^{i-1}}{(s+\alpha_0)^{i-1}} \frac{\alpha_0^{k-i}}{(s+\alpha_0)^{k-i}}$$

Because

$$\left\| \frac{\alpha_0^i}{(s+\alpha_0)^i} \right\|_{\infty\delta} = \left(\left\| \frac{\alpha_0}{s+\alpha_0} \right\|_{\infty\delta} \right)^i = \left(\frac{2\alpha_0}{2\alpha_0 - \delta} \right)^i \leq 2^i, \quad i \geq 1$$

and

$$\left\| \frac{1}{s+\alpha_0} \right\|_{\infty\delta} = \frac{2}{2\alpha_0 - \delta} \leq \frac{2}{\alpha_0}, \quad \left\| \frac{s^{i-1}}{(s+\alpha_0)^{i-1}} \right\|_{\infty\delta} = 1, \quad i \geq 1$$

we have

$$\begin{aligned}
\|F_1(s, \alpha_0)\|_{\infty\delta} &\leq \left\| \frac{1}{s+\alpha_0} \right\|_{\infty\delta} \sum_{i=1}^k C_k^i \left\| \frac{s^{i-1}}{(s+\alpha_0)^{i-1}} \right\|_{\infty\delta} \left\| \frac{\alpha_0^{k-i}}{(s+\alpha_0)^{k-i}} \right\|_{\infty\delta} \\
&= \frac{\sum_{i=1}^k C_k^i 2^{k-i+1}}{2\alpha_0 - \delta} \\
&\leq \frac{c}{\alpha_0}
\end{aligned}$$

where $c \triangleq \sum_{i=1}^k C_k^i 2^{k-i+1}$ is a constant independent of α_0. The above inequalities are established by using $\alpha_0 > \delta$. □

In the stability proofs of indirect adaptive schemes, we need to manipulate polynomial operators which have time-varying coefficients. We should

A. SWAPPING LEMMAS

note that given $a(t) \in \mathcal{R}^{n+1}$, and $\alpha_n(s) \triangleq [s^n, s^{n-1}, \ldots, s, 1]^\top$, $A(s,t), \bar{A}(s,t)$ defined as

$$A(s,t) \triangleq a_n(t)s^n + a_{n-1}(t)s^{n-1} + \cdots + a_1(t)s + a_0(t) = a^\top(t)\alpha_n(s)$$

$$\bar{A}(s,t) \triangleq s^n a_n(t) + s^{n-1} a_{n-1}(t) + \cdots + s a_1(t) + a_0(t) = \alpha_n^\top(s)a(t)$$

refer to two different operators because of the time-varying nature of $a(t)$. For the same reason, two polynomial operators with time-varying parameters are not commutable, i.e.,

$$A(s,t)B(s,t) \neq B(s,t)A(s,t), \quad \bar{A}(s,t)B(s,t) \neq \bar{B}(s,t)A(s,t)$$

The following two swapping lemmas can be used to interchange the sequence of time-varying polynomial operators that appear in the proof of indirect adaptive control schemes.

Lemma A.3 *(Swapping Lemma A.3).* Let

$$A(s,t) = a_n(t)s^n + a_{n-1}(t)s^{n-1} + \cdots a_1(t)s + a_0(t) = a^\top(t)\alpha_n(s)$$

$$B(s,t) = b_m(t)s^m + b_{m-1}(t)s^{m-1} + \cdots b_1(t)s + b_0(t) = b^\top(t)\alpha_m(s)$$

be any two polynomials with differentiable time-varying coefficients, where $a = [a_n, a_{n-1}, \ldots, a_0]^\top$, $b = [b_m, b_{m-1}, \ldots, b_0]^\top$, $a(t) \in \mathcal{R}^{n+1}$, $b(t) \in \mathcal{R}^{m+1}$ *and* $\alpha_i(s) = [s^i, s^{i-1}, \cdots, s, 1]$. *Let*

$$C(s,t) \triangleq A(s,t) \cdot B(s,t) = B(s,t) \cdot A(s,t)$$

be the algebraic product of $A(s,t)$ *and* $B(s,t)$. *Then for any function* $f(t)$ *such that* $C(s,t)f$, $A(s,t)(B(s,t)f)$, $B(s,t)(A(s,t)f)$ *are well defined, we have*

$$\text{(i)} \quad A(s,t)(B(s,t)f) = C(s,t)f + a^\top(t)D_{n-1}(s)\left[\alpha_{n-1}(s)(\alpha_m^\top(s)f)\dot{b}\right] \tag{A.5}$$

and

(ii) $A(s,t)(B(s,t)f) = B(s,t)(A(s,t)f) + G(s,t)\left((H(s)f)\begin{bmatrix}\dot{a}\\\dot{b}\end{bmatrix}\right)$

(A.6)

where
$$G(s,t) \triangleq [a^\top D_{n-1}(s), b^\top D_{m-1}(s)]$$

$D_i(s), H(s)$ are matrices of dimension $(i+2) \times (i+1)$ and $(n+m) \times (n+1+m+1)$ respectively, defined as

$$D_i(s) \triangleq \begin{bmatrix} 1 & s & s^2 & \cdots & s^i \\ 0 & 1 & s & \cdots & s^{i-1} \\ 0 & 0 & \ddots & \ddots & \vdots \\ \vdots & & \ddots & 1 & s \\ 0 & \cdots & & 0 & 1 \\ 0 & \cdots & & 0 & 0 \end{bmatrix}$$

$$H(s) \triangleq \begin{bmatrix} 0 & \alpha_{n-1}(s)\alpha_m^\top(s) \\ -\alpha_{m-1}(s)\alpha_n^\top(s) & 0 \end{bmatrix}$$

Proof (i) Using the relation $s(f_1 f_2) = f_1 s f_2 + (sf_1)f_2$, we can write

$$A(s,t)(B(s,t)f) = a_0 B(s,t)f + \sum_{i=1}^{n} a_i \sum_{j=0}^{m} s^i (b_j s^j f)$$

$$= a_0 B(s,t)f + \sum_{i=1}^{n} a_i \sum_{j=0}^{m} s^{i-1}\left(b_j s^{j+1} f + \dot{b}_j s^j f\right)$$

$$= a_0 B(s,t)f + \sum_{i=1}^{n} a_i \sum_{j=0}^{m} \left[s^{i-2}\left(b_j s^{j+2} f + \dot{b}_j s^{j+1} f\right) + s^{i-1}\dot{b}_j s^j f\right]$$

$$\vdots$$

(A.7)

$$= a_0 B(s,t)f + \sum_{i=1}^{n} a_i \sum_{j=0}^{m} \left(b_j s^{j+i} f + \sum_{k=0}^{i-1} s^k \left(\dot{b}_j s^{j+i-1-k} f\right)\right)$$

$$= \sum_{i=0}^{n} a_i \sum_{j=0}^{m} b_j s^{j+i} f + \sum_{i=1}^{n}\sum_{j=0}^{m}\sum_{k=0}^{i-1} s^k \left(\dot{b}_j s^{j+i-1-k} f\right) = C(s,t)f + r(t)$$

where
$$r(t) \triangleq \sum_{i=1}^{n} a_i \sum_{k=0}^{i-1} s^k \sum_{j=0}^{m} \left(\dot{b}_j s^{j+i-1-k} f\right)$$

A. SWAPPING LEMMAS

and the variable t in a, b, f is omitted for the sake of simplicity.

From the expression of $r(t)$, one can verify through simple algebra that

$$r(t) = a^\mathsf{T} D_{n-1}(s) r_b(t) \qquad (A.8)$$

where

$$r_b(t) \triangleq \begin{bmatrix} \sum_{j=0}^m \dot{b}_j s^{j+n-1} f \\ \sum_{j=0}^m \dot{b}_j s^{j+n-2} f \\ \vdots \\ \sum_{j=0}^m \dot{b}_j s^{j+1} f \\ \sum_{j=0}^m \dot{b}_j s^j f \end{bmatrix}$$

To simplify the expression further, we write

$$\begin{aligned} r_b(t) &= \left(\begin{bmatrix} s^{n-1+m} & s^{n-1+m-1} & \cdots & s^{n-1} \\ s^{n-2+m} & s^{n-2+m-1} & \cdots & s^{n-2} \\ \vdots & \vdots & & \vdots \\ s^m & s^{m-1} & \cdots & 1 \end{bmatrix} f \right) \begin{bmatrix} \dot{b}_m \\ \dot{b}_{m-1} \\ \vdots \\ \dot{b}_0 \end{bmatrix} \\ &= \left(\alpha_{n-1}(s) \alpha_m^\mathsf{T}(s) f \right) \dot{b} \qquad (A.9) \end{aligned}$$

Using (A.8) and (A.9) in (A.7), the result of Lemma A.3 (i) follows immediately.

(ii) Using the result (i), we have

$$A(s,t)(B(s,t)f) = C(s,t)f + a^\mathsf{T} D_{n-1}(s) \left(\left[\alpha_{n-1}(s) \alpha_m^\mathsf{T}(s) f \right] \dot{b} \right) \qquad (A.10)$$

$$B(s,t)(A(s,t)f) = C(s,t)f + b^\mathsf{T} D_{m-1}(s) \left(\left[\alpha_{m-1}(s) \alpha_n^\mathsf{T}(s) f \right] \dot{a} \right) \qquad (A.11)$$

Subtracting (A.11) from (A.10), we obtain

$$A(s,t)(B(s,t)f) - B(s,t)(A(s,t)f) \qquad (A.12)$$
$$= a^\mathsf{T} D_{n-1}(s) \left(\left[\alpha_{n-1}(s) \alpha_m^\mathsf{T}(s) f \right] \dot{b} \right) - b^\mathsf{T} D_{m-1}(s) \left(\left[\alpha_{m-1}(s) \alpha_n^\mathsf{T}(s) f \right] \dot{a} \right)$$

The right-hand side of (A.12) can be expressed, using block matrix manipulation, as

$$[a^\mathsf{T} D_{n-1}(s), b^\mathsf{T} D_{m-1}(s)] \left(\begin{bmatrix} 0 & \alpha_{n-1}(s) \alpha_m^\mathsf{T}(s) \\ -\alpha_{m-1}(s) \alpha_n^\mathsf{T}(s) & 0 \end{bmatrix} \begin{bmatrix} \dot{a} \\ \dot{b} \end{bmatrix} \right)$$

and the proof is complete. $\square$

Lemma A.4 (Swapping Lemma A.4) *Let*

$$A(s,t) \triangleq a_n(t)s^n + a_{n-1}(t)s^{n-1} + \cdots a_1(t)s + a_0(t) = a^\top(t)\alpha_n(s)$$
$$\bar{A}(s,t) \triangleq s^n a_n(t) + s^{n-1}a_{n-1}(t) + \cdots sa_1(t) + a_0(t) = \alpha_n^\top(s)a(t)$$
$$B(s,t) = b_m(t)s^m + b_{m-1}(t)s^{m-1} + \cdots b_1(t)s + b_0(t) = b^\top(t)\alpha_m(s)$$
$$\bar{B}(s,t) \triangleq s^m b_m(t) + s^{m-1}b_{m-1}(t) + \cdots sb_1(t) + b_0(t) = \alpha_m^\top(s)b(t)$$

where $s \triangleq \frac{d}{dt}$, $a(t) \in \mathcal{R}^{n+1}, b(t) \in \mathcal{R}^{m+1}$ *and* $a,b \in \mathcal{L}_\infty$, $\alpha_i = [s^i, \cdots, s, 1]^\top$. *Let*

$$A(s,t) \cdot B(s,t) \triangleq \sum_{i=0}^n \sum_{j=0}^m a_i b_j s^{i+j}, \quad \overline{A(s,t) \cdot B(s,t)} \triangleq \sum_{i=0}^n \sum_{j=0}^m s^{i+j} a_i b_j$$

be the algebraic product of $A(s,t)$, $B(s,t)$ *and* $\bar{A}(s,t), \bar{B}(s,t)$, *respectively. Then for any function* f *for which* $\overline{A(s,t) \cdot B(s,t)}f$ *and* $\bar{A}(s,t)(B(s,t)f)$ *and* $\bar{B}(s,t)(A(s,t)f)$ *are defined, we have*

(i) $\quad \bar{A}(s,t)(B(s,t)f) = \bar{B}(s,t)(A(s,t)f) + \alpha_{\bar{n}}^\top(s)F(a,b)\alpha_{\bar{n}}(s)f \quad$ (A.13)

where $\bar{n} = \max\{n,m\} - 1$ *and* $F(a,b)$ *satisfies* $\|F(a,b)\| \leq c_1|\dot{a}| + c_2|\dot{b}|$ *for some constants* c_1, c_2

(ii) $\bar{A}(s,t)\left(B(s,t)\frac{1}{\Lambda_0(s)}f\right) = \frac{1}{\Lambda_0(s)}\overline{A(s,t) \cdot B(s,t)}f + \alpha_n^\top(s)G(s,f,a,b)$ (A.14)

for any Hurwitz polynomial $\Lambda_0(s)$ *of order greater or equal to* m, *where* $G(s,f,a,b)$ *is defined as*

$$G(s,f,a,b) \triangleq [g_n, \ldots, g_1, g_0], g_j = -\sum_{j=0}^m W_{jc}(s)\left((W_{jb}(s)f)(\dot{a}_i b_j + a_i \dot{b}_j)\right)$$

and W_{jc}, W_{jb} *are strictly proper transfer functions that have the same poles as* $\frac{1}{\Lambda_0(s)}$.

Proof (i) From the definition of $\bar{A}(s,t), A(s,t), \bar{B}(s,t), B(s,t)$, we have

$$\bar{A}(s,t)(B(s,t)f) = \sum_{i=0}^n \sum_{j=0}^m s^i(a_i b_j s^j f) \quad (A.15)$$

A. SWAPPING LEMMAS

$$\bar{B}(s,t)(A(s,t)f) = \sum_{i=0}^{n}\sum_{j=0}^{m} s^j(a_i b_j s^i f) \qquad (A.16)$$

Now we consider $s^i(a_i b_j s^j f)$ and treat the three cases $i > j$, $i < j$ and $i = j$ separately: First, we suppose $i > j$ and write

$$s^i(a_i b_j s^j f) = s^j\left(s^{i-j}(a_i b_j s^j f)\right)$$

Using the identity $s(a_i b_j s^l f) = a_i b_j s^{l+1} f + (\dot{a}_i b_j + a_i \dot{b}_j) s^l f$, where l is any integer, for $i - j$ times, we can swap the operator s^{i-j} with $a_i b_j$ and obtain

$$s^{i-j}(a_i b_j s^j f) = a_i b_j s^i f + \sum_{k=1}^{i-j} s^{i-j-k}(\dot{a}_i b_j + a_i \dot{b}_j) s^{j-1+k} f$$

and, therefore,

$$s^i(a_i b_j s^j f) = s^j(a_i b_j s^i f) + \sum_{k=1}^{i-j} s^{i-k}(\dot{a}_i b_j + a_i \dot{b}_j) s^{j-1+k} f, \quad i > j \qquad (A.17)$$

Similarly, for $i < j$, we write

$$s^i(a_i b_j s^j f) = s^i((a_i b_j s^{j-i}) s^i f)$$

Now using $a_i b_j s^l f = s(a_i b_j) s^{l-1} f - (\dot{a}_i b_j + a_i \dot{b}_j) s^{l-1} f$, where l is any integer, for $j - i$ times, we have

$$(a_i b_j s^{j-i}) s^i f = s^{j-i}(a_i b_j) s^i f - \sum_{k=1}^{j-i} s^{j-i-k}(\dot{a}_i b_j + a_i \dot{b}_j) s^{i-1+k} f$$

and, therefore,

$$s^i(a_i b_j s^j f) = s^j(a_i b_j s^i f) - \sum_{k=1}^{j-i} s^{j-k}(\dot{a}_i b_j + a_i \dot{b}_j) s^{i-1+k} f, \quad i < j \qquad (A.18)$$

For $i = j$, it is obvious that

$$s^i(a_i b_j s^j f) = s^j(a_i b_j s^i f), \quad i = j \qquad (A.19)$$

Combining (A.17), (A.18) and (A.19), we have

$$\bar{A}(s,t)(B(s,t)f) = \sum_{i=0}^{n}\sum_{j=0}^{m} s^j(a_i b_j s^i f) + r_1 = \bar{B}(s,t)(A(s,t)f) + r_1 \qquad (A.20)$$

where

$$r_1 \triangleq \sum_{i=0}^{n}\sum_{\substack{j=0 \\ j<i}}^{m}\sum_{k=1}^{i-j} s^{i-k}(\dot{a}_i b_j + a_i \dot{b}_j)s^{j-1+k}f - \sum_{i=0}^{n}\sum_{\substack{j=0 \\ j>i}}^{m}\sum_{k=1}^{j-i} s^{j-k}(\dot{a}_i b_j + a_i \dot{b}_j)s^{i-1+k}f$$

Note that for $0 \leq i \leq n, 0 \leq j \leq m$ and $1 \leq k \leq |i-j|$, we have

$$j \leq i-k, \quad j+k-1 \leq i-1, \quad \text{if } i \geq j$$
$$i \leq j-k, \quad i+k-1 \leq j-1, \quad \text{if } i < j$$

Therefore all the s-terms in r_1 (before and after the term $\dot{a}_i b_j + a_i \dot{b}_j$) have order less than $\max\{n,m\}-1$, and r_1 can be expressed as

$$r_1 = \alpha_{\bar{n}}^\top(s) F(a,b) \alpha_{\bar{n}}(s) f$$

where $\bar{n} = \max\{n,m\}-1$ and $F(a,b) \in \mathcal{R}^{\bar{n}\times\bar{n}}$ is a time-varying matrix whose elements are linear combinations of $\dot{a}_i b_j + a_i \dot{b}_j$. Because $a,b \in \mathcal{L}_\infty$, it follows from the definition of $F(a,b)$ that

$$\|F(a,b)\| \leq c_1 |\dot{a}| + c_2 |\dot{b}|$$

(ii) Applying Lemma A.1 with $W(s) = \frac{s^j}{\Lambda_0(s)}$, we have

$$a_i b_j \frac{s^j}{\Lambda_0(s)} f = \frac{s^j}{\Lambda_0(s)} a_i b_j f - W_{jc}((W_{jb}f)(\dot{a}_i b_j + a_i \dot{b}_j))$$

where $W_{jc}(s), W_{jb}(s)$ are strictly proper transfer functions, which have the same poles as $\frac{1}{\Lambda_0(s)}$. Therefore

$$\bar{A}(s,t)\left(B(s,t)\frac{1}{\Lambda_0(s)}f\right) = \sum_{i=0}^{n}\sum_{j=0}^{m} s^i \left(a_i b_j \frac{s^j}{\Lambda_0(s)}f\right)$$

$$= \sum_{i=0}^{n}\sum_{j=0}^{m} \frac{s^{i+j}}{\Lambda_0(s)}(a_i b_j f) - \sum_{i=0}^{n}\sum_{j=0}^{m} s^i W_{jc}((W_{jb}f)(\dot{a}_i b_j + a_i \dot{b}_j))$$

$$= \frac{1}{\Lambda_0(s)} \overline{A(s,t) \cdot B(s,t)} f + r_2 \qquad (A.21)$$

where

$$r_2 \triangleq -\sum_{i=0}^{n}\sum_{j=0}^{m} s^i W_{jc}((W_{jb}f)(\dot{a}_i b_j + a_i \dot{b}_j))$$

From the definition of r_2, we can write

$$r_2 = \alpha_n^\top(s) G(s,f,a,b)$$

by defining

$$G(s,f,a,b) \triangleq [g_n,,\ldots,g_1,g_0], \quad g_j = -\sum_{j=0}^{m} W_{jc}(s)\left((W_{jb}f)(\dot{a}_i b_j + a_i \dot{b}_j)\right)$$

□

B Optimization Techniques

An important part of every adaptive control scheme is the on-line estimator or adaptive law used to provide an estimate of the plant or controller parameters at each time t. Most of these adaptive laws are derived by minimizing certain cost functions with respect to the estimated parameters. The type of the cost function and method of minimization determines the properties of the resulting adaptive law as well as the overall performance of the adaptive scheme.

In this section we introduce some simple optimization techniques that include the method of *steepest descent*, referred to as the *gradient method*, *Newton's method* and the *gradient projection method* for constrained minimization problems.

B.1 Notation and Mathematical Background

A real-valued function $f : \mathcal{R}^n \mapsto \mathcal{R}$ is said to be continuously differentiable if the partial derivatives $\frac{\partial f(x)}{\partial x_1}, \ldots, \frac{\partial f(x)}{\partial x_n}$ exist for each $x \in \mathcal{R}^n$ and are continuous functions of x. In this case, we write $f \in \mathcal{C}^1$. More generally, we write $f \in \mathcal{C}^m$ if all partial derivatives of order m exist and are continuous functions of x.

If $f \in \mathcal{C}^1$, the **gradient** of f at a point $x \in \mathcal{R}^n$ is defined to be the column vector

$$\nabla f(x) \triangleq \begin{bmatrix} \frac{\partial f(x)}{\partial x_1} \\ \vdots \\ \frac{\partial f(x)}{\partial x_n} \end{bmatrix}$$

If $f \in \mathcal{C}^2$, the **Hessian** of f at x is defined to be the symmetric $n \times n$ matrix having $\partial^2 f(x)/\partial x_i \partial x_j$ as the ijth element, i.e.,

$$\nabla^2 f(x) \triangleq \left[\frac{\partial^2 f(x)}{\partial x_i \partial y_j}\right]_{n \times n}$$

A subset $\mathcal{S}$ of $\mathcal{R}^n$ is said to be **convex** if for every $x, y \in \mathcal{S}$ and $\alpha \in [0, 1]$, we have $\alpha x + (1 - \alpha)y \in \mathcal{S}$.

A function $f : \mathcal{S} \mapsto \mathcal{R}$ is said to be *convex over the convex set* $\mathcal{S}$ if for every $x, y \in \mathcal{S}$ and $\alpha \in [0, 1]$ we have

$$f(\alpha x + (1 - \alpha)y) \leq \alpha f(x) + (1 - \alpha)f(y)$$

Let $f \in \mathcal{C}^1$ over an open convex set $\mathcal{S}$, then f is convex over $\mathcal{S}$ iff

$$f(y) \geq f(x) + (\nabla f(x))^\mathsf{T}(y - x), \quad \forall x, y \in \mathcal{S} \tag{B.1}$$

If $f \in \mathcal{C}^2$ over $\mathcal{S}$ and $\nabla^2 f(x) \geq 0 \ \forall x \in \mathcal{S}$, then f is convex over $\mathcal{S}$.

Let us now consider the following unconstrained minimization problem

$$\begin{aligned} \text{minimize } & J(\theta) \\ \text{subject to } & \theta \in \mathcal{R}^n \end{aligned} \tag{B.2}$$

where $J : \mathcal{R}^n \mapsto \mathcal{R}$ is a given function. We say that the vector θ^* is a global minimum for (B.2) if

$$J(\theta^*) \leq J(\theta) \quad \forall \theta \in \mathcal{R}^n$$

A necessary and sufficient condition satisfied by the global minimum θ^* is given by the following lemma.

Lemma B.1 *Assume that $J \in \mathcal{C}^1$ and is convex over $\mathcal{R}^n$. Then θ^* is a global minimum for (B.2) iff*

$$\nabla J(\theta^*) = 0$$

The proof of Lemma B.1 can be found in [132, 196].

A vector $\bar{\theta}$ is called a *regular point of the surface* $S_\theta = \{\theta \in \mathcal{R}^n \,|\, g(\theta) = 0\}$ if $\nabla g(\bar{\theta}) \neq 0$. At a regular point $\bar{\theta}$, the set

$$M(\bar{\theta}) = \left\{\theta \in \mathcal{R}^n \,\big|\, \theta^\mathsf{T} \nabla g(\bar{\theta}) = 0\right\}$$

is called the *tangent plane* of g at $\bar{\theta}$.

B.2 The Method of Steepest Descent (Gradient Method)

This is one of the oldest and most widely known methods for solving the unconstrained minimization problem (B.2). It is also one of the simplest for which a satisfactory analysis exists. More sophisticated methods are often motivated by an attempt to modify the basic steepest descent technique for better convergence properties [21, 132, 196]. The method of steepest descent proceeds from an initial approximation θ_0 for the minimum θ^* to successive points $\theta_1, \theta_2, \cdots$ in $\mathcal{R}^n$ in an iterative manner until some stopping condition is satisfied. Given the current point θ_k, the point θ_{k+1} is obtained by a linear search in the direction d_k where

$$d_k = -\nabla J(\theta_k)$$

It can be shown [196] that d_k is the direction from θ_k in which the initial rate of decrease of $J(\theta)$ is the greatest. Therefore, the sequence $\{\theta_k\}$ is defined by

$$\theta_{k+1} = \theta_k + \lambda_k d_k = \theta_k - \lambda_k \nabla J(\theta_k), \quad (k = 0, 1, 2, \cdots) \quad (B.3)$$

where θ_0 is given and λ_k, known as the *step size* or *step length*, is determined by the linear search method, so that θ_{k+1} minimizes $J(\theta)$ in the direction d_k from θ_k. A simpler expression for θ_{k+1} can be obtained by setting $\lambda_k = \lambda \ \forall k$, i.e.,

$$\theta_{k+1} = \theta_k - \lambda \nabla J(\theta_k) \quad (B.4)$$

In this case, the linear search for λ_k is not required, though the choice of the step length λ is a compromise between accuracy and efficiency.

Considering infinitesimally small step lengths, (B.4) can be converted to the continuous-time differential equation

$$\dot{\theta} = -\nabla J(\theta(t)), \quad \theta(t_0) = \theta_0 \quad (B.5)$$

whose solution $\theta(t)$ is the descent path in the time domain starting from $t = t_0$.

The direction of steepest descent $d = -\nabla J$ can be scaled by a constant positive definite matrix $\Gamma = \Gamma^\top$ as follows: We let $\Gamma = \Gamma_1 \Gamma_1^\top$ where Γ_1 is an $n \times n$ nonsingular matrix and consider the vector $\bar{\theta} \in \mathcal{R}^n$ given by

$$\Gamma_1 \bar{\theta} = \theta$$

Then the minimization problem (B.2) is equivalent to

$$\text{minimize } \bar{J}(\bar{\theta}) \triangleq J(\Gamma_1 \bar{\theta})$$
$$\text{subject to } \bar{\theta} \in \mathcal{R}^n \quad \text{(B.6)}$$

If $\bar{\theta}^*$ is a minimum of $\bar{J}$, the vector $\theta^* = \Gamma_1 \bar{\theta}^*$ is a minimum of J. The steepest descent for (B.6) is given by

$$\bar{\theta}_{k+1} = \bar{\theta}_k - \lambda \nabla \bar{J}(\bar{\theta}_k) \quad \text{(B.7)}$$

Because $\nabla \bar{J}(\bar{\theta}) = \frac{\partial J(\Gamma_1 \bar{\theta})}{\partial \bar{\theta}} = \Gamma_1^T \nabla J(\theta)$ and $\Gamma_1 \bar{\theta} = \theta$ it follows from (B.7) that

$$\theta_{k+1} = \theta_k - \lambda \Gamma_1 \Gamma_1^T \nabla J(\theta_k)$$

Setting $\Gamma = \Gamma_1 \Gamma_1^T$, we obtain the scaled version for the steepest descent algorithm

$$\theta_{k+1} = \theta_k - \lambda \Gamma \nabla J(\theta_k) \quad \text{(B.8)}$$

The continuous-time version of (B.8) is now given by

$$\dot{\theta} = -\Gamma \nabla J(\theta) \quad \text{(B.9)}$$

The convergence properties of (B.3), (B.4), (B.8) for different step lengths are given in any standard book on optimization such as [132, 196]. The algorithms (B.5), (B.9) for various cost functions $J(\theta)$ are used in Chapters 4 to 9 where the design and analysis of adaptive laws is considered.

B.3 Newton's Method

Let us consider the minimization problem (B.2) and assume that $J(\theta)$ is convex over $\mathcal{R}^n$. Then according to Lemma B.1, any global minimum θ^* should satisfy $\nabla J(\theta^*) = 0$. Usually $\nabla J(\theta^*) = 0$ gives a set of nonlinear algebraic equations whose solution θ^* may be found by solving these equations using a series of successive approximations known as the Newton's method.

Let θ_k be the estimate of θ^* at instant k. Then $\nabla J(\theta)$ for θ close to θ_k may be approximated by the linear portion of the Taylor's series expansion

$$\nabla J(\theta) \simeq \nabla J(\theta_k) + \frac{\partial}{\partial \theta} \nabla J(\theta) |_{\theta=\theta_k} (\theta - \theta_k) \quad \text{(B.10)}$$

B. OPTIMIZATION TECHNIQUES

The estimate θ_{k+1} of θ^* at the $k+1$ iteration can now be generated from (B.10) by setting its right-hand side equal to zero and solving for $\theta = \theta_{k+1}$, i.e.,

$$\theta_{k+1} = \theta_k - H^{-1}(\theta_k)\nabla J(\theta_k), \quad k = 0, 1, 2, \ldots \quad (B.11)$$

where $H(\theta_k) = \frac{\partial}{\partial \theta}\nabla J(\theta_k) = \nabla^2 J(\theta_k)$ is the Hessian matrix whose inverse is assumed to exist at each iteration k.

A continuous time version of (B.11) can also be developed by constructing a differential equation whose solution $\theta(t)$ converges to the root of $\nabla J(\theta) = 0$ as $t \to \infty$. Treating $\theta(t)$ as a smooth function of time we have

$$\frac{d}{dt}\nabla J(\theta(t)) = \frac{\partial}{\partial \theta}\nabla J(\theta)\dot{\theta} = H(\theta)\dot{\theta} \quad (B.12)$$

Choosing

$$\dot{\theta} = -\beta H^{-1}(\theta)\nabla J(\theta), \quad \theta(t_0) = \theta_0 \quad (B.13)$$

for some scalar $\beta > 0$, we have

$$\frac{d}{dt}\nabla J(\theta(t)) = -\beta \nabla J(\theta)$$

or

$$\nabla J(\theta(t)) = e^{-\beta(t-t_0)}\nabla J(\theta_0) \quad (B.14)$$

It is therefore clear that if a solution $\theta(t)$ of (B.13) exists $\forall t \geq t_0$, then equation (B.14) implies that this solution will converge to a root of $\nabla J = 0$ as $t \to \infty$.

When the cost J depends explicitly on the time t, that is $J = J(\theta, t)$ and is convex for each time t, then any global minimum θ^* should satisfy

$$\nabla J(\theta^*, t) = 0, \quad \forall t \geq t_0 \geq 0$$

In this case, (B.12) becomes

$$\frac{d}{dt}\nabla J(\theta, t) = \frac{\partial}{\partial \theta}\nabla J(\theta, t)\dot{\theta} + \frac{\partial}{\partial t}\nabla J(\theta, t)$$

Therefore, if we choose

$$\dot{\theta} = -H^{-1}(\theta)\left(\beta \nabla J(\theta, t) + \frac{\partial}{\partial t}\nabla J(\theta, t)\right) \quad (B.15)$$

for some scalar $\beta > 0$, where $H(\theta, t) = \frac{\partial}{\partial \theta} \nabla J(\theta, t)$ we have

$$\frac{d}{dt} \nabla J(\theta, t) = -\beta \nabla J(\theta, t)$$

or

$$\nabla J(\theta(t), t) = e^{-\beta(t-t_0)} \nabla J(\theta(t_0), t_0) \qquad (B.16)$$

If (B.15) has a solution, then (B.16) implies that such solution will converge to a root of $\nabla J = 0$ as $t \to \infty$.

Newton's method is very attractive in terms of its convergence properties, but suffers from the drawback of requiring the existence of the inverse of the Hessian matrix $H(\theta)$ at each instant of time (k or t). It has to be modified in order to avoid the costly and often impractical evaluation of the inverse of H. Various modified Newton's methods use an approximation of the inverse Hessian. The form of the approximation ranges from the simplest where H^{-1} remains fixed throughout the iterative process, to the more advanced where improved approximations are obtained at each step on the basis of information gathered during the descent process [132]. It is worth mentioning that in contrast to the steepest descent method, the Newton's method is "scale free" in the sense that it cannot be affected by a change in the coordinate system.

B.4 Gradient Projection Method

In sections B.2 and B.3, the search for the minimum of the function $J(\theta)$ given in (B.2) was carried out for all $\theta \in \mathcal{R}^n$. In some cases, θ is constrained to belong to a certain convex set

$$\mathcal{S} \triangleq \{\theta \in \mathcal{R}^n \,|\, g(\theta) \leq 0\} \qquad (B.17)$$

in $\mathcal{R}^n$ where $g(\cdot)$ is a scalar-valued function if there is only one constraint and a vector-valued function if there are more than one constraints. In this case, the search for the minimum is restricted to the convex set defined by (B.17) instead of $\mathcal{R}^n$.

Let us first consider the simple case where we have an equality constraint, that is, we

$$\begin{array}{ll} \text{minimize} & J(\theta) \\ \text{subject to} & g(\theta) = 0 \end{array} \qquad (B.18)$$

B. OPTIMIZATION TECHNIQUES

where $g(\theta)$ is a scalar-valued function. One of the most common techniques for handling constraints is to use a descent method in which the direction of descent is chosen to reduce the function $J(\theta)$ by remaining within the constrained region. Such method is usually referred to as the **gradient projection method**.

We start with a point θ_0 satisfying the constraint, i.e., $g(\theta_0) = 0$. To obtain an improved vector θ_1, we project the negative gradient of J at θ_0 i.e., $-\nabla J(\theta_0)$ onto the tangent plane $\mathcal{M}(\theta_0) = \left\{\theta \in \mathcal{R}^n \,\middle|\, \nabla g^\mathsf{T}(\theta_0)\theta = 0\right\}$ obtaining the direction vector $\text{Pr}(\theta_0)$. Then θ_1 is taken as $\theta_0 + \lambda_0 \text{Pr}(\theta_0)$ where λ_0 is chosen to minimize $J(\theta_1)$. The general form of this iteration is given by

$$\theta_{k+1} = \theta_k + \lambda_k \text{Pr}(\theta_k) \tag{B.19}$$

where λ_k is chosen to minimize $J(\theta_k)$ and $\text{Pr}(\theta_k)$ is the new direction vector after projecting $-\nabla J(\theta_k)$ onto $\mathcal{M}(\theta_k)$. The explicit expression for $\text{Pr}(\theta_k)$ can be obtained as follows: The vector $-\nabla J(\theta_k)$ can be expressed as a linear combination of the vector $\text{Pr}(\theta_k)$ and the normal vector $N(\theta_k) = \nabla g(\theta_k)$ to the tangent plane $\mathcal{M}(\theta_k)$ at θ_k, i.e.,

$$-\nabla J(\theta_k) = \alpha \nabla g(\theta_k) + \text{Pr}(\theta_k) \tag{B.20}$$

for some constant α. Because $\text{Pr}(\theta_k)$ lies on the tangent plane $\mathcal{M}(\theta_k)$, we also have $\nabla g^\mathsf{T}(\theta_k)\text{Pr}(\theta_k) = 0$ which together with (B.20) implies that

$$-\nabla g^\mathsf{T} \nabla J = \alpha \nabla g^\mathsf{T} \nabla g$$

i.e.,

$$\alpha = -(\nabla g^\mathsf{T} \nabla g)^{-1} \nabla g^\mathsf{T} \nabla J$$

Hence, from (B.20), we obtain

$$\text{Pr}(\theta_k) = -\left[I - \nabla g(\nabla g^\mathsf{T} \nabla g)^{-1} \nabla g^\mathsf{T}\right] \nabla J \tag{B.21}$$

We refer to $\text{Pr}(\theta_k)$ as the projected direction onto the tangent plant $\mathcal{M}(\theta_k)$. The gradient projection method is illustrated in Figure B.1.

It is clear from Figure B.1 that when $g(\theta)$ is not a linear function of θ, the new vector θ_{k+1} given by (B.19) may not satisfy the constraint, so it must be modified. There are several successive approximation techniques that can be employed to move θ_{k+1} from $\mathcal{M}(\theta_k)$ to the constraint surface $g(\theta) = 0$

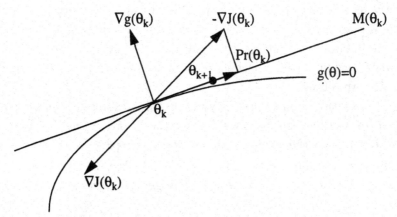

Figure B.1 Gradient projection method.

[132, 196]. One special case, which is often encountered in adaptive control applications, is when θ is constrained to stay inside a ball with a given center and radius, i.e., $g(\theta) = (\theta - \theta_0)^T(\theta - \theta_0) - M^2$ where θ_0 is a fixed constant vector and $M > 0$ is a scalar. In this case, the discrete projection algorithm which guarantees that $\theta_k \in \mathcal{S} \ \forall k$ is

$$\bar{\theta}_{k+1} = \theta_k + \lambda_k \nabla J$$
$$\theta_{k+1} = \begin{cases} \bar{\theta}_{k+1} & \text{if } |\bar{\theta}_{k+1} - \theta_0| \leq M \\ \theta_0 + \frac{\bar{\theta}_{k+1} - \theta_0}{|\bar{\theta}_{k+1} - \theta_0|} M & \text{if } |\bar{\theta}_{k+1} - \theta_0| > M \end{cases} \quad (B.22)$$

Letting the step length λ_k become infinitesimally small, we obtain the continuous-time version of (B.19), i.e.,

$$\dot{\theta} = \text{Pr}(\theta) = -\left[I - \nabla g (\nabla g^T \nabla g)^{-1} \nabla g^T \right] \nabla J \quad (B.23)$$

Because of the sufficiently small step length, the trajectory $\theta(t)$, if it exists, will satisfy $g(\theta(t)) = 0 \ \forall t \geq 0$ provided $\theta(0) = \theta_0$ satisfies $g(\theta_0) = 0$.

The scaled version of the gradient projection method can be obtained by using the change of coordinates $\Gamma_1 \bar{\theta} = \theta$ where Γ_1 is a nonsingular matrix that satisfies $\Gamma = \Gamma_1 \Gamma_1^T$ and Γ is the scaling positive definite constant matrix. Following a similar approach as in section B.2, the scaled version of (B.23) is given by

$$\dot{\theta} = \bar{\text{Pr}}(\theta)$$

B. OPTIMIZATION TECHNIQUES

where

$$\bar{Pr}(\theta) = -\left[I - \Gamma \nabla g (\nabla g^\mathsf{T} \Gamma \nabla g)^{-1} \nabla g^\mathsf{T}\right] \Gamma \nabla J \qquad \text{(B.24)}$$

The minimization problem (B.18) can now be extended to

$$\begin{aligned} & \text{minimize } J(\theta) \\ & \text{subject to } g(\theta) \leq 0 \end{aligned} \qquad \text{(B.25)}$$

where $\mathcal{S} = \{\theta \in \mathcal{R}^n \,|\, g(\theta) \leq 0\}$ is a convex subset of $\mathcal{R}^n$.

The solution to (B.25) follows directly from that of the unconstrained problem and (B.18). We start from an initial point $\theta_0 \in \mathcal{S}$. If the current point is in the interior of $\mathcal{S}$, defined as $\mathcal{S}_0 \triangleq \{\theta \in \mathcal{R}^n \,|\, g(\theta) < 0\}$, then the unconstrained algorithm is used. If the current point is on the boundary of $\mathcal{S}$, defined as $\delta(\mathcal{S}) \triangleq \{\theta \in \mathcal{R}^n \,|\, g(\theta) = 0\}$ and the direction of search given by the unconstrained algorithm is pointing away from $\mathcal{S}$, then we use the gradient projection algorithm. If the direction of search is pointing inside $\mathcal{S}$ then we keep the unconstrained algorithm. In view of the above, the solution to the constrained optimization problem (B.25) is given by

$$\dot{\theta} = \begin{cases} -\nabla J(\theta) & \text{if } \theta \in \mathcal{S}_0 \text{ or } \theta \in \delta(\mathcal{S}) \text{ and } -\nabla J^\mathsf{T} \nabla g \leq 0 \\ -\nabla J + \frac{\nabla g \nabla g^\mathsf{T}}{\nabla g^\mathsf{T} \nabla g} \nabla J & \text{otherwise} \end{cases} \qquad \text{(B.26)}$$

where $\theta(0) \in \mathcal{S}$ or with the scaling matrix

$$\dot{\theta} = \begin{cases} -\Gamma \nabla J(\theta) & \text{if } \theta \in \mathcal{S}_0 \\ & \text{or } \theta \in \delta(\mathcal{S}) \text{ and } -(\Gamma \nabla J)^\mathsf{T} \nabla g \leq 0 \\ -\Gamma \nabla J + \Gamma \frac{\nabla g \nabla g^\mathsf{T}}{\nabla g^\mathsf{T} \Gamma \nabla g} \Gamma \nabla J & \text{otherwise} \end{cases} \qquad \text{(B.27)}$$

B.5 Example

Consider the scalar time varying equation

$$y(t) = \theta^* u(t) \qquad \text{(B.28)}$$

where $y, u : \mathcal{R}^+ \mapsto \mathcal{R}$ are bounded uniformly continuous functions of time and $\theta^* \in \mathcal{R}$ is a constant. We would like to obtain an estimate of θ^* at each time t from the measurements of $y(\tau), u(\tau), 0 \leq \tau \leq t$. Let $\theta(t)$ be the estimate of θ^* at time t, then

$$\hat{y}(t) = \theta(t) u(t)$$

is the estimate of $y(t)$ at time t and

$$\epsilon(t) = y - \hat{y} = -(\theta(t) - \theta^*)u \qquad (B.29)$$

is the resulting estimation error due to $\theta(t) \neq \theta^*$. We would like to generate a trajectory $\theta(t)$ so that $\epsilon(t) \to 0$ and $\theta(t) \to \theta^*$ as $t \to \infty$ by minimizing certain cost functions of $\epsilon(t)$ w.r.t. $\theta(t)$.

Let us first choose the simple quadratic cost

$$J(\theta, t) = \frac{1}{2}\varepsilon^2(t) = \frac{(\theta - \theta^*)^2 u^2(t)}{2} \qquad (B.30)$$

and minimize it w.r.t. θ. For each given t, $J(\theta, t)$ is convex over $\mathcal{R}$ and $\theta = \theta^*$ is a minimum for all $t \geq 0$. $J(\theta, t)$, however, can reach its minimum value, i.e., zero when $\theta \neq \theta^*$.

The gradient and Hessian of J are

$$\nabla J(\theta, t) = (\theta - \theta^*) u^2(t) = -\epsilon u \qquad (B.31)$$

$$\nabla^2 J(\theta, t) = H(\theta, t) = u^2(t) \geq 0 \qquad (B.32)$$

Using the gradient method we have

$$\dot{\theta} = \gamma \epsilon u, \quad \theta(0) = \theta_0 \qquad (B.33)$$

for some $\gamma > 0$. Using the Newton's method, i.e., (B.15), we obtain

$$\dot{\theta} = -\frac{1}{u^2}\left[-\epsilon u + \frac{\partial}{\partial t}\nabla J\right] = \frac{\epsilon}{u^2}[u + 2\dot{u}] \qquad (B.34)$$

which is valid provided $u^2 > 0$ and $\dot{u}$ exists.

For the sake of simplicity, let us assume that $u^2 \geq c_0 > 0$ for some constant c_0 and $\dot{u} \in \mathcal{L}_\infty$ and analyze (B.33), (B.34) using a Lyapunov function approach.

We first consider (B.33) and choose the function

$$V(\tilde{\theta}) = \frac{\tilde{\theta}^2}{2\gamma} \qquad (B.35)$$

B. OPTIMIZATION TECHNIQUES

where $\tilde{\theta}(t) = \theta(t) - \theta^*$ is the parameter error. Because θ^* is a constant, we have $\dot{\tilde{\theta}} = \dot{\theta} = \gamma \epsilon u$ and from (B.29) $\epsilon = -\tilde{\theta} u$. Therefore, along any solution of (B.33) we have

$$\dot{V} = \frac{\tilde{\theta}\dot{\tilde{\theta}}}{\gamma} = \tilde{\theta}\epsilon u = -\tilde{\theta}^2 u^2 \leq -c_0 \tilde{\theta}^2 < 0$$

which implies that the equilibrium $\tilde{\theta}_e = 0$, i.e, $\theta_e = \theta^*$ is e.s.

Similarly, for (B.34) we choose

$$V(\tilde{\theta}) = \frac{\tilde{\theta}^2 u^4}{2}$$

Because u^2 is bounded from above and below by $c_0 > 0$, $V(\tilde{\theta})$ is positive definite, radially unbounded and decrescent. The time derivative $\dot{V}$ of $V(\tilde{\theta})$ along the solution of (B.34) is given by

$$\dot{V} = u^4 \tilde{\theta}\dot{\tilde{\theta}} + 2\tilde{\theta}^2 u^3 \dot{u} = u^2 \epsilon \tilde{\theta} u + 2u^2 \epsilon \tilde{\theta} \dot{u} + 2\tilde{\theta}^2 u^3 \dot{u}$$

Using $\epsilon = -\tilde{\theta}u$ we have

$$\dot{V} = -\tilde{\theta}^2 u^4 - 2\tilde{\theta}^2 u^3 \dot{u} + 2\tilde{\theta}^2 u^3 \dot{u} = -\tilde{\theta}^2 u^4 \leq -c_0^2 \tilde{\theta}^2$$

which implies that the equilibrium $\tilde{\theta}_e = 0$ i.e., $\theta_e = \theta^*$ of (B.34) is e.s.

Let us now assume that an upper bound for θ^* is known, i.e., $|\theta^*| \leq c$ for some constant $c > 0$. If instead of (B.30) we consider the minimization of

$$J(\theta, t) = \frac{(\theta - \theta^*)^2}{2} u^2$$

$$\text{subject to } g(\theta) = \theta^2 - c^2 \leq 0$$

the gradient projection algorithm will give us

$$\dot{\theta} = \begin{cases} \gamma \epsilon u & \text{if } \theta^2 < c^2 \\ & \text{or if } \theta^2 = c^2 \text{ and } \epsilon u \theta \leq 0 \\ 0 & \text{otherwise} \end{cases} \quad (B.36)$$

where $\theta(0)$ is chosen so that $\theta^2(0) \leq c^2$. We analyze (B.36) by considering the function given by (B.35). Along the trajectory of (B.36), we have

$$\dot{V} = -\tilde{\theta}^2 u^2 \leq -2\gamma c_0 V(\tilde{\theta}) \quad \text{if } \theta^2 < c^2 \text{ or if } \theta^2 = c^2 \text{ and } \epsilon u \theta \leq 0$$

and
$$\dot{V} = 0 \text{ if } \theta^2 = c^2 \text{ and } \epsilon u \theta > 0.$$

Now $\theta^2 = c^2$ and $\epsilon u \theta > 0 \Rightarrow \theta \theta^* > c^2$. Since $|\theta^*| \leq c$ and $\theta^2 = c^2$, the inequality $\theta \theta^* > c^2$ is not possible which implies that for $\theta^2 = c^2, \epsilon u \theta \leq 0$ and no switching takes place in (B.36). Hence,

$$\dot{V}(\tilde{\theta}) \leq -2\gamma c_0 V(\tilde{\theta}) \quad \forall t \geq 0$$

i.e., $V(\tilde{\theta})$ and therefore $\tilde{\theta}$ converges exponentially to zero.

Bibliography

[1] Anderson, B.D.O., " Exponential Stability of Linear Equations Arising in Adaptive Identification," *IEEE Transactions on Automatic Control*, Vol. 22, no. 2, pp. 83-88, 1977.

[2] Anderson, B.D.O., "Adaptive Systems, Lack of Persistency of Excitation and Bursting Phenomena," *Automatica*, Vol. 21, pp. 247-258, 1985.

[3] Anderson, B.D.O., R.R. Bitmead, C.R. Johnson, P.V. Kokotović, R.L. Kosut, I. Mareels, L. Praly and B. Riedle, *Stability of Adaptive Systems*, M.I.T. Press, Cambridge, Massachusetts, 1986.

[4] Anderson, B.D.O. and R.M. Johnstone, "Adaptive Systems and Time Varying Plants," *Int.. J. of Control*, Vol. 37, no.2, pp. 367-377, 1983.

[5] Anderson, B.D.O., and R.M. Johnstone, "Global Adaptive Pole Positioning," *IEEE Transactions on Automatic Control*, Vol. 30, no. 1, pp. 11-22, 1985.

[6] Anderson, B.D.O. and J.B. Moore, *Optimal Control: Linear Quadratic Methods*, Prentice Hall, Englewood Cliffs, New Jersey, 1990.

[7] Anderson, B.D.O. and S. Vongpanitlerd, *Network Analysis and Synthesis: A Modern Systems Theory Approach*, Prentice Hall, Englewood Cliffs, New Jersey, 1973.

[8] Andreiev, N., "A Process Controller that Adapts to Signal and Process Conditions," *Control Engineering*, Vol. 38, 1977.

[9] Annaswamy, A.M. and K.S. Narendra, "Adaptive Control of Simple Time-Varying Systems," *Proceedings of the 28th IEEE Conference on Decision and Control*, Tampa, Florida, December 1989.

[10] Aseltine, J.A., A.R. Mancini and C.W. Sartune, "A Survey of Adaptive Control Systems," *IRE Transactions on Automatic Control*, Vol. 3, no. 6, pp. 102-108, 1958.

[11] Åström, K.J., "Theory and Applications of Adaptive Control–A Survey," *Automatica*, Vol. 19, no. 5, pp. 471-486, 1983.

[12] Åström, K.J. and T. Bohlin, "Numerical Identification of Linear Dynamic Systems from Normal Operating Records," in P.H. Hammod (Ed.) *Theory of Self-Adaptive Control Systems*, Plenum Press, New York, pp. 96-111, 1966.

[13] Åström, K.J. and P. Eykhoff, "System Identification–A Survey," *Automatica*, Vol. 7, pp. 123, 1971.

[14] Åström, K.J., P. Hagander, and J. Sternby, "Zeros of Sampled Systems," *Automatica*, Vol. 20, no. 1, pp. 21-38, 1984.

[15] Åström, K.J. and B. Wittenmark, *Adaptive Control*, Addison-Wesley Publishing Company, Reading, Massachusetts, 1989.

[16] Athans, M. and P.L. Falb, *Optimal Control*, McGraw-Hill, New York, 1966.

[17] Bai, E.W. and S. Sastry, "Global Stability Proofs for Continuous-time Indirect Adaptive Control Schemes," *IEEE Transactions on Automatic Control*, Vol. 32, no. 4, pp. 537-543, 1987.

[18] Barkana, I., "Adaptive Control: A Simplified Approach," in C. T. Leondes (Ed.), *Advances in Control and Dynamics*, Vol. 25, Academic Press, New York, 1987.

[19] Bellman, R.E., *Dynamic Programming*, Princeton University Press, Princeton, New Jersey, 1957.

[20] Bellman, R.E. *Adaptive Control Processes–A Guided Tour*, Princeton University Press, Princeton, New Jersey, 1961.

[21] Bertsekas, D.P., *Dynamic Programming*, Prentice Hall, Englewood Cliffs, New Jersey, 1987.

[22] Bitmead, R.R. "Persistence of Excitation Conditions and the Convergence of Adaptive Schemes," *IEEE Transactions on Information Theory*, Vol. 30, no. 3, pp. 183-191, 1984.

[23] Bitmead, R.R., M. Gevers and V. Wertz, *Adaptive Optimal Control*, Prentice Hall, Englewood Cliffs, New Jersey, 1990.

[24] Boyd, S. and S. Sastry, "Necessary and Sufficient Conditions for Parameter Convergence in Adaptive Control," *Automatica*, Vol. 22, no. 6, pp. 629-639, 1986.

[25] Brockett, R.W., *Finite Dimensional Linear Systems*, Wiley, New York, 1970.

[26] Broussard, J. and M. O'Brien, "Feedforward Control to Track the Output of a Forced Model," *Proceedings of the 17th Conference on Decision and Control*, pp. 1144-1155, 1979.

[27] Caldwell, W.I., *Control System with Automatic Response Adjustment.* American patent, 2,517,081. Filed 25, April 1947, 1950.

[28] Carroll, R. and D.P. Lindorff, "An Adaptive Observer for Single-Input Single-Output Linear Systems," *IEEE Transactions on Automatic Control*, Vol.18, pp. 428-435, 1973.

[29] Chalam, V.V., *Adaptive Control Systems: Techniques and Applications*, Marcel Dekker, New York, 1987.

[30] Chen, C.T., *Introduction to Linear System Theory*, Holt, Rinehart and Winston Inc., New York, 1970.

[31] Chen, C.T. and P.E. Caines, " On the Adaptive Control of Stochastic Systems with Random Parameters," in *Proceedings of the 23rd IEEE Conference on Decision and Control*, pp. 33, 1984.

[32] Coppel, W.A., *Stability and Asymptotic Behavior of Differential Equations*, Heath and Company, Boston, Massachusetts, 1965.

[33] Cristi, R., "Internal Persistency of Excitation in Indirect Adaptive Control," *IEEE Transactions on Automatic Control*, Vol. 32, no. 12, 1987.

[34] Cruz, Jr, J.B. *System Sensitivity Analysis*, Dowden, Hutchinson & Ross Inc., Stroudsburg, Pennsylvania, 1973.

[35] Dasgupta, S. and B.D.O. Anderson, "Priori Information and Persistent Excitation," Technical Report, Australian National University, 1987.

[36] Datta, A., "Transient Performance Improvement in Continuous-Time Model Reference Adaptive Control: An L_1 Formulation," *Proceedings of the 1993 American Control Conference*, San Francisco, California, pp. 294-299, June 1993.

[37] Datta, A. and M.T. Ho, "Systematic Design of Model Reference Adaptive Controllers with Improved Transient Performance," Department of Electrical Engineering, Texas A& M Univ., College Station, Tech. Report, TAMU-ECE92-12, Nov. 1992.

[38] Datta, A. and P.A. Ioannou, "Decentralized Adaptive Control," in C.T. Leondes (Ed.), *Advances in Control and Dynamic Systems*, Vol. XXXVI, 1989.

[39] Datta, A. and P.A. Ioannou, "Performance Improvement Versus Robust Stability in Model Reference Adaptive Control," *Proceedings of the 30th IEEE Conference on Decision and Control*, pp. 1082-1087, December, 1991.

[40] De Larminat, Ph., "On the Stabilizability Condition in Indirect Adaptive Control," *Automatica*, Vol. 20, pp. 793-795, 1984.

[41] Desoer, C.A., *Notes for a Second Course on Linear Systems*, Van Nostrand Reinhold, New York, 1970.

[42] Desoer, C.A. and M. Vidyasagar, *Feedback Systems: Input-Output Properties*, Academic Press Inc., New York, 1975.

[43] Dion, J.M., Dugard, L. and Carrillo, J., "Interactor and Multivariable Adaptive Model Matching," *IEEE Transactions on Automatic Control*, Vol. 33, no. 4, pp. 399-401, 1988.

[44] Dorf, R.C. *Modern Control Systems*, 6th Edition, Addison-Wesley Publishing Company, Reading, Massachusetts, 1991.

[45] Doyle, J., B.A. Francis and A.R. Tannenbaum, *Feedback Control Theory*, MacMillan, New York, 1992.

[46] Duarte, M.A. and K.S. Narendra, "Combined Direct and Indirect Adaptive Control of Plants with Relative Degree Greater Than One," Technical Report, No. 8715, Center for System Science, Yale University, New Haven, Connecticut, 1987.

[47] Dugard, L. and G.C. Goodwin, "Global Convergence of Landau's "Output Error with Adjustable compensator" Adaptive Algorithm", *IEEE Transactions on Automatic Control*, Vol. 30. no. 6, pp. 593-595, 1985.

[48] Egardt, B., *Stability of Adaptive Controllers*, Lecture Notes in Control and Information Sciences, Vol. 20, Springer-Verlag, Berlin, 1979.

[49] Elliott, H., Cristi, R. and Das, M. "Global Stability of Adaptive Pole Placement Algorithms," *IEEE Transactions on Automatic Control*, Vol. 30, no. 4, pp. 348-356, 1985.

[50] Elliott, H. and Wolovich, W.A., "A Parameter Adaptive Control Structure for Linear Multivariable Systems," *IEEE Transactions on Automatic Control*, Vol. 27, no. 2, pp. 340-352, 1982.

[51] Elliott, H. and W.A. Wolovich, "Parametrization Issues in Multivariable Adaptive Control," *Automatica*, Vol. 20, no. 5, pp. 533-545, 1984.

[52] Eykhoff, P. *System Identification: Parameter and State Estimation*, John Wiley & Sons, Inc., New York, 1974.

[53] Fel'dbaum, A.A., *Optimal Control Systems*, Academic Press, New York, 1965.

[54] Feuer, A. and A.S. Morse, "Adaptive Control of Single-Input, Single-Output Linear Systems," *IEEE Transactions on Automatic Control*, Vol. 23 pp. 557-569, 1978.

[55] Fradkov, A.L., *Adaptive Control in Large Scale Systems* (in Russian), M, Nauka, Phys & Math, 1990.

[56] Fradkov, A.L., "Continuous-Time Model Reference Adaptive Systems– An East-West Review," *Proceedings of the IFAC Symposium on Adaptive Control and Signal Processing*, Grenoble, France, July 1992.

[57] Franklin, G.F., J.D. Powell and A. Emami-Naeini, *Feedback Control of Dynamic Systems*, 2nd Edition, Addison-Wesley, Reading, Massachusetts, 1991.

[58] Fu, M. and B.R. Barmish, "Adaptive Stabilization of Linear Systems via Switching Control," *IEEE Transactions on Automatic Control*, Vol. 31, no. 12, pp. 1097-1103, 1986.

[59] Gavel, D.T. and D.D. Siljak, "Decentralized Adaptive Control: Structural Conditions for Stability," *IEEE Transactions on Automatic Control*, Vol. 34, no. 4, pp. 413-425, 1989.

[60] Gawthrop, P.J. "Hybrid Self-Tuning Control," *IEE Proceedings, Part D*, Vol. 127, pp. 229-336, 1980.

[61] Gawthrop, P.J., *Continuous-Time Self-Tuning Control: Volume I– Design*. Research Studies Press LtD., John Wiley & Sons Inc., New York, 1987.

[62] Gilbert, W.J., *Modern Algebra with Applications*, John Wiley & Sons Inc., New York, 1976.

[63] Giri, F., F. Ahmed-Zaid and P.A. Ioannou, "Stable Indirect Adaptive Control of Continuous-time Systems with No Apriori Knowledge about the Parameters," *Proceedings of the IFAC International Symposium on Adaptive Systems in Control and Signal Processing*, Grenoble, France, July 1992.

[64] Giri, F., J.M. Dion, M. M'Saad and L. Dugard, "A Globally Convergent Pole Placement Indirect Adaptive Controller," *Proceedings of the 26th IEEE Conference on Decision and Control*, Los Angeles, California, pp. 1-6, December 1987.

[65] Giri, F., J. M. Dion, L. Dugard and M. M'Saad, "Parameter Estimation Aspects in Adaptive Control," *Proceedings of IFAC Symposium on Identification and System Parameter Estimation*, Beijing, China, 1988.

[66] Giri, F., M. M'Saad, L. Dugard and J. M. Dion, "Pole Placement Direct Adaptive Control for Time-Varying Ill-Modeled Plants," *IEEE Transactions on Automatic Control*, Vol. 35, no. 6, pp. 723-726, 1990.

[67] Glover, K., "All Optimal Hankel-Norm Approximations of Linear Multivariable Systems and Their L^∞-Error Bounds," *International Journal of Control*, Vol. 39, no. 6, pp. 1115-1193, 1984.

[68] Golden, M.P. and B.E. Ydstie, "Chaos and Strange Attractors in Adaptive Control Systems, *Proceedings of the 10th IFAC World Congress*, Vol. 10, pp. 127-132, Munich, 1987.

[69] Gomart, O. and P. Caines, "On the Extension of Robust Global Adaptive Control Results to Unstructured Time-Varying Systems," *IEEE Transactions on Automatic Control*, Vol. 31, no. 4, pp. 370, 1986.

[70] Goodwin, G.C., D.J. Hill and X. Xianya, " Stochastic Adaptive Control for Exponentially Convergent Time-Varying Systems," in *Proceedings of the 23rd IEEE Conference on Decision and Control*, pp. 39, 1984.

[71] Goodwin, G.C. and D.Q. Mayne, "A Parameter Estimation Perspective of Continuous Time Adaptive Control," *Automatica*, Vol. 23, 1987.

[72] Goodwin, G.C., P.J. Ramadge and P.E. Caines, "Discrete-Time Multivariable Adaptive Control," *IEEE Transactions on Automatic Control*, Vol. 25, no. 3, pp. 449-456, 1980.

[73] Goodwin, G.C. and K.C. Sin, *Adaptive Filtering Prediction and Control*, Prentice Hall, Englewood Cliffs, New Jersey, 1984.

[74] Goodwin, G.C. and E. K. Teoh, " Adaptive Control of a class of Linear Time Varying Systems," *Proceedings of the IFAC Workshop on Adaptive Systems in Control and Signal Processing*, San Francisco, June 1983.

[75] Goodwin, G.C. and E.K. Teoh, "Persistency of Excitation in the Presence of Possibly Unbounded Signals," *IEEE Transactions on Automatic Control*, Vol. 30, no. 6, pp. 595-597, 1985.

[76] Grayson, L.P., "Design via Lyapunov's Second Method," *Proceedings of the 4th JACC*, Minneapolis, Minnesota, 1963.

[77] Gupta, M.M. (Ed.), *Adaptive Methods for Control System Design*, IEEE Press, New York, 1986.

[78] Hahn, W. *Theory and Application of Lyapunov's Direct Method.* Prentice Hall, Englewood Cliffs, New Jersey, 1963.

[79] Hale, J. K., *Ordinary Differential Equations*, Wiley, New York, 1969.

[80] Harris, C.J. and S.A. Billings (Eds), *Self-Tuning and Adaptive Control: Theory and Applications*, Peter Peregrinus, London, 1981.

[81] Hsu, L. and R.R. Costa, "Bursting Phenomena in Continuous-Time Adaptive Systems with a σ-Modification," *IEEE Transactions on Automatic Control*, Vol. 32, no. 1, pp. 84-86, 1987.

[82] Ioannou, P.A., "Decentralized Adaptive Control of Interconnected Systems", *IEEE Transactions on Automatic Control*, Vol. 31, no. 3, pp. 291-298, 1986.

[83] Ioannou, P.A. and A. Datta, "Robust Adaptive Control: A Unified Approach," *Proceedings of the IEEE*, Vol. 79, no. 12, pp. 1735-1768, 1991.

[84] Ioannou, P.A. and A. Datta, "Robust Adaptive Control: Design, Analysis and Robustness Bounds," in P.V. Kokotović (Ed.), *Grainger Lectures: Foundations of Adaptive Control*, Springer-Verlag, New York, 1991.

[85] Ioannou, P.A. and P.V. Kokotović, *Adaptive Systems with Reduced Models*, Lecture Notes in Control and Information Sciences, Vol. 47, Springer-Verlag, New York, 1983.

[86] Ioannou, P.A. and P.V. Kokotović, "Instability Analysis and Improvement of Robustness of Adaptive Control," *Automatica*, Vol. 20, no. 5, pp. 583-594, 1984.

[87] Ioannou, P.A. and J. Sun, "Theory and Design of Robust Direct and Indirect Adaptive Control Schemes," *Int. Journal of Control*, Vol. 47, no. 3, pp. 775-813, 1988.

[88] Ioannou, P.A. and G. Tao, "Model Reference Adaptive Control of Plants with Known Unstable Zeros," USC Report 85-06-01, University of Southern California, Los Angeles, California, June 1985.

[89] Ioannou, P.A. and G. Tao, "Frequency Domain Conditions for Strictly Positive Real Functions," *IEEE Transactions on Automatic Control*, Vol. 32, no. 1, pp. 53-54, 1987.

[90] Ioannou, P.A. and G. Tao, "Dominant Richness and Improvement of Performance of Robust Adaptive Control," *Automatica*, Vol. 25, no. 2, pp. 287-291, 1989.

[91] Ioannou, P.A. and K.S. Tsakalis, "A Robust Direct Adaptive Controller," *IEEE Transactions on Automatic Control*, Vol. 31, no. 11, pp. 1033-1043, 1986.

[92] Ioannou, P.A. and T.Z. Xu, "Throttle and Brake Control Systems for Automatic Vehicle Following," *IVHS Journal*, Vol. 1, no. 4, pp. 345-377, 1994.

[93] James, D.J., "Stability of a Model Reference Control System," *AIAA Journal*, Vol. 9, no. 5, 1971.

[94] Johnson, Jr., C.R., *Lectures on Adaptive Parameter Estimation*, Prentice Hall, Englewood Cliffs, New Jersey, 1988.

[95] Kailath, T., *Linear Systems*, Prentice Hall, Englewood Cliffs, New Jersey, 1980.

[96] Kalman, R.E. "Design of a Self Optimizing Control System," *Transaction of the ASME*, Vol. 80, pp. 468-478, 1958.

[97] Kalman, R. E., and J. E. Bertram, "Control Systems Analysis and Design via the 'Second Method' of Lyapunov," *Journal of Basic Engineering* Vol. 82, pp. 371-392, 1960.

[98] Kanellakopoulos, I., *Adaptive Control of Nonlinear Systems*, Ph.D Thesis, Report No. UIUC-ENG-91-2244, DC-134, University of Illinois at Urbana-Champaign, Coordinated Science Lab., Urbana, IL 61801.

[99] Kanellakopoulos, I., P.V. Kokotović and A.S. Morse, "Systematic Design of Adaptive Controllers for Feedback Linearizable Systems," *IEEE Transactions on Automatic Control*, Vol. 36, pp. 1241-1253, 1991.

[100] Kaufman, H. and G. Neat, "Asymptotically Stable Multi-input Multi-output Direct Model Reference Adaptive Controller for Processes Not Necessarily Satisfying a Positive Real Constraint," *International Journal of Control*, Vol. 58, no. 5, pp. 1011-1031, 1993.

[101] Khalil, H.K., *Nonlinear Systems*, MacMillan, New York, 1992.

[102] Khalil, H.K. and A. Saberi, "Adaptive Stabilization of a Class of Nonlinear Systems Using High Gain Feedback," *IEEE Transactions on Automatic Control*, Vol. 32, no. 11, pp. 1031-1035, 1987.

[103] Kim, C., *Convergence Studies for an Improved Adaptive Observer*, Ph. D thesis, University of Connecticut, 1975.

[104] Kokotović, P.V. "Method of Sensitivity Points in the Investigation and Optimization of Linear Control Systems," *Automation and Remote Control*, Vol. 25, pp. 1512-1518, 1964.

[105] Kokotović, P.V. (Ed.), *Foundations of Adaptive Control*, Springer-Verlag, New York, 1991.

[106] Kokotović, P.V., H.K. Khalil and J. O'Reilly, *Singular Perturbation Methods in Control: Analysis and Design*, Academic Press, New York, 1986.

[107] Krasovskii, N.N. *Stability of Motion: Application of Lyapunov's Second Method to Differential Systems and Equations with Delay*, Stanford University Press, Stanford, California, 1963.

[108] Kreisselmeier, G. "Adaptive Observers with Exponential Rate of Convergence," *IEEE Transactions on Automatic Control*, Vol. 22, no. 1, pp. 2-8, 1977.

[109] Kreisselmeier, G., "An Approach to Stable Indirect Adaptive Control," *Automatica*, Vol. 21, no. 4, pp. 425-431, 1985.

[110] Kreisselmeier, G., "Adaptive Control of a Class of Slow Time-Varying Plants," *Systems and Control Letters*, Vol. 8, no. 2, pp. 97-103, 1986.

[111] Kreisselmeier, G., "A Robust Indirect Adaptive Control Approach," *International Journal of Control*, Vol. 43, no. 1, pp. 161-175, 1986.

[112] Kreisselmeier, G., "An indirect adaptive controller with a self-excitation capability," *IEEE Transactions on Automatic Control*, Vol. 34, no. 5, pp. 524-528, 1989.

[113] Kreisselmeier, G. and B.D.O. Anderson, "Robust Model Reference Adaptive Control," *IEEE Transactions on Automatic Control*, Vol. 31, no. 2, pp. 127-133, 1986.

[114] Kreisselmeier, G. and G. Rietze-Augst, "Richness and Excitation on an Interval–with Application to Continuous-time Adaptive Control," *IEEE Transactions on Automatic Control*, Vol. 35, no. 2, 1990.

[115] Kronauer, R. E. and P. G. Drew, "Design of the Adaptive Feedback Loop in Parameter-Perturbation Adaptive Controls," *Proceedings of the IFAC Symposium on the Theory of Self-Adaptive Control Systems*, Teddington, England, September 1965.

[116] Krstić, M., I. Kanellakopoulos and P.V. Kokotović, "A New Generation of Adaptive Controllers for Linear Systems," *Proceedings of the 31st IEEE Conference on Decision and Control*, Tucson, Arizona, December 1992.

[117] Krstić, M., I. Kanellakopoulos and P.V. Kokotović, "Passivity and Parametric Robustness of a New Class of Adaptive Systems," Report CCEC-92-1016, Center for Control Engineering and Computation, University of California, Santa Barbara, California, 1992.

[118] Krstić, M., I. Kanellakopoulos and P.V. Kokotović, "Nonlinear Design of Adaptive Controllers for Linear Systems," Report CCEC-92-0526, Center for Control Engineering and Computation, University of California, Santa Barbara, California, 1992.

[119] Krstić, M., P. V. Kokotović and I. Kanellakopoulos, "Transient Performance Improvement with a New Class of Adaptive Controllers," *Systems & Control Letters*, Vol. 21, pp. 451-461, 1993.

[120] Kudva, P. and K. S. Narendra, "Synthesis of an Adaptive Observer Using Lyapunov's Direct Method," *Int. Journal of Control*, Vol. 18, pp. 1201-1210, 1973.

[121] Kuo, B. C. *Automatic Control Systems*, 6th Edition, Prentice Hall, Englewood Cliffs, New Jersey, 1991.

[122] Kwakernaak, H. and R. Sivan, *Linear Optimal Control Systems*, Wiley Interscience, New York, 1972.

[123] Landau, I.D. *Adaptive Control: The Model Reference Approach*, Marcel Dekker, Inc., New York, 1979.

[124] LaSalle, J.P., and S. Lefschetz, *Stability by Lyapunov's Direct Method with Application*, Academic Press, New York, 1961.

[125] LaSalle, J.P., "Some Extensions of Lyapunov's Second Method." *IRE Transactions on Circuit Theory*, pp. 520-527, December 1960.

[126] Lefschetz, S., *Stability of Nonlinear Control Systems*, Academic Press, New York, 1963

[127] Ljung L. and T. Soderstrom, *Theory and Practice of Recursive Identification*, MIT Press, Cambridge, Massachusetts, 1983.

[128] Lozano-Leal, R., "Robust Adaptive Regulation without Persistent Excitation," *IEEE Transactions on Automatic Control*, Vol. 34, no. 12, pp. 1260–1267, 1989.

[129] Lozano-Leal, R. and G.C. Goodwin, "A Globally Convergent Adaptive Pole Placement Algorithm Without a Persistency of Excitation

Requirement," *IEEE Transactions on Automatic Control*, Vol. 30, no. 8, pp. 795-797, 1985.

[130] Luders, G. and K.S. Narendra, "An Adaptive Observer and Identifier for a Linear System," *IEEE Transactions on Automatic Control*, Vol. 18, no. 5, pp. 496-499, 1973.

[131] Luders, G and K.S. Narendra, "A New Canonical Form for an Adaptive Observer", *IEEE Transactions on Automatic Control*, Vol. 19, no. 2, pp. 117-119, 1974.

[132] Luenberger, D.G. *Optimization by Vector Space Methods*, John Wiley & Sons, Inc., New York, 1969.

[133] Lyapunov, A.M. "The General Problem of Motion Stability" (1892) In Russian. Translated to English, *Ann. Math. Study*, no. 17, 1949, Princeton University Press, 1947.

[134] Malkin, I.G., "Theory of Stability of Motion," Technical Report Tr. 3352, U.S. Atomic Energy Commission, English Ed., 1958.

[135] Mareels, I.M.Y., B.D.O. Anderson, R.R. Bitmead, M. Bodson, and S.S. Sastry, "Revisiting the MIT Rule for Adaptive Control," *Proceedings of the 2nd IFAC Workshop on Adaptive Systems in Control and Signal Processing*, Lund, Sweden, 1986.

[136] Mareels, I.M.Y. and R.R. Bitmead, "Nonlinear Dynamics in Adaptive Control: Chaotic and Periodic Stabilization," *Automatica*, Vol. 22, pp. 641-655, 1986.

[137] Martensson, B. "The Order of Any Stabilizing Regulator is Sufficient Information for Adaptive Stabilization," *System & Control Letters*, Vol. 6, pp. 299-305, 1986.

[138] Martin-Sanchéz, J.M., "Adaptive Control for Time-Varying Process," in *Proceedings of 1985 American Control Conference*, pp. 1260, 1985.

[139] Massera, J.L., "Contributions to Stability Theory," *Annals of Mathematics*, Vol. 64, pp. 182-206, 1956.

[140] McRuer, D., I. Ashkenas and D. Graham, *Aircraft Dynamics and Automatic Control*, Princeton University Press, Princeton, New Jersey, 1973.

[141] Mendel, J. M., *Discrete Techniques of Parameter Estimation: The Equation Error Formulation*, Marcel Dekker, Inc., New York, 1973.

[142] Meyer, K. R., "On the Existence of Lyapunov Functions for the Problem on Lur'e," *SIAM Journal of Control*, Vol. 3, pp. 373-383, 1965.

[143] Michel, A. and R.K. Miller, *Ordinary Differential Equations*, Academic Press, New York, 1982.

[144] Middleton, R.H and G.C. Goodwin, *Digital Control and Estimation: A Unified Approach*, Prentice Hall, Englewood Cliffs, New Jersey, 1990.

[145] Middleton, R.H. and G. C. Goodwin, "Adaptive Control of Time-Varying Linear Systems," *IEEE Transactions on Automatic Control*, Vol 33, No. 2, pp. 150-155, Feb. 1988.

[146] Middleton, R.H., G.C. Goodwin, D.J. Hill and D.Q. Mayne, "Design Issues in Adaptive Control," *IEEE Transactions on Automatic Control*, Vol. 33, no. 1, pp. 50-58, 1988.

[147] Miller D.E. and E.J. Davison, "An Adaptive Controller which Provides Lyapunov Stability," *IEEE Transactions on Automatic Control*, Vol. 34, no. 6, pp. 599-609, 1989.

[148] Miller D.E. and E.J. Davison, "An Adaptive Control Which Provides an Arbitrarily Good Transient and Steady-State Response," *IEEE Transactions on Automatic Control*, Vol. 36, no. 1, pp. 68-81, 1991.

[149] Monopoli, R.V., "Lyapunov's Method for Adaptive Control Design," *IEEE Transactions on Automatic Control*, Vol. 12, no. 3, pp. 334-335, 1967.

[150] Monopoli, R.V., "Model Reference Adaptive Control with an Augmented Error Signal," *IEEE Transactions on Automatic Control*, Vol. 19, pp. 474-484, 1974.

[151] Monopoli, R.V. and C.C. Hsing, "Parameter Adaptive Control of Multivariable Systems," *International Journal of Control*, Vol. 22, no. 3, pp. 313-327, 1975.

[152] Morari, M. and E. Zafiriou, *Robust Process Control*, Prentice Hall, Englewood Cliffs, New Jersey, 1989.

[153] Morse, A.S., "Global Stability of Parameter Adaptive Control Systems," *IEEE Transactions on Automatic Control*, Vol. 25, pp. 433-439, 1980.

[154] Morse, A.S., "Parametrizations for Multivariable Adaptive Control," *Proceedings of the 20th IEEE Conference on Decision and Control*, pp. 970-972, San Diego, California, 1981.

[155] Morse, A.S., "Recent Problems in Parameter Adaptive Control," in I. D. Landau (Ed.), *Outils et modeles mathematiques pour l'automatique l'analyse de Systemes et le traitment du signal*, CNRS, Paris, 1983.

[156] Morse, A.S. "An Adaptive Control for Globally Stabilizing Linear Systems with Unknown High-Frequency Gains," *Proceedings of the 6th International Conference on Analysis and Optimization of Systems*, Nice, France, 1984.

[157] Morse, A.S. "A Model Reference Controller for the Adaptive Stabilization of any Strictly Proper, Minimum Phase Linear System with Relative Degree not Exceeding Two," *Proceedings of the 1985 MTNS Conference*, Stockholm, Sweden, 1985.

[158] Morse, A.S., "A $4(n+1)$-dimensional Model Reference Adaptive Stabilizer for any Relative Degree One or Two Minimum Phase System of Dimension n or Less," *Automatica*, Vol. 23, no. 1, pp. 123-125, 1987.

[159] Morse, A.S., "High Gain Adaptive Stabilization," *Proceedings of the Carl Kranz Course*, Munich, Germany, 1987.

[160] Morse, A.S., "High Gain Feedback Algorithms for Adaptive Stabilization," *Proceedings of the Fifth Yale Workshop on Applications of Adaptive Control Systems Theory*, New Haven, Connecticut, 1987.

[161] Morse, A.S., "Towards a Unified Theory of Parameter Adaptive Control: Tunability," *IEEE Transactions on Automatic Control*, Vol. 35, no. 9, pp. 1002-1012, 1990.

[162] Morse, A.S., "A Comparative Study of Normalized and Unnormalized Tuning Errors in Parameter-Adaptive Control," *Proceedings of the 30th IEEE Conference on Decision and Control*, Brighton, England, December 1991.

[163] Morse, A.S., "Towards a Unified Theory of Parameter Adaptive Control-Part II: Certainty Equivalence and Input Tuning," *IEEE Transactions on Automatic Control*, Vol. 37, no. 1, pp. 15-29, 1992.

[164] Morse, A.S., "High-Order Parameter Tuners for the Adaptive Control of Linear and Nonlinear Systems," *Proceedings of the U.S.-Italy Joint Seminar on Systems, Models, and Feedback: Theory and Applications*, Capri, Italy, 1992.

[165] Morse, A.S., D.Q. Mayne and G.C. Goodwin, "Applications of Hysteresis Switching in Parameter Adaptive Control," *IEEE Transactions on Automatic Control*, Vol. 37, no. 9, pp. 1343-1354, 1992.

[166] Morse, A.S. and F.M. Pait, "MIMO Design Models and Internal Regulators for Cyclicly-Switched Parameter-Adaptive Control Systems," *Proceedings of 1993 American Control Conference*, San Francisco, California, pp. 1349-1353, June 1993.

[167] Mudgett, D.R. and A.S. Morse, "Adaptive Stabilization of Linear Systems with Unknown High Frequency Gains," *IEEE Transactions on Automatic Control*, Vol. 30, no. 6, pp. 549-554, 1985.

[168] Naik, S.M., P.R. Kumar and B.E. Ydstie, "Robust Continuous Time Adaptive Control by Parameter Projection," *IEEE Transactions on Automatic Control*, Vol. 37, no. 2, pp. 182-198, 1992.

[169] Narendra, K.S. (Ed.), *Adaptive and Learning Systems: Theory and Applications*, Plenum Press, New York, 1986.

[170] Narendra, K.S. and Annaswamy, A.M., "Robust Adaptive Control in the Presence of Bounded Disturbances," *IEEE Transactions on Automatic Control*, Vol. 31, no. 4, pp. 306-315, 1986.

[171] Narendra, K. S. and A.M. Annaswamy, "Persistent Excitation of Adaptive Systems," *International Journal of Control*, Vol. 45, pp. 127-160, 1987.

[172] Narendra, K.S. and A.M. Annaswamy, *Stable Adaptive Systems*, Prentice Hall, Englewood Cliffs, New Jersey, 1989.

[173] Narendra, K.S., I.H. Khalifa and A.M. Annaswamy, "Error Models for Stable Hybrid Adaptive Systems," *IEEE Transactions on Automatic Control*, Vol. 30, no. 4, pp. 339-347, 1985.

[174] Narendra, K.S., Y.H. Lin and L.S. Valavani, "Stable Adaptive Controller Design, Part II: Proof of Stability," *IEEE Transactions on Automatic Control*, Vol. 25, no. 3, pp. 440-448, 1980.

[175] Narendra, K.S. and L. E. McBride, Jr., "Multivariable Self-Optimizing Systems Using Correlation Techniques," *IRE Transactions on Automatic Control*, Vol. 9, no. 1, pp. 31-38, 1964.

[176] Narendra, K.S. and R.V. Monopoli (Eds.), *Applications of Adaptive Control*, Academic Press, New York, 1980.

[177] Narendra, K.S. and J.H. Taylor, *Frequency Domain Criteria for Absolute Stability*, Academic Press, New York, 1973.

[178] Narendra, K.S. and L.S. Valavani, "Stable Adaptive Controller Design— Direct Control", *IEEE Transactions on Automatic Control*, Vol. 23, pp. 570-583, 1978.

[179] Nussbaum, R.D. "Some Remarks on a Conjecture in Parameter Adaptive Control," *System & Control Letters*, Vol. 3, pp. 243-246, 1983.

[180] Ogata, K. *Modern Control Engineering*, 2nd Edition, Prentice Hall, Englewood Cliffs, New Jersey, 1990.

[181] Ohkawa, F., "A Model Reference Adaptive Control System for a Class of Discrete Linear Time Varying Systems with Time Delay," *International Journal of Control*, Vol. 42, no. 5, pp. 1227, 1985.

[182] Ortega, R., "On Morse's New Adaptive Controller: Parameter Convergence and Transient Performance," *IEEE Transactions on Automatic Control*, Vol. 38, no. 8, pp. 1191-1202, 1993.

[183] Ortega, R. and T. Yu, "Robustness of Adaptive Controllers: A Survey," *Automatica*, Vol. 25, no. 5, pp. 651-678, 1989.

[184] Osburn, P.V., A.P. Whitaker and A. Kezer, "New Developments in the Design of Model Reference Adaptive Control Systems," Paper No. 61-39, Institute of the Aerospace Sciences, 1961.

[185] Ossman, K.A., "Indirect Adaptive Control for Interconnected Systems," *IEEE Transactions on Automatic Control*, Vol. 34, no. 8, pp. 908-911, 1989.

[186] Papoulis, A., *Probability, Random Variables, and Stochastic Processes*, Third Edition, Prentice Hall, Englewood Cliffs, New Jersey, 1991.

[187] Parks, P.C. "Lyapunov Redesign of Model Reference Adaptive Control Systems," *IEEE Transactions on Automatic Control*, Vol. 11, pp. 362-367, 1966.

[188] Phillipson, P.H., "Design Methods for Model Reference Adaptive Systems," *Proc. Inst. Mech. Engrs.*, Vol. 183, no. 35, pp. 695-700, 1969.

[189] Polderman, J.W., "Adaptive Control and Identification: Conflict or Conflux?" Ph.D thesis, Rijksuniversiteit, Groningen, November 1987.

[190] Polderman, J.W., "A State Space Approach to the Problem of Adaptive Pole Placement Assignment," *Math Control Signals Systems*, Vol. 2, pp. 71-94, 1989.

[191] Polycarpou, M. and P.A. Ioannou, "On the Existence and Uniqueness of Solutions in Adaptive Control Systems," *IEEE Transactions on Automatic Control*, Vol. 38, 1993.

[192] Popov, V. M., *Hyperstability of Control Systems*, Springer-Verlag, New York, 1973.

[193] Praly, L., "Robust Model Reference Adaptive Controller–Part I: Stability Analysis," *Proceedings of the 23rd IEEE Conference on Decision and Control*, 1984.

[194] Praly, L., S.T. Hung, and D.S. Rhode, "Towards a Direct Adaptive Scheme for a Discrete-time Control of Minimum Phase Continuous-time System," *Proceedings of the 24th IEEE Conference on Decision and Control*, pp. 1199-1191, Ft. Lauderdale, Florida, December 1985.

[195] Praly, L. and E. Trulsson, "Decentralized Indirect Adaptive Control," *Commande des systemes complexes technologiques*, pp. 295-315, 1986.

[196] Rao, S.S., *Optimization: Theory and Application*, Wiley Eastern LtD, 1984.

[197] Rohrs, C.E., L. Valavani, M. Athans, and G. Stein, "Robustness of Continuous-time Adaptive Control Algorithms in the Presence of Unmodeled Dynamics," *IEEE Transactions on Automatic Control*, Vol. 30, no. 9, pp. 881-889, 1985.

[198] Rosenbrock, H.H., *State-space and Multivariable Theory*, Wiley-Interscience, New York, 1970.

[199] Royden, H.L., *Real Analysis*, The Macmillan Company, New York, 1963.

[200] Rudin, W., *Real and Complex Analysis*, 2nd Edition, McGraw-Hill, New York, 1974.

[201] Sastry, S. and M. Bodson, *Adaptive Control: Stability, Convergence and Robustness*, Prentice Hall, Englewood Cliffs, New Jersey, 1989.

[202] Shackcloth, B. and R.L. Butchart, "Synthesis of Model Reference Adaptive Systems by Lyapunov's Second Method," *Proc. of the 2nd IFAC Symposium on the Theory of Self-Adaptive Control Systems*, Teddington, England, 1965.

[203] Sethares, W.A., C.R. Johnson, Jr. and C.E. Rohrs, "Bursting in Adaptive Hybrids," *IEEE Transactions on Communication* Vol. 37, no. 8, pp. 791-799.

[204] Šiljak, D.D. "New Algebraic Criteria for Positive Realness", *J. Franklin Inst.*, Vol. 291, no. 2, pp. 109-120, 1971.

[205] Singh, R.P. and K.S. Narendra, "Priori Information in the Design of Multivariable Adaptive Controllers," *IEEE Transactions on Automatic Control*, Vol. 29, no. 12, pp. 1108-1111, 1984.

[206] Slotine, J.J. and W. Li, *Applied Nonlinear Control*, Prentice Hall, Englewood Cliffs, New Jersey, 1991.

[207] Sobel, K., *Model Reference Adaptive Control for Multi-input, Multi-output Systems*, Ph.D. thesis, Rensselaer Polytechnic Institute, Troy, New York, June 1980.

[208] Sobel, K., H. Kaufman, and L. Mabius, "Implicit Adaptive Control Systems for a Class of Multi-input, Multi-output Systems," *IEEE Transactions Aerospace Electronic Systems*, Vol. 18, no. 5, pp. 576-590, 1982.

[209] Sondhi, M.M. and D. Mitra," New Results on the Performance of a Well-Known Class of Adaptive Filters," *Proceedings of the IEEE*, Vol. 64, no. 11, pp. 1583-1597, 1976.

[210] Stein, G., "Adaptive Flight Control–A Pragmatic View," in K.S. Narendra and R.V. Monopoli (Eds.), *Applications of Adaptive Control*, Academic Press, New York, 1980.

[211] Sun, J., "A Modified Model Reference Adaptive Control Scheme for Improved Transient Performance," *Proceedings of 1991 American Control Conference*, pp. 150-155, June 1991.

[212] Sun, J., A.W. Olbrot and M.P. Polis, "Robust Stabilization and Robust Performance Using Model Reference Control and Modeling Error Compensation," *Proceedings of 1991 American Control Conference*, June 1991.

[213] Tao, G., "Model Reference Adaptive Control: Stability, Robustness and Performance Improvement," Ph.D. thesis, EE-Systems, University of Southern California, August 1989.

[214] Tao, G., " Model Reference Adaptive Control of Multivariable Plants with Unknown Interactor Matrix," *Proceedings of the 29th IEEE Conference on Decision and Control*, Honolulu, Hawaii, pp. 2730-2735, December 1990.

[215] Tao, G. and P.A. Ioannou, " Strictly Positive Real Matrices and the Lefschetz-Kalman-Yakubovich Lemma," *IEEE Transactions on Automatic Control*, Vol. 33, no. 9, 1988.

[216] Tao, G. and P.A. Ioannou, "Robust Model Reference Adaptive Control for Multivariable Plants," *International Journal of Adaptive Control and Signal Processing*, Vol. 2, no. 3, pp. 217-248, 1988.

[217] Tao, G. and P.A. Ioannou, "Model Reference Adaptive Control for Plants with Unknown Relative Degree," *Proceedings of the 1989 American Control Conference*, pp. 2297-2302, 1989.

[218] Tao, G. and P.A. Ioannou, "Robust Stability and Performance Improvement of Multivariable Adaptive Control Systems," *Int. J. of Control*, Vol. 50, no. 5, pp. 1835-1855, 1989.

[219] Taylor, L.W. and E.J. Adkins, "Adaptive Control and the X-15," *Proceedings of Princeton University Conference on Aircraft Flying Qualities*, Princeton University Press, Princeton, New Jersey, 1965.

[220] Truxal, J.G. "Adaptive Control," *Proceedings of the 2nd IFAC*, Basle, Switzerland, 1963.

[221] Tsakalis, K.S., "Robustness of Model Reference Adaptive Controllers: An Input-Output Approach," *IEEE Transactions on Automatic Control*, Vol. 37, no. 5, pp. 556-565, 1992.

[222] Tsakalis, K.S. and P.A. Ioannou, "Adaptive Control of Linear Time-Varying Plants," *Automatica*, Vol. 23, no. 4, pp. 459-468, July 1987.

[223] Tsakalis, K.S. and P.A. Ioannou, "Adaptive Control of Linear Time-Varying Plants: A New Model Reference Controller Structure," *IEEE Transactions on Automatic Control*, Vol. 34, no. 10, pp. 1038-1046, 1989.

[224] Tsakalis, K.S. and Ioannou, P.A., "A New Indirect Adaptive Control Scheme for Time-Varying Plants," *IEEE Transactions on Automatic Control*, Vol. 35, no. 6, 1990.

[225] Tsakalis, K.S. and P.A. Ioannou, "Model Reference Adaptive Control of Linear Time-Varying Plants: The Case of 'Jump' Parameter Variations," *International Journal of Control*, Vol. 56, no. 6, pp. 1299-1345, 1992.

[226] Tsakalis K.S. and P.A. Ioannou, *Linear Time Varying Systems: Control and Adaptation*, Prentice Hall, Englewood Cliffs, New Jersey, 1993.

[227] Tsakalis, K.S. and S. Limanond, "Asymptotic Performance Guarantees in Adaptive Control," *International Journal of Adaptive Control and Signal Processing*, Vol. 8, pp. 173-199, 1994.

[228] Tsiligiannis, C.A. and S.A. Svoronos, "Multivariable Self-Tuning Control via Right Interactor Matrix," *IEEE Transactions on Automatic Control*, Vol. 31., no. 4, pp. 987-989, 1986.

[229] Tsypkin, Y.Z., *Adaptation and Learning in Automatic Systems*, Academic Press, New York, 1971.

[230] Unbehauen, H., (Ed.) *Methods and Applications in Adaptive Control*, Springer-Verlag, Berlin, 1980.

[231] Vidyasagar, M., *Control System Synthesis – A Factorization Approach*, M.I.T. Press, Cambridge, Massachusetts, 1985.

[232] Vidyasagar, M., *Nonlinear Systems Analysis*, 2nd Edition, Prentice Hall, Englewood Cliffs, New Jersey, 1993.

[233] Wen, J., *Direct Adaptive Control in Hilbert Space*, Ph.D thesis, Elextrical, Computer and Systems Engineering Department, Rensselaer Polytechnic Institute, Troy, New York, June 1985.

[234] Wen, J. and M. Balas, "Robust Adaptive Control in Hilbert Space," *Journal of Mathematical Analysis and Applications*, Vol. 143, pp. 1-26, 1989.

[235] Whitaker, H.P., J. Yamron, and A. Kezer, "Design of Model Reference Adaptive Control Systems for Aircraft," Report R-164, Instrumentation Laboratory, M. I. T. Press, Cambridge, Massachusetts, 1958.

[236] Willems, J.C. and C.I. Byrnes, " Global Adaptive Stabilization in the Absence of Information on the Sign of the High Frequency Gain," *Proceedings of INRIA Conference on Analysis and Optimization of Systems*, Vol. 62, pp. 49-57, 1984.

[237] Wolovich, W.A., *Linear Multivariable systems*, Springer-Verlag, New York, 1974.

[238] Wonham, W.M., *Linear Multivariable Control: A Geometric Approach*, 3rd Edition, Springer-Verlag, New York, 1985.

[239] Ydstie, B.E., "Bifurcation and Complex Dynamics in Adaptive Control Systems," *Proceedings of the 25th IEEE Conference on Decision and Control*, Athens, Greece, 1986.

[240] Ydstie, B.E., "Stability of Discrete Model Reference Adaptive Control — Revisited," *Syst. Contr. Lett.*, Vol. 13, pp. 1429-438, 1989.

[241] Ydstie, B.E., "Transient Performance and Robustness of Direct Adaptive Control," *IEEE Transactions on Automatic Control*, Vol. 37, no. 8, pp. 1091-1105, 1992.

[242] Yuan, J.S-C and W.M. Wonham, "Probing Signals for Model Reference Identification," *IEEE Transactions on Automatic Control*, Vol. 22, pp. 530-538, 1977.

Index

Ackermann's formula, 520
Adaptive control, 5, 313, 434
    decentralized, 735
    direct (implicit), 8
    explicit (explicit), 8
    model reference, 12, 313
    multivariable, 734
    pole placement, 14, 434, 709
    proportional and integral, 444́
    time-varying, 732
Adaptive observer, 250, 267
    hybrid, 276
    Luenberger, 269
    nonminimum, 287
    with auxiliary input, 279
Admissibility, 475, 498
Asymptotic stability, 106
Autocovariance, 256
Averaging, 665

Barbălat's Lemma, 76
Bellman-Gronwall Lemma, 101
Bezout identity, 40
Bilinear parametric model, 58, 208
Bursting, 693

Certainty equivalence, 10, 372
Characteristic equation, 35

Class $\mathcal{K}$ function, 109
Command generator tracker, 736
Complementary sensitivity
    function, 136
Controllability, 29
Convex set, 784
Coprime polynomials, 40
Covariance resetting, 196
Cyclic switching, 505

Dead zone, 607
Decentralized adaptive control, 737
Decrescent function, 110
differential operator, 37
Diophantine equation, 42
Dominant richness, 642
Dynamic normalization, 571, 584

Equation error method, 152
ϵ-modification, 601
Equilibrium state, 105
Estimation, 144
Exponential stability, 106

Forgetting factor, 198

Gain scheduling, 7
Gradient algorithm, 180, 786

Hessian matrix, 784
High frequency instability, 548
High-gain instability, 551
Hölder's inequality, 71
Hurwitz polynomial, 40
Hybrid adaptive law, 217, 617

Instability, 545
Internal model principle, 137
Input/output model, 34
Input/output stability, 79
Internal stability, 135
Invariant set, 112

KYP Lemma, 129

Leakage, 556
    ϵ-modification, 600
    σ-modification, 586
    switching-σ modification, 588
Least squares estimation, 192
Linear parametric model, 49
Linear quadratic adaptive control, 486, 731
Linear quadratic control, 459
Linear systems, 27
Lipschitz condition, 73
LKY Lemma, 129
Luenberger observer, 267
Lyapunov matrix equation, 123
Lyapunov stability, 105
    direct method, 108
    indirect method, 118
    Lyapunov-like function, 117

Minimum phase transfer function, 40
Minkowski inequality, 71
MIT rule, 18
MKY Lemma, 129
Model reference adaptive control, 12, 313, 651, 694
    assumptions, 332, 412
    direct, 14, 372, 657, 667
    indirect, 14, 396, 689
    modified MRAC, 699
Model reference control, 330
Multivariable adaptive control, 736

Negative definite function, 109
Norm, 67
    extended $\mathcal{L}_p$ norm, 70
    induced norm, 67
    matrix norm, 71
Normalization, 162, 571
Nussbaum gain, 215

Observability, 30
Observability grammian, 277
Observer, 250
    adaptive observer, 250, 636
    Luenberger observer, 267
Output-error method, 151
Output injection, 221

Parallel estimation model, 151
Parameter convergence, 220, 297
    partial convergence, 265
Parameter drift, 545
Parameter estimation, 144
    gradient method, 180
    least squares algorithm, 191

INDEX

projection, 202
SPR-Lyapunov design, 170
Parameter identifier, 250
Parametric model, 47
  bilinear, 58, 208
  linear, 49
Persistent excitation, 150, 176, 225
Pole placement adaptive control, 14, 437, 711
  direct APPC, 15
  indirect APPC, 15, 466
Pole placement control, 447
Positive definite function, 109
Positive real, 126
Projection, 202, 604, 789
Proper transfer function, 35
Pure least-squares, 194

Radially unbounded function, 110
Region of attraction, 106
Relative degree, 35
Riccati equation, 460
Richness of signals, 255, 643
  dominant, 643
  sufficient, 255
Robust adaptive control, 635
Robust adaptive law, 575
Robust adaptive observer, 649
Robust parameter identifier, 644

Schwartz inequality, 71
Self-tuning regulator, 16, 435
Semiglobal stability, 661
Sensitivity function, 16, 136
Series-parallel model, 152
Singular perturbation, 535

Small gain theorem, 96
Small in m.s.s., 83
SPR-Lyapunov design, 170
Stability, 66, 105
  BIBO stability, 80
  internal stability, 135
  I/O stability, 79
Stabilizability, 498
State space representation, 27
State transition matrix, 28
Strictly positive real, 126
Sufficient richness, 255
Swapping lemma, 775
Switched excitation, 506
Sylvester Theorem, 44
  Sylvester matrix, 44
  Sylvester resultant, 45

Time-varying systems, 734
Transfer function, 34
  minimum phase, 40
  relative degree, 35
Tunability, 415

Uniform stability, 106
Uniformly complete observability (UCO), 90, 221
Uniformly ultimately bounded, 107
Unnormalized MRAC, 343
Unstructured uncertainty, 632

A CATALOG OF SELECTED
DOVER BOOKS
IN SCIENCE AND MATHEMATICS

CATALOG OF DOVER BOOKS

Mathematics–Bestsellers

HANDBOOK OF MATHEMATICAL FUNCTIONS: with Formulas, Graphs, and Mathematical Tables, Edited by Milton Abramowitz and Irene A. Stegun. A classic resource for working with special functions, standard trig, and exponential logarithmic definitions and extensions, it features 29 sets of tables, some to as high as 20 places. 1046pp. 8 x 10 1/2. 0-486-61272-4

ABSTRACT AND CONCRETE CATEGORIES: The Joy of Cats, Jiri Adamek, Horst Herrlich, and George E. Strecker. This up-to-date introductory treatment employs category theory to explore the theory of structures. Its unique approach stresses concrete categories and presents a systematic view of factorization structures. Numerous examples. 1990 edition, updated 2004. 528pp. 6 1/8 x 9 1/4. 0-486-46934-4

MATHEMATICS: Its Content, Methods and Meaning, A. D. Aleksandrov, A. N. Kolmogorov, and M. A. Lavrent'ev. Major survey offers comprehensive, coherent discussions of analytic geometry, algebra, differential equations, calculus of variations, functions of a complex variable, prime numbers, linear and non-Euclidean geometry, topology, functional analysis, more. 1963 edition. 1120pp. 5 3/8 x 8 1/2. 0-486-40916-3

INTRODUCTION TO VECTORS AND TENSORS: Second Edition–Two Volumes Bound as One, Ray M. Bowen and C.-C. Wang. Convenient single-volume compilation of two texts offers both introduction and in-depth survey. Geared toward engineering and science students rather than mathematicians, it focuses on physics and engineering applications. 1976 edition. 560pp. 6 1/2 x 9 1/4. 0-486-46914-X

AN INTRODUCTION TO ORTHOGONAL POLYNOMIALS, Theodore S. Chihara. Concise introduction covers general elementary theory, including the representation theorem and distribution functions, continued fractions and chain sequences, the recurrence formula, special functions, and some specific systems. 1978 edition. 272pp. 5 3/8 x 8 1/2.
0-486-47929-3

ADVANCED MATHEMATICS FOR ENGINEERS AND SCIENTISTS, Paul DuChateau. This primary text and supplemental reference focuses on linear algebra, calculus, and ordinary differential equations. Additional topics include partial differential equations and approximation methods. Includes solved problems. 1992 edition. 400pp. 7 1/2 x 9 1/4. 0-486-47930-7

PARTIAL DIFFERENTIAL EQUATIONS FOR SCIENTISTS AND ENGINEERS, Stanley J. Farlow. Practical text shows how to formulate and solve partial differential equations. Coverage of diffusion-type problems, hyperbolic-type problems, elliptic-type problems, numerical and approximate methods. Solution guide available upon request. 1982 edition. 414pp. 6 1/8 x 9 1/4. 0-486-67620-X

VARIATIONAL PRINCIPLES AND FREE-BOUNDARY PROBLEMS, Avner Friedman. Advanced graduate-level text examines variational methods in partial differential equations and illustrates their applications to free-boundary problems. Features detailed statements of standard theory of elliptic and parabolic operators. 1982 edition. 720pp. 6 1/8 x 9 1/4. 0-486-47853-X

LINEAR ANALYSIS AND REPRESENTATION THEORY, Steven A. Gaal. Unified treatment covers topics from the theory of operators and operator algebras on Hilbert spaces; integration and representation theory for topological groups; and the theory of Lie algebras, Lie groups, and transform groups. 1973 edition. 704pp. 6 1/8 x 9 1/4.
0-486-47851-3

Browse over 9,000 books at www.doverpublications.com

CATALOG OF DOVER BOOKS

A SURVEY OF INDUSTRIAL MATHEMATICS, Charles R. MacCluer. Students learn how to solve problems they'll encounter in their professional lives with this concise single-volume treatment. It employs MATLAB and other strategies to explore typical industrial problems. 2000 edition. 384pp. 5 3/8 x 8 1/2. 0-486-47702-9

NUMBER SYSTEMS AND THE FOUNDATIONS OF ANALYSIS, Elliott Mendelson. Geared toward undergraduate and beginning graduate students, this study explores natural numbers, integers, rational numbers, real numbers, and complex numbers. Numerous exercises and appendixes supplement the text. 1973 edition. 368pp. 5 3/8 x 8 1/2. 0-486-45792-3

A FIRST LOOK AT NUMERICAL FUNCTIONAL ANALYSIS, W. W. Sawyer. Text by renowned educator shows how problems in numerical analysis lead to concepts of functional analysis. Topics include Banach and Hilbert spaces, contraction mappings, convergence, differentiation and integration, and Euclidean space. 1978 edition. 208pp. 5 3/8 x 8 1/2. 0-486-47882-3

FRACTALS, CHAOS, POWER LAWS: Minutes from an Infinite Paradise, Manfred Schroeder. A fascinating exploration of the connections between chaos theory, physics, biology, and mathematics, this book abounds in award-winning computer graphics, optical illusions, and games that clarify memorable insights into self-similarity. 1992 edition. 448pp. 6 1/8 x 9 1/4. 0-486-47204-3

SET THEORY AND THE CONTINUUM PROBLEM, Raymond M. Smullyan and Melvin Fitting. A lucid, elegant, and complete survey of set theory, this three-part treatment explores axiomatic set theory, the consistency of the continuum hypothesis, and forcing and independence results. 1996 edition. 336pp. 6 x 9. 0-486-47484-4

DYNAMICAL SYSTEMS, Shlomo Sternberg. A pioneer in the field of dynamical systems discusses one-dimensional dynamics, differential equations, random walks, iterated function systems, symbolic dynamics, and Markov chains. Supplementary materials include PowerPoint slides and MATLAB exercises. 2010 edition. 272pp. 6 1/8 x 9 1/4. 0-486-47705-3

ORDINARY DIFFERENTIAL EQUATIONS, Morris Tenenbaum and Harry Pollard. Skillfully organized introductory text examines origin of differential equations, then defines basic terms and outlines general solution of a differential equation. Explores integrating factors; dilution and accretion problems; Laplace Transforms; Newton's Interpolation Formulas, more. 818pp. 5 3/8 x 8 1/2. 0-486-64940-7

MATROID THEORY, D. J. A. Welsh. Text by a noted expert describes standard examples and investigation results, using elementary proofs to develop basic matroid properties before advancing to a more sophisticated treatment. Includes numerous exercises. 1976 edition. 448pp. 5 3/8 x 8 1/2. 0-486-47439-9

THE CONCEPT OF A RIEMANN SURFACE, Hermann Weyl. This classic on the general history of functions combines function theory and geometry, forming the basis of the modern approach to analysis, geometry, and topology. 1955 edition. 208pp. 5 3/8 x 8 1/2. 0-486-47004-0

THE LAPLACE TRANSFORM, David Vernon Widder. This volume focuses on the Laplace and Stieltjes transforms, offering a highly theoretical treatment. Topics include fundamental formulas, the moment problem, monotonic functions, and Tauberian theorems. 1941 edition. 416pp. 5 3/8 x 8 1/2. 0-486-47755-X

Browse over 9,000 books at www.doverpublications.com

CATALOG OF DOVER BOOKS

Mathematics–Logic and Problem Solving

PERPLEXING PUZZLES AND TANTALIZING TEASERS, Martin Gardner. Ninety-three riddles, mazes, illusions, tricky questions, word and picture puzzles, and other challenges offer hours of entertainment for youngsters. Filled with rib-tickling drawings. Solutions. 224pp. 5 3/8 x 8 1/2. 0-486-25637-5

MY BEST MATHEMATICAL AND LOGIC PUZZLES, Martin Gardner. The noted expert selects 70 of his favorite "short" puzzles. Includes The Returning Explorer, The Mutilated Chessboard, Scrambled Box Tops, and dozens more. Complete solutions included. 96pp. 5 3/8 x 8 1/2. 0-486-28152-3

THE LADY OR THE TIGER?: and Other Logic Puzzles, Raymond M. Smullyan. Created by a renowned puzzle master, these whimsically themed challenges involve paradoxes about probability, time, and change; metapuzzles; and self-referentiality. Nineteen chapters advance in difficulty from relatively simple to highly complex. 1982 edition. 240pp. 5 3/8 x 8 1/2. 0-486-47027-X

SATAN, CANTOR AND INFINITY: Mind-Boggling Puzzles, Raymond M. Smullyan. A renowned mathematician tells stories of knights and knaves in an entertaining look at the logical precepts behind infinity, probability, time, and change. Requires a strong background in mathematics. Complete solutions. 288pp. 5 3/8 x 8 1/2. 0-486-47036-9

THE RED BOOK OF MATHEMATICAL PROBLEMS, Kenneth S. Williams and Kenneth Hardy. Handy compilation of 100 practice problems, hints and solutions indispensable for students preparing for the William Lowell Putnam and other mathematical competitions. Preface to the First Edition. Sources. 1988 edition. 192pp. 5 3/8 x 8 1/2. 0-486-69415-1

KING ARTHUR IN SEARCH OF HIS DOG AND OTHER CURIOUS PUZZLES, Raymond M. Smullyan. This fanciful, original collection for readers of all ages features arithmetic puzzles, logic problems related to crime detection, and logic and arithmetic puzzles involving King Arthur and his Dogs of the Round Table. 160pp. 5 3/8 x 8 1/2. 0-486-47435-6

UNDECIDABLE THEORIES: Studies in Logic and the Foundation of Mathematics, Alfred Tarski in collaboration with Andrzej Mostowski and Raphael M. Robinson. This well-known book by the famed logician consists of three treatises: "A General Method in Proofs of Undecidability," "Undecidability and Essential Undecidability in Mathematics," and "Undecidability of the Elementary Theory of Groups." 1953 edition. 112pp. 5 3/8 x 8 1/2. 0-486-47703-7

LOGIC FOR MATHEMATICIANS, J. Barkley Rosser. Examination of essential topics and theorems assumes no background in logic. "Undoubtedly a major addition to the literature of mathematical logic." – *Bulletin of the American Mathematical Society.* 1978 edition. 592pp. 6 1/8 x 9 1/4. 0-486-46898-4

INTRODUCTION TO PROOF IN ABSTRACT MATHEMATICS, Andrew Wohlgemuth. This undergraduate text teaches students what constitutes an acceptable proof, and it develops their ability to do proofs of routine problems as well as those requiring creative insights. 1990 edition. 384pp. 6 1/2 x 9 1/4. 0-486-47854-8

FIRST COURSE IN MATHEMATICAL LOGIC, Patrick Suppes and Shirley Hill. Rigorous introduction is simple enough in presentation and context for wide range of students. Symbolizing sentences; logical inference; truth and validity; truth tables; terms, predicates, universal quantifiers; universal specification and laws of identity; more. 288pp. 5 3/8 x 8 1/2. 0-486-42259-3

Browse over 9,000 books at www.doverpublications.com

CATALOG OF DOVER BOOKS

Mathematics–Algebra and Calculus

VECTOR CALCULUS, Peter Baxandall and Hans Liebeck. This introductory text offers a rigorous, comprehensive treatment. Classical theorems of vector calculus are amply illustrated with figures, worked examples, physical applications, and exercises with hints and answers. 1986 edition. 560pp. 5 3/8 x 8 1/2. 0-486-46620-5

ADVANCED CALCULUS: An Introduction to Classical Analysis, Louis Brand. A course in analysis that focuses on the functions of a real variable, this text introduces the basic concepts in their simplest setting and illustrates its teachings with numerous examples, theorems, and proofs. 1955 edition. 592pp. 5 3/8 x 8 1/2. 0-486-44548-8

ADVANCED CALCULUS, Avner Friedman. Intended for students who have already completed a one-year course in elementary calculus, this two-part treatment advances from functions of one variable to those of several variables. Solutions. 1971 edition. 432pp. 5 3/8 x 8 1/2. 0-486-45795-8

METHODS OF MATHEMATICS APPLIED TO CALCULUS, PROBABILITY, AND STATISTICS, Richard W. Hamming. This 4-part treatment begins with algebra and analytic geometry and proceeds to an exploration of the calculus of algebraic functions and transcendental functions and applications. 1985 edition. Includes 310 figures and 18 tables. 880pp. 6 1/2 x 9 1/4. 0-486-43945-3

BASIC ALGEBRA I: Second Edition, Nathan Jacobson. A classic text and standard reference for a generation, this volume covers all undergraduate algebra topics, including groups, rings, modules, Galois theory, polynomials, linear algebra, and associative algebra. 1985 edition. 528pp. 6 1/8 x 9 1/4. 0-486-47189-6

BASIC ALGEBRA II: Second Edition, Nathan Jacobson. This classic text and standard reference comprises all subjects of a first-year graduate-level course, including in-depth coverage of groups and polynomials and extensive use of categories and functors. 1989 edition. 704pp. 6 1/8 x 9 1/4. 0-486-47187-X

CALCULUS: An Intuitive and Physical Approach (Second Edition), Morris Kline. Application-oriented introduction relates the subject as closely as possible to science with explorations of the derivative; differentiation and integration of the powers of x; theorems on differentiation, antidifferentiation; the chain rule; trigonometric functions; more. Examples. 1967 edition. 960pp. 6 1/2 x 9 1/4. 0-486-40453-6

ABSTRACT ALGEBRA AND SOLUTION BY RADICALS, John E. Maxfield and Margaret W. Maxfield. Accessible advanced undergraduate-level text starts with groups, rings, fields, and polynomials and advances to Galois theory, radicals and roots of unity, and solution by radicals. Numerous examples, illustrations, exercises, appendixes. 1971 edition. 224pp. 6 1/8 x 9 1/4. 0-486-47723-1

AN INTRODUCTION TO THE THEORY OF LINEAR SPACES, Georgi E. Shilov. Translated by Richard A. Silverman. Introductory treatment offers a clear exposition of algebra, geometry, and analysis as parts of an integrated whole rather than separate subjects. Numerous examples illustrate many different fields, and problems include hints or answers. 1961 edition. 320pp. 5 3/8 x 8 1/2. 0-486-63070-6

LINEAR ALGEBRA, Georgi E. Shilov. Covers determinants, linear spaces, systems of linear equations, linear functions of a vector argument, coordinate transformations, the canonical form of the matrix of a linear operator, bilinear and quadratic forms, and more. 387pp. 5 3/8 x 8 1/2. 0-486-63518-X

Browse over 9,000 books at www.doverpublications.com